# The Principles and Practice of Medical Education

Shaping the Future of Healthcare

# The Principles and Practice of Medical Education

## Shaping the Future of Healthcare

**Sameer Mohommed Khan**
Department of Physiology, Kasturba Medical College, Manipal Academy of Higher Education, Manipal, Karnataka, India

ELSEVIER
ACADEMIC PRESS
An imprint of Elsevier

Academic Press is an imprint of Elsevier
125 London Wall, London EC2Y 5AS, United Kingdom
50 Hampshire Street, 5th Floor, Cambridge, MA 02139, United States



ISBN: 978-0-443-44197-4

For Information on all Academic Press publications
visit our website at https://www.elsevier.com/books-and-journals

*Publisher:* Megan Ball
*Acquisitions Editor:* Elizabeth Brown
*Editorial Project Manager:* Priyamvada Jha
*Production Project Manager:* Kiruthigadevi
*Cover Designer: Arumugam Kothandan*

Typeset by MPS Limited, Chennai, India

# Dedication

*This book is dedicated to*

*My students who have inspired me to acquire and proliferate knowledge and skills that made me what I am today*

*My wife Dr. Anisa Sameer Khan for her understanding, encouragement, and valuable guidance*

*My daughters, Samaira and Ridaa, for their patience and support throughout the process*

# Contents

# Part 4
# Assessment and feedback in medical education

# Part 5
# Quality in medical education

# Foreword

Medical education is evolving from "teaching content" to developing clinicians who can improve patient outcomes, work effectively in teams, and keep learning across an entire career. This future-facing shift is guided by well-established educational principles (how adults learn, how competence develops, and how quality is assured) and by practical methods such as competency-based curricula, authentic workplace learning, and programmatic assessment. Through six distinctive sections, this monograph addresses evolving ideas of how fundamentals of medical education can and should be addressed.

The core principle of medical education, inherent in the introduction (Part 1), is that medical learners are adults: they learn best when education is relevant, problem-centered, active, and connected to real clinical responsibilities rather than passive memorization. Thus, an evidence-based curriculum (Part 2), which clearly states the objectives, outcomes, and competencies, is the scaffold on which the curriculum framework is built—a framework that covers the challenges posed by the hidden curriculum, effective evaluation, and the intelligent use of learning management systems. Frameworks such as Can MEDS illustrate what "competence" includes beyond medical expertise—communication, collaboration, advocacy, scholarship, leadership/management, and professionalism—so graduates can meet modern health system demands.

As is clear in the text, instructional methodologies (Part 3) should be outcomes-oriented—so curricula begin by defining the abilities graduates must demonstrate, then design learning and assessment backward from those outcomes. These methodologies, for example, problems-based learning, microteaching, workplace learning, etc., are being challenged by daily advances in artificial intelligence and explored in the text. Assessment and feedback (Part 4)—written, clinical, and workplace measures are the touchstones of medical education because they determine the capability of a student to finally enter practice. Assessment shapes what learners prioritize, so modern programs (as shown in this text) emphasize frequent, low-stakes assessment aligned to milestones, combined into defensible decisions about progression (often called "programmatic assessment"). Assessment and evaluation are fundamental measures in program evaluation, which determines whether or not a medical program should be accredited. They are closely bound to quality in medical education (Part 5), that is, quality improvement and accountability. Global standards for basic medical education emphasize benchmarked continuous quality improvement (CQI), aligned with quality assurance and quality control, while still allowing local adaptation to context.

The principles and practices of medical education, as set out throughout this text, clearly illustrate that effective medical education blends foundational science, clinical skills, and behavioral/social sciences in order that learners can make safe decisions and care for diverse patients in context. It is also clear from the text that the future of healthcare relies on medical education based in authentic clinical participation—learning in the workplace with mentoring — because many competencies (teamwork, clinical reasoning under uncertainty, and professional judgment) develop through supervised practice and reflection. Research in medical education (Part 6) enjoys a long and distinguished history, which has affected profound changes in not only the education of physicians but also of many other health and social care professionals. This part of the monograph is an excellent introduction to both quantitative and qualitative research methods and a challenge to continue moving that work forward in response to theoretical and technological advances across healthcare.

When the principles and practices set out in this monograph are implemented well, medical education becomes a lever for health system improvement: graduates are prepared not only to diagnose and treat but also to communicate effectively, collaborate across professions, and advocate for patients and communities. A principles-based approach to standards, as envisioned in this text, encourages medical schools to innovate responsibly, adapting curricula to local population needs while meeting globally recognized expectations for quality. The long-term impact of applying these principles is a workforce better equipped for rapid scientific change, increasing complexity, and the everyday realities of team-based, patient-centered care. This articulate monograph is the most welcome roadmap for the planning of future medical education.

**John H.V. Gilbert, CM, PhD, LLD (honoris causa), FCAH**
*Professor Emeritus, University of British Columbia*

# Preface

Dear Friends,

As an author and medical educationist, I am of the firm opinion that medical education is always evolving in order to achieve the best outcomes in practice, research, and instruction. In this regard, many of the subtleties and nuances are still unknown.

The significance of medical education in the healthcare sector cannot be understated because it has overcome numerous obstacles and transformed the methods of instruction and learning.

In this context, I came to see how crucial it is for all healthcare professionals to understand medical education and its constituent parts in order to grasp the different ideas and tenets upon which medical education is based and how this can help students develop a holistic mindset. The significance of medical education in establishing the high standards of quality healthcare and education that our patients, students, and the community at large are expected to receive cannot be overstated.

My journey towards medical education started as a student of the fellowship program at the prestigious Manipal University. This experience was life changing as it ignited my passion and desire to delve deeply into the many facets of medical education and investigate the enormous potential that medical education had in shaping the student's holistic development. In order to enhance my passion with an aim to spread the knowledge and experience that I gained here, I took the initiative write a textbook that would benefit students from various backgrounds as medical education is the core of all health professionals.

I found the process of creating this book to be both enjoyable and challenging because it allowed me to examine the various aspects of medical education. It also provided me with an opportunity to comprehend and value how the field of medical education has, in fact, altered our approach to the teaching and learning process in medicine.

Since it has something for everyone, this book is primarily intended for medical students, medical educators, medical researchers, and practicing physicians.

My main motivation for writing this book is to provide the audience with an introduction to the foundational ideas of medical education.

I've attempted to cover a variety of subjects in this book, from the principles of teaching and learning process to the more recent ones like potential applications of artificial intelligence in medical education. Across many fields, this book provides a comprehensive and panoptic introduction to medical education and its various facets in an understandable manner. The textbook is divided into six sections that cover a wide range of subjects. Each section is thoughtfully chosen and written to meet the demands of healthcare professionals while taking their level of awareness into consideration. The text's organization offers a methodical approach to comprehending and appreciating the complexities of medical education.

This book's key characteristics that make it an essential resource are:

- A comprehensive and current text that incorporates the various facets of medical education
- Easy to read and understand
- Reaching a broad audience
- simplifying the subject
- methodically structuring the content to make it easier to understand

This book should be used as a guide to help you understand and use medical education in your future pursuits. I wish you luck in your future professional endeavors.

Comments and suggestions for improving the book from teachers and students are always valued and duly recorded.

With regards,

**Dr. Sameer Mohommed Khan**

# Acknowledgments

First of all, I thank the almighty for giving me the courage, constantly guiding me in the right way, and enabling me to turn my dream into reality.

I express my sincere gratitude to all the reviewers of this book for believing in me and my work and rendering me the much-needed support and encouragement.

My sincere gratitude to Prof. John H.V. Gilbert, who has been a wonderful mentor and a very kind gentleman. His opinions regarding the book's foreword have improved its quality.

It's my pleasure to acknowledge the cooperation and support that I got from the staff of Elsevier publications who have been kind and generous towards me and also for their tremendous support and faith in me and my work.

I am delighted to express my profound appreciation to my dear wife, Dr. Anisa Sameer Khan, for motivating me, supporting me, and encouraging me throughout the entire process as we accomplish our objective.

# PART 1

## INTRODUCTION TO MEDICAL EDUCATION

Chapter 1

# Fundamentals of medical education

*"Medical education is not just a program for building knowledge and skills in its recipients...it is also an experience which creates attitudes and expectations."*

—Abraham Flexner

## 1.1 Introduction

Since ages medicine has been regarded as one of the finest and noblest professions as people who practiced medicine were highly respected and revered in the society. What began as just the transfer of knowledge has now transformed into one of the most sophisticated and rapidly growing field. Whether it was to challenge the religious beliefs or to prove the various scientific inventions and discoveries medical education has indeed stood the test of times. Although it appears to be a straightforward process of teaching the learner new information and skills, it actually involves a number of elements that not only help the learner become more skilled in their line of work but also affect how the learner views the quality of healthcare. It includes basic training at medical school, specialized training in the form of a residency or fellowship, and ongoing continuous education throughout a practitioner's career to ensure clinical competency and to guarantee high-quality patient care. This ensures that the healthcare professionals have a suitable platform to demonstrate their knowledge and skills while also staying current. Despite starting off as a knowledge-and-skill-based approach to health and illness, it has progressively grown into a broad field that includes every component that goes into providing high-quality healthcare.

Medical education's ultimate goal is to provide society with a workforce of professionals who are informed, skilled, prioritize patient care, and who exhibit a lifelong dedication to acquiring new skills and honing their current ones. Additionally, medical education is focused on producing compassionate healthcare professionals that can support societal well-being in addition to being persistent in information-gathering and solution-providing. The evolution of medical education from antiquity to the present reflects humanity's relentless pursuit of knowledge and progress in healthcare. Furthermore, the swift change in learners' focus from simply gaining knowledge and skills to becoming skilled healthcare providers due to a more modern, holistic approach to medical education that reflects society's values and attitudes concerning healthcare has added more momentum (Ackerknecht E 1984).

## 1.2 Major events that shaped medical education

Medicine has been seen as a synthesis of art and science since the time of the itinerant barber-surgeon. Despite advances in medical theory, the primary focus of medical education has been to provide a competent physician who is capable enough to take care of the health of the society. The competent physician should act as a change agent who is in a position to uplift healthcare and strive for the betterment of the community at large. Medicine has evolved significantly throughout time to accommodate shifts in scientific knowledge, societal expectations, and instructional methodologies. The long and transformative path from traditional practices to modern medical education has enabled advancements in healthcare. Medical education has seen significant change over the years. Indeed, it has made a name for itself from its modest origins in ancient Greece to the cutting-edge, technologically advanced educational system of today (Norman G 2012). If some of the significant groundbreaking events hadn't occurred, this would not have been possible. These occasions not only increased awareness of medical education but also established the benchmark that allowed for the realization of the goal of becoming a competent and qualified healthcare professional.

Being aware of the fundamentals of medical education is essential because it helps the student comprehend the various aspects of medical education and how it will benefit him in his professional life. Acquiring knowledge and understanding of the fundamental ideas of medical education will also allow the learner to put the many theoretical ideas into practice,

**The Principles and Practice of Medical Education. DOI:** https://doi.org/10.1016/B978-0-443-44197-4.00011-4

which will not only provide the learner an advantage in terms of becoming a better healthcare professional so that they can provide the best possible service to the community. Additionally, the learner will be better prepared for effective lifelong learning, improved patient care and communication, the capacity to instruct others, a deeper comprehension of the healthcare system, and their own professional development if they have a solid understanding of the complexities of medical education and its components. At the same time, it is crucial for educators to understand the complex aspects of contemporary medical education and how it is evolving. Given how quickly medical education is developing, having an understanding of its many subtleties will give them the skills they need to be a ray of hope for students and assist them in achieving their personal and professional objectives. Additionally, in order to improve student learning outcomes, increase teaching efficacy, and ultimately improve patient care, teachers must be knowledgeable with the foundations of medical education (Fig. 1.1).

### 1.2.1 The Greek origin

In pre-Hippocratic Greece, religious and supernatural beliefs had an impact on medical education and practice. Before the emergence of Hippocrates and his school of Kos, healing was frequently associated with religion and mythology, and early schooling was focused on Asclepions, which were temples dedicated to Asclepius, the Greek god of medicine (Papageorgiou, 2022). Established in the sixth century B.C., Asclepions functioned as mystic medical schools for a small group of "godly blessed" priests. The healing process known as incubation, or temple slumber, was used to rehabilitate patients who came to these hallowed places. Asclepius and his daughters Hygeia and Panacea, as well as one of his priests, visited them in their dreams while they were sleeping in the abaton, or dormitory, and they received divine guidance and inspiration. Over time, the impact of religion and superstition on medicine gradually diminished until Hippocrates' rational medicine arrived, drawing a line between magic and reason.

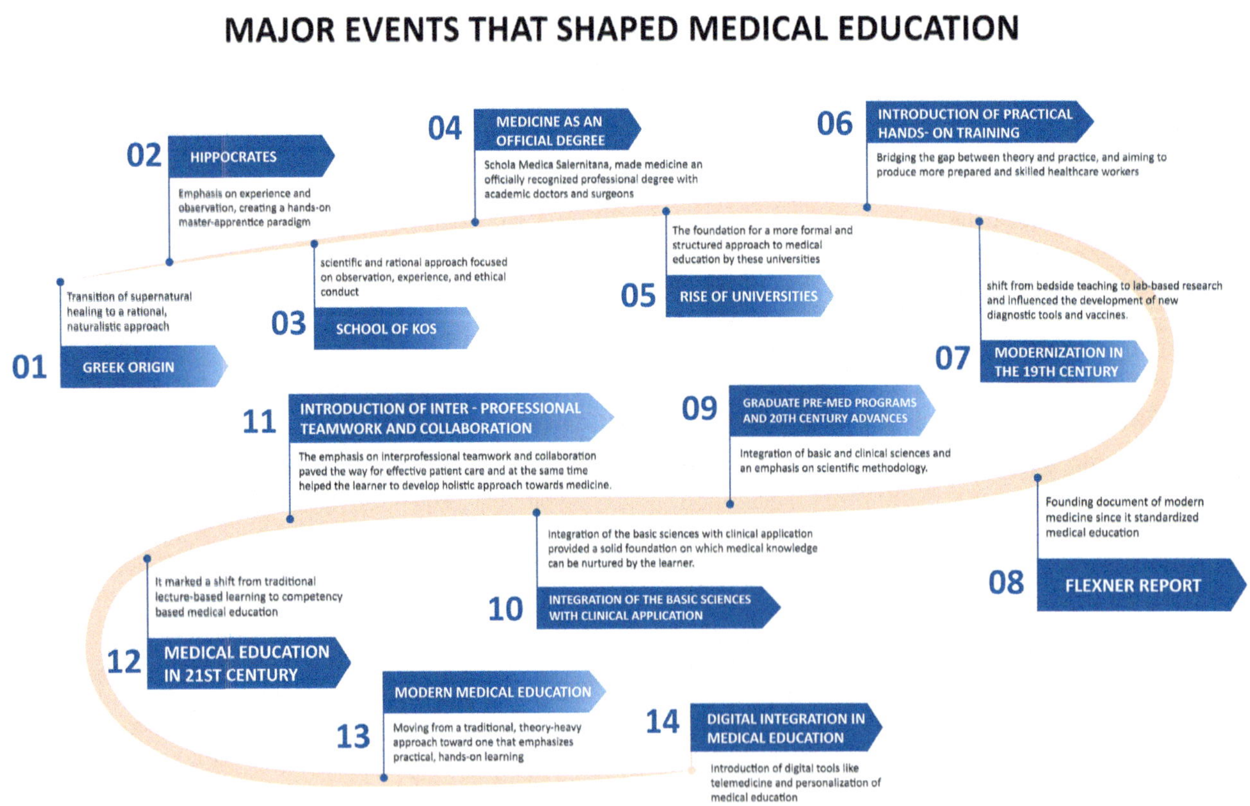


FIGURE 1.1 **Major events that shaped medical education.** This figure depicts the major events that shaped medical education.

### 1.2.2 Hippocrates

Modern medicine, which is based on the observation of clinical indications and logical conclusions rather than religious or magical beliefs, is often credited to Hippocrates, an ancient Greek physician who flourished during Greece's classical period. Hippocrates was successful in shifting the emphasis to experience and observation, creating a hands-on master-apprentice paradigm in which students learned how to gather thorough histories, make logical assumptions, and treat patients by close observation and help (Suvajdzic, Ljiljana & Djendic 2016). With an emphasis on moral behavior and a comprehensive approach to health, this signaled the start of organized medical education. The Hippocratic principle of "to help, or at least to do no harm" remains relevant even after 2500 years. In the modern context, it serves as a deontological necessity, compelling physicians to weigh the risks associated with any attempt at healing.

### 1.2.3 The school of Kos

Kos' medical school is linked to Hippocratic medicine, the first inventive period of contemporary medicine. The Hippocratic school of Kos revolutionized medical education by introducing a scientific and rational approach focused on observation, experience, and ethical conduct, shifting from divine explanations to natural disease processes and patient-centered care (Yapijakis, 2009). The Hippocratic oath's ethical framework and a thorough understanding of clinical observation, disease diagnosis, and dietary and lifestyle treatment are important contributions of this approach. This foundational work established the basis for modern medical education, emphasizing evidence-based knowledge and practical, hands-on training for physicians. The Hippocratic diagnostic approach, which was founded on observation and logical reasoning, provided the fundamental concepts of medical practice. In addition to emphasizing the notion that every illness has a natural cause and that the doctor's job is to be at the patient's bedside, Kos' medical school had a profound understanding of human suffering (Kostakopoulos, 2024). Hippocrates rejected supernatural explanations and separated the profession of medicine from superstition and magic.

### 1.2.4 Medicine as an official degree

Emperor Frederick II's Constitution of Melfi in 1231, which regulated the practice of medicine by requiring the Licentia Medendi (license to practice) granted by the Schola Medica Salernitana, made medicine an officially recognized professional degree with legal sanction by the 13th century. This paved the way for the eventual formalization of medical degrees as we know them today by laying the groundwork for professional standards and formal medical education. From approximately AD 1100 to the middle of the 19th century, there were two categories of medical professionals academic doctors who were scholarly gentlemen with just theoretical training, possibly with the exception of understanding how to produce medications and surgeons with hands-on training. However, there was no schedule for academic medical education. Universities provided lecture programs in place of predetermined curricula. In the early part of the 1800s, there was a gradual but substantial change in the medical education system.

### 1.2.5 Medieval contributions and the rise of universities

Significant advancements in medical education took place as the world entered the medieval age. The first European universities, such as the university of Paris in France and the University of Bologna in Italy, were established at this time before graduate premed programs became publicly accessible. The foundation for a more formal and structured approach to medical education was laid by these institutions' creation of structured medical curricula. This period saw a significant departure from solely religious and superstitious conceptions of health and illness in favor of a more scientific approach to medicine. However, the prevailing cultural and religious constraints placed some restraints on the advancement of medical research. The church, which at the period controlled medical education, often prohibited the study and dissection of human bodies, which hindered a more thorough understanding of human anatomy and physiology.

Galen, a Greek physician, writer, and philosopher who lived in Rome from 129 to 216 CE, had a major impact on medical theory and practice until the middle of the 17th century. Galen revolutionized medical education by establishing anatomy and experimental physiology as core disciplines, emphasizing clinical observation like pulse diagnosis and patient-centered narratives, and writing extensively, making his works the foundation for western medicine (Hajar, 2021). He published hundreds of medical treatises, including volumes on pharmacology, anatomy, physiology, and therapy. He published and talked extensively about the anatomy of the body, highlighting the functions of the heart, brain, and blood, using ideas from his anatomical dissections.

### 1.2.6 Introduction of practical hands-on training

The idea of practical, hands-on instruction in medicine has long existed. Early bedside teaching methods date back to the middle of the sixteenth century, and by the 1880s, they had greatly expanded to include practicums and laboratory dissections. In medical education, practical hands-on training entails applying theoretical information through practical exercises and techniques (Bleakley, 2008). By bridging the gap between theory and practice, this strategy aims to produce more prepared and skilled healthcare workers by giving students hands-on experience and confidence before they engage with actual patients.

### 1.2.7 Modernization in the 19th century

The 19th century saw a dramatic shift in medical education. Some of the major events that shaped medical education in this period were the rise of bacteriology and the germ theory of disease, which established a scientific basis for medicine. This led to a shift from bedside teaching to lab-based research and influenced the development of new diagnostic tools and vaccines. Furthermore, emphasis was being given to practical, hands-on experience with patient care and surgery, the establishment of university-affiliated hospitals, and the introduction of standardized curricula emphasizing scientific knowledge rather than traditional methods. Reforms were driven by both state authorities and medical organizations, moving medicine from a theoretical, guild-based system to a more academic and scientifically grounded profession. Additionally, medical students started to receive more demanding clinical training around this time. Students got more involved in in-patient care and surgery as medical education started to place a greater emphasis on hands-on experience rather than only on theoretical knowledge. More recent systematic approaches to practical training are shown by modern innovations like formal fieldwork in the 1970s and introduction of preregistration home appointments in 1948. Compared to the primarily theoretical education of earlier times, this hands-on method of teaching represented a dramatic change in acquiring the much needed skills and paved the way for the medical practitioner to hone their skills so that they can contribute to the betterment of the society.

#### *1.2.7.1 Significant changes in medical education during this period*

##### 1.2.7.1.1 Transition from apprenticeship to clinical training

One of the notable changes was that instead of learning from a single master, medical students now received more structured instruction in hospitals and operating rooms. This was a very important step as it laid the proper learning environment in which the medical student could exhibit his knowledge and skills and also develop to be a more competent healthcare provider. During the transition from medical apprenticeship to clinical training a substantial shift from theory to practice occurs necessitating both professional and emotional adjustment. Medical schools used techniques like cognitive apprenticeship, which makes use of expert modeling to make thinking visible. It also created a more realistic learning opportunities through hands-on patient care. A flexible curriculum that bridges the gap between fundamental knowledge and clinical reality, mentorship, and support networks are essential for the successful transition from apprenticeship to clinical training.

##### 1.2.7.1.2 Introduction to normative courses

The introduction of the normative courses by the universities proved to be of great benefit as they laid the foundation for future medical education. Universities began to offer defined curricula that covered both scientific theory and practical application, in contrast to the less structured, theoretical instruction of the past. By establishing uniform guidelines for curriculum, examinations, and certification, normative courses in medical education strived to create a standardized, high-quality training. These "normative" courses also referred to as foundational courses play a key role in inculcating the learner with the fundamental knowledge and abilities they will require for their future employment, covering both ethical behavior and scientific ideas. The premed courses conducted by various universities across the globe like the Bachelor of Science in biomedical sciences in United States or individual courses in Biology, Chemistry, and Physics offered by various universities are a great example of the foundational courses.

##### 1.2.7.1.3 Prominence on useful skills

Practical experience and skills such as providing patient care, performing surgeries, and assisting with autopsies were highly valued during training. Clinical reasoning, technical proficiency, and critical "soft skills" like communication, empathy, teamwork, leadership, and resilience were given much more importance in medical education. Producing well-rounded doctors who can deliver excellent, patient-centered care requires a focus on both technical proficiency and these interpersonal skills, which were given priority.

#### 1.2.7.1.4 Creation of university-affiliated hospitals

As university-affiliated hospitals grew, they became a hub for modern medical education by providing spaces for clinical training and research. University-affiliated hospitals were created to combine patient care with medical education and research while giving the institution complete control over medical education. This concept gained popularity after the Flexner Report in 1910. The prime focus of these hospitals was to establish proficiency in clinical training and research. Furthermore, they also focused on education integration, standardization, and quality of healthcare. It also enabled the universities the power to control the standard of clinical settings and training.

#### 1.2.7.1.5 Professionalization of medical education

The need to improve standards, regulate organizations, elevate medicine to a more professional level, and regulate medical organizations gave rise to the emergence of professional organizations. To maintain a high standard of medical care and public trust, regulation and professionalization in medical education entailed both internal development of professional values and external control. While professionalization is established within educational institutions through curricula, role modeling, and the introduction of soft skills to instill in future physicians a sense of ethics, responsibility, and patient-centeredness, external regulation is controlled by disciplinary and legal organizations that establish standards.

#### 1.2.7.1.6 Emergence of scientific specializations

The period saw the beginning of formalization in domains like hospital-based medicine and specialist surgery in response to new medical knowledge. It was believed that the medical specialties emerged in the nineteenth century as a result of the shift to pathological anatomy, the emergence of medical technologies, and due to urbanization. The demand for more in-depth knowledge, the advent of hospital-based, science-focused medicine, and administrative efficiency were also some of the 19th-century factors that fueled the creation of medical specializations. With the advent of lab sciences and clinical clerkships, this evolution changed medical education from a generalist, lecture-based paradigm to a more specialized, scientific area.

### 1.2.8 Flexner Report: 1910

Abraham Flexner's scathing assessment on American medical schools, which was funded by the Carnegie Foundation had its roots in education, revolutionized medical education and, by extension, the medical industry as a whole (Spinelli, 2022). The total absence of guidelines for clinical practice and medical education in US led to the publication of the Flexner Report in 1910.

The Flexner Report is regarded as the founding document of modern medicine since it standardized medical education and it did so on the basis of scientific concepts and verifiable facts. The Flexner Report, which embraced scientific knowledge and its expansion as the defining ethos of a modern physician, transformed the structure and methodology of medical education (Duffy, Thomas 2011). Abraham Flexner provided recommendations for a revamped curriculum after assessing the situation at various institutions. Above all, he suggested that curricula should be simplified and that the number of medical degree-granting schools be limited. Both the ideal and current situations of medical education were included in his findings. Additionally, Flexner envisioned clinical teaching in academically oriented institutions, where students would be trained to do research motivated by the concerns that arose during patient care by reflecting psychiatrists and doctors (Ludmerer, 2011).

### 1.2.9 Graduate premed programs and 20th-century advances

The 20th century saw a major shift in medical education, largely defined by the 1910 Flexner Report, which advocated for university-based medical schools with rigorous science, standardized curricula, and hands-on clinical training. A focus on scientific methods and the merging of basic and clinical sciences were two major advances during this period. One of the major advancements is the emergence of research in medical education as a distinct field in the 1950s, driven by the explosion of medical knowledge, calls for public accountability, increased funding, and evolving healthcare systems, moving beyond traditional apprenticeships to a more scientific, evidence-based approach to understanding learning, teaching, and assessment in medicine. The increasing importance of research, curricular innovations in the late 20th century erupted due to the growing demands for biomedical knowledge significantly impacted the medical curriculum. (Stevens & Goulbourne, 2012)

At the same time, a need to integrate basic sciences with clinical applications earlier in the medical education process was also a commonly voiced concern. The rise of problem-based learning in the last quarter of the century pioneered at McMaster University, Canada, in the year 1969 turned out to be a blessing in disguise, as it transformed the way medical subjects were being taught and learnt. Owing to these developments, more importance was now being given to collaborative and active learning.

### 1.2.10 Integrating of the basic sciences with clinical application

Due in large part to the Flexner Report in 1910, the idea of combining the basic sciences with clinical application in medicine was greatly developed and established in the early 20th century. A more methodical approach to connecting scientific knowledge with clinical practice was introduced by this report, which suggested reorganizing medical education to include a stronger scientific foundation. The integration of the basic sciences with clinical application proved to be very beneficial as it provided a solid foundation on which medical knowledge can be nurtured by the learner. On the other hand, it made the learner to be more confident in approaching the patient and understanding the various issues that might have to be tackled. The success of this integration is exemplified by well-known "2+2" curriculum model, which consists of basic sciences first, followed by clinical sciences.

### 1.2.11 Introduction of interprofessional teamwork and collaboration

From the early days, medical education was visualized to be a teamwork in which different teams through their knowledge and expertise contributed to the delivery of highest quality medical care. Early efforts to codify interprofessional teamwork and collaboration in medicine began in the United States in the 1940s and gained considerable traction in the United Kingdom in the late 1960s. These concepts emerged in the second half of the 20th century with an aim to enhance the quality of medical education (Mohammed, 2021). The emphasis on interprofessional teamwork and collaboration paved the way for effective patient care and at the same time helped the learner to develop holistic approach towards medicine. In order to address the complexity of patient care, the Institute of Medicine first brought attention to the idea in the 1970s and then again in 2001. In the 2000s, the World Health Organization (WHO) also promoted it as a worldwide approach to enhance patient care and health systems, underscoring its significance.

### 1.2.12 Medical education in the 21st century

Modernization of 21st-century medical education involves a shift from traditional lecture-based learning to competency-based education. It also emphasized the need for graduates to be well versed with technological advancements (Birden, Hudson & Page, Sue. 2007). It integrates digital tools, promotes interprofessional teamwork, and fosters soft skills like emotional intelligence, alongside a greater focus on patient safety, quality improvement, and medical ethics within the curriculum. Technology advancements, curriculum changes, and the necessity to train doctors who are more adept at navigating intricate, quickly changing healthcare systems are some of the main forces behind this shift. These developments dramatically changed medical education from the traditional classroom, lab, and bedside methods that predominated until the late 20th century to a more modern system wherein the focus was slowly shifting towards acquiring the competencies. One facet of contemporary medical education is the learning of knowledge and its application to patient care. Current medical education models incorporate a wide range of elements, including effective teamwork, collaborative practice, ethical values, attributes such as professionalism, as well as advanced technologies for teaching both hard and soft skills.

Medical education is undergoing a paradigm change right now. The topic of whether to use the tried-and-true old methods or adopt a completely new avatar is unavoidably on the minds of legislators, educators, and students alike. The COVID-19 pandemic forced the previously rigorous traditional classroom instruction to be supplanted with an unusual and sometimes unnerving fully online option. In order to revive medical education, all stakeholders now have the opportunity to pause, rethink, rebalance, and move forward. Recognizing the challenges that will unavoidably come up is crucial as we prepare to transform the medical education system. There is an urgent need to concentrate on the development of foundational skills. These skills will only be beneficial if they can be used by all students in real-life situations and benefit humanity, regardless of whether an undergraduate is learning how to try to elicit a heart murmur, or a resident is learning various techniques, like palpating for hepatomegaly.

### 1.2.13 Modern medical education

#### *1.2.13.1 Moving from theory to practice*

Modern medical education is definitively moving from a traditional, theory-heavy approach toward one that emphasizes practical, hands-on learning. This shift is a response to the need for graduates to be better prepared for the complexities of modern healthcare and to bridge the long-standing gap between theoretical knowledge and real-world clinical practice. One of the major factors that is influencing this transition is competency-based education, which ensures that graduates have the necessary competencies in order to deliver effective patient care and at the same time uphold the quality of medical education. Additionally, in the realm for teaching and learning, the active learning methodologies have made a major impact as the traditional lectures, characterized by passive information reception, are being supplemented or replaced by student-centered methods that require active participation. It is also to be kept in mind that medical education theories which form the core of the medical education immensely influence the practice of modern medicine by developing the mindset of the teacher and the learner on how medical education is taught and practiced. This transition aids in shaping physicians' thinking, communication, and approach to patient care.

#### *1.2.13.2 Evidence based medical education*

Evidence-based medical education (EBME) is an approach where educators use a systematic process to improve their teaching by applying the best available research evidence to their decisions. It involves critically evaluating the quality of educational research and evidence to make informed changes to teaching methods, rather than relying on tradition or assumption (Rajashekhar & Kodkany 2002). The core of EBME is to create a culture where teachers actively use evidence to inform their practice, similar to how evidence-based medicine guides patient care. Some of the important aspects of evidence-based medical education are critical thinking where teachers are encouraged to think critically about their teaching methods and question their effectiveness. It also includes evidence-driven decisions which emphasizes that instead of making decisions based on what has always been done, educators use research findings to guide their choices about educational strategies.

### 1.2.14 Digital integration in medical education

Understanding the basic medical sciences and their relationships to patient care are increasingly essential components of modern medical education. Contemporary medical education is no exception to the quickly changing global educational scene. Innovative approaches and technologies are being used by medical schools all around the world to improve student learning and equip the upcoming generation of healthcare professionals for whatever obstacle they may face (Alrashed, 2024). Healthcare and medical education are two important areas of medicine that are being impacted by rapid technology breakthroughs. Since the medical field is dynamic and necessitates the integration of newer technologies, medical education must be able to prepare future doctors in accordance with changing trends in practice patterns, the role of medicine in disease diagnosis and treatment, and innovations and advancements in medical science.

Future medical education seems likely to be increasingly interdisciplinary and collaborative. The emerging fields of artificial intelligence (AI), genomics, and other cutting-edge technologies are poised to bring about yet another revolution in medical professional training (Khan sameer 2025). It is expected that as the population ages, the focus will shift toward a more comprehensive and integrated strategy, with a focus on preventative care and the management of chronic illnesses. The goal is to provide future medical professionals with the technical expertise and adaptability they need to cope with rapidly changing technological environments, so they can deliver effective, customized patient care in the years to come. Digital tools like telemedicine and electronic health records are included into modern medical curricula, illustrating the field's continuous progress. The idea of personalized medicine, a medical paradigm that encourages healthcare personalization with decisions and treatments catered to specific patients, is also becoming more and more well liked. In order to prepare upcoming medical professionals to manage complicated and varied health issues in a worldwide setting, this integration is essential (Timotheou, 2023).

## 1.3 Major components of medical education

A curriculum that covers the fundamental sciences and clinical specialties, the development of practical clinical skills through practical hospital experience, and the development of professional traits like ethics, communication, and empathy are all important components of medical education. In order to maintain competence throughout a physician's career, the education process is a continuum that is frequently separated into basic education, postgraduate specialist training, and continuous professional development (Sukrajh, Verona & Adefolalu, Adegoke. 2021). (Fig. 1.2).

### 1.3.1 Curriculum

The curriculum is the thorough, organized strategy for learning in medical education that encompasses various aspects such as learning objectives, content, teaching methodologies, and evaluation techniques intended to direct students toward particular knowledge, abilities, and attitudes. It is a dynamic, ever-changing entity that necessitates a methodical, needs-based approach to ensure the achievement of learning outcomes. It successfully integrates educational concepts into practice and continues to be pertinent to the needs of community health.

### 1.3.2 Instructional methods

Instructional methods are crucial in medical education because they are essential for delivering the educational content to the learners. As teachers, it is important that we utilize instructional methods wisely so that they serve the purpose that is intended of them. The use of instructional methods depends on the activity, level of the students, and the skills that are intended to be imparted. A range of instructional techniques are employed to meet teaching and learning goals and accommodate different student types. The use of technology and the development of modern teaching methods have caused a significant change in educational practices in recent years. The success of applying the instructional methods lies in the fact that the topic and instructional methods are aligned with each other.

### 1.3.3 Assessment and feedback

Assessment and feedback are two crucial elements of teaching and learning. Assessment not only helps to evaluate the learner but provides an opportunity for the teacher to understand the way in which students learn and understand the subject. Teachers can also use assessment to determine the best ways to extend, support, and intervene with their students. Additionally, assessment offers a means of knowing whether the learning outcomes are achieved.

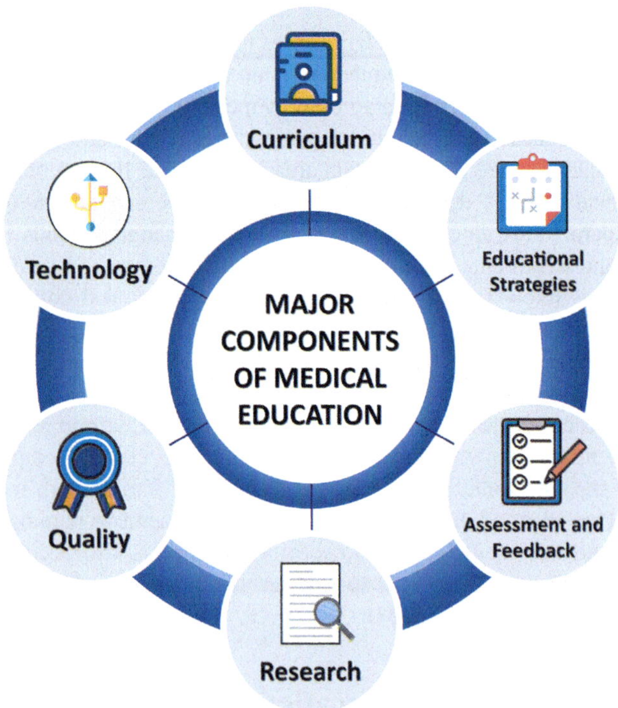


FIGURE 1.2 **Major components of medical education.** This figure reflects the major components of medical education.

Feedback is central to medical education in promoting learning and ensuring the meeting of standards. Feedback is the key to unlocking fresh learning. Feedback is an essential tool for the transition from student to practitioner in medical education, because it helps close the gap between current and desired performance, encourages self-evaluation, directs the acquisition of skills in terms that can be clinical, communication, or professional, and ensures patient safety by improving competency. It improves learning outcomes by offering particular insights for action planning, calibration, and improvement that are frequently absent or insufficient in traditional training.

### 1.3.4 Research

In order to enhance teaching, learning, and evaluation strategies, medical education research entails the methodical examination of the procedures and results of training health professionals. It produces data to guide the creation of curricula, improve pedagogical approaches, and guarantee top-notch instruction that satisfies changing healthcare demands. In addition to advancing healthcare through discovery, enhancing teaching strategies, and preparing flexible, lifelong learners who can critically evaluate literature and improve patient care, research in medical education helps students by developing critical thinking, evidence-based medicine skills, and stronger residency applications. It develops critical abilities that are necessary for a medical industry that is always changing, such as data analysis, problem-solving, communication, and teamwork. Research in healthcare also plays a pivotal role in identifying the most effective treatments and, more specifically, which medicines work best for individual patients. It may provide important information on the effectiveness of a medical procedure and any possible adverse effects.

### 1.3.5 Quality

A critical component of medical education is quality, which includes a demanding curriculum, excellent faculty, high standards of evaluation, sufficient learning materials, efficient procedures for ensuring quality, and accreditation to guarantee that graduates are qualified and able to address the health demands of society. Through set standards, accreditation, and cycles of quality improvement, medical education quality aims to enhance patient outcomes and produce qualified healthcare workers.

Apart from this, taking regular feedback from the peers and students helps us to assess the strengths and the areas of weakness in quality assurance. To guarantee that medical schools fulfill international standards, international organizations such as the WHO and World Federation of Medical Education (WFME) promote standardized quality assurance and accreditation.

### 1.3.6 Technology

Medical education technology has changed significantly over time, mostly as a result of technological improvements. In order to satisfy the demands of students in a fast-changing healthcare environment, medical education has changed from traditional lecture-based instruction to the use of multimedia technology and virtual simulations. Through simulation and e-learning platforms, technology in medical education enhances skill acquisition and knowledge retention by integrating tools like virtual reality, augmented reality, and AI to create immersive, personalized, and interactive learning experiences. By integrating telemedicine and digital health tools into curriculum, this method improves accessibility, encourages self-directed learning, and equips aspiring medical professionals for a digitally linked healthcare environment. Improved clinical performance, flexible resource availability, and safe practice settings for intricate operations are some of the main advantages.

## 1.4 Art of medical education

The art of medical education involves finding a perfect amalgamation between traditional and innovative practices and at the same time developing educational programs that are in par with the current societal needs and expectations. Finding the ideal curriculum and teaching strategies is driven by several important factors, such as interests of the learner, skills to be learnt, and the population to be served. Additionally, the art of medical education combines humanistic abilities with scientific rigor, employing humanities, visual arts, and narrative to develop professionalism, empathy, keen observation, and critical thinking. In the end, this fosters patient-centered care and turns students into self-aware, compassionate healers who possess more than just technical knowledge. In order to better connect with patients and manage the intricacies of sickness, it places a strong emphasis on reflective practice, comprehending human experience, and enhancing communication (Scheele, Fedde, 2012), At the same time, it is important to keep in mind that by skillfully using the strengths of the teacher and the learner, one is able to achieve the intended graduate attributes and contribute to the personal and professional development of the learner. A quality improvement program will be the pinnacle of the effort to attain mastery in the art of medical education. Even if the art of medical teaching appears to be quite difficult, the rewards are ultimately worth the effort.

## 1.5 Principles of medical education

A learner-centered, competency-based approach that combines scientific knowledge with clinical skills, professionalism, and reflective practice is the cornerstone of medical education principles. It emphasizes lifelong learning, adaptability to societal demands, and continuous improvement through feedback, technology, and robust teacher development. Active learning, relevance to community health issues, combining fundamental sciences with clinical application, and ethical foundation are some of the essential components. Some of the core principles of medical education include the following.

### 1.5.1 Learner-centered approach

A fundamental tenet of contemporary medical education is the learner-centered approach, which shifts the emphasis from teacher-directed instruction to empowering students as active participants, encouraging self-directed learning, critical thinking, and lifelong learning skills by customizing instruction to each student's needs, encouraging problem-solving, and turning educators into facilitators rather than merely information providers, ultimately producing more capable and flexible doctors.

### 1.5.2 Competency and skill-based medical education

Competency and skill-based medical education is a core principle shifting focus from time-based learning to demonstrating specific abilities, ensuring graduates possess defined levels of proficiency in essential areas like patient care, medical knowledge, communication, professionalism, teamwork, and systems-based practice, using frameworks like Entrustable Professional Activities and milestones for personalized, outcome-focused training (Frank, 2010).

### 1.5.3 Reflective practice

It is commonly acknowledged that reflective practice is an essential professional need and a cornerstone of contemporary medical education. It is a methodical procedure that helps doctors and students evaluate their experiences in order to increase their comprehension, sharpen their judgment, and enhance both the standard of patient care and their own professional development.

### 1.5.4 Relevance to society

Relevance to society is a fundamental principle in modern medical education, shaping curricula to meet community health needs and improve health outcomes. Medical education guarantees that future physicians are not only scientifically knowledgeable but also socially aware, community-focused, and able to significantly improve public health by incorporating societal relevance.

### 1.5.5 Professionalism and ethics

Professionalism and ethics are foundational principles in medical education, focusing on developing virtuous physicians through core ethical tenets like autonomy, beneficence, nonmaleficence, justice, and professional attributes, which include honesty, compassion, accountability, integrity, lifelong learning to build patient trust, ensure high-quality care, and navigate complex medical challenges. These principles guide decision-making, shaping a doctor's commitment to patient well-being, societal responsibility, and continuous self-improvement, moving beyond just knowledge to encompass a strong professional identity.

### 1.5.6 Lifelong learning

A key tenet of medical education is lifelong learning, which is an ethical obligation that guarantees physicians stay up to date with quickly changing knowledge in order to deliver safe, excellent, evidence-based patient care. It also fosters adaptability, excellence, and patient trust, going beyond simple certification to include professional identity, innovation, and managing intricate healthcare systems.

## 1.6 Key stakeholders in medical education

Key stakeholders in medical education include medical students, patients, healthcare providers such as doctors, nurses, and staff, educational institutions such as medical schools and universities, policymakers and government bodies, accreditation agencies, funding bodies, and researchers. Other important groups are community members, patient advocates, professional societies, and even technology developers, as their input influences curriculum, patient care standards, and overall quality of healthcare.

### 1.6.1 Groups directly involved in the education process

#### *1.6.1.1 Medical students and residents*

Medical students and residents are crucial stakeholders in medical education, influencing curriculum design, learning environments, and patient care quality, with their involvement essential for creating relevant, learner-centered programs that address real-world needs, foster skills like leadership, and improve overall healthcare outcomes through collaborative approaches like cocreation and Residents-as-Teachers models. Their perspectives ensure educational practices align with evolving healthcare demands, promoting learner empowerment, better transition to practice, and stronger community connection. Moreover, the interaction between medical students and residents is crucial in that the residents not only provide medical students hands-on clinical training but also as a source of knowledge and support so that the medical students feel comfortable in the hospital environment. In order to prepare for their future careers as educators, medical students help out by mentoring their peers, supporting research, and honing their own teaching abilities through positions like teaching assistants.

#### *1.6.1.2 Faculty and educators*

The perspectives and involvement of faculty and educators are crucial for curriculum creation, program evaluation, and the production of qualified healthcare professionals, making them important internal stakeholders in medical education. In the domains of curriculum creation and delivery, where faculty members are the main forces behind curriculum delivery and translate educational objectives into practice, the role and significance of faculty as stakeholders are widely recognized. Their advice is essential for creating and executing innovative teaching strategies and making sure the curriculum is pertinent to present and future healthcare requirements. They also actively participate in mentoring and role modeling, serving as mentors and important role models for students, impacting their clinical skills, ethical comprehension, and professional growth.

#### *1.6.1.3 Medical schools and universities*

These organizations provide the framework, resources, and oversight required to guarantee the success of medical education initiatives. In order to prepare students for residency and practice, medical schools and universities are in charge of organizing and delivering medical education, from clinical training to foundational sciences. In addition to developing and overseeing the curriculum, conducting research, evaluating student performance, and conducting supervision to guarantee patient safety, they also offer continuing medical education to working physicians. Universities also make sure that their programs adhere to national accreditation requirements and are dedicated to enhancing their quality.

#### *1.6.1.4 Program administrators and leadership*

Department heads, deans, and other professionals responsible for monitoring and enforcing educational rules are also important stakeholders of medical education. For medical education, program administrators and leaders are essential. Administrators handle day-to-day operations and support, while leaders supervise quality and strategic vision. While leaders define objectives, promote the program, and handle intricate connections, their responsibilities also include creating curricula, guaranteeing quality and accrediting criteria, offering faculty development, assisting trainees, and creating a good learning environment. They are jointly in charge of the thorough training program that produces knowledgeable and capable medical professionals (Fig. 1.3).

FIGURE 1.3 **Groups directly involved in the education process.** This figure shows groups directly involved in the education process.

### 1.6.2 External groups and organizations

#### *1.6.2.1 Patients and family members*

Patients and families are important stakeholders in medical education as they stand to gain the most. They offer practical insights that help shape curricula, improve patient-centered care, improve communication skills, and guarantee that training meets the real needs of those receiving care, ultimately improving health outcomes and system accountability. Their participation adds real-world experience, encourages empathy, and creates a more efficient, pertinent, and patient-centered learning environment for aspiring medical professionals.

#### 1.6.2.2 Governmental organizations and policymakers

The direction, funding, quality, and capacity of medical education to address public health needs are all significantly influenced by governmental organizations and policymakers, who are important stakeholders in the field. They also influence the management and funding of the healthcare and education sectors, and establish national policy. Their participation guarantees that public safety regulations, workforce demands, and national health objectives are all in line with medical education plans. By defining the objectives of the healthcare system, allocating resources, and enacting laws, they have an impact on medical education, which in turn shapes the kinds of doctors who are trained and the services they may offer.

#### 1.6.2.3 Accreditation bodies

These groups ensure the quality and uniformity of medical education programs by establishing standards. Accreditation bodies act as stakeholders in medical education by setting quality standards, evaluating institutions against those standards, and ensuring the public that graduates are competent and safe. They fulfill this role through a process of periodic review, often including self-study and a peer review site visit, which helps ensure patient safety and contributes to the improvement of medical schools and the overall quality of medical education.

#### 1.6.2.4 Professional societies and associations

These are the organizations that support professional standards and development that represent certain medical specialties or the whole healthcare workforce. Professional societies and associations like WFME or any national medical association like the National Medical Commission in India and the American Medical Association act as stakeholders in medical education by influencing policy, developing curricula, providing continuing education, and ensuring the competency of faculty and practitioners through accreditation and setting standards. They use their collective voice to represent the interests of their members to policymakers, educators, and the public, and they often have formal roles in committees and on accreditation bodies.

#### 1.6.2.5 Payers and funding bodies

Organizations that offer financial support for medical education, affecting program design and research agendas are also important stakeholders. Funding bodies may include government agencies like Medicare, Medicaid, and the NIH or professional organizations such as the AAMC and AMEE, private foundations, and the pharmaceutical and healthcare industries. Payers and funding bodies are key stakeholders in medical education because they provide significant funding, influence the direction of training through financial incentives, and shape curriculum by determining which services and interventions are valued. By controlling the flow of financial resources, they can promote or hinder the development of specific medical fields and specialties, impacting what and how future physicians are taught. They also influence the practical application of medical knowledge by setting standards for care and demanding evidence-based practices (Fig. 1.4).

### 1.6.3 Broader societal stakeholders

#### 1.6.3.1 Healthcare institutions and providers

The quality of recent medical graduates' education affects clinics, hospitals, and other healthcare facilities that employ them. Since they offer the training environment, give expertise, and assume responsibility for preparing future professionals to satisfy the demands of patients and the healthcare system, healthcare institutions and providers are important stakeholder in medical education. Additionally, the freshly graduated student may get employment at these institutions. They provide students with clinical experiences, guarantee that training is in line with practical application, and assist in the creation of curricula and continuing professional development programs that equip healthcare professionals to provide high-quality treatment.

#### 1.6.3.2 Community organizations and leaders

Local organizations that support the health needs of the community and offer information on how educational initiatives benefit their constituents are also important components of medical education. Community organizations and leaders act as stakeholders in medical education by providing insights into community needs, fostering trust between the community and healthcare institutions, and ensuring educational programs are relevant and socially accountable. They achieve this

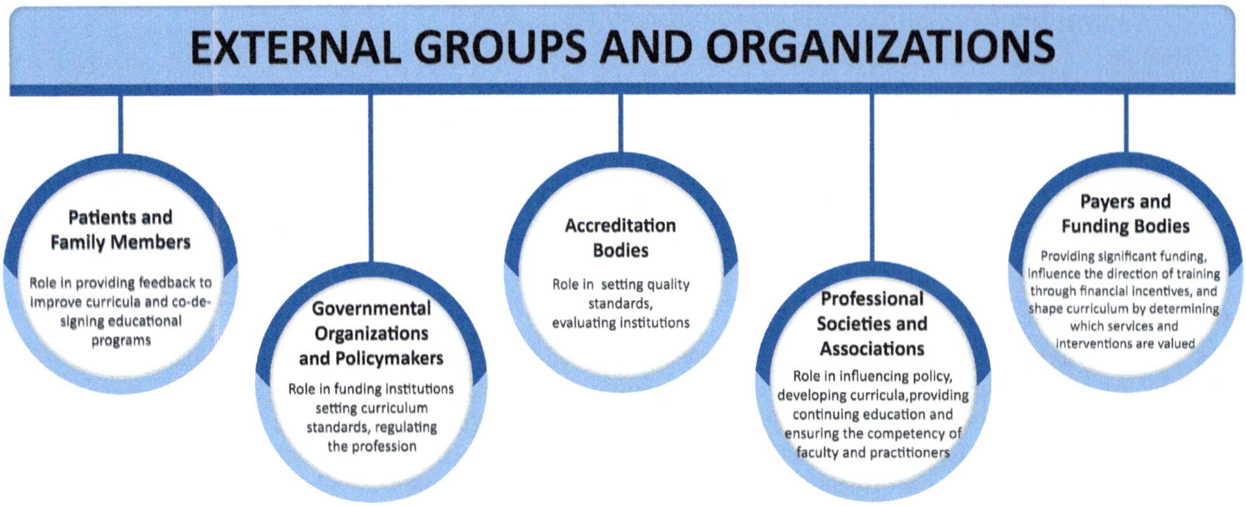


FIGURE 1.4 **External groups and organizations involved in medical education.** This figure shows the external groups and organizations involved in medical education.

through a variety of actions, such as participating in curriculum development, influencing policy, and ensuring the long-term success of health-related initiatives within their community. Furthermore, these community organizations serve as a source of valuable feedback that helps the medical institution to provide the educational and clinical services in the best possible way.

### *1.6.3.3 Researchers and developers of health information technology*

These individuals carry out research on medical education and create instruments such as Meded tools and AI that have the potential to revolutionize the way education is provided. Researchers and developers of health information technology act as stakeholders in medical education by shaping the curriculum to ensure future clinicians are prepared for digital health and also are capable of integrating technology into teaching and learning process. They are able to impact education in a variety of ways by utilizing the information gained by several medical education research and the input received from diverse sources. By focusing on the development of skills like accessing electronic health records, managing health data, and interpreting data science, these technologists support medical education and assist healthcare professionals in being ready for future technological breakthroughs. Their participation guarantees that medical education keeps up with technology developments and equips students to use new health information systems with competence (Fig. 1.5).

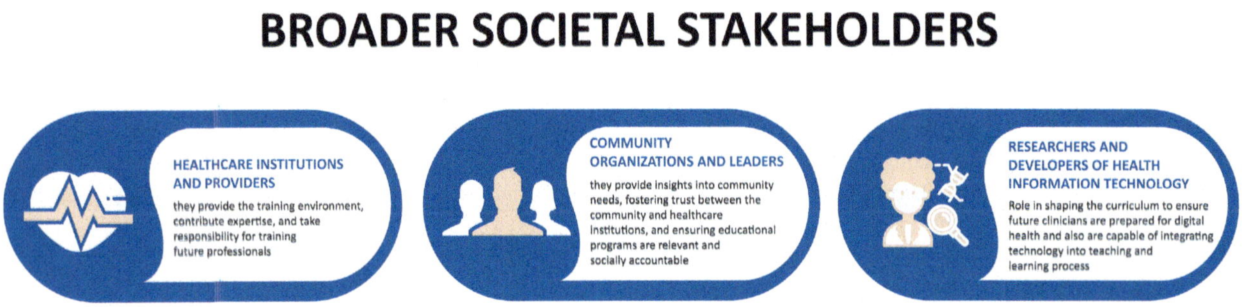


FIGURE 1.5 **Broader societal stakeholders involved in medical education.** This figure shows the broader societal stakeholders involved in medical education.

## 1.7 Determinants of quality medical education

Determinants of quality medical education include the educational program's design and content, including its goals, structure, and curriculum. Essential factors also include human and material resources, such as qualified faculty and adequate facilities. Furthermore, a positive learning environment, effective teaching and assessment methods, robust quality assurance processes, and the capacity for innovation and technology integration are added benefits. Additionally, the overall student experience, the clinical learning environment, the competencies and well-being of the students themselves are critical in upholding the essence of quality medical education (Fig. 1.6).

### 1.7.1 Changing graduate attributes

The advanced knowledge, abilities, attitudes, and values that universities feel their graduates should have upon finishing a program of study are known as graduate attributes. The change in graduate attributes is being credited to the incorporation of skills like teamwork, communication, and critical thinking, which has indeed brought in resilience into the curriculum. Through this medical education is fostering more flexible, patient-centered, and creative doctors. This change improves clinical practice, healthcare delivery, and the profession's overall commitment to the community by bringing education into line with societal demands. In order to develop lifelong, moral, and compassionate professionals who are equipped for the changing healthcare environment, medical schools must modernize their teaching strategies to actively foster these qualities, going beyond the traditional learning of knowledge.

### 1.7.2 Teacher–learner relationship

The teacher–learner relationship in medical education is crucial for fostering professional growth, improving learning outcomes, and shaping future physicians by building trust, providing mentorship, ensuring patient safety, modeling empathy, and influencing career choices, moving beyond mere knowledge transfer to encompass emotional support,

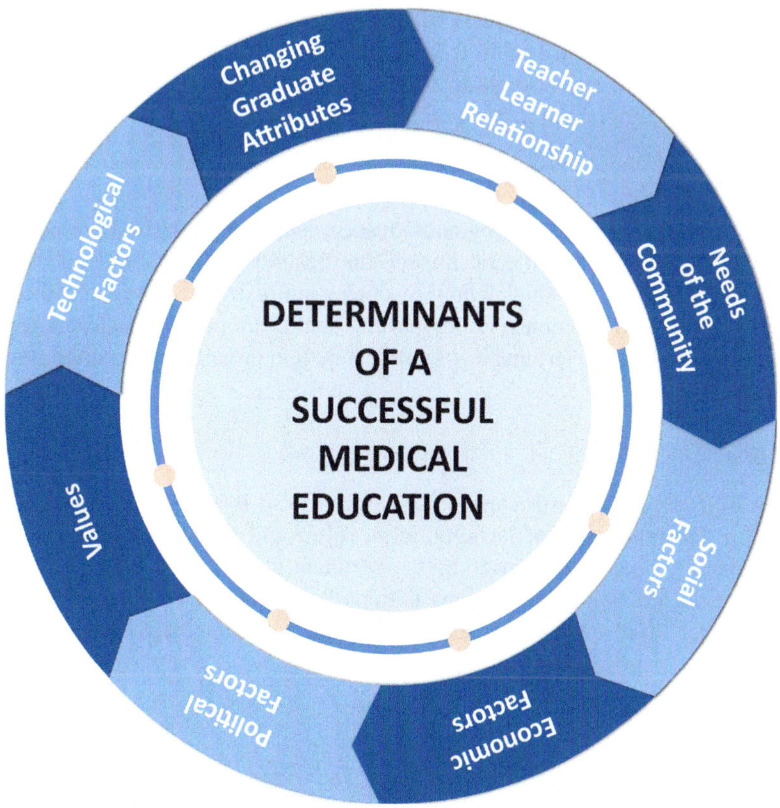


FIGURE 1.6 **Determinants of a successful medical education.** This figure shows the determinants of a successful medical education.

ethical development, and self-directed learning in a high-stress environment. Teachers build relationships with their students when they treat them with respect, trust, and care for their needs. It has been demonstrated that constructive educational connections promote students' growth and development, enhance the learning environment, improve students' academic attitudes, raise student happiness, reduce disciplinary difficulties, and generate helpful feedback.

### 1.7.3 Needs of the community

Healthcare stands out as a nonnegotiable basis among the many developmental pillars and societal necessities, crucial to any community's long-term prosperity and well-being. It is impossible to overestimate the significance of easily available, effective, and complete healthcare because it has a direct impact on people's longevity, productivity, and quality of life. Maintaining and enhancing population health is the major objective of healthcare. This includes taking preventative steps to lessen the likelihood of illnesses as well as treating injuries. The foundation of a strong workforce is a healthy populace, which boosts economic growth and productivity.

Patients also want improved management efficiency, health and outcome equity, social and financial risk protection, and responsiveness. In order to address and reduce health disparities among various population segments, a well-structured curriculum and program outcomes which are based on the needs assessment are essential.

Geographical locations, racial or ethnic variances, and socioeconomic variables are frequently the causes of disparities. All people, regardless of background, have the chance to live longer and healthier when they have access to high-quality healthcare. To ensure that everyone has an equal opportunity at health and well-being, it is important to level the playing field.

### 1.7.4 Social factors

The nonmedical elements that affect health outcomes are known as social determinants of health (SDOH). These circumstances encompass a broad range of forces and systems that influence day-to-day existence, including political systems, social norms, development agendas, economic policies, and social policies.

Reducing disparities in SDOH speeds up the process of achieving health equality, which is a condition where everyone has the chance to reach their optimal level of health. It has been demonstrated that SDOH affects health more than either genetics or access to medical care. In order to attain higher-quality healthcare, it is crucial that elements like social responsibility and accountability be ingrained in the brains of the students.

### 1.7.5 Economic factors

Economic factors, such as the status of the economy and advancements, particularly those pertaining to the founding of medical schools and colleges, have a significant impact on healthcare in society and the nation at large. The multidisciplinary topic of health economics studies how resources are distributed, how healthcare systems are financed, and how economic issues affect health outcomes. Policymakers, healthcare professionals, and academicians must have a thorough understanding of the economic determinants of public health in order to create strategies that effectively improve community well-being.

### 1.7.6 Political factors

It is not only essential to have a good infrastructure or resources, but it is also important to try to make the healthcare faculty feel accountable for the importance of laws, policies, rules, and guidelines. In this way, medical education is greatly influenced and shaped by the decisions made by the pertinent policymakers and regulatory agencies. It is a truth that the frameworks in which doctors work change along with political landscapes. Changes in healthcare policies impact everything from patient treatment to resource allocation, demonstrating the interdependence of political and public health frameworks.

### 1.7.7 Values

Medical education is changing dramatically in the modern era. The introduction of value-based medical education is one of the most significant advances. It is crucial that the learner understands the importance of values from the outset, and that attempts be made to help them internalize them through a variety of activities, such as role-playing and simulations. As the fundamental elements that direct decision-making, foster trust between patients and caregivers, and establish the standard of patient care, values are seen as an essential part of any healthcare facility that actively serves the community. Since it aids in the learner's personal and professional development, the integration of values into medical education has become increasingly important.

### 1.7.8 Technological factor

Among the goals of incorporating technology into medical education include improving decision-making, increasing perceptual variance, improving skill coordination, practicing for significant or unusual events, learning team training, and developing psychomotor abilities. Providing high-quality medical, educational, and health services to all segments of the community depends on all of these aspects. Furthermore, modern medical education has changed and is now accessible to students all over the world owing to the development of online learning and information and communication technology.

## 1.8 Medical education and quality of healthcare

Competent and skilled healthcare professionals are the foundation of high-quality healthcare. The goal of medical education programs is to give aspiring medical professionals the information and abilities they need to identify, manage, and avoid a variety of illnesses. In order to guarantee that doctors are adequately equipped to handle the intricate demands of healthcare, these programs place a strong focus on demanding instruction, critical thinking, and practical experience (Fig. 1.7).

### 1.8.1 Contribution of medical education to healthcare quality

#### *1.8.1.1 Promoting innovation*

Medical education promotes creativity and research in addition to disseminating preexisting information. Academic institutions and medical schools offer a rich environment for cutting-edge research, with teachers and students collaborating to find novel medical technologies, cures, and treatments. The projects that are being taken up by medical educationists and researchers focus on the objective of better healthcare outcomes, and the discoveries that arise from medical research have the potential to completely transform healthcare and enhance patient outcomes.

#### *1.8.1.2 Addressing inequalities in healthcare*

Medical education is also crucial to addressing healthcare disparities. By creating a diverse workforce of healthcare professionals who understand and are sensitive to the unique needs of various communities, medical education helps to close the gaps in healthcare quality and access. Focusing on SDOH and providing culturally competent care are important components of medical education that ensure that healthcare providers can treat every patient equally.

#### *1.8.1.3 Adjusting to new developments in technology*

The medical field is rapidly changing as a result of technological breakthroughs. AI, telemedicine, and medical education tools are some of the innovative ways that healthcare is evolving. Medical education programs must equip the next generation of

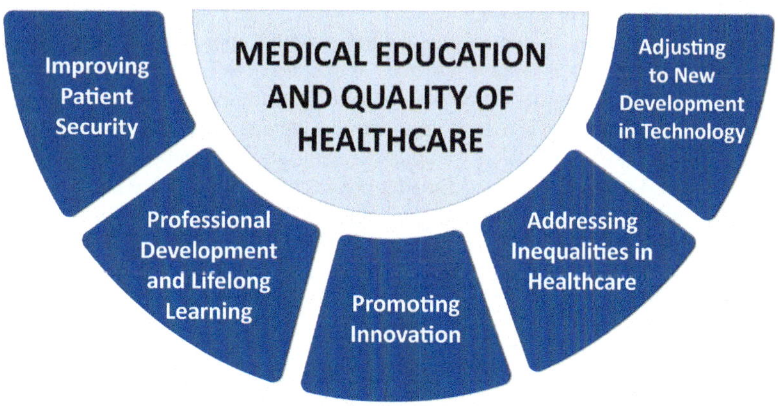


FIGURE 1.7 **Medical education and quality of healthcare.** This figure reflects medical education and quality of healthcare.

healthcare professionals to embrace and benefit from these technological breakthroughs. One can bring about the much-needed transition towards integrating technology into education by encouraging medical students to have a lifelong learning mentality and to be curious. To maintain interest in this rapidly evolving sector of technology, it is also essential that regular updates be introduced at both the faculty and student levels through faculty development programs or hands-on sessions.

#### 1.8.1.4 Professional development and lifelong learning

Professional development and lifelong learning are essential, continuous processes in medical education that assist healthcare professionals in acquiring new skills, knowledge, and attributes required to provide high-quality, evidence-based patient care in a rapidly evolving medical sector. Research, technology advancements, and changing patient care standards are the driving forces behind activities like continued professional development and continuing medical education, which are essential components of this process. The healthcare professional may always succeed in the workplace by including abilities like leadership, teamwork, and interprofessional collaboration. Improved patient outcomes, career promotion, maintaining professional competency, and adjusting to new difficulties are just a few of the significant advantages that can result from this.

#### 1.8.1.5 Improving patient safety

Medical education plays a crucial role in guaranteeing patient safety, which is of utmost importance in the healthcare industry. Healthcare workers receive education on error prevention techniques, patient safety procedures, and the value of good communication in multidisciplinary healthcare teams. Reducing medical errors and enhancing patient outcomes require these abilities and expertise.

To put it briefly, the foundation of contemporary healthcare is medical education. In addition to encouraging innovation, addressing healthcare disparities, embracing technological breakthroughs, fostering lifelong learning, and improving patient safety, it gives healthcare personnel the information, abilities, and mindset they need to deliver competent and compassionate care.

**Summary**

The advances in modern medical education symbolizes one of the greatest achievements of humanity. The lessons from the past, when combined with recent advancements, give promise for a future where medical education will keep evolving, adapting, and improving, ultimately leading to better patient care and health outcomes. The future of medicine is greatly influenced by medical education, which is at the forefront of healthcare innovations. It serves as the cornerstone around which healthcare workers construct their expertise, knowledge, and abilities, which in turn affects patient care and community health in general. It is impossible to overestimate the significance of strong medical education programs as healthcare continues to change. They represent a commitment to enhancing the health and well-being of people and communities around the world, in addition to serving as a route to becoming a healthcare professional.

**Key point**

**Future perspective: modernizing the medical education**

Future medical education will focus on a humanistic, patient-centered approach, enhanced by technology like AI for personalized learning and data analysis. It is imperative that, as medical educators, all our efforts and resources will be channelized in that direction. Leveraging digital technology, expanding study opportunities, and offering career guidance to create an environment where students can excel across various fields in alignment with their interests and abilities should become central objectives of medical education. Curricula will emphasize early and continuous clinical experience, interdisciplinary collaboration, and community-based learning to foster a broader perspective on healthcare needs. Key future competencies will include integrating technological literacy with strong critical thinking, communication, and ethical reasoning to prepare physicians for a dynamic and complex global healthcare system, which should be the focus if we want learners to excel and be successful in their careers.

## References

Ackerknecht, E.H., 1984. From barber-surgeon to modern doctor. Bulletin of the History of Medicine 58 (4), 545–553.

Alrashed, F.A., Ahmad, T., Almurdi, M.M., Alderaa, A.A., Alhammad, S.A., Serajuddin, M., et al., 2024. Incorporating technology adoption in medical education: a qualitative study of medical students' perspectives. Advances in Medical Education and Practice 15, 615–625.

Birden, H., Page, S., 2007. 21st century medical education. Australian Health Review: A Publication of the Australian Hospital Association 31 (3), 341–350. Available at https://doi.org/10.1071/AH070341.

Bleakley, A., Bligh, J., 2008. Students learning from patients: let's get real in medical education. Advances in Health Sciences Education 13 (1), 89–107. Available at https://doi.org/10.1007/s10459-006-9028-0.

Duffy, T.P., 2011. The flexner report - 100 years later. Yale Journal of Biology and Medicine 84 (3), 269–276.

Frank, JR., Snell, LS., Cate, OT., Holmboe, ES., Carraccio, C., Swing, SR., et al., 2010. Competency-based medical education: theory to practice. Medical Teacher 32(8), 638-45. Available at https://doi.org/10.3109/0142159X.2010.501190. PMID: 20662574.

Hajar, R., 2021. Medicine from Galen to the Present: A Short History. Heart Views 22 (4), 307–308. Available at https://doi.org/10.4103/heartviews.heartviews_125_21 Epub 2022 Feb 11. PMID: 35330648; PMCID: PMC8939383.

Kostakopoulos, N.A., Bellos, T.C., Katsimperis, S., Tzelves, L., 2024. Hippocrates of Kos (460-377 BC): The Founder and Pioneer of Clinical Medicine. Cureus 16 (10), e70602. Available at https://doi.org/10.7759/cureus.70602 PMID: 39483540; PMCID: PMC11526839.

Khan, S., 2025. Fundamentals of AI for medical education research and Practice. January 20, 2025. Academic press B978-0-443-33584-6.00007-4.

Ludmerer, K.M., 2011. Abraham flexner and medical education. Perspectives in Biology and Medicine 54 (1), 8–16. Available at https://doi.org/10.1353/pbm.2011.0009.

Mohammed, C.A., Anand, R., Saleena Ummer, V., 2021. Interprofessional Education (IPE): A framework for introducing teamwork and collaboration in health professions curriculum. Med J Armed Forces India 77 (Suppl 1), S16–S21. Available at https://doi.org/10.1016/j.mjafi.2021.01.012 Epub 2021 Feb 2. PMID: 33612927; PMCID: PMC7873741.

Norman, G., 2012. Medical education: past, present and future. Perspectives on Medical Education 1 (1), 6–14. Available at https://doi.org/10.1007/s40037-012-0002-7.

Papageorgiou, C., Konstantinou, G., Lambrinoudakis, V., Papageorgiou, C., Yiannopoulou, K.G., 2022. Asclepion of Epidaurus: the application of a historical perspective in medical education. Philos Ethics Humanit Med 17 (1), 7. Available at https://doi.org/10.1186/s13010-022-00120-6 PMID: 35440062; PMCID: PMC9020009.

Rajashekhar, H.B., Kodkany, B.S., Naik, V.A., Kotur, P.F., Goudar, S.S., 2002. Evidence Based Medicine and its impact on Medical. Indian Journal of Anaesthesia 46.

Scheele, F., 2012. The art of medical education. Facts, Views & Vision in ObGyn 4, 266–269.

Spinelli, H., 2022. Abraham flexner: trayectoria de vida de un educador. Salud Colectiva 18, e4053. Available at https://doi.org/10.18294/sc.2022.4053.

Suvajdzic, L., Djendic, A., Sakac, V., Canak, G., Dankuc, D., 2016. Hippocrates-The father of modern medicine. Military Medical and Pharmaceutical Journal of Serbia 73 (12), 1181–1186. Available at https://doi.org/10.2298/vsp150212131s.

Stevens, F.C.J., Goulbourne, J.D.S., 2012. Globalization and the modernization of medical education. Medical Teacher 34 (10e684). Available at https://doi.org/10.3109/0142159X.2012.687487.

Sukrajh, V., Adefolalu, A.O., 2021. Understanding learning and the components of the learning process in medical education: a review of the literature. European Journal of Education and Pedagogy 2 (1), 69–72. Available at https://doi.org/10.24018/ejedu.2021.2.1.53.

Timotheou, S., Miliou, O., Dimitriadis, Y., Sobrino, S.V., Giannoutsou, N., Cachia, R., et al., 2023. Impacts of digital technologies on education and factors influencing schools' digital capacity and transformation: a literature review. Education and Information Technologies 28 (6), 6695–6726. Available at https://doi.org/10.1007/s10639-022-11431-8.

Yapijakis, C., 2009. Hippocrates of Kos, the father of clinical medicine, and asclepiades of Bithynia, the father of molecular medicine. In Vivo (Athens, Greece) 23 (4), 507–514.

Chapter 2

# Principles of learning and learning cycle

*"Learning is the process whereby knowledge is created through the transformation of experience."*

—David Kolb

## 2.1 Introduction

Medical education plays a vital role in advancing healthcare by producing trained professionals, improving patient outcomes, and adapting to ongoing medical advancements through new research, technology, and therapies. By utilizing their knowledge and talents, members of the medical fraternity can stay up to date, pursue their interests, and significantly impact their communities. This will provide them the tools they need to thrive in a rapidly changing environment. It also ensures adherence to ethical standards, fosters professionalism, enhances clinical reasoning, and provides the workforce needed to meet complex healthcare demands.

In addition to describing how students learn, research-based assertions known as "principles of learning" provide teachers with essential instructions for improving instruction and student achievement. These guidelines evaluate a student's mental and physical readiness to understand certain medical ideas. Additionally, by enabling the teacher to incorporate different elements like feedback, social and collaborative learning environments, and active participation, these principles serve as a guide for improving the learning process and making it relevant and meaningful for the student.

## 2.2 Characteristics of a good learner

A good learner is considered to be one who is curious, persistent and self-aware, able to comprehend the knowledge provided to him or her, and apply it in the best possible way so that it serves to fulfill the needs of the society. Qualities like empathy, communication skills, discipline, adaptability, intellectual curiosity, ethical responsibility, and a strong commitment to patient-centered care are all signs of an excellent medical student. These qualities support both academic achievement and the growth of a kind, capable, and reliable doctor (Khawar, 2022). Some of the key qualities that define a good learner are as follows.

### 2.2.1 Professional and personal attributes

#### *2.2.1.1 Empathy and compassion*

A genuine care for the well-being of patients coupled with an effort to understand and alleviate the pain of others is one of the most vital characteristics to be developed by the learner. For medical students, empathy and compassion are essential because they enhance patient care, result in better health outcomes, and fortify the bond between the patient and the doctor. With the help of these abilities, students can comprehend the viewpoint of a patient, which can result in more precise diagnoses, more treatment compliance, and patient satisfaction. A student's personal professional well-being and career longevity depend on a compassionate approach towards patients.

#### *2.2.1.2 Integrity and ethics*

Integrity, dependability, and a dedication to moral ideals are critical attributes that must be ingrained in the student. Integrity and ethics are crucial for medical students because they are the foundations of patient trust, professional accountability, and informed decision-making. Aspiring doctors can better manage difficult situations, maintain the reputation of the medical community, and provide patient-centered treatment by upholding these ideals from the start of their careers. It is important to make an effort to include these attributes into the current curriculum. Along with being

**The Principles and Practice of Medical Education. DOI:** https://doi.org/10.1016/B978-0-443-44197-4.00008-4

taught as part of a variety of teaching and learning activities, such as case-based discussions, role-playing difficult ethical problems, experiential learning through clinical exposure, and bedside teaching, these attributes should be exhibited by the teacher. These principles will, in fact, greatly influence the learner's development as a better professional.

#### 2.2.1.3 Respect

Being polite and decent to patients, colleagues, and the medical community at large is one of the primary qualities that must be exhibited by medical graduate. For medical students, having respect towards their teacher, colleagues, and patients are critical, as it creates a positive learning atmosphere, increases trust, instills confidence, and molds them into capable and empathetic doctors. Respect fosters collaboration, creativity, and resilience. Respect is ultimately a fundamental ethical and professional obligation that affects future physicians' personal and professional growth. Innovative teaching methods like problem-based learning, team-based learning, and other interactive activities thrive on this quality. The ability of a learner to respect the opinions and suggestions of his peers as well as the teacher will not only foster a positive learning environment but also shape the learners' personality.

#### 2.2.1.4 Resilience

Medicine is regarded as one of the toughest and demanding careers to pick up. Medical students are expected to demonstrate excellent state of both physical and mental health. The ability to manage medical school's demands and challenges without compromising one's health is indeed a herculean task. In order for the medical students to shows skills of organizing their study, time management, manage stress, overcome hardship, and preserve their mental health in the face of the rigorous coursework and clinical setting, resilience is essential. The development of resilient skills should be encouraged as a part of teaching-learning process. Setting up deadlines for assignments or projects and closely monitoring their attendance and grades are some of the ways that can be used to assess resilience amongst the students. It helps students to overcome obstacles, grow from mistakes, and avoid burnout, which benefits their personal health and future capacity to deliver high-quality patient care.

#### 2.2.1.5 Self-awareness

It is true that self-awareness is crucial for medical students. Building trust and confidence with coworkers, professors, and patients requires an understanding of one's own emotions and how they affect social connections. Being self-aware is essential for doctors because it lowers the chance of medical errors and provides more empathetic, objective treatment for patients. By ensuring that all treatment alternatives are fairly presented, it can also result in improved clinical decision-making. Lastly, improving relationships with patients and coworkers might result from being conscious of one's communication style. Professionals who are self-aware are better able to take proactive measures to preserve their own well-being by identifying their own emotional triggers and constraints. Self-awareness helps students understand their own emotional responses, values, and limitations. It also helps doctors to make informed decisions, improve connections with patients and coworkers, and maintain well-being in a stressful profession. To teach self-awareness to medical students, it is best to use a mix of direct and indirect methods, such as facilitated discussions on emotional clinical situations or reflective journaling which gives the learner a chance to gauge his or her reaction and emotions towards a particular challenging situation. It is always better if educators' model self-aware behavior by being open about their own feelings and reflections, which shows students that self-awareness is a professional expectation.

### 2.2.2 Cognitive and academic skills

#### 2.2.2.1 Intellectual curiosity and dedication

The learner's personal and professional development will be greatly influenced by exhibiting the traits of intellectual curiosity and commitment to the subject. The learner can stay up to date with medical breakthroughs by cultivating interest to learn and not relying solely on recollection to completely comprehend things. It strengthens critical thinking, promotes lifelong learning, and enhances problem-solving skills. Dedication towards learning medicine and a commitment towards patient care are essential qualities to be learnt. By being dedicated to the subject, one is able to do justice to what has been learnt. Dedication is also crucial for medical students as it drives perseverance through challenges, ensures high-quality patient care by fostering trust and continuity, and contributes to success in both their studies and future careers. It leads to better clinical performance, professional advancement, and the development of a strong sense of purpose beyond academics. When combined, these attributes result in enhanced memory and recall, better patient care, and a stronger capacity to adjust to the quickly evolving field of medicine.

#### 2.2.2.2 *Critical thinking*

Critical thinking is the ability to analyze information, synthesize knowledge, and draw informed conclusions. For medical students, critical thinking is essential because it results in more precise diagnoses, wiser treatment choices, and better patient outcomes. The fundamental skill in this regard include analysis, evaluation, and problem-solving, which are applied through processes like questioning assumptions, identifying logical connections between ideas, and discerning fact from misinformation. By using this vital skill, doctors prevent medical errors, solve difficulty issues, assess complex circumstances, and adjust to changing medical procedures. It is a fundamental component of contemporary medical education, pushing students to develop a deeper comprehension of clinical reasoning in place of rote memorization. The inclusion of the critical skills into the medical curriculum is highly regarded. It is imperative that as educators we must work towards inculcating this into the learner. Teaching-learning activities such as problem-based learning and case-based learning can be used to teach these skills to the students.

#### 2.2.2.3 *Observation*

The importance of observation skill, especially in medicine, cannot be overstated as it forms the very basis of learning. Students can grow both emotionally and professionally by becoming good observers. Students watch their teachers constantly and attempt to imitate them as best they can, not just when they are demonstrating a process or a skill. To comprehend people's emotions and behaviors and respond appropriately, the ability to observe is also essential. In this sense, it is critical for students to observe their coworkers and educators and behave in a way that doesn't insult them. Observation skills are vital for medical students because they are necessary for accurate diagnosis, effective treatment monitoring, and the development of good clinical reasoning and patient care abilities. By thoroughly examining patient signs, symptoms, and interactions, students can develop procedural skills, enhance their empathy and communication abilities, and better comprehend clinical material. These skills improve both technical competence and the humanistic aspects of medicine.

#### 2.2.2.4 *Evidence-based and technologically competent*

One of the most notable aspects of contemporary medical education is the learner's capacity to be directed toward evidence-based medical education and their commitment to embrace medical education based on facts and figures. In addition to acquiring the necessary knowledge and abilities related to their field of expertise, students are also keen to investigate the information's original source using a variety of resources and feel satisfied that they are engaging in authentic learning. These abilities help both the teacher and the student because they make the material more genuine, which fosters mutual trust.

#### 2.2.2.5 *Committed to lifelong learning*

A medical student's ability to value lifelong learning is one of the main things that sets them apart. Students who are dedicated to lifelong learning are better equipped to comprehend the complexities and subtleties of medical education. Additionally, a dedication to lifelong learning will allow the student to stay up to date on the most recent advancements, treatments, technologies, and knowledge in the field of medical education, enabling them to provide the best patient care and uphold the standard of healthcare.

### 2.2.3 Interpersonal and communication skills

#### 2.2.3.1 *Effective communication*

The ability to clearly explain information, listen closely, and understand other people's perspectives is the essence of communication. For medical students, effective communication is essential because it gives them a chance to voice out their opinions, likings, and dislikings about the various aspects of medical education process. Being able to communicate clearly to the patient fosters trust, instills confidence, and improves health outcomes by increasing patient knowledge and compliance. Having good communication skills is also one of the important pillars of collaborative practice and interprofessional education. In this regard, it is known to improve collaboration between medical staff, aid in precise diagnosis and treatment planning, and facilitate the handling of delicate situations like bad news or involving patients and their families during the discussion of critical illness. Building excellent doctor–patient connections during a medical profession requires early development of strong communication skills.

#### 2.2.3.2 *Teamwork*

Being a good team player in the medical field is of immense importance. It has to be kept in mind that gone are the days where medical education used to be taught in silo. Medical educators across the globe now concentrate on how to instill a sense of teamwork and collaborative learning among the learners. As it promotes improved communication, lowers medical errors, and increases patient safety through shared accountability and coordination, teamwork is essential for medical students. The skill of being a team player is dependent on the personal characteristics of the individual. For instance, students with good communication skills and adaptability are considered better in a team than those students who lack these skills. As medical educationists, it is our responsibility to observe these skills and encourage them to develop into great team players. By teaching students how to be a team player, we are making them ready for the collaborative nature of contemporary healthcare.

#### 2.2.3.3 *Interpersonal skills*

In modern medical education, interpersonal skills have gained a lot of importance as they are directly linked to improved patient outcomes and a more effective healthcare system. In addition to medical knowledge, strong interpersonal skills are now regarded as an essential component of a physician's expertise. Medical education has no doubt moved from in silo learning to a collaborative and interprofessional approach. Owing to this shift, it has become imperative that the students develop these skills and apply them in daily learning process. Components of interprofessional skills like active listening and teamwork not only are helpful to a medical student to effectively carry-on daily activities but at the same time help the learner to build a strong bond between their own peers and teachers. This will not only enhance learning experience but also bring out the best out of the learner. Moreover, skills like conflict resolution play a critical role in team-based activities and also in clinical setting as they help solve major issues that may arise both at personal and professional level. A quick and effective conflict resolution will also contribute to the decision-making process and overall quality of healthcare.

### 2.2.4 Behavioral and practical skills

#### 2.2.4.1 *Adaptability*

The capacity to adjust to changing conditions and environments is known as adaptability and is a vital medical skill that the learner needs to develop. It is a well-known fact that apart from imparting knowledge, medical colleges also serve as a melting point of different backgrounds and cultures as students from different regions converge at a single place. A medical student needs to adapt to all this and still be able to perform to the best of their abilities. Moreover, adapting to the learning environment, clinical setting, learning styles, technological advancements, and innovative curriculum as early as possible is very advantageous for the learner. In the clinical setting, the ability to adapt early will help the doctor to deliver high-quality treatment, handle unforeseen patient scenarios at the earliest, and embrace lifelong learning. As teachers, we have to encourage and support the learner so that he or she can undergo this process smoothly. Teaching and learning activities such as role-playing and simulated patient interactions, combined with structured small-group sessions, can be effectively used to demonstrate the essence of interpersonal skills to the students.

#### 2.2.4.2 *Professionalism*

In the current scenario of medical education, professionalism is now seen as an essential competency that has to be ingrained in a medical student. One of the best ways to learn professionalism is by observing the teachers who serve as role models. By exhibiting professionalism, students are in a position to understand and respect their colleagues as well as conduct themselves ethically, demonstrating integrity, competence, and commitment to core medical values. All these will culminate into the learner developing into an ethically competent professional. It will also enable the learner to serve as a role model for the next generation. Fundamental values like respect, integrity, and ethical behavior are also some of the important components of professionalism. Students can learn about medical professionalism through a variety of methods, including bedside instruction, role-modeling, community activities, and scenario-based discussions.

#### 2.2.4.3 *Time management*

One of the hardest things for a medical student to master is time management. It's critical for medical students to maintain organization and time management skills because they have a lot of tasks to complete during their profession while juggling clinical and academic obligations. Effective time management lowers stress, boosts productivity, and prevents burnout since it helps students to stay focused, organized, and prepared to achieve their objectives. To develop time management skills, medical students must learn how to prioritize and organize their study time, break down large tasks, minimize distractions, create regular routines, and create and use a comprehensive schedule.

## 2.3 Learning environment

The physical, social, and psychological contexts in which students learn from classrooms to clinical settings are all included in the learning environment in medical education. Important elements include safe, welcoming, abuse-free environments, realistic patient experiences, and supportive peer and instructor connections (Bassaw, 2003). A good learning environment not only enhances academic accomplishment by increasing focus, motivation, and engagement, but it also promotes social and emotional well-being by cultivating resilience, confidence, and a sense of belonging. While a bad learning environment can impede learning and lead to fatigue, a positive one fosters student well-being and engagement.

### 2.3.1 Essential elements of a productive learning environment

#### *2.3.1.1 Psychological safety*

It is important to understand that medical students undergo a lot of stress in the initial days of their study. More often than not the stress is psychological in nature. This stress can arise due to many factors like new learning environment, new place, or simply homesickness. In these situations, it is important to understand that the student needs to be made comfortable so that they can achieve their goals. One of the important aspects of this is to have a soothing and welcoming learning environment. A good learning environment should be such that the student should feel free to express their concerns, pose inquiries, and offer criticism without fear of abuse or reprisal. In this aspect, the task of orientating the newly enrolled students to the medical college carries great importance as it gives a chance for the student to understand the learning environment (Othman, Mohammad & Rukban, 2010).

#### *2.3.1.2 Supportive relationships*

Supportive relationships are crucial in the medical field since a student's personality is greatly shaped by the interactions they have with their teachers and colleagues. Strong emotional support is built on these relationships. Supportive relationships are essential in medical education because they create a courteous, secure environment that fosters learning, well-being, and professional development. These relationships, which are built on empathy and trust among peers, professors, and students, improve academic performance, foster a sense of community, and help students build vital humanistic and therapeutic skills for their future careers.

#### *2.3.1.3 Open communication*

Open communication in medical education is crucial for fostering empathy, patient-centered care, and cooperation because it builds trust, empowers patients to make decisions, facilitates clear information flow among healthcare providers, and reduces errors in hierarchical settings. These factors improve patient outcomes, safety, and satisfaction. In the end, it raises the standard of care by fostering a culture where junior staff members feel comfortable speaking up, improving treatment adherence, and developing critical clinical skills. It is advantageous to the student because it fosters the development of key competencies, such as active listening, empathy, and cultural sensitivity. All of these are necessary for forming therapeutic partnerships. Additionally, it promotes teamwork by minimizing mistakes, enhancing coordination, and promoting open communication, particularly in high-stress situations (Steinmair, Dagmar & Zervos, 2022).

#### *2.3.1.4 Proper physical space*

A learning environment must have comfortable, well-furnished classrooms, well-spaced labs, and functional instruments. Each of these has a direct impact on students' capacity for focus, teamwork, and comfort. The best qualities of the learner are also highlighted when there are enough room and structure to facilitate activities. By creating a safe, comfortable, and organized environment, enough physical space encourages academic success. Additionally, it helps students by improving their emotional health, attentiveness, and involvement. Proper physical space also refers to the design and layout of learning environments to maximize educational effectiveness, promote collaboration, and ensure psychological safety. This includes flexible, well-lit, and acoustically optimized spaces that align with modern pedagogical methods.

#### 2.3.1.5 *Effective mentoring and teaching*

Knowledgeable, passionate, and up-to-date teachers who provide clear guidance and chances for professional development are also an essential part of the learning environment. Since they facilitate both formal and informal learning, direct professional and personal development, and enhance patient care results, effective mentorship and teaching are crucial in medical education. In order to help trainees navigate their careers, improve their abilities, and comprehend the "implicit curriculum" of medicine, which include leadership and communication, mentors serve as coaches and role models. In order to create a productive learning environment that benefits the mentee, mentor, and institution, strong mentoring relationships are based on two-way communication, trust, and the mentor's capacity to adjust to the mentee's requirements.

#### 2.3.1.6 *Equity and respect*

When all students, regardless of background, are treated with respect and are not the target of harassment, humiliation, or discrimination, equity and respect are established in the learning environment. Additionally, respect and equity create a learning atmosphere where all students feel appreciated, have equal opportunities, and can acquire the skills needed to deliver patient-centered care. This entails ensuring that everyone has fair access to resources and information, fostering a respectful environment between students and instructors, and using teaching and evaluation strategies that reduce bias. Improved student performance, more creativity, and a more varied and competent future healthcare staff are all associated with fostering an inclusive atmosphere.

## 2.4 Principles of adult learning

For a student, one of the greatest transitions of learning happens when they get exposed to medical education. In this situation, the medical students have to adapt to various components like learning environment, learning styles, and teaching strategies. On the other hand, teaching faculty should understand and appreciate these difficulties faced by the learner and offer help and support. They need to understand that as adults the learning styles and preferences have changed. Therefore, the teaching and learning process has to be modified in such a way that it matches with the needs and expectations of the adult learner. In this regard, the principles of adult learning come in handy as they provide an opportunity to design and facilitate learning that considers the nuances and complexity of the learning environment as well as the development of skills that help adult learners succeed in the more uncertain and changing environment (Hutchinson, 2003). Each of the six guiding concepts listed below are essential and does, in fact, constitute the core of education. Furthermore, to a greater or lesser extent, each of these should be included in the design of a module, course, or program since they are all related to and necessary.

### 2.4.1 Authenticity

It refers to the fact that the content taught to the learner should be authentic and should provide all the needed knowledge and skill which will prepare the individual for real-world work settings. Moreover, if the content taught is a reflection of real-world scenarios, there is a huge possibility that it will boost the learning and confidence of the student. To introduce authenticity into medical education activities, using real-world problems, patient-centered case studies, and real-life applications of a certain instrument or tool is highly beneficial.

### 2.4.2 Alignment

It is important to keep in mind that the curricular content is always built in alignment with the needs and expectations of the learner. Hence it is imperative that the learning objectives, outcomes, learning activities, assessment design, and the learning environment are aligned and work in concert to guarantee educational success. For example, in any system-based course, the course learning objectives, topic-specific learning objectives, teaching tactics, and assessment techniques must all be in line with each other. A true success can only be achieved when students feel that their needs are being satisfied during the teaching and learning process in terms of achieving the objectives of the subject being taught.

### 2.4.3 Holistic

In order to promote self-transformation and progress, the holistic principle of adult learning places an emphasis on helping learners understand new ideas in a meaningful context by involving them intellectually, emotionally, socially, and physically. It treats knowledge as a dynamic interplay between explicit, implicit, and emancipatory aspects, acknowledging the connection of an individual with their experiences, other people, and their surroundings. Thus, learning is seen as a process that integrates various facets of knowledge and experience into a cohesive whole, nourishing the individual and their potential. The role of a teacher here is to encourage the learner to think out of the box and come up with novel ideas so that they can contribute to the betterment of medical education and in turn to the quality of healthcare.

### 2.4.4 Feedback

Feedback in adult learning is one of the crucial aspects as it not only reflects the true understanding of the topic learnt but at the same time provides room for improvement. As teachers it is vital to understand that regular constructive feedback is essential to facilitate adult learning. These feedback process can be formal or informal sessions. For the feedback to be effective, it is essential to recognize and value the learner's abilities. In order to counteract any detrimental elements in the student's learning process, care must also be taken to draw attention to the weaknesses and offer rational remedies. Adult learners value solutions that are supported by reason and which come from their own teachers. Students must actively participate in providing and receiving feedback from professors and peers in order to evaluate their own performance. Teachers who give students regular, thorough, and helpful feedback can give them the self-assurance and encouragement they need to take control of their education. By providing constructive feedback, teachers can help students develop a growth mindset and become more resilient learners. The practice of conducting formative assessment sessions on a regular basis gives a chance to the learner to obtain the feedback and reflect on his or her performance. Feedback given during the mentoring sessions gives the chance to discuss and reflect the hurdles he or she might face during the course of study.

### 2.4.5 Judgment

It is an essential part of the learning and assessment processes since the development and use of judgment are vital in assisting students to understand their own work and how they are performing in relation to what is needed or expected. Students should be taught how they can apply the skill of reasoning and critical thinking to their own work. The process of judgment not only instills a sense of satisfaction as they can gauge their work and work upon their strengths and improve their weakness, but it also gives them an opportunity to actively take part in the learning process. Feedback and judgment are intertwined processes, as both require students to take an active role in their education.

### 2.4.6 Future-orientedness

It refers to the ability of learner to deal with unfamiliar situations and new challenges that arise outside of the immediate course or training. It emphasizes the ability to address new or unusual problems. Many of the so-called 21st-century skills, like creativity, critical thinking, and digital literacy, are necessary for being future-oriented. To solve problems effectively in the unknown, one must have a solid understanding of a discipline or process. The students should be made aware of what challenges they might face in their career and what skills they might require to solve those issues. Being future-oriented also refers to how best the learner can apply the skills that are being taught to them in medical school. Developing inquiry skills, knowing what questions to ask, and knowing where and how to gather facts to assist overcome challenges are all components of being future-oriented (Fig. 2.1).

## 2.5 Active learning

An approach to teaching that actively involves pupils in the learning process is known as "active learning." This method, as opposed to passive methods like didactic lectures, focuses more responsibility on the student in medical education. All active learning exercises still require the professional medical educator's direction, though. Active learning has been expanded in the medical curriculum to encompass any activity involving the collection of information and the solution of problems, whether done alone or in a group. Active learning techniques that emphasize interactive, student-centered strategies to improve knowledge retention, clinical reasoning, and professional development are gradually replacing the conventional lecture-based model of medical education. Active learning techniques such as problem-based learning, team-based learning, case-based learning, and simulation-based learning are gaining a lot of importance due to their ability to enhance clinical reasoning, foster the development of interpersonal skills necessary for healthcare practice, and advance a deeper comprehension and application of medical knowledge.

From the perspective of the learner, active learning is also proving to be a real game changer as it gives a chance for the learner to be actively involved in the learning process. Students should be encouraged to contribute to active learning by engaging in discussions, solving problems, and explaining concepts to peers. Some of the strategies that can be applied

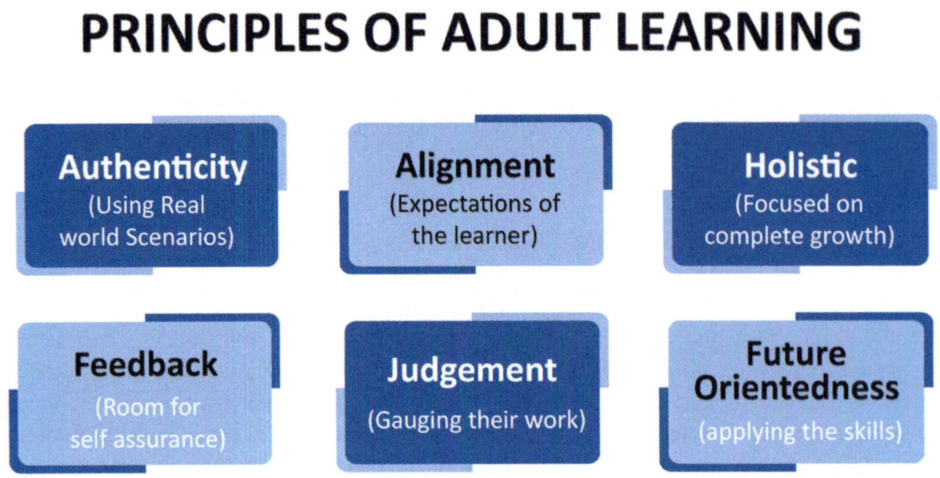


FIGURE 2.1 **Principles of adult learning.** This figure shows the principles of adult learning.

include participating in group work and problem-solving, reflecting on and summarizing material in their own words, and teaching concepts to others. Additionally, active learning helps students become more prepared for contemporary medical practice by bridging the gap between theoretical knowledge and real-world application. Active learning offers a promising future for medical education, despite certain obstacles, such as the strain on resources and the requirement for faculty training. It has been shown to be highly advantageous to use active learning techniques like retrieval practice and spaced studying, use visual aids and mnemonics for remembering, engage in peer-to-peer discussions, and use collaborative learning. In order to make learning in medical education effective, strategies such as simulation-based learning offer a variety of chances for practical clinical experience with proper patient consent and an emphasis on patient-centered care.

## 2.6 Learning styles

Learning styles refer to individual preferences for how students best absorb and process complex medical information. When instruction is tailored to each student's preferred method of learning, there is a greater chance that they will understand the material and retain it for longer time. This leads to increased confidence, better performance, more motivation, and enhanced staff involvement. It is important to understand these styles and make an effort to inculcate these in teaching and learning process. It is always better to mix the different learning styles in order to cater to different learners (Pashler, 2009) (Fig. 2.2).

### 2.6.1 Read/write

One of the most popular approaches used in medical education is the read-or-write strategy. This approach is perfect for students who are more likely to read or write the subject to be learned. This category includes using educational materials such as textbooks, articles, written instructions, and notes to accomplish a thorough understanding of the topic. These students' written communication and comprehension skills are exceptional. They can efficiently arrange information through their writing and are excellent note-takers. This type of learning is highly valued in medical education since it enhances students' comprehension and writing skills, which are crucial for exam responses. These students can improve their recollection by using acronyms, mnemonic devices, practice photos, taking notes in class, and revisiting references.

### 2.6.2 Visual

Visual learning in medical education leverages sight for understanding through diagrams, videos, and imagery, significantly boosting retention and clinical reasoning by making complex concepts like anatomy or pathology clearer than text alone, helping students recognize patterns, and improving performance in skill acquisition, with tools like radiology images enhancing critical thinking and establishing confidence in patient care. Medical students benefit greatly from this method, which enhances their capacity to understand complex imaging, identify patterns in illnesses, and hone

FIGURE 2.2 **Learning styles.** This figure shows the various learning styles.

clinical reasoning abilities. This kind of learning proves to be advantages where there is a need to remember diagrams related to anatomy, pathology, microbiology and also while remembering X-rays, ECG recordings, or CT images.

### 2.6.3 Auditory

The auditory learning style in medical education entails learning material by hearing, favoring talks, lectures, and spoken instructions over written or visual materials, and frequently processing ideas through verbalization and repetition. These students, who frequently do well at remembering spoken information and obeying spoken instructions, benefit from techniques like group discussions, audio recordings, reading aloud, and mnemonics using rhymes. They should practice verbal repetition, record and listen to lectures, discuss concepts with a study partner, and employ strategies like reading aloud or making rhythmic mnemonic devices in order to maximize their learning. Listening to lectures and podcasts about difficult subjects like genetics or the pathophysiology of diseases that are recorded and replayed is a typical example of this type of learning. These learners may have trouble with dense material, silent reading, and highly visual presentations, among other things.

### 2.6.4 Kinesthetic

Kinesthetic learners have an ability to learn best when information is delivered in a tactile style. The student learns best when the task includes practical sessions or skill lab activities because they are hands-on learners. Instead of just reading or passively listening, these learners would rather do the task themselves. These students are better able to remember and apply the material when this method is used. We can improve students' learning abilities by asking them to create large-scale, life-size models, like a large heart model where they can simulate blood flow, or by using simulations or manikins to practice procedures, such as drawing blood or performing surgery, which provides essential physical repetition.

### 2.6.5 Logical

In contrast to hands-on kinesthetic or intuitive learning styles, logical learning styles in medical education concentrate on comprehending systems, patterns, and reasoning. Students that use this kind of instruction emphasize theory, facts, and statistics. They are often seen addressing analytical problems, such as detecting illnesses or understanding complex physiological processes. Although many medical students use multimodal approaches, medical students with this predilection succeed with structured information, logical sequences, and evidence-based approaches. They also benefit from case studies, data analysis, and understanding the "why" behind clinical decisions. For instance, students are given a patient's history, signs, symptoms, and lab results during small group exercises like case-based learning and problem-based learning. In these situations, deductive thinking and methodical problem-solving are necessary for them to collaborate in order to examine the data, determine the differential diagnosis, and plan for investigations and management.

### 2.6.6 Social

Social learners, also known as interpersonal learners, enjoy working in groups. These students typically thrive in study groups where they can benefit from one other's expertise. This enhances their understanding of the material and aids in their retention. Social learning theories, such as Bandura's theories, which suggest that people acquire new behaviors by observing and imitating others, lend credence to this strategy. Activities that support medical students' development of social learning skills include case-based learning, simulation exercises, peer teaching, gamification, and community outreach programs. Future doctors can better prepare for the emotional demands of teamwork and healthcare by incorporating social learning.

### 2.6.7 Solitary

In medical education, the solitary learning style, also known as intrapersonal learning, describes a predilection for solitary study, introspection, and calm settings for processing complicated knowledge. Given the enormous quantity of material that medical students must absorb, which frequently necessitates substantial self-directed learning, this approach is especially pertinent. This kind of learner could feel overburdened or distracted rather than flourishing in the group. The finest outcomes are anticipated when this kind of learner operates independently. Setting up a peaceful, private area at home or in another convenient location is one strategy for studying alone. Students gain from solitary learning because it lets them study at their own speed, fosters focused attention by lowering distractions, and helps them become independent and self-reliant. Using online resources, video lectures, and e-learning platforms like MOOCs are teaching practices that promote autonomous learning among medical students (Buşan,2009).

## 2.7 Learning theories

Learning theory seeks to explain how individuals acquire, process, retain, and recall knowledge during the process of learning. Environmental, cognitive, and emotional influences, along with prior experiences, play a vital role in comprehending, acquiring, and retaining skills or knowledge. On the other hand, motivation plays an important role in enabling the process of learning and is said to be the driving force where activity is started and sustained to achieve a target. The knowledge of how the students learn and the methods they follow is highly beneficial to medical educationists as it plays a vital role in developing successful curricula and providing effective education. In addition to helping students absorb the information being taught, learning theories can help teachers manage their behavior (Badyal, Dinesh & Singh, Tejinder, 2017). The field of education has seen the emergence of numerous learning theories that have impacted its evolution. However, rather than being purely theoretical or academic ideas, these theories are practical teaching methods that teachers may utilize to enhance learning for students of all ages.

Five of the most prominent learning theories: behaviorism, cognitivism, connectivism, constructivism, humanism, and adult learning theory (andragogy; Fig. 2.3).

### 2.7.1 Behaviorism

Behaviorism theory which was developed by John B. Watson and is commonly associated with Ivan Pavlov and B.F. Skinner is characterized by students learning mainly by observing the actions of others, whether those individuals be peers of their own age or adults in positions of authority. According to the theory, learning is a shift in behavior in the intended direction. Reward and encouragement for proper behavior, repetition, feedback, and reinforcement for necessary behavioral corrections to ensure they are made and maintained are some of the strategies used to bring about this transformation. It is based on the idea that behaviors are acquired through conditioning, which is a process of reinforcement and punishment. For instance, the instructor will use either positive or negative reinforcement, which entails adding something positive or taking something negative to encourage the desired behavior or discourage an undesirable behavior as the case may be. Rewarding or praising a behavior, such as volunteering during group discussions, is an

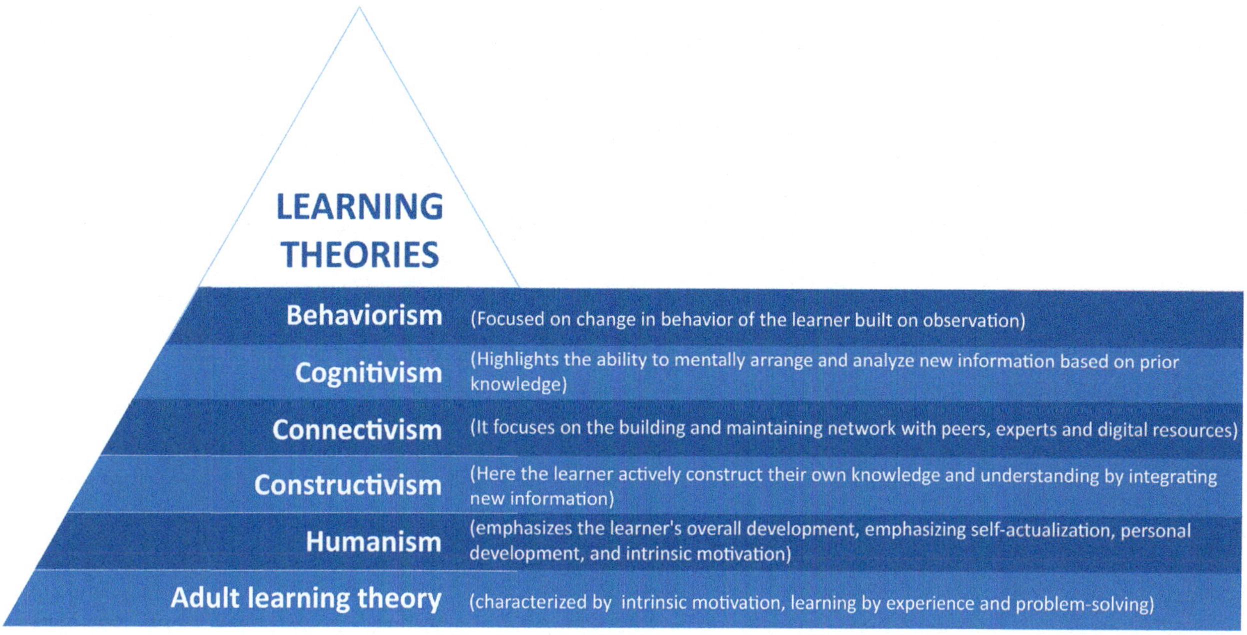


FIGURE 2.3 **Learning theories.** This figure shows the various learning theories.

example of positive reinforcement. This theory's main advantages are that it offers structure, precise expectations, quantifiable results, efficient classroom management, and powerful incentive through reinforcement.

### 2.7.2 Cognitivism

Cognitivism theory was introduced by Jean Piaget, a Swiss psychologist who pioneered cognitive development theory with an emphasis on the concept of inquiry and exploration. The internal environment and cognitive structures of the learners are the main focus of learning in the cognitivist approach. According to this theory, the learner assigns meaning to experiences using cognitive skills like insight, information processing, perceptions, and memory. Unlike behaviorism, cognitivism maintains that learning occurs mainly when students attempt to organize and evaluate new knowledge in their minds. As cognition is one of the important aspects of learning in medicine, being able to analyze and process information will always be advantageous to the students. The teacher's role in this framework is to facilitate cognitive processing by helping the learner "learn how to learn." To encourage cognitive learning in medical students, active learning strategies like case studies, debates, and quizzes to promote deeper processing and link new information to prior knowledge are used (Dong, Lio, Sherer, Jiang, 2021).

### 2.7.3 Connectivism

Connectivism has been described as a 21st-century theory of learning. Medical education today is heavily influenced by technology. In this digital age where people have come to rely on technology to seek answers and find information, it is clear that technology is changing how students learn in and out of the classroom. Students can now ask questions on their smartphones or type them into Google instead of using traditional learning tools like references as they did in the past. Today's students use laptops and smartphones as information hubs instead of learning from teachers and textbooks. Connectivism also places a high importance on the capacity to find and organize information in order to conduct reliable research. The increasing use of technology as an educational tool has changed the learning landscape. Technology has not only made learning easy but at the same time it has given a chance to the learner to actively participate in the process of learning. Another important aspect of this is that technology gives the learner to connect with others on a regular basis and much more effectively. These are some of the reasons that makes this theory more relevant and acceptable with the present generation of learners. To incorporate connectivism into medical education, emphasizing the creation and upkeep of networks with colleagues, professionals, and digital resources through the use of social media, collaborative platforms, and individual learning settings is considered to the important. This entails utilizing interactive platforms like simulations to support deep, networked learning, encouraging students to critically assess and connect information from a variety of online sources, and promoting teamwork. Knowing this theory as a teacher can be beneficial as they can use more resources and methods to establish a learning environment that helps the students succeed.

### 2.7.4 Constructivism

In medicine, having prior knowledge has always been beneficial to understand, apply, and to gain further knowledge and expertise about a particular topic. This is one of the important aspects of constructivism which emphasizes the fact that students build upon, or construct, prior knowledge in order to acquire new information. Constructivism asserts that learners actively construct their own knowledge and understanding by integrating new information with their existing experiences and beliefs. This is an active, social, and contextual process where prior knowledge forms the foundation for new learning, and the learner is a participant rather than a passive recipient of information. Some of the important aspects of this theory are that the knowledge is constructed rather than innate or passively absorbed, thereby making learning as an active process. Teachers and educators can reap the benefit of this approach effectively by helping students better understand concepts they have already learned, relating them to their everyday experiences to help new lessons resonate.

### 2.7.5 Humanist learning theory

Founded by pioneers like Carl Rogers, James F.T. Bugental, and Abraham Maslow, humanist learning theory is a learner-centered approach to education. Humanism is a theory of learning that emphasizes the learner's overall development, emphasizing self-actualization, personal development, and intrinsic motivation. The focus here is on the individual learner because it is believed that learning is a personal act viewed through the lens of the individual's emotions, needs, values, and experiences. It is a student-centered method that emphasizes not only cognitive growth but also student choice, creativity, and emotional health. Humanist learning theory lays a strong emphasis on attending to each student's specific emotional and academic requirements in order to empower them to assume more responsibility for their own education. This method is predicated on the notion that people are kind and will act appropriately if their basic needs are met. By incorporating empathy and respect into the curriculum through techniques like role-modeling, patient-centered case studies, and reflection exercises, the humanistic approach can be used in the teaching and learning process. This entails establishing a nurturing learning atmosphere, fostering empathy through patient narratives, and helping students understand the human aspects of care in clinical settings and in their relationships with coworkers.

### 2.7.6 Adult learning theory (andragogy)

Malcolm Knowles created the adult learning theory, often known as andragogy, which holds that adults learn differently from children and stresses their self-directedness, dependence on life experience, and problem-based learning style. According to this theory, adult learners might be intrinsically motivated, need to understand the "why" behind learning, and learn best through experience and problem-solving, all of which contribute to more successful, engaging, and pertinent educational programs. The principles of andragogy, teaching, and learning in higher education are analyzed and synthesized with consideration for the particular requirements of adult learners. The unique characteristics of adult learning were emphasized by Knowles' theory, which was first put forth in 1980 and then improved in 1984 with the addition of the fifth principle (Knapke, 2024). The five underlying presumptions of this thesis on the traits of adult learners are as follows.

#### *2.7.6.1 Self-concept*

A person's views, attitudes, and perceptions about themselves are collectively referred to as their self-concept. In essence, it identifies the person and provides an answer to one of the most crucial concerns regarding his identity. Numerous individuals and the environment have an impact on this aspect of self-concept. It is seen that a person's self-concept shifts from one of a dependent personality to one of a self-directed human being as they get older and mature. In medical education, self-concept is crucial because it facilitates the shift from dependency to independence as individuals grow from a dependent to a self-directed position in which they assume responsibility for their education.

#### *2.7.6.2 Adult learner experience*

An adult's experience base grows as they become older, making them a more valuable resource for learning. This will be advantageous since self-direction, problem-centered learning, and a foundation of life experience are characteristics of the adult learner experience. Adults are inherently driven and look for education that directly relates to their personal or professional lives so they may incorporate new information with what they already know. The adult experience is one of ownership, application, and personal development, in contrast to children's subject-centered and externally motivated learning.

#### *2.7.6.3 Readiness to learn*

According to adult learning theory, adults are motivated to learn when they sense a need or see a clear, immediate application for the knowledge or skills in their lives, such as for their profession, personal development, or to address a practical problem. It is connected to new social duties and obligations, and learning works best when it is timely, relevant, and useful. The fact that people are willing to learn when they feel the need to do so in order to function better in a particular aspect of their lives serves as an example of their readiness to learn. The topic's relevance to them has a significant impact on their readiness to learn. This implies that they must recognize the importance of what they are learning in light of their future professional and personal objectives. For instance, a third-year resident would benefit more from a lecture on a certain surgical procedure than a first-year resident. As adults enjoy greater control over their learning process and might be motivated to learn when they have a choice in what, when, and how they learn, the readiness to learn also represents the feature of self-directed learning.

#### *2.7.6.4 Orientation to learning*

The development of self-directed learners who are driven by internal objectives and the urge to solve issues is the main focus of adult education. This approach to learning highlights how individuals use their past experiences as a resource, need to understand why they are learning, want to be actively involved in their education, and want to apply new information to pressing, real-world issues. Instead of being the only "transmitter of knowledge," medical educators should take on the role of facilitators, helping students apply their knowledge to clinical settings. Adults learn best when the material is instantly pertinent to their life and when they can use the new information to solve issues. This is why adults have a problem-based and immediate relevance orientation to learning. In this sense, "why" a subject is significant and how it directly relates to patient care should be the main focus of medical education. To assist students relate to patient experiences right away, the Competency-Based Medical Education paradigm incorporates early clinical exposure. Adults, on the other hand, like to learn for immediate, practical use, in contrast to youngsters who frequently learn for future application. They must recognize the worth and applicability of what they are currently learning since they are goal-oriented.

#### *2.7.6.5 Motivation to learn*

The theory's capacity to improve learning for doctors by matching their innate traits such as being self-directed, experienced, and problem-oriented is what drives medical educators to study adult learning theory. The need to see immediate, practical application, enhance clinical practice through problem-based learning, and cultivate a sense of autonomy and respect for their prior expertise are important motivators. By putting these ideas into practice, patient care is eventually improved through increased engagement, knowledge retention, and skill application. One of the main things that propels adult learning is internal motivation. As a result, internal motivators like career advancement, self-improvement, and a desire for personal interest are more potent than external rewards like grades when it comes to motivating medical professionals. These innate motivations are accessed through the use of adult learning theory. Furthermore, people are motivated when they have a choice in what, why, and how they learn, and they like to be in charge of their education. The approach places a strong emphasis on self-directed learning, which is consistent with adult learners' independence and need for control. They exhibit the characteristics of independence and self-directed learning in this way.

## 2.8 Learning cycle

A learning cycle is an organized, multiphase process for picking up new information and abilities (Sukrajh, Verona & Adefolalu, Adegoke, 2021).There are several types of learning cycles, including the 5E Instructional Model and David Kolb's Experiential Learning Cycle. A development through various phases to allow deep comprehension, a focus on active student interaction and reflection for effective, meaningful learning, and a cyclical nature where learning influences future experiences are some of the key characteristics of learning cycles.

### 2.8.1 Purpose of learning cycles

#### 2.8.1.1 Deeper understanding

As they move through distinct phases, students acquire a greater understanding of the contents that they are supposed to learn and apply. The learning cycle's sequential stages of inquiry, concept development, and application help learners to go beyond simple rote learning to a zone where they can inculcate the habit of deep understanding of the concepts and learning experiences. In an attempt to cultivate the habit of deep understanding, learners make connections between new and existing information by actively investigating a concept, gaining a personal understanding of it, and then using it in novel contexts. A deeper understanding helps the learner in effective application of medical knowledge, improve clinical reasoning, increase student engagement and motivation, and foster the development of interpersonal skills essential for healthcare practice.

#### 2.8.1.2 Active engagement

One of the advantages of the learning cycles is that they encourage students to participate actively in the process through inquiry and discovery. The learning cycle's main aim is to foster active engagement amongst the learners since it shifts students from passive listening to active participation which improves comprehension, retention, and the growth of vital abilities like teamwork and problem-solving. Instead of merely absorbing knowledge, students actively explore, explain, and apply what they learn within a learning cycle. Active engagement technics in medical education can be inculcated by using a variety of interactive teaching methods like problem-based learning, case-based learning, flipped classrooms, and small-group discussions, while also providing opportunities for direct clinical involvement.

#### 2.8.1.3 Knowledge application

As is well known, the secret to a medical student's professional success is the application of their knowledge. One of the most crucial abilities a clinician should have is the ability to apply their expertise. By translating abstract concepts into real-world applications, knowledge application strengthens comprehension and promotes learning through the cycle of application, reflection, and adaptation. This stage allows students to evaluate their comprehension by solving problems, making connections, and developing transferable abilities, ensuring that learning is about more than just gaining knowledge. Keeping these things in mind, it is vital to focus on early and frequent clinical exposure, incorporate interactive learning strategies like simulation-based education and problem-based learning, give regular, helpful feedback, and motivate students to participate actively through research projects and case-based discussions in order to instill knowledge application in medical students.

#### 2.8.1.4 Teacher framework

Learning cycles have shown to be incredibly helpful for teachers as well as students since they provide educators with a framework for organizing lessons that promote learning and are focused on the needs of each individual student. A teacher framework gives the learning cycle a purpose by supporting teachers in an ongoing process of analysis, planning, instruction, and evaluation to improve student learning outcomes. This framework serves as a tool for effective, student-centered instruction rather than just content delivery by assisting in ensuring that instruction is in line with the needs and goals of the students. The goal is to create a cycle of continuous growth and reflection with a focus on students' achievement.

### 2.8.2 Experiential learning

Experiential learning also known as learning by doing is the process of learning via firsthand experience and introspection. In medical education, the idea of experiential learning has shown to be particularly helpful since it includes practical exercises that help students apply theories they have learned in the classroom to real-world scenarios, improve their skills, and hone their critical thinking and problem-solving abilities. Structured initiatives like faculty development, curriculum creation, and longitudinal outcome assessments can help achieve this. Clinical rotations, simulations, and community-based projects are examples of specific techniques that are intended to lead students through the learning cycle of active experimentation, abstract conceptualization, reflective observation, and concrete experience (Yardley, Sarah & Teunissen, Pim & Dornan, Tim, 2012). Some of the common learning cycles are described below.

### 2.8.3 David Kolb's experiential or reflective cycle

David Kolb was an eminent American educational theorist whose interests and publications focused on experiential learning. David Kolb created the Kolb learning cycle or the reflective cycle as it is also known in 1984, which demonstrates how to create an experiential learning process that works. According to this learning cycle, it is believed that effective learning occurs through a continuous cycle of experience, reflection, conceptualization, and experimentation (Kolb, David, 1984). Kolb's theory focuses a lot on the internal cognitive processes of the learner. According to Kolb, learning is gaining abstract ideas that are adaptable to a variety of circumstances and novel experiences which in turn serve as the catalyst for the creation of new ideas. Kolb firmly held the following fundamental ideas about application and learning.

#### *2.8.3.1 Learning is cyclical*

Kolb's model emphasizes that learning is a continuous cycle of experience, introspection, conceptualization, and experimentation rather than a linear process. Because learning is a continuous process that includes gaining knowledge, applying it, and then reflecting on the outcomes to guide future attempts, it is circular. This cycle progresses from experience and perplexity to clarity, which is subsequently strengthened or improved by additional contemplation and testing. Concrete experience, reflective observation, abstract conceptualization, and active experimentation are the four phases of Kolb's experiential learning cycle.

#### *2.8.3.2 Experience is the basis of learning*

The theory emphasizes the importance of active engagement in the learning process and holds that knowledge is created via transformative experiences. The foundation of learning is experience, as it grounds knowledge in practical application and individual engagement through the "learning by doing" process. It entails actively interacting with a subject through practical exercises, which are subsequently processed through critical analysis and introspection to produce significant findings and provide a link between theory and practice. Compared to passive methods alone, this makes learning more effective, memorable, and personalized.

#### *2.8.3.3 Kolb's cycle: four steps*

1. The concrete learning experience
2. Reflective observation
3. Abstract conceptualization
4. Active experimentation

### 2.8.4 Concrete learning experience (feeling)

Concrete learning experience forms the basis of experiential learning. Through this concept, the student gets a chance to have a hands-on experience either of a new situation or event. In this regard, the learner has also a chance to think about his previous experiences and correct them. Concrete learning experiences aid in learning by directly applying abstract concepts to real-world situations, which helps them become more memorable and intelligible. Concrete learning in medicine is applied through practical experiences like patient encounters, clinical rotations, and simulations, followed by reflection, analysis, and active experimentation. By making students active participants rather than passive recipients, this method improves motivation, strengthens retention of the knowledge, and creates a personal connection to the subject matter. This style of learning also assists in developing key life skills like problem-solving, critical thinking, and flexibility.

### 2.8.5 Reflective observation of the new experience (watching)

As a part of the reflective observation of new experience, learners engage themselves to think about the new experience using what they already know. Any discrepancies that occur between understanding and experience are particularly important and are to be noted. By assisting students in critically analyzing their experiences and drawing connections between previously learned material and newly acquired knowledge, reflective observation promotes learning. This is a crucial phase in Kolb's experiential learning cycle as it promotes deeper understanding by getting students to reflect on what transpired, how they felt, what went well, and how it may have been done differently. This approach develops analytical and critical thinking abilities, which enhances performance in the future and results in a deeper comprehension of difficult subjects. For instance, recalling a clinical encounter in detail by focusing on the critical aspects, like what happened, who was involved, what was the setting of this incident, and reflecting on your internal response to the experience are all part of this process.

### 2.8.6 Abstract conceptualization (thinking)

Abstract conceptualization in medical education refers to the idea where students synthesize practical experiences into theoretical frameworks, synthesizing insights to create new ideas and theories. One may change an existing abstract concept or come up with a new one through careful introspection. Abstract conceptualization facilitates learning by enabling students to go beyond concrete experiences and produce more expanded concepts, general laws, and logical models. Activities like idea mapping, case-based brainstorming, and guided reflection to assist students make the connection between theoretical knowledge and real-world experiences can be used when teaching abstract conceptualization in medical education. The goal is to move students from reflecting on a clinical event to forming new principles or adjusting existing ones, often by framing the experience and explaining the underlying logic. This stage requires structured activities, such as facilitated discussions, to guide learners in making intellectual connections and adjusting their perspective.

### 2.8.7 Active experimentation (doing)

Active experimentation in medical education is a stage where students apply new knowledge and skills to real-world or simulated situations to test theories and refine their abilities. Activities include performing procedures on simulated patients, participating in case studies, or engaging in community-based fieldwork, with a focus on applying what has been learned to solve problems and improve future outcomes. The learner apply their thoughts or ideas to the surroundings to observe what happens. Active experimentation enables students to test theories, modify their behavior, and gain a deeper, more practical understanding by providing them with the chance to use recently learned skills and knowledge in simulated or real-world situations. This experiential phase enhances learning by providing new experiences that can then be taken into consideration and form the basis of a new learning cycle. It goes beyond theory by facilitating practical application, problem-solving, and the development of critical thinking skills through trial and error (Fig. 2.4).

## 2.9 Learning styles

Kolb's learning cycle identifies four learning styles based on how individuals prefer to process experiences, namely, diverging, assimilating, converging, and accommodating. These styles emerge from combining preferences for concrete experience, abstract conceptualization, reflective observation, and active experimentation within a continuous cycle of learning. The four learning styles that Kolb identified are as follows.

### 2.9.1 Diverging

A diverging learning technique strikes a balance between hands-on experience and introspective observation. It is characterized by the ability to view concrete situations from multiple perspectives and valuing feelings and imagination over action. In medical education, diverging learners combine concrete experience and reflective observation. They are imaginative and creative, preferring to gather information and assess experiences from various viewpoints rather than focusing on a single solution. One of the strengths of these type of learners is that they are good at brainstorming, generating ideas, and working collaboratively in groups.

### 2.9.2 Assimilating

A balance between abstract thought and introspective observation is achieved by the assimilation learning strategy. This kind of learner values theory and factual precision over intuition or practical experience, and they place a high value on the logical and methodical integration of new material into their preexisting understanding. This method, which actively connects new material to a student's preexisting cognitive structures rather than having it memorized without context, is thought to be essential to effective learning. These students prefer to review the material and assess the hands-on experience. Assimilators choose for tutorial-led research, self-directed activities, and traditional lectures supported by books and videos. They do best when they have time to think and analyze the information.

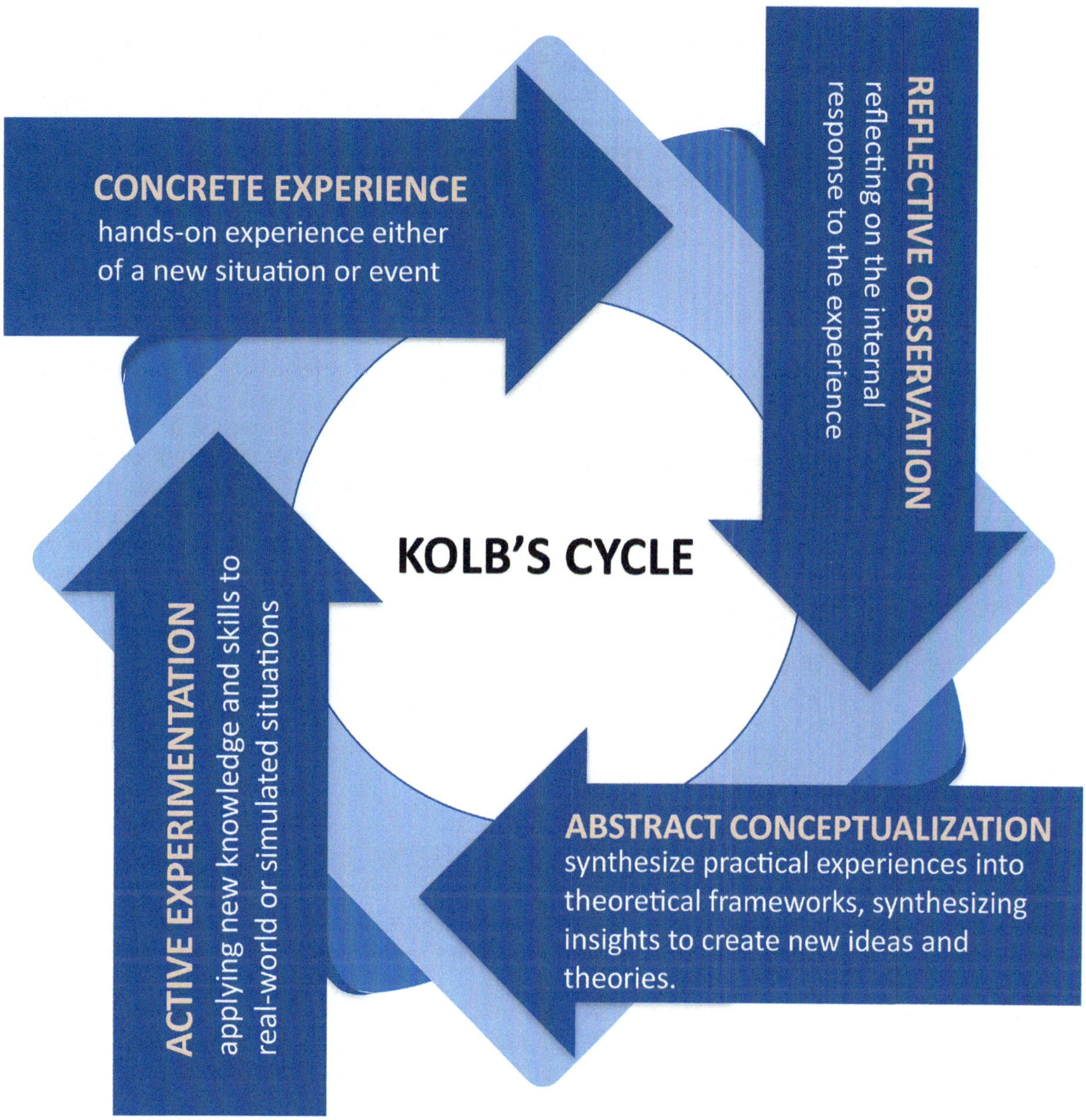


FIGURE 2.4 **Components of Kolb's experiential learning cycle.** This figure shows the components of Kolb's experiential learning cycle.

### 2.9.3 Converging

Convergent learning, sometimes referred to as "think and do," strikes a balance between active inquiry and abstract thought. Learners in this process frequently combine the "thinking" and "doing" learning styles. As they thrive at solving problems and are drawn to real-world applications of their education, the students here are regarded as problem solvers. Students here are considered as experimenters as they learn best by trial and error and applying concepts in real-world situations. Here, the learner approaches learning with a problem-solving focus and searches for technical answers. Worksheets, computer-based tasks, and interactive activities that require problem-solving are all valued by the learners. Simulation-based learning for practical skills, case-based learning where students apply theory to solve patient problems, and oral examinations where they must use deductive reasoning to respond to hypothetical scenarios are examples of convergent learning.

### 2.9.4 Accommodating

To accommodate diverse learning styles in medical education, a variety of instructional strategies that work for different types of learners must be used. Teachers may foster an inclusive atmosphere that meets a range of needs by implementing multimodal teaching practices. This will improve learning outcomes and produce more qualified healthcare workers. From kinesthetic learners who thrive on hands-on experience to those who prefer visual or auditory input, this method helps all students understand and remember knowledge more efficiently. This can be achieved by combining a number of strategies that cater to multimodal learners and by utilizing flexible teaching approaches like problem-based learning, which combines theory with practical application. Ultimately, a flexible approach can improve student engagement and learning outcomes.

### 2.9.5 The 5E learning model

It is a truth that in a time of evolving technology, dwindling attention spans, and increased expectations on educators, it is becoming more and more challenging to engage students in the learning process. A group at the Biological Science Curriculum Study under the direction of Rodger W. Bybee created the 5E Model of Instruction in 1987 with a view to promote active learning. Beyond reading and listening, a good student should also be able to ask questions, observe, model, analyze, explain, make conclusions, argue from evidence, and discuss their own understanding.(Putra, Febrianto & Kholifah 2018) The primary objective of the 5E model is to engage the student in the classroom through phased lessons. Inquiry-based learning is the defining characteristic of this approach. Students are encouraged to find information on their own rather than having it imposed upon them when inquiry is used as a teaching technique. By incorporating students into the learning process, the 5E model essentially promotes student participation.

### 2.9.6 Five stages of the 5E model

1. Engagement,
2. Exploration,
3. Explanation,
4. Elaboration,
5. Evaluation (Fig. 2.5).

### 2.9.7 Engagement

Engagement is the initial stage of the 5E learning cycle. Teachers engage students in dialogue and encourage them to explore their ideas in order to learn more about their interests and prior knowledge as well as what they now understand. In this stage, teachers give students assignments, questions, or issues that arouse their curiosity and encourage critical thinking. The involvement should ideally offer the "need to know" that motivates students and fosters tenacity. A good teacher should be able to engage the students through active learning strategies like case-based learning, debates, and simulations. Connecting basic science to clinical medicine, providing real-world scenarios, and involving students in daily clinical tasks will develop a sense of responsibility. The teacher should also focus on encouraging collaboration through group work and peer teaching, while also fostering a positive learning environment through mentorship, respectful communication, and incorporating principles of diversity, equity, and inclusion.

### 2.9.8 Exploration

The second phase of the 5E learning cycle, known as exploration, gives students the opportunity to pose and respond to their own questions. Students can experiment, explore, make mistakes, and learn in a low-stakes setting during this trial-and-error phase. Students may carry out experiments and make inferences from data and observations during exploration activities. From case studies and problem-based learning to direct patient involvement and volunteer work, these activities require students to accomplish tasks while reflecting on their actions. Building skills, encouraging critical thinking, investigating job options, and cultivating professionalism and empathy are the objectives. Clinical rotations, humanistic or community-based experiences, and active learning techniques are examples of exploration activities in medical education.

### 2.9.9 Explanation

During the explanation stage, students have the opportunity to share their ideas, explain how they understand concepts, and describe the subject or idea they are studying. By putting concepts into their own words, students can improve their

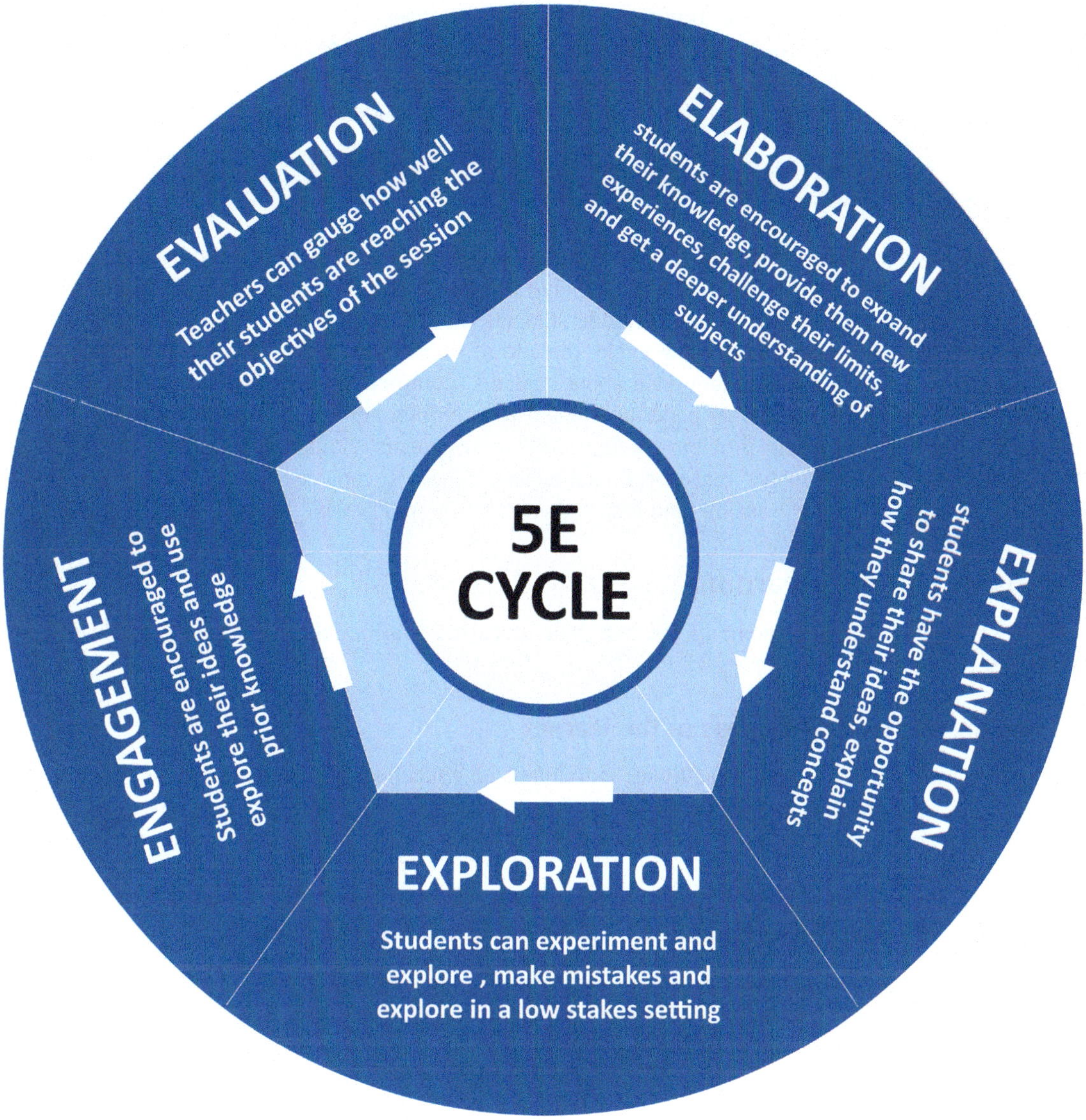


FIGURE 2.5 **Components of the 5E cycle.** This figure shows the components of the 5E cycle.

own understanding. During the explanation process, students must make their own decisions about what should be said and the best way to convey their views. Students typically present their designs to one another in the classroom or in a public place, such as exhibitions or online discussion board, during the explanation phase of a maker-centered lesson. Students can demonstrate their understanding of specific scientific topics, present their work, and talk about how they came up with it. Students receive explanations and comments as needed from the instructor and their peers. As in all 5E phases, the instructor helps students learn by making recommendations, asking relevant questions, and overlaying student ideas with academic terminology. Students' explanations of their work and the ideas they emphasize also provide excellent chances for formative assessment.

## 2.9.10 Elaboration

The elaboration phase in the 5E learning cycle is where students apply their new knowledge to new related situations, deepening their understanding and reinforcing learning through real-world connections and extensions, often involving problem-solving, new investigations, or projects to practice transferring concepts beyond the initial lesson. Teachers facilitate this phase by asking probing questions and challenging students to use new vocabulary and reasoning, promoting deeper conceptual understanding and the transfer of learning. Teachers also encourage students to expand their knowledge, provide them new experiences, challenge their limits, and get a deeper understanding of subjects during elaboration. Students can use the knowledge they have gained from the early stages of creation and presentation to enhance their work during the elaboration stage. Students can now revise their presentations, include fresh concepts or ideas, or consider other uses.

### 2.9.11 Evaluation

At the final assessment stage of the 5E learning cycle, students get the chance to assess themselves. This stage also aims to provide teachers with information that they can use to gauge how well their students are reaching the objectives of the session. It can be a powerful educational tool to provide students the chance to think about and evaluate their own and their peers' ideas. If 5E is considered a learning cycle, the end of one cycle also marks the start of a new one. In this method, teachers can use the information they gain about students' comprehension in one iteration to guide subsequent learning cycles. Instead of using traditional paper-and-pencil tests, which are insufficient to measure all of a learner's competences, including creativity and problem-solving skills, assessments should be based on our capacity to finish assignments, produce work, and solve real-world challenges. Teachers can assess students' grasp of disciplinary principles, acquire insight into their learning, and assess their mastery of scientific process skills using this method.

## 2.10 Steps to enhance learning

In this section, we try to summarize some of the important steps that can be taken to enhance the learning process in medical education.

### 2.10.1 Knowing the level and needs of the learner

Understanding a learner's needs and level facilitates individualized education, which raises motivation, engagement, and retention. This knowledge enables teachers to spot obstacles, offer focused assistance, and design a more productive learning environment that builds on prior knowledge and takes into account each student's unique learning preferences, strengths, and mental health.

### 2.10.2 Tailoring the content to be learnt

Learning content that is tailored improves retention, increases engagement, and makes learning more relevant. Giving students authority increases motivation, accommodates a range of skill levels and learning preferences, and promotes deeper comprehension as opposed to rote memorizing.

### 2.10.3 Encouraging active participation

By boosting engagement, enhancing knowledge retention, and fostering the development of vital abilities like communication and problem-solving, active participation improves learning. Students embrace confidence, gain a deeper knowledge of subjects, and become more motivated to learn when they actively participate in activities like group projects, discussions, and practical assignments.

### 2.10.4 Making learning interactive

In addition to facilitating collaboration and accommodating a variety of learning styles, making learning interactive improves engagement, retention, and critical thinking. Passive listening is turned into an active process where students apply their information, get rapid feedback, and learn from their peers through interactive techniques including group projects, quizzes, and simulations.

### 2.10.5 Initiating team work and collaboration

By fostering important skills like communication and problem-solving, boosting motivation and engagement, and deepening comprehension through peer-to-peer explanation, teamwork and cooperation improve learning. Collaborating with others pushes people to question presumptions, consider different viewpoints, and develop a sense of shared responsibility, all of which improve learning outcomes and get them ready for future collaborative settings (Richardson, 2015).

### 2.10.6 Using real-life experience

Real-world experience enhances learning since it becomes more applicable, remembered, and beneficial. This experiential learning approach not only improves comprehension and engagement but also cultivates critical skills like creativity, problem-solving, and resilience, all of which are essential for success in the classroom and in the future career.

### 2.10.7 Technology-based learning

By encouraging collaboration, enabling personalized learning, and increasing engagement, technology-based learning improves education. It makes learning more dynamic with tools like games and simulations, gives students access to a greater variety of resources, and helps get them ready for a digital future. It also helps teachers give quick feedback, offers flexible learning alternatives, and supports a variety of learning styles.

**Summary**

Medical knowledge acquisition is a lifetime necessity for doctors and a critical ability for medical students. By emphasizing experiential and team-based learning to better reflect real-world patient care and by incorporating evidence-based techniques like spaced repetition and active recall to enhance information retention, principles of learning can revolutionize medical education. In order to cultivate critical thinking, empathy, and the capacity to traverse intricate healthcare systems, this change also entails promoting transformative learning.

**Key point**

**Future perspectives: the era of collaborative learning**

Technology like AI, virtual reality, and mobile learning will influence the learning process in the future, resulting in more individualized, flexible, and immersive learning environments. In order to educate doctors for a dynamic and technologically advanced healthcare environment, it will also place a strong emphasis on lifelong, self-directed, competency-based learning that incorporates clinical experience, humanistic skills, and interprofessional teamwork. The integration of these into the learning process will be a real game changer. In order to work in a healthcare system that prioritizes preventive, patient-centered care, and interdisciplinary collaboration, future medical students will be expected to combine traditional clinical abilities with technology competency and adaptability. Strong communication, empathy, and ethical reasoning will be prioritized, along with digital health literacy, data analytics, and the capacity to collaborate with diverse teams.

## References

Badyal, D.K., Singh, T., 2017. Learning theories: the basics to learn in medical education. International Journal of Applied and Basic Medical Research 7 (5), 1.

Bassaw, B., Roff, S., McAleer, S., Roopnarinesingh, S., de Lisle, J., Teelucksingh, S., et al., 2003. Students' perspectives on the educational environment, Faculty of Medical Sciences, Trinidad. Medical Teacher 25 (5), 522–526. Available at https://doi.org/10.1080/0142159031000137409.

Buşan, A.M., 2014. Learning styles of medical students - implications in education. Current Health Sciences Journal 40 (2), 104–11010 .12865/CHSJ.40.02.04.

Dong, H., Lio, J., Sherer, R., Jiang, I., 2021. Some Learning Theories for Medical Educators. Med Sci Educ 31 (3), 1157–1172. Available at https://doi.org/10.1007/s40670-021-01270-6 PMID: 34457959; PMCID: PMC8368150.

Hutchinson, L., 2003. ABC of learning and teaching. Educational Environment. BMJ 326, 12.

Khawar, A., Frederiks, F., Nasori, M., Mak, M., Visser, M., van Etten-Jamaludin, F., et al., 2022. What are the characteristics of excellent physicians and residents in the clinical workplace? A systematic review. BMJ Open 12 (9), e065333. Available at https://doi.org/10.1136/bmjopen-2022-065333 PMID: 36127103; PMCID: PMC9490566.

Knapke, J.M., Hildreth, L., Molano, J.R., Schuckman, S.M., Blackard, J.T., Johnstone, M., et al., 2024. Andragogy in Practice: Applying a Theoretical Framework to Team Science Training in Biomedical Research. Br J Biomed Sci 81, 12651. Available at https://doi.org/10.3389/bjbs.2024.12651 PMID: 38605981; PMCID: PMC11008574.

Kolb, D.A., 1984. Experiential Learning: Experience as the Source of Learning and Development. Prentice Hall, Englewood Cliffs, NJ.

Othman, M., Rukban, A., Khalil, M., Al-zalabani, A., 2010. Learning environment in medical schools adopting different educational strategies. Educational Research and Reviews 5, 126–129.

Pashler, H., McDaniel, M., Rohrer, D., Bjork, R., 2008. Learning Styles: Concepts and Evidence. Psychological Science in the Public Interest: a journal of the American Psychological Society 9 (3), 105–119. Available at https://doi.org/10.1111/j.1539-6053.2009.01038.x.

Putra, F., Kholifah, I., Subali, B., Rusilowati, A., 2018. 5E-Learning Cycle Strategy: Increasing Conceptual Understanding and Learning Motivation. Jurnal Ilmiah Pendidikan Fisika Al-Biruni 7, 171. Available at https://doi.org/10.24042/jipfalbiruni.v7i2.2898.

Richardson, O., Chan, M.-K., Lang, E.S., Harvey, B.J., 2015. Scholar. CanMEDS 2015 Physician Competency Framework.

Steinmair, D., Zervos, K., Wong, G., Loeffler-Stastka, H., 2022. Importance of communication in medical practice and medical education: An emphasis on empathy and attitudes and their possible influences. World Journal of Psychiatry 12, 323–337. Available at https://doi.org/10.5498/wjp.v12.i2.323.

Sukrajh, V., Adefolalu, A., 2021. Understanding Learning and the Components of the Learning Process in Medical Education: A Review of the Literature. European Journal of Education and Pedagogy 2, 69–72. Available at https://doi.org/10.24018/ejedu.2021.2.1.53.

Yardley, S., Teunissen, P., Dornan, T., 2012. Experiential learning: AMEE guide No. 63. Medical teacher 34, e102–15. Available at https://doi.org/10.3109/0142159X.2012.650741.

Chapter 3

# Teaching and learning process

*"Teaching is more than imparting knowledge; it is inspiring change. Learning is more than absorbing facts; it is acquiring understanding."*

—William Arthur Ward

## 3.1 Introduction

The teaching-learning process is a cyclical, systematic interaction between teachers and learners to achieve educational goals, involving stages like identifying needs, planning learning, facilitating instruction, assessing understanding, and providing feedback. A teacher–learner relationship is a connection characterized by mutual trust, respect, and collaboration to achieve educational goals. Owing to this, it is imperative to comprehend this idea and investigate the processes of the teacher–student relationship, the different components and tenets, its advantages and goals, and above all its breadth and significance in the current educational environment.

The primary goals of the teaching-learning process are to facilitate an effective education by covering a wide variety of comprehensive objectives and results. The learner's intellectual, cognitive, analytical, logical, ethical, emotional, and social growth is intended to be improved by these goals. These goals center on the development of skills, attitudes, values, and knowledge and comprehension. Key elements include the teacher, learner, content, learning environment, and various teaching methods, all working together in a relationship designed to transfer knowledge, develop skills, and cultivate attitudes (Arslan, Polat, 2016).

### 3.1.1 Changing roles of a teacher: from mystic gurus to lifelong learner

Since ancient times, teachers are often revered as gurus, sages, philosophers, or priests, who imparted wisdom and knowledge to their disciples or followers. Ancient teachers were portrayed as mystic gurus because they were viewed as spiritual guides who imparted knowledge beyond the ordinary, sometimes acting as a conduit for divine wisdom and possessing qualities of enlightenment. The teaching and learning process has undergone a paradigm shift over time. The passive, teacher-centered approach to teaching has given way to an active, student-centered one that emphasizes abilities like teamwork and critical thinking. This has affected how educators must be ready to embrace the difficulties brought about by this shift.

The end of the 20th century and turn of the 21st century saw the emergence of new technologies in learning. Educators would now harness the power of rapidly evolving web-based platforms and devices for learning. The advent of online learning became widespread and challenged teachers to teach online. This disruption continues to make the teaching profession more intricate to this day. In recent years, there has been an increased focus on the emotional aspects of learning. Teachers have used coaching, positive psychology, and mindfulness techniques in their work. Since they are flexible, tech-savvy, cooperative, and have strong communication, critical thinking, and problem-solving skills, the 21st-century educators are thought of as lifelong learners (Hathur, Kulkarni, 2024).

## 3.2 Teacher–learner relationship

According to conventional wisdom, a teacher is someone who transfers information, skills, and wisdom to his students. Conversely, a learner is a person who is eager to pick up new skills and information. Although the terms "teacher" and "student" have separate definitions, they are in fact closely related. The bond between a teacher and a student has long been regarded as one of the most holy and valuable relationships. In order for the students to grow holistically and be able to perform to the best of their abilities, it is thought that the teacher not only instructs the students but also instills in them

**The Principles and Practice of Medical Education. DOI:** https://doi.org/10.1016/B978-0-443-44197-4.00017-5

the necessary knowledge, skills, and attitude, making them a successful healthcare professional but more importantly a better human being. One of the most important aspects of learning is the interaction between a teacher and his students. This is an important factor that influences students' conduct, motivation, academic engagement, and progress. These days, the instructor and the student are seen as more than just knowledge providers and seekers. The relationship between teachers and students is highly valued nowadays because they work together as a team, understand one another, celebrate each other's accomplishments, and recognize each other's strengths and weaknesses. In the process, they also have an impact on how education should be shaped to promote quality (Mitchell, Kensler, Tschannen-Moran, 2018).

### 3.2.1 Benefits of positive teacher–student relationship

At all levels of an educational institution, both within the classroom and throughout the entire learning environment, strong and pleasant interactions between teachers and students can be beneficial. By setting clear objectives for each student, recognizing their strengths and shortcomings, providing equitable opportunities for participation in class discussions, and encouraging them to believe in their own abilities, teachers may help students succeed academically. The following are some advantages of having a good relationship between the teacher and the students:

**1. Respecting the individuality**

A great teacher–student relationship depends on respecting each student's uniqueness because it builds trust, boosts academic performance, and establishes a welcoming and inclusive learning environment. Teachers can customize their instruction, create real relationships, and inspire students to participate, take chances, and feel appreciated by recognizing each student's distinct interests, strengths, and needs. A healthy connection requires trust, which is built when educators genuinely care about the lives and backgrounds of their students. Respecting individuality has several advantages, including raising student motivation and engagement, establishing a friendly and safe environment, and helping them feel valuable (Plaut, Baker, 2011).

**2. Building trust**

Everyone is aware that relationships are built on trust. The bond between a teacher and a student is among the most sacred. By fostering a safe, secure, and encouraging environment where students feel free to take chances, ask questions, and participate in class, developing trust in this area improves teacher–student interactions. The relationship is greatly impacted by crucial components such as knowing the learner, communicating with empathy, actively listening, and demonstrating real concern by getting to know students' interests and lives. Mutual respect, which is fostered by this trust, increases academic engagement and promotes social-emotional development.

**3. Understanding the learner**

Relationships between teachers and students depend on an understanding of the learner, as it builds trust, increases engagement, and produces a more individualized and productive learning environment. The teacher needs to be cognizant of the students' diverse personal and professional demands. It is important to recognize that students come from a variety of cultural and educational backgrounds. For a student in a medical school, the instructor will be their first point of contact or the person they turn to when they believe they can resolve any problem. Teachers can better personalize their support, establish a closer relationship, and motivate students when they are aware of their strengths, needs, interests, and emotions. This improves academic achievement and gives students a sense of security and confidence in the classroom (Wubbels, Brekelmans, 2005).

**4. Appreciate the learner style**

As it promotes more effective and engaging learning, greater student confidence, and improved academic success, understanding a learner's style is essential for a healthy teacher–student connection. Teachers can modify classes to fit each student's needs when they recognize, value, and adjust to various learning styles, needs, and expectations. This helps students retain material and creates a more encouraging and welcoming learning environment.

**5. Promotes academic achievement**

Learners who build strong bonds with their teachers fare better than those who have unpleasant relationships. Positive teacher–student relationships increase academic achievement by fostering a positive learning environment, increasing student motivation and engagement, and improving overall welfare. Strong relationships provide students more self-assurance and emotional stability, which motivates them to participate more in class, demonstrate abilities like teamwork, and conquer challenges. This is achieved by creating an atmosphere in which students feel valued, understood, and supported, a sentiment that may be fostered through frequent, one-on-one, and open communication.

**6. Encourages self-esteem**

Students' low self-esteem is frequently the result of a confluence of unfavorable events, including subpar academic results, trouble acclimating to a new setting, and obstacles related to language, culture, and other factors. Both the teacher and the student must work together to overcome these obstacles. Self-esteem can be built by cultivating positive teacher–student relationships and fostering a supportive environment where students feel valued, accepted, and secure. A student's self-esteem will also be greatly increased by praising their efforts and highlighting their strengths. This is accomplished by open communication, empathy, and high standards, which encourages the learner to take intellectual and social risks, develop resilience, and have confidence in their own abilities. Teachers can encourage this by showing genuine attention, offering helpful feedback, and creating a safe learning atmosphere.

**7. Professional development**

One of the key benefits of a good student–teacher connection is that teachers work hard to improve their professional and interpersonal skills. Focusing on these relationships can enhance professional development since they are associated with higher levels of student motivation, achievement, and self-regulation. Teachers may improve this relationship through professional development that emphasizes building trust, communicating effectively, showing empathy, and being consistent. As a result of this, all students will benefit from a more positive learning atmosphere and better teaching results.

**8. Benefits for the educators**

A good teacher–learner relationship is a true example of reciprocal regard. One of the essential qualities of the teacher is that the teacher should learn to unlearn. The advantage of having a good teacher–learner relationship is that it provides the teacher a room for improvement. A positive relationship allows teachers to better understand their students' needs, leading to more personalized support, which in turn creates a calmer, more productive classroom and higher levels of teacher well-being. It not only provides a strong sense of increased job satisfaction but also gives a feeling that we as teachers are indeed change agents that have a tremendous potential to influence the learning process.

## 3.3 Teaching principles

Teaching principles in medical education emphasize active, learner-centered approaches, incorporating adult learning theory, reflective practice, and effective feedback. Teaching principles are designed to provide a holistic approach to the teaching-learning process. These principles take into consideration some of the vital aspects such as accountability, transparency and upholding values to name a few. Teachers should plan to enhance their interactions with students in order to facilitate high-quality learning.

### 3.3.1 Accountability

As per the World Health Organization, accountability in medical education is defined as a commitment in guiding education, research, and medical services in line with the priorities of community, region, and world health needs, based on the values of quality, equity, and cost-effectiveness. Accountability is one of the main goals of the health system and is of great importance among policymakers and planners of all organizations, especially the health system. Therefore, health policymakers are seeking measures to make the system more accountable to patients and the community. Accountability is also regarded as a crucial component of teaching skills because it drives professional growth, improves student outcomes, and builds trust between teachers, students, and the wider community. It ensures teachers are dedicated to high-quality instruction, adapt to student needs, and are responsible for both academic and personal development. A teacher can display accountability in medical education by clearly defining and communicating expected learning outcomes, using transparent and competency-based assessments, and fostering a curriculum that responds to community health needs. Other methods include creating partnerships with community stakeholders, promoting specialties needed by the community, and ensuring their own instructional practices are effective.

### 3.3.2 Transparency

Transparency in medical education is indeed a vital component of the teaching-learning process as it gives clear view to what is expected from the students. Transparency in medical education means providing clear and open information about the curriculum, its development, and its outcomes to students, patients, and the public. This includes disclosing how students will be evaluated, the structure of the program, and the reasons behind curriculum choices, enabling students to prepare effectively and fostering trust among all stakeholders. It also involves being forthright about the processes and data that inform educational decisions to ensure that the curriculum remains responsive and accountable.

### 3.3.3 Holistic approach

The holistic approach in medicine has indeed carved a niche for itself, and its effect on teaching and learning process cannot be overstated. As a teacher, it is every important to understand that in order to reap the benefits of medicine, it has to be taught holistically. Then and only then will it cater to the different needs of the community. Some of the benefits that the holistic approach in medical education has that it leads to the development of well-rounded, resilient physicians by supporting students' mental, social, and emotional well-being alongside their academic growth. This approach prepares them for the complexities of modern healthcare, improves patient outcomes, and enhances communication and collaboration skills essential for a successful career.

### 3.3.4 Upholding values

Values in medical education form one of the vital cornerstones of the teaching-learning process. It is important for a teacher to uphold values to create a positive learning environment, foster students' ethical development, and prepare them to be responsible citizens. By modeling values like respect, fairness, and integrity, teachers build trust, encourage students' personal growth, and ensure a structured, supportive classroom where students can thrive academically and socially. This goes beyond simply imparting knowledge, as it shapes students' character and equips them with the skills to make sound, ethical decisions throughout their lives.

### 3.3.5 Familiarize with the educational process

As educators we need to be familiarized about the expected outcomes. Our ability to understand the educational process will make the teaching process smooth. At the same time, we will be in a better position to make students understand the topic. Learning activities should generally be organized to produce a coherent curriculum. One of the important aspects of this is that the teacher should be able to take part in not only teaching the outcomes but must be involved in all the aspects of the teaching-learning process in order to achieve the expected results. The lecturer should be aware of how and where his subject fits within the medical school curriculum.

### 3.3.6 Effective use of prior knowledge

The effective use of prior knowledge in teaching process serves as a great start in order to comprehend and plan various activities both by the teachers and the students. These are the materials created for the session that are meant to act as the starting point for learning. Learners' prior knowledge and experience are essential and a great asset in new learning situations. Knowles asserts that personal experiences are a great learning resource for adult learners. An example of a learning stimulus is providing students with handouts that include case scenarios, questions to answer, or blank spaces to fill up before the session. Students should ideally read over the material, think about it, and create questions about it before the class begins. Students should be encouraged to explain relevant basic ideas and concepts both before and during the session, such as the anatomy and physiology of the system or the clinical topic that will be covered. Learners with well-organized prior knowledge of the subject behave more strategically, make better decisions, and assimilate new information more fully than those with less prior knowledge. A pretest is a smart way to determine how well students have understood earlier sessions.

### 3.3.7 Establishing relevance in medical education

Meaningful learning is believed to take place when students choose what, how, and when they wish to learn. As medical educators, it is important we understand and appreciate the needs of the students and fully support them in their choices. When the material being learned has personal significance, learning outcomes improve. Therefore, the goal is to engage students in real-world tasks in order to enhance their skills. Furthermore, when the learning process is relevant to the learner's interests and workplace, information is more likely to be kept and implemented, which may lead to a change in behavior and practice. In this regard, the teacher is at the best position to guide the student as to how the relevance is established. One of the best examples of relevance is integrating basic sciences with clinical medicine. This will not only make the topic more interesting but at the same time make it more relevant as the expectations of the students are fulfilled.

### 3.3.8 Stimulate curiosity and interest among students

For a student, curiosity and interest are the major driving forces which enable learning. It has to be understood that the curiosity and interests of each student are different, and it is the duty of the teacher to aptly recognize that. Deep learning and better understanding of the topic can be achieved only if these things are taken care of. Curiosity is fostered by a variety of factors, such as learning stimuli, a secure environment, and constructive feedback for students. In order to instill a sense of curiosity and interest amongst the students, it is impertinent that the teacher applies various modes of teaching. To instill curiosity, encourage students to ask "why" questions, provide hands-on and open-ended activities, foster a safe environment for mistakes, and model your own curiosity through enthusiasm and exploration. Creating opportunities for new experiences, reflecting on new ideas, and reading widely are also effective strategies to promote a curious mindset.

### 3.3.9 Encourage students to actively contribute to the educational process

In modern medical education where all the aspects are targeted towards making learning process more effective and meaningful, the contributions from the perspective of students are of immense value. Students are no longer just passive listeners, but they are involved in all the important aspects of medical education. Whether it is the formulation of the learning outcomes, designing the instruction methods or methods of evaluation the role of students cannot be ignored. Moreover, by encouraging the students to actively participate in the classroom activities, receive peer feedbacks, and reflect on their achievement, we as educators are helping the students to actively participate in the educational process. Innovative learning modalities like problem-based learning, team-based learning, or group discussion all stress on collaborative learning and teamwork which strengthens the concept of active involvement of the learner.

### 3.3.10 Provide a safe learning environment

The provision of safe learning environment is one of the prime requisites of learning process. The student has to be made safe and comfortable by designing the learning climate in a way to facilitate active learning and fostering a positive atmosphere. The role of teachers in creation of this environment is vital as they are in a best position to understand the needs of the students and act accordingly. In this regard, the teacher also acts as a bridge between the students and the administration.

### 3.3.11 Providing student with ample opportunities

The teacher should ensure that students have several opportunities to practice new skills like communication, teamwork, and collaboration. These can be practiced by incorporating activities such as interactive lecturing, case-based discussion, or working in community-oriented projects which gives the student ample opportunity to practice and hone their soft skills. Moreover, by practicing with each other, standardized patients, manikins, role-playing, and simulations in the clinical skills lab, students can get a certain level of proficiency in the basic psychomotor and other abilities. These methods provide hands-on learning opportunities for the development of clinical skills, which are believed to be essential to ensure patient safety and effective care. The practice should be backed by appropriate assessment and feedback before students engage with real patients.

### 3.3.12 Establish a collaborative and team-based learning environment

With the advent of collaborative learning, the idea of working in silo has taken a backseat. It is now become imperative that students collaborate and express their thoughts. Medical professionals of the future will be learning from and with others. A collaborative setting where all students are encouraged to understand and even contribute to the development of varied ideas will be beneficial for long-term learning. A collaborative learning environment will not only encourage the student to perform better, but at the same time, it has the capacity to develop the learner in a way that he or she can imbibe the skills of decision-making, communication, trust building, and conflict-resolution. Positive interdependence occurs when students work together in small groups to improve learning for all participants, share resources, encourage one another, and recognize successes. Individual responsibility can be demonstrated by assigning each student a test, having them explain what they have learned, or having them watch and document the work of other students. In this regard, the role of the instructor is vital to create an environment which enhances the teamwork and collaboration amongst the learner. In activities such as group projects, case discussions may be used in this regard.

### 3.3.13 Implement effective assessment procedures

Assessment is a well-known, potent stimulus and motivator for learning and, ideally, for studying. Having said that the assessment methods should be carefully chosen keeping in mind the activity to be assessed and also the benefits that can be obtained by the students. Both formative and summative assessments are a great way to exhibit the students' abilities in terms of knowledge, skills, and attitude. Assessment should not only be done at the end of the course rather it should be done periodically to gauge students' performance, thereby giving them chance for improvement.

### 3.3.14 Incorporate frequent and constructive feedback

The essence of feedback in medical education cannot be overstated. Feedback has become one of the important pillars of educational process as it serves many purposes. It is not only an indication of how the student has performed a certain task, but it is also a chance for the learner to understand his or her strengths and weakness. Feedback has the potential to be an effective teaching tool when handled appropriately. In its absence, mistakes go uncorrected, exceptional work goes unrewarded, and clinical competency is either achieved by trial and error or not at all. The type of feedback should be carefully chosen and must be tailored to individual student. If wrongly executed, students' confidence and sense of safety will be damaged. The feedback session works best when it is done quickly, preferably just after observation, and with the trainee's consent. One of the important aspect of the feedback process is that the student should be given a choice to reflect his achievement. The student then offers suggestions for how to improve his or her performance as well as areas that require improvement. The teacher then lists what has to be changed and how, avoiding judgmental comments and adjectives.

### 3.3.15 Invite feedback on teaching from teachers and students

Teachers frequently take criticism and evaluation of their teaching in addition to providing feedback to their peers and students. Feedback and reflection should be an integral part of our teaching strategies because they are associated with improvement and transformation. Asking students for comments, using a simple checklist or questionnaire, or having a colleague observe and assess teaching methods are all ways to assess a teaching session, whether it is didactic or clinical. A lesson may also be videotaped and viewed at a later time by the instructor or a colleague. This approach to evaluating the efficacy of instruction can be quite educational.

### 3.3.16 Provide your students with opportunities to reflect on their practice

Students should be given the opportunity to reflect on their practice, which includes generating new ideas and perspectives as well as assessing and analyzing their own work. Teachers can support students' introspection by using a structured logbook or a portfolio. A portfolio that includes both reflections on the experience and evidence of achievement in a specific learning difficulty might be used. Think-aloud modeling is one of the methods used in medical education to enhance the reflective practice. It is a metacognitive strategy where experts verbalize their thought processes like clinical reasoning, decision-making, or diagnostic steps while solving problems, making implicit knowledge explicit for learners. In this regard, it can be used by the physician to perform a difficult task such as resolving a clinical problem so that all the students can observe the process. It is mainly used to make the implicit explicit as it aids in clarifying the steps, strategies, and rationale involved in complex medical problem-solving, moving from conscious thought to reflection. In order to facilitate the student's cognitive process, it allows them to express their ideas aloud. This helps them organize their thoughts by creating connections between what they already know and what is available, as well as by identifying any gaps in their knowledge and data.

### 3.3.17 Develop lifelong learning and study skills

As we are aware that studying medicine is a lifelong process, it is important for the learner to understand only reading and understanding various concepts in medicine will not make him or her a good doctor. Along with that, skills like teamwork and others are also needed. Teachers can foster lifelong learning by cultivating a growth mindset and curiosity, promoting self-directed and inquiry-based learning, encouraging the use of technology, and building critical thinking and goal-setting skills. A supportive classroom environment that encourages exploration and views mistakes as learning opportunities is also crucial.

### 3.3.18 Present yourself as a role model

One of the greatest gifts the teacher can give to his student is by being his or her role model. The faculty is considered as living example of the institute's principles and beliefs. They demonstrate commendable professional qualities and articulate informed opinions, both of which are essential for students' growth on the personal and professional fronts. It is being observed that the teachers who set an example for their students will be more effective in shaping the life and career of the student. Regular faculty–student connection, both within and outside of the classroom, has a big impact on student motivation and involvement. A teacher must always exhibit professionalism and etiquette so that a positive influence is cast upon the learner. Students consider their teachers to be their role model only if they see something special in them, someone whom they can relate to and someone with whom they can share their problems and experiences.

## 3.4 Teaching-learning process

The teaching-learning process is a multistep process in which a teacher measures and analyzes students' learning needs, creates learning goals, and uses innovative teaching-learning techniques to transfer knowledge. This relationship is not only limited to the transfer of knowledge but also is responsible for creating a unique bond that when nurtured well has endless possibilities. Since this procedure is similar to a two-way "circular" communication paradigm, it can only be successful in the presence of a two-way equal stream of communication and discussion (Susan Peña, 2025). The several facets of the teaching-learning process are discussed in the subsequent paragraphs

### 3.4.1 Objectives of the teaching-learning process

In order to promote an effective education, the teaching-learning process' objectives cover a wide variety of comprehensive goals and outputs. These goals are intended to support students' intellectual, cognitive, analytical, logical, moral, social, and emotional growth. Since these goals serve as the cornerstone upon which a solid foundation for intellectual growth is erected, only when all of them are accomplished can the entire span of the teaching-learning process be observed. Below are a list of the many goals and their fundamental principles (Fig. 3.1).

#### *3.4.1.1 Knowledge acquisition and understanding*

The first objective in an educational process is to develop cognitive and intellectual reasoning, along with absorbing preexisting, factual information. The teacher isn't just imparting knowledge, but the student is also understanding and applying it. These are the basic principles for learning new concepts. It should be noted that the process of knowledge acquisition and understanding should be a two-way process, thereby encouraging the student to participate in the learning process. Knowledge acquisition entails the following four processes (Fig. 3.2).

**1. Acquiring factual knowledge**

Acquiring factual knowledge is one of the first steps of learning medicine. Students learn the fundamental ideas, facts, and concepts related to a particular subject. Acquiring factual knowledge is crucial in medical education because it forms the

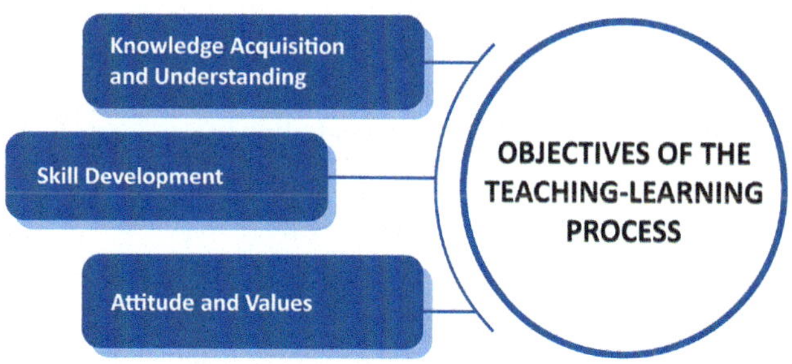


FIGURE 3.1 **Objectives of the teaching-learning process.** This figure depicts objectives of the teaching-learning process.

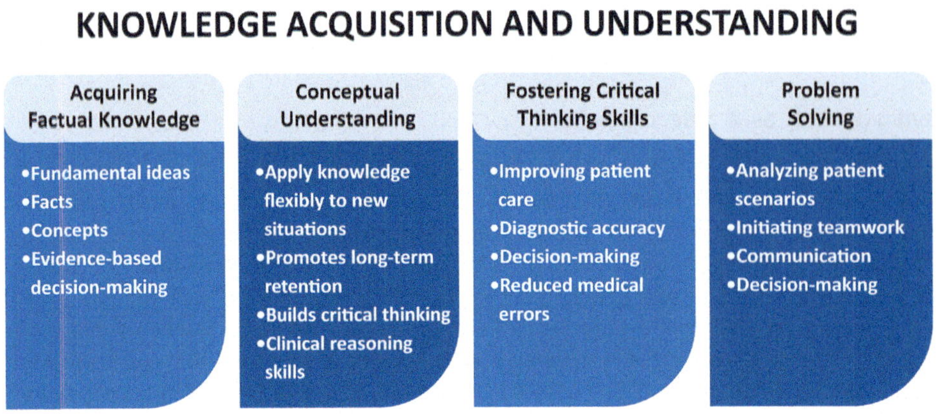


FIGURE 3.2 **Knowledge acquisition and understanding.** This figure shows the knowledge acquisition and understanding.

foundation for critical thinking, clinical skills, and evidence-based decision-making. From the teachers' point of view, the knowledge imparted should be relevant, evidence based, and applicable to ensure that it will allow future physicians to effectively solve problems, apply knowledge to patient care, and maintain highest quality of healthcare. Activities such as questions and answers sessions and using multiple choice questions for assessment are great methods to uplift the essence of factual knowledge.

**2. Conceptual understanding**

Conceptual understanding is vital in medical education because it allows students to apply knowledge flexibly to new situations, promotes long-term retention, and builds critical thinking and clinical reasoning skills. Instead of just memorizing protocols, students understand the "why" behind treatments and procedures, enabling them to adapt care for complex, atypical cases and make more informed decisions. This deeper understanding helps bridge the gap between basic sciences and clinical practice, fostering the development of expertise and preventing reliance on rote memorization.

**3. Fostering critical thinking skills**

Students gain the capacity to assess arguments, examine data, and make their own decisions based on their comprehension. Fostering critical thinking in medical education is crucial for improving patient care, diagnostic accuracy, and decision-making. It equips future physicians with the ability to analyze complex information, evaluate evidence, and solve problems effectively, leading to better patient outcomes, reduced medical errors, and a more adaptable healthcare workforce. Activities that encourage critical thinking in medical students include problem-based learning, case-based learning, and simulation exercises. Other effective methods involving classroom discussions and debates, flipped classroom models, peer review, and creative projects like designing their own cases enables the development of critical skills amongst the students.

**4. Problem-solving**

Problem-solving is crucial in medical education as it prepares students for real-world clinical challenges by enhancing critical thinking, decision-making, and communication skills. By inculcating teaching methodologies like problem-based learning and team-based learning, students have a great chance to learn and exhibit skills like analyzing patient scenarios, initiating teamwork, fostering the quality of communication, and applying theoretical knowledge to complex, dynamic situations, ultimately leading to better patient outcomes.

### *3.4.1.2 Skill development*

It is important to understand that the professional growth of the student is not only measured in terms of intellect and cognition. However, at the same time, it is vital that we as teachers appreciate student's holistic growth as well. One of the most crucial elements of the teaching-learning process is development of skills. Students have the opportunity to develop a range of new skills that will help them advance both personally and professionally because of their insightful teachers. Among the abilities that students must acquire are the following (Fig. 3.3):

FIGURE 3.3 **Skill development.** This figure shows the skill development.

**1. Communication skills**

Communication is an integral part of medical education process. It has gained a lot of importance as it forms the basis on which healthcare operates. It is only through proper and effective communication that one can achieve the best possible result in terms of patient care and quality of healthcare. For medical students, communication skills are essential because they facilitate improved rapport, trust building, and sharing of ideas within their peers or the healthcare team. Inculcating these skills is imperative on the part of the teacher as it not only improves the learner but also prepares him or her to exhibit highest quality of patient care and become a lifelong learner.

**2. Collaboration skills**

For medical students, collaboration skills are essential because they facilitate interprofessional communication and teamwork, which enhance patient care, safety, and results. These abilities improve students' readiness for the contemporary, team-based healthcare environment, lower medical errors, and increase job satisfaction for healthcare professionals. In the end, more coordinated, effective, and comprehensive patient care results from a collaborative approach.

**3. Technology skills**

Obtaining proficiency using digital tools and technology for study, research, and practice has become the need of the hour. It is one of those skills that is being expected by every healthcare professional. For medical students, technology skills are essential because they increase learning through interactive platforms like virtual or augmented reality and simulations, help them make better decisions, and get them ready for the current healthcare industry, where technology plays a key role in patient care and diagnosis. These abilities are necessary for practicing in a secure setting, gaining access to a wealth of resources, and adjusting to a future in which the medical industry is tech-enabled and data-driven. It should be understood that the use of technology needs to be closely monitored by the teacher so that it does not override the basic knowledge and skills of the learner

**4. Life skills**

For medical students, life skills are vital as they mold the learner to be a better person. Inculcating life skills such as teamwork, communication, problem-solving, stress management, and critical thinking have proved to be an integral part of medical education and practice. In this regard, the role of teacher is vital as they are in a best position to share their experience based on what benefits are obtained by developing these above-mentioned skills. The teaching of life skills should not be done as a separate activity, but rather it should be a part of the day-to-day activities.

#### *3.4.1.3 Attitude and values*

Finally, with the correct mindset, learning can continue well beyond adulthood through intellectual and holistic development. The teaching-learning process is pointless if the learner lacks curiosity, the capacity to distinguish between good and bad, and an awareness of their civic duties. In this regard, no matter how good the instruction is, it will all be ignored. Some essential attitudes and values that should be inculcated in students are as follows (Fig. 3.4):

**1. Curiosity of learning**

Teachers should establish a secure learning atmosphere, use inquiry-based and active learning strategies, promote questioning and reflection, and serve as passionate role models in order to stimulate medical students' interest. Using interactive education, making resources available, gamification, and encouraging a mindset that views mistakes as chances for learning rather than failures are some specific tactics that can be used to enhance the curiosity.

**2. Open-mindedness and respect**

It is vital to imbibe a sense of open-mindedness and respect amongst the learners from the very beginning of their career. Students should be able to appreciate and accommodate their colleagues who come from various regional and cultural backgrounds. By learning to be open minded and respectful of others, the learner develops himself to be a better individual. Being open minded and respectful will be very beneficial to the learner as it provides a secure and encouraging learning atmosphere. It also enables the learner to address various viewpoints, incorporate experiential learning through community service, and employ role-playing and simulations to develop empathy and communication skills. Activities involving the role of interprofessional set up such as collaborative group projects or a collaborative healthcare should be encouraged to emphasize these skills.

**3. Ethical responsibility**

Teachers should instill in their students a sense of moral behavior, social responsibility, and global awareness. As a result, they will grow up to be more morally upright citizens. Using a mix of case-based and interactive learning, formal ethics instruction, and creating an atmosphere of ethical role modeling are all ways to instill ethical responsibility in medical students. This involves making ethics applicable to clinical settings and practical through techniques including role-playing, problem-based learning, and case discussions. Students can also acquire the required moral compass through mentorship, reflective practice, and open discussion of ethical topics like patient autonomy, confidentiality, and professional conduct.

**4. Personal growth**

To support their holistic development, teachers should encourage their students to imbibe qualities of self-awareness, confidence, and sense of purpose. The early development of a strong sense of self will assist students. Encouraging medical students to have a growth mindset by prioritizing effort over skill, redefining failures as teaching moments, and asking for helpful criticism in order to support their personal and professional development will go on a long way to modify the personal and professional life of the learner. The role of a teacher or a mentor towards instilling the quality of personal growth in the learner is of prime importance. The art of setting goals, time management skills, management of stress, appreciating the strengths of the learner, and giving constructive feedback all contribute to personal growth.

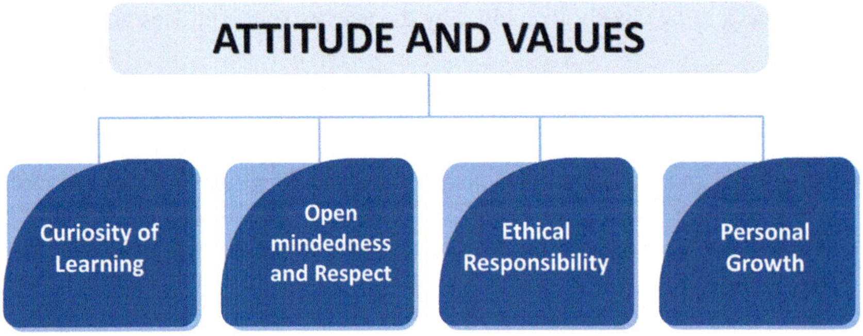


FIGURE 3.4 **Attitude and values.** This figure shows the attitude and values.

### 3.4.2 Elements of the teaching-learning process

The teaching-learning process consists of numerous essential components. Together, these essential components which may appear insignificant on their own form a well-oiled mechanism that facilitates effective teaching and learning. All of these elements can serve as useful recommendations to enhance the teaching-learning procedure. A holistic teaching-learning plan may not be as effective overall if even a few components are missing. Each of the below mentioned elements are crucial to attain success in terms of quality of medical education as well as patient care (Sukrajh, Adefolalu, 2021).

#### 3.4.2.1 *Teachers*

In modern medical education, there has been a tremendous shift in the role of a teacher. Teachers who organize, carry out, and assist in the teaching-learning process also play a critical role in all aspects of medical education. They are knowledge providers, facilitators, role models, or mentors, who have the capacity to influence the behavior of the learner. Apart from that, teachers strive hard to promote a positive mindset amongst their students so that they apply their knowledge, skills, and attitude towards achieving their goals.

#### 3.4.2.2 *Learners*

Students actively sought to expand their knowledge and abilities. Learners in the medical teaching-learning process are active participants who identify their own needs, establish objectives, locate resources, and critically assess information rather than passive recipients. Their work entails deliberate participation through practical exercises, deliberate observation, and critical thought, which enhances comprehension and aids in the development of their professional identities as doctors.

#### 3.4.2.3 *Curriculum*

The curriculum serves as the main framework for medical education, directing the learning process by establishing goals, subject matter, instructional strategies, and evaluation procedures in order to generate qualified medical professionals. It integrates several educational components like clinical practice, professional identity development, and basic sciences, and it guarantees a logical evolution of information and abilities. It also adjusts to changing demands. An organized and thorough educational experience is produced by a well-designed curriculum, guaranteeing that students acquire the necessary skills and are ready for their future positions in the healthcare industry.

#### 3.4.2.4 *Instructional methods*

Many strategies are used to spread knowledge and encourage learning. These can include interactive conversations with professionals, educational excursions and field trips, application-based stimulating activities, research opportunities such as projects and essays, and visual and auditory aids such as films and videos. In order to make learning dynamic, applicable, and effective in medical education, instructional approaches must go beyond passive lectures to include clinical problems, use modern tools, and accommodate different learning styles. They assist students in developing clinical skills and preparing for the rigors of professional practice by applying their knowledge of the basic sciences to complex, real-world patient care scenarios.

#### 3.4.2.5 *Assessment and evaluation*

Assessment and evaluation are of prime importance in medical education. Assessment both formative and summative conducted in the forms of exams, group projects, and feedback are just a few of the many methods used to evaluate the learner. These assessments when done properly have the ability to inspire students and assist them in identifying their strengths and limitations. It is also used as a platform from where it is possible to gauge learning objectives and guide curriculum development. They help teachers modify their teaching style, understand the needs of the learner, and help them to perform to the best of their abilities.

#### 3.4.2.6 *Learning resources*

The learning resources are referred to all those things that support the teaching and learning process. Learning resources include things like textbooks, research papers, journals, interactive smart classroom aids, online databases, simulation software, and online reference tools. Learning materials are crucial to the teaching-learning process in medical education as they make learning engaging and dynamic, provide up-to-date information, and accommodate different learning styles.

They promote self-directed learning for lifetime competency, improve understanding and retention, and aid in the development of clinical reasoning through practical experience.

#### 3.4.2.7 Classroom environment

It is the physical and mental environment in which learning takes place. A good, supportive learning environment promotes growth, the development of a healthy curiosity, and a positive attitude toward learning and knowledge absorption. Since a supportive environment encourages teamwork, participation, and in-depth learning, the classroom setting plays a significant role in medical education. It affects students' desire and capacity to cope with the demands of medical school, which in turn affects their conduct, academic performance, and general well-being. The way the curriculum is delivered, the institutional culture, and the caliber of interactions between students and staff are important considerations.

#### 3.4.2.8 Technology

Technology has indeed become an integral part of medical education. As critical components of medical education, it is vital that both the learner and the teacher should be aware of the common technologies that are used in medical education. The teaching-learning process in medical education is greatly aided by technology, which offers interactive and easily accessible learning opportunities through resources like electronic devices, virtual reality, and simulations. It facilitates remote learning and cooperation, enhances decision-making and skill acquisition, and permits practice in secure settings.

#### 3.4.2.9 Feedback and reflection

Feedback is an essential component of the teaching-learning process for continuous improvement. A positive and constructive feedback promotes effective assessment and reflection, which results in continuous development. As they encourage self-regulated learning, enhance clinical competency, and cultivate lifelong learning habits, feedback and reflection are indispensable part of medical education. Reflection allows students to critically examine their experiences, incorporate new information, and hone professional skills, while feedback offers an outside evaluation to pinpoint performance gaps and strengths. When given in the right way, feedback and reflection have the ability to boost the confidence, enhance the experience, and bring about a positive change in the individual.

#### 3.4.2.10 Communication

Teachers and students should always be able to communicate clearly and effectively. It is important to understand that communication is a two-way process between the student and teacher. It is the duty of the teacher to encourage communication and develop those skills in the student in order for the student to absorb and grow appropriately. It not only enables the teacher and the learner to share ideas but also enables the learner to execute quality patient care in the best possible way.

#### 3.4.2.11 Motivation

This speaks to the factors, both internal and external, that affect students' engagement and excitement. The teaching-learning process is made much more worthwhile when students are motivated because they are more likely to be involved and excited about learning. Since motivation promotes student engagement, in-depth learning, and academic achievement, it is essential to comprehend the various factors that affect the learner and strive to work in that direction for the best results. It affects a student's perseverance, health, and capacity to become a skilled physician. Regular motivation from teachers and peers go a long way in attainment of academic success and at the same time help the individual to gain much needed confidence and inculcate the habit of lifelong learner.

#### 3.4.2.12 Adaptability

Adaptability is essential in medical education because it gives students the resilience and flexibility required for a changing sector. It enhances academic achievement, fosters a lifelong learning mindset that is crucial for patient care, and helps students handle novel or uncertain situations. A teaching-learning process should be adaptable in its implementation to meet the diverse needs of learner. Finding each student's individual strengths and weaknesses, learning capacity, intellectual abilities, and areas of aptitude is crucial to ensuring a smooth learning experience. After that, the curriculum and instructional techniques ought to be adjusted appropriately.

#### *3.4.2.13 Learning style assessment*

Different learning styles and preferences among students should be properly recognized and accommodated. As an educator, the teacher should be flexible enough to experiment the different learning styles and find out the best possible one which can be applied to the student. The learning style should be tailored to the student. A truly inclusive learning environment can be created by evaluating and modifying the lesson plan to meet the needs of kinesthetic, visual, and auditory learners, among others. Each diversity should be valued and acknowledged, and the teaching-learning process should be modified to accommodate them. A crucial part of medical education is learning style evaluation, which benefits both educators and learners by revealing personal preferences and improving instruction and learning results. Through the use of various teaching techniques, it enables teachers to adapt their lessons to a range of learning preferences while enabling students to employ efficient study techniques and develop their capacity for self-directed learning. By incorporating these elements, educators can create a dynamic and effective teaching-learning environment to facilitate meaningful learning experiences for students.

## 3.5 Teaching cycle

This systematic approach to education ensures that the learner receives a well-rounded learning experience. It is a guiding concept that promotes efficient conduction of all the processes in medical education. Every aspect of the teaching cycle is important as it significantly contributes to the success of the medical program (Patel, Kalkat, 2019). This cycle comprises five essential stages. Every step is the teaching cycle showcases active involvement of various elements like the teacher, learner, and the administration (Fig. 3.5).

### 3.5.1 Identifying needs

The first and most crucial phase in the teaching cycle is determining what your students need. A medical educator should be able to recognize the different demands of the student. It is important to understand what is expected from the learner out of the program and whether their needs are justified by the teaching process. A needs assessment is a best way to find out and understand the various needs of the learner. The results of the initial evaluation will likely determine whether your learner has any special needs or whether you agree to personalized learning plans and needs. For instance, you must determine the educational background of the learner and determine if he or she needs special assistance. During this phase of the teaching cycle, you must make sure that every student can meet the objectives and make progress in their education. The demands of your student and assisting them in moving on to the next phase are the main focus of this starting point. A teacher should always be ready to provide students with as much support and direction as possible.

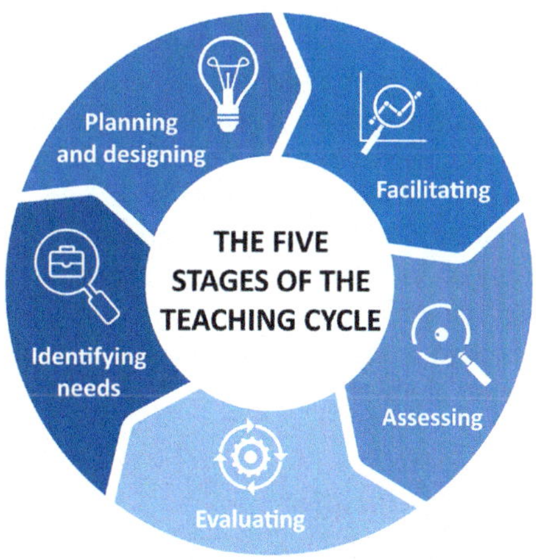


FIGURE 3.5 **The five stages of the teaching cycle.** This figure shows the five stages of the teaching cycle.

### 3.5.2 Planning and designing

Planning and designing is the next phase of the teaching cycle. It is good to revise the course content and the teaching and learning methods and to update them every time. The course reports and the feedback forms, if any, should be discussed in the department to smoothen the process of teaching and learning. It is imperative to make sure that all the objectives and criteria of that program are met. The designing of the lesson plan or the blue print should always be done in alignment with the course learning outcomes. The selection of the instructional methods and the assessment methods are very critical for the successful outcome and this should be kept in mind while planning and designing the topic.

### 3.5.3 Facilitating

The delivery of content is the focus of this phase of the teaching cycle. Delivering the content in the right way so that it meets the expectations of the learner is important. It is vital that the learner is inspired and benefitted from the topic. Additionally, we must ensure that the students are aware of and comprehend the lesson's objectives. It is crucial that students understand what is expected of them after they leave the classroom. Facilitating a teaching-learning process is indeed an art that has to be mastered by teachers. As every instructional method is different and requires a different set of skills, it is imperative that the teacher is ready to adapt to the need of the situation. At the same time, successful conduction of any activity depends on the reaction given by the students to the efforts taken by the teacher.

### 3.5.4 Assessing

As educators, we are in charge of the learners' ongoing evaluation and growth. It's crucial to employ a variety of assessment techniques to make sure that students are engaged and comprehend the lesson. The use of both formative or summative are beneficial in this regard. The goal of formative assessment is to track students' progress. This gives you continuous feedback that you may utilize to enhance your instruction. Students can use this to determine their strengths and shortcomings. As a teacher or instructor, we should take keen interests in the growth of the student and encourage them by highlighting their positive aspects. A good teacher is also able to understand the difficulties that the learner might be facing and motivates him or her to achieve better results. A summative assessment, on the other hand, is used to gauge how well the students have learned and understood the concepts. This would occur at the conclusion of a course, after which it could be compared to a benchmark or standard. When thinking about the next activities or courses for your learner, this will help you focus your efforts on what needs to be done and how can the learner improve his grades. Both of these are intended to give the teacher and the student a clear picture of the learner's development and level of knowledge. For you to be able to offer assistance or make adjustments to enhance students' success, every evaluation you employ must be trustworthy and consistent.

### 3.5.5 Evaluating

Assessing the overall efficacy of the teaching and learning process is the last step in the teaching cycle. To determine what went well and what needs improvement, educators evaluate their instruction, course materials, and evaluation techniques. Getting input from other people, like your students and coworkers, is an important part of the assessment phase. Annual reports from the curriculum committee or course reports from the course coordinators are a great way to look at what worked and what didn't and to list down the steps to be taken for improvement. The evaluation process should be done on a regular basis to attend the problem as early as possible. Surveys, conversations with students, panel discussions, and peer observations are some ways that teachers can get input about the course's advantages and shortcomings. Teachers can modify their future teaching methods as needed in light of the evaluation results. This procedure optimizes the chances of student achievement and guarantees ongoing enhancements to the quality of instruction.

### 3.5.6 The teaching cycle as a guiding principle

The teaching cycle is still a vital framework for anyone looking to improve their teaching and learning methods and maximize the learning experience for their students in medical education, even as education continues to change and advance. By embracing every phase of the teaching cycle, educators may enable their students to develop into lifelong learners and provide them the abilities and traits necessary to prosper in a rapidly changing and dynamic world.

## 3.6 Role of the teacher in teaching-learning process

A teacher's role in the teaching-learning process is vital and complex, involving a range of duties and tasks. The teacher is essentially at the center of the teaching-learning process, and throughout the learning process, he or she wears multiple hats. Students won't even gain the skills they need to remember material if the teacher doesn't create a comprehensive learning environment (Murati, 2015). A few key roles of the teacher are described below (Fig. 3.6):

**1. Facilitator of learning**

As a facilitator in medical education, a teacher helps students become self-directed learners by establishing safe spaces, posing thought-provoking questions, encouraging critical thinking, and modeling clinical reasoning. The emphasis shifts from being the only source of knowledge to enabling students to create their own understanding, manage their learning, and develop lifelong skills through active engagement, reflection, and constructive feedback, as well as by serving as a mentor, coach, and role model (Maniago, Lumapenet, Balayanan, 2023).

**2. Knowledge source**

Teachers are a primary source of information and expertise in their subject areas. They should convey content effectively and organize it, ensuring that students have a firm foundation of understanding. The teacher's function in contemporary medical education has significantly changed from being the main "sage on the stage" that is single transmitter of knowledge to a "guide on the side," whereby the teacher represents a facilitator of learning. Teachers still serve as informational resources, but their capacity to promote active learning, mentorship, and the growth of clinical reasoning determines their effectiveness in the modern era of medical education.

**3. Mentor and role model**

Teachers in medical education serve as mentors who offer individualized guidance, support career development, encourage critical thinking, and assist in navigating complex challenges, bridging the gap between knowledge and wisdom to cultivate compassionate, competent physicians. They also act as role models by exhibiting professional behavior, clinical skills, ethics, and attitudes, greatly influencing students' identities. Their impact goes beyond textbooks as they impart vital professional principles through their attitude and behavior, thereby upholding role modeling as an essential component of students' personal and professional growth (Wijnen-Meijer, 2022).

**4. Curriculum developer**

In order to ensure that the curriculum is pertinent, useful, and satisfies changing patient care needs, teachers in medical education play a critical role as curriculum developers by offering their expertise to design learning objectives, choose

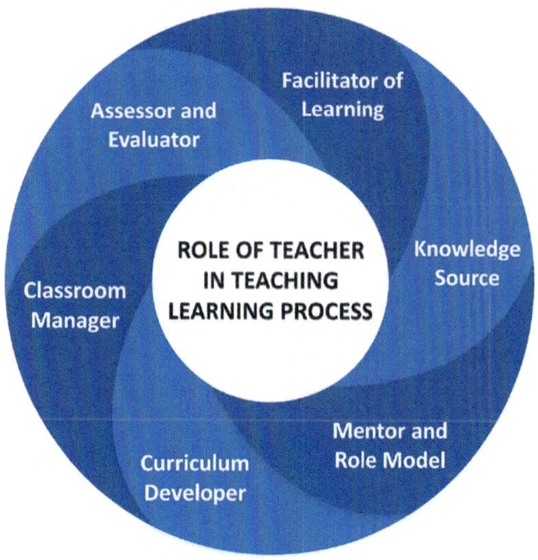


FIGURE 3.6 **Role of teacher in teaching-learning process.** This figure shows role of teacher in teaching-learning process.

content, and develop effective strategies. They go beyond simply imparting knowledge to become facilitators, mentors, and change agents. In order to close the gap between theory and practice, they need continual pedagogical training. Their engagement guarantees alignment with real-world clinical demands, cultivates critical thinking, and changes professional attitudes through role modeling.

**5. Classroom manager**

A teacher serving as a classroom manager in medical education is in charge of establishing a secure, orderly, and engaging atmosphere that promotes professional, cognitive, and clinical development in addition to upholding discipline. To increase student engagement and close the gap between theoretical knowledge and clinical application, they oversee the learning environment, whether it is a lecture hall, a small group discussion, or a clinical bedside setting.

**6. Assessor and evaluator**

The assessment of the teaching-learning process falls under the purview of teachers. Teachers should create customized tests to measure students' learning and development. In order to help students grow, they must also provide them feedback and adjust their teaching strategies to meet their requirements. They must exercise their discretion and determine the best academic course of action for their students. In essence, a teacher's role in the teaching-learning process is dynamic and goes beyond simply dispensing knowledge. Good instructors foster their students' social, emotional, and ethical development in addition to their academic progress, making a substantial contribution to their overall development. There are no bad students, as we have all discovered. And by wearing all of these hats and embodying these roles as educators, we can become better teachers.

## 3.7 Improving the process of teaching and learning

Enhancing the teaching-learning process is a complex issue that affects both educators and students. A greater portion of this duty falls on teachers, who act as the facilitators in this situation (Tripathi, 2018). The following are some tactics that can be used to enhance the teaching and learning process:

**1. Create a positive learning environment**

As a teacher, it is essential that we strive towards creating a welcoming and inclusive environment where students can freely express their opinions. To encourage a polite and cooperative learning atmosphere, it is necessary that the teacher must set clear expectations and guidelines from the start. Students will genuinely develop into strong-willed individuals who aren't afraid to voice their opinions and be authentic when they feel at ease and accepted in the classroom.

**2. Stay updated on educational trends**

As educationists, it is important that we seek for opportunities for ongoing professional development to keep up with the latest developments in technology, teaching strategies, and educational research. By maintaining flexibility and adaptability in our approach, we as educators can contribute to the best of our abilities to the success of the program. Moreover, staying updated proves to be of great benefit for the student as they can have full benefit from our knowledge and expertise. Attending specialized seminars, educational retreats, research conferences, and expert sessions will help you continue to advance as mentors and educators. At the same time, faculty development programs at the departmental or college level for the teaching fraternity is also important as it gives the faculty a chance to stay updated and hone their teaching skills.

**3. Promote critical thinking**

It is vital that we as educators impart the skill of critical thinking in our students. It is crucial to design tasks and exercises that go beyond simple memory and application of factual knowledge. It is necessary to motivate the student to evaluate, analyze, and apply what they have learnt. Giving students the ability to think critically helps them approach problems correctly and find the best solution.

**4. Establish real-world connections**

In today's world, learning and education cannot be confined to classroom. Whenever there is a chance, the teacher should be able to create opportunities so that the student is able to learn with, from, and each other. By encouraging to learn in an interprofessional way, the students have a chance to connect with the outside world. Students can be better equipped to address actual societal concerns by learning about real-world situations. It is essential to include subjects that call for real-world applications in order to assist students in understanding the practical significance of what they are studying. The connection between theoretical knowledge and its practical applications is strengthened by field trips, real-world case studies, and guest lectures.

**5. Cultivate a growth mindset**

Instead of concentrating only on grades, encourage students to have a growth attitude by recognizing their efforts, tenacity, and resilience. Students are more receptive to learning when they feel that their efforts are valued. Students should be taught that hard work and dedication may enhance and refine their innate talents, intelligence, and abilities. Students should be encouraged to identify their area of strength and develop it for their future pursuits.

**6. Incorporate technology**

Instead of shunning technology, it is vital that the present-day teachers embrace the rapid advances in educational technology. Thc days of a classroom with just a blackboard and chalk are long gone. Several educational technology solutions are available to broaden the scope of the teaching and learning process and improve the smoothness and engagement of the classroom experience. In the 21st century, it is highly recommended to use technologies such as interactive whiteboards and displays, online assessments and analysis tools for easier testing, collaborative learning platforms for a peer-based learning environment, and augmented and virtual reality tools for a more interactive classroom (Kofi, Ocansey, 2021).

**Summary**

Teaching-learning process is the heart of education. It is the most powerful instrument of education to bring about desired changes in the behavior of students. The importance of the teaching-learning process in medical education lies in developing highly competent and effective healthcare professionals by equipping them with essential knowledge, skills, and attitudes, fostering lifelong learning, and ensuring patient-centered care. It ensures that students learn effectively in clinical settings, understand professional values, become excellent communicators, and can adapt to the complexities of medicine, leading to better patient outcomes and a more robust healthcare system.

**Key point**

**Future perspectives: transforming teaching via technology**

In order to produce more efficient and flexible training, medical education teaching methods of the future will combine cutting-edge technology with individualized and active learning. The widespread use of artificial intelligence and virtual/augmented reality for personalized, simulation-based learning, as well as a stronger focus on competency-based education, team-based learning, and flipped classrooms to enhance clinical skills and patient outcomes, are important trends that have the potential to transform medical school education. Incorporating modern technologies into the teaching process should be prioritized in a process to move into the future. At the same time, more stress should also be given to regularize the training of the faculty in order to make them ready to counter any new challenges that they might encounter.

## References

Arslan, Y., Polat, S., 2016. The relationship between teachers' trust in students and classroom discipline Beliefs. International Education Studies 9 (12), 81. Available at https://doi.org/10.5539/ies.v9n12p81.

Bidabadi, N.S., Isfahani, A.N., Rouhollahi, A., Khalili, R., 2016. Effective Teaching Methods In Higher Education: Requirements And Barriers j. Adv Med Educ Prof 4 (4), 170–178.

Hathur, B., Kulkarni, P., 2024. Changing Roles of Medical Teachers in the Era of Competency-based Medical Education. APIK Journal of Internal Medicine 12 (1), 1–3. Available at https://doi.org/10.4103/ajim.ajim 100 23.

Kofi, D., Ocansey, W., Xu, Z., Zhang, X., Xu, W., Qian, H., Mao, F., Dickson, K.W.O., 2021. Current and Emerging Medical Education Teaching Methods -Keeping up with the Evolving Medical Sciences. Revista de Educaci6n (Madrid) 392, 52–104.

Maniago, K., Lumapenet, H., Balayanan, H., 2023. Teachers' role towards teaching and learning process. International Journal of Advance Research And Innovative Ideas In Education 9, 4813–4815.

Mitchell, R.M., Kensler, L., Tschannen-Moran, M., 2018. Student trust in teachers and student perceptions of safety: positive predictors of student identification with school. International Journal of Leadership in Education 21 (2), 135–154. Available at https://doi.org/10.1080/13603124.2016.1157211.

Murati, R., 2015. The role of the teacher in the educational process. New Horizons in Education 5 (2).

Patel, A., Kalkat, H., 2019. Teaching models in the clinical years of medical education. Advances in Medical Education and Practice 10, 309–310. Available at https://doi.org/10.2147/AMEP.S186141.

Peña, S., Martinez-Santander, C., Cantos-Reyes, V., Rojas, M.F., 2025. Active methodologies and didactic techniques in medical education: A systematic review. Educación Médica 26 (4), 101040.

Plaut, S., Baker, D., 2011. Teacher-student relationships in medical education: Boundary considerations. Medical Teacher 33, 828–833. Available at https://doi.org/10.3109/0142159X.2010.541536.

Sukrajh, V., Adefolalu, A.O., 2021. Understanding learning and the components of the learning process in medical education: a review of the literature. European Journal of Education and Pedagogy 2 (1), 69–72. Available at https://doi.org/10.24018/ejedu.2021.2.1.53.

Tripathi, R., 2018. Importance and improvements in teaching-learning process through effective evaluation methodologies. ESSENCE International Journal for Environmental Rehabilitation and Conservation 9 (2), 7–16. Available at https://doi.org/10.31786/09756272.18.9.2.202.

Wijnen-Meijer, M., 2022. Learning to teach in medical education. GMS J Med Educ 39 (1), Doc14. Available at https://doi.org/10.3205/zma001535 PMID: 35368834; PMCID: PMC8953194.

Wubbels, T., Brekelmans, M., 2005. Two decades of research on teacher–student relationships in class. International Journal of Educational Research 43 (1-2), 6–24. Available at https://doi.org/10.1016/j.ijer.2006.03.003.

Chapter 4

# Values in medical education

*"The aim of education is the knowledge, not of facts, but of values."*

—William Ralph Inge

## 4.1 Introduction

Medicine and humanity are deeply intertwined, with medicine's scientific and technical aspects complemented by its humanitarian "art." Humanity is expressed through the practice of medicine in the physician–patient relationship, which requires compassion, empathy, and respect, leading to better patient care and outcomes. The field of medical humanities specifically focuses on this relationship by studying the human aspects of illness through the arts and social sciences, which helps healthcare professionals connect with patients on a deeper level and combat burnout. The development of capable and moral healthcare workers depends heavily on the values taught in medical school. In this regard, the teacher plays a pivotal role in not only inculcating values amongst the students but at the same time be a role model and demonstrate highest degree of values to the students. These values are heavily embedded into our curriculum and every effort is made to teach them in the best possible way so that everyone stands to gain maximum benefit.

In order to develop professional identity, enhance decision-making in challenging circumstances, and advance moral and compassionate patient care, values are essential. By directing a doctor's behavior toward fundamental values like justice, integrity, and respect for others, they serve as the cornerstone of professionalism and, in the end, strengthen the healthcare system. According to Micheal Porter, values in healthcare is defined as the achievement of "outcomes that matter to patients" relative to costs.

## 4.2 Need for value-based education

In modern medical education, the need to shift to value-based education has gained importance as it moves medicine beyond a volume-based to one that rewards providers for delivering high-quality care and achieving good patient health outcomes. Moreover, it ensures that medical students develop critical soft skills like empathy and ethical decision-making, which are essential for building trust and effectively partnering with patients. For the physicians, the value-based system is beneficial as it trains them to be mindful of costs while still delivering high-quality care and to use data and analytics to make informed decisions that improve system efficiency (Nafea, Ebtihaj, 2024).

## 4.3 Societal expectations

Understanding and appreciating what society expects of a doctor is crucial before moving forward as health professional educators. Only then will we be able to defend our educational approach and benefit the community as a whole. A doctor must meet societal standards for competence, morality, and accountability as part of a social contract built on trust and a dedication to the welfare of patients. Clinical knowledge, sympathetic and caring communication, and openness in describing diagnosis and treatment strategies are important requirements. Additionally, doctors are supposed to respect patient autonomy, actively listen, and advocate for both individual and community health. A very robust value-based education system is required to demonstrate all of these to the greatest extent possible.

## 4.4 Value system in medical education

Medical education has long been associated with imparting knowledge, skills, and a positive attitude to the student. The value system in medical education, however, has been getting a lot of traction lately. It highlights how important it is for a

**The Principles and Practice of Medical Education. DOI:** https://doi.org/10.1016/B978-0-443-44197-4.00026-6

healthcare professional to inculcate values in his personality and effectively use them when making critical judgments. In medical schools, values are taught as through a combination of structured curriculum, clinical experiences, and mentorship to foster professional formation and ethical reasoning (Sturmberg J P, Taher S, 2025).

## 4.5 Key values in medical education

Key values in medical education are the ethical and professional principles that guide the practice and behavior of healthcare professionals. Imparting fundamental values and qualities such as empathy, ethics, honesty, and respect not only gives a chance for the healthcare professional to exhibit highest quality of patient care but at the same time helps them to build trust and to have an excellent doctor–patient relationship. In order to guarantee an excellent patient care and safety, a value-rich educational experience is very much needed that will push the students to comprehend and combine their personal beliefs with professional behavior.

In addition to humanism, empathy, humility, professional integrity, and a dedication to high-quality patient care and lifelong learning, medical education values also place a strong emphasis on autonomy, beneficence, nonmaleficence, and justice. These principles direct moral behavior, support career advancement, encourage civil and cooperative settings, and mold the attitudes and actions of aspiring doctors. Education and values are closely related to each other. The latter forms the foundation of the educational philosophy from which the curriculum, program, and values that must be emphasized in the curriculum are derived. Medical students are expected to embody core values like integrity, honesty, empathy, and respect to maintain public trust in the profession. They are also expected to demonstrate qualities such as dedication, discipline, and a commitment to lifelong learning to handle the demanding nature of medical training and practice. In order to incorporate values in medical education and to uphold the importance in students' personal and professional growth, many countries have taken active steps to change the medical curricula so that values can become an active part of the teaching and learning process. In an effort to bring humanistic concepts back into medical practice, a global partnership created the International Charter for Human Values in Healthcare, which was founded in early 2011 and serves as a foundation of fundamental values for all healthcare interactions. The goal of this charter was to improve the human aspects of healthcare interactions, including those between physicians and patients, interprofessional and other teams, across departments, and with coworkers and other stakeholders (Fig. 4.1).

### 4.5.1 Key ethical principles

#### *4.5.1.1 Autonomy*

One of the key ethical principles of medical education is autonomy, which incorporates a student's sense of self-governance in their education and empowers them to take charge of their education, make educated decisions, and mold their educational journey. In order to encourage the ethical principle of autonomy, it is vital that the student should be involved in decision-making for class rules, allowing them to choose topics or project formats and guiding them to set

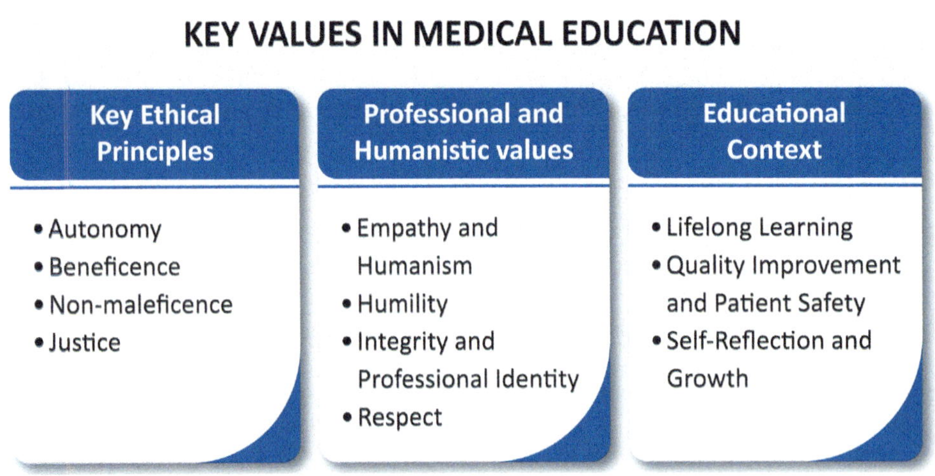


FIGURE 4.1 Key values in medical education.

goals and monitor their own progress. Medical teachers should regularly reflect on their teaching approach as it will determine their students' motivation. Autonomy-supportive teaching enhances intrinsic motivation, which means doing something because it is interesting and enjoyable.

The role of a teacher is vital for the successful implementation of the autonomy. In this regard, the teacher should be able to make a shift from authoritative to a more flexible approach. This is because a controlling style of teaching enhances extrinsic motivation, which means doing something because of outside pressures. By using simulation-based learning rather than lectures to teach certain topics like examination of cardiovascular system learning may be made more interesting for students. Additionally, by outlining the significance of the subjects and allowing students to choose their own learning strategies and evaluation dates, a sense of autonomy and increase intrinsic motivation is fostered amongst the students because they will feel accountable for their education. With the backing of self-determination theory, it encourages a sense of agency and alignment with personal beliefs in the classroom, which in turn promotes internal motivation, deeper learning, psychological well-being, engagement, and the growth of a professional identity. It suggests that when healthcare providers support patients' feelings of autonomy, competence, and relatedness, it fosters autonomous motivation for health behaviors, leading to better adherence and outcomes.

#### 4.5.1.2 Beneficence

The idea that doctors have an obligation to act in their patients' best interests and try to "do good" by advancing their welfare and interests is known as beneficence in medical education (Purwati, Ani & Ruslita, Gita, 2025). This means actively assisting patients, ensuring their safety, removing hazardous situations, and speaking out for their needs. In order to enhance health outcomes and preserve the dignity of people they serve, medical school students are taught to prioritize patient welfare, build trust, and apply moral principles like beneficence. This can be taught by integrating beneficence into a longitudinal curriculum that addresses ethics, professional competence, and clinical reasoning. All these aspects are critical for fostering a strong commitment to patient welfare. A healthcare provider can put the beneficence principle into practice in a clinical context by offering excellent, patient-centered care that involves advancing patients' needs, preventing harm, and promoting their well-being. This is accomplished by honoring patient autonomy, collaborating with other medical professionals, and paying close attention to a patient's physical and mental well-being.

#### 4.5.1.3 Nonmaleficence

It is critical to recognize that the concept of nonmaleficence, or "to do no harm," is fundamental to the success of medical education and the achievement of students' career goals. The final goal is to make sure that the patient's potential advantages outweigh any risks or discomfort. In this regard, a strong emphasis is placed on minimizing intentional harm and lowering patient risks by balancing potential advantages and disadvantages. In order to inculcate the "do no harm" ethic in medical students, nonmaleficence education combines classroom instruction with real-world application. Instead of merely teaching medical students about ethical rules through lecture, cutting-edge teaching strategies like problem-based learning not only promote collaborative, student-centered, and self-directed learning, but they also inspire them to learn how to analyze the facts, reason, and think critically, which will enable them to make complex ethical decisions. This will not only ensure that the decisions taken are in the best interests of the patient but also at the same time certify that no harm has been caused to the patient.

#### 4.5.1.4 Justice

The capacity to fairly represent all parties involved is one of the hallmarks of high-quality medical education and care. Making sure that everyone has equitable access to healthcare services, regardless of their socioeconomic status, race, ethnicity, or other traits, is at the heart of this principle. Students need to be taught more than only skills, information, and attitudes in order to develop critical awareness, or consciousness of oneself and others. Understanding and appreciating the value of justice and how it applies to day-to-day activities is crucial for learners. Students have the opportunity to demonstrate the quality of justice by participation in role-playing or group activities. In order to safeguard patients' rights, prevent discrimination, and ensure that resources and benefits are allocated equitably within the healthcare system, it is critical to respect the principles of justice when providing healthcare. One of the best ways to teach the value of justice, according to medical educators and faculty involved in the teaching and learning process, is to incorporate social justice into the curriculum and teach ethical principles.

## 4.5.2 Professional and humanistic values

### 4.5.2.1 Empathy and humanism

Recognizing the intrinsic value and dignity of every person, as well as sharing and comprehending the emotions of others, is the core principle of empathy and humanism. Acknowledging the qualities of empathy and humanism is important for a medical student as it forms the basis of medical practice. In the context of medicine, compassion is the ability of medical personnel to identify, comprehend, and address a patient's suffering with empathy and a wish to lessen their suffering. It is to be understood that apart from treating patients, it is also important for a healthcare professional to develop a human connection with them, supporting them, and earning their trust. Positive patient outcomes, patient happiness, and the development of a robust healthcare system all depend on compassion. In order to acknowledge the human experience of illness and suffering, this value goes beyond a strictly clinical approach and emphasizes the significance of comprehending and sharing patients' sentiments. Inculcating this principle in medical students involves diverse educational methods like role-playing with standardized patients, bedside teaching, or while history taking and examination (Prabhu, 2011).

### 4.5.2.2 Humility

It has been well appreciated that humility fosters personal growth and stronger relationships by encouraging learning, empathy, and a willingness to admit mistakes. It also contributes to better problem-solving skills, improved decision-making, and greater emotional intelligence, while also helping individuals to handle stress more effectively. Cultivating humility allows for continuous improvement and can make someone more open to feedback and new perspectives. As a healthcare professional or one who is actively involved in patient care humility is one quality that should be at the top of the list. Practicing the quality of humility is the prime responsibility of a medical student. By being humble, a medical student is able to better understand, respect, and value the opinions of others. In the clinical scenario, being humble goes a long way in establishing a positive and strong relationship with peers, team members, colleagues, and more importantly with patients.

### 4.5.2.3 Integrity and professional identity

For a medical practitioner, exhibiting integrity, moral conduct, and a strong sense of self are some of the vital components of clinical practice. This value emphasizes how crucial integrity, openness, and adherence to moral standards are in all facets of medical practice, including teaching and patient care. The maintenance of integrity and professional identity forms the basis of both the patient–physician interaction and the healthcare system as a whole. It includes being truthful, reliable, and committed to acting in the patients' best interests. Upholding integrity in medical education involves being honest and ethical in academic and clinical settings, which includes fully engaging with coursework, maintaining patient confidentiality, respecting professional boundaries, and being truthful about one's own qualifications. Medical students should be encouraged to uphold these principles in their learning environment by being honest and truthful about their work, upholding academic policies, and following the rules and regulations they are intended to follow. On the other hand, professional identity can be exhibited by integrating professional values, behaviors, and knowledge into one's own identity. Activities that demonstrate professional identity in medical education include writing reflections, experiential learning in clinical settings, and mentorship. These activities allow students to develop self-awareness and internalize professional values, leading them to think, act, and feel like a physician (Desai, 2022).

### 4.5.2.4 Respect

Treating everyone with respect and decency, including oneself, coworkers, and patients, is one of the vital duties of a healthcare professional. By demonstrating the quality of respect, one wholeheartedly acknowledges each person's intrinsic autonomy, dignity, and worth. It is not only a professional duty but also a moral need that influences moral behavior and fosters trust between healthcare providers and their patient. Inculcating the quality of mutual respect amongst the students and teaching them to be appreciative and respectful towards their peers is one of the important responsibilities of the teacher. Compassionate care, recognition of patient choices, and guaranteeing their active involvement in health-related decisions are ways to show this regard. Activities that promote teamwork by encouraging communication, collaboration, and mutual respect among students and other healthcare professionals or those activities that require the person to provide and receive constructive feedback enforce the person to appreciate the quality of respect.

### 4.5.3 Educational context

#### *4.5.3.1 Lifelong learning*

Learning medicine is indeed a dynamic and ongoing process as it requires one to be continuously reading, updating, and looking for solutions all throughout the career of a medical professional. Owing to this, a medical student becomes a lifelong learner. Lifelong learning has also established itself as a fundamental tenet in medical education and practice. By imbibing the quality of lifelong learning, medical students are able to stay current, offer the best care possible, and enhance quality of healthcare. This habit of lifelong learning is not a one-day transition, rather it is a slow and continuous process which the student is able to achieve through pure dedication and perseverance towards medical education. As medical teachers, it is important to appreciate the strengths of the students and to instill confidence amongst the learner so that they are able to go through the challenges and adapt to this important quality. From the learners' point of view, he or she should be constantly self-motivated in order to enjoy this process. Attending various workshops, conferences, and seminars will enable them to learn new things and also make them more confident about the latest developments in the field of medical education.

#### *4.5.3.2 Quality improvement and patient safety*

Adherence to highest levels of quality and safety is one of the cornerstones of medical education. Teaching the basics of quality improvement process to medical students has gained importance as it enables the learner to deliver high-quality care to the patients by maintaining the highest degree of patient safety. Medical students with a sound knowledge of quality better appreciate the essence of academic achievement. It also prepares them for a more successful professional career. These learners are in a better position to exhibit their competency by providing excellent medical care to their patients. It is a well-known fact that poor quality care can be inherently unsafe and has to be addressed urgently. Quality improvement and patient safety are intimately related to each other because a showcase of high standard of patient care is a reflection of the quality standards that has been practiced at that clinical setting. To teach quality improvement and patient safety in medical school, it is necessary to introduce these concepts as elective courses. For better understanding and to highlight the importance of these topics in clinical practice, strategies like experiential learning or interprofessional approach should be used as a part of the teaching process.

#### *4.5.3.3 Self-reflection and growth*

Promoting the concept of self-reflection in order to help healthcare professionals strengthen their values, make better decisions, and feel more satisfied and psychologically healthy is an important part of medical education and clinical practice. Self-awareness, critical thinking, and lifelong learning are an essential component for professional development and better patient care. As we are all aware, reflection in medical education reflects the capacity of individuals to self-analyze themselves and understand their own strengths and weakness. It also gives the learner to think about the possible solutions which can be applied to a particular problem. Additionally, it enables the learner to learn from previous experiences, avoid the same mistakes, and at the same time apply those lessons to future actions. By doing a critical analysis of the situation, the learners are in a better position to make meaningful decisions, chalk their own path, and establish clear goals which will enable them to achieve both personal and professional milestones. In order to develop flexible, resilient, and self-reliant practitioners who can handle changing healthcare concerns, it is imperative that the habits of self-reflection and growth be ingrained in the students. Activities like portfolio writing, group discussions, multisource feedback, and mentoring practices can help the students to appreciate the importance of reflection in medicine.

## 4.6 Value-based medical education

An approach to medical education known as value-based medical education (VBME) places a strong emphasis on the value of patient-centered treatment, moral decision-making, and economical utilization of healthcare resources. It moves the emphasis from just providing healthcare services to optimizing patient value, taking their preferences into account, and enhancing health outcomes while controlling expenses. In addition to clinical expertise, this method highlights the significance of ethical, cultural, and communicative abilities. Moreover, it seeks to develop healthcare workers who are sympathetic, culturally aware, and dedicated to ongoing patient outcome improvement. In order to instill values in medical education, various strategies are used, such as updating the curriculum to incorporate the essential values of medicine, regularly training the faculty by emphasizing the development of their knowledge and skills in ethical decision-making, using interactive teaching methods, and designing the curriculum to cultivate a new generation of doctors who are patient-centered. At the same time, it is equally important to involve the learners in quality improvement initiatives (Supe A, Shah H, 2021).

### 4.6.1 Key aspects of value-based medical education

The goal of VBME is to produce doctors who prioritize the health of their patients, understand the economy, and appreciate the morality. In order to improve population health and support the well-being of the healthcare workforce, it is crucial to teach students how to prioritize patient needs and preferences, minimize waste, comprehend the financial implications of treatments, and work in integrated teams to address a patient's holistic health needs. It emphasizes the importance of patient-centered care, clinical competence, and professional responsibility, shifting the focus from simply delivering medical services to ensuring optimal patient outcomes. Some of the important aspects that are highlighted by VBME are as follows.

#### *4.6.1.1 Patient-centered care*

In clinical decision-making, VBME places a strong emphasis on comprehending and honoring patient preferences, values, and objectives. VBME and patient care put an emphasis on delivering high-quality, reasonably priced healthcare that puts the needs and outcomes of the patient first rather than just the volume of services rendered. In order to enhance health outcomes and cut down on needless expenses, this strategy places a strong emphasis on integrated care, communication, ethics, and the application of evidence-based approaches.

#### *4.6.1.2 Ethics and professionalism*

It encourages students to make morally good decisions in clinical practice by incorporating professionalism and ethical values into the curriculum. VBME places importance on incorporating professionalism and ethical concepts into medical education, producing doctors who are not only skilled but also have high moral standards and a dedication to the welfare of their patients. This approach aims to foster trust, integrity, and human values in healthcare interactions in order to strengthen the doctor–patient bond and improve the quality of care (Hilton SR & Slotnick HB, 2005).

#### *4.6.1.3 Interprofessional collaboration*

To enhance patient care and expedite procedures, VBME encourages cooperation and teamwork among medical experts from various specialties. Interprofessional education and VBME are related strategies that emphasize moral principles, collaboration, and patient-centered care in an effort to enhance healthcare. The significance of ethics, compassion, and patient preferences in medical practice is emphasized in VBME. Furthermore, interprofessional collaboration entails students from various health professions learning alongside one another to improve teamwork and communication in healthcare settings. When combined, they support a comprehensive strategy for healthcare that puts the health of the patient, moral behavior, and productive teamwork first (Gonzalo, 2017).

#### *4.6.1.4 Cost-effectiveness*

It emphasizes how crucial it is to make prudent use of healthcare resources and refrain from needless procedures or treatments. Cost-effectiveness and VBME are related ideas since the former seeks to give aspiring medical professionals the know-how to provide high-value treatment, while the latter offers a framework for evaluating the worth of medical interventions. In order to prepare a generation of doctors who can handle the transition to patient-centered and cost-effective healthcare systems, it is imperative that value-based care principles be incorporated into medical school.

#### *4.6.1.5 Continuous learning and improvement*

VBME promotes a dedication to evidence-based practice, lifelong learning, and ongoing healthcare delivery improvement. VBME places a strong prominence on the necessity of integrating humanistic, ethical, and economical concepts into healthcare education in addition to ongoing learning and development. This approach aims to prepare future doctors to deliver patient-centered care, make the most use of the resources at their disposal, and improve overall health outcomes. Continuous learning ensures that medical personnel stay knowledgeable about the latest advancements throughout their careers.

## 4.7 Role of a teacher in value-based medical education

A teacher's job in VBME is to model and emphasize professional principles and ethical behavior. Through conversation and introspection, the instructor also serves as a facilitator, helping students to establish and maintain their own principles. Teachers in a value-based education system incorporate values into the curriculum and regular classroom interactions while acting as mentors, role models, and facilitators. In addition to actively using techniques like role-playing, discussions, and community service to help students build a strong moral compass, empathy, and a sense of responsibility, they must live up to the ideals they teach (Awan, Amara & Ali, 2025). The various roles adapted by a teacher in VBME are illustrated in Fig. 4.2

### 4.7.1 Role model

Teachers' actions, attitudes, and behavior have a big impact on students' values. They should demonstrate the professional and ethical behavior they expect from their students, as well as act as a true role model in inculcating what they teach. Teachers serve as role models in VBME by modeling professional ethics, values, and behaviors for students to follow. By being a good role models, teachers have a big impact on students' attitudes and professional growth (Ho, Maria & Tani, Massimiliano, 2007).

### 4.7.2 Facilitator

The duties and responsibilities of a teacher in modern medical education goes beyond just giving lectures. Modern teachers promote discussions and active engagement that lead to deep understanding and reflection. They are always keen to help students integrate theory and practice, think critically about their experiences, and create their own sense of values. Instead of merely imparting knowledge, a teacher in VBME serves as a facilitator by demonstrating and helping students to build strong ethical reasoning and professionalism (Bashir A and McTaggart, 2011). Additionally, teachers can instill a sense of value-based learning in medical students by fostering a welcoming, inclusive environment and employing

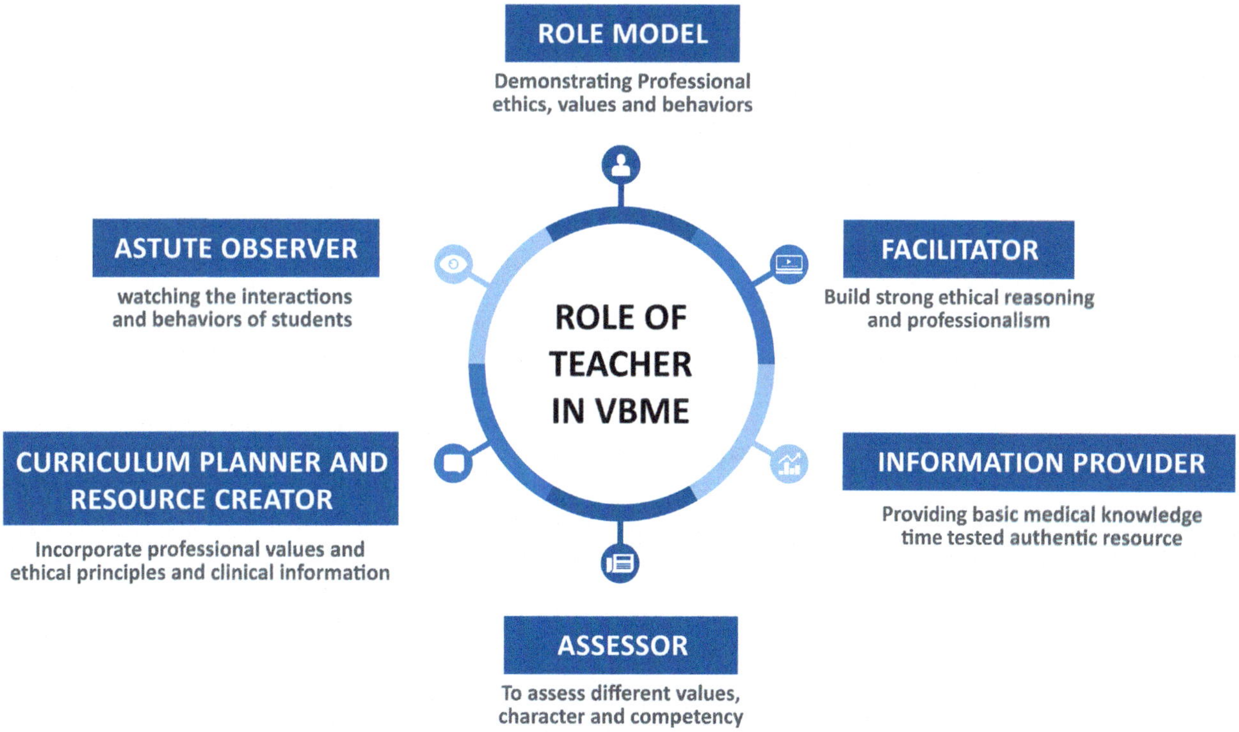


FIGURE 4.2 Role of teacher in value-based medical education (VBME).

strategies that go beyond direct instruction, like active listening, open-ended questions, and role-playing exercises that assist students in exploring and internalizing values.

### 4.7.3 Information provider

In order to appreciate the essence of VBME, it is imperative on the part of the learner to learn from the right resources. The resources should be time tested and authentic. In this regard, the teacher serves as the go-to person. Teachers not only transmit basic medical knowledge but also teach the students concepts such as value-based care so that students are in a position to understand the importance of providing high-quality, cost-effective patient-centered care. In addition to using formal techniques like conferences, case studies, and simulations, teachers also impart knowledge through real-world, hands-on experiences during clinical rotations. Modeling value-based care, teaching financial literacy and resource management, and fostering conversations that link clinical judgments to value principles so that students can apply them to patient care are some of the main duties of the teacher. These varied roles of a teacher change their position from one of a lone information provider to one of a facilitator who delivers information in a new context.

### 4.7.4 Assessor

The assessment of values amongst the learner is indeed challenging and a critical task which has to be completed by the teacher. In order to assess different values in a student, the teacher has to look beyond knowledge and skills. They have to take into account the student's professionalism and ethical development. By using the right assessment tools, the teacher will be in the best position to assist students to improve and align their behavior with professional standards. In VBME, a teacher acts as an assessor by assisting students in internalizing professional and ethical values in addition to medical knowledge. This function goes beyond simple grading to include serving as a facilitator, role model, and assessor of character and competency. Through assessment, the instructor demonstrates what matters, inspires students to acquire these values, and offers formative feedback to aid in their professional growth.

### 4.7.5 Curriculum planner and resource creator

The role of teachers in creating resources and curricula that integrate values education with clinical and scientific training is very vital for the successful implementation of VBME. This ensures that value-based topics are taught in a systematic way. A teacher's responsibilities as a curriculum planner and resource developer in VBME include forming the educational process to incorporate professional values and ethical principles in addition to clinical information. It is also important that the curriculum should be planned in such a way that it is aligned with the specific learning outcomes and the assessment methods. To develop a value-based curriculum, it is important for the teacher to identify core values through community input, then infuse these values across all subjects with integrated learning activities like role-playing and community service.

### 4.7.6 Astute observer

The teacher should also be wise enough to observe and assess every factor that could influence the learning process. As a perceptive observer, a teacher in VBME carefully watches and considers the interactions and behaviors of students in order to spot and mold the growth of qualities like professionalism, empathy, and compassion. Teachers must therefore understand these significant issues and assist students in understanding and resolving them. In addition to providing traditional education, this position involves mentoring, role modeling, and fostering conversations that link clinical practice to fundamental principles. The instructor serves as a mentor, using observations to offer focused criticism, encourage introspection, and make sure the hidden curriculum aligns with the intended ethical standards. Furthermore, by practicing evidence-based medical education and respecting the individuality of the students goes a long way to facilitate the VBME.

## 4.8 Value-based curriculum framework

It has to be appreciated that the value-based curriculum teaches more than just technical knowledge and skills. This method of teaching incorporates fundamental principles into each facet of the educational process. The goal of this kind of education is to create well-rounded, ethically conscious and capable leaders. Essentially, a value-based curriculum framework incorporates fundamental principles throughout the educational process. It emphasizes how people develop holistically by prioritizing all the three aspects which are social, emotional, and intellectual development. A value-based curriculum links learning to ethical dilemmas and real-world difficulties that the learner may face in their roles, preparing them to handle complexity with a balanced viewpoint. It aims to build students' character and equip them to make constructive contributions to society, going beyond academic success ( Khalil, 2025).

Goal-setting for patient-centered outcomes, content that incorporates ethical quandaries and patient preferences, and instructional strategies like clinical simulations and role modeling that foster professional judgment and empathy are all components of a value-based curriculum in medical education. In order to assess a student's judgment and commitment to values, assessment and evaluation techniques that go beyond factual knowledge are also essential. When combined, these are the fundamental components of a medical education program that is founded on values. Models that incorporate values through needs assessments, the creation of a hierarchy of values, and the creation of value-based teaching strategies and content are examples of value-based curriculum frameworks in medical education. Some of the important examples of curricular frameworks that incorporate values into medical education are described below (Fig. 4.3).

### 4.8.1 Value-based ethical framework

An approach to decision-making that places more emphasis on an individual's or organization's basic values such as integrity, equity, and respect rather than just rigorous regulations or penalties is known as a value-based ethical framework. Students learn how to reason and make fair decisions in challenging, real-world circumstances using a value-based approach. One of the ways to teach about a value-based ethical framework is to conduct activities that include role-plays and simulated scenarios. By using these realistic scenarios and simulated patients, the student can be taught about how to obtain informed consent, break bad news, or handle resource allocation conflicts. Another good example would be to conduct team-based learning where there is an opportunity for interprofessional education. By having students work alongside other healthcare disciplines, students are able to understand that different professions may hold different values and assumptions, promoting collaboration and better patient outcomes Almatrafi, 2024).

### 4.8.2 Value-based practice framework

In medical education, a value-based practice framework is one that incorporates a person's own values with the best available data and expertise to guide clinical judgment. By encouraging respect for one another, value awareness, and communication with patients, families, and coworkers, it serves as a counterpoint to evidence-based treatment and offers a method for professionals to negotiate complex and contradictory values. To stress on the importance of value-based practice framework, clinical case studies that highlight value-based conflicts can be developed and can be taught as small group discussions sessions (Patel, 2012).

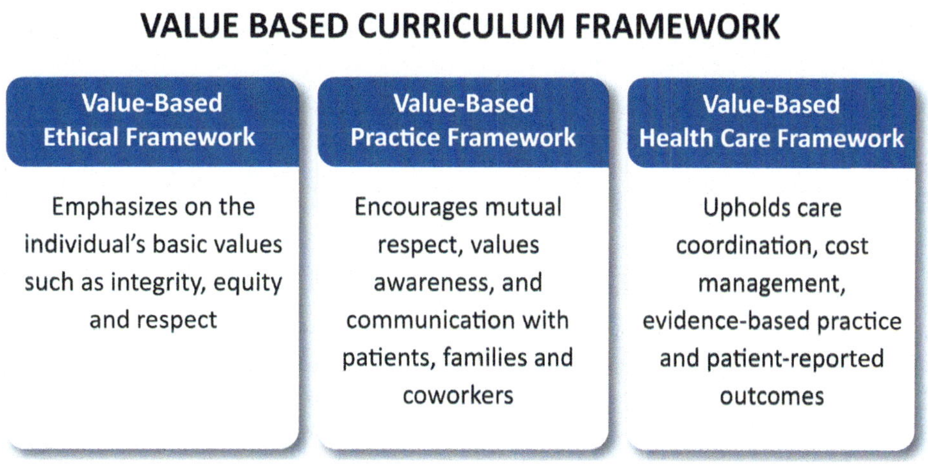


FIGURE 4.3 Value-based curriculum framework.

### 4.8.3 Value-based health care framework

In medical education, a value-based health care paradigm teaches aspiring doctors to prioritize patient outcomes in relation to costs over service volume. By doing this, the fee-for-service model of education is replaced with one that prioritizes population health, health equity, patient experience, and a well-supported healthcare staff. In order to prepare physicians for this new approach, medical education must now include instruction in care coordination, cost management, evidence-based practice, and patient-reported outcomes. By using multiple teaching methods like structured teaching which include conferences, simulations, journal clubs, and clinical teaching, which include rounds and bedside discussions, the importance of this framework can be highlighted.

## 4.9 How value-based medical education differs from traditional medical education

VBME places an emphasis on patient-centered treatment, quality, and outcomes, whereas traditional medical education is a lecture-style, content-heavy model that focuses on preclinical and clinical phases and teaches a wide range of basic sciences. The primary distinctions are seen in the curriculum's emphasis wherein value-based models incorporate concepts like patient preferences, integrated care, and evidence-based practice to enhance overall patient efficiency and health, while traditional models are frequently knowledge-based and sequential. The typical teaching approach is mainly lecture-based and frequently includes separate preclinical and clinical phases (Shrivastava, 2024). Furthermore, VBME incorporates value-based care principles to teach students how to manage a patient's entire health. To encourage participation and deeper comprehension, VBME frequently employs active learning strategies including case-based learning and simulation. Additionally, the VBME promotes the integration of information from several fields, including the social sciences, ethics, and medicine.

### 4.9.1 Inculcating values amongst the learners

Medical educators must have a solid understanding of professional principles in order to instruct and support their students' growth in these areas, which in turn helps them navigate their careers and emphasizes the significance of a patient-centered approach in healthcare (Holtzman, 2020). Role-plays, conversations, case studies, reflections, group exercises, project debates, and volunteer work are all ways to teach values. A healthy concealed curriculum should be given just as much attention. Since faculty members are the best role models, they might need training. Some of the strategies for inculcating values include the following.

#### *4.9.1.1 Establish defined learning objectives*

Establishing clear learning objectives, designing realistic and difficult scenarios, and assigning students to play different roles like doctor, patient, and spectator with an aim to instill values through role-playing in medical education is one of the main ways to inculcate the importance of values. In order to promote active learning, the development of empathy, and the reinforcement of professional values like as communication, integrity, and compassion, the sessions should be structured to include preparation, role-playing, and a debriefing with feedback and reflection (Colldén, C., & Hellström,2018).

#### *4.9.1.2 Promote empathy and reflection*

To stress on the values of empathy and reflection, it is important to incorporate early and continuous patient interaction, employ reflective activities like journaling and group discussions, and include perspective-taking exercises like patient narratives. Other successful tactics include promoting social service, utilizing technology such as virtual reality, and offering guidance and criticism.

#### *4.9.1.3 Integrate values into the curriculum*

To integrate values into a medical education curriculum, there should be a right blend of teaching methods like case studies, reflection, and role-playing with faculty training, institutional support, and assessments that go beyond cognitive skills. This calls for a methodical, step-by-step process in which faculty members model moral behavior and incorporate subjects like empathy, ethics, and cultural sensitivity into the current curricular framework.

#### *4.9.1.4 Foster a supportive environment*

To integrate values in medical education through a supportive environment, educators should act as role models, foster psychological safety by setting clear expectations, encourage open communication, and create a sense of community through mentorship and peer interaction. Implementing these strategies helps cultivate professionalism, empathy, and ethical conduct in students, which are essential for patient-centered care.

#### *4.9.1.5 Offer practical experiences*

It is possible to include values into medical education through hands-on experiences by promoting team-based care, incorporating patient perspectives into the curriculum, and providing early and thorough clinical exposure. Students can develop values like patient-centered care, empathy, and professional identity by actively engaging in real-world situations, participating in quality improvement projects, and collaborating with interprofessional teams. The use of simulation and virtual reality, along with a focus on introspection, are important strategies for bridging theory to practice and promoting a deeper understanding of medical ethics and high-value care.

#### *4.9.1.6 Reward positive behavior*

In order to incorporate values into medical education, teachers should model professional values themselves and employ constant, instantaneous, and targeted positive reinforcement, such as verbal praise, acknowledgment, and privileges. Clear expectations must be set, honest and nonjudgmental criticism must be given, and rewards must be relevant to the students and catered to their specific needs.

## 4.10 Benefits of value-based medical education

VBME benefits both patients and providers by encouraging patient-centered care, improving outcomes through prevention and individualized treatment, improving care coordination, encouraging data-driven decisions, and raising satisfaction by matching care with patient values and goals. Ultimately, this results in more compassionate, effective, and efficient healthcare. In order to transition from disease-focused care to holistic well-being, it prepares aspiring physicians to handle socioeconomic determinants of health and prudently manage resources. Some of the major benefits of VBME and care are listed below.

### 4.10.1 Better patient outcomes

VBME can help patients achieve better health outcomes since it emphasizes on patient-centered care and evidence-based treatment. VBME improves patient outcomes by connecting provider incentives with patient health, prioritizing preventive and personalized care, boosting patient satisfaction through a patient-centered approach, and leveraging data to support better decision-making. In contrast to a fee-for-service strategy, a value-based approach encourages patients to maintain their health rather than undergo further operations, which ultimately leads to improved health, more patient satisfaction, and lower costs.

### 4.10.2 Higher patient satisfaction

When patients feel valued, heard, and actively participated in their care, they are more likely to be happy with their overall healthcare experience. By teaching healthcare professionals to emphasize a comprehensive, patient-centered approach that values the whole person, individualized care plans, and proactive prevention, VBME helps to increase patient satisfaction. Better communication, better health outcomes, and patient empowerment to take an active role in their own treatment are all facilitated by this education.

### 4.10.3 Lower healthcare expenses

VBME can assist in controlling healthcare costs by focusing on cost-effectiveness and minimizing pointless procedures. VBME reduces healthcare costs by increasing patient health, reducing unnecessary treatments, and optimizing the use of existing resources by teaching physicians to prioritize outcomes over procedures. Attending medical schools that prioritize preventative care, care coordination, and the underlying causes of disease would better prepare future doctors to manage chronic illnesses and create a more sustainable healthcare system.

### 4.10.4 Increased physician satisfaction

Physicians are more likely to be satisfied with their careers when they believe they are offering high-quality, value-driven care. Through enhancing patient relationships, lowering burnout, developing skills for improved patient outcomes, and aligning with their professional mission, VBME helps to increase physician satisfaction (El Fadel, 2025). VBME gives doctors more time for holistic care, encourages ongoing professional growth, and links them to the inherent satisfaction of better health by prioritizing quality and patient well-being over high patient volume.

### 4.10.5 Better equipped medical personnel

Aspiring medical professionals are prepared by VBME to manage the complexities of the healthcare system and adapt to evolving treatment models. VBME equips employees with skills in data analysis, integrated care, and cost-conscious decision-making by emphasizing patient outcomes and cost-effectiveness. This approach equips medical staff to deliver more customized, efficient, and superior care by prioritizing the quality of results over the volume of services.

### 4.10.6 Enhanced professionalism

VBME encourages the growth of professionalism, which includes moral behavior, clear communication, and a dedication to ongoing development. A doctor's professional identity and behavior are shaped by their values, which also affect how they interact with patients and coworkers. VBME improves professionalism by fostering core values like patient-centeredness, integrity, and accountability through real-world experiences and discussions. It moves beyond purely cognitive skills to develop noncognitive attributes such as empathy, compassion, and collaboration by linking values to specific, observable behaviors in a safe environment. This educational approach prepares future physicians to navigate complex healthcare systems and meet societal needs with a strong ethical foundation.

### 4.10.7 Improves health care

A healthier society is a result of ethical and responsible healthcare workers. VBME improves healthcare by training healthcare providers to focus on patient outcomes, cost-effectiveness, and integrated care, which leads to higher quality, more equitable, and more efficient care. This approach fosters a shift from volume-based to value-based payment models, incentivizing providers to deliver comprehensive, patient-centered services that maximize positive health results while minimizing waste and improving patient experience.

### 4.10.8 Overcoming obstacles

Values support responsible decision-making under trying circumstances and aid in navigating difficult ethical conundrums. VBME improves overcoming obstacles by equipping future physicians with the skills to navigate systemic inefficiencies and focus on patient outcomes, cost-effectiveness, and quality of care. This includes understanding the financial and social barriers patients face, emphasizing preventative care and holistic, whole-person care, and fostering a system-based approach to problem-solving that integrates individual learning with systems improvement. In essence, VBME aims to produce healthcare professionals who are not only clinically competent but also compassionate, ethical, and committed to providing the best possible care for their patients.

## 4.11 Challenges of value-based medical education

Resistance to cultural changes, disjointed healthcare systems, a lack of information technology (IT) infrastructure and data standardization, financial incentives from traditional fee-for-service models that are not aligned, and the challenge of creating integrated, team-based care models are some of the obstacles facing VBME. Effective procedures, a sizable upfront technological investment, and continual faculty training to operate new data collecting and outcome assessment technologies are all necessary for VBME implementation.

### 4.11.1 Cultural and systemic barriers

#### *4.11.1.1 Resistance to cultural shifts*

A fundamental challenge is changing long-held organizational culture and habits within healthcare institutions and among faculty. Resistance to cultural shifts acts as a barrier in VBME by hindering the development of skills needed for effective cross-cultural care, leading to patient mistrust and suboptimal outcomes. This resistance can stem from a lack of diversity in leadership, outdated systems, and ingrained biases, making it difficult for trainees to learn about and practice culturally sensitive care. Without addressing these cultural barriers, the quality of care for diverse patient populations suffers, as differences in beliefs, communication styles, and expectations are not adequately understood or respected.

#### 4.11.1.2 *Fragmented systems*

Healthcare systems are often siloed, hindering the interdisciplinary, collaborative care that value-based model demands. Fragmented healthcare systems are a barrier to VBME because they prevent the teaching and application of a truly integrated, team-based approach, which is central to value-based care. This fragmentation limits students' ability to learn in a coordinated environment, as they are often siloed in specialty-focused training that lacks the necessary interdisciplinary collaboration, communication, and comprehensive patient data needed to understand holistic, patient-centered care. Consequently, education is hindered, and graduates are less prepared to deliver the coordinated, high-value care that the system requires.

#### 4.11.1.3 *Institutional buy-in*

Implementing major changes in curriculum or residency programs requires strong, time-consuming institutional support that can be difficult to achieve. Institutional buy-in is a barrier to VBME due to the resistance to change from traditional, fragmented systems, lack of resources, and difficulty integrating new curricula that emphasize collaboration, patient-centered outcomes, and cost-effectiveness. Without institutional support, a value-based approach struggles to gain traction, as it requires a significant shift in culture, infrastructure, and teaching methods to move away from a siloed, specialty-focused model.

### 4.11.2 Curriculum and training needs

#### 4.11.2.1 *Developing new paradigms*

Medical education curricula need to be updated to produce graduates equipped to address the health problems of the 21st century. Developing new paradigms is a barrier to VBME because it clashes with established, pathology-focused curricula and requires a significant shift in mindset, resources, and teaching methods. This resistance to change, often referred to as academic inertia, makes it difficult for medical schools and residency programs to quickly adopt and integrate the new concepts of patient-centered care, data-driven decision-making, and a focus on outcomes and cost-effectiveness into their training.

#### 4.11.2.2 *Faculty training*

Educators need training to understand and implement value-based principles, manage new technologies, and adapt to changing educational needs. Faculty training is a barrier to VBME due to a lack of formal training, limited time and resources, and faculty discomfort with new concepts and teaching methods. Insufficient training leaves faculty ill-equipped to teach value-based principles, while heavy workloads and other institutional barriers limit opportunities for faculty development, making it difficult for them to adapt from traditional, pathology-focused curricula to a new, value-centered one.

### 4.11.3 Technical and data challenges

#### 4.11.3.1 *Inadequate information technology infrastructure*

Healthcare systems often lack the necessary technological infrastructure to effectively collect, analyze, and share the vast amounts of data needed for value-based care. Inadequate IT infrastructure is a barrier in VBME by hindering the collection, analysis, and use of patient outcome data and by limiting access to digital learning tools and resources. It prevents the tracking of performance measures, the integration of patient-reported outcomes, and the implementation of value-based models that require robust data-driven decision-making.

#### 4.11.3.2 *Data standardization and interoperability*

A lack of standardized electronic health records and interoperability between different platforms creates significant hurdles for data exchange and analysis. Data standardization and interoperability are barriers in VBME because they hinder the ability to collect, share, and use comprehensive patient data across different systems. This creates data silos, leading to incomplete patient records, inconsistent data, inefficient care coordination, and challenges in accurately tracking outcomes and value. Without standardized formats and interoperable systems, it is difficult to effectively integrate data for teaching and learning purposes, slowing down progress in digital health initiatives.

#### 4.11.3.3 Measuring outcomes

Developing reliable and standardized methods for measuring patient outcomes and demonstrating the financial sustainability of value-based approaches remains difficult. Measuring outcomes is a barrier in VBME due to the complexity of defining and standardizing measures, issues with data collection and privacy, and the difficulty of including nonquantifiable aspects like patient well-being and the humanistic side of care. Disagreements on what constitutes value and aligning different stakeholder perspectives also present significant challenges.

### 4.11.4 Financial and resource constraints

#### 4.11.4.1 Financial incentives

The continued use of fee-for-service payment schemes, in which the quantity of services rendered determines the cost rather than the caliber or results of the treatment, is one of the major hurdles in the implementation of value-based health care. This has drawn criticism since it may result in overuse of services, increased total expenses, and fragmented care because of inadequate provider coordination. Financial incentives can be a barrier in VBME because they can have unanticipated consequences, such as lowering intrinsic drive and putting quantitative results ahead of quality care.

#### 4.11.4.2 Upfront investment

VBME requires significant upfront investment in technology, staff training, and new financial modeling to manage financial risks associated with accountable care. Upfront investment acts as a barrier to VBME because it requires significant initial spending on technology, data infrastructure, and training for both organizations and staff to shift from fee-for-service models. Practices must be able to manage the costs of new software and systems for data collection and analysis to track performance and patient outcomes, which can be particularly difficult for smaller practices with limited financial reserves. This financial risk and uncertainty discourage investment in new models of care that take time to become profitable, especially given the potential for frequent changes in value-based care programs and reimbursement policies.

#### 4.11.4.3 Resource limitations

Inadequate resources and inefficient internal workflows can overburden staff and impede the adoption of value-based care principles. Resource limitations are a major barrier to VBME because they hinder the necessary infrastructure, technology, and staffing required for the model to function effectively. Without adequate funding, insufficient staffing, and outdated technology, it is difficult for educators and institutions to implement strategies like comprehensive patient care, data collection, and outcome tracking that are central to value-based learning. This makes it hard to teach and practice the model sustainably and successfully (Fig. 4.4).

## 4.12 Value-based healthcare

Value in healthcare is the ratio of the patient's attained health outcomes to the total cost of attaining those results, taking into account a wide range of elements such as the patient's experience, the influence on society, and the fair use of resources. It is a multifaceted idea that takes into account the patient's top priorities, the sustainability of care, and the effective use of resources within the healthcare system (Garcés-Ferrer, Jorge, 2024) Value-based health care is a novel paradigm for rethinking the way care is delivered in order to improve patient outcomes and save costs; nevertheless, medical educators face the difficulty of giving students the knowledge and skills necessary to comprehend and interact with these concepts. Healthcare providers, such as hospitals and doctors, are paid according to the health results of their patients under a VBHC paradigm. Value-based care agreements reward clinicians who use evidence-based medicine to assist their patients to live healthier lives, promote healthy patient outcomes, and lessen the burden of chronic diseases. VBHC programs are vital to a larger quality strategy to reform how healthcare is delivered and paid for (Almeida, 2023).

## 4.13 Impact of values on principles and practice of medical education

Values have a significant influence on the principles and practice of medical education by defining professionalism, guiding ethical behavior, creating the curriculum, and influencing the hidden curriculum and learning environment. They serve as the primary moral compass for both educators and students. Professionalism is one of the fundamental ideals ingrained in medical education. In this sense, the goal of medical education is to inculcate virtues like honesty, decency,

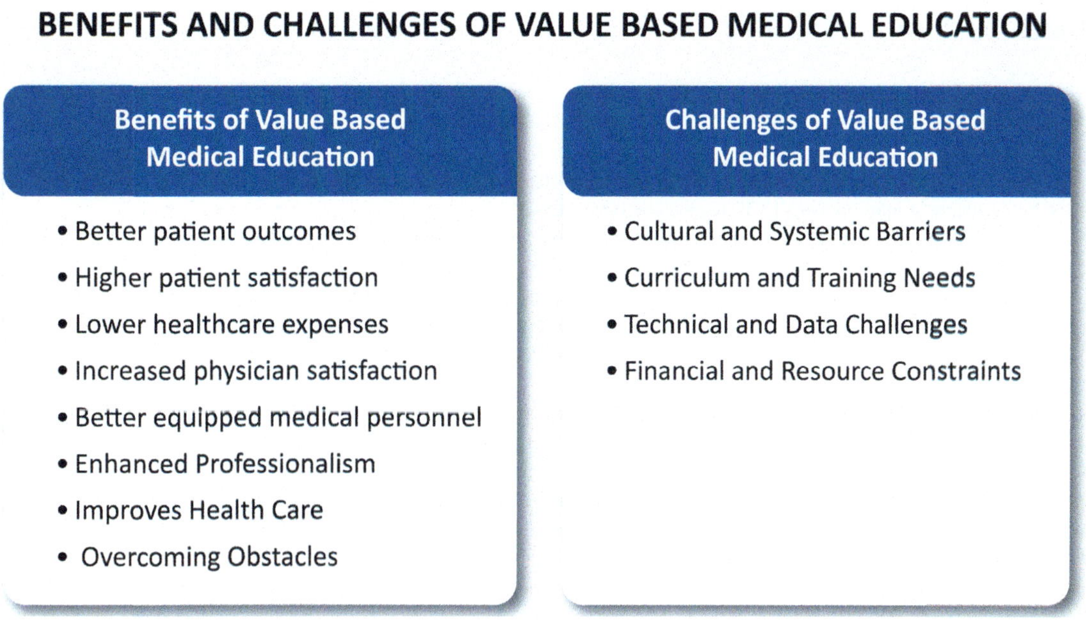


FIGURE 4.4 Benefits and challenges of value-based medical education.

compassion, selflessness, and responsibility. During their training and practice, students are encouraged to establish these ideals as the cornerstone of their medical professional identities.

However, these fundamental principles are directly reflected in the actual application of medical education, such as role modeling, in which administrators and teachers impart values to students not only through formal instruction but also through their behavior and attitudes. Students' professional behavior and ethical compass are significantly shaped by the informal and frequently unstated "hidden curriculum" based on the ideals they witness in their mentors.

**Summary**

In conclusion, it is anticipated that medical schools will equip the upcoming generation of healthcare professionals to provide patient-centered care by delivering value-based medical education. Given that value-driven teaching will greatly enhance patient outcomes and well-being, it is imperative that a multifaceted strategy be used to guarantee that it is incorporated into the current medical education delivery model. It is critical to recognize that the global trend of moving from volume-based to value-based healthcare models that put patient outcomes and cost-effectiveness first aligns medical education with the healthcare system transition. Additionally, it equips a new generation of doctors with the know-how to succeed in changing healthcare environments. It improves care by making sure that resources are spent effectively and that therapies are customized to each patient's specific needs.

**Key point**

**Future perspectives: values as a catalyst in medical education**

Future perspectivity on values in medical education place a strong emphasis on the development of professional resilience and ethical leadership, a humanistic approach to medicine combined with technology improvements, and a stronger emphasis on public health. In this context, values act as a catalyst in medical education as it transforms a human being to being human. The incorporation of values in the students upholds its importance as one of the main curricular components. Important ideas like emphasizing compassion and ethics in addition to technical skills, creating a more comprehensive and student-centered learning environment, and modifying curricula to reflect a proactive and prevention-oriented model of care must be prioritized because they not only help students become good professionals but also instill integrity and ethical behavior in both patients and colleagues.

## References


Almatrafi, F., Ammar, A., 2024. Values-Based Medicine Is an Ethical Concept for Implementing the Ethical Principles in Daily Practice. Adv Tech Stand Neurosurg 50, 335–346. Available at https://doi.org/10.1007/978-3-031-53578-9_12 PMID: 38592537.

Almeida, A.A.R., Da Mota, M.H.C., Cruz, Sde S., Massa, B.F., Ferraz, R.R.N., De Liberal, M.M.C., et al., 2023. The impact of healthcare workers' values and perception of organizational culture on hospital performance: a case study. Contribuciones A Las Ciencias Sociales 16 (11), 25621–25636. Available at https://doi.org/10.55905/revconv.16n.11-051.

Awan, A.H., Ali, S., Siddique, U., Salman, R., Khan, Y., Ayub, H., 2025. Role of a teacher in medical education: a faculty's perspective from HBS. Pakistan Journal of Health Sciences 153–158. Available at https://doi.org/10.54393/pjhs.v6i5.3090.

Bashir, A., McTaggart, I.J., 2022. Importance of faculty role modelling for teaching professionalism to medical students: individual versus institutional responsibility. Journal of Taibah University Medical Sciences 17 (1), 112–119. Available at https://doi.org/10.1016/j.jtumed.2021.06.009.

Colldén, C., Hellström, A., 2018. Value-based healthcare translated: a complementary view of implementation. BMC Health Services Research 18 (1). Available at https://doi.org/10.1186/s12913-018-3488-9.

Desai, M.K., Kapadia, J.D., 2022. Medical professionalism and ethics. Journal of Pharmacology and Pharmacotherapeutics 13.

El Fadel, O., Goldberg, Z.N., Jain, A., et al., 2024. Integrating Choosing Wisely, Value-Based Care Principles, Into Undergraduate Medical Education: A Pilot Study. Cureus 16 (3), e56912. Available at https://doi.org/10.7759/cureus.56912.

Fernández-Salido, M., Alhambra-Borrás, T., Casanova, G., Garcés-Ferrer, J., 2024. Value-based healthcare delivery: a scoping review. International Journal of Environmental Research and Public Health 21 (2), 134. Available at https://doi.org/10.3390/ijerph21020134.

Gonzalo, J.D., Dekhtyar, M., Hawkins, R.E., Wolpaw, D.R., 2017. How can medical students add value? Identifying roles, barriers, and strategies to advance the value of undergraduate medical education to patient care and the health system. Academic Medicine 92 (9), 1294–1301. Available at https://doi.org/10.1097/acm.0000000000001662.

Hilton, S.R., Slotnick, H.B., 2005. Proto-professionalism: how professionalisation occurs across the continuum of medical education. Medical Education 39 (1), 58–65. Available at https://doi.org/10.1111/j.1365-2929.2004.02033.x.

Holtzman, J.N., Deshpande, B.R., Stuart, J.C., Feeley, T.W., Witkowski, M., Hundert, E.M., et al., 2020. Value-based health care in undergraduate medical education. Academic Medicine 95 (5), 740–743. Available at https://doi.org/10.1097/ACM.0000000000003150.

Ho, M.T., Tani, M., 2007. What medical students value from their teachers. Australian Health Review: A Publication of the Australian Hospital Association 31 (3), 358–361.

Khalil, H., Ameen, M., Davies, C., Liu, C., 2025. Implementing value-based healthcare: a scoping review of key elements, outcomes, and challenges for sustainable healthcare systems. Frontiers in Public Health 13. Available at https://doi.org/10.3389/fpubh.2025.1514098.

Nafea, E.T., 2024. Professional values and ethics in medical education. In: Amarin, Z.O. (Ed.), Advances in Medical Education and Training. IntechOpen. Available at https://doi.org/10.5772/intechopen.114233.

Patel, M.S., Davis, M.M., Lypson, M.L., 2012. The VALUE Framework: training residents to provide value-based care for their patients. J Gen Intern Med 27 (9), 1210–1214. Available at https://doi.org/10.1007/s11606-012-2076-7 PMID: 22573146; PMCID: PMC3514984.

Prabhu, N.K., 2011. Human values in education: reflecting on the core. International Research Journal 2 (12), 1727–1732.

Purwati, Ani, Ruslita, Gita, 2025. The Principle of Beneficence of Doctor-Patient Communication Regarding the Implementation of the Hospital Code of Ethics in Health Services. Journal of Law, Politic and Humanities 5, 4606–4619. Available at https://doi.org/10.38035/jlph.v5i6.1867.

Shrivastava, S.R.B.L., Shrivastava, P.S., Tiwade, Y., 2024. Shifting from conventional curriculum to value-based medical curriculum: narrative review. Journal of Pharmacy and Bioallied Sciences 16 (Suppl 2), S1084. Available at https://doi.org/10.4103/jpbs.jpbs_1232_23.

Sturmberg, J.P., Taher, S., 2025. What we mean by 'values in healthcare': the importance of reaching a consensus. Cureus 17 (2). Available at https://doi.org/10.7759/cureus.79034.

Supe, A., Shah, H., 2021. Value based medical education. Medical Journal Armed Forces India 77, S8. Available at https://doi.org/10.1016/j.mjafi.2020.12.001.

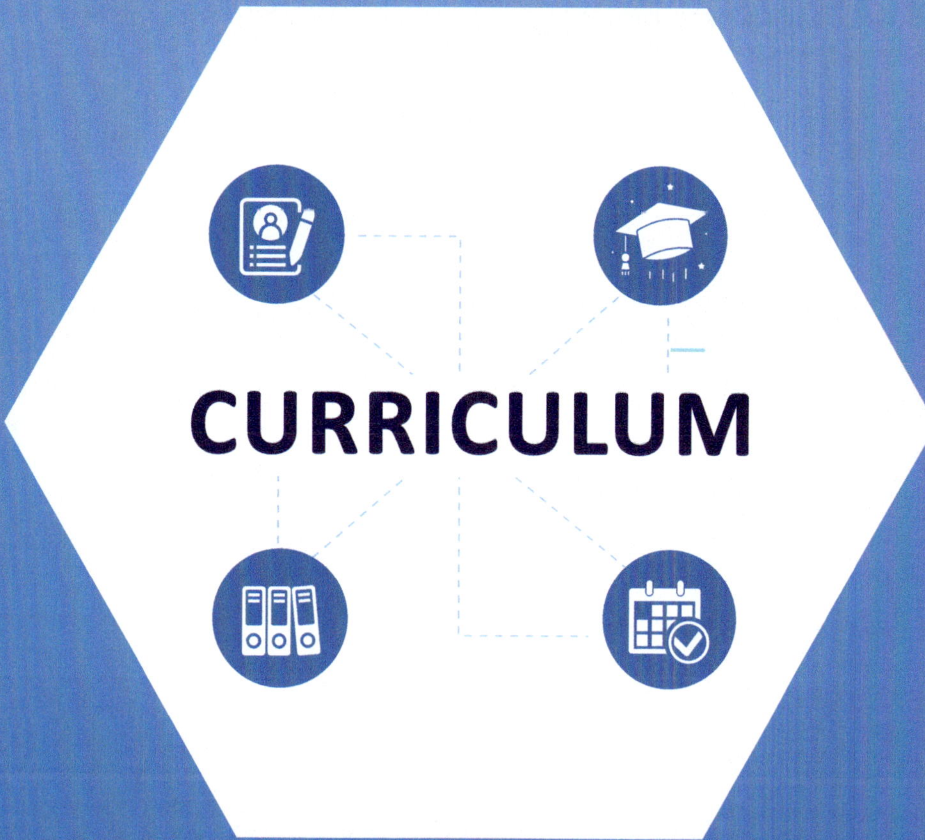


# PART 2

## CURRICULUM

Chapter 5

# Educational objectives, outcomes, and competencies

*"Setting goals is the first step in turning the invisible into the visible."*

—Tony Robbins

## 5.1 Introduction

It is a well-known fact that medical education is undergoing a continuous transformation. Whether in terms of planning the curricula, teaching methods, assessment, or introduction of new technologies like artificial intelligence (AI), there is always a tremendous amount of information to be taught, learnt, and applied continuously. In order to successfully achieve these, it is imperative that a clear-cut strategy is being applied at all levels. In this regard, educational objectives or goals serve as a beacon.

Educational objectives, learning objectives, and learning outcomes are fundamental components of the teaching and learning process, providing a clear direction and framework for both educators and learners. For any teaching and learning process to be effective, it is imperative that the educational objectives must be laid down and they should follow a certain standard. These objectives serve as guiding principles, shaping curriculum design, instructional methods, and assessment strategies. They articulate what students should know, understand, and be able to do upon completion of a course, program, or educational endeavor. They also outline the specific knowledge, skills, and attitudes that learners should acquire through a course or program.

## 5.2 Educational objective or goals

Educational objectives also called as goals are at the top of the hierarchy because they represent a larger idea or a goal that is meant to be accomplished. These objectives are established at the university level. In addition to reflecting the university's goals, it also directs the development of the program's learning objectives. Educational objectives in medical education are broad in scope and reflect the desired outcomes of a medical program, guiding curriculum, and instruction to produce competent, patient-centered physicians who are lifelong learners. These goals encompass various aspects, including patient care, medical knowledge, communication skills, professionalism, practice-based learning, systems-based practice, and interprofessional collaboration. They generally have one or more objectives that need to be achieved. For instance, educational goals in medical education aim to develop competent, ethical, and compassionate physicians. They also focus on personal and professional development ensuring that the healthcare professional is moving towards lifelong learning and are able to advocate for their patients (Table 5.1).

## 5.3 Significance of a well-formulated goal

### 5.3.1 For the students

A well-formulated goal in medical education acts as a vital compass, providing clear direction and motivation for students and instructors alike. By defining specific and measurable outcomes, these goals ensure that learning is intentional, efficient, and aligned with the core competencies required for a future medical career. The goals also give a chance to develop a world class academic environment which enables medical education to transcend beyond boundaries. It also upholds the needs of academic excellence and global competitiveness to be embodied into the learner. The significance of a well-formulated goal lies in the fact that it contributes to various facets of medical education which can be described as follows.

**The Principles and Practice of Medical Education.** **DOI:** https://doi.org/10.1016/B978-0-443-44197-4.00033-3

**TABLE 5.1 Key features of educational goals.**

| | |
|---|---|
| 1 | Educational goals are the general aims or purposes of the program and its curriculum. |
| 2 | Educational goals should be effective, meaningful, achievable, and be assessable. |
| 3 | Educational goals describe broad learning outcomes/competences and concepts that speak about what you want students to learn and are expressed in general terms. |
| 4 | Educational goal should provide a broad framework for determining the more specific learning objectives of a program. |
| 5 | Educational goal should be consistent with the mission of the program and the institution. |
| 6 | A single learning goal may include many specific subordinate learning objectives. |

#### 5.3.1.1 *Defines physician competency*

Goals ensure the development of the knowledge, skills, and attitudes needed to be a competent physician, including the ability to handle difficult cases and understand basic medical mechanisms. Moreover, educational goal provides a framework of specific, measurable outcomes that medical students must achieve by the time of graduation. In addition to this, the educational goals helps the learner to develop practical patient skills by enforcing the curriculum to focus on what a learner can do rather than focusing on what to teach. This prepares doctors to be immediately capable of managing patient care. Similarly, it helps to reinforce core professional behaviors like integrity, empathy, and respect for diversity, which are then integrated into teaching and assessment. By fostering the quality of lifelong learning, an educational goal enables the physician to demonstrate the capacity for continuous self-evaluation and appraisal of new scientific evidence. This instills the habit of ongoing improvement throughout a medical career.

#### 5.3.1.2 *Drives curriculum and teaching*

A well-formulated goal guides the design of curriculum, teaching methods, and assessments, ensuring that the educational experience is focused on achieving specific outcomes. Educational goals drive curriculum and teaching by serving as the foundational blueprint for the entire learning process. By defining the desired knowledge, skills, and values that students should acquire, goals provide a clear roadmap for designing coherent curricula and selecting effective teaching strategies. Having a well-formulated educational goal will help in the establishment of a structured framework by translating a program's vision into a structured plan. For instance, a goal of fostering critical thinking skills leads curriculum developers to design learning experiences that require analysis, evaluation, and problem-solving and not just memorization. It also helps educator in the selection of course content, instructional materials, and resources. If the goal is to prepare students for a diverse, interconnected world, the curriculum should integrate to fulfill the needs and demands of a society. For both the learner and the teacher, educational goals act as a reference in guiding the sequence and scope of the topics to be covered in the course or a program. Goals help the teacher to determine the order in which topics are introduced and the depth to which they are covered. A well-planned curriculum ensures that concepts build logically on prior knowledge, allowing students to progress toward the ultimate learning goals. The educational goals also set the stage for assessment as they provide a basis for measuring learning.

#### 5.3.1.3 *Promotes patient safety*

Educational goals promote patient safety by equipping current and future healthcare professionals with the knowledge, skills, and attitudes necessary to minimize errors and foster a robust culture of safety within healthcare systems. Additionally, empowering patients to take an active role in their own treatment, encouraging a culture of accountability and continuous improvement, and developing core competencies in healthcare personnel all contribute to patient safety. For instance, a medical student who notices that a wound on a patient's limb has not been cleaned properly and alerts the nursing team, thereby preventing a potential infection, is said to be well aware of the components of patient safety.

#### 5.3.1.4 *Fosters lifelong learning*

In order to stay up to date with developments and the changing demands of the healthcare system, goals highlight the necessity for doctors to pursue further education throughout their careers. Learning objectives support lifelong learning in medical education by fostering basic abilities like metacognition and self-directed learning, which are crucial for keeping up with the rapid changes in medicine. By encouraging students to set their own objectives, look into problems, and evaluate their progress, medical education can transition from passive knowledge absorption to a self-driven process that is essential for providing high-quality patient care throughout a physician's career.

#### 5.3.1.5 *Encourages professional development*

Educational goals set expectations for professional behavior, including compassion, teamwork, and self-care, which are crucial for a physician's overall effectiveness and well-being. Learning goals encourage professional development in medical education by providing a framework for focused learning, self-regulated learning, and continuous improvement through feedback. Medical professionals and students can better focus their efforts on skill mastery, personal development, and lifelong learning by establishing clear goals, which will ultimately increase their competence and confidence.

#### 5.3.1.6 *Motivates students*

Well-defined educational goals give students a "road map" for their learning journey. They can see what they are expected to accomplish, which allows them to track their own progress, focus their study time, and better prepare for assessments. This road map also plays a critical role in increasing their motivation and engagement by setting clear expectations for their performance and progress. Learning goals motivate medical students by providing direction and meaning, fostering self-efficacy and resilience, and encouraging deeper, more autonomous learning. They help students connect their education to their long-term "life goals" of becoming a certain type of doctor, increasing intrinsic motivation and the use of effective study strategies.

#### 5.3.1.7 *Enables systemic improvement*

Medical education can produce doctors who are equipped to treat individual patients as well as enhance the healthcare system by emphasizing objectives linked to system-based care, public health concepts, and quality improvement. By coordinating the curriculum, evaluations, and instructional strategies to produce a cohesive and efficient system, learning objectives propel systemic progress in medical education. Improved and prepared doctors along with improved patient outcomes result from this process, which establishes a feedback loop for ongoing quality improvement.

### 5.3.2 For the medical educators

For the medical educator, a well-formulated goal gives clear direction whereby the teacher can plan his strategies in a way to obtain maximum benefit. At the same time, it enables the teacher with a vision in which all their efforts should be directed in order to fulfill the needs of the learner and the program at large. The teacher can also stand to gain enhancement in lesson and curriculum design as clear goal serves as the foundation for instructional decisions, from selecting materials and structuring lectures to guiding the choice of relevant activities and assessments. It provides a blueprint for what students will do and how their success will be measured, ensuring fair grading and cohesive instruction.

## 5.4 Learning objectives

### 5.4.1 Definition

Learning objectives are defined as specific goals that delineate what educators intend students to learn as a result of a teaching interaction, guiding the selection of teaching approaches, learning activities, and evaluation strategies. These are much more specific than goals and are also easier to measure. It helps to think of them as steps you take to achieve your goal (Ruiz et al., 2009). Learning objectives are succinct, straightforward statements that outline the intended learning outcomes of education. These outcomes include certain abilities, attitudes, and values that students should possess in order to represent the larger goals. In terms of the knowledge or abilities that students will be able to exhibit following teaching, these statements describe the expected results of a curriculum, course, lesson, or activity. Learning objectives lay the groundwork for evaluation by outlining the goals and anticipated outcomes of instructional activities. Learning objectives outline the topics that instructors will cover in a course or program. They are generally less broad than goals and broader than student learning outcomes (Swick, 2010). These learning objectives can be classified based on types of objectives, domains, and the person (Fig. 5.1).

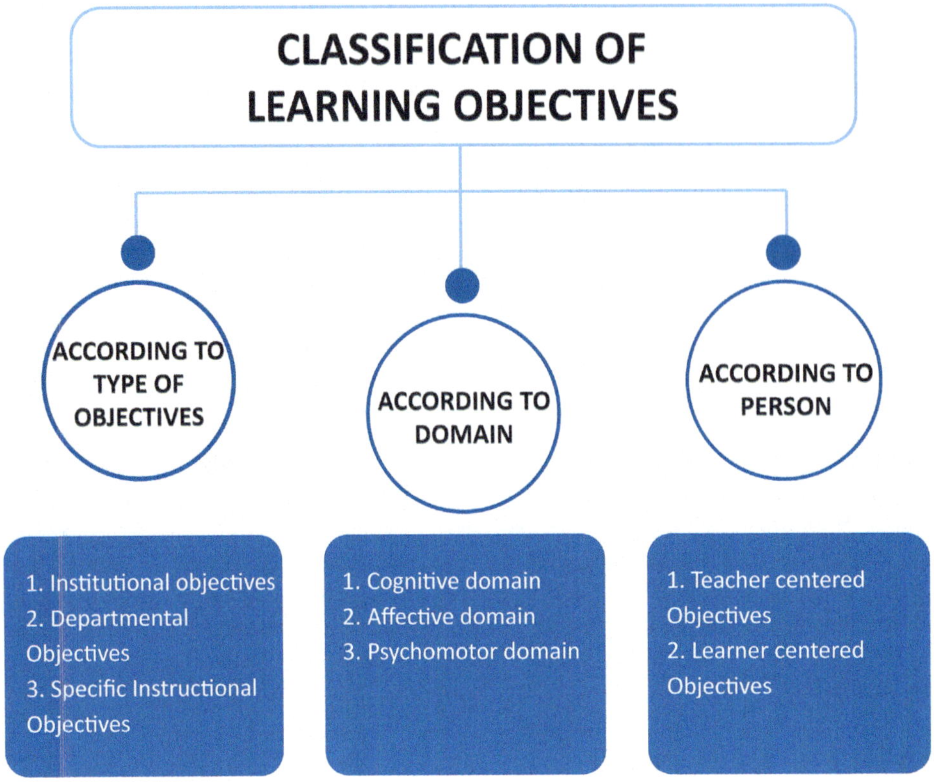


FIGURE 5.1 Classification of learning objectives.

## 5.4.2 According to type of objectives

### *5.4.2.1 Institutional objectives*

Institutional objectives in medical education are the overarching goals and measurable outcomes a medical institution aims to achieve, focusing on producing competent, ethical, and compassionate physicians who can provide high-quality, patient-centered care. For instance, developing strong patient care and medical knowledge, fostering professionalism and ethical conduct, improving communication and interpersonal skills, and demonstrating systems-based practice to serve diverse communities are some of the key components of institutional objectives. These objectives are to be strongly ingrained into the learner so that they are always reminded of what they are supposed to achieve during their medical career.

### *5.4.2.2 Departmental objectives*

In medical education, departmental objectives are goals established by a specific department to direct the educational process. These objectives often fall into one of many categories, such as enhancing student competency, promoting faculty development, or advancing medical education research. Training students in patient care, medical knowledge, and professional behaviors, as well as maintaining the caliber of instruction, introducing new educational technology, and creating curricula, are important components in this respect. It is to be kept in mind that in order to achieve the departmental objectives, it is vital that the faculty understand the goals of the institution. The process of evaluating departmental objectives is critical and involves systematically assessing the goals set by a department to determine their relevance, effectiveness, and alignment with organizational priorities. This process ensures that objectives are not only achievable but also contribute to the broader mission and vision of the organization.

### *5.4.2.3 Specific instructional objectives*

Specific instructional objectives in medical education are characterized by being measurable, observable, and action-oriented statements that define what students should be able to do after a lesson, course, or program. These objectives focus on the immediate skills or knowledge that students should acquire. Essentially, they are short-term targets that guide day-to-day teaching activities. For instance, if the topic is about mechanics of breathing, an instructional objective might be: "Students will be able to identify the muscles of breathing." This is something a teacher wants their students to accomplish during the lesson. By formulating a specific instructional objective which in this case is learning about the muscles of breathing, students will be benefited in the sense that they will develop a habit of focused study as these objectives help students prioritize their time and focus on the most important information, making studying more efficient. This in turn guides them towards a deeper understanding and clinical application. The key characteristic of instructional objectives is their specificity and measurability.

## 5.4.3 According to domain

### 5.4.3.1 *Cognitive domain*

In medical education, the objectives in cognitive domain refers to intellectual skills like knowledge, problem-solving, and critical thinking, typically organized by Bloom's taxonomy into levels from simple knowledge recall to complex evaluation. The objectives play a critical role in defining how the learner is able to achieve outcomes about knowing, thinking, and processing information in a structured way, with higher levels building upon the skills of lower levels. For example, learning about the structure of the heart along with its blood supply represents the factual knowledge that the students must learn before they can apply this information to understand the pathophysiology of cardiac disease.

### 5.4.3.2 *Affective domain*

Affective domain objectives in medical education focus on the development of values, attitudes, and emotions, cultivating professional and compassionate behavior rather than just knowledge or skills. These objectives are categorized hierarchically from a basic level of awareness to a high level of internalized professional values, ensuring that students develop the emotional intelligence and ethical framework needed for patient care. Affective domains like values and attitude form a critical part in the personal and professional development of the learner as they are crucial for developing qualities like empathy, integrity, and communication skills, which are essential for patient-centered care. The inclusion of the affective domain is vital to be incorporated into the curriculum so as to inculcate these values into the learner.

### 5.4.3.3 *Psychomotor domain*

In medical education, psychomotor objectives have a special place in the teaching and learning process as it puts the learner into spotlight. It focuses on the development of skills and coordination of the learner. These objectives are crucial because they require the integration of cognitive understanding with physical actions, and they are measured by the speed, precision, and correct execution of motor skills. The psychomotor domain, especially in the senior students, has to be given a lot of importance, and efforts should be taken to prioritize these skills in day-to-day teaching and learning activities. Examples include performing a physical examination or using medical instruments accurately.

## 5.4.4 According to person

### 5.4.4.1 *Teacher-centered objectives*

Teacher-centered objectives in medical education are goals set by the instructor that focus on the teacher's control over content delivery and the students' passive reception of knowledge. These objectives often revolve around ensuring that the students master foundational knowledge through lectures and demonstrations, with the teacher being the primary authority and responsible for all aspects of the learning process, including assessment. The teacher-centered objectives primarily focus on content and instruction to be presented by the teacher in the class and center around what is to be taught to the students. For instance, the objective can be framed as "the objective of this session is to revive the mechanism of regulation of blood pressure." These objectives also focus on what the teacher will present during a session. For instance, the objective can be constructed as "to describe the etiology and pathophysiology of Parkinson's disease."

### 5.4.4.2 *Learner-centered objectives*

Learner-centered objectives in medical education are learning goals focused on what the student will be able to do as a result of an educational activity, rather than what the instructor will teach. The core component of a learner-centered approach is one where the teacher acts as a facilitator to guide students who take responsibility for their own learning, develop self-directed skills, and apply knowledge to real-world scenarios. These objectives focus on the student's ability to apply knowledge, analyze a case, and make informed decisions, rather than simply recalling facts. For instance, learner-centered objectives like "Develop a differential diagnosis and initial management plan for pneumonia" needs the learner to apply knowledge, analyze, and critically think about the various conditions which can mimic pneumonia as in this case. It also needs the learner to develop a management plan based on their understanding of the disease (Chatterjee and Corral, 2017).

## 5.4.5 Learning objectives can also be classified on the basis of areas of focus

Learning objectives are more than just lofty aspirations for a course. Learning objectives can also be classified into different sorts in curriculum creation based on their level of cognitive complexity and specificity. The following classification is more often used in the teaching and learning process.

### *5.4.5.1 Knowledge objectives*

These objectives emphasize the acquisition of factual information, concepts, and theories. They typically include activities such as memorizing material, describing essential phrases, and explaining fundamental concepts. The emphasis is on the preservation and transmission of foundational knowledge. These kinds of objectives are more important in the initial years, that is, years 1 and 2 of the medical program, wherein it is necessary for the learner to understand, remember, and recall various facts related to the topic. The specific learning objectives are designed for the topic, and the assessment methods are designed to cater to the knowledge-based objectives as they carry maximum weightage rather than understanding and application.

### *5.4.5.2 Understanding objectives*

These objectives involve understanding and interpreting information. Learners are expected to demonstrate their comprehension by summarizing, paraphrasing, and explaining ideas in their own words. The goal is to ensure that students grasp the meaning and significance of the material they have learned. These objectives enable the learner to have a deeper understanding of the topic. As the specific learning objectives of comprehension are learnt in more detail, this will give a chance for the learner to apply this knowledge and understanding to enhance their practical skills.

### *5.4.5.3 Application objectives*

These objectives require learners to apply their knowledge and understanding in real-world or novel situations. They may involve solving problems, analyzing case studies, or making connections between concepts. The focus is on the practical application of acquired knowledge and skills. Application objectives are given more importance in final-year students as the subjects they study in this period are based on practical application of knowledge and understanding. It should be taken care that while designing the specific learning outcomes of the topic, more weightage should be given to sharpen the practical or hands-on experience of the learner rather than just reinforcing knowledge and understanding.

### *5.4.5.4 Analysis objectives*

These objectives concentrate on breaking down complex information into its parts and identifying patterns or relationships. Learners are expected to compare, contrast, and evaluate different perspectives. The emphasis is on critical thinking and the ability to deconstruct and examine information in depth. Here, the learner is not only expected to understand and apply the knowledge they have, but at the same time, they must be able to analyze a particular situation or a case in order to achieve the specific learning objectives. For instance, activities like case-based learning give ample opportunities for the students to exhibit their analytical skills where they try to figure out various aspects of the case like diagnosis and management through thorough analysis of the specific signs and symptoms.

### *5.4.5.5 Synthesis objectives*

These objectives entail combining concepts from several sources to form a new information. Fostering creativity and the capacity to blend various ideas in novel ways is the aim of this objective. In order to produce new knowledge, a distinctive interpretation, or a new understanding, the learner is expected to collect, assess, and integrate data from many sources. One of the important ways to achieve these objectives is to practice synthesis activities like drafting patient notes, such as Subjective, Objective, Assessment, and Plan (SOAP). Healthcare professionals can record patient data in an easy-to-understand and structured manner using this format, which enhances clinical decision-making, communication, and continuity of treatment. Before summarizing their opinions, assessments, and treatment recommendations, clinicians must collect and examine a patient's medical history, lab results, and other information. However, disjointed electronic health records that do not adequately allow information synthesis can hinder this process.

#### 5.4.5.6 *Evaluation objectives*

These objectives require learners to make judgments and determine the value or quality of ideas. They may involve critiquing arguments, assessing evidence, or proposing improvements. The focus is on developing the ability to make informed decisions and evaluations based on specific criteria. These might involve activities or various processes like assessing the learner achievement in order to determine if students have imbibed the intended knowledge, skills, and abilities and can apply them in practice. The role of a learner in achieving these objectives is exemplified by the fact that they can provide feedback to educators and guide improvements to teaching methods and the overall curriculum to meet learner needs and societal demands with an aim to improve teaching and curriculum. Even for the process of program evaluation, the objectives pertaining to evaluation become important as they give a chance to ensure program quality by providing a system for quality assurance and accountability, ensuring educational standards are maintained and helping the curriculum evolve over time.

## 5.5 Taxonomy of learning objectives

### 5.5.1 Definition

Taxonomies in medical education are classification systems based on an organizational scheme. It is a set of carefully defined terms, organized from simple to complex and from concrete to abstract, wherein they provide a framework of categories into which one may classify educational goals.

### 5.5.2 Advantages of taxonomy

Writing learning objectives, creating assessments, and designing instruction are all made easier with the use of a taxonomy for educational objectives. Its benefits include ensuring a balance between various cognitive levels, fostering progressive learning by advancing students from fundamental to complicated abilities, and producing more interesting and quantifiable learning experiences. In the end, the framework fosters deeper comprehension and critical thinking by ensuring that curriculum, assessments, and teaching methods are in line with the intended learning outcomes. It provides precise and measurable goals because it uses a hierarchical structure and specific verbs to indicate what students should know and be able to accomplish following a lesson or course. As a result, goals become observable and quantifiable.

### 5.5.3 Bloom's taxonomy

Benjamin Bloom was an educational psychologist working at the University of Chicago. He proposed the taxonomy in 1956 which fostered a common vocabulary for thinking about learning goals. Bloom's taxonomy engendered a way to align educational goals, curricula, and assessments that are used in medical schools. It also structured the breadth and depth of the instructional activities and curriculum that teachers provide for students. Bloom's taxonomy is a model that describes the cognitive processes of learning and developing mastery of subject. As it offers a hierarchical framework for categorizing educational goals and competencies, Bloom's taxonomy has gained a lot of importance. In addition, it enables educators to create more efficient and coordinated learning activities and evaluations. By directing the creation of lesson plans and assessments that gradually increase student knowledge, it goes beyond simple memorization to promote higher-order thinking skills like analysis, evaluation, and creation. Deeper comprehension and the growth of practical abilities are encouraged by this. Furthermore, it assists the teachers in designing the course and the curriculum by helping them set clear, measurable learning objectives at appropriate levels of complexity for the students they are teaching (Salas et al., 2024).

Most commonly envisioned as a pyramid, with more basic cognitive activities at the base and higher processes like synthesis and evaluation at the top, it showcases that knowledge of the topic forms the base on which all the other domains are built. The original model was centered around static knowledge and abstract abilities, while the revised model which emerged in 2001 focuses on measurable activities. In this version, the original base of the pyramid, "Knowledge," was replaced with "Remember," while "Application" was replaced with the more active "Applying" (Fig. 5.2).

### 5.5.4 Bloom's revised taxonomy

The revised Bloom's Taxonomy emphasizes students' learning outcomes through the use of refined terms. The revised taxonomy is a refreshed take on Bloom's Taxonomy from 1956, which examined cognitive skills and learning behavior. In order to highlight cognitive processes as active acts, the new edition included significant changes, such as switching from nouns to verbs. For example, knowledge was substituted with remembering. Additionally, there was a change in the

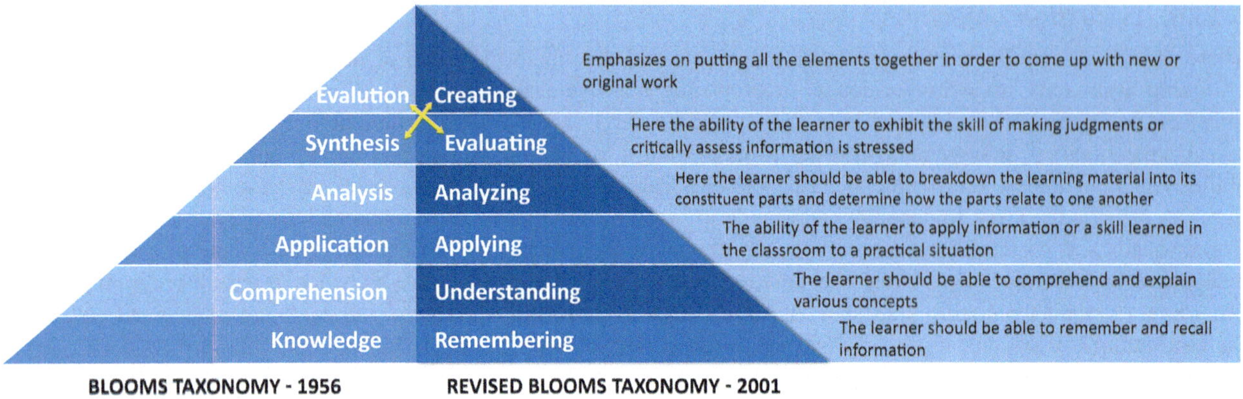


FIGURE 5.2 Bloom's and revised Bloom's taxonomies.

hierarchy, with "Creation" replacing "Evaluation" at the top level. A wider audience and a more genuine instrument for curriculum planning, instruction, and assessment were the goals of this revision (Anderson and Krathwohl, 2001).

### 5.5.5 Levels of revised Bloom's taxonomy

The revised taxonomy focuses on six levels. The six levels are remembering, understanding, applying, analyzing, evaluating, and creating. Bloom's action verbs are also used when developing learning outcomes. Each level is conceptually different.

#### *5.5.5.1 Remembering*

Medicine is indeed a very vast subject that requires a constant reading and remembering information. The process of remembering necessitates that the learner should be able to remember and recall information from long-term memory. For instance, remembering the branches of the brachial plexus or the stages of a nerve action potential are critical to the understanding of the various functional aspects or clinical medicine related to those topics. It is indeed imperative that learners are made aware of this fact and encouraged to remember and recall the important factual information as often as they can. It is also important to keep in mind that the learning outcomes intended to target the ability to remember must incorporate the verbs such cite, define, list, match, name, outline, quote, recall, report, reproduce, retrieve, show, state, tabulate, and tell. For instance, the learning outcome "List the key findings on an ECG that are indicative of heart blocks" is an example of a learning outcome in the knowledge domain.

#### *5.5.5.2 Understanding*

A student's comprehension of a certain subject opens the door to a more in-depth learning process. This comprehension capacity is demonstrated by one or more explanations. Understanding entails demonstrating knowledge of a certain idea or information, such as describing the pathophysiology of a disease or how a hormone functions. Bloom's taxonomy lists classify, describe, explain, interpret, summarize, and translate as action verbs for "understand" that entail exhibiting an understanding of concepts and ideas. These verbs emphasize reiterating or clarifying something in your own words, drawing connections between ideas, and showcasing your understanding of the subject. For instance, the learning outcome "Describe the pathophysiology of Parkinson's disease" is an example of a learning outcome in the understanding domain.

#### *5.5.5.3 Applying*

The domain of applying showcases the ability of the learner to apply information or a skill learned in the classroom to a practical situation such as in demonstration of examination of the cranial nerves or the measurement of blood pressure. In this, the student recollects what he or she has learnt from their teachers and try to demonstrate the skills either in the skill lab or during clinical examinations. For the application level of Bloom's taxonomy, action verbs focus on using knowledge in new situations, including verbs like apply, calculate, demonstrate, solve, use, illustrate, perform, model, choose, complete, operate, and implement, showing learners can use concepts to tackle problems or complete tasks where it is intended to examine medical data, such as test results and patient history, to understand a condition or assess where the main purpose is to know the ability of the learner to evaluate a patient's health status to determine a course of action.

For instance, the learning outcome "Apply the ABCs (Airway, Breathing, Circulation) to a simulated patient in a cardiac arrest scenario" is an example of a learning outcome of the applying domain.

#### *5.5.5.4 Analyzing*

In medical education, the skill of analyzing the information provided and applying it to various situations is indeed a game changer as far as scoring top grades in medical school or achieving professional success is concerned. This skill requires the learner to breakdown the learning material into its constituent parts and determine how the parts relate to one another or to an overall structure or purpose. One of the important skills in achieving the analyzing domain for the learner is critical reasoning to determine patterns and connections between different sources of information. The action verbs that suits the most for this domain include analyze, arrange, classify, compare, and differentiate. For instance, the learning outcome "Analyze the steps of a thyroidectomy and explain the purpose of each step" is an example of a learning outcome in the analyzing domain.

#### *5.5.5.5 Evaluating*

The domain of evaluation in medical education has gained traction due to the fact that it covers what a student should know, feel, and do to become a competent physician implicating that the domains of cognition, affect, and psychomotor are well covered. The evaluation domains go way beyond knowledge and understanding as it emphasizes that the learner should be able to exhibit the skill of making judgments by being able to critically assess information, compare different options, and make reasoned judgments. It also stimulates the higher-order thinking amongst the learner as it goes beyond simply recalling or understanding information to actively using it to make decisions in their day-to-day clinical practice. The action verbs that are intended to be used for this domain are appraise, assess, and critique. For instance, "Following a simulation, the trainee will be able to appraise their own clinical performance and reflect on areas for improvement, using a structured debriefing tool."

#### *5.5.5.6 Creating*

The creating level represents the final level of the revised Bloom's taxonomy model, which emphasizes on putting separate elements together to form a detailed final product. This domain is regarded as the highest level as it is not only the final level but at the same time is the most complex stage in the cognitive hierarchy following levels like remembering, understanding, applying, analyzing, and evaluating. One of the main features of this domain is that the learner here focuses on maintaining the originality that is on generating something unique and novel. In the realm of medical education, the domain of creating is exhibited in a physician's ability to integrate vast amounts of information to create a new diagnosis, treatment plan, or research proposal or using advanced skills where students can move beyond the four other levels of cognitive process to use their knowledge and skills in a truly innovative way such as creating an original lesson plan or presentation to explain a difficult medical concept to peers, effectively creating new educational material based on their understanding. Appropriate action verbs for this level include design, construct, produce, formulate, and invent. For instance, the learning outcome of "Given a patient's complete medical history, present a comprehensive, patient-specific treatment plan, detailing the rationale for each intervention" is a good example to understand the creating domain.

## 5.6 Using Bloom's revised taxonomy in assessment

The application of Bloom's taxonomy in assessment has shown to be highly advantageous since it provides a framework for crafting questions based on the domain that is meant to be measured. It guarantees that tests are in line with learning goals, encouraging a deeper comprehension of the material and assisting in the identification of certain areas in which students might have difficulty. It promotes the growth of higher-order cognitive abilities by going beyond simple memory (Shumway and Harden, 2003).

### 5.6.1 Modernizing the Bloom's taxonomy in the era of artificial intelligence

Incorporating AI-specific competences and emphasizing the distinct human cognitive abilities that AI cannot imitate, like moral reasoning and profound empathy, are key components to modernizing Bloom's taxonomy in the AI era. The original architecture is still applicable, but its use has changed as AI technologies can help with higher-level tasks like creating and analyzing as well as many lower-level ones like remembering and understanding. Bloom's Taxonomy is being reexamined in the era of AI to move the emphasis from lower-level tasks that AI can readily complete to higher-order abilities like evaluation, creativity, and critical thinking. The framework is used to create AI-powered learning experiences, but in order to guarantee the effective use of AI in medical education, it is important to be aware of the numerous challenges that the learner might get involved in like AI bias, comprehending its limitations, and using AI as a tool to foster more complex cognitive processes rather than depending on it for straightforward answers. In this aspect, the teachers should be able to modify their teaching and assessment strategies so as to guarantee that AI is used in a more professional and ethical way (Hmoud and Ali, 2024).

AI can be used to exploit lower-level activities because it can effectively handle jobs like remembering and understanding, thereby freeing up educators to concentrate on higher-level skills. These are some benefits of integrating AI into Bloom's taxonomy. AI can be used by students to swiftly locate material or obtain summaries, but the focus then switches to what they do with that information. AI can prove to be a double-edged sword. While it can be helpful in readily providing information to the learner, it is also harmful if the information is not used in the right way.

### 5.6.2 Formulation of a learning objective

Learning objectives should start with a clear, action-oriented verb and describe what the student will be able to do after a topic is learnt. They should be aligned with the needs and expectations of the learner and the society at large. It should be noted that both Bloom's taxonomy and SMART formula (specific, measurable, attainable, realistic, and time-bound) are necessary for writing effective learning objectives (Ogbeiwi, 2017). Learning objectives are the bedrock of any educational institution and serve as the guiding light for a lesson, curriculum, activity, or method. In short, it is a roadmap for both students and teachers. It helps students to understand the "what" and "why" of the lesson. It allows students to improve their engagements and develop metacognitive skills. For the teachers, it proves to be beneficial in structuring the curriculum in appropriate methods. It allows teachers to find out which areas students may be struggling with. Developing clear and well-defined learning objectives is essential for creating effective educational ventures (Bovend'Eerdt et al., 2009) (Fig. 5.3).

Using the SMART formula is one of the most common ways used to formulate educational objectives. It is generally accepted that the SMART acronym was first written down in November 1981 in Spokane, Washington, by George T. Doran,

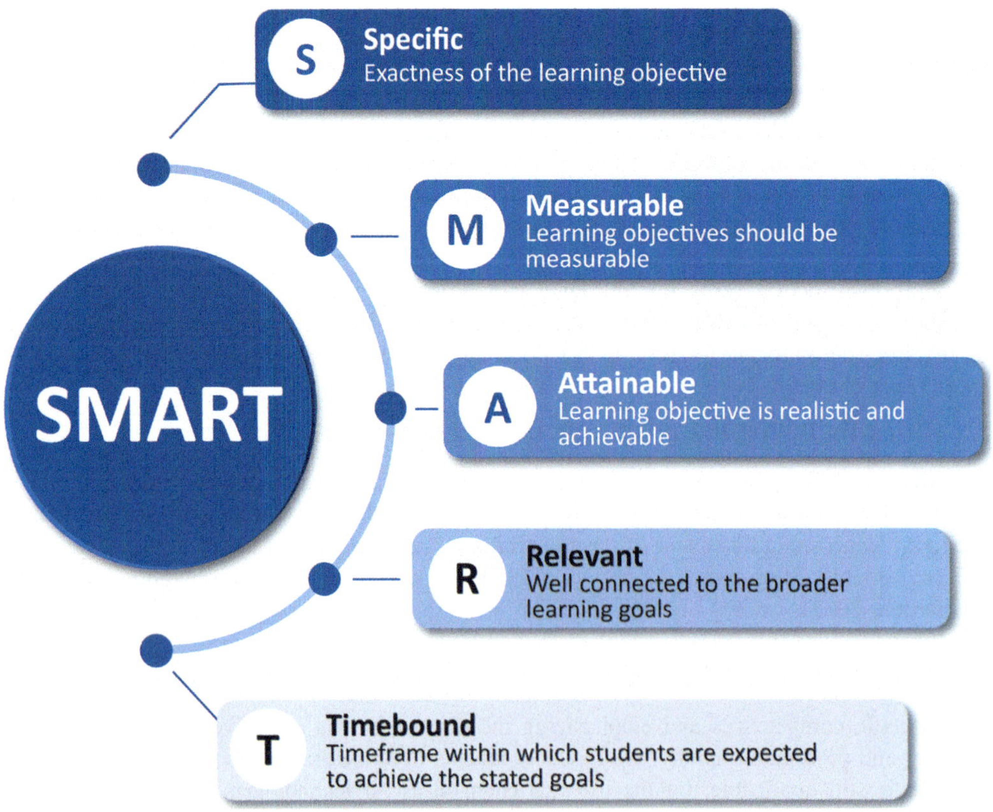


FIGURE 5.3 SMART (specific, measurable, attainable, realistic, and time-bound) formula.

a consultant and former director of Corporate Planning for Washington Water Power Company. The need for formulating effective learning objectives was felt when educationists realized that keeping the contents as specific as possible is crucial for improving the quality and effectiveness of medical education, patient care, and healthcare management. The structured framework helps both learners and clinicians move from vague intentions to concrete, actionable plans with clear criteria for success. For the medical educator, the formulation of SMART objectives helps to align curriculum and assessment aspects by providing a clear blueprint for curriculum development by specifying what students should be able to do at the end of a course or rotation. This ensures that teaching methods, learning activities, and assessment tools all support the same educational outcome, preventing disjointed instruction. Furthermore, having the smart way of objectives helps in streamlining staff performance and training as they provide a structured approach for this process. This ensures staff development goals are purposeful, measurable, and aligned with organizational priorities, which in turn leads to better teaching and learning process and also in delivery of effective patient care (Sottas, 2011).

## 5.6.3 Components of the specific, measurable, attainable, realistic, and time-bound framework

### 5.6.3.1 Specific

The term "specific" denotes the exactness of the learning objective. It is necessary that the learning objectives in curriculum development must be well-defined and precise. Educators should strive to identify the specific outcomes that students will attain upon successfully completing the course or lesson, rather than establishing a vague or general goal. An alternative approach to this is to think about what the learner will be able to comprehend or perform after completing the course that they were unable to perform or comprehend previously. The approach of making the learning objective specific is very critical to the teaching and learning process. For the student, it guides to achieve their goals by increasing motivation and focus as the learners know exactly what they are expected to accomplish and why it is important. Furthermore, they are more motivated to engage with the material. On the other hand, for the educator, the specific learning objective is advantageous because it streamlines curriculum planning by helping them to design more precious teaching materials. By defining what needs to be achieved, they can select and organize content, activities, and resources more effectively and avoid trying to cover too much.

### 5.6.3.2 Measurable

It is a well-known fact that in medical education, assessment drives the learning process. In this regard, it is vital to design the learning objective which is measurable. Only then we will be able to justify the educational process. Making the learning objectives measurable will provide clarity for both instructors and learners, direct instruction and assessment, and at the same time promote accountability. For the instructor, it will act as a guideline to effectively frame the instructional design as these objectives serve as a guiding light for developing course material, learning activities, and instructional strategies. They help educators focus on what is important, avoid extraneous content, and ensure all parts of the curriculum are aligned and assessed. At the same time, for the learner, measurable objective will provide clarification of their expectations. These objectives help students understand exactly what is expected of them and what they should be able to do by the end of a lesson or course. This clarity reduces uncertainty and allows them to focus their study efforts effectively.

### 5.6.3.3 Attainable

It should be kept in mind that the learning objectives developed should always be realistic and achievable within the timeframe of the lesson or unit. Formulating attainable learning objectives is crucial for both educators and learners. For the learner, the formulation of an attainable objective contributes in the reduction of stress, as an attainable objective with a reasonable timeline helps reduce anxiety. Knowing they have the resources and time to succeed allows learners to concentrate on mastering the material. In addition, the attainable learning outcome promotes ownership of learning in the sense that when learners are aware of what is expected, they can track their own progress and take responsibility for their learning. This helps build self-directed learning skills, such as planning and self-assessment. For the educator, a well-formulated attainable objective helps to enhance the effectiveness of teaching process. Realistic objectives enable the educators to design more focused and purposeful learning experiences. This allows for better curriculum planning and the selection of relevant instructional materials and activities.

### 5.6.3.4 Relevant

One of the important aspects to be aware of is to make certain that the objectives which are formulated are well connected to the broader learning goals of the curriculum and are appropriate for the target audience. Making learning objectives relevant offers numerous advantages for both students and instructors, significantly enhancing motivation, engagement, and the long-term effectiveness of learning. For the learner, the creation of relevant objectives will help to enhance knowledge retention and transfer. This is because connecting new information to real-world applications will help students retain concepts more deeply. This contextual learning promotes the ability in the learner to flexibly apply knowledge to different situations, which is a key indicator of deep learning. For the educator, the formulation of attainable objective will guide assessment design by helping the instructors to create assessments that are truly aligned with what is being taught. This ensures that tests and assignments accurately measure whether students have met the intended goals.

#### 5.6.3.5 *Time-bound*

It is necessary to formulate the learning objectives with a certain timeframe within which students are expected to achieve the stated goals. By establishing a timeframe for a learning objective, a clear structure and urgency is provided, which leads to increased focus, motivation, and effective resource management. The inclusion of a deadline is a key component of the popular SMART goal framework and is crucial for translating an idea into an actionable plan. For the learner, it helps to improvise the time management skills as it teaches the learners how to plan and prioritize their tasks effectively. Breaking a large project into smaller, time-bound milestones, for example, helps prevent learners from feeling overwhelmed and trains them to manage their workload. At the same time, for the educator, it enables to optimize teaching strategies by observing how the learners manage their workload around deadlines. Owing to this, instructors can gain insight into student behavior and adjust their teaching methods accordingly.

Some of the important steps to be considered while crafting an effective learning objectives are described below.

### 5.6.4 Step 1: determine desired learning outcomes

In medical education, determining desired learning outcomes provides a clear framework for curriculum design, teaching methods, and assessment, ultimately leading to more competent and prepared physicians. This ensures that all educational activities are aligned with the specific knowledge, skills, and professional behaviors that students must master to succeed in their careers. Once the aims of a program are delineated, program planners are to determine the learning outcomes needed to meet such aims. For the learner, determining the desired learning outcomes ensures career readiness as it focuses on the competencies required for practice. Moreover, the formulation of defined outcomes ensure that medical students are equipped with the practical knowledge and skills which are needed to be a successful healthcare professional. For the educator, the determination of the desired learning outcomes will provide a clear roadmap for making course design decisions, from selecting content to structuring learning activities and determining the sequence of topics.

### 5.6.5 Step 2: start with action verbs

Learning objectives should be centered around observable and measurable actions. By using action verbs when designing educational outcomes, clarity, measurability, and a clear focus can be provided both for the instructors and students. These verbs define observable and assessable student behaviors, moving beyond vague terminologies like know or understand. Using action words while formulating objectives is beneficial as it saves time and effort. By writing clear learning outcomes with action verbs, instructors can plan their teaching more efficiently from the start, ensuring that they focus on what is truly needed for students to succeed. For the learners, the use of a specific action verb provides more clarity enabling them to know exactly what is expected of them by the end of a topic or course. This clarity fosters engagement and helps them concentrate their efforts on developing specific skills and knowledge.

### 5.6.6 Step 3: specify the conditions and criteria

Specifying the conditions and criteria when designing educational outcomes helps to provide a clear framework for students, instructors, and assessors, leading to a more focused and effective learning experience. This process is crucial for achieving transparency, consistency, and measurability in education. The conditions may include resources, tools, or constraints that the learners will encounter during the learning process or assessment. For the learner, specifying the conditions and criteria provides clear expectations as it gives a chance for the students to precisely comprehend what they are expected to achieve, which reduces ambiguity and helps them direct their efforts toward the most important concepts and skills. This clarity can alleviate unnecessary stress and anxiety. Similarly, for the educator, specifying conditions and criteria helps instructors to streamline curriculum planning and design a more focused and purposeful curriculum, ensuring all content, activities, and resources align with the desired outcomes. This avoids the inefficiency of covering too much or irrelevant material.

### 5.6.7 Step 4: incorporate Bloom's taxonomy

By aligning the learning objectives with the various components such as remembering, understanding, applying, analyzing, evaluating, and creating of Bloom's taxonomy, which is one of the most widely recognized framework for categorizing cognitive processes, the teacher can ensure that learners engage in higher-order thinking skills and develop a deeper understanding of the subject matter. By providing a clear and measurable objectives, the taxonomy's hierarchical structure and associated action verbs help create specific, clear, and measurable learning outcomes that students can easily understand. These features in turn enable the medical educators to create a balanced curriculum wherein the instructors are encouraged to balance their curriculum by including a variety of cognitive skills, rather than focusing primarily on rote memorization.

### 5.6.8 Step 5: align with the program objectives

It is imperative that the specific learning objectives are well aligned with the program learning objectives. It is important to have a clear understanding of the program goals before formulating the specific learning objectives. Aligning the specific learning objectives with the program objectives is vital for effective educational design, quality assurance, and student success. The alignment has many advantages, both for the learner and educator. For the learner, this alignment is very crucial as it helps them to develop a sense of deeper, more meaningful learning because students can see how individual courses and assignments connect to the overarching program objectives. This fosters a growth mindset towards education. Meanwhile, for the educator, this alignment acts as a boon because it provides an opportunity to identify the learning gaps where certain skills or knowledge are under-addressed, or areas of over-concentration where they are redundantly covered.

### 5.6.9 Step 6: involve stakeholders and subject experts

Involving stakeholders and subject matter experts is crucial for developing learning objectives that are relevant, accurate, and effective. Their participation ensures that the learning outcomes align with real-world needs and organizational goals, which increases learner motivation and the overall success of the educational program. By involving the stakeholders, medical educators have an opportunity to gather diverse perspectives, whereby a clearer, more holistic picture of needs and potential challenges is depicted. On the other hand, the involvement of subject matter experts into the process of formulating the objective ensures the accuracy and credibility as the subject expert guarantees that the content and objectives are factually correct, up to date, and aligned with healthcare standards and best practices. Without a subject matter experts, there is a risk of providing learners with outdated or inaccurate information.

### 5.6.10 Step 7: review and refine

One of the best things about the learning objectives is that the objectives are always open for review and refinement. It is important that they should be revisited and refined as and when needed. Reviewing and refining learning objectives is crucial for ensuring a course is focused, effective, and aligned with student needs. This process helps instructors design relevant assessments and activities, guides students by setting clear expectations, and ensures that objectives remain achievable and meaningful throughout the learning process. Regularly evaluating and updating objectives prevents gaps in learning and keeps the course relevant. At the same time, it is vital that constant feedback be taken from learners and instructors which will provide opportunities to make adjustments and ensure that the objectives remain relevant, clear, and achievable. It is also important to collaborate with subject matter experts, instructional designers, and other stakeholders to review and refine the learning objectives.

## 5.7 Assessing the learning objectives

The assessment of the learning objectives forms one of the critical steps in the teaching and learning process. It not only helps the teachers to understand how much the student has learnt a particular topic or concept, but it also helps them to identify the gaps or redundancies in the process of content delivery. In this regard, the major steps to be followed while assessing the learning objectives are enumerated below.

### 5.7.1 Review your learning objectives

Before beginning to assess the learning objectives, it is important to review them and make sure they are well-written and aligned with your program goals. A useful framework for writing learning objectives is the SMART criteria. This goes to say that the learning objectives should state exactly what the learners will be able to do, how realistic and attainable they are, how they relate to the program goals and the needs of the learners, and when they will be accomplished. To accomplish the aim of reviewing the learning objectives, a checklist or a rubric which would help to evaluate your learning objectives has to be used.

### 5.7.2 Choose appropriate assessment methods

Choosing an appropriate assessment method to evaluate the objective is a very vital step that should be done with utmost care. Different types of assessments can be used depending on the nature and level of the learning objectives. For instance, the use of formative assessments to monitor the learners' progress and provide feedback along the way is one of the best ways to assess. One can also use summative assessments to evaluate the learners' achievement and outcomes at the end of the program, such as tests, projects, portfolios, or presentations. It is essential to select the assessment methods that best match the learning objectives and provide valid and reliable evidence of the learners' performance (Bara and Bara, 2023).

### 5.7.3 Collect and analyze data

Following the process of choosing the appropriate assessment methods, the next step would be to collect and analyze the data that they generate. This would include the collection, storage, and organization of data and what tools or software will be used to analyze it. The criteria or standards which will be used to judge learners' performance and the way the results are reported have to be clearly specified. The process of collecting and analyzing data for specific learning objectives is indeed a collaborative effort which involves various individuals. For instance, the teachers in the classroom are in the best position to collect and analyze student performance data to improve their teaching and provide personalized learning. Teachers can use either formative or summative assessments or informal observations and student self-assessments.

### 5.7.4 Interpret and communicate results

The final step of assessing the learning objectives is to interpret and communicate the results of the data analysis. The results obtained are a reflection of the effectiveness of the learning objectives and the program. It is vital that the results are communicated to all the stakeholders and explain what they mean and what implications they have for the program. Interpreting and communicating the results also gives a chance for the stakeholders to put forth recommendations or suggestions for improvement based on the achieved results and feedback (Fig. 5.4).

### 5.7.5 Significance of learning objectives

By concentrating on particular, quantifiable, and observable physician competencies for better practice, learning objectives are essential in medical education because they offer a framework for curriculum design, direct teaching, and assessment, align expectations for students and faculty, and guarantee that education translates into improved patient care. They make clear the desired results, maximize instructional time, and are frequently necessary for certification, explicitly connecting education to patient safety and professional standards. The important significances of learning objectives are summarized below.

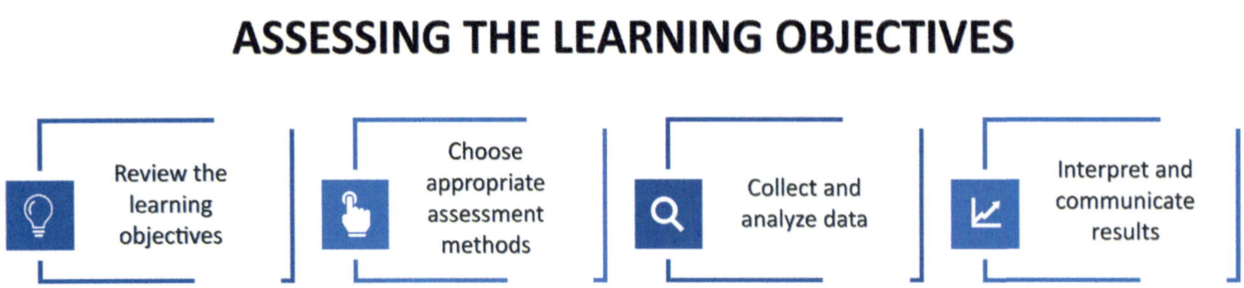


**FIGURE 5.4** Assessing the learning objectives.

#### 5.7.5.1 Clarity and focus

The learning objectives serve as a strategy that links every step of the learning process. They can be used by teachers to plan their lessons, select appropriate teaching methods and resources, and maintain focus. By establishing specific learning objectives, educators may ensure that their classes remain focused and purposeful.

#### 5.7.5.2 Increased student engagement

When students understand the goals and purpose of a course, they are more dedicated to their study. Learning objectives let students set goals for themselves, monitor their progress, and comprehend how each class increases their overall mastery of the subject. The transparency of the learning process increases students' motivation and engagement levels.

#### 5.7.5.3 Efficient evaluation

Learning objectives provide a clear framework for developing assessments that accurately measure students' development. By matching specific learning objectives with student learning evaluations, teachers can identify the precise skills and knowledge that students need to demonstrate competency. Evaluations are ensured to be accurate and trustworthy measures of students' understanding by employing a targeted approach.

#### 5.7.5.4 Data-driven instruction

By comparing student performance to learning objectives, teachers can identify areas where students may be struggling. Teachers can better adapt their classes to the needs of their students and create a more dynamic and responsive learning environment by employing a data-driven approach.

#### 5.7.5.5 Communication and collaboration

Clear communication and cooperation between teachers, students, and other stakeholders are encouraged by learning objectives. If everyone involved in the educational process is aware of the expectations and learning objectives, they may work together more effectively.

## 5.8 Learning outcomes

As against the objectives, learning outcomes are the evaluation of an activity performed against the intended result. Learning outcomes are statements of expectations written in measurable terms that express what a student is expected to know, think, or able to do at the end of the learning experience. In other words, learning outcomes are the measurable results of a student's learning, which define the knowledge and skills they have gained from an educational experience (Amin et al., 2007).

### 5.8.1 Levels of learning outcomes

There are three levels of learning outcomes.

#### 5.8.1.1 Institutional learning outcomes

Institutional learning outcomes (ILOs) are significant because they define the core knowledge, skills, and abilities that all students should acquire by the end of their learning experience, guiding curriculum design, program evaluation, and student support. They provide a framework for institutions to ensure consistent quality across programs, facilitate accreditation, improve instruction, and help students understand the overarching goals of their education and develop core competencies like critical thinking and communication. For instance, the institutional learning outcomes for a medical university might focus on professional competence, ethical practice, and the ability to care for patients in a collaborative, compassionate, and culturally sensitive way.

#### 5.8.1.2 Program learning outcomes

Program learning outcomes in medical education specify the fundamental knowledge, abilities, and professional behaviors that graduates must exhibit. These include clinical competence such as taking a patient's history, diagnosing them, and managing them, scientific comprehension such as anatomy, physiology, pathology, and research, effective communication, interprofessional teamwork, ethical practice, leadership, and a dedication to lifelong learning, all of which are focused on patient-centered care and community health. They prepare physicians to ethically progress medical knowledge and adjust to complex healthcare systems by bridging the gap between fundamental sciences and practical application. "Demonstrating a patient-centered approach to care, applying basic and clinical sciences to solve patient problems, communicating effectively with diverse patients and colleagues, and performing clinical procedures safely" are a few examples of the program's learning outcomes.

#### *5.8.1.3 Course learning outcomes*

Course learning outcomes in medical education define the fundamental knowledge, skills, and professional attitudes graduates must exhibit with an emphasis on applying fundamental sciences to clinical reasoning, patient care, professional ethics, and effective communication. They go beyond merely content knowledge to quantifiable competencies like clinical skills, evidence-based practice, and patient-centered care. These outcomes, which are frequently evaluated through practical tests and real-world clinical scenarios, direct curriculum design, guaranteeing that the graduates can operate as competent, safe, and ethical practitioners from the outset. They are essential for defining a course's objective, guiding instruction and assessment, and setting clear expectations for students. For example, the course learning outcome for a cardiovascular system course might be written as "Demonstrate knowledge and understanding of the structure and function of the cardiovascular system in normal and disease state."

#### *5.8.1.4 Specific learning objectives*

In medical education, specific learning outcomes specify quantifiable, observable behaviors and competencies that graduates must attain. These include clinical skills like taking a patient's history, performing examinations and procedures, professional behaviors like ethics, communication, and teamwork, foundational science application like pathophysiology, and public health in terms of prevention and health promotion, going beyond memorization to show evidence-based practice, critical thinking, and holistic patient care. They are specialized goals that contribute to more general program objectives, guaranteeing that physicians are capable of carrying out particular duties like handling crises or working well in health teams.

### 5.8.2 Learning objectives versus outcomes

Learning objectives and outcomes, while related, are distinct concepts. Objectives define what students should learn during a lesson or course, while outcomes represent what students can demonstrate as a result of that learning. In simpler terms, objectives are the aims of a lesson or course, while outcomes are the proof that those aims have been achieved. For example, a learning objective might be to understand a concept, while the learning outcome would be to demonstrate that understanding through a specific task like a quiz or presentation. In essence, objectives are the goals, and outcomes are the evidence of achieving those goals (Harden, 2002).

The key differences between learning objectives and outcomes are summarized below (Table 5.2).

## 5.9 Competencies and their role in medical education

### 5.9.1 Definition

Educational competency in medical education refers to the ability of the learner to successfully perform tasks and achieve outcomes that are essential for patient care, by using a combination of knowledge, skills, and attitudes. The ability of the

**TABLE 5.2** Difference between the learning objectives and learning outcomes.

| Feature | Learning objectives | Learning outcomes |
|---|---|---|
| Focus | Teaching: What students should learn | Learning: What students can demonstrate |
| Nature | Broad, general statements | Specific, measurable statements |
| Example | "Students will be able to identify the different parts of a cell." | "Students will correctly label all parts of a cell diagram." |
| Perspective | Future-oriented | Past-oriented |

learner to exhibit these skills forms the basis of competency-based medical education (CBME), an outcome-based approach where the emphasis is on the acquisition of specific, observable competencies needed to meet patient and societal needs, and uses assessments to track progress, allowing learners to advance at their own pace (McCrory, 2025).

## 5.9.2 Core competencies

The core competencies in medical education, as defined by organizations like the ACGME and AAMC, are medical knowledge, patient care, practice-based learning and improvement, interpersonal and communication skills, professionalism, and systems-based practice. These core areas ensure physicians have the skills and ethical grounding to provide high-quality, safe, and compassionate care to diverse patients by understanding scientific principles, effectively communicating, continuously improving, and working within the healthcare system (Kavic, 2002) (Fig. 5.5).

### *5.9.2.1 Medical knowledge*

The ability to understand and apply biomedical, clinical, epidemiological, and social and behavioral sciences to patient care is the prime requirement expected from a student. Medical knowledge is a competency because it is the demonstrated ability to apply scientific understanding to patient care, going beyond simple recall to include the continuous learning and adaptation required in medicine. Competency in this area means a healthcare professional can use their knowledge of biomedical, clinical, and social-behavioral sciences to effectively diagnose, treat, and promote the health of patients, which is essential for performing patient-centered care.

### *5.9.2.2 Patient care*

Patient care is referred to as the capacity to provide compassionate, appropriate, and effective care, including performing essential medical, diagnostic, and surgical procedures. Patient care is a competency because it combines essential knowledge, skills, and professional behaviors necessary to provide safe and effective treatment for patients. It is a core requirement for healthcare professionals to demonstrate competence in areas like diagnosing and treating illness, communicating with patients and families, and maintaining ethical, compassionate, and professional conduct.

### *5.9.2.3 Practice-based learning and improvement*

The skill to investigate and evaluate patient care, appraise scientific evidence, and continuously improve performance through self-evaluation and lifelong learning is known as practice-based learning and improvement. Practice-based

FIGURE 5.5 Core competencies in medical education.

learning and improvement is a competency because it requires professionals to continuously evaluate their own practice, assimilate evidence to improve patient care, and engage in lifelong learning to close knowledge and skill gaps. It is a core competency in fields like medicine that focuses on self-improvement through active analysis, reflection, and the application of new knowledge in real-world situations, rather than just acquiring theoretical knowledge (Whitcomb, 2003).

#### *5.9.2.4 Interpersonal and communication skills*

The ability to effectively exchange information and collaborate with patients, families, and other healthcare professionals to improve patient satisfaction and treatment adherence are the components of interpersonal and communication skills. Interpersonal and communication skills are a competency because they are a set of observable and improvable abilities that are crucial for effectively interacting with others and achieving goals in personal and professional settings. This competency is a combination of both the performance of tasks such as communication skills and the impact on others like interpersonal skills, including active listening, empathy, and conflict resolution. They are essential for building relationships, fostering teamwork, and driving success.

#### *5.9.2.5 Professionalism*

Professionalism refers to the demonstration of ethical principles, integrity, compassion, reliability, respect for patients, and a commitment to self-improvement by the health professional. Professionalism is a core competency in medical education, encompassing a commitment to ethical principles, professional responsibilities, and patient well-being. It is a foundational element that includes qualities like integrity, honesty, and accountability, and is assessed through a combination of observed behavior, self-reflection, and training to ensure physicians provide high-quality care while maintaining trust with patients and society. Recent advancements also emphasize incorporating equity and anti-oppression into this competency.

#### *5.9.2.6 Interprofessional collaboration*

Interprofessional collaboration as a competency involves professionals from different fields working together effectively to achieve common goals, such as improving patient care. This competency requires skills like communication, mutual respect, and accountability, and is developed through education and training to improve patient safety, quality, and outcomes. Interprofessional collaboration is a core competency in medical education, preparing students to work with other health professionals to improve patient care. This competency is built on key frameworks, such as the Interprofessional Education Collaborative (IPEC)'s four core competencies which include Values and Ethics, Roles and Responsibilities, Communication, and Teams and Teamwork. Developing these skills is essential for the modern, team-based healthcare landscape

#### *5.9.2.7 Systems-based practice*

The ability to understand and utilize the healthcare system to provide patient care that is safe, effective, and efficient, including being aware of social determinants of health and resources, is known as systems-based practice. Systems-based practice is a core competency in medical education that requires physicians to understand and navigate the complex healthcare system to provide optimal patient care. It involves being aware of the larger context of healthcare, using system resources effectively, and contributing to quality improvement, patient safety, and cost-effective care. Systems-based practice focuses on how a physician's work connects to the system as a whole and emphasizes the roles of multidisciplinary teams, cost containment, and health equity.

## 5.10 Emerging competencies

Emerging competencies in medical education include digital health, such as using AI and data analytics, health systems science to navigate complex healthcare environments, and advanced skills like precision education and personalized coaching. These new skills are being integrated alongside traditional competencies like patient care, medical knowledge, and communication (Kofi et al., 2021).

### 5.10.1 Digital health

Digital health is an emerging competency in medicine as it gives aspiring physicians the tools they need to leverage cutting-edge technology like wearables, telemedicine, remote patient monitoring, and AI to deliver better, more effective patient care. A new curriculum must incorporate digital health competencies to guarantee that doctors can manage data efficiently, deliver remote treatment, interact with patients via digital platforms, and critically assess digital tools. Without this training, there's a chance that the workforce won't be ready for the growing digital revolution of healthcare, which could result in poor decision-making and detrimental effects on patient care.

### 5.10.2 Artificial intelligence and data analytics

It is impossible to overlook the rise of AI and its impact on healthcare. Due to their rapid transformation of healthcare delivery and the need for future physicians to be technologically and ethically literate, AI and data analytics are developing capabilities in medical education. By incorporating these abilities into medical curricula, students are better prepared to use AI technologies efficiently, assess their results critically, and negotiate the moral challenges of data-driven medicine (Musick, 2005).

### 5.10.3 Personalized education

Because it combines with the outcome-based methodology of CBME, uses adaptive technology to develop unique learning pathways, and creates self-directed, lifelong learners better suited for contemporary healthcare, personalized education is an emerging competency in medical education. With this change, the "one-size-fits-all" teaching approach is abandoned in favor of empowering learners according to their individual needs, interests, and developmental stages.

### 5.10.4 Impact of the competencies on the learner

Introducing competencies into the curriculum is a significant shift from traditional models, focusing on students' demonstration of mastery of skills and knowledge. The primary significance lies in better preparing students for real-world demands, promoting personalized learning, and fostering a growth mindset. The introduction of competencies into curriculum benefits both the learner and the educator. From the learner perspective, the important benefits are that it provides a chance to foster deeper learning and mastery as it emphasizes on having a thorough knowledge of the subject failing which the students cannot progress from one level to another in the medical program. It also prepares the learner to have an enhanced real-world readiness as it emphasizes on practical application and workplace-relevant skills such as communication, critical thinking, creativity, and problem-solving. This approach ensures graduates are better equipped for the modern workforce (Yeoh, 2019). Moreover, it promotes personalized learning as it acknowledges that students learn at different paces and in different ways. It allows for flexible learning pathways and provides individualized support, which particularly benefits nontraditional learners and students with disabilities.

For the educator, it provides a chance for a clearer articulation of the learning expectations, which in turn culminate in significance for educators and institutions focused teaching and learning. It also enables for the authentic assessment as the assessment process shifts from traditional tests to more meaningful and authentic tasks, such as projects, presentations, and portfolios. This provides a more comprehensive picture of a student's abilities.

## 5.11 Impact on principles and practice of medical education

Learning objectives and competencies have completely changed medical education by moving the emphasis from time-based training and factual knowledge acquisition to outcome-based, learner-centered education that stresses observable abilities and practical skills required for safe and effective patient care. As the clearly defined skills improve accountability and alignment with society and patient requirements, they make medical training standards apparent to students, faculty, and the general public. Since learning objectives and competencies helps in identifying the essential knowledge, abilities, and attitudes needed for practice, they direct curriculum creation. This guarantees that the curriculum is updated, in line with the demands of modern healthcare, and prevents unintentional duplication or removal of essential material.

**Summary**

Objectives are crucial in medical education because they guide student learning, align curriculum with learning outcomes, and ensure accountability. By strategically incorporating learning objectives into curriculum design, educators can enhance the quality of teaching and learning experiences, promote student engagement and motivation, and facilitate personalized and adaptive learning pathways. For the faculty, having a well-defined objective provides clear expectations for what students should know or be able to do, structure teaching and assessment, and help faculty focus instruction on measurable skills and competencies. Ultimately, distinct objectives contribute to better-trained, more competent physicians and improved patient safety. Embracing the versatility and flexibility of learning objectives empowers educators to create dynamic and impactful educational experiences that fulfill the various requirements and preferences of today's learners.

**Key point**

**Future perspectives: key shifts in educational objectives**

Understanding and appreciating the necessity of upgrading educational objectives is crucial, as medical education is evolving into a future arena due to the explosion of digital technology and the rise of artificial intelligence. Future perspectives on educational objectives center on digital competencies, data analysis, and AI literacy, along with essential human-centric abilities like communication, empathy, and teamwork. Key aspects that require focus are fostering critical thinking, adaptability, and social-emotional abilities like curiosity and resilience in order to equip students for a world that is rapidly changing. This means a shift to learning possibilities that are more student-centered, flexible, and connected to real-world problem-solving, with a greater emphasis on lifelong learning and creativity.

## References


Amin, Z., Khoo, H., Chong, Y.-S., 2007. Outcomes in Medical Education. Annals of the Academy of Medicine, Singapore 36, 711. Available at https://doi.org/10.47102/annals-acadmedsg.V36N9p711.

Anderson, L.W., Krathwohl, D.R., 2001. A Taxonomy for Learning, Teaching, and Assessing: A Revision of Bloom's Taxonomy of Educational Objectives. Addison Wesley Longman, New York.

Bara, E., Bara, G., 2023. The connection between assessment and learning outcomes. Balkan Journal of Interdisciplinary Research 9, 51–58. Available at https://doi.org/10.2478/bjir-2023-0006.

Bovend'Eerdt, T.J.H., Botell, R.E., Wade, D.T., 2009. Writing SMART rehabilitation goals and achieving goal attainment scaling: a practical guide. Clinical Rehabilitation 23 (4), 352–361. Available at https://doi.org/10.1177/0269215508101741.

Chatterjee, D., Corral, J., 2017. How to write well-defined learning objectives. Journal of Education in Perioperative Medicine 19 (4), E610.

Harden, R.M., 2002. Learning outcomes and instructional objectives: is there a difference? Medical Teacher 24 (2), 151–155. Available at https://doi.org/10.1080/0142159022020687.

Hmoud, M., Ali, S., 2024. AIEd Bloom's Taxonomy: A Proposed Model for Enhancing Educational Efficiency and Effectiveness in the Artificial Intelligence Era. The International Journal of Technologies in Learning 31, 111–128. Available at https://doi.org/10.18848/2327-0144/CGP/v31i02/111-128.

Kavic, M.S., 2002. Competency and the six core competencies. JSLS: Journal of the Society of Laparoendoscopic Surgeons/Society of Laparoendoscopic Surgeons 6 (2), 95–97.

Kofi, D., Ocansey, W., Xu, Z., Zhang, X., Xu, W., Qian, H., Mao, F., Kofi, W., Ocansey, D., 2021. Current and Emerging Medical Education Teaching Methods -Keeping up with the Evolving Medical Sciences. Revista de Educación (Madrid) 392. pp. 52–104.

McCrory, K.E., 2025. Core competencies, milestones, and entrustable professional activities. In: Kellerman, R., Irwin, G. (Eds.), Graduate Medical Education in Family Medicine. Excellence in Medical Education, vol 2. Springer, Cham, pp. 107–119. Available at https://doi.org/10.1007/978-3-031-70741-4_12.

Musick, D.W., 2005. Emerging trends in medical education: what are they? And why are they important? North Carolina Medical Journal 66 (3), 244–248. Available at https://doi.org/10.18043/ncm.66.3.244.

Ogbeiwi, O., 2017. Why written objectives need to be really SMART. British Journal of Healthcare Management 23 (7), 324–336. Available at https://doi.org/10.12968/bjhc.2017.23.7.324.

Ruiz, J.G., Mintzer, M.J., Issenberg, S.B., 2009. Learning objects in medical education. Medical Teacher 28 (7), 599–605. Available at https://doi.org/10.1080/01421590601039893.

Salas, A., Tan, M., Andrienko, S., Cengiz, K., Wisco, J.J., 2024. Reaching the top of Bloom's Taxonomy: an innovative pilot program for preclinical undergraduate and medical school students to create curricula for STEMM outreach/service-learning programs. Frontiers in Education 9. Available at https://doi.org/10.3389/feduc.2024.1446513.

Shumway, J.M., Harden, R.M., 2003. AMEE guide no. 25: the assessment of learning outcomes for the competent and reflective physician. Medical Teacher 25 (6), 569–584. Available at https://doi.org/10.1080/0142159032000151907.

Sottas, B., 2011. Learning outcomes for health professions: the concept of the swiss competencies framework. GMS Z Med Ausbild 28 (1), Doc11. Available at https://doi.org/10.3205/zma000723.

Swick, H.M., 2010. Educational objectives in medicine—some special problems. Programmed Learning and Educational Technology 16 (4), 322–327. Available at https://doi.org/10.1080/0033039790160410.

Whitcomb, M.E., 2003. Putting patients first: the need to reform graduate medical education. Academic Medicine 78 (9), 851–852. Available at https://doi.org/10.1097/00001888-200309000-00001.

Yeoh, K.-G., 2019. The future of medical education. Singapore Medical Journal 60, 3–8. Available at https://doi.org/10.11622/smedj.2019003.

Chapter 6

# Curriculum framework

*"Tell me and I forget, teach me and I may remember, involve me and I learn."*

—Benjamin Franklin

## 6.1 Introduction

In medical education, the curriculum is the blueprint for developing competent, ethical, and adaptable healthcare professionals who can meet the needs of patients and society. A curriculum is defined as a sophisticated blend of educational strategies, course content, learning outcomes, educational experiences, assessment, the educational environment, and the individual students' learning style. The curriculum is important as it communicates to stakeholders including students, teachers, patients, and the public more generally on what can be expected of the education program. It also provides a benchmark for program evaluation and for accreditation (El-Astal, 2023). Curricular frameworks are crucial in medical education because they provide a structured "roadmap" for learning, ensuring a consistent and reliable pathway to competency by aligning curriculum content with defined learning outcomes (Tolentino et al., 2024). They guide educators on what to teach and what students should know and be able to do, enhance reliability in assessment, and facilitate integration across subjects, leading to a more comprehensive understanding for learners.

The importance of these frameworks lies in the fact that they clearly outline what students should know, what skills they should develop, and how their progress will be assessed. Moreover, having a curricular framework ensures alignment with the needs of the healthcare workforce and society, ensuring graduates are relevant and competent. It also contributes to the facilitation of integration by guiding the process of curriculum mapping and by eliminating gaps or redundancies in the content. Curriculum frameworks have the capacity to influence curriculum design, teaching methodologies, and assessment strategies (Harden & Yıldız, 2024). They can be tailored to different levels of medical education like undergraduate or postgraduate and specific areas like global health or artificial intelligence (AI) in medicine. However, the curriculum framework is not a set of detailed lesson plans, detailed scope and sequence of curriculum objectives, or a mandate for specific methodologies or programs.

## 6.2 Types of curriculum

In medical education, the classification of curriculum can be approached in several ways based on educational philosophy, organization and structure, content, mode of delivery, innovation, and other types. Below is a detailed overview of the main types and classifications used in medical education.

1. Based on educational philosophy,
2. Based on organization and structure,
3. Based on content integration,
4. Based on mode of delivery,
5. Based on innovation or modern trends,
6. Other types of curricula (Fig. 6.1).

### 6.2.1 Based on educational philosophy

A medical education curriculum based on an educational philosophy moves beyond a traditional, lecture-based model to incorporate theories like constructivism and social learning. Instead of just memorizing facts, students engage in student-centered methods like problem-based learning (PBL), team-based learning, and case-based learning (CBL), where they

**The Principles and Practice of Medical Education. DOI:** https://doi.org/10.1016/B978-0-443-44197-4.00015-1

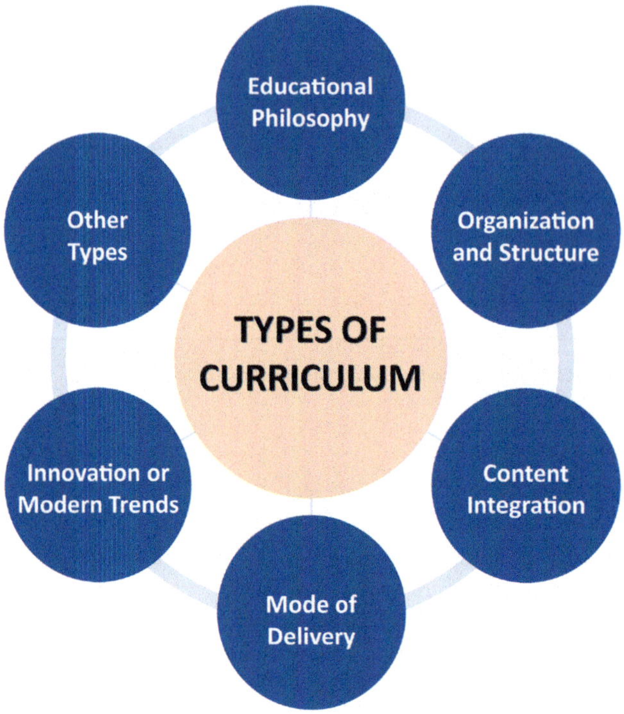


FIGURE 6.1 Types of curriculum.

learn through active problem-solving, collaboration, and applying knowledge to real-world scenarios. The goal is to create lifelong learners who are ethical, critical-thinking physicians prepared for any specialty.

#### 6.2.1.1 *Subject-centered curriculum (perennialism and essentialism)*

The subject-centered curriculum is based on two educational philosophies, perennialism and essentialism, which emphasize the importance of a core body of knowledge and the need for students to master it, forming the foundation of a subject-centered curriculum. The main objective of a subject-centered curriculum is to increase students' knowledge and comprehension of certain subjects or disciplines (Tozer, 2014) This kind of curriculum is frequently organized around a number of subjects and competencies within a specific field. This is also referred to as the traditional curriculum or discipline-based curriculum as in this the students are taught different disciplines as entirely separate, stand-alone courses. A subject-centered curriculum is considered a "focused" curriculum because its primary goal is to concentrate on a specific subject matter, such as anatomy or physiology rather than the needs of the student. For instance, students in the preclinical years learn about the curriculum dedicated to anatomy, physiology, and biochemistry separately. This approach prioritizes the mastery of content, skills, and knowledge within a discipline, often using traditional methods like direct instruction and standardized assessments. This type of curriculum concentrates on the systematic and organized way of teaching as well. It also emphasizes on skill practice, as it often includes a lot of drill and practice in specific skills to ensure mastery. Furthermore, it is teacher-directed, as in this, the instruction typically involves the teacher delivering content directly to students, who are expected to absorb and understand the material.

#### 6.2.1.2 *Learner-centered curriculum (needs, interests, and learning preferences)*

The learner-centered curriculum adapts content to each student's unique needs and interests. This curriculum is focused on the student. The needs, interests, and learning preferences of each individual student are the main focus of a student-centered program. It gives students greater influence over their education by emphasizing active learning and student interaction. Some of the main characteristic features is that it encourages the active participation from the students and ensures they take part in the learning process through problem-solving, inquiry-based learning, and discussions (Spencer and Jordan, 1999). At the same time, it also acknowledges that the students have unique learning styles and paces, allowing them to progress in a way that best suits them. It also gives the students choice and autonomy whereby the students often have a say in what and how they learn, with choices in topics, projects, and how they demonstrate their knowledge. Examples of student-centered curricula in medicine include PBL, where students work in groups to solve clinical cases, CBL, and simulation-based learning which uses realistic scenarios to develop skills.

#### 6.2.1.3 *Problem-centered curriculum (problem focused)*

The goal of a problem-centered curriculum is to educate students how to solve real-world issues. Instead of using lectures to steer learning, a problem-centered medical curriculum known as PBL employs clinical situations. Students analyze real-world patient issues in groups, pinpoint their knowledge gaps, and gather the data needed to comprehend the situation. This approach integrates knowledge from various disciplines to solve a specific, often complex issue. This method places a strong emphasis on critical thinking, self-directed learning, and using fundamental scientific concepts in a clinical setting (Lim, 2023). A problem-centered curriculum is a focused curriculum because it concentrates on real-world problems as the central theme for learning, rather than traditional subjects. This type of the curriculum is structured around a specific problem which becomes the focal point for all learning activities. It is also advantageous as it promotes integrated learning whereby it breaks down subject silos by requiring students to draw on knowledge and skills from multiple disciplines to address the problem.

#### 6.2.1.4 *Experience-centered curriculum (learning by doing)*

An experience-centered curriculum is an educational approach that prioritizes learning through real-world experiences, practical activities, and reflection rather than solely through traditional instruction. It centers on students' interests and actively involves them in concrete activities, such as community projects or scientific observations. It is built on the premise that doctors acquire the majority of their essential skills "on the job" and that reflection on these real-world experiences is crucial for developing clinical capability and professional identity. The goal is for students to connect what they are learning to the world around them and to be able to articulate the purpose and outcome of their experiences. One of the key features of the experience-centered curriculum is that it is learner-centered as it ensures that the student is an active participant in their education, with learning built around their interests, needs, and the challenges they face in practice. Moreover, the experiential learning emphasizes on "learning by doing" through real patient encounters, simulations, and community engagement, rather than passive knowledge acquisition from textbooks or lectures alone (Fig. 6.2).

### 6.2.2 Based on organization and structure

#### 6.2.2.1 *Traditional (discipline-based) curriculum*

The traditional-based curriculum is structured around the basic science disciplines like anatomy, physiology, biochemistry in the preclinical years, followed by clinical subjects like medicine, surgery, pediatrics, etc. This type of curriculum is teacher-centered and is discipline-based, where the subjects are taught as separate, distinct courses. In addition, this type of curriculum mainly focuses on lectures as the primary teaching method and recommends the use of textbook-based learning. Another feature of this curriculum is that content is organized in a sequential manner, completing one set of subjects before moving on to the next. The main aim here is to build a deep and comprehensive foundation in the scientific principles of medicine. For instance, the "2 + 3" model (2 years preclinical + 3 years clinical) is a good example of this type of curriculum.

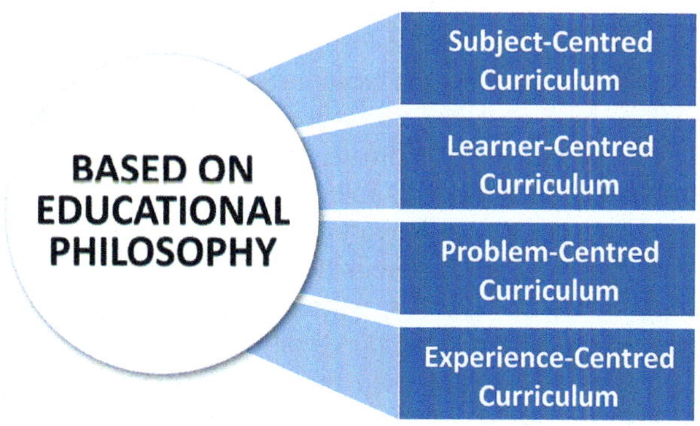


FIGURE 6.2 Based on educational philosophy.

### 6.2.2.2 *Integrated curriculum*

The integrated curriculum is structured to combine basic and clinical sciences around organ systems or themes. An integrated curriculum in medicine is a teaching method that combines different subjects, like basic sciences and clinical medicine, to provide a holistic understanding of patient care. Instead of learning disciplines in isolation, students learn how they interrelate, often by applying knowledge from basic sciences to real-world clinical scenarios from an early stage. This approach promotes better knowledge retention and skill development and is becoming increasingly popular worldwide. This type of curriculum provides a holistic learning experience as it integrates basic sciences, clinical medicine, and patient care to build a cohesive understanding of health and disease (Bandiera et al., 2013). In addition, it also gives a chance for the learner to have early clinical exposure wherein the students can engage with clinical environments, patient care, and case studies from the beginning of their medical education. It is also characterized by contextualized learning wherein it presents scientific concepts within a clinical context, which helps students see the relevance and application of what they are learning. One of the main features of this type of curricula is that it focuses on the interdisciplinary approach as it breaks down traditional barriers between subjects to create a more connected and meaningful learning experience.

There are two types of integration done in this curriculum which are horizontal and vertical integration as explained below:

#### 6.2.2.2.1 Horizontal integration

Horizontal integration in medical curriculum involves bringing together different subjects from the same phase of study to create a more connected and holistic learning experience, such as combining anatomy, physiology, and biochemistry to study a specific body system. This method breaks down traditional departmental barriers and provides a more meaningful, interconnected view of the material for students. For example, it links the disciplines taught at the same stage, for instance, anatomy and physiology of the cardiovascular system.

#### 6.2.2.2.2 Vertical integration

Vertical integration of a medical curriculum is an educational approach that combines basic science with clinical practice from the start of medical school, rather than separating them into distinct preclinical and clinical phases. This model introduces clinical context early, linking foundational knowledge to real-world application throughout the entire program, which can improve student understanding, engagement, and transition to professional practice. It helps the student link basic sciences with clinical applications early in training.

### 6.2.2.3 *Problem-based learning (PBL) curriculum*

PBL plays a crucial role in medical education by shifting away from traditional lecture-based techniques and organizing the whole curriculum around addressing challenging and real-world clinical scenarios. A PBL curriculum is structured around authentic, open-ended problems with students actively working in groups to identify learning needs, research, and develop solutions, making the problem the central driver for acquiring knowledge and skills, shifting the teacher's role to facilitator. The curriculum unfolds through problem-solving phases which include understanding the problem, defining learning objectives, self-directed study, and regrouping to synthesize solutions, building deep understanding and relevant skills like critical thinking and collaboration.

### 6.2.2.4 *Community-oriented/community-based curriculum*

This type of curriculum is structured or designed around health needs of the community. A community-based curriculum in medicine is a learning strategy where medical students train in community settings to address real-world health problems and learn about social determinants of health. This approach involves collaboration with community stakeholders and provides an authentic learning experience where students follow patients in their home and community contexts, moving beyond the hospital setting to learn about healthcare delivery in the community. In this, the students learn in community settings, such as clinics, primary health centers, etc., wherein the focus is on prevention, primary care, and public health. It inculcates a sense of collaboration amongst the learners, where it requires the learner to exhibit the qualities of active collaboration with community members, healthcare providers, and other stakeholders. There is a focus on the real-world where the curriculum is designed to address real-world health problems and the specific needs of the community it serves. It also enforces the patient-centered learning, wherein the students follow patients across different settings, including their homes, to understand their health challenges holistically.

#### 6.2.2.5 *Competency-based curriculum (CBC)*

It is structured around defined outcomes or competencies students must demonstrate. A competency-based curriculum in medicine is an educational approach that focuses on a student's demonstration of specific skills, knowledge, and attitudes needed for real-world medical practice, rather than just accumulating knowledge. It requires a shift from traditional, time-based models to one that emphasizes continuous assessment and allows students to advance at their own pace once they have mastered a competency. Some of the vital elements include entrustable professional activities, explicit learning milestones, and ongoing, longitudinal assessment to ensure graduates are "job-ready." The main features of this curriculum are that it focuses on outcomes as the curriculum is organized around a set of expected competencies, such as clinical reasoning, communication, and evidence-based decision-making. The curriculum also enforces that the students take more responsibility for their own learning through self-directed learning, reflection, and practice, rather than being passive recipients of information.

#### 6.2.2.6 *Outcome-based education curriculum*

This curriculum is structured similar to competency-based curriculum, wherein it begins with identifying outcomes such as knowledge, skills, and attitudes and has a backward learning design. An outcome-based education curriculum in medicine focuses on the competencies a graduate must possess that is the "product" rather than a rigid, time-based process. This approach redesigns the entire curriculum, from content and teaching methods to assessment, to ensure students achieve specific learning outcomes, leading to a more flexible, competency-centered, and individualized learning experience. In this, the focus is on the outcomes where the primary goal is to define and measure the final competencies of medical graduates, such as being able to care for patients ethically and competently. There is also a central role of assessment, since here assessment is not an afterthought but the core of the system, used to evaluate student achievement and inform the curriculum. It is often formative, providing ongoing feedback to students. It is also flexible and adaptable in that it allows for variable timeframes, as students can progress at their own pace once they achieve the required outcomes, rather than adhering to a fixed schedule (Fig. 6.3).

### 6.2.3 Based on content integration

A curriculum based on content integration in medical education unifies basic sciences and clinical medicine since it dismantles traditional silos and teaching subjects contextually through themes like a disease or systems. It frequently uses techniques like PBL and spiral or vertical/horizontal models to improve comprehension, application, and retention for comprehensive patient care. Horizontal integration, vertical integration, and spiral integration are important strategies for delivering relevant, practical learning.

FIGURE 6.3 Based on organization and structure.

Harden's "Integration Ladder" describes 11 steps of content integration in a curriculum, ranging from minimal integration (isolation) to full integration (transdisciplinary). The higher up the ladder, the less emphasis there is on individual disciplines and the more on integrated themes and real-world application, with students taking increasing responsibility for their own learning. The steps in this curriculum include isolation or fragmentation, awareness, harmonization or consultation, nesting or infusion, temporal coordination or parallel teaching, sharing or joint teaching, correlation or concomitant program, complementary or mixed program, multidisciplinary or webbed, interdisciplinary or monolithic and transdisciplinary or authentic. This type of curriculum acts as a planning tool, wherein the ladder acts as a useful framework for curriculum planners to evaluate the existing level of integration and decide on the desired level for different parts of the curriculum. In addition, it provides the flexibility, because here it may not be necessary or desirable to achieve the highest level of integration throughout an entire program. In this, the appropriate level can vary by module or stage. In order to move up the ladder, there is a need for effective collaboration and communication among staff, as well as a more centralized organizational structure. There are higher levels of integration which place more responsibility on students to synthesize information, promoting critical thinking and real-world application of knowledge.

### 6.2.4 Based on mode of delivery

#### *6.2.4.1 Face-to-face (traditional classroom or bedside teaching)*

A face-to-face curriculum involves direct, in-person instruction where students and instructors are in the same physical location. This format facilitates real-time communication, hands-on activities, collaborative learning, and immediate feedback, allowing educators to gauge understanding and adapt their teaching style in real-time. The curriculum is delivered through structured in-person sessions in a classroom or training room. It is characterized by direct interaction between the students and instructors who are physically present, enabling immediate and direct communication.

#### *6.2.4.2 Blended (mix of in-person and online learning)*

A blended curriculum, also known as hybrid learning, combines online educational materials with traditional face-to-face classroom instruction. This approach uses a mix of digital resources and in-person teaching to create a more flexible and personalized learning experience that caters to different learning styles and paces. In this, there is a combination of online learning which includes online lectures, interactive exercises, videos, digital textbooks, and other web-based resources that students can access at their own pace. It also involves the traditional classroom learning which involves in-person, instructor-led activities, discussions, and hands-on work, allowing for direct interaction between students and teachers.

#### *6.2.4.3 Fully online (virtual curriculum)*

A fully online curriculum, also known as a virtual curriculum, is an educational program delivered entirely through the internet using digital media. This format involves online instruction that can be either self-paced or asynchronous, or real-time or synchronous, where students and instructors interact via video or audio through platforms like Zoom or Google Meet. This type of curriculum is based on the internet-based delivery, wherein all the instructional content, communication, and progress tracking occur online. In this type, there is flexibility provided whereby the students can often complete coursework at a time and place that is convenient for them (Fig. 6.4).

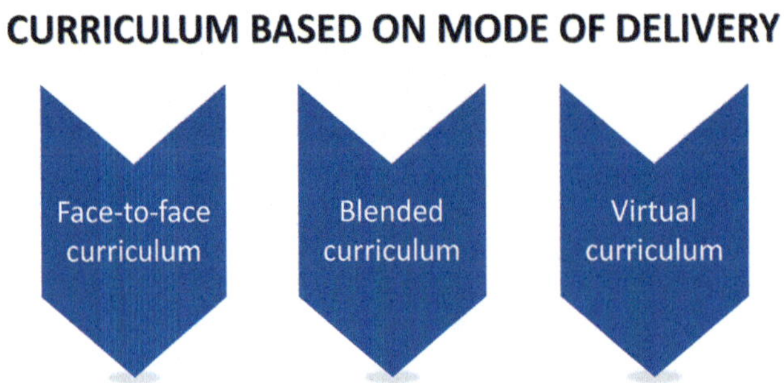


FIGURE 6.4 Based on mode of delivery.

### 6.2.5 Based on innovation or modern trends

#### 6.2.5.1 *Spiral curriculum*

In this, topics are revisited with increasing complexity. A spiral curriculum is an educational approach that involves revisiting key topics repeatedly throughout a course, with each encounter building on prior knowledge and adding greater complexity. This iterative method, developed by Jerome Bruner, reinforces learning by continually returning to concepts at a deeper, more abstract level and connecting new material to what students already know. One of the most important characteristic features of spiral curriculum is that it emphasizes on reinforcement as repeated exposure to concepts helps solidify them in the student's memory. It allows for the logical progression from simple ideas to more complex ones.

#### 6.2.5.2 *Hybrid curriculum*

It is a combination of traditional- and problem-based or integrated approaches. A hybrid curriculum blends in-person and online learning, integrating both to create a single, cohesive learning experience. It intentionally combines face-to-face instruction with online activities, allowing some learning to take place remotely and often asynchronously, while in-person time is reduced but not eliminated. This approach uses online resources to enhance classroom discussions and activities and can increase flexibility for students. One of the features of the hybrid curriculum is that it combines a lot of learning modalities.

#### 6.2.5.3 *Simulation-based curriculum*

A simulation-based curriculum is an educational approach that uses realistic, interactive scenarios to teach students skills and knowledge in a safe, controlled environment. It uses simulated patients or virtual environments for learning. One of the characteristic features of simulation-based curriculum is that the scenarios created using the simulation mimics real-world tasks and challenges to provide context for learning. This method involves creating mock situations to allow learners to practice and apply their skills without real-world consequences, promoting experiential learning through a cycle of action, reflection, and application. For example, by providing a simulation of cardiopulmonary resuscitation, it gives a chance for the student to practice and learn the process in a safe environment before they can actually perform the procedure in a live patient (Fig. 6.5).

### 6.2.6 Other types of curriculum

#### 6.2.6.1 *Written curriculum*

It is the officially recorded, planned curriculum. The official, documented curriculum created by educational authorities and specialists is called the written curriculum, sometimes referred to as the explicit or overt curriculum. Teachers are obliged to adhere to certain subjects, material, learning objectives, and instructional practices in the classroom. For example, the written curriculum clearly talks about the goals and objectives of the course in the form of clear, measurable statements describing what learners will be able to do upon completion. It also specifies the teaching strategies that will be used in the course like lectures, PBL, small group discussions, simulation, clinical skills labs, and clinical rotations.

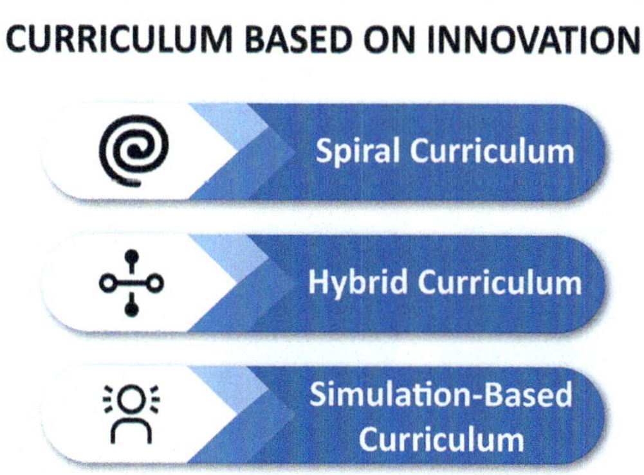


FIGURE 6.5 Based on innovation.

### 6.2.6.2 *Taught curriculum*

The represents the curriculum as taught by the instructor in the classroom. The actual instruction and learning opportunities that teachers provide to students in the classroom on a daily basis are referred to as the "taught curriculum." It is how classes, lectures, group projects, and other activities bring the written curriculum to life. It is impacted by the teacher's methods, ideology, and the needs of their students. It can differ from teacher to teacher and is less standardized than the written curriculum. This curriculum signifies the way in which the educational content is taught. The curriculum in medicine is generally divided into preclinical sciences like anatomy, physiology, and clinical sciences like medicine and surgery.

### 6.2.6.3 *Hidden curriculum*

The hidden curriculum refers to the inadvertent lessons like social norms and expectations learned in the educational setting. The lessons, values, and viewpoints that students pick up indirectly through the medical school environment, social interactions, and culture are all important components of the hidden curriculum. Although they are not specifically mentioned in the official written curriculum, they have a big influence on how the learners behave and grow. Examples of the hidden curriculum in medicine include implicitly teaching professionalism in practice such as seeing senior doctors handle complex situations with accountability, empathy, and conflict resolution.

### 6.2.6.4 *Assessed curriculum*

It refers to the portion of the curriculum that is assessed and tested. In medicine, assessed curricula use a wide range of techniques to assess students' clinical abilities, knowledge, professional conduct, and practical application of what they have learned. An assessed curriculum, also known as a tested curriculum, is the portion of a curriculum that is evaluated through various assessments to measure student learning and understanding. It includes all formal and informal methods used to assess students, such as quizzes, tests, projects, portfolios, and presentations, and is used to inform future instruction and evaluate the effectiveness of the written and taught curriculum.

### 6.2.6.5 *Null curriculum*

This type of curriculum includes topics, concepts, and substance that are purposefully left out of the official curriculum. These omissions may be the result of practical limitations, societal values, or educational priorities. The null curriculum is the content that medical schools intentionally or unintentionally do not teach, encompassing the subjects, skills, and perspectives that are omitted from the formal curriculum. It includes topics deliberately excluded due to perceived unimportance or controversy, leading to a lack of exposure to certain concepts or practices. This "absent curriculum" can have a significant educational effect, as students may learn that the omitted topics are not essential or worthy of consideration (Fig. 6.6).

## 6.3 Curriculum versus syllabus

The significance of curriculum in medical education lies in its role as a comprehensive framework for guiding learning, ensuring quality, and adapting to the evolving needs of healthcare. It outlines the intended learning outcomes, such as knowledge, skills, and attitudes, and provides the structure for teaching, assessment, and evaluation to ensure graduates are competent, professional, and ready to serve communities. A well-designed curriculum aligns with health workforce needs, incorporates new evidence-based practices, and includes mechanisms for continuous improvement through feedback from various stakeholders (Woods et al., 2010). For instance, the curriculum in most medical schools across the globe is 5 years long and is divided into a preclinical year (year 1 and 2) and clinical years (years 3, 4 &5). The first 2 years are divided into a system-oriented block that covers multiple disciplines such as anatomy, physiology, biochemistry,

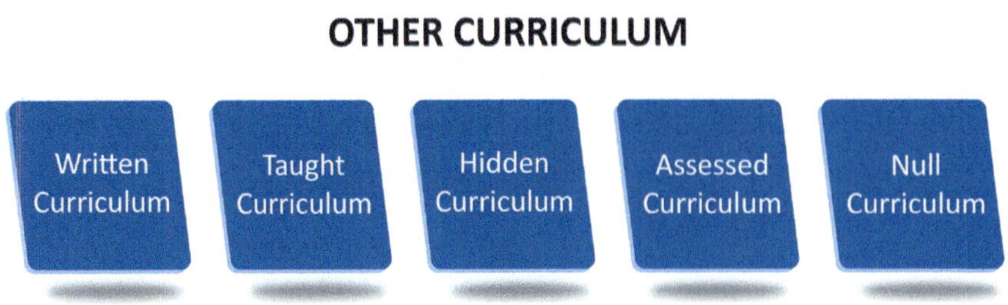


FIGURE 6.6 Other types of curriculum.

pharmacology, and pathology. During the next 3 years, the students are introduced to basic clinical skills as part of the vertical integration of the curriculum where the main focus is on studying clinical subjects like medicine, surgery, pediatrics, and obstetrics and gynecology.

On the other hand, a syllabus is a descriptive list of subjects to be taught in a class. The entire educational program encompasses multiple courses, subjects, and interrelations. It focuses on a single course or subject within the broader curriculum. A medical school syllabus is significant as it serves as a roadmap for students and faculty, outlining the course's expectations, learning outcomes, and grading policies. It acts as a contract between instructor and student, ensures alignment with accreditation standards, and provides a framework for integrating foundational science with clinical application. Ultimately, a well-designed syllabus is a critical tool for structuring the curriculum, ensuring educational goals are met, and preparing future physicians for practice. For instance, the syllabus in physiology would refer to all the topics and its contents like the topic of general physiology would highlight on the components of the cell, transport mechanism, or homeostasis. It also gives a clear idea what the student will learn in that particular subject or course (Burton and McDonald, 2001) (Table 6.1).

## 6.4 Curricular frameworks

A curriculum framework serves as a blueprint for consistency and quality while permitting the development of particular topic. It is a guiding structure that describes the fundamental elements of a medical school's curriculum, defining what students should learn, such as content, skills, and values, and providing guidelines for how it is taught and assessed. Without being comprehensive lesson plans, it ensures coherence throughout an educational system by establishing standards for learning outcomes, curriculum areas, goals, and occasionally instructional methodologies (Zhao et al., 2023).

## 6.5 Key aspects of curriculum frameworks in medical education

### 6.5.1 Defining learning outcomes

Frameworks specify what students should know and be able to do upon completion of a course or program. Defining learning outcomes is crucial for curriculum frameworks because they clarify what students will know, be able to do, and value upon completion of a course or program. This clarity benefits instructors by guiding them in designing aligned content, activities, and assessments, and helps students by setting clear expectations and increasing their motivation. Overall, learning outcomes create a student-centered approach ensuring that the curriculum effectively serves educational goals and standards.

**TABLE 6.1** Difference between curriculum and syllabus.

| Features | Curriculum | Syllabus |
|---|---|---|
| Definition | A curriculum is defined as a sophisticated blend of educational strategies | Syllabus is a descriptive list of subjects to be taught in a class. |
| Purpose | To provide a structured plan for education, serving as a roadmap for both students and teachers | To serve as a foundational contract between an instructor and students |
| Scope | Wide | Narrow |
| Nature | Prescriptive | Descriptive |
| Created by | University | Subject expert |
| Term | Long term | Until the specific subject is finished |
| Flexibility | No room for alterations | Highly flexible |
| Available for students | By request | Always |

### 6.5.2 Guiding curriculum development

They provide a structure for organizing content, sequencing learning experiences, and selecting appropriate teaching and assessment methods. Guiding curriculum development within a curricular framework is crucial because it ensures a consistent, high-quality, and aligned educational experience for students by providing a roadmap for what and how to teach. This guidance is essential for establishing a foundation of knowledge and skills, creating a structure for instructional planning and assessment, promoting equity, and adapting education to meet the evolving needs of society.

### 6.5.3 Promoting integration

Frameworks can encourage the integration of basic sciences with clinical training, as well as interprofessional collaboration. Promoting integration in a curriculum framework is essential for fostering holistic student development and preparing learners for a complex, interconnected world. An integrated approach connects different subjects and life experiences, which increases student engagement and enhances skills like critical thinking, problem-solving, and creativity.

### 6.5.4 Addressing societal needs

Addressing societal needs gives the educator a chance to incorporate principles of social accountability, focusing on issues like refugee and migrant health and addressing health inequities. Addressing societal needs in the curriculum is crucial for preparing students to be active, informed, and responsible citizens in an ever-evolving world. By aligning educational goals with real-world issues, a curriculum becomes relevant and empowers students to create positive social change.

### 6.5.5 Adapting to emerging technologies

In recent era of technological growth, curricular frameworks are evolving to include the integration of AI in medical education. The integration of digital technologies into education represents a significant evolution in the pedagogical landscape, with the potential to enhance accessibility, engagement, and personalization in learning. Adapting the curriculum to emerging technologies involves strategically integrating tools like AI, virtual reality, and gamification to enhance teaching and learning. This transformation aims to make learning more personalized, engaging, and relevant for 21st-century students.

## 6.6 Types of curriculum frameworks

### 6.6.1 Competency-based frameworks

A competency-based framework is defined as a basic conceptual structure, including relevant physician's duties and obligations around which core competencies can be built. It is important to be aware that in an era of greater public accountability, medical curricula must ensure that all graduates are competent in all essential domains. However, not all curricula currently in use explicitly define and address all desired outcomes. Furthermore, not all current curricula assess or ensure that graduates have acquired all the necessary abilities (Nikolov et al., 2014). On the other hand, the competency-based medical education (CBME) frameworks focus on assessing and developing specific, observable abilities or competencies required for effective medical practice, rather than simply accumulating time in training. To embrace CBME and prepare medical students for practice, medical educators usually utilize an organized national or international competency framework that describes the abilities that physicians must possess to meet the needs of patients and society.

### 6.6.2 Frameworks for medical education competencies

Common frameworks for medical education competencies include the ACGME's six core competencies (patient care, medical knowledge, practice-based learning and improvement, interpersonal and communication skills, professionalism, and systems-based practice) and frameworks emphasizing CBME, which focuses on learner outcomes and observable abilities rather than time spent in training (McDougal et al., 2023). Other frameworks, such as the ones for entering medical students, focus on specific personal and professional qualities like empathy, ethical responsibility, and communication skills.

### 6.6.3 Competency-based medical education (CBME)

The competency framework outlines the specific knowledge, skills, attitudes, and behaviors required, while CBME is the method of designing, implementing, and assessing the curriculum to ensure learners achieve these competencies and are "ready for autonomous practice." Essentially, the framework lists the competencies and CBME serves the educational process to achieve them. CBME has emerged as a transformative approach to medical training, revolutionizing traditional models of education. CBME is an outcome-based approach to the design, implementation, and evaluation of education programs and to the assessment of learners across the continuum that uses competencies or observable abilities (Frank et al., 2010). The goal of CBME is to ensure that all learners achieve the desired patient-centered outcomes during their training. This approach shifts the focus from a time-based curriculum to one based on demonstrated, observable abilities. The strengths of CBME are its emphasis on outcomes, tailored learning experiences, and assessment-driven feedback. On the other hand, the vital challenges include faculty development, resource allocation, standardization, and resistance to change (Albanese et al., 2008).

## 6.7 Key components of competency-based medical education

### 6.7.1 Observable abilities

Observable abilities, or competencies, in CBME refer to all those demonstrable skills, knowledge, values, and attitudes a healthcare professional needs to practice medicine effectively. They are observable and measurable, meaning that they can be seen and assessed in real-world situations, allowing for progress to be tracked and for students to advance when they are ready, rather than at a set time. The examples in this regard include performing a physical examination, communicating with a patient, or making ethical decisions.

### 6.7.2 Integration of knowledge, skills, and attitudes

They are a blend of what a learner knows, what they can do, and their professional behaviors and attitudes. In other words, the integration of knowledge, skills, and attitudes in CBME means combining the "knowing," "doing," and "being" skills to successfully perform professional tasks in a medical context. This integration is the core principle of competency, where a physician must apply scientific understanding, clinical abilities, and a professional mindset to navigate complex patient care situations.

### 6.7.3 Focus on patient-centered outcomes

Focusing on patient-centered outcomes is one of the prime purposes of the competency-based medical education. In this, the educational goals are based on the health needs of patients and the population, rather than a time-based curriculum. This shift requires customizing the learning experience to individual needs, assessing trainees on their ability to meet clearly specified outcomes.

### 6.7.4 Shift from time-based to outcome-based learning

CBME emphasizes achieving specific standards of performance rather than simply completing a set amount of time in training. The shift from time-based to outcome-based learning in CBME moves the focus from a fixed duration of training to achieving specific, demonstrable skills and knowledge. In a time-based model, a set amount of time is assumed to be sufficient for learning, but the actual learning outcomes can vary significantly between individuals. Conversely, in the outcome-based approach, the learning outcomes are fixed, while the time it takes for a learner to achieve them is flexible, allowing individuals to progress at their own pace based on mastery.

### 6.7.5 Continuous assessment and feedback

Learning is supported by frequent, formative assessments and built-in feedback to help learners reach the desired standard. Continuous assessment and feedback in CBME are iterative processes that use frequent, formative evaluations to track a learner's progress toward achieving specific competencies, providing timely, specific feedback to guide improvement and support their development. This differs from traditional models by focusing on demonstrable skills and knowledge rather than time-based completion, with assessments serving to assist learning and identify gaps early, not just to deliver a final grade.

### 6.7.6 Real-life practice

Assessment and training focus on the skills needed for real-life medical practice, including aspects beyond just medical knowledge, such as communication, teamwork, and quality improvement (QI). Real-life practice is crucial in CBME, because it allows students to bridge the gap between theory and practice, build hands-on skills, and develop patient-centered competencies in a real-world context. This experience is vital for preparing graduates to be "practice-ready," enhancing their critical thinking, clinical decision-making, and communication abilities in a manner that traditional learning cannot replicate.

## 6.8 Teaching the competencies

The key competencies which are being taught to medical students are clinical skills like clinical reasoning and patient care, communication and professionalism such as teamwork and ethical conduct, educational and organizational skills like planning, assessment, and providing feedback, and personal development like lifelong learning and self-awareness. Teaching activities to build medical student competencies include hands-on practice like simulation and clinical skills labs, active learning methods such as case-based and PBL, and technology-enhanced approaches like e-learning and game-based learning. These activities are designed to develop a range of competencies, from clinical reasoning and procedural skills to teamwork and communication. For instance, in activities like CBL students work in groups to analyze realistic clinical cases, discussing differential diagnoses and management plans to improve clinical reasoning and teamwork.

## 6.9 Assessing the competencies

A competency assessment can help educators make informed decisions and ensure that the learner is skilled, knowledgeable, competent, and is well prepared to encounter the various challenges that they might face both personally and professionally. By using a CBME approach that focuses on observable abilities, one can aptly assess the various competencies. Competencies like patient care, medical knowledge, communication skills, ethics, professionalism, and systems-based practice are assessed through a variety of methods such as observation, testing, interviews, and self-assessment. The ultimate goal is to foster expertise, not just competence, and requires continuous, multifaceted evaluation throughout a physician's career.

## 6.10 Methods for assessment

### 6.10.1 Observation-based assessment

An example of observation-based assessment in medicine is watching and recording a surgical resident's performance of a physical exam during a patient consultation, including their communication style, technical skills, and clinical decision-making. This observation made by the examiner or a senior colleague will readily point out whether the student is exhibiting the desired competency which is suitable to that particular task. Moreover, it provides a holistic view as by observing performance of the learner in the given task, evaluators can gain a more comprehensive understanding of a student's strengths and areas for improvement that might be missed with traditional tests.

### 6.10.2 Test-based assessment

Test-based assessment in medicine evaluates a student's knowledge, skills, and judgment through various structured testing methods. The test-based methods are one of the time-tested methods to evaluate the student's ability to apply knowledge and skills to real-world situations. Assessment methods like practical examinations or objective structured clinical examination are a great way to undertake this type of assessment.

### 6.10.3 Interview-based assessment

Interview-based assessments in medicine evaluates competencies like patient care, communication, professionalism, and problem-solving through structured questions, often using the STAR (Situation, Task, Action, Result) method. These assessments can be structured or behavioral, with interviewers asking specific questions about past experiences or hypothetical scenarios, and then scoring answers against predetermined criteria to ensure objectivity and consistency. The multiple mini-interviews is a modern, structured interview format used by many medical schools to assess a wide range of nonacademic skills. Instead of a single, long interview, candidates rotate through a series of short, independent stations.

### 6.10.4 The 360-degree feedback assessment

The 360-degree feedback in medical education is a comprehensive assessment tool that gathers performance evaluations from multiple sources, including supervisors, peers, subordinates, and even patients. This method which is based on multisource feedback provides a more holistic view of a medical trainee's competencies, including communication and interpersonal skills that traditional evaluations often miss, and can be used to identify areas for targeted improvement. Confidentiality and careful implementation are key to ensuring honest and constructive feedback (Fig. 6.7).

### 6.10.5 Learning outcome-based framework

An outcome-based framework in medical education is a system where the curriculum and assessment are designed to achieve predefined learning outcomes, focusing on what a student can do as a doctor rather than just the process of learning. One of the most common models which provides a framework of educational strategies that can be used to develop or evaluate a curriculum with a strong outcome-based approach is the SPICES model where (S: student-centered, P: patient-centered, I: integrated, C: community-based, E: elective, and S: systematic). Many medical schools have successfully applied the SPICES model within an outcome-based education framework to ensure their graduates achieve a set of specific competencies. This framework emphasizes defining clear learning outcomes and designing the curriculum to ensure students achieve those outcomes. The outcome-based framework focuses on outcomes wherein the central principle lies in defining the desired competencies of the graduate first, then building the curriculum, teaching methods, and assessments around these outcomes. Moreover, there is an emphasis on learning, not teaching as the framework prioritizes student learning over the act of teaching, allowing for variable time spent on learning different topics as needed to achieve mastery.

The whole curriculum, including the content, instructional strategies, and assessments, is created backward from the intended learning outcomes in the outcome-based framework. An outcome-based curriculum has a number of challenges, such as the difficulty of precisely defining and evaluating learning goals, the requirement for substantial faculty training and a cultural shift, the increased workload for instructors in planning and evaluation, and problems with curriculum alignment and resource allocation. The CANMEDS framework is a well-known illustration of an outcome-based curriculum in medical education. The CANMEDS framework, which is a kind of medical competency framework, identifies and characterizes the skills doctors need to successfully address the healthcare requirements of the patients they treat. It also signifies that the medical practitioner is entitled to play a variety of roles like medical expertise, communicator, collaborator, leader, health advocate, scholar, and professional (Frank, 2015) (Fig. 6.8).

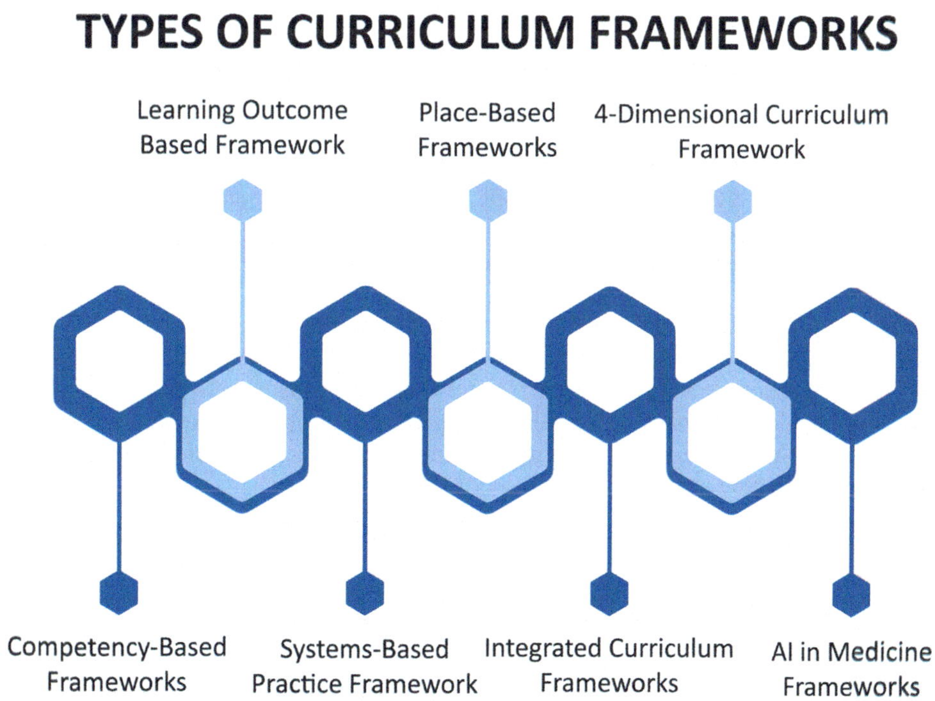


FIGURE 6.7 Types of curriculum frameworks.

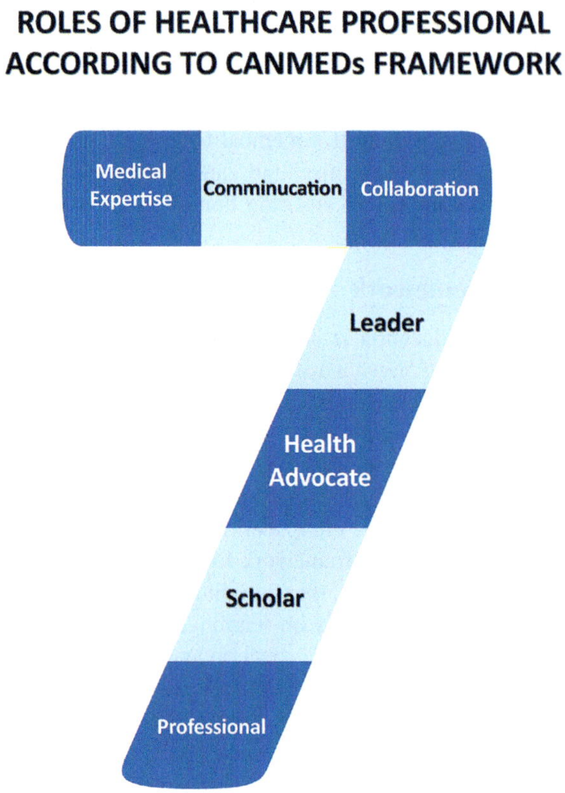


FIGURE 6.8 The seven roles of a healthcare professional identified by the CanMEDS framework.

### 6.10.6 Systems-based practice framework

The systems-based practice framework in medical education is a core competency that teaches physicians to understand and work within complex healthcare systems to provide optimal patient care. It requires physicians to be aware of the larger context of healthcare, including patient safety, QI, cost-effectiveness, and the social determinants of health, and to utilize the system's resources effectively. This competency moves beyond a single-patient focus to recognize how the system itself impacts care quality and safety. This framework primarily focuses on understanding the system, wherein it entails that the physicians must recognize how the various parts of the healthcare system interact to deliver care. Importance is also given to optimizing care by equipping the physicians with the skills to use system resources efficiently and effectively for the benefit of the patient. Furthermore, it highlights the importance of patient safety as it involves understanding the principles of patient safety and how to mitigate risks within the system. In this process, the physicians learn to identify and implement changes to improve the quality of care and patient outcomes, leading to overall QI (Hogan et al., 2024).

Inadequate faculty expertise and training, a lack of standardized curricula and assessment tools, difficulty defining and implementing the competency, and workload compression that leaves little time for system-based practice activities are some of the major obstacles of this framework. These problems lead to a mismatch between educational objectives and the demands of the health system, uneven training, inconsistent evaluation, and a possible knowledge and competence gap for residents. Examples of the systems-based practice framework in medical education include QI projects, multidisciplinary rounds, and patient safety initiatives.

### 6.10.7 Place-based frameworks

A place-based education refers collectively to the pedagogical approaches that emphasize experiential, community-based, contextual, or ecological learning in order to foster stronger connectivity to local settings, cultures, and surroundings. When creating curricula, these frameworks take into account the particular traits and requirements of particular groups. In medical education, a place-based framework places a strong emphasis on studying and practicing medicine in a particular community, making use of its special features and resources to meet local health needs. This strategy seeks to improve healthcare delivery, boost medical personnel' efficacy, and increase community well-being in general. The key aspects of a place-based framework in medical education includes understanding the local context such as focusing on understanding the specific social, economic, environmental, and cultural factors of a particular location that influence health. Furthermore, the importance is also given to community engagement wherein the students and faculty actively engage with the community, collaborating with local residents, healthcare providers, and organizations to identify health priorities and develop tailored solutions.

Healthcare providers can deliver more appropriate and effective care by cultivating a relationship with the community they serve. Since graduates with experience in a particular region are more likely to be interested in practicing there, the place-based framework also helps with recruiting and retention, which helps with workforce shortages in underprivileged and rural areas. Obtaining cooperation from medical professionals, paying them for their time, and providing adequate physical support for student activities are some of the challenges associated with a place-based framework in medical education. A lack of student comprehension of community-based practice, the difficulty of community engagement, and problems with long-term finance and stability are further obstacles. Since many programs lack long-term follow-up, ensuring continuous, high-quality evaluation is another major challenge.

### 6.10.8 Integrated curriculum frameworks

An integrated curriculum framework in medical education aims to connect and blend different subjects and disciplines to provide a more holistic and relevant learning experience. It moves away from traditional, siloed teaching of subjects towards a more interconnected approach, helping students see how different areas of knowledge relate to each other and to real-world clinical practice. The presentation of basic sciences in the context of clinical practice helps the students to appreciate the immediate application of their learning, making it more meaningful and easier to retain. This establishes the significance of integrated curricular framework in medical education. Furthermore, it promotes the development of a holistic approach to healthcare by assisting students in viewing and treating patients in a more comprehensive manner rather than as a collection of interconnected systems. Additionally, by promoting the repetition and gradual development of concepts, it helps students retain information better. This framework also promotes teamwork, which can improve student happiness and provide a more positive, stress-free learning environment. Faculty members who are accustomed to and schooled in the old, compartmentalized curriculum may provide a serious obstacle. Besides, careful planning, coordination, and faculty development are essential for the effective administration and execution of an integrated curriculum.

### 6.10.9 Four-dimensional curriculum framework

The four-dimensional curriculum framework (4DF) is conceptualized as a dynamic, multidimensional tool to guide curriculum design, development, and reform. Its fundamental aim is to "help health professional educators link educational practice to health policy, workforce, and professional practices in a coherent and reflexive way (Steketee et al., 2024)." The four dimensions are interrelated as follows.

#### *6.10.9.1 Dimension 1 (D1)*

This dimension concentrates on how the healthcare practices will be oriented in the future, taking into account both historical and modern practices, changing workplace demands, and future orientations.

#### *6.10.9.2 Dimension 2 (D2)*

This dimension covers the standards, knowledge, abilities, attitudes, and professional ideals expressed in professional practice. This dimension connects to dimension 1 by acknowledging that practice and skill development must be in line with evolving practice priorities and settings as well as real-world demands. As a professional, it focuses on the dynamic interaction between "knowing, being, and doing."

#### *6.10.9.3 Dimension 3 (D3)*

The fundamental educational activities that support the growth of the skills and abilities covered in dimension 2 are addressed in dimension 3. Workplace practices, changing healthcare policy, and patient demands all have an impact on how learning, teaching, and assessment are designed, planned, and delivered.

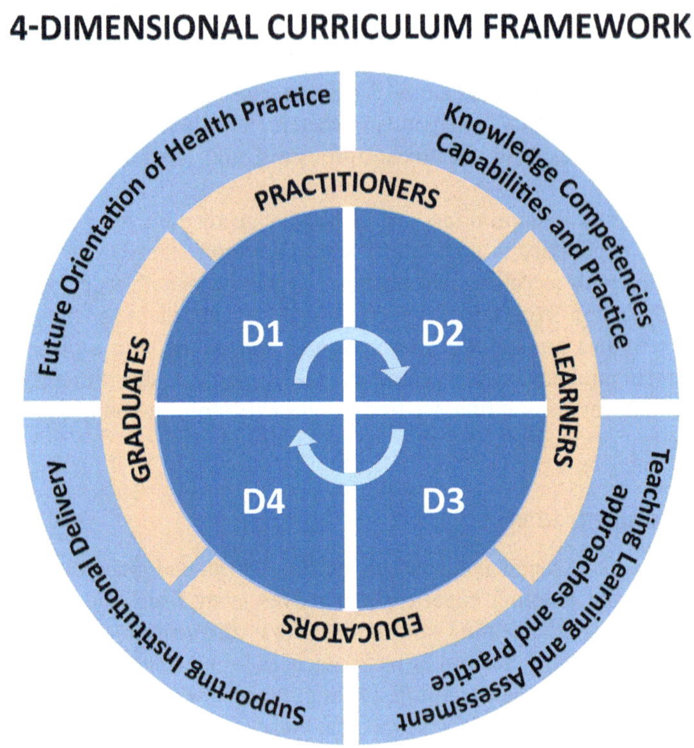


FIGURE 6.9 Four-dimensional curriculum model.

### *6.10.9.4 Dimension 4 (D4)*

The organizational governance and administrative practices unique to the educational setting in which the learning program is being delivered are taken into account in the fourth dimension. It shows how the university views itself, where it is located, and how it interacts with the medical field (Fig. 6.9).

## 6.11 Implementing the four-dimensional model for teaching and learning

To implement a four-dimensional teaching and learning model, educators can use strategies like interdisciplinary projects, active learning, and reflection to develop knowledge, skills, character, and metacognition. For example, a project could require students to learn a new topic, collaborate and communicate effectively, consider ethical implications, and reflect on their learning process. This requires shifting from a teacher-centered approach to a more holistic, learner-centered one that focuses on preparing students for real-world challenges. Implementing this model not only helps to foster knowledge and skills as it focuses on core academic knowledge but also integrate 21st-century skills like critical thinking, communication, and creativity through interdisciplinary projects and PBL. In addition, it also fosters the development of character and social-emotional learning by integrating activities that build character traits like resilience, empathy, and values through discussions, ethical considerations, and social-emotional learning strategies.

The challenges associated with the four-dimensional model occur due the fact that it lays over-emphasis on the physical dimension and neglects the other dimensions like psychological, social, and spiritual dimensions of care. This requires a fundamental rebalancing of the curriculum by integrating with biological and biomedical aspects of health which medical students often feel ill-prepared to discuss with patients.

### 6.11.1 Artificial intelligence in medicine frameworks

As healthcare moves to a more digital environment, there is a growing need to train future physicians on the clinical uses of AI. However, clinical training in AI has often been inconsistent or lacking. Given the increasing significance and potential impact of AI technology on health care delivery, there is an increasing demand to integrate AI into medical school curricula. AI in medicine frameworks are designed to include AI into the curriculum with an aim to integrate AI principles and applications into medical education, preparing future healthcare professionals for an increasingly

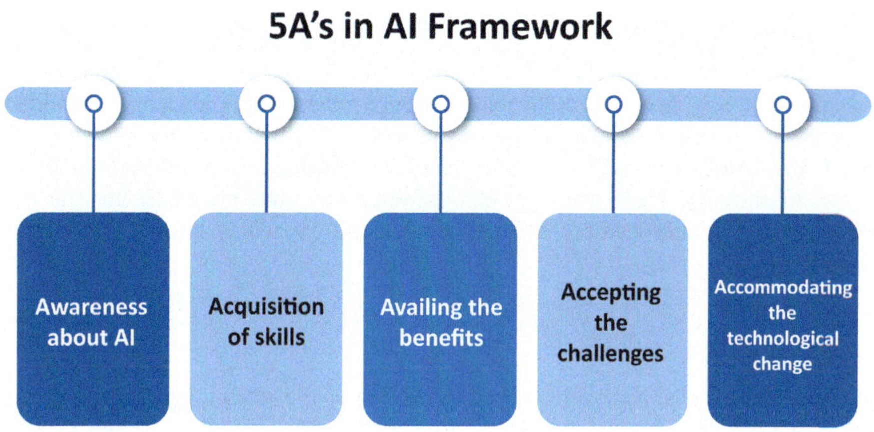


FIGURE 6.10 The 5A's in artificial intelligence (AI) framework.

AI-driven field. This framework typically involves incorporating AI-related topics into existing medical curricula, developing specialized AI courses, and utilizing AI-powered tools for learning and assessment. In addition, there needs to be an inclusion of AI competencies as well as sub competencies within each domain. Some of the important competencies that are essential for medical graduates to achieve can be described as 5A's:

**A—Awareness about AI**: Understanding the basics of AI and how it impacts medical education and health care.

**A—Acquisition of skills**: Gaining skills like communication, teamwork, collaboration, and adaptability.

**A—Availing the benefits**: Reaping the benefits that AI has brought in the world of medical education and healthcare.

**A—Accepting the challenges**: Realizing the various challenges that are encountered by implementing the AI curriculum framework and working towards combating them.

**A—Accommodating the technological change**: Understanding that AI is here to stay and acknowledging the technological revolution (Fig. 6.10).

For instance, the AIFM-ed Curriculum Framework for Postgraduate Family Medicine Education on Artificial Intelligence was developed for AI training in postgraduate family medicine education. This framework serves as a structured guide for integrating AI competencies into medical education, ensuring that future family physicians are equipped with the necessary skills to use AI effectively in their clinical practice (Tolentino et al., 2025). These efforts are led by national medical associations such as the UK National Health Service, the US American Medical Association, and Canada's Royal College of Physicians and Surgeons. They have released documents recommending policies for integrating AI within their respective medical educational institutions. Furthermore, institutions also are encouraged to consider adapting the AIFM-ed framework within their own programs, tailoring it to meet the specific needs of their trainees and health care environments.

## 6.12 Key components of an artificial intelligence-based medical curriculum framework

### 6.12.1 Foundational artificial intelligence knowledge

An essential part of integrating artificial intelligence into medical education is incorporating fundamental AI expertise, such as machine learning, data analysis, and the ethical implications of AI in healthcare, into the core medical curriculum. The first step in educating future doctors to work in an AI-integrated healthcare system is to introduce basic AI expertise. This information can be used to construct curricula in a number of domains, teaching students how to use AI tools efficiently, morally, and responsibly through a variety of pedagogical approaches.

### 6.12.2 Specialized artificial intelligence modules

Developing dedicated courses or modules focused on specific AI applications in medicine, such as AI-assisted diagnostics, drug discovery, personalized medicine, and robotic surgery, is necessary to facilitate a deeper understanding of the various aspects of AI. Moreover, the knowledge of how these modules are created and can be applied to clinical practice should be prioritized for students in the final year of medicine.

### 6.12.3 Artificial intelligence–powered learning tools

If AI-driven tools for automated assessment, simulation-based training, and personalized learning are incorporated into an AI-based curriculum, it will significantly improve student engagement and knowledge retention and give students a competitive edge in preparing for the future. By automating processes like creating content and assessments, customizing learning routes, and offering data-driven insights for ongoing development, AI-powered technologies are being developed to assist in the creation of curricula. These resources can save teachers time, guarantee that the content is current, and produce more engaging and adaptable learning experiences that are catered to the unique requirements and learning preferences of each student.

### 6.12.4 Ethical considerations

One of the essential elements that must be covered in the curriculum is the ethical, legal, and societal ramifications of AI in healthcare, including data privacy, bias, and algorithmic transparency. Additionally, in order for learners to comprehend and value both the positive and negative elements of these concerns, the challenges related to them must be addressed as part of the curriculum.

### 6.12.5 Interprofessional collaboration

One of the key components for the successful integration of AI into the curriculum is encouraging cooperation between computer scientists, medical practitioners, and AI specialists to guarantee the integration of AI in healthcare. This ensures that the AI tools are both socially and technically sound. This method entails developing curriculum frameworks that incorporate AI into certain professional contexts, like the healthcare industry, and emphasizing both the useful application of AI tools and the critical evaluation of their results. In addition to enhancing teacher knowledge, collaboration can give students the ethical awareness and skills they need to use AI successfully in their future employment.

## 6.13 Developing and implementation of curriculum framework

Some of the key aspects that should be considered while starting the process of curriculum development is to make sure that the well-defined objectives pinpoint the essential skills and knowledge that students must master (Masters, 2020). After objectives have been established, active teaching strategies come into play, delivering content in ways that captivate learners, while cross-disciplinary content ensures that lessons remain relevant and engaging for learners from all backgrounds. Equally important, assessments serve as the final piece, gauging progress and confirming whether students are meeting established standards. Some of the important aspects of developing curriculum are as follows.

### 6.13.1 Needs assessment and stakeholder collaboration

Understanding and valuing stakeholder demands is the first step in creating excellent curriculum frameworks. The goal of the needs assessment process is to find gaps and rank the techniques that could be useful in resolving any problems that may come up during implementation. Adoption and implementation are significantly improved when educators work together to create curriculum frameworks. In order to determine the next steps to be taken in terms of curriculum implementation, stakeholders' cooperation during the curriculum development process is crucial. The stakeholders contribute ideas, knowledge, and assistance to make the learning process pertinent and successful.

### 6.13.2 Designing and structuring the framework

One of the best examples of the medical educationists' teamwork is the framework's design and organization. Involving all stakeholders is the key to a well-designed curriculum as it not only encompasses the views and feedback of all the stakeholders but also gives them an opportunity to actively participate in this vital process. The curriculum must satisfy the requirements and expectations of the students and be adaptable to changes. All four crucial components such as the content, teaching and learning methodologies, assessment procedures, and evaluation procedures will be considered in a well-designed curriculum. One-size-fits-all thinking is challenged by this collaborative approach to curriculum design and structure, which produces adaptable, transparent frameworks that function in a variety of contexts.

### 6.13.3 Implementation and continuous review

Implementing a medical education curriculum involves preparing a teaching plan, delivering instruction using various strategies. This in turn is done by determining the content and structure of the curriculum and ensuring it is aligned with the objectives. At the same time, it is vital to choose the appropriate teaching strategies that match the objectives intended to be achieved. After having successfully designed, structured, and implemented, one of the vital questions to be pondered about is How do we make sure that the curricular frameworks stay relevant? This is achievable by the process of undertaking continuous review. Regular updates, driven by stakeholder input, helps the medical educator to keep a check on the implementation process. This cycle of review and refinement ensures frameworks evolve with educational needs, staying practical and impactful.

### 6.13.4 Best practices for using curriculum frameworks

As in any other process, the practice of undertaking the best practices for using curricular frameworks guarantees that the strategies, methods, and techniques that are used in this process are effective, efficient, and reliably lead to desired educational and clinical outcomes. In this process, educators have the opportunity to align lessons with objectives, weave in assessments, and pursue ongoing training. It is vital to involve the learners to join in, whereby they can use these frameworks as guides to discuss the progress with teachers. These practices boost engagement and outcomes. Teachers can observe a sense of student understanding and mastery through specific behaviors and actions that are aligned with curriculum objectives. In addition, walkthroughs by administrators evaluate alignment, creating a culture of continuous improvement. Some of the vital steps involve forming a well-qualified team for the development of the framework. At the same time, it is vital to define the clear learning outcomes and use the framework as a guide for creating lesson plans, teacher training, and assessment strategies.

### 6.13.5 Integrating assessments and feedback

Integrating the assessment with the other components of the curricular is a very critical step. The primary aim of integrating assessment with the curricular process is to make sure that the students have learnt and are able to display the skills expected of them. In this process, deep and meaningful learning is encouraged and promoted in the learner. This integration also ensures that a competent, practice-ready physicians are produced in this process. In the process of integration, it is to be made sure that assessments accurately measure whether learning objectives are met, guide students toward what is important, and prepare graduates for modern healthcare demands. Similarly, integrating feedback in the process of curriculum development is always important as it is crucial to keep the medical education relevant and at the same time preserve the quality of medical education. It also guarantees that the curriculum objectives are met through continuous improvement loops involving students, faculty, and other stakeholders.

## 6.14 Role of stakeholders in developing the curricular framework

Stakeholders are the key players in developing the curricular framework as their contributions have a major impact on all the aspects of the curriculum. The major stakeholders involved in this process are the learners, teachers, institutional education committees like the academic affairs committee, quality committee, and the medical education committee, the college administration, and patients. Each of these stakeholders have a vital role to play towards the successful development and implementation of the curriculum. For instance, the learners provide the much-needed input and feedback to ensure the curriculum is relevant, engaging, and meets their learning needs. The role of the teacher lies in the development, implementation, and adaptation of the curriculum. By this process, the teacher is in the best position to guide the learner through the learning process and make sure that they achieve their career goals.

The role of committees such as the academic committees or the quality committee is to warrant that there is a smooth implementation of the curriculum and to address any issues pertaining to the quality of teaching and learning process or the assessment process which form the major pillars of the curriculum. Patients contribute to curriculum development by providing real-world perspectives to ensure training is patient-centered, helping to shape educational content, and participating in the development of learning materials and assessments. Their involvement ensures medical education focuses on the actual needs and expectations of patients, leading to more empathetic and effective future healthcare professionals. Patients can also contribute as experts by sharing their experiences and expertise in a collaborative and constructive way.

## 6.15 Impact of the curricular frameworks on the teaching-learning process

By offering structure and objectives, directing lesson planning, and raising student engagement, curricular frameworks have an impact on the teaching-learning process. By relating the material to real-world situations, they can enhance relevance, help teachers link activities and evaluations, and specify what students should learn. In the end, a well-crafted curriculum contributes to a more efficient, interesting, and meaningful learning environment that equips students for success in the future. Curriculum frameworks have an impact on the teaching process because they provide teachers with a blueprint that outlines the skills, information, and learning objectives to be taught. Simultaneously, it improves lesson planning by assisting teachers in organizing their subjects and exercises in a way that satisfies both the learning objectives and the needs and expectations of the students. Additionally, the curricular framework ensures that assessments are in line with learning objectives, offering a means of gauging student development and guiding instructional tactics. The curriculum framework by its impact on the teaching-learning process ensures that the learning becomes more meaningful and interesting for students as real-world events, practical exercises, and pertinent content are incorporated into the learning process. Simultaneously, it works to enhance the results, which raises student accomplishment, particularly when they are updated and incorporate new educational strategies. Additionally, the students gain from the curriculum's clear structure, which makes it easier for them to comprehend how their learning progresses and what is expected of them.

## 6.16 Evaluating the curricular framework

A curriculum framework's effectiveness, relevance, and quality in achieving its goals are assessed by evaluating the curricular framework which is done through a methodical approach. Setting precise evaluation standards, collecting information on elements like content and student performance, and getting input from stakeholders are all part of this process. The results are then utilized to pinpoint the curriculum's advantages and disadvantages, make adjustments, and guide future choices to guarantee that the learners are adequately prepared (Spiel et al., 2006). Medical education curriculum frameworks should be reviewed on a regular basis to determine their efficacy, fix any issues, and make sure they are still relevant to changing demands. This includes summative assessments at important milestones to evaluate overall results and make more substantial modifications, as well as formative assessments during implementation to direct urgent improvements. Essential evaluation triggers, including certification requirements or essential stakeholder comments, are also crucial.

### 6.16.1 Key steps in evaluating a curriculum framework

#### *6.16.1.1 Define purpose and objectives*

Determine the objectives and desired learning results that the curriculum aims to accomplish is a vital step. The goal of reviewing a curriculum framework is to methodically ascertain its overall value, efficacy, and relevance. It is a crucial, continuous process that offers the information required to make well-informed decisions on whether to accept, change, or remove curriculum components.

#### *6.16.1.2 Gather data*

In order to evaluate a curriculum frameworks efficacy, relevance, and overall impact, data must be gathered from a variety of sources using a multiple techniques. The evaluation's objectives which can be focused on formative, summative, or diagnostic assessment and the context of the curriculum will determine the precise data and gathering methods. The process will be more authentic if data is gathered from a variety of sources to assess the curriculum's components and their impact, including student performance data and learning outcomes, teacher and student feedback, analysis of instructional materials, assessments, and teaching methods, and evaluation of resources like facilities and technology.

#### *6.16.1.3 Involve stakeholders*

A comprehensive, relevant, and transparent approach is ensured by involving stakeholders in the assessment of a curriculum framework. By bringing together a range of perspectives, the curriculum evaluation process can be reinforced, made more dependable, and made more beneficial. Feedback from students, teachers, administrators, parents, patients, and community members is crucial for a fair and effective curriculum review since it provides a range of viewpoints.

#### 6.16.1.4 Analyze findings

Here, patterns, trends, strengths, and weaknesses are found by comparing the gathered data with the predetermined criteria. Analyzing the results of a curricular framework evaluation include evaluating the information gathered to ascertain the impact, efficacy, and relevance of the curriculum. This procedure offers practical insights for enhancing the curriculum, influencing future educational policy, and guaranteeing consistency with targeted educational objectives.

#### 6.16.1.5 Implement changes

Creating a strategy for improvement based on the analysis, which can entail changing the course material, instructional techniques, or evaluation methods is an important part of curriculum evaluation. A curriculum evaluation framework must be changed by first defining new objectives, then developing and piloting the new framework with training and clear communication, followed by continuous monitoring and iterative improvement. Building staff awareness of new procedures, incorporating formative assessment for ongoing development during implementation, and making sure the new evaluation's goals are pertinent and attainable are all helpful first steps.

#### 6.16.1.6 Monitor and document

One of the primary responsibilities in this step is to constantly monitor the impacts of any changes and create a comprehensive report that describes the entire procedure, including conclusions and actions. A systematic process of monitoring and documenting a curriculum's assessment ensures its effectiveness, relevance, and alignment with learning goals. Monitoring focuses on ongoing implementation, whereas evaluation is a periodic review of outcomes and overall impact. The process of documentation offers a comprehensive record for accountability and continuous improvement.

### 6.16.2 Common evaluation models and approaches

#### 6.16.2.1 Tyler's objectives-centered model

Tyler's objectives-centered model, which offers a four-step, methodical approach centered on guaranteeing consistency between objectives, learning experiences, and outcomes, is used to assess medical education courses (Cruickshank, 2018). The steps in the process are as follows:

1. Clearly defining educational objectives,
2. Identifying and choosing learning experiences to match those objectives,
3. Logically organizing those experiences,
4. Assessing their efficacy to make necessary modifications.

Although this model is commended for its structure and emphasis on quantifiable results, it is critiqued for possibly being overly strict and for having trouble accounting for sophisticated cognitive abilities or inventiveness.

#### 6.16.2.2 Context, input, process, product (CIPP) model

The context, input, process, product or the CIPP model evaluates a medical education curriculum framework by systematically assessing the program's environment, resources, implementation, and outcomes. This comprehensive approach helps make decisions, identify strengths and weaknesses, and improve the curriculum's effectiveness and relevance through formative and summative evaluation (Toosi et al., 2021).

### 6.16.3 Components of the context, input, process, product (CIPP) model

#### 6.16.3.1 Context

This step determines needs, objectives, and priorities by assessing the program's surroundings. This entails analyzing the particular requirements of the medical education program as well as the issues it seeks to resolve. It also evaluates the program's overall goals and the resources that are available.

#### 6.16.3.2 Input

This assesses the tools and tactics used to carry out the program. This entails figuring out workable ways to accomplish the program's objectives while taking into account the resources needed, such as faculty, facilities, and funding. It also involves evaluating the viability of various strategies from an economic standpoint.

#### 6.16.3.3 Process

This provides formative input by keeping an eye on the program's implementation. This entails evaluating the program's execution in relation to the plan, taking into account the budget, schedule, and activities. Additionally, it assesses stakeholder satisfaction and pinpoints any implementation-related problems.

#### 6.16.3.4 Product

It assesses the curriculum's results and outcomes. This involves evaluating the overall quality of the outcomes (effectiveness) and the extent to which program objectives were met. It also entails assessing the program's durability, transferability, and long-term effects.

**Summary**

It is common knowledge that curriculum frameworks are essential to medical education since they offer a methodical path that guarantees dependability, stability, and a thorough educational experience. They direct curriculum development and evaluation, guaranteeing that students fulfill particular competencies and that material is integrated across disciplines, enhancing both the efficacy of instruction and the standard of patient care. It is true that a number of factors interact to ensure the curriculum is designed and implemented successfully. The primary goal of curriculum frameworks in medical education is to give students a well-organized plan for what they should know and be able to accomplish, guaranteeing an excellent, consistent, and pertinent educational experience.

**Key point**

**Future perspectives: from competence to quality**

A curricular framework that is future-focused should be adaptive, flexible, and skill-based. It should also have integrated assessments, defined learning objectives, and a strong focus on teacher and student development. In order to prepare the next generation of medical students for the challenges of contemporary medical education, it is crucial to ensure that future skills like critical thinking, problem-solving, interprofessional learning, collaboration, and digital literacy are integrated and practiced.

Additionally, it's critical to instill in learners a growth mindset so they can accept difficulties and view errors as teaching moments. In order to make the teaching and learning process more robust and efficient, the curriculum framework of the future should be able to address the diverse demands of the teachers as well as overcome obstacles such a heavy workload. Productivity and teaching quality can be improved by offering facilities, resources, and a safe workplace. On the other side, teachers can be equipped to successfully mentor students in a changing environment by encouraging active learning and empowering them through faculty development and ongoing professional development.

## References

Albanese, M.A., Mejicano, G., Mullan, P., Kokotailo, P., Gruppen, L., 2008. Defining characteristics of educational competencies. Medical Education 42 (3), 248–255. Available at https://doi.org/10.1111/j.1365-2923.2007.02996.x.

Bandiera, G., Boucher, A., Neville, A., Kuper, A., Hodges, B., 2013. Integration and timing of basic and clinical sciences education. Medical Teacher 35 (5), 381–387. Available at https://doi.org/10.3109/0142159X.2013.769674.

Burton, J.L., McDonald, S., 2001. Curriculum or syllabus: which are we reforming? Medical Teacher 23 (2), 187–191. Available at https://doi.org/10.1080/01421590020031110 PMID: 11371297.

Cruickshank, V., 2018. Considering Tyler's curriculum model in health and physical education. Journal of Education and Educational Development 5, 207. Available at https://doi.org/10.22555/joeed.v5i1.1443.

El-Astal, M., 2023. What is curriculum? Building a broader understanding of the term. Journal of Curriculum and Teaching 12, 188. Available at https://doi.org/10.5430/jct.v12n6p188.

Frank, J., 2015. CanMEDS 2015 Physician Competency Framework, Royal College of Physicians and Surgeons of Canada, Canada.

Frank, J., Snell, L., ten Cate, O., Holmboe, E., Carraccio, C., Swing, S., et al., 2010. Competency-based medical education: Theory to practice. Medical Teacher 32, 638–645. Available at https://doi.org/10.3109/0142159X.2010.501190.

Harden, R, Yıldız, S, 2024. Medical Teacher 46.

Hogan, S., Md, T., Edgar, L., Holmboe, E., 2024. Systems-based practice 20 years on: Navigating the system for better care. Academic Medicine. Available at https://doi.org/10.1097/ACM.0000000000005640.

Lim, W.K., 2023. Problem Based Learning in Medical Education: Handling Objections and Sustainable Implementation. Advances in Medical Education and Practice 14, 1453–1460. Available at https://doi.org/10.2147/AMEP.S444566 PMID: 38164409; PMCID: PMC10758192.

Masters, K., 2020. Artificial intelligence developments in medical education: a conceptual and practical framework. MedEdPublish 9 (1). Available at https://doi.org/10.15694/mep.2020.000239.1.

McDougal, L., Johnson, P.F., Kirk, L., Weiss, K.B., 2023. The ACGME framework for medically underserved areas and populations and graduate medical education. Journal of Graduate Medical Education 15 (2), 272–275. Available at https://doi.org/10.4300/jgme-d-23-00119.1.

Nikolov, R., Shoikova, E., Kovatcheva, E., 2014. Competence Based Framework for Curriculum Development. Available at https://doi.org/10.13140/RG.2.1.4201.6086.

Spencer, J.A., Jordan, R.K., 1999. Learner centred approaches in medical education. BMJ (Clinical Research ed.) 318 (7193), 1280–1283. Available at https://doi.org/10.1136/bmj.318.7193.1280 PMID: 10231266; PMCID: PMC1115656.

Spiel, C., Schober, B., Reimann, R., 2006. Evaluation of curricula in higher education: challenges for evaluators. Evaluation Review 30 (4), 430–450. Available at https://doi.org/10.1177/0193841X05285077.

Steketee, C., Monica, M., Brewer, M., Rogers, G., 2024. The four-dimensional curriculum framework: 10 years on. Focus on Health Professional Education: A Multi-Professional Journal 25 (2), 98–109. Available at https://doi.org/10.11157/fohpe.v25i2.722.

Toosi, M., Modarres, M., Amini, M., Geranmayeh, M., 2021. Context, input, process, and product evaluation model in medical education: A systematic review. Journal of Education and Health Promotion 10 (1), 199. Available at https://doi.org/10.4103/jehp.jehp_1115_20 PMID: 34250133; PMCID: PMC8249974.

Tolentino, R., Baradaran, A., Gore, G., Pluye, P., Abbasgholizadeh-Rahimi, S., 2024. Curriculum Frameworks and Educational Programs in AI for Medical Students, Residents, and Practicing Physicians: Scoping Review. JMIR Medical Education 10, e54793. Available at https://doi.org/10.2196/54793 PMID: 39023999; PMCID: PMC11294785.

Tolentino, R., Hersson-Edery, F., Yaffe, M., Abbasgholizadeh-Rahimi, S., 2025. AIFM-ed curriculum framework for postgraduate family medicine education on artificial intelligence: mixed methods study. JMIR Medical Education 11, e66828. Available at https://doi.org/10.2196/66828.

Tozer, S., 2014. Encyclopedia of educational theory and philosophy. SAGE Publications, United States.

Woods, A., Luke, A., Weir, K., 2010. Curriculum and syllabus design. International Encyclopedia of Education. Elsevier Ltd., Australia, pp. 362–367. Available at http://doi.org/10.1016/B978-0-08-044894-7.00058-0, http://www.sciencedirect.com/science/referenceworks/9780080448947.

Zhao, L., Fan, W., Wang, S., 2023. Education Research International, pp. 1–9. Available at http://doi.org/10.1155/2023/8728567 Conception, Features, and Framework of Curriculum and Instruction Integration2023.

Chapter 7

# Curriculum planning and implementation

*"The curriculum of the future will be what one might call the humanistic curriculum."*

—John Goodlad

## 7.1 Introduction

Curriculum development is the systematic process of planning, designing, implementing, and evaluating educational programs to create meaningful learning experiences in order to produce a functional, effective, and adaptable educational program. It is the methodical process of creating a learning roadmap for students, outlining the knowledge and skills they should acquire. It also makes an effort to describe how the students will be taught and at what period of time the students will learn the content to achieve particular learning goals. This process aligns with standards and institution's mission to guarantee a coherent, successful educational experience. It shifts from general yearly objectives to targeted daily lessons, including topic sequencing, strategy selection, and flexible planning, to accommodate a range of student requirements.

Curriculum planning, which highlights on what to teach focusing on the standards, goals, and content, is the fundamental first step in the larger process of curriculum development. In addition to ensuring that the teaching and learning process runs smoothly, a well-designed and executed curriculum will direct both the teacher and the student toward the goals and objectives that are supposed to be accomplished at the conclusion of the program. Curriculum planning ensures that teaching is intentional, coherent, and effective, guiding educators in what to teach, how to teach it, and how to know if students have learned it (Harden et al., 2024).

## 7.2 Curriculum development and curriculum planning

The methodical process of creating, implementing, and assessing educational programs to produce qualified, well-rounded doctors is known as curriculum development in medical education. This process includes goal-setting, content selection, teaching method selection, and outcome assessment, all while adjusting to new innovations like competency-based education and technology. It's a systematic framework that guarantees deep learning, critical thinking, professionalism, and lifelong learning for aspiring physicians, not just a list of topics (Tagawa et al., 2014). In medical education, curriculum planning is the systematic process of creating, developing, and putting into practice a thorough training plan for future physicians. It focuses on desired competencies, knowledge, skills, and professional attitudes and uses techniques like problem-based learning and assessment to make sure graduates meet changing medical standards and societal demands. To develop an efficient, integrated, and adaptable learning experience, it entails defining precise objectives, choosing content and instructional strategies like simulation, bedside teaching, and e-learning, and setting up evaluation procedures. In essence, development designs, implements, and refines the real curriculum based on the blueprint created by planning, making sure that it is in line with the objectives and needs of the students.

## 7.3 Phases and steps in curriculum development

The important phases in curriculum development includes key steps like planning, relevance of content, chalking down the implementation process, and the also determine the evaluation and reporting (Thomas et al., 2015). It also shows the interaction and relationships of the various essential phases of the curriculum development process:

1. Assessment of the needs
2. Goals and objectives
3. Curriculum designing

**The Principles and Practice of Medical Education. DOI:** https://doi.org/10.1016/B978-0-443-44197-4.00007-2

4. Implementation
5. Evaluation

Each phase has several steps or tasks to complete in logical sequence. These steps are not always separate and distinct but may overlap and occur concurrently. It would make no sense to design learning activities before learner outcomes and content are described and identified. Before the start of the actual process, it is imperative that some of the preliminary steps are to be completed. Every step of the curriculum development process should be carefully planned keeping the stakeholders in mind and also attempt should be made to involve them in every process. For example, the curriculum team members, who have direct knowledge of the target audience, should be involved in conducting the needs assessment.

## 7.3.1 Phase I: needs assessment

The needs assessment is an essential component of the planning as it lays the foundation for all of the curriculum development steps. The steps in this phase include the following:

### *7.3.1.1 Identify the needs of the learner*

Determining the needs of the students is a crucial first step in curriculum planning. Among the crucial actions done at this stage include determining current knowledge and skills, understanding goals and motivations, and considering contextual factors. Finding out what specific skills or information they currently lack that are required for the desired outcome is one of the most important questions to be answered during this step. This will help determine the holes that need to be filled in the curriculum. Another important question that is to be answered is what misconceptions do students have regarding the subject? This will allow the instructor to directly address the misconceptions and make the necessary corrections throughout the course. By considering how their present proficiency levels stack up against the requirements for this course, it is possible to determine the learner's level of preparedness for the subject matter. Identifying the needs of the learner must be data-driven. This can be verified by inquiring about the statistics, work examples, and performance reports that bolster our understanding of their current abilities.

### *7.3.1.2 Form curriculum development team*

Once the nature and scope of the issue have been broadly defined, the members of the curriculum development team can be selected. The members selected should represent people who are serving in the curriculum committee, student assessment committee, and the administrative committee. Here, the objective is to build an efficient team by gaining the team members' knowledge in the subjects covered by the curriculum. This will not only ensure that the process of curriculum development is well formulated but at the same time will ensure that the quality is maintained. Incorporating a student representation into the committee is also a good idea because it not only brings a fresh viewpoint to the process but also helps us understand what the students may expect from it. Emphasizing the values of cooperation and teamwork is also crucial to clearly identify the roles and responsibilities of team members.

### *7.3.1.3 Conduct needs assessment and analysis*

#### 7.3.1.3.1 Needs assessment

As mentioned earlier, the needs assessment stage requires gathering information from various sources such as students, teachers, parents, and other stakeholders to understand their needs, goals, and objectives. Needs assessment may involve conducting surveys, focus groups, interviews, or observations to collect data about the learners and the context in which the curriculum will be implemented. The advantages of this needs assessment process are that it is able to identify the gaps between what the learner knows and what they need to know. It also tells us if the scope of the problem is clarified and defined. The results may prompt decision makers to allocate resources for a curriculum development team to prepare curriculum materials. The second step in this needs assessment process is analysis, which outlines methods for utilizing the collected data and information. This analysis aids in prioritizing needs, identifying knowledge and practice gaps, and determining the characteristics of the target audience.

## 7.3.2 Phase II: setting goals and objectives

Based on the information gathered from the needs assessment stage, the goals and objectives of the curriculum are set. These should be specific, measurable, attainable, relevant, and time-bound (SMART). Setting goals and objectives helps to ensure that the curriculum is designed to meet the needs and expectations of the learners and the community. In curriculum creation, establishing goals and objectives guarantees that activities, content, and assessments are in line with intended results and offers a clear learning path. Students gain a feeling of purpose and direction from this process, which also increases their motivation and aids in the development of critical abilities like self-control and time management. In the end, clear goals and objectives result in more successful and interesting learning opportunities. Establishing goals and objectives will also guarantee that assessments, long-term learning, and the everyday teaching and learning process all contribute to the same educational outcome. Additionally, it will give them a goal to strive for, which will help them prioritize their work and better manage their time.

## 7.3.3 Phase III: designing the curriculum

Once the goals and objectives have been set, the curriculum is designed. This stage involves developing a framework for the curriculum, including the content, instructional strategies, assessment methods, and materials. The content should be aligned with the goals and objectives of the curriculum and should be organized in a logical and meaningful way. Instructional strategies should be selected based on the needs of the learners and the goals of the curriculum. Assessment methods should be aligned with the learning objectives and should provide meaningful feedback to learners. Phase III determines intended outcomes referring to what learners will be able to do after participation in curriculum activities, the content referring what will be taught, and the methods indicating how it will be taught. The steps included in this process are as follows:

### *7.3.3.1 State intended outcomes*

In curriculum development, this stage of stating intended outcomes entails specifying specific, quantifiable activities that students will take to demonstrate learning. It also emphasizes on what students do in terms of their skills, knowledge, and values rather than what teachers teach. Strong, action verbs from taxonomies should be used to indicate the depth of learning. These outcomes also serve as a guide for evaluation and activities, thereby guaranteeing alignment.

### *7.3.3.2 Selection of the content*

The next challenge in the curriculum development process is selecting content that will make a real difference in the lives of the learner and ultimately society as a whole. The following are the main questions that need to be answered at this stage: "What will the learner need to know if the intended outcome is to be attained? What attitudes, behaviors, abilities, and information will need to be learned and practiced?." When choosing curriculum content, criteria such as relevance, importance, learnability, and feasibility are used to select content that is practical to teach, fulfills student requirements, and matches with goals. This ensures balance, articulation, and logical sequencing for meaningful learning experiences. Assessing learner needs, taking real-world applications into account, making sure the content is up to date, and logically arranging it for continuity and integration are important tasks.

### *7.3.3.3 Design the teaching-learning activities*

The next stage will be to logically arrange the content so that it builds upon itself and moves from simple to sophisticated once the particular subjects and themes required to meet the learning objectives have been chosen. To guarantee balance and breadth, the content must also be divided into more manageable, smaller pieces or subject schemes. Finding and compiling excellent resources, including lesson plans, materials, and exercises, is crucial for every idea.

## 7.3.4 Phase IV: implementation

In this stage, the curriculum is implemented in the classroom. Teachers and other educators use the curriculum to guide their instruction and enable the students to engage with the content and activities. It is a good practice to use a backward design approach when creating teaching-learning activities. It is good to start with clearly defining learning objectives followed by planning assessments to measure them and finally take up the task of creating engaging activities, content, and resources that actively develop students' knowledge and skills while guaranteeing alignment and sequencing for successful learning. Needs analysis, goal-setting, active versus passive method selection, content sequencing, assessment planning, and time allocation are all crucial processes that prioritize student involvement and practical relevance.

### 7.3.5 Phase V: evaluation and reporting

#### *7.3.5.1 Design evaluation strategies*

In the curriculum development paradigm, evaluation acts as both a phase and a particular step. Curriculum evaluation techniques include data collection through surveys, interviews, and observations, as well as student performance analysis through portfolios, tests, and other metrics. The process of curriculum evaluation encompasses both formative and summative evaluation methods. The primary aim here is to find out and fix issues pertaining to various aspects of curriculum as early as possible. Doing this will also enable continuous changes. The evaluation strategies when implemented on a regular basis provide useful information regarding the effectiveness of the curriculum and at the same time will help us to work on weakness if there are any.

#### *7.3.5.2 Reporting*

Getting the results into the hands of those who can use them is the last step in an evaluation approach. The reporting can be completed in the form of formal, extensive written reports, which usually consist of an executive summary, introduction, methodology, findings, discussion (interpretation), and recommendations that can be put into practice. Targeted presentations, such as conferences, seminars, or in-person meetings, can also be used to engage stakeholders directly. These presentations can be tailored to the interests of a particular audience and frequently highlight important findings with the use of visual aids like graphs or charts.

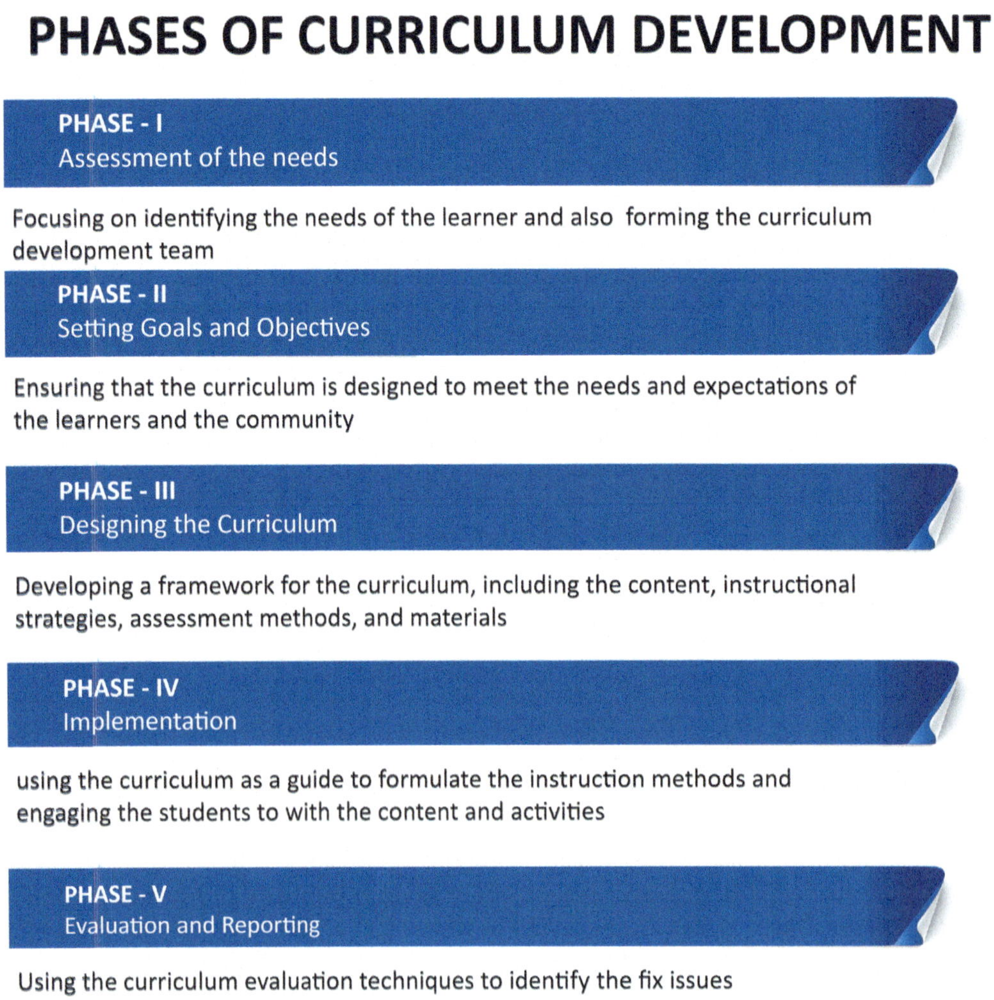


FIGURE 7.1 Phases of curriculum development.

#### 7.3.5.3 Evaluation

Once the curriculum has been implemented, it is evaluated to determine its effectiveness. In order to make informed changes, evaluating a curriculum entails a methodical process of collecting data on its efficacy, relevance, and alignment with goals using models like CIPP or Tyler's, concentrating on student outcomes, teaching strategies, and resources and involving stakeholders which include teachers, students, and parents through formative assessment which occurs during development and summative assessment that takes place after implementation. Setting objectives, gathering a variety of data from both the performance and feedback, comparing it to standards, and developing action plans for ongoing improvement are important elements.

#### 7.3.5.4 Revision

Based on the evaluation, the curriculum is revised and updated to improve its effectiveness. It refers to the process which is involved in updating a medical curriculum, wherein the focus is on analyzing present advantages and disadvantages which are brought into light by carrying out the process of needs assessment and obtaining feedback. In this regard, it is also important to compare how the obtained results of evaluation align with the vision and goals of the university and the institution. In order to ensure stakeholder engagement and ongoing quality improvement, organizing themes, prioritizing up-to-date information, and integrating basic science with clinical application may also be needed.

#### 7.3.5.5 Implementation of revised curriculum

The implementation of the revised curriculum is the last step which takes place once the revisions have been made. The revised curriculum is implemented in the classroom and it is regularly observed if there are any changes to be made in terms of all aspects of teaching and learning process. The evaluation and revision cycle is a continuous process as the curriculum is continuously monitored and updated to ensure that it remains effective (Fig. 7.1).

## 7.4 Frameworks in curriculum development

### 7.4.1 Kern's framework for curriculum development

A systematic approach, such as Kern's six-step model, offers a structured pathway for curriculum design. While these steps are discussed in sequence below, it is important to keep in mind that the steps may occur in parallel as well as out of sequence. Moreover it should be noted that the various steps are intertwined and are dependent on one another (Bodrick et al., 2025).

#### 7.4.1.1 Problem identification and general needs assessment

Understanding what inspires a teacher to start creating a curriculum is the goal of Kern's framework for curriculum development. Is the goal to impart new knowledge or techniques, or is it more a reflection of hospital administration or accrediting agency requirements? It's important to assess whether the time and effort required to create and administer the curriculum would ultimately improve a healthcare provider's knowledge base and, subsequently, a health outcome when evaluating the proposal or demand for a new curriculum. An essential step in this process is carrying out a general requirements assessment, often known as problem identification. The identification of the problem and the general needs assessments can be made using a variety of methods, such as group consensus, expert opinion, or already published data. Since it promotes wider generalizability, this stage is crucial for justifying the later distribution of the newly created curriculum.

#### 7.4.1.2 Targeted needs assessment

The general needs assessment from step one is applied to the real learning environment and students in this step. Involving stakeholders such as teachers, students, and administrators is essential to properly defining the problem since a focused needs assessment is the stage where an ideal method is superimposed upon the subject of interest. This is where baseline data, such as what are their acknowledged weaknesses and knowledge base? what drives these students? which learning styles do they prefer?, about the learner can be gathered. Numerous techniques for assessment can be used, such as direct observation, focus groups, interviews, informal discussions, and tests. Markers of validity and correctness should ideally be used when gathering this data.

#### 7.4.1.3 Goals and objectives

A key component of curriculum building is the creation of compelling goals and objectives. They set the curriculum's expectations and direct the rest of the developmental process. Goals are general aspirations to achieve in this phase, whereas targets are precise and quantifiable. The intended outcome of a curriculum's execution is known as an objective, and it must be specified before implementation in order to achieve the desired outcome. Crucially, for a goal to be effective, it must be well-written. The educational process, the learners themselves, or particular systems-level results can all be the focus of objectives. The SMART framework should be used for creating objectives. This framework guides the educator in the development of objectives that are Specific, Measurable, Achievable, Realistic, and Time-based. Bloom's taxonomy, which is the most popular model for objective development, emphasizes the use of action verbs to describe how students utilize their information cognitively. Higher-level learning takes place as objectives that are higher on the established framework are generated using this strategy, which is structured as a hierarchy. The lowest level of learning objectives which is remember is placed at the base of Bloom's taxonomy, which is frequently in the form of pyramid comprises of action verbs like remember, understand, apply, analyze, evaluate, and create in ascending order. A list of aims and objectives that is too long can overwhelm teachers and students, stifle innovation, and restrict learner-centered education.

#### 7.4.1.4 Educational strategies

A curriculum developer's role is to support learning, but there are several ways to do this. The method by which students learn is becoming more and more known in the late 20th and early 21st centuries. Once learning objectives are set, curricular content flows naturally from them. In order to increase student interest and investment, Kern advocates using a range of different strategies for reinforcement of a specific topic. This is because students will eventually realize their ideal learning methods reflected in the curriculum. A variety of teaching strategies should be employed and selected in accordance with practicality and the resources at hand. For instance, if the teachers don't have the required tools, a simulation session shouldn't be created. Similarly, educators ought to employ literature-based strategies that work best for adult learners. Due to its significance in continuing medical education, self-directed learning has recently gained more recognition as a crucial component of medical education curriculum. Given the significance of the skill in lifelong learning, it should be included as a teaching approach in designed curricula. Since not all students are proficient in this type of learning, it is necessary to provide them with particular techniques to maximize their self-directed learning habits.

#### 7.4.1.5 Implementation

The curriculum is eventually prepared for implementation following the time, effort, and resources used in the first four processes mentioned above. Four stages should ideally take place for a curriculum to be implemented successfully:

1. Garnering support from stakeholders, such as students, teachers, and pertinent administrators.
2. Planning for the implementation of the desired change, including identifying resources like staff, time, facilities, and funding.
3. Operationalizing the intervention, ideally planning for a pilot period, phase-in period, and full curriculum implementation.
4. Ensuring the viability and longevity of the curriculum by involving multiple parties so the curriculum does not rely on the efforts of a single person.

#### 7.4.1.6 Evaluation and feedback

In the end, the new curriculum's achievements and contributions ought to be documented in an evaluation system. Assessments based on an individual's level can indicate increased knowledge or skill. Individual assessments can be utilized to evaluate the instructional program when taken collectively. The evaluation's goal determines what it includes. For example, is the program's success being assessed, and if so, by what standard? A successful curriculum is always evolving. Effective curricula are dynamic and adapt to the changing needs of students in response to evaluation and feedback data. Lastly, program evaluations should be publicized for wider distribution after they are completed (Fig. 7.2).

### 7.4.2 Models of curricular development

There are several models for developing a curriculum. Here are a few popular ones:

#### 7.4.2.1 Tyler's model of curriculum development

Tyler's model was developed by American educator Ralph W. Tyler in 1949. It is also known as the objective model and is among the most widely used curriculum design models. It is based on the notion that the objective of education is to

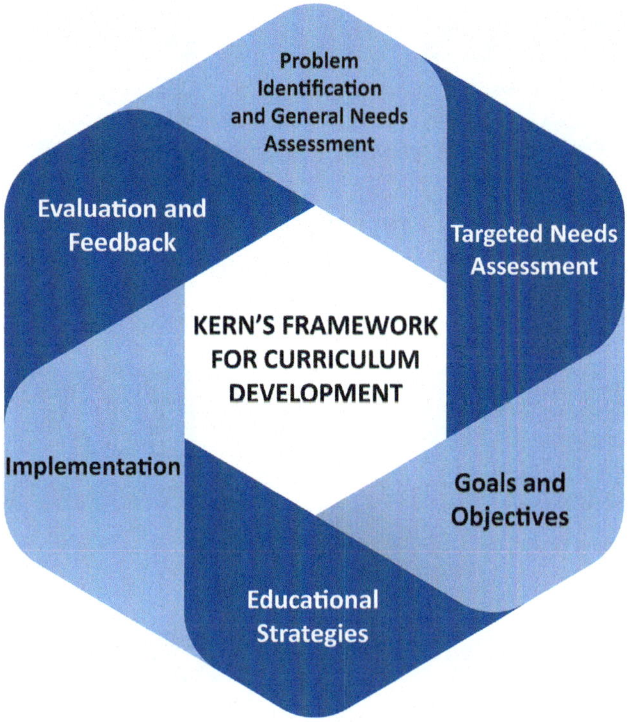


FIGURE 7.2 Kern's framework for curriculum development.

develop learning that is meaningful and useful to learners. Moreover, by concentrating on precisely defined, quantifiable goals that are derived from society and student requirements, this objective-focused paradigm guides curriculum design and stresses consistency across goals, activities, and outcomes. It is a linear model that considers four fundamental aspects of developing a curriculum which include the following:

1. The purpose or objective of education to be achieved,
2. The learning experience required to attain the defined objective,
3. Organizing educational activities for effective learning experiences,
4. Assessing the learning experiences.

The strength of this model lies in its concise structure and systematic approach. At the same time, it allows for flexibility. This concept makes it simple for educators to create curricula for every topic area and educational level. Tyler's model allows educators to evaluate the efficacy of their lesson plans and implement necessary modifications since it emphasizes on curriculum evaluation. However, it has a few shortcomings. It does not give much importance to social learning. Its specific learning objectives don't address critical thinking or emotional development. As such, this model may not meet the diverse requirements of students (Fig. 7.3).

### *7.4.2.2 Taba's model of curriculum development*

Taba's model emphasizes the planning of instructional strategies. Also known as the interactive model, it was designed in 1962 by Hilda Taba, an Estonian-American curriculum theorist, reformer, and teacher. It follows a grassroots or down-top approach and promotes a major role for teachers (Portillo et al., 2020). The model consists of seven stages of the learning and teaching system, which are mutually interactive. The various stages are describe as follows (Fig. 7.4).

#### 7.4.2.2.1 Diagnosis of learners' needs

The first and most important step in the Taba model of curriculum development is the analysis of learners' needs. In this step, educators must determine each student's unique learning needs, interests, and skills while also taking societal and cultural demands into account. In order to collect information and determine what students need to achieve, teachers employ techniques like surveys, interviews, and assessments. This information then influences the next stages of the curriculum design process.

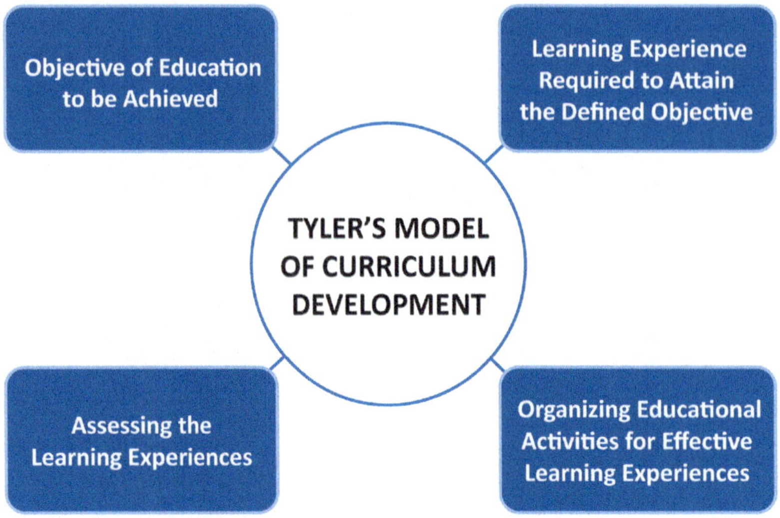


FIGURE 7.3 Tyler's model of curriculum development.

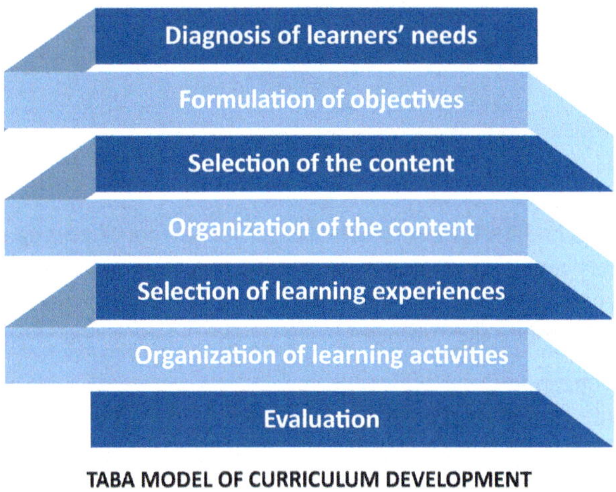


FIGURE 7.4 Taba models of curriculum development.

Identifying the needs of the students, which entails compiling information on what they need to succeed in their educational journey, is one of the crucial aspects in this process. The student's background must be taken into account in this regard. Understanding students' cultures, backgrounds, and social learning styles should be prioritized. In order to decide what students should learn, the diagnosis stage should also take into account the needs of society and culture.

### 7.4.2.2.2 Formulation of objectives

Following the diagnosis of learner needs, this stage is aimed at setting specific goals to address those needs. The selection and arrangement of learning activities and content are based on the objectives. Determining what students should be able to learn and do based on their needs, skills, and interests is the step's main goal. In terms of substance, the goals frequently cover things like knowledge gain, cognitive process development, and the development of a positive personality. The primary goal of formulating the objectives is to make it easier to choose acceptable learning experiences and to choose and organize content after they are chosen.

### 7.4.2.2.3 Selection of the content

The third step is content selection, which primarily entails selecting topics that directly relate to the previously stated learning objectives. In addition to taking into account the students' needs, interests, and academic performance, this phase includes making sure that the information is legitimate, important, and aligns with the goals. Finding information that directly satisfies the learning objectives set in the preceding stage and aligning that with the objectives is the main goal of this step. Additionally, this method of selection guarantees validity and significance by having curriculum designers assess the content's accuracy, value, and relevance.

#### 7.4.2.2.4 Organization of the content

The fourth main step in a seven-stage sequential and linear process is content organization. The content is organized according to the aforementioned procedure, which mandates that the chosen content be arranged in a specific order. The learners' academic accomplishments and maturity levels are important factors to take into account at this stage. At this point, concepts like "known to unknown," "simple to complex," and "concrete to abstract" are used to create order and guarantee cumulative learning.

#### 7.4.2.2.5 Selection of learning experiences

The stage of selection of learning experience entails picking teaching strategies and exercises that will actively engage students with the material and support them in meeting the learning objectives. To encourage student participation and higher-order thinking abilities, teachers use strategies including group projects, problem-solving activities, and assignments. Choosing the teaching strategies that will allow students to engage with the material and meet the goals is the primary goal of this phase. At this point, the teacher chooses suitable and interesting exercises based on the structured material and the traits of the students.

#### 7.4.2.2.6 Organization of learning activities

The sixth phase in a seven-step sequential and linear process involves organizing the learning activities. The chosen learning experiences must be arranged and sequenced as part of the organization of learning activities. Both the learners' traits and the order of the content determine the organization. To guarantee efficient and cumulative learning, the teacher must organize how these activities will proceed in the classroom, potentially creating vertical and horizontal integration.

#### 7.4.2.2.7 Evaluation

This is the seventh and last phase in the seven-step process, which focuses on determining if the desired learning outcomes have been achieved. The evaluation step is a formal assessment that establishes whether the curriculum creation process was successful. Its function is to verify whether the goals and objectives established in the previous stages have been achieved. The evaluation's results will help assess the overall learning outcome and guide future curriculum development. Instead of emphasizing correct or incorrect answers, this model encourages open-ended questioning. One of the main advantages is that it provides the opportunity to use critical thinking abilities to go further into an idea.

## 7.4.3 John Goodlad's model of curriculum development

Goodlad's model was developed by Canadian theorist and educational researcher John Goodlad. The curriculum creation model developed by John Goodlad stresses an all-encompassing and integrative approach that incorporates four main data sources, five decision-making levels, and ongoing evaluation throughout the process. The concept emphasizes striking a balance between society demands and personal education, considering education to have a moral component based on democratic principles (Goodlad, 1981). One of the important aspects of this models is that Goodlad thought of values as data sources. Goodlad concentrates on four data sources in his curriculum development methodology, which are as follows:

1. Values
2. Funded knowledge
3. Conventional wisdom
4. The learner's needs and interests

### *7.4.3.1 Key differences from other models*

Goodlad's model is unique because it introduces organizing centers as learning opportunities, emphasizes continual evaluation, and uses values as a key data source. The paradigm relies on explicit value declarations regarding data sources. This model suggests ongoing assessment throughout the whole curriculum creation process.

### 7.4.4 Wheeler's model of curriculum development

Curriculum design is seen as an ongoing, dynamic process in D.K. Wheeler's cyclical, prescriptive model of curriculum development. It highlights the interconnectedness of curriculum parts and the use of evaluation feedback to guide continuing revision (Prihantoro, 2021). It was proposed in 1967 as an upgrade to linear models such as Ralph Tyler's. This model designs and implements a curriculum using an organized and methodical procedure. It seeks to assist teachers in developing and implementing a dynamic curriculum that is successful, efficient, and relevant. The five stages of curriculum development included in Wheeler's model are as follows:

1. Selection of aims, goals, and objectives
2. Selection of learning experiences
3. Selection of content or subject matter
4. Organization and integration of learning experiences and content
5. Evaluation and revision of curriculum

This model's cyclical nature, which makes it a continuous loop rather than a one-time event, is one of its key features. The assessment results lead back to the objectives phase. Because each of the five stages is connected to and influences the others, the model also illustrates the interdependence of elements. In order to make wise judgments and support situational analysis, the model places a strong emphasis on researching the particular context or setting in which the curriculum will be applied. Since it can incorporate new information and make adjustments based on feedback and social changes, it is characterized by flexibility and reactivity. It emphasizes how the different curricular components are related to one another. Curriculum designers can start working on it at any point in this model. Situational analysis is another area of emphasis in this model. However, putting this model into practice could be challenging and time-consuming.

### 7.4.5 Kerr's model of curriculum development

Kerr's model of curriculum development was proposed by John Kerr, a British curriculum specialist. Several features resemble those in Tyler's and Wheeler's models, but the difference lies in the emphasis on the interrelatedness of the various components that are a part of this model (Yudhistir & Jugessur, 2018). This interrelatedness can be either direct or indirect and is achieved by the flow of the data between the components. John Kerr proposed four components for his curriculum development model:

1. Objectives
2. Knowledge
3. School learning experiences
4. Evaluation

Kerr's model recommends sorting objectives as cognitive, psychomotor, and effective. It also underlines that knowledge needs to be organized, integrated, sequenced, and reinforced. This model pays attention to the needs and interests of students, the influence of social conditions and school, and the maturity of the learners. However, one drawback is that it doesn't focus on teacher input.

## 7.5 Curriculum planning

### 7.5.1 Definition of curriculum planning

Curriculum planning is the process of creating a structured framework that outlines what students will learn, how they will learn it, and how their progress will be assessed. It involves identifying learning outcomes, selecting appropriate content and teaching strategies, and designing assessments, all while aligning with state or district standards and considering individual student needs.

#### *7.5.1.1 Key aspects of curriculum planning*

##### 7.5.1.1.1 Defining learning outcomes

By defining the learning outcomes, the medical educator will be able to identify the specific knowledge, skills, and competencies students should acquire during the course of their study. Learning outcomes in curriculum development are measurable statements that specify what students will be able to know, do, or value upon completing a learning experience. They are written from the student's perspective and serve as a guide for designing curriculum, aligning activities and assessments, and communicating clear expectations to learners. These written outcomes can be at the level of a topic, course, or entire program, connecting learning across different stages.

#### 7.5.1.1.2 Selecting content

Choosing the subject matter, topics, and materials that will be included in the curriculum is one of the vital steps as it determines how far can we justify the outcomes that have been identified. Selecting content for a curriculum involves ensuring it is valid, relevant, useful, learnable, and feasible by applying a range of criteria and procedures. The process of selecting content should be the responsibility of the teachers and the subject experts as they are in the best position to select the contents and also align them with the learning outcomes. Fundamental aspects include aligning content with educational objectives, considering the needs and interests of learners, and ensuring the content is accurate, practical, and developmentally appropriate and aligned with the learning objectives. The selection of the content should be based on judgment by the subject experts, data obtained and subsequent analysis done, and consensus.

#### 7.5.1.1.3 Determining teaching methods

Deciding on the instructional strategies, activities, and approaches that will be used to deliver the content is one of the most vital steps as it influences the learning process to a great extent. Determining teaching methods in curriculum development involves considering the curriculum's purpose, learning outcomes, and content, as well as student needs, subject matter complexity, and available resources like classroom and facilities. Effective methods are chosen by balancing these factors to create an engaging and effective learning experience, often using a combination of teacher-centered and student-centered approaches.

#### 7.5.1.1.4 Designing assessments

Developing methods to evaluate student learning and progress is the main purpose of this step. Designing assessments for curriculum development involves starting with clear learning objectives and aligning all assessments with these outcomes. This process includes selecting a variety of assessment types which suit the teaching methods and the outcomes to be assessed. Effective assessment design is an ongoing, cyclical process that uses data to inform and improve teaching and learning.

#### 7.5.1.1.5 Organizing the learning experience

The curriculum should be planned in such a way that it will not only inculcate the knowledge, skills, and attitude in the learner, but at the same time, it should positively enhance the learning experience. Organizing the learning experience in curriculum development involves aligning teaching methods, activities, and resources with specific learning goals to create a purposeful, sequential, and engaging experience for students. It also includes checking prior knowledge, setting clear goals, structuring experiences logically with continuity and sequence, using diverse methods like hands-on activities and discussions, and providing opportunities for practice and application. The process aims to ensure students gain skills, understanding, and desired behavioral changes in a way that is relevant and motivating for their age and level.

## 7.5.2 Importance of curriculum planning

### *7.5.2.1 Ensures alignment*

Curriculum planning ensures alignment by connecting learning objectives, content, and assessments to educational standards and goals. This is achieved by first analyzing standards to create specific, measurable learning objectives, then mapping the curriculum to ensure content and activities directly support those objectives. Finally, assessments are designed to accurately measure student progress toward mastery of the objectives. It should be guaranteed that the teaching activities undertaken to deliver the content also guarantees that the alignment is ensured (Atwa and Gouda, 2014).

### *7.5.2.2 Maximizes learning outcomes*

By carefully considering learning objectives, content, and teaching methods, curriculum planning can improve student learning and achievement. Curriculum planning maximizes learning outcomes by creating a structured, coherent, and engaging learning experience that is aligned with clear goals. It ensures lessons are logically sequenced, build on prior knowledge, and incorporate modern teaching methodologies, while also allowing for personalization to meet diverse student needs and promoting critical thinking. A well-planned curriculum provides a roadmap for teachers, enabling focused instruction and consistent, high-quality delivery that supports both academic achievement and the development of higher-order thinking skills.

#### 7.5.2.3 Promotes effective teaching

Curriculum planning provides teachers with a clear framework for their instruction, helping them to organize their lessons and deliver content effectively so that they meet the expectations of the learners. Curriculum planning promotes effective teaching by providing structure and purpose, aligning instruction with goals, and facilitating adaptability. It helps the instructor to break down complex subjects into manageable parts, ensures the specific learning outcomes are meant, and allows for differentiation to meet diverse student needs. This leads to focused learning and better outcomes by creating a clear, cohesive learning path for both teachers and students.

#### 7.5.2.4 Facilitates evaluation

Curriculum planning provides a basis for evaluating the effectiveness of the curriculum and making adjustments as needed. Curriculum planning facilitates evaluation by providing a clear, structured, and intentional framework against which performance and outcomes can be measured. In addition to this, curriculum evaluation also helps to provide the clear learning outcomes as these explicit outcomes become the primary benchmarks used during the evaluation phase to determine if students have achieved the intended knowledge and skills. Furthermore, the curriculum planning helps to integrate assessment strategies. Curriculum planning involves the deliberate design and integration of appropriate assessment methods, such as quizzes, projects, and exams, that directly align with the learning objectives. This ensures that the tools used for evaluation are valid and reliable measures of what the curriculum intends to teach.

#### 7.5.2.5 Supports student growth

A well-planned curriculum can help students develop the knowledge, skills, and competencies they need to succeed in their academic and personal lives. In essence, curriculum planning is a crucial process for creating effective and meaningful learning experiences for all students. Curriculum planning supports student growth by providing a structured roadmap that guides learning, ensuring students master essential concepts and skills. It helps by setting a clear pace, aligning education with standards, and creating a consistent and engaging learning environment tailored to diverse student needs. This structure allows teachers to focus on effective instruction and assessment, while students receive clear learning goals and opportunities to build deeper understanding.

#### 7.5.2.6 Guarantees clearer learning objective

The precise knowledge and abilities that students should acquire by the end of the course are defined with the aid of curriculum design. By assisting in the alignment of instruction, resources, and assessments with predetermined goals, curriculum design supports learning objectives. It offers a methodical framework for precisely defining the knowledge and skills that students should possess. In order to ensure a coherent and successful learning experience, a defined learning aim will also make it easier to choose the content, instructional methodologies, and evaluation techniques.

#### 7.5.2.7 Content is organized

It ensures that the sequence of lessons is logical and sees to that the upcoming topics are built upon previous knowledge. Curriculum planning helps the learner to shift from basic to complex topics, ensuring that concepts build upon one another. It defines a clear scope, determining the depth and breadth of topics covered, and uses pacing guides to schedule when topics will be taught. This structure ensures content is coherent, builds mastery, and aligns with educational goals and standards.

#### 7.5.2.8 Teaching methods are appropriate

The plan includes strategies for delivering content, whether through lectures, discussions, multimedia, or hands-on activities. Curriculum planning ensures appropriate teaching methods by establishing clear learning objectives that guide the selection of methods, creating a logical sequence for instruction, and incorporating assessment to gauge student progress and adapt strategies. It also includes considering students' prior knowledge and needs to ensure methods are inclusive and effective for all learners.

#### 7.5.2.9 *Assessments align with goals*

Curriculum planning ensures assessment alignment with goals by starting with clear learning objectives, designing assessments that directly measure those objectives, and using a feedback loop to inform and improve both instruction and assessment. This process involves mapping assessments to the specific skills and knowledge students are expected to acquire, using tools like alignment matrices, and continuously evaluating results to make necessary adjustments. All these elements work together to create a learning experience that is both effective and engaging.

### 7.5.3 Understanding curriculum planning

Before the curriculum is planned, it is crucial to fully comprehend it from every angle to ensure that all of its components have been treated fairly. It is important to realize that curriculum planning is the process of creating and arranging an educational program's structure, content, and delivery. Creating a thorough framework that directs students' learning is more important than just identifying subjects or topics to be taught. Ensuring that students gain the necessary knowledge, skills, and competences in an effective and interesting way is the aim of curriculum development. In order to create a structured framework that optimizes learning results, it is necessary to match educational objectives with instructional tactics and evaluation techniques. In this regard, the important aspects that need to be clarified include what knowledge and abilities the learners must acquire, referring to the curriculum; why they should acquire it, suggesting the justification or purpose; and how it will be taught, indicating the various teaching methods and tactics. What level is the curriculum aimed at, pertaining to whether it is at the national, academic, or medical school level? Are all the essential components—that is, the content, instructional strategies, and evaluation—sufficiently covered? Which curricular approach will be used? Will it be subject-centered, learner-centered, or competency-based, and how will learning be assessed and evaluated? It basically boils down to matching learning opportunities with desired outcomes.

#### 7.5.3.1 *Content*

The knowledge, ideas, and abilities that students must learn are referred to as the curriculum's content. This could involve theoretical ideas, real-world applications, or both. Teachers must make sure that students actively participate in the learning process, comprehend the importance of what they are learning, and have opportunity to apply their knowledge in meaningful ways; simply imparting knowledge is insufficient. To meet the various demands of students, the content must be properly chosen and arranged. Teachers must take into account both the material to be taught and the method by which students will interact with it while creating a curriculum. This entails combining teaching strategies with learning goals in a way that promotes a smooth learning process. Presenting this material in a way that is both accessible and entertaining while also satisfying the learner's needs and expectations is the difficult part.

#### 7.5.3.2 *Process of learning*

Curriculum planning also focuses on the process of chalking out how the content will be delivered and how learners will interact with it. It plays a central role in shaping how learning takes place. When done thoughtfully, it creates meaningful, engaging, and effective learning experiences. On the other hand, if it is weak or unstructured, it can lead to confusion, gaps in knowledge, or disengaged students. For instance, it helps to provide clear learning direction as it defines what students should learn in terms of learning goals and objectives and also at the same time it provides the much-needed clarity to aid both the teachers and learners to understand the significance of each lesson and how it connects to broader learning outcomes. When students know why they are learning something, they're more motivated and engaged. At the same, it guarantees the coherence and continuity of the topic by organizing the learning process so that it is presented more logically and progressively.

## 7.6 Role of stakeholders in curriculum planning

Curriculum planning is not a solitary endeavor. It involves collaboration between various stakeholders, each of whom brings a unique perspective and expertise to the table. Stakeholders like students, teachers, administrators, and community members are essential to curriculum planning by providing diverse perspectives, ensuring the curriculum is relevant and practical, and offering necessary resources and support. Learners are the primary focus, while teachers are crucial for design and implementation, administrators oversee resources, and the community offer support and local context. This collaborative approach ensures the curriculum meets the needs of students and is successfully brought to life in an educational setting (Foster, 2023).

### 7.6.1 Educators: designing and delivering the curriculum

Educators are the primary architects of the curriculum. They are responsible for selecting content, creating learning objectives, and designing the teaching methods. Moreover, educators need to ensure that the curriculum is accessible and inclusive. Teachers are essential to curriculum planning as they provide practical insights, help align curriculum with student needs, create detailed lesson plans, and implement and evaluate the curriculum in the classroom. Their role extends from selecting and customizing content to designing engaging learning experiences and adapting the curriculum to best suit their students. The teacher must also consider the diverse backgrounds, learning styles, and technological resources of students. Providing practical insights is one of the main responsibilities of teachers as they use their classroom experience to offer valuable input on learning goals, content selection, and instructional methods, ensuring the curriculum is relevant and effective. They also contribute towards customizing and designing as they select and adapt curricula to meet the specific needs of their students, design engaging learning experiences, and create detailed lesson plans based on the curriculum. The role of a teacher is also vital in the process of implementation and evaluation of the curriculum as they are responsible for delivering the curriculum, using effective teaching strategies, and adapting content to meet the diverse learning styles and progress of their students. The teacher is in the best position to collaborate and provide feedback as they can easily collaborate with curriculum developers, administrators, and other educators to provide feedback on the curriculum's effectiveness and suggest improvements.

### 7.6.2 Learners: active participants of the curriculum

Learners are active participants in the learning process as well as in curriculum planning by providing feedback, identifying interests, and helping to shape the learning experience. By involving students, educators can create more relevant and engaging curricula that foster a sense of ownership, increase motivation, and lead to better outcomes. Their feedback is essential in refining and improving the curriculum. Learners also play a role in shaping the learning environment. Through their engagement with course materials, participation in discussions, and collaboration with peers, they help create a dynamic and interactive learning experience. By taking part in the curriculum development process, the learner is able to provide insights on what topics interest them, what they find challenging, and where they need more practice or knowledge. Moreover, they can identify interests and talents and can highlight areas of personal interest and specific talents they feel should be included in the curriculum. The role of a learner in the process of content selection is vital as they can suggest specific activities or areas of study they want to be include and help define what success looks like in the curriculum. By participating in the curriculum development process and by providing feedback on what works and what doesn't, students help educators refine teaching strategies and materials to better suit their needs and learning styles. This ultimately shapes the learning experience.

### 7.6.3 Administrators: supporting the curriculum's implementation

Administrators provide the organizational framework that enables curriculum planning to succeed. Administrators are central to curriculum planning by providing leadership, allocating resources, and ensuring alignment with educational standards. Their role includes setting a vision, guiding the curriculum development process, and supporting teachers through professional development by ensuring that they have the skills to implement the new curriculum effectively. They also help in handling resources by managing the budget and allocating financial, human, and material resources for curriculum initiatives. They also oversee the procurement of necessary instructional materials and resources such as technology tools, learning management systems, and administrative support. They also help align the curriculum with institutional goals and ensure compliance with educational standards and regulations.

### 7.6.4 Policymakers: setting guidelines and ensuring quality

Policymakers play a crucial role in curriculum planning by setting the overall vision, goals, and standards for education. They establish frameworks, determine resource allocation, engage with stakeholders, and oversee implementation to ensure that the curriculum aligns with societal needs and promotes student success. Their work includes defining priorities, creating guidelines, and implementing accountability measures while also needing to allow for flexibility and adaptation at the local level. They have a role in setting the vision and standards by defining the high-level goals, standards, and priorities that guide curriculum development and innovation. These stakeholders are also involved in creating the framework as they provide a set of guidelines that medical schools and educators must follow when designing or implementing a curriculum. The allocation of resources is also one of the responsibilities of policymakers as they influence curriculum development by determining the resources and funding for various educational programs, which can impact how subjects are prioritized. Policymakers also are in the best position to engage stakeholders including educators and community members to gather input and ensure that the curriculum meets societal and student needs.

## 7.7 Curriculum implementation and its strategies

Curriculum implementation is the process of putting a planned curriculum into practice within an institutional setting. It involves translating the curriculum's objectives and resources into actionable lesson plans, teaching strategies, and learning experiences. Effective implementation hinges on teachers understanding the curriculum, adapting it to student needs, and utilizing appropriate resources and teaching methods. Curriculum implementation also aids to turning theoretical plans into practical learning experiences for students. It involves not just delivering content but also integrating various teaching and learning methods, ensuring alignment with learning objectives, and evaluating the effectiveness of the implemented curriculum (Mutalik et al., 2022).

### 7.7.1 Strategies to implementation of curriculum

Effective curriculum implementation requires a strategic plan that includes professional development, stakeholder engagement, resource allocation, and ongoing assessment. Key strategies involve creating a supportive environment, using a phased rollout, adapting the curriculum to diverse needs, and consistently gathering feedback to make necessary adjustments (Fasinro, 2024).

#### *7.7.1.1 Planning and preparation*

Creating a strategic plan is the initial stage of planning and preparing. In order to guarantee that learning objectives are in line with educational goals, this involves procedures like establishing a clear vision, precisely defining the goals, outlining the timelines, and allocating the resources. To promote a common goal and guarantee its implementation, a stakeholder team of educators, administrators, and designers must also be assembled. Understanding the curricular structure and all of its complex nuances will ensure that a strong foundation has been established for subsequent implementation.

#### *7.7.1.2 Professional development and support*

Teachers must be ready to take on this crucial responsibility if the implementation process is to be successful. In this sense, it is crucial to offer training and targeted professional development to give them the abilities and information required to successfully implement the new curriculum. To keep instructors informed about any modifications made to the core curriculum, professional development should be conducted on a regular basis.

#### *7.7.1.3 Implementation and adaptation*

It is always preferable to test the curriculum in pilot projects before progressively increasing its use. To accommodate and satisfy the requirements of all students, including those with disabilities or from diverse cultural and linguistic backgrounds, the curriculum should also be tailored to each individual. The curriculum's learning activities ought to be planned so that they are logically sequenced, scaffolded, and presented in a relaxed, learner-friendly setting. Making sure that the curriculum's resources are contextualized to meet the unique needs of the student is also crucial.

#### *7.7.1.4 Monitoring and feedback*

Establishing a feedback loop is an excellent practice for the monitoring and feedback process since it allows for the regular gathering of input from parents, students, and teachers, which helps to improve teaching methods and maintain the curriculum's relevance. A procedure for tracking implementation should also be in place, such as gathering information on fidelity and staff training to pinpoint areas in need of more assistance. The information obtained from the feedback have to be regularly reviewed and appraised. To assess the curriculum's efficacy and make the required adjustments, it is important to set up a number of checkpoints (Fig. 7.5).

### 7.7.2 Key aspects of curriculum implementation

#### *7.7.2.1 Translating the curriculum*

This step involves taking the written curriculum document and developing it into concrete lesson plans, schedules, and activities. Translating the curriculum is a key aspect of implementation because it transforms a planned document into

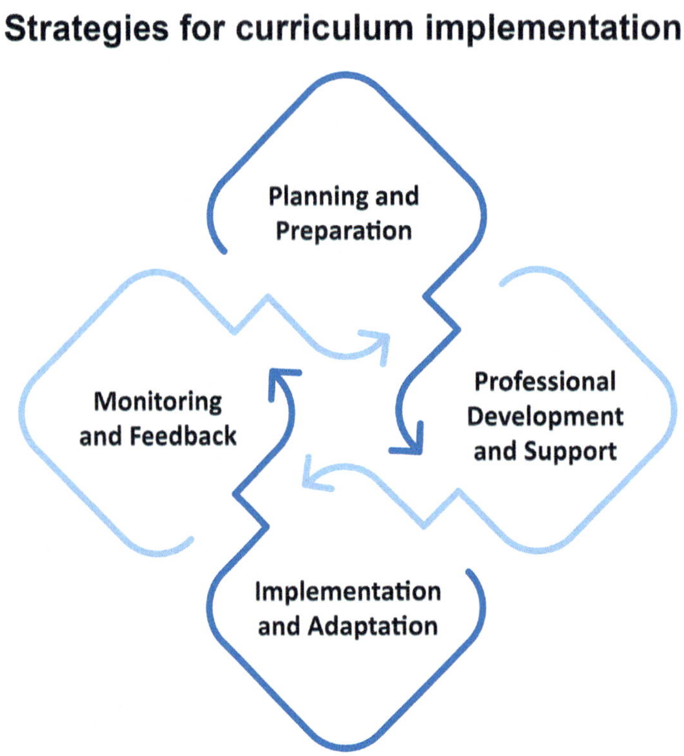


FIGURE 7.5 Strategies for curriculum implementation.

practical, relevant learning experiences for students. This process goes beyond a literal word-for-word conversion to include cultural adaptation, ensuring the content is meaningful and accessible to students from different backgrounds and contexts. By adapting the curriculum, teachers can improve student engagement, comprehension, and performance.

### 7.7.2.2 *Teaching and learning*

It is the daily interaction between teachers and students as they engage with the curriculum's content and learning experiences. Teaching and learning are key aspects of curriculum implementation because they are the practical application of the curriculum's plans, turning learning goals into classroom realities through effective instruction and student engagement. Teachers use the curriculum to guide their lesson planning, structure content delivery, and assess student understanding, while students actively learn through the teacher's methods. Successful implementation relies on this dynamic interaction, where teachers adapt curriculum to diverse student needs and utilize appropriate resources and strategies.

### 7.7.2.3 *Instructional strategies*

Instructional strategies are key to curriculum implementation because they provide the methods teachers use to actively engage students and help them meet learning goals. This step enables the teachers to utilize various methods and approach to deliver the curriculum, adapting their strategies to different learning styles and needs. They are the practical application of a curriculum, transforming planned objectives into a dynamic learning experience by selecting techniques like lectures, case studies, or group work that best suit the content, learners, and outcomes. Without effective strategies, a curriculum remains a theoretical plan rather than a tool for student learning and development.

### 7.7.2.4 *Resource utilization*

Resource utilization in curriculum implementation refers to the usage of textbooks, technology, and other materials to support student learning. It is considered as a key aspect of curriculum implementation because the availability and effective use of human, physical, financial, and technological resources directly determine the quality of the learning environment and student outcomes. It contributes to enhance teaching and learning by adequately providing the teaching materials, such as textbooks, lab equipment, and digital tools, which will enable teachers to present subject matter in more engaging and innovative ways, thereby stimulating student interest and motivation and increasing the overall quality of instruction.

#### *7.7.2.5 Assessment*

Evaluating student learning and the effectiveness of the curriculum is an integral part of the implementation process. Assessment is a key aspect of curriculum implementation as it provides feedback to improve teaching and learning, measures student progress toward learning objectives, and ensures the curriculum is effectively implemented. By continuously assessing student understanding, educators can identify strengths and weaknesses, provide necessary support, and make data-driven adjustments to instructional strategies and curricular content. This feedback loop ensures that the curriculum is responsive to student needs and leads to the desired learning outcomes.

#### *7.7.2.6 Teacher's role*

Teachers are considered pivotal in curriculum implementation, as they interpret, adapt, and deliver the curriculum to students. In addition, a teacher's role is crucial to curriculum implementation because they are the primary interpreters and adapters of the curriculum in the classroom, translating it into actionable teaching strategies that meet the specific needs of their students. They provide valuable input for curriculum development, assess its effectiveness through student outcomes, and adjust the curriculum to be more engaging and effective.

#### *7.7.2.7 Student needs*

Curriculum implementation needs to be responsive to the diverse backgrounds, learning styles, and needs of students. Student needs are central to curriculum implementation as they drive engagement and motivation, allowing for personalized learning experiences through differentiated instruction and tailored content. A student-centered approach ensures that teaching strategies and learning activities are relevant, accessible, and challenging for all learners, leading to greater success and retention. By focusing on individual strengths and preferences, curriculum implementation becomes a dynamic process of adapting to ensure every student can reach their full potential.

## 7.8 Levels of curriculum implementation

### 7.8.1 National level

At the highest level, the national curriculum is formulated, outlining broad educational goals, standards, and policies. This level sets the overall direction for the education system and ensures alignment with national priorities. It is also responsible for national assessments, teacher training frameworks, and resource allocation. A national curriculum is a common set of learning standards and subjects for medical schools, mandated by a national government to ensure consistent education across the country. It outlines what subjects are taught and the skills students are expected to learn (Sharma et al., 2022).

A national curriculum for medical education typically outlines a consistent framework, including core subjects, learning phases like preclinical and clinical, and competency standards, while allowing for institutional flexibility. For instance, in Saudi Arabia, the national medical program often employs integrated, system-oriented, student-centered, and problem-based learning strategies. The curriculum is typically divided into basic science years and clinical years, with a specific focus on community-oriented care and professional development through frameworks like CanMEDS (Fleming, 1989).

### 7.8.2 At the level of medical school

A medical school curriculum is typically structured into two main phases. These being preclinical and clinical. The preclinical phase involves classroom-based study of basic sciences like anatomy and physiology, while the clinical phase consists of supervised, hands-on training in hospitals and clinics, covering patient history taking, physical examinations, and specific medical specialties. Modern curricula often integrate these phases and emphasize active learning and professional development to prepare students for residency and practice. Teachers are responsible for translating the curriculum into engaging and effective learning experiences for students.

One of the special features of these curriculum lies in its interconnectedness. It's important to note that these levels are interconnected and work together to achieve educational goals. Effective curriculum implementation requires strong communication, collaboration, and coordination among all stakeholders at each level. For example, national policies and standards will continuously guide the work of medical institutions.

### 7.8.3 Importance of curriculum implementation

#### *7.8.3.1 Achieving curriculum goals*

Effective implementation ensures that the intended learning outcomes and objectives of the curriculum are achieved. Achieving curriculum goals is the primary importance of curriculum implementation, as it provides a systematic roadmap for students and teachers to align with educational standards and develop necessary skills for future success. Effective implementation ensures learning is relevant, engaging, and consistent by providing a framework for lesson planning, resource allocation, and teacher preparedness, ultimately leading to desired outcomes (Ludvigsson, 1999).

#### *7.8.3.2 Enhancing student learning*

When properly applied, the curriculum can help children develop positive attitudes, better skills, and deeper comprehension. Because it gives teachers an organized roadmap, synchronizes instruction with educational objectives, and promotes the development of critical skills, effective curriculum implementation is key for improving student learning. In order to prepare students for success, a well-implemented curriculum makes sure that lessons are taught methodically and logically, includes a range of activities to engage different learning styles, and offers a framework for evaluating student progress.

#### *7.8.3.3 Providing a framework for teaching*

Curriculum implementation provides teachers with a clear guidance for instruction, ensuring consistency and coherence. A framework for teaching the importance of curriculum implementation involves phases like engage, enact, and embed which guide educators through understanding the new curriculum, applying it in their teaching, and finally integrating it into their practice. This process helps teachers, administrators, and students understand the curriculum's role in providing a roadmap for learning, ensuring consistency, and aligning educational goals with current standards.

#### *7.8.3.4 Facilitating improvement*

The process of curriculum implementation can reveal areas where the curriculum or teaching practices can be improved. Effective curriculum implementation is crucial for improvement because it acts as a pathway for teaching, aligns educational goals, and enhances student outcomes. It drives improvement by providing a framework for teachers, ensuring consistency, promoting relevance and engagement, and creating a mechanism for continuous monitoring and adjustment. Success depends on leadership support, adequate resources, and feedback loops that lead to refinement and better results for students.

#### *7.8.3.5 Driving educational change*

Successful implementation can lead to positive changes in teaching practices, student learning, and overall institutional culture. Effective curriculum implementation is crucial for driving educational change because it is the process that translates planned educational visions and reforms into tangible classroom practices and improved student outcomes. Without effective implementation, even the best-designed curriculum remains a mere written document, leading to an "implementation gap" where desired changes fail to materialize in the classroom.

## 7.9 Role of curriculum planning towards the principles and practice of medical education

Curriculum planning plays a pivotal role in shaping medical education by translating educational principles into practical learning experiences. It serves as the bridge between theory and practice, ensuring that training programs are aligned with contemporary healthcare needs and educational best practices. Effective curriculum planning underpins several core principles of modern medical education. In terms of competency-based education, planning defines the specific competencies such as patient care, medical knowledge, and professionalism that graduates must attain. It structures the curriculum to ensure measurable outcomes are achieved rather than focusing solely on time spent in training (Betelmal and Alnasfi, 2024). In practice, curriculum planning involves systematic design and implementation steps that bring these principles to life. For instance, in terms of needs, assessment planning begins by identifying the healthcare needs of the community and the competencies required by future physicians. This ensures the curriculum remains relevant and responsive to societal demands.

**Summary**

Curriculum planning is a complex and collaborative process that involves much more than just outlining what should be taught. It requires a strategic approach that integrates content and process, involves key stakeholders, and adapts to the specific needs of learners in various contexts. The principles, processes, and practical considerations involved in effective curriculum planning and implementation in medical education are also important. It emphasizes that a well-designed curriculum is central to producing competent and patient-centered medical graduates who can meet societal health needs. By aligning teaching methods, resources, and assessments with clear learning objectives, educators can create a learning experience that empowers students to succeed in their academic and professional pursuits. In discussing curriculum implementation, the chapter highlights the importance of planning, preparation, monitoring, and feedback. It is also vital to understand the role of stakeholders in the implementation process.

**Key point**

**Future perspectives: from knowledge to competencies**

Future curriculum development will focus on dynamic, adaptable, and student-centered learning, moving away from static, textbook-based models to incorporate technology, interdisciplinary studies, and skills for the 21st century. Future medical education curricula will integrate advanced technology like AI and telemedicine, shift towards patient-centered and public health approaches, and incorporate new competencies such as interdisciplinary collaboration, cultural competence, and lifelong learning. Key trends include moving away from content overload to a more dynamic and flexible model, incorporating ethics, and redesigning assessments to measure a broader range of skills relevant to future healthcare challenges.

## References

Atwa, H.S., Gouda, E.M., 2014. Curriculum integration in medical education: A theoretical review. Intellectual Property Rights 2, 113. Available at https://doi.org/10.4172/2375-4516.1000113.

Betelmal, I., Alnasfi, S., 2024. Implementation of curriculum strategies for medical education. Benghazi University Medical Journal 1, 75–80. Available at https://doi.org/10.37376/benunivmedj.v1i1.7143.

Bodrick, M., Aljuffali, L., Alrasi, M., Aldahash, A., Arifi, S., Almutairi, M., Alhabib, I., Obaid, A., 2025. Versatility of the Kern six-step curriculum approach across health sciences education - from simplicity to complexity. Journal of Medical and Health Studies 6, 133–140. Available at https://doi.org/10.32996/jmhs.2025.6.2.20.

Fasinro, K., 2024. Curriculum implementation: challenges and the prospect of education resource centres to aid effective implementation. African Educational Research Journal 12 (1), 1–5. Available at https://doi.org/10.30918/aerj.121.23.102.

Fleming, W., 1989. The national curriculum and medical education. BMJ (Clinical Research ed.) 298 (6667), 135. Available at https://doi.org/10.1136/bmj.298.6667.135.

Foster, K., 2023. Involving stakeholders in re-imagining a medical curriculum. The Asia Pacific Scholar 8 (1), 43–46. Available at https://doi.org/10.29060/taps.2023-8-1/sc2807.

Goodlad, J.I., 1981. Curriculum development beyond 1980. Educational Evaluation and Policy Analysis 3 (5), 49–54. Available at https://doi.org/10.3102/01623737003005049.

Harden, R.M., Lilley, P., Yıldız, S., 2024. The curriculum: the heart and soul of a medical school. Medical Teacher 46 (12), 1524–1528. Available at https://doi.org/10.1080/0142159X.2024.2416307, http://www.tandfonline.com/loi/imte20#.VufR2bdf2DY.

Ludvigsson, J., 1999. A curriculum should meet future demands. Medical Teacher 21 (2), 127–129. Available at https://doi.org/10.1080/01421599979734.

Mutalik, A., Ramanathan, R., Seetharaman, N., Shanmugam, J., Subramanian, B., Mariappan, K., Ramalingam, V., Mani, V., Shanmugham, T., 2022. Evaluating the effectiveness of curriculum implementation and supplementation program in a private medical college in south India. International Journal of Community Medicine And Public Health 9, 3753. Available at https://doi.org/10.18203/2394-6040.ijcmph20222567.

Portillo, E., Look, K., Mott, D., Breslow, R., Kieser, M., Gallimore, C., 2020. Intentional application of the Taba curriculum model to develop a rural pharmacy practice course. Innovations in Pharmacy 11 (1). Available at https://doi.org/10.24926/iip.v11i1.2089. PMID: 34017632; PMCID: PMC8132528.

Prihantoro, C.R., 2021. Examining the use of wheeler-model based curriculum development in a learning management system for vocational study program. International Journal of Education and Practice 9 (3), 507–519. Available at https://doi.org/10.18488/journal.61.2021.93.507.519, http://www.conscientiabeam.com/pdf-files/art/61/IJEP20219(3)507-519.pdf.

Sharma, M., Murphy, R., Doody, G.A., 2022. Barriers and facilitators for implementation of a national recommended specialty core-curriculum across UK medical schools: a cross-sectional study using an online questionnaire. BMJ Open 12, e053565. Available at https://doi.org/10.1136/bmjopen-2021-053565.

Tagawa, M., Saiki, T., & Nishigori, H., 2014. Curriculum development in medical education. 25–35. https://doi.org/10.11307/mededjapan.45.1_25.

Thomas, P.A., Kern, D.E., Hughes, M.T., Chen, B.Y., 2015. Curriculum Development for Medical Education: A Six-step Approach Curriculum Development for Medical Education: A Six-step Approach, Third Edition. Johns Hopkins University Press, United States, pp. 1–300. Available at https://muse.jhu.edu/book/44600/.

Yudhistir S.M.F., & Jugessur, 2018. Use of Bronfenbrenner's Ecological model as Pillars of Educational Research for hypothesis formulation. International Journal of Humanities and Social Science Invention (IJHSSI), 12 (4), 105–107. ISSN (Online): 2319–7722, ISSN (Print): 2319–7714. Available at https://www.ijhssi.org.

Chapter 8

# Curricular approaches

*"If you can't describe what you are doing as a process, you don't know what you are doing."*

—Ed Deming

## 8.1 Introduction

The entire strategy and framework for how students learn, including the subject matter, teaching methods, classroom environment, and assessment strategies, is known as a curricular approach. It essentially describes the entire educational process and the interrelationships between different curricular components. It is important to remember that curriculum methods usually stem from and are influenced by curriculum ideas (Mohammed, 2013). One of the important aspects which influence the curricular approach is the relationship it has with the curricular principles. The curriculum principles are the fundamental ideas, attitudes, and rules that influence how an educational program should be created, implemented, and assessed. Some of the important principles being coherence, relevance, integration, learner-centeredness, competency orientation, flexibility, and alignment between outcomes, instruction, and assessment (Chhatria, 2024).

Curricular approaches are the specific frameworks or methods used to organize and implement the curriculum. The importance of competency and lifelong learning are examples of curriculum concepts in medical education, and one strategy would be to arrange the learning activities in a way that upholds such values. The curricular approach ensures that our work is consistent with the institutional mission while guiding the development of learning goal narratives, outcomes, and strategies through the utilization of academic literature and national trends. The viewpoints of curriculum design and development, as well as the responsibilities that educators, students, and curriculum specialists play in curriculum preparation, are all illustrated by a curriculum approach. It also includes the goals and objectives of the program. A variety of approaches, each with a distinct focus and philosophy, are used to generate competent and adaptable healthcare professionals. The relation between the curricular principles and curricular approaches is vital as the curriculum principles define the educational philosophy, and curricular approaches are the strategic models used to realize that philosophy in action.

## 8.2 Curriculum principles

Through learner-centered, flexible, and inclusive design, medical education curriculum concepts prioritize continuous improvement and lifetime learning. Important principles include meeting community and healthcare system requirements, fostering reflective practice, integrating basic sciences with clinical competence, and developing professional attributes like ethics and communication (Remesh, 2017). Another element of effective courses that advances students is a robust mechanism for assessment and feedback.

## 8.3 Core principles

### 8.3.1 Flexible and learner-centered

As a curriculum principle, flexibility entails creating instruction that can accommodate a wide range of student needs, interests, and speeds going beyond inflexible structures to provide individualized pathways, choices in learning methods in terms of time, place, content and outcomes also encouraging critical thinking, engagement, and lifelong learning by giving students autonomy and relevance rather than merely content delivery. A curriculum based on learner-centeredness and flexibility focuses on the individual needs, interests, and pace of each student, modifying education to accommodate different learning styles and preferences. This approach promotes active engagement, boosts motivation, raises retention

The Principles and Practice of Medical Education. DOI: https://doi.org/10.1016/B978-0-443-44197-4.00003-5

rates, and develops skills like critical thinking and problem-solving by allowing students to make decisions about their education.

### 8.3.2 Relevance

Relevance as a curricular principle is the concept that curriculum should connect with students' lives, interests, and future goals to make learning more meaningful and engaging. This principle involves linking classroom content to the real world, real-world challenges, and students' personal experiences to improve motivation, participation, and retention. It moves away from rote memorization towards a deeper understanding and a sense of ownership over the learning process. The main aspects of the curriculum relevance are that its connection with students. In order to assist students to see how their study relates to their future activities, the curriculum should take into account their future-oriented goals and focus on their plans and aspirations.

### 8.3.3 Coherence

Coherence as a curricular principle means that the parts of an educational program learning objectives, content, teaching methods, and assessments are logically and consistently aligned. This ensures that all components work together to create a unified and connected learning experience for students, where new ideas build upon previous ones and learning is not fragmented. A coherent curriculum helps students see the connections between subjects and makes learning more meaningful. One of the important points to be considered with regard to coherence is alignment. The alignment between learning objectives, content, instructional methods, and assessments is vital for the success of the curriculum. For example, a lesson on a specific topic should use teaching methods and assessments that are consistent with the learning goals for that topic. It is also important that a logical sequencing is followed, wherein the content is organized in a logical progression, with each new idea building on what has already been learned. This requires careful sequencing of both content and teaching approaches to ensure smooth transitions for students.

### 8.3.4 Progression

Progression as a curricular principle means that a curriculum is structured to systematically develop knowledge and skills over time, building on prior learning from simple to complex. It involves a coherent sequence of learning experiences that increases in effectiveness, breadth, and depth, leading to more sophisticated understanding and application of skills. The main aspects in this regard are that the progression should be smooth and should cater to building on prior knowledge such that each new concept builds logically on what a student has already learned. It should also move smoothly between complexities, wherein the learning will be developed from the basic skills to complex tasks, ensuring students are not overwhelmed.

### 8.3.5 Integration

Basic sciences should be integrated with clinical sciences and patient care to provide a more comprehensive and practical education. A fundamental tenet of curriculum is integration, which entails tying together disparate topic areas to produce a more comprehensive and significant educational experience. It goes beyond conventional, compartmentalized topics to help students see the connections between ideas like basic sciences and clinical importance, encouraging critical thinking and problem-solving abilities. By structuring instruction around themes or real-world issues, this method enables students to draw connections and apply what they have learned to new situations.

### 8.3.6 Ethics and professionalism

The curriculum must explicitly address the development of professionalism, communication skills, ethics, and a commitment to the student code of ethics. Ethics and professionalism are core, mandatory curriculum principles in modern medical education, universally recognized as essential for maintaining public trust and ensuring high-quality, patient-centered care. They form the basis of medicine's social contract with society. They are essential for fostering public confidence, guaranteeing responsibility, and equipping students to deal with difficult moral conundrums in the workplace. Medical education integrates ethics and professionalism throughout the entire training program, not as isolated subjects but as a pervasive theme. This is achieved through a multidimensional approach through the formal curriculum in the form of explicit courses, interactive lectures, and problem-based learning (PBL) sessions using case studies to teach the theoretical foundations and decision-making processes for ethical dilemmas. It can also be incorporated through the informal and hidden curriculum, wherein the students also learn from observation and real-life experiences, making the ethical conduct of faculty members and clinical supervisors as role models.

### 8.3.7 Equity and inclusivity

Equity and inclusivity as core curriculum principles in medical education empower health professionals to be advocates for their patients, leading to better health outcomes and more culturally and socially humble practitioners. Implementing these principles requires a systemic, institution-wide approach rather than isolated interventions. Some of the key strategies for integration is through embedding the inclusive design from the start. Rather than adding equity and inclusivity as a standalone topic, incorporate it throughout all curriculum design processes, from learning outcomes to assessment methods (Bhat, Krishna, 2025). The Universal Design for Learning framework offers a useful blueprint for creating flexible learning environments that anticipate diverse needs. On the other hand, diversifying and decolonizing the content by actively reviewing the curriculum materials for biases and stereotypes and ensuring that the content is not solely related to only privileged class or the higher ethic groups but also includes case studies featuring historically marginalized communities such as various ethnic groups is of vital importance to uphold the principle of equity and inclusivity (Doyle, Claire, 2025).

### 8.3.8 Practicing reflection

Medical instructors and students should be encouraged to engage in reflective practice in order to identify successful teaching techniques and consistently improve learning opportunities. As a fundamental component of curriculum, reflective practice is an ongoing, methodical process that enables teachers and students to evaluate experiences critically, question presumptions, and make well-informed modifications to enhance instruction and learning results. It fosters lifelong learning skills by changing education from a passive process of receiving information to a dynamic process of inquiry and personal growth.

### 8.3.9 Evaluation and feedback

Evaluation and feedback are core, interconnected principles in medical education, serving as the backbone for a continuous improvement cycle in curriculum design, teaching methods, and student performance. They are crucial for developing clinically competent, self-aware, and professional physicians. Evaluation is the systematic process of collecting and reviewing information to determine the extent to which an educational program aligns with its objectives and to make informed decisions for improvement. In the realm of curriculum development evaluation, results provide essential qualitative data that helps faculty enhance courses, adapt teaching methodologies, and ensure the curriculum remains relevant to current biomedical knowledge, technology, and community needs. It also contributes to quality improvement process in healthcare education as it helps to solve problems and build the knowledge base for best practices.

### 8.3.10 Relationship between the curriculum principle and curricular approach

The relationship between curriculum principles and approaches is close and interdependent, wherein the principles provide the foundation or philosophy, while the approach represents the practical way those principles are put into action in designing and delivering the curriculum. Principles influence why and what should be taught, whereas the approaches determine how it is taught. Together they create a complete framework for planning and teaching.

For instance, curriculum principle can be of learner-centeredness where education meets learners' needs and interests. On the other hand, curricular approach will be learner-centered approach when the students take an active role in learning. In other words, the principles drive the approach by determining which approach is most suitable for achieving them. For example, a principle focused on achieving clear, measurable outcomes will lead to an approach that is more technical and results-oriented. The curricular approach, on the other hand, operationalizes principles where the chosen approach provides the specific steps and structure needed to make the principles a reality in the classroom. It's the bridge between a broad educational vision and specific learning experiences.

## 8.4 Curricular approach

A curricular approach demonstrates the perspectives of curriculum design and development, as well as the roles of teachers, students, and curriculum experts in curriculum preparation. It also contains the curriculum's aims and objectives. Our expertise, values, and perspectives are reflected in the way we approach curricula. A curriculum approach, which represents a meta orientation or holistic position, includes curriculum domains that are common, important knowledge within the field, curriculum theory and practice, and curriculum foundations that highlight an individual's philosophy, view of psychology and learning theory, and view of social issues. A curriculum's creation and design, the role of the learner, teacher, and curriculum specialist in curriculum planning, the curriculum's objectives, and the crucial issues that need to be looked at are all expressed in an approach. A curriculum approach reflects our perspectives on education and society. It is possible to determine whether our professional viewpoint and the formal organizational vision are at odds by comparing our curricular approach to that of our institution.

## 8.5 Core components of curricular approach

The main elements of a curricular approach are usually aims, goals, and objectives, content, learning experiences, learning environment, and evaluation. Together, these components direct the process of teaching and learning. These five elements are interrelated and have an impact on one another. Below is a description of each of the essential elements (Yamani, Masoumeh, 2016) (Fig. 8.1).

### 8.5.1 Aims, goals, and objectives

Different from specific, quantifiable objectives, aims are the broad, overarching goals or purposes of training that concentrate on producing knowledgeable, competent, ethical professionals who serve society by promoting health and managing illness, ultimately preparing them for lifelong learning and patient-centered care. Instead of being merely duties, they serve as a "mission statement" for the entire educational process, incorporating the desired traits, values, and overall impact on people and the community. They represent an ideal of what students ought to become as a result of their education. Aims signify a broad purpose as they define the core goals of education, such as cultivating critical thinkers, encouraging lifelong learning, or preparing students for the workforce. Since they give the "why" that determines the "what" which represents the material, "how" which are the learning experiences, and "how well" that is the assessment, goals serve as the cornerstone around which the rest of the curriculum is constructed. On the other hand, goals are more detailed declarations that offer guidance for accomplishing the objectives. They serve as the overarching vision for what students should accomplish and are the cornerstone of a curricular strategy. They specify the curriculum's goal, which directs the content selection, learning experience design, and assessment techniques. In addition, goals offer focus and consistency since they convert general educational objectives into precise, long-term hopes for student development. These offer guidance by defining the curriculum's overarching goal and endpoint, guaranteeing a distinct and coherent educational route. Additionally, it directs experiences and material. Goals determine the content that will be taught as well as the exercises and methods that will be employed to support learning by outlining what students should know or be capable of. One of the most important aspects of goals is that they serve as the foundation for assessments by enabling the measurement of the curriculum's efficacy by offering a standard for determining whether or not student learning has been attained.

The most detailed descriptions of what students should know or be able to achieve after completing a particular learning experience are called objectives. Objectives are the cornerstones of a curricular approach because they outline the desired outcomes and act as a guide for the whole program. They serve the curriculum by directing the choice of materials,

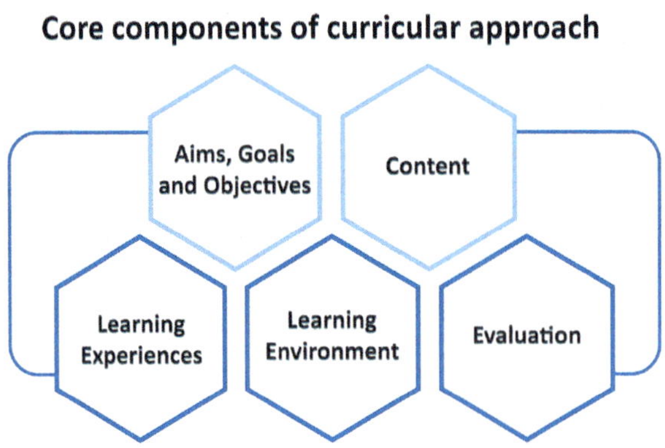


FIGURE 8.1 Core components of curricular approach.
This figure shows the core components of curricular approach.

educational activities, and evaluation techniques to guarantee that they are all in line with the overarching objectives of education. In medical education, objectives serve as core principles by defining the expected knowledge, skills, and attitudes for graduates, ensuring a standardized and high-quality learning experience. They are structured around key competencies like patient care, medical knowledge, and professionalism, guiding curriculum design and assessment to ensure a comprehensive and competency-based education (CBE). Objectives help institutions to clearly articulate learning outcomes, making the curriculum more systematic and effective for students, faculty, and ultimately, patients. Clear assessment and evaluation of student learning are made possible by well-defined objectives, which are said to be SMART (specific, measurable, achievable, relevant, and time-bound).

### 8.5.2 Content

The word content in medical education is used in reference to the material, subjects, and themes that will be covered in the curriculum. It also covers the attitudes, abilities, and knowledge that students are supposed to learn and apply in their daily lives. In a content-focused approach, content is chosen according to standards such as validity, usefulness, and relevance to learners, and it is logically arranged to guarantee a clear, organized, and productive learning process that aids students in expanding on their existing knowledge. In medical education, content is a fundamental component of the curricular approach, defining the body of knowledge, skills, attitudes, and values students are expected to learn. The role of the content is of prime importance as it forms the foundation of learning and provides the essential subject matter such as anatomy, physiology, and pathology that forms the backbone of a medical student's knowledge base. It also contributes to the achievement of outcomes. While modern approaches often emphasize competencies and intended learning outcomes, the content is the vehicle through which these outcomes are achieved. Content must also align with the specific skills and behaviors a future physician needs to exhibit. Since the content-centered approach mainly concentrates on the transmission of knowledge through particular teaching methods and the selection of important subject matter, the role of content in the curricular approach revolves around knowledge transmission. Furthermore, the selection and arrangement of information are guided by the existence of certain objectives. The material gives students the fundamental knowledge and abilities they need to build more sophisticated comprehensions and competences.

### 8.5.3 Learning experiences

These are the teaching methods, exercises, and materials that educators employ to promote learning. They assist students in interacting with the information and acquiring the required skills by bridging the gap between the content and the intended learning goals. Since learning experiences are the real activities and interactions that students utilize to accomplish specific learning objectives, they are an essential part of a curriculum. These experiences, which can range from lectures, skill lab sessions, practical sessions, internships to group projects and conversations, must be purposefully planned, developmentally sequenced, and in line with the goals, subject matter, and evaluation techniques of the curriculum. Coherent, meaningful, and productive learning journeys are created for students through the utilization of learning experiences in a successful curriculum. It should be mentioned that the learning activities are not haphazard but rather, they are carefully planned and organized to assist students in achieving particular learning objectives and results. Every learning experience must be in line with the curriculum's content and overall educational goals. Building on students' past knowledge and skills, learning activities should always be designed to be developmentally appropriate.

### 8.5.4 Learning environment

The learning environment is a core and influential component of the curricular approach, acting as both a critical factor in successful curriculum implementation and a direct determinant of student outcomes. Curriculum theorists consider the environment as one of the essential components of education, alongside the learner, teacher, and subject matter. The significance of learning environment in the curricular approach lies in the fact that it influences outcomes as it impacts student motivation, engagement, well-being, and academic achievement, often as much as the content itself. A positive environment is a worthwhile outcome in its own right and a means to improved learning outcomes. At the same time, it guides pedagogy and assessment as the design of the learning environment dictates the feasibility and effectiveness of different teaching methods and assessment strategies. For example, a curriculum emphasizing collaboration requires a community-centered environment that provides opportunities for peer-to-peer interaction and group work.

### 8.5.5 Evaluation

In order to ascertain whether the goals have been met, this component entails evaluating student learning. It contains a range of tools and techniques for assessment that are used to gauge both the success of the curriculum and the progress of the students. Evaluation is an essential, continuous, and rigorous component of the curricular approach because it provides the data and insights needed to determine the effectiveness of the curriculum, make informed decisions for improvement, and ensure conformance with educational objectives and social demands. It acts as a feedback system that links the other curricular components that include objectives, content, and technique in a continuous cycle of planning, implementation, and improvement. The main purpose of evaluation is to ascertain how well students have met the curriculum's goals and planned learning outcomes. Teachers, administrators, and curriculum creators are among the stakeholders who rely on evaluation for crucial, data-driven insights when making decisions about whether to accept, change, expand, or discontinue a program. Additionally, formative and summative assessments assist in determining the advantages and disadvantages of the curriculum, teaching strategies, and instructional resources, enabling prompt and essential modifications and enhancements.

## 8.6 Types of curricula approach

Ornstein and Hunkins identify six major curriculum approaches which are behavioral, managerial, systems, academic, humanistic, and reconceptualist approaches. These approaches reflect different beliefs about the purpose of education, the nature of knowledge, and the roles of students and teachers. The different types of curriculum approaches are described below (Fig. 8.2).

### 8.6.1 Behavioral approach

The most traditional and widely used method to curriculum is the behavioral approach. It is rational and prescriptive, based on scientific and technical concepts, and incorporates models, paradigms, and methodical approaches to curriculum development. The behavioral method is based on a blueprint with well-defined goals and objectives. One of the unique features of this approach is that all the content and activities are sequenced and aligned with predetermined objectives. Furthermore, the initial goals and objectives are used to evaluate the learning outcomes. This approach relies on technical and scientific principles. It includes paradigms, model, and step-by-step strategies. The method has also been referred to as technocratic, rational-scientific, experientialist, logical, and conceptual-empiricist (Isaac, Rief, 2009). The behavioral approach to curriculum development is also a methodical, planned strategy that emphasizes measurable, visible changes in student behavior as the main learning result. It places a strong emphasis on setting clear, precise, and quantifiable learning goals early on and then creating the curriculum in terms of material and activities to support those goals. After then, an

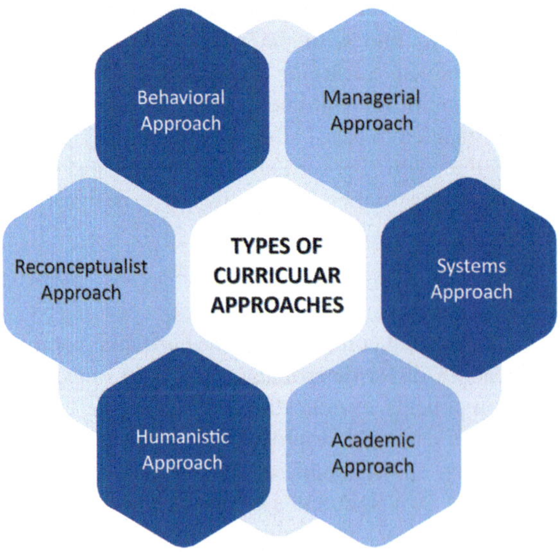


FIGURE 8.2 Types of curricular approaches.
This figure depicts the types of curricular approaches.

evaluation is carried out to see whether the desired behavioral changes have taken place. The behavioral approach's emphasis on the following elements is one of its defining characteristics.

#### *8.6.1.1 Focus on observable behavior*

Instead of concentrating on internal mental processes or subjective sensations, this method gives priority to activities that can be directly observed and measured. By organizing teachings on quantifiable behaviors and particular results rather than internal ideas or emotions, the behavioral curriculum approach emphasizes observable behavior. Learning is characterized as a shift in observable behavior brought about by the student's contact with the environment, which is controlled by a system of rewards, penalties, and stimuli. The curriculum is structured to provide for the explicit definition, instruction, and assessment of desirable behaviors through student behavior.

#### *8.6.1.2 Objectives-driven*

The objective-driven approach in medical education is an instructional method that uses clearly defined, measurable learning objectives to guide the entire educational process, including curriculum design, teaching methods, and student assessment. This approach focuses on the specific outcomes such as what a student can do or know at the end of the program rather than just the content covered which are the inputs. Setting SMART (specific, measurable, achievable, relevant, and time-bound) learning objectives is the first step in developing a curriculum. As the behavioral curricular approach is based on the idea that educational goals must be specified as distinct, observable, quantifiable behavioral outcomes, it is intrinsically goal-driven. This method offers a methodical framework in which every curricular element is clearly in line with accomplishing these predetermined goals.

#### *8.6.1.3 Structured and sequential*

A structured and sequential approach in medical education is an educational methodology where the curriculum is organized in a logical, linear progression, with each stage or topic building systematically upon the previous one. This contrasts with more fragmented or parallel learning models by ensuring a foundational understanding before advancing to more complex concepts and skills. The exercises and information are thoughtfully arranged and sequenced to lead students step-by-step toward the intended behavioral outcomes. By emphasizing observable, quantifiable results and dissecting difficult abilities into a number of smaller, more manageable steps, the behavioral curricular approach naturally facilitates structured and sequential learning. Students will have a clear, predictable, and effective learning route thanks to its methodical design.

#### *8.6.1.4 Reinforcement and feedback*

The method encourages and shapes desirable actions by using feedback and positive reinforcement. Behavioral techniques involve short, repeated practice opportunities, frequently in a safe, simulated environment, to support controlled practice and repetition. This allows for constant application and skill enhancement. Through habituation, this repetition fortifies the desired actions and offers numerous chances for prompt reinforcement and feedback. In order to establish a strong correlation between an action and its outcome, the behavioral method also promotes prompt and consistent feedback that is given right after an observed performance. Effective operant conditioning depends on this immediacy, which enables students to quickly correct errors or reinforce positive behaviors.

#### *8.6.1.5 Systematic evaluation*

The term systematic evaluation is a planned, continuous, and objective process of collecting and analyzing data to determine whether educational objectives are achieved, the effectiveness of teaching–learning strategies, and the areas for curriculum improvement. In the behavioral approach, evaluation is tightly aligned with predefined learning objectives, often framed as competencies or behaviors. The degree to which students exhibit the preset behavioral objectives is the basis for evaluating learning results. The behavioral curriculum method emphasizes well-defined, observable, and quantifiable learning objectives in order to facilitate systematic evaluation in medical education. This emphasis makes it possible to evaluate student achievement and the efficacy of the program in an organized and impartial manner. In addition to helping students learn and get ready for the test, a well-defined purpose motivates them to do better.

The Kirkpatrick model of program evaluation is the most widely recognized and used framework for systematic evaluation in the behavioral approach to curriculum in medical education. This approach is rooted in the idea that educational outcomes should be observable and measurable. The behavioral approach itself emphasizes clearly specified, observable, and measurable learning objectives, which align well with the structure of the Kirkpatrick model, particularly focusing on observable behavioral changes.

## 8.6.2 Managerial approach

The managerial approach, which is reminiscent of organizational theory, views the medical institutions as a social system where administrators, instructors, curriculum experts, and students interact. Teachers who use this method organize the curriculum in terms of staff, programs, timetables, space, resources, and equipment. This method promotes the selection, coordination, supervision, and communication of those involved in curriculum decisions. Committee and group procedures, interpersonal interactions, leadership approaches and personalities, and decision-making are all taken into account. The managerial approach depends on a plan, logical steps, and rational concepts. It frequently concentrates on the administrative and supervisory facets of curriculum, particularly the planning and execution process. The dean's leadership role in organizing, planning, and executing the curriculum is highlighted by the managerial approach to curriculum creation. It emphasizes successful implementation and resource allocation, viewing the curriculum as a system with inputs, processes, and outputs. In order to promote medical institutions improvement, this strategy also emphasizes the significance of managing people, data, and procedures. In this method, in addition to the dean or the principal, a curriculum supervisor or specialist carries out a number of crucial duties within the organization, including assisting in the development of the educational objectives of the medical institutions or community, planning the curriculum with students, parents, teachers, and support staff, coordinating or evaluating a survey of student needs, designing the programs of study by grade level and/or subject, scheduling classes, and even creating the academic calendar.

The organization and execution of this strategy are more important to the administrators as they are more interested in enhancing curriculum in light of goals, policies, and individuals on a systemic level than they are in subject matter, techniques, and resources. For example, implementing strategic planning for program development, utilizing resource allocation frameworks for departmental budgeting, incorporating change management principles during curriculum reform, and giving students real-world decision-making responsibility in clinical settings under supervision. The important characteristic features of this type of approach are as follows:

### *8.6.2.1 Head of the institution as curriculum leader*

The dean is regarded as the primary figure in developing the curriculum, establishing guidelines, and guaranteeing its successful execution and is regarded as a curriculum leader in a medical institution. To assure effective and compliance curriculum delivery, the dean or a senior academic officer focuses primarily on the administrative, resource, and systems management components. For example, introducing a new curriculum and continuously monitoring whether adequate facilities and resources are provided for the conduction of the courses are all responsibilities of the dean and the supportive administrative staff.

### *8.6.2.2 Systems thinking*

Systems thinking in medical education is a managerial approach that views the medical education system as interconnected parts, rather than isolated issues. This managerial method focuses on understanding the relationships between different elements, like curriculum, faculty, students, and the broader healthcare system, to identify how to create sustainable, long-term improvements and achieve desired educational outcomes. It moves beyond focusing on symptoms to address the root causes of problems by analyzing how different parts of the system interact and influence each other. For example, in integrated curriculum design, a systems thinking approach creates an integrated, interprofessional curriculum that clearly connects disciplines like anatomy, physiology, and pathology rather than managing them in silos. This aids students in comprehending the big picture of patient care, which requires integrating information from several disciplines.

### *8.6.2.3 Efficiency and effectiveness*

In a managerial approach to medical education, effectiveness is defined as achieving the desired goals and producing the intended results such as doing the right things, while efficiency is the ability to do so without wasting resources, time, or energy like doing things right. Effectiveness in medical education focuses on the outcomes and whether the educational program successfully produces competent healthcare professionals who can deliver quality patient care. For example, redundancies and gaps are found and removed during the process of curriculum mapping and integration, which involves methodically mapping the entire curriculum and integrating basic and clinical sciences across different stages that is vertical integration and across disciplines at the same stage which is horizontal integration. This saves teacher and student time by preventing the same information from being taught repeatedly in several courses. Efficiency concerns the process of education and how well resources such as time, money, staff, and facilities are utilized to achieve those results. Here the focus is on maximizing productivity and minimizing waste. It's about optimizing the input-to-output ratio. The measurement is concerned with assessing the optimal use of resources, using metrics like cost per student, time taken per task, resource utilization rates, or the reduction of administrative waste.

#### 8.6.2.4 Focus on processes

It highlights the procedures that go into delivering the curriculum, such as classroom management, instructional time, and assessment. In a medical curriculum that emphasizes process in a managerial approach, the attention is not only on the final results or outcomes but also on the particular activities, steps, and methods employed in healthcare delivery and education. This strategy seeks to guarantee that established, quality-controlled processes are regularly adhered to, producing dependable results.

#### 8.6.2.5 Teacher as a manager

As coordinators and supervisors of students' learning, teachers are viewed as managers of the educational process. In order to maximize student learning and resource allocation, a teacher in a medical curriculum manages the learning environment by using management techniques like organizing, planning, coordinating, and controlling. In a medical curriculum, for instance, a clinical teacher in a hospital ward successfully manages time constraints, patient availability, and physical resources to guarantee that every student receives sufficient practical experience and supervision throughout their rotations.

#### 8.6.2.6 Stakeholder involvement

The strategy frequently includes stakeholders including teachers, administrators, and occasionally students in the planning process, even if the dean takes the lead. Involving stakeholders in the managerial approach to medical curriculum is crucial to maintaining the education's relevance, responsiveness to healthcare demands, and eventual improvement of health outcomes. Throughout the whole curriculum lifecycle which includes planning, design, delivery, and evaluation effective management entails identifying important stakeholders, assessing their impact and interests, and putting a variety of engagement tactics into practice.

#### 8.6.2.7 Emphasis on implementation

In this, the practical issues of implementing the curriculum are included in addition to curriculum design. Improving patient outcomes and converting theoretical knowledge into practical, effective healthcare practices require a focus on implementation within a managerial perspective in the medical curriculum. It is to be noted that an implementation-focused managerial approach stresses the actual deployment of evidence-based therapies into normal practice. This guarantees that practical experiences may be successfully applied in challenging clinical settings by future doctors. When compared to the behavioral approach where the focus is on particular learning goals and quantifiable results, the managerial approach is more comprehensive and covers the full curriculum delivery system. For the success of the managerial approach, it is important that clear goals are set and job responsibilities be described to each member of the committee so that the task can be carried out smoothy with no hassles or conflicts. One of the best examples of the managerial activities in curriculum and teaching is in planning and preparing the academic calendar. This process is a collective effort of various group of individuals and committees. The initial draft is prepared by the members of the curriculum committee with the help of departmental heads and the program coordinator. The program coordinator is responsible to recheck the document taking into consideration the scheduled credit hours and contact hours of the individual course. During this process, it is even important to ensure that there is an effective use of time and resources. This is then approved by the vice dean and the dean.

### 8.6.3 Systems approach

The systems approach is a method of planning, implementing, and evaluating a curriculum by viewing it as an integrated and interdependent system. In medical education, it ensures that all elements of the curriculum work together to produce competent healthcare professionals who meet societal needs. An emphasis on structuring education into a system emerged from a managerial perspective that places a strong emphasis on organizing people and policies. This approach views the curriculum as an interconnected system with components such as objectives, learning methods, assessment, faculty, and the learning environment working together to accomplish particular learning objectives. For instance, the outcome structured curriculum connects basic sciences horizontally to the organ systems they relate to, linking them to clinical reasoning and patient scenarios to enhance a student's ability to handle complex situations. In other words, the systems approach is a structured, organized method of curriculum development that views curriculum as a system consisting of inputs, processes, outputs, and feedback, all interacting to achieve clearly defined educational outcomes.

Furthermore, instead of seeing the educational process as a collection of discrete components, the systems approach to curriculum development sees it as a complex, linked system. Through component analysis, change design and implementation, and ongoing process evaluation and improvement, this method seeks to optimize the overall system. Of late, a lot of medical institutions employ a systems approach called total quality management, which is based on Ed Deming's 14 ideas for enhancing the workplace. This strategy, which is also based on industry, is a paradigm shift that emphasizes the importance of students, thorough data collection and analysis, self-monitoring and inspection, teamwork, communication, cooperation, and team accountability. Some of the key aspects of the systems approach in curriculum development are as follows:

#### *8.6.3.1 Holistic perspective*

The systems approach takes the curriculum as a whole into account, recognizing that modifications to one component of the system may have an impact on other components. It acknowledges how different components of the educational environment such as administration, counseling, instruction, and evaluation are interrelated. Instead of concentrating only on isolated symptoms or specific bodily parts, a holistic perspective within a systems approach in a medical curriculum teaches aspiring medical professionals to see patients and healthcare issues as part of a complex, interrelated whole (Ahn, Phillips, 2006). To comprehend the intricate relationships that affect health outcomes, it goes beyond a strictly reductionist approach, which dissects issues into their component parts.

#### *8.6.3.2 Systematic process*

The strategy entails a methodical planning, development, execution, and evaluation procedure to guarantee that every element functions in unison. It highlights the significance of well-defined goals and objectives, suitable teaching techniques, and efficient assessment techniques. The systematic process of the systems approach in a medical curriculum is a cyclical model built on the fundamental elements of input, process, output, and ongoing evaluation and feedback. It views medical education as a dynamic, integrated system with the goal of generating qualified medical graduates who address the health requirements of society.

#### *8.6.3.3 Input, process, and output*

The educational process is modeled by the systems approach as a system having inputs like students, resources like teaching and learning, and outputs like student learning, skills, and knowledge. The strategy seeks to improve the quality of outcomes by streamlining the process and inputs. The resources that enter the system, such as students, instructors, and curriculum content, are referred to as input. The process encompasses the actions that alter these inputs, such as instructional strategies and educational opportunities, and the output is the outcome of that transformation, which transforms the students into competent, skilled medical graduates. This framework facilitates the methodical development and assessment of the medical education program's efficacy.

### 8.6.4 Academic approach

The academic approach, sometimes known as the classical, encyclopedic, synoptic, intellectual, or knowledge-oriented approach, aims to study and synthesize key curriculum perspectives, trends, and concepts (Gamulin, 2004). This method is more often philosophical or historical than social or pragmatic. Curriculum development is typically discussed in a scholarly, theoretical manner and covers a wide range of topics related to education. This method, which gained popularity in the 1930s and persisted through the 1950s, is based on the writings of Boyd Bode, Henry Morrison, and John Dewey. The influx of new topics related to curriculum during this period expanded the field to include many trends and issues and led to the integration of various instructional, teaching, learning, guidance, evaluation, supervision, and administrative procedures. The academic approach to curriculum building places a strong emphasis on intellectual rigor and subject matter expertise, concentrating on the development of cognitive skills within academic disciplines and the transmission of known information. It emphasizes in-depth knowledge in disciplines like anatomy, physiology, and biochemistry by frequently utilizing philosophical foundations. This method seeks to develop critical thinking and problem-solving skills and is based on classic educational principles. As contemporary curricularists approach curriculum from a postmodern academic perspective, the academic approach has partially returned the focus on the nature and structure of knowledge. Some of the key characteristic features of this approach are as follows:

#### 8.6.4.1 Knowledge-centered

The methodical delivery and mastering of academic material form the foundation of the program. In a medical curriculum, a knowledge-centered academic approach is a curriculum design concept that emphasizes the sequential mastering, systematic organization, and in-depth study of fundamental scientific and medical knowledge. It highlights that building the clinical reasoning and problem-solving abilities necessary for successful medical practice requires a strong, integrated knowledge base. One of the examples of the knowledge-centered approach is that the students learn about the structure of the muscles of breathing and the relation of pressure and volume in the respiratory system before they understand the mechanics of breathing and its clinical application. This process is also supported by teachers through the interactive lectures where the teachers, being experts in their discipline, often use systematic lectures and direct instruction to communicate complex subject matter efficiently.

#### 8.6.4.2 Disciplinary focus

The curriculum is structured on well-known academic fields including anatomy, physiology, biochemistry and clinical sciences. In a medical curriculum, a disciplinary focus in academic approach which is also referred to as a traditional or subject-based curriculum is distinguished by a compartmentalized structure in which each medical specialty is taught independently as a separate subject with little integration between them. A classic example of a disciplinary focus in medical education is the preclinical curriculum of many traditional 4-year medical school programs, often called the "discipline-based model." In this, the first 2 years of medical school are concentrated mainly on teaching the basic science subjects such as anatomy, physiology, biochemistry, and pathology. On the other hand, the next 2 years are concentrated on the clinical subjects like medicine, surgery pediatrics, and obstetrics.

#### 8.6.4.3 Intellectual rigor

It is a well-known fact that the development of critical thinking, analytical abilities, and in-depth comprehension is of immense importance for a student in order to carry out a successful clinical practice. In order to generate competent, analytical, and moral doctors, medical curricula must be intellectually rigorous by emphasizing higher-order thinking, relevant material, and appropriate degrees of challenge and support. This emphasizes in-depth comprehension, critical thinking, and the application of knowledge to challenging clinical issues rather than just memorizing facts.

#### 8.6.4.4 Transmission of knowledge

The strategy frequently concentrates on communicating accepted theories and information in the pertinent fields. Knowledge transmission in medical curricula has changed from traditional, teacher-centered techniques like passive lectures to a range of student-centered, active learning strategies intended to improve clinical reasoning, long-term retention, and the capacity to apply knowledge in practical situations. Fostering learning transfer, such as applying acquired knowledge to novel and challenging clinical scenarios, is the main objective.

#### 8.6.4.5 Potential for depth

Although it may be criticized for possibly ignoring real-world application or student-centered learning, it might result in a solid foundation in some topics. The possibility for depth in the academic approach within the medical curriculum is a topic of continuous debate and is typically viewed as striking a balance between the need for deep understanding and the breadth of material. Many contemporary curricular changes seek to promote a deeper, more analytical approach, but unfortunately, the sheer amount of content frequently compels students to embrace surface-level, memorization-based learning.

### 8.6.5 Humanistic approach

The previous techniques, according to some curriculum leaders, are overly restrictive and technocratic. They argue that curriculum designers who strive to be rational and scientific overlook the sociopsychological dynamics of classrooms and medical institutions, the personal and social aspects of curriculum and instruction, the artistic, physical, and cultural aspects of subject matter, and the necessity of learners' self-actualization and self-reflection. The humanistic approach to curriculum development emphasizes personal development, self-discovery, and the realization of individual potential while focusing on the learner as a whole. It creates a welcoming and compassionate learning atmosphere by putting the needs, interests, and experiences of the learner first (Biruny, 2021). In order to achieve self-actualization, this strategy seeks to foster intrinsic motivation, self-directed learning, and a positive self-concept. In contrast to competitive, teacher-dominated, large-group learning, curriculum professionals who support this approach typically place more faith in cooperative learning, individual learning, small-group learning, and social activities. Every student contributes significantly to the curriculum and collaborates with teachers, parents, and curriculum experts to develop lessons in the classroom. The concepts of professional collegiality and mentor systems are more prominent in medical institutions that use this method, and curriculum leaders and supervisors typically allow professors more influence into curriculum decisions. Students are frequently invited to participate in curriculum meetings to voice their opinions because curriculum committees are bottom-up rather than top-down. Among its fundamental tenets are the following:

#### *8.6.5.1 Student-centered*

With their individual needs and goals directing the learning process, the learner is at the center of the curriculum. In order to actively involve students in their learning process and promote empathy, professional ethics, and the capacity to care for patients, a student-centered, humanistic approach in a medical curriculum entail changing the role of the instructor to that of a facilitator. This is accomplished by combining a variety of active learning techniques, including role-playing, simulation, and group projects, with a focus on the professional and personal growth of the student through introspection, feedback, and the incorporation of humanistic principles into therapeutic practice.

#### *8.6.5.2 Holistic development*

In order to create well-rounded people, the humanistic approach concentrates on the cognitive, affective (emotional), and psychomotor (physical) areas of learning. Holistic development in a humanistic medical curriculum refers to training a "whole person" by methodically promoting their intellectual, physical, emotional, social, and spiritual development in addition to their academic or technical abilities (Benatar, 2014). This strategy is based on the idea that a well-rounded doctor is better able to offer thorough, sympathetic, and patient-centered treatment because all facets of a person are interconnected.

#### *8.6.5.3 Intrinsic motivation*

This highlights the value of encouraging sincere curiosity and enthusiasm for learning as opposed to depending on rewards or penalties from outside sources. Intrinsic motivation, according to a humanistic approach to a medical curriculum, is the internal drive of medical students to study, develop, and care for others for the intrinsic interest, fulfillment, and happiness it brings rather than for extrinsic incentives like status or grades. Self-determination theory, which holds that people are growth-oriented by nature and flourish when their fundamental psychological needs are satisfied, serves as a major foundation for this idea.

#### *8.6.5.4 Self-actualization*

This strives to assist students in realizing their greatest potential and developing into independent, self-directed learners. The idea of a healthcare worker realizing their full potential via the development of traits like empathy, acceptance, creativity, and a strong sense of purpose is known as self-actualization in a humanistic approach to the medical curriculum. It goes beyond only curing illness to support the development of the healthcare professional as a whole, inspiring them to become genuine, creative, and deeply aware of both themselves and their patients.

#### *8.6.5.5 Positive learning environment*

It establishes a secure, encouraging, and sympathetic environment where the learner feels appreciated, respected, and inspired to express themselves. A positive learning environment in a humanistic medical curriculum is one that encourages students' personal development, self-directed learning, and empathy. It is safe, respectful, and encouraging. It places emphasis on the learner as a whole person, acknowledging that successful learning and the development of a professional identity depend on emotional health. By using the "patient-as-teacher" paradigm, in which patients share their experiences to help students develop empathy, and having clinical professors serve as role models who exhibit and talk about patient-centered care and emotional reactions are two instances of a humanistic approach in medical education. Other examples include developing supportive learning settings, incorporating humanistic methods into all facets of the curriculum, including volunteer work and skill development, and using role-playing and simulations with debriefing sessions to develop self-awareness.

### 8.6.6 Reconceptualist or postmodern approach

The humanistic approach to curriculum is considerably extended by the postmodern, or reconceptualist approach. Postmodern curriculum theorists concentrate on the more significant ideological problems in education. They look into and have an impact on the political, social, and economic institutions of society. Theory is more important to postmodernists than real-world applications. The reconceptualist approach to curriculum development views education as a dynamic and developing process rather than a set of fixed objectives, emphasizing social justice, critical thinking, and reflection on individual and societal settings. It encourages educators and students to critically analyze the cultural values ingrained in the curriculum, challenge established power structures, and supports student-centered learning (Karki, 2020). Critical thinking is one of the key tenets of this method. According to reconceptualist, the curriculum should encourage students to challenge presumptions and examine data from different angles. Furthermore, social justice is one of the vital components of this approach since this strategy aims to create a fairer learning environment by highlighting issues of power, inequality, and educational access. Moreover, conceptualists stress the value of reflecting on educational experiences and relating them to one's own life and larger societal contexts.

The problem-posing education (PPE) is one method of demonstrating the reconceptualist approach in clinical practice. PPE asks students to critically analyze the sociopolitical factors that lead to health inequities, whereas traditional PBL approaches patient cases as biomedical puzzles to be solved. When talking about a diabetic patient who has trouble adhering to treatment, for example, a PPE approach would encourage students to look into systemic problems such as the patient's housing situation, access to healthcare, or food security instead of just concentrating on the biological consequences of inadequate glycemic control. This encourages aspiring doctors to perceive themselves as advocates who can address the systemic causes of illness. In addition to the above mentioned approaches, the following are the two most popular approaches followed in medical education.

## 8.7 Backward design approach

Backward design in medical education is a three-stage process that begins with the desired learning outcomes, such as specific skills and competencies, and works backward to plan the curriculum and assessments. This approach ensures that all teaching and assessment activities are directly aligned with the program's goals and the competencies future healthcare professionals need (Kailey Rinaldi, 2025). The three stages of the backward design approach are described below (Fig. 8.3).

### 8.7.1 Identifying desired results

It starts by defining the specific competencies and learning outcomes students should achieve by the end of the program or course. Identifying desired results in backward design is the first stage and focuses on determining what students should know, understand, and be able to do by the end of a unit or course. This is achieved by starting with the end in mind, considering standards and "big ideas" to establish enduring understandings and essential questions that guide the rest of the design process. In medical education, this can involve using competency-based frameworks like the ACGME Milestones to define the expected level of performance for each competency.

### 8.7.2 Determining acceptable evidence

This includes designing assessments that will provide clear evidence of whether students have met the desired results. These assessments should be designed to measure the specific skills and knowledge identified in the first step. This means defining what constitutes a successful demonstration of understanding, knowledge, and skills by considering a range of evidence, such as performance tasks, projects, and traditional quizzes, and using criteria or rubrics to evaluate student work based on the desired results. Examples include objective structured clinical examinations, patient simulations, or portfolio reviews.

### 8.7.3 Planning learning experiences and instruction

This phase involves developing learning activities, content, and teaching methods that will guide students toward achieving the desired outcomes and passing the assessments. This step involves selecting or creating resources and

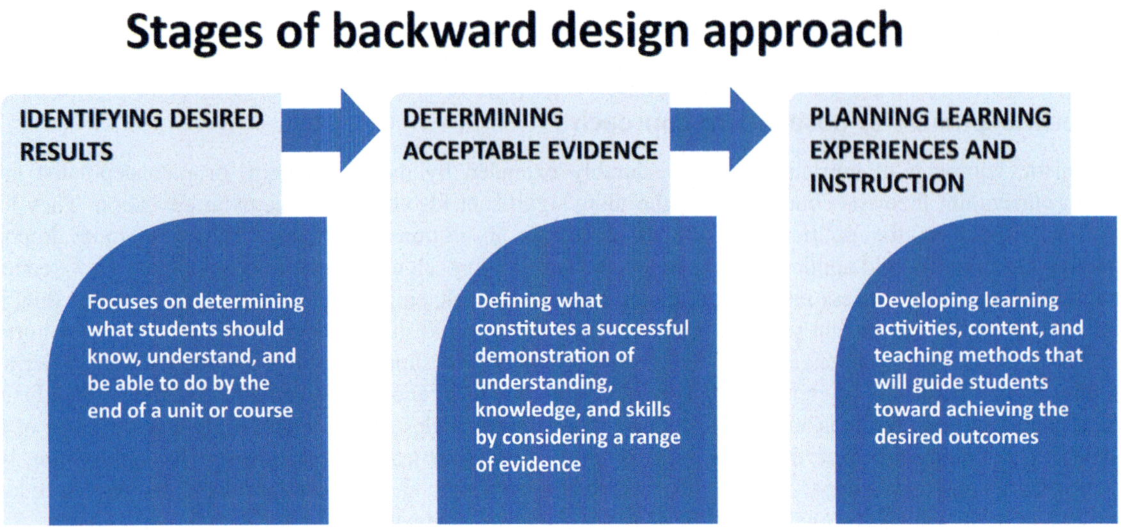


FIGURE 8.3 Stages of backward design approach.
This figure shows the stages of backward design approach.

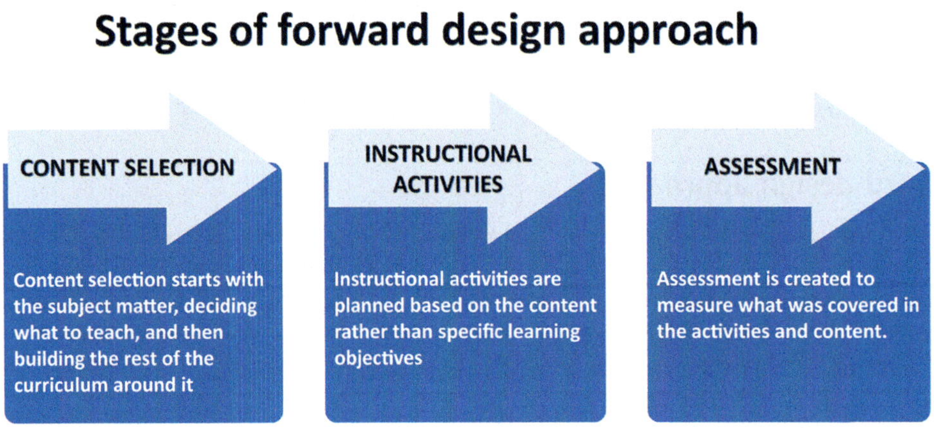


FIGURE 8.4 Stages of forward design approach.
This figure shows the stages of forward design approach.

instructional strategies that will best help students master the required knowledge and skills. Unlike traditional models, the "backward design" approach prioritizes aligning activities with the goals and assessments, rather than simply "covering content."

## 8.8 Forward design approach

The forward design in medical education is the traditional, content-focused approach to curriculum development, where the process begins with determining the content and moves chronologically to instructional activities and, finally, assessment. The traditional forward design model follows a linear, "waterfall" approach with a primary focus on the teaching process rather than the learning outcomes. The processes involved in the forward design approach are as follows (Fig. 8.4).

### 8.8.1 Content selection

This involves deciding the subject matter and putting together a syllabus based on the educator's knowledge. In forward design, content selection involves starting with the subject matter, deciding what to teach, and then building the rest of the curriculum around it. This is the first step, where the curriculum developer determines the specific knowledge, skills, and concepts to include in a syllabus, which then dictates the methodology and assessment methods in a sequential, linear fashion. This approach, often called the "waterfall model," prioritizes content over outcomes and is common in situations with standardized materials or general learning objectives.

### 8.8.2 Instructional activities

This step involves choosing teaching methods like didactic lectures to impart the selected content. In forward design, instructional activities are planned based on the content rather than specific learning objectives. This approach begins by identifying the subject matter, then developing lessons, activities, and materials, and finally creating assessments to measure what was taught. For example, a teacher might first decide to teach communication skills to the student, then create role-playing activities and handouts, and finally develop assessments like a simulation or role-play to gauge the trainees' performance.

### 8.8.3 Assessment

This refers to the designing of assessments at the end to measure student performance relative to the content taught, often based on normalized performance. In forward design, assessment is the last step after content and methodology are planned. The process begins with the syllabus and teaching activities, and the assessment is then created to measure what was covered in the activities and content.

This approach is often compared to the "waterfall model," where each stage depends on the completion of the previous one. One of the prominent limitations of this approach often results in passive learning and may not guarantee that learning objectives are met, as the focus is on information delivery rather than skill acquisition and competency development.

## 8.9 Selecting the right approach

How we choose the best curricular approach is one of the most important components of curriculum. This is crucial because medical education necessitates a thorough approach that takes into account program objectives, staff and student needs, available resources, and the local healthcare context for choosing the best curriculum approach in medical education. In order to generate graduates who are capable and prepared for the workforce, it is crucial that educational goals and instructional techniques be in harmony.

## 8.10 Teaching strategies

Teaching strategies are the general methods, approaches, and tactics that teachers employ to support learning, engage students, and assist them in comprehending and retaining information. These teaching strategies can be adjusted according to various subjects, student needs, and learning objectives in order to establish productive and encouraging learning environments. They are the adaptable lesson plans that direct training, from lectures for the entire class to practical projects, encouraging in-depth understanding and skill application. The different teaching strategies used in medical education are described below (Fig. 8.5).

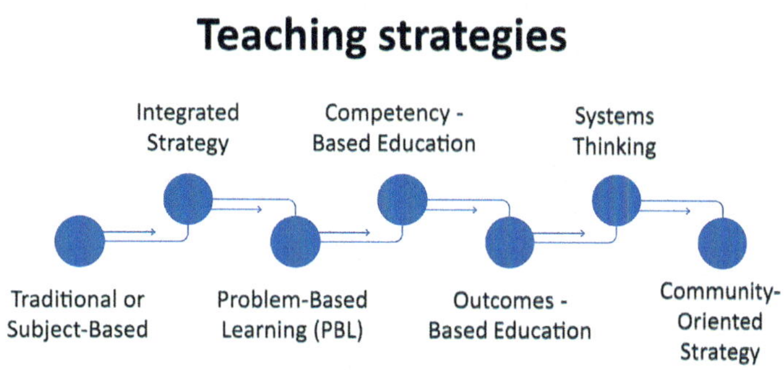


FIGURE 8.5 Teaching strategies.
This figure shows the various teaching strategies.

### 8.10.1 Traditional/subject-based

A compartmentalized approach, frequently with distinct courses for basic and clinical sciences, and an emphasis on fact memorization are characteristics of the subject-based method. In medical education, subject-based curricula are conventional, segmented methods in which basic sciences such as anatomy, physiology, and pathology are taught in a particular order. Regarding the framework of the student-centered curriculum, each subject is taught separately and is organized around conventional academic fields. Clear learning objectives, a reasonable structure, and the ability for teachers to teach in their areas of expertise are some of the subject-based approach's key advantages. Additionally, it makes it possible to comprehend each subject matter thoroughly. One of the main drawbacks of this subject-centered strategy is that it results in a fragmented body of knowledge due to poor subject integration. It may be less successful in making the connection between clinical application and fundamental scientific theory. Lectures, for example, are an excellent illustration of this tactic since they are a type of instruction in which the teacher serves as the main source of knowledge and provides students with information in an organized manner. Although contemporary approaches frequently include active learning practices to boost engagement, this one-way flow of information is effective for providing a lot of stuff to many students at once.

### 8.10.2 Integrated strategy

In medical education, integration is a commonly utilized teaching approach that unifies several subjects and disciplines to increase the effectiveness and relevance of learning. By dismantling conventional boundaries between courses like anatomy and physiology, this method aids students in making connections between the fundamental sciences and clinical practice, improves their retention of information, and fosters a more comprehensive grasp of medical concepts (Atwa, Gouda, 2014). The integration technique is effective in medical education because it brings together similar subjects that are often taught separately, like the anatomy and physiology of the kidneys. However, by integrating several topic areas, it fosters critical thinking and encourages students to apply knowledge in a more comprehensive manner. Additionally, because students can observe how the fundamental sciences are directly used in clinical settings, learning becomes more engaging and promotes knowledge retention. In the end, it also makes it possible for students to acquire critical abilities, including analytical skills and a basis for successful communication and comprehensive patient care. In contrast to subject-based teaching, integrated curricula are frequently organized around systems or problems and combine disciplines to demonstrate their links. Although they can be difficult to put into reality, integrated approaches are thought to better prepare students for lifetime learning by linking theory and practice, whereas subject-based strategies offer an organized and thorough understanding of certain fields. A case-based learning (CBL) module on diabetes mellitus is an illustration of an integrated teaching strategy in medical education. It provides a comprehensive understanding of the disease by integrating the contributions of several departments, including biochemistry, pathology, pharmacology, medicine, and community medicine.

### 8.10.3 Problem-based learning (PBL)

A popular teaching method in medical education, PBL strategy puts students at the center of their education to foster critical thinking and problem-solving abilities. It mimics the difficulties of actual medical practice by using challenging, real-world clinical dilemmas to help students as they attempt to comprehend and solve them. Curriculum development is influenced by the PBL approach, which has become a significant component of many medical institutions curricula and has been adopted as the main educational philosophy by several institutions (Wood, 2003). Student-centered learning, which promotes active and collaborative learning by having students work on problems in small groups, is one of the key components of PBL. However, since this approach is intended to improve critical thinking, analysis, and problem-solving abilities, all of which are essential for clinical practice, it fosters skill development. Additionally, compared to traditional techniques, it offers a more cohesive and integrated learning experience by teaching medical students how to manage various variables and complicated clinical concerns. This is indeed a learner-centered method that encourages critical thinking and self-directed learning by having students solve real-world medical problems.

### 8.10.4 Competency-based education (CBE)

In contrast to traditional time-based learning, CBE, a popular and developing teaching approach in medical education, emphasizes a student's proven practical talents and skill mastery. It acts as a foundation for curriculum development, instruction, and evaluation, guaranteeing that aspiring doctors have the fundamental information and abilities needed for clinical practice (Frank, 2010). One of the main characteristics of the CBE in medical education is its emphasis on performance and skills. Rather than merely accruing credit hours, the CBE requires students to show that they have mastered particular competencies, such as the capacity to conduct a focused physical examination or manage an inpatient with a new acute problem. However, this technique provides for more specialized and adaptable training because students can go through the curriculum at any moment once they have mastered the necessary abilities. As a result, it is flexible and offers a chance for personalized learning. Additionally, there is a heavy emphasis on deep comprehension that is displayed by action, as learning objectives are established through application. The development of certain, quantifiable skills necessary for practicing medicine is heavily emphasized, with a focus on graduates' functional capacities.

### 8.10.5 Outcomes-based education

It is comparable to competency-based learning, but it places more emphasis on the overall learning objectives and how they relate to patient and societal demands. The outcomes-based education concentrates on the teaching process and emphasizes what students should be able to accomplish at the conclusion of a learning experience. This learner-centered, performance-based approach bases curricular content, instructional practices, and assessment techniques on the established learning outcomes. Because the outcome-based approach is learner-centered, with teachers serving as consultants and guides, it has proven beneficial (Gruppen, 2012). The curriculum is created backward from well-defined learning outcomes that specify what a graduate should be able to achieve because the focus is on the outputs. Additionally, it improves learning by permitting greater flexibility in the learning schedule and approaches, which can facilitate individualized learning. To improve student initiative and clinical competency, it is also combined with other tactics like the flipped classroom model. Furthermore, it offers space for ongoing development based on performance data and feedback.

### 8.10.6 Systems thinking

In medical education, systems thinking is a teaching approach that aids students in comprehending the connections and influences between various components of the healthcare system, such as patients, providers, and policies. By employing techniques like interprofessional education (IPE), CBL, and technological integration, it transcends a single-patient focus to a more comprehensive understanding of care. This method is utilized to enhance patient outcomes, problem-solving abilities, and the capacity to negotiate challenging healthcare settings (Norris Grohs, 2026). By bringing students from various healthcare professions together to learn and collaborate as a team, the system thinking approach has been beneficial in fostering IPE. By incorporating systems thinking ideas into the curriculum, sometimes through specialized courses, it also provides an opportunity for curricular integration, assisting students to see the broad picture and address the underlying causes or issues. It encourages experiential learning by providing students with practical experience with complicated systems using techniques including CBL, simulation games, and reflective diaries. It is essentially a comprehensive strategy that takes into account how different curriculum components are related to one another as well as the larger context of healthcare practice.

### 8.10.7 Community-oriented/community-based curriculum

Community-based education has emerged as a teaching approach in medical education with an aim to address real-world health issues and build practical skills. This method places a strong emphasis on working together with stakeholders and members of the community, offers real-world learning opportunities, and supports students' development of socio-behavioral abilities including cooperation and communication (Hays, 2007). One of this strategy's most notable aspects is that it allows for authentic learning, in which students interact with patients and families in their normal social and community settings. In addition, in this local healthcare, organizations and providers work together to address the unique health needs of the community. In order to develop graduates who can effectively serve communities, particularly marginalized ones, it also emphasizes social accountability. This has proven to be beneficial wherein the students comprehend healthcare inequities and fosters social responsibility. In addition, students develop critical abilities like leadership, teamwork, communication, and health literacy while gaining practical experience by undertaking this type of approach.

## 8.11 Relation between curricular approach and teaching strategies

Teaching strategies are the particular methods and techniques utilized to deliver the curriculum and accomplish its desired learning outcomes, whereas the curricular approach offers the general framework and direction for a medical education program. To enhance student learning, the two must be compatible because they are intrinsically related. The relationship between curricular approaches and teaching strategies is dynamic and complementary to each other. The curricular approach provides the overall framework or philosophy of teaching and learning, while teaching strategies are the specific methods and techniques used to implement that approach in the classroom. Teaching strategies operationalize the curricular approach. The curricular approach signifies the "what" and "why" of teaching (educational philosophy, goals, and content organization). On the other hand, the teaching strategies uphold the concept of the "how" of teaching (methods to help students learn effectively). The curricular approach determines which teaching strategies are most suitable. Conversely, effective teaching strategies ensure that the intended curriculum is successfully delivered. The techniques, protocols, and procedures teachers employ to present course material, engage students, and accomplish learning objectives are known as teaching strategies. These are the overarching plans that direct instruction, supporting a variety of learning requirements and fostering a connection between students and the subject matter. Among the crucial elements of this partnership are the following:

## 8.12 Curricular approach as the blueprint

The curricular approach serves as the "blueprint" that establishes the program's overall structure, content, and philosophy. It provides answers to the "what" and "why" of education. It also involves a systematic process of designing and organizing content into manageable units, courses, and programs. This includes identifying specific learning objectives, desired outcomes, and the scope and sequence of topics, providing a clear roadmap for all stakeholders.

## 8.13 Teaching strategies as the delivery method

The operational techniques and processes employed during instruction to carry out the curriculum are known as teaching strategies which take into consideration the details provided by the blueprint. They are chosen to fit the goals established by the curricular approach and represent the "how" of education. Lectures, PBL, team-based learning (TBL), CBL, small-group discussions, standardized patient encounters and simulation, bedside teaching, and role modeling are examples of teaching techniques.

## 8.14 The interplay and relationship

A student-centered, integrated curriculum, for instance, requires active learning strategies like PBL and TBL, where the teacher serves as a facilitator. This is an example of how the two are highly interconnected, with the curriculum dictating strategy. Conversely, a traditional, subject-based curriculum frequently places a greater emphasis on passive, lecture-based education. Besides, it makes curriculum delivery possible by emphasizing that putting the curriculum's educational concepts into practice requires the successful use of the selected teaching tactics. Additionally, it permits flexibility and development because modern medical education places a strong emphasis on the transition to outcome-based, competency-driven curricula that combine active learning techniques to develop critical thinking and practical problem-solving abilities. The curriculum ought to offer a structure that permits the appropriate use and development of different teaching methods. Curriculum and instruction must be compatible in order to promote this relationship more effectively, since teaching methods that are out of sync can compromise the objectives of the curriculum model. Essentially, the teaching tactics are the vehicles and driving techniques utilized to successfully reach the objective, while the curriculum approach provides the destination and map.

## 8.15 Importance of a curricular approach

A curricular approach is crucial as it offers a methodical, goal-oriented framework for learning that improves student achievement, engagement, and retention. It facilitates individualized learning experiences, directs education, and fosters critical thinking and problem-solving abilities. For educators and institutions, it fosters better coordination, enhances staff collaboration, and allows for more effective use of resources by creating clear outcomes and aligning daily actions with broader institutional goals. The curricular approaches prove to be beneficial to both the students and the teacher. For the students, the following are the benefits provided by having a good curricular approach

### 8.15.1 Improved academic and personal development

A well-designed curricular approach improves academic and personal development by providing a structured, relevant, and engaging learning framework that fosters essential skills and caters to individual needs. As a result, it offers an opportunity for better academic performance while also fostering the learner's personal growth. By engaging students in real-world challenges and inquiry-based learning, curricular approach foster critical thinking, reasoning, and decision-making skills that are vital for academic success and future careers. It also encourages skill building amongst the students as the holistic curriculum goes beyond academics to nurture essential life skills such as communication, collaboration, time management, and leadership through group work, discussions, and projects

### 8.15.2 Increased persistence and completion

Students' perseverance, retention, and completion rates are known to increase with a well-defined, organized curriculum. Curricular methods boost student persistence and completion rates since they help in creating clear, pertinent, and captivating educational pathways that actively assist students' academic and social integration. Increased perseverance and competition support transparent, well-organized routes where students can effectively traverse their academic journeys through well-defined plans and course sequences, minimizing confusion and needless credit accumulation. Students are more driven to persevere when they see how their classes relate to their long-term professional objectives. By going beyond typical lectures to include project-based, problem-based, and experiential learning, it also promotes active and applied learning. This makes the content more interesting and applicable to real-world scenarios. This increases comprehension, enhances retention of information, and maintains students' interest in their studies.

### 8.15.3 Personalized learning

It enables teachers to modify their methods to fit a variety of student demands and learning styles. Instead of employing a one-size-fits-all approach, personalized learning within a curricular approach entail adjusting curriculum, speed, and instruction to each student's unique requirements, talents, and interests. In order to foster ownership and involvement, this entails students actively participating in goal-setting and progress monitoring. In order to assist each student in learning the same standards in their own unique way, teachers use data to give differentiated support, either in small groups or one-on-one sessions. In this, the emphasis is moved from a teacher-led lecture to a student-centered setting where each student has a customized learning plan based on their own starting position, skills, and preferred learning style, and personalized learning supports student-centered learning. Additionally, it guarantees that students are actively participating, wherein students take an active role in their own education by establishing objectives, organizing their courses, and deciding how to show what they have learned.

### 8.15.4 Greater sense of belonging

The curricular approach facilitates a greater chance of student retention as the students feel more included and accepted. A curricular approach increases students' sense of belonging by making learning more relevant and validating their identities. In addition, as it incorporates students' varied experiences and backgrounds into the curriculum and instructional strategies, students are more likely to feel appreciated, respected, and a part of the medical institutions community. This also makes them content, as they see their cultures, histories, and viewpoints represented in what they learn. A more inclusive curriculum with diverse examples from a range of cultural, ethnic, linguistic, and socioeconomic origins results from a stronger sense of belonging. This prevents feelings of alienation and fosters a healthy self-concept by guaranteeing that all students see themselves and their experiences represented.

### 8.15.5 Clearer understanding

Learning is made easier by breaking down difficult concepts into smaller, more accessible chunks. A curricular approach's main benefit is that, by creating an organized and purposeful learning route, it gives both students and teachers a clearer, more methodical understanding. Students really benefit from this clarity since it establishes clear expectations, allowing them to understand precisely what they are supposed to learn and the desired results. This eliminates uncertainty, lowers tension, and enables people to efficiently concentrate their efforts. Additionally, a well-designed curriculum ensures that core concepts are learned before moving on to more difficult ideas, which supports the study's logical progression. Over time, a deeper, more comprehensive understanding of the subject is made possible by this methodical accumulation of knowledge, which also minimizes comprehension gaps.

### 8.15.6 Enhanced effectiveness

It improves overall efficacy and aids in schedule planning by offering a clear framework for instruction and learning. A curricular approach increases effectiveness since it offers an organized, purposeful, and coherent framework for learning that leads to higher student outcomes, increased engagement, better use of resources, and a more proactive approach to student support. A well-planned curriculum clearly outlines what students should know and be able to do, ensuring that all instructional strategies and assessments are in line with precise, quantifiable learning outcomes. This is one of the key factors that contributes to increased effectiveness. This makes learning more concentrated and effective by giving teachers and students a clear roadmap. Additionally, it guarantees that the subject is taught in an organized, sequential manner that builds on previously learned ideas and abilities. This continuity enables a deeper and more thorough comprehension of difficult subjects and helps close learning gaps, which raises overall student accomplishment.

### 8.15.7 Better collaboration

Better collaboration between instructors and students is naturally fostered by a curricular approach, especially one that promotes cross-curricular or integrated learning and incorporates collaborative curriculum development. There are many benefits to this collaboration, including improved curriculum development as teachers from various subject areas or disciplines collaborate to create a more coherent, thorough, and well-aligned curriculum that avoids knowledge gaps and reinforces important concepts across disciplines. By giving educators an organized forum to share materials, creative teaching techniques, and successful teaching strategies, it also enables them to share best practices and knowledge. Students benefit from a richer and more varied learning experience as a result of this pooling of collective intelligence.

### 8.15.8 More proactive approach

In order to lessen the need for reactive actions and possibly save time and resources, staff members can be more driven in attending to the requirements of students. Furthermore, a curricular approach encourages a more enterprising approach to education since it focuses on prevention, structure, and skill development rather than just responding to issues after they arise. This planning aids in anticipating any behavioral problems or learning gaps and resolves them before they worsen. On the other hand, it reduces interruptions and off-task behavior by creating clear expectations and consistent routines that make the learning environment more predictable and safer for the learner. As a result, there is more time for instruction and a more seamless, concentrated learning environment.

### 8.15.9 Alignment with goals

The establishment of a cohesive, integrated, and efficient educational experience where all elements like learning objectives, content, instruction, and assessment cooperate to accomplish clearly stated results is the main benefit of the curricular approach with relation to goal alignment. Students receive a clear grasp of what is expected of them and why their learning is important by enforcing the alignment with goals, which establishes a clarity in learning objectives. Additionally, it promotes a coherent and consistent learning route, which reduces learning gaps and guarantees that students acquire the requisite knowledge and skills which have been demonstrated to promote academic performance and achievement.

### 8.15.10 Effective resource allocation

Effective resource allocation is one of the benefits of curricular approaches since it is believed that if the curricular approach is well organized and meaningful, resources including staff time can be used more efficiently. In addition since it offers a planned, data-driven framework that matches resources with specific educational goals and student requirements, a curricular approach improves efficient resource allocation while reducing waste. This is accomplished through a strategic alignment in which the curriculum begins with well-defined learning objectives and results. Owing to this type of clarity, educational leaders may strategically allocate resources such as material, human, and financial to the areas that most directly support these goals, ensuring that resources are employed where they have the greatest influence on student learning. Facilitating the regular assessment and review of the curriculum, which produce statistics on student performance and resource efficacy, is one of the ways by which curricular approach promotes data-driven decision-making. Instead of depending on presumptions or conventional methods, this evidence-based approach allows administrators to make well-informed decisions, giving funding and support priority for programs that are performing well and modifying allocations for those that are not.

**Summary**

The curricular approach is a deliberate, learning-centered strategy used in educational settings, frequently in student affairs and higher education, to create experiences that support students' growth and learning. The curricular method, which is based on research, theory, and institutional mission, structures learning around well-defined outcomes, in contrast to traditional programming models that depend on discrete events or activities. Assessment-driven practice ensures that activities are regularly assessed and improved based on proof of student learning, and learning experiences build upon one another. Its performance is mostly dependent on departmental cooperation, regular training for teachers, and integration with students' academic and extracurricular surroundings. In general, the curricular approach changes the emphasis from providing programs to encouraging significant, quantifiable learning, giving students a comprehensive and well-rounded educational experience.

**Key point**

**Future perspectives: from scientific approach to self-actualization**

The curriculum's development and execution, as well as its future, are very challenging jobs. This is because students, teachers, and other stakeholders have ever-changing needs and expectations. To handle these constantly evolving scenarios, it is imperative to consider a multitude of factors. The curriculum must be integrated across time and disciplines and should be outcome-driven. Additionally, it ought to be competency- and patient-centered. Evidence should support this strategy. Additionally, the curriculum method should be regularly evaluated and improved. To improve 21st-century curriculum ideas, educators should place a strong emphasis on critical thinking, creativity, and teamwork. In order to monitor student progress, spot gaps, and customize learning experiences, the curriculum approach of the future will mostly rely on sophisticated data systems and learning analytics. Teachers will be able to improve learning outcomes and interventions in real time with the help of adaptive platforms and AI tools.

## References

Ahn, A.C., Tewari, M., Poon, C.S., Phillips, R.S., Ahn, A.C., Tewari, M., et al., 2006. The clinical applications of system approach. PLoS Medicine 3 (7), e209.

Atwa, H.S., Gouda, E.M., 2014. Curriculum integration in medical education: a theoretical review. Intellectual Property Rights: Open Access 2 (2). Available at https://doi.org/10.4172/2375-4516.1000113.

Benatar, S., 2014. The humanistic side of medical education. South African Medical Journal 105 (1), 3. Available at https://doi.org/10.7196/SAMJ.9043.

Bhat, Krishna, 2025. IP International Journal of Medical Paediatrics and Oncology 11.

Biruny, D.H., Salsabila, I., 2021. Application of a humanistic approach to health care and education. AMCA Journal of Community Development 1 (1), 9–12. Available at https://doi.org/10.51773/ajcd.v1i1.22.

Chhatria, S., Pani, B., Digal, A., Sangeeta, M., Chhatria, S., Pani, B., et al., 2024. Principles and Approaches to Curriculum Development 24, 51–64.

Doyle, Claire, 2025. Advances in Simulation 10.

Frank, J., 2010. Medical teacher 32.

Gamulin, S., 2004. Academic approach to academic medicine. Croatian Medical Journal 45 (3), 245–247.

Gruppen, L., 2012. Korean Journal of Medical Education 24.

Hays, R., 2007. Teaching and Teacher Education 23.

Isaac, M., Rief, W., 2009. Role of behavioural and social sciences in medical education. Current Opinion in Psychiatry 22 (2), 184–187. Available at https://doi.org/10.1097/yco.0b013e328327b3b6.

Karki, B.K., 2022. Postmodern perspective towards human health and illness. KMC Research Journal 4 (4), 29–36. Available at https://doi.org/10.3126/kmcrj.v4i4.46476.

Kailey, Chan, 2025. Journal of Professional Nursing 58.

Mohammed, G., 2013. Curriculum: approaches and theories. Iraqi Association for Education and Psychology Journal 96.

Norris, M.B., Grohs, J.R., 2026. BMC Medical Education 26.

Remesh, A., 2017. Curriculum design principles for developing a module in medical education. Progress in Medical Sciences 1 (1), 34. Available at https://doi.org/10.5455/pms.20171223064024.

Wood, D.F., 2003. BMJ 326.

Yamani, N., Rahimi, M., 2016. The core curriculum and integration in medical education. Research and Development in Medical Education 5 (2), 50–54. Available at https://doi.org/10.15171/rdme.2016.011.

Chapter 9

# Hidden curriculum

*"Being a role model is the most powerful form of educating."*

—John Wooden

## 9.1 Introduction

Introduced in the 1960s and applied to medical education in 1994, hidden curriculum encompasses the unspoken norms, values, and behaviors that students absorb, often without realizing it. The term hidden curriculum was first defined by Frederic Hafferty as the "set of influences that function at the level of organizational structure and culture." He further extended the meaning of the hidden curriculum grossly as the "understandings, customs, rituals, and taken-for granted aspects of what goes on in the life-space we call medical education" and viewed the hidden curriculum in medical education as an institutional-level concept best visible in sectors like policy development, evaluation, resource allocation, and institutional slang (Hafferty FW 1998). It refers to the implicit lessons and values that students absorb during their training, beyond the formal curriculum taught in lectures and textbooks. These lessons, often conveyed through observation, interactions, and institutional practices, shape a student's professional identity, behaviors, and attitudes. While not explicitly taught, the hidden curriculum can significantly influence how future physicians approach their work, impacting their empathy, professionalism, and overall approach to patient care.

While the formal curriculum consists of the courses, lessons, and learning activities that the students participate in, and the knowledge and skills educators intentionally teach to students, the hidden curriculum also represents a set of influences that function at the level of the organizational structure and culture that affect learning. It refers to the unwritten, unofficial, and often unintended lessons, values, and perspectives that are communicated to the learners. The hidden curriculum in medical education is a powerful but often unnoticed force that shapes students' professional identities and ethical behavior (Nyquist JG.2016). The hidden curriculum is more than simple transmission of knowledge and skills. It is also considered as a socialization process. Wittingly or unwittingly, norms and values transmitted to future physicians often undermine the formal messages of the declared curriculum. The hidden curriculum consists of what is implicitly taught by example on a day-to-day basis, not the explicit teaching of lectures, grand rounds, and seminars. It has to be noted that the hidden curriculum is not limited to choice of specialty. It also undermines the student as caring and ethical professionals. In the realm of hidden curriculum, collegiality, patient-centered care, and ethical practice are often subordinated for factual knowledge or are brushed aside by practical realities (Das, Suprava 2024).

## 9.2 Features of a hidden curriculum

The important features of hidden curriculum are socialization, unspoken rules and values, role modeling, institutional culture, and impact on empathy (Fig. 9.1).

**The Principles and Practice of Medical Education. DOI:** https://doi.org/10.1016/B978-0-443-44197-4.00034-5

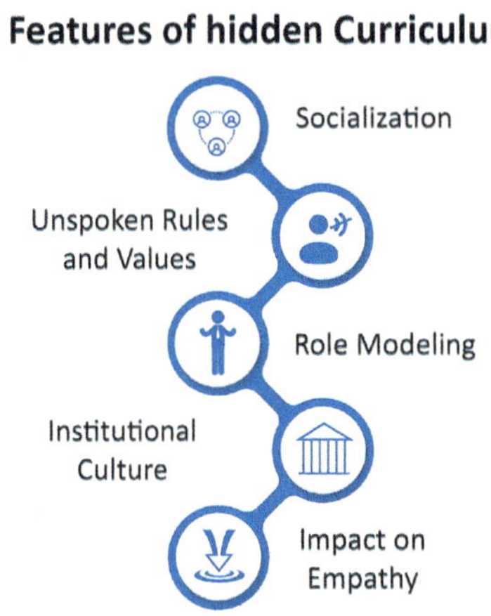


FIGURE 9.1 Features of hidden Curriculum.

## 9.2.1 Socialization

The hidden curriculum is a crucial part of the socialization process, helping students transition into the medical profession and understand its norms and values. Socialization is a core, often unintentional, feature of the hidden curriculum in medical institutions. It refers to the process by which students absorb implicit lessons, values, attitudes, and social norms from the institution environment, teacher behaviors, and peer interactions, rather than from the formal, explicit academic curriculum. This process prepares students to function in society and the workplace. Key aspects of socialization reflected via the hidden curriculum would include the transmission of social norms and values as medical institutions serve as primary agents of secondary socialization, where students internalize societal expectations without formal acknowledgment. The important things addressed are issues like punctuality and time management as rigid class schedules and deadlines implicitly teach the value of being on time, a crucial skill for the modern workforce. The competition and cooperation are other aspects which emphasize individual achievement and competition, while group projects teach collaboration and teamwork skills. In terms of classroom behavior the conformity of dress codes, behavior expectations, for example, asking permission before speaking and other unwritten rules encourage conformity to established standards.

## 9.2.2 Unspoken rules and values

It includes the unwritten rules, expectations, and beliefs that govern behavior in clinical settings. Unspoken rules and values are a core, defining feature of the hidden curriculum, encompassing the implicit lessons, norms, and expectations that students learn in an educational environment outside of the formal, explicit curriculum. These informal teachings significantly influence student behavior, attitudes, and socialization, often mirroring and reinforcing broader societal structures (Matorevhu, Alois 2022). Some of the key aspects of the unspoken rules and values are that these are implicit and unwritten as these rules are rarely explicitly stated or written down in any curriculum guide or mission statement. They are absorbed by students subconsciously through daily observation and interaction. In addition, the learning through experience enables the students learn these rules by participating in college activities, observing how teachers and peers behave, and experiencing the consequences of their own actions.

## 9.2.3 Role modeling

Students learn from observing how experienced physicians interact with patients, colleagues, and the healthcare system. These role models, both positive and negative, significantly impact student learning. Role modeling is a primary mechanism through which the hidden curriculum is transmitted to students. Role modeling as a feature of the hidden curriculum is critical in enabling the learners to learn by observation as students, particularly in fields like medicine, constantly observe the actions, interactions, and professional conduct of their teachers, senior doctors, and peers. This observation, which can be both conscious and unconscious, leads to imitation and the internalization of values and norms that shape their own professional identity and behavior. Furthermore, there is implicit transmission of values, as while the formal curriculum may explicitly teach professional ethics, the hidden curriculum reveals what truly matters in daily practice through the behaviors that are modeled. For example, if a teacher demonstrates empathy in a classroom discussion but is dismissive of patients in a clinical setting, students receive a hidden message that efficiency or a lack of emotional engagement is acceptable in practice.

### 9.2.4 Institutional culture

The hidden curriculum reflects the overall culture of the medical institution, including its priorities, values, and power structures. Institutional culture is a primary source and a key feature of the hidden curriculum. The institutional culture in a hidden curriculum is important as it upholds the unspoken rules and expectations which include the daily operations and informal expectations of an institution, whereby it teaches the students lessons beyond the formal subjects. The important issues addressed and taught to the students are respect for authority in which the students are expected to interact with faculty and senior staff conveys messages about hierarchy and power dynamics (Fortoul van der 2013). In addition, traits like punctuality and discipline are highly regarded and emphasis is given to being on time and adhering to dress codes and schedules which ensure that the values about conformity and responsibility that prepare students for the workforce are instilled amongst the learner. It is impertinent that with reference to the institutional culture, a healthy balance between competition versus collaboration is maintained. For instance, an institutional culture that heavily prioritizes individual awards and high grades may foster competition, whereas an environment that emphasizes teamwork and shared resources encourages collaboration.

### 9.2.5 Impact on empathy

The hidden curriculum encompasses the underlying values, norms, and behaviors conveyed indirectly through educational environments, significantly shaping medical students' professional adaptability. It should be noted that the hidden curriculum has both significant positive and negative influences on students. The positive aspects of the hidden curriculum generally center on socialization and the development of essential life and professional skills not explicitly taught in the formal curriculum. Owing to this, the social skills and interaction are highly prioritized, wherein the students learn how to interact with peers and authority figures, fostering cooperation, communication, and social awareness. Additionally, it promotes personal growth by supporting the development of qualities like responsibility, accountability, self-control, patience, perseverance, and a strong work ethic (Martimianakis M 2015). Conversely, it has been noted that the hidden curriculum can negatively impact medical students' empathy levels, potentially leading to cynicism and emotional distance from patients. The negative impacts often involve the reinforcement of societal inequalities and the transmission of undesirable values and behaviors. This negative impact of hidden curriculum is pronounced particularly in stressful environments, wherein values like efficiency override compassion through role model and workplace pressures. One of the important things that students often learn is by observing behaviors of their role models and if they see role models who are unemphatic or cynical, they may adopt those same attitudes. The transition to unempathetic behavior leads to stress and burnout amongst the students, as high-stress environments, common in many educational settings like medical institution, are linked to increased burnout and decreased empathy, while a positive interventions like early patient exposure and explicit empathy training can create an empathic hidden curriculum to counteract these effects.

## 9.3 Need for hidden curriculum

It is a well-known fact that modern medical education is not based on only knowledge and skills of the learner. In addition to these important competencies, it has become necessary for a medical student to understand and exhibit those skills which will not only enable them to connect with their peers but also with their patients. The adaptation of these skills like socializing or maintaining their integrity play a colossal role in guaranteeing that quality and holistic healthcare is delivered to every patient (Lempp, 2004). The hidden curriculum is needed in modern medical education because it is an unavoidable force that shapes professional identity, medical professionalism, and humanism in ways that traditional learning cannot (Sarikhani, Y 2020). It is essential for socializing students into medicine by teaching unspoken values, norms, and behaviors through observation and experience, which helps them develop professional identity, ethical conduct, and critical skills necessary for navigating the complexities of clinical practice. Some of the important reasons as to why we need hidden curriculum are as follows:

### 9.3.1 Professional identity formation

It is a powerful tool for socialization, helping trainees understand what it means to be a doctor by observing role models' behavior, professionalism, and values in real-world clinical settings. Professional identity formation is a need of the hidden curriculum because the hidden curriculum transmits implicit professional values, ethics, and norms through social interactions, role modeling, and institutional culture, which are critical for shaping a professional's identity. This informal learning process influences professional development both positively and negatively by teaching students how to act, what to value, and how to navigate professional situations outside of formal instruction. Managing the hidden curriculum is essential for educational institutions to ensure a more positive and holistic professional identity for students. The hidden curriculum shapes the professional identity by upholding the aspect of role modeling where the students learn by observing the behaviors of instructors, senior professionals, and peers. This includes both positive role models and negative examples, which can lead to emotional dissonance or stress. In addition to that, it also ensures that the organizational culture which is characterized by the unspoken rules, hierarchies, and the overall culture of an institution convey powerful messages about how a professional should behave and what is valued.

### 9.3.2 Transmission of unspoken norms

Medical education inherently involves subtle messages about expected behaviors and attitudes that are not taught in classrooms but are absorbed through observation and interaction with peers, faculty, and patients. The transmission of unspoken norms is a core function and an inevitable part of the hidden curriculum in education. This process occurs through implicit learning and serves several key purposes, including socialization into society's expectations, preparing students for the workforce, and reinforcing existing social structures. The unspoken norms are not explicitly taught in formal lessons but rather are absorbed by students through the overall institution environment, daily interactions, and institutional culture. These aspects are heavily reflected through teacher behaviors, wherein the way in which educators showcase themselves with respect to their communication styles, body language, and their interaction with the learners, especially while praising or giving feedback, subtly convey messages about appropriate conduct and behavior. By laying emphasis on punctuality, dress codes, and adherence to schedules which endorse the rules and policies of the institution, it teaches students about conformity and respect for rules, preparing them for the demands of the adult working world.

### 9.3.3 Real-world application of ethics

It bridges the gap between theoretical knowledge of ethical principles and their practical application in complex clinical situations, shaping how students handle moral dilemmas. The real-world application of ethics through the hidden curriculum primarily occurs via role modeling, organizational culture, and informal social interactions. In many professional and educational settings, implicit lessons often have a more significant impact on ethical behavior than formal ethics classes. The medical field is a prime example where the hidden curriculum heavily influences professional ethics. For instance, the students observe the behavior of senior doctors and instructors who they consider as their role models. When a senior physician treats nurses or junior staff with respect, students learn the value of teamwork and dignity in the workplace. Conversely, witnessing blame-shifting or a lack of respect can teach students that such behaviors are not acceptable in the profession. Moreover, with respect to patient consent and rights, the unwritten rule might be to skip formal patient consent for nonsurgical procedures due to time constraints in a busy public hospital. This implicitly teaches students that efficiency can take precedence over patient autonomy, which is indeed a clear ethical violation.

### 9.3.4 Development of noncognitive skills

The hidden curriculum provides ongoing opportunities to develop crucial noncognitive skills like empathy, communication, and teamwork, which are essential for patient-centered care. The hidden curriculum develops noncognitive skills, such as cooperation, perseverance, and integrity, through the unspoken norms and social environment of an institution, which complements the formal curriculum. This is a necessary function as these skills are crucial for academic and life success, sometimes even more so than cognitive ability alone. Medical institutions foster this development through their structure, daily interactions, and opportunities for students to take on roles like leadership or setting group goals. Hidden curriculum develops noncognitive skills in teaching the students implicit values like politeness, honesty, and hard work through daily routines and interactions, not just through formal lessons. In addition, the institution's social environment is a powerful tool for developing skills like cooperation, tolerance, and responsibility. This ensures that the quality of socialization is displayed by the students at all times.

### 9.3.5 Socialization into the medical community

It helps students understand the complexities of the physician–patient relationship, the interprofessional team dynamics, and the broader community context in which they will practice. The hidden curriculum serves as the primary mechanism for socializing medical students into the medical community by implicitly conveying the unwritten rules, norms, values, and behaviors necessary to function as a professional. In this aspect, the hidden curriculum fulfills the need for socialization by helping in building the learner's professional identity. It plays a critical role in shaping a student's professional identity beyond the formal curriculum. This involves internalizing a sense of responsibility, duty, and selflessness, often through observing the actions of senior doctors and peers. The socialization into the medical community also provides an opportunity to students, whereby they learn the practical, unwritten rules of the medical world, such as navigating the professional hierarchy, understanding appropriate communication styles, and mastering the process of presenting a patient's story in case discussions.

## 9.4 Components of hidden curriculum

The key components of hidden curriculum in medical education deeply influences professional identity formation, ethics, and behavior and are classified into the following categories:

1. Structural factors
2. Organizational culture
3. Educational factors
4. Cultural factors
5. Social factors (Fig. 9.2).

### 9.4.1 Structural factors

Structural factors influencing the hidden curriculum include institutional characteristics like power hierarchies, policies of the medical institution and the universities, and the physical environment. These structures transmit implicit lessons about social norms, values, and expectations through aspects like organizational structure, teacher–student and student–student interactions, and the overall institution culture, which can perpetuate social inequalities or reinforce dominant ideologies. These factors have emerged as the major driving force of hidden curriculum in medical education. Factors like power dynamics clearly display that the hierarchies are reinforced by how decisions are made. For example, when a student council primarily plans a community service and has no real influence on policy, or when administrators lead meetings where student and teacher input is not the central to the decision-making process, it teaches students about the limits of

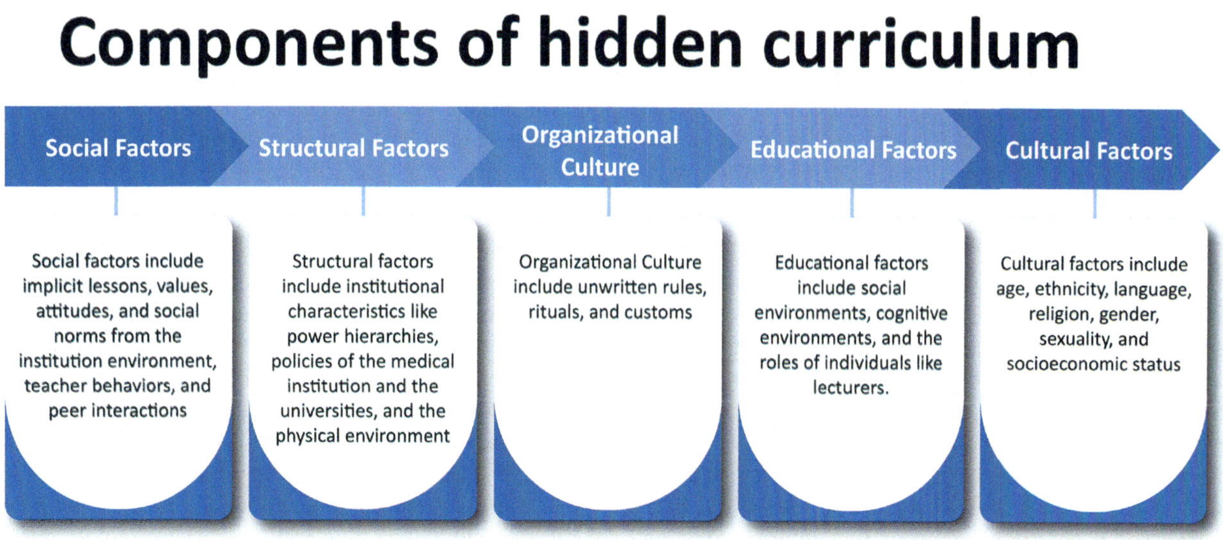


FIGURE 9.2 Components of hidden curriculum.

their power. It has to be noted that the policies and routines are a very vital component as the hidden curriculum is transmitted through routines like class schedules, disciplinary procedures, and grading systems, which are part of the formal structure but carry implicit messages. Other factors like the physical environment of the classroom showcasing how the students sitting in rows facing the teacher implicitly reflects that the teacher is the primary source of knowledge and authority, and that students should be passive learners.

From the social and cultural perspective, the upholding and transmitting of social norms and values by the medical institution may help cultivate an individualistic mindset, promote cooperation, and also foster community and support. From the perspective of interpersonal and pedagogical structure, the way in which the teacher and the student interact and differential treatment provided by teachers based on behavior or performance strongly influences the student's perception of social hierarchies and can reveal biases. Good peer interaction which is built up by involving the students in group projects and interactions with peers can teach about social development, teamwork, and how to navigate social roles and conflict. From the context of physical and environmental factors, the overall physical environment, including the layout of classrooms, clinics, or labs, can send implicit messages about hierarchy and how students should behave. Finally with respect to social and institutional factors which include the administrative and social structures, which encompasses the hierarchy and decision-making processes, are crucial components that teach students about institutional culture. It has to be kept in mind that enforcement on student-centeredness and empowerment is highly valued as the extent to which an institution is student-centered and provides opportunities for student empowerment, enabling the student to grow both personally and professionally. Moreover, the students also appreciate the qualities of accountability and equal opportunities wherein the students and faculty are held accountable and equal opportunities are provided to both of them. This indeed shapes the learning environment to a great extent.

### 9.4.2 Organizational culture

Organizational culture also tremendously influences the hidden curriculum where it has been observed that the learning of medical students is largely influenced by the dominant culture of educational organization. Unwritten rules, rituals, and customs are the most important aspects of an organization's culture. Moreover, it has great impact in professional development, clinical competency, ethical practice, and educational outcomes. These implicit lessons that are transmitted through elements like organizational structure, dominant cultural attitudes, rituals, and the way resources are allocated ultimately shape the professional identity and behavior of the learner. The hidden curriculum, fueled by the organization's culture, influences the students as much as, or sometimes more than, the formal curriculum. One of the ways that the organizational culture manifests in the hidden curriculum is by implicating rules and norms wherein the students learn unstated expectations for behavior, attitudes, and professional conduct that are not explicitly taught in the formal curriculum. On the other hand, by the way of culture web which is a model used to analyze the hidden curriculum by identifying its various components, such as symbols, rituals, power structures and routines, educators are in a position to recognize and understand the unspoken norms, values, and beliefs transmitted to students alongside the formal curriculum. This professional identity is also influenced by observing and navigating the organizational culture, wherein students develop their professional identity as they "fit in" with the community of practice they are joining. At the same time, it is vital to keep in mind that the cultural context of educational environments has inevitable positive or negative influences on the learning outcomes. It is essential to identify the undesirable aspects of organizational culture and modify them to deal with negative.

### 9.4.3 Educational factors

Educational factors including teaching methods and evaluation approaches are also identified as an independent component of hidden curriculum in medical education. Interactive teaching methods have the nature of inductive learning and can enhance future working competencies among medical students. Assessment of students based on problem-solving and decision-making skills far beyond just passing the exams is associated with more active learning and, as a result, more efficient clinical practice. Moreover, it seems that teaching and evaluation methods, especially in clinical settings, may have a role modeling function with high impacts on learning and practice. One important aspect of educational factors is that these components of hidden curriculum are associated with development of some key values of the medical professionalism such as excellence and leadership. Educational factors that are components of the hidden curriculum include social environments, cognitive environments, and the roles of individuals like lecturers. These factors operate beyond the formal curriculum, influencing students through implicit lessons on values, norms, and skills gained through interactions, teaching methods, and the broader institutional culture. Some of the main educational factors of the hidden curriculum that may have an impact on the learner is the social environment which include the social dynamics, relationships, and interactions among students and their teachers. On the other hand, the presence of a cognitive environment which encompasses the way knowledge is presented and assessed, teaching methods, evaluation processes, the types of questions asked, the way content is explained, and the emphasis on rote memorization versus critical thinking are all important contributors to this factor.

### 9.4.4 Cultural factors

The hidden curriculum, comprising unwritten social rules and cultural norms, is significantly influenced by cultural factors. These factors, such as age, ethnicity, language, religion, gender, sexuality, and socioeconomic status, shape how students perceive and interact within educational settings. The hidden curriculum can reinforce or challenge dominant cultural values, impacting students' learning experiences and social development. Cultural factors are key components of the hidden curriculum which are conveyed through the dominant culture of the educational environment, the behaviors of teachers and peers, and the implicit messages in curriculum materials. For instance, learning how to get along with peers, how to live up to the expectations of their teachers, and the cultural biases that may be presented in textbooks. The cultural factors operate within the hidden curriculum mainly focusing on the institutional culture and environment, as the overall atmosphere, organizational structure, and daily rituals of an institution are powerful carriers of cultural messages. Moreover, the presence of various interpersonal interactions showcasing the way in which the teachers and students interact, including their attitudes and behaviors, teaches the students about social conduct, appropriate professional conduct, and how to navigate social hierarchies.

Some of the important aspects that are influenced by the cultural factors are in terms of language and communication where there is a dominance of certain languages like English which can create a hidden curriculum that favors the cultural norms of native speakers. This can marginalize students from different linguistic and cultural backgrounds. The other aspects that have to be considered here is the social class and socioeconomic status, wherein the presence of students from different socioeconomic backgrounds may encounter varying expectations and opportunities within the hidden curriculum. Some may face challenges in navigating institution culture due to cultural differences or lack of resources. Similarly, the issues of gender and sexuality, where the hidden curriculum can perpetuate gender stereotypes or reinforce heteronormative expectations, can potentially affect students' self-esteem and social interactions with respect to the ethnicity and race. Here institutions may inadvertently favor certain cultural perspectives or values, leading to a hidden curriculum that may not be inclusive of all students.

The cultural factors of the hidden curriculum can lead to both positive and negative effects, such as teaching teamwork and resilience, but also potentially promoting stereotypes, academic pressure, and a disconnect between students' home cultures and the institution's dominant culture. Ultimately, it influences how students navigate authority, social hierarchies, and their own sense of self in relation to the wider world. The cultural factors of the hidden curriculum impact students by shaping their social norms, values, and identity through unspoken institution rules and interactions. Some of the major impacts of the hidden curriculum are in terms of socialization, wherein the hidden curriculum plays a crucial role in students' social development and identity formation. It shapes their understanding of social norms, appropriate behavior, and their place within the medical institution community. Furthermore, it also impacts the academic achievement as the cultural factors embedded in the hidden curriculum can affect students' engagement, motivation, and academic performance. It also enhances the professional identity of the learner which are of much significant in a profession like medicine. The hidden curriculum significantly influences how students develop their professional identities, including their understanding of roles, responsibilities, and ethical considerations.

### 9.4.5 Social factors

The implicit lessons transmitted through the social interactions, institution culture, and organizational structures significantly influence students' social skills, attitudes, and behaviors. Social factors play a crucial role in shaping the hidden curriculum, impacting how students learn and interact within the institution environment. These factors include the social dynamics between students and teachers, peer interactions, institutional policies, and cultural and economic influences, which shape attitudes, self-perceptions, and social skills. For example, social hierarchies and cultural biases can become embedded in the hidden curriculum, sometimes perpetuating inequality while other aspects can promote positive social development, such as fostering collaboration and respect. Social factors play a vital role in developing student's personality by inculcating in them a sense of social dynamics in a number of ways. One of the most important ways is the teacher–student interactions. The way teachers interact with different students can convey messages about authority and respect. For instance, a teacher's way of emphasizing on certain values like obedience or creativity or displaying professionalism and ethical behavior towards their colleagues and with their students can make the student to have a greater understanding on the behavior and alter their personality so that they can display the traits expected of them in front of their colleagues and patients. On the other hand, peer dynamics that majorly include the student-to-student interactions, including conversations, friendships, and social cliques, are a powerful part of the hidden curriculum. Students learn how to behave in a social context by observing and participating in these dynamics. Some of the major impacts of social factors on the hidden curriculum are seen in terms of the teachers' ability to manage conflict resolution, wherein the teachers' approaches to resolving conflicts between students or to handle any delicate issues amongst the students can teach students about problem-solving and conflict management. In addition, by displaying the qualities of collaboration and teamwork, teachers or the institutions can set an example of how these aspects can indeed lead to the betterment of the learner and help them to achieve their personal and professional goals. Medical institutions with a strong emphasis on teamwork and collaboration can foster these skills in students, while those with a more competitive environment might discourage collaboration.

## 9.5 Impact of the hidden curriculum on medical education

The hidden curriculum can have both positive and negative consequences. On one hand, it can foster positive social skills, promote personal development, and encourage a sense of belonging. However, on the other hand, it can lead to prejudice, social division, and perpetuate inequalities. Understanding the hidden curriculum is crucial for educators to create a positive and equitable learning environment for all students (Lee CA 2023). It is well known that the hidden curriculum significantly impacts the teaching and learning process by shaping students' understanding of unspoken rules, social norms, and values, which can affect academic performance, social skills, and self-concept. Students who can navigate these implicit expectations may gain an advantage, while those who struggle may face lower self-efficacy and poorer academic outcomes. Educators can mitigate negative impacts by making expectations explicit and fostering a more equitable environment.

### 9.5.1 Impact on students

The impact of the hidden curriculum on the part of the student is mainly in terms of academic achievement, wherein the success often depends on a student's ability to adapt to the hidden curriculum. Those who can decode unstated rules and teacher expectations may perform better academically. In addition, it is also concerned with the social and emotional development of the learner as the hidden curriculum teaches important social skills like politeness and responsibility. It can influence students' ability to collaborate, their social skills, and their tolerance for others. At the same time, one has to be careful as it can also promote negative traits like competitiveness and perfectionism. Moreover, it has an impact on the self in the sense that it upholds the implicit messages about worth and potential that can lead students to internalize positive or negative views of themselves, impacting their self-esteem and sense of identity. The hidden curriculum shapes students' understanding of future roles, including career paths and professional identities.

### 9.5.2 Impact on the teaching and learning process

The hidden curriculum emerges through institution culture, teacher attitudes, peer interactions, classroom organization, and institutional norms. The impact of the hidden curriculum on the teaching and learning process is in terms of socialization and value formation, wherein on the positive front, it can helps students learn social norms, respect, punctuality, cooperation, and responsibility skills essential for life beyond institution. On the negative front, it may reinforce conformity, discourage critical thinking, or promote passive acceptance of authority if not balanced with open discussion. With respect to the teacher student relationship, it can help teachers to communicate values and expectations through tone, attention, and fairness, thereby encouraging equitable treatment that fosters confidence and motivation. One must also be careful with regard to the teacher–student relationship, as any biases or unequal treatment can demotivate students and perpetuate stereotypes. In terms of reinforcement of social inequality, one must be aware that hidden curriculum can reproduce societal hierarchies. For example, medical institutions in affluent areas often promote leadership and independence, while those in lower-income areas may emphasize obedience and discipline. This shapes students' self-concepts and career aspirations differently, perpetuating inequality.

A strong influence on learning motivation and behavior is also felt in terms of classroom management and reward systems. The development of critical thinking and attitudes happens when the teachers model curiosity, openness, and questioning. Conversely, a rigid, authoritarian environment may discourage students from challenging ideas or expressing opinions. The hidden curriculum also exhibits the cultural and gender bias by reflecting only the dominant cultural or gender norms and ignoring minority contributions or reinforce stereotypes that can marginalize students. Inclusive practices, on the other hand, can empower underrepresented groups. The institution climate and emotional development is influenced by hidden curriculum, wherein the emotional tone of a classroom, for instance, whether it is supportive or stressful, deeply affects students' willingness to participate and take intellectual risks. A positive hidden curriculum nurtures self-esteem, while a negative one will lead to anxiety or alienation.

### 9.5.3 Role of teacher in addressing the hidden curriculum

The role of a teacher in combating the various facets of hidden curriculum is of prime importance. As teachers form an important bridge between the students and the administration and also the society, they are in the best position to intervene and correct a particular situation and at the same time uphold the thought and expectations of the students in front of decision makers and policy builders (Mosalanejad 2013). Some of the important roles of teachers in this regard are described below (Fig. 9.3).

### 9.5.4 Reflecting on personal biases and teaching practices

Encouraging teachers to become more aware of the hidden curriculum within their classrooms is the first step toward mitigating effects of personal bias. This involves self-reflection and making adjustments to promote fairness and inclusivity. To address the issues of self-reflection and bias teachers can keep a reflective journal to record their interactions with students, noting any patterns in their behavior that might indicate unconscious favoritism or bias. Regular bias training workshops can help teachers identify their implicit biases and understand how they influence classroom dynamics. On the other hand, reviewing classroom management strategies can help the teachers to review how discipline is maintained in the class and to see if certain groups of students are disproportionately disciplined and adjust their approach to ensure fairness. Adopting restorative justice practices can also help shift from punitive discipline to conflict resolution and relationship-building, promoting equity.

### 9.5.5 Promoting inclusivity by using diverse examples and perspectives

Creating an inclusive and equitable environment means actively challenging the hidden curriculum's tendency to perpetuate inequality. This requires changes at the classroom, institution, and policy levels. The incorporation of diverse curricula which incorporates authors, historical figures from different backgrounds, and perspectives in the curriculum helps all students see themselves represented and valued. For instance, while talking on the topic of history of medicine or latest advances in the field of medicine, it is always wise to have global perspectives and lesser-known narratives from marginalized communities. At the same time, importance should be given to ensure equitable access to resources. This can be done by first conducting an equity audit to assess disparities in access to advanced classes, extracurricular activities, and technology. Based on the findings, the institution can call for implementation of targeted interventions.

### 9.5.6 Encouraging critical discussion about societal norms and fairness

Teaching students to recognize and challenge the hidden curriculum empowers them to become active participants in their own education and encourages them to think critically about societal norms and power structures. One of the important things to be remembered is that you can separate the social issues from the realm of medicine. The overall success of implementing medical strategies especially in the field of public health is always connected to topics like power dynamics, privilege, and social norms. In this context, encouraging critical reflection and discussion is a vital step. Teachers can

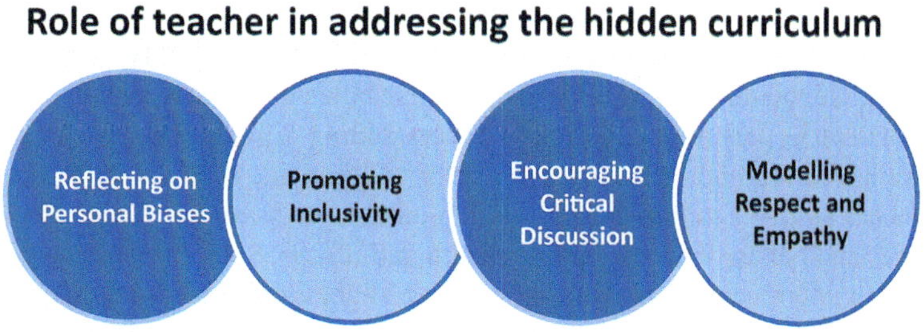


FIGURE 9.3 Role of teacher in addressing the hidden curriculum.

initiate classroom activities that encourage students to reflect on their experiences in institution. For example, students might be asked to write reflective essays on the unspoken rules they observe in their institution and how these rules impact their behavior. Discussions on topics like privilege, power dynamics, and social norms can help students recognize the hidden curriculum and develop a critical understanding of how it shapes their view. In addition, by facilitating classroom debates and projects on current social issues, it allows students to explore different perspectives and question societal norms, further helping them see how the hidden curriculum reflects broader social dynamics. Collaborative projects focused on social justice, like for instance researching about the social determinants of health and health inequalities, encourage students to think critically about equity and fairness in their environment.

### 9.5.7 Modeling the respect and empathy in daily interactions

The intentional choice of which values or virtues to highlight with students may shape the direction of character formation in students. This intentional choice encompasses the focus of the classroom culture. In an act to model the respect and empathy in daily interactions, the teachers must ask themselves as to who do they want the students to be and how do they want the students to understand virtues or values. This can be achieved within the classroom, through enforcing appropriate procedures and praising their students. The educator can be intentional about recognizing positive virtues, such as respect and empathy. Additionally, through discussions in lessons or literature, the teacher can intentionally emphasize values and how the individuals or characters demonstrate those. Students should partake in such discussions and share ideas to enhance their understanding of virtues, or even challenge the ideas of others. By doing all these activities, the teacher is able to uphold the democratic classroom practices and exhibit the skills of shared decision-making, open dialogue, and cooperation. It should be noted that the hidden curriculum powerfully shapes learners' attitudes, behaviors, and identities. While it can transmit valuable social skills, it can also perpetuate inequality or discourage critical thought. Teachers who are aware of it can intentionally shape the hidden curriculum to support equity, empowerment, and deeper learning.

## 9.6 Medical education and sociology

Before we move on with the different frameworks of hidden curriculum, it is vital for us to first be cognizant with the strong relationship that exist between the field of medical education and sociology. As we all are aware, medical education is not only about the teaching and learning of the knowledge, skills, and attitudes necessary to diagnose, treat, prevent illness, and promote health, but at the same time, it is also concerned with the needs and expectations of the society. The study of how social variables, including class, gender, culture, and socioeconomic status, affect health, illness, and healthcare systems is known as medical sociology, or sociology in medicine. It examines how social structures, disparities, and behaviors affect health outcomes, as well as how society handles illness through its healthcare institutions and practices. The social determinants of health, the doctor–patient interaction, the experience of sickness, and health policy are some of the crucial research topics (Jenkins 2021). Sociology examines the medical education as a social system, by analyzing the process of medical socialization and the impact of inequality within the profession.

### 9.6.1 How sociology influences medical education

The amalgamation of sociology and medical education has proven to be very beneficial in order to understand health disparities, whereby sociology highlights how social stratification on the basis of socioeconomic status, race, gender affects health outcomes. It also ensures that a good assessment of these disparities will enable medical education to address these prominent health inequalities. Similarly, it also contributes to improve patient care as it gives future doctors the ability to connect with a diverse patient population by understanding different social, cultural, and economic contexts, which can ameliorate communication and trust between doctor and patient. It also upholds the principle of sociological imagination, wherein it teaches students to look beyond individual and focus on the bigger picture and make an attempt to understand how social forces and behaviors contribute to health and illness. Sociology influences the hidden curriculum in medicine by framing it as an interpersonal–social process that shapes professional identity through interactions with faculty and patients, impacting values, behaviors, and professionalism. Furthermore, the elements of sociology enable us to study medical education as a social process, examining how students are socialized into their professional roles, the dynamics of power, inequality, and stress within the medical field.

## 9.6.2 Frameworks of hidden curriculum

As a teacher and the medical educationists, it is vital to figure out as to how the conceptual approaches with regard to the unwritten rules, values, and social lessons of an institution are conveyed. These frameworks primarily focus on the concepts of the institutional and organizational aspects, on function and also on active engagement by taking into account various aspects like the structure of the institution and policies that shape student learning, the interpersonal–social dimensions, examining teacher–student relationship, peer interactions and its contextual–cultural aspects, societal and cultural influences, and also the motivational–psychological aspects which look at the psychological impact on student motivation and behavior. Here are some ways to conceptualize and analyze the hidden curriculum.

### *9.6.2.1 Conceptual frameworks*

These frameworks view the hidden curriculum through the lens of different academic disciplines. The conceptual framework outlines the interplay of different elements on the hidden curriculum, thereby demonstrating the relational and dynamic nature of the hidden curriculum (Lawrence C 2018). A conceptual framework of the hidden curriculum often developed to better identify, understand, and manage the hidden curriculum's various components, such as institutional values, lecturer influence, and the impact of the world of work on student development. The framework is the analytical tool, while the hidden curriculum is the phenomenon being studied. Here the focus is on the following aspects.

#### 9.6.2.1.1 Institutional–organizational

In a conceptual framework, an "institution" refers to the established systems, rules, and norms that shape behavior, while an "organization" is a structured entity operating within those institutional contexts. "Institutional–organizational" is a widely recognized and fundamental component in conceptual frameworks for understanding the hidden curriculum. It refers to the implicit rules, structures, and cultural norms embedded within the organization that influence student learning and behavior outside of the formal, explicit curriculum. It also encompasses the structural factors such as the official rules, regulations, administrative methods, and the physical setting of the institution. These elements convey implicit messages about what the institution values. One of the prime elements in the realm of institutional and organizational concept is the institutional culture, including shared values and beliefs. These are being transmitted through daily practices, power dynamics, and social interactions among staff and students. For the student, the concept of institutional organization significantly affects learning by shaping the teaching-learning environment, social integration, resource allocation, and overall clarity of purpose. It provides the much-needed clarity and predictability by defining the roles, responsibilities, and decision-making processes, which translates into a more organized and predictable learning environment for students. Students in such environments are more likely to feel comfortable, engaged, and able to focus on academic achievement because they understand expectations and procedures. Furthermore, it enables the student to know how the information flows within the institution. Decentralized and integrated structures with open communication channels facilitate the rapid and transparent transfer of knowledge and ideas among staff and students, which is essential for a high-quality learning environment.

#### 9.6.2.1.2 Interpersonal–social

Interpersonal–social is a widely recognized and significant component or conceptual dimension within frameworks used to study the hidden curriculum, particularly in educational and professional settings like medical education. This dimension focuses on the informal social interactions and relationships that transmit values, norms, and behaviors outside the formal, explicit curriculum. In addition, it focuses on the social dynamics of the classroom and institution, such as interactions between students, teachers, and staff. Some of the important aspects of the interpersonal and social concept is the various informal processes such as day-to-day interactions, conversations, and collegial relationships among students, faculty, residents, staff, and patients. The interpersonal–social concept critically influences role modeling wherein students observe and learn from the behavior of their mentors and peers, which can profoundly shape their professional identity and actions, often more powerfully than explicit instruction. For instance, by gaining knowledge about the interpersonal–social concepts, the skill of active listening and empathy is being enforced wherein the students can be trained to listen attentively to a patient's concerns and respond with empathy, building trust. Furthermore, by understanding the interpersonal–social concepts, the competencies like teamwork and collaboration can be enhanced so that a health professional works respectfully and effectively with colleagues from various professions such as nurses, social workers, etc., in an interprofessional healthcare team. On the other hand, by upholding the interpersonal and social concepts, constructive relationship can be enhanced and at the same time the healthcare professional can be trained in conflict management skills.

#### 9.6.2.1.3 Contextual–cultural

It considers the broader cultural context, including social norms, values, and power structures that influence what is taught implicitly. In a conceptual framework of the hidden curriculum, the contextual–cultural component refers to the idea that the implicit values, norms, and beliefs transmitted to students are deeply influenced by the specific context and culture of the educational institution and its surrounding society. One of the key aspects of the contextual cultural concept is that it believes that the hidden curriculum is not a universal, fixed set of lessons, rather it is subjective and situational. The contextual–cultural component highlights that the specific lessons learned implicitly in one institution or culture may differ significantly from those in another. For instance, a student from India has a different set of cultural values than a student who has studied in Australia. The cultural concept also signifies that the local and broader cultural environment, including societal norms regarding race, gender, class, tolerance, and professional ethics, serves as a primary source for the hidden curriculum's content. These elements are often absorbed indirectly through social interactions, institutional practices, and a general "climate" of the learning environment. A key example of how contextual–cultural aspects are integrated is within a framework for culturally competent healthcare provision is noticed in its core principle. In this, the framework emphasizes that effective medical practice requires an understanding of how one's own cultural background and biases, as well as the patient's, influence healthcare interactions. From the perspective of the student, a good application of the contextual–cultural causes increased student engagement and motivation. When students see their own cultural experiences and identities reflected in the curriculum and classroom environment, they are more likely to be motivated, engaged, and invested in their learning.

#### 9.6.2.1.4 Motivational–psychological

It investigates the psychological effects on students, including how it fosters or hinders motivation, self-belief, and social-emotional learning. The motivational–psychological component is a key dimension in the conceptual framework of the hidden curriculum. This framework helps researchers and educators systematically identify and understand the implicit lessons students learn in educational settings. The role of motivational and psychological component focuses on how the implicit aspects of the institutional environment and social interactions affect students' internal states, including their motivation and psychological well-being. For a medical student these two concepts form the basis on which learning is built and carried forward. It is seen that more than the external motivation it is the intrinsic motivation that drives learning. This leads to a sense of satisfaction from academic success. On the other hand, factors like psychological well-being and self-perception can profoundly impact students' self-esteem, self-concept, identity formation, sense of belonging, and overall well-being. Motivational and psychological factors are crucial in the teaching-learning process because they directly influence a student's behavior, effort, and success. It is a fact that motivation drives students to be more positive, energetic, and persistent when facing challenges in student behavior and performance. Additionally, psychological factors determine the goals students set for themselves and how intensely they work to achieve them. Essentially, the conceptual framework impacts medical education in many ways such as by providing a structure for curriculum design, research, and assessment, which helps guide educators in understanding complex phenomena and evaluating outcomes. They act as a lens to focus research questions, design effective teaching methods, and interpret results by highlighting key variables and their relationships, such as those involving the teacher, learner, and learning environment.

### *9.6.2.2 Concept of functionalism*

Some frameworks focus on the purpose or function of the hidden curriculum. Frameworks that analyze the function of the hidden curriculum are largely based on sociological theories, primarily the functionalist perspective and the conflict perspective. Functionalism views the hidden curriculum as serving a positive, necessary function in society like socialization. It argues that the hidden curriculum prepares students for future roles in the workforce and society by teaching valuable skills and norms. Some of the important features of concept of functionalism are mentioned below:

#### 9.6.2.2.1 Social control

Within the functionalist framework, the hidden curriculum serves the function of social control by subtly transmitting to students the dominant social norms, values, and beliefs that help maintain social order and stability in society. At the same time, it also acknowledges how the hidden curriculum can be used to teach obedience, docility, and conformity, which serves as a form of social control. The implication of the social control in the hidden curriculum leads to the inculcation of values, wherein a common set of values and beliefs, such as respect for authority, punctuality, diligence, and conformity, are being instilled. These are often unstated but consistently reinforced through daily routines, rules, and informal interactions. It is seen that the components of the institutional structures, such as classroom organization, discipline activities, and the power dynamic between teachers and students, prepare individuals to accept hierarchical relationships. This process is seen as training for future roles in the workplace and other societal institutions, where obedience to authority is expected.

#### 9.6.2.2.2 Functionalism versus conflict

The perpetuation of inequality analyzes how the hidden curriculum can reinforce existing social inequalities by providing different opportunities or expectations for students of different social classes. In general, it draws a comparison between the functionalist framework and the conflict theory.

Functionalist framework: This framework views the hidden curriculum, and education generally, as serving necessary and positive functions for society, such as socialization into shared norms and values, social integration, and sorting students according to their talents for appropriate roles. In these, functionalists view education as a vital institution that serves necessary functions for social order and stability. In a positive sense, the hidden curriculum is viewed positively, considering it a "latent function" that transmits essential norms, values, and skills like punctuality, respect for authority, and cooperation necessary for a well-functioning society. Additionally, it also upholds the perspective of meritocracy as the system is believed to operate for sorting individuals into appropriate roles based on their abilities and efforts. They argue that inequality in outcomes is a natural and necessary result of ensuring that the most qualified individuals fill the most important positions, which benefits society as a whole.

Conflict perspective: This framework argues that the hidden curriculum is a tool for maintaining existing social inequalities and power structures. It suggests that the implicit lessons taught in institution, such as docility, obedience to authority, and an acceptance of hierarchy, prepare students of different social classes for their respective roles in a stratified workforce. Conflict theorists explicitly argue that the hidden curriculum serves to maintain and perpetuate social inequalities across class, race, and gender lines. In support of this framework, key theorists like Pierre Bourdieu, Samuel Bowles, and Herbert Gintis argue that the hidden curriculum reinforces the values and norms of the dominant social class, rewarding students from privileged backgrounds who possess the necessary cultural capital. Working-class students may face barriers, which limits their social mobility. Furthermore, the framework is of the opinion that the institution mirrors the workplace to create a docile, obedient workforce willing to accept hierarchical structures. This prepares students from lower socioeconomic backgrounds for lower-tier jobs, while preparing privileged students for positions of power and leadership.

#### 9.6.2.2.3 Development of skills and competencies

The hidden curriculum develops skills and competencies as a component of function-based frameworks by imparting unwritten lessons through institutional structures, culture, and interactions, such as teamwork, adaptability, and professional integrity, which are crucial for employability and social integration beyond the formal curriculum. The framework supports the transmission of values and norms since it teaches social and professional values, such as honesty, respect, and empathy, through interactions and the institutional environment. Moreover, it shapes professional identity by exposing students to implicit aspects of the role, such as resilience and teamwork, which are critical for professional socialization. The hidden curriculum develops skills and competencies through implicit norms, values, and social interactions, not from formal lessons. It fosters skills like punctuality, discipline, and cooperation, along with professional competencies such as collaboration, resilience, and ethical conduct, which are crucial for employability and social integration. Educators can deliberately harness the hidden curriculum to supplement formal education, guiding students to develop crucial social-emotional and professional competencies.

### *9.6.2.3 Active engagement*

Active engagement within the hidden curriculum in medical education does not rely on a single, formal framework, but rather a comprehensive approach involving multiple integrated strategies that aim to make the implicit explicit. One of the main principles involved here is the democratic education. In medical education, "democratic education" within the context of the hidden curriculum refers to the implicit transmission of values related to equality, shared decision-making, inclusivity, and student autonomy in the learning environment. Democratic education, through active engagement of students, provides a framework for identifying, critically analyzing, and potentially reshaping the powerful, often unwritten, lessons of the hidden curriculum. It actively seeks to minimize, make explicit, or redirect the hidden curriculum to have a positive developmental impact on students. Discussion with students is a key way to explore the hidden curriculum's impact. Furthermore, theorists like John Dewey argued that medical institutions should help students build the social and emotional skills necessary to engage in democratic life. This is achieved through the hidden curriculum's daily practices, which can foster or hinder these skills.

Democratic education, which values student voice, self-determination, and active participation in decision-making, offers a means to address these issues. Active engagement is central to this approach. By exposing and articulating the hidden curriculum and also by fostering environments where students can voice their experiences without fear of judgment, institutions can bring the implicit curriculum to the surface. Democratic education acts as a component of the active engagement in the hidden curriculum by making democratic values a tangible, lived experience rather than just a formal lesson. For the learner, the democratic approach proves to be of immense benefit as democratic education positively affects learners by increasing intrinsic motivation, improving critical thinking, and enhancing social-emotional skills through active participation and shared decision-making. This approach fosters student ownership over their learning, leading to deeper engagement, higher achievement, better attendance, and the development of valuable citizenship skills needed for a democratic society. Learners gain critical skills like communication, collaboration, and adaptability, along with a deeper understanding of societal norms, responsibility, and cultural awareness. This active participation helps build character, promotes empathy, and makes students more ready to navigate the complexities of the real world.

## 9.7 Teaching the hidden curriculum

To teach the hidden curriculum in medical education, it is vital to foster reflective practice and psychological safety, create structures for students to reflect on their environment, promote role modeling of desired behaviors, and encourage frank discussions about professional values and challenges. In order to achieve these, the various strategies employed include journaling, group discussions, and examining institutional policies, but the most powerful lessons often come from direct, critical observation and experience within the clinical environment. Activities to teach the hidden curriculum in medicine include facilitating reflective practice through journaling and debriefings, promoting observational learning by analyzing interactions and social cues, and intentionally fostering a positive learning environment through mentorship, teamwork, and open communication (Basyiruddin 2020). Educators can also use a structured approach of priming, noticing, processing, and choosing to guide students in critically examining their experiences and professional behaviors.

## 9.8 Assessment of the hidden curriculum

### 9.8.1 Qualitative

The most frequently used method to assess or explore the hidden curriculum involved is by conducting individual or focus group interviews which would be recorded and transcribed for subsequent qualitative analysis. Beyond the use of interviews and focus groups, another common qualitative assessment method is a thorough review of written reflections or narratives written by trainees, often mandatory as part of a reflective course, or as part of an end of term assessment. Qualitative assessment of the hidden curriculum also involves using methods like observations and thematic analysis to understand students' and faculty's experiences and perceptions (Li GJ 2024). This approach is used to explore how implicit learning through informal interactions, role modeling, and institutional culture shapes the professional identity, ethics, and clinical skills, with findings revealing both positive and negative influences. The qualitative methods such as semi-structured interviews with students and faculty, as well as direct observation, to collect rich, descriptive data are very helpful in this regard. In addition to this, the collected data, often from interviews, is transcribed and analyzed using techniques like thematic analysis to identify patterns, themes, and codes. The focus here is to understand the lived experience of students, thereby gathering insights into the informal, unwritten rules and values they absorb in the clinical environment.

### 9.8.2 Quantitative

The quantitative assessment method is usually in the form of cross-sectional surveys. Quantitative assessment of the hidden curriculum in medical education are the methods used to measure the frequency and impact of its positive and negative aspects, such as professional identity formation or the reinforcement of medical hierarchies. The tools used in quantitative assessment are the surveys and questionnaires often using a 5-point Likert scale to ask students about their experiences with the hidden curriculum. This enables analysis of how the hidden curriculum influences student attitudes, values, and professional development. The Hidden Informal Curriculum Assessment Tool (HICAT) tool also helps in quantifying the frequency and impact factor of positive and negative aspects of the hidden curriculum for medical students (Mcgurgan 2015). In this method, the assessment items are frequently categorized into domains based on literature reviews, such as role modeling, workplace environment, and social interactions. It is also observed that some studies use a mixed-methods approach, combining quantitative survey data with qualitative data from student reflections or interviews to provide a more complete picture.

## 9.9 Impact of the hidden curriculum on the principles and practice of medical education

The hidden curriculum profoundly impacts medical education and practice by transmitting implicit values and behaviors that can either reinforce or undermine formal teachings of professionalism, ethics, and humanism. It is an unavoidable force that primarily works through role modeling and organizational culture. The hidden curriculum significantly shapes the core principles instilled in future physicians such as professional identity formation, wherein the students develop their professional identity by observing senior clinicians and institutional norms. This can lead to either a positive consolidation of their identity or emotional distress and dissonance if the observed behaviors conflict with their personal values. It also contributes to the ethical and moral reasoning, whereby the students learn actual ethical conduct from the day-to-day actions of their mentors. Moreover, observing integrity and patient respect reinforces formal lessons, whereas witnessing dishonesty or a lack of transparency can lead to the internalization of unethical practices. In terms of medical practice, the hidden curriculum can strengthen patient trust by implicitly teaching respectful and transparent communication, or it can damage relationships by fostering a detached or dismissive approach to patient concerns. In addition, it helps the student to observe rigid hierarchies or mistreatment among staff, reinforcing these as acceptable cultural patterns, which can inhibit open communication and effective collaboration within multidisciplinary teams.

## 9.10 Implications of hidden curriculum

### 9.10.1 Positive implications

The positive implications of the hidden curriculum include the development of social skills, a strong work ethic, and professional identity. It helps students learn societal norms, develop values like responsibility and empathy, and prepare for future roles by promoting skills such as teamwork, conflict resolution, and adaptability (Meyer R 2025). The hidden curriculum can foster personal development, boost self-confidence, and create a more holistic learning environment by shaping students' perceptions and behaviors beyond the formal curriculum. Some of the important positive implications are towards the personal and social development, wherein students learn and practice important social skills like navigating relationships, collaborating with peers, and resolving conflicts. Moreover, through their medical institution experiences, students internalize societal expectations for behavior, attitudes, and future roles (Brown MEL 2020). In addition to this, the hidden curriculum supports the professional and skill-based development and promotes the work ethic, as it helps in building a strong work ethic by teaching concepts like punctuality, discipline, and professionalism. At the same time, it upholds the professional identity, as in a field like medicine, it is crucial for developing a professional identity and acquiring values like empathy, compassion, and ethical conduct through observation and interaction.

### 9.10.2 Negative implications

Negative implications of the hidden curriculum include perpetuating social inequalities, reinforcing gender roles, and creating a detrimental learning environment that can lead to stress, burnout, and unprofessional behavior. It can negatively affect students' self-esteem, mental health, and professionalism by promoting competitiveness, fostering trust issues, and pressuring students to conform by abandoning their identities (Ullah H 2025). Existing evaluator procedures may be unjust and invalid since not all students are born into environments that readily enable them to acquire language practices that are perceived as norms leading to shared academic assessment standards. If these vulnerable student populations are disproportionately disadvantaged by institutional policies and practices, and educators unconsciously align to the hidden curriculum, this reflects implicit bias.

**Summary**

The hidden curriculum refers to the unspoken, unofficial lessons, values, and expectations that the institutions communicate to students alongside the formal curriculum. While subjects like anatomy, physiology, and biochemistry are taught explicitly, institutions also impart social norms, behavioral expectations, and cultural messages that are not written in any curriculum. This hidden curriculum shapes students' attitudes, identities, and social development. It influences how they understand authority, competition, cooperation, gender roles, class expectations, and what behaviors are rewarded or discouraged. For example, punctuality, obedience, respect for hierarchy, and conformity often become part of what students learn simply by participating in institution routines. Although the hidden curriculum can help students develop important life skills, it can also reinforce social inequalities. Students from different backgrounds may interpret or experience these unspoken rules differently, leading to advantages for some and disadvantages for others. Educators and policymakers therefore emphasize the importance of recognizing the hidden curriculum to create more equitable and inclusive educational environments.


**Key point**

Future perspectives: bridging the gap between medicine and sociology

Teachers will play a critical role in purposefully shaping the hidden curriculum in the future by serving as moral and social guides, teaching time management skills, and modeling positive values and behaviors. This will result in the explicit development of skills. In order to promote holistic student development and emphasize critical thinking, ethical decision-making, and adaptability for a complex society, their future job will entail combining the hidden curriculum with formal and informal learning. The teachers' primary responsibilities go beyond merely serving role models. It also extends to purposefully exhibiting virtues, social standards, and professional conduct that are essential for students' growth and to encouraging deliberate modeling. The teachers help in fostering greater awareness and transparency amongst the learners. The future of the hidden curriculum will be in the integration of digital citizenship, wherein technology will shape the learning process. It is a fact that the hidden curriculum will extend to online spaces. Students will absorb unspoken rules about digital behavior, privacy, online ethics, and global communication, necessitating conscious guidance from educators.

# References


Basyiruddin, M., Rukayah, Roemintoyo, 2020. Teaching Strategies as a Powerful Hidden Curriculum: A Review Study. https://doi.org/10.2991/assehr.k.200129.096.

Brown, M.E.L., Coker, O., Heybourne, A., Finn, G.M., 2020. Exploring the hidden curriculum's impact on medical students: professionalism, identity formation and the need for transparency. Medical Science Educator 30 (3), 1107–1121. Available at https://doi.org/10.1007/s40670-020-01021-z, springer.com/education+&+language/journal/40670.

Das, S., Sadikan, M.Z., Oo, K.T., 2024. The perspective of hidden curriculum in the field of medical education. International Journal of Transformative Health Professions Education 2 (1), 109–115. Available at https://doi.org/10.71354/ijthpe.02.01.28.

Fortoul van der Goes, T.I., Núñez-Fortoul, A., 2013. What we say and what we do, incongruence in the teaching of good habits: the hidden curriculum. Investigative and Educational Medicine 2 (7), 119–121. Available at 10.1016/S2007-5057(13)72699-1.

Hafferty, F.W., 1998. Beyond curriculum reform: confronting medicine's hidden curriculum. Academic Medicine 73 (4), 403–407. Available at https://doi.org/10.1097/00001888-199804000-00013, http://www.lww.com/product/?1040-2446.

Jenkins, T.M., Underman, K., Vinson, A.H., Olsen, L.D., E. Hirshfield, L., 2021. The resurgence of medical education in sociology: a return to our roots and an agenda for the future. Journal of Health and Social Behavior 62 (3), 255–270. Available at https://doi.org/10.1177/0022146521996275.

Lawrence, C., Mhlaba, T., Stewart, K.A., Moletsane, R., Gaede, B., Moshabela, M., 2018. The hidden curricula of medical education: a scoping review. Academic Medicine 93 (4), 648–656. Available at https://doi.org/10.1097/ACM.0000000000002004, http://www.lww.com/product/?1040-2446.

Lee, C.A., Wilkinson, T.J., Timmermans, J.A., Ali, A.N., Anakin, M.G., 2023. Revealing the impact of the hidden curriculum on faculty teaching: a qualitative study. Medical Education 57 (8), 761–769. Available at https://doi.org/10.1111/medu.15026, http://onlinelibrary.wiley.com/journal/10.1111/(ISSN)1365-2923.

Lempp, Clive, 2004. BMJ (Clinical research ed.) 329.

Li, G.J., Sherwood, M., Bezjak, A., Tsao, M., 2024. Assessing the hidden curriculum in medical education: a scoping review and residency program's reflection. Canadian Medical Education Journal 15 (5). Available at https://doi.org/10.36834/cmej.78841.

Martimianakis, M.A., Michalec, B., Lam, J., Cartmill, C., Taylor, J.S., Hafferty, F.W., 2015. Humanism, the hidden curriculum, and educational reform: a scoping review and thematic analysis. Academic Medicine 90 (11), S5. Available at https://doi.org/10.1097/ACM.0000000000000894, http://www.lww.com/product/?1040-2446.

Matorevhu, Alois, 2022. Journal of Research in Instructional 2.

Mcgurgan, Alexandra, 2015. MedEdPublish 10.

Meyer, R, 2025. Med Sci Educ 35.
Mosalanejad, Bahar, 2013. Strides Dev Med Educ 10.
Nyquist, J.G., 2016. The hidden curriculum in health professional education. Journal of Chiropractic Education 30 (1), 48–49. Available at https://doi.org/10.7899/jce-15-20.
Sarikhani, Y., Shojaei, P., Rafiee, M., Delavari, S., 2020. Analyzing the interaction of main components of hidden curriculum in medical education using interpretive structural modeling method. BMC Medical Education 20 (1). Available at https://doi.org/10.1186/s12909-020-02094-5.
Ullah, H, 2025. Med Educ Pract 13.

Chapter 10

# Curriculum mapping

*"A curriculum map should inspire possibilities instead of thinking inside of the box."*

—John Spencer

## 10.1 Introduction

Curriculum mapping is the systematic process of documenting and assessing the medical education curriculum to ensure that learning objectives, instructional activities, and assessments are all in line. Curriculum mapping enhances the curriculum's coherence and efficacy by helping to identify gaps, duplications, and inconsistencies. A curriculum map can be used as a planning tool to decide where the learning objectives will be taught and assessed during the initial stages of curriculum design (Al-Eyd G 2018). Curriculum mapping is a powerful and strategic tool for ensuring coherence, alignment, and effectiveness in educational planning and delivery. Using the curriculum map enables the medical educators to find gaps, redundancies, and areas for development by methodically recording and evaluating a curriculum's scope and sequence. This enables them to enhance their methods and better serve their students' needs. The following are the major roles of curriculum maps in medical education:

1. Curriculum mapping assist in making sure that the curriculum appropriately addresses each program learning outcome.
2. The curriculum map plays a role in locating any curricular structural issues.
3. The curriculum map assists in identifying structural issues and determining how to address them.
4. The curriculum map facilitates the recording of the subjects and locations covered.

The biggest benefit of curriculum mapping is its ability to improve the relationships between curriculum, assessment, and instruction. However, while mapping the curriculum, it should be noted that the educators must also assess how their activity aligns with college and institutional regulations. Curriculum mapping allows teachers to keep track of how many of the required standards, content, and abilities they have covered and what still has to be covered. Teachers are actively involved in all the processes including organizing, analyzing, and evaluating the courses to find gaps and repetitions in the application of learning standards. With a well-developed and customized curriculum plan made especially for the program, teachers might find this process simpler and even be able to go deeper. Curriculum mapping, which shows the relationships between all of its components, is essential to accomplishing any curriculum's objectives and results. It makes it easy to see how the various curricular components relate to one another. The map illustrates how learning objectives are reached through certain learning events and evaluated and serves as a visual representation of alignment. Additionally, it displays the timeline and sequence in which it depicts the specifics of the curriculum's topics and activities. In terms of resources and tools, it emphasizes the supplies and equipment required for any educational task. Additionally, it aids in curriculum visualization by painting a comprehensive image of the program and emphasizing the relationships between its various components.

## 10.2 Need for curriculum mapping

In medical education, curriculum mapping is necessary to make sure that the curriculum is clear, cohesive, and successful in meeting learning goals (Shrivastava 2021). It facilitates the identification and correction of gaps or redundancies, the alignment of the proclaimed curriculum with what is taught and tested, and the clarification of the relationships between various learning outcomes and components. It also encourages continuous quality improvement and enhances transparency and communication. The relationships between learning objectives, content, instructional techniques, and assessments are also highlighted through the curriculum map. In this sense, it is also advantageous to match program

**The Principles and Practice of Medical Education. DOI:** https://doi.org/10.1016/B978-0-443-44197-4.00013-8

objectives and accrediting requirements with instruction. In the end, it offers a framework for accreditation, encourages better student results, and supports curricular improvement.

## 10.3 Components of a curriculum map

Learning expectations, which are in the form of learning objectives or outcomes at different levels; learning event details, such as course codes, duration, and location; pedagogy, such as instructional techniques; and evaluation, which include both formative and summative, are the elements of curriculum mapping in medical education. Curriculum mapping also includes assessment data, standard compliance, and a clear scope and sequencing. Among the potential benefits for both teachers and students are improved instructional coherence, consistency with standards, and a better understanding of the learning process.

### 10.3.1 Learning objectives

Learning objectives, a crucial component of curriculum mapping, are used to demonstrate and assess how they are introduced, developed, and mastered over a program's courses. These outline what students should know and be able to do after completing a unit or course. They must also align with broader program learning objectives. By mapping objectives, teachers may ensure coherence, identify gaps or overlap, and align instruction with the desired student outcomes (Kopera-Frye 2008). The learning outcomes play an important role in alignment where the curriculum mapping uses a visual tool such as a grid to show which learning objectives each course is meant to achieve. This helps ensure that the curriculum aligns with the program's overall goals. Since teachers can assess whether the learners are being taught at the appropriate level of complexity across the curriculum, mapping the learning objectives also provides a way to ensure instructional scaffolding.

### 10.3.2 Learning event information

Learning event information includes details like course codes, duration, and venue which are also vital components of the curriculum map as they record what is taught, when it is taught, and how it is taught in a medical curriculum. This data, which is used to find duplications and misalignments, comprises particular learning objectives, material, instructional strategies, and assessments related to each event. Curriculum mapping is the act of dividing the entire curriculum into distinct years, semesters, and levels using a visual representation, usually a grid or table. It also makes it easier for us to see the courses that will be offered in each of these years, along with the credit hours and course codes. It may also specify how long the courses will be taught. The program's regular and elective courses will also be shown on the curriculum map. This curriculum map development process ensures comprehensive coverage and provides a foundation for topic and assessment design. Some of the information tracked in learning events are the nature of the learner's activity such as receiving information, practicing, creating and the corresponding teacher activity like transmitting, guiding, and evaluating. In addition, the alignment level in terms of the contribution of the learning activity which is often marked as introduced, reinforced or developed, or mastered are also indicated. It is also possible to know whether the event includes formative feedback or summative assessments such as quizzes, projects, or clinical observations to measure progress. The curriculum map also serves as a guide for what students should learn over time, guaranteeing that proper growth and sequencing are adhered to. It accomplishes this by logically arranging content to expand on prior knowledge. Additionally, mapping gives teachers the chance to determine what is being taught, where it is being taught, and whether any subjects are being taught more than once or ignored. This offers the teachers assurance that the gaps and duplications are adequately recognized.

### 10.3.3 Instructional strategies

Instructional techniques are essential for the planning and structuring of curriculum mapping. These teaching techniques are used to get students interested in the subject matter. It is important that the instructional strategies be varied and appropriate for the learning objectives and the needs of the students. Instructional strategies, which outline how content will be presented in compliance with standards and learning objectives, are a crucial component of curriculum mapping. Curriculum maps include instructional strategies for preparing a range of learners, connecting concepts, and ensuring a coherent and logical learning progression across the academic year. This integration helps teachers manage the "how" of learning to ensure effective delivery and student comprehension by going beyond just the "what" and "when" of education (Harden RM 2001). These tactics, which include lectures, group projects, and experiential learning, are all integrated into curriculum maps to outline the specific teaching methods that will be used for each unit or lesson. Additionally, curriculum maps ensure that the teaching strategies chosen align with the content, learning objectives, and academic standards for the semester, level, or subject. The purpose of instructional strategies in curriculum mapping is to make sure that, at the institutional, program, course, and session levels, the curriculum map clearly connects instructional strategies to certain learning outcomes. Furthermore, by documenting strategies, the map guarantees that instruction is scaffolded, providing a natural learning progression that gradually removes assistance as students become proficient.

### 10.3.4 Assessments

As it offers proof of student learning and guarantees that learning objectives are in line with instructional strategies and assessment techniques, assessment is an essential part of curricular mapping in medical education. Teachers can create a thorough method for assessing students' knowledge, abilities, and attitudes by incorporating assessment plans into curricular maps. This helps to improve and refine the curriculum. Assessment plays a crucial part in curricular mapping by providing data that demonstrates that students have accomplished learning objectives and developed the requisite knowledge, abilities, and attitudes. Curriculum maps also allow instructors to ensure that the assessments they choose are appropriate for the intended learning outcomes, and that they are distributed across the curriculum in a way that aligns learning and evaluation. Additionally, when assessments are mapped against the curriculum, it makes it easier to find and fix flaws like areas where students are not being assessed or where content is not being taught correctly, which helps detect gaps and redundancies. It helps to improve the curriculum since assessment results offer both qualitative and quantitative information that can be utilized to make data-driven decisions concerning curriculum, teaching strategies, and assessment instruments. In the end, it promotes balanced integration since, in integrated medical curricula, mapping assessments ensures a balanced approach to content evaluation across many disciplines and program phases.

### 10.3.5 Alignment with standards

One of the most important aspects of curriculum mapping in medical education is standard alignment. A well-created curriculum map helps to guarantee that learning outcomes satisfy criteria by ensuring that the curriculum is in line with particular standards, such as those from accreditation agencies. Curriculum mapping alignment with standards is used as a quality improvement strategy to guarantee that a program satisfies set standards and legal requirements. It guarantees that students will have the chance to acquire the knowledge and abilities specified in those standards. This procedure entails methodically connecting evaluations, learning objectives, and course content to program objectives and outside standards. As a result, a clear framework is created that demonstrates how each course contributes to the overarching goals of education and compliance. This ensures a consistent, top-notch education covering all essential knowledge and abilities. This phase is essential because it ensures consistency in what is taught and assessed, regardless of the instructor or classroom. It also assists instructors in closing learning gaps by identifying and addressing information and skill deficiencies that could otherwise go undiscovered.

### 10.3.6 Skills

Curriculum maps should address topic knowledge as well as the development of specific abilities. A crucial part of the medical curriculum map is skills. Throughout the course of the educational program, a medical curriculum map clearly indicates where and how particular knowledge, abilities, and attitudes are taught and evaluated. The curriculum map, which emphasizes the development and mastery of both clinical and professional skills, is an essential instrument for guaranteeing the coherence and alignment of medical education. The learning outcomes, which specify skills within the quantifiable learning objectives that students are expected to accomplish upon completion of a course or program, are important components of skill integration in a medical curriculum map. In order to bridge the gap between "knowing" and "doing," the curriculum mapping places a strong emphasis on the application of academic information to practical, real-world clinical skills. This approach ensures coherence across courses and semester levels by tracking specific skills like critical thinking and problem-solving in addition to information to avoid gaps and overlaps. A skills-based curriculum map shows how a program builds essential skills to prepare students for future success. One of the main roles in curriculum mapping is the identification of skills, which is the process by which basic skills like communication, teamwork, and critical thinking are recognized. At the conclusion of a program, it is also anticipated that the students are matured. It also aids in skill sequencing since it demonstrates how abilities are introduced, reinforced, and ultimately mastered, often moving from lower-level to higher-level courses.

### 10.3.7 Resources

The term curriculum resources refer to materials that serve as the primary source for a teacher to prepare and plan for the delivery of curriculum standards, as well as materials that students regularly interact with to gain the necessary knowledge and skills. Resources are a crucial part of curriculum mapping since they are the instruments and materials required to achieve learning objectives. They include textbooks, handouts, and reference articles and are in accordance with the standards and tests. Moreover, the curriculum mapping technologies generate digital libraries of shared resources that educators may access and use. Curriculum maps can also function as a repository for new resources, linking them directly to the standards they support and encouraging teacher collaboration. This can involve using learning management systems and reading assignments and reference materials. Learning environments, such as lecture halls, skill laboratories, or practical labs, and teachers and personnel participating in delivery are also considered as resources. In addition to helping students plan their studies and access pertinent materials, resource mapping ensures that the assistance required for the curriculum to be delivered effectively is available.

## 10.4 Types of curriculum maps

Curriculum mapping is a crucial technique in medical education for arranging, visualizing, and coordinating course material with assessments, competencies, and learning outcomes. Depending on the area of the curriculum being studied or overseen, there are various kinds of curriculum maps, each with a specific function. Here's a structured overview of the main types of curriculum maps used in medical education (Fig. 10.1).

### 10.4.1 Outcome-based curriculum map

An outcome-based curriculum map is a technique used in medical education that matches specific curriculum content, instructional strategies, and assessments with learning outcomes to ensure students gain the requisite skills. It is a systematic, cyclical process that helps define the educational path, identify gaps, and demonstrate that the curriculum meets standards. This approach, which is crucial to competency-based medical education, places more emphasis on students mastering higher-order knowledge than merely collecting credits (Alam, Tanweer 2022). Here, the primary focus is on the alignment of assessments, instructional activities, and learning outcomes. The main objective of this map is to ensure that each educational task is designed to help students develop specific skills. This map starts by defining the exact, measurable outcomes that graduates must achieve, such as certain skills, knowledge, and attitudes. These outcomes are then matched to the official written curriculum, which include course materials, learning activities, and assessments like clinical rotations and examinations. The map ensures both horizontal alignment, which provides a consistent learning experience throughout all courses in the subsequent stage, and vertical alignment, which links program-level outcomes to

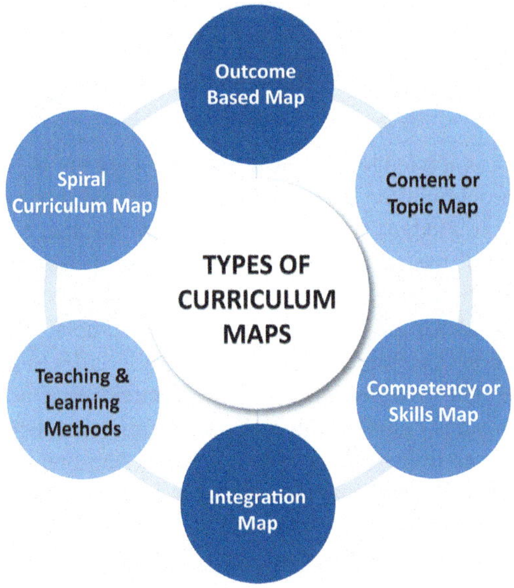


FIGURE 10.1 **Types of curriculum maps.** This figure depicts the types of curriculum maps.

particular course outcomes. Creating a continuous loop that involves mapping, aggregating, reviewing, identifying gaps, and implementing adjustments is also helpful in improving the curriculum. A curriculum map might, for instance, illustrate how course learning outcomes like conduct a patient history and physical examination address institutional learning outcomes like demonstrate patient-centered communication, and then be evaluated in a session using a formative assessment like an objective structured clinical examination.

### 10.4.2 Content or topic map

A content or subject map is one of the most often used and beneficial forms of curriculum mapping in medical education. It concentrates on the concepts, principles, and issues covered in a medical curriculum. It focuses on the actual subject matter being taught. In medical education, a content curriculum map is a planning and communication tool that shows the connections between various curriculum elements, including learning objectives, instructional strategies, and evaluations. It acts as a road map that shows how information is integrated both horizontally, that is, across disciplines, and vertically, that is, across different phases, to guarantee transparency and consistency for students, instructors, and administrators. In order to enable more integrated, outcome-based educational models, find gaps, remove duplications, and comply with accreditation requirements, these maps are crucial. Additionally, it helps with curriculum design by offering a comprehensive perspective to guarantee that all learning objectives and required competences are satisfied through suitable teaching methods and evaluations. By demonstrating what is taught, how it is taught, and how it is evaluated, it also increases transparency by making the educational process understandable to all parties involved. For example, in the basic sciences, the topic of the circulatory system includes the anatomy of the heart, cardiac physiology, and cardiac enzymes. Similarly, the clinical science years cover heart conditions such coronary artery disease, heart failure, and medication which are all well connected through outcomes, instructional strategies, and assessment. Further examples of content or subject maps include auscultation, ECG interpretation, and case-based discussion as part of the skills or assessment.

### 10.4.3 Competency or skills map

A competency or skills map is a framework used in medical education that outlines the knowledge, skills, and attitudes required from medical professionals. Curriculum development and evaluation are common uses for it. It serves as a guide for learning and assessment by breaking down complex competencies into smaller, measurable components, guaranteeing curriculum coherence and preparing students for clinical practice. Here, the main focus is on clinical, procedural, communication, and professional skills and competencies. The main purpose of this map is to track each competency's introduction, reinforcement, and evaluation. The map also helps define core competencies, which describe the essential abilities like patient care, medical knowledge, professionalism, communication, and systems-based practice that the student should acquire. It also facilitates a more thorough approach to learning and assessment by breaking down complex procedural skills into a hierarchical list of required competencies. Additionally, it guides curriculum design because the map ensures that all necessary learning objectives are included in the curriculum and that learning develops logically, which is often demonstrated by curriculum mapping. A curriculum map might, for example, indicate that a medical student is expected to exhibit advanced patient interviewing abilities in the third and fourth years after learning basic history taking in the first year and being evaluated on patient interviews in the second year.

### 10.4.4 Integration map

A graphic depiction of how several disciplines interact and evolve to help students accomplish learning objectives is called an integration curricular map. It is a tool for coordinating education across disciplines, recording what is taught and when, and guaranteeing a more thorough and integrated learning experience (Waldvogel TEE 2025). Integration curricular maps often link information thematically and can be used to show the relationships between programs, courses, and learning objectives. The combination of basic and clinical sciences is its main focus. In medical education, curriculum mapping facilitates integration through several strategies, including horizontal, longitudinal, and spiral integration, which tries to tear down barriers between basic and clinical sciences. The integration can be done in three ways. For instance, physiology, anatomy, and biochemistry for the digestive system are examples of horizontal integration. Conversely, longitudinal and spiral integration refers to revisiting and gradually expanding upon subjects over a number of years. The three types are described below.

#### 10.4.4.1 Horizontal integration

Using a horizontal integration strategy, many basic science fields are connected within a particular topic or organ system. Instead of teaching the anatomy, physiology, and biochemistry of the heart as distinct courses, this curriculum would combine them into a single block. For example, the characteristics of cardiac tissue and myocardial metabolism are linked to the structure of the heart.

#### 10.4.4.2 Longitudinal integration

Using the longitudinal integration approach, a particular subject or idea is revisited at various stages of the curriculum. Over time, it aids students in understanding the relationship between fundamental science and clinical practice. For example, in the final year, where the students are taught clinical medicine, the tutor repeats or revisits the anatomy of the heart that was taught in the initial years before moving forward.

#### 10.4.4.3 Spiral integration

In spiral integration which is considered as a more sophisticated type of longitudinal integration, a particular topic is reviewed at progressively higher levels of complexity. In the first year, for instance, a student might learn the fundamentals of a structure and function of a particular organ or system; in the second year, they might study the diseases and clinical symptoms related to that; and in a subsequent clinical year, they might learn about the investigations, treatment, and management. The spiral integration map's primary feature is that it compels students to return to the topics because they frequently encounter the same themes, topics, or subjects throughout a course. The map also depicts rising complexity, which is the process by which every new encounter with a subject becomes more intricate and expands upon prior understanding, making it more profound (Bruinsma M 2007). The map also highlights the value of expanding on previously taught content, since it makes it easier to gain a deeper understanding by relating and restating what has already been learnt.

### 10.4.5 Teaching and learning methods map

A curriculum map for teaching and learning methods is a strategy document that synchronizes teaching strategies with program objectives, student learning outcomes, content, and assessments throughout time. It works as a long-term, big-picture plan that differs from a short-term lesson plan by providing a more comprehensive overview of how teaching and learning activities are arranged to achieve certain goals. It creates a visual link between the learning objectives that students are expected to achieve and the content and learning activities that teachers are teaching. It also offers a thorough synopsis of the whole program, demonstrating how various courses help achieve program-level objectives. Seeing how various teaching strategies are applied to evaluate student learning, and whether those evaluations are successfully capturing the intended results, is also beneficial. By visually organizing the curriculum, teachers may ensure a logical progression of skills, identify and correct gaps or redundancies, and improve the program's overall quality. This type of map concentrates on methods of delivery such lectures, simulations, problem-based learning, and clinical placements. The main objective of this map is to ensure that a variety of methods are applied appropriately and in accordance with outcomes. To illustrate how the learning objective "understand cardiac physiology" is addressed through didactic lectures as in year 1 and subsequently reinforced with case-based learning and physical examination as in year 4, a map can employ a matrix. Over time, this method helps find any gaps or duplications in teaching methods and guarantees alignment.

## 10.5 Key strategies for curriculum mapping

Establishing a clear set of learning objectives, encouraging teacher collaboration, and aligning both vertically and horizontally are important tactics for medical education curriculum mapping. In addition, the utilization of data and proof should be emphasized. Maintaining a dynamic curriculum is also equally important. Assuring logical development and coherence, including feedback and routine evaluation, and locating and resolving gaps or redundancies are additional crucial tactics.

### 10.5.1 Establish clear goals and frameworks

Setting clear goals and a common framework, which ensures coherence, standard alignment, and relevant instruction across courses and semester levels, is a crucial part of curricular mapping. This initial phase provides a road map that guides all ensuing planning, teaching, and assessment, enhancing the students' educational experience. Clear goals, sometimes referred to as learning objectives and outcomes, outline precisely what students should know, understand, or be capable of at the conclusion of a course or program. This goal-oriented approach, which often employs "backward planning" concepts, ensures that all efforts are focused on specific, measurable objectives and steers clear of aimless, activity-oriented education. This also ensures coherence and alignment because a clear framework provides a basis for aligning the entire curriculum with institutional mission statements, overall objectives, and external academic standards or accreditation requirements. This alignment ensures that every curricular component works together to provide a coherent educational experience by progressively developing skills from one level to the next and promoting consistency across multiple teachers of the same topic.

### 10.5.2 Visualize the curriculum

This produces a visual depiction, like a spreadsheet or map, that illustrates how various curriculum components are spread out across time. A crucial tactic in medical education curriculum mapping is curriculum visualization, which offers a clear, comprehensive picture of the complete program and facilitates the identification of gaps, duplications, and misalignments. Coherence and alignment between the desired learning outcomes, instructional strategies, and assessments are ensured by this spatial representation. The visual maps make the curriculum clear to all stakeholders, including students, faculty, and administrators, which facilitates better communication about what, when, and how topics are taught. This is one of the advantages of visualizing the curriculum. By providing the whole curriculum in one location, visualization also aids in the discovery of gaps and redundancies. Teachers may quickly see areas where crucial content may be lacking or where subjects are needlessly duplicated across several courses. Concurrently, it guarantees constructive alignment, which is the act of ensuring that learning objectives are well matched with instructional methods and assessment methodologies.

### 10.5.3 Focus on vertical and horizontal alignment

Curricular mapping guarantees a coherent learning route by connecting standards both horizontally within the same semester level or program and vertically across semester levels. Vertical alignment results in a natural growth of content and skills from one semester to the next, whereas horizontal alignment ensures consistency and collaboration among teachers at the same semester level or within a specific program. When combined, they create a curriculum that provides students with a smooth, continuous education in every topic. It ensures that basic knowledge and abilities are necessary for later, more complicated topics by connecting learning objectives, instruction, and assessments from a lower semester or course to the next level up. Vertical alignment also helps students progress their knowledge and abilities from one course or semester level to the next. Regardless of their teacher or classroom, all students in the same semester or enrolled in the same course will get instruction of a similar caliber and be exposed to the same learning needs due to the horizontal alignment.

### 10.5.4 Emphasize evidence and data use

One of the key strategies of the curricular map is its emphasis on the use of data and evidence. By highlighting the use of data and evidence, curriculum mapping can be converted from a descriptive activity into a powerful, data-driven tool for continuous development. The data collected in the form of student outcomes and feedback is very useful for evaluating the curriculum's effectiveness. This approach uses a range of data types to validate, evaluate, and enhance the curriculum, ensuring that it aligns with learning objectives and student needs. This also helps detect gaps and redundancy in order to discover areas of over-coverage, under-coverage, or misalignment with learning objectives. Additionally, based on indications of patterns in student learning and achievement, the data is used to update or enhance maps. Making informed decisions on program-level change initiatives is aided by the data, which provides precise evidence to support meaningful, collaborative discussions and decisions. Instead of depending only on judgment, decisions are now based on what has been proven to be effective. Additionally, it allows for the measurement of student achievement since curriculum maps that include assessment data facilitate teachers' monitoring of students' progress toward specific learning objectives. This information helps assess how successfully courses yield the intended outcomes.

### 10.5.5 Keep the curriculum dynamic

The process of considering the curriculum map as a "living document" that is constantly evaluated, analyzed, and improved in response to input, data, and changing educational requirements is one of the critical elements in curriculum mapping. It shifts from a one-time documentation work to a continuous improvement cycle. A dynamic curriculum strategy's primary feature is that it functions as a continuous improvement loop, with the curriculum map's frequent review and updating acting as its focal point. This means implementing the necessary changes in response to data from student performance, teacher feedback, and evolving institutional norms. It is clear that a dynamic curriculum is inherently flexible, allowing teachers to deviate from initial plans as necessary to meet the specific needs and developmental stages of their students. It encourages adaptability in teaching methods and resource integration, including incorporating new technology or current events.

### 10.5.6 Leverage technology tools

Leveraging technology for curriculum mapping refers to the use of digital technologies for curriculum development, administration, and analysis. This method facilitates data-driven improvements, improves teamwork, and streamlines processes (Balzer F 2016). Technology can be utilized to connect courses to learning objectives, track progress, identify gaps in assessment data, and provide teachers with online resources. This data can also be used to determine whether courses are producing the intended outcomes and where assessments should be placed. This approach increases mapping effectiveness and could lead to a more adaptable and responsive curriculum. Furthermore, artificial intelligence is now a fundamental structural element in medical education curriculum mapping rather than just an experimental addition. The intricate process of matching specific session goals with more general institutional and international norms is increasingly automated by artificial intelligence technologies. However, sophisticated natural language processing finds overlaps and hidden content gaps in medical curricula that conventional evaluations could overlook. To guarantee a balanced growth of competencies, artificial intelligence systems examine skill development patterns and keyword density.

### 10.5.7 Align assessments with learning goals

Matching assessments with learning objectives is a curriculum mapping strategy that ensures exams are trustworthy and accurately measure student performance. This approach starts with the creation of learning objectives, which are then followed by evaluation tasks and standards that establish whether or not students have achieved the objectives. It ensures that there is a clear and logical connection between what is taught, what is assessed, and what students are expected to know and do. It's usually a good idea to start with the backward design approach, which starts with desired outcomes and aligns learning activities and assessments accordingly. It is useful to keep in mind that before creating any assessments, precise, measurable learning objectives must be set using the appropriate action verbs. The assessment activities are designed to allow students to demonstrate the knowledge and skills outlined in the learning goals. For example, if professional communication is the goal, the evaluation should be a done in a way that assesses professional communication rather than any other competency. A curriculum map must also be created to show how each learning objective is taught and assessed throughout a course or program. This makes it easier to identify any gaps or duplications.

### 10.5.8 Provide training and support

Providing support and training is a crucial strategy for the successful implementation and long-term sustainability of curriculum mapping initiatives. It ensures that educators have the skills, expertise, and confidence necessary to successfully contribute to the process, which enhances curriculum quality and student outcomes (Bayer CR 2019). One of the most crucial aspects of training and assistance is that it enables all employees to transition from potential resistance to active participation by assisting them in understanding the purpose, benefits, and mechanisms of curriculum mapping. It also supports the creation of professional development programs that give educators the resources they need to use mapping technologies, establish precise learning objectives, align instruction and assessments with standards, and analyze map data. Additionally, because the training helps create a consistent vocabulary and standards for mapping across numerous departments or semester levels which are both essential for creating a curriculum that is comprehensive and coherent, it ensures quality and consistency.

### 10.5.9 Communication and transparency

Communication and transparency are a crucial curriculum mapping components that fosters trust, alignment, and responsibility among all stakeholders, including teachers, administrators, and students. This transparency will enable everyone to understand the learning objectives, topic content, exams, and expected learning outcomes. Transparent communication can improve collaboration between educators and administrators. Teachers can uncover chances for interdisciplinary linkages, exchange best practices, and provide a unified learning environment by openly sharing what is taught in different courses and semesters. Transparency avoids subject gaps and unnecessary repetition by ensuring that instruction is in line with established educational standards and institution-wide objectives, which also facilitates alignment and coherence.

## 10.6 The curriculum mapping process

Creating a curriculum map is like making a plan for a successful educational endeavor. It's a systematic procedure with several key components that come together to offer a comprehensive and well-rounded educational experience (Watson 2020). The following steps are involved in the curriculum mapping process (Fig. 10.2):

### 10.6.1 Identify learning objectives

Curriculum mapping is centered on learning objectives. These objectives outline the information and abilities that students should learn. They serve as the compass, guiding the entire process of learning. Determining learning objectives is a fundamental step in medical education curriculum mapping that offers a clear road map for what students should know or be able to do at different program levels. Determining learning objectives serves as a basis for design and alignment. The entire curriculum is designed in accordance with the well-defined learning objectives, which guarantee that the course material, teaching strategies, and evaluation procedures all contribute to the achievement of particular, intended results. Additionally, it ensures coherence and integration, with the objectives acting as clear connections between various curriculum elements, including sessions, courses, and the program as a whole, encouraging both horizontal and vertical integration of information and skills.

### 10.6.2 Sequence content

After determining the learning objectives, the following stage is to construct a rational path that arranges the content logically. This means arranging skills and subjects in a way that ensures a smooth progression in complexity and difficulty. By helping students build a solid foundation before moving on to increasingly difficult concepts, this sequencing produces a scaffolded learning path. A number of crucial processes are included in the iterative, cyclical process of curriculum mapping for medical education, which includes the sequencing of information. This procedure guarantees a logical learning progression, prevents gaps and duplications, and complies with accreditation requirements. Determining the objectives is one of the most important elements in the sequencing of medical education content. Setting specific learning goals at the institutional, program, course, and session levels must come first. These objectives outline the essential knowledge, skills, and attitudes that students must acquire. Locating and compiling preexisting content is also essential. It is impertinent to compile thorough information about the subjects taught, instructional methodologies, learning resources, and assessment forms.

### 10.6.3 Select assessment methods

The assessment techniques employed determine how educational outcomes are evaluated. Instructors choose assessments that effectively gauge students' comprehension and align with the learning goals. This may include tests, assignments, quizzes, and presentations. Effective assessment methods provide insights into students' growth and guide curriculum modifications. In order to ensure that assessments are in line with learning objectives and program outcomes, choosing assessment methods is an essential part of curricular mapping in medical education. Criteria including validity, reliability, feasibility, and alignment with the desired learning outcomes serve as a guide for this approach. The assessment's primary function in curricular mapping is to promote learning. Alignment is essential because assessments have a significant impact on what and how students learn.

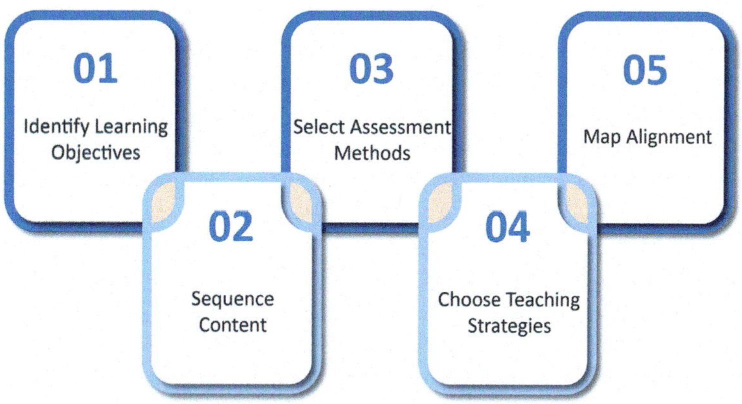


FIGURE 10.2 **Curriculum mapping process.** This figure depicts the curriculum mapping process.

### 10.6.4 Choose teaching strategies

To engage students and convey information, teachers use instructional strategies. These strategies should align with the learning objectives and take into account various learning styles. Teaching tactics, also known as pedagogy, educational methods, or instructional strategies, are typically chosen during the step of planning of learning experiences and instruction or the design stage of curriculum mapping. The choice of these tactics is essential for coordinating the delivery of the material with the desired learning goals and evaluation techniques. Finding the best methods to support student learning and guarantee that the intended outcomes are achieved is the role of choosing teaching strategies. This entails connecting instructional strategies to particular learning goals. Coherence between the curriculum, instructional strategies, and evaluations must be maintained concurrently. It also aids instructors in creating a common, coherent course or program plan.

### 10.6.5 Map alignment

The process of linking learning objectives and outcomes at different levels such as the program, course, and session and with assessments to make sure the curriculum is effective and cohesive is known as map alignment. To find gaps, duplications, and misalignments, it entails making a "map" that graphically depicts the connections between what is taught, learned, and evaluated. Some of the key aspects of map alignment are that it links the learning objectives, wherein the mapping process connects program learning outcomes with course learning outcomes and specific learning objectives to ensure they are all working toward the same goals. Map alignment guarantees that the material and teaching strategies are in line with the specified learning objectives and are presented in a way that will assist students in achieving those objectives. Furthermore, it is essential in establishing a direct connection between the assessment techniques and the learning objectives and sessions. This guarantees that students are evaluated based on the knowledge they are supposed to acquire (Hege I 2010). Additionally, it confirms that each component of the curriculum map works together to determine whether coherence and consistency are upheld, a process known as mapping alignment. This means making sure that instructional strategies support learning objectives and content, and that assessment practices align with learning objectives. This alignment ensures the curriculum map's purpose, consistency, and cohesiveness. Curriculum connections, the paper-based approach, and other methods are used to enter data linked to curriculum mapping and create reports using the information gathered. Excel spreadsheets, document files, and online survey tools can also be used.

### 10.6.6 When to design a curriculum map

It should be kept in mind that it is not only important to create a curriculum map, but it is equally important to revise it on a regular basis in order to find gaps, duplications, and to make sure that a curriculum is in line with learning objectives. In this regard, a curriculum map should be created at the beginning of a new curriculum's development or when assessing an existing one (McLean M 2010). Teachers can use it as a tool to plan and monitor their lessons prior to the start of a term. There are three stages at which the curriculum map can be created or revised, which are illustrated in Fig. 10.3.

#### *10.6.6.1 During initial curriculum development*

This makes it easier to prepare with the "end in mind," guaranteeing that the curriculum makes sense and that learning objectives are integrated from the outset. Creating a curriculum map during the early stages of curriculum development is a methodical, multistep process that progresses from high-level vision to in-depth planning and continuous revision. This procedure entails defining the program's or curriculum's overarching goal and high-level vision. This entails figuring out the desired effect, a broad course, and the central idea that will lead the creation of the curriculum. Establishing program learning outcomes by converting the vision into precise, quantifiable statements is also crucial. What students should be able to accomplish after finishing the program is outlined in these statements. Listing all required courses and essential learning opportunities, such as fieldwork or internships, that each student will participate in is equally important. These make up the curriculum map's framework.

#### *10.6.6.2 To evaluate an existing curriculum*

A curriculum map can be evaluated by comparing it to learning outcomes, analyzing student performance data, and conducting stakeholder reviews to find any gaps, duplications, or alignment problems. It is always good to start by examining the map to determine whether learning objectives are taught in a sequential manner and adequately assessed. Subsequently, data and feedback is used to verify whether students are reaching those objectives. In order to ascertain whether all program learning outcomes are being taught and evaluated at the proper level and in the proper order across

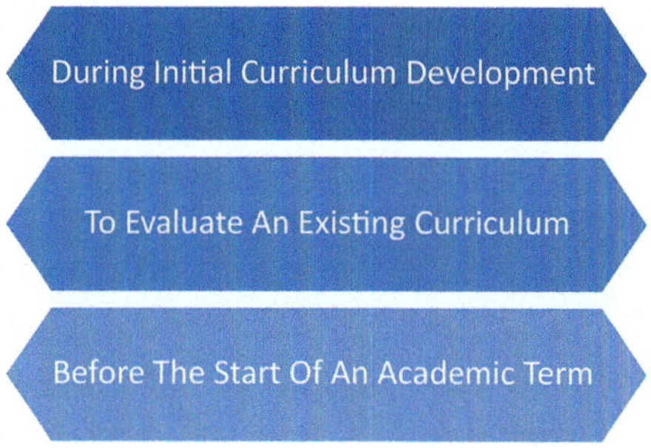


FIGURE 10.3 **When to design a curriculum map.** This figure shows when to design a curriculum map.

courses, it is necessary to study and analyze the curriculum map. In order to make sure that every core course supports at least one program learning outcome and to identify any core courses that do not, it is crucial to simultaneously check for course support. Additionally, it's important to assess skill development and search for learning outcomes that have been excessively repeated. This is helpful for ensuring alignment with standards and outcomes, as well as allowing a post-hoc study of what is currently being taught to identify any gaps or redundancies.

#### *10.6.6.3 Before the start of an academic term*

Preparing a curriculum map in advance of a term is essential since it serves as a road map for education, guaranteeing that students' learning progresses in a coherent and straightforward manner. In the end, it improves student learning and outcomes by assisting in the identification and removal of gaps or redundancies, aligning learning outcomes with assessments, and enabling teachers to maintain pace targets. The curriculum map serves as a guide for the entire term, offering teachers a clear path and lowering uncertainty about what to teach next, which is one of the process's advantages. Additionally, it helps teachers stay on track to accomplish their term's deadlines and instructional goals, which enhances pacing. In order to make sure they cover the necessary content and are in line with the overall curriculum, teachers can utilize the map for daily planning to check what has been taught and what is planned for the future.

### 10.6.7 Role of stakeholders in curriculum mapping process

Stakeholders including learners, teachers, administrators, parents, and the community are vital to the curriculum mapping process by giving input, feedback, and support at different phases. To maximize the curriculum's efficacy, learners are the main focus, instructors are its implementers, administrators make sure resources are available, and parents and community members provide insightful viewpoints, local knowledge, and support.

#### *10.6.7.1 Learners*

Since they offer a distinct, firsthand perspective on what is really being taught and understood, learners play a critical role in curriculum mapping by helping to discover gaps, overlaps, and areas for improvement from their point of view. By making learning routes clearer and facilitating the application of knowledge in practical settings, their involvement can also result in more motivation and engagement. Students can also use the curricular map to better understand expectations and how courses relate to one another. Students participate in the process by offering their own viewpoint. Students can offer a genuine perspective of what is really taught and learnt by engaging with the curriculum, which may deviate from the scheduled content. Taking an active part in the curriculum benefits the students as they gain an increased sense of motivation and involvement. Students' motivation and engagement can increase when they perceive a clear, cohesive learning pathway throughout their degree.

#### *10.6.7.2 Teachers*

They are important developers and implementers who offer firsthand input on what works and what doesn't in the classroom. They add to the curriculum's practical design and enrichment. As the implementers who find gaps and duplications, guarantee vertical and horizontal alignment, and offer suggestions for improvement, teachers play a crucial role in curricular mapping. Their responsibilities include working with other educators to develop a common vision, tying program results to topic outcomes, and recording actual classroom activities to guide future changes. Teachers play a crucial role in curriculum mapping by connecting learning by aligning their particular topic content with the larger program and course-level targets. This ensures that students' abilities and knowledge move logically across semester levels and disciplines. As students collaborate with colleagues to comprehend how their disciplines contribute to shared program goals, it also promotes cooperation and a shared vision.

#### *10.6.7.3 Administrators*

They manage the process from an institutional standpoint, supervise the curriculum's execution, and guarantee that it is adequately supported by resources. By offering direction, resources, and supervision, administrators are in charge of starting, assisting, and supporting the curriculum mapping process. They are responsible for collecting data, creating a cooperative atmosphere, and making sure the finished maps are in line with the objectives of the entire institution. In order to ensure that the curriculum is implemented successfully, they also examine and approve the maps and provide professional development support to teachers. As they start the process and make sure it is in line with the institution vision and objectives, administrators play a crucial role as visionaries and initiators. In order to direct the process, they also collect and distribute data on standards, philosophies, and trends. In their capacity as facilitators, they establish a cooperative setting where educators can collaborate successfully using techniques like roundtables and brainstorming. Additionally, they offer the structures and resources required for the entire procedure, including safe online platforms.

#### *10.6.7.4 Community members*

Community people are essential to the medical curriculum mapping process because they can contribute local expertise, act as resources or guest speakers, and connect the curriculum to outside possibilities and needs that enhance its relevance. Additionally, they make sure that the instruction is culturally sensitive, socially responsible, and pertinent to the health needs of the community. Through their involvement, academic institutions and the society they serve are better connected. Community members play a crucial role in identifying and prioritizing health needs because they offer invaluable insight into the real health issues, priorities, and difficulties that the local population faces, including the social determinants of health, such as the circumstances surrounding a person's birth, growth, living, employment, and aging. The curriculum can stay current and adaptable to the shifting requirements of the community thanks to this input. They also contribute to the design and content of the curriculum by giving medical education a real-world context that might not be available in a conventional hospital setting.

#### *10.6.7.5 External stakeholders*

By guaranteeing that the curriculum is pertinent, up to date, and in line with the real needs of the healthcare system and society, external stakeholders play a crucial role in medical curriculum mapping. These include governmental bodies, professional associations, and clinical partners who can have an impact on the curriculum by establishing guidelines and standards. Their contributions help create graduates who can handle issues in the real world and inform the process of ongoing quality improvement. Ensuring relevance and competency is one of the main functions and contributions of external stakeholders. In order to ensure that the curriculum develops the information and skills required for effective performance in a dynamic context, clinical partners such as practitioners, hospital managers, and directors of residency programs offer insights into patient complexity and the difficulties of contemporary healthcare settings. In order to close the "transition gap" between undergraduate and graduate medical education, they can assess students' readiness and offer suggestions for enhancements. However, they also ensure that the curriculum satisfies the demands of society. Patients, families, local communities, and public health organizations provide input that enables medical school programs to reflect population health needs and lived experiences, improving healthcare and educating graduates to be champions for equity and health improvement.

## 10.7 Importance of curriculum mapping

As it links courses to standards and facilitates a seamless flow of learning across disciplines and semester levels, curriculum mapping is crucial for ensuring educational alignment. It improves student performance and creates a more cohesive learning environment by assisting in the identification and closure of information gaps or overlaps. It also helps teachers by promoting teamwork, facilitating data-driven curricular enhancements, and assisting them in comprehending the structure of the program. The significant advantages of curriculum mapping are shown in Fig. 10.4.

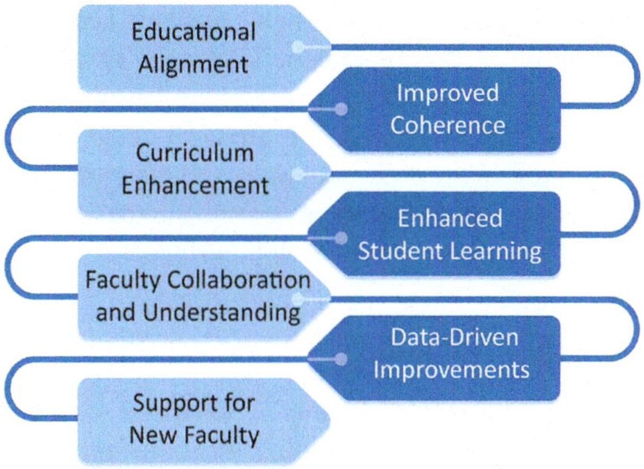


FIGURE 10.4 **Importance of curriculum mapping.** This figure shows the importance of curriculum mapping.

### 10.7.1 Educational alignment

Curriculum mapping is crucial for educational alignment because it ensures that learning objectives, instructional methodologies, and assessments are consistent and in accordance with program objectives and standards (Herrmann T 2019). It generates a visual summary that indicates any gaps or overlaps in the material, prevents duplication, and aids students in understanding the logical flow of their learning. This process helps teachers' continuous development and promotes consistency across courses, semester and levels. Curriculum mapping improves educational alignment by ensuring that instruction meets program objectives and external standards, such as state or national learning outcomes, by tying what is taught to what is assessed.

### 10.7.2 Improved coherence

A coherent learning process is a connected, sequential educational journey in which all components such as curriculum, instruction, assessment, and goals work together harmoniously, building on past knowledge and reinforcing one another to produce a cohesive, meaningful, and effective learning experience free from pointless details that strain working memory. Students can build on prior information without experiencing significant gaps or unnecessary repetition when a consistent learning sequence is provided. Curriculum mapping creates a steady flow of knowledge throughout courses and grades by aligning learning objectives with what is taught and assessed (Plaza CM 2007). In addition, it gives both teachers and students a clear, structured path since it makes sure that students don't repeat knowledge needlessly or overlook crucial topics. For learning to be more efficient and seamless, this alignment and organization are essential. Coherence enhances both vertical and horizontal alignment as it ensures that knowledge and skills grow logically over time and that there is consistency among courses at the same level. Since mapping provides educators with a comprehensive view, it also helps them discover any locations where content might be missing or duplicated, preventing gaps and overlaps. It also allows for adjustments to increase efficiency and ensure that all specifications are met.

### 10.7.3 Curriculum enhancement

Curriculum maps assist teachers in identifying areas for improvement by giving them a visual depiction of what is taught. This results in stronger curricula and more successful teaching techniques. Curricular mapping improves curricula since it guarantees adherence to standards, enhances coherence, and identifies gaps or duplications for ongoing improvement (Gulbis, K 2021). It provides educators with a framework for organizing lessons, collaborating effectively, and utilizing data to enhance teaching strategies to better meet the needs of their students. It also improves instructional practice by helping teachers with class preparation, progress tracking, and timeliness. It also helps teachers understand what learners have already learned and what they will learn in the future, which leads to improved education.

### 10.7.4 Enhanced student learning

Learners are better able to understand where they are in their learning process, which aids in memory retention and test preparation. Curriculum mapping enhances student learning by providing students with a more defined and dependable learning path by ensuring congruence between learning outcomes, instruction, and assessment. It steers clear of repeated material, promotes consistency across semester or levels, and helps teachers keep track of students' progress so they may modify their lessons as needed. This approach helps students build on their prior knowledge, become more engaged, and understand the purpose of their education by establishing a scaffolded learning environment. It provides students with clarity and focus, which improves their learning. When students are clear about what they must learn, why it is important, and how it will be evaluated, they are more interested and successful. A curriculum map guarantees that all students, regardless of their teacher, have a cohesive and consistent learning experience while also ensuring continuity and consistency by guaranteeing that the key concepts are taught without gaps or unnecessary repetition.

### 10.7.5 Faculty collaboration and understanding

Teachers can work together and communicate across departments, semesters, or levels by using maps as a common point of reference. Additionally, this aids educators in comprehending what students have learnt thus far and what they will study going forward. Curriculum mapping enhances faculty collaboration and comprehension by creating a shared framework that clarifies the purpose of each course, identifies any gaps or duplications, and promotes the sharing of resources and best practices. This process ensures that students have a cohesive learning experience, encourages teacher interaction, and allows for rigorous curriculum change. By offering a clear and visual framework that aligns courses with program objectives and student learning outcomes, curriculum mapping guarantees that all faculty members are on the same page. By encouraging shared knowledge, this improves faculty collaboration. Additionally, by serving as a conduit for communication between educators, it promotes collaborative discussions about curriculum, skills, and assessments.

### 10.7.6 Data-driven improvements

Data driven improvement methods offer an evidence-based approach to curriculum review and modification by utilizing assessment data to measure how well students are meeting learning objectives. Curriculum mapping is crucial for data-driven improvements because it provides teachers with an open, methodical framework for collecting, evaluating, and using student performance data to make informed decisions about instruction and content. Teachers can easily identify any gaps or repetitions in the curriculum by graphically comparing the "delivered curriculum" that is what is actually taught with the "intended curriculum" which refer to the standards and learning objectives. Data may then verify whether these locations are connected to particular student challenges. Aligning with standards and assessments is beneficial because curriculum maps facilitate the matching of teaching materials and assessments with institutional or national requirements. By ensuring that assessment data appropriately reflects students' progress toward predetermined learning outcomes, this alignment increases the data's usefulness for attempts to improve.

### 10.7.7 Support for new faculty

A centralized curriculum map can be a helpful tool for new teachers to understand the content and structure of the curriculum. Curriculum mapping is crucial for new teachers because it provides a road map that clarifies what should be taught and when, aligns courses with standards, and lessens the workload. It creates structure, increases confidence, and streamlines the process of creating daily courses by giving a clear image of the scope and sequence of the full year. This allows new teachers to focus more on the needs of their students rather than always creating plans from scratch. Curriculum mapping provides organization and clarity, with a clear, visual layout of the curriculum that makes it plain what has been taught, what comes next, and how particular courses link to more general subjects. These are the key benefits of curriculum mapping for new instructors. This clarity is quite beneficial for new teachers acclimating to a new system. It also aids in standardizing alignment and assisting new teachers in coordinating their education with official outcomes by guaranteeing that assessments and lessons are created to satisfy institutional or national regulations.

## 10.8 Challenges in creating the curriculum mapping

Obtaining teacher support, the time and resources needed, the fragmented and inconsistent curriculum design across several courses or instructors, and issues with technical tools are all obstacles to creating a curriculum map (Chakrabarti, R 2021). Dealing with misinterpretations of learning outcomes, maintaining vertical and horizontal alignment between various courses, and managing the process with clear leadership and communication are additional difficulties. Among the typical difficulties are the following:

### 10.8.1 Time consumption

Since it requires a substantial, frequently excessive amount of work from the medical educators, time consumption is a major obstacle in the curriculum mapping process in medical education. This significant time investment may result in a number of issues, such as data management challenges and teacher alienation. The procedure of collecting extensive data as part of the data gathering for a medical curriculum is usually very time-consuming, complex, and dense. This frequently entails gathering data from a large number of faculty members from various departments and locations, each of which has its own teaching strategies and learning goals. Additionally, a lot of medical schools first compile their curricula using manual techniques like many spreadsheets. Due to the sluggish and error-prone nature of data entry and maintenance, this procedure is extremely inefficient and adds greatly to the time burden (Shilling, Tamara, 2013).

### 10.8.2 Lack of teacher acknowledgment and collaboration

A major obstacle in medical curriculum mapping is the lack of teacher support and cooperation, which is caused by things like worries about course autonomy, excessive workloads, and a lack of knowledge about the goals and advantages of the mapping. Accurate recording and ongoing curriculum improvement may be hampered by this reluctance. The faculty members' perception that autonomy is at danger is one of the causes of this difficulty. Curriculum mapping may be seen by faculty members who are used to individual teaching approaches as a challenge to their authority over their classes and methods of instruction. The time-consuming process of curriculum mapping is an additional strain for medical educators, who frequently have to juggle the demanding needs of teaching, research, and clinical care. Furthermore, one of the main obstacles in this regard is a lack of knowledge and training. Teachers might not have the required expertise in curriculum mapping software and technique, and they might not completely comprehend the "big picture" of the entire curriculum or how their particular course fits into it.

### 10.8.3 Insufficient training and support

A major obstacle to the medical curriculum mapping process is inadequate training and support, which frequently results in faculty resistance, unreliable data, and an inability to use the map effectively for accreditation and quality improvement. Faculty resistance and lack of engagement are some of the main effects of inadequate training and support. When faculty members are unfamiliar with curriculum mapping concepts and the tools involved, they may oppose the process or fail to finish the task successfully because they do not understand its value and purpose. Additionally, different departments and individuals use different terminology to tag comparable content due to a lack of standardized training in medical education and the usage of curriculum mapping tools. This "inconsistent labeling" makes it impossible to match with national competency frameworks and conduct meaningful analysis.

### 10.8.4 Maintaining and updating regularly

In medical education, maintaining and updating a curriculum map on a regular basis is a major and well-known difficulty. Although a curriculum map is a "living document," practical challenges frequently make it difficult to maintain it up to date. Slow feedback loops, where curricular modifications frequently need several levels of review and committee sign-offs, are one of the reasons behind this. Faculty members may become disengaged and students may not be able to benefit from critical improvements as a result of these delayed processes. Furthermore, there is a lack of coordination and ownership where the mapping project may eventually lose pace and accuracy if there is no designated coordinator in charge of communication and content updates.

### 10.8.5 Resource constraints

The curriculum mapping process in medical education is severely hampered by resource limitations, which mainly affect the availability of crucial components like faculty time, committed staff, funding, and suitable technology infrastructure. These restrictions may make the mapping effort less accurate, thorough, and long-lasting. The absence of specialized curriculum committees or experts in curriculum design, alignment, and mapping could hinder the process. Instructors may not have the necessary guidance and support to use mapping tools and connect course material with objectives and assessments if they are not properly trained. Adoption of contemporary, centralized curriculum management systems or software may also be hampered by budgetary restrictions. Version control instability, fragmented data, and inefficiency can result from many institutions' reliance on manual techniques like spreadsheets or isolated platforms.

### 10.8.6 Technology integration issues

The medical curriculum mapping process has a number of difficulties with technology integration, such as faculty members' lack of technological skills, insufficient infrastructure and funding, and educators' opposition to change. Deficits in faculty expertise and training are the main obstacles to technology integration. Since many medical educators lack the formal training and technical know-how needed to use and incorporate digital tools into curriculum mapping and instruction. This lack of expertise may result in uneven application and discourage academics from using new technology. A major obstacle, especially in environments with limited resources, is the absence of a strong technical infrastructure, which includes inadequate hardware, software, internet connectivity, and backup systems. Another issue is the high upfront expenditures of deploying and maintaining cutting-edge equipment.

### 10.8.7 Ensuring coherence and alignment

Due to the complexity of medical information, the wide range of varied stakeholders and teaching locations, and possible faculty resistance or lack of training, ensuring coherence and alignment in medical curriculum mapping is a major problem. The complexity and volume of content are some of the main obstacles in this regard. The volume and constantly changing nature of medical knowledge make it challenging to arrange content logically and guarantee that fundamental ideas are suitably reinforced at various training stages. Additionally, there are many departments, faculty members, and affiliated hospitals involved in medical education, with each having its own teaching philosophies and areas of interest. It can be difficult to organize these disparate groups to work toward a single curricular goal since faculty members may operate independently and make different decisions about how to present and evaluate the material. Educators and administrators can create methods, such as starting small and setting aside time for preparation, to create more sustainable and successful curriculum mapping procedures by being aware of these typical obstacles.

**Summary**

The methodical process of documenting, analyzing, and aligning a curriculum's components to make sure that instruction and learning are consistent, cohesive, and in line with intended results is known as curriculum mapping. It usually entails mapping out what is taught, how it is taught, and how learning is evaluated across programs, courses, or grades. Learning objectives or standards, instructional methodologies, material, assessments, and the order or chronology of instruction are important components that are frequently included in a curriculum map. Teachers can find curriculum gaps, duplications, and misalignments by organizing this data graphically or in a systematic manner. Curriculum mapping guarantees that students receive a comprehensive and rationally organized educational curriculum, facilitates better instructional preparation, and encourages teacher collaboration. Additionally, it encourages continuous reflection and improvement of teaching methods and assists medical schools in matching curriculum to standards.

**Key point**

**Future perspectives: from identifying gaps to real-time collaboration**

In the future, medical educators will play a critical role in curriculum mapping by acting as designers and data analysts. They will do more than just impart knowledge as they will be able to strategically organize classes and use data to adapt them to students' needs. They will collaborate with peers, integrate both new and existing material, and use technology to enhance efficiency and personalization in order to create flexible, student-first learning pathways. Teachers will actively create, adapt, and modify the curriculum based on student needs and learning objectives rather than just following a premade plan. They will create lesson plans, identify connections between several subjects, and ensure that instruction is appropriately scheduled and structured. Curriculum mapping will depend more and more on cutting-edge digital technologies and AI-powered platforms in the future to automate data collecting, identify gaps, and make recommendations for enhancements. These tools have the ability to assess student performance and match curricular changes with current data. Personalized learning routes are anticipated to be included in future curriculum maps. With the increased flexibility of maps, teachers will be able to modify the content and pace according to the requirements, interests, and readiness levels of their students.

## References

Al-Eyd, G., Achike, F., Agarwal, M., Atamna, H., Atapattu, D.N., Castro, L., et al., 2018. Curriculum mapping as a tool to facilitate curriculum development: a new school of medicine experience. BMC Medical Education 18 (1). Available at https://doi.org/10.1186/s12909-018-1289-9.

Alam, Tanveer, 2022. Informatica. 46.

Balzer, F., Hautz, W.E., Spies, C., Bietenbeck, A., Dittmar, M., Sugiharto, F., et al., 2016. Development and alignment of undergraduate medical curricula in a web-based, dynamic learning opportunities, objectives and outcome platform (LOOOP). Medical Teacher 38 (4), 369–377. Available at https://doi.org/10.3109/0142159X.2015.1035054, http://www.tandfonline.com/loi/imte20#.VufR2bdf2DY.

Bayer, C.R., Respress, E., Willock, R.J., Heiman, H.J., 2019. CurriCulum mapping and needs assessment to inform the training of health policy leaders. Ethnicity and Disease 29, 413–420. Available at https://doi.org/10.18865/ED.29.S2.413, https://www.ethndis.org/edonline/index.php/ethndis/article/view/1006.

Bruinsma, M., Jansen, E.P.W.A., 2007. Curriculum mapping: integrating multiple perspectives on the curriculum. Curriculum and Teaching 22 (1), 25–45. Available at https://doi.org/10.7459/ct/22.1.03.

Chakrabarti, R., Wardle, K., Wright, T., Bennie, T., Gishen, F., 2021. Approaching an undergraduate medical curriculum map: challenges and expectations. BMC Medical Education 21 (1). Available at https://doi.org/10.1186/s12909-021-02778-6.

Gulbis, K., Krüger, T., Dittmar, M., Peters, H., 2021. Approaches to mapping an undergraduate medical curriculum to a national competency-based catalogue of learning outcomes. Medical Teacher 43 (4), 439–447. Available at https://doi.org/10.1080/0142159X.2020.1854704, http://www.tandfonline.com/loi/imte20#.VufR2bdf2DY.

Harden, R.M., 2001. AMEE Guide No. 21: Curriculum mapping: a tool for transaprent and authentic teaching and learning. Medical Teacher 23 (2), 123–137. Available at https://doi.org/10.1080/01421590120036547.

Hege, I., Nowak, D., Kolb, S., Fischer, M.R., Radon, K., 2010. Developing and analysing a curriculum map in occupational- and environmental medicine. BMC Medical Education 10 (1). Available at https://doi.org/10.1186/1472-6920-10-60.

Herrmann, T., Leggett, T., Herrmann, T., Leggett, T., 2019. Curriculum mapping: aligning content and design radiol technol. 2019;90:530–3 cited here | PubMed | Google scholar. Radiologic Technology 90 (5), 530–533. Available at http://www.radiologictechnology.org/content/90/5/530.full.pdf#page=1&view=FitH.

Kopera, Frye, 2008. Collected Essays on Learning and Teaching. 1.

McLean, M., Gibbs, T., 2010. Twelve tips to designing and implementing a learner-centred curriculum: prevention is better than cure. Medical Teacher 32 (3), 225–230. Available at https://doi.org/10.3109/01421591003621663.

Plaza, C.M., Draugalis, J.L.R., Slack, M.K., Skrepnek, G.H., Sauer, K.A., Plaza, C.M., et al., 2007. Curriculum mapping in program assessment and evaluation Am J Pharm Educ. 2007;71:20 Cited Here | PubMed | CrossRef | Google Scholar. American Journal of Pharmaceutical Education 71 (2), 20. Available at https://doi.org/10.1016/s0002-9459(24)03309-6.

Shilling, T., 2013. Opportunities and challenges of curriculum mapping implementation in one school setting: considerations for school leaders. Journal of Curriculum and Instruction 7 (2). Available at https://doi.org/10.3776/joci.2013.v7n2p20-37.

Shrivastava, S.R.B.L., Shrivastava, P.S., 2021. Scope and need of curriculum mapping in accomplishing learning outcomes in the field of medical education. Indian Journal of Health Sciences and Biomedical Research (KLEU) 14 (2), 178. Available at https://doi.org/10.4103/kleuhsj.kleuhsj_388_20.

Waldvogel, TEE, 2025. BMC Med Educ. 25.

Watson, Wendy, 2020. Focus on Health Professional Education. 21.

# Chapter 11

# Evaluation of the curriculum

*"Evaluation's most important purpose is not to prove but to improve."*

—Daniel Stufflebeam

## 11.1 Introduction

The quest of excellence has become more than just a goal in the constantly changing global medical educational scene. The vital responsibility of developing the next generation of leaders, thinkers, and innovators is placed on educational institutions worldwide. Medical schools must constantly innovate, adapt, and improve their programs and curricula in order to fulfill this enormous obligation. In this context, curriculum evaluation refers to the methodical process of gathering and evaluating data in order to assess the value, efficacy, and quality of an educational program, as well as whether it satisfies standards and accomplishes its objectives (Coles & Grant, 1985). It entails evaluating objectives, content, learning opportunities, and results using a variety of techniques in order to offer input for continuous enhancements, direct decision-making regarding modifications and guarantee conformity with student requirements and standards.

## 11.2 Purpose of curriculum evaluation

### 11.2.1 Determine the strengths and weaknesses of the curriculum

Curriculum evaluation helps identify strengths in terms of effective content, engaging methods, good outcomes, and weaknesses in terms of knowledge gaps, irrelevant material, and poor alignment by methodically collecting data from students, teachers, and administrators and employing formative and summative methods to ensure goals are met, enhance instruction, and direct future development for better student learning. It entails evaluating student performance, obtaining input from stakeholders, comparing results to standards, and making data-driven decisions for ongoing development.

### 11.2.2 Establishing the student achievement of the curriculum goals

Curriculum evaluation's purpose of assessing student achievement of goals is to determine if the curriculum is effective in helping students meet learning objectives, which then guides improvements to the curriculum, teaching methods, and instructional approaches. This process provides data to judge the overall quality of the program, identify areas needing revision, and ensure the curriculum is developing students as intended. Ultimately, it helps to guarantee that the students receive the best possible education by confirming that teaching meets their needs and that the curriculum leads to desired outcomes.

### 11.2.3 To monitor students' learning experiences while in the program

One of the goals of curriculum evaluation is to track students' learning experiences since it gives teachers direct feedback on how the curriculum impacts students, enabling them to pinpoint areas that need improvement. Through both informal

**The Principles and Practice of Medical Education. DOI:** https://doi.org/10.1016/B978-0-443-44197-4.00002-3

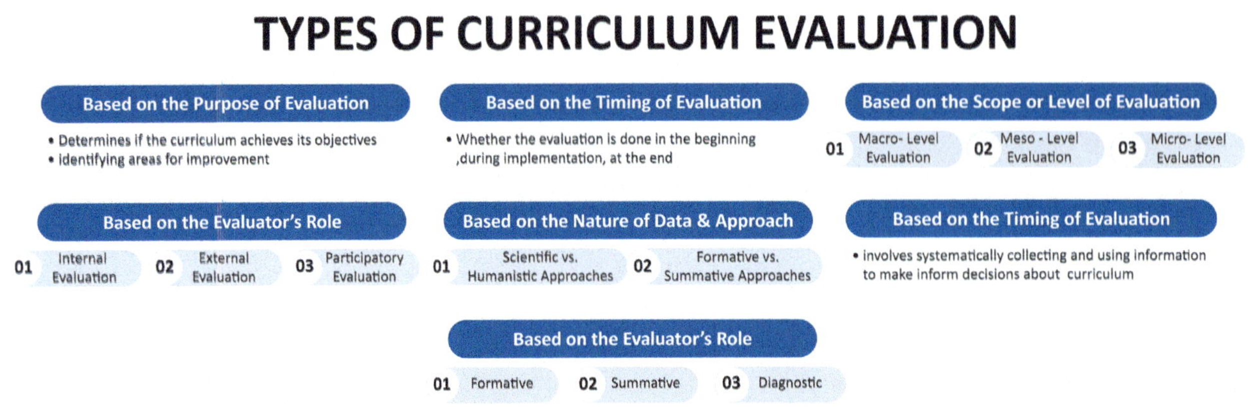


FIGURE 11.1 **Types of curriculum evaluation.** This figure shows the types of curriculum evaluation.

and formal means, this process helps to ensure that students are reaching learning objectives and permits adjustments to improve their entire learning experience.

### 11.2.4 Review, modification, and reorganization of the curriculum

Curriculum assessment fulfills the fundamental goals of review, adaptation, and rearrangement since it aids in methodical collection of information on the efficacy and efficiency of an educational program. It also helps identify whether a curriculum needs to be modified, enhanced, or replaced by evaluating its effectiveness. This procedure supports ongoing educational improvement by ensuring that the curriculum stays effective, current, and in line with its objectives. Essentially, it is a quality control mechanism to assure that the curriculum, its courses, the processes undertaken, teaching strategies, and student achievement of goals are meeting the required standards (Ullah et al., 2024).

## 11.3 Types of curriculum evaluation

The types of curriculum evaluation are illustrated in Fig. 11.1.

### 11.3.1 Based on the purpose of evaluation

Curriculum evaluation can be based on its purpose, which includes determining if a curriculum achieves its objectives, identifying areas for improvement, and informing decisions about its continuation, modification, or elimination. Other purposes involve assessing a curriculum's effectiveness against national standards, ensuring it meets the needs of the digital age, or aligning it with modern educational research. Ultimately, the purpose of the evaluation dictates the focus, data collection methods, and the ultimate goal of the assessment process.

### 11.3.2 Based on the timing of evaluation

The main types of curriculum evaluation based on the timing are formative, summative, and diagnostic, which differ by their time of conduction.

#### *11.3.2.1 Formative evaluation*

Formative evaluation is an ongoing process undertaken during learning or program development with an aim to gather feedback and make improvements in real-time, rather than waiting until the end. Its goal is to monitor progress, identify strengths and weaknesses, provide timely feedback for adjustment, and enhance effectiveness, contrasting with summative evaluation, which judges final outcomes.

#### *11.3.2.2 Summative evaluation*

Unlike formative assessment, which directs learning during the process, summative evaluation is a formal assessment carried out at the conclusion of a learning period, such as a unit, course, or program, to measure overall achievement against set standards and determine the "worth" or final outcome of the instruction.

#### 11.3.2.3 Diagnostic evaluation

A diagnostic evaluation is a comprehensive assessment to identify the root cause of a problem in medical education such as finding a disease or assessing learning gaps, or other fields by gathering detailed information to understand strengths, weaknesses, and specific needs, guiding targeted interventions rather than just general classification.

### 11.3.3 Based on the scope or level of evaluation

This classification is based on the degree of remoteness from the direct teaching and learning experiences of the students and the scale of decision-making involved. Curriculum evaluation, when classified by its scope or level, generally falls into three main categories:

**1. Macro-level evaluation**

Macro-level curriculum evaluation assesses the broad framework of a curriculum, focusing on overall goals, policies, and the balance of subjects, rather than individual lessons. It involves a holistic review of the entire program to ensure it aligns with societal and educational objectives, often using methods like surveys, analysis of policy documents, and student progress data across the institution.

**2. Meso-level evaluation**

Meso-level curriculum evaluation involves assessing the curriculum at the institutional level, focusing on how the curriculum is planned, implemented, and managed within the institutional context. This evaluation examines how the curriculum aligns with broader policies, the effectiveness of teaching strategies across grade levels, and the coordination of resources. It is a bridge between national policies which represent macro level and classroom activities which are considered as micro level.

**3. Micro-level evaluation**

Micro-level curriculum evaluation focuses on the classroom-level implementation of a specific, often smaller, section of a curriculum. It involves assessing and adjusting teaching strategies, learning resources, and activities in real-time or through a cycle of observation and feedback, without changing the curriculum's core structure. This evaluation uses tools like observations, feedback, and assessments to refine the delivery of specific topics to meet learning objectives.

### 11.3.4 Based on the evaluator's role

#### 11.3.4.1 Internal evaluation

It is done by teachers and administrators often involved in formative evaluation during curriculum development and implementation. Their role is to provide ongoing feedback, identify strengths and weaknesses, and make timely adjustments to teaching methods and materials to improve student learning.

#### 11.3.4.2 External evaluation

This type of curriculum evaluation typically involves an independent individual or group hired for summative evaluation. They provide an objective, unbiased assessment of the final curriculum's overall effectiveness, often for public accountability or policy decisions for instance taking the decision whether to adopt a curriculum on a large scale or discontinue it.

#### 11.3.4.3 Participatory evaluation

Participatory curriculum evaluation is an approach that involves all stakeholders, including students, teachers, and community members, in the evaluation process to assess and improve the curriculum. This method emphasizes collaboration, using the perspectives of primary users to enhance the utilization of evaluation findings and create more relevant and effective programs.

### 11.3.5 Based on the nature of data or approach

Based on the nature of data or the approach used, curriculum evaluation can be broadly categorized into the following types:

**1. Scientific versus humanistic approaches**

The scientific approach to curriculum evaluation uses quantitative data and objective, experimental methods to measure outcomes, focusing on systematic organization and achievement of specific goals. In contrast, the humanistic approach uses qualitative data and a holistic perspective to understand the experiences, interpretations, and social interactions of all stakeholders, emphasizing flexibility and learner-centered participation. The choice of approach depends on the context, but a blended approach is often recommended for a comprehensive view (Adebayo, 2018).

**2. Formative versus summative approaches**

Formative curriculum evaluation happens during development to fix flaws and improve the process, while summative evaluation occurs after completion to judge the final curriculum's success. Formative approach is aimed at providing continuous feedback to make improvements, while summative approach is aimed at learning, providing a final judgment on a student's or program's achievement.

### 11.3.6 Based on the decision-making function (Stufflebeam's context, input, process, product [CIPP] model)

Curriculum evaluation is fundamentally a decision-making process that systematically collects and uses information to determine the value, effectiveness, and quality of an educational program and guide future improvements. The function of evaluation is to provide data-driven insights to help stakeholders make informed decisions. Curriculum evaluation based on a decision-making function involves systematically collecting and using information to inform decisions about whether to continue, stop, modify, or disseminate a curriculum. This approach moves beyond simply measuring outcomes and integrating data from various assessments to guide practical improvements and strategic planning (Warju, 2016). Several curriculum evaluation models are designed specifically to facilitate different types of decision-making include Stufflebeam's context, input, process, product (CIPP) model and Cronbach's model.

### 11.3.7 Based on evaluation focus in medical education

Based on evaluation focus, curriculum evaluation in medical education can be categorized as formative such as in during development, summative which refers to the final product, and diagnostic wherein the attempt identifying causes of deficits. Other important evaluation types include process evaluating implementation and outcome evaluating results. These can be further structured through evaluation models like CIPP (context, input, process, product) or Kirkpatrick's four-level model.

## 11.4 Components of curriculum evaluation

The main components of curriculum evaluation include evaluating the curriculum's aims and objectives, its content, the learning experiences provided, and the evaluation methods themselves. These components are part of a larger process that includes defining the evaluation's purpose, collecting and analyzing data, and using the findings to improve the curriculum.

### 11.4.1 Curriculum goals

When planning the evaluation of curriculum goals, faculties want to determine if the curriculum goals are appropriate and reasonable. More specifically, faculties are interested in the extent to which the curriculum goals reflect the practice and standards of the educational institution. Curriculum evaluation assesses whether the curriculum is achieving its stated goals, which are the broad, long-term aims of the program, by examining student outcomes, content, and teaching methods. This allows for objective judgments on the curriculum's quality and helps determine if revisions are needed to meet educational objectives. The roles of the curriculum in evaluation lies in the fact that the goals establish the benchmark as they define the ultimate purpose and expected achievements of the curriculum, such as preparing students for responsible citizenship. Curriculum evaluation is then designed to systematically determine if these goals are being met. At the same time, they guide the evaluation design. The goals inform the evaluation process by helping to establish criteria and develop specific questions for evaluation. For example, if a goal is to develop critical thinking skills, the evaluation will focus on assessing student performance in areas that demonstrate critical thinking.

### 11.4.2 Curriculum design

When planning evaluation of the curriculum design, the scope of this component becomes evident. The curriculum design encompasses the curriculum goals and the components of the program such as courses, their sequence, alignment, and mode of delivery. As well, it includes teacher and student activities, and policies governing the curriculum. Curriculum design is a component of evaluation because it creates the blueprint that evaluation assesses. The curriculum design specifies the curriculum's aims, content, and learning experiences, while evaluation measures how well these elements were achieved and identifies areas for improvement. Evaluation provides data-driven insights that inform future curriculum design revisions to better meet student needs and educational goals. The design process defines what the curriculum is supposed to achieve, including specific objectives, content, and desired learning experiences. These elements become the benchmarks against which success will be measured during evaluation. The curriculum design also includes the assessment plan as it is a vital part of the evaluation process. This involves creating the specific tools and methods that will be used to assess student outcomes and the overall effectiveness of the curriculum.

### 11.4.3 Teaching strategies

Teaching strategies are a crucial component of curriculum evaluation because they are the methods used to implement the curriculum, and their effectiveness directly shows how well the curriculum is meeting its goals. Evaluating teaching strategies helps determine if the methods are stimulating learning, achieving objectives, and if changes are needed for continuous improvement. The teaching strategies are used in curriculum evaluation to connect the objectives to outcomes as the teaching strategies translate curriculum objectives into action. Evaluating how well a strategy works shows if students are meeting the intended learning outcomes. It also contributes to assessing effectiveness as by evaluating the success of specific teaching strategies, evaluators can determine the overall effectiveness of the curriculum and provide feedback for improvement. When planning evaluation of teaching strategies, faculty can be guided by literature that describes effective teaching and lists desirable teacher competencies, behaviors, or characteristics for classroom and clinical courses. These ideas from the literature can be adopted, adapted, or extended to suit the curriculum.

### 11.4.4 Courses

Evaluation of courses is, in some measure, is a microcosm of the evaluation of curriculum goals, design, and outcomes in which all aspects of course design and implementation are considered. Courses are a key component of curriculum evaluation because they embody the curriculum's structure, content, and learning experiences. Evaluating courses involves assessing how well they achieve their objectives, examining the course's structure such as the topic selection, organization, content, and the learning experiences provided, and using this information to make improvements to the curriculum as a whole. Evaluating a course's effectiveness directly reflects how well it has translated the curriculum's goals into actual content and learning activities. One of the prime focuses of evaluation is the way the courses are structured and organized including the selection and sequencing of topics. It is vital that the organization must be logical and support the learning objectives.

### 11.4.5 Human and physical resources

An important dimension of curriculum evaluation is ascertaining if suitable and sufficient human and physical resources are present. Human and physical resources are essential components of curriculum evaluation because their adequacy, quality, and management directly determine the feasibility and effectiveness of curriculum implementation and the achievement of educational objectives. Evaluation of the curriculum must assess the availability and proper utilization of these resources to identify strengths, weaknesses, and areas for improvement. The importance of human resources in the context of education primarily refers to teachers, administrative staff, supervisors, and external evaluators. Their role in curriculum evaluation involves several aspects such as in implementation and delivery of the curriculum in order to uphold the quality of the curriculum which largely depends on the competence and professional development of the teachers who implement it. Evaluation assesses if teachers have the necessary skills, knowledge, and training to deliver the content effectively. Additionally, as active participants in evaluation, teachers and staff provide crucial feedback on the practical challenges and successes of the curriculum in the classroom. They are the primary source of data on student engagement, learning barriers, and the feasibility of instructional strategies.

### 11.4.6 Learning climate

The learning climate is the social, emotional, and intellectual atmosphere that exists within the institution. It influences the quality of life of students, faculty, and staff. The learning climate is important regardless of the medium through which the curriculum is offered. Learning climate is a major and essential component of curriculum evaluation, serving as both a worthwhile outcome in its own right and a significant determinant of student academic achievement. It is widely used as a key criterion for assessing the effectiveness of educational programs and instructional methods. Curriculum theorists consider the learning environment and instruction as the two major codeterminants of student outcomes. Evaluating the learning climate helps to assess the effectiveness and determine if a new curriculum or teaching strategy is successfully creating the intended positive environment, for instance, more student-centered, innovative, or collaborative. It is a fact that a new curriculum often creates a measurably different learning climate even when traditional outcome measures cannot distinguish between curricula. At the same time, it can help in identifying the strengths and weaknesses as it pinpoints the specific elements of the educational program that need improvement, such as inadequate teacher support or a poor social atmosphere.

### 11.4.7 Polices

Curriculum policies are intended to support students' achievement of curriculum goals while ensuring that academic standards are maintained. Therefore, in reviewing and evaluating curriculum policies, it is important to consider whether the policies are appropriate, reasonable, understood by faculty and students, and applied consistently. Evaluators might also wish to ascertain if there have been situations that might indicate a need for new policies. Policies are an integral and foundational component of curriculum evaluation. They provide the mandatory framework, guidelines, and standards against which the curriculum's effectiveness, relevance, and alignment with legal and societal requirements are assessed. Policies function as a component of curriculum evaluation by contributing towards the establishment of aims and objectives. The policies, often set at national or state levels, define the overarching aims, goals, and expected outcomes of an educational system. Curriculum evaluation uses these policy statements as the primary benchmark to determine if the curriculum is meeting its intended objectives. Moreover, the policies assure that alignment and compliance is maintained as the evaluation checks that the curriculum design, content, instructional strategies, and assessment methods are in line with the established policies and statutory guidelines. This includes compliance with legal requirements such as those related to special education needs, safeguarding, equity, and inclusion.

### 11.4.8 Learning outcomes

In the domain of learning outcome, it is essential to determine if current students are progressing toward this outcome and if graduates are successful as they begin practice. Evaluation of student learning outcomes is viewed by some as the most important aspect of curriculum evaluation. Learning outcomes are a fundamental and measurable component of curriculum evaluation, serving as the primary evidence of whether an educational program is achieving its stated goals. They shift the focus from teaching intentions to what the student is expected to know, understand, and be able to demonstrate upon completion of a course or program. The major function of learning outcomes in curriculum evaluation is to give tangible, observable, and quantifiable proof of student achievement, which directly demonstrates the efficacy of the curriculum. Additionally, it aids in directing assessment as a clearly defined learning result and informs the creation of assessment tasks such as tests, portfolios, etc., that directly gauge students' mastery of particular abilities and information. This guarantees that evaluations are in line with the desired learning objectives.

## 11.5 Approaches to curriculum evaluation

### 11.5.1 Scientific approach

The scientific approach to curriculum evaluation uses quantitative methods like test scores, statistical analysis, and pretests or posttests to measure objective, measurable outcomes such as student achievement and learning gains. It is a systematic process that compares results against a standard to determine success, often overlooking qualitative data like student feelings or the diversity of individual circumstances. The core principles of the scientific approach are as described below:

**1. Quantitative data**

It relies on numerical data like test scores, percentages, and statistical analysis to provide objective results. Quantitative data is a core principle of the scientific approach to curriculum evaluation because it provides objective, measurable evidence of a curriculum's effectiveness, which is essential for systematic analysis and evidence-based decision-making. It has to be kept in mind that the scientific method emphasizes objectivity and the reduction of bias. Quantitative data, by nature, is numerical and collected through standardized tools like standardized tests, surveys with fixed-response options, and attendance records which minimizes subjective interpretation in the evaluation process. The scientific approach also focuses on the measurement of outcomes and on using numerical data generated by learners to measure specific outcomes and performance. Examples include test scores, grades, completion rates, and attendance percentages, which provide clear, quantifiable metrics of student achievement and progress.

**2. Measurable outcomes**

Measurable outcomes focuses on quantifiable results, such as learning gains or achievement levels, to determine effectiveness. In the scientific approach to curriculum evaluation, measurable outcomes are a core principle because they enable objectivity, data-driven decision-making, and empirical verification of effectiveness. This approach views evaluation as a systematic process of objectively establishing facts through testing and data analysis. The prime roles of measurable outcomes are in upholding the objectivity and accountability as the they provide clear, specific, and observable criteria for success, reducing subjectivity and personal bias in evaluation. This ensures that the evaluation is based on tangible evidence of student learning and instructional effectiveness, fostering accountability among educators and institutions. In addition, these contribute to the empirical data collection as the scientific approach relies heavily on quantitative data and statistical analysis to determine if the intended goals have been met. Measurable outcomes facilitate this by allowing student performance to be quantified and compared against established benchmarks or intended results.

**3. Systematic comparison**

Systematic comparison is a core principle of the scientific approach in curriculum evaluation because it provides a rigorous, evidence-based framework for making objective, data-driven decisions about the curriculum's effectiveness, quality, and future development. It helps in identification of strengths and weaknesses by systematically comparing a curriculum's performance, for example, by comparing student outcomes against predetermined objectives, external standards, or alternative programs, evaluators can clearly identify specific areas of strength and weakness. In addition to this, it leads to data-driven decision-making by comparing the necessary evidence to inform decision-making. For instance, comparing the results of students in a new program to a control group in a conventional program provides data on the effectiveness of the new approach.

**4. Objectivity and precision**

In curriculum evaluation, objectivity and precision are core principles of scientific approach which help to ensure that the process is free from personal bias, is reliable, and produces verifiable, evidence-based conclusions. Objectivity in curriculum evaluation means the process and results are not influenced by the personal beliefs, values, or prejudices of the evaluator. This leads to elimination of bias because the scientific approach uses structured methodologies to minimize the influence of individual opinions, ensuring that the evaluation genuinely reflects the curriculum's effectiveness rather than the evaluator's preconceptions. It also ensures that there is a standardized measurement. On the other hand, the precision in a scientific approach to curriculum evaluation involves using rigorous, data-driven, and systematic methods to obtain consistent and objective measurements of a curriculum's effectiveness. This approach is often contrasted with more humanistic, qualitative approaches and focuses on measurable outcomes to make evidence-based decisions.

## 11.5.2 Humanistic approach

The humanistic approach to curriculum evaluation is a qualitative method that focuses on the individual's personal experience, emotions, and holistic development rather than on quantitative data. It uses methods like interviews, case studies, and observations to understand the social-emotional and cognitive aspects of learning, emphasizing the learner's perspective and growth. The goal is to understand the complexity of the educational environment from multiple viewpoints and adapt the curriculum to meet students' needs. There is also a focus on the individual wherein this approach centers on the personal experiences of learners and teachers, valuing each individual's development and sense of self-worth. The key principles of this approach can be described as follows:

**1. Qualitative data**

It relies on qualitative methods such as interviews, open discussions, and observations to gather rich, narrative data about feelings, emotions, and experiences. Qualitative data is one of the key principles of the humanistic approach to curriculum evaluation. The humanistic approach emphasizes understanding the subjective experiences, personal growth, and overall well-being of individuals, which cannot be adequately captured through quantitative data alone. One of the important features related to the reliance on qualitative data is that it focuses on the whole person. In addition, the humanistic education values the integration of the cognitive and affective domains of learning. Qualitative methods, such as interviews, observations, and journals, allow evaluators to gather rich, in-depth information about this complex, intertwined aspects of a student's development. It also ensures that the subjective experience and personal meaning is provided as the humanistic approach believes that reality is subjective and that personal meaning is crucial to learning. Qualitative data collection methods are designed to explore how individuals perceive and make sense of their educational experiences, allowing participants to express their thoughts, emotions, and experiences in their own words.

**2. Holistic learning**

The evaluation considers the whole person intellectually, emotionally, and socially and assesses the curriculum's impact on overall well-being and personal growth. Holistic learning is a key principle of the humanistic approach to curriculum evaluation. Both the humanistic learning theory and the philosophy of holistic education are closely intertwined, emphasizing the development of the "whole person" beyond mere academic or intellectual growth. The role of holistic learning in humanistic curriculum evaluation is that it is integrated (Winarko and Budiwati, 2024). Holistic learning views all subjects and life experiences as interconnected. Humanistic curriculum evaluation assesses how effectively the curriculum fosters these connections and helps students find meaning and purpose in their learning and life, rather than focusing on fragmented, isolated subjects.

**3. Multiple perspectives**

It acknowledges that a curriculum's success can be understood in multiple ways, and it seeks to capture these diverse viewpoints. In the humanistic approach to curriculum evaluation, the emphasis is on the individual learner's experience, personal growth, and self-actualization rather than just objective, measurable outcomes. The principle of incorporating multiple perspectives is a natural extension of this, ensuring that the evaluation process is holistic, learner-centered, and respects the subjective reality of all participants. One of the main features of multiple perspectives is student-centeredness, which emphasizes the significance of comprehending students' own perspectives on their education and personal development. Since only self-evaluation is seen genuinely significant in this technique, students are pushed to assess their own development and level of satisfaction. Furthermore, it strengthens the holistic perspective of the student, which is in line with humanism's goal of educating the "whole person," incorporating social, emotional, cognitive, and physical development. Therefore, evaluation must take into account not only academic performance but also these numerous elements from multiple perspectives.

**4. Flexibility and adaptation**

Unlike more rigid approaches, humanistic evaluation is adaptable and can be used to actively involve students in modifying the curriculum to better suit their needs. In a humanistic approach to curriculum evaluation, flexibility and adaptation are core principles that ensure the educational experience is relevant to each student's unique needs, interests, and personal growth. This approach moves beyond rigid, content-focused evaluation to focus on the holistic development of the individual. Flexibility in the humanistic curriculum is a strategy that recognizes and embraces students' diverse needs, learning styles, and interests. The flexible curriculum upholds student's autonomy and choice by allowing learners a significant freedom and autonomy in choosing their learning paths, content, and methods. This empowers students to take ownership of their education, which is a central tenet of humanistic theory. It also enables flexible pacing by taking into account each student's unique learning pace and recognizing that not all students learn at the same rate. This indicates that an evaluation procedure takes into account individual development rather than depending on a single, uniform schedule.

## 11.5.3 Modernistic approach

The modernist approach to curriculum evaluation is a scientific, objective, and data-driven method that focuses on cause-and-effect relationships between curriculum and outcomes. It emphasizes measurable objectives and uses quantitative data to determine a curriculum's success. Key aspects include using formative and summative assessments to generate data for statistical analysis and making decisions based on empirical evidence. The core principles of the modernistic approach are as described below:

### 1. Objective and empirical

The modernistic approach to curriculum evaluation is founded on the core principle of objectivity and empiricism, primarily through the application of the scientific method. This paradigm views the institution as a "factory" and focuses on a systematic, product-oriented, and data-driven process. This principle aims to demonstrate a clear connection between curriculum and outcomes, and it is based on the idea that there is a quantifiable, objective reality. The emphasis on empirical evidence, wherein the knowledge and judgments on the efficacy of a curriculum must be gained from sensory experience, systematic observation, experimentation, and the measurement of behavior or outcomes, forms one of the main traits that characterize this principle. However, it also emphasizes objectivity as an ideal phenomenon since the approach aims to reduce personal bias, emotions, and preconceptions in the appraisal process. To guarantee that the results are unbiased and trustworthy across various assessors, standardized procedures, quantitative data, and statistical analysis are used.

### 2. Focus on outcomes

The primary goal is to see if specific instructional strategies and curricula lead to defined learning outcomes. In the modernistic approach to curriculum evaluation, outcomes serve as the core principle, fundamentally shifting the focus from educational inputs that is what the teacher plans to teach to the actual results and achievements of the students that is what the students can demonstrably do. This is encapsulated in the concept of outcome-based education. The prime aspects of the outcomes-based approach are that it provides clarity and prioritization as the process begins with the clear specification of intended learning outcomes or competencies such as knowledge, skills, and attitudes that students must attain by the end of a course or program. This clarity helps align expectations for all stakeholders, including students, teachers, and employers. It also uses the backward design as the desired outcomes drive the entire curriculum design. Teaching methods, learning activities, and assessments are designed and aligned to ensure that they effectively facilitate the achievement of these predetermined outcomes. This is a significant departure from traditional models where content delivery might take precedence over measurable results.

### 3. Quantitative data

It relies heavily on data that can be statistically analyzed, such as test scores, to measure effectiveness. Quantitative data is a core principle of the modernistic approach to curriculum evaluation due to its emphasis on objectivity, measurement, and empirical evidence. This approach is rooted in the belief that educational outcomes can and should be measured numerically to make informed, evidence-based decisions about program effectiveness. The central role of quantitative data is to seek an unbiased view of program outcomes as it facilitates objectivity and reduces bias. Quantitative data, collected through standardized tools like tests and surveys with numerical scales, minimizes the influence of subjective judgment, which is a major benefit over purely qualitative methods. The other aspect is that this approach assumes that the constructs under study, for example, the student performance, learning gains, and program impact can be measured and expressed in numerical form, such as test scores, grades, attendance rates, and completion rates.

### 4. Measurable objectives

Modern curriculum design emphasizes setting quantifiable objectives that can be measured to verify that students are meeting them. In the modernistic approach to curriculum evaluation, measurable objectives are the core principle because they provide a systematic, data-driven, and objective framework for determining the effectiveness and accountability of an educational program. In addition, it provides clarity and focus, as the measurable objectives often following the SMART criteria—specific, measurable, achievable, relevant, time-bound—offer a clarity for both educators and students. They explicitly define the desired outcomes, ensuring that all instructional activities and content are aligned toward achieving those specific goals. Moreover, the objective assessment as a part of the modernistic approach, rooted in the work of medical educationists like Ralph Tyler, views learning as an observable change in behavior. Measurable objectives make it possible to use quantitative data and standardized assessments like achievement tests or observation forms to objectively gauge student performance and determine if the intended behavioral changes have occurred.

## 11.5.4 Postmodernistic approach

A postmodern approach to curriculum evaluation is decentralized, subjective, and dynamic, focusing on diverse, local, and individual experiences rather than predetermined, universal standards. It values qualitative data, such as student narratives and experiences, to understand how the curriculum impacts students' construction of knowledge and identity. This method critiques power structures within the curriculum, challenging the narratives given by those in authority and emphasizes self-organization and transformative learning (Boboc, 2012). The key features of the postmodern approach to evaluation are mentioned below:

**1. Subjectivity and diversity**

In a postmodern approach to curriculum evaluation, subjectivity and diversity are core principles that directly challenge the traditional modernist emphasis on objectivity, universal truths, and standardized outcomes. In terms of subjectivity, postmodernism argues that reality and knowledge are subjective, or based on individual interpretation and experience, rather than an objective, external reality. These principles in turn recognizes that there is no single "truth" and that learning is unique, contradictory, and influenced by individual, cultural, and historical contexts. Evaluation focuses on understanding these diverse, local perspectives. Postmodern approach rejects the idea of a single, one-size-fits-all curriculum and views it as insufficient. Instead, it focuses on diversity, individualism, context, and a plurality of perspectives. Moreover, there is an emphasis on lived experience in which evaluation must account for the unique lived experiences, histories, and perspectives of each student. The individual's construction of meaning from personal experience is central to the learning process and consequently, to its evaluation. On the other hand, the diversity in the postmodernism embraces pluralism and multiplicity, asserting that diverse perspectives, cultures, and knowledge systems should be valued and integrated into the educational experience. On the other hand, it includes a key ethical responsibility, which is to allow for and elevate the voices and experiences of those who have been historically subjugated or excluded due to race, gender, class, or cultural background.

**2. Critique of power**

With regard to power, it actively seeks to decenter and challenge narratives about truth, knowledge, and power within the curriculum. In postmodernist approaches to medical education, a critique of power is a central feature. It refers to questioning how power operates in the production of medical knowledge, educational structures, professional identities, and clinical practice. Some of the important aspects of the critique of power in postmodern medical education is that it questions authority and hierarchies, wherein postmodernism challenges the traditional top-down hierarchies in medical education such as teacher over student or doctor over patient with an aim to define truth in medicine and to ascertain as to whose voices are to be excluded. For instance, many a times, the opinions of senior clinician are often treated as unquestionable, while students' or patients' perspectives are marginalized. Postmodernism through the approach of critique of power also rejects the idea that medical knowledge is purely objective or neutral. Instead, it argues that knowledge is socially constructed and shaped by institutional, cultural, political, and economic forces. In addition to this, it addresses to the hidden curriculum, professional identity, patient voice, and shared decision-making. In essence, the critiques of power aims to promote awareness of one's own power, democratize learning, value multiple forms of knowledge, and finally foster ethical, socially responsive physicians.

**3. Dynamic and fluid evaluation**

The postmodernistic approach views curriculum aims, goals, and outcomes as fluid and generative rather than fixed and predetermined. In this regard, dynamic and fluid evaluation form one of the core principles of the postmodernist approach to curriculum evaluation. This approach fundamentally rejects the fixed, objective, and universal standards of modernist evaluation models, such as the Tyler's model, which relies on predetermined objectives and standardized outcomes. The key aspects of this dynamic and fluid approach are that here the evaluation is viewed as dynamic, contextual, and continuously evolving rather than fixed, standardized, and purely objective. It challenges traditional assessment systems that rely heavily on rigid exams and numerical scores. Moreover, the postmodernism views knowledge and truth as subjective, multiple, and context-dependent. Therefore, evaluation cannot be based on a single narrative of what constitutes success or a single set of universal standards. It is also to be kept in mind that evaluation is tied to specific local contexts, cultures, and individual experiences. It acknowledges that what is valuable in one situation may not be in another, and thus, the criteria for evaluation must be flexible and responsive to the immediate environment and the people involved.

**4. Holistic and ecological perspective**

The holistic and ecological perspectives dissolve the artificial boundary between the classroom and the outside community, valuing the interconnectedness of knowledge, experience, and the natural world. Holistic and ecological perspectives are core principles of a postmodern approach to curriculum and its evaluation. The holistic perspective in a postmodern curriculum emphasizes the development of the whole person and the interconnectedness of all things. This approach challenges the fragmented and reductionist assumptions of traditional, modern education often referred to as the

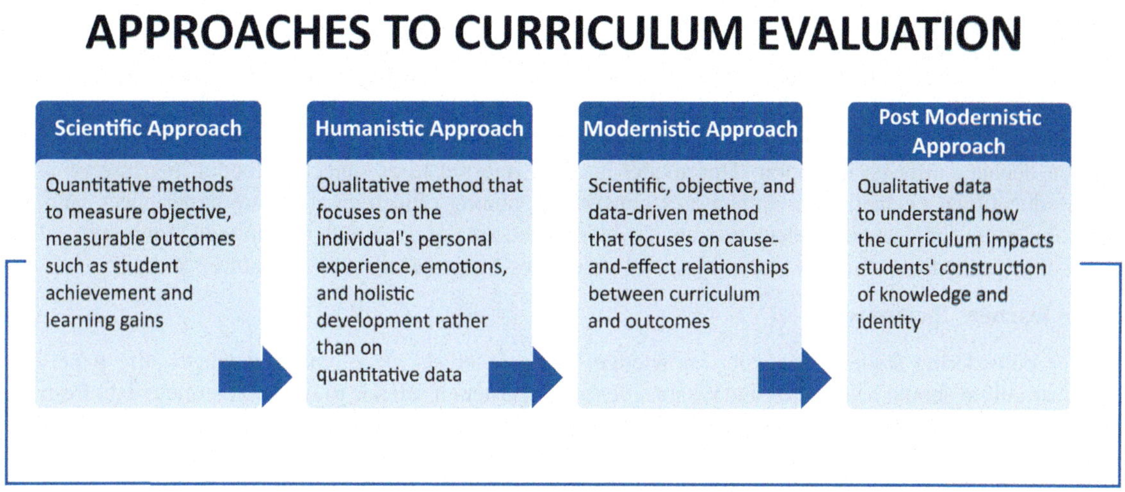


FIGURE 11.2 **Approaches to curriculum evaluation.** This figure shows the various approaches to curriculum evaluation.

mechanistic worldview. In the realm of curriculum evaluation, considering the emotional, social, ethical, and academic needs of students in an integrated format, rather than separate, isolated components is a vital feature. In addition, the holistic approach prioritizes meaningful experiences over the simple transmission of a predetermined, standardized curriculum.

**5. Focus on identity and self-esteem**

This sees education as a process of identity construction rather than discovery, where evaluation aims to understand how individuals and groups are empowered to attain their own chosen goals. In a postmodernist approach to curriculum evaluation, identity and self-esteem are not viewed as fixed, inherent traits but as dynamic, socially, and culturally constructed elements. The core principles focus on pluralism, diversity, individual experience, and empowerment. The prime factors regarding postmodernism is that it challenges the traditional idea of a single, stable identity and instead views identity as fluid, fragmented, and context-dependent. Curriculum evaluation must therefore consider how the curriculum interacts with these multiple, evolving identities. This also focuses on the celebrating diversity and pluralism, as this approach advocates for multicultural and multilingual education that acknowledges and values diverse voices, backgrounds, and experiences, particularly those of marginalized groups. Evaluation focuses on ensuring the curriculum is inclusive and avoids promoting a single, dominant culture.

**6. Emphasis on qualitative data**

Emphasis on qualitative data is a core principle of the postmodernist approach in curriculum evaluation, as qualitative methods such as analyzing student narratives, discussions, and autobiographical histories provide insight into the learning experience, complementing or replacing the quantitative data emphasized in traditional approaches. Postmodernism fundamentally challenges the "grand narratives" and claims of universal truth associated with modernism and its preference for objective, quantitative methods. The shift to qualitative data in a postmodern framework is based on several core principles like subjectivity of reality in which the postmodernism posits that reality is socially constructed and subjective, not objective and universally measurable. Qualitative data, gathered through methods like interviews, observations, and personal narratives, is better suited to capturing these diverse, lived experiences and multiple perspectives (Fig. 11.2).

## 11.6 Models for curriculum evaluation

Some of the most common models of curriculum evaluation include the following:

1. Tylers model,
2. CIPP,
3. Stakes congruence,
4. Illuminative evaluation model.

### 11.6.1 The Tyler's model

The process of curriculum development and evaluation has been significantly shaped by a few seminal works, among which Ralph W. Tyler's model holds a place of particular prominence. Understanding this model is crucial for educators, curriculum designers, and anyone involved in the educational process, as its principles continue to resonate and inform practice, even decades after its inception. This model is often referred to as objective model, because of its objective approach to educational evaluation. It emphasizes consistency among objectives, learning experiences, and outcomes. Curriculum objectives indicate both behaviors to be developed and area of content to be applied (Mobit et al., 2024). Tyler recommends that curriculum planners identify general objectives by gathering data from three sources:

**1. From the learners themselves**

This involves considering students' existing knowledge, skills, interests, developmental stages, life experiences, and needs. The curriculum should be relevant and responsive to the learners it intends to serve. To collect data from learners, a variety of methods, including both quantitative data collection such as quizzes, tests, and surveys and qualitative data collection such as observations, interviews, and work samples are used. It is important to define what data is needed, choose appropriate methods, and collect data consistently to help understand student progress and inform teaching strategies.

**2. From outside the medical institution**

The collection of the data by the curriculum developers to analyze the demands of society, including current social, cultural, and economic issues, as well as the values and aims that society upholds, is critical. It not only gives an idea about the needs and expectations of the society but at the same time makes the stakeholders aware as to whether the curriculum is in alignment with the goals as planned or there is any need to take any measures to rectify it. This will have great benefits on the student as well as they will be prepared for effective participation in life beyond the walls of medical institutions.

**3. From subject matter specialists**

Experts in various academic disciplines provide insights into the essential knowledge, concepts, skills, and methods of evaluation that are considered valuable and usable within their fields. This ensures that the curriculum is academically sound and reflects important disciplinary content (Fig. 11.3).

Through the Tyler's model, the four vital components of curriculum are being addressed. These being the following:

**1. Objectives**

In concern with the objectives one of the vital things to be considered is to determine what educational purposes should the institution seek to attain? This question forms the cornerstone of the Tyler's model, as it emphasizes the critical importance of clearly defining educational purposes, or objectives, before any other curriculum planning occurs. Tyler himself stated that objectives are "the most critical criteria for guiding all the other activities of the curriculum maker." These objectives should articulate what students are expected to know, understand, or be able to do as a result of the curriculum, often stated in terms of observable and assessable behaviors. A crucial aspect is that these objectives must align with the

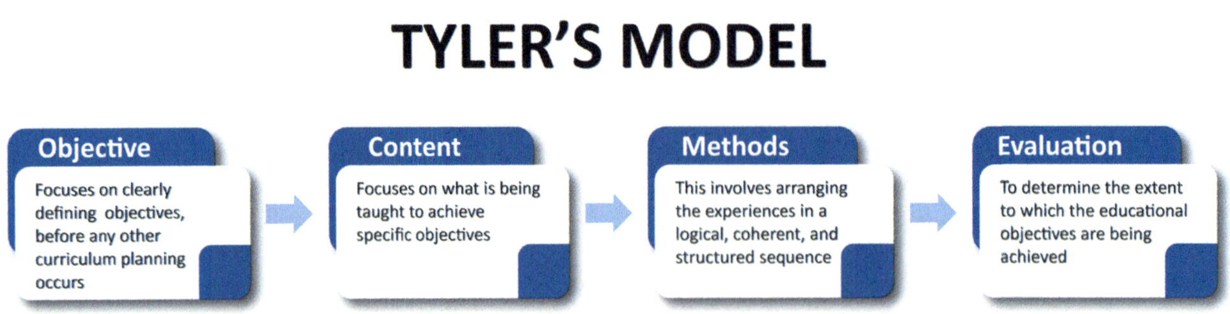


FIGURE 11.3 **Tyler's model.** This figure depicts the Tyler's model.

overarching philosophy of the educational institution. This step is about defining desired learning outcomes that will guide the entire curriculum design.

**2. Content**

In Ralph Tyler's model of curriculum evaluation, content is crucial because it's the "what" that is being taught and is the subject matter used to achieve specific objectives. Content must be selected based on the needs of learners and society, then organized logically and aligned with the learning experiences and evaluation methods to ensure the curriculum is effective. The importance of content in Tyler's model is that it forms the primary source for developing educational objectives, which are the foundation of the entire curriculum. In addition, it provides the substance for learning. As we are aware that the content is the actual material like knowledge, skills, and values that students are expected to learn. It also guides the selection of learning experiences. Once the objectives are set, the content informs the types of learning experiences and instructional strategies educators will use to help students meet those objectives.

**3. Methods**

This step focuses on how educational experiences can be effectively organized. After selecting appropriate learning experiences, the third step focuses on their effective organization. This involves arranging the experiences in a logical, coherent, and structured sequence to maximize their cumulative effect and ensure a smooth progression of learning. An effective organization aims for synergy, where the interconnectedness of experiences leads to deeper and more lasting learning than isolated activities could achieve. Tyler highlighted three key criteria for organizing learning experiences:

**a. Continuity**

This refers to the vertical reiteration of major curriculum elements. Important concepts, skills, and values are revisited at different points in the curriculum, allowing for reinforcement and deeper understanding over time.

**b. Sequence**

This principle emphasizes that each successive learning experience should build upon the preceding one. Content and skills are arranged in an order of increasing complexity, breadth, or depth, ensuring that learning is progressive.

**c. Integration**

This relates to the horizontal relationship of curriculum experiences. It involves linking concepts and skills from different subject areas or connecting learning to students' broader life experiences, helping them to see the relationships between different fields of knowledge and to unify their learning. For example, vertical integration ensures that skills learned in earlier years are built upon in later years, while horizontal integration might involve connecting topics across the same level.

**d. Evaluation**

This step focuses on how we can determine whether these purposes are being attained. The final aspect in the Tyler rationale addresses the evaluation of the curriculum. This is an ongoing process designed to determine the extent to which the educational objectives, defined in the first step, are being realized by the program of curriculum and instruction. Evaluation involves assessing student learning outcomes, analyzing the effectiveness of the selected and organized learning experiences including instructional methods and materials and gathering feedback from various stakeholders, such as students and teachers (Table 11.1).

**TABLE 11.1** Tyler's model—strengths versus weaknesses.

| Strengths | Weaknesses |
|---|---|
| Clear and simple structure | Perceived as linear and rigid, not reflecting the dynamic reality of curriculum work |
| Systematic and logical approach to planning | Overemphasis on narrow, behavioral objectives |
| Strong focus on objectives and outcomes | Potential neglect of creativity, critical thinking, and affective learning |
| Provides clear direction for instruction | Limited student involvement in the curriculum development process |
| Encourages teacher reflection and evaluation | May reduce teacher autonomy and flexibility |
| Promotes objectivity in evaluation | Lacks contextual sensitivity for diverse learners and settings |
| Facilitates accountability | Constructing behavioral objectives can be difficult and time-consuming |

### 11.6.2 Context, input, process, product (CIPP) evaluation model (Stufflebeam)

The CIPP evaluation model is a decision-oriented framework for evaluating programs, projects, and organizations, based on four components which include Context, Input, Process, and Product. Developed by Daniel Stufflebeam, its primary goal is to provide information for improvement and accountability rather than just proving worth. This model is often used to inform planning, structure, and execution, and it can be applied to both formative evaluation which is improvement-focused and summative which is judgment-focused evaluation. The four components of the CIPP model are as mentioned below:

**1. Context (C)**

This evaluates the environment to identify needs and problems and to set goals. It answers the question, "What needs to be done?" In the CIPP model, the evaluation component assesses the environment to define goals and priorities. This stage involves identifying the needs, problems, assets, and opportunities within a program's or project's target area to establish a rationale for intervention. It is essentially a needs assessment that helps ensure a program's objectives align with the reality of the situation.

**2. Input (I)**

This involves assessing the resources, alternatives, and plans that could be used to achieve the goals. It answers the question, "How should it be done?" In the CIPP model, input evaluation assesses the resources and plans needed to achieve program goals, including factors like budget, staffing, and facilities. It is a preprogram stage that helps determine the most effective approach for addressing the needs identified in the context evaluation by evaluating competing strategies and their associated work plans.

**3. Process (P)**

This assesses the implementation of the program to determine if it is being carried out as planned. The process as in the CIPP model involves evaluating the implementation of a program to see how well the plan is being carried out and whether it is being done correctly. It monitors and documents the program's activities, strategies, and methodologies to identify what is working and what needs refinement for continuous improvement.

**4. Product (P)**

This evaluates the outcomes to determine success and whether goals were met. It answers the question, "Did it succeed?" In the CIPP model, Product (P) evaluation measures the outcomes and achievements of a project or program, including both intended and unintended results. This final stage assesses the program's overall success by determining if it met its objectives and delivered value to stakeholders, and it informs decisions about continuing, modifying, or terminating the effort (Fig. 11.4).

### 11.6.3 Stake's model (countenance model)

Stake's countenance model, also known as the responsive evaluation model, is a framework for curriculum evaluation in medical education that emphasizes both descriptive and judgmental aspects. It involves creating a matrix to organize data based on the program's intended and observed antecedents, transactions, and outcomes, and then using this information to make judgments about congruence, that is, how well intentions match outcomes, and contingencies, that is, how outcomes relate to conditions. This approach is flexible and stakeholder-inclusive, gathering feedback from students, faculty, and staff to understand the curriculum's effectiveness from multiple perspectives (Stake, 1967). The key components of the model include the following:

**1. Intention**

In Robert Stake's countenance model of curriculum evaluation, the "intention component" refers to the intended aspects or objectives of the curriculum as perceived by its developers and stakeholders. It is one of the two main columns in Stake's evaluation matrix, the other being "observations." The intention component plays a critical role in the evaluation process as it helps to define the ideal curriculum, wherein the intention aspect outlines what the curriculum aims to achieve or what the developers hope will happen. This includes the planned context, processes, and desired results. It also focuses on the basis for congruence analysis, wherein the primary use of the "intentions" data is for comparison with the observations.

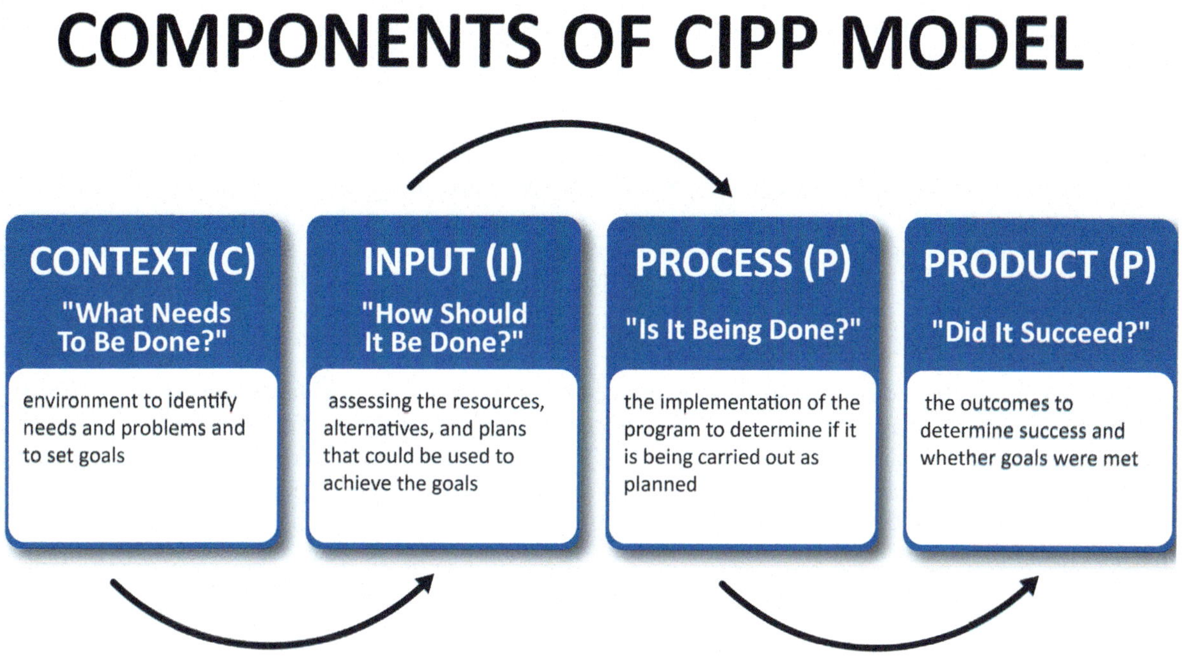


FIGURE 11.4 **CIPP (context, input, process, and product) model.** This figure shows the CIPP model.

**2. Observation**

In Robert Stake's countenance model of curriculum evaluation, observation is a crucial component as the observations done here are used to record the actual conditions, activities, and outcomes of a program. This data is then compared to the intended aspects of the curriculum to identify discrepancies. The primary function of observation in Stake's model is to provide empirical data about what is genuinely happening in the educational setting, as opposed to what was planned or expected. Stake emphasizes collecting rich, qualitative data from natural settings to build a comprehensive understanding of the program. During this stage, the observation data is collected across three specific phases of the educational program. These being the following:

**a. Antecedents (input)**

This step involves observing the conditions existing before instruction begins that may relate to outcomes. This could include the physical environment, available resources such as infrastructure, and the background or prior experiences of students and teachers.

**b. Transactions (process)**

This step involves and focuses on observing the dynamic and continuous interactions and activities that form the process of instruction. Evaluators watch for the actual teaching methods, student engagement, and communication patterns in the classroom.

**c. Outcomes (output)**

This involves observing the effects or results of the instructional program. This goes beyond test scores to include observed student behaviors, attitudes, and skills.

**3. Congruence**

The degree of correspondence between the intended and observed aspects of the curriculum is given by congruence. Congruence is the comparison of what was intended versus what was actually observed and achieved during the curriculum's implementation. It measures the alignment between the program's stated intents which are antecedents and process intents and its observations which are the actual outcomes serving as a key component for judging the curriculum's success. Congruence is determined by examining data to see if intended outcomes were realized, and this comparison is a primary way to make an evaluation judgment about the curriculum. Congruence looks at the overlap between the intended outcomes and the actual results. It asks the fundamental question: "Did what was intended actually happen?"

**4. Contingency**

The relationships between the antecedents, transactions, and outcomes are the highlights of contingency. Contingency refers to the relationship between the different components of a curriculum: antecedents, transactions, and outcomes. It is a descriptive analysis that looks for the empirical connections between what existed before the curriculum began that is antecedents, what happened during its implementation referring to transactions, and the results it produced which are the outcomes. It is one of the two principal ways to process descriptive evaluation data, alongside congruence.

## 11.6.4 Illuminative evaluation model

The illuminative model of curriculum evaluation in medical education is a qualitative approach that focuses on understanding a program's context, process, and outcomes rather than just measuring results. It uses methods like focus groups and interviews to "illuminate" the complex reality of the program, providing a deep, descriptive understanding that helps to refine the curriculum and enhance future teaching and learning experiences (Alderman, 2015). The illuminative model of curriculum evaluation was developed by Malcolm Parlett and David Hamilton. The key principles of the illuminative model can be described as follows:

**1. Focus on description and interpretation**

Instead of simply measuring with numbers, it aims to describe what is happening in the program and interpret the findings within their specific context. In the illuminative model of curriculum evaluation, description and interpretation are the core functions that replace traditional quantitative measurement and prediction. The goal is to "illuminate" the complex reality of the educational program within its natural context, focusing on understanding processes and experiences rather than just measuring outcomes against predefined objectives. In terms of description, it involves the comprehensive and detailed documentation of the educational program as it actually operates in a real-world setting.
In this, the evaluator seeks to understand the "learning milieu" referring to the wider context, environment, and culture in which the program functions and the "instructional system" which include the formal plans, syllabus, techniques, and equipment. In addition to that, it facilitates the gathering of rich data by using qualitative methods for data collection such as observation, interviewing, and document review to capture the actual experiences and perceptions of the participants, who in this case are teachers and students. It also captures the program's significant features, recurring events, and critical processes. Interpretation follows the descriptive phase and involves making sense of the gathered information to explain "how it operates" and "how it is influenced by the various situations in which it is applied." Interpretation plays a role by trying to understanding the meaning, whereby the evaluator interprets the data to understand the meaning participants attach to their actions, decisions, and beliefs within the program's social world. As a result of this, it provides insightful knowledge, whereby the aim is to generate a deeper knowledge and understanding about the program's dynamics, rather than a simple judgment of success or failure.

**2. Emphasis on context**

It examines the wider environment in which the curriculum is implemented, including the goals, background, and resources involved. The illuminative model of curriculum evaluation places a central and profound emphasis on the role of context. It views a curriculum innovation as inseparable from the unique cultural, social, and institutional environment in which it operates. The emphasis on the context can be appreciated by the fact that the model is rooted in a social anthropology paradigm. Here, the focus is on comprehending the procedures and intricate connections between the curriculum's content, structure, and context. This stands in stark contrast to conventional models, which frequently isolate variables and concentrate solely on quantifiable results or outputs.

**3. Uses qualitative methods**

It relies heavily on qualitative data collection methods, such as interviews and focus groups, to gather rich, detailed information about participants' experiences. Qualitative methods are central to the illuminative model of curriculum evaluation, providing the in-depth understanding of the "how" and "why" a curriculum operates within its specific context. These methods allow evaluators to focus on the lived experiences of participants and the complex social dynamics, revealing insights that quantitative data alone cannot provide. This involves using techniques like interviews, informal observation, and document analysis to study how a program is implemented, influenced by its environment, and perceived by teachers and students. The role of qualitative methods here is to have a thorough understanding of the "how" and "why" wherein the qualitative data helps explain the processes behind outcomes, moving beyond simple measurement to explore how the curriculum actually works in practice. It also contributes to explore lived experiences as it captures the perspectives of those involved, such as teachers and students, to understand their unique experiences and perceptions of the program.

**4. Goal is to "illuminate"**

The main goal is to provide a comprehensive understanding of the program's complexities, helping to clarify issues and raise the level of sophistication of the debate around it. In the illuminative model of curriculum evaluation developed by Parlett and Hamilton, the term "illuminate" means to shed light on, clarify, and interpret the complex realities, processes, and contexts of an educational program or innovation. The goal is to provide a deep, insightful understanding of how the program actually operates in its natural setting rather than simply measuring predefined outcomes or effectiveness against specific metrics. The "illumination" process involves understanding the context by examining the wider social and cultural environment in which the program functions. At the same time, importance has been given to focusing on processes by investigating the interactions, issues, and significant features that emerge during the implementation of the curriculum.

**5. Data-driven but context-rich**

While data is gathered, the evaluation is not solely focused on decision-making instead, it also concentrates on providing deep insight that informs future improvements. The illuminative model prioritizes content rich description and interpretation over mere measurement. Its key features contribute significantly to the context-rich requirement. At the same time, it examines the wider social, psychological, and material environment in which teaching and learning occur, providing deep insights into the real-world conditions affecting the curriculum. Instead of a standardized methodological package, it uses a general research strategy where the problem defines the methods. This flexibility allows evaluators to use diverse methods like observation, interviews, and case studies tailored to the specific context, uncovering unique and unpredicted phenomena. All this will uphold the adaptable and eclectic strategy.

## 11.7 Modernizing curriculum evaluation process through artificial intelligence

Modernizing the curriculum evaluation process through AI involves leveraging data analytics, automation, and machine learning to create a more personalized, efficient, and data-driven educational experience. AI transforms traditional, static evaluation methods into a dynamic and responsive system of continuous improvement. Some of the essential ways by which AI modernizes curriculum evaluation are by upholding technology to foster automated grading and feedback, wherein AI tools can automatically grade objective assessments such as multiple-choice, true or false questions, etc., and even evaluate written responses using Natural Language Processing (Tozsin et al., 2024). This provides students with instant, unbiased feedback, allowing for immediate correction and freeing up educators' time for more complex tasks and one-on-one student interaction. On the other hand, by the optimum usage of data-driven insights and predictive analytics, AI platforms can analyze vast amounts of student performance and engagement data to identify trends and patterns that human evaluators might miss. This helps educators and administrators to identify knowledge gaps at both individual and cohort levels, predict student outcomes, and pinpoint those at risk of falling behind, enabling early intervention and support and at the same time evaluate the effectiveness of specific teaching methods or course materials, guiding data-informed curriculum adjustments. In addition to all these, adaptive assessments and personalized learning paths enable the creation of adaptive learning systems that adjust the difficulty and type of questions in real-time based on a student's performance, ensuring they are consistently challenged at an appropriate level. This moves the evaluation from a one-size-fits-all approach to a tailored experience that meets individual needs and learning styles.

## 11.8 Importance of curriculum evaluation

Education has always been on the move as it is in the process of continuous evolution influenced by advancements in knowledge, shifts in societal needs, and changes in pedagogical approaches. To stay at the forefront of education and meet the needs of the students and stakeholders, it is necessary that the curricula must be regularly evaluated.

### 11.8.1 Adaptation to a changing world

Curriculum evaluation is crucial for adapting to a changing world by ensuring education remains relevant and effective in the face of new societal, technological, and workforce demands. It provides a structured process to assess the strengths and weaknesses of a curriculum, enabling educators to update content, improve teaching methods, and align educational goals with evolving needs to prepare students for the future. Curriculum evaluation is important for adaptation because it ensures relevance as a changing world requires a curriculum that is adaptable and dynamic. Evaluation helps identify if the current curriculum is preparing students with the necessary skills, such as critical thinking, adaptability, and lifelong learning (Reed et al., 2023). In addition, it responds to societal shifts, wherein the rapid changes like globalization, climate change, and technological advancements necessitate a curriculum that reflects these new realities. Evaluation provides a mechanism to assess and incorporate these changes.

### 11.8.2 Meeting accreditation standards

Curriculum evaluation is crucial for meeting accreditation standards because it provides the evidence that an institution is delivering a quality, relevant, and effective education. Accreditation bodies use evaluation data to verify that an institution's programs meet predefined benchmarks and are committed to continuous improvement (Youker, 2018). The vital roles of curriculum evaluation in accreditation are in demonstrating quality and accountability, wherein the accreditation serves as a quality assurance mechanism, and the evaluation process generates the data necessary to prove that an institution's educational offerings meet or exceed established standards. This demonstrates accountability to students, parents, employers, and the community. In addition to that, it ensures alignment with standards and objectives, whereby the evaluation process systematically assesses whether the curriculum's design, content, and learning outcomes align with the specific criteria set by accrediting bodies, as well as with institutions and professional standards.

### 11.8.3 Enhancing educational quality

Curriculum evaluation enhances educational quality by providing data-driven feedback to improve teaching, assess student outcomes, and ensure the curriculum remains relevant and effective. This process allows educators to identify strengths and weaknesses, align objectives with real-world needs, and make informed decisions about curriculum design, content, and methodology for continuous improvement (Zhang et al., 2023). Curriculum evaluation enhances educational quality by contributing to the teaching and learning, wherein the evaluation provides feedback to teachers on their instructional practices and helps identify areas where students need support. This allows for adjustments to teaching methods and the development of more effective learning experiences. It also ensures that there is a proper alignment with goals. It verifies whether the curriculum is meeting its educational objectives and if students are learning in the most effective way possible.

### 11.8.4 Data-driven decision-making

Data-driven decision-making is crucial for curriculum evaluation because it moves from assumptions to evidence-based improvements, ensuring the curriculum effectively meets student needs and adapts to a changing world. This approach uses student performance data, test scores, and other feedback to pinpoint areas of the curriculum that need reinforcement, modify programs to reflect industry demands, and ultimately improve student outcomes and learning experiences. Data-driven decision-making improves curriculum evaluation by enhancing the accuracy and reliability. This is done by analyzing data like grades and test results in which curriculum specialists can accurately assess if students are achieving learning objectives and identify specific curriculum areas that may be underperforming. It also promotes responsiveness to the needs of healthcare field as the data from the job market can be used to ensure the curriculum is relevant and equip students with the necessary skills for emerging fields.

### 11.8.5 Demonstrating accountability

Curriculum evaluation demonstrates accountability by systematically gathering data to determine the effectiveness and worth of educational programs, ensuring that institutions are answerable to various stakeholders for their performance, use of resources, and the quality of education provided (Kodagoda et al., 2024). In addition, it promptly reports outcomes to stakeholders as the evaluation provides concrete evidence of how well the curriculum is meeting its intended goals and objectives. This data is then reported to stakeholders, including parents, students, administrators, and the public, assuring them that the institution is delivering on its promises. It also ensures that there is an alignment with standards by checking if the curriculum aligns with established local, state, and national educational standards and benchmarks. This ensures that the institution is meeting external requirements and is held responsible for maintaining a certain level of educational quality.

### 11.8.6 Fostering innovation

Curriculum evaluation fosters innovation by identifying strengths and weaknesses, which then provides data-driven insights to inform and guide improvements. This process is crucial for developing more creative, engaging, and effective curricula that use new methods like project-based learning to better prepare students for a rapidly changing world. By regularly assessing and adapting a curriculum, educators can move away from rigid, traditional methods toward ones that promote critical thinking, problem-solving, and adaptability. Innovation is fostered by curriculum evaluation by the provision of data-driven feedback, wherein evaluation gives objective data on what is and isn't working, which is essential for making informed, evidence-based decisions about curriculum changes. On the other hand, it identifies opportunities for new approaches as it reveals the need to shift from outdated teaching methods to innovative ones, such as incorporating technology, project-based learning, and real-world problem-solving.

## 11.9 Impact on teaching-learning process

Curriculum evaluation positively impacts the teaching-learning process by providing data to improve teaching methods, align objectives with student outcomes, and ensure the curriculum remains relevant and effective. It identifies strengths and weaknesses, guides the modification of content and materials, and ultimately leads to a more effective and adaptable learning environment for students. (Levy-Feldman, 2025). The major outcomes of curriculum evaluation on the teaching-learning process are that it improves teaching and learning, wherein evaluation identifies areas where students may be struggling, which can prompt teachers to adjust their teaching methods and strategies. It also helps ensure that the curriculum is aligned with learning objectives and student needs. In addition, it also guides curriculum development as the results of an evaluation provide data-driven insights for curriculum designers and educators. This helps them make informed decisions on what needs to be modified, such as content, teaching approaches, and learning materials. From the students' perspective, it enhances student outcomes by assessing the effectiveness of the curriculum. Evaluation helps to determine if students are achieving the intended knowledge, skills, and attitudes. This feedback loop is crucial for improving student learning experiences and results (Lupina et al., 2023).

## 11.10 Impact of curriculum evaluation on the principles and practice of medical education

Curriculum review has a significant impact on the principles and practices of medical education since it ensures that educational programs continue to be successful, current, and in line with the changing demands of society and the healthcare industry. It serves as an ongoing feedback loop that propels the necessary modifications to satisfy accreditation requirements and enhance learning results. Some of the key influences of the curriculum evaluation process are in the following areas:

**1. Evidence-based decision-making**

Curriculum evaluation provides the data necessary to determine which teaching methods, assessments, and content areas are working and which are not. This moves the practice of medical education away from tradition-based decisions towards evidence-based strategies, ensuring resources are used effectively. Evidence-based decision-making in curriculum evaluation refers to the process of evaluating the efficacy of educational programs utilizing high-quality data, research, and professional experience. It relies on scientific data to create, alter, or sustain advances in education, going beyond intuition or tradition.

**2. Quality assurance and accreditation**

Medical schools must meet stringent national and international accreditation standards. Regular and rigorous curriculum evaluation is essential to demonstrate compliance and maintain accreditation, which validates the quality and rigor of the education provided. Both quality assurance and accreditation are interrelated procedures in curriculum evaluation that guarantee educational programs fulfill particular requirements for efficacy, relevance, and quality. While accreditation is a formal, external assessment that verifies certain requirements, quality assurance is a continuous, internal process of evaluating and enhancing instruction and curriculum. When combined, they promote accountability, public trust, and ongoing improvement of educational quality (Afolabi et al., 2015).

**3. Alignment with professional competencies**

Evaluation helps ensure that the curriculum effectively develops the core competencies required for medical practice, such as clinical skills, professionalism, communication, and systems-based practice. This ensures graduates are prepared for the demands of residency and clinical work. The alignment with professional competencies in curriculum evaluation is a process of ensuring that the knowledge, skills, and attitudes taught and evaluated in an educational program directly correspond to the established standards, necessary skills, and ethical practices demanded by a particular profession. It ensures graduates are prepared for the workforce by bridging the gap between academic learning and the practical, real-world demands of a vocation.

**4. Continuous quality improvement (CQI)**

Continuous quality improvement is a continuous, methodical, data-driven process that actively evaluates, improves, and refines educational initiatives. Instead of waiting for failure, it employs repeated feedback loops such as data collection, strengths or weakness analysis, change implementation, and outcome measurement in order to constantly enhance teaching and learning. The cornerstone of ongoing quality improvement in medical education is curriculum evaluation. By identifying gaps or weaknesses, such as those related to out-of-date material or inefficient teaching methods, it enables educators to make targeted improvements and promote a culture of excellence and flexibility.

**5. Stakeholder feedback integration**

In general, stakeholder feedback integration refers to the process of collecting, evaluating, and acting upon feedback from people or groups interested in or impacted by an educational program in order to enhance its efficacy, relevance, and quality. In the context of medical education, curriculum evaluation incorporates feedback from a diverse range of stakeholders, including students, faculty, patients, and healthcare employers. This ensures that the curriculum is responsive to the needs and expectations of all parties, enhancing its relevance and acceptance. In order to promote shared ownership, enhance accountability, and guarantee that the curriculum satisfies practical demands, it goes beyond simple consultation and actively involves stakeholders which include students, instructors, and administration in the evaluation process.

**6. Innovation and adaptability**

Curriculum evaluation plays a crucial role in fostering innovation and flexibility in education since it offers a systematic, evidence-based procedure for revising teaching strategies, content, and structures to meet changing demands. Evaluation turns the curriculum from a rigid, static blueprint into a dynamic, "future-proof" instrument that enhances learning outcomes and gets students ready for a world that is changing quickly. In essence, curriculum evaluation is not merely a post-program review, in fact, it is an integrated, ongoing process that shapes the fundamental principles and day-to-day practices of medical education, ensuring the production of competent and effective physicians.

**Summary**

Curriculum evaluation is the systematic process of assessing and determining the effectiveness and quality of a curriculum. The purpose of curriculum evaluation is to ensure that educational programs meet their objectives, support student learning, and adapt to changing needs and contexts. This chapter outlines the fundamental principles, methods, and types of curriculum evaluation, as well as its importance in the educational process. Curriculum evaluation is a vital component of the educational process. It provides crucial feedback that helps improve teaching and learning, ensures accountability, and supports informed decision-making. By employing various evaluation methods and involving diverse stakeholders, curriculum evaluation contributes to the continuous enhancement of educational programs.

**Key point**

**Future perspectives: embracing the humanistic and modern aspect of the curriculum**

The future perspective of curriculum evaluation is evolving in response to changes in educational needs, advancements in technology, and the growing emphasis on personalized learning. As the world becomes more interconnected and education becomes more dynamic, curriculum evaluation will increasingly need to adapt to new challenges and opportunities. The future of curriculum evaluation is dynamic, multifaceted, and increasingly tied to technology, personalized learning, and global

**(cont'd)**

challenges. Curriculum evaluation will evolve to not only assess traditional academic outcomes but also support the development of essential life skills, social-emotional learning, and competencies that will prepare students for the complexities of the future. By way of embracing these new perspectives, curriculum evaluation will play a critical role in shaping educational systems that are more responsive, inclusive, and effective in meeting the needs of all learners.

## References


Adebayo, A., 2014. Perspectives in Education. HIS LINEAGE PUBLISHING HOUSE, Nigeria.

Adebayo, A. 2018. A Review of Scientific and Humanistic Approaches in Curriculum Evaluation.

Afolabi, F.O., Adesope, O.M., Ogunwuyi, A.O., 2015. Curriculum Evaluation: A Veritable Tool for Quality Assurance in Education. Journal of Education Research & Behavioral Sciences 4 (6), 161–167.

Alderman, L., 2015. Illuminative evaluation as a method applied to Australian government policy borrowing and implementation in higher education. Evaluation Journal of Australasia 15 (1), 4–14. Available at https://doi.org/10.1177/1035719x1501500102.

Boboc, M., 2012. The postmodern curriculum in a modern classroom. International Journal of Education 4 (1). Available at https://doi.org/10.5296/ije.v4i1.1167.

Coles, C.R., Grant, J.G., 1985. Curriculum evaluation in medical and health-care education. Medical Education 19 (5), 405–422. Available at https://doi.org/10.1111/j.1365-2923.1985.tb01345.x.

Kodagoda, G.S., Jowsey, T., Bishop, J., Forbes, M., Grant, L.J., Green, P., et al., 2024. Internal evaluation of medical programs is more than housework: A scoping review. PLoS One 19 (10), e0305996. Available at https://doi.org/10.1371/journal.pone.0305996 PMID: 39453962; PMCID: PMC11508059.

Levy-Feldman, I., 2025. The role of assessment in improving education and promoting educational equity. Education Sciences 15 (2), 224. Available at https://doi.org/10.3390/educsci15020224.

Lupina, S.P., Ganzon, W. Jr. D., Pangandoyon, R.T., Mandin, M.O., 2023. Strategy evaluation in teaching strategies: a systematic literature review. International Journal of Scientific and Research Publications 13 (5), 302–308. Available at https://doi.org/10.29322/ijsrp.13.05.2023.p13739.

Mobit, M.O., Elit, L., Palmer, D., Palmer, N., Fanfon, T., 2024. Curriculum mapping evaluation of a Tyler model designed physiotherapy curriculum of the Baptist institute of health science in Cameroon. Cogent Education 11. Available at https://doi.org/10.1080/2331186X.2024.2329367.

Reed, S., Picca, A., Lockwood, L., Thoreson, L., Burke, A.E., Mahan, J.D., et al., 2023. A qualitative analysis of assessment and evaluation practices within the individualized curriculum. Academic Pediatrics 23 (5), 1001–1010. Available at https://doi.org/10.1016/j.acap.2023.04.007, http://www.elsevier.com/wps/find/authored_newsitem.cws_home/companynews05_01099.

Stake, R.E., 1967. The Countenance of Educational Evaluation. Teachers College Record: The Voice of Scholarship in Education 68, 1–15. Available at https://doi.org/10.1177/016146816706800707O.

Tozsin, A., Ucmak, H., Soytürk, S., Aydın, A., Gozen, A., Fahim, M., et al., 2024. The role of artificial intelligence in medical education: A systematic review. Surgical Innovation 31, 415–423. Available at https://doi.org/10.1177/15533506241248239.

Ullah, H., Huma, S., Yasin, G., Ashraf, M., Tahir-ud-Din, Q., Shabana, H., et al., 2024. Curriculum and program evaluation in medical education: a short systematic literature review. Annals of Medicine & Surgery 86 (10), 5988–5994. Available at https://doi.org/10.1097/ms9.0000000000002518.

Winarko, B., Budiwati, N., 2024. Humanistic Pedagogy: Approaches to Enhancing Individual Development in Modern Education. EDUTEC: Journal of Education and Technology 8. Available at https://doi.org/10.29062/edu.v8i2.1047.

Warju, W., 2016. Educational program evaluation using CIPP model. Innovation of Vocational Technology Education 12 (1). Available at https://doi.org/10.17509/invotec.v12i1.4502.

Winarko, B., Budiwati, N., 2024. Humanistic pedagogy: Approaches to enhancing individual development in modern education. EDUTEC : Journal of Education and Technology 8. Available at https://doi.org/10.29062/edu.v8i2.1047.

Youker, B., 2018. Internal evaluation. The SAGE encyclopedia of educational research, measurement, and evaluation Vol. 4. SAGE Publications, pp. 857–858.

Zhang, Y., Wu, X., Zhu, C., Zhou, J., 2023. Classroom quantitative evaluation: a method of both formative and summative evaluation. Sustainability 15 (3), 1783. Available at https://doi.org/10.3390/su15031783.

Chapter 12

# Learning management systems

*"Coming together is a beginning; keeping together is progress; working together is success."*

—Henry Ford



## 12.1 Introduction

A learning management system is a software application or platform used to administer, document, track, report on, and deliver educational courses, training programs, and other learning materials. It acts as a central hub for learners and educators to access content, complete assignments, receive feedback, and interact through features like virtual classrooms and forums. Learning management system platforms can be cloud-based for accessibility or self-hosted for more control, and they are used across various sectors, including schools, universities, and businesses, to streamline the learning process. Learning management systems are widely used in medical education to supplement traditional teaching by centralizing content, facilitating flexible and self-paced learning, managing assessments, and ensuring compliance with accreditation standards (Afridi, Lubna, 2025). They enhance the learning experience for both undergraduate and postgraduate students. While the learning management system is a powerful supplemental tool, it cannot fully replace the hands-on clinical training and direct patient interaction that are integral to medical education. It is most effective when integrated into a blended learning approach that combines digital resources with in-person teaching.

## 12.2 Key features of learning management system

### 12.2.1 Content delivery

Content delivery is a core, fundamental component of a learning management system in medical education, acting as the primary mechanism for distributing essential educational materials to students and professionals. The role and importance of content delivery in medical learning management system is to provide a centralized, accessible, and flexible platform for all learning resources, supporting continuous and complex medical training. In addition to this, it contributes to the centralization and organization, wherein the learning management system serves as a single hub for vast amounts of medical information, including lecture notes, textbooks, images, and videos of surgeries, preventing fragmentation of resources. It also contributes to the accessibility and flexibility by giving a chance to the medical professionals, who often have demanding, irregular schedules, an opportunity to access content 24/7 from various devices, enabling self-paced and on-demand learning that fits into their demanding work lives. It also maintains the consistency and standardization by

**The Principles and Practice of Medical Education. DOI:** https://doi.org/10.1016/B978-0-443-44197-4.00025-4

ensuring that all learners receive the same, up-to-date, and evidence-based information, which is crucial for maintaining consistent and high standards of patient care and regulatory compliance such as Health Insurance Portability and Accountability Act (HIPAA) and Accreditation Council for Continuing Medical Education (Chahal, Kavita, 2021).

### 12.2.2 Compliance management

Compliance management within a learning management system in medical education is a crucial, automated, and streamlined process for ensuring that all professionals adhere to the complex array of healthcare laws, regulations, and accreditation standards. The vital role of compliance management in medical education is also highlighted in the area of automated tracking and reporting. Here the learning management system automatically monitors the completion status of required courses and certifications, providing real-time data on the organization's compliance status. This replaces manual, error-prone tracking methods. It also contributes to certification and license management as the learning management system tracks the status and expiration dates of licenses and certifications, sending automated notifications and reminders to both learners and administrators when renewals are due.

### 12.2.3 Customizable learning paths

One of the important features of the learning management systems in medical education is that it offers personalized learning experiences tailored to different roles and skill levels within a healthcare setting, such as students, residents, healthcare professionals, doctors, and administrative staff. This customizable learning paths in a medical education which is unique and beneficial to the learner provide a tailored, self-paced, and flexible educational experience for healthcare professionals, which is crucial for competency development and keeping pace with rapidly evolving medical practices. Some of the essential features of the customizable learning paths is that they provide a role-based customization in which the learning paths can be automatically assigned or designed based on a user's specific role, ensuring content relevance to their responsibilities. This also helps in the implementation of the adaptive learning algorithms where advanced learning management system platforms use artificial intelligence to analyze a learner's progress, skill level, and performance data in real-time, dynamically adjusting the content and difficulty to fill knowledge gaps effectively. Moreover, complex medical topics and procedures are broken down into short, bite-sized modules, allowing busy professionals to complete training during short breaks or between shifts without disrupting workflow contributing to the modular content delivery or microlearning.

### 12.2.4 Collaboration

One of the prime roles of the learning management system in medical education is to facilitate communication and collaboration among learners, potentially from different locations or specialties as it fosters essential skills like critical thinking, communication, problem-solving, and teamwork which are vital for effective healthcare practice. Collaboration features within learning management system facilitate active learning and knowledge sharing among students and between students and instructors, ultimately leading to improved patient care. This feature of collaboration has some important benefits in medical education in that it contributes to development of essential skills amongst the medical professionals The development of these skills is important as the healthcare professionals work in interdisciplinary teams. In this regard, early exposure to collaborative environments in a learning management system helps them develop crucial interpersonal, leadership, and teamwork skills. In addition, it facilitates enhanced knowledge acquisition and retention which occurs when the students discuss, brainstorm, and teach peers through online platforms. Through this, they deepen their understanding of complex medical concepts and improve knowledge retention.

## 12.3 Basic principle and working of learning management system

A learning management system is a software platform designed to manage, deliver, and track educational content and training programs. Its basic principle is to provide a centralized, accessible, and automated environment that connects administrators and instructors and learners, facilitating the entire learning lifecycle efficiently and cost-effectively. The core principles guiding a learning management system's design and function include centralization as all learning materials, user data, and reports are stored in one central location, making them easy to manage, access, and update. It also facilitates the accessibility as it provides the content which can be accessed by learners anytime, anywhere, and from various devices, removing geographical and time barriers to learning. It also automates the routine administrative tasks such as user enrollment, grading, notifications, and certification, freeing up time for instructors and administrators to focus on content quality and interaction. A learning management system works by enabling administrators to manage and upload content, then assign it to users. Learners log in to access their assigned courses, complete modules, take assessments, and track their progress through a centralized platform. The system provides tools for tracking user performance, generating reports, and facilitating communication between users.

### 12.3.1 Working of learning management system

At its core, a learning management system operates as a comprehensive platform that enables the creation, delivery, and management of learning content and activities. The general functioning of a learning management system is described below (Fig. 12.1).

### 12.3.2 Content creation

Using authoring tools or content libraries, instructors, or administrators create educational materials, including courses, modules, and multimedia resources. In terms of content, the instructor can upload and arrange instructional resources into courses, including presentations, videos, and tests. In turn, whether a user is a teacher or a student, the system enables the creation and maintenance of user accounts, which are frequently arranged into groups or roles.

### 12.3.3 User management

Administrators of learning management systems oversee user accounts, roles, permissions, and access levels to ensure proper material delivery and user interaction. An essential part of a learning management system is user administration, which enables administrators to set up, maintain, and create user accounts with particular roles and permissions. It permits the customization of learning paths and development plans for various users, including administrators, instructors, and learners, and it guarantees data integrity and security by limiting access to courses and features. The key functions of user management in a learning management system are in the user creation and management, wherein the administrators can add new users individually or in bulk and manage their profiles, which may include details like name, email, and other user properties. It also supports role assignment wherein the system supports different user roles such as administrator, instructor, and learner, each with unique tasks and permissions to control what they can see and do on the platform.

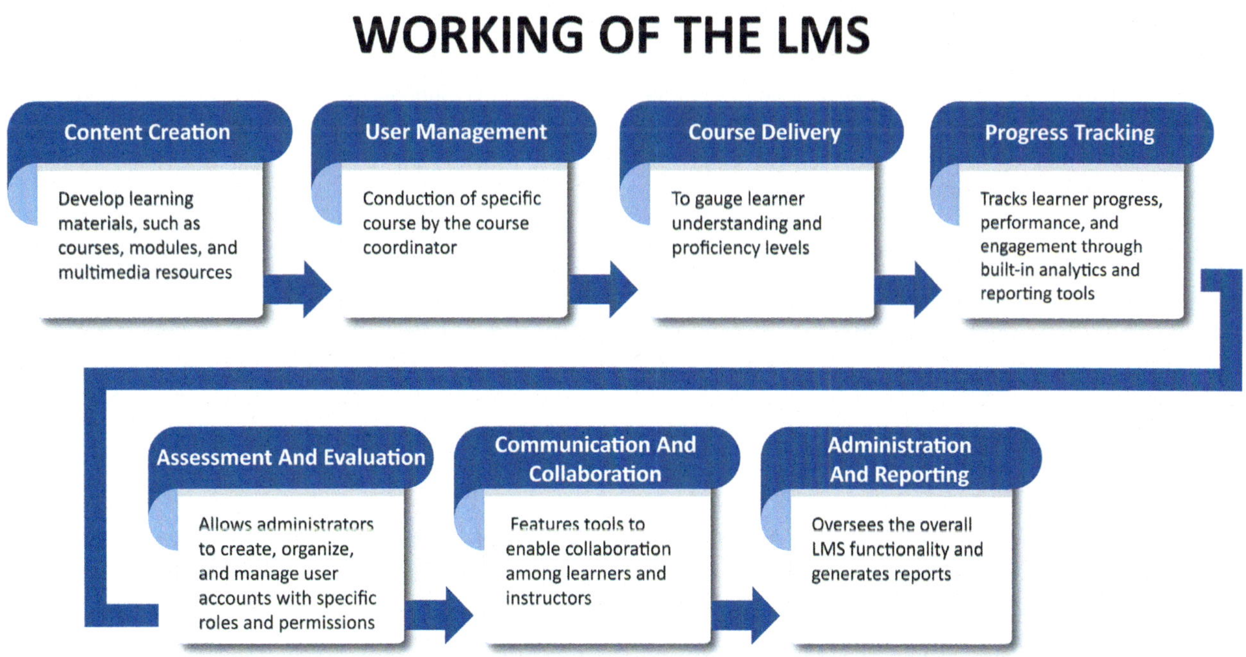


FIGURE 12.1 Working of the learning management system.

### 12.3.4 Course delivery

Learning management systems deliver medical education through online content, using interactive modules like e-PBL, surgical demos, assessments, and discussion forums, supporting both undergraduate and postgraduate training, blended learning, and CME credits, offering features for content management, progress tracking, and compliance reporting to enhance efficiency and skill development. Key functions include sharing resources, managing tasks, providing feedback, and facilitating continuous learning for updated standards and better patient care, often using platforms like Blackboard (John, Kevin, 2022). From the perspective of learners, it is important for the learner to login to access enrolled courses, navigate modules, consume materials, and participate in activities at their own pace.

### 12.3.5 Progress tracking

The learning management system tracks learner progress, performance, and engagement through built-in analytics and reporting tools. The process of tracking and reporting the progress can be accomplished as the learning management system has the capability to track each user's progress, such as course completion and assessment scores. Further to complete the process of reporting the system generates reports on user performance and training effectiveness, providing data for administrators. It enables timely identification of knowledge gaps and facilitates personalized learning paths to ensure medical professionals meet specific competency standards. One of the main characteristics of progress tracking in a medical education learning management system is that it helps with thorough reporting and analytics by producing in-depth data on student engagement, assessment scores, and course completion rates. It also ensures the real-time monitoring by allowing the instructors and administrators to monitor student activity, such as time spent on modules and performance in assessment in real-time. In addition, it forms a vital part of tracking the development of core competencies and skills, which are often linked to the direct observation data entered by preceptors or clinical instructors.

### 12.3.6 Assessment and evaluation

Learning management system implements assessment activities, including quizzes, assignments, and surveys, to gauge learner understanding and proficiency levels. In an learning management system, assessment gathers student data to gauge learning, while evaluation uses that data to judge if learning objectives are met and to improve teaching or systems, with the learning management systems streamlining both via automated grading, analytics, and diverse tools for real-time feedback and better outcomes.

### 12.3.7 Communication and collaboration

Discussion boards, chat, video conferencing, and shared project spaces are examples of communication and collaboration tools found in learning management systems. These tools facilitate peer learning, real-time feedback, and group work through features like calendars, to-do lists, and integrated platforms, which improve engagement and comprehension in both online and blended learning environments.

### 12.3.8 Administration and reporting

Administration and reporting in a learning management systems involves managing users, courses, content, and system settings, while in learning management systems, reporting tracks and analyzes learning data in terms of progress, performance, and engagement to provide insights for better decision-making, ensuring compliance, and proving training effectiveness, with both functions relying on automation and data to streamline learning and development. Some of the most popular learning management system systems used in medical education are as follows:

### 12.3.9 Moodle

Moodle learning management system is a free, open-source, and highly customizable online platform for creating, delivering, and managing educational courses and training programs, used globally by schools, universities, businesses, and individuals to build personalized learning environments with interactive content, assessments, and communication tools. It's a modular system, allowing organizations to adapt it to their specific learning goals, supporting various methods like blended learning, distance education, and flipped classrooms (Fediv, Vasyl, 2021). Moodle in medical education is an open-source learning management system used to create flexible, interactive online learning environments for students and professionals, offering features like course delivery, assessments such as quizzes, peer reviews, discussion forums, and progress tracking to support blended learning, compliance training, and continuous professional development. It allows medical institutions to build personalized, collaborative platforms for delivering digital content, managing training, and fostering community among learners and instructors, enhancing education beyond traditional methods. The advantages provided by the Moodle learning management system in medical education involve both students and healthcare professionals.

### 12.3.9.1 *For medical students*

#### 12.3.9.1.1 Blended learning

Moodle improves blended learning in medical education by centralizing resources, enabling interactive content like videos, tests, or simulations, encouraging collaboration through forums that allow for flexible access, automating assessments, tracking progress with analytics, and supporting a variety of approaches like flipped classrooms. In the end, this improves engagement, comprehension, and performance for both theoretical knowledge and practical skills.

#### 12.3.9.1.2 Resource hub

Moodle serves as a central, easily accessible resource hub in medical education since it gives educators, a flexible, multimedia platform to post, arrange, and administer a wide range of educational resources for students and healthcare professionals. It serves as a central repository by offering a single, safe place to arrange and keep a variety of educational resources, thereby eliminating resource sharing. This makes it possible for instructors and students to access crucial information from anywhere at any time. Additionally, the platform enables educators to update content on a regular basis to reflect the most recent developments, best practices, and standards in the quickly changing medical sector.

#### 12.3.9.1.3 Interactive learning

Interactive learning through the Moodle platform transforms passive students into active participants by integrating engaging elements like quizzes, simulations, discussions, and gamification, fostering deeper understanding, better retention, and critical thinking through direct engagement with content, instructors, and peers, rather than just static material. Learning management system platforms facilitate this by offering features for multimedia, collaborative spaces, personalized paths, and immediate feedback, making learning dynamic and rewarding.

#### 12.3.9.1.4 Assessment and feedback

Assessment and feedback in learning management system are crucial for tracking progress and improving learning, involving tools for quizzes, assignments, and exams that offer formative and summative evaluation, providing immediate or detailed comments, and supporting self-directed learning through features like automated grading and analytics, ultimately boosting student performance and engagement by guiding them towards mastery. Moodle greatly affects evaluation and feedback by making them a more automated, adaptable, and accessible process. It makes it possible to use a variety of interactive, data-driven evaluation techniques like workshops, quizzes, and assignments that improve feedback, encourage student participation, and support both formative and summative assessment, frequently leading to better performance and engagement than traditional methods.

#### 12.3.9.1.5 Adaptive learning

Adaptive learning in a learning management system uses artificial intelligence and data analytics to create personalized learning paths, adjusting content, difficulty, and pace in real-time based on a user's performance, engagement, and knowledge gaps, moving beyond one-size-fits-all training to deliver exactly what each learner needs. The main way that the Moodle platform influences adaptive learning is by providing educators with a flexible, open-source framework that enables them to design individualized, student-centered, and self-paced learning experiences. In addition, Moodle offers a wide range of tools, such as branched lessons, conditional activities, and sophisticated plugins, that facilitate manual or, more recently, AI driven adaptive learning.

### 12.3.9.2 *For healthcare professionals*

#### 12.3.9.2.1 On-demand training

A learning management system is extensively used in medical education to provide flexible, on-demand training that supplements traditional methods and addresses the rigorous requirements of the field. It is an essential tool for delivering continuous, scalable, and accessible educational content to busy professionals and students. The primary advantage of a learning management system is in terms of its accessibility and flexibility where the learning management system offers a continuous access to training materials from any location and device, which is crucial for busy medical professionals with unpredictable schedules. This supports self-paced, asynchronous learning that fits into packed shifts. In this context, on-demand training is significantly transformed by the Moodle platform, mainly by replacing required, planned sessions with flexible, user-driven learning opportunities, 24/7 accessible, and customized learning paths that increase retention and engagement.

#### 12.3.9.2.2 Competency tracking

Learning management systems is also advantageous as it helps us to track competencies by centralizing resources, delivering standardized training, enabling continuous assessment via quizzes or simulations, and providing real-time analytics to monitor progress against required skills, ensuring compliance and personalized learning paths for students and professionals, improving patient care through better-qualified staff. The Moodle platform facilitates a transition from time-based learning to competency-based learning, whereas learners advance based on their mastery of certain abilities rather than seat time, thereby enhancing competency monitoring. This is accomplished using Moodle's structured architecture, which allows for automated, real-time monitoring of learner performance by directly mapping learning activities (quizzes, assignments, SCORM packages) to specified competencies.

#### 12.3.9.2.3 Role-specific training

Role-specific training is provided by the Moodle platform, as it offers a flexible, centralized, and configurable platform that adapts information to the particular needs of various healthcare professionals. As a learning management system, it is highly effective in medical education. Role-specific training is greatly impacted by the Moodle platform, particularly through Moodle Workplace, which makes highly customized, automated, and focused learning experiences possible. It replaces the "one-size-fits-all" approach to training with an organized paradigm in which the information is customized for particular departments, job functions, or skill levels.

#### 12.3.9.2.4 Efficiency and scale

The Moodle platform has an impact on medical education by facilitating cost-effective, high-capacity, asynchronous teaching for thousands of users and by significantly increasing productivity through automation and optimized workflows. Medical personnel can complete training at their own pace while upholding high standards thanks to the widespread conversion of traditional methods into blended or entirely online learning.

## 12.3.10 Canvas

With features for assignments, grading, video integration, discussion boards, and tracking student progress, the Canvas learning management system is a digital platform used by hospitals and universities in medical education to centralize, deliver, and manage all course content. This allows medical professionals and students to learn at any time and from any location (Kamath, 2025). The distribution of lecture notes, PowerPoints, videos, PDFs, and references for topics like anatomy, pharmacology, and clinical skills is one of the key characteristics of this learning management system. Additionally, it controls assignments and due dates, divides the curriculum into modules, and gives teachers analytics to track student participation. Like the Moodle platform, it provides busy learners with round-the-clock access to resources using mobile apps or online browsers. It even makes it easier to turn in assignments digitally and offers prompt feedback.

## 12.3.11 Blackboard

Blackboard learning management systems in medical education serves as a central online hub for course materials, communication, and assessments, supporting blended learning by delivering content, facilitating discussions and managing assignments. It enhances engagement through features like content sharing and virtual classrooms (Sampayan, 2024). One of the positive features of this learning management system is that it has a well-organized and professionally designed interface. This makes it more user friendly for the students and teachers. Some of the major applications of Blackboard in medical education is in the following areas:

### *12.3.11.1 Content management and delivery*

Blackboard facilitates content management and distribution since it offers a centralized learning management system that allows teachers to arrange, store, and distribute a variety of course materials to students,. Its features make it simple to create material, make it accessible globally, and integrate a variety of learning resources. The ability for students to access course materials, announcements, and assignments on their terms using a variety of devices at any time and from any location is one of the primary benefits of Blackboard. Teachers have the flexibility to set rules to release content based on particular conditions, like a date, time, or a student's performance on prior items, which allows for a personalized learning experience. This allows for adaptive release and personalization of the contents.

#### *12.3.11.2 Virtual classrooms and communication*

Blackboard also gives the instructor the facility of hosting virtual classrooms via the Blackboard Collaborate, wherein the content can be delivered into the students anytime and anywhere. Through its features like announcements, virtual chats, and email, it creates a seamless instructor–student and student–student interaction. In essence, Blackboard serves as a powerful complementary tool, offering structure and interaction to medical education, making it more adaptable and robust.

#### *12.3.11.3 Assessment and evaluation*

Blackboard's assessment and evaluation features allow instructors to create various tests or assignments for formative or summative feedback, manage submissions, and evaluate outcomes, while students submit work and receive grades, all tracked within the learning management system for improved teaching and learning. Instructors use features like rubrics for assignments and manage submissions within content areas, with options for manual grading or automated checks.

#### *12.3.11.4 Collaborative learning*

Through its integrated tool, Blackboard Collaborate, which offers real-time virtual classrooms with video, audio, chat, and shared whiteboards or applications for interactive group work, discussions, and content sharing, Blackboard enables synchronous and recorded collaborative activities like group projects, virtual labs, and discussions to increase engagement, comprehension, and social presence in online courses. One of Blackboard's drawbacks is that, in order to transform health professional training as effectively as possible, it necessitates continual training for teachers and ways to bridge digital gaps and maintain focused student involvement.

### 12.3.12 TOPYX

TOPYX is a flexible learning management system for medical education with features like remote training, compliance tracking, certification management, and patient education. It supports a variety of content and encourages social learning for better staff competency and patient care through efficient, safe online delivery. It has demonstrated its value to the teacher by assisting with content and course administration through the uploading, arranging, and delivery of training materials, courses, and assignments to students (Sánchez, 2023). However, since it tracks student performance and development using dashboards, statistics, and analytics to determine who needs more help, it aids in progress tracking. It maximizes flexibility for the learner by allowing them to access the course materials on any mobile device at any time and from any location. By enabling them to take part in a range of educational activities, such as assignments, tests, and interactive social learning tools like forums and real-time chat, it also keeps the learner interested.

## 12.4 Implementing learning management system

Establishing objectives, choosing a platform, migrating content, training users, and continuing evaluation are all part of the strategic, multistep process of implementing a learning management system (Tibyampansha, 2017). This ensures that the system meets organizational learning objectives for efficiency, improved communication, and tracking and integrates with current systems. A committed team, strong content creation, extensive testing, well-thought-out planning, stakeholder support, rigorous training, and an emphasis on user experience to promote adoption are all necessary for a successful implementation. Sequential approach, beginning with clearly defining objectives, evaluating needs, and collaborating with stakeholders to create the implementation strategy and continuous evaluation of the system's performance to satisfy the changing needs of medical education are all essential components of the implementation process.

### 12.4.1 Steps for implementation

The various steps in implementation of learning management systems are described below.

#### 12.4.1.1 Define goals and assess needs

Establishing objectives and evaluating needs is a fundamental step in medical education that comes before and directs the efficient use of a learning management system. It guarantees that the installation of the learning management system is strategically in line with the particular needs of medical professionals and the ultimate objective of enhancing patient care and practice outcomes. The broad, long-term objectives of the curriculum or training program, such as generating qualified doctors or guaranteeing ongoing professional development, are established by the goals in medical education. These overarching aims are converted into precise, quantifiable learning objectives via the learning management system component, which organizes the course material.

#### 12.4.1.2 Create an implementation plan

For the implementation of the learning management system, it is crucial to create a thorough schedule and strategic plan that assigns roles and tasks to the members of the implementation team. For the successful implementation of the learning management system, it is important to ensure that there is a smooth integration with the curriculum, align with educational goals, and encourage user acceptance.

#### 12.4.1.3 Develop and curate content

In order to facilitate learning in the cognitive, affective, and psychomotor domains, it's critical to develop original content and compile preexisting resources to create effective and captivating online learning environments. The new material should be produced by understanding the learner's requirements and expectations, as well as establishing specific learning objectives are always essential first steps (Kumar, Girish, 2022). In order to maintain the content current and relevant, it is crucial that it is rationally arranged, accessible, and interactive. Since it outlines exactly what students should know or be able to perform after finishing the course, writing the content helps define learning objectives. By emphasizing interactivity, utilizing a range of media, and avoiding lengthy text passages, it also aids in the creation of captivating content.

#### 12.4.1.4 Set up and integrate the learning management system

The process of integrating the learning management system necessitates a methodical strategy that focuses on particular institutional requirements, regulatory compliance, and thorough user training(Bozoğlu, 2016). With features like content repositories, online tests, and feedback tools, integrating a learning management system into medical education centralizes course delivery, improves flexible, self-paced learning and enhances communication and assessment. This ultimately increases student engagement, knowledge retention, and clinical preparedness, especially with blended learning models that combine digital resources with hands-on training.

#### 12.4.1.5 Launch and announce

In order to promote uptake and engagement, this step entails formally introducing the learning management system and making sure that all users are properly informed. In a medical school setting, launching and announcing a learning management system necessitates a methodical strategy that highlights its unique advantages for teachers and students, such as improved accessibility, customized learning routes, and strong compliance tracking. It would cover the stages of preparation, planning, and content announcement.

#### 12.4.1.6 Monitor and evaluate

This step entails regularly monitoring student progress, collecting feedback, and assessing the system's efficacy in relation to the original objectives. For optimization, postimplementation evaluations are essential. Assessing student satisfaction and learning outcomes, analyzing system utilization and technical performance and assessing the content's compatibility with curricular goals are all important aspects of monitoring and evaluating a learning management system in medical education. Using established evaluation frameworks and concentrating on technical functionality, content quality, and overall effectiveness, this process incorporates both quantitative data from system logs and qualitative comments from students and educators (Fig. 12.2).

**FIGURE 12.2** Steps in implementation of learning management system.

## 12.5 Benefits of a learning management system in medical education

A learning management system in medical education offers organized and easily accessible training, guaranteeing consistent education, compliance, and better patient outcomes with features like individualized learning, progress tracking, and chances for ongoing education (Back, 2016). Some of the significant advantages of the learning management systems are in the areas of curriculum development and implementation, teaching and learning process, and assessments. In addition, these systems are useful for automating administrative activities, delivering reliable and affordable training, and supplying useful data for quality improvement and performance analysis.

### 12.5.1 Curriculum planning and implementation

A learning management system plays a crucial role in curriculum planning and implementation by providing a centralized digital platform that supports organization, delivery, tracking, and evaluation of educational content. Some of the vital roles that learning management systems play in terms of curriculum planning are as follows:

#### *12.5.1.1 Content organization*

Learning management systems helps to centralize structures and delivers digital learning materials into coherent courses, enabling easy access, progress tracking, role-based customization, and consistent updates for learners and administrators, improving efficiency, engagement, and accessibility for varied learning needs. It moves beyond simple storage to create organized learning paths with modules, lessons, and topics, ensuring learners get relevant information directly, saving time and boosting skill development. In addition, the learning management system helps in the process of content delivery and accessibility and aids in delivering the course materials to students anytime, anywhere, and on any device with an internet connection, supporting flexible and remote learning models such as online, blended, or hybrid.

#### *12.5.1.2 Curriculum mapping and design*

A learning management system is a powerful digital platform that centralizes the processes of curriculum mapping, design, delivery, and evaluation. One of the features of the learning management system is that it aids is in the alignment with standards by allowing educators to explicitly tag and align course activities and assessments with specific institutional, state, or national learning standards, which is crucial for accreditation processes. The learning management systems also contributes in designing learning pathways as it enables the educators to structure flexible and personalized learning paths, where content can be released based on student progress or mastery of prior concepts.

### *12.5.1.3 Alignment of the content and assessments*

Learning management systems greatly simplify the alignment of content and assessments by offering a centralized platform that combines instructional materials, activities, and assessment tools. The integration of learning management system through clearly mapping learning objectives to particular materials and assessments will enable the educators to guarantee a seamless educational experience. Instructors can specify and save particular learning objectives and competencies using capabilities included in the majority of contemporary learning management systems platforms. The learning outcomes that each activity addresses can then be clearly identified by directly connecting these objectives to assignments, tests, discussion boards, and other content items. In the realm of assessment and evaluation, the learning management system provides tools for creating, administering, and automatically grading quizzes and assignments, offering immediate feedback to students and instructors.

### *12.5.1.4 Facilitating collaboration*

Learning management systems have grown in importance for both teachers and students due to their numerous features and user-friendliness. It is true that the formal channels of communication between students and learners are now learning management systems like Blackboard. A learning management system significantly improves student–teacher collaboration through a variety of integrated tools and capabilities. Collaboration outside of the classroom is made possible, for instance, by tools like email, message boards, and discussion boards. Teachers are able to ask questions, provide detailed remarks, and answer inquiries, while students are free to ask questions and receive clarification at their own pace.

### *12.5.1.5 Personalization and customization*

Learning management systems also play a part in personalization and customization because they allow teachers to create different versions of courses, create individualized learning journeys, and provide targeted content based on student needs and learning styles, all of which are essential components of successful curriculum planning. A learning management system's role in personalization and customization has changed as medical education moves into the era of artificial intelligence. Customization enables administrators to brand the look, integrate with other systems, and customize dashboards, paths, and rules for particular roles, ultimately increasing engagement, relevance, and outcomes by moving beyond static content to dynamic, learner-centric experiences. Data and AI are used to create unique learning journeys, adapting content, pace, and format to individual needs.

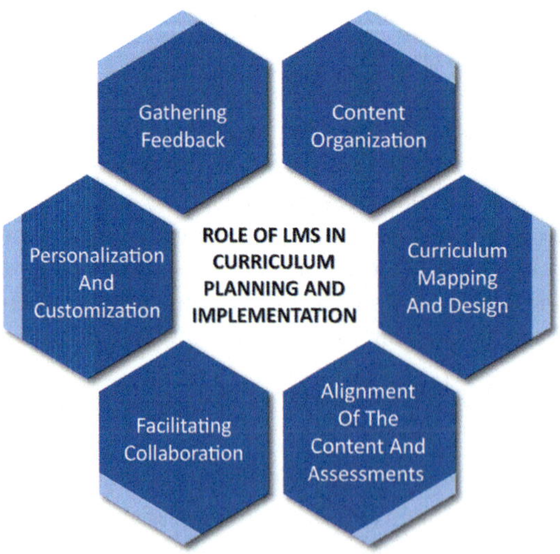


**FIGURE 12.3** Role of learning management system in curriculum planning and implementation.

#### 12.5.1.6 *Gathering feedback*

At the same time, learning management systems helps in gathering feedback for improvement as the built-in tools for surveys and feedback allow curriculum developers to collect student and faculty input on the effectiveness of course materials, which informs a process of continuous improvement and refinement (Fig. 12.3).

### 12.5.2 Teaching and learning methods

A learning management system performs a number of crucial functions in the teaching and learning process by centralizing content, simplifying administrative duties, boosting engagement, and facilitating data-driven insights. Since the learning management system functions as a digital hub where instructors may upload, organize, and manage a variety of learning materials, including documents, videos, quizzes, and presentations, it assists instructors with content management and delivery (Alamgir, Ashar, 2025). This supports a consistent teaching style by guaranteeing that all content is in one easily accessible location and can be updated with ease. Because the system automates numerous time-consuming administrative chores, like tracking attendance, managing user enrollment, and grading assessments like automated quizzes, it also aids in streamlined administration. Teachers are able to spend more time instructing and providing insightful feedback as a result. In terms of assessment and evaluation, learning management systems enable educators to create and administer a range of online examinations and utilize integrated grading tools and rubrics. This provides a systematic way to monitor student progress and facilitates the development of a trustworthy assessment system.

### 12.5.3 Assessment

A learning management system streamlines assessment by providing a central platform for creating, delivering, and evaluating various forms of assessments, from automated quizzes to instructor-graded assignments. Learning management systems enhance assessment through automated grading, detailed progress tracking and reporting, instant feedback for students, and data-driven insights for personalization. This allows educators to efficiently monitor student performance, identify learning gaps, provide targeted support, and ultimately improve learning outcomes in online, blended, and self-paced learning environments. Following are the ways in which the learning management systems facilitate assessment:

#### 12.5.3.1 *Content and assessment creation*

Learning management systems provide an organized method of assessing students' comprehension by offering tools for creating a variety of assessment kinds, such as tests, quizzes, and assignments. In a learning management system, content and assessment production include developing and arranging educational resources, such as documents and videos, and then developing instruments, such as examinations, assignments, and quizzes, to gauge student progress. Drag-and-drop interfaces, templates, AI-powered tools for productivity, and the capacity to create certifications are some of the key features. While content creation tools provide a variety of learning experiences, assessments are used to measure knowledge retention, pinpoint areas of weakness, and offer feedback.

#### 12.5.3.2 *Automated grading*

Automated grading in learning management systems uses algorithms and AI to instantly score various student work, saving time for the teachers, ensuring consistency, and providing immediate student feedback for faster learning, though complex subjective assessments still need human oversight for nuance like creativity. These systems boost efficiency, scalability, and fairness by freeing teachers for deeper engagement, transforming assessment from a bottleneck to a supportive tool for improved learning outcomes. Automated grading in a learning management system is a computer-based process that uses technology like artificial intelligence to evaluate student work, such as multiple-choice tests and essays. This system provides faster, more consistent grading, which saves educators time and provides students with immediate feedback, enhancing the learning experience.

#### 12.5.3.3 *Tracking and reporting*

Learning management systems make it possible to track student performance, progress, and engagement in great detail. This data is useful for spotting patterns and problem areas. The process of gathering, evaluating, and presenting information on student development, engagement, and course efficacy is known as tracking and reporting in a learning management system. Making educated decisions to enhance training, spot skill gaps, and guarantee compliance is made possible by this data. Real-time progress tracking, thorough assessment analysis, and the creation of reports that may be customized for different stakeholders are some of the important features of these systems.

#### 12.5.3.4 *Instant feedback*

Students can receive immediate feedback on their performance, fostering a continuous improvement cycle and a better understanding of their strengths and weaknesses. Learning management systems provide instant feedback through automated grading, in-the-moment correction messages, and real-time analytics, which significantly enhances student motivation, knowledge retention, and the ability to correct misconceptions immediately. The mechanisms for instant feedback in a learning management system are in the realm of the automated grading, for instance, in the objective question types such as multiple-choice, the learning management system can automatically grade responses and immediately display the results to the student. Another aspect of the role of learning management systems in assessment is in specific answer feedback as instructors can prewrite specific explanations for why each potential answer option is right or wrong. This feedback is triggered and displayed as soon as the student makes a selection.

#### 12.5.3.5 *Accessibility and consistency*

Learning management systems offer a single, easily accessible site for educational resources and tests, guaranteeing that all students receive the same content. While consistency guarantees equity and dependable outcomes for all students, accessibility in a learning management system guarantees that all students, including those with impairments, may access and participate in assessments. Standardizing the assessment process and objectives for every student, assessment consistency helps in promoting justice and fairness.

#### 12.5.3.6 *Enhanced efficiency*

A learning management system frees up teachers to concentrate more on instruction and less on paperwork by reducing their administrative responsibilities. Learning management system assessments are more efficient when tasks like grading and feedback are automated, saving teachers time, and when timely and comprehensive results are provided, assisting educators and students in identifying areas for growth. The assessment process is streamlined by automated elements, which improve student progress tracking and enable more frequent and consistent feedback. By automating grading and assessment generation, the learning management system reduces administrative work for instructors, freeing up time for instruction and individualized support. It facilitates effective processes for students by streamlining the entire assessment process, from preparation and delivery to tracking and reporting. Furthermore, it facilitates rapid feedback by providing prompt or immediate answers on tests, which is crucial for learning and aids students in identifying areas that need improvement. It also makes it easier for students to identify their own progress and areas for development, which helps them fill in specific knowledge and skill gaps more effectively.

#### 12.5.3.7 *Flexibility*

Flexibility in using a learning management system for assessment provides significant benefits by allowing for adaptability in time, location, format, and weighting, which caters to diverse learner needs and streamlines administrative tasks for educators (Camilleri, M. A., & Camilleri, 2023). The ability for learners to access information at any time and from any location is one of the advantages of flexibility in learning management system evaluations. Students may fit their studies around their personal and professional obligations because they have round-the-clock access to learning materials and examinations through a variety of devices. Additionally, it facilitates self-paced learning for the learning management system by enabling asynchronous learning, which lets students advance through tests at their own speed, spending more time on difficult ideas and swiftly going through well-known ones.

### 12.5.4 Role of learning management system in skill development

A learning management system plays a crucial role in skill development by centralizing training, personalizing learning paths, tracking progress, and offering flexible and engaging content. It acts as a platform for skill-based training, allowing organizations to deliver targeted courses, monitor performance, and issue certifications upon completion (Faisal Rahim, 2022). This helps learners continuously update their skills, align with career goals, and build proficiency in a flexible, accessible, and modern format. It provides the opportunity for centralized and flexible learning by acting as a single source of truth for all learning, housing a wide variety of training resources, courses, and materials in one accessible location. The learners can access materials, participate in training, and complete courses at their own convenience, from multiple devices and locations, promoting continuous learning. In addition to this, it contributes to the enhanced engagement and motivation by providing learning management system platforms which can incorporate interactive elements like gamification, multimedia, and group projects to make the learning process more engaging and motivating. It also assists in the process of social learning as many systems include social learning features like discussion forums and live chats, which foster peer-to-peer collaboration and knowledge sharing.

### 12.5.5 Data-informed decision-making

Learner management systems give teachers access to analytics that help them better understand the behavior and performance of their students. This allows them to improve their teaching methods and help students succeed. One of the main advantages of adopting a learning management system for assessment is data-informed decision-making, which offers evidence-based insights for ongoing course and learning experience improvement. Teachers can increase student engagement and results by identifying knowledge gaps, customizing learning routes, and improving curriculum, instructional strategies, and content through the analysis of assessment data. Decision-making is influenced by learning management system assessment data. A learning management system can assist in identifying a student's strengths and weaknesses in terms of tailored learning paths, which are made possible by monitoring individual performance. This enables the recommendation of certain assignments or content to meet each student's particular needs. Additionally, it aids in curriculum and content improvement since analytics can identify areas in which students struggle. This information can be used to make decisions about upgrading course materials, enhancing teaching tactics, or adding additional resources to address frequent learning issues.

## 12.6 Contribution towards principles and practice of medical education

Learning management systems make a substantial contribution to the concepts and practice of medical education by improving accessibility, standardizing content distribution, encouraging active learning, and simplifying administration. The concepts of medical education support accessibility and self-directed learning. Additionally, learning management systems enable students to learn at their own pace and convenience, accommodating a variety of learning styles and schedules, by offering a centralized, round-the-clock access point to materials, modules, and assessments. Additionally, in reality, learning management systems make course management more effective for instructors by replacing diverse approaches like email or physical handouts with a single, well-organized repository for lectures or any other information.

## 12.7 Challenges of using learning management system in medical education

Challenges of using learning management system in medical education include user adoption issues due to a lack of technical skills, a steep learning curve, and resistance to new technology (Geada, 2022). Other significant challenges include data privacy concerns, the cost of implementation, lack of interactivity, difficulties in engaging users, and the inability to replicate hands-on clinical experience. The challenges can incorporate three aspects:

### 12.7.1 User-related challenges

#### *12.7.1.1 Technical skills and training*

One of the main obstacles to implementing a learning management system in medical education is that both teachers and students lack the necessary technical expertise and training, which makes implementing a new system challenging and annoying. Underutilization or discontent with the platform is frequently caused by this talent gap, inadequate infrastructure, and unwillingness to change.

#### *12.7.1.2 User adoption and engagement*

One of the biggest challenges is getting staff and students to regularly utilize and interact with the platform, particularly if they find it difficult to use or believe it increases their workload. When utilizing a learning management system in medical education, user acceptance and engagement are major obstacles that frequently result from a mix of subpar design, inadequate training and support, pedagogical mismatch with conventional techniques, and innate resistance to change among students and teachers.

#### *12.7.1.3 Motivation*

One of the important hurdles of the learning management systems is maintaining the motivation of the learner. If the material is uninteresting or the learning management system's goal is unclear to them, learners may become demotivated. When utilizing a learning management system in medical education, motivation is a major problem that frequently results from poor system usability, perceived irrelevance, and a lack of individualized interaction.

## 12.7.2 Content and curriculum challenges

### *12.7.2.1 Lack of interactivity*

Simulations and other interactive, customized learning experiences that are essential for medical training may be absent from standard learning management system capabilities. A major obstacle to the use of learning management systems in medical education is their lack of interactivity, which jeopardizes important in-person interactions, impedes the development of clinical skills, and causes social isolation and student disengagement.

### *12.7.2.2 Inability to replicate hands-on experience*

Learning management system platforms cannot completely replace the essential practical, hands-on experience needed for clinical skills especially in disciplines like surgery. One well-known and important issue with employing learning management systems in medical education is the inability to duplicate practical clinical practice. Moreover, complex psychomotor skills, like completing physical examinations, suturing, and surgical operations, as well as soft skills like patient communication, cannot be taught through the learning management system. Furthermore, since a learning management system mainly allows content delivery and online interaction, there is a lack of in-person connection. The tactile, multisensory setting of a genuine clinic or lab, where students engage with real equipment, simulated patients, or cadavers, is difficult to replicate.

### *12.7.2.3 Content creation*

For educators, producing current, interesting, and high-quality content for a learning management system can be a difficult and time-consuming endeavor. When utilizing a learning management system in medical education, content production is a major problem because of the time and technical knowledge needed, the requirement to make difficult material interactive and engaging, and the need to make sure the content is current, relevant, and consistent. Medical faculty frequently have severe workloads with clinical obligations, leaving insufficient time for teaching and learning activities. In addition, one of the main issues of content creation is that it results in time-intensive development since producing high-quality, effective digital material takes time.

## 12.7.3 Technical and administrative challenges

### *12.7.3.1 Data privacy*

Protecting sensitive patient and user data is a critical concern that requires robust security measures. Data privacy is a significant challenge for using learning management systems in medical education due to the highly sensitive nature of health information and strict regulatory requirements like HIPAA and General Data Protection Regulation. The key data privacy challenges encompass handling sensitive health data. As medical education involves handling highly confidential health data such as patient records, it requires a higher standard of protection than typical academic data. Standard learning management system platforms are often not designed for storing this "special category" data, increasing the risk of a breach.

### *12.7.3.2 Cost*

Effective learning management system implementation and upkeep can be costly, especially for smaller schools with tighter budgets. A major obstacle in medical education is the expensive cost of setting up and maintaining a learning management system, especially given the range of direct and indirect costs involved. Purchasing, developing, software upgrades, security patches, bug fixes, general maintenance, and setting up a learning management system can all have high upfront costs. These costs can vary depending on their complexity and custom-built systems. However, implementation is not the end of the ongoing process of maintenance and other costs.

### *12.7.3.3 Technical issues*

Learning management systems can be affected from technical glitches, downtime, and compatibility problems that disrupt learning. The common technical issues of using learning management systems in medical education primarily involve usability and navigation problems, compatibility, and integration difficulties. Navigation problems can make resources, assignments, and essential information hard to locate, especially if course sites are poorly structured or filled with dense information. Moreover, both medical students and educators often find learning management system interfaces confusing and difficult to use, leading to dissatisfaction and hindering effective studying or teaching.

#### 12.7.3.4 Customization

Additionally, some learning management systems may not have the customizable features required to satisfy the particular requirements of a medical curriculum. In medical education, customizing a learning management system entail modifying administrative features, user interfaces, and content to satisfy particular institutional requirements, boost participation, and increase training efficacy. This entails developing competency-specific modules, customizing learning pathways for various positions, and integrating the system with other platforms (Lyall, 2024).

**Summary**

The function of learning management systems as fundamental digital infrastructures in modern medical education is examined in this chapter. It demonstrates how learning management system platforms facilitate medical curriculum delivery, organization, assessment, and evaluation in undergraduate, graduate, and continuing professional development contexts. The learning management system is offered as a pedagogical enabler that fosters active learning, introspection, teamwork, and the development of long-term competencies rather than only as a technological tool. Faculty participation, learner acceptance, technological infrastructure, integration with other educational and healthcare systems, and concerns about data security and privacy are among the implementation problems that are discussed. In order to maximize the educational value of learning management system deployment, the chapter emphasizes the significance of institutional strategy, faculty development, and instructional design assistance.

**Key point**

Future perspectives: the advent of learning experience platforms

The future of learning management systems in medical education is characterized by the integration of advanced technologies like artificial intelligence, virtual reality, and augmented reality to create personalized, immersive, and data-driven learning experiences. Future directions for learning management systems in medical education also include integration with simulation technologies and learning experience platforms. It concludes that while learning management systems platforms are not sufficient on their own to ensure educational quality, they are essential tools that, when thoughtfully implemented and pedagogically aligned, can significantly enhance teaching, learning, and assessment in medical education. On the other hand, the role of student awareness on the basics of learning management system and the applications is critical to the successful usage of the learning management system. Similarly, the role of faculty training and regular workshops cannot be understated. It is vital that the institution and the departments work in this direction so as to ensure that medical education is delivered in the right way.

## References

Afridi, S., Zain, K., Mahsood, N., Noor, N., Noori, M., Kashif, L., 2025. Learning management systems (LMS) in medical education: insights from undergraduate medical students in Peshawar. Journal of Medical Sciences 33 (2), 72–76. Available at https://doi.org/10.52764/jms.25.33.2.3.

Alamgir, A. (2025). Learning Management System (LMS) Curriculum 2025 RMU. 10.13140/RG.2.2.22672.72963.

Back, D.A., Behringer, F., Haberstroh, N., Ehlers, J.P., Sostmann, K., Peters, H., 2016. Learning management system and e-learning tools: an experience of medical students' usage and expectations. International Journal of Medical Education 7, 267–273. Available at https://doi.org/10.5116/ijme.57a5.f0f5. Available fromhttps://www.ijme.net/archive/7/learning-management-system-and-e-learning-students-expectations.pdf.

Bozoğlu, O., Armağan, S., Erdönmez, Ç., 2016. Integrating a learning management system into the instructional program and learning process: challenges and advantages. British Journal of Education, Society & Behavioural Science 18 (4), 1–7.

Camilleri, M.A., Camilleri, A.C., 2023. Learning from anywhere, anytime: utilitarian motivations and facilitating conditions for mobile learning. Technology, Knowledge and Learning 28 (4), 1687–1705. Available at https://doi.org/10.1007/s10758-022-09608-8, https://www.springer.com/journal/10758.

Chahal, K., 2021. Features of learning management systems (LMS) for improving teaching and learning. Journal of Interdisciplinary Cycle Research 13, 192–206.

Faisal Rahim, M., Ahmed Skaikh, M., Ul Haque, S., Jabeen, S., Ansari, T., 2022. Role of a Learning Management System in Medical Curriculum: Role of a Learning Management Sys tem in Medical Curriculum. Pakistan Journal of Health Sciences 3 (05). Available at https://doi.org/10.54393/pjhs.v3i05.246.

Fediv, O., Buzdugan, I., Vivsianyk, V., Prysiazhniuk, I., Prysyazhnyuk, V., 2021. The Role of Moodle Software among Medical Students During Distance Learning. Scientific Bulletin of Mukachevo State University Series "Pedagogy and Psychology" 7, 63–69. Available at https://doi.org/10.52534/msu-pp.7(2).2021.63-69.

Geada, N., Jamil, G.L. (Eds.), 2022. Digital Active Methodologies for Educative Learning Management. IGI Global Scientific Publishing. Available at https://doi.org/10.4018/978-1-6684-4706-2.

John, K., 2022. Learning Management System and Its Role in the Effective Delivery of Medical Education. SBV Journal of Basic, Clinical and Applied Health Science 5, 105–106. Available at https://doi.org/10.5005/jp-journals-10082-03164.

Kamath, S.G., Nayak, K.R., Nayak, V., Verma, S., 2025. Leveraging learning management systems in medical education: a scoping review of use, outcomes, and improvement pathways. Medical Education Online 30 (1).

Kumar, S.A., Niveditha, G., Seema, P., Bengalorkar, G.M., 2022. Usage of learning management system to implement online teaching methodology in undergraduate medical students. Clinical Dermatology Review 6 (2), 149. Available at https://doi.org/10.4103/cdr.cdr_71_22.

Lyall, J., Freund, K., Webb, A., 2024. Challenges and tensions in the role of the LMS for medical education: Time for the "next generation LMS"? ASCILITE Publications 304–311. Available at https://doi.org/10.14742/apubs.2017.783.

Sampayan, E., 2024. Blackboard Learning Management System - An Artificial Intelligence Approach: Challenges and Prospects in Nursing Education. International Journal of Science and Healthcare Research 9, 230–234. Available at https://doi.org/10.52403/ijshr.20240129.

Sánchez, L., Peñarreta, D., Soria Poma, X. (2023). Learning Management Systems for Higher Education: A Brief Comparison. 10.36227/techrxiv.23615523.

Tibyampansha, D., Ibrahim, G., Kapanda, G., et al., 2017. Imple-mentation of a learning management system for medical students: a case study of Kilimanjaro Christian Medical University College [version 1]. MedEdPublish 6, 50.

# INSTRUCTIONAL METHODOLOGIES

# PART 3

## INSTRUCTIONAL METHODOLOGIES

Chapter 13

# Traditional teaching methods

*"The greatest danger of traditional education is that it might remain purely verbal."*

—Miraa Komarovsky

## 13.1 Introduction

Traditional teaching in medical education centers on teacher-led, didactic methods like lectures, textbooks, and rote memorization, focusing on foundational scientific theory before clinical application, with key techniques including large-group lectures, small-group tutorials, direct patient observation, and structured case reviews, emphasizing factual knowledge delivery and instructor control. While effective for broad knowledge, these methods are increasingly complemented by student-centered approaches like problem-based learning and case-based learning to enhance critical thinking.

## 13.2 Teaching methodologies in medical education

In addition to more contemporary techniques like problem-based learning, case-based learning, and simulation-based learning, classic lectures and clinical instruction are also used in medical education (Mitchell, 2024). While some innovative approaches emphasize active learning through flipped classrooms and team-based learning, others use technology like augmented and virtual reality and e-learning platforms. The particular learning objectives and the student's preferred learning style frequently influence the method selection. Medical education teaching methods can be broadly categorized as either traditional, such as lecture-based, or modern, which are primarily learner-centered. Modern approaches, such as problem-based learning, case-based learning, team-based learning, simulation-based learning, e-learning, and flipped classrooms, emphasize problem-solving, collaboration, and technology, all of which are meant to foster critical thinking and clinical skills (Challa, 2021).

1. Traditional and interactive models
    1. Lectures
    2. Demonstration
    3. Discussions
2. Modern and student-centered methods
    1. PBL (problem-based learning)
    2. TBL (team-based learning)
    3. CBL (case-based learning)
    4. Flipped classroom
3. Technology-based models
    1. E-learning
    2. Simulation-based learning
    3. Massive open online courses

**The Principles and Practice of Medical Education. DOI:** https://doi.org/10.1016/B978-0-443-44197-4.00004-7

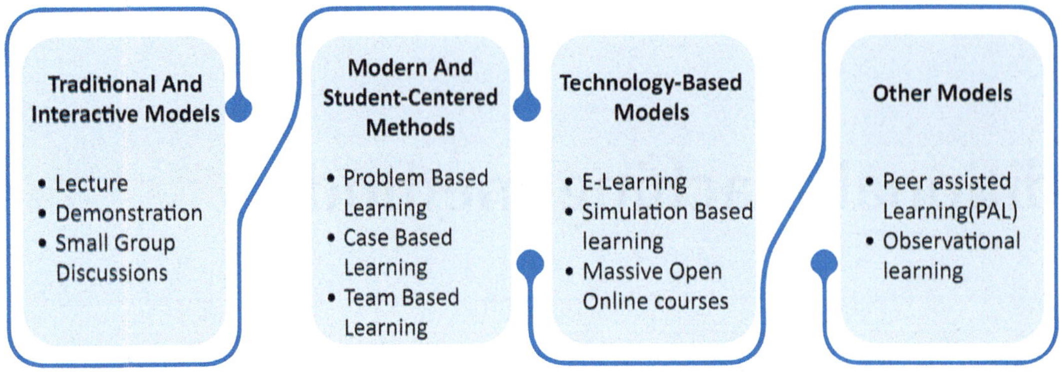


FIGURE 13.1 Classification of teaching methods.

4. Other models
   1. Peer-assisted learning
   2. Observational learning

The classification of teaching methods is illustrated in Fig. 13.1

Out of all the above-mentioned methods of teaching, the current chapter mainly focuses on the traditional and interactive approach of teaching in medical education. The rest of the methods are dealt in detail in the subsequent chapters.

## 13.2.1 Traditional and interactive models

Traditional teaching techniques are one of the most widely used in medical education. In this teaching method, the instructor or teacher is seen as a source of information. The traditional method of teaching medicine is teacher-centered and mostly uses lectures, textbooks, and two-dimensional illustrations to communicate factual knowledge. Even while it provides a foundational knowledge base, this approach is frequently critiqued for being less effective than more modern, interactive ways at developing deep comprehension, spatial reasoning, and clinical ability. However, one of the disadvantages of the traditional method is that it has very minimal teacher–student connection. Some of the important characteristics of traditional teaching methods are described below.

### *13.2.1.1 Teacher-centered*

The teacher-centered approach in traditional teaching places the teacher as the primary source of knowledge, delivering information via lectures and direct instruction, with students passively receiving it through listening and note-taking, focusing on content mastery, classroom order, and assessment through tests (Luscombe, 2016). The main characteristic features of the teacher-centered approach include the teacher controlling pace, content, and activities, while students demonstrate learning by recalling facts, often leading to efficient material coverage but potentially limiting critical thinking and active engagement. This method views the instructor as the primary source of knowledge and as in charge of the classroom. Some of the key benefits of the teacher-centered method is that the instructor is involved in the efficient content coverage, whereby the enormous volumes of material may be covered rapidly. Since they are in control of this approach, teachers are highly effective at managing the classroom since they help to keep things organized and structured.

### *13.2.1.2 Didactic teaching*

Didactic teaching is a conventional, teacher-centered approach that emphasizes the direct transfer of knowledge from teacher to student, usually through lectures, lesson plans, and demonstrations. It places a strong emphasis on authority, memorization, and passive learning. Although it is frequently criticized for restricting critical thinking, it is nonetheless prevalent for core courses.

### 13.2.1.3 *Reliance on static materials*

One of the main characteristics of traditional medical education is the heavy reliance on static, two-dimensional instructional materials, which often convey knowledge in an isolated, one-way fashion. This contrasts with modern methods that employ dynamic, interactive, and technology-driven resources. One of the primary features of relying on static resources is the reliance on didactic lectures and textbooks, which are a significant part of traditional medical education. In order to visualize complex systems, such as those found in anatomy, students mostly employ diagrams and illustrations from textbooks and static presentations.

### 13.2.1.4 *Focus on memorization*

It is true that in medical education a lot of information has to be memorized. Information pertaining to branches of the arteries, physiological values or the enzymes in the rate limiting steps of crucial biochemical reactions require the student to remember and recall it at all times. This is indeed one of the primary aspects of traditional medical education in which a heavy reliance on rote recall, or rote learning, is given. This approach heavily emphasizes the passive acquisition of a vast amount of factual knowledge, which students are expected to understand through repeated study sessions and memory-based assessments. The emphasis on memorizing to facilitate retention of vast factual knowledge, such as detailed physiology, pharmacology, and pathology, rather than the development of critical thinking or clinical reasoning skills is one of the major drawbacks of traditional teaching. The advantages of rote memorization as in building a foundational knowledge base or facilitating quick recall and accuracy have made it an indispensable part of the learning process in medical education.

## 13.2.2 Phases of the delivery of content

As far as the delivery of content in traditional teaching method is concerned, it can be grouped into three stages which are before class, in class, and after class. Before the conduction of the class, the students will be given a sneak peek of the contents of the topic. The teachers don't engage with the students throughout class; instead, they spend much of the time teaching about the topic. After class, the students are expected to read the reference provided by the tutor and complete any assignments if needed (Fig. 13.2).

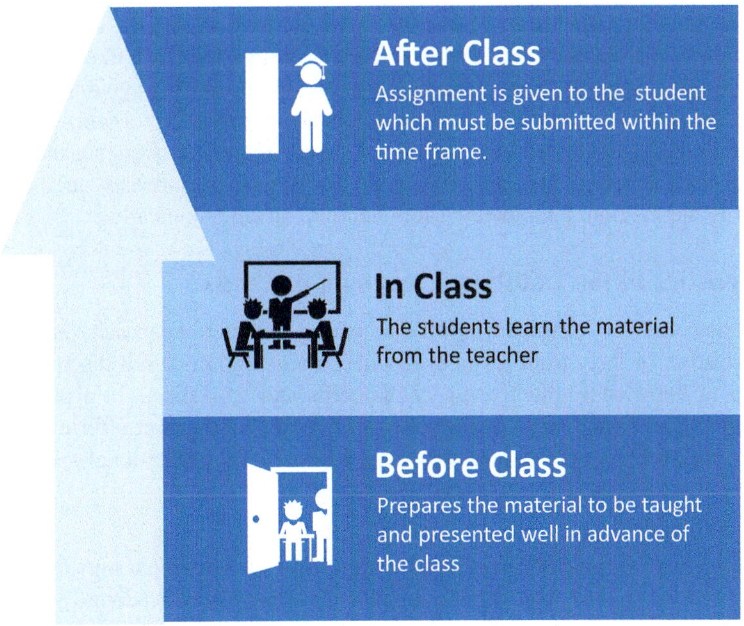


FIGURE 13.2 This figure shows the phases of delivery of content.

#### 13.2.2.1 Before class

In a traditional lecture, the tutor prepares the content to be taught and presented well in advance of the class, making sure that it aligns with the specific learning objectives of the subject. At this point, students are very important because they evaluate the course materials that the instructor has provided them before starting a particular course or topic. This course document, also known as the study guide, is a useful tool since it gives students information about the material to be taught so that they may prepare for it in advance. Additionally, the student knows exactly what will be taught and discussed. It is imperative that the reference material is also given to the students before class so as to facilitate their study process. The same references that were given to the students should be used when preparing for the lecture, and any additional references should be made apparent.

#### 13.2.2.2 In class

In a traditional classroom, students listen passively as the teacher presents the material on the subject using power point slides or the chalk-and-talk method. In the classroom, there isn't much interaction between the teacher and the students. The primary responsibility of the student will be to record all of the information provided by the instructor and to take notes on it in order to utilize it for tests. One of the main benefits for the tutor is that the students learn the material and the teacher covers all the key topics related to the subject as planned. The drawback is that unless the teacher asks a question, the student is unable to engage in class. One of the biggest challenges is delivering all the information in the allocated time, which is usually 50 minutes for the lecture. In this sense, the lecture needs to be carefully organized to include all pertinent elements and align with the intended learning objectives.

#### 13.2.2.3 After class

Along with in-class activities, the tutor designs assignments related to the lesson's subject matter that the students must finish. The students are expected to complete these assignments within the allotted time. After receiving clear instructions from the instructor, the student is expected to reread the reference, complete the assignment, and verify that they have comprehended everything. At this stage, there is minimal interaction between the teacher and the students because they are expected to work independently. This will boost the student's confidence and competence in finding and using materials on their own, in addition to providing them with the opportunity to work independently.

## 13.3 Frameworks for the traditional teaching

Frameworks for traditional teaching center on a teacher-led, content-focused model, emphasizing structured curriculum, rote memorization and direct instruction, often using deductive methods within fixed classroom settings, aiming for standardized academic knowledge and discipline. The Kolb's experiential learning cycle, Peyton's four-step approach, and Bloom's taxonomy are frameworks for traditional medical education that offer organized techniques for teaching and evaluation even in conventional, lecture-based environments. These frameworks enable incorporating active learning to increase engagement and the caliber of learning, but they also place an emphasis on the planned progression from fundamental information to application, guaranteeing that learners advance logically.

### 13.3.1 Key characteristics of the traditional teaching methods

Traditional teaching frameworks are characterized by a teacher-centered approach, emphasizing direct instruction, lectures, and rote memorization of facts from textbooks, with a focus on structured discipline, standardized testing, and passive student reception of knowledge, often using chalkboards and worksheets in a physical classroom setting. The curriculum follows a predetermined pace, ensuring systematic content coverage but offering little flexibility for individual learning styles or critical thinking. Some of the characteristic features of the traditional teaching methods are as follows:

#### 13.3.1.1 Structured curriculum

A key and distinguishing feature of conventional teaching approaches is structured curriculum, which provides teachers and students with a comprehensive road map for the duration of a course or academic year. Order, predictability, and mastery of fundamental information are its main features. Since the material is arranged logically and step-by-step, one of the main features of the structured curriculum is that it helps with the sequential and linear progression. To ensure a methodical acquisition of knowledge, essential concepts are presented initially, and following sessions immediately build upon previously understood content. At the same time, it also highlights that the clear and specific learning goals are established in advance, and these objectives dictate what students are expected to know or be able to do by the end of a unit or course, providing a clear target for instruction and assessment. In addition, it also places a strong emphasis on subjects like anatomy, physiology, and pharmacology which forms the backbone of clinical medicine (Papanna, 2012).

#### 13.3.1.2 Hierarchical knowledge transfer

In traditional teaching methods, the term "hierarchical nature of knowledge" refers to a systematic, top-down approach in which learning proceeds step-by-step and knowledge is arranged into fundamental and advanced levels. In order to impart this structured knowledge, the traditional approach is usually teacher-centered and uses lectures and rote memorization. Additionally, since the content is typically organized from basic, foundational concepts to more complex, higher-order skills, the core information is given priority here. This ensures that before moving on to more difficult topics, students have the requisite knowledge. The teacher is viewed as the authority in this structure because they are the primary source and authority of knowledge and because they arrange and disseminate information to the learners in a top-down manner. The instructor controls the learning process by deciding which topics to cover and in what order.

#### 13.3.1.3 Assessment and evaluation

Conventional teaching approaches use objective, paper-based assessments like multiple-choice and essay type of questions for assessment and evaluation, which mainly concentrate on memorization and recall. These techniques are typically summative, offering a definitive assessment of a student's knowledge and are simple to grade. They may, however, overlook higher-order cognitive abilities like creativity and problem-solving. Traditional methods of assessment primarily concentrate on obtaining data regarding students' learning, frequently via standardized tests. The main goal is to provide a snapshot of a student's performance on an exam by measuring their knowledge at a certain moment in time. Some of the main strategies employed here are written tests, quizzes, and exams that typically ask students to describe a certain concept or select the appropriate response. Lower-level cognitive skills including information recognition, definition, and memorization are highlighted here. In contrast, evaluation in traditional methods refers to the process of evaluating the data obtained from assessments to determine the value or caliber of student performance. Assigning a final grade based on test and exam scores is one of the approaches used in this regard. Here, the emphasis is on evaluating a student's learning rather than offering specific suggestions for development.

#### 13.3.1.4 Professionalism and ethics

The development of professional habits, beliefs, and a robust ethical framework inside a hierarchical system is a key component of the traditional paradigm. In traditional teaching methods, professionalism and ethical principles are primarily taught through lectures, assigned readings from textbooks or professional codes of ethics, and role modeling by educators. These methods generally focus on knowledge acquisition and adherence to established rules, rather than encouraging critical thinking, open-ended discussion, or the development of individual ethical reasoning skills, which are more common in modern, student-centered approaches. For example, instructors use lectures and conversations to explain the significance of professional conduct and the ethical principles in medical profession, clarify important vocabulary, and present ethical theories. Large-group or small-group discussions are frequently held after this to examine various viewpoints and elucidate ideas. To expand on the understanding of professionalism and ethics, students evaluate ethical conundrums and discuss possible solutions through studies and scenarios, which are a type of primary traditional technique. Although these are typically less interactive than more recent implementations, they still assist students in applying broad ideas to particular, realistic circumstances (Zinski, 2017).

#### 13.3.1.5 Faculty as authority figures

As authorities and specialists, the faculty play a critical role in forming the learners' professional identities and exemplifying the fundamental principles of the field. As the traditional teaching methods are teacher-centered, students are expected to passively absorb knowledge. Additionally, the faculty is viewed as the only source of knowledge, with teachers being regarded as experts who possess all relevant knowledge that is subsequently imparted to students, frequently through lectures or direct teaching. Additionally, the faculty is seen as the controller and decision maker, having authority over the curriculum's content, class pace, learning resources, and methods of assessment. They are in charge of upholding discipline and order in the learning environment.

#### 13.3.1.6 Focus on mastery of content

Traditional teaching methods, primarily characterized by lecture-based, teacher-centered instruction, treat time as a constant and student achievement as a variable. One of the key characteristics of the traditional approach is that the material is taught at a set pace, with the teacher delivering a predetermined curriculum within a predetermined timeframe, such as a semester or week, and all students moving on to the next topic at the same time, regardless of whether they have fully understood the current material (Peña, 2025). Here, the emphasis is on the knowledge-based approach, which prioritizes ensuring that students have acquired a substantial amount of clinical and factual knowledge before they may practice.

### 13.3.2 Key traditional frameworks and their application

#### 13.3.2.1 Kolb's experiential learning cycle

Kolb's experiential learning cycle explains a cycle of learning through experience, which can be used to traditional instruction by breaking up a lecture into stages such as concrete experience, reflective observation, abstract conceptualization, and active exploration.

#### 13.3.2.2 Role of Kolb's experiential learning cycle in traditional learning

Kolb's experiential learning cycle improves traditional medical education by bridging the gap between theoretical knowledge and practical application through a four-stage cycle. This cycle helps students to integrate what they learn in the classroom with clinical practice in traditional settings by using patient scenarios as experiences, analyzing them to develop diagnoses and treatment plans, and then applying these to new cases. This approach makes learning more dynamic and effective, especially in competence-based medical education, by enhancing clinical competency, critical thinking, and patient management skills.

Traditional medical education is incorporated into Kolb's cycle in several ways. Students can directly experience things like taking a patient's history, doing a physical examination, or handling a simulated cardiac arrest through the concrete experience. Using the reflective observation component, students think about what happened during the interaction, including reviewing the initial evaluation and creating a differential diagnosis. Through abstract conceptualization, students develop abstract concepts like a working hypothesis and a final diagnosis. Through active experimentation, which in this case involves implementing a management approach and completing a follow-up, students apply their recently learned information to unique settings. The Kolb's experiential learning cycle has the advantage of helping students apply what they have learned in real-world or simulated patient scenarios. This is because the cycle explicitly integrates theoretical knowledge with practical experience, which is essential for competency-based medical education. Additionally, by shifting from passive learning to active engagement and practical application, the cycle enhances

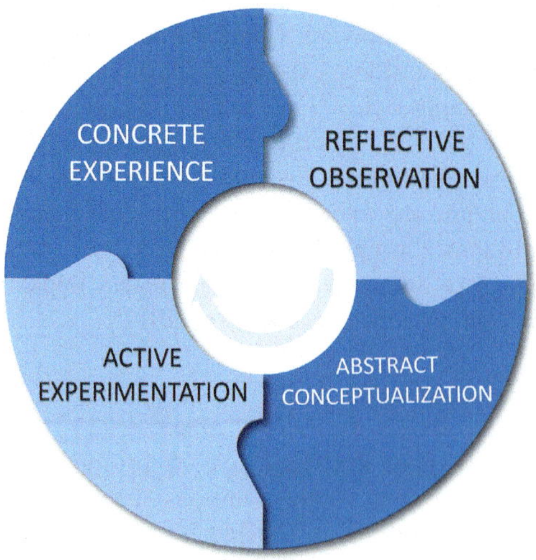


FIGURE 13.3 Kolb's experiential learning cycle.

knowledge retention by deepening the embedding of information in long-term memory and increasing its retrievability (Fig. 13.3).

### 13.3.3 Peyton's four-step approach

Peyton's four-step approach, which incorporates guided practice, independent practice, assessment, and demonstration, is used to teach clinical skills (Cenaj, 2025). This method enhances skill acquisition, confidence, and retention by breaking down complex procedures into manageable steps, moving from passive observation to active, supervised execution.

#### *13.3.3.1 Step 1: demonstration*

The first step in Peyton's classic four-part approach to teaching practical skills is demonstration. In this step, the teacher demonstrates the entire skill at a normal rate without providing any additional comments or explanation while the student observes to obtain a clear overall picture of the correct way. The demonstration helps to confirm the student's original findings as they closely watch the skill's smooth flow and anticipated result in a real-world situation. Since it gives a clear mental image of the entire process, which is essential for the next learning stages, it also helps set the benchmark. Additionally, it offers a basis for learning. The Peyton's four-step approach is in line with the social cognitive model of sequential skill acquisition which states that the learning begins with seeing an expert perform a task, which is followed by imitation and, ultimately, self-regulated execution.

#### *13.3.3.2 Step 2: deconstruction*

Deconstruction is the second step in Peyton's four-step method for teaching practical abilities. The instructor repeats a demonstration of the skill, this time dissecting it into smaller, discrete sub-steps and giving a thorough vocal explanation for each. The skill is performed by the instructor once more, but this time in slow motion and with thorough spoken explanations of each step.

#### *13.3.3.3 Step 3: comprehension*

According to Peyton's four-step method, the learner vocally describes each step of a skill in the comprehension stage. The instructor then follows the student's as they carry out the activity, verifying understanding and pointing out any gaps before the student completes it on their own. As a result, the student must explain the process to the teacher, who serves as the performer, shifting the cognitive load.

#### *13.3.3.4 Step 4: performance*

The performance stage also known as "Trainee does" is the last and most important part in Peyton's four-step method for teaching procedural skills, where the student independently performs the complete skill under the instructor's observation. This is in contrast to "traditional learning" methods, which were less structured and frequently resulted in insufficient skill acquisition. These methods typically involved a straightforward "see one, do one" approach. The student executes the skill on their own. Following that, there is usually an evaluation and feedback session where the learner's performance is assessed and comments are provided. The performance stage's goal is to ensure an autonomous application, when the student performs the skill on their own without the instructor's verbal or physical support at each sub-step at that exact moment. By allowing the learner to move from an instructed performance to an autonomous one by combining the knowledge and motor abilities acquired in the previous three levels, it also helps with skill consolidation. It also enables

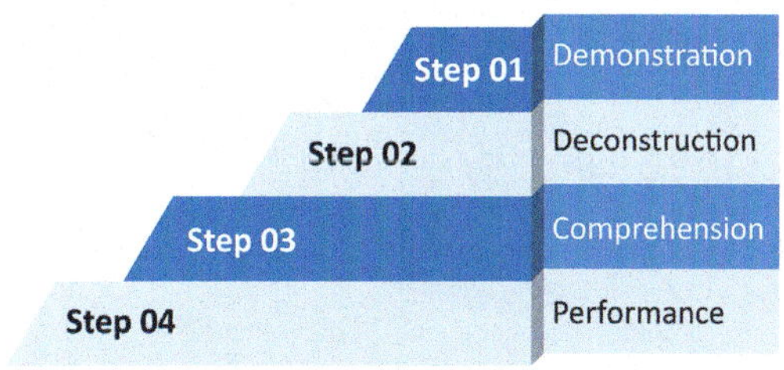


FIGURE 13.4 Peyton's four-step approach.

the instructor to observe this self-directed performance and provide timely feedback and modifications when needed, which is crucial for enhancing abilities and avoiding errors (Fig. 13.4).

#### 13.3.3.5 Advantages of using the Peyton's four-step approach

In traditional classroom environments, Peyton's four-step approach provides benefits including quick and efficient skill learning, increased confidence, and deeper comprehension. Peyton's four-step approach can be utilized in conjunction with conventional teaching techniques, especially as a highly efficient and organized substitute for conventional, less engaging "see one, do one, teach one" presentations. Through the use of demonstration, deconstruction, comprehension, and performance in small groups, it improves the acquisition of skills. Additionally, because this approach goes beyond simple observation, it promotes a more thorough grasp of a subject, which aids in deeper comprehension. In order to improve their knowledge base, students are asked to break down the skill and recollect procedures.

### 13.3.4 Bloom's taxonomy

In order to ensure that doctors develop deep clinical reasoning beyond memorization for better patient care, Bloom's taxonomy offers a hierarchical framework for organizing medical education. This framework moves learners from basic recall, represented by the term remembering, to complex skills, such as critical thinking and innovation, represented by the term creating. It helps teachers connect learning activities to cognitive levels and directs curriculum design, objective setting, and assessment. From the traditional teaching point of view, it often starts with lower levels like remembering facts but aims to progress students up the pyramid for deeper understanding, using questions and activities specific to each level (Fig. 13.5).

### 13.3.5 Advantages of using the Bloom's taxonomy in traditional teaching

#### 13.3.5.1 Focusing on lower-level skills

One of the characteristics of conventional teaching methods is that they frequently emphasize memorization and understanding, concentrating on lower-level cognitive skills such as "Remember" and "Understand." Both the teacher and the student benefit from concentrating on the lower-level skills. As there are several definitions and concepts that can only be learned by repeated reading and memorizing, it helps the learner understand the significance of memorization. For example, the definitions and normal values of the physiological parameters. It is also beneficial for the teacher because conventional methods are excellent at methodically presenting a lot of fundamental material in a logical order. This methodical, well-defined approach aids in guaranteeing thorough covering of crucial content.

#### 13.3.5.2 Hierarchical learning

By offering a systematic, sequential learning path that breaks down difficult subjects into simpler, prerequisite concepts, hierarchical learning supports conventional teaching approaches. This method ensures that learners have a solid foundation before moving on by guiding teachers in the order of basic to advanced material. Additionally, before students are asked to explain an idea based on their comprehension, they may be requested to memorize important terminology.

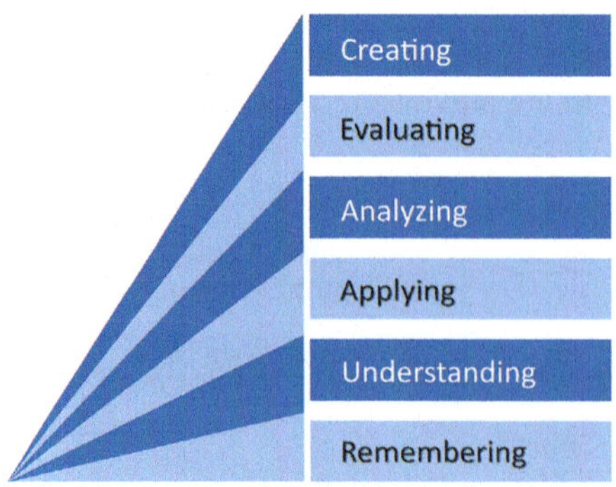


FIGURE 13.5 Bloom's taxonomy.

Traditional approaches also stress how important it is to have a solid foundation of knowledge before advancing to more sophisticated abilities. Additionally, it offers the convenience of assessment because evaluating basic comprehension and factual memory is sometimes simpler and easier than evaluating higher-order thinking abilities. This also supports the teacher-led model of learning, in which the instructor serves as the main source of knowledge and students are expected to assimilate and replicate it.

## 13.4 LEPO framework (learning environment, learning processes, and learning outcomes)

The LEPO (learning environment, learning processes, and learning outcomes) framework can be used to analyze traditional teaching methods by examining the relationships between the learning environment which include as classroom layout and teacher presence, the learning processes such as rote memorization and lectures, and the learning outcomes such as knowledge recall. A teacher-centric setting, passive learning activities, and results that are mainly concerned with information retention are characteristics of traditional techniques (Phillips, 2010). Since the framework takes into consideration every aspect of the learning process, it offers an organized method for assessing these conventional components and comprehending their limitations. The LEPO framework functions by focusing on the three concepts (Fig. 13.6).

### 13.4.1 Learning environment

The learning environment in a traditional paradigm is frequently static and controlled by the teacher. Delivering information from a fixed position, the teacher serves as the center authority. This structure is reinforced by the physical arrangement of the classroom, such as the rows of desks facing the front, which makes it difficult for the learners to sit in groups and debate the subject.

### 13.4.2 Learning processes

According to the LEPO framework, learning in medical education is a dynamic interaction where learning processes are the essential activities that students participate in, such as practice questions, concept mapping, discussion, and using third-party resources to develop knowledge and skills. These activities are influenced by the environment and result in both knowledge and skill acquisition outcomes. In this sense, debates, self-directed study with outside resources, and active engagement with the material are all important processes that eventually influence how students internalize medical knowledge outside of lectures (Wong, 2021).

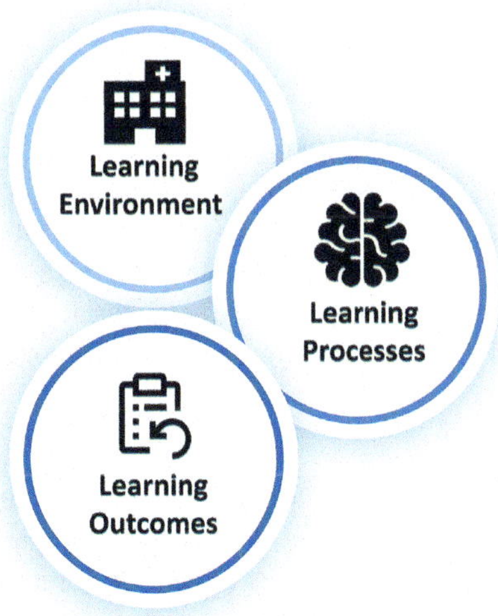


FIGURE 13.6 Learning environment, learning processes, and learning outcomes (LEPO) framework.

### 13.4.3 Learning outcomes

Learning outcomes encompass the measurable results, such as knowledge, skills, behaviors, or understanding. Students frequently exhibit this once they have mastered the task. In contrast to broad teaching objectives, the emphasis here is on what graduates can accomplish, such as using scientific foundations, communicating, performing clinical skills, and addressing community health. These results are attained by efficient learning settings and procedures, such as interactive sessions, visual aids, and condensed lectures. They are essential for creating assessments and enhancing curricula, guaranteeing that future physicians are capable and caring professionals.

## 13.5 Traditional teaching strategies

In medical education, teacher-centered techniques like lectures, direct instruction, and bedside observation are considered traditional teaching methodologies. Clinical skills are frequently acquired through supervised patient practice following didactic sessions, and these approaches mostly rely on the instructor imparting knowledge and students recalling facts through repetition. This strategy has been critiqued for being less student-centered than more recent approaches, even though it is still widely used. In these environments, students receive instruction from teachers in person at predetermined times while adhering to conventional classroom procedures. The traditional teaching techniques that are frequently employed in medical education are the following:

1. Lecture-based instruction
2. Practical sessions
3. Bedside instruction
4. Small group teaching
5. Tutorials

### 13.5.1 Lecture-based learning

Lecture-based learning is a traditional, teacher-centered teaching method where instructors verbally transmit large amounts of information to students, who primarily listen and take notes, often using visual aids like projectors or whiteboards for structured delivery, focusing on foundational knowledge. Lecture-based learning aims to effectively impart fundamental knowledge to a large number of students, provide context and information synthesis, pique students' interest and motivation, model critical thinking and thought processes, and enable the instructor to assess comprehension through in-the-moment interaction. This methodology has drawbacks when it comes to encouraging independent thought and applying knowledge to clinical practice, yet it is frequently used to swiftly cover a large amount of theoretical information from textbooks (Peter Cantillon, 2003). Although lectures are primarily intended to impart knowledge, they can also foster comprehension and encourage introspection when they are created in an interactive manner, allowing students to personalize the material.

### 13.5.2 Frameworks of lecture

The conventional, teacher-centered method, which emphasizes an instructor imparting knowledge to a passive audience, is one framework for lecture-based learning. More contemporary frameworks, such as the interactive lecture, which include activities, discussions, and questions to keep students interested, are also commonly used along with the innovative teaching methods. A concise introduction, a body that presents the material, and a conclusion that provides a summary and comments are all common components of effective lecture design frameworks.

#### *13.5.2.1 Traditional lecture framework*

It is the most traditional and popular lecture technique. Since it is a teacher-centered method, the lesson's content, pace, and flow are all under the teacher's control. In this, the role of the students is passive as their participation is only limited to actively listening, processing information, and taking notes. There is minimal or no engagement of the student as the lecturer imparts knowledge to the listener in a one-way fashion (Panjehpour, 2012).

### 13.5.3 Structure of the traditional lecture framework

The format of a traditional lecture is usually linear and consists of three phases:

#### 13.5.3.1 Introduction

The first step is introduction, in which the instructor presents the subject, outlines the lecture's goals, draws on existing knowledge, and provides a synopsis of what will be addressed. The primary goal at this stage is to mentally prepare students for new knowledge.

#### 13.5.3.2 Presentation

The presentation, which is the primary body of the lecture, is the second step. The instructor frequently uses a white board or a power point presentation to thoroughly explain the subject utilizing charts and pictures. One benefit of this phase is that the content is organized logically and chronologically, making it simple for the student to understand. The students pay close attention to the teacher's presentation during the entire process. This step's primary goal is to successfully impart knowledge to the student.

#### 13.5.3.3 Conclusion

The final step is devoted to summarizing or concluding the topic, in which the instructor provides a summary of the key points and concluding remarks. In lecture-based learning, the summarizing and concluding step is an essential last stage that serves to bring the audience to a close, reinforce important points, and consolidate information. It aids students in remembering the most important details for long-term memory. Bullet points, which primarily concentrate on the key ideas and important details related to the subject being taught, are typically used for this. The lecture-based method efficiently delivers large amounts of foundational information to many students at once, provides structure, allows instructors to share expertise and passion, and can be recorded for future access, making it cost-effective for covering broad topics or introducing new subjects. It's useful for setting context, demonstrating professional thinking, and catering to auditory learners, while also developing student skills in listening and note-taking. It also allows the subject matter experts to share unique insights, passion, and model disciplinary thinking, making complex topics relatable. As a one-way communication paradigm that frequently fails to accommodate a variety of learning styles or foster critical thinking and problem-solving abilities, the lecture-based approach has drawbacks such as encouraging passive learning, generating boredom, and decreasing interest. It frequently lacks involvement and real-world application, and its efficacy greatly depends on the instructor's skill. It can overwhelm students with too much knowledge too rapidly, resulting in poor memory. Since this approach mostly depends on the professor's presentation abilities, a lecturer with poor expertise and presentation skills might ruin the experience, which is a significant disadvantage in terms of efficacy.

### 13.5.4 Demonstrative lecturing

The lecture-demonstration method, also known as demonstrative lecturing, is a teaching approach used in medical education that blends a live or recorded demonstration of a practical skill or process with conventional verbal explanation. Since it bridges the gap between theory and real-world application, this integrated method is especially beneficial for teaching clinical and technical skills. This approach effectively links theory and practice by providing a useful and visual example, helping students comprehend concepts that are difficult to grasp through explanation alone. Through observation and subsequent practice in a controlled situation, such as a skills lab or simulated setting, the main objective is to assist students in learning practical skills. In contrast, listening and taking notes are the main methods of learning in a purely didactic presentation. Demonstrative lecturing in medical education primarily addresses competencies related to procedural skills, patient care, practice-based learning, interpersonal and communication skills, and teamwork. The effectiveness of demonstrative methods is enhanced when combined with active student participation, observation, assistance, and performance, leading to improved skill acquisition and retention compared to lecture-only methods. An effective demonstrative lecture follows a structured approach. The various steps included in the conduction of the demonstrative lecture are described below:

#### 13.5.4.1 Planning and preparation

A good demonstrative lecture can be transformed from a simple presentation into an engaging, interactive learning experience with careful planning and preparation. To make sure the demonstration goes well, the instructor must meticulously prepare it, establish clear learning objectives, collect all required supplies and equipment, and practice the procedure beforehand. Additionally, it is critical that the instructor comprehends the audience. In addition to this the teacher should have a good knowledge about the students' level, past knowledge, learning preferences, and any special needs. This makes the course relevant and interesting by adjusting the examples and content to their requirements.

#### 13.5.4.2 Introduction

To effectively introduce a topic during a demonstration lecture in medical education, the lecturer should grab the learners' attention, connect the new material to prior knowledge and real-world relevance, define learning objectives, and establish a safe, interactive learning environment. Additionally, it should set a clear expectation for the demonstration by providing an overview of the items to be utilized, defining any appropriate language, and briefly presenting pertinent background information. It is important that the tutor confidently greet the class and introduce themselves. It is always good to start with a captivating element to pique interest and stimulate thought as it helps to hook the student to the topic.

#### 13.5.4.3 Demonstration with commentary

The integration of demonstration with commentary is a highly effective method for teaching practical skills, bridging the gap between theory and practice by combining visual, hands-on learning with verbal explanation. This combination of spoken and visual education improves comprehension. Here, the instructor completes the experiment or assignment step-by-step while methodically and clearly outlining each step. The instructor demonstrates the method by continuously explaining verbally what they are doing and why. This commentary is essential because it gives background information, clarifies decision-making, and expresses the reasoning behind each action in real time. In medical education, demonstration with commentary is a potent teaching technique that is frequently used to educate psychomotor skills as well as communication or thinking abilities. While carrying out a process or interaction, the educator explains to the students the underlying principles and their own cognitive processes. The instructor's demonstration of the proper hand-washing technique is one of the most common examples of this, which explains the significance of each step such as duration, coverage areas, and not touching the faucet with clean hands to ensure proper infection control.

#### 13.5.4.4 Student observation and interaction

Student interaction and observation are essential elements of a demonstrative lecture that turn a passive presentation into an engaging learning environment. These components serve as a link between theoretical ideas and real-world implementation. This method's main teaching tool is observation, which enables students to see the "how" and "why" of a process. Additionally, it facilitates students' understanding of challenging or intangible ideas like chemical processes or medical operations. In this case, the students mainly watch the procedure. The instructor asks questions either during or after the performance to clear up any confusion and encourage critical thinking.

#### 13.5.4.5 Follow-up and practice

Following the teacher's demonstration, students could have the chance to practice the skill on their own or participate in an activity or conversation that strengthens what they have learned. The instructor assesses the students' progress and gives criticism. Effective follow-up and practice in a demonstrative lecture are essential for solidifying learning and ensuring students can apply the concepts or skills demonstrated. Immediate application, guided discussion, and reflective activities are some of the key strategies in this regard.

### 13.5.5 Advantages of the demonstrative lecture

The successful integration of theory with practical application and the enhancement of the development of critical psychomotor and clinical skills are the two main advantages of demonstrative lectures in medicine. It provides an enhanced method of instruction when compared to standard verbal-only lectures. By enabling students to observe an expert conduct a technique step-by-step before attempting it themselves, the demonstrative lecture helps students acquire practical skills more successfully. This is due to the fact that seeing is often simpler to recall than hearing. Combining visual, auditory, and frequently kinesthetic learning through subsequent practice engages multiple senses, resulting in a deeper understanding and better long-term retention of complex medical information and procedures. This also helps to improve comprehension and retention. Additionally, it aims to bridge the gap between theory and practice by directly connecting the theoretical knowledge acquired in lectures or textbooks to its practical application.

### 13.5.6 Practical sessions

Practical sessions are a key component of traditional medical education, emphasizing the application of academic information in simulated or real-world settings. Practical sessions are characterized by being organized, hands-on, teacher-supervised activities including skill-based exercises, lab experiments, and demonstrations, which aim to close the gap between clinical practice and classroom lectures. Here students learn by observing, imitating, and executing a method or process under the teacher's close supervision. As a conventional approach, practical sessions have the following features:

#### 13.5.6.1 Teacher-centered

In medical education, practical sessions have traditionally been characterized by a teacher-centered approach, where the instructor demonstrates a skill or procedure and students passively observe. This model, rooted in the conventional structure of basic and clinical sciences, prioritizes the efficient transmission of a large volume of information and fundamental skills.

#### 13.5.6.2 Structured and repetitive

Students are frequently given detailed, step-by-step directions to follow throughout the sessions, which may require repetition in order to perfect a skill. The structured and repetitive nature of practical sessions is a deliberate pedagogical approach essential for building core competencies, ensuring consistency in training quality, and ultimately enhancing patient safety. This approach is rooted in learning theories, such as deliberate practice and spaced repetition, which are crucial for mastering complex cognitive and psychomotor skills.

#### 13.5.6.3 Emphasis on physical skills

The focus on physical skills during practical sessions in medical education is vital for cultivating key competencies, ensuring patient safety, and enhancing diagnostic precision and treatment results. These experiential sessions connect theoretical learning with practical clinical experience, fostering crucial confidence and skillfulness in aspiring doctors. A significant benefit of prioritizing physical skills is the enhancement of patient outcomes since practical training in physical assessments, surgical methods, and various procedures equips physicians with the hands-on abilities necessary to elevate the quality of care and patient results.

#### 13.5.6.4 Complementary to lectures

Theoretical knowledge acquired through lectures is frequently reinforced through practical exercises. Practical sessions act as complementary to the lectures as they provide a means of putting what has been learnt into practice. While lectures offer a fundamental understanding, practical sessions serve as a supplement by assisting in the application of theoretical knowledge. One of the important advantages here is that the practical sessions offer hands-on experience to reinforce concepts presented in lectures and significantly contribute in greater understanding of the topic. The role of practical sessions as a complementary to lectures is appreciated in subjects like anatomy and pathology wherein the introduction of practical sessions acts as boon for the students to gain knowledge and practical experience about the topic and apply it to clinical medicine.

#### 13.5.6.5 Limited student autonomy

Limited student autonomy in practical sessions is often a byproduct of traditional teacher-led instruction, institutional constraints, and a lack of specific training for both students and educators. In many practical settings, students remain dependent on instructors for step-by-step guidance rather than leading their own learning process. This can often result from the nature of traditional practical sessions, wherein traditional approaches often focus on the transfer of knowledge from the teacher to the student, leaving little room for students to make decisions about their methods or objectives. Moreover, the presence of fixed curricula and rigid assessments, such as predefined syllabi and standardized tests, can restrict a teacher's ability to offer choices in how practical tasks are performed.

#### 13.5.6.6 Focus on memorization and application

The main objective is for the students to precisely duplicate a certain procedure or ability, which may require them to attentively focus on the procedures so that they are able to memorize them in order to use them effectively. Combining active recall with practical application techniques is an effective way to enhance memorization and application during practical sessions. The practical sessions greatly improve traditional teaching by offering hands-on experience that results in deeper understanding, better knowledge retention, and the development of critical real-world skills. This is one of the practical sessions' advantages over the traditional teaching method. Additionally, it helps close the gap between abstract theory and real-world situations by offering a greater comprehension of the practical application of principles, which makes difficult topics more relatable and understandable. Additionally, it improves knowledge retention by fostering stronger memory associations through active participation in classes and the use of many senses like touch, sight, and sound, which results in longer-lasting knowledge retention than passive listening or reading alone. It also provides an opportunity for the student to acquire essential practical skills like problem-solving, critical thinking, creativity, and flexibility as it negotiates real-world obstacles and making mistakes in a controlled context.

### 13.5.7 Bedside teaching

Bedside teaching is a cornerstone of clinical medical education, offering an ideal, hands-on method for students to learn history taking, physical exams, communication, empathy, and professionalism directly with patients, serving as a bridge between classroom theory and real-world practice, though facing challenges like time constraints and patient acuity, requiring skilled preceptors to role-model holistic care (Peters and Ten Cate, 2022). Some of the unique features of bedside teaching are described as follows:

#### *13.5.7.1 Direct patient interaction*

Direct patient interaction in bedside teaching is a vital part of clinical education where students learn history taking, physical examination and communication skills directly with real patients under supervision, fostering empathy, professionalism, and bridging theory with practice for more holistic, patient-centered care. It involves a triad of teacher, learner, and patient, focusing on the human side of medicine, though requiring careful management of patient consent, privacy, and clear communication to balance learning needs with patient comfort, often by discussing complex theory away from the bedside.

#### *13.5.7.2 Skill acquisition*

Bedside teaching plays a crucial role in fostering skill acquisition in a student as skills involved in history taking, physical examination, clinical reasoning, communication, empathy, and professionalism in real-time patient care can only be learnt by having a close contact with patient themselves. Bedside teaching builds both essential humanistic skills and diagnostic abilities by immersing learners in authentic clinical scenarios.

#### *13.5.7.3 Role modeling*

Role modeling in bedside teaching is a crucial teaching strategy where educators demonstrate ideal professional behaviors, communication, and clinical skills through observation and imitation, shaping learners' professional identity by showing compassion, empathy, and effective patient interaction in real clinical situations. It involves more than just showing clinical competence as it requires actively demonstrating patient-centered care, honesty, positive attitude, and respect, integrating both tangible skills and intangible attitudes that are hard to teach didactically. Bedside teaching offers many benefits over more contemporary, theoretical approaches since it places students directly in patient care settings. By creating a safe and supervised setting, it gives an opportunity for the student to directly interact with the patient where they can practice communication skills, obtain histories, and perform physical examinations on real patient. Additionally, bedside teaching offers significant advantages by integrating theory with practice, enhancing clinical skills like history taking, physical exams, clinical reasoning, and developing crucial professionalism, communication, and empathy through direct patient interaction and role modeling, benefiting students, clinicians, and patients by improving care, making learning relevant, and fostering a patient-centered approach. In addition to these, from the teacher's point of view, it gives them room to provide the students a quick, tailored feedback on their performance in real time, rectifying any flaws in their methods or logic. This feedback is essential and works better than general comments due to individuality of the response as against the general feedback provided to the whole class.

#### *13.5.7.4 Rote learning and repetition*

Rote learning and repetition is a very vital component of medical education in order to build a strong foundation in terms of acquiring knowledge and remembering facts. This method proves to be of immense benefit when the students attempt to remember important definitions, facts or physiological values. This kind of learning builds foundational knowledge, especially in anatomy, biochemistry, and pharmacology, but fall short in fostering deep understanding, critical thinking, and problem-solving needed for complex clinical application, necessitating integration with evidence-based strategies like active recall and spaced repetition for effective, durable learning. In order to foster deeper comprehension and application, more successful medical education approaches frequently combine rote memorization with strategies like active learning, spaced repetition, and elaboration. In addition, rote learning and repetition helps in memory consolidation as it helps transfer information from short-term to long-term memory, making recall reliable. This is very important for a student who is attempting to answer questions which require recall of factual knowledge such as while writing a long essay question or answering multiple-choice questions.

### 13.5.8 Small-group discussion

Although small-group instruction is frequently student-centered, it can also be used in a teacher-centered manner, particularly by transitioning traditional educators, in which the teacher serves as a facilitator while maintaining control over the structure and content delivery, such as directing discussion and observation to guarantee that students learn particular outcomes, emphasizing active engagement within a structured, teacher-defined framework rather than purely independent exploration. By considering small-group discussions as organized, required, and faculty-led events that connect fundamental research with clinical application, they can be incorporated into conventional, lecture-heavy medical curriculum.

### 13.5.9 Tutorials

A tutorial is a class or brief series of classes where a small group of students receives in-depth education on a particular subject from one or more instructors. In essence, tutorials are a fundamental, traditional medical education approach that uses rigorous instruction in small groups of six to twelve students. This allows for deeper dives into lecture subjects, practice solving problems, concept clarification, and self-directed learning, frequently under the guidance of a facilitator. These instructor-led sessions serve as a link between lengthy lectures and clinical application by emphasizing the development of critical thinking, problem-solving, and collaboration skills. They are also excellent for establishing foundational science understanding. Similar to lectures but in a more intimate environment, the tutorials are frequently expert-led, with a tutor who is typically a faculty member leads the discussions and imparts knowledge. The tutorial session is structured, with the teacher setting the agenda, posing questions, and guiding the flow of information, particularly in the beginning or when dealing with typical problems.

## 13.6 Role of teacher in traditional approach

In traditional teaching, the teacher is the central authority, acting as the primary source of information and "sage on the stage" who directly transmits fixed knowledge to passive students through lectures and rote memorization, controlling content, pace, and environment to ensure discipline and the efficient transfer of foundational skills (Wang, Qianjiao, 2019). Their role is to impart information and guide students through structured, textbook-focused lessons, with assessment often relying on standardized tests to measure knowledge recall. Some of the key roles of a teacher in a traditional teaching method include the following:

### 13.6.1 Key roles of a teacher in a traditional teaching method include

#### *13.6.1.1 Authority and source of knowledge*

In the classroom, the teacher has complete power and is the main source of knowledge and information. The authority of the teacher in the class is due to the fact that the teacher being a subject expert has specialized knowledge, professional competence, and teaching skills. Students respect and respond to this authority when they perceive the teacher as a knowledgeable instructor who can effectively help them learn the subject matter. As a facilitator of learning, teachers act less as the sole "knowledge holders" and more as facilitators who guide students to think critically, ask questions, and explore information for themselves. They help students navigate vast amounts of information and develop independent learning skills.

#### *13.6.1.2 Lecturer and demonstrator*

Teachers directly impart knowledge to their students by conducting lectures, demonstrations, or by using other methods. In the traditional teaching approach, a teacher either serves as a demonstrator, by combining lectures with visual and practical exercises, or as a lecturer, imparting knowledge in a one-way, teacher-centered manner. Gaining practical information or skills in conducting an experiment or performing a clinical treatment is best accomplished through hands-on demonstrations that allow for increased student participation. In this sense, a teacher's function as a subject matter expert and demonstrator can greatly increase students' knowledge and comprehension of the subject since they can readily connect the theoretical ideas with the topic's practical application. In a traditional teaching approach, lecturers serve as information providers who manage the curriculum and surroundings while students passively absorb knowledge, frequently by memorization. In both authoritarian and teacher-controlled models, the teacher is the dominant authority figure who creates the learning environment.

#### 13.6.1.3 Controller of the learning environment

A teacher as a controller of the learning environment actively manages the classroom, setting clear rules, guiding student behavior, structuring activities, and maintaining authority to ensure an organized, focused, and productive space for learning. As the only authority and main source of information in the traditional teaching approach, the teacher serves as the principal controller of the learning environment. This role is vital when introducing new, complex topics to the class which the students are expected to pay full attention. In essence, the teacher as a controller ensures order and effective delivery, creating a foundation for learning, while understanding the need to transition to other roles, like facilitator, as students become more independent.

#### 13.6.1.4 Director of learning

In order to ensure that students adhere to a set curriculum and learning framework, the instructor guides and leads the teaching activities. In traditional teaching methods, the teacher functions as the director of learning by serving as the primary source of knowledge and the central authority in the classroom. In this, the instructor assumes multiple responsibilities, such as curriculum dictation, in which the teacher decides the subjects, topics, and sequence of learning activities. They primarily deliver information through lectures and direct instruction, with students expected to absorb and retain this content. They maintain strict control over the learning environment, including student behavior, pace, and activities. The teacher designs and grades assessments like tests, quizzes, assignments to measure student comprehension against predefined standards proving to be authority of assessment. Essentially, the teacher orchestrates the entire educational process, guiding students through a structured, often linear, path of information acquisition.

#### 13.6.1.5 Focus on rote memorization

The emphasis on rote memory in a traditional teaching approach entails a teacher guiding students through repetitions and recitations until the material is retained. A teacher's role in rote memorization involves facilitating repetition for foundational facts while balancing it with meaningful learning, connecting rote facts to real-world applications, using varied methods for different learners, and breaking down large information into chunks for quick recall, ultimately using rote learning as a foundation to support deeper understanding and problem-solving. This approach is criticized for perhaps impeding deeper comprehension, critical thinking, and creativity by emphasizing memorization over conceptual knowledge and application, even though it might be useful for laying a foundation of facts.

#### 13.6.1.6 Facilitator of passive learning

According to this approach, students are passive learners who are supposed to pay attention, take notes, and absorb what the teacher says. When using traditional teaching approaches, the instructor frequently unintentionally facilitates passive learning. This happens because the traditional approach is mostly teacher-centered, with the instructor serving as the main information source and the students acting as passive consumers of knowledge. Additionally, this method provides very with few chances for discussion, group projects, or student-led conversations.

### 13.6.2 Advantages of traditional teaching methods

Some of the advantages of traditional teaching methods include structured learning, direct instructor interaction for immediate feedback, stronger teacher–student relationships, better development of foundational knowledge for exams, and fostering accountability, discipline, and hands-on experience through in-person settings, which some students find more effective than online methods for core clinical skills and understanding.

#### 13.6.2.1 Enhanced interaction

While modern, interactive teaching methods are generally designed to enhance interaction, certain traditional methods, such as bedside teaching and small-group clinical instruction, inherently prioritize and facilitate high levels of direct, face-to-face interaction between the instructor and students, which is challenging to fully replicate with purely online or technology-based approaches. Additionally, by encouraging direct, in-person contacts between students and teachers, traditional techniques can result in a more encouraging learning atmosphere and a more in-depth, individualized comprehension of the subject matter (Sukkurwalla, 2024).

#### 13.6.2.2 Immediate feedback

In a typical classroom, teachers can answer questions right away and give immediate feedback, which keeps students interested and helps them better understand difficult subjects. Immediate feedback also helps to clarify mistakes, reinforce proper practices, and increase long-term knowledge retention by immediately connecting actions to consequences in real time. It makes it easier for students to make quick corrections because they can do so as soon as they receive feedback for example during demonstration of a skill or a regular bedside teaching session. Additionally, the promptness of the feedback makes the learning process more dynamic and pertinent, which in turn increases student motivation and engagement. Students are motivated to establish greater learning goals because they are given an instant sense of where they stand and what they need to improve.

#### 13.6.2.3 Structured learning environment

Students' time management skills can be enhanced and learning can be organized with the support of a traditional setup, which offers a familiar and consistent atmosphere with a defined structure. One benefit of traditional teaching in medical education is a structured learning environment, which offers a clear framework with established procedures, effective information transmission, and instant feedback. When it comes to being ready for standardized tests, this methodical approach guarantees that learners acquire knowledge in a logical order, with clearly defined objectives and a consistent format. This structured nature of the traditional teaching method helps the student to easily adapt to the sequence in which the activities are conducted during the course and also enables them to prepare for the exams in a better way.

#### 13.6.2.4 Skill-based and hands-on experience

The hands-on experience learning required for medical training, such as understanding clinical skills and procedures, is best provided by traditional techniques, particularly when combined with practical sessions. By combining practical, skills-based instruction through techniques like laboratory practice with simulators and bedside teaching, which enable direct application of theory in a controlled setting, traditional medical education methods offer an edge. Although more recent approaches frequently add simulation for greater realism and patient contact practice, this strategy aids students in developing practical abilities and adapting their skills for real-life circumstances.

#### 13.6.2.5 Social and emotional development

The physical classroom environment encourages social interaction among students, which is vital for the development of social acceptability, teamwork, and communication skills which are a critical component of providing effective patient care. Traditional teaching approaches naturally support social and emotional development through essential components like structured personal interaction, role modeling, and the holistic development that takes place in clinical settings like ward rounds. The close interaction between the student and the teacher helps them to mutually understand each other and appreciate mutual qualities so that the students can gain maximum benefit.

#### 13.6.2.6 Access to facilities

For thorough training, traditional medical education frequently offers direct access to crucial physical resources including labs, libraries, and clinical skills simulation centers. While the lecture itself is a teaching method, the traditional format requires students and instructors to be physically present in a specific place such as a lecture hall, thereby allowing the teacher and the student to interact on almost a daily basis. Direct access to various laboratories such as histology lab, dissection rooms, pathology labs which require specialized equipment and spaces gives an opportunity for the learner to practice and gain perfection in various practical skills which will prepare them for future clinical practice. Gaining practical, hands-on experience in hospitals, clinics, and other healthcare settings is a cornerstone of medical education and requires direct access to these real-world facilities.

#### 13.6.2.7 Support for standardized testing

Many medical students are required to take standardized tests, which can fit in nicely with the regimented nature of traditional learning, which frequently depends on lectures and textbooks. One of the main benefits of traditional teaching methods in medical education is that they efficiently educate students for the objective and uniform structure of standardized examinations used for licensure exams, progress tracking, and admissions.

#### 13.6.2.8 *Experiential learning and strong academic foundation*

Traditional teaching methods can support experiential learning and clinical exposure by providing the foundational knowledge necessary to engage with real-world scenarios. Traditional methods, such as lectures by establishing foundational knowledge, are effective at conveying the core theories, principles, and facts of a subject. In this regard, the traditional teaching methodologies acts as a strong anchor for the modern medical education. This provides students with the essential knowledge base needed to interpret and analyze the situations they encounter during hands-on experiences or clinical rotations. In addition, lectures also provide the theoretical framework that helps students understand the "why" behind the "how" in practical settings. For example, understanding the pathophysiology of a disease in a classroom setting is crucial before observing or participating in its treatment in a clinical environment.

### 13.6.3 Disadvantages of traditional teaching methods

Due to passive learning, a "one-size-fits-all" approach that ignores various learning styles, and an emphasis on memorizing over problem-solving, traditional teaching approaches might result in low student engagement. Additionally, because students may not be encouraged to think independently, these approaches may lead to a teacher-centric dynamic that hinders student-led inquiry and limits innovation. The following are some of the main drawbacks of conventional teaching methods:

#### 13.6.3.1 *Limited engagement and passivity*

Through lectures, students are frequently expected to passively absorb material, which can result in disinterest, boredom, and a lack of desire. Traditional teaching methods often lead to limited student engagement and passivity due to their teacher-centric, lecture-based approach that primarily emphasizes the memorization of facts rather than active participation or critical thinking. Due to the very nature of the traditional teaching method, students are essentially expected to listen, take notes, and absorb knowledge rather than actively engage in the learning process. Additionally, in this method, there is a dearth of active participation because there are frequently little possibilities for debate, discussion, and cooperative endeavors. This hinders the development of communication, critical thinking, and problem-solving abilities amongst the students.

#### 13.6.3.2 *One-size-fits-all approach*

Individual learning demands, styles, and speeds are rarely taken into account by traditional approaches. Some students may fall behind as a result, and others may not be appropriately challenged. One of the main drawbacks of traditional teaching is the "one-size-fits-all" approach, which ignores the fact that students have different learning needs, styles, and speeds. Over a period of time, this approach proves to be detrimental to the learner. In addition to failing to accommodate students with learning disabilities or those who require alternative approaches to succeed, this can cause slower learners to struggle to keep up and rapid learners to feel held behind. As the emphasis switches from individual student advancement to rote memorization and standardized curriculum coverage, the outcome is frequently disengagement and a lack of deep knowledge. While many students benefit from other approaches, such as visual, kinesthetic, or auditory methods, traditional methods assume that all students learn at the same rate and in the same way.

#### 13.6.3.3 *Poor development of critical thinking*

Traditional teaching methods often hinder critical thinking development by prioritizing passive memorization over active analysis, questioning, and problem-solving, leading to students who can recall facts but struggle to apply knowledge, evaluate information, or form independent judgments, creating a disconnect between learning and real-world application. This approach fosters a teacher-centric environment where students receive information rather than constructing understanding, which limits engagement and hinders the development of essential cognitive skills needed for complex challenges.

#### 13.6.3.4 *Teacher-centric dynamic*

The teacher-centric dynamic in traditional teaching is disadvantageous to the students by fostering passive learning, reducing engagement, stifling creativity, and hindering critical thinking, as it focuses on rote memorization and limits collaboration, ultimately failing to prepare students with the adaptable, problem-solving skills needed for modern careers. This method treats students as empty vessels, making them disengaged and less motivated, with limited ownership over their learning journey.

#### 13.6.3.5 Lack of creativity and innovation

An excessive focus on fact memorization can hinder students' capacity to think creatively and independently. Since they mainly focus on rote memorization memorizing facts and processes rather than thoroughly understanding concepts or applying knowledge in novel ways, traditional teaching techniques are frequently accused for suppressing creativity and innovation.

#### 13.6.3.6 Limited use of technology

These approaches may not keep up with technology developments, which could leave students unprepared for today's digital demands. One drawback of the limited use of technology in conventional teaching techniques is that it limits students' access to contemporary educational resources, which can impede the development of vital digital skills that will be needed in the future. Additionally, it prohibits the use of cutting-edge tools that can improve comprehension and engagement, such as interactive simulations, online research, and individualized learning platforms.

#### 13.6.3.7 Dependence on physical resources

Students in remote or underprivileged locations may face obstacles because traditional techniques frequently call for actual classrooms and resources. Traditional teaching's reliance on physical resources is a drawback since it might restrict access for distant learners. Furthermore, by having limited access to information, students' knowledge is often restricted to the resources immediately available at the institution, such as specific books in a library or physical handouts. This can hinder exploration beyond the available materials. Moreover, sharing and distributing physical materials among a large number of students can be inefficient, often leading to scarcity of essential texts or the need for students to share resources.

## 13.7 Enhancing the effectiveness of tradition-based teaching

To enhance traditional teaching, the use of modern resources and methods to assist direct instruction, such as technology, active learning, and cooperative strategies, has proved to very beneficial. Traditional teaching methods can be made more effective and student-centered by stressing on student engagement, integrating different approaches, and creating a supportive and exciting environment. In addition to all these, the integration of technologies such as blended learning, artificial intelligence (AI), and interactive platforms; the incorporation of personalized learning with adaptive tools; a focus on critical 21st-century skills; the fostering of positive learning environments; and the encouragement of real-time engagement and peer interaction will prepare students for the various challenges of modern medical education and clinical practice. However, when the material appears relevant, students learn more effectively by making connections to real-world scenarios. For example, teaching the disease's signs and symptoms with a real patient and including relevant historical events, like discoveries or improvements, could enhance traditional teaching methods.

## 13.8 Traditional versus innovative approach

Traditional teaching is teacher-centered, focusing on lectures, rote memorization, and standardized tests for knowledge transmission, while innovative approaches are student-centered, emphasizing on active engagement, technology, critical thinking, and personalized, collaborative learning to foster deeper understanding and real-world skills. The best approach often blends both, using traditional structure for foundational knowledge and modern methods for complex skills, adapting to student needs, subject matter, and environment.

## 13.9 Need for shift from tradition to innovative teaching

In order to move from traditional to innovative teaching, it is important for educators to focus on creating active, student-centered learning by embracing blended learning approaches, using technology like augmented or virtual reality and AI, and adding hands-on experiences (Shrivastava, 2022)). The goals include developing critical thinking and problem-solving skills, customizing instruction, and better preparing students for a rapidly changing world. It is also important to acknowledge the need for change, keeping the students' requirements in mind and carefully blend new tactics with effective traditional methods rather than abandoning them entirely. The need for shift from tradition to innovative technologies is determined by the following factors:

### 13.9.1 Outdated methods

It is widely held that traditional lecture-based teaching approaches are inadequate for fostering critical thinking, problem-solving abilities, and student involvement in today's classrooms. Traditional teaching methods sometimes lead to passive learning and disengagement because they fail to foster critical thinking, problem-solving, and collaborative skills. Traditional teacher-centered approaches that focus on memorization and lectures are not always advantageous in a rapidly progressing world of medical education. Innovative methods, on the other hand, are more effective because they are student-centered and participative, fostering the skills necessary for success in the twenty-first century.

### 13.9.2 Need for 21st-century skills

Since traditional approaches, which emphasize rote memorization and passive learning, are insufficient for developing the dynamic competencies required in a modern, technology-driven, and globally interconnected world. The demand for 21st-century skills necessitate a shift from traditional to innovative teaching methods. Innovative approaches are more adapted to foster teamwork, flexibility, and digital literacy that modern healthcare and a globalized society demand.

### 13.9.3 Technology integration

The shift from traditional to innovative teaching methods, which fundamentally changes how education is delivered and received, depends heavily on technology integration. The motivations underlying this shift include improving learning outcomes, getting students ready for a digital world, and offering more flexible, engaging, and personalized educational experiences. Technological advancements in virtual reality, AI, and internet platforms offer new ways to improve education's accessibility, customization, and engagement.

### 13.9.4 Diverse student needs

Innovative teaching is required since traditional "one-size-fits-all" approaches are unable to satisfy the varied demands of students. Innovative approaches, such as differentiated instruction and culturally responsive teaching, can offer more individualized learning routes and accommodate a variety of learning styles, increasing the inclusivity and equity of education as they facilitate individualized support, a variety of instructional formats, and accessible learning experiences.

## 13.10 Relevance of traditional teaching methods in the era of artificial intelligence

Traditional education approaches are still useful in the age of AI as these methods offer structure, discipline, and vital social and interpersonal skills that AI cannot match. The future of education in modern medicine should incorporate the best aspects of both, using AI for efficiency and personalization while utilizing conventional techniques for social-emotional learning, mentoring, and the development of critical thinking and creativity (Ramsamooj, 2025). Traditional teaching methods are still important in today's medical school because they give students structure and discipline and teach them how to respect authority, obey rules, and successfully manage their time (Li, 2025). Additionally, they promote social and interpersonal skills since in-person interactions with peers and teachers are essential for the development of empathy, communication, and teamwork. Since teachers guide students instead of just imparting knowledge, these methods also offer a chance to enhance emotional support, context, and depth that AI lacks. The traditional approaches also have the benefit of emphasizing higher-order thinking, which promotes meaningful, in-depth discussion and introspection—both of which are essential for the development of critical thinking abilities that go beyond memorization.

**Summary**

This chapter highlights the long-standing role of traditional teaching methods in formal education by examining their fundamental traits, philosophies, and uses. It describes how classroom methods have been influenced for generations by teacher-centered instruction, which is based on organized lesson delivery, straightforward explanation, and repetition. Important techniques like lectures, demonstrations, rote learning are described, emphasizing their advantages in providing coherent, well-organized information and laying the groundwork. The chapter also discusses the responsibilities of instructors and students in traditional classrooms, where the teacher is the primary source of authority and knowledge and students are expected to listen, absorb information, and demonstrate their understanding through standardized examinations. Overall, the chapter also acknowledges the fact that while traditional teaching methods may seem rigid or outdated to some, they nevertheless offer a helpful framework, especially when combined with more modern, student-centered approaches.

**Key point**

**Future perspectives: traditional teaching methods as an anchor**

Technology integration will be the main emphasis of traditional medical school teaching methods in the future, and it is anticipated that existing methods will change rather than vanish. Didactic lectures and other traditional teaching techniques are losing ground to active, simulation-based, and digital learning as they transition from primary instruction to a blended, "high-yield" role. Pure lecturing is still essential for clinical training, as it serves as an anchor for modernization of medical education. Integrating these and utilizing technology to improve rather than replace crucial human contact is the way of the future. Teachers will increasingly use digital tools including interactive platforms, smart boards, AI tutoring systems, and multimedia materials in addition to lectures, demonstrations, and organized classes. Additionally, although creativity, teamwork, and problem-solving are highly appreciated in today's medical education, old approaches may be revived to promote fundamental abilities like memorization and disciplined study habits that offer stability in rapidly changing classrooms.

## References


But, V., Bulboaca, A., Bolundut, D., Boarescu, P.-M., Bulboaca, A., 2020. Medical education - from traditional learning methods to e- learning methods. Applied Medical Informatics 42, 1–8.

Cenaj, D. Schulte-Uentrop, L. Laura Kröger, L.F. Küllmei, J. Haus, J.M. Moll-Khosrawi, P. 2025. The Effectiveness of Peyton's 4-Step Approach to Teach Resuscitation Skills: A Randomized Controlled Clarification Study Journal of Medical Education and Curricular Development 24 (12). Available at https://doi:10.1177/23821205251358090. PMID: 40735144; PMCID: PMC12304607.

Challa, K.T., Sayed, A., Acharya, Y., 2021. Modern techniques of teaching and learning in medical education: a descriptive literature review. MedEdPublish 10 (1). Available at https://doi.org/10.15694/mep.2021.000018.1.

Luscombe, C., Montgomery, J., 2016. Exploring medical student learning in the large group teaching environment: examining current practice to inform curricular development. BMC Medical Education 16 (1). Available at https://doi.org/10.1186/s12909-016-0698-x, http://www.biomedcentral.com/bmcmededuc/.

Li, H., 2025. AI foundations in China's medical physiology education: pedagogical practices and systemic challenges. Advances in Medical Education and Practice 16, 1439–1453. Available at https://doi.org/10.2147/AMEP.S532951, www.dovepress.com/advances-in-medical-education-and-practice-journal.

Mitchell, S., Sehlbach, C., Franssen, G.H.L., Janczukowicz, J., Guttormsen, S., 2024. Taxonomy of teaching methods and their use in health professions education: a scoping review protocol. BMJ Open 14 (1), e077282. Available at https://doi.org/10.1136/bmjopen-2023-077282.

Papanna, K.M., Kulkarni, V., Tanvi, D., Lakshmi, V., Kriti, L., Unnikrishnan, B., et al., 2013. Perceptions and preferences of medical students regarding teaching methods in a Medical College, Mangalore India. African health sciences 13 (3), 808–813. Available at https://doi.org/10.4314/ahs.v13i3.41India, http://www.ajol.info/index.php/ahs/article/download/93730/83155.

Panjehpour, M., Ataee, N., 2012. Comparison the effectiveness of problem solving method with lecturebased method in the teaching of metabolic biochemistry. Iranian Journal of Medical Education 11 (9), 1318–1325.

Peter Cantillon, Linda Hutchinson, Diana Wood 2003. ABC of learning and teaching in medicine 2003 BMJ Publishing Group ISBN 07279 16785.

Peters, M., Ten Cate, O., 2022. Bedside teaching in medical education: a literature review. Perspectives on Medical Education 3 (2), 76–88. Available at https://doi.org/10.1007/s40037-013-0083-y.

Peña, S., Martinez-Santander, C., Cantos-Reyes, V., Rojas, M.F. 2025. Active methodologies and didactic techniques in medical education: A systematic review, Educación Médica 26 (4), 101040, ISSN 1575-1813. Available at, https://doi.org/10.1016/j.edumed.2025.101040.

Phillips, R., McNaught, C., Kennedy, G., 2010. Towards a Generalised Conceptual Framework for Learning: The Learning Environment, Learning Processes and Learning Outcomes (LEPO) framework 2495–2504.

Ramsamooj, A., Ibrahim, S.M., Gerriets, V.A., Cusick, J.K., Ramsamooj, R., 2025. Artificial intelligence versus traditional learning in a medical school setting. Cureus 17 (6). Available at https://doi.org/10.7759/cureus.85262.

Sukkurwalla, A., Zaidi, S.J.A., Taqi, M., Waqar, Z., Qureshi, A., 2024. Exploring medical educators' perspectives on teaching effectiveness and student learning. BMC Medical Education 24 (1). Available at https://doi.org/10.1186/s12909-024-06465-0, https://bmcmededuc.biomedcentral.com/.

Shrivastava, S.R.B.L., Shrivastava, P.S., 2022. Lecture at crossroads in medical education: is it time to say goodbye or introduce specific strategies to enhance their effectiveness? Current Medical Issues 20 (4), 259–261. Available at https://doi.org/10.4103/cmi.cmi_50_22.

Wong, K., Kapoor, V., Tso, A., OConnor, M., Convissar, D., Kothari, N., et al., 2021. Environments, processes, and outcomes - using the LEPO framework to examine medical student learning preferences with traditional and electronic resources. Medical Education Online 26 (1). Available at https://doi.org/10.1080/10872981.2021.1876316.

Wang, Q., 2019. Research on the Role of Teachers in Traditional Teaching Mode and "MOOC Plus Flipped Classroom" Teaching Mode*. Available at https://doi.org/10.2991/iccese-19.2019.224.

Zinski, K.T.C.P.W., Blackwell, F.M., Belue, W.S., Brooks, 2017. Is lecture dead? A preliminary study of medical students' evaluation of teaching methods in the preclinical curriculum. International Journal of Medical Education 8, 326–333. Available at https://doi.org/10.5116/ijme.59b9.5f40.

Chapter 14

# Small-group teaching

*"Small groups provide students an opportunity to practice much needed social-emotional-learning skills."*
—*Laura Smith and Christina Krantz*

## 14.1 Introduction

Small-group teaching is an active learning strategy where a teacher works with a small cluster of students for focused instruction, often after a whole-class lesson, to provide targeted support, reinforce concepts, and customize learning to individual needs, fostering deeper engagement and skill development (Rajput M, Brahmbhatt, S 2022). It allows for personalized feedback, addresses specific learning gaps, and encourages critical thinking and collaboration in a more intimate setting than a large lecture. Small-group teaching is an interactive pedagogical approach in medical education that involves a facilitator and a small number of students to foster active learning, critical thinking, and collaboration.

## 14.2 Definition of small-group teaching

Small-group teaching is an instructional strategy in which a teacher guides a small group of students as they collaborate on a subject. The instructor serves as a facilitator rather than a lecturer, and the students participate in peer-to-peer learning, problem-solving, and conversations to increase their comprehension (Walton, H 1997). This approach is student-centered and places a strong emphasis on active engagement to enhance critical thinking and teamwork abilities. Although small-group teaching is not defined by any one person the approach has been influenced by a number of influential individuals and philosophies. Small group teaching, which is defined as an instructional technique focused on communication, teamwork, and active involvement rather than just group size, is strongly associated with contemporary educational thought. One of the important figures in the concept of small group teaching are Mills and Alexander (2013), who emphasized teamwork over numbers.

## 14.3 Prerequisites of small-group teaching

The prerequisites for effective small-group teaching include clear learning objectives, a well-structured lesson plan, and a supportive learning environment that fosters collaboration and communication. Teachers need to plan for student engagement and manage the classroom effectively, while students must have some foundational knowledge and be willing to participate actively in discussions and activities. The prerequisites of small-group teaching involve both the teachers and the student. From the teachers' point of view, a crucial prerequisite is the creation of precise learning objectives that guarantee the class is concentrated and effective. Additionally, it's critical to develop a thorough lesson plan that includes interesting activities, precise directions, and a method for giving focused feedback. A crucial prerequisite is the existence of efficient classroom management, which creates procedures and guidelines that foster safety and respect while facilitating smooth transitions and concentrated work. In order to customize the lesson to each student's unique needs, it is also critical for the teacher to comprehend the prior knowledge of your students.

The responsibility of the teacher lies in fostering a positive environment by creating a psychologically safe space where students feel comfortable taking risks, asking questions, and vocalizing their thoughts without fear of embarrassment. The facilitator should also be prepared to guide discussions, challenge students to think critically, and encourage active participation from all members (Ahmed, MU 2014). From the students' perspective, they should have a fundamental comprehension of the subject. They should also be willing to actively participate in conversations, activities, and group

**The Principles and Practice of Medical Education. DOI:** https://doi.org/10.1016/B978-0-443-44197-4.00029-1

projects. The presentation of collaboration abilities is one of the elements that is emphasized here, since students must be ready to collaborate with peers, pay attention to their viewpoints, and react positively. The students' willingness to participate in the class is crucial because they should be prepared to ask questions and share their thoughts, even if it seems difficult.

## 14.4 Role of group dynamics

Group dynamics are the underlying forces that shape how a group functions and how its members interact. In a small teaching group, these dynamics are influenced by the individuals' backgrounds, skills, communication styles, and attitudes, as well as the teacher's role and presence. Understanding group dynamics in small-group teaching means recognizing how student relationships, interactions, and communication within the group influence the learning environment and outcomes. Group dynamics are essential in small-group teaching because they promote active participation, collaboration, and higher-order thinking skills. By managing the interactions within a group, educators can enhance student motivation and communication, ensure equitable participation, and foster a supportive learning environment where students learn from each other and develop key skills like teamwork and problem-solving. This group dynamics facilitates active participation and collaboration. Teaching in small groups depends on dynamics that motivate everyone to participate. To guarantee that everyone takes part, roles can be established, which lessens the possibility that one person would dominate or that others would feel excluded. Similarly, group dynamics improve cooperation and communication, thereby enhancing their communication skills in a real-world context which will help the students in clinical practice. As students collaborate to solve issues or finish assignments, it also strengthens their potential for collaboration. Some of the key elements of these dynamics include group roles like a leader, group norms, shared expectations for behavior, group relationships the degree of task and social cohesion, and the communication patterns that occur. Managing these factors, will enable the educators can foster a positive and productive learning atmosphere, improving student engagement, well-being, and academic achievement.

## 14.5 Key features of small-group teaching

### 14.5.1 Active participation

Students are actively involved in the learning process rather than passively receiving information. Active participation is a key feature of small-group teaching because it fosters a student-centered, interactive, and collaborative learning environment. Students are actively involved in discussing, applying, and reflecting on new knowledge, rather than passively receiving it from the instructor. This approach increases student engagement, motivation, and confidence, and also helps develop critical thinking and communication skills through peer-to-peer interaction. Active participation has an effect on small-group teaching as it facilitates the student-led discussion as small-group size allows for every student to have the opportunity to contribute to discussions, which is often difficult in a larger class. It also enhances the deep learning as students are encouraged to apply and transfer new knowledge through in-depth discussions, collaboration, and reflection, moving beyond surface-level understanding.

### 14.5.2 Interactive learning

Small-group teaching promotes discussion, collaboration, and reflection among students and between students and the facilitator. Interactive learning is a key feature of small-group teaching because it fosters active student participation through discussions, problem-solving, and hands-on activities, leading to a deeper understanding of the material. This approach encourages collaboration among peers, provides immediate feedback, and allows for more individualized and differentiated instruction, as the teacher can act as a facilitator rather than just a lecturer (Jones, RW 2007). Interactive learning works in small groups as it promotes active engagement, wherein the students are not passive recipients of information but actively participate by discussing, collaborating, and applying new knowledge. It also promotes collaboration and peer-to-peer learning in which the students work together, share ideas, and learn from one another's perspectives, which enhances communication and teamwork skills.

### 14.5.3 Skill development

The interactive nature of small-group instruction promotes skill development by going beyond simple knowledge transmission to the active practice and improvement of critical thinking, problem-solving, teamwork, and communication.

One of the main elements that promotes skill development is that it encourages active engagement because each student has an opportunity to express their thoughts, pose questions, and participate in conversations in a smaller setting, all of which directly improve communication skills. Instant feedback is also beneficial because it allows teachers to provide students quick, tailored feedback on their performance, which helps them improve their approach to tasks and develop their critical thinking and problem-solving skills.

### 14.5.4 Deep learning

Small-group teaching promotes deep learning by encouraging active engagement, cooperative problem-solving, critical thinking, and knowledge retention. Students' active participation in their education, the growth of their collaboration and self-directed learning abilities, and their capacity to relate ideas to novel and practical problems are all important aspects. Compared to standard lectures, small groups provide a more dynamic and engaging setting that boosts student interest and the possibility of long-term retention. Beyond rote memorization, deep learning fosters higher-order thinking by encouraging critical thinking, analysis, and the application of information to novel problems in small groups.

### 14.5.5 Facilitator role

The role of a facilitator is a key feature of small-group teaching, as they guide student-led learning by creating a structured yet inclusive environment. Setting clear objectives, controlling group dynamics, encouraging active involvement from all members, asking open-ended questions to spark critical thinking, and offering support and feedback without taking over the conversation are all important components. This approach empowers students to take ownership of their learning while developing collaborative skills. The role of the facilitator in small-group teaching is that the facilitator sets the session's objectives and ensures a logical flow of content. They provide a framework and timeline but provide the freedom to the group to take the lead within it. Moreover, the facilitator guides the discussion wherein instead of lecturing, the facilitator asks questions that encourage students to think, debate, and problem-solve. They can ask clarifying questions or prompt students to explain their reasoning to deepen understanding (Fig. 14.1).

### 14.5.6 Small-group teaching strategies

Clear objectives, adaptable and flexible grouping, focused and interesting training, continuous evaluation and feedback, and the use of a variety of cooperative activities like think-aloud, debates, PBL, and jigsaws are all effective small-group teaching techniques (Meo, SA 2013). The main tenets are to keep sessions brief and targeted, encourage active engagement, and adjust group composition and education to students' changing requirements in a flexible and responsive manner. Some of the important steps to be carried out before conducting a small-group teaching sessions is planning and preparation, which include the following:

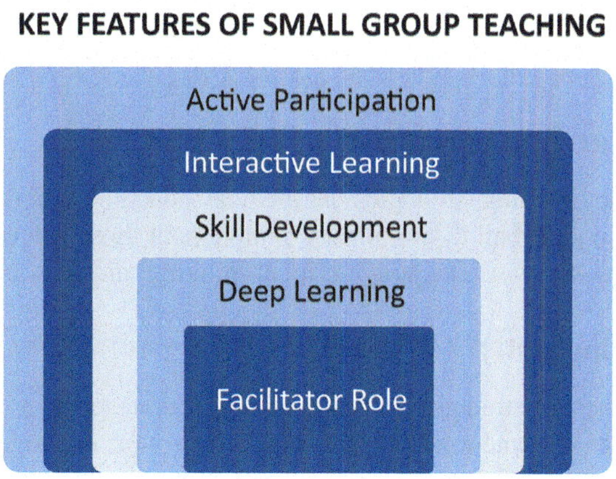


FIGURE 14.1 Key features of small-group teaching.

#### 14.5.6.1 *Assess student needs*

Assessing student needs for small-group teaching involves using regular formative and diagnostic assessments to understand individual and group learning gaps, strengths, and preferences. These assessments provide data to form targeted groups for remediation or enrichment, and to adjust content, process, and products to meet diverse needs. To make sure the groups and tactics are successful and to adjust as students advance, ongoing observation and assessment are crucial. Formative and diagnostic assessments, which use tools like pretests, quizzes, and exit tickets to test understanding while instruction is taking place, are some of the techniques for determining the requirements of students. This aids in spotting false beliefs before they take hold. However, in order to guide your initial planning, the diagnostic tests assist us in understanding the learner's current knowledge and ability sets (Burgess, A 2020).

#### 14.5.6.2 *Set clear goals*

Setting clear goals is a foundational element of effective small-group teaching. This is crucial in small-group teaching because it provides focus and direction, enhances student motivation, and improves learning outcomes. Clear goals help students understand what is expected, how to prioritize their efforts, and how to gauge their own progress, making every minute of the session purposeful and productive. This approach not only boosts academic performance but also develops essential skills like time management and self-discipline. The importance of clear goals is that it provides focus and direction as the clear goals give students a target to aim for, which reduces feelings of overwhelm and helps them prioritize tasks within the group. It ensures that both the teacher and students are working towards the same objective. It also boosts motivation as when students have clear goals, they can see what success looks like and feel a greater sense of accomplishment when they achieve smaller, manageable steps (Steinert, Y. 1996). Students are motivated to persevere and take on more difficult tasks by this encouraging feedback. Additionally, by breaking down objectives into smaller, more manageable phases, it helps students better organize and plan their time and resources. They gain the ability to plan tasks to achieve goals, which results in a more effective use of time.

#### 14.5.6.3 *Plan targeted lessons*

After determining the needs of the students and establishing clear, quantifiable learning objectives, focused small-group classes are planned. Subsequently, it's critical to create activities that fit these objectives and the skills of the students. This should be followed by a lesson plan and continuous formative evaluation to track advancement. Lastly, it's critical to reflect on the meeting in order to improve future planning. Short, differentiated tasks and activities that directly address the stated learning objectives and take into account the individual skills of the group must be created (David Newbie, 2001).

#### 14.5.6.4 *Maintain flexibly in student groups*

It is important to organize students into groups based on their skills, needs, or interests. To maintain flexibility in student groups for small-group teaching, a regular assessment of the student needs to be done through formative assessments and observations to form and adjust groups based on current learning objectives. Using a variety of methods for grouping, such as grouping by skill, ability, or to provide peer support, and ensuring that the groups are short-term and fluid which allows for quick shifts as students' needs evolve. Providing clear instructions and opportunities for collaboration also supports this dynamic approach. Permanent grouping should be avoided to better support student growth.

## 14.6 Methods of small-group teaching

Some of the effective small-group learning strategies in medical education include case-based learning, PBL, and techniques like think–pair–share and snowball discussions. To implement these, educators should act as facilitators rather than lecturers, using open-ended questions to encourage critical thinking and reasoning.

### 14.6.1 Problem-based learning (PBL)

PBL is one of the most common teaching strategies in which small-group teaching is highlighted. In this, the students are divided into the small groups of 8–10, wherein they are engaged in working together to solve real-world problems, fostering collaboration and critical thinking. PBL in medical education transforms small-group teaching by shifting the focus from teacher-led lectures to student-centered, active learning driven by clinical cases. This approach promotes deeper understanding and retention by requiring students to collaborate, critically think, and self-direct their learning, while the facilitator acts as a guide rather than a source of all answers (Karimi, R 2011).

PBL influences on small-group teaching as it shifts the role of the teacher from being a knowledgeable lecturer to a facilitator who guides the students. They are responsible for providing structure without giving away answers, promoting metacognitive skills, and managing group dynamics to ensure all members participate. It also promotes the active, student-centered learning wherein the PBL makes small groups the central hub for learning, with students actively working together to analyze case scenarios. This promotes active engagement and deepens understanding through collaboration and discussion rather than passive listening.

### 14.6.2 Team-based learning (TBL)

Team-based learning (TBL) influences small-group teaching by transforming it from traditional, passive methods to an active, collaborative, and problem-solving-focused model. It enhances student engagement, improves academic performance, and develops crucial clinical skills like critical thinking and communication through structured group activities, immediate feedback, and peer assessment(Dnyanesh, 2024). By emphasizing student-driven conversation and problem-solving instead of teacher-led lectures, TBL transforms small-group instruction and creates a more dynamic and engaged learning environment. Additionally, because it can result in greater academic success, it enhances academic performance. Moreover, since TBL is structured, it fosters critical thinking and problem-solving skills by encouraging students to apply core information to solve complicated, clinically oriented challenges.

### 14.6.3 Think–pair–share (TPS)

Think–pair–share is a straightforward, successful method in which students think on their own, discuss their ideas with a partner, and then present their group's ideas to the entire class. It is a popular, structured collaborative learning technique that serves as a teaching method for small groups, particularly in pairs. Students are engaged by having them consider a question on their own, discuss it with a partner to gain confidence, and then present their views to the class or a larger group. Think–pair–share has an impact on small-group instruction in medical education by boosting student engagement and participation, encouraging critical thinking and teamwork, and enhancing the caliber of conversations and the learning process as a whole (Guenther, AR 2024). It gives students an organized method to think on their own first, then discuss their ideas with a partner before presenting them to the group as a whole. This can increase confidence and result in a more equal distribution of contributions. By dividing the task into digestible pieces, it invites more students to participate, which boosts interaction and involvement in small-group instruction.

The initial individual "think" phase allows all students to process information, and the "pair" phase creates a more comfortable setting for discussion before the "share" phase. It also promotes critical thinking and collaboration as this strategy cultivates critical thinking by requiring students to articulate their thoughts to a peer, which helps them clarify and build upon their ideas. It also fosters a collaborative environment where students can learn from diverse perspectives and provide feedback to one another. Moreover, it improves the quality of discussions by allowing students to discuss and refine their ideas in pairs. The "share" phase often results in higher-quality contributions to the larger group discussion. It helps students move from simple regurgitation of facts towards deeper understanding and application of concepts.

### 14.6.4 Snow ball technique

In medical education, the snowball method is an active learning strategy that encourages critical thinking, participation, and peer teaching in a low-pressure setting. Students begin working independently, then form pairs to discuss ideas, then merge into groups of four, eight, and so on, gradually developing consensus and a deeper understanding of a clinical case or topic before sharing with the entire class. Since the snowball technique starts with introspection before conversations, it ensures that every student participates immediately and encourages active engagement. This encourages participation from all students, even those who would be reluctant to speak in front of a big audience, and eliminates the passivity that is sometimes associated with traditional lectures(Gudeli, (2024). It also enhances comprehension and retention by building knowledge gradually through progressive group discussions wherein the students can clarify ideas, consider different viewpoints, and achieve a better understanding of complex topics.

### 14.6.5 Fishbowl

The fishbowl is a small-group teaching method in medical education where an inner circle discusses a topic often a patient interaction role-play while an outer circle observes, taking notes on content and process, then providing structured feedback during a debrief, fostering active listening, critical thinking, and communication skills for both speakers and observers. It effectively teaches clinical skills like interviewing, allows for peer learning in a safe setting, and involves all students, with roles often switching(Mutalik, 2016). One of the main advantages of this method in medical education is that it's great for practicing medical professionalism, communication, and interviewing. Additionally, it encourages participation from both speakers and listeners. In addition, it fosters group dynamics and subject analysis and provides a safe space for students to learn from their peers' achievements and failures (Gandham, 2016). This method can be applied in medical education in a number of situations, such as discussing any concept application, such as in physiology or biochemistry, assessing ethical concerns in patient care, or rehearsing patient interviews for breaking bad news.

## 14.7 Conduction of small-group teaching

In medical education, small-group teaching entails a facilitator using interactive techniques like case-based learning, PBL, and TBL to help a small group of students accomplish particular learning goals. For conduction to be effective, the facilitator must serve as a guide rather than a lecturer, promoting critical thinking, active engagement, and the growth of teamwork and communication abilities. Setting explicit goals, organizing the meeting, controlling group dynamics, and giving feedback are some of the crucial actions in this regard. A few guidelines for teaching in small groups.

### 14.7.1 Define objectives

To make sure that the conversation stays on topic and is effective, it is critical to clearly outline the learning objectives prior to the session. In medical education, setting specific goals for small-group instruction is essential as it gives the teacher and the students direction, organization, and focus. These objectives contribute to enhance student comprehension and skill development by making sure that teaching and learning activities align with the intended outcomes and the broader curriculum. The formulation of the smart objectives enhances engagement and efficiency by keeping the session focused on specific goals as objectives promote efficient use of limited teaching time and foster deeper discussion and critical thinking relevant to the topic at hand. Moreover, these objectives ensure that small-group discussions contribute meaningfully to the overall goals of the medical curriculum, preventing repetition and ensuring comprehensive coverage of necessary topics.

### 14.7.2 Adopt a facilitative role

Instead of lecturing, the instructor's job is to facilitate. They should direct the conversation, pose thought-provoking questions, and make sure every student contributes to the session, and there is no room for domination by any student. In small-group medical education, taking on a facilitative role is key because it encourages active, student-centered learning, which is more successful than traditional lectures in fostering the development of critical clinical skills like problem-solving, critical thinking, and teamwork (Grijpma, JW 2025). Instead of merely imparting knowledge, the facilitator directs the learning process. One advantage of the tutor's facilitative function is that it promotes active engagement by asking open-ended questions that invite all students to share their opinions. This contrasts with the passive reception of information in a lecture setting. It also develops critical thinking and problem-solving by guiding students through case discussions and problem-based scenarios without giving direct answers, facilitators prompt learners to analyze evidence, synthesize information, and develop their own solutions, fostering deeper understanding and clinical reasoning skills.

### 14.7.3 Encourage active learning

Improving critical thinking, clinical reasoning, problem-solving abilities, and knowledge retention in medical school requires small-group instruction that promotes active learning. Additionally, it promotes self-directed learning, teamwork, and communication skills and the ability to incorporate different viewpoints from peers. Active learning in small groups is advantageous because it fosters improved critical thinking and reasoning skills amongst the students, which are necessary for problem-solving in clinical practice. Furthermore, it offers a means of enhancing information retention because, in contrast to traditional lecturing, active learning fosters deeper comprehension and greater retention of difficult medical topics due to its interactive and collaborative aspect.

### 14.7.4 Promote collaboration and communication

Designing activities that require students to work together, listening to different perspectives, and developing teamwork and communication skills are all essential components for healthcare settings. Promoting collaboration and communication during small-group teaching in medical education is crucial as it enhances diagnostic skills and improves team readiness among future medical professionals. Fostering these skills has gained importance as they lead to improved clinical reasoning and diagnosis, wherein discussing cases in a group encourages students to verbalize their thought processes, consider diverse perspectives, and refine their diagnostic reasoning, mirroring the collaborative nature of real-world medical practice. In addition, it inculcates a sense of teamwork and interprofessional readiness. It has to be noted that the modern healthcare setup is inherently team-based. Small-group activities simulate this environment, helping students develop the interpersonal skills necessary to work effectively with colleagues from various disciplines, which is essential for patient safety and quality care.

### 14.7.5 Manage group dynamics

Managing group dynamics is critical in ensuring that participation of the students is balanced and that they feel comfortable sharing their ideas. It is also vital for enhancing learning outcomes, improving communication skills, fostering a supportive environment, and ensuring effective knowledge acquisition in medical education. Effective group management directly impacts student engagement and the overall quality of the educational experience. It also optimizes learning outcomes as well-managed dynamics ensure that all students participate actively. This prevents dominant students from monopolizing discussions and encourages quieter students to contribute, leading to a more comprehensive understanding of the topic. Managing group dynamics also helps to improving learner satisfaction and motivation as a positive and respectful group atmosphere significantly boosts student motivation and satisfaction with the course, making the learning process more enjoyable and effective.

### 14.7.6 Provide feedback

Providing feedback to students in small groups is essential as it helps close the performance gap between current and expected performance, reinforces best practices, and encourages motivation and self-reflection. Additionally, it aids students in honing their knowledge and abilities. Quick, organized comments can boost students' self-esteem, direct them toward development, and create a more productive learning environment. Providing feedback helps the students to improves their performance and understand where they stand and what specific areas need improvement to close the gap between actual and desired performance. It also promotes self-reflection by encouraging students to evaluate their own performance, identify their strengths and weaknesses, and develop a plan for improvement. It is important that the method of peer assessment is followed in this regard to gain maximum benefit.

### 14.7.7 Be prepared for challenges

It is critical that the tutor understands the possible difficulties with the small-group teaching approach. Maintaining an effective learning environment, making sure that time and resources are used efficiently, and supporting the students' learning journey as well as the facilitator's professional efficacy all depend on being ready for obstacles while teaching small groups. This prior preparation tends to be highly helpful in creating an effective learning environment, ensuring the session remains productive and conducive to learning as it allows the facilitator to anticipate potential disruptions or student concerns. This proactive strategy maintains student engagement and helps control classroom tensions. One of the challenges that the facilitator experiences is time management. Having a structured plan helps facilitators stay on the topic and deliver it within the allotted time. When unexpected challenges arise, a well-prepared facilitator can address them quickly and return to the session's objectives without significant delays (Fig. 14.2).

## 14.8 Role of a student in small-group teaching methods

In small-group teaching, a student's role shifts towards active participation. This may include taking responsibility for their own learning, collaborating with peers, and engaging in discussions to clarify understanding and build higher-order skills like problem-solving and critical thinking. Students are expected to contribute to the group's work, articulate their ideas, ask questions, and work with others to achieve specific goals.

### 14.8.1 Active participation and responsibility

In this regard, the student should be an active participant who is responsible for their own learning, not just a passive recipient of information. The student should be prepared and attend the session will all seriousness so as to gain maximum benefit out of it. They should be ready to engage and complete any required prework and being prepared to contribute. At the same time students must be willing to contribute to discussions share thoughts, ask questions, and provide their own perspectives to the group.

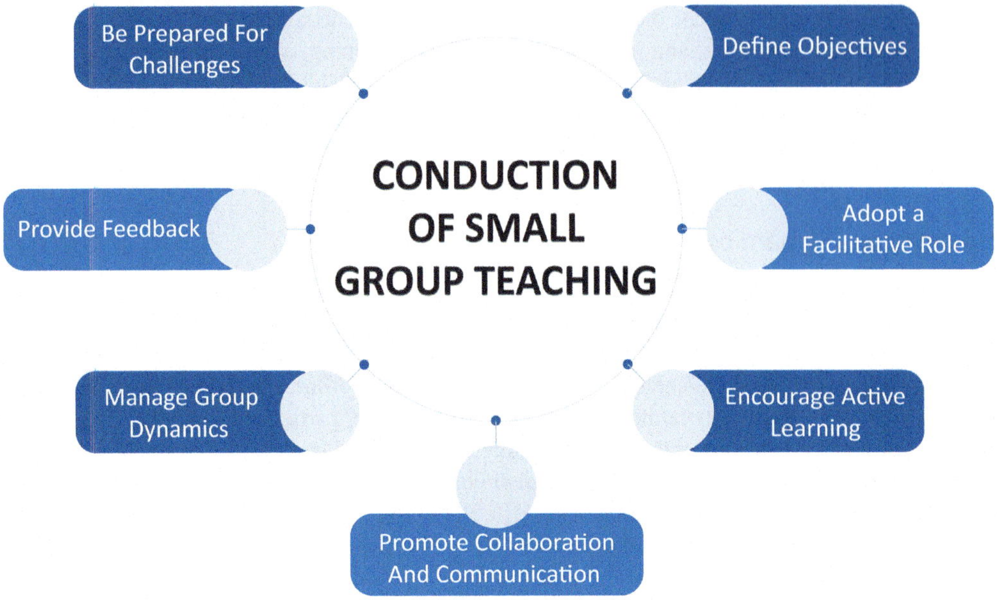


FIGURE 14.2 Conduction of small-group teaching.

### 14.8.2 Collaboration and peer-to-peer learning

The small-group teaching methods give a very good opportunity to work with peers and actively engage with other group members to explore topics, solve problems, and make connections. This is indeed a great opportunity for the student to showcase their collaboration and teamwork skills. It also provides them a chance to learn from others. The support and learn that the students get from their peers helps them to clarify confusing points and offer different perspectives. It is also a chance for the students to develop social skills by learning to draw out quieter members, manage dominant speakers, and ensure that everyone in the group contributes.

### 14.8.3 Intellectual engagement

Since it provides a forum for students to express their opinions and comprehend those of others in the group, small-group instruction also promotes to articulate comprehension of the subject. Additionally, it clears up any confusion that may have been bothering the learner. The learner can have a better understanding of the subject and clear up any doubt by presenting pertinent questions and contrasting their opinions with those of their peers. As students practice and use higher-level thinking abilities including reasoning, problem-solving, and critical thinking through group projects, small-group instruction aids in the development of higher-order skills.

## 14.9 Factors affecting small-group teaching

Factors affecting small-group teaching include student-related elements like their prior knowledge and learning styles, tutor-related factors such as their communication skills and subject-matter expertise, and contextual factors like the learning environment, available resources, and the difficulty of the material. The group's dynamic, including interpersonal challenges and socio-emotional well-being, and the quality of the instructional materials also play a crucial role. These factors can be grouped as follows:

### 14.9.1 Student-related factors

#### *14.9.1.1 Prior knowledge and experience*

Student's prior knowledge and experience significantly impact small-group teaching by influencing engagement and participation, learning engagement, the effectiveness of instruction and the need for different levels of support. Accurate prior knowledge helps students make connections, lowers cognitive load, and allows them to contribute more effectively to discussions. However, inaccurate or insufficient prior knowledge can hinder learning, require more teacher support, and necessitate adapting teaching to address gaps and misconceptions. Ultimately, teachers must assess and leverage students' existing knowledge to tailor small-group instruction effectively. Some of the impacts of accurate prior knowledge is that it enhances learning and retention as prior knowledge provides a framework for new information, helping students to better retain and organize it. At the same time, it reduces cognitive load. When students have a strong foundation, they experience lower cognitive load, freeing up mental resources for deeper learning and self-regulated learning strategies.

#### *14.9.1.2 Learning styles and needs*

It is seen that different learners require different levels of structure and autonomy, and they have varied academic levels, interests, and strengths. Learning styles and needs significantly affect small-group teaching by necessitating a varied and adaptable approach. Understanding a group's diverse learning preferences such as visual, auditory or kinesthetic allows educators to use different methods like visual aids, discussions, or hands-on activities to meet each student's needs. A more inclusive and productive environment where students can flourish at their own pace is created when learning styles and needs are prioritized. If they are not, teaching methods may be ineffective, resulting in disengagement and a failure to grasp content. The facilitator can match a student's preferred learning style and inspire motivation for the subject by being aware of their learning preferences and styles. On the other hand, a mismatch might result in apathy, particularly when it comes to tasks that call for higher-order thinking abilities. Here, the facilitator's job is to mentor and assist every student, including those who lack motivation or need more direction. Knowledge of individual needs helps the teacher to provide the most appropriate support and interventions.

#### *14.9.1.3 Motivation and engagement*

Academic accomplishment, knowledge retention, critical thinking, and the development of social skills are all directly impacted by motivation and involvement in small-group instruction. Students that are highly interested and driven are more likely to actively participate, persevere through difficulties, and produce superior results. The topic's nature also affects motivation. Students are more likely to attend and participate if the subject is engaging and pertinent to them. On the other hand, motivation and engagement fosters deeper learning and knowledge retention wherein by active participation the student can move beyond superficial memorization to deeper-level processing of information. When students are motivated, they pay more attention and try to understand material more thoroughly, leading to better long-term retention and the ability to apply knowledge in real-world contexts.

#### *14.9.1.4 Communication and cognitive skills*

The development of communication and cognitive skills for the students are critical in the small-group teaching as it leads to development of a better rapport with the teacher as well as with the entire group. Effective communication and strong cognitive skills are crucial for small-group teaching because they enhance student engagement, improve learning outcomes, and build collaboration. Communication skills allow students to explain their reasoning and articulate ideas, which stimulates deeper cognitive processing, while cognitive skills like meta-cognition enable students to monitor their own learning and contribute more effectively to the group. Teachers can facilitate this by providing clear instructions, probing questions, and structured support to promote meaningful interaction and collaborative knowledge-building. It also gives a chance for the student to develop interpersonal skills because the students develop stronger verbal communication and interpersonal skills by working collaboratively and negotiating meaning, which are valuable for future careers.

### 14.9.2 Tutor-related factors

#### *14.9.2.1 Communication and facilitation skills*

The ability of a tutor to ask questions, listen, respond, and encourage group dynamics is critical to success of the small-group teaching. A tutor's communication and facilitation skills directly affect small-group instruction by fostering an engaging, collaborative, and productive learning environment. These skills allow the tutor to guide the group, encourage participation, evaluate understanding, and manage group dynamics. These abilities are particularly beneficial since better communication results in clarity and understanding. The instructor can more successfully communicate the learning objectives and expectations for the group's participation by being succinct and straightforward. Additionally, the tutor can assess each student's unique learning needs by consistently asking focused questions, which encourages active participation.

#### 14.9.2.2 *Subject-matter expertise*

The tutor's knowledge of the subject matter is crucial for guiding student achievement. Subject-matter expertise in a tutor significantly impacts small-group teaching by enabling deeper explanations, more effective problem-solving guidance, and the ability to address misconceptions accurately. While expert tutors may focus more on directing content, nonexpert tutors might rely more on process facilitation, though the most effective tutors possess both subject knowledge and strong facilitation skills. Expertise can make learning more enjoyable and engaging, though interpersonal skills like social congruence also have a substantial impact on learning outcomes. A tutor with in-depth expertise can give comprehensive, precise explanations and promptly clear up students' misconceptions before they become ingrained misconceptions. Tutors can also help students solve problems by using their conceptual knowledge to direct them toward successful problem-solving techniques rather than only demonstrating the procedures.

#### 14.9.2.3 *Personal approach*

It is vital on the part of the tutor to create a supportive, personal educational approach that focuses on creating a positive atmosphere. A tutor's personal approach significantly impacts small-group teaching by fostering a safe, engaging, and interactive learning environment through qualities like empathy, clear communication, and the ability to encourage peer-to-peer learning. A good approach moves beyond direct instruction to guide students, ask questions, and facilitate discussion, which can increase student confidence and lead to deeper learning and collaboration. Conversely, a poor approach can result in less student engagement and a less productive atmosphere.

#### 14.9.2.4 *Flexibility*

Flexibility on the parts of the tutor is very important as it gives an opportunity for the tutors to modify their teaching methods in real time in response to the demands of the group. In small-group instruction, a tutor's adaptability enables customized, student-centered instruction that adjusts to the students' varied and evolving needs, interests, and learning preferences. Although it necessitates careful preparation to handle the change in material and pace, this adaptability increases engagement, fosters collaborative skills, and can result in enhanced academic and social-emotional outcomes. It is challenging to accomplish this in a whole-class situation, but flexibility allows the tutor to adjust lessons in real time to suit particular learning gaps or strengths. Additionally, this guarantees that each student gets the help they need to succeed without feeling singled out.

#### 14.9.2.5 *Prior experience with small-group teaching methods*

Prior experience with small-group teaching methods positively affects future small-group teaching by improving an instructor's ability to plan, facilitate, and manage sessions, leading to a more effective learning environment for students. Experience helps educators understand group dynamics, set clear objectives, manage classroom logistics, and adapt strategies based on student needs and prior learning, resulting in enhanced student confidence, critical thinking, and collaboration. This prior experience can come in handy as it enhances small-group teaching. It can lead to an improved planning and facilitation as experienced instructors have a better understanding of how to design and structure a session, including setting ground rules, creating stimulus questions, and balancing individual student needs with the overall group task. Prior experience helps educators manage challenges, such as ensuring all students participate while working with a small group, preventing personality clashes from disrupting learning, and maintaining focus on the learning objectives.

### 14.9.3 Contextual factors

#### 14.9.3.1 *Learning environment*

A supportive and well-structured learning environment directly enhances small-group teaching by increasing engagement, allowing for personalized feedback, and fostering a sense of psychological safety. When the environment is positive and comfortable, students are more likely to actively participate, take academic risks, and take ownership of their learning, while teachers can more effectively tailor instruction to individual needs. Small groups naturally provide students more chances to engage with the teacher and each other, which increases student engagement. In addition, a safe learning atmosphere will help students feel at ease and supported, which will allow them to express uncertainty, ask questions, and take chances without worrying about being judged.

#### 14.9.3.2 Institutional support

Institutional-level decisions, such as the availability of resources and suitable spaces, are important foundational factors. Institutional support significantly affects small-group teaching by providing the necessary resources, infrastructure, and policies that enable or hinder its effective implementation and quality. Strong support enhances the learning environment and outcomes, while a lack of support creates barriers for both teachers and students. Two important things are to be noted in this regard. The first one is concerned with the provision of institutional support wherein a robust institutional support towards small-group teaching can positively affect its conduction. Adequate financial support and the allocation of tangible resources like classrooms with flexible seating arrangements are very essential for collaborative work. The second important aspect is to encourage teacher training and development thereby providing opportunities for continuous professional development in modern pedagogical techniques, such as facilitating small groups, improve teacher competencies and confidence.

#### 14.9.3.3 Instructional materials and problem quality

The difficulty level of the topics and the quality of the problems presented are critical influences on learning outcomes. High-quality instructional materials and problem quality profoundly affect small-group teaching by directly influencing student engagement, comprehension, motivation, and overall academic performance. High-quality instructional materials are content-rich resources aligned with academic standards and evidence-based practices. In small-group settings, their quality has a significant positive impact in terms of enhanced understanding and retention. If the problem or the case scenario is constructed well, then it will help to break down complex, abstract concepts, making them clearer and more practical to students. The use of diverse media such as visuals, audio, real-world examples, etc., accommodate various learning styles, which is particularly effective in a small group where individual needs can be closely addressed.

#### 14.9.3.4 Allocation of time for preparation

Allocation of sufficient time for preparation significantly impacts the effectiveness of small-group teaching by allowing the instructor to design relevant activities, structure sessions effectively, prepare resources, and provide tailored feedback. This preparation time ensures the learning objectives are met, maximizes student engagement, and facilitates deeper learning. The impact of having sufficient time for preparation is seen in the fact that it helps to tailor the content and activities. Adequate preparation also allows instructors to research the students' prior knowledge and learning needs, enabling them to create highly relevant case studies, discussion questions, and problem-solving exercises that directly address the group's specific context and learning objectives. In addition, time for preparation ensures a clear session structure, including well-defined introductions, main activities, and summaries. This organization helps students understand the flow, manage their time effectively during the session, and consolidate their learning later.

### 14.9.4 Group dynamic factors

#### 14.9.4.1 Group size and composition

The size and mix of personalities within a group can impact dynamics. Group size and composition significantly impact small-group teaching by influencing student engagement, learning outcomes, and social dynamics. Smaller groups usually consists of around 3–10 students often lead to greater participation and personalized attention, while composition affects learning through factors like mixed abilities, personality, and social connectedness. Group composition is a crucial factor in designing effective instruction because it impacts how students interact and learn from each other. It is advantageous to have smaller groups as they provide more opportunities for each student to speak and engage with the material, leading to increased participation and a more active learning environment. Moreover, with fewer students, teachers can provide more targeted feedback and instruction tailored to individual learning needs.

#### 14.9.4.2 Interpersonal dynamics

Personality clashes, communication style differences, and conflict management within the group can either hinder or foster progress. Interpersonal dynamics significantly affect small-group teaching by influencing collaboration, communication, and overall learning outcomes. Positive dynamics foster open communication and problem-solving, while negative dynamics can lead to conflict, social loafing, and reduced participation. The teacher's own attitude and management style are crucial in shaping a positive and productive group climate.

#### *14.9.4.3 Socio-emotional well-being*

The emotional well-being of group members is important for the group to function effectively and succeed at its task. Socio-emotional well-being positively affects small-group teaching by improving teacher-student relationships, classroom management, and student motivation, while negative well-being can lead to burnout, reduced empathy, and poorer student outcomes. Teachers with high socio-emotional competence are better equipped to manage emotions and create supportive interactions, leading to a more effective and positive learning environment for students. Conversely, a lack of well-being can create a negative cycle where stress hinders empathy and patience, disrupting the group dynamic and student performance (Fig. 14.3).

### 14.9.5 Small-group teaching as an innovative approach

Small-group teaching is an innovative approach that transforms passive learning into active, collaborative, and interactive experiences through guided discussions and problem-solving. This method is innovative because it shifts the instructor's role from a lecturer to a facilitator, fostering higher-order thinking skills like critical thinking, problem-solving, and communication, which are often underdeveloped in large lecture settings. It benefits students by addressing knowledge gaps, encouraging self-directed learning, improving knowledge retention, and developing crucial teamwork and leadership skills (Euliano, TY 2001).

## 14.10 Small-group versus large-group teaching

Small-group teaching is more learner-centered, interactive, and often results in better academic performance, while large-group teaching is typically teacher-centered and can be less interactive. Small groups allow for more personalized attention and peer collaboration, leading to higher engagement and knowledge retention, whereas large groups can be less intimate and require more effort to ensure all students are engaged.

## 14.11 Difference between small- and large-group teaching

The difference between small-group and large-group teaching is presented in Table 14.1.

## 14.12 Enhancing the small-group teaching

To enhance small-group teaching in medical education, set clear goals and use engaging activities like case studies or role-playing. Facilitators should act as guides, encourage active participation, and manage group dynamics effectively by creating a safe environment and managing time. Providing constructive feedback and integrating technology like simulators can further improve learning outcomes. In addition to the above the tutor should take care that all the important components of small-group teaching like structure and planning in terms of defining clear learning outcomes and preparing the students, facilitation techniques in terms of managing group dynamics and encourage participation, assessment and feedback in terms of provide constructive feedback and at last environmental and technological support by create a comfortable environment and integrating technology should be addressed adequately.

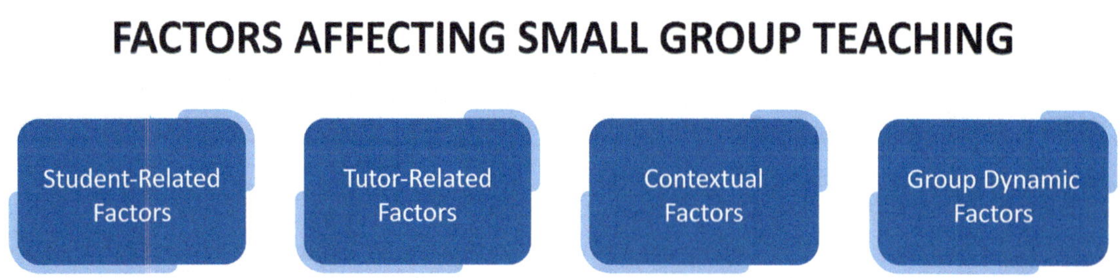


FIGURE 14.3 Factors affecting small-group teaching.

**TABLE 14.1** Difference between small-group and large-group teaching.

| Sl no. | Feature | Small-group teaching | Large-group teaching |
|---|---|---|---|
| 1 | Role of student | Active participant, involves discussion, problem-solving, and collaboration | Often passive, primarily a one-way delivery of information through lectures |
| 2 | Role of teacher | Facilitator who guides discussion and provides individual feedback | Instructor who delivers content to a large audience |
| 3 | Interaction | High degree of interaction between students and with the teacher | Low to no student-to-student interaction. The interaction is typically between the teacher and individual students |
| 4 | Learning Focus | Deep understanding, concept development, and critical thinking through debate and social interaction | Exposure to new concepts and competition |
| 5 | Advantages | Higher engagement, better knowledge retention, and improved academic outcomes | Can be more efficient for delivering foundational information to large numbers of students |
| 6 | Disadvantages | Requires significant resources and personnel; may not cover as much content as large lectures | Can be less effective for developing deeper conceptual understanding and can lead to passive learning |
| 7 | Example | Problem-based learning | Lecture |

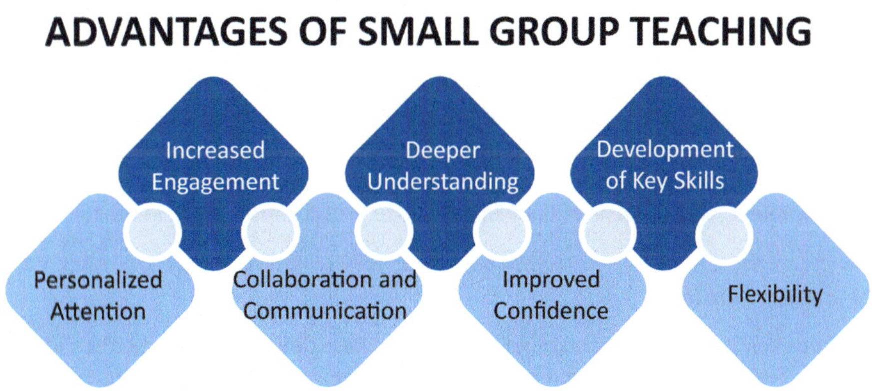


FIGURE 14.4 Advantages of small-group teaching.

## 14.13 Advantages and disadvantages of small-group teaching

Small-group teaching enhances engagement, collaboration, personalized feedback, and skill development, allowing for differentiated instruction and deeper learning. However, disadvantages can include challenges in managing inconsistent participation, assessment difficulties, increased teacher workload, potential for unequal contribution among students, and the need for significant planning and preparation by the instructor.

## 14.14 Advantages of small-group teaching

The advantages of small-group teaching are illustrated in Fig. 14.4

### 14.14.1 Personalized attention

The smaller size allows teachers to provide more individualized support, monitor progress, and tailor instruction to meet students' specific needs. Personalized attention is a key advantage of small-group teaching because it allows teachers to tailor instruction to each student's unique strengths, weaknesses, and learning pace. This focused approach enables teachers to provide targeted support, immediate feedback, and to identify and address learning gaps early, which can improve academic performance and build student confidence. It facilitates tailored instruction wherein the teachers can adapt lessons, materials, and strategies to match the specific needs of the group, ensuring no student is left behind and that advanced learners are appropriately challenged. It also encourages targeted support as with fewer students, a teacher can focus on specific challenges, providing extra help and more in-depth explanations when needed.

### 14.14.2 Increased engagement

Students are more likely to participate and engage in discussions and activities in a less intimidating setting. Increased engagement is a key advantage of small-group teaching because the intimate setting allows for more opportunities for active participation, collaboration, and individualized attention from the teacher. Students feel more comfortable contributing, asking questions, and taking risks in a less intimidating environment, which leads to a deeper understanding of the material. This also fosters a sense of ownership and responsibility for their own learning. Small groups facilitate increased engagement because with fewer students, there are more chances for each individual to speak up in discussions, share ideas, and get their questions answered. It also fosters increased comfort as the smaller, more intimate setting makes students feel more seen and heard, which gives them the psychological safety needed to participate without fear of being wrong in front of a large group (Lama, (2015).

### 14.14.3 Enhanced collaboration and communication skills

Students learn to work together, share ideas, and develop essential social and communication skills through peer interaction. Small-group teaching enhances collaboration and communication skills by encouraging students to actively share ideas, listen to peers, and work together on problems. This environment builds stronger interpersonal skills through discussions and debates, prepares students for real-life teamwork, and improves their ability to articulate their thoughts clearly and respectfully. Enhanced collaboration leads to shared problem-solving where the students can tackle more complex problems together than they could individually, pooling their knowledge and skills to find solutions. It also gives a chance to the student to understand and appreciate the diverse perspectives as by working in small groups exposes students to different viewpoints, which deepens their understanding of a topic and helps them learn from one another.

### 14.14.4 Deeper understanding

Working in smaller groups can lead to deeper learning as students clarify information, articulate ideas, and learn from one another. Small-group teaching promotes deeper understanding by creating an environment for increased collaboration, immediate feedback, and personalized instruction. Students can work together to explain concepts to each other, receive instant feedback from both peers and teachers, and get targeted help on their specific needs, leading to better comprehension and retention. It leads to collaborative learning and peer teaching in which the students are in a position to explain concepts to one another, which reinforces their own understanding and helps their peers.

### 14.14.5 Improved confidence

The intimate setting can empower shyer students to speak up, ask questions, and build confidence. Improved confidence is a key advantage of small-group teaching because the intimate setting fosters a safer environment for students to participate, take intellectual risks, and share their ideas without the intimidation of a large class. This confidence is built through more personalized attention from the teacher, more frequent opportunities for verbal and nonverbal contributions, and a sense of community that encourages students to be more engaged and vocal. Confidence is improved by small-group teaching as it creates a safe environment to participate. In this setting, students are less intimidated to speak up or ask questions in a small group than in a large lecture setting. It also gives a chance to the tutor to provide more immediate and tailored feedback, helping students correct misconceptions and build a stronger understanding of the material, which leads to greater confidence.

### 14.14.6 Development of key skills

Small-group activities foster teamwork, problem-solving, and argumentation, which are crucial for academic and life success. Small-group teaching develops key skills like communication, critical thinking, problem-solving, teamwork, and self-directed learning. The interactive and collaborative environment allows students to share ideas, test hypotheses, and learn from each other, while the smaller setting provides more opportunities for personalized support and increased confidence. This approach fosters both cognitive and social skills that are beneficial for academic and professional success.

### 14.14.7 Flexibility

Small groups can be more mobile and can adapt their agenda to address specific difficulties or explore topics more deeply. Flexibility is a key advantage of small-group teaching, as it allows teachers to adapt lessons instantly to student needs, interests, and learning styles. This approach makes it easier to provide targeted support for those who are struggling and offer challenges to those who are ahead. The ability to change groups frequently prevents students from being labeled, fosters more inclusive environments, and promotes better social and emotional development through diverse collaboration. The teacher also has a chance to make real-time adjustments to change the pace and focus of a lesson based on a group's performance, rather than following a rigid, preplanned schedule for the entire class (Fig. 14.5).

## 14.15 Disadvantages of small-group teaching

### 14.15.1 Unequal participation

Unequal participation is a disadvantage of small-group teaching because some students may dominate the group while others become passive, leading to a lack of engagement for those who contribute less. This can result from a few members taking over the workload, while others become non participatory which can be demotivating for the entire group. Other related issues include a lack of productivity, potential for conflict, and unfair assessment if the whole group receives the same grade despite differing contributions. The main concerns here are the dominant students who make take control of

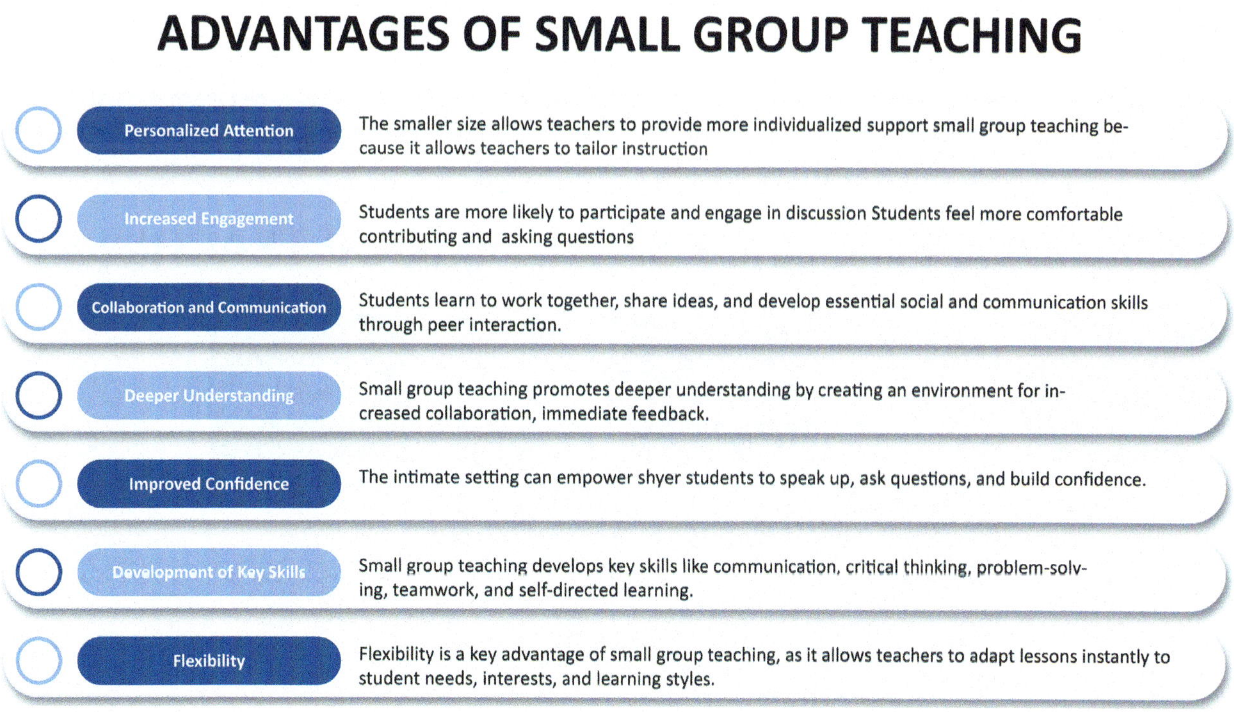


FIGURE 14.5 Importance of advantages of small-group teaching.

the task, leaving little room for others to contribute. On the other hand, the presence of passive students either by choice or because they feel their input is not valued or needed can lead to them feeling a low sense of agency.

### 14.15.2 Assessment challenges

It can be challenging for teachers to effectively assess the learning and individual contributions of each student within the group. Assessment in small-group teaching presents challenges such as difficulty in evaluating individual contributions, potential for unequal workload distribution, and the risk of group dynamics hindering performance. It can be hard to tell if the final product reflects everyone's learning, and the need to balance individual validity with group collaboration can be difficult for teachers. Furthermore, assessment strategies may need to shift to include practical skills and ethics alongside knowledge, which requires careful planning. The assessment challenges in small-group teaching lie in evaluating individual learning as it becomes difficult to determine the specific contribution of each student to a group project, as the final product may not accurately represent all members' understanding. Moreover, a group's success doesn't always correlate with individual student performance on subsequent exams. In addition to this, there is an unequal participation and workload as some students may dominate the group's work while others contribute very little or nothing, a phenomenon known as "free-riding." This makes it unfair to grade all members the same.

### 14.15.3 Time management

Planning and managing multiple small groups can be time-consuming for teachers, requiring extensive preparation and coordination. Time management is a disadvantage of small-group teaching because it requires more planning, resources, and constant monitoring from the teacher to ensure each group is on task, which can lead to a heavy workload and difficulty in managing multiple groups simultaneously. Teachers must juggle different materials, time constraints, and the potential for student off-task behavior in each group, making it more demanding than a traditional, large-group lesson.

### 14.15.4 Teacher workload

While it allows for personalized attention, it also increases the number of groups the teacher needs to manage and support simultaneously. One of the main disadvantages of small-group teaching is the significant increase in teacher workload due to the intensive demands of planning, managing, and assessing multiple groups simultaneously. This is more pronounced when there is a shortage of faculty members. Furthermore, the time consumed in the planning and preparation phase of conduction also adds to the stress and the teacher workload.

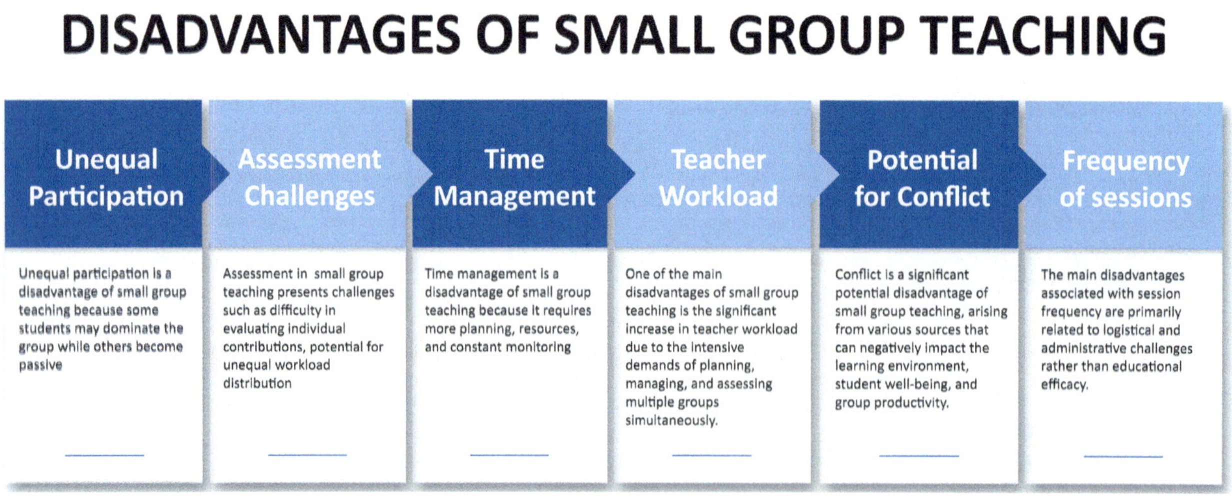


FIGURE 14.6 Disadvantages of small-group teaching.

### 14.15.5 Potential for conflict

Group dynamics can sometimes lead to negative interactions, conflicts, or a sense of tension among students. Conflict is a significant potential disadvantage of small-group teaching, arising from various sources that can negatively impact the learning environment, student well-being, and group productivity. Differing opinions and communication styles can be the main cause of conflict as when individuals with diverse perspectives and communication styles work together, disagreements are natural. Some students may struggle to accept criticism or new ideas, leading to friction.

### 14.15.6 Frequency of sessions

The main disadvantages associated with session frequency are primarily related to logistical and administrative challenges rather than educational efficacy. The main challenges pertain to coordinating multiple frequent sessions which can be complex, as aligning the availability of all students and the facilitator/instructor can be challenging. Moreover, it causes increased workload for instructors as the planning, preparing for, and conducting numerous small-group sessions adds a significant administrative and teaching burden on faculty compared to preparing for a single, large lecture (Fig. 14.6).

**Summary**

Small-group teaching is a learner-centered instructional approach widely used in medical education to promote active engagement, critical thinking, and professional skill development. Unlike large lectures, small groups create an environment where students can interact closely with peers and facilitators, apply knowledge, and receive personalized feedback. Small-group formats help students bridge theory and practice by encouraging discussion, problem-solving, and reflection. They provide opportunities to develop communication skills, teamwork, and clinical reasoning competencies essential for effective medical practice. The format also supports deeper learning, as learners must articulate their understanding and respond to diverse viewpoints. Small-group teaching is an essential component of modern medical education, fostering deeper understanding, skill acquisition, and professional competency. When well-facilitated, it transforms learners from passive recipients of information into active, reflective participants capable of engaging in lifelong learning and collaborative clinical practice.

**Key point**

Future perspectives: fostering a culture of collaborative learning

The future of small-group teaching in medical education is dynamic and technology-enhanced, with an increasing focus on interprofessional collaboration, competency-based assessment, personalization, and inclusivity. As medical training continues to adapt to evolving healthcare needs, small-group learning will remain a vital modality which is shaped by innovation but grounded in its core strengths of interaction, reflection, and shared inquiry. The main focus areas will be to integrate technology-enhanced learning, wherein the advances in digital education will significantly augment small-group learning. Moreover, there will be a greater emphasis on interprofessional education as small-group format can encourage a collaborative learning amongst a variety of health professions in order to uphold the quality of medical education and healthcare.

## References

Abdelkhalek, N., Hussein, A., Gibbs, T., Hamdy, H., 2010. Using team-based learning to prepare medical students for future problem-based learning. Medical Teacher 32 (2), 123–129. Available at https://doi.org/10.3109/01421590903548539.

Ahmed, M.U., Roy, S., 2014. Small group teaching—an update. Bangladesh Journal of Medical Microbiology 8 (2), 28–31 no date; Jones, 2007; Meo, 2013; Rajput and Brahmbhatt, 2022a; Burgess et al., 2020; Steinert, 1996.

Burgess, A., van Diggele, C., Roberts, C., Mellis, C., 2020. Facilitating small group learning in the health professions. BMC Medical Education 20 Available from. Available at https://doi.org/10.1186/s12909-020-02282-3, http://www.biomedcentral.com/bmcmededuc/.

David, Newbie, 2001. A Handbook of Medical Teacher. Kluwer Academic Publishers, Netherlands.

Dnyanesh, S., Shirahatti, A., Bhimalli, S.M., Yadav, S.K., Dnyanesh, K., 2024. Team-based learning (TBL) in medical education: fostering collaboration and enhancing learning outcomes. Medical Journal of Dr. D.Y. Patil Vidyapeeth. 17 (5), 1036–1042. Available at https://doi.org/10.4103/mjdrdypu.mjdrdypu_908_23.

Euliano, T.Y., 2001. Small group teaching: clinical correlation with a human patient simulator. American Journal of Physiology - Advances in Physiology Education 25 (1-4), 36–43. Available from. Available at https://doi.org/10.1152/advances.2001.25.1.36, http://advan.physiology.org/contents-by-date.0.shtml.

Grijpma, J.W., De la croix, A., Dolmans, D.H.J.M., Meeter, M., Grooten, R.H.J.M., Kusurkar, R.A., 2025. Preparing medical teachers for small-group active learning: a design-based research study. Perspectives on Medical Education 14 (1), 578–589. Available from. Available at https://doi.org/10.5334/pme.1759, https://pmejournal.org/articles/1759/files/68cbf4f09410c.pdf.

Jones, R.W., 2007. Learning and teaching in small groups: characteristics, benefits, problems and approaches. Anaesthesia and Intensive Care 35 (4), 587–592. Available at https://doi.org/10.1177/0310057x0703500420.

Karimi, R., 2011. Interface between problem-based learning and a learner-centered paradigm. Advances in Medical Education and Practice 2, 117. Available at https://doi.org/10.2147/AMEP.S12794.

Lama, P., Kulkarni, J., Binod, T., Pranoti, S., 2015. The impact and significance of small and large group teaching and learning in medical curriculum. Journal of the Anatomical Society of India 2. Available at https://doi.org/10.1016/j.jasi.2018.06.085 2349-1604.

Meo, S.A., 2013. Basic steps in establishing effective small group teaching sessions in medical schools. Pakistan Journal of Medical Sciences 29 (4) Available from. Available at https://doi.org/10.12669/pjms.294.3609, http://www.pjms.com.pk/index.php/pjms/article/view/3609/1547.

Mutalik, M., 2016. Student perceptions and learning outcome on a fishbowl strategy-based pharmacology seminar on drug dependence. International Journal of Basic and Clinical Pharmacology 879–883. Available at https://doi.org/10.18203/2319-2003.ijbcp20161539.

Mills D, Alexander P. Small group teaching: a toolkit for learning. 2013. Retrieved from: https://www.heacademy.ac.uk/system/files/res sources/small_group_teaching_1.pdf.

Rajput, M., Brahmbhatt, S., 2022. Small group teaching. International Journal of Health Sciences 6622–6626. Available at https://doi.org/10.53730/ijhs.v6ns4.11276.

Steinert, Y., 1996. Twelve tips for effective small-group teaching in the health professions. Medical Teacher 18 (3), 203–207. Available at https://doi.org/10.3109/01421599609034161. Available from: http://www.tandfonline.com/loi/imte20#.VufR2bdf2DY.

Walton, H., 1997. Small group methods in medical teaching. Medical Education 31 (6), 459–464. Available at https://doi.org/10.1046/j.1365-2923.1997.00703.x.

Chapter 15

# Innovative teaching methods

*"Individualizing education and starting with empathy for those we serve is where innovative teaching and learning begins."*

—George Couros

## 15.1 Introduction

As educators work to prepare future clinicians for an increasingly complicated healthcare environment, medical education is undergoing a significant transition. The clinical reasoning, flexibility, and humanistic abilities required by contemporary practice cannot be developed solely through traditional didactic approaches, which were formerly thought to be the cornerstone of medical education. Innovative teaching strategies that emphasize active involvement, authenticity, and ongoing skill development have become more popular as a result of advances in science education, quick technological advancements, and moves toward competency-based training (Kalyani and Rajasekaran, 2018).

Cutting-edge methods like flipped classrooms, problem-based learning (PBL), team-based learning (TBL), and simulation-based training are transforming how students acquire and apply medical information. These approaches promote better learning by enabling students to integrate theory and practice, collaborate effectively in groups, and critically assess their decisions in safe but useful environments.

By facilitating immersive, repeated encounters that connect classroom education and clinical exposure, technologies like virtual and augmented reality significantly increase learning potential. Medical educators are also adopting personalized and adaptive learning systems that take into account each student's talents, learning preferences, and rate of advancement. By providing more adaptable and customized pathways, this change not only improves competency attainment but also promotes learner well-being.

Innovative teaching methods were introduced in medical education gradually, with key early examples being the "case method" in 1912 and the rise of PBL in the late 1960s and 1970s. More modern interactive and technology-based methods like simulation-based training and case-based learning emerged more recently to promote active, hands-on learning and better prepare students for clinical situations.

## 15.2 Goals of innovative teaching

### 15.2.1 Bridging the theory–practice gap

Innovative teaching in medical education seeks to reduce the theory–practice gap by attempting to establish a link between theoretical knowledge and the practical skills and decision-making abilities needed in clinical settings. Connecting classroom knowledge with practical clinical skills is necessary to bridge the theory–practice gap in medical education. This is mainly accomplished through cutting-edge teaching techniques like PBL, TBL, simulation-based learning, and virtual reality, which provide safe, controlled environments where students can practice challenging procedures and build clinical competence and confidence that directly affect patient care.

### 15.2.2 Developing essential skills

In order to prepare medical professionals to deal with the many subtleties of the complex healthcare contexts, helping them to develop fundamental skills is a crucial part of medical education. Innovative teaching strategies in medical education aim to develop critical thinking, problem-solving, and clinical competencies through hands-on, collaborative, and technology-enhanced approaches like simulation and PBL. This goes beyond rote memorization.

**The Principles and Practice of Medical Education. DOI:** https://doi.org/10.1016/B978-0-443-44197-4.00031-X

By teaching students how to work in teams, apply information to complicated circumstances, and think logically in a safe atmosphere before treating patients, these strategies are essential for educating students to adapt to the changing healthcare landscape (Soni et al., 2024). By allowing students to practice procedures and manage actual events in a secure, controlled setting, innovative ways help them learn critical skills and gain competence and confidence prior to patient interaction. Additionally, it aids in the development of critical thinking and problem-solving skills since methods like evidence-based medicine, case-based learning, and PBL teach students how to evaluate complicated medical data, draw logical conclusions, and solve patient issues.

### 15.2.3 Increasing engagement and retention

Enhancing information retention and comprehension through the development of more effective and captivating learning experiences is one of the goals of the innovative teaching methods. This is done by changing the learning process from passive knowledge transmission to active, application-based learning. Innovative teaching strategies in medical education seek to improve student engagement and retention.

The innovative teaching techniques also help to increase engagement by promoting active learning, such as case-based learning, PBL, and simulations, which require students to actively participate in problem-solving as opposed to passively absorbing knowledge. Because of this practical approach, students continue to be cognitively engaged with the material. Furthermore, it provides the setting and context in which state-of-the-art methods often bridge the gap between theoretical comprehension and clinical application, making the material seem immediately applicable to students' future careers. This perceived relevance naturally raises interest and engagement.

### 15.2.4 Preparing for the future of healthcare

Giving the next generation of medical professionals the technological know-how and skills they need to successfully navigate the evolving healthcare landscape is one of the aims of the innovative teaching methods. In addition, innovative teaching methods give future healthcare workers the critical thinking, adaptability, and technical skills needed for a changing healthcare environment by shifting from passive learning to active, collaborative, and technology-integrated approaches. This involves encouraging cooperation and communication, developing practical skills through simulation and virtual reality, encouraging ongoing education and problem-solving, and getting ready for the integration of digital health technologies like telemedicine and artificial intelligence. Furthermore, it cultivates essential abilities for a dynamic environment where innovative methods like PBL and case-based learning motivate students to actively analyze circumstances rather than passively take in information, enabling them to successfully apply their knowledge in clinical settings. It also encourages adaptation by letting the student move away from traditional lectures. These methods place a strong emphasis on adaptation and lifelong learning, which are crucial for managing the quickly evolving healthcare sector (Fig. 15.1).

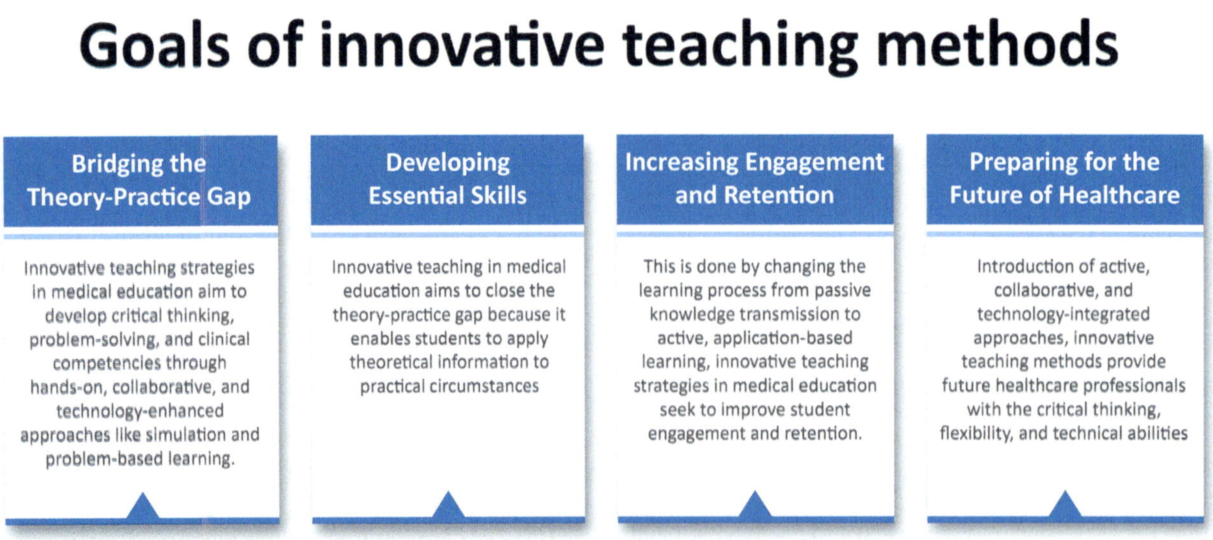


FIGURE 15.1 **Goals of innovative teaching methods.** This figure shows the goals of innovative teaching methods.

## 15.3 Need for innovative teaching

A more dynamic and adaptable approach to teaching and learning is required due to the complexity of contemporary education. When used separately, neither modern nor conventional approaches can adequately meet the many needs of today's students. Because of their organized, teacher-centered approach, traditional teaching methods are good at imparting fundamental knowledge and guaranteeing consistency. However, they frequently lack the flexibility to accommodate various learning styles, creativity, and critical thinking abilities, all of which are becoming more and more important in the fast-paced world of today.

Innovative teaching is crucial in medical education to prepare for rapidly changing healthcare, moving beyond lectures to engaging, technology-enhanced methods like simulation, PBL, and flipped classrooms, fostering critical thinking, teamwork, and adaptability for complex, patient-centered care. The need stems from dynamic healthcare environments, diverse patient needs, and the integration of new technologies, requiring educators to create active, experiential, and collaborative learning that mirrors real-world practice and promotes lifelong learning. Innovative teaching methods are needed to address the following issues in modern medical education.

### 15.3.1 Evolving healthcare

Teachers must adapt innovative strategies to prepare students for a dynamic profession because medical treatments, processes, and technologies are always evolving. By radically changing the way care is delivered to be more preventative, individualized, predictive, and accessible, innovative approaches address the changing nature of healthcare.

These advancements are intended to improve patient outcomes, increase productivity, and lower growing expenses. Their primary sources of inspiration for the expansion of medical education techniques are changes in healthcare practices and technological breakthroughs. Large-scale datasets are analyzed by technologies like artificial intelligence and big data to support diagnosis, forecast results, and improve treatment strategies. Additionally, in order to avoid depending on outdated approaches, it is crucial for both educators and students to have an innate curiosity about new methods, procedures, and research. This will enable them to actively engage in student-centered teaching strategies that transcend standard lectures and help them both accept and value contemporary learning practices.

### 15.3.2 Fostering key skills

Fostering key skills like critical thinking, problem-solving, teamwork, and communication skills have time and again proved to be very effective in delivering healthcare and raising the quality of medical education. These skills can easily be taught through the innovative teaching approaches which move teaching strategies beyond passive learning. These innovative strategies encourage critical thinking, creativity, teamwork, and flexibility by putting students in real-world scenarios, providing customized feedback, and encouraging self-directed exploration of new tools and environments. Activities like collaborative projects, group discussions, and peer-teaching activities require students to work effectively in teams, articulate their ideas clearly, listen to diverse perspectives, and negotiate to reach common goals. The use of simulations and scenario-based learning also improve cognitive abilities for comprehending difficult ideas. Additionally, it improves teamwork and communication by preparing students for collaborative settings through group assignments and the usage of digital platforms.

### 15.3.3 Student engagement

Students' enthusiasm, curiosity, and lifetime love of learning can be enhanced by contemporary techniques like case-based discussions and simulations. By moving from passive lectures to active participation, innovative teaching approaches increase student engagement by making learning more individualized, relevant, and engaging. Project-based learning, gamification, and technological integration are examples of strategies that foster curiosity, problem-solving, and teamwork, which results in increased comprehension and retention. This change from passively absorbing information to actively creating it increases students' motivation and aids in the development of critical abilities. Furthermore, it facilitates collaborative learning by requiring students to actively participate in group projects and debates, share ideas, and contribute to collective learning. This boosts students' motivation and sense of connection to the subject matter.

## 15.4 Innovative teaching methods

Student-centered strategies like PBL and simulation-based learning, which encourage active engagement and critical thinking, are examples of innovative teaching techniques in medical education. Virtual reality, gamification, and e-learning are examples of technological tools that improve engagement and offer realistic, immersive teaching. TBL, flipped classrooms, and case-based learning are further successful strategies that encourage cooperation, problem-solving, and a deeper comprehension of intricate medical ideas.

## 15.4.1 Types of innovative teaching methods

1. **Active and collaborative learning models**
   a. Problem-based learning (PBL)
   b. Case-based learning (CBL)
   c. Team-based learning (TBL)
   d. Flipped classroom
2. **Simulation-based learning**
3. **Technology-enhanced methods**
   a. E-learning
   b. Virtual reality (VR)
   c. Gamification

The types of innovative teaching methods are illustrated in Fig. 15.2.

### *15.4.1.1 Active and collaborative learning models*

#### 15.4.1.1.1 Problem-based learning (PBL)

PBL in medical education is a student-centered method where learning starts with a complex clinical problem, not lectures, prompting students to self-direct research in small groups, develop hypotheses, and apply knowledge to solve the case, fostering critical thinking, better retention, and real-world skills like teamwork and communication, mimicking actual clinical practice (Shah et al., 2024). It integrates basic and clinical sciences contextually, making learning more engaging and effective for long-term knowledge application compared to traditional didactic methods.

The goal of PBL, which is a student-centered teaching approach, is to solve challenging, real-world situations (Sun et al., 2017). In this method, as a facilitator, the instructor helps students determine what they already know, what they still need to learn, and how to locate the data needed to answer the open-ended problem. Instead of directly presenting facts and concepts to students, PBL uses challenging real-world issues as a means of fostering their understanding of concepts and principles.

### *15.4.1.2 Definition*

PBL as defined by the founders at McMaster University was described as learning that results from the process of working toward the understanding or resolution of a problem. The problem is encountered first in the learning process and serves as

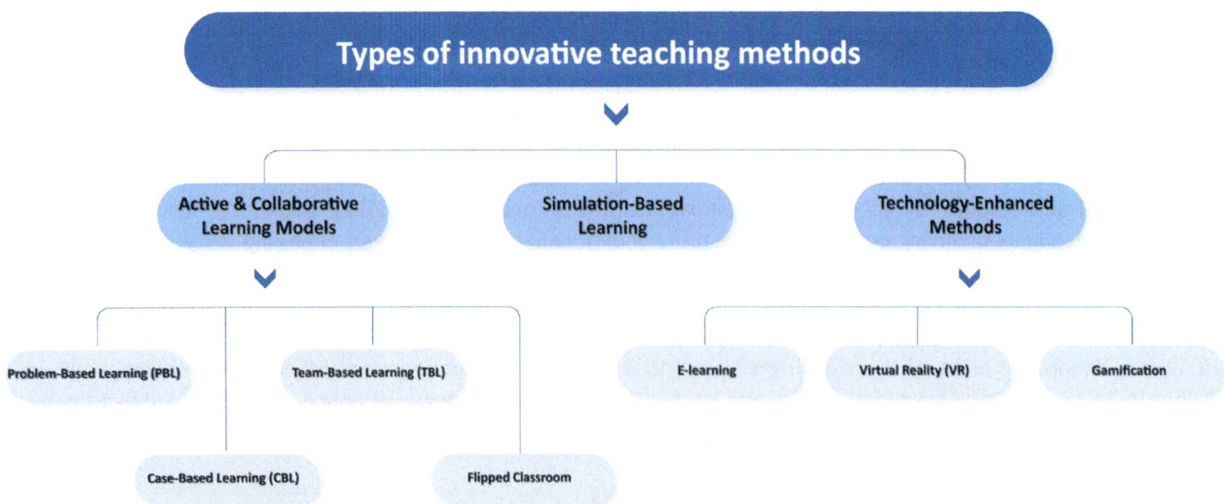


FIGURE 15.2 **Types of innovative teaching methods.** This figure shows the types of innovative teaching methods.

a focus or stimulus for the application of problem-solving or reasoning skills, as well as for the search for or study of information or knowledge needed to understand the mechanisms responsible for the problem and how it might be resolved (Servant-Miklos, 2019).

### *15.4.1.3 Key characteristics*

#### 15.4.1.3.1 Student-centered

One of the key characteristics of PBL is that it is student-centered, shifting the focus from the teacher to the learner and enabling them to take control of their own education by actively participating in difficult and practical problems. By taking charge of their education through small-group conversations and cooperative problem-solving, students actively investigate, investigate, and integrate basic science with clinical concepts, making learning more relevant and interesting. Additionally, by encouraging self-directed learning, it helps students become autonomous learners by helping them recognize their own learning objectives and needs.

#### 15.4.1.3.2 Real-world problems

Since problems are frequently drawn from real-world situations, students find the learning to be interesting and relevant. One of the main features of PBL is the utilization of real-world issues, which give students a genuine context in which to acquire concepts and principles rather than just information. Students are inspired to actively explore, collaborate, and apply their knowledge by taking on challenging, open-ended problems that resemble obstacles in both personal and professional life. This fosters the development of critical thinking, self-direction, and problem-solving abilities (Mansur et al., 2012). The real-world issues give students perspective and relevance, enabling them to relate newly learned material to their own experiences and recognize the usefulness of what they are learning. Because these challenges are frequently unstructured and lack a single correct answer, it also offers an opportunity for students to acquire authentic skills by requiring them to apply the same research, analysis, and critical thinking techniques they would in a professional setting.

#### 15.4.1.3.3 Collaborative learning

In order to foster cooperation and communication, PBL is typically conducted in small groups. Collaborative learning is an essential part of PBL because it stresses cooperation, peer-to-peer learning, and the co-construction of knowledge by using groups to actively address real-world problems. This approach helps students acquire critical thinking, communication, and problem-solving skills that are essential in the workplace by changing them from passive recipients to active problem-solvers. Collaborative learning involves students working in small groups to solve complex, open-ended problems rather than receiving information directly from a facilitator. The group develops the trait of shared responsibility as it assigns duties according to members' strengths and strives toward a common goal.

#### 15.4.1.3.4 Skill-focused

By participating in the PBL exercise, students can enhance their critical thinking, research, and problem-solving skills in addition to their material understanding. One of the main and distinguishing features of PBL is its skill-focused approach. The foundation of the entire PBL method is the idea that students should acquire a variety of transferable, practical skills in addition to factual information. PBL helps with skill emphasis because it pushes students to actively develop important 21st-century skills through challenging, open-ended, and real-world problems. It teaches students how to identify, analyze, and characterize issues, develop workable solutions, evaluate their effectiveness, and select the best course of action in order to imitate how experts deal with challenges in the field.

#### 15.4.1.3.5 Facilitative role of teacher

Teachers serve as facilitators in PBL, offering resources and assistance without disclosing the solutions. A key feature of PBL is the teacher's facilitative role, which is crucial for leading students through the learning process as opposed to merely imparting knowledge. In this regard, the role of the educator shifts from the "sage on the stage" to the "guide on the side." In PBL, the teacher's role shifts from lecturer to facilitator or coach, guiding students to develop clinical reasoning, teamwork, and self-directed learning by asking probing questions, monitoring group dynamics, providing resources strategically, and ensuring logical progress through the problem, rather than just dispensing content knowledge. They help structure inquiry, challenge assumptions, foster critical thinking, and encourage students to become independent learners and problem-solvers.

### 15.4.1.4 Core values of problem-based learning

PBL is built on four learning principles derived from the constructivist learning theory, which are constructive, contextual, collaborative, and self-directed learning.

#### 15.4.1.4.1 Constructive learning

Through constructive learning, students create their own knowledge networks by taking into account what they already know, finding, analyzing, and discussing a range of source materials, and adding new viewpoints. Pupils are instructed to utilize logical reasoning, check opposing information, ask critical questions, and look into alternative points of view. A fundamental tenet of PBL is constructive learning, in which students actively create their own knowledge by collaborating with peers and a facilitator to solve real-world challenges. Students use their past knowledge to construct new understanding rather than passively absorbing material, which is subsequently polished through group debate and independent study. In order to support the learning process and scaffold the acquisition of new knowledge and skills, the teacher's role changes from lecturer to guide.

#### 15.4.1.4.2 Contextual learning

Students apply their knowledge in a relevant setting through contextual learning. They accomplish this by examining real-world examples from professional practice, hotly contested subjects in scientific circles, or social issues that call for a scholarly perspective. By connecting the problems presented to individual experiences, brainstorming about what people already know about the subject enhances contextuality. A fundamental principle of PBL is contextual learning which emphasizes the importance of understanding information and skills within a specific context or real-world setting. By connecting abstract ideas to actual circumstances, this method helps students absorb new information and recognize its applicability. Students become more motivated, learn how to apply their information, and hone their critical thinking and problem-solving abilities when they work on relevant tasks.

#### 15.4.1.4.3 Collaborative learning

The fundamental approach of PBL is collaborative learning, in which students collaborate in groups to solve real-world issues, hone their communication and critical thinking abilities, and gain knowledge from one another. In this process students actively participate in collaborative learning by sharing ideas and criticism in order to jointly develop knowledge. Within the group, it is the students who discuss and solve a problem by creating learning objectives, critically assessing each other's ideas, undertaking independent research, and debating the results. In essence, collaborative learning in PBL shifts focus from instructor-led facts to active, peer-to-peer knowledge construction, using shared inquiry, discussion, and mutual support to develop deep understanding and essential 21st-century competencies like decision-making and coordination.

#### 15.4.1.4.4 Self-directed learning

Self-directed learning is central to PBL, empowering students to take initiative by identifying learning needs, setting goals, finding resources, and evaluating outcomes as they tackle complex problems, fostering autonomy, critical thinking, and lifelong learning skills, making PBL a key driver for developing self-directed learners who actively manage their knowledge acquisition rather than passively receiving it. Self-directed learning allows students to take control of their education by finding and organizing resources and information as well as planning, monitoring, and evaluating their comprehension.

### 15.4.1.5 Problem-based learning teaching strategies

Complex problems are presented as part of PBL teaching methodologies to foster critical thinking and teamwork. In order to provide meaningful learning that meets the needs of the community, educators facilitate brainstorming, discussion, and research. They do this by connecting theory to practice through case studies and simulations (Elshama, 2019).

### 15.4.1.6 Conduction of the problem-based learning

#### 15.4.1.6.1 Prerequisites for conduction of problem-based learning

One of the main prerequisites for the conduction is that the teachers must design challenging, real-world issues, provide a collaborative learning environment with clear ground rules, and make sure students have access to the resources they need in order to implement PBL. To effectively participate in the PBL, students need to have a solid foundation in active learning, critical thinking, teamwork, and self-directed study.

Essential components for the learning environment, the students, and the teachers are the conditions for implementing PBL.

### *15.4.1.7 For the learning environment and curriculum*

#### 15.4.1.7.1 Authentic, open-ended problems

The fundamental component of PBL is a scenario that is related to a real-world event or problem that does require a set of competences to comprehend and solve. The problem should also be in line with the learning objectives and curriculum requirements.

#### 15.4.1.7.2 Resource availability

In order to properly explore the issue, students must also have access to a variety of learning materials, including reference books, journals, and online information. One of the roles of the facilitator is to help the students to get access to these educational materials.

#### 15.4.1.7.3 Collaborative setting

Since PBL is group oriented, its successful conduction and implementation require an atmosphere that fosters small-group learning and interaction.

#### 15.4.1.7.4 Assessment alignment

Instead of focusing only on traditional subject recall through standardized tests, the assessment techniques should be in line with the intended results of PBL, such as critical thinking, teamwork, and self-directed learning abilities.

### *15.4.1.8 For students for problem-based learning*

#### 15.4.1.8.1 Foundational knowledge

In PBL, foundational knowledge isn't taught first. Instead, students discover it as needed while working on an authentic problem, identifying gaps, researching, and constructing understanding collaboratively, building core concepts and critical thinking skills within context, contrasting with traditional learning where basics are delivered up front.

#### 15.4.1.8.2 Critical thinking and problem-solving

The inherent ability of the student to think analytically, evaluate information, and problem-solving is always beneficial to the students. PBL requires students to use critical thinking and problem-solving skills to tackle complex, real-world problems, and in doing so, they further develop and strengthen these exact skills through the learning process. The act of engaging with a problem analyzing it, researching it, and working towards a solution is the process by which these skills are honed.

#### 15.4.1.8.3 Metacognitive skills

The capacity to keep an eye on their own education, spot knowledge gaps, and plan how to get the information and resources they need. Instead, PBL is a successful learning approach that actively cultivates and advances students' metacognitive abilities. Students must practice metacognition through PBL key activities, such as organizing and tracking their problem-solving process, which gradually strengthens these abilities.

#### 15.4.1.8.4 Collaboration and communication

One of the key elements of PBL is the ability to work well in a group, express ideas clearly, and provide peers constructive criticism. PBL focuses on students collaborating in groups to address difficult topics or solve intricate, real-world situations. True collaboration entails a shared process of invention, problem-solving, and discovery rather than merely assigning duties.

#### 15.4.1.8.5 Self-directed learning

One of the key elements of PBL is the drive and capacity to take responsibility for their education, carry out independent research, and efficiently manage their time in order to achieve learning objectives. PBL necessitates self-directed learning since PBL calls on students to take charge, recognize their own learning needs, and oversee their own learning process.

#### 15.4.1.8.6 Flexibility and adaptability

One of the most important traits of a student in PBL is an openness to new knowledge and experiences, the capacity to modify their approach, and the willingness to learn from failures. Flexibility and adaptability are not so much prerequisites for PBL as they are essential abilities acquired during the process and a prerequisite for both students and teachers to successfully execute it. As PBL is open-ended and dynamic, participants must welcome change and modify their strategies as they solve challenging, real-world issues.

#### *15.4.1.9 For the facilitator in the problem-based learning*

##### 15.4.1.9.1 Problem design expert

A facilitator acting as a problem design expert in PBL guides the process of creating effective, engaging, and realistic problems that drive student learning and critical thinking. The expert structures the problem to be appropriately challenging without being overwhelming. They ensure that complexity is introduced in stages, allowing students to build foundational knowledge before tackling more intricate details.

##### 15.4.1.9.2 Group facilitation skills

The facilitator acts as an expert in guiding discussions, encouraging participation, managing group dynamics, and fostering a constructive learning environment without directly providing solutions. The facilitator merely observes the group and notes down the way in which the session is conducted by the leader of the group.

##### 15.4.1.9.3 Curriculum knowledge

A deep understanding of the curriculum to ensure that the chosen problems effectively integrate key concepts and align with desired learning outcomes. Basic initial knowledge is an essential prerequisite for students to effectively engage in and benefit from PBL. By possessing the deep knowledge of curriculum, the teachers must be able to align PBL activities with specific curriculum standards and learning outcomes. Without this knowledge, projects might not cover the necessary material, making it difficult to measure their effectiveness and potentially leaving gaps in student learning.

##### 15.4.1.9.4 Assessment skills

The ability to design and implement assessments that evaluate the learning process, taking into account the skills of the student is one of the vital requirements for the facilitator.

## 15.5 Steps of problem-based learning conduction

The seven steps of PBL, often called the "Maastricht seven-jump process," are as follows (Fig. 15.3):

**1. Clarifying terms in the problem scenario**

In order to ensure that everyone is familiar with the problem situation, it is crucial to clarify terms used in the problem scenario. In this sense, any new terms, ideas, or jargon are discussed and explained by the group. The group

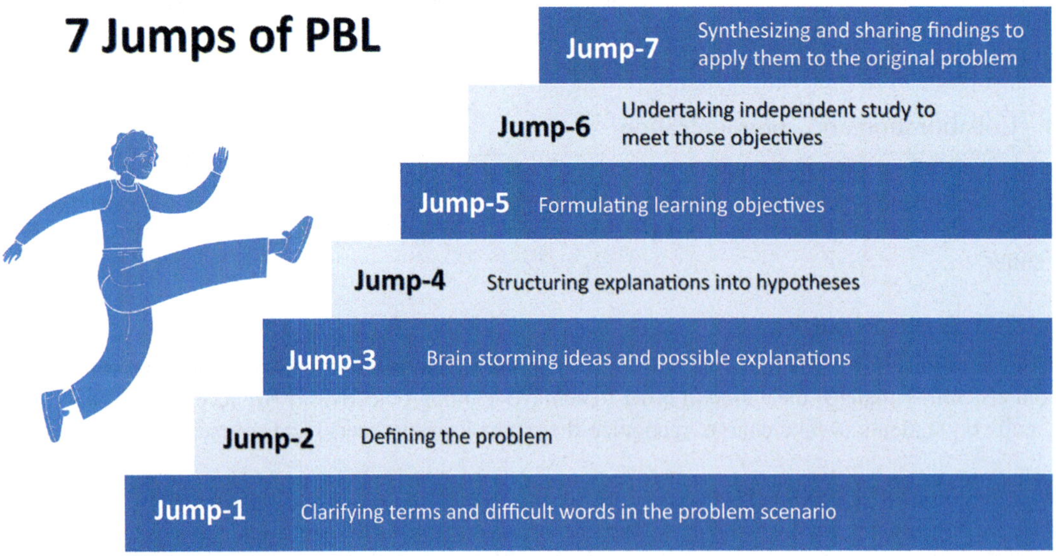


FIGURE 15.3 **Seven jumps of problem-based learning (PBL).** This figure shows the seven jumps of PBL.

must first decide on the meaning of any unfamiliar phrases, terminology, and concepts that are offered in the problem scenario in order to guarantee that everyone has a consistent starting point and knowledge before proceeding.

**2. Defining the problem**

Based on the first conversation, the group identifies and precisely defines the primary problem or question at hand in this step. In the Maastricht "seven jumps" approach to PBL, identifying the issue is typically the second stage. The group must share their perspectives and knowledge in this step in order to create a common understanding of the issue and determine which phenomena need to be explained in the form of one or more questions.

**3. Brainstorming ideas and possible explanations**

In PBL, brainstorming is an essential early-stage activity where students come up with preliminary concepts and potential solutions to comprehend a challenging, real-world issue before carrying out formal study. In PBL, brainstorming is used to generate a variety of concepts, ideas, and tenable hypotheses related to the problem situation by utilizing the group's accumulated prior knowledge.

**4. Structuring explanations into hypotheses**

In this step, the group arranges and shapes the ideas that were generated into tentative explanations or hypotheses that might solve the issue in the fourth phase of organizing those explanations into hypotheses. Explanations are organized into hypotheses in the seven jumps of PBL. This stage comes after brainstorming and entails grouping the group's preliminary concepts and past knowledge into provisional fixes or justifications for the issue.

**5. Formulating learning objectives**

In the fifth phase of the seven-jump PBL approach, the group determines what new information is required to solve the problem by creating learning objectives. In this phase, the group leverages the gaps in their collective knowledge to develop particular learning objectives that will be investigated independently in the following step, depending on the triggers found in the problem after brainstorming and organizing possible solutions. This step's goal is to pinpoint particular issues or subjects that the group must research in order to comprehend the issue and come up with a solution. Here, the process is formulating these goals by talking about what they have previously come up with and figuring out what details are lacking. These objectives act as a roadmap for the upcoming independent study phase.

**6. Undertaking independent study to meet those objectives**

A crucial part of a seven-step procedure frequently employed in PBL is conducting independent study to meet those objectives. In order to acquire the information and comprehension required to address the given issue, students participate in self-directed learning throughout this phase. They use a variety of resources to conduct research and locate knowledge that will help them meet the learning objectives. This step which usually follows the analysis and creation of learning objectives, students conduct independent research on the subjects that have been identified as learning objectives using a variety of resources, including academic journals, textbooks, online databases, and library materials. Additionally, it helps students synthesize knowledge by having them critically assess the data they have collected and apply it to the particular facets of the problem they are attempting to solve.

**7. Synthesizing and sharing findings to apply them to the original problem**

The last step (step 7) involves consolidating, evaluating, and applying the knowledge gained from individual self-directed study to the initial problem situation by synthesizing and sharing findings. To fulfill this commitment, the group gets back together to discuss their discoveries, synthesize the new information, and apply it to the initial issue, which results in a better comprehension and a possible solution. This step's goal is to turn disparate facts into a cohesive understanding. Reporting back, in which students share the data they collected from their individual research with their group members, is one of the most important tasks in this step. During this phase, the team collaborates to compile the individual findings, discuss them, and combine the new information with what they already know and the theories they came up with earlier. Furthermore, the main objective is to connect the newly learned information to the initial problem description. In order to explain the phenomena in the problem, the group tests their initial theories and assesses what they have learned.

### 15.5.1 Role of student in problem-based learning

By working in groups to investigate and solve difficult real-world problems, students who actively engage in PBL take control of their learning. They are in charge of describing the problem, locating knowledge gaps, conducting independent research, and developing evidence-based solutions. Important student duties include critical thinking, effective communication, teamwork, self-evaluation, and reflecting on their growth to improve self-directed learning skills. They also conduct independent research and actively consider what they already know and what they still need to learn in order to solve the issue. Here, students work together in small groups to solve challenges by sharing thoughts, knowledge, and perspectives.

### 15.5.2 Role of teacher in problem-based learning

With PBL, the emphasis is shifted from subject delivery to student-centered learning, with the teacher serving as a guide and facilitator rather than a lecturer. Creating challenging, real-world challenges, guiding students through the inquiry process with insightful questions, and encouraging teamwork are important responsibilities. Along with giving comments and assisting students in developing abilities like critical thinking and self-direction, the instructor also evaluates the students' comprehension and problem-solving techniques. One of the main responsibilities of a teacher in PBL is to generate and present real, unstructured situations that motivate students to learn. Additionally, the instructor plays the role of a facilitator, guiding students as they solve problems, posing questions to encourage critical thinking, and assisting them in overcoming challenges without offering concrete solutions.

### 15.5.3 Competencies covered

Among the skills covered by PBL include communication, critical thinking, and problem-solving. It also promotes the growth of skills including digital literacy, information synthesis, self-directed learning, and collaboration. In order to acquire these competencies, students actively work to address real-world challenges, which calls for analysis, investigation, and presentation of solutions.

### 15.5.4 Advantages

Developing critical thinking and problem-solving abilities, fostering a deeper and more enduring comprehension of the material, and boosting student motivation and engagement through real-world application are just a few benefits of PBL. Additionally, it promotes critical professional abilities including communication, cooperation, collaboration, and self-directed learning (Dolmans and Schmidt, 1996).

#### *15.5.4.1 Skill development*

In terms of skill development, it promotes critical thinking and problem-solving by teaching students how to assess information, analyze complicated problems, and come up with logical answers. Additionally, it fosters collaboration and communication skills among students by teaching them how to solve problems in groups, which improves their interpersonal communication, teamwork, active listening, and negotiating abilities. Additionally, it fosters lifelong learning habits by empowering students to take charge of their education by identifying their own learning needs, locating resources, and taking ownership of their education. Students learn how to ask critical questions, compile and assess data from many sources, and verify its accuracy through PBL.

#### *15.5.4.2 Learning outcomes*

By applying concepts to real-world challenges, the PBL platform helps learners develop a deeper and more enduring comprehension than rote memorization. Additionally, it boosts motivation and engagement since students are more likely to be inspired and involved when they perceive a clear goal for their education and are able to focus on issues that are pertinent to them. PBL also helps students get ready for future challenges by giving them practice using their knowledge and skills in real-world scenarios.

#### *15.5.4.3 Teacher and classroom benefits*

PBL contributes to the creation of a positive learning environment by encouraging a more student-centered, adaptable, and resilient classroom culture. Better relationships between students and teachers are also fostered, which leads to a shift in the role of the teacher to that of a facilitator and a more positive dynamic between the two.

### 15.5.5 Disadvantages of problem-based learning

The disadvantages of PBL include the fact that it takes a lot of time for both teachers and students, may cover less content than traditional methods, and presents difficulties with assessment. Additional drawbacks include the potential for unfair group dynamics and social loafing, anxiety in some students, and a high demand for experienced teachers and sufficient resources.

#### *15.5.5.1 For students*

Sometimes, students find PBL to be a time-consuming activity since it requires them to spend more time on research and discussion, which may conflict with other subjects. Some students may feel anxious and unsatisfied since they are not used to self-directed learning. Students may suffer from cognitive overload if they are unable to handle the exercises' open-ended nature. Learning may be superficial if PBL is not properly organized. As a result, students might not cover the required depth or breadth of information.

#### *15.5.5.2 For teachers and the learning environment*

It has been observed that highly skilled teachers are necessary for PBL since they need specialized training to formulate difficult problems and act as facilitators rather than lecturers. Furthermore, employing a single rubric to assess teamwork, creativity, and process management in addition to topic knowledge makes grading PBL a challenging task. Furthermore, it's possible that certain students could dominate discussions while others don't contribute much, or that some participants will indulge in "social loafing" and work less hard. The focus on challenges, as opposed to traditional methods, may cause content to be overlooked, which could be problematic for standardized assessments.

## 15.6 Case-based learning

Case-based learning is a student-centered, discussion-based approach in medical education that uses real-world clinical cases to teach medical knowledge and reasoning. Its characteristics include integrating theory and practice, encouraging critical thinking and professional skills including diagnosis and treatment planning, and encouraging active learning through small-group discussions (Kaur, 2014). The instructor's main responsibility is to serve as a facilitator, helping the students work together to solve problems that often don't have a single right answer.

### 15.6.1 Characteristics of case-based learning

Some of the key characteristics of case-based learning are the following:

**1. Student-centered**

One of the main features of case-based learning is student-centeredness, which puts students at the center of the learning process and changes the teacher's position from lecturer to facilitator. Instead of having students passively absorb knowledge from a teacher, the emphasis here is on active student interaction and exploration. Through inquiry, discussion, and problem-solving, students collaborate to examine real-world issues, build their own knowledge, and develop practical skills. This method places a strong emphasis on active student participation.

**2. Discussion-driven**

One of the main and most important aspects of case-based learning is discussion. It is the process by which students actively interact with the subject matter, apply their theoretical understanding to practical situations, and hone their critical thinking abilities. Students in small groups engage in rigorous, cooperative learning in this setting.

**3. Real-world cases**

The utilization of real-world cases, which provide students with real-world scenarios to analyze, collaborate on, and resolve, is a key component of case-based learning. These instances are presented as narratives with complexity, ambiguity, and multiple points of view in order to mirror real-life decision-making. This approach seeks to go beyond rote memorization and promote practical expertise, critical thinking, and problem-solving skills by placing students in real-world, interactive, and context-rich scenarios. In order to make the lecture more applicable and memorable, the students apply their knowledge to real, in-depth patient scenarios.

**4. Collaborative problem-solving**

A key component of case-based learning is collaborative problem-solving, where students analyze real-world events in groups. This method fosters critical thinking, teamwork, communication, and problem-solving abilities through rigorous group interaction in which students develop knowledge and strive toward a common solution to a challenging issue. In this, the groups collaborate to evaluate a case, pinpoint learning objectives, put forth theories, and create a diagnostic and treatment plan.

**5. Facilitator-led**

A facilitator or instructor leads the conversation, demonstrates clinical reasoning, and promotes involvement without offering all the solutions. A fundamental feature of case-based learning is facilitator-led learning, in which the teacher serves as a guide to facilitate the student learning process. In this capacity, the facilitator is also responsible for organizing the classroom and helping the students to solve problems without just offering solutions.

**6. Integrates theory and practice**

By teaching students how to apply theoretical information to real-world scenarios, case-based learning helps close the gap between basic science and clinical practice. One of the main features of case-based learning is the integration of theory and practice. In this method, students apply theoretical knowledge to real-world or fictitious case scenarios, bridging the gap between abstract ideas and practical problem-solving (Tsekhmister, 2023).

**7. Promotes key skills**

It is well known that case-based learning fosters a number of critical abilities that extend beyond just knowledge acquisition. Critical thinking, clinical reasoning, decision-making, and teamwork are important abilities that are promoted by case-based learning.

## 15.6.2 Role of student in case-based learning

Students actively participate in case-based learning, analyzing real-world situations and using their prior knowledge to discover solutions. Working with peers, identifying issues, coming up with solutions, conducting research, and presenting their conclusions and logic to the group are all part of their job description. Additionally, it is the responsibility of students to think critically, ask questions, and reflect on their learning experience and possible areas for growth.

## 15.6.3 Role of teacher in case-based learning

Instead of giving lectures, the instructor in case-based learning serves as a guide or facilitator. By fostering a psychologically safe setting, posing insightful questions, and directing conversations toward learning goals, they steer the learning process and motivate students to collaborate in order to solve challenging, open-ended problems.

## 15.6.4 Competencies addressed by case-based learning

By involving students in real-world situations, case-based learning mainly addresses a variety of core competencies that are necessary for professional practice, such as critical thinking and problem-solving, clinical or practical reasoning, communication, collaboration and teamwork, decision-making and judgment, and professional behaviors and ethics. These abilities encompass practical, interpersonal, and self-directed skills in addition to content knowledge.

## 15.6.5 Advantages of case-based learning

By applying information to real-world situations, case-based learning provides benefits like improved critical thinking and problem-solving abilities (Srinivasan et al., 2007). Additionally, it enhances student collaboration and communication, encourages long-term memory and deeper comprehension, and develops professional abilities like empathy and introspection.

### *15.6.5.1 Skill-based advantages*

Critical thinking and problem-solving are two of the skill-based benefits. Case-based learning sharpens students' analytical and evaluative skills by requiring them to examine difficult situations, identify important issues, and generate creative solutions. Additionally, it makes learning more relevant and significant by enabling students to apply theoretical information to real-world issues. Additionally, by encouraging students to go beyond rote memory and engage in deeper analysis, thought, and the capacity to discern between important and irrelevant aspects, this approach fosters higher-order thinking.

#### *15.6.5.2 Collaborative and communicative advantages*

It provides collaborative and communicative advantages in addition to enhancing empathy and communication. The student-led conversations improve oral communication abilities and the ability to understand complex problems from multiple perspectives. Students who work together on cases also develop closer relationships with their peers, which could be beneficial for future networking and group support. Additionally, by giving students the opportunity to interact with and learn directly from seasoned experts, this enhances career connections and provides invaluable networking and educational opportunities.

### 15.6.6 Disadvantage of case-based learning

Case-based learning has drawbacks, such as students feeling overburdened by the open-ended nature of the cases, instructors having to devote a lot of time and resources, the possibility of inefficient use of class time if discussions are poorly managed, and the limited generalizability of findings from a single case. Additionally, it might be challenging for teachers to develop suitable situations and for students to adopt the approach without enough assistance.

#### *15.6.6.1 For the students*

Students may sometimes feel overwhelmed by the open-ended questions and lack of definitive answers in case-based learning, which has been shown to be detrimental to them. This can cause irritation and make it difficult for them to participate, particularly if they have no prior knowledge. Without sufficient guidance, the approach may feel unstructured, which could result in ineffective use of class time. As some students may dominate the conversation while others remain silent, it might result in unproductive participation, underscoring the necessity of cautious facilitation.

#### *15.6.6.2 For instructors and institutions*

From the instructor's point of view, case-based learning is a drawback since it requires a lot of time and resources because creating cases, getting ready for discussions, and handling student answers can all take a lot of time. Furthermore, it can be difficult to create situations that are neither too easy nor too complex, and instructors may oppose case-based learning because they are inexperienced with it. Additionally, it presents evaluation issues, making it challenging to measure student learning and assess higher-level thinking skills. Additionally, the teachers might not be comfortable with the methodology and might rather use more conventional lecture-based methods.

#### *15.6.6.3 Team-based learning (TBL)*

"An active learning and small group instructional strategy that provides students with opportunities to apply conceptual knowledge through a sequence of activities that include individual work, teamwork, and immediate feedback" is the definition of TBL. Professor Larry Michaelsen in 1980 created TBL which is an active, collaborative learning approach to foster critical thinking and deep understanding by involving students in preclass course concepts, in-class readiness assessments, and application-focused team activities. This approach is becoming more and more popular in higher education, professional development, and even medical education. It develops abilities like teamwork and problem-solving and turns lectures into lively conversations. TBL's main learning goal is to go beyond merely teaching material and concentrate on giving students the chance to experience applying course principles to solve problems (Punja et al., 2014).

## 15.7 Key characteristics of team-based learning

Permanent, diverse teams, individual and group accountability, a readiness assurance procedure with instant feedback, and the application of concepts through cooperative problem-solving are all essential elements of TBL. TBL flips the conventional classroom model and promotes critical thinking and communication skills by having students prepare and work on topics outside of class and using class time for application and discussion.

## 15.7.1 Team formation and accountability

### *15.7.1.1 Diverse and long-term teams*

To facilitate the establishment of high-performing units, students are assigned to teams that stay together throughout the course. These teams are purposefully made up of people with different experiences and viewpoints.

### *15.7.1.2 Personal and collective responsibility*

Every student has an obligation to contribute to the team and prepare the content of the topic for discussion. TBL also requires the whole team to perform as a whole as it exhibits the collective responsibility and collaborative effort of the team.

### *15.7.1.3 Peer review and assessment*

Peer review and assessment is a characteristic and crucial feature of TBL, serving to motivate student preparation, encourage accountability within the team, and provide constructive feedback essential for both individual and collective learning.

### *15.7.1.4 Readiness assurance process: individual and team testing*

The readiness assurance process is a core component of TBL, featuring sequential individual and team testing to ensure student preparation and accountability, followed by group application exercises.

### *15.7.1.5 Quick feedback*

Quick feedback is a defining and essential characteristic of TBL. It serves a pivotal role in the learning cycle by immediately reinforcing correct understanding, identifying misconceptions, and allowing for timely correction during the learning process, not just at the end. It gives the team a chance to address any inconsistencies and clear up any misunderstandings, while the information is still fresh in their minds.

### *15.7.1.6 Explanation*

In the process of conducting the TBL activity, explanation plays an important part as this gives a chance to the instructors to initiate a class-wide discussion and clarify or address any unanswered questions at the conclusion of the process.

## 15.7.2 Application and learning activities

### *15.7.2.1 Application exercises*

Application exercises are essential team activities in TBL, where groups use newly learned concepts to solve challenging, real-world issues, going beyond mere recollection to deeper knowledge and critical thinking. In this case, the majority of the activity's time is spent applying concepts through tasks like solving significant, real-world problems. Through organized exercises like case studies, critiques, or creative problem-solving, the application exercise promotes deep learning and group decision-making.

### *15.7.2.2 Solving problems*

Solving problems in TBL involves a structured, active process where students prepare individually, then collaborate in teams to apply knowledge to complex scenarios using a readiness assurance process, followed by application activities, class discussion, and peer evaluation, building critical thinking, teamwork, and deep understanding.

### *15.7.2.3 Collaborative analysis*

In TBL, collaborative analysis entails small groups of students working together to analyze complex problems, discuss concepts, and apply knowledge after individual preparation. It goes beyond simple content delivery to promote critical thinking, communication, and problem-solving through structured activities like application exercises and readiness assurance tests, with peer feedback and transparent group processing.

## 15.8 Steps in conduction of team-based learning

TBL follows a structured cycle that emphasizes preparation, engagement, and application:

**1. Initial preparation phase**

In order to obtain a basic comprehension of the subject, in this step, the students independently read the allocated materials before class as part of the first preparation phase of TBL. This preparatory work, which may consist of readings references, books, or any study material, guarantees that students are ready for the ensuing evaluation and application stages. It is good if every student has a foundational understanding of the main ideas in order to facilitate more fruitful class discussions and problem-solving.

**2. Readiness assurance process**

A readiness assurance test in TBL is a two-part assessment process that starts with an individual test (iRAT) and is immediately followed by a team test (tRAT) on the same content. The individual test ensures students prepare individually, while the team test allows them to discuss and apply their knowledge collaboratively using tools like IF-AT scratch cards to reach a consensus. This process holds students accountable for preparation and fosters peer teaching and deeper understanding. As part of the individual test and team test, the tutor must construct questions for the students to answer by reading the provided reference. Typically, there are two short answer questions and five multiple-choice questions.

**a. Individual readiness test (iRAT)**

Students demonstrate their individual mastery of the topic by taking a closed-book, multiple-choice test after arriving at class. After the allotted seven to ten minutes, the tutor picks up the papers, and the students move on to the next phase.

**b. Team readiness test (tRAT)**

A tRAT in TBL is a crucial step where teams retake the same quiz (iRAT) they just completed individually, using scratch-off cards (IF-ATs) or online platforms for immediate feedback, forcing consensus, revealing knowledge gaps, fostering deep learning, and ensuring everyone comes to class prepared for application activities. It transforms individual learning into a collaborative, high-energy event, solidifying understanding and building crucial teamwork skills.

**3. Minilecture**

In TBL, a minilecture is a brief, targeted presentation by the instructor that explains ideas that students found difficult during the readiness assurance process. It usually lasts no longer than ten to fifteen minutes. It addresses certain misunderstandings or queries that the teams discovered after the individual and team examinations and the team's appeals. Before students go to application exercises, the goal is to effectively define the most challenging aspects of the subject rather than to teach the complete thing.

**4. Application activities**

In TBL, an application question is a case study or problem that calls for a group to use the information and abilities acquired during the preclass phase to resolve a practical issue. Teams collaborate to reach a resolution, frequently using multiple-choice or free-response questions. They then have to present their solution to the class in order to defend it. This method places a strong emphasis on communication, problem-solving, and critical thinking abilities. Teams solve challenging, real-world challenges by using their conceptual understanding. Using the knowledge they have gained from the other steps, students collaborate and engage in debate (Fig. 15.4).

### 15.8.1 Role of the tutor in team-based learning

In TBL, a tutor acts as a facilitator rather than a traditional lecturer, helping students with the readiness assurance process, monitoring discussions, making comments, and ensuring broad participation. Additionally, it is the tutor's duty to establish a safe learning atmosphere and support students in developing a deeper comprehension by offering questions and challenging concepts. Their primary goal is to help students become proficient, independent learners by helping them grasp the subject through teamwork.

### 15.8.2 Role of student in team-based learning

Students participate actively and responsibly in TBL by preparing on their own before class, working together in stable teams to apply knowledge and solve problems during sessions, and contributing to a more in-depth, interactive learning

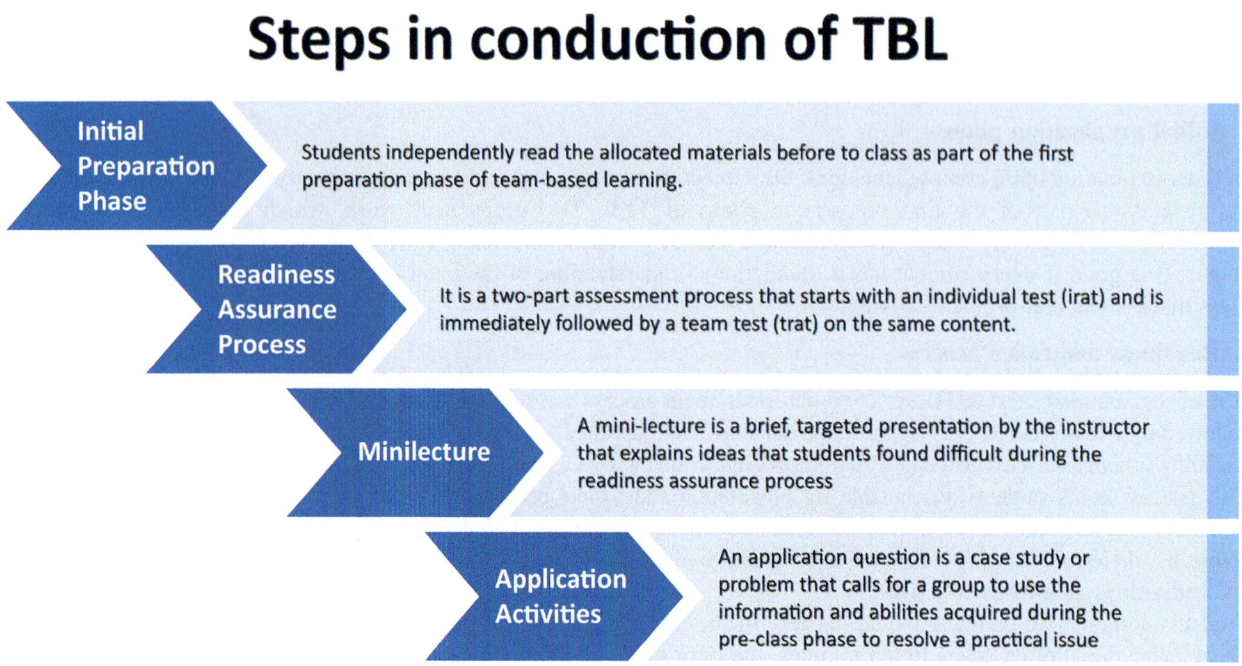


FIGURE 15.4 **Steps in conduction of team-based learning (TBL).** This figure shows the various steps in conduction of TBL.

experience through discussion, peer support, and simultaneous reporting of answers. They must take responsibility for their education, hold themselves and their teams accountable through peer evaluation and readiness assurance testing, and actively engage in challenging assignments to hone their critical thinking and communication skills.

### 15.8.3 Advantages of team-based learning

With TBL, students actively participate in conceptual discussion and application, transforming passive learning into an active process. By encouraging improved information acquisition, improved teamwork, and the development of higher-order thinking skills, it also improves learning outcomes (Burgess et al., 2017). Furthermore, critical thinking, problem-solving, communication, and teamwork are all developed through TBL. Since the combination of individual and team testing encourages students to come to class prepared and contribute significantly to their teams, students who participate in the TBL session feel more accountable. It also frees up educators to focus on problem-solving and instruction rather than just knowledge transfer.

### 15.8.4 Disadvantages of team-based learning

One of the disadvantages of workload and involvement is uneven participation, where some students may dominate while others contribute very little. Additionally, some students could rely on the team to complete the task; this is referred to as "free-riding," and it causes resentment among the team members who bear the burden. Concerns about working inefficiently also exist since disagreements and discussions may slow progress, making individual work more effective for simpler tasks. Arguments that can result in conflict can hinder learning and cause stress, which is one of the group dynamics and learning disadvantages. Furthermore, slower learners may hurry or faster learners may slow down due to the group's tempo, failing to meet individual need. It is also possible that the needs of the group will be prioritized over those of the individual, with the advancement and expectations of the group possibly overshadowing the individual's learning needs.

The ineffective group construction that necessitates careful evaluation of student characteristics like ability level and learning style is one of the implementation and structure-related drawbacks. There is also potential for poor implementation if the learning activities are not structured correctly, resulting in a lack of deep learning.

### 15.8.5 Flipped classroom

In a flipped classroom, students learn new content at home through texts or videos while active learning, discussions, and problem-solving occur during class. With this approach, the learning environment is reversed, with the teacher acting as a facilitator during class to assist students in applying the material in increasingly complex ways while they learn at their own pace outside of it (Ozdamli and Aşıksoy, 2016). Some of the key characteristics of the flipped classroom are the following:

**1. Inverted learning environment**

One of the main features of a flipped classroom is an inverted learning environment, in which the conventional roles of classroom and home learning are reversed. In order to free up class time for active learning, problem-solving, discussion, and hands-on activities under teacher supervision, the teacher's lecture content is relocated outside of the classroom. With this approach, the traditional lecture period is shifted outside of the classroom, while problem-solving and application are brought within. Additionally, students can actively participate in their education by first assimilating the material at their own pace, which makes it possible to place a greater emphasis on applying and extending knowledge during class.

**2. Flexible pacing**

Flexible pacing, which allows students to study at their own pace utilizing pre-class materials, is one of the key components of flipped classrooms and allows for unique learning experiences. By enabling students to pause, rewind, or rewatch lectures and access content whenever and wherever they choose, this gives them the time they need to understand a subject before moving on. Then, during in-person class time, students can engage in more challenging activities like problem-solving, allowing the instructor to provide more individualized support.

**3. Active learning in class**

Active learning, which defies the traditional method by having students concentrate on foundational information at home and use class time for practical activities, problem-solving, and in-depth discussion under the teacher's supervision, is a crucial component of a flipped classroom. In this, class time is devoted to active engagement, such as group projects and hands-on experiments, rather than only listening to lectures. This shift from passive listening to active participation enables higher-level learning, personalized feedback, and enhanced student–teacher engagement during class sessions.

**4. Teacher as facilitator**

The instructor becomes a guide or facilitator who works directly with students to solve their obstacles and provides more individualized assistance, rather than being the exclusive source of information. In a flipped classroom, the teacher's role shifts from that of a typical lecturer to that of a facilitator who guides active, in-class learning. This allows students to engage in more sophisticated thinking activities, such discussions and problem-solving, while the teacher provides one-on-one support, answers questions, and promotes cooperation. The facilitator creates a supportive environment where students can work through the content at their own pace with their classmates.

**5. Increased student responsibility**

One of the main features of the flipped classroom is increased student responsibility because it puts the student at the center of the learning process. In order to use class time for active, hands-on activities like problem-solving and teacher-led debates, this model demands students to take charge of their education by engaging with new content prior to class. Autonomy, greater comprehension, and better learning outcomes are all encouraged by this active involvement.

**6. Higher-order learning**

One of the main features of flipped classrooms is higher-order learning, which frees up class time for higher-order thinking abilities like analysis, problem-solving, and creation by moving lower-level learning, such memorizing, to extracurricular activities. Students can use class time to develop higher-level thinking abilities, such concept analysis and application, using the underlying knowledge they have learned at home. Through exercises like problem-solving, case studies, and group discussions, this in-class focus enables more direct teacher assistance, peer cooperation, and knowledge application (Hurtubise et al., 2015).

**7. Collaboration**

Collaboration is one of the key components of the flipped classroom since it converts class time from passive lectures to active, group projects. Students use class time for group projects, debates, problem-solving, and peer-to-peer interaction while learning basic concepts through recorded materials at home. In addition to learning from one another and receiving more direct feedback from the teacher, students can practice advanced skills like communication and critical thinking.

## 15.8.6 Conduction of the flipped classroom

The flipped classroom reverses typical teaching methods by having students learn new information on their own time, frequently through texts or videos, and using class time for active, group projects, discussions, and problem-solving with teacher aid. The implementation of a flipped classroom involves three main steps:

**1. Before class**

Acquiring knowledge is the primary objective of this stage. During this phase, students are expected to engage with teacher-prepared instructional materials in order to get a basic comprehension of the new subject. The main goals of this session are to view recorded video lectures, complete assigned readings, take notes, and take brief online exams to assess basic knowledge. Here, it is the teacher's responsibility to create engaging, concise, and clear materials that also clearly outline the prerequisites for students' preparedness.

**2. During class**

The application and strengthening of knowledge are characteristics of this level. In the subsequent phase, the instructor serves as a facilitator or guide while the class time changes from passive listening to active involvement and knowledge application. Discussions, problem-solving exercises, case studies, team-based projects, group activities, and experiments are the primary activities that take place here. Here, the teacher's primary responsibilities are to direct learning, offer prompt feedback and individualized assistance, deal with each student's unique challenges, and promote peer-to-peer cooperation.

**3. After class**

Consolidation and knowledge expansion are the hallmarks of this stage. At this point, students assess their comprehension and expand their knowledge by additional practice and introspection. Completing any assignments, reflecting further, or taking formative and summative tests are some of the activities. Here, the teacher's job is to assess students' development and use their input to modify subsequent classes.

## 15.8.7 Role of the student in flipped classroom

In a flipped classroom, the student's role changes from that of a passive recipient to that of an active participant who must prepare for class by studying new material at home and then using it in class through exercises and group projects. Students must watch lectures, finish preclass assignments, take responsibility for their education, and participate fully in class discussions and problem-solving.

## 15.8.8 Role of teacher in flipped classroom

In a flipped classroom, the instructor becomes a facilitator who supports and directs student-led learning instead of a traditional lecturer. Among the primary responsibilities include creating in-class activities that promote teamwork and problem-solving, preparing and selecting movies and other out-of-class information, and giving students personalized support and prompt feedback. During class, the instructor takes on the role of a mentor, monitoring students' development, responding to their inquiries, and assisting them in gaining a deeper comprehension.

## 15.8.9 Advantages of the flipped classroom

Personalized learning, higher student engagement, and the growth of critical abilities like self-directed learning and teamwork are the primary benefits of a flipped classroom. Teachers can use class time for active learning, one-on-one assistance, and group projects while students can learn at their own speed by moving content delivery outside of the classroom. Additionally, this method enhances academic achievement and results in a deeper comprehension of the subject.

## 15.8.10 Disadvantages of the flipped classroom

The flipped classroom has drawbacks, such as depending on students' motivation for independent preclass study, requiring teachers to devote a substantial amount of time up front to creating materials, and possibly causing unfairness owing to unequal access to technology and the internet. Some students may oppose the concept since they believe it is the teacher's responsibility to lecture, and the change may provide difficulties for both educators and educational institutions, such as making sure students have finished their assignments outside of class.

#### *15.8.10.1 Simulation-based education (SBE)*

In medical education, simulation-based learning uses realistic simulations to train healthcare workers in a risk-free setting, enabling them to practice both technical and non-technical abilities without endangering actual patients. It entails simulating clinical events, from routine procedures to uncommon emergencies, utilizing technologies such as virtual reality, high-fidelity mannequins, or standardized patients (Elendu et al., 2024). Repetitive practice, thorough feedback, and the growth of abilities like cooperation, communication, and decision-making are all made possible by simulation-based learning, which eventually improves patient care and safety.

One of the main benefits of simulation-based learning is that it offers a low-stakes environment where trainees can make mistakes, grow from them, and gain confidence without facing repercussions for actual patients. Additionally, it improves the learner's skill development because it has been shown to be successful in practicing both "soft skills" like professional behavior, teamwork, and communication as well as technical skills like cardiopulmonary resuscitation and surgery. It can mimic everyday scenarios as well as uncommon, dangerous occurrences that trainees might not come across in conventional training (Al-Elq, 2010).

One of the benefits of simulation-based learning is that it provides a secure setting for skill development, increasing engagement, and fostering critical thinking without real-world repercussions. By lowering the hazards and resource requirements associated with conventional approaches, it can be more economical. It also delivers tailored and standardized training and realistic scenarios.

### 15.8.11 Technology-enhanced methods

Technology-enhanced learning (TEL) in medical education integrates digital tools and methods to improve teaching and learning, making it more flexible, interactive, and personalized by using techniques like blended learning, virtual or augmented reality simulations, mobile apps, and e-learning platforms. TEL goes beyond traditional lectures to improve knowledge, skills like clinical reasoning and communication, and professional attitudes for future doctors and healthcare professionals.

#### *15.8.11.1 E-learning*

E-learning in medical education refers to the use of electronic technology, primarily the internet, to provide and enhance education for students and healthcare professionals. It is made up of interactive platforms, multimedia, and online resources that allow for flexible, self-paced learning outside of a traditional classroom. Information is taught, skills are practiced through simulation, and assessments are carried out via e-learning. It can be applied in place of or in addition to conventional methods.

Easy access to instructional resources, including videos, audio, question banks, and virtual patients, at any time and from any location is one of the key characteristics of e-learning. Additionally, it makes studying enjoyable by giving students the freedom to choose their own study time, pace, and order, which helps to accommodate various learning styles and hectic schedules. Additionally, it offers interactive information that goes beyond static text to incorporate virtual reality, interactive simulations, and other multimedia that offer more in-depth educational opportunities.

E-learning, which offers customizable, student-centered learning experiences, is used in medical education using a variety of digital modalities, including videos, applications, and virtual patients. It can involve synchronous or asynchronous sessions and is supported by systems like learning management system. Teachers must transition from material distributors to facilitators during the conduction, which calls for consideration of issues like resource quality and technical obstacles.

#### *15.8.11.2 Virtual reality*

In medical education, virtual reality employs computer-generated, immersive environments to provide healthcare professionals with interactive, realistic, and safe training. It has many uses, from honing nontechnical abilities like critical thinking and communication to practicing intricate surgical techniques and anatomical exploration (Lampropoulos et al., 2025). By offering constant, excellent learning opportunities, this technology can standardize training and enable repeating practice of uncommon or dangerous situations.

Its main features are that it provides a platform for surgical and procedural training where students and medical professionals can carry out challenging procedures like laparoscopic surgery or trauma response without putting real patients in danger. Students may visually examine and modify internal organs and systems from different angles thanks to technology, which enables the creation of complex 3D models of the human body. This makes studying anatomy and physiology more realistic. This technology offers a scalable, flexible, and engaging alternative or supplement to traditional methods like cadavers by enhancing understanding of human systems and providing opportunities for remote, self-paced learning and decision-making in real-world settings like operating rooms or emergency rooms.

#### *15.8.11.3 Gamification*

The use of game-design components like points, badges, and leaderboards in nongaming educational situations to increase learning effectiveness and engagement is known as gamification in medical education. This method enhances students' motivation, clinical reasoning abilities, and knowledge retention using mechanics, including problem-solving challenges, simulations, and instant feedback (Ahmed et al., 2015). A sense of accomplishment and competition are fostered by the points, badges, and leaderboards, which promote motivation and ongoing education. Additionally, students can practice clinical skills, problem-solving, and decision-making in a risk-free setting thanks to interactive scenarios, simulations, and virtual patient cases.

Additionally, it is distinguished by challenges and quests where students are guided through learning objectives more actively than with traditional techniques through planned assignments and problem-solving exercises. This creative method creates dynamic, risk-free environments where students may practice difficult clinical skills, diagnose virtual patients, and learn from their mistakes without endangering real patients by using interactive simulations and game-like challenges.

## 15.9 Traditional versus innovative teaching

The difference between traditional and innovative teaching methods is presented in Table 15.1.

### 15.9.1 Blending traditional and modern approaches

A balanced strategy that can meet the changing demands of education is provided by combining the structure of conventional methods with the application of contemporary approaches. The basis for methodical learning, discipline, and mastery of fundamental concepts is provided by traditional methods. These techniques guarantee that students learn fundamental information and abilities in a controlled setting. The learning process can be infused with creativity, adaptability, and critical thinking through the application of contemporary techniques, including interactive learning, technology integration, and collaborative approaches. These methods not only increase student engagement but also make it possible to tailor instruction to each student's needs, resulting in more individualized learning opportunities. By combining these two approaches, teachers can provide students with a well-rounded education that develops their brains and creativity. While modern methods promote engagement, creativity, and problem-solving, the strict discipline of earlier systems ensures that students have a strong academic foundation. This hybrid approach better prepares students for the problems of the real world while assisting them in understanding the subject matter more thoroughly and meaningfully.

**TABLE 15.1** Difference between traditional and innovative teaching methods.

| Sl no. | Feature | Traditional teaching method | Innovative teaching method |
|---|---|---|---|
| 1 | Characteristic feature | Memorization, teacher-centeredness, and information transfer | Active participation, skill-based learning, and student-centered |
| 2 | Teacher's role | Principal source of information and authority | Facilitator and guide |
| 3 | Student's role | Information recipient who is passive | Collaborator and active participant |
| 4 | Methods | Lectures, hands-on training, and memorization | Flipped classrooms, problem-based learning, and team-based learning |
| 5 | Technology | Very little to no use of technology | Integrated use of software, internet resources, and smartboards |
| 6 | Learning environment | Physical classroom, set schedule | Adaptable, mixed, or online |
| 7 | Assessment | Mostly conventional examinations and tests | A combination of conventional tests, lectures, and hands-on |

**Summary**

In this chapter, the focus is on innovative teaching methods that have emerged as educational paradigms shift to meet the needs of diverse learners in the 21st century. Traditional teaching methods, often characterized by lecture-based instruction and passive learning, are being replaced by approaches that emphasize active engagement, personalized learning, and the integration of technology. The chapter emphasizes that innovation in teaching is not about adopting a single method but about creating a flexible, adaptive approach that meets the needs of students in an increasingly complex and fast-changing world. Effective teachers are those who are willing to experiment with new techniques, incorporate technology, and create learning environments that are engaging, inclusive, and supportive. Through these innovative approaches, educators can inspire a love of learning and help students develop the critical skills needed for success in the future.

**Key point**

**Future perspectives: redefining learning**

The future of innovative teaching methods is closely tied to the evolving landscape of education, driven by technological advancements, societal changes, and a growing understanding of how people learn best. As the educational environment continues to evolve, we can anticipate several key trends and perspectives that will shape the future of teaching and learning. There will be increased integration of artificial intelligence, where it will play an increasingly central role in personalizing education.

Adaptive learning platforms powered by AI will analyze students' strengths and weaknesses in real-time, adjusting content and pacing to optimize learning. The future of teaching will likely see a greater emphasis on personalized learning, where each student's educational journey is tailored to their needs, interests, and abilities. This shift will allow students to take more control over their learning, fostering independence and a sense of ownership.

The future of innovative teaching methods holds tremendous promise for transforming the educational experience, making learning more personalized, engaging, and accessible. By embracing technology, fostering collaboration, and promoting lifelong learning, educators will empower students to succeed in an increasingly complex, fast-changing world. These shifts will not only redefine how students learn but also how society views the role of education in preparing individuals for the future. As such, the evolution of teaching methods will continue to play a pivotal role in shaping the future of work, technology, and global citizenship.

# References

Ahmed, M., Sherwani, Y., Al-Jibury, O., Najim, M., Rabee, R., Ashraf, M., 2015. Gamification in medical education. Medical Education Online 20, 29536. Available at https://doi.org/10.3402/meo.v20.29536.

Al-Elq, Abdulmohsen H., 2010. Simulation-based medical teaching and learning. Journal of Family and Community Medicine 17 (1), 35. Available at https://doi.org/10.4103/1319-1683.68787.

Burgess, A., Bleasel, J., Haq, I., Roberts, C., Garsia, R., Robertson, T., et al., 2017. Team-based learning (TBL) in the medical curriculum: better than PBL? BMC Medical Education 17 (1). Available at https://doi.org/10.1186/s12909-017-1068-z.

Dolmans, D., Schmidt, H., 1996. The advantages of problem-based curricula. Postgraduate Medical Journal 72 (851), 535–538. Available at https://doi.org/10.1136/pgmj.72.851.535.

Elendu, C., Amaechi, D.C., Okatta, A.U., Amaechi, E.C., Elendu, T.C., Ezeh, C.P., et al., 2024. The impact of simulation-based training in medical education: a review. Medicine (United States) 103 (27), e38813. Available at https://doi.org/10.1097/MD.0000000000038813, https://journals.lww.com/md-journal/pages/default.aspx.

Elshama, S., 2019. How to Apply Problem-Based Learning in Medical Education? A Critical Review. Iberoamerican journal of medicine 2, 14–18. Available at https://doi.org/10.5281/zenodo.3585108.

Hurtubise, L., Hall, E., Sheridan, L., Han, H., 2015. The flipped classroom in medical education: engaging students to build competency. Journal of Medical Education and Curricular Development 2. Available at https://doi.org/10.4137/JMECD.S23895. PMID: 35187252.

Kalyani, D., Rajasekaran, K., 2018. Innovative teaching and learning. Journal of Applied and Advanced Research S23. Available at https://doi.org/10.21839/jaar.2018.v3is1.162.

Kaur, R., 2014. Case based learning as an innovative teaching tool. International Journal of Basic & Clinical Pharmacology 3 (2), 1. Available at https://doi.org/10.5455/2319-2003.ijbcp20140428.

Lampropoulos, G., del Bosque, A., Fernández-Arias, P., Vergara, D., 2025. Virtual reality in medical education, healthcare education, and nursing education: an overview. Multimodal Technologies and Interaction 9 (7), 75. Available at https://doi.org/10.3390/mti9070075.

Mansur, D.I., Kayastha, S.R., Makaju, R., Dongol, M., 2012. Problem based learning in medical education. Kathmandu University Medical Journal 10 (40), 78–82. Available at https://doi.org/10.3126/kumj.v10i4.11002, http://www.kumj.com.np/issue/40/78-82.pdf.

Ozdamli, F., Aşıksoy, G., 2016. Flipped classroom approach. World Journal on Educational Technology 8, 98. Available at https://doi.org/10.18844/wjet.v8i2.640.

Punja, D., Kalludi, S.N., Pai, K.M., Rao, R.K., Dhar, M., 2014. Team-based learning as a teaching strategy for first-year medical students. Australasian Medical Journal 7 (12), 490–499. Available at https://doi.org/10.4066/AMJ.2014.2244. PMID: 25646125.

Servant-Miklos, V.F.C., 2019. A revolution in its own right: how maastricht university reinvented problem-based learning. Health Professions Education 5 (4), 283–293. Available at https://doi.org/10.1016/j.hpe.2018.12.005.

Shah, S.S.H., Syed, A., Othman, R.Q.A., Chatha, C.W.A., Rasool, G., Agarwal, A., 2024. Problem based learning (PBL): an analysis of medical students' reflection. Gomal Journal of Medical Sciences 22 (3), 188–192. Available at https://doi.org/10.46903/gjms/22.03.1486, https://gjms.com.pk/index.php/journal/article/download/1486/pdf.

Soni, I., Mane, M., Singh, S.K., Upadhye, V.J., Deo, R.C., Saini, P., et al., 2024. Innovative teaching methods in healthcare education: a comparative analysis. Health Leadership and Quality of Life 3. Available at https://doi.org/10.56294/hl2024.394, https://hl.ageditor.ar/index.php/hl/article/view/394/966.

Srinivasan, M., Wilkes, M., Stevenson, F., Nguyen, T., Slavin, S., 2007. Comparing problem-based learning with case-based learning: effects of a major curricular shift at two institutions. Academic Medicine 82 (1), 74–82. Available at https://doi.org/10.1097/01.acm.0000249963.93776.aa.

Sun, J., Chen, Yu-J., Wu, Yi-Z., 2017. Application and evaluation of problem-based learning in undergraduate clinical education in 2014-2018. MedEdPublish 7, 228. Available at https://doi.org/10.15694/mep.2018.0000228.1.

Tsekhmister, Y., 2023. Effectiveness of case-based learning in medical and pharmacy education: a meta-analysis. Electronic Journal of General Medicine 20 (5), em515. Available at https://doi.org/10.29333/ejgm/13315.

Chapter 16

# Microteaching

*"A teacher who loves learning earns the right and the ability to help others learn."*

—Ruth Beechick

## 16.1 Introduction

One cutting-edge method for teacher preparation is microteaching. Microteaching is an innovative teacher training tool. It was developed and shown for the first time in 1960 at Stanford University in the United States of America. Dwight W. Allen created the term "Microteaching" in 1963 at the same university.

Teachers in particular can use microteaching as a training method to hone their abilities. The fact that microteaching is an essential component of professional development for educators, enabling them to hone their teaching abilities and consistently enhance instructional practices, demonstrates the significance of microteaching.(Cavanaugh 2022) In medical education, microteaching is a condensed teaching approach in which teachers prepare, deliver, and get feedback on a brief, targeted lecture for a small number of students. This session is frequently captured on video. In a low-risk setting, this regulated, repeating procedure aids in the improvement of particular teaching abilities like lesson design, presentation, and feedback delivery (Ghatge 2023). Microteaching provides healthcare professionals with constructive criticism and boosts their confidence and teaching skills by breaking down complex material into smaller, more manageable chunks (Allen, D. W. 1968).

## 16.2 Characteristics of microteaching

Microteaching is distinguished by its brief duration which usually lasts for 5–10 minutes, with a small class size preferably of 5–10 students, and focusing on a single teaching technique at a time, like explaining or asking questions. It gives trainees instant feedback, enabling reteaching and skill improvement in a controlled, reduced setting. The important characteristics of microteaching include the following:

### 16.2.1 Scaled-down teaching

One of the main features of microteaching is scaled-down teaching, which makes real-world instruction easier by lowering the complexity of the class size, session length, and number of skills practiced at once. It is teaching a single idea to a small number of students for a brief amount of time, which facilitates the teacher's ability to concentrate, get feedback, and improve (Rachel 2016).

### 16.2.2 Focused skill development

A key feature of microteaching is focused skill development, which divides difficult instruction into smaller, more manageable chunks so that teachers can practice and become proficient in certain skills like questioning, explaining, or using teaching aids in a brief, controlled session. Repetition and instant feedback allow for targeted development, which is then followed by reteaching until the skill is learned. In order to pinpoint areas for growth, the trainee receives prompt, helpful feedback from peers and supervisors following each session.

**The Principles and Practice of Medical Education. DOI:** https://doi.org/10.1016/B978-0-443-44197-4.00010-2

### 16.2.3 Repetition and refinement

Repetition and refinement are core principles of microteaching's cyclical process which allows teacher trainees to practice and master specific instructional skills in a controlled, low-pressure environment. The structured cycle of repeated practice and targeted feedback is crucial for skill development and building teacher confidence.

### 16.2.4 Controlled environment

A controlled environment is a crucial component of microteaching. The complexities of a real classroom setting are simplified so that a teacher can focus on improving certain skills in a friendly, low-pressure situation. It is also distinguished by the short lesson length, with teaching sessions typically lasting only 5–10 minutes. This helps to isolate and concentrate on particular abilities, like employing reinforcement or questioning strategies. Additionally, rather than handling every facet of a complicated lesson at once, it concentrates on a single ability, with the main goal being to practice and enhance one particular teaching technique at a time.

### 16.2.5 Objective assessment

A key component of microteaching is objective assessment, which is carried out using precise standards and instant feedback to assist teacher candidates in mastering particular abilities. Instead of focusing on overall student outcomes, the assessment procedure examines observable instructional actions in a controlled environment. In this case, the teaching abilities are assessed based on predetermined standards, and instruments such as video recordings allow for self-evaluation.

### 16.2.6 Individualized training

The learner receives one-on-one attention to concentrate on specific teaching skills with this highly tailored training technique. For aspiring and experienced teachers, microteaching is a well-known, highly tailored training approach. Additionally, by offering a low-stakes, secure, and controlled environment, microteaching enables trainees to gradually gain confidence in their particular talents before using the combined skills in a real, complex classroom context. The method encourages self-directed learning and evaluation, which also helps the teacher with self-reflection, since students may watch their own performance, for example, through video recording, and actively engage in evaluating their strengths and weaknesses (Fig. 16.1).

## 16.3 Purpose of microteaching

### 16.3.1 Skill development

The main goal of microteaching is skill development since it enables teachers to concentrate on and practice particular, feasible teaching techniques like questioning, explaining, or classroom management in a safe setting. Teachers can learn specific skills through practice, get quick feedback, and improve their methods before implementing them in a real classroom (Ambili, R. 2013). By breaking the intricate process of teaching into smaller, more manageable chunks, microteaching allows teachers to focus on one skill at a time, such as utilizing probing questions or providing clear instructions. This increases focused practice and aids in skill development. Additionally, it promotes a healthy learning

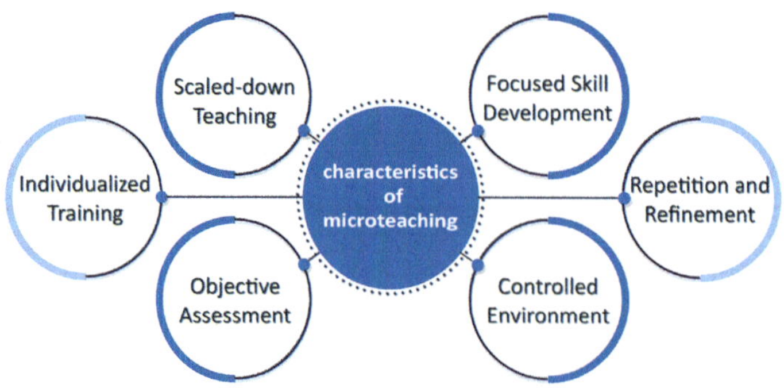


FIGURE 16.1 **Characteristics of microteaching.** This figure shows the characteristics of microteaching.

environment by giving teachers a low-risk, safe atmosphere to try out new methods and approaches without having to worry about overseeing a class full of students.

### 16.3.2 Safe practice environment

One of the main features of microteaching is to give teachers a secure, regulated practice setting where they can hone their teaching abilities. They can try out new techniques, make errors, and get helpful criticism in this low-stakes environment without the stress or possible detrimental effects on students in an actual classroom. One of the main advantages of a safe practice environment is that it lowers anxiety and boosts confidence. This is because practicing in a small, encouraging group of peers or a few students helps both new and seasoned teachers develop their self-confidence and lowers the anxiety that comes with performing in a full classroom. Additionally, it permits experimentation since the controlled setting enables teachers to test novel teaching methods that they might be reluctant to use in a high-stakes situation, such as particular questioning styles or the use of visual aids and curriculum materials.

### 16.3.3 Constructive feedback

One of the main traits of microteaching is to provide constructive feedback, which gives a teacher-in-training a quick and organized way to pinpoint their strengths and shortcomings. This procedure is giving a brief lecture to mentors or peers who provide detailed, useful feedback on abilities like lesson design, delivery, and classroom management. The objective is to hone instructional strategies in a low-stakes setting before implementing them for a wider audience, resulting in ongoing development and introspection.

Constructive feedback is effective in microteaching because it provides prompt, targeted feedback following a practice session, enabling in-the-moment reflection and improvement. Furthermore, because the evaluation is based on observations of particular teaching abilities, such clarity, tempo, and student engagement, it is both objective and specific. It assists teachers in determining what was successful and what requires improvement.

### 16.3.4 Increased retention

Microteaching significantly increases retention by delivering content in small, focused segments that align with how the brain processes information and by incorporating spaced repetition and active learning techniques. One of the important components in this regard is the "teach, critique, reteach" cycle. This technique highlights educators' continuous development by focusing on preparing, practicing, receiving feedback, and replanning. The "teach, critique, reteach" approach offers a chance for real-world implementation and prompt, helpful feedback. Errors are eliminated and more robust, long-lasting teaching skills are developed through this cyclical process of practice and correction.

Additionally, microteaching reduces cognitive overload and enables mastery of individual components, making them easier to remember and apply later, by focusing on one particular teaching skill at a time, such as questioning strategies, the use of visual aids, reinforcement, or presenting clarity.

### 16.3.5 Collaboration and reflection

It fosters cooperation between instructors, encouraging the exchange of information and skills to improve group practice. Microteaching's main goals are collaboration and reflection, which enable teachers to hone their abilities through class planning, instruction, and analysis with peers. Through evaluating one's own instruction and learning from others, this method promotes self-awareness, creates professional connections, and offers chances for constructive criticism.

Teachers take turns teaching and giving feedback to one another during microteaching sessions, which improves peer feedback. Peers can identify each other's strengths and shortcomings by working together to provide insightful criticism on instructional methods. In the end, this results in community building, where the cooperative character of the activity fosters professional connections and a sense of community among educators.

## 16.4 Types of microteaching skills

Microteaching skills include a variety of teaching techniques practiced in a controlled, scaled-down environment to develop a teacher's competence. Microteaching focuses on specific aspects of teaching, such as explaining, questioning, reinforcing student responses, demonstrating concepts, managing the classroom, and effectively introducing and closing lessons. These skills are practiced in a cycle of planning, teaching, getting feedback, and reteaching to improve teaching performance. These skills can be divided into two categories:

## 16.4.1 Foundational teaching skills

### *16.4.1.1 Explaining*

Explaining is a microteaching ability that entails giving a small group a clear explanation of a concept in a brief, targeted lesson in short amount of time. To make sure students grasp the subject, it needs certain elements including a concise introduction, well-organized ideas, examples, and a summary. In order to enhance their capacity to effectively explain topics, teachers exercise this skill in a controlled environment and get rapid feedback.

### *16.4.1.2 Questioning*

Questioning is a basic microteaching strategy that supports critical thinking development, comprehension evaluation, and student involvement. In order to ask effective questions, it's important to prepare relevant questions, allow enough time for pauses, address each student by name, and provide meaningful comments. Other crucial elements include using a variety of question kinds, such as open-ended, probing, and higher-order, as well as avoiding posing unclear or contradictory questions.

### *16.4.1.3 Reinforcement*

Reinforcement is a microteaching technique that uses both positive and negative cues to promote engagement and learning by strengthening the connection between a student's reaction and the teacher's feedback. In order to effectively discourage erroneous replies, negative reinforcement should be used sparingly and carefully in conjunction with verbal praise or with nonverbal cues like smiling or nodding, and other types of positive reinforcement for correct answers. Motivating learners, boosting their engagement, and fostering confidence in a nurturing learning atmosphere are the main objectives of this technique.

### *16.4.1.4 Classroom management*

In microteaching, classroom management is a crucial skill used to create a focused, low-pressure learning environment. Teachers can try particular management strategies with a small number of students during a microteaching session, such as setting rules, rerouting behavior, and utilizing nonverbal clues, and they can get quick feedback to improve their strategy. This helps teachers become more adaptable and confident in managing real classrooms by highlighting skills like classroom organization, managing student behavior, and motivating students (Bajaj 2014).

## 16.4.2 Related and practical skills

### *16.4.2.1 Lesson planning*

A key component of microteaching is lesson planning, which ensures clarity, organization, and a greater likelihood of success by offering a focused, planned roadmap for a brief, practice-based lesson. In a condensed amount of time, which is usually 5–10 minutes, this entails identifying precise learning objectives, choosing succinct content, organizing particular teaching activities, and creating a straightforward assessment technique. A well-crafted microteaching lesson plan facilitates efficient practice, revision, and feedback while assisting the instructor in concentrating on particular abilities, such as questioning or explanation.

### *16.4.2.2 Presentation*

Presenting a brief, targeted lesson to a small group in order to hone particular abilities like oral presentation, organization, and the use of teaching aids is a fundamental microteaching ability for educators. With the ultimate goal of enhancing overall teaching competency, the process necessitates lesson design, instruction, feedback, and reteaching to improve performance in areas including clarity, engagement, and visual aids.

### *16.4.2.3 Feedback*

A key component of microteaching is feedback, which consists of organized and helpful criticism from peers, supervisors, and video recordings to assist teachers in honing certain teaching techniques. Teachers can discover strengths, rectify shortcomings, and practice a skill in isolation before applying it to entire lessons due to this feedback loop, which includes self-evaluation and can involve students. In order to enhance performance, the instructor replans the lesson, gets feedback, and retrains, thereby signifying the cyclic nature of the process.

#### 16.4.2.4 *Introduction and summary*

In order to grab students' interest, establish rapport, and inspire them for the class, the introduction is a microteaching technique that links new information to what they already know. Important elements include using teaching aids, clearly defining the lesson's objectives and topic, and assessing existing knowledge with pertinent questions. A seamless transfer from prior knowledge to the new material is ensured by effective introductions, which also avoid confusion and generate curiosity amongst the students.

#### 16.4.2.5 *Using teaching aids*

In a controlled setting, trainee instructors can make concepts come to life through the use of charts, audio, and images. Before being used in the classroom, this improves skill development, strengthens learning, and offers concrete feedback. In microteaching, the use of instructional aids is crucial for honing and showcasing particular abilities like engagement and presentation. Focusing on a single skill, like employing aids, will help the teachers to practice a certain teaching technique, get feedback, and enhance their ability to correctly prepare a class and explain concepts in a controlled context (Thangaraju P 2023).

#### 16.4.2.6 *Time management*

For teachers to properly organize, present, and reflect on a brief session by organizing activities, maintaining focus, and allotting time for each section, time management is an essential microteaching ability. To make sure the class goes well and achieves its learning objectives, it entails defining clear goals, prioritizing work, employing timers, and managing unforeseen disturbances.

#### 16.4.2.7 *Skill variety*

The intentional alteration of a teacher's actions and behaviors to capture and maintain students' attention and interest during a class is known as skill diversity or stimulus variation in microteaching. This ability is crucial for boosting student interest and avoiding boredom. To grab and hold students' attention, the instructor could, for example, move purposefully throughout the classroom. Additionally, they can accentuate ideas or arguments by employing suitable and natural body language, hand gestures, and facial expressions.

## 16.5 Core components of microteaching

### 16.5.1 Focus on a single skill

Teachers focus on learning a single skill, such as applying questioning strategies, giving concise explanations, or efficiently running the classroom, rather than teaching a whole subject. The goal of microteaching is to isolate and improve certain teaching competencies. Instead of attempting to combine several difficult aspects at once, the focus here is on a single skill. This makes it possible to focus on that particular ability through practice, observation, and feedback. The justification for concentrating on a single skill is that it makes complexity easier to understand. It is challenging to master all of the talents at once because teaching is a complex process that calls for the teacher to concentrate on numerous skills. These abilities should be divided into smaller, more manageable parts so that a teacher can become proficient in each one before integrating them. Additionally, when observers and mentors are evaluating a single behavior or skill, they can offer more accurate and helpful criticism.

### 16.5.2 Controlled practice

A key feature of microteaching, which is essentially a system of training that simplifies the real classroom setting to assist teachers in mastering particular skills in a safe, encouraging setting, is controlled practice. Controlled practice in microteaching is a core element where trainee teachers focus on a single, specific teaching skill like questioning or explaining in a scaled-down, safe environment involving a small group within a short time to build proficiency, get immediate feedback, and refine performance before applying it in real classrooms, making it a systematic, cyclical process for skill mastery (Cavanaugh 2022). This aspect of microteaching is essential because it enables trainee teachers to focus on one specific teaching behavior or skill at a time, such asking questions or giving explanations, rather than managing every facet of a full class at once.

### 16.5.3 Immediate feedback

After the lesson, peers and supervisors provide the teacher quick, helpful feedback that helps them see their strengths and shortcomings. Immediate feedback is widely acknowledged as a fundamental and essential part of the microteaching cycle for improving a teacher's overall instructional competency (Bell A 2008). Because of its promptness, trainee instructors can successfully change their teaching methods, praise their strengths, and quickly detect and remedy their faults. Immediate feedback plays a crucial part in changing behavior since it helps to clear up misunderstandings and correct mistakes as soon as they are committed, which stops bad habits from being reinforced. It also facilitates the development and enhancement of skills by focusing on one specific ability at a time, such as by using instructional aids or asking questions. Compared to delayed input, immediate feedback offers focused advice on that specific skill, resulting in quicker mastery.

### 16.5.4 Replanning and reteaching

Using replanning and reteaching, the instructor modifies their lesson plan and teaches the skill again until the student masters it. Replanning and reteaching are essential elements of the microteaching cycle that serve as the main means of improving and refining skills in a safe, encouraging setting. Trainee instructors can incorporate criticism and practice a particular ability until they reach a level of proficiency thanks to this cyclical process. Replanning and reteaching play a crucial role since they promote progress. The learner can assess their shortcomings and adjust their lesson plan and methodology based on the helpful criticism they received following the first teaching session. Additionally, it permits a trial-and-error method, as the secure, smaller setting makes it possible to try out novel tactics and learn from errors without the stress of a huge classroom. Additionally, it boosts confidence since the teacher's self-confidence is bolstered by effectively implementing the comments in the reteach session prior to entering a real classroom.

### 16.5.5 Recording and self-evaluation

Teachers can evaluate their own performance and learn more about their teaching approach thanks to the frequent videotaping of sessions. The fundamental, crucial elements of microteaching are recording and self-evaluation, which are effective instruments for professional growth, self-awareness, and skill improvement for both inexperienced and seasoned teachers (Ajayi-Dopemu 1986). One important microteaching tool that helps provide an objective record of a teaching session for later examination is the use of audio and video recording equipment. These recordings offer an objective perspective of a teacher's performance, enabling them to see themselves from the perspective of their pupils and take note of their own application of techniques, body language, and material usage. This aids in reflection as well because the recording is an essential tool for the teacher to methodically evaluate their performance and pinpoint particular strengths and shortcomings that could otherwise go unnoticed (Fig. 16.2).

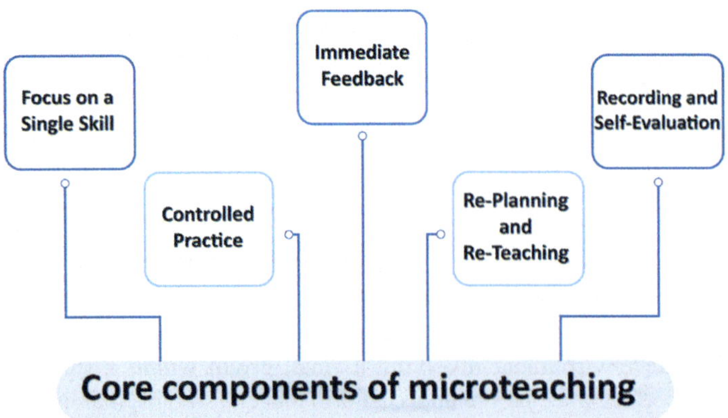


FIGURE 16.2 **Core components of microteaching.** This figure shows the core components of microteaching.

## 16.6 Microteaching cycle

The microteaching cycle is a systematic approach that helps teachers improve their teaching skills through regular practice and feedback. The six crucial stages of the cycle which include plan, teach, feedback, replan, reteach, and refeedback all support the growth of the educator's overall competency.

### 16.6.1 Plan

A teacher candidate creates a lesson plan to concentrate on a particular teaching ability during the planning phase of the microteaching cycle. To effectively teach the concept and practice the skill, this entails choosing a topic, determining learning objectives, and organizing exercises and questions. During the planning stage, the instructor chooses a particular skill or goal to concentrate on and creates a thorough lesson plan. This plan ensures that the lesson is organized and goal-oriented by outlining the teaching strategies, resources, and techniques.

### 16.6.2 Teach

The "Teach" stage in the microteaching cycle is the core performance phase, where the teacher trainee delivers a short, focused lesson which lasts for around 5–10 minutes to a small group, applying a specific skill like questioning or explaining. During this step, the instructor uses the tactics that they have studied and prepared for during the planning phase in a genuine, but in a scaled-down, classroom setting. The instructor presents the lesson to a small group of students or peers during the teaching phase. This step is also characterized by observation, wherein the teacher educator and peers observe and take notes on the performance, gathering data for feedback (Remesh A. 2013).

### 16.6.3 Feedback

The feedback phase of the microteaching cycle is getting helpful criticism from instructors or peers during a quick training session. This information, which can be provided through direct discussion or video review, enables the learner to pinpoint their areas of strength and growth in order to improve their teaching abilities. Setting the framework for the learner to replan, reteach, and seek further feedback is a crucial step for improvement. The feedback's objective is to give details on the trainee's performance in a secure setting. The feedback enables the learner to assess their performance and determine whether their attempt to mimic particular instructional practices was successful. Furthermore, the feedback will be constructive, highlighting both the positive aspects and the areas that want development. This makes it possible to comprehend the issue more thoroughly and to evaluate oneself. Feedback can be given verbally, in writing, or through a video recording of the lesson.

### 16.6.4 Replan

The instructor reviews the course plan after getting feedback and makes any required changes in light of the findings. The teacher can improve their strategy during the replanning process by addressing any areas that were found to need improvement during the observation. The initial lesson plan is revised during the microteaching cycle's replanning phase in response to feedback gathered following the first teaching session. In this stage, the instructor evaluates the input to pinpoint areas that need work and adjusts the lesson's goals, tactics, and resources to fix any issues before the next teaching session. Through repeated practice and feedback, this iterative method enables the improvement of instructional techniques. In order to determine what succeeded and what needs improvement, the instructor analyzes feedback from peers, supervisors, or recorded sessions during the replanning phase. Additionally, it will assist in recognizing and pinpointing specific shortcomings in the class, such as a lack of student participation, an unclear explanation, or poor questioning.

### 16.6.5 Reteach

During the reteach phase, the revised lesson is presented to a different group or the same audience using the modifications made during the replanning phase. This stage assesses how well the modifications worked and aids the instructor in strengthening their abilities. In the microteaching cycle's reteaching phase, the lesson is taught once more after being replanned in light of the initial teaching session's feedback. In order to practice and enhance the skills found during the first teaching and feedback stages, the instructor implements the recommended modifications when instructing the same or a different group. Applying the modifications made during the replanning phase and evaluating their efficacy are the goals of the reteaching phase. In this case, either the original group or a new set of peers are given the updated lesson by the teacher.

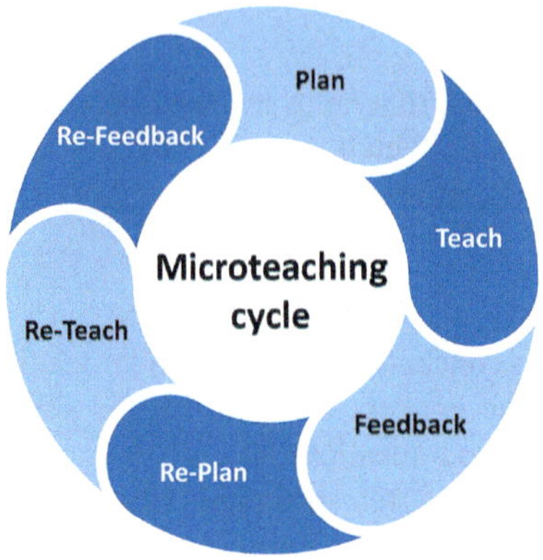


FIGURE 16.3 **Microteaching cycle.** This figure shows the microteaching cycle.

### 16.6.6 Refeedback

In the last stage of refeedback, the teacher gets input on the updated lesson. This input is essential for comprehending the effects of the changes and offers suggestions for future development. To guarantee ongoing professional progress, the cycle may be repeated as necessary. The supervisor gives feedback on the "reteach" lesson during the refeedback phase of the microteaching cycle, which enables the teacher candidate to further develop their abilities. Behavioral transformation requires this additional feedback loop, which is repeated until the trainee has adequately mastered the teaching skill which occurs usually after multiple cycles of planning, teaching, feedback, replanning, and reteaching. It improves the previous session since the refeedback is given after the trainee has had a chance to reteach the lesson in light of the initial feedback. It also provides constructive criticism in the form of insightful comments and explanations to improve the teacher candidate's skills (Fig. 16.3).

## 16.7 Microteaching techniques

Microteaching techniques train teachers by focusing on single skills in short, recorded lessons with small groups, using video for self-review and feedback to refine abilities like questioning, explaining, and reinforcing, all within a low-stakes cycle of planning, teaching, feedback, and reteaching until mastery. Key techniques involve specific skills like questioning, closure, reinforcement, specific methods like problem-based, inquiry and structured processes like the "Five R's": recording, reviewing, responding, refining, and redoing. The common microteaching techniques in medical education are as follows:

### 16.7.1 Questioning technique

In microteaching, effective questioning strategies are used to encourage critical thinking, gauge comprehension, and engage students. These strategies include the ability to formulate clear questions, use a variety of question types such as factual, open-ended, probing, ensuring good distribution, using wait time, and probing deeper with prompts like seeking further information or redirecting, all within a brief teaching segment to practice and hone these crucial teaching skills. Giving students time to consider after asking a question is essential for promoting more deliberate responses and critical thinking, thus it's a skill to master.

### 16.7.2 Clinical demonstration

The clinical demonstration technique in microteaching involves a teacher showing a specific clinical skill or a procedure in a short session, allowing for focused practice, observation, and immediate, specific feedback to develop psychomotor and communication skills in a controlled setting, bridging theory to practice efficiently. The key steps in this regard include planning, teacher demonstration, student observation, return demonstration, and analysis or feedback, often recorded for review, preparing educators to teach complex clinical tasks effectively. This technique ensures that the student has mastered the skills of presentation and explanation along with the procedural skill.

### 16.7.3 Case-based teaching

Case-based teaching in microteaching involves using real-life scenarios to guide short, focused teaching practice sessions, helping preservice teachers apply theory to complex situations, develop critical thinking, and improve skills like problem-solving, decision-making, and communication within a simulated, reflective environment, often enhanced by technology for recording and peer feedback. It moves beyond traditional lecturing by centering learning on authentic problems, allowing teachers to practice and refine their methods in a safe space before real classrooms.

### 16.7.4 Bedside teaching simulation

Simulated teaching is a teacher training technique where aspiring teachers practice in a controlled, artificial environment like role-playing or virtual scenarios to develop and refine skills like classroom management, lesson delivery, and communication, providing a low-risk space for mistakes and feedback before entering real classrooms. It bridges theory and practice by letting trainees experience realistic situations, build confidence, and receive immediate constructive criticism from peers and instructors. Bedside teaching can be effectively integrated with simulation techniques and structured through a microteaching framework to provide a safe, controlled, and effective learning environment for healthcare students. This combined approach helps develop clinical skills, communication, and professionalism without compromising patient safety.

### 16.7.5 Feedback-giving technique

The feedback and reflection phase is crucial because it acts as an eye opener for the trainee. Peers or mentors provide instant feedback that is constructive and supportive, including topics like engagement strategies, classroom management, communication, and subject delivery. Additionally, it provides an opportunity for the instructor to evaluate their own performance and pinpoint their areas of strength and growth. Even the development of self-awareness and skill improvement are aided by this process.

### 16.7.6 Use of teaching aids

Using teaching aids in microteaching involves integrating tools like visuals, models, handouts, or multimedia to make short lessons more engaging, clarify concepts, and practice specific skills like presentation or explanation, focusing on how aids enhance engagement, understanding, and skill mastery in a controlled setting. Teachers prepare and use these aids to support a single skill within a brief lesson, then receive feedback to refine their use before reteaching, making abstract ideas concrete and improving teaching effectiveness.

### 16.7.7 Microteaching methods

There are several different methods of microteaching, each with its own strengths and weaknesses.

#### *16.7.7.1 Modeling method*

The modeling method in microteaching is a technique where a trainer or expert teacher demonstrates a specific teaching skill or technique, which the trainees then observe, discuss, and practice themselves. This method is an integral part of the initial knowledge acquisition phase of the microteaching cycle. The proper use of the skill is observed by trainees. After the expert demonstrates the talent, the trainees prepare a brief microlesson for a small group of peers or students that lasts 5–10 minutes and focuses on the observed skill. Later, under the supervision of peers and a supervisor, the trainee instructs a small microclass, usually 3–10 participants in the microlesson. A video recording of the session is frequently made. Peers and the supervisor provide constructive comments and recommendations to the trainee right after the session. Later, the trainee modifies their lesson plan to include the recommendations for enhancement based on the feedback. This kind of microteaching works well for instructing trainees how to complete a certain task. It may, however, be less successful in assisting them in comprehending the rationale behind a particular approach or strategy.

#### *16.7.7.2 Lab method*

In the lab approach of microteaching, aspiring or in-service faculty members can practice, refine, and receive feedback on specific teaching techniques in a controlled, repeated laboratory setting. The complexities of typical classroom learning are broken down into manageable, focused sessions using this strategy. The methodology is based on a systematic, iterative cycle of practice and feedback to boost teaching ability and confidence.

Similar to the model technique mentioned above, each session usually consists of a limited number of participants, either peers or a small group of students, and a brief lesson. Instead of a full lecture, focus on skills like clinical communication, explaining complex conditions to patients—for example, complications of diabetes mellitus—or demonstrating a psychomotor skill. This technique also makes use of using audio or video recording for immediate self-review and peer or mentor feedback.

#### 16.7.7.3 *Clinic method*

The clinical method of microteaching in medical education involves "scaling down" teaching into short, focused sessions often using the one-minute preceptor model to develop specific clinical teaching skills. It focuses on discrete micro skills for example, probing for within brief, real-world patient encounters, using feedback often video-recorded to refine techniques in a low-stakes environment, bridging theory with practical patient care.

The development of useful teaching behaviors for hectic clinical settings is one of the key advantages of employing these strategies. Additionally, it gives educators like faculty and residents more self-assurance. Moreover, it also increases the effectiveness and efficiency of teaching on the run, which emphasizes the significance of preparing healthcare professionals to become more capable clinical instructors, supervisors, and assessors in hectic clinical settings by concentrating on the practical skills for instructing peers, junior staff, or students while providing patient care. Furthermore, it helps to close the gap between theoretical understanding and patient care.

#### 16.7.7.4 *Minilecture method*

In microteaching, the minilecture technique entails a teacher giving a brief, targeted lesson to a small group on a particular skill, concept, or strategy in a controlled setting for practice and quick, helpful feedback. This allows for skill improvement and confidence building prior to actual classroom application. It is a fundamental component of the microteaching cycle and focuses on acquiring a single teaching skill in a low-stakes environment, such as questioning, explaining, or introducing a topic. One of the most important aspects of minilectures is their conciseness and focus, in which the lessons are brief and concentrate on a single skill or topic, such as presenting a lesson or elucidating a difficult concept. It is also characterized by a controlled environment and a small audience.

## 16.8 Conduction of microteaching

The conduction of a microteaching session follows a structured, cyclical process designed to help teachers practice and refine specific teaching skills in a controlled, low-pressure environment.

### 16.8.1 Phase 1: knowledge acquisition (plan)

Before any practice begins, the teacher trainee must understand the theory and components of the specific teaching skill they will focus on.

#### 16.8.1.1 *Study the skill*

The trainee learns about a specific skill (e.g., "skill of questioning" or "skill of stimulus variation") through lectures, discussions, and observing demonstrations by experts or watching video recordings.

#### 16.8.1.2 *Planning the session*

A successful minilesson for microteaching in medical education should concentrate on a single, targeted teaching skill and a succinct subject while adhering to an organized schedule of introduction, content delivery, engagement activity, and closure in a brief amount of time (usually 5–10 minutes). A precise, quantifiable learning goal is determined during this stage. It is beneficial to concentrate on one essential teaching skill or a single, succinct therapeutic notion rather than presenting a wide range of topics. Knowing your audience beforehand and adjusting your examples and content to your peers' or learners' current level of expertise is a good technique. It is crucial to schedule the session's structure and scheduling by allotting sufficient time for an introduction, the primary activity or topic, and a conclusion or assessment. A typical format is 3–5 minutes for the introduction, 10–15 minutes for the substance, and 2–3 minutes for the end. Learning the procedure is improved by using the proper case studies or visual aids. Practicing your presentation will help you become more confident and efficiently manage your time.

### 16.8.2 Phase 2: skill acquisition

This is the practical, hands-on phase where the teacher trainee actively practices the target skill in a controlled, "scaled-down" environment. This phase is driven by a continuous cycle of planning, teaching, and receiving feedback.

#### *16.8.2.1 Choosing a skill to be learnt or practiced*

The first stage is to decide which particular teaching technique you wish to practice, such as engaging students, asking questions, or providing clear explanations of subjects. It is best to pick one particular teaching ability for microteaching in medical education so that it can be practiced and improved through concentrated, iterative sessions. The ability should be something the student wishes to develop and pertinent to the setting of medical education.

#### *16.8.2.2 Presenting the minilesson*

In a controlled setting, a small group of peers who are your learners are given the minilesson. A successful minilesson in medical education microteaching requires careful preparation, captivating delivery of a single concept or skill, and active integration of helpful criticism. A small group of peers or students are given a brief (5–15 minutes) session in a secure, replicated setting. It includes the following phases. To assist your class, it's crucial to collect or prepare any required teaching resources, such as handouts, visual aids, or digital presentations. A 2–3-minute introduction to the subject should be included in the minilesson, along with a clear statement of the learning objectives to set expectations. Ask a question or present a real-world clinical scenario to get the students interested right away. The primary material, which lasts 5–7 minutes and contains the essential details of the selected teaching technique, comes next. To check for comprehension and promote involvement, a brief, interactive exercise or quick evaluation should be scheduled later. Examples of such activities include a rapid-fire question round, a one-question quiz, or asking participants to describe a significant point. The summary, which includes the key takeaways and connects the learning to prior knowledge or potential future applications in clinical practice, comes last.

#### *16.8.2.3 Record the session*

To evaluate your own performance at a later time, record the lesson on video. In order to record a successful minilesson for microteaching in medical education, it is important to use appropriate audiovisual aids, incorporate self-reflection and peer feedback, plan clearly, and offer a single teaching skill.

#### *16.8.2.4 Review the recording*

The teachers should themselves watch the class video to see how well they used the selected skill, used their body language, and their teaching style. There are many benefits to reviewing recordings in microteaching, chief among them being that it gives participants a tangible instrument for objective self-reflection and precise, useful feedback from peers and supervisors in medical education. A medical educator can objectively assess their own teaching performance by watching a recording, spotting nonverbal clues, communication patterns, and areas that could be overlooked in a live session. It also makes it easier for mentors and peers to provide extremely detailed, fact-based comments. Feedback might be focused on specific instances in the tape rather than broad remarks.

#### *16.8.2.5 Get constructive feedback*

Talk about your lesson with your classmates or a mentor, who can offer constructive criticism that highlights both your areas of strength and growth. Receiving constructive criticism as part of microteaching in medical education has many benefits, with the main ones being skill development, self-awareness, and readiness for practical clinical practice. By identifying particular areas for improvement in their teaching methods, communication style, and content delivery, medical educators can become more successful educators. By getting feedback, participants learn about their strengths and shortcomings and frequently become aware of facets of their teaching approach that they were not previously aware of, encouraging a reflective practice.

#### *16.8.2.6 Reflect on performance*

The instructor can pinpoint particular methods for enhancing instruction by using the comments and their own observations. A key component of successful microteaching in medical education is performance reflection, which offers a number of benefits that improve learning and skill development. Medical educators can immediately pinpoint areas for improvement in communication, instructional strategies, and feedback procedures by critically analyzing how they deliver their lessons. Deeper awareness of one's own teaching style, prejudices, and strong points is fostered by this approach, which results in more genuine and productive interactions between educators and students.

#### *16.8.2.7 Replan, reteach, and refeedback*

##### 16.8.2.7.1 Replan

Rearrange the lesson to target the areas that need improvement based on the comments. In medical education, refining the lesson plan as part of the microteaching cycle has several benefits, such as improved teaching abilities, better content organization and clarity, and higher teacher confidence and self-efficacy. One of the main benefits of topic refinement is that it enables methodical skill development because microteaching enables teachers to concentrate on learning certain, discrete teaching techniques one at a time. By fine-tuning the lesson plan, the session may be tailored to these particular skills, guaranteeing a concentrated and productive practice. The planning and refinement process also places a strong emphasis on content relevancy, logical order, and conciseness. This aids medical educators in simplifying difficult medical ideas into manageable chunks, improving students' understanding and memory of the material.

##### 16.8.2.7.2 Reteach

This method involves teaching the same subject or an updated version to the same group in order to put the modifications into practice and continue honing the ability. A key component of the feedback loop is the reteaching aspect of microteaching, which enables medical educators to quickly implement constructive comments, correct mistakes, and hone their teaching techniques in a safe, low-stakes setting.

One of the main benefits of the reteaching phase is that it makes it easier for trainees to apply their skills right away. By incorporating the criticism they got during the critique session into a revised lesson plan and reteaching, they can consolidate new behaviors and remove mistakes. Additionally, it results in behavior modification. The cycle of "teach, critique, reteach, and refeedback" is a useful method for changing particular teaching behaviors and skills in a way that may not be achievable in a single, conventional teaching session.

##### 16.8.2.7.3 Refeedback

In the microteaching cycle, the "refeedback" phase is the critical second round of feedback that takes place after the trainee has replanned and retaught a lesson based on the initial feedback. The goal of this phase is to further refine the specific teaching skill until mastery is attained, with an emphasis on behavior modification and confidence building for actual classroom use.

### 16.8.3 Continuous cycle

The continuous cycle in microteaching signifies the repetition of the process of replanning, reteaching, and getting feedback until mastery of the particular teaching ability. The key advantage of repeating the process in a microteaching cycle is skill mastery.

### 16.8.4 Phase 3: transfer phase

In the last and crucial phase of the microteaching cycle, teacher trainees applies the skills they have individually mastered such as questioning and explaining to real, full-scale classroom instruction, adapting and applying them to actual student needs. This is the culmination of practice in a low-stakes setting to real-world application. It bridges the gap between controlled microlessons and the real classroom environment.

## 16.9 Integration of microteaching as a part of the curriculum

Integrating microteaching into the curriculum involves making this practical, skill-focused training method a structured, formal part of a teacher education program, not just an add-on activity. This process aims to bridge the gap between educational theory and real-world teaching practice, helping prospective and in-service educators develop specific skills in a controlled, low-stakes environment. A structured, multistep plan is necessary for successful integration, which includes steps like defining clear objectives. It's crucial to specify the teaching skills or competencies that the microteaching sessions are meant to develop, such as effective use of technology, encouraging inquiry-based learning, and assessment skills. Additionally, the module or course must be designed to incorporate microteaching as a formal course component, such as a microteaching course with certain time and learning objectives that complement the program's overall objectives. Providing sufficient training facilities and resources, such as materials, observation tools, and possibly technology like video recording for self-evaluation and peer review, is equally crucial. Finally, it is crucial to organize the sessions.

## 16.10 Technology-enhanced microteaching

The use of digital tools, platforms, and technologies in microteaching to enhance instruction and learning is known as "technology-enhanced microteaching." This method seeks to produce more dynamic, participatory, and successful microteaching sessions by utilizing contemporary educational tools (S. Ledger 2020). The following is a discussion of some of the main characteristics of technology-enhanced microteaching:

### 16.10.1 Use of multimedia

By using multimedia, teachers can easily convey difficult medical concepts like anatomy, physiology, or clinical procedures using multimedia components including instructional films, animations, and podcasts. To improve comprehension and retention, these might be included in microteaching sessions. Additionally, more captivating and interactive presentations with images, diagrams, and even brief video clips are made possible by using programs like PowerPoint, Prezi, or Google Slides. These presentations are particularly helpful for teaching medical procedures or elucidating medical disorders.

### 16.10.2 Simulations and virtual learning environments

Simulations and virtual learning environments enhance microteaching by providing a controlled, repeatable space for preservice teachers to practice skills, receive feedback, and develop confidence without the pressure of a real classroom. These tools offer realistic scenarios, reduce complexity, and allow for practice that bridges the gap between theory and real-world application. The use of technology, such as virtual reality, further increases immersion and interactivity through avatars, enabling teachers to repeatedly practice and refine their performance.

### 16.10.3 Interactive tools for feedback

During the microteaching session, real-time tests and polling are made possible using platforms such as Kahoot, Quizlet, or Socrative. These resources can be utilized in a medical setting to assess student comprehension and retention by administering knowledge tests following each lesson. Additionally, real-time comments and conversations are made possible by digital platforms like Teams, Zoom, and Google Classroom. Peer reviews of microteaching sessions, where instructors or peers can offer comments via chat, voice, or direct annotations on shared papers or presentations, can benefit greatly from this.

### 16.10.4 Recordings and self-reflection

Self-reflection is a crucial element of microteaching. Teachers can assess their communication, presentation, and style of instruction by leveraging technology to record microteaching sessions by using applications like Zoom, OBS Studio, or even mobile applications. This is particularly helpful for evaluating the efficacy of teaching strategies and the lucidity of medical explanations in medical education. In order to support ongoing professional growth, instructors can examine the recording of the session, evaluate their performance, and pinpoint areas for improvement.

### 16.10.5 Collaborative learning tools

A range of collaborative learning tools can greatly improve microteaching, which is giving brief courses to small groups and getting feedback. Whether the process is fully virtual or conducted in person, these tools make communication, content exchange, and peer evaluation easier. The Collaborative Learning Annotation System (CLAS), a platform for annotating and providing structured feedback on educational activities, or WeVu, which enables the recording and uploading of videos to a shared space where peers and instructors can offer comments and annotations, are examples of tools that can be used for both video recording and feedback.

However, when people are unable to meet in person, they can use tools for communication and discussion like Zoom Video and Microsoft Teams, which are conferencing programs that allow for virtual meetings, conversations, and screen sharing.

### 16.10.6 Online learning platforms and learning management systems

Microteaching can be integrated with more comprehensive online courses thanks to learning management systems like Moodle, Canvas, and Blackboard. Teachers can use technology-enhanced microteaching tactics to improve individual microteaching sessions while uploading instructional content, assigning tasks, starting conversations, and monitoring student progress (Owidi 2025). Before taking part in synchronous sessions or in-class discussions, students can examine recorded microteaching sessions using self-paced courses offered by these platforms.

### 16.10.7 Data analytics and performance tracking

In microteaching, data analytics and performance monitoring entails employing methodical data collection and analysis to give trainee teachers unbiased, fact-based feedback that improves skill development and instructional decision-making. A crucial element of successful microteaching is the ability of instructors to offer customized feedback that targets particular areas for development through the analysis of performance data. Improved teaching abilities, better classroom management, more teacher confidence, and a more seamless transfer to real-world teaching contexts are all results of incorporating data analytics into microteaching. It offers teacher education programs a solid, data-driven framework to guarantee that aspiring educators are prepared to handle the challenges of the contemporary classroom.

## 16.11 Advantages of microteaching

By strengthening teaching abilities, boosting instructor confidence, and improving student learning through a methodical, feedback-driven process, microteaching offers substantial benefits in medical education. Microteaching has been shown to be advantageous in several ways:

### 16.11.1 Improved teaching practice

Microteaching improves teaching practice by providing a controlled environment for focusing on and refining specific skills, such as questioning or lesson planning, through a process of teaching, getting fast feedback, and reteaching. This method helps teachers become more confident, try out new strategies, and identify areas that need improvement through reflection and lesson analysis. Microteaching improves teaching practice by isolating specific abilities, like set induction, questioning, or classroom management, and allowing teachers to practice and refine them in a focused way. In a low-stakes setting, teachers can swiftly identify and correct errors and shortcomings by getting prompt, helpful feedback from peers, supervisors, or even video recordings (Dayanindhi, V.K 2018).

### 16.11.2 Encourages self-assessment

Microteaching encourages self-evaluation by providing a systematic, safe, and cyclical process that fosters critical self-reflection and the recognition of one's own strengths and weaknesses. The practice of filming and reviewing microteaching sessions allows teacher candidates to watch and assess their own performance objectively. This playback serves as a "mirror," providing them with an unhindered, clear view of their interactions with pupils, body language, and teaching methods that they might not otherwise be aware of. Furthermore, microteaching provides a structured reflection through a particular "feedback" phase in which participants are especially urged to think on how they presented the lesson, how effectively their selected strategies worked, and how the students reacted. This rigorous technique encourages a deeper, more critical self-awareness of their work.

### 16.11.3 Builds confidence

Microteaching builds confidence by providing a safe, stress-free environment for practicing and refining specific teaching methods through a structured cycle of peer review, reflection, and repetition. This process systematically reduces stress and increases confidence before a teacher meets a full classroom. By using quick lessons, often lasting 5–10 minutes, with a small group of peers or students in a safe learning setting, microteaching simplifies traditional teaching. In this secure, supportive setting, new teachers can experiment and make mistakes as learning opportunities, which lessens their fear of failing or suffering negative outcomes. It also allows for concentrated skill development because it focuses on learning one specific skill at a time rather than juggling all aspects of education at once. When teachers successfully acquire and apply these particular skills, they become more competent, which immediately leads to increased confidence.

### 16.11.4 Identifies strength areas

Teachers can find their talents and use them into their teaching strategies by using microteaching. Microteaching helps uncover a teacher's areas of strength by providing a controlled, low-stakes environment for practice, peer and expert observation, and focused feedback on specific abilities. This targeted strategy makes strengths easier to see for both the teacher and their observers. Through concentrated skill practice, where teachers practice one skill at a time, observers are able to pinpoint specific skills in which a teacher naturally excels. Furthermore, prompt and comprehensive feedback from peers and a supervisor reveals the teacher's effective tactics, promoting positive behaviors and confirming what works. Additionally, by seeing a recording of their own performance, educators can objectively assess which elements of their lesson went well and where their delivery was particularly effective.

### 16.11.5 Supports continuous professional growth

It guarantees that educators stay up to date on the most recent teaching techniques by promoting ongoing professional development. Microteaching allows teachers to hone specific skills, experiment with new methods in a low-stakes environment, and continuously refine their teaching style by creating a structured cycle of practice, feedback, and reflection. When teachers effectively practice and refine their skills in a controlled setting and get positive feedback, their self-confidence and self-efficacy significantly increase. This increased confidence leads to more confident and successful performance in actual classrooms. Microteaching, which creates a professional learning community where educators may share ideas, learn from each other's experiences, and create a supportive atmosphere, frequently includes peer observation and evaluation.

### 16.11.6 Facilitates experimentation with new strategies

Before implementing new teaching techniques in an actual classroom, educators can hone them in a low-risk environment. The main way that microteaching encourages experimentation with new tactics is by offering a secure, regulated "laboratory" where teachers can test new approaches without the dangers of a traditional classroom. This is followed by a planned cycle of instant feedback and iterative reteaching. As microteaching occurs in a virtual environment with a limited number of students or peers, it lessens the stress and anxiety that come with a full, real classroom. Teachers are encouraged to venture outside of their comfort zones and try out novel or cutting-edge teaching techniques and resources because of this psychological safety.

### 16.11.7 Develop specific teaching skills

Enhancing certain skills like effective communication, lesson planning, and classroom management is the aim of microteaching. Microteaching improves specific teaching skills using a "teach, critique, reteach" method that focuses on one skill at a time in a controlled environment. It allows teachers to refine skills like lesson planning, presentation, and classroom management before teaching a full class, receive prompt constructive feedback from peers or mentors, and refine their approaches in a laid-back setting. This focused practice boosts confidence and expertise in particular skills, which are then used to more difficult teaching situations. By using the skill-focused method, which focuses on a specific ability rather than presenting a whole subject, teachers can develop skills such as giving clear instructions, using questioning techniques, or managing classroom behavior.

### 16.11.8 Benefits of microteaching to the students

Microteaching helps students by creating more dynamic, productive, and engaging learning environments. It also helps teachers improve their skills in areas like classroom management, communication, and subject delivery through practice and feedback. By allowing teachers to tailor lessons to each student's requirements, this boosts student engagement, motivation, and achievement. Students gain from teachers who are more confident and adept at communicating concepts. Here is a list of some noteworthy benefits:

#### *16.11.8.1 More engaging learning*

A more upbeat and effective classroom environment might result from teachers practicing and perfecting attention-grabbing techniques. By reducing difficult subjects, adding interactive and practical exercises, and utilizing contemporary, diversified formats like quizzes and simulations, microteaching increases student engagement. It enables teachers to hone their skills in a targeted manner, resulting in more self-assured and dynamic instruction that meets the needs of each individual student and offers prompt feedback.

#### *16.11.8.2 Personalized learning*

Students may become more engaged and motivated as a result of a teacher who exudes confidence and effectiveness. Microteaching increases student engagement and produces more effective and engaging sessions by giving teachers the chance to refine skills like classroom management and tailored instruction in a controlled setting. With the help of microteaching methods, teachers can gain self-assurance, experiment with different strategies, and create a more dynamic, student-centered classroom where students take more ownership of their education.

#### *16.11.8.3 Increased student engagement*

Students may become more engaged and motivated as a result of a teacher who exudes confidence and effectiveness. By allowing teachers to hone skills like classroom management and individualized instruction in a controlled environment, microteaching improves student engagement and results in more successful and captivating sessions. Through this approach, educators can develop their confidence, try out new tactics, and foster a more engaging, student-centered classroom where students take greater responsibility for their education.

#### *16.11.8.4 Enhanced teacher effectiveness*

Teachers become more skilled at delivering content by refining certain abilities in a controlled environment, which has a direct impact on student learning outcomes. Microteaching increases teacher effectiveness for students by providing teachers with the opportunity to practice specific skills in a secure environment, receive prompt feedback, and acquire confidence. Better lesson delivery, student engagement, and overall learning results result from this. This process helps teachers refine their methods before putting ideas like lesson planning and classroom management into practice in real classrooms.

#### *16.11.8.5 Better classroom management*

Through the development of classroom management skills, microteaching enables aspiring educators to provide students with a more organized and productive learning environment. Microteaching improves classroom management by providing teachers with the opportunity to practice certain skills in a controlled, safe environment. This approach boosts confidence and provides immediate feedback while assisting teachers in developing strategies for setting norms, rerouting behavior, and creating a positive environment. Regular use of these tactics helps teachers become more adept and comfortable teaching real classrooms.

#### *16.11.8.6 Improved problem-solving and critical thinking*

Teachers who have received microteaching training are frequently better able to pose questions and inspire students to use critical thinking and independent problem-solving skills. Microteaching helps students develop their critical thinking and problem-solving skills by providing a low-stakes environment where they may practice teaching specific skills, receive prompt and helpful feedback, and continuously improve their techniques. This cycle of practice, feedback, and replanning allows students to analyze problems, test solutions, and develop more effective teaching strategies through introspection and observation (Fig. 16.4).

## 16.12 Disadvantages of microteaching

The primary drawbacks of microteaching include the time-consuming nature of the preparation and sessions, as well as the artificial and unrealistic atmosphere that might not adequately prepare teachers for real classrooms. Additionally, by emphasizing individual skills, it might result in a fragmented picture of education, and peer audiences' comments may not be genuine. Several significant drawbacks of microteaching can be described as below:

### 16.12.1 Lack of realism

A dynamic classroom with a wide range of student requirements and unforeseen difficulties cannot be accurately replicated in the controlled environment. Because the controlled, artificial setting does not adequately prepare aspiring teachers for the complexity, variety of student requirements, and unpredictable dynamics of a realistic classroom,

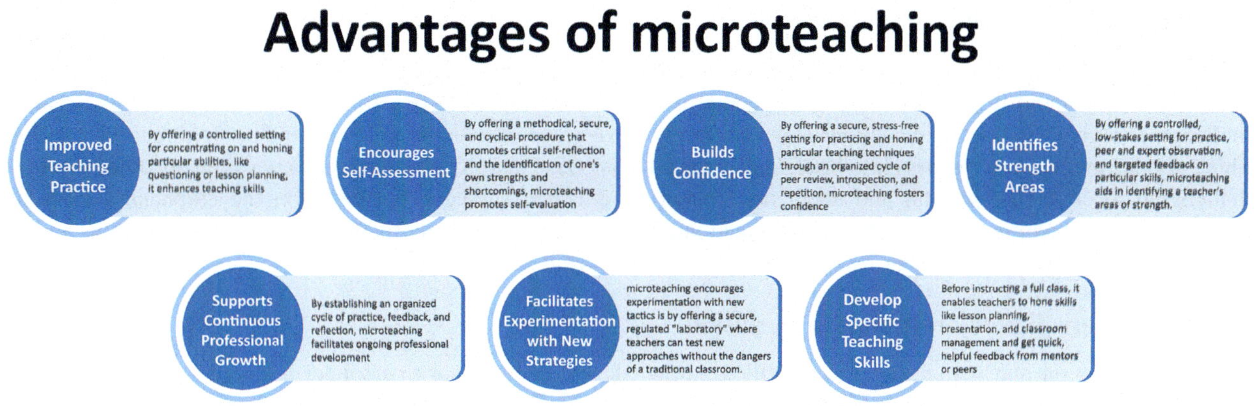


FIGURE 16.4 **Advantages of microteaching.** This figure reflects the advantages of microteaching.

microteaching's lack of realism is a major drawback. One major disadvantage is the existence of an artificial environment. Microteaching differs from managing a big class of thirty or more students since it occurs in a highly controlled environment, frequently with a tiny, unrepresentative group of students or peers acting as students.

### 16.12.2 Time-consuming

Microteaching takes a lot of time because the cycle is recurrent and involves planning, teaching, obtaining feedback, replanning, and reteaching. Microteaching sessions can be costly and time-consuming to plan and conduct. This approach may be time-consuming for each learner, especially if there are many trainees or if they need to complete the cycle multiple times to improve. The time also includes material preparation and any logistical delays like scheduling and video filming.

### 16.12.3 Fragmented skills

Concentrating on a single skill at a time can provide a limited perspective of the comprehensive teaching process, and the skills that are exercised may not be readily transferable to other contexts. The fragmentation of abilities is a significant disadvantage of microteaching since it provides a limited, artificial, and fragmented perspective of the entire teaching process. This overemphasis on individual components may make it difficult for trainees to successfully integrate these skills in a real, complex classroom situation. Furthermore, it may be challenging for trainees to incorporate the several abilities they have performed into a cohesive, fluid teaching approach. The many skills may even overlap in a regular classroom, making their specific application less practical.

### 16.12.4 Limited student engagement

The brief time prevents meaningful engagement, and peer audiences might not offer the same genuine responses or difficulties as real students. One drawback of microteaching is low student participation because the tiny group size is not representative of a typical classroom, which may result in a lack of experience in handling larger groups and a variety of behaviors. Additionally, when the emphasis is on discrete skills rather than comprehensive instruction, students may feel less engaged, and the artificial setting may not adequately prepare them for the problems of teaching in the real world.

### 16.12.5 Potential for boredom

Both the trainee and the "students" may become weary of the format's repetition. One known drawback of microteaching is the possibility of boredom. Both the students watching and taking part in the sessions and the preservice instructors practicing may find the program monotonous. Over time, the cyclical nature of the microteaching process can become tedious. Furthermore, the small group size and controlled setting frequently lack the lively and impromptu character of a genuine classroom, which might make the experience feel less interesting or significant.

### 16.12.6 Unsuitable for complex material

It can be challenging to manage big groups or sufficiently cover complicated subjects in the little time available. Microteaching's inability to effectively teach complex or difficult topics is one of its main drawbacks. It might be challenging to give thorough explanations and a comprehensive grasp of complex subjects due to the method's structure, which divides the teaching process into brief, skill-focused sessions. The microteaching session's time constraints make it challenging to cover the breadth and depth needed for complicated subjects. Furthermore, the emphasis on discrete abilities in microteaching might result in a disjointed understanding of the subject matter and teaching process as a whole rather than the integrated, thorough lesson required for complicated subjects.

### 16.12.7 Overreliance on trained supervision

The overreliance on and requirement for a skilled, knowledgeable supervisory personnel is one possible drawback of microteaching because the quality of the feedback and direction given has a significant impact on how effective microteaching is. This calls for supervisors (faculty members, seasoned mentors, etc.) who have received specialized training in the methods of data collection, observation, and constructive criticism. The execution of a microteaching program is essentially ineffective without appropriately trained and educated personnel.

**Summary**

Through the use of microteaching, medical educators can hone their teaching techniques in a short-format, controlled environment. Microteaching, which usually consists of short 10–15-minute lessons, concentrates on certain teaching techniques such describing clinical procedures, interviewing patients, or explaining difficult medical topics. Peer or mentor feedback follows these sessions, which are frequently videotaped for introspection. Microteaching is particularly useful in medical education because it allows teachers to enhance their efficacy without the stress of a lengthy lecture or clinical session. In a low-risk setting, instructors can improve their communication, presentation, and teaching skills via concentrated, repetitive practice.

**Key point**

Future perspectives: teaching how to teach

Microteaching will probably become more important in producing successful instructors as medical education develops. Microteaching in medical education has a bright future thanks to technological developments, tailored learning, and feedback systems. Microteaching is a useful, effective, and adaptable way to enhance teaching methods as medical educators deal with the growing need to be both proficient clinicians and successful educators. Microteaching will continue to be a crucial instrument in developing the next generation of medical educators and enhancing patient care through improved instruction, whether through the incorporation of state-of-the-art technology, the growth of international collaboration, or the emphasis on soft skills.

## References

Ajayi-dopemu, Y., Talabi, J.K., 1986. The effects of videotape recording on microteaching training techniques for education students. Journal of Educational Television 12 (1), 39–44. Available at https://doi.org/10.1080/0260741860120105.

Allen, D.W., Eve, A.W., 1968. Microteaching. Theory into Practice 7 (5), 181–185. Available at https://doi.org/10.1080/00405846809542153.

Ambili, R., 2013. Micro teaching, an efficient technique for learning effective teaching. Journal of Research in Medical Sciences 18 (2), 158–163. Retrieved from. Available at http://www.ncbi.nim.nih.gov/pmc/articles/pmc3724377.

Bajaj, P., Patil, M.S., Almale, B., 2014. Microteaching in medical education. MVP Journal of Medical Sciences 1 (2), 84. Available at https://doi.org/10.18311/mvpjms/2014/v1/i2/822.

Bell, A., Mladenovic, R., 2008. The benefits of peer observation of teaching for tutor development. Higher Education 55 (6), 735–752. Available at https://doi.org/10.1007/s10734-007-9093-1.

Cavanaugh, S., 2022. Microteaching: theoretical origins and practice. Educational Practice and Theory 44 (1), 23–40. Available at https://doi.org/10.7459/ept/44.1.03.

Cavanaugh, S. & Rouech, K.. (2023). Utilizing a Microteaching Learning Cycle with Student Teachers: A Practice-Based Pedagogical Tool. 2023. 42-56.

Dayanindhi, H., 2018. Effectiveness of microteaching as a method of developing teaching competence among in-service medical teachers. Journal of Advances in Medical Education & Professionalism 6, 155.

Ghatge, R., Desai, S.A., Pawar, R., Mokalikar, U., 2024. Microteaching, is there any need for medical education? Medical Journal of Dr. D.Y. Patil Vidyapeeth. 17 (6), 1298–1299. Available at https://doi.org/10.4103/mjdrdypu.mjdrdypu_918_23.

Ledger, S., Fischetti, J., 2020. Micro-teaching 2.0: technology as the classroom. Australasian Journal of Educational Technology 36 (1), 37–54. Available at https://doi.org/10.14742/ajet.4561, https://ajet.org.au/index.php/AJET/article/download/4561/1603.

Owidi, S., Lyanda, J., 2025. Integrating artificial intelligence in micro teaching: the role of ChatGPT for customized feedback and interactive learning. International Journal of Research in Social Sciences and Humanities 12, 1–10. Available at https://doi.org/10.5281/zenodo.15130275.

Rachel, Koross, 2016. IRA International Journal of Education and Multidisciplinary Studies 4.

Remesh, A., 2013. Microteaching, an efficient technique for learning effective teaching. Journal of Research in Medical Sciences 18 (2), 158–163 PMID: 23914219; PMCID: PMC3724377.

Thangaraju, P., Medhi, B., 2023. Microteaching: overview and examination evaluation. Indian Journal of Pharmacology 55 (4), 257–262. Available at https://doi.org/10.4103/ijp.ijp_912_21, http://www.ijp-online.com/.

Chapter 17

# Workplace-based learning

*"All genuine learning comes through experience."*

—John Dewey

## 17.1 Introduction

Workplace-based learning is an educational approach that connects classroom knowledge with real-world application to improve employability by allowing students to acquire academic and technical skills through practical experience in an actual workplace. It encompasses a variety of activities, including job shadowing, internships, apprenticeships, and on-the-job training, all of which are intended to prepare people for future professions and give them insight into various positions and corporate cultures.

In medical education, workplace-based learning is essential for bridging theory and practice. It uses real patient encounters for authentic skill development, feedback, and professional identity formation. It involves structured activities like mini-CEX and direct observation, but its successful implementation necessitates supportive environments, dedicated trainers, protected time, and addressing issues like heavy workloads. As the foundation of competency-based medical education, it is an essential component of training where students gain knowledge by actively participating in patient care, progressing from observation to autonomous performance with structured feedback. (Sajjad, 2015) Workplace-based learning aids students in choosing a profession and increases their employability by giving them real-world skills, verifiable experience, networking opportunities, and concrete proof of their success. Workplace-based learning significantly helps medical residents choose a job by giving them practical experience, assisting in the development of their professional identities, and allowing them to assess the actual requirements and culture of a specific workplace or specialty. By bridging the gap between academic study and professional practice, workplace-based learning gives candidates the verified experience and practical abilities that medical employers greatly value.

## 17.2 Workplace as a learning environment

In medical education, the workplace is an essential learning environment that puts theory into practice through actual patient interactions. However, it's a complex ecosystem with opportunities like practical experience and identity formation, as well as challenges like heavy workloads, resource problems, and striking a balance between service and learning. (Spencer, 2003) Constructive feedback, decent student-teacher ratios, a strong team culture, sufficient resources, and leadership that prioritizes education alongside patient care are all necessary for effective learning and necessitate deliberate effort to create a favorable environment.

### 17.2.1 Skill acquisition

Workplace learning is frequently informal, real, and immersed in particular work activities and social interactions, in contrast to formal classroom settings where learning may be abstract and context-independent. Skill acquisition in medical workplace-based learning involves experiential learning through authentic patient care, guided by mentorship, structured scaffolding like task analysis, and reflection, moving trainees from novice to expert by integrating knowledge with practical application, communication, and professionalism. (Dornan, 2009) Key elements include hands-on practice, feedback, problem-solving, and building patient relationships, all crucial for developing competent, adaptable physicians. Some of the challenges in this regard include time constraints and ensuring a supportive learning culture.

**The Principles and Practice of Medical Education.** **DOI:** https://doi.org/10.1016/B978-0-443-44197-4.00012-6

### 17.2.2 Development of situation-specific competencies

Situation-specific competences in the workplace refer to a person's capacity to use their knowledge, abilities, and behaviors to carry out particular duties in a given setting. These skills go beyond generic or theoretical knowledge and are crucial for success in real-world, dynamic situations. The development of situation-specific competencies in medical education is intrinsically linked to workplace-based learning which is primarily informed by theories of situated learning and social constructivism. This approach shifts the focus from generalized, theoretical knowledge acquisition in classrooms to the mastery of practical skills and tacit knowledge within authentic clinical contexts.

### 17.2.3 Opportunities for upskilling and reskilling

In medicine, upskilling involves learning new skills to improve performance in a current role, such as a resident learning to operate a new medical device. Reskilling involves acquiring entirely new skills to transition into a different medical role, such as a lab technician learning the skills needed for patient education. Both are crucial for a hospital or clinic setting as they contribute to the organizational flexibility in the face of swift technological advancements changing patient needs as well as for the professional development of employees. Opportunities for on-the-job training are prevalent in the workplace because this practical approach enables students to learn new procedure or abilities under actual working situations, frequently under the supervision of an experienced peer or supervisor. Additionally, upskilling and reskilling in medical education through workplace-based learning offers significant opportunities to address skills gaps, improve patient care, and enhance employee retention. These opportunities are driven by rapid technological advancements, evolving patient expectations, and persistent workforce shortages. (Ahmad, 2025) Organizing frequent workshops and seminars is crucial since these in-depth, frequently expert-led gatherings are perfect for networking, sharing ideas, and improving specific skills. Additionally, formal mentorship and peer learning programs match students with mentors or peers to offer customized career counseling, hands-on training, and knowledge sharing, promoting a culture of collaboration and professional development.

### 17.2.4 Informal learning and social cues

In the workplace, informal learning is an unplanned, impromptu, and social process where workers pick up knowledge and abilities via regular tasks and interactions, greatly impacted by social cues that promote cooperation and trust. Social cues in the workplace have a significant impact on informal learning in medical education because both novice and seasoned doctors use these unspoken cues to build their professional identities, negotiate professional relationships, and learn context-specific information. To self-monitor and modify their behavior, particularly communication styles, doctors, for example, are aware of and utilize a range of informal indicators, such as patient emotions, the ambiance of a consultation, nonverbal communication, or a colleague's tone (Fig. 17.1).

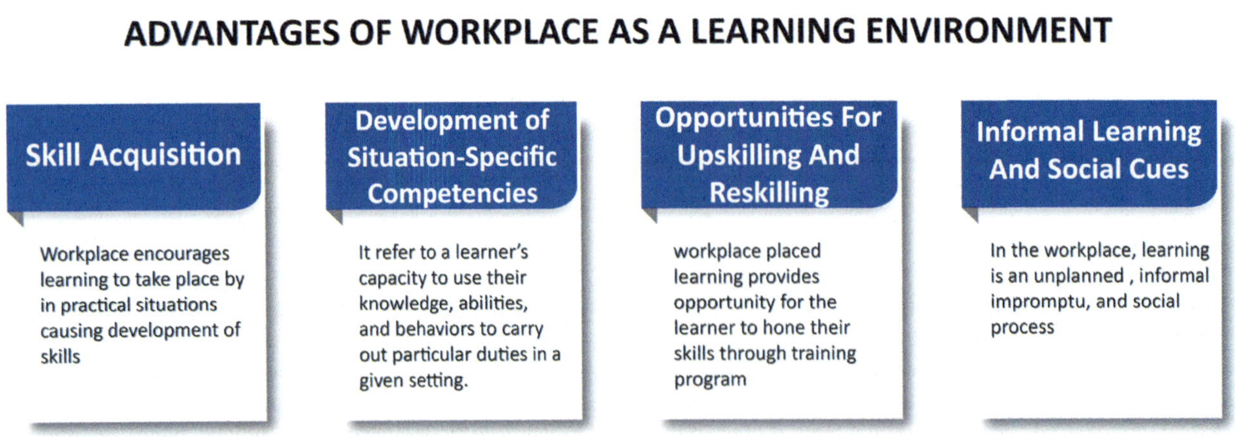


FIGURE 17.1 Advantages of workplace as a learning environment.

## 17.3 Key features of workplace-based learning

In the medical field, workplace-based learning is an organized teaching method where students use their knowledge and skills in actual clinical environments, such as clinics and hospitals. Direct patient engagement, supervised professional responsibilities, and support from a triad of patients, students, and doctor-teachers make it an essential component of medical education. The task, context, and learner are important elements, and the implementation of educational theories and a supportive learning environment are necessary for efficacy.

### 17.3.1 Real-world experience

Workplace-based learning in medical education offers vital, hands-on experience, moving beyond textbooks to develop clinical skills, professional identity, and patient-centered care, often through clerkships, apprenticeships, and real patient encounters in diverse settings like hospitals, clinics, or community practices. Constructive criticism, introspection, supervision, interprofessional cooperation, and the use of tools like mini-CEX and entrustable professional activities are important components, but obstacles like heavy workloads, time constraints, and making sure learning environments are supportive continue to be critical for success. Technical expertise, problem-solving abilities, and soft skills like cooperation and communication can all fall under this category. Real-world workplace-based learning provides essential hands-on experience, bridging theory and practice through clinical clerkships, internships, and simulations, fostering professional identity, communication, and complex clinical skill development in authentic settings like hospitals or community clinics, supported by supervision, feedback, and reflective practice. Some of the key elements in this regard include supportive environments, appropriate patient mix, supervision, and integrating learning activities with real patient care to prepare competent, reflective clinicians.

### 17.3.2 Learning from practice

A key component of a workplace-based setting is learning from practice, which frequently happens informally through daily chores, social interactions, and reflection on real-world difficulties. By immersing people in real-world job experiences, this method goes beyond typical classroom instruction and allows for the development of both clinical and "soft" skills. The fact that learning is most successful when it is placed in the real situation where it will be applied is one of the key components of learning from practice. (Attenborough, 2019) Workplace-based learning in medical education uses real clinical settings for supervised, hands-on experience, bridging theory and practice through direct patient interaction, observation, and problem-solving, fostering skills, professional identity, and evidence-based practice. Some of the important components include a supportive environment, skilled supervisors providing feedback, collaborative learning, and linking practice with classroom theory, making it a vital, yet challenging, part of training that requires faculty development and strong systems.

### 17.3.3 Assessment for learning

In the workplace, assessment for learning is the ongoing process of obtaining and analyzing data regarding employees' development in order to modify training or teaching methods and enhance performance. It places more emphasis on formative assessment than summative assessment, emphasizing continuous feedback and improvement over a final mark or assessment. Additionally, people receive timely, constructive feedback on a frequent basis while working on their assignments. Assessment for learning uses real-time feedback and formative processes within clinical settings to guide skill development, bridging theory and practice, and is crucial for competency-based education by providing structured learning and supporting progression, often through tools like direct observation of procedural skills (DOPS) or mini clinical exercises focusing on growth rather than just final judgment. Effective implementation requires integrating teaching, learning, and assessment, emphasizing reflection, and addressing tensions between formative and summative goals to maximize learning impact and ensure patient safety.

### 17.3.4 Holistic development

Workplace-based learning naturally fosters holistic growth by cultivating a broad variety of cognitive, practical, and interpersonal competencies in a real-world situation. This approach allows students to develop socially, professionally, and personally by fusing education with practical application. In this way, workplace-based development settings require the concurrent use of a blend of technical expertise, critical thinking, problem-solving, and decision-making abilities. Versatility is one of the features expected from a healthcare practitioner. In regard to this, holistic development in workplace-based learning in medical education involves training well-rounded physicians by integrating traditional medical knowledge and technical skills with essential personal and professional competencies like empathy, ethics, communication, and teamwork. This approach moves beyond the biomedical model to address the entire person, including their psychological, social, and emotional needs.

### 17.3.5 Structured learning

A key element of successful workplace-based learning is structured learning, which turns routine, unstructured work duties into purposeful learning opportunities. Structured learning in workplace-based medical education brings formal frameworks, like Peyton's four-step model or specific feedback cycles, to authentic clinical settings, transforming routine tasks into deliberate learning opportunities through supervision, structured feedback, and goal setting, bridging theory and practice for improved skill acquisition and competency, despite challenges with time constraints and inconsistent supervision quality. Success hinges on balancing work demands with protected learning time, providing constructive feedback, ensuring varied patient exposure, and strong organizational support for educators (Fig. 17.2).

## 17.4 Types of workplace-based learning

Clinical rotations, internships, shadowing, and direct patient care under supervision are examples of workplace-based learning in the medical field. These can be organized using formal techniques like case-based discussions and supervised clinical procedures, or informally through daily contacts and observation. Important elements also include assessments of learning, such as through portfolios, and assessments for learning, such as direct observation.

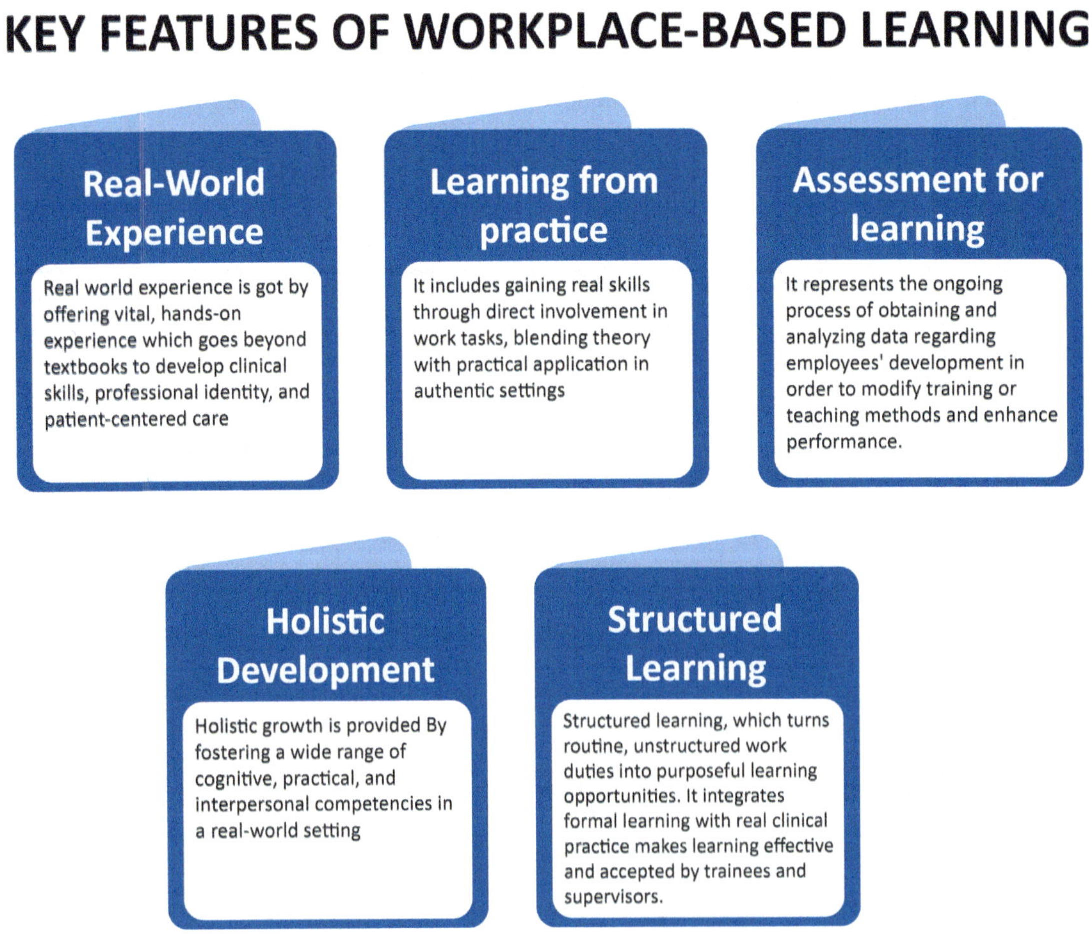


FIGURE 17.2 Key features of workplace-based learning.

## 17.4.1 Formal workplace-based learning

### *17.4.1.1 Internships*

Internships are a crucial and indispensable method of workplace-based learning in medical education, designed to bridge the gap between theoretical knowledge acquired in medical school and the practical application of skills in authentic healthcare settings. As they practice medicine under the close supervision of their instructors and senior staff, internships are an extremely important time in a healthcare professional's career. The concept of experiential learning plays a vital role in this regard as the interns learn by doing and applying their knowledge in real-world scenarios, which helps connect theory to practice. This hands-on experience, involving patient history taking, physical examinations, counseling, and exposure to various medical and surgical specialties, is vital for skill acquisition and building confidence.

### *17.4.1.2 Clinical rotations*

Clinical rotations are a core workplace-based learning method in medical education, immersing students in real clinical settings to integrate theory with practice through hands-on patient care, supervised tasks, teamwork, and experiential learning, fostering clinical reasoning and professional development under the guidance of experienced staff. They transition learning from classroom theory to authentic practice, developing essential skills like diagnosis, patient management, communication, and interprofessional collaboration within the dynamic healthcare environment. Clinical reasoning, diagnostic skills, and patient management are all enhanced by this immersive approach. Additionally, as clinical rotations are conducted in real-world clinical settings, such as clinics and hospitals, where medical professionals work on a regular basis, students gain familiarity with the reality of healthcare delivery through this exposure.

### *17.4.1.3 Direct observation and procedural checklists*

"Direct observation and procedural checklists" refer to a structured method, most notably the DOPS is a workplace-based medical learning approach that involves direct observation with procedural checklists. It entails a supervisor watching a trainee carry out a therapeutic operation, assessing performance using a systematic checklist, and giving prompt, helpful feedback to enhance abilities. This strategy is beneficial for formative learning and offers authentic assessment because it helps trainees see their inadequacies and build practical skills in a real clinical situation. (Hassan, 2011) With this method, a trained supervisor watches a trainee perform a technique in a real-world clinical situation. Preprocedural preparation, technique, and postprocedural management are only a few of the areas of the procedure that the supervisor assesses using an organized, standardized checklist. Following the process, the supervisor gives prompt, detailed comments based on the observations noted on the checklist. This method's feedback is meant to be formative, assisting the learner in identifying areas of strength and growth prior to subsequent evaluations or the actual application of the skill on a patient.

### *17.4.1.4 Case-based discussions*

Case-based discussions are a powerful workplace-based learning method, especially in medicine, involving a structured, private talk between a trainee and supervisor about a real patient case the trainee managed, focusing on clinical reasoning, decision-making, professionalism, and application of knowledge to assess skills and provide feedback for development, rather than just hypothetical situations. (Wu, 2025) Case-based discussions go beyond memorization and make learning memorable and applicable by immersing students in real-world problems and promoting active engagement, critical thinking, and collaborative problem-solving within their professional context.

## 17.4.2 Informal and self-directed learning

### *17.4.2.1 Medical shadowing*

Medical shadowing is a foundational form of workplace-based learning in healthcare education, rooted in the historical apprenticeship model. It involves observing a healthcare professional in an authentic clinical setting to gain insight into the daily realities of the profession, patient interactions, and the healthcare environment. In this, the students or prospective professionals follow a healthcare provider to understand the daily realities, workflows, and professional interactions within a clinical context. This is basically an opportunity for experiential, observable learning. (McKavanagh, 2012) Learning by observation, rather than providing direct patient care or carrying out clinical responsibilities, is the main feature of shadowing. Being a passive observer enables the person to concentrate on how medical personnel interact with the healthcare team, communicate with patients, and make choices. It enhances classroom learning by offering beneficial, first-hand exposure to the real practice of medicine. This gives students insight into the moral and psychological difficulties of the field and enables them to relate academic knowledge to actual patient situations.

#### *17.4.2.2 Implicit learning*

In a healthcare context, it has been shown that knowledge acquisition frequently happens unconsciously through information intake and passive observation. Implicit learning is a type of workplace-based learning in the medical field that involves learning without conscious awareness or intent. It is an informal, unstructured, and often opportunistic style of learning that takes place as part of regular therapeutic practice, in contrast to explicit, formal instruction. One of the best examples of implicit learning in medical practice is when a physician learns to recognize tiny patterns in a patient's symptoms or body language that indicate a specific illness without being aware of every single cue they are processing. Another example is when trainees and medical students observe more experienced colleagues and learn professional behaviors, moral principles, and communication techniques that influence their own professional identities.

#### *17.4.2.3 Reactive and deliberative learning*

In the medical field, reactive learning is a type of unstructured, on-the-job learning that happens naturally in reaction to urgent clinical circumstances, issues, or mishaps. With little time for formal planning or introspection during the actual action, it is a deliberate attempt to meet an immediate need. A particular, real-world problem, incident, or query that comes up while a medical professional is performing their job is what starts the reactive learning process. Reactive learning occurs "in action" while completing tasks in a hectic clinical setting, frequently without specific time made out for formal instruction, making it contextual and instantaneous. In medical education, an example of reactive learning occurs when a medical resident manages a patient's suddenly deteriorating condition on the spot and learns from that high-pressure, real-time experience. On the other hand, deliberative learning in the field of medical education is acknowledged as a particular, deliberate form of workplace-based learning in which learning activities are proactively designed to meet predetermined objectives for professional growth. This is in contrast to implicit learning that occurs more casually while working. In contrast to reactive learning, it is distinguished by structured, repetitive practice on specific tasks with immediate, expert feedback, aimed at moving beyond mere experience to true expertise. An example of deliberative learning is of procedural training in small groups, wherein students practice specific, step-by-step techniques on standardized models. Here medical students might practice trauma procedures in small groups of 4–6, receiving individualized, immediate feedback from an instructor after each session.

#### *17.4.2.4 Self-regulated and self-directed learning*

Self-regulated learning is crucial in medical education's workplace, enabling trainees to independently set goals, use strategies, monitor progress, and reflect within clinical settings, transforming from structured preclinical learning to autonomous growth by leveraging supervisor feedback, peer interaction, and self-assessment, essential for lifelong learning and handling complex patient situations. While traditionally supported by lectures, self-regulated learning in the clinic involves deliberate reflection, seeking feedback, and adjusting strategies based on patient context and supervisor guidance, with mentors playing a vital role in fostering this growth towards independent practice. On the other hand, self-directed learning in medical education transforms the workplace, such as hospitals and clinics, into a primary learning environment, empowering students to identify their own needs, set goals, find resources, and evaluate outcomes, fostering crucial lifelong learning skills through active participation rather than passive reception, though it requires structure, guidance, and addressing challenges like uncertainty to build confidence and competence effectively. It integrates with experiential learning and competencies, moving beyond traditional lectures to develop professional autonomy, critical thinking, and adaptability essential in medicine (Fig. 17.3).

## 17.5 Theories supporting workplace-based learning

Social learning, constructivism, experiential learning, adult learning theory, and self-efficacy theory are some of the key ideas that support workplace-based learning. (Shrivastava, 2024) These theories place an emphasis on learning through social interaction, observation, building on past knowledge, and real-world experiences. Behaviorism, cognitive theories, and connectivism, which emphasize reinforcement, memory, and networked learning, respectively, are further pertinent theories.

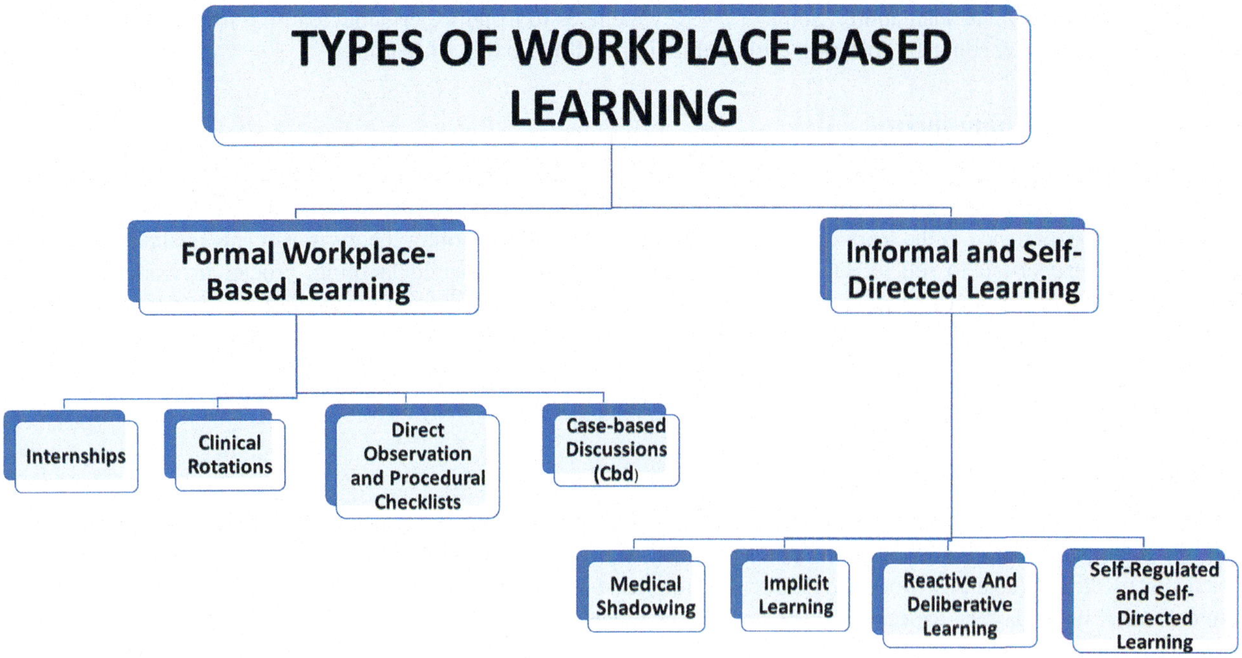


FIGURE 17.3 Types of workplace-based learning.

### 17.5.1 Social learning theory

Social learning theory in medical workplace learning emphasizes learning through observation, imitation, and modeling senior clinicians, alongside reinforcement and social interaction, explaining how students develop skills and professional behaviors in real clinical settings, moving beyond just formal instruction to active participation in a community of practice. It explains how learners internalize complex clinical reasoning, communication, and attitudes by watching experts, practicing, getting feedback, and building self-efficacy, making clinical rotations, mentorship, and team meetings crucial for skill acquisition. (Mukhalalati, 2022) In the realm of social learning theory, a lot of importance has been given to clinical rotations, mentorship, and team meetings as they give an opportunity for the trainee to imbibe all the features essential for skill development.

### 17.5.2 Constructivism

According to this theory, students actively create their own knowledge and comprehension of the world by having experiences and thinking back on them. Building new knowledge on top of prior knowledge and experience becomes the process of workplace learning. Constructivism in medical workplace learning posits that doctors actively build knowledge through experience, reflection, and social interaction, and not just passive reception. It is applied by using methods like problem-based learning, simulation, and interprofessional dialogue to connect new clinical situations with existing knowledge, fostering deeper understanding, critical thinking, and better application of skills in real settings, like using Vygotsky's ideas for guided participation. This approach goes beyond passive memorization and prepares students for real-world patient encounters by emphasizing problem-solving, critical thinking, and reflection on real-world scenarios. Additionally, it changes the instructor's position from lecturer to facilitator by guiding students via interactive exercises and conversation, enabling them to take responsibility for their own development.

### 17.5.3 Experiential learning

Experiential learning in medical workplace-based learning transforms theory into practice through direct patient experiences, reflection, and applying knowledge, often guided by Kolb's cycle, which includes concrete experience, reflective observation, abstract conceptualization, and active experimentation. It moves beyond rote memorization to develop clinical skills, ethical understanding, and self-directed learning in authentic healthcare settings like rotations, improving competency and adaptability. Some of the Key methods to foster experiential learning involve simulation, role-play, and real patient encounters, fostering critical thinking and ownership of learning.

### 17.5.4 Adult learning theory

According to this hypothesis, which is most famously linked to Malcolm Knowles, adults and children learn in different ways. It implies that adult learners are driven by both personal and professional objectives, self-directed, and bring a richness of life experience to the learning process. (Abela, 2009) Adult learning theory transforms medical education by shifting from teacher-led to self-directed, experience-based, problem-centered learning, crucial in workplace settings where adult doctors learn best when learning is relevant, immediate, and builds on existing patient experiences, fostering lifelong skill development through intrinsic motivation and collaborative practice. Some of the important principles include understanding the "why," utilizing real-life patient cases, encouraging autonomy, and valuing prior clinical experience for effective skill acquisition in authentic settings.

### 17.5.5 Self-efficacy theory

This theory, which was created by Albert Bandura, centers on a person's confidence in their capacity to achieve in a specific circumstance or undertaking. Self-efficacy, which emphasizes the concept of belief in one's ability to succeed, is crucial in medical education's workplace-based learning as it predicts effort, persistence, and performance in complex clinical tasks, with mastery experiences, feedback, peer modeling, and clear goals enhancing a student's confidence to apply skills and learn from real-world scenarios, ultimately improving skill acquisition and adaptive expertise.

## 17.6 Designing a workplace learning environment

In medical education, creating a workplace learning environment necessitates taking into account social, psychological, and physical aspects, making sure there are sufficient resources, cultivating a supportive culture through stakeholder partnerships and constructive criticism and offering protected learning time. (Verhees, 2024) Leveraging current clinical settings, maintaining a supportive organizational culture with administrative engagement, making good use of space to promote learning, and involving committed individuals as a bridge between practice and education are important components. Effective learning also requires striking a balance between the demands of service delivery and education, as well as encouraging active reflection and teamwork. When designing the workplace, some essential factors that must be taken into account are the following:

### 17.6.1 Supportive culture and leadership

Strong leadership and a supportive culture are crucial for the successful design and implementation of workplace-based learning in medicine. Leadership sets the tone for a positive learning environment, while a supportive culture built on trust and shared values fosters psychological safety, open communication, and engagement, all of which eventually led to greater patient care and professional development. In order to ensure psychological safety and foster an environment where trainees and staff feel comfortable asking questions, owning up to mistakes, and taking interpersonal risks without fear of judgment or criticism, a supportive culture within a medical workplace is crucial for workplace-based learning to flourish. Conversely, it emphasizes the value of constructive criticism because a culture that values and welcomes frequent, insightful, and constructive criticism enables people to pinpoint their areas of weakness and promotes ongoing development.

### 17.6.2 Physical environment

The physical environment has a significant impact on the effectiveness of workplace-based learning in medicine because it influences learners' cognitive load and shapes opportunities for observation, interaction, and reflection. When designing the physical environment for medical workplace-based learning, it's crucial to make sure that it supports the knowledge and skills that are supposed to be learned. For example, the physical environments of operating rooms, wards, and outpatient departments differ and facilitate the acquisition of distinct abilities. Verify that the setting is appropriate for achieving the intended learning goals. In this regard, the physical layout should be intentionally created to facilitate learning rather than obstruct it.

### 17.6.3 Psychological and social environment

For workplace-based learning in the medical education to be successful, a conducive social and psychological environment must be created. These components are crucial for promoting psychological safety, social integration into the healthcare team, and interpersonal risk-taking. It should be emphasized that psychological safety is essential because it provides the framework, enabling students to take interpersonal risks without fear of rejection, embarrassment, or detrimental impacts on their careers. Furthermore, it assists in ensuring the learners' emotional support and wellbeing, so they can cope with workplace stress, heavy workloads, and challenging situations.

### 17.6.4 Learner-centered and experiential design

When developing successful workplace-based learning in the medical field, learner-centered and experiential design are essential strategies that emphasize the requirements of the individual and active engagement in real-world situations. It is crucial that the design be learner-centered. This approach tailors the educational experience for every student by considering their unique prior knowledge, learning preferences, and career goals. Additionally, because training is based on a tailored learning plan developed in conjunction with a supervisor rather than a one-size-fits-all curriculum, it highlights individualized learning strategies. The learner's present proficiency and potential growth areas are taken into account in this strategy. The relevance of learning objectives to the learner's particular professional goals boosts engagement and motivation. Furthermore, experiential design places a strong emphasis on learning by doing, thinking back on the experience, and using that knowledge in new contexts. It also emphasizes how learning is a crucial component of real-world clinical practice. This could entail overseeing procedures, handling patient cases, or taking part in team handovers.

### 17.6.5 Structure and strategy

Designing effective workplace-based learning in medicine requires a structured approach that integrates formal learning into authentic practice, supported by a conducive environment and specific educational strategies. It also involves careful planning, a supportive environment, and specific design principles. In addition, it involves clear roles, integration of theory and practice, and a focus on active, reflective learning. Creating a supportive learning environment, assigning real-world tasks to students, and offering structured feedback are all important components of designing successful workplace-based learning in the medical field. Establishing a positive organizational philosophy that values learning, making sure that the physical and professional environment is favorable with sufficient resources and manageable workloads, and incorporating active reflection and constructive feedback from clinicians are all important tactics that are crucial in this regard.

## 17.7 Integrating workplace-based learning into the curriculum

The first step in integrating workplace-based learning into a curriculum is developing a strategy that includes stakeholders, establishes learning objectives, and aligns them with students' career ambitions and classroom curricula. Equally important in this regard are finding eligible employers and making all participants students, employers, and educators ready for the experience. The next phase is to implement the workplace-based learning activity, which includes scheduled activities and student reflection. Following this is the evaluation of the program and recognition of everyone's contributions. The steps involved in integrating workplace-based learning into the curriculum are described below.

### 17.7.1 Planning and design

#### *17.7.1.1 Engage stakeholders*

A planned and continuous strategy with open communication, collaborative goal-setting, and clearly defined responsibilities is necessary to engage stakeholders in the integration of workplace-based learning. Students are considered crucial stakeholders and main benefactors, and their involvement guarantees that the curriculum is interesting, relevant, and fulfills their learning needs and professional goals. Additionally, educators or faculty members play a significant role as they ensure that academic material matches with workplace-based learning experiences in terms of curriculum design, implementation, and evaluation. In addition, the administrators, in their capacity as stakeholders, supervise execution, assign resources which include money, personnel, and supplies and guarantee alignment with institutional objectives. Lastly, employers who participate as stakeholders help to ensure that the curriculum satisfies industry and societal demands by offering job placements, mentorship, and real-world problems.

#### 17.7.1.2 Defining objectives

Work-based learning objectives are clear, measurable, and doable goals that direct the process and outline the knowledge and skills that students will gain during their work placement. They emphasize the development of both technical and general employability skills, such as professionalism, communication, and critical thinking, and are designed to link classroom education with real-world application. The objectives should also be designed to help students develop important job abilities like communication, teamwork, and an understanding of professional culture. It should also encourage students to grow their professional networks, learn about industry-specific topics outside of the classroom, and understand expectations at work.

#### 17.7.1.3 Identifying resources

Identifying resources for workplace-based learning involves recognizing people, such as mentors and peers; structured programs, such as internships, apprenticeships, and job shadowing; digital tools, such as software, online courses, and platforms; physical assets, such as equipment and facilities; and informational materials, such as manuals, industry standards, and data to create authentic, hands-on experiences that link learning to career goals and organizational needs, utilizing both internal company assets and external community or government support.

#### 17.7.1.4 Creating a framework

It is necessary to develop a clear framework for incorporating workplace-based learning into a curriculum. This framework should include essential elements such as alignment with curriculum goals, specific learning objectives, cooperation with partners in the healthcare industry, and an organized series of experiences ranging from awareness to preparation. Along with student preparation and reflection, the framework should include learning assessment that connects academic material to practical activities and abilities.

#### 17.7.1.5 Ensuring alignment

Making sure the curriculum, workplace learning activities, and evaluation are all linked to achieve shared objectives and learning outcomes is known as alignment. This entails connecting academic material to practical work duties, offering mentorship in the classroom and in the workplace, and including reflection to help students make connections between their experiences and acquire essential skills. When theory and practice are successfully integrated into a cohesive learning environment, students learn more effectively and meaningfully.

### 17.7.2 Implementation

#### 17.7.2.1 Recruiting employers

Recruiting employers for workplace learning in medical education necessitates strategic collaboration between academic institutions and healthcare companies, with a focus on mutual benefits and open communication, in order to produce a workforce that is competent and ready. In order to create a clear understanding of the objectives and advantages of the partnership, such as tackling local worker shortages or enhancing patient care, it is important to develop a shared vision and plan. To ensure alignment with service and educational needs, it is also necessary to include leaders who are both clinical and nonclinical. Building strong linkages with academic institutions is also essential so that healthcare organizations may work with colleges and universities to develop relevant curriculum, offer internships, and offer possibilities for on-the-job training.

#### 17.7.2.2 Preparing participants

To guarantee that participants can successfully integrate classroom learning with real-world application, work-based learning preparation entails a planned approach that includes skill development, expectation management, and administrative preparation. In this aspect, fostering self and career exploration assist students in identifying their interests, strengths, and professional objectives. Additionally, employing career assessments and one-on-one discussions with mentors to assist students in understanding potential career alternatives and how workplace-based learning fits into their future aspirations is an essential step. It is also crucial to receive training in important soft skills and workplace norms. This includes teamwork, problem-solving, communication, professionalism, and time management. Through role-playing exercises that simulate real-world scenarios, they can learn appropriate workplace behavior and etiquette.

#### *17.7.2.3 Conducting the activity*

It is crucial to conduct workplace learning activities in medical education using a range of formal and informal techniques, such as peer-based groups, shadowing, and internships, with an emphasis on direct patient contact, active participation, and feedback from seasoned professionals. The objectives of carrying out a successful activity can be met by emphasizing experiential learning that fosters problem-solving, teamwork, and gradual independence in a secure setting. Students are given the chance to collaborate in official or informal peer groups through activities such as peer learning. This helps with task clarification, role navigation, stress management, and learning from each other's experiences with patients and supervisors. On the other hand, activities that support direct patient care allow students to actively engage in performing professional tasks, including direct patient engagement, in order to promote in-depth learning and skill development.

#### *17.7.2.4 Facilitating reflection*

Reflection must be developed as a fundamental, continuous, and systematic process that clearly connects theory to practice and promotes critical self-analysis in order to be successfully incorporated into a workplace-based learning curriculum. Since it connects theory to practice, promotes clinical decision-making, and supports professional development, reflection is an essential part of workplace-based learning in medical education. It enables medical practitioners to evaluate events critically, draw lessons from both successes and failures, develop their professional identities, and ultimately deliver better patient care. As mentioned earlier, reflection helps with the practical application of medical knowledge by assessing experiences in light of what is already known or still needs to be learned. Additionally, it improves clinical decision-making by helping professionals and students analyze cases more effectively and integrate their knowledge and experience to make better diagnosis and treatment choices.

### 17.7.3 Evaluation and recognition

#### *17.7.3.1 Gathering feedback*

A new workplace-based learning curriculum needs a robust feedback system for both program improvement and student development. Effective feedback in a workplace-based learning setting is a dynamic, three-way process involving the student, the workplace supervisor, and the academic institution. These assessments could be formal, planned exams like direct observation of practical skills or case-based conversations, with time allotted for feedback discussions. In order to assist students while they are still learning, it can also be organized informally, such as through "just-in-time" discussions that focus on certain assignments.

#### *17.7.3.2 Continuous improvement*

Continuous improvement in workplace-based learning is an ongoing, systematic approach to enhance processes, skills, and overall performance by integrating learning into daily workflows and organizational culture. It shifts the focus from one-off training events to a sustained commitment to growth and adaptability. Feedback should be timely, precise, actionable, and presented as a two-way dialogue with well-defined objectives in order to be used for workplace learning. In order to develop a clear action plan for improvement, it is also important to promote a culture that solicits and welcomes feedback from a variety of sources.

#### *17.7.3.3 Recognizing contributions*

Improved student results, increased curricular and institutional relevance, and more employer or community participation are some of the major benefits of incorporating workplace-based learning into a curriculum. Because of this, it is crucial to thank and acknowledge all involved parties, particularly the employers who donate their time and resources. For instance, by enabling the application of theory to practice, students are able to apply their academic knowledge and technical abilities to real-world problems and projects, which enhances their comprehension of the subject. Workplace-based learning experiences serve as a primary venue for the development of highly valued soft skills, such as professionalism, communication, teamwork, critical thinking, and problem-solving, which are challenging to teach in a traditional classroom setting.

## 17.8 Benefits of integrating the workplace-based learning into the curriculum

In order to connect theoretical knowledge with real-world application and promote skill development and professional advancement, workplace-based learning integration in medical education entails matching formal curriculum with real-world clinical settings. In order to prepare students for real-world medical practice, this integration necessitates recognizing the workplace as a learning environment, using structured supervision, feedback, and reflection, and including all members of the healthcare team in defining competencies and structuring learning experiences. Some of the benefits obtained from this integration are described as follows:

### 17.8.1 Authentic learning environments

Authentic learning environments are advantageous for workplace-based learning, as they develop problem-solving skills through challenging, pertinent activities, boost motivation by directly tying learning to employees' careers, and improve skills through application to real-world circumstances. This method improves knowledge retention and increases employees' flexibility and readiness for difficulties in the workplace. Through the authentic learning environments, students acquire and master both technical and soft skills, like problem-solving, teamwork, and communication, in a real-world context. (Aryasomayajula, 2018) Additionally, by taking on real-world problems, students improve their critical thinking and creative problem-solving abilities because the activities are directly related to their jobs. In the end, it increases motivation since learning becomes more effective and interesting when it is immediately applied to their current and future careers.

### 17.8.2 Integrated learning

The learners gain from integrated learning in the workplace by developing their practical skills, being more prepared for the workforce, and having a clearer career. Additionally, it improves soft skills like cooperation and communication, increases employee engagement, and offers beneficial networking possibilities. Workplace-based learning in medical education promotes integrated learning by bridging theory and practice, allowing students to apply classroom knowledge to real patient scenarios, develop clinical reasoning, receive immediate feedback, and gain essential professional competencies, ultimately preparing them better for complex healthcare realities through contextual, hands-on experience.

### 17.8.3 Structured supervision and feedback

Enhancing the transfer of learning and developing skill sets requires putting procedures in place for supervision, giving feedback, and encouraging reflection following clinical practice. In workplace-based learning, structured supervision and feedback are helpful because they create a positive atmosphere, improve performance through focused improvement, boost motivation, and support professional growth. Through an organized, conversational process of reflection and assistance, they assist students in understanding their progress, identifying opportunities for improvement, and developing critical skills. Structured supervision and criticism are beneficial because they help students see their strengths and shortcomings, which directly improves performance in areas like reports and presentations. Additionally, it gives learners explicit assistance by providing clear information about their progress, educating them about their learning needs, and recommending resources that will help them achieve their goals.

### 17.8.4 Interprofessional involvement

In order to mediate the learning process and guarantee thorough professional development, it is essential to involve all members of the healthcare team, including peers, other health professionals, and experienced practitioners. Significant advantages of interprofessional participation in workplace-based learning include better patient outcomes, teamwork, and communication, as well as increased job satisfaction and professional growth. The main objective of interprofessional collaboration, which is to improve the standard of care and services rendered, is upheld by the existence of interprofessional involvement in the workplace. Interprofessional teams can deliver more thorough and all-encompassing treatment by combining a variety of viewpoints and expertise, which has been demonstrated to lower medical errors, avoid complications, and enhance general health outcomes. In order to develop critical soft skills like active listening, collaborative decision-making, and conflict resolution, learning with, from, and about different professions also helps to improve communication and teamwork abilities. This promotes respect and a sense of common purpose, both of which are essential for productive teamwork and efficient workflows.

### 17.8.5 Fostering a supportive learning culture

Building a supportive learning culture in workplace-based learning benefits both the individual and the organization by promoting continuous skill development, improving performance, and raising employee engagement. Some of the key benefits are increased adaptability and problem-solving skills, enhanced morale and retention, and a better fit between employee competences and business goals. In this sense, learners gain from the opportunity for ongoing skill development since a supportive culture promotes lifelong learning, guaranteeing that staff members keep current with changes in the sector and develop new, transferable abilities. Additionally, it has been noted that receiving constructive criticism on a regular basis helps students recognize their strengths and shortcomings, which improves performance and reduces surprises during official evaluations (Fig. 17.4).

## 17.9 Role of stakeholders in workplace-based learning

### 17.9.1 Administration

Administration is essential to workplace-based learning in medical education because it provides the infrastructure, resources, and strategic direction needed to support effective training programs. Ensuring that educational goals align with institutional capabilities and legal obligations is the responsibility of administration. Allocating resources is one of the most significant administrative duties. Administrators are responsible for securing and allocating adequate funding, staffing, and facilities and technology needed to support workplace-based learning activities. As they manage operations and guarantee that workplace-based learning is an organized and integrated component of the formal education program rather than an ad hoc activity, they also participate in the process of curriculum integration and support. This entails assuring compliance with accrediting requirements and integrating educational goals into routine work procedures. One of the administration's most important responsibilities is policy and governance, which includes creating, implementing, and overseeing regulatory frameworks and regulations pertaining to ethical behavior, quality and safety, and training. They make sure that all applicable laws and professional standards are followed.

### 17.9.2 Role of the teacher

Teachers play important roles in workplace-based learning in medical education by playing a variety of different roles. For instance, as facilitators, they lead students through clinical experiences; as coaches, they offer timely and constructive feedback to enhance skills; and as role models, they exhibit professionalism and integrate knowledge with practice. In order to develop successful learning opportunities in the clinical setting, they must also serve as curriculum innovators and evaluators.

#### *17.9.2.1 Facilitator and mentor*

In workplace-based learning, a teacher's role changes from that of a typical instructor to that of a mentor and facilitator who encourages and tests students as they apply their knowledge in practical settings. Instead of only imparting knowledge, a teacher's main responsibility as a facilitator is to manage resources, foster a positive learning environment, and direct the learning process. One of a teacher's primary duties is to guide students' discovery by asking insightful

FIGURE 17.4 Benefits of integrating the workplace-based learning into the curriculum.

questions and encouraging them to find answers on their own, rather than giving them direct answers. Along with regulating access to essential tools, equipment, and experiences, the teacher is also in charge of organizing the office environment to guarantee a variety of learning opportunities.

#### 17.9.2.2 Feedback provider

In workplace-based learning, a teacher serves as a feedback provider by providing prompt, targeted, and encouraging feedback that helps students bridge the gap between their actual and ideal performance. To help the student acquire new abilities and meet learning objectives, this entails highlighting strengths, pinpointing areas for improvement with specific tactics, and speaking in a constructive, growth-oriented manner. Instead of merely evaluating, the responsibility is to support learning and growth, making sure the feedback is comprehensible and useful. Instead of drawing generalizations, the teacher should give feedback as near to the session as feasible by describing conduct in precise, unambiguous terms.

#### 17.9.2.3 Role model

In workplace-based learning contexts, teachers play a critical role as role models by exhibiting professional behavior, ethical conduct, and a persistent dedication to learning. Teachers bridge the gap between academic knowledge and practical application by setting an example for students to follow via their actions. By exhibiting workplace qualities, including punctuality, a strong work ethic, accountability, and respect for coworkers and clients, teachers exhibit professionalism. They provide students with concrete examples of appropriate behavior in the workplace by demonstrating how to dress professionally and uphold professional boundaries. Teachers show the value of honesty, integrity, and moral judgment in challenging work environments. They teach students how to respect ethical values in their future employment by openly resolving conflicts and upholding professional norms, which is crucial for establishing credibility and confidence.

#### 17.9.2.4 Promoter of reflection

A teacher who encourages reflection in the workplace creates an atmosphere where learning is more than just completing assignments. To increase comprehension, improve performance, and promote ongoing development, they lead staff members through a methodical process of self-evaluation, critical thinking, and experience analysis. It is the teacher's responsibility to provide a safe setting where people feel comfortable asking questions, owning up to mistakes, and openly discussing their experiences. For genuine introspection, this psychological safety is essential. Additionally, by candidly sharing their own experiences, difficulties, and lessons learned, the instructor exhibits their own reflective process. Others are inspired to follow suit by this transparency.

#### 17.9.2.5 Designer of learning opportunities

In workplace-based learning, a teacher creates learning opportunities by organizing real-world experiences, serving as a facilitator, and tying the curriculum to the working world. Their responsibilities include designing deliberate learning pathways, coordinating them with particular career objectives, offering direction, and assisting students in acquiring the abilities and viewpoint required for success in the workplace. In order to accomplish this, the instructor must take on the role of a coach and set clear standards for success and skill growth.

### 17.9.3 Role of the student or trainee

The student is not just a passive information consumer but also a key and active participant in workplace-based learning in medical education. In addition to actively participating in patient care and education, they also contribute to curriculum creation, offer input, and cultivate their professional identities.

#### 17.9.3.1 Active learner

In order to develop practical skills and a deeper comprehension of their profession, students participating in workplace-based learning must be proactive, actively participate in tasks, and reflect on their experiences. Here students are required to go beyond passive observation and actively participate by posing questions, collaborating with others, resolving issues, and taking responsibility for their education. Finding information, completing assignments, acting professionally, and applying newly learned knowledge to connect classroom theory to real-world application are all crucial responsibilities of the student in this situation.

#### 17.9.3.2 *Patient-centered engagement*

Beyond a passive, observant role, a student's main responsibility in patient-centered involvement in workplace-based learning is to be an engaged, courteous, and sympathetic partner in the patient's care and learning process. This method assists students in acquiring critical professional skills like clinical reasoning, empathy, and communication abilities. For example, it is impolite for the students to communicate openly and honestly with patients, paying attention to their worries, values, beliefs, and life circumstances. To promote mutual understanding and foster trust, they should steer clear of medical jargon and describe diseases and treatments in simple terms.

#### 17.9.3.3 *Seek and utilize feedback*

In workplace-based learning, a student's role is to be an active learner who initiates, listens to, interprets, and acts upon feedback to enhance performance and professional development. It is also important to have a two-way conversation with peers and supervisors, and use this input to reflect, fix errors, promote best practices, and direct future learning. In this context, a student's primary responsibility is to actively seek out feedback, particularly in unexpected work contexts. This demonstrates a proactive approach to education. At the same time, students should be very responsive and listen to criticism with an open mind.

#### 17.9.3.4 *Professional reflection*

In workplace-based learning, a student's job as a professional reflector is to consciously examine their experiences in order to enhance abilities, make connections between theory and practice, and direct future growth. This entails assessing performance, pinpointing areas for improvement, and explaining how their professional practice and understanding have evolved to improve both individual and organizational performance. Continuous learning and professional development are greatly aided by reflection, which converts implicit learning into explicit knowledge.

#### 17.9.3.5 *Proactive in learning*

In workplace-based learning, a proactive student is a responsible, involved person who takes the initiative to further their personal growth and benefit the institution. This position entails actively looking for learning opportunities, posing questions, organizing their own education, working with colleagues, and applying newly acquired knowledge to projects and challenges in the real world. Students can enhance their ability to learn new skills, stay motivated, retain information, and solve problems by taking initiative.

## 17.10 Benefits of workplace-based learning

Medical students can benefit from workplace-based learning in a number of ways, including deeper, contextual learning, better teamwork and communication skills, increased professionalism, and the formation of lifelong learning habits through practical experience and feedback. This approach improves students' self-esteem and increases their employability by bridging the knowledge gap between clinical practice and classroom education.

### 17.10.1 Application of knowledge

Students benefit from workplace-based learning in medical school because it enables them to apply theoretical information in practical clinical situations, which enhances critical thinking and problem-solving abilities. As students interact with patients and a healthcare team, this hands-on application strengthens their learning, closes the knowledge gap between theory and practice, and fosters their professionalism, self-assurance, and communication skills. Workplace-based learning offers a clear opportunity to apply classroom knowledge to actual patient situations, strengthening comprehension and promoting better clinical judgment.

### 17.10.2 Skill development

Workplace-based learning in medical education is a powerful tool for skill development, offering hands-on experience that improves both technical and soft skills like clinical competency, communication, problem-solving, and teamwork. This type of learning allows students to apply theoretical knowledge in real-world settings, leading to better retention, professional identity development, and readiness for a career after graduation. (Gillespie, 2020).

### 17.10.3 Active engagement

Students are actively involved in the learning process, moving beyond passive observation to become participatory members of the healthcare team, leading to deeper learning. Active engagement is a key benefit of workplace-based learning in medical education because it promotes deeper understanding, critical thinking, and skill application. (Mertens, 2018) Workplace-based learning increases students' motivation, fosters the development of critical interpersonal and problem-solving skills, and improves knowledge retention through hands-on experience by having them actively participate in real-world clinical scenarios.

### 17.10.4 Reflection and feedback

Reflection and feedback are key benefits of workplace-based learning in medical education because they help bridge the gap between theory and practice, leading to deeper understanding, improved clinical skills, and self-regulated learning. Trainees can improve their performance and apply their knowledge to real-world scenarios by routinely reflecting on their experiences and getting timely, organized feedback from peers and supervisors. By integrating feedback into the learning process, this combination enhances clinical reasoning and promotes a lifelong learning habit.

### 17.10.5 Preparation for challenges

Workplace-based learning is a cornerstone of medical education, providing significant benefits in preparing future medical professionals for real-world challenges. This experiential approach helps trainees develop key skills in several crucial areas. For instance, it helps in developing resilience and stress management, wherein workplace-based learning immerses learners in the demanding and often unpredictable environment of clinical practice. This exposure to high-pressure situations, emotional challenges, and patient interactions helps them build resilience and develop effective coping mechanisms. Trainees learn how to handle stress in a professional setting by handling challenging conversations, handling demanding workloads, and personally witnessing the emotional toll of patient care.

### 17.10.6 Fostering lifelong learning

Workplace-based learning in medical education fosters lifelong learning by integrating education directly into the clinical environment, leveraging several mechanisms to develop practitioners who continuously seek knowledge. Workplace-based learning inherently provides immediate opportunities to apply theoretical knowledge to real patient scenarios. The continuous, formative feedback from senior clinicians helps trainees identify knowledge gaps in real-time, encouraging self-directed study and a mindset of continuous improvement.

### 17.10.7 Bridging the gap between education and work

Workplace-based learning bridges the gap between education and work by providing hands-on experience where students can apply theoretical knowledge in real-world settings, develop crucial professional skills like communication and teamwork, and gain a better understanding of workplace demands. This approach prepares graduates for the professional environment by making them more work-ready upon graduation. Providing early exposure to the dynamics of a professional healthcare setting, including patient interactions, ethical considerations, and the pace of clinical work, helps to reduce the shock of transitioning from student to practitioner.

### 17.10.8 Enhancing the employability

Workplace-based learning enhances employability in medical education by bridging the gap between academic knowledge and real-world clinical practice, leading to better patient care. This hands-on experience develops essential skills, including technical proficiency, critical thinking, and communication, while also fostering professionalism and building confidence for future careers. Employers benefit from graduates who are better prepared, requiring less onboarding and integration into the workplace.

## 17.11 Challenges of workplace-based curriculum

### 17.11.1 Heavy workload and staff shortages

Heavy workloads and staff shortages are major challenges to workplace-based learning in medical education, as they lead to time pressure, burnout, and a reduced ability of busy staff to provide adequate training, feedback, and a supportive learning environment. Staff shortages and heavy workloads can prevent the integration of educational opportunities into the workday and lead to a focus on immediate patient care over teaching.

### 17.11.2 Inadequate training and motivation

Inadequate supervisor training and motivation are significant challenges to effective workplace-based learning in medical education, impacting both the quality of the learning environment and trainee development. Inadequate training of the trainer leads to the ineffective feedback delivery where the trainers may lack the skills to provide constructive, actionable, and timely feedback, a cornerstone of effective workplace-based learning. This can lead to trainees being unaware of their specific weaknesses or how to improve. It also leads to missed teaching opportunities as without the proper training, the trainers may fail to identify the opportunities of teaching in the busy clinical setting, often prioritizing clinical tasks over educational interactions. Additionally, the institution's promotion or compensation structure may not sufficiently value or reward educational efforts due to a lack of recognition and benefits. This may lessen the desire to invest time and effort in education. In the end, these difficulties result in a less-than-ideal learning environment where students could feel abandoned, receive inadequate direction, and fall short of their full potential. Systemic adjustments are needed to address these problems, such as offering organized programs for faculty development and revamping institutional incentive structures to place teaching on an equal footing with clinical care and research.

### 17.11.3 Unsupportive learning environment

An unsupportive learning environment in medical education is a challenge to workplace-based learning because it can include negative social and emotional dynamics, such as poor instructor–student relationships, a lack of psychological safety, and high stress, which hinders a student's motivation, engagement, and ability to learn effectively. This contrasts with a supportive environment where students feel safe to ask questions, make mistakes, and apply their knowledge in realistic scenarios, which is crucial for problem-based learning.

### 17.11.4 Balancing care and education

Balancing patient care and education in medical workplaces is a challenge due to the time constraints of clinical work, the risk of educational needs being overlooked during busy periods, and the need to maintain a focus on patient safety. Effectively managing time and prioritizing tasks, ensuring that learning objectives are met, and mitigating burnout among both students and supervising staff are key to overcoming this challenge.

### 17.11.5 Inability to identify of learning opportunities

The inability to identify learning opportunities is a challenge in workplace-based learning because learners may lack the experience to recognize them, while educators may have difficulty finding and creating them within a busy clinical environment. This can result from factors like a learner's inexperience, the time constraints of the workplace, and a lack of formal training for educators on how to integrate learning into daily clinical duties. This issue can prevent learners from achieving their full potential in workplace-based learning settings.

### 17.11.6 Lack of support and resources

Lack of support and resources is a significant challenge in workplace-based learning in medical education, stemming from a shortage of trained educators, insufficient time and high workloads for both trainees and supervisors, and inadequate material resources and infrastructure. This can manifest as poor coordination between clinical and academic staff, insufficient clinical tools and facilities, and a lack of dedicated time for training, which hinders the effectiveness of the learning environment.

**Summary**

Workplace-based learning is a cornerstone of contemporary medical education, emphasizing the acquisition of clinical knowledge, skills, and professional behaviors within authentic healthcare environments. Unlike classroom-based instruction, workplace-based learning immerses learners in real patient care settings where they observe, participate, and reflect on clinical activities under supervision. Overall, it can be said that the workplace-based learning is indispensable for preparing competent, reflective, and adaptable physicians. When well-structured and supported, it fosters authentic skill development, nurtures professional identity, and bridges the gap between theoretical knowledge and clinical practice.

**Key point**

Future perspectives: inculcating competencies amongst the learner

The future of workplace-based learning in medical education will integrate competency-based models, distributed healthcare settings, and technology, leading to more practical and effective training. Key trends include a greater focus on workplace-based learning in primary care and other distributed settings, the use of AI-driven personalization and data analytics to tailor learning, and stronger collaboration between educational institutions and clinical environments to create a more seamless learning experience. In addition, faculty development programs are essential for the future of workplace-based learning in medical education by equipping educators with the necessary skills for evolving teaching methods, like competency-based and interprofessional learning. It provides training in new techniques, assessment strategies, and the integration of technology, enabling faculty to effectively mentor students in real-world clinical environments and support institutional goals for higher-quality healthcare. The ultimate goal is to bridge the gap between academic knowledge and real-world practice to produce more prepared and confident healthcare professionals.

## References

Abela, J., 2009. Adult learning theories and medical education: a review. Malta Medical Journal 21 (1), 11–18. Available at http://www.um.edu.mt/umms/mmj/showpdf.php?article=234.

Ahmad, A., Abd Rahim, A.F., Wan Mohammad, W.M.Z., Mat Nor, M.Z., 2025. A scoping review on experiential learning in workplace-based: clinical placements. Education in Medicine Journal 17 (1), 55–77. Available at https://doi.org/10.21315/eimj2025.17.1.6, http://eduimed.usm.my/current.html.

Aryasomayajula, S., Raithatha, A., Haywood, M., Jobanputra, R., Roplekar, R., Acharya, V., 2018. Hands-on teaching, shadowing, and supported learning through acute clinics to help improve the confidence of and meet training needs for junior doctors working in ear, nose, and throat surgery. Advances in Medical Education and Practice 9, 827–835. Available at https://doi.org/10.2147/AMEP.S161187, https://www.dovepress.com/getfile.php?fileID=46169.

Attenborough, J., Abbott, S., Brook, J., Knight, R.A., 2019. Everywhere and nowhere: work-based learning in healthcare education. Nurse Education in Practice 36, 132–138. Available at https://doi.org/10.1016/j.nepr.2019.03.004, http://www.elsevier-international.com/journals/nepr/.

Dornan, T., Scherpbier, A., Boshuizen, H., 2009. Supporting medical students' workplace learning: experience-based learning (ExBL). The Clinical Teacher 6 (3), 167–171. Available at https://doi.org/10.1111/j.1743-498x.2009.00305.x.

Gillespie, H., Findlay White, F., Kennedy, N., Dornan, T., 2020. Enhancing workplace learning at the transition into practice: lessons from a pandemic. Medical Education 54 (12), 1186–1187. Available at https://doi.org/10.1111/medu.14240.

Hassan, S., 2011. Faculty development: DOPS as workplace-based assessment. Education in Medicine Journal 3 (1). Available at https://doi.org/10.5959/eimj.3.1.2011.or4.

McKavanagh, P., Boohan, M., Savage, M., McCluskey, D., McKeown, P., 2012. Evaluation of a final year work-shadowing attachment. The Ulster Medical Journal 81 (2), 83–88. Erratum in: Ulster Med J. 2014 Sep;83(3):184. Kavanagh, Peter [corrected to McKavanagh, Peter]. PMID: 23526851.

Mertens, F., de Groot, E., Meijer, L., Wens, J., Gemma Cherry, M., Deveugele, M., et al., 2018. Workplace learning through collaboration in primary healthcare: a BEME realist review of what works, for whom and in what circumstances: BEME Guide No. 46. Medical Teacher 40 (2), 117–134. Available at https://doi.org/10.1080/0142159x.2017.1390216.

Mukhalalati, B., Elshami, S., Eljaam, M., Hussain, F.N., Bishawi, A.H., 2022. Applications of social theories of learning in health professions education programs: a scoping review. Frontiers in Medicine 9. Available at https://doi.org/10.3389/fmed.2022.912751, journal.frontiersin.org/journal/medicine.

Sajjad, M, Mahboob, U, 2015. Improving workplace-based learning for undergraduate medical students. Pakistan Journal of Medical Sciences 31 (5), 1272–1274. Available at https://doi.org/10.12669/pjms.315.7687.

Shrivastava, S.R.B.L., Shrivastava, P.S., 2024. Maximize workplace-based learning through appliscation of learning theories. Journal of the Scientific Society 51 (2), 125–128. Available at https://doi.org/10.4103/jss.jss_226_22.

Spencer, J., 2003. Learning and teaching in the clinical environment. BMJ (Clinical Research ed.) 326 (7389). Available at https://doi.org/10.1136/bmj.326.7389.591.

Verhees, M.J.M., Landstra, A.M., Engbers, R., Van De Pol, M.H.J., Huijbregts, R., Van Meggelen, R.A., et al., 2024. Designing a workplace-based learning environment for learning health promotion: a design-based research. Perspectives on Medical Education 13 (1), 357–367. Available at https://doi.org/10.5334/pme.1203, https://storage.googleapis.com/jnl-up-j-pme-files/journals/1/articles/1203/667d4cfa2d07d.pdf.

Wu, H., Varshney, K., McArthur, L., Stocks, N., Walters, L.. 2025. Utility of case-based discussions as workplace based assessments in medical training: a scoping review. https://doi.org/10.21203/rs.3.rs-8083958/v1.

Chapter 18

# Clinical teaching methodologies

*"Medicines cure diseases, but only a doctor can cure patients."*

—Carl Jung

## 18.1 Introduction

Clinical teaching methodologies include methods like bedside clinics, case studies, group discussions, and simulation, which vary in direct patient involvement and focus on developing clinical reasoning, critical thinking, and practical skills. These methods aim to bridge the gap between theoretical knowledge and practical application in a supervised clinical setting to better prepare students for independent patient care. According to the American Physical Therapy Association, clinical education is "a formal supervised experiential learning, focused on development and application of patient or client-centered skills and professional behaviors. It is designed so that students gain substantial, relevant clinical experience and skills, engage in contemporary practice, and demonstrate competence before beginning entry level practice."

Clinical education is one of the best educational opportunities for students in medical programs. Through experiential learning, students can put their skills, knowledge, and attitudes to use in the workplace where they will eventually be expected to perform. Clinical education is also beneficial for programmatic improvement. Based on how effectively students do in the clinical setting, the program receives important feedback about how well prepared they are to assist people of the community. Clinical learning forms the core of medical education. The effectiveness of the clinical teaching is being influenced by numerous elements. These elements include the availability of a supportive atmosphere, learning in real-world clinical settings, exposure to a wide range of clinical experiences, and self-directed learning (Spencer, 2003). In order for clinical education to be effective, clinical teachers must assist students, promote introspection, and give regular, helpful feedback. It is important to organize and include early and frequent clinical experiences into curricula. Giving undergraduate medical students these experiences is linked to the development of positive attitudes as well as the acquisition of practical and diagnostic abilities.

## 18.2 Aims of clinical teaching

Clinical teaching seeks to close the knowledge gap between professional practice and theory by offering supervised, practical learning experiences. The following are the main objectives of clinical teaching methods:

### 18.2.1 Integration of theory and practice

The goal of integrating theory and practice in clinical education is to close the knowledge gap between the classroom and practical application, which is essential for patient safety, professional confidence, and competency. It entails applying theoretical principles directly to patient care through the use of simulation, reflective journaling, evidence-based practice, and strong mentor support.

### 18.2.2 Skill acquisition and mastery

Through organized, intentional, simulation-based practice, skill acquisition and mastery in clinical education seek to advance trainees from beginner to expert, guaranteeing competence prior to patient care. In the end, this strategy reduces medical errors and improves patient outcomes by emphasizing high standards, thorough assessment, and prompt feedback.

**The Principles and Practice of Medical Education. DOI:** https://doi.org/10.1016/B978-0-443-44197-4.00018-7

### 18.2.3 Clinical reasoning and problem-solving

The goal of clinical reasoning and problem-solving in clinical education is to help students become more adept at gathering cues, processing information, and making fast, accurate judgments that improve patient care and lower errors. In order to promote clinical thinking, it is vital to teach the students how to relate theoretical knowledge to real-world, context-specific situations using tools like simulation and structured, reflective feedback.

### 18.2.4 Professionalism and role modeling

Professionalism and role modeling in clinical education seek to mold students' professional identities as they have the capability of exhibiting crucial behaviors, abilities, and moral principles. Students learn qualities like empathy, integrity, and clinical competence from watching professionals. While negative modeling emphasizes the necessity of deliberate, conscious education, effective modeling reinforces appropriate behavior.

### 18.2.5 Holistic, patient-centered care

The goal of holistic, patient-centered care in clinical education is to prepare future practitioners to treat the whole person, thereby addressing the social, emotional, mental, spiritual, and physical needs in addition to symptoms. Instead of concentrating only on biomedical models, this strategy promotes empathy, trust, collaborative decision-making, and personalized care.

### 18.2.6 Collaborative practice

Collaborative practice has gained immense importance in clinical education as it encourages communication, shared decision-making, and mutual trust among varied professions, thereby educating healthcare students for real-world, team-based treatment. Through coordinated, interprofessional teamwork, this strategy improves patient safety, lowers medical errors, and maximizes the quality of care. It entails teaching students how to collaborate with patients and their families to achieve shared objectives.

### 18.2.7 Confidence building

Confidence development in clinical education aims to transform students into competent, self-assured practitioners by fostering a low-stress, supportive setting that encourages skill acquisition. It links theory and practice through early exposure, simulation, and constructive criticism, reducing anxiety and increasing students' preparedness to manage difficult patient care.

### 18.2.8 Self-directed learning

Clinical teaching enhances self-directed learning by turning students from passive observers into active participants who manage their own educational path. In this way, clinical educators are increasingly working as facilitators rather than primary knowledge providers, utilizing specific models and tools to encourage student autonomy and lifetime learning practices.

### 18.2.9 Upholding principles of medical education

Clinical teaching approaches, when used effectively, uphold fundamental ideals of medical education by bridging the gap between theory and practice, developing critical skills, and preparing students for professional practice. These approaches are frequently regarded as the cornerstone of education for health professionals. However, the quality of implementation, clinical educators' training, and the handling of practical issues like patient access and time limits all have a significant impact on how well these principles are kept.

## 18.3 Different methods of clinical teaching

While there are many clinical teaching approaches, this chapter will concentrate on the most popular ones. Clinical teaching approaches are mainly divided into two categories, namely classical and modern methods. Clinical rounds, case presentations, and bedside instruction are examples of classical clinical approaches. Virtual reality (VR) and augmented reality (AR), e-learning, and simulation-based learning are examples of modern clinical education techniques. This also includes techniques like case-based learning (CBL) and problem-based learning (PBL). These methods aim to foster critical thinking, practical skills, and a comprehensive understanding of patient care (Fig. 18.1).

**A. Classical clinical teaching methodologies**
1. Bedside teaching,
2. Case presentations,
3. Clinical round.

**B. Modern clinical teaching methodologies**
1. Simulation-based learning,
2. E-learning,
3. Virtual and augmented reality,
4. Problem-based learning,
5. Case-based learning.

## 18.3.1 Classical clinical teaching methodologies

### *18.3.1.1 Bedside teaching*

Bedside teaching is an essential, patient-centered approach to medical education in which medical professionals instruct learners, whether students or residents, at the patient's bedside in order to foster professional, clinical, and communication skills. Physical examinations, patient-learner interaction, and collaborative care planning are all part of it, and they all help to develop expert clinical reasoning and empathy (Peters, 2013). Effective bedside teaching necessitates precise objectives, courteous patient interaction, and targeted advice. It offers learners instant feedback, professional role modeling, vital practical experience, and patient involvement in their own care. Bedside education is still an essential method for producing skilled and caring doctors, despite obstacles.

## 18.3.2 Key aspects of bedside teaching

### *18.3.2.1 Authentic learning*

It is the main method for learning how to take a patient's history, perform a physical examination, use clinical reasoning, and communicate with real patients as the clinical setting with actual patients and their complicated issues offers a really authentic learning experience that cannot be entirely recreated in a classroom. This method involves students in relevant,

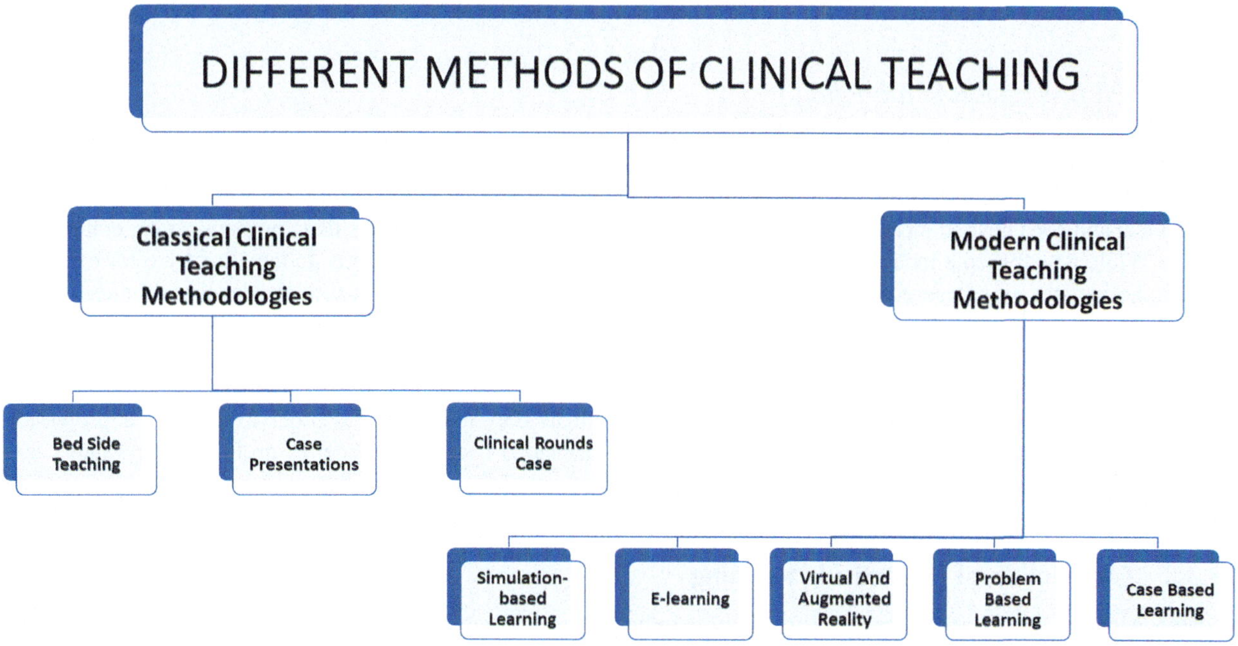


FIGURE 18.1 **Different methods of clinical teaching.** This figure shows the different methods of clinical teaching.

practical exercises that reflect the intricacies of real-world medical practice. Since it allows students to interact with genuine patient cases mirroring the realities of clinical practice, authentic learning in bedside teaching exposes students to real-world relevance (Gonzalo, 2010). In contrast to typical textbook circumstances, this boosts motivation because the knowledge is directly applicable to their clinical practice. Additionally, it helps students bridge the gap between theory and practice by applying theoretical knowledge and practicing critical skills including taking a history, doing a physical examination, and communicating in a real-world context. Clinical reasoning and problem-solving skills are developed through actual learning because the dynamic nature of patient encounters necessitates critical thinking, the formulation of differential diagnoses, and the justification of clinical decisions based on patient data and observations.

#### *18.3.2.2 Patient-centered approach*

Effective bedside teaching requires a patient-centered approach, which enables healthcare students to witness and practice professionalism, empathy, and communication in a real-world clinical setting. Instead of seeing the patient as a passive subject, this method views them as an active participant in both their treatment and the teaching process. Transforming medical students from technical practitioners into compassionate clinicians is said to depend on teaching a patient-centered approach during bedside teaching. The chance of follow-through and favorable health outcomes increases when students learn to include patients in shared decision-making, resulting in treatment regimens that are in line with the patient's life.

#### *18.3.2.3 Integrated skills*

Bedside teaching makes it possible to develop and evaluate several abilities simultaneously as a cohesive process rather than separately. Bedside teaching can successfully incorporate a variety of integrated skills to improve medical education, despite its conventional concentration on clinical examination and history taking. This method goes beyond merely diagnosing illnesses to include the comprehensive skills needed for efficient patient care. Since bedside teaching offers a perfect environment for modeling and practicing efficient patient–doctor communication, communication skills are among the crucial abilities that should be integrated among the learners. This entails actively listening, resolving patient concerns, developing treatment strategies, and clearly and sympathetically communicating complicated medical information. Students can also get instant feedback on their communication abilities at the same time. Students' clinical thinking and decision-making skills are also enhanced by involving them in the bedside decision-making process. Teachers can also demonstrate how to integrate information from the patient's history, physical examination, and investigations to arrive at a diagnosis and treatment plan. Consequently, abstract reasoning processes are made plain and observable.

#### *18.3.2.4 Professional role modeling*

Teachers make a significant contribution by acting as role models for students, exhibiting professionalism, leadership, and productive teamwork. A key element of bedside teaching is professional role modeling, which entails using imitation and observation to exhibit professional values, clinical abilities, and attitudes. It helps mold a learner's professional identity and conduct and goes beyond merely demonstrating clinical skill. It encompasses traits like communication, empathy, and passion. While negative role models can instill bad habits, positive role models inspire students to strive for excellence. This emphasizes the significance of educators being cognizant of their influence and actively developing their own role modeling abilities. It should be noted that the majority of learning in clinics occurs through the process of observation and imitation, in which students closely watch the activities, relationships, and decision-making processes of their instructors and subsequently mimic these behaviors. Additionally, the clinical learning process supports the social learning process, in which students are impacted by their environment and individuals they look up to in order to build their professional identity and competence. It also affects the learner's career decisions because a student's professional attitude, career direction, and level of satisfaction with their residency program can all be greatly influenced by role models.

### 18.3.3 Conduction of the bedside teaching

Bedside teaching is a crucial pedagogical method in medical education where instructors teach students and residents at the patient's bedside. Effective conduction of a session requires careful planning and execution of several components. Some of the key steps for conducting effective bedside teaching include the following:

### *18.3.3.1 Preparation (preround)*

For bedside instruction to be effective, the preround or preparation period is essential. In order to guarantee that learners have a seamless, fruitful experience and that patient care is disrupted as little as possible, the session must be planned and organized. The following are the main tasks that must be completed during the preround phase:

#### 18.3.3.1.1 Define learning objectives

In this step, the instructor should ascertain the learning objectives and educational requirements beforehand. Since each level calls for a different set of abilities to be taught, this aids in adjusting the teaching points to the learner's level. For instance, teaching the junior students how to take a clinical history and the senior students' differential diagnoses.

#### 18.3.3.1.2 Select suitable patients

As the most important elements of bedside teaching, choosing appropriate patients with pertinent clinical symptoms or instances that align with the learning objectives is crucial. Patients who are willing and medically stable enough to take part in a training session should always be chosen.

#### 18.3.3.1.3 Obtain patient consent

The patient must be informed about the goal of the session and who will be present, and their verbal or written consent must be obtained without compulsion. Making sure they are aware that they can end the meeting at any moment is a good practice. This demonstrates the doctor's professionalism and moral obligation while also confirming the patient's readiness to engage in the training process.

#### 18.3.3.1.4 Review patient information

It is crucial that educators and students go over the patient's history, clinical findings, and test results. It makes it possible to stay informed about the case and comprehend any new, noteworthy developments in the disease's or treatment's clinical course. Additionally, it guarantees that the session is more effective and concentrated on the patient's main and present problems.

#### 18.3.3.1.5 Orient the learners

Gathering the team away from the patient's bedside to discuss the patient in brief, establish clear expectations for the session, assign specific roles to each learner, for example, who will conduct the physical examination, or present the case and set ground rules for bedside etiquette, such as communication style and jargon usage, are all good practices and exhibit the learner's professionalism and etiquette.

### *18.3.3.2 Introduction (at the bedside)*

Bedside teaching's introduction or "at the bedside" phase entails crucial actions to create a courteous, fruitful, and instructive experience for the patient, students, and instructor. Among the crucial actions that must be taken during this stage are the following:

#### 18.3.3.2.1 Introduce the team and learners

Introducing the patient to everyone in attendance is the first step. Avoiding technical jargon and speaking in a language that the patient is comfortable with are crucial. It should be mentioned that patients have a right to know who is at their bedside and they normally desire to know.

#### 18.3.3.2.2 Orient the patient

This step involves briefly explaining the purpose and objectives of the visit from the patient's perspective. Letting the patient know what to expect from you with regard to what you will be doing is very important. Statements such as "We are here today so the students can practice their examination skills, which will help them become better doctors" by the students is helpful. This will allow the patient to be comfortable and cooperate with the student.

#### 18.3.3.2.3 Obtain explicit consent

It is advised that the learners confirm the patient's willingness to participate in the teaching session. It is important to remember that patient participation is voluntary and all the details of the consent are well understood by the patient in the language which they understand.

#### 18.3.3.2.4 Set the "rules of engagement" and time limit

It is important to set the ground rules for bedside etiquette and clarify the time commitment. For instance, letting the patient know that you will be with them for about 10 minutes is a good practice as it will make the patient comfortable and minimize anxiety.

#### 18.3.3.2.5 Assign learner roles and state expectations

At this point, the learners can briefly outline the specific tasks they will perform or the areas of the examination they will focus on. This gives learners a clear purpose and helps structure the session.

#### 18.3.3.2.6 Ensure patient comfort

It is vital that the learners continuously pay attention to the physical and mental comfort of the patient. This is to make sure that the patient is stable and comfortable during the entire session. Any abnormality or discomfort from the patient side should be dealt with utmost seriousness.

#### 18.3.3.2.7 Briefly review the case

If necessary, the learner can quickly go over the patient's relevant demographic data and brief medical history, often at the patient's bedside but in a way that respects their primary importance to the session.

#### 18.3.3.2.8 Set the stage for active learning

It is ideal to employ advanced organizers of any kind to provide the students a framework, such as outlining what to anticipate or what important details to pay attention to throughout the interaction. To activate past knowledge and reflect on new learning, learners can, for example, rapidly complete a KNOW, WANT TO KNOW, LEARNED (KWL) chart both before and after the session. In this instance, the instructor asks students to briefly write down what they know and what they would like to know about heart failure before they enter the class. Following a debriefing, the students fill in the learned column.

### *18.3.3.3 Execution (the teaching encounter)*

The teaching encounter phase of bedside teaching involves the following key steps, which often build on preparatory actions and lead to evaluative activities. Some of the essential features of this phase are as follows:

#### 18.3.3.3.1 Active teaching and learning

This is the core component of the encounter. It involves facilitating hands-on learning experiences for the student with the patient. In this phase, the teacher should use open-ended questions and promote active engagement and clinical reasoning rather than just demonstrating.

#### 18.3.3.3.2 Graduated responsibility

In bedside teaching, as medical students, interns, and residents show competence, experience, and dependability, they are given increasing degrees of autonomy, authority, and independence in patient care through graduated responsibility, also known as graded responsibility.

#### 18.3.3.3.3 Observation

In this step, the instructor carefully observes the learner's performance during history taking, physical examination, or specific procedures to assess their skills and identify areas for improvement. It is also beneficial to observe for any particular skill in the students. For example, some students have a very good communication skill while some have excellent skills of observation. These things should be kept in mind.

#### 18.3.3.3.4 Patient involvement

During the entire session of the bedside learning, it is the responsibility of the team to keep the learning environment lively by involving the patient and encouraging them to contribute, correct, and ask questions throughout the session, making them feel like an active participant rather than a passive subject.

#### 18.3.3.3.5 Role modeling

During the session, the teacher should demonstrate high level of professionalism, ethical behavior, empathy, and effective communication skills. These skills will not only justify the role of a teacher as a good leader but at the same time set a benchmark against which the learners will be graded by the patient. In this regard, the teacher will also serve as a role model for their learners.

#### 18.3.3.3.6 Immediate feedback

It is the responsibility of the teacher to provide brief, specific, and real-time feedback during the encounter to correct errors or reinforce good practices immediately, allowing the learner to modify their behavior "in action."

#### 18.3.3.3.7 Refinement of learning issues

The teacher uses the observations from the encounter to refine the specific learning needs and focus for any follow-up teaching, often conducted in a separate location away from the patient.

### *18.3.3.4 Feedback and closure (postround or at the bedside)*

Bedside teaching's feedback and closure phase entails organized debriefing, introspection, and the development of an action plan for future learning. These should take place right after leaving the patient's room. One of its main characteristics of the feedback phase is the creation of a safe environment to reduce fear and potential embarrassment. It is always a good idea to make sure the feedback is given in a quiet, encouraging environment outside of the patient's room. Additionally, there should always be space for self-evaluation, in which the student is asked to evaluate their own performance in terms of how they did and what went well. This promotes self-reflection and reveals their insight. In addition, it is necessary to acknowledge and reinforce specific, observed positive behaviors or skills with concrete examples. For instance, appreciating the explanation of the treatment plan given by the student to the patient can reinforce effective communication.

In terms of closure, it's critical to highlight essential points where the primary learning objectives and "take-home messages" from the interaction are consolidated. This enables strengthening of learning and retention process. Before leaving the patient's bedside, the learner should always make sure the patient receives a summary of their care plan, a chance to ask any last questions, and a thank you for attending the lesson. In order to address the learner's identified areas for improvement, it will be beneficial if the team can work together to create a clear, agreed-upon action plan.

### *18.3.3.5 Follow-up (postround)*

After the team leaves the patient's bedside, the follow-up phase of bedside teaching is essential for assessment, feedback, and learning consolidation. Debriefing is one of the crucial tasks that must be completed in this situation. The team should go to a quiet space, like a conference room or hallway, to discuss sensitive material or matters that are inappropriate to discuss in front of the patient. This permits a thorough clinical conversation while guaranteeing patient anonymity (Fig. 18.2).

## 18.3.4 Benefits of bedside teaching

### *18.3.4.1 For learners*

Bedside teaching benefits learners by improving communication and clinical skills through direct patient interaction, fostering empathy and professionalism via role modeling, and enhancing critical thinking and reasoning in a real-world setting. It provides hands-on experience with physical examinations and patient histories that cannot be simulated, allowing for immediate feedback and more confident skill development. It also promotes a patient-centered approach to care and helps learners understand the human impact of illness. These benefits can target the following three aspects.

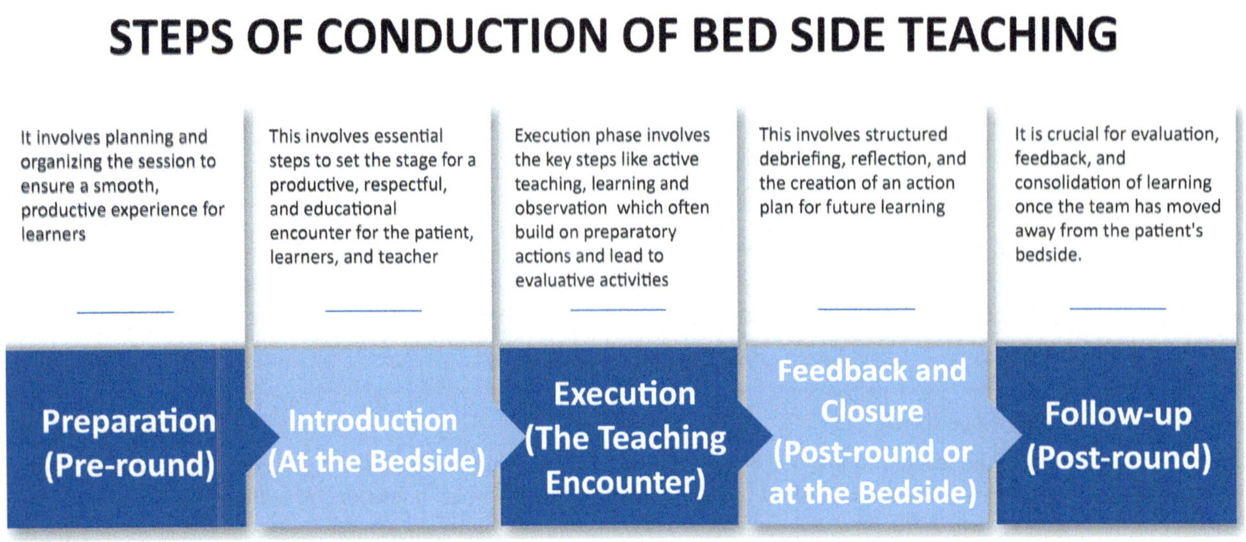


FIGURE 18.2 **Steps of conduction of bedside teaching.** This figure depicts the steps of conduction of bedside teaching.

#### 18.3.4.1.1 Skill development

Bedside teaching helps to enhance the clinical skills as it allows the learners to practice physical examinations, interpret clinical signs, and gain tactile experience by doing palpation, percussion as a part of the diagnosing specific conditions like organomegaly or ascites. In addition, it encourages the student to hone some of the vital skills like communication by facilitating the chance to practice history taking, building rapport with patients, and communicating with both patients and the healthcare team. It also focuses on developing critical thinking by encouraging learners to develop their reasoning skills by using clinical information to form diagnoses and treatment plans.

#### 18.3.4.1.2 Professional and personal growth

Bedside teaching is one of the best platforms to instill professionalism amongst the learners as it provides opportunities for professional role modeling, demonstrating compassion, empathy, and respectful communication with patients. The students are also given a glimpse of patient-centered care where they are taught to view the patient as an individual rather than a disease, involving them in their own care plan and respecting their time and perspective. All these will have a positive impact on the students' personality by increasing their confidence as the students become more confident in speaking to and examining real patients, providing the much needed reassurance to the patient.

#### 18.3.4.1.3 Learning and feedback

As far as the learning and feedback is concerned, bedside teaching promotes active learning, wherein the learners are engaged to actively take part in the session by putting across their views and opinions rather than only passively observing the session. Moreover, it gives an opportunity for educators to observe the learner's skills and reasoning process and offer immediate, direct, and specific feedback. Ultimately it gives a chance for the learner to understand illness in a holistic way by not only targeting the disease but also to appreciate the personal and human impact of illness on a patient's life.

### *18.3.4.2 For patients and families*

Bedside teaching benefits patients and families by fostering patient-centered care, increasing communication, and improving understanding and involvement in the treatment plan. It allows patients to be active partners in their care, ask questions directly to the healthcare team, and feel more involved, which can lead to better health outcomes and adherence to treatment. For families, this involvement can reduce anxiety and build trust in the healthcare providers (Ramirez, 2016). There are several benefits that are being provided in this regard, such as improved patient-centered care, as the patients' values, preferences, and perspectives are directly integrated into the care plan, promoting a more holistic approach. In addition, a direct interaction with the healthcare team, including students and residents, allows for clearer communication and the opportunity for patients to ask questions and express concerns. Moreover, there are increased involvement and empowerment by the patients and families which gives them a chance to be directly involved in discussions and decisions about their care, making them feel more in control and less like passive recipients of care.

#### *18.3.4.3 For educators*

Bedside teaching provides educators with benefits like the opportunity to serve as professional role models, give immediate and direct feedback to learners, and observe clinical skills firsthand. This method also allows educators to foster teamwork and communication skills, reinforce patient-centered care principles, and improve the quality of care through real-time correction of misinformation and observation. The bedside teaching proves to be of benefit for the educator in the sense that it sets the stage where educators can demonstrate desired behaviors, including compassion, communication, and teamwork, in a real-world context. There is also a chance for the educator to observe a learner's clinical skills in real-time and provide specific, direct feedback for improvement.

### 18.3.5 Challenges of bedside teaching

Challenges of bedside teaching include time constraints, high workloads, and constant interruptions, often exacerbated by system issues like the emphasis on technology and electronic health records, which can reduce hands-on patient interaction. Additional challenges involve dealing with a diverse range of student learning styles, patient-related concerns, and teacher-related issues such as a lack of preparation or skills. These challenges can be divided into two categories:

#### *18.3.5.1 Teacher and system challenges*

##### 18.3.5.1.1 Time and workload

Time constraints and heavy clinical workloads are widely recognized as primary barriers to effective bedside teaching. These factors contribute significantly to the decline of bedside teaching as physicians struggle to balance patient care demands, administrative duties, and educational responsibilities. Some of the important factors in this regard are the competing responsibilities where the clinician-teachers face with regard to their time, including direct patient care, documentation, research, and administrative tasks, making it difficult to dedicate protected time for teaching. The issue of increased patient turnover owing to the shorter patient hospital stays in modern healthcare settings is also a vital factor as there is less time to identify suitable patients with specific clinical findings for teaching purposes and to conduct comprehensive teaching sessions.

##### 18.3.5.1.2 Technological shifts

Technological shifts present several challenges to traditional bedside teaching, primarily by changing the environment, diverting attention, and creating a skill gap between faculty and students. The presence of physical barriers owing to the increasing reliance on electronic health records means that clinical data is often accessed on computers, not at the patient's bedside. This can physically distance the teacher and student from the patient, as they focus on screens rather than physical examination or direct interaction. Furthermore, the time consumed in data entry and in navigating complex clinical software reduces the time available for direct patient engagement and spontaneous teaching moments at the bedside.

##### 18.3.5.1.3 System priorities

System priorities present a major challenge to effective bedside teaching by creating a hospital environment focused on efficiency and patient turnover over the time-intensive nature of education. These demands put significant pressure on clinicians and teachers and limit dedicated teaching time. Issues like time constraints where the clinicians face increased demands on their time, including higher caseloads, extensive electronic health record documentation, and conflicting administrative responsibilities, contribute to this limitation. This leaves less time for the unhurried, focused interactions necessary for quality bedside teaching. In addition, clinical teaching activities, which can take longer, may be viewed as inefficient or a strain on workflow due to the emphasis on hospital efficiency, cost-effectiveness, and shorter patient stays.

##### 18.3.5.1.4 Lack of training and experience

Effective bedside teaching is severely hampered by faculty members' lack of training and experience, which frequently results in disorganized sessions, discomfort for the faculty, and lost learning opportunities for medical students. The fact that many clinical faculty members lack formal training in educational principles or particular bedside teaching strategies is one of the significant factors in this regard. They frequently impart knowledge in a manner that may not be in line with contemporary pedagogical best practices. Sometimes faculty discomfort are exhibited by the teachers who lack confidence in their teaching abilities may feel uncomfortable conducting operations or having patient conversations in front of students and patients. As a result, they may end up using classroom-style instruction outside of the patient's room. Time constraints can also be a barrier since, in the hectic clinical setting, faculty members may put clinical responsibilities ahead of teaching or believe they don't have enough time to complete a comprehensive teaching session at the bedside.

##### 18.3.5.1.5 Organizational issues

Organizational issues present significant challenges to effective bedside teaching by impacting resource allocation, scheduling, and overall system management. The presence of poor planning and coordination can result in several groups of students in the same area, uncoordinated intern shifts, or educational rounds clashing with other hospital activities like meal times or patient tests. The lack of a clear, structured timetable makes it difficult to ensure consistent educational opportunities. In addition, having inadequate resources and facilities such as shortages of essential physical and human resources pose a major challenge. This includes insufficient numbers of faculty members and personnel, small or crowded patient rooms, and a lack of necessary equipment like stethoscopes or diagnostic tools at the bedside. Added to this, the presence of a high student-to-teacher ratio makes personalized, one-on-one teaching difficult and can be overwhelming for both patients and instructors.

#### *18.3.5.2 Student and patient challenges*

##### 18.3.5.2.1 Diverse learning styles of students

Diverse learning styles present a significant challenge in bedside teaching, requiring clinical educators to adapt their instructional methods to meet individual trainee needs effectively (Crumlish, 2009). Since trainees absorb information in various ways, a one-size-fits-all approach often fails to engage all learners equally or optimize knowledge acquisition and retention. Situations created due to mismatch of styles further add to the problem as traditional bedside teaching often relies heavily on auditory and observational learning. This approach can be a disadvantage to residents and medical students who learn primarily by kinesthetic or visual styles.

##### 18.3.5.2.2 Patient-related issues

Patient-related difficulties pose a considerable barrier to bedside teaching as they add unpredictability and thus require a careful balance between teaching objectives, patient well-being, and clinical responsibilities. The patient may feel vulnerable, nervous, or exhausted, and they may not want to be used as a teaching subject, among other important factors. Obtaining informed permission can be challenging when people are ill or in pain, despite the fact that it is essential.

### 18.3.6 Ways to improve bedside teaching

Teachers should choose cases with high educational value and prepare both the patient and the learner in order to enhance bedside teaching. In this sense, using an organized framework such as the One Minute Preceptor or SNAPPS model is also beneficial. Focusing on patient-centeredness throughout the session by encouraging patient engagement, communicating clearly, and giving targeted feedback to foster critical thinking are all important while improving bedside teaching.

#### *18.3.6.1 Structured approach*

A structured approach improves bedside teaching by using a clear framework that includes preparation, delivery, and reflection. This involves setting specific learning objectives based on the learner and patient, using tools like the One Minute Preceptor or SNAPPS, and incorporating a structured feedback process after observing the learner's skills. Some of the vital elements in this regard are preparing learners and patients, guiding the clinical interaction, and providing specific, constructive feedback on both skills and clinical reasoning.

#### *18.3.6.2 Focus on goals*

Focusing on clear goals is crucial for improving bedside teaching by setting a clear roadmap for both the educator and the learner. This involves establishing specific learning objectives beforehand, orienting learners to the session's purpose and their roles, and then tailoring the content to these goals to ensure maximum satisfaction and knowledge gain for the student. By having defined goals, educators can structure the session effectively, manage time, and ensure learning is both relevant and efficient.

#### *18.3.6.3 Promote reflection*

To promote reflection for better bedside teaching, educators should create a safe environment, guide learners to reflect during the encounter with questions, and facilitate postsession debriefing. This can be achieved by encouraging self-assessment, linking new experiences to prior knowledge, and using structured methods like the "what, so what, now what" mode.

#### *18.3.6.4 Provide feedback*

To improve bedside teaching through feedback, educators should solicit feedback from learners and peers, while also providing timely, specific, and constructive feedback to learners on their performance. This involves creating a supportive environment, setting clear objectives, and using the bedside encounter as a direct opportunity for both learning and improvement.

#### *18.3.6.5 Case presentation*

Case presentation is a clinical teaching methodology that uses a patient's story to teach and evaluate clinical reasoning, problem-solving, and diagnostic skills. It requires students to organize and present patient information, while enabling educators to assess the student's clinical thinking process and provide targeted feedback.

### 18.3.7 Components of a case presentation

A case presentation includes identifying patient information which includes patient's history, physical examination findings, laboratory and results, diagnosis, a treatment plan, and the outcome. For a clinical case, the presentation should chronologically detail the patient's journey from their chief complaint to their final status, while a nonmedical case may have components like a problem statement, findings, and a solution (Onishi, 2008). Clinical case presentation includes addressing various aspects of the patient and the clinical signs and symptoms. It starts with identifying personal information about the patient such as name, age, gender, etc., and the reason for admission or consultation. Later on, a complete clinical history and clinical examination is conducted to find out more about the signs and symptoms pertaining to the suspected clinical condition. This is always backed up by evidence in the form of laboratory tests, imaging, and other investigations. Finally, a treatment and management plan is proposed based on the gathered information.

### 18.3.8 Role of case presentation in clinical teaching

Case presentation is a crucial clinical teaching method that enhances students' clinical reasoning and diagnostic skills through active engagement with patient scenarios. It allows for the assessment of a learner's knowledge and evaluation of their clinical thought process in a real-world context (McLean, 2016). Beyond assessment, it provides a reflective opportunity for students to learn from their experiences and helps educators identify areas for improvement.

### 18.3.9 Methods of case presentation

Clinical case presentation techniques include interactive strategies like the signpost method, formal frameworks like the SNAPPS method, and conventional storytelling. Presentations usually follow a chronological narrative that includes patient information, history, physical examination, investigation results, assessment, and management plan, regardless of the method. Clinical case presentation methods can be categorized as in Fig. 18.3.

#### *18.3.9.1 Traditional/narrative methods*

##### 18.3.9.1.1 Storytelling

This approach takes the listener through the patient's journey from the primary complaint to the resolution by presenting the case as a narrative. In order to make complicated medical information more interesting and approachable, storytelling techniques for clinical case presentations entail organizing a narrative with a beginning, middle, and finish. Introducing the patient's history and problem, outlining the creative solution, and discussing the result and its implications are all important strategies. This approach may be applied in a variety of mediums, including digital storytelling and spoken presentations, and it helps humanize the experience, promote empathy, and improve communication.

**METHODS OF CASE PRESENTATION**

**Traditional/Narrative Methods**

- **Storytelling:** Here the case is presented as a narrative, guiding the audience through the patient's journey from the chief complaint to the final outcome
- **Chronological order:** In this the cases are presented in the order it occurred, making it easier to follow the patient's progress over time

**Structured/Framework-Based Methods**

- **SNAPPS:** It is designed to foster clinical reasoning and shift the role of the preceptor from expert to facilitator, while the learner takes a more active role
- **Problem-oriented:** Here the patient information is organized around the patient's specific health problems, rather than by the source of the data
- **Signpost method:** Here the trainee provides a concise overview of a patient's case and then "signposts" to the presenter the different aspects they have considered

**Digital Formats**

- **Slide Presentations:** This process uses slides to outline the case in a structured manner
- **Electronic Posters:** These are typically single-page digital documents often in the form of PDF that display the case information and can be viewed on screens

FIGURE 18.3 **Methods of case presentation.** This figure reflects the methods of case presentation.

#### 18.3.9.1.2 Chronological order

This approach makes it simpler to track the patient's development throughout time by presenting the cases in the chronological order of occurrence. The clinical team can easily follow the diagnostic rationale and management choices as the presentation progresses because the chronological arrangement of a clinical case follows a regular, logical sequence that reflects the patient's path through care. For example, the general introduction or patient identification should come first, followed by the patient's medical history, current sickness history, and physical examination.

### *18.3.9.2 Structured/framework-based methods*

#### 18.3.9.2.1 SNAPPS

The six-step, learner-centered SNAPPS clinical teaching approach is intended to improve clinical reasoning and change the emphasis from passive reporting to active, evidence-based discussion. It works especially well in hectic outpatient settings, allowing preceptors to assist effectively while allowing trainees to communicate their clinical thinking rather than just facts. It is frequently utilized in outpatient settings to improve clinical reasoning, encourage self-directed learning, and streamline instruction particularly when time is of the essence. The components of SNAPPS are listed below:

***S—Summarize briefly the history and findings*** This is the initial stage in which the student provides a succinct, organized summary of the patient's medical history and important physical findings. An active, patient-centered, and effective "problem representation" is intended to replace a thorough, passive, chronological recitation of information.

***N—Narrow the differential to two or three relevant possibilities*** The "N" in the SNAPPS teaching technique refers for "Narrow the differential," a crucial stage in clinical reasoning where the student narrows down a lengthy list of potential diagnoses to the two or three that are most likely. This part pushes students to actively prioritize options depending on the patient's presentation rather than merely listing them.

***A—Analyze the differential by comparing and contrasting the possibilities*** In this third and crucial phase in the SNAPPS teaching technique, student goes beyond merely listing possible diagnoses to actively compare and contrast them using the patient's data. It acts as a link between recognizing opportunities and creating a tangible management strategy, enabling the preceptor to assess the learner's clinical reasoning in real time.

***P—Probe the preceptor by asking questions about uncertainties*** The probe the preceptor component, which is the fourth phase in the SNAPPS teaching technique, serves as an essential learner-driven tool for promoting conversation, filling in knowledge gaps, and enabling prompt feedback. This step gives the student the ability to start a conversation by expressing doubts, outlining challenges, or proposing different strategies based on the patient case they just presented, in contrast to traditional techniques where the teacher poses the questions.

***P—Plan management for the patient's medical issues*** The fifth element of the SNAPPS teaching approach is "Plan management," which is a crucial stage where the student moves from case analysis to starting a tangible management plan. This stage places a strong emphasis on learner autonomy by encouraging the student to take charge of patient care while utilizing the preceptor as a resource for advice.

***S—Select a case-related issue for self-directed learning*** Choosing a suitable patient case is an essential initial stage in the SNAPPS clinical teaching technique that allows the student to take the lead in the discussion and practice clinical reasoning in a condensed amount of time. A case that presents two to three reasonable differential diagnoses and permits a concise, 3-minute overview is optimal.

***Benefits of using SNAPPS as a clinical teaching method*** The SNAPPS clinical teaching approach improves medical education as it encourages active, learner-centered case presentations that strengthen critical thinking, diagnostic reasoning, and self-directed learning. It boosts students' confidence in making clinical decisions by encouraging them to express their doubts and question their preceptors (Wolpaw, 2003). In outpatient settings, the organized six-step method works well without appreciably lengthening presentation times.

#### 18.3.9.2.2 Problem-oriented method

The structured approach, known as the "problem-oriented method" of clinical case presentation, arranges patient data according to the patient's particular health issues rather than the data's source. In order to support a methodical, problem-solving approach to patient treatment, it entails compiling a thorough list of the patient's issues, which may include symptoms, diagnoses, and aberrant results. The complete medical record is then arranged around these issues. This approach frequently records progress notes for every issue using the SOAP format (Subjective, Objective, Assessment, Plan).

#### 18.3.9.2.3 Signpost method

The signpost method is a clinical case presentation technique in which the trainee gives a brief summary of a patient's case before pointing out to the presenter the various factors they have taken into consideration, including the history, physical examination, investigations, differential diagnosis, and management plan. This method allows for a more focused and productive discussion centered on clinical reasoning rather than basic facts by directing the preceptor to the trainee's priorities and thought process (Brunton, 2022).

### *18.3.9.3 Digital formats*

Clinical case presentations in digital formats use a variety of mediums and platforms, mostly for publication, education, and discussion. These formats make use of technology to improve information sharing, clarity, and engagement between students and healthcare providers. Slide shows and electronic posters are two examples of often utilized digital formats.

#### 18.3.9.3.1 Slide presentations

These include some of the most popular formats, such as PowerPoint, Google Slides, and Canva, where slides are used to present the patient's medical history, physical examination, investigations, diagnosis, course of treatment, and results in an organized manner. Visual aids like as radiographs, charts, graphs, and histology pictures are frequently incorporated in the slide presentation.

#### 18.3.9.3.2 Electronic posters (e-posters)

E-posters are used in conferences and academic settings. These are typically single-page digital documents, often in the form of a PDF in landscape orientation, that display the case information and can be viewed on screens. E-posters promote active and deep learning as students are actively involved in gathering information, organizing it, and creating the e-poster, which encourages critical and reflective thinking and leads to better knowledge retention than passive learning methods. It also helps in enhancing communication and presentation skills as this process provides a platform for students to practice explaining their research or topic clearly and concisely to peers and faculty, which helps build confidence for future professional presentations.

## 18.3.10 Benefits of case presentation as a clinical methodology

Case presentation is an effective clinical teaching method that improves clinical reasoning, communication, and problem-solving skills by applying theoretical knowledge to real-world patient scenarios. It enhances clinical reasoning by requiring abstraction of information and stimulates informal discussion, leading to a deeper understanding of patient care. Some of the major benefits obtained by case presentations are as follows:

### *18.3.10.1 Enhances clinical reasoning*

Case presentation enhances clinical reasoning by requiring presenters to organize information, practice abstraction, and reflect on the diagnostic process. This process forces trainees to summarize, narrow the differential diagnosis, analyze options, and plan management, while also improving their communication skills and ability to think on their feet. For the other team members, it provides learning opportunities from diverse cases and allows them to actively engage and discuss the material with their peers and facilitators, which improves their own diagnostic abilities.

### *18.3.10.2 Improves problem-solving skills*

Case presentation improves problem-solving skills by allowing the learners to apply theoretical knowledge to complex, real-world scenarios, which strengthens their critical thinking, analytical skills, and clinical reasoning. This process encourages active participation, collaboration, and the development of a more holistic understanding of a problem, ultimately building confidence in decision-making abilities.

### *18.3.10.3 Develops communication skills*

Case presentations develop communication skills by forcing presenters to structure complex information logically, communicate clearly and concisely, and adapt their message to a specific audience. This process improves both verbal and nonverbal communication, including active listening, storytelling, and confidence, and prepares individuals for professional interactions in their field.

### *18.3.10.4 Promotes active learning*

Case presentation promotes active learning by shifting the focus from passive listening to active engagement with real-world problems. It fosters critical thinking and problem-solving skills as students must analyze scenarios, apply theoretical knowledge, and develop solutions. Students' comprehension is deepened, engagement is increased, and course topics can be used practically when they are required to synthesize material and explain their reasoning.

### *18.3.10.5 Encourages reflection and analysis*

Case presentation encourages reflection and analysis by providing a structured framework that requires individuals to systematically review, interpret, and contextualize information, bridging the gap between theory and real-world practice. Here the students get to reflect on the case, justify their reasoning, and often receive feedback on their management plan, which can lead to improved patient outcomes in the future.

### *18.3.10.6 Fosters collaboration*

These presentations often spark informal interactions and discussions among students and faculty, creating a more dynamic and collaborative learning environment. Case presentation fosters collaboration by creating a structured environment for knowledge sharing, diverse perspectives, and shared problem-solving. It provides a common platform for professionals from various backgrounds to engage in meaningful dialogue, which ultimately leads to improved outcomes.

#### *18.3.10.7 Teaches empathy*

Students can develop more empathy for their patients by examining a patient's viewpoint and the complexity of their circumstances. It provides a chance for medical students or professionals to put themselves in the patient's position, comprehend their particular experience of sickness, and form an emotional bond that transcends the biological facts.

### 18.3.11 Clinical rounds

Clinical rounds are a widely used teaching methodology in healthcare, where a team discusses clinical cases to teach clinical reasoning, communication, and patient care in a real-world setting (Bassaw, 2011). Effective teaching involves preparation before rounds, active participation during rounds with a focus on teaching, and a debriefing after rounds to consolidate learning. Making learning visible, combining patient care and teaching, giving both immediate and thoughtful feedback, and creating a supportive learning atmosphere are all important tactics of clinical rounds. Clinical rounds and bedside teaching are related but separate tasks in medical education that frequently overlap in practice. The main distinction is that bedside teaching concentrates on learner education, whereas clinical rounds concentrate on patient management (Shaik, 2023).

### 18.3.12 Conduction of clinical rounds

Clinical rounds as a teaching methodology typically follow a structured process that includes preround preparation, bedside rounds for case presentation and examination, and postround discussion and evaluation. The steps involving the conduction of clinical rounds can be divided into three stages such as the following:

#### *18.3.12.1 Preround*

##### 18.3.12.1.1 Plan and prepare

Effective planning and preparation are crucial for productive clinical rounds, ensuring patient safety, efficient communication, and valuable learning opportunities. In the preround phase before seeing patients, the healthcare team meets together to set the agenda and roles. Then the patients are selected by typically focusing on a balance between clinical care, education, and patient well-being. The selection process is influenced by the type of round, the team's objectives, and the patient's individual needs and preferences. It is also important for each member of the team to review the patient information with a focus on patient's case history, recent progress, and investigation results.

##### 18.3.12.1.2 Notify patients

This step involves notifying the patients, family members, and staff about clinical rounds effectively by using a multifaceted approach that combines direct communication with standardized tools, ensuring all parties are aware of the schedule and their role in the discussion. It is important to explain the purpose of rounds and how the patient and family can be involved.

##### 18.3.12.1.3 Obtain consent

Obtaining consent during clinical rounds involves an ongoing, respectful, and transparent communication process with the patient, rather than a single event. The goal is to ensure the patients are fully informed and are voluntarily participating in their own care decisions. It is vital to take care that the patients' autonomy is maintained at all times. The patient should be provided with sufficient information to make an informed choice. This includes the diagnosis, nature and purpose of recommended interventions, potential benefits, material risks, and alternatives including the option of no treatment.

#### *18.3.12.2 Bedside rounds*

##### 18.3.12.2.1 Patient introduction

This includes giving a brief introduction of the patient to the group and explain the purpose of the rounds. A patient introduction during clinical rounds is a structured oral presentation to a medical team that follows an established format, such as the SOAP (Subjective, Objective, Assessment, Plan) or a comprehensive history and physical examination structure. The presentation should be confident, clear, and focused on relevant details to guide patient management.

#### 18.3.12.2.2 Case presentation

A case presentation during clinical rounds is a structured way for a healthcare provider to share a patient's relevant clinical information with the rest of the team to facilitate discussion and patient management. The presentation typically includes the patient's history, physical examination findings, investigations, and management plan. The case presentation should be concise and focus on the information that is most relevant to the patient's current problem. A lengthy recitation of all available facts should preferably be avoided.

#### 18.3.12.2.3 Demonstration of clinical skill

In this stage, the instructor might demonstrate relevant clinical skills such as a clinical sign or symptom pertaining to the disease. The students should be encouraged to ask questions to probe their understanding and the instructor's reasoning.

#### 18.3.12.2.4 Arriving at a differential diagnosis

At this stage, after having gone through the details of the clinical history and clinical examination findings, an attempt should be made to arrive at a differential diagnosis. The initial differential diagnoses should be based on the most likely conditions or the red flags picked up during the clinical discussion. Later the choice of the differential diagnosis has to be discussed with a proper clinical reasoning.

### *18.3.12.3 Postround*

#### 18.3.12.3.1 Summarize and reflect

Summarizing and reflecting are essential components of clinical rounds, serving to monitor patient progress, coordinate care plans, and facilitate medical education and team communication. On the other hand, reflective practices help the team to consider different perspectives, improve compassionate care, and address nonclinical issues that may impact the patient–caregiver relationship, ultimately leading to a more holistic approach to patient well-being.

#### 18.3.12.3.2 Provide feedback

Providing feedback to the learner after the clinical rounds is important. The focus of the feedback should be on creating a supportive, respectful environment. The feedback should be a conversation aimed at improvement, not a judgment of the person. This will not only encourage the student to do better in the next clinical round session, but it also boosts their confidence and desire to learn.

#### 18.3.12.3.3 Assign self-directed learning

During this stage, it is important that the instructor identifies a specific, case-related issue for the student to research further and present at a later time. Assigning a self-directed learning activity gives a chance for the student to explore the case-related issues and come up with a new or innovative approach towards the case. It will also help the student to stay relevant and updated on the latest discoveries and research work pertaining to the case discussed.

#### 18.3.12.3.4 Assess achievement

Formative feedback and direct observation are the main methods used to evaluate clinical rounds' performance. Structured methods and assessment instruments are used to evaluate particular abilities in real time. Direct observation of procedural skills (DOPS), in which an assessor watches a trainee carrying out clinical duties with patients and the team during rounds, is a popular technique for evaluating achievement. A structured feedback model, such as Pendleton's feedback model, which begins with the learner identifying what went well before moving on to areas for progress, can also be used for the evaluation. A crucial component of the assessment procedure should be to provide a targeted, helpful feedback.

## 18.3.13 Benefits of the clinical rounds as a teaching methodology

Clinical rounds, especially those conducted at the bedside, offer significant benefits as a teaching methodology in medical education by bridging the gap between theory and practice and fostering essential clinical competencies (Hulland, 2017). Some of the important benefits obtained by clinical rounds include the following:

##### *18.3.13.1 For students*

###### 18.3.13.1.1 Application of knowledge

Clinical rounds offer a real-time, patient-centered learning environment and are an essential link between theoretical knowledge and practical implementation. This method helps trainees go from beginner to expert as it enables the students to use, hone, and expand their knowledge of clinical medicine in the setting of direct patient care.

###### 18.3.13.1.2 Development of clinical skills

By giving a chance to repeatedly practice and refine their core skills in clinical medicine like history taking, physical examinations, interpreting diagnostic tests, performing procedures, and developing clinical judgment and decision-making abilities, clinical rounds become a very important aspect of teaching methodology in medical education.

###### 18.3.13.1.3 Exposure to real-world scenarios

At the same time, by attending the clinical rounds, the learners encounter diverse patient populations, various medical conditions, and the dynamic, unpredictable nature of real healthcare settings, which classroom simulations cannot fully replicate. This will also prove to be very exciting as the students have a chance to come out of the textbooks and explore the real-world of medicine in all its facets.

###### 18.3.13.1.4 Interprofessional collaboration

The interprofessional collaboration that happens during the clinical rounds is indeed a boon for the students. Here the students have an opportunity to interact with a full healthcare team including nurses, pharmacists, therapists, social workers, etc., and learn about the comprehensive nature of patient care, fostering teamwork, and improving communication skills across disciplines.

###### 18.3.13.1.5 Professional role modeling and ethics

Clinical rounds not only help the learner to apply the theoretical knowledge to clinical practice, but it also gives a platform where the learners can observe their senior physicians and other professionals acting as role models, implicitly acquiring behaviors related to professionalism, communication skills, empathy, and ethical decision-making in patient care.

###### 18.3.13.1.6 Increased self-efficacy and confidence

Clinical rounds are widely acknowledged as an essential teaching strategy for boosting healthcare professionals' and students' self-efficacy and confidence. These rounds enable students to transition from academic theory to competent clinical practice, lowering anxiety and enhancing their capacity to handle challenging patient situations by offering an organized, engaging, and frequently secure learning environment.

##### *18.3.13.2 For patients*

###### 18.3.13.2.1 Enhanced understanding

One of the main patient-centered advantages of clinical rounds, especially when done at the bedside, is improved comprehension of a patient. The professional team obtains a more comprehensive, sympathetic, and accurate understanding of the patient's illness, social environment, and personal preferences by shifting conversations from a conference room to the patient's room.

###### 18.3.13.2.2 Improved communication and trust

Clinical rounds are a fundamental component of patient-centered care that immediately improve communication and trust, especially when they are performed at the bedside. They change the patient experience from passively receiving care to actively participating in it, which improves clinical outcomes and increases patient satisfaction.

###### 18.3.13.2.3 Greater involvement

One of the main advantages of clinical rounds is increased patient involvement. Clinical rounds change the patient from a passive recipient of care to an active participant by moving conversations from conference rooms to the patient's room. This improves the patient's comprehension of their condition, boosts their level of satisfaction, and increases the accuracy of their care plan.

#### *18.3.13.3 For the healthcare system*

##### 18.3.13.3.1 Improved quality of care

Through improved patient-centeredness, teamwork, and communication, clinical rounds raise the standard of treatment. By involving a multidisciplinary team, they result in better-coordinated care plans, fewer adverse events, and a more thorough understanding of the patient's condition. They can increase patient satisfaction, trust, and treatment adherence by including the patient and family.

##### 18.3.13.3.2 Efficiency

Clinical rounds improve efficiency in healthcare settings by promoting collaboration among interdisciplinary teams, expediting decision-making, and simplifying communication. When properly organized, they speed up patient discharge procedures, cut down on needless interruptions, and save time spent on administrative duties.

## 18.4 Modern clinical teaching methodologies

### 18.4.1 Simulation-based learning

Simulation-based learning in medical education is a method that uses realistic, simulated environments and scenarios to train students and clinicians without involving real patients. This approach allows for hands-on practice of both technical and nontechnical skills in a safe, controlled setting, enabling repeated practice and immediate feedback to improve performance and patient safety. Simulation-based learning is valuable for training in procedures, crisis intervention, communication, and teamwork, and for experiencing rare or complex cases that are not frequently encountered in clinical settings (Al-Elq, 2010).

### 18.4.2 Conduction of simulation-based learning

The steps for conducting simulation-based learning in medical education involve a systematic process that can be broken down into preparation, the simulation activity itself, and postsimulation phases, with debriefing being the most critical component for effective learning.

#### *18.4.2.1 Planning and preparation phase*

The planning and preparation phase ensures that the simulation is well-designed and resource-ready. Some of the important steps to be completed in this phase are as follows:

##### 18.4.2.1.1 Identify learning objectives

The first step is to clearly define the purpose and specific learning outcomes for the session like knowledge acquisition, specific technical skills, teamwork, or ethical considerations. It should also be kept in mind whether the use of simulation method is suited to the topic to be learnt. For example, simulation-based learning is more suitable where there are clinical oriented topics such as in cardiovascular system or respiratory system rather than topics which are purely theoretical based.

##### 18.4.2.1.2 Target the audience

Determining the appropriate year or level of learners is important to decide on the content to be taught. For instance, if the learner is from year 2 or 3, a more basic approach to clinical teaching is preferred wherein the target will be to teach history and physical examination, whereas for the years 4 and 5, the aim will be to teach the clinical features and the diagnosis. It is also important to determine whether the student is a novice resident or belongs to an interprofessional team in order to design the case accordingly.

##### 18.4.2.1.3 Design the scenario

The clinical scenario developed should be realistic and relevant that aligns with the learning objectives. As mentioned above, the level and year of the student should be kept in mind in designing the case scenario. In addition to the alignment with the learning objectives, the scenario should challenge learners and encourage critical thinking.

##### 18.4.2.1.4 Select simulation type and resources

It is important to decide and choose the appropriate fidelity simulation, such as low-fidelity task trainer, standardized patient, or high-fidelity manikin, and gather necessary materials, equipment, and a suitable location. For instance, low-fidelity task trainers are used for teaching fundamental, hands-on skills, especially those that require repetition, like performing a medical procedure such as fetal scalp blood sampling. Their primary uses include building confidence, allowing practice at one's own pace in a safe environment, and confirming mastery of basic psychomotor skills before advancing to more complex training or real-world situations. They are cost-effective, require less setup, and focus on isolated skills without the complexity of high-fidelity simulators.

##### 18.4.2.1.5 Prepare learners and facilitators

It is important to provide presession materials or prior knowledge instruction so that the learners have a foundational understanding before the simulation begins. At the same time, the instructors or facilitators should be trained so as to ensure that they can guide the session effectively, provide real-time tutoring if needed, and lead the debriefing session.

##### 18.4.2.1.6 Running a trail

It is a good practice to run a trial of the scenario with staff to refine the process and ensure everything runs smoothly before the official implementation. This also will ensure familiarization as it will allow the instructor to become fully comfortable with all the supporting materials, mechanics, and potential outcomes of the simulation. In addition, it will ensure that the troubleshooting is done as it will allow the instructors to identify potential technical issues or logistical problems before the simulation is run with students. It will also help in anticipating ways the simulation might go wrong, allowing them to prepare solutions or adapt their presimulation discussions with the class.

### 18.4.3 Implementation phase (the simulation)

The implementation phase is the hands-on phase where learners engage with the scenario.

#### *18.4.3.1 Introduction*

In this stage, it is important to orient the learners to the environment, the equipment, the ground rules, and the expected roles and responsibilities. The students should be clearly made aware of advantages and disadvantages of the selected simulation methods to be used. It should be noted that the introduction should set a safe and supportive tone.

#### *18.4.3.2 Run the scenario*

At this stage, the students will actually run the session where they are given a chance to handle the challenge and perform the required tasks and procedures such as mastering basic lifesaving skill like cardiopulmonary resuscitation or invasive procedures which include central line placement. During the simulation, the facilitator observes the students' actions, ensuring they stay on task and are appropriately challenged. If the students go in an unexpected but valuable direction, the facilitator may allow the scenario to adapt so that it guides them back to the learning objectives.

#### *18.4.3.3 Close the scene*

In this stage, a conclusion of the simulation session is done by the facilitator either once the learning objectives have been met or the scenario naturally concludes.

### 18.4.4 Postsimulation phase (debriefing and evaluation)

The postsimulation phase is characterized by debriefing and evaluation. This is where the most significant learning occurs.

#### *18.4.4.1 Debriefing and feedback*

Immediately following the simulation, a structured debriefing session is conducted by a trained facilitator. A debriefing session in simulation teaching is a guided, reflective discussion that occurs after a simulated event to help participants analyze their actions, thoughts, and decisions. It is a crucial learning component that solidifies knowledge, improves future performance by identifying and addressing gaps, and connects theoretical knowledge to practical application. A skilled facilitator guides the process through open-ended questioning, creating a safe environment for learners to self-assess and critically reflect on their performance.

#### 18.4.4.2 Evaluation and assessment

A multifaceted approach is used in simulation teaching evaluation and assessment to measure student performance, knowledge, and abilities in a secure setting. Assessment, which collects information about a participant's performance, and evaluation, which uses that information to make judgments, are essential elements. Checklists, rubrics, and debriefing sessions are used to evaluate the cognitive, emotional, and psychomotor domains. These can be summative, measuring ultimate achievement, or formative, providing feedback and continuous progress.

### 18.4.5 Benefits of simulation-based learning

Simulation-based learning offers a risk-free environment for practicing skills, improving decision-making, and boosting confidence. It provides realistic, immersive experiences that bridge the gap between theory and practice, leading to better knowledge retention, increased engagement, and the ability to repeat scenarios for mastery (Elendu, 2024). Some of the key benefits of simulation-based learning are depicted in the following figure (Fig. 18.4).

#### 18.4.5.1 Safe and risk-free practice

Students can experiment and make mistakes without facing real-world repercussions, which is particularly helpful for patient safety in industries like healthcare. The primary benefit of simulation-based learning is that it offers a risk-free, secure setting where participants can practice and make mistakes without facing repercussions in the real world. This enables students to acquire experience and proficiency in a controlled environment prior to dealing with real patients or handling a crisis. Additionally, it results in zero injury because it enables medical professionals to test emergency responses and carry out intricate treatments on high-fidelity manikins without endangering the patient. Repeated practice in a secure environment fosters proficiency and confidence. Professionals are better equipped to manage high-pressure circumstances in real life when they have confidence in their abilities.

#### 18.4.5.2 Enhanced decision-making and problem-solving

Students are forced to think critically and apply their knowledge in dynamic situations via realistic, difficult scenarios. Simulation-based learning enhances decision-making and problem-solving abilities by providing a safe environment for practicing complex scenarios, experimenting with strategies, and learning from the results of decisions without real-world ramifications. It forces active learning, critical thinking, and the development of higher-order skills by allowing users to

FIGURE 18.4 **Benefits of simulation-based learning.** This figure reflects the benefits of simulation-based learning.

analyze situations, test different solutions, and see the immediate and long-term outcomes. Students gain a stronger understanding of theoretical concepts and their real-world applications through this cyclical "claim, evidence, reasoning" process.

#### 18.4.5.3 Improved skills and confidence

Faster onboarding and improved work readiness can result from learners gaining competency and confidence in their skills through realistic task practice. Among the many benefits of simulation-based learning is the enhancement of abilities and self-assurance. It speeds up skill acquisition by offering a safe practice environment where complex surgeries or investigations can be undertaken repeatedly without the real-world repercussions of failure. Instantaneous, objective performance feedback from simulators enables learners to spot and fix mistakes in real time, resulting in quicker and more efficient learning than traditional methods. One of the key advantages is that students can practice uncommon or crucial situations that they might not come across during regular on-the-job training, making sure they are ready for a greater variety of circumstances.

#### 18.4.5.4 Increased engagement and motivation

Simulations are more engaging and motivating than traditional approaches because they are interactive and immersive. The main benefits of simulation-based learning include increased motivation and engagement because interactive, hands-on activities are more pleasurable and pertinent, which improves retention. Simulations make learning more dynamic and interesting by allowing students to "learn by doing," explore in a safe environment, and receive immediate feedback.

Simulations are more engaging than passive learning since they produce realistic and interactive experiences that increase learners' attention and enjoyment. As a result of this, an enhancement of motivation and engagement is seen amongst the learner. Additionally, simulations allow students to actively practice skills and make judgments by placing them in realistic settings.

#### 18.4.5.5 Opportunities for repetition

Self-directed learning and the capacity to learn from errors are made possible by the ability for learners to practice and master scenarios via repetition. Simulation-based learning capitalizes on the potential for deliberate and infinite repetition by allowing students to practice complex tasks as often as needed to become proficient. Muscle memory and automaticity are developed through this repetitive process, which is essential in high-stakes professions like medicine. Additionally, a safe setting promotes repetition since it lets students make mistakes and grow from them without facing repercussions in the real world. This lack of risk promotes resilience and exploration. Additionally, every learner can experience the same scenario repeatedly through simulations, ensuring continuous exposure to certain learning objectives.

#### 18.4.5.6 Immediate and personalized feedback

In order to help students better comprehend their decisions and develop their judgment, simulations can offer real-time feedback and facilitate debriefing sessions. The ability of simulation-based learning to deliver quick, tailored feedback is one of its main benefits. This direct and timely feedback technique significantly accelerates skill acquisition and performance improvement because simulations allow students to see the immediate results of their actions in a safe, controlled setting. The system or instructor can promptly identify the error and facilitate prompt intervention if a decision has a detrimental impact.

#### 18.4.5.7 Cost and resource efficiency

As they eliminate the need for pricey tools or supplies, simulations might be more affordable than in-person instruction. Simulation-based learning offers significant advantages in terms of cost and resource efficiency by reducing the expense of physical resources, reducing the likelihood of expensive equipment being broken, and concentrating the need for skilled instructors. Training using physical equipment is often expensive due to the cost of the equipment itself, maintenance, supplies, and facilities. By replacing these physical requirements with virtual environments, simulators remove many recurring costs.

#### 18.4.5.8 Realistic experience

Simulations can help close the knowledge gap between theory and practice as they offer practical experience that is directly applicable to real-world problems. The realistic experience offered by the simulation-based learning enables students to practice and develop abilities in a secure setting that replicates real-world circumstances. This method of learning by doing improves decision-making abilities without the danger of real-world repercussions, increases knowledge retention, and bridges the gap between theory and practice.

### 18.4.6 Disadvantages of simulation-based learning

Disadvantages of simulation-based learning include high costs, the need for technical expertise, and the potential for simulations to lack realism or transferability to real-world situations. Other drawbacks include challenges in assessment, potential learner skepticism, and the significant time and resources required for development and maintenance. In addition, simulations may not perfectly replicate the complexities and unpredictability of real-world scenarios, which can lead to a gap between simulated performance and actual ability. Simulated environments provide the artificial environment which may not fully capture the dynamics of real-world situations, such as the stress of a genuine emergency or the nuances of a classroom setting.

### 18.4.7 E-learning

E-learning in clinical medical education provides flexible, accessible, and personalized training through digital tools like videos, interactive case studies, and quizzes, preparing students for real-world practice (Doherty, 2010). It can supplement or replace traditional lectures without compromising quality, but challenges like limited access to technology and infrastructure in some regions must be addressed. E-learning platforms help with knowledge retention, skills development, and offer features such as self-assessment tools to monitor progress.

### 18.4.8 Conduction of e-learning

E-learning is a popular and successful clinical teaching strategy that is frequently used in conjunction with traditional methods in a blended learning approach. In order to promote communication and knowledge transfer in a flexible, learner-centered setting, it is conducted using technology and strong educational concepts. E-learning requires methodical planning, development, execution, and evaluation when used as a clinical teaching approach. The fundamental procedures guarantee that the online format efficiently facilitates the growth of clinical knowledge, logic, and useful abilities. Clinical skills courses, which focus on fundamental procedures like vital sign assessment, phlebotomy, infection control such as handwashing, and basic life support, or medical terminology courses, which teach the fundamental vocabulary used in healthcare settings, are examples of typical e-learning courses (Delungahawatta, 2022). However, it also includes advanced and specialized topics like critical care, which is taught as an online master's program covering the management of critically ill people, including trauma, organ failure, and infection. The following are the steps of conduction of e-learning.

#### *18.4.8.1 Planning and design*

##### 18.4.8.1.1 Identify learner needs and objectives

Understanding the target audience's existing knowledge, skill gaps, and learning preferences is the first step in the process. In order to set specific, measurable, attainable, relevant, and time-bound goals, it is crucial to assess the learner's existing knowledge, abilities, and experience. After conducting a needs analysis, it is important to determine precise, quantifiable, and pertinent learning goals for the e-learning program or module keeping in mind the learning needs and objectives. This process can be accomplished by using frameworks such as Bloom's taxonomy. It is vital to make sure that these objectives are measurable and directly meet the needs that have been identified.

##### 18.4.8.1.2 Design the course structure

Outlining the course flow, which divides the course's content into digestible modules or units, is a crucial part of creating the course structure. A logical development that expands on earlier knowledge is crucial in this regard. A clear introduction with learning objectives, modularized content divided into manageable courses with interactive elements, and a summary and assessment at the end are all components of a good e-learning course structure. Prior to developing, it is crucial to establish the goals of the course and the demands of the target audience, draft a thorough outline, develop and test the material, and then apply and assess the progress of the learners.

#### 18.4.8.1.3 Select appropriate methods and tools

The learning objectives and the approaches must coincide. A thorough examination of your learning objectives, target audience, and available resources is necessary before choosing the best e-learning techniques and tools. Combining different techniques and choosing resources that facilitate smooth delivery and engagement is the optimal strategy. Selecting an appropriate learning management system or platform to hold the contents is crucial, as is deciding how the content will be presented, such as through readings, videos, case studies, simulations, and discussion forums.

### *18.4.8.2 Content development*

#### 18.4.8.2.1 Create engaging content

Clear learning objectives and a modular framework should be the first steps in creating interesting e-learning content. Interactivity, storytelling, and real-world scenarios should then be added. Using a variety of strategies to keep students engaged and motivated throughout the course by including as strategies and tools such as frequent feedback, gamification, and multimedia are helpful in this regard.

#### 18.4.8.2.2 Incorporate interactive elements and assessments

In order to engage the learner and make the process more engaging, it is crucial to incorporate interactive aspects. To measure comprehension and advancement, it's crucial to combine formative and summative evaluations. It is important to remember that assessments should be in line with learning objectives.

#### 18.4.8.2.3 Obtain necessary approvals

It has to be made sure that every piece of content used in the process of e-learning is correct, authentic, current, and verified by subject matter experts. Getting the curriculum committee's permission for the course design and making sure it complies with the organization's overall learning objectives and quality requirements are considered as good practices.

### *18.4.8.3 Implementation and facilitation*

#### 18.4.8.3.1 Launch the course

Prior to launching an e-learning course, the content must be finalized and made to be of a high caliber. Another crucial element is getting the platform ready for dissemination by publishing it to an internal system or a learning management system. Prior to launch, quality checks must be carried out. After that, a strategic communication plan must be put into place, which should involve introducing the course, emphasizing its advantages, and facilitating learners' access.

#### 18.4.8.3.2 Facilitate interaction and mentoring

To facilitate effective interaction and mentoring in e-learning, it is vital to design collaborative activities, leverage diverse communication tools, and establish a supportive learning environment. The goal is to build a sense of community that counters the isolation often associated with online learning. In addition, encouraging regular communication between students and faculty through discussion forums or virtual sessions is also important. Mentoring and support are essential in an online environment to reduce learner isolation and promote engagement.

#### 18.4.8.3.3 Ensure accessibility and support

To ensure accessibility and support in e-learning, it is important to design content with multiple formats, clear navigation, and compatibility with assistive technologies. Providing support through consistent training for instructors, alternative assessment options, and a clear feedback mechanism for users to report issues is also necessary to ensure the success of the e-learning process.

### *18.4.8.4 Evaluation and refinement*

#### 18.4.8.4.1 Gather feedback

To gather feedback in e-learning, methods like anonymous surveys, focus groups, and in-course questions to collect direct feedback can be used. In addition, gathering indirect feedback from student performance on assessments and engagement with gamification elements is also helpful. Effective feedback should be specific, timely, and lead to actionable improvements in the course.

#### 18.4.8.4.2 Analyze performance data

Analyzing e-learning performance involves collecting and analyzing data to understand learner progress, engagement, and outcomes, which then informs improvements to the course and learning experience. The process typically includes setting goals, gathering data from various sources like learning management systems and surveys, and analyzing this data to identify strengths and weaknesses. Effective analysis allows for personalized learning, optimized content, and overall enhancement of the e-learning program's effectiveness.

#### 18.4.8.4.3 Iterate and improve

In e-learning, iteration and improvement entail an ongoing cycle of design, development, feedback collection, and improvement. This strategy, which is frequently informed by vigorous instructional design approaches or iterative design, guarantees that courses continue to be efficient, interesting, and pertinent to the needs of students.

## 18.4.9 Benefits of e-learning in clinical teaching

E-learning in clinical teaching offers significant benefits including enhanced flexibility and accessibility, standardized content delivery, cost-effectiveness, and the ability to foster self-directed learning. It complements traditional methods and is particularly valuable for medical professionals with demanding schedules. Some of the key benefits of e-learning in clinical teaching include the following:

### *18.4.9.1 Flexibility and accessibility*

E-learning transcends geographical and time barriers, allowing both undergraduate and postgraduate medical trainees to access educational materials and complete training at their convenience, fitting study around demanding clinical schedules and personal commitments.

### *18.4.9.2 Standardized and consistent training*

Digital platforms provide standardized training modules, ensuring all healthcare staff receive the same high-quality education and exposure to evidence-based content, which is crucial for maintaining consistent standards of patient care across different locations or institutions.

### *18.4.9.3 Cost-effectiveness*

E-learning can reduce costs associated with traditional classroom training, such as travel, accommodation, and physical infrastructure such as lecture halls and printed materials. This makes education more affordable for both institutions and students.

### *18.4.9.4 Support for self-directed learning*

It encourages students to take greater control over their education by learning at their own pace and revisiting complex topics or video demonstrations as needed to achieve mastery.

### *18.4.9.5 Enhanced knowledge retention and performance*

A well-designed e-learning, especially when using interactive and multimedia elements like videos, animations, case studies, and quizzes can lead to improved knowledge acquisition, better knowledge retention, and higher scores in assessments compared to traditional methods alone.

### *18.4.9.6 Opportunities for specialization and collaboration*

E-learning platforms provide a wealth of chances for specialization and collaboration since it embraces interactive, community-driven, and focused educational approaches instead of passive, self-paced learning. Through resources like virtual classrooms, peer-to-peer mentoring, and cooperative, project-based work, these platforms allow users to acquire specific skills and participate in social learning.

### *18.4.9.7 Safe practice environments*

The use of VR and simulations in e-learning allows students to practice complex diagnostic and therapeutic skills on virtual patients in a safe, controlled environment, without the risk of harming actual patients. However, e-learning is most effective when integrated into a blended learning approach that combines digital education with essential hands-on clinical experience to ensure students develop necessary practical and communication skills.

### 18.4.10 Challenges of e-learning in clinical teaching

E-learning in clinical teaching faces several challenges, primarily concerning the difficulty in developing practical skills online, technical barriers, and a lack of interpersonal interaction. These issues affect both students and faculty members. The key challenges of e-learning include the following:

#### *18.4.10.1 Difficulty in acquiring clinical or practical skills*

The inability to effectively teach and evaluate practical, clinical, and laboratory skills, such as physical examinations and surgical procedures, without in-person, patient-based interaction and supervision, is a significant obstacle in e-learning.

#### *18.4.10.2 Technical issues and infrastructure*

Reliance on technology means that issues like unstable internet connectivity, lack of access to appropriate devices such as computers or iPads, and outdated hardware can disrupt the learning process, particularly in resource-limited areas.

#### *18.4.10.3 Lack of face-to-face interaction*

The absence of in-person interaction with instructors and peers can lead to feelings of isolation and disconnection. This also makes it difficult to receive immediate feedback and clarification, which are crucial for effective learning and communication skills development in a clinical setting.

#### *18.4.10.4 Student motivation and self-discipline*

E-learning demands a high degree of self-motivation and effective time management. Without the structured environment of a traditional classroom, some students may struggle to stay focused, manage distractions like social media, and meet deadlines.

#### *18.4.10.5 Faculty preparedness and resources*

Instructors may lack the necessary technical skills and training to design engaging, high-quality e-learning content effectively. Developing robust, interactive online modules is often time-consuming and requires significant institutional support and resources.

#### *18.4.10.6 Assessment and academic integrity*

Ensuring the authenticity of student work and assessing complex learning outcomes beyond basic knowledge testing can be more challenging in an online environment compared to supervised, in-person examinations.

#### *18.4.10.7 Resistance to change*

Some educators and students may have a negative attitude toward e-learning, preferring traditional teaching methods and being resistant to adopting new technologies or pedagogical approaches.

### 18.4.11 Virtual and augmented reality

VR creates a fully immersive, simulated environment that replaces the real world, while AR overlays digital information onto the real world without replacing it. The main difference is that VR disconnects you from your physical surroundings with a headset, whereas AR uses devices like smartphones or AR glasses to add to what you already see. In medical education, VR and AR are used to provide immersive and interactive training for hands-on procedures, anatomical study, and practicing patient interactions in a safe, risk-free environment (Hsieh, 2018). The first step to use VR for clinical teaching is to conduct a needs assessment in order to identify learning objectives and technical requirements. Then a VR program is implemented by setting up the necessary hardware and software, creating or choosing a suitable virtual environment, and integrating VR sessions into the curriculum to provide immersive, risk-free practice opportunities. The final step would be to evaluate the program's effectiveness and use the feedback to improve future iterations. The conduction of clinical teaching using VR includes the following steps.

#### *18.4.11.1 Plan and design the virtual reality program*

##### 18.4.11.1.1 Conduct a needs assessment

A needs assessment is conducted by involving various stakeholders, such as educators, students, and technical experts, to identify specific problems and learning needs that VR can address. In this, it is important to consider factors like accessibility, assessment methods, and required personnel.

##### 18.4.11.1.2 Set learning objectives

The next step is to clearly define what skills or knowledge students should gain. For example, a VR experience could focus on patient assessment, diagnosis, or surgical techniques.

##### 18.4.11.1.3 Design the virtual environment

It is important to first choose or create a realistic and interactive 3D virtual setting based on the requirement of the topic to be taught. For example, a virtual setting would include an operating room, an emergency department, or a clinical lab. It is important to integrate interactive elements such as the ability to perform procedures, examine anatomical models, or respond to emergencies.

#### *18.4.11.2 Implement and integrate the virtual reality program*

After clearly defining the learning objectives and designing the virtual environment, the next step would be to set up the technology which includes acquiring and preparing the necessary hardware such as headsets, controllers, computers, and software simulation platforms. It is also vital to train users to ensure they are comfortable with the technology and then integrate the VR training into the curriculum, using it as a supplement to traditional methods.

#### *18.4.11.3 Integrate into the curriculum*

The third step would be to integrate the VR sessions with traditional teaching methods like lectures and hands-on labs to create a comprehensive learning experience. In order to do this, it is important that a dedicated time is scheduled for students to use VR, either in lab or in a more flexible environment. VR is primarily used for both repetitive practice and to bridge the gap between theory and practical application in a safe, simulated environment. It also helps in providing support and training as it offers guidance on how to use the virtual reality equipment and software to ensure a smooth learning experience.

#### *18.4.11.4 Evaluate and iterate*

Evaluating and iterating VR for clinical teaching involves a structured, multiphase approach that utilizes specific frameworks and metrics to ensure the technology is effective, usable, and integrated with existing curricula. This can be done by monitoring progress where the way in which students are using the VR modules can be monitored and their performance can be assessed. It is also important to collect feedback from both students and instructors on what is working well and what can be improved. The data collected should be used to make necessary adjustments to the VR content, learning objectives, or overall implementation strategy.

### 18.4.12 Benefits of virtual reality in clinical teaching

VR and AR are increasingly being integrated into medical education, offering numerous benefits for clinical teaching. Some of the important benefits of using VR and AR are listed below:

#### *18.4.12.1 Enhanced visualization of complex concepts*

VR allows students to immerse themselves in learning basic sciences such as learning human anatomy via the 3D models or learning about the molecular structure in biochemistry, enabling them to explore body systems, organs, and tissues from every angle. This deep, interactive visualization can lead to the better understanding of complex structures and physiological processes. On the other hand, AR can superimpose anatomical images or 3D models onto the body of a patient or a mannequin, enhancing understanding during physical examination.

#### 18.4.12.2 Safe, risk-free practice

VR offers a safe environment for students to practice procedures and clinical scenarios without the risk of causing harm to real patients. This is especially valuable for high-risk procedures like surgeries, where practicing on virtual patients can build skills and confidence before hands-on experience. AR, on the other hand, can be used to guide medical students through real-life procedures by providing real-time feedback and instructions on the patient's body, reducing errors during clinical practice.

#### 18.4.12.3 Simulated scenarios and decision making

VR can create highly realistic clinical scenarios, such as emergency situations or rare diseases, which are difficult to simulate in real life. These simulations help students develop critical thinking and decision-making skills in a controlled, repeatable environment. Apart from this, students can practice diagnostic reasoning and treatment plans in a variety of situations, allowing them to experience a wide range of possible scenarios they might not otherwise encounter in a traditional clinical setting.

#### 18.4.12.4 Remote learning and accessibility

VR and AR technologies can facilitate remote learning by enabling students to participate in simulations and virtual labs from anywhere in the world. This is particularly beneficial for students in geographically isolated areas or for medical institutions that want to reach a wider audience. Students can access virtual cadavers, surgical procedures, or clinical simulations, helping them stay engaged and continue learning even when they are not physically present in the classroom or clinical setting.

#### 18.4.12.5 Hands-on experience without the need for physical resources

VR and AR eliminate the logistical constraints of physical resources like cadavers, mannequins, or medical equipment. Students can practice as many times as needed without requiring additional physical materials, and instructors can create various clinical scenarios on-demand. Virtual simulations can reduce the need for physical materials like cadavers, mannequins, and medical supplies. VR and AR technologies also save on travel and accommodation costs for students and faculty involved in clinical rotations and practical workshops.

#### 18.4.12.6 Improved retention and engagement

The immersive and interactive qualities of VR and AR can boost student engagement and retention of information. When students actively participate in their education, they learn more effectively, and these technologies offer a more experiential, hands-on approach than traditional lecture-based methods. Because they are actively participating and seeing scenarios rather than merely reading about them, students are more likely to remember clinical knowledge and procedural skills taught in an immersive setting.

#### 18.4.12.7 Personalized learning

VR and AR platforms can adapt to the learner's pace and needs. For example, if a student struggles with a particular procedure or concept, they can repeat the simulation until they achieve mastery. Instructors can monitor performance in real-time and provide targeted feedback.

#### 18.4.12.8 Assessment and feedback

VR and AR can be used for continuous assessment, as platforms can track a student's performance during simulated exercises. This allows instructors to identify areas where students may need further practice and provide immediate feedback. The data collected from simulations can also be used to assess competency in various procedures or clinical scenarios.

### 18.4.13 Challenges in using the virtual and augmented reality as clinical teaching methodologies

#### 18.4.13.1 High cost and resource requirements

The initial investment in VR headsets, AR devices, haptic simulators, software licenses, and maintenance is often expensive. Moreover, the technical infrastructure in the form of powerful computers, dedicated rooms, and wireless networks adds additional cost. Smaller or resource-limited institutions may struggle to adopt these technologies.

#### 18.4.13.2 Technical limitations and reliability issues

The technical limitations are that the VR and AR systems may experience software glitches, calibration problems, or hardware malfunctions, interrupting learning sessions. On the other hand, limited field of view, lag, or low-resolution graphics can reduce realism and affect the learning experience.

##### 18.4.13.2.1 Lack of standardization

One of the important drawbacks is that there is no universal curriculum or standard for VR- and AR-based clinical training. There is a lot of variations in content quality, fidelity, and instructional design, which make it difficult to ensure consistent learning outcomes. Even accreditation bodies have not fully integrated VR/AR competencies into standard medical curricula.

##### 18.4.13.2.2 Integration challenges in traditional curriculum

Redesigning the curriculum, modifying the schedule, and closely aligning it with learning objectives are all necessary when integrating VR and AR. Implementation may also be slowed by faculty resistance to technological change. Additionally, it is challenging to incorporate VR and AR into clinical programs without overburdening students due to time constraints.

##### 18.4.13.2.3 Dependence on technology

There is also over-reliance or dependence on simulations, which may reduce exposure to real patient variability. The drawback here is that the students may become proficient in controlled digital environments but less adaptable to real-world uncertainty.

### 18.4.14 Problem-based learning (PBL)

PBL in medical education is a student-centered approach that uses real-world clinical cases to develop critical thinking, problem-solving, and self-directed learning skills. Instead of passive lecturing, students work in small groups to analyze complex cases, identify learning issues, and research solutions, which improves knowledge integration, application, and retention (Bridges, 2012).

### 18.4.15 Problem-based learning as a clinical teaching methodology

PBL is an effective clinical teaching methodology because it promotes active, student-centered learning and develops critical thinking skills by having students work collaboratively on real-world clinical cases. It has been shown to enhance knowledge retention, improve critical thinking abilities, and increase student satisfaction compared to traditional methods, though some research suggests inconsistent results regarding critical thinking improvement.

### 18.4.16 Advantages of problem-based learning in clinical teaching

#### 18.4.16.1 Integrates theory and practice

PBL uses clinical scenarios, such as patient cases, to provide a direct link between theoretical knowledge and its real-world application in patient care. It helps to integrate theory and practice by having students solve complex, real-world problems using a framework that requires them to apply theoretical knowledge to a practical situation. Students are prompted to identify knowledge gaps, research relevant theories, and then collaborate to implement strategies and evaluate the outcomes, thereby creating a cycle of application and reflection that builds professional competence and critical thinking skills.

#### 18.4.16.2 Develops critical thinking and diagnostic skills

PBL enhances the development of critical thinking and diagnostic skills as it facilitates the students to gain a fundamental clinical skill such as the ability to formulate hypotheses, develop a differential diagnosis, and choose suitable investigations and treatment. Furthermore, exposing students to real-world problems that are authentic, complex, and open-ended leads to development of critical thinking and diagnostic skills. Students must actively analyze the situation, identify knowledge gaps, conduct independent research, work with peers, and apply their findings to develop a solution.

#### 18.4.16.3 Enhances self-directed learning

PBL promotes self-directed learning by requiring students to take the initiative in diagnosing their learning needs, setting goals, and finding resources to solve complex problems. In PBL, students work collaboratively to first identify what they need to learn and then independently research and study the relevant topics before returning to the group to apply their new knowledge. This process, guided by a teacher acting as a facilitator, develops self-directed learning skills by giving students ownership over their learning journey and teaching them to become lifelong learners.

#### 18.4.16.4 Promotes teamwork and communication

PBL promotes teamwork and communication by requiring students to collaborate on complex, real-world problems, which necessitates active listening, sharing ideas, negotiating roles, and providing constructive feedback to reach a shared solution. This process fosters essential skills like negotiation, compromise, active listening, and social skills as students work together to analyze a problem and develop a solution.

#### 18.4.16.5 Improves patient interaction skills

Through the analysis of genuine, patient-centered case studies, PBL enhances students' capacity to connect with patients and promotes critical thinking, teamwork, and communication. This method goes beyond rote memorization to teach students how to apply knowledge, work in groups, and develop empathy by highlighting the patient's humanistic requirements in a clinical setting.

#### 18.4.16.6 Boosts motivation and engagement

PBL makes learning more relevant and student-centered and increases enthusiasm and motivation, as students see the immediate value of the knowledge they are acquiring. It also boosts motivation and engagement by using real-world problems that make learning relevant. Students become more engaged because they are actively involved in a collaborative, hands-on approach that develops critical thinking, problem-solving skills, and a sense of accomplishment.

#### 18.4.16.7 Encourages reflection

PBL encourages reflection by embedding it into the learning cycle, requiring students to analyze the problem, identify knowledge gaps, and evaluate their solutions. Reflection is not a single event but an ongoing process that includes reflecting before, during, and after the problem-solving process, using methods like self-assessment, group discussion, and revisiting prior steps to improve future learning.

### 18.4.17 Challenges of problem-based learning as a clinical teaching methodology

The substantial time and resources needed for implementation, the possibility of insufficient content coverage, the heavy workload for both students and instructors, issues with group dynamics, and difficulties with assessment are some of the challenges associated with using PBL as a clinical teaching method. Additionally, it calls for well-crafted cases, in-depth tutoring, and a change in student-directed learning practices, all of which can be initially taxing for students. The following are some significant disadvantages of employing PBL as a clinical teaching strategy:

#### 18.4.17.1 High demand on educators

The high demand on educators is a significant challenge in using PBL as a clinical teaching method, since it requires more time for preparation, facilitation, and skill development compared to traditional methods. Educators need more time for planning and engagement, and they often lack the specific training and experience required to be effective facilitators, leading to inconsistencies in their guidance and feedback. This high demand stems from the shift from a traditional lecture format to a small-group, student-centered approach that necessitates a different set of skills from the educator.

#### 18.4.17.2 Lack of hands-on experience

One acknowledged and important issue in applying PBL in clinical teaching is the absence of practical clinical experience. The main disadvantage of a completely PBL paradigm or one with inadequate practical integration is the possibility of a gap between theoretical knowledge and practical application, even though it is excellent at developing these cognitive and analytical skills.

#### 18.4.17.3 Student-directed learning

Student-directed learning becomes a challenge in PBL, as it stems from students needing to take initiative, identify their own learning needs, and manage their own learning process. This is because students may struggle with lacking the necessary skills for self-directedness, such as goal setting, critical analysis, and time management, which are essential for success in PBL. It requires a shift in mindset from passive reception to active, independent learning, which can be a difficult transition for students.

#### 18.4.17.4 Uneven content coverage

As PBL is student-centered and self-directed, it can result in the removal of important topics and varying levels of comprehension within student groups. This presents a significant obstacle when adopting PBL as a clinical teaching method. If the process is not well controlled, there is a chance that the curriculum coverage will be less thorough than with traditional approaches.

#### 18.4.17.5 Student workload

One major issue with employing PBL as a clinical teaching approach is increased student workload. Since the majority of PBL is focused on students' independent study, they frequently encounter a significant rise in their overall workload, which can cause stress and a sense of overload. Additionally, it can take the students a long time to become used to this new, student-centered approach, which could cause tension and anxiety.

#### 18.4.17.6 Poor group dynamics

Poor group dynamics present a significant challenge in PBL within a clinical setting, potentially leading to reduced student motivation, unequal knowledge acquisition, and negative learning outcomes. These issues stem from various interpersonal and structural factors. In addition to this, learning can be negatively affected if the group dynamics are dysfunctional, for example, if one student dominates the discussion or if there is a lack of constructive participation.

#### 18.4.17.7 Subjective assessment

Since subjective assessment may affect the validity and impartiality of student evaluations, it poses a serious obstacle to the use of PBL in clinical education. Standardized assessment can be difficult in PBL because on-the-spot evaluations may be inconsistent and student performance may be subjectively assessed.

### 18.4.18 Case-based Learning (CBL)

In medical education, CBL is a student-centered teaching approach that helps students apply theoretical knowledge to real-world clinical problems. This approach improves critical thinking, clinical reasoning, and problem-solving skills by having students assess a case, discuss a differential diagnosis, and choose a treatment plan in a supervised, small-group setting. Unlike PBL, CBL focuses on applying past knowledge to solve a problem rather than identifying the problem.

### 18.4.19 Case-based learning as a clinical teaching methodology

CBL is a clinical teaching methodology that uses real-world clinical cases to improve medical students' problem-solving, critical thinking, and clinical reasoning skills by bridging the gap between theoretical knowledge and practical application. Through collaborative, student-centered discussions of patient scenarios, students learn to analyze problems, apply basic science knowledge, and develop diagnostic and management skills in a more interactive and engaging environment than traditional lectures (Zhao, 2022). PBL and CBL share many similarities due to their shared pedagogical foundation in constructivist learning theory, which emphasizes that students actively create knowledge rather than passively absorb it. Moving away from conventional lecture-based approaches, both of these approaches promote active participation, group debate, critical analysis, teamwork, and problem-solving abilities. In both of these methodologies, actual or authentic clinical case situations are used to bridge the gap between theory and practice and make learning relevant. Instead of being a traditional lecturer who imparts all the knowledge, the instructor in both methods largely serves as a guide or facilitator.

**Summary**

Clinical education approaches aim to improve students' clinical reasoning, psychomotor skills, communication abilities, and professional attitudes by putting them in real-world practice settings. Unlike classroom learning, clinical training must balance patient safety and learner development while adapting to changing healthcare environments. Clinical instructors play a variety of roles, such as educator, supervisor, facilitator, evaluator, and role model. Effective clinical teaching requires a purposeful blend of instructional strategies, learner-centered approaches, and supportive learning environments. Employing these strategies will enable the creation of competent, confident, and compassionate healthcare professionals.

**Key point**

**Future perspectives: a shift towards the hybrid model**

Future clinical teaching methods will be influenced by rapidly advancing technology, evolving healthcare demands, and a growing emphasis on competency-based education. It promises a dynamic blend of technology, individualized instruction, interdisciplinary collaboration, and humanistic care. In addition to being technically skilled, the future generation of clinical educators will be adaptable, compassionate, and ready to use innovation while maintaining core therapeutic principles in order to meet the evolving needs of modern healthcare.

# References

Al-Elq, A.H., 2010. Simulation-based medical teaching and learning. Journal of Family and Community Medicine 17 (1), 35. Available at https://doi.org/10.4103/1319-1683.68787.

Bassaw, B., Naraynsingh, V., 2011. Ward-rounds: role in clinical teaching and learning in contemporary medicine. The West Indian Medical Journal 60, 601–603.

Bridges, S., McGrath, C., Whitehill, T.L., Problem-Based, 2012. Learning in Clinical Education: The Next Generation. Springer, Netherlands. Available at http://doi.org/10.1007/978-94-007-2515-7.

Brunton, L., Tazzyman, A., Ferguson, J., Hodgson, D., Nelson, P.A., 2022. The challenges of integrating signposting into general practice: qualitative stakeholder perspectives on care navigation and social prescribing in primary care. BMC Primary Care 23 (1), 27314553. Available at https://doi.org/10.1186/s12875-022-01669-z, https://bmcfampract.biomedcentral.com/.

Crumlish, C.M., Yialamas, M.A., McMahon, G.T., 2009. Quantification of bedside teaching by an academic hospitalist group. Journal of Hospital Medicine 4 (5), 304–307. Available at https://doi.org/10.1002/jhm.540, http://www3.interscience.wiley.com/cgi-bin/fulltext/122440051/PDFSTART.

Delungahawatta, T., Dunne, S.S., Hyde, S., Halpenny, L., McGrath, D., O'Regan, A., et al., 2022. Advances in e-learning in undergraduate clinical medicine: a systematic review. BMC Medical Education 22 (1). Available at https://doi.org/10.1186/s12909-022-03773-1.

Doherty, I., McKimm, J., 2010. E-learning in clinical teaching. British Journal of Hospital Medicine 71 (1), 44–47. Available at https://doi.org/10.12968/hmed.2010.71.1.45973.

Elendu, C., Amaechi, D.C., Okatta, A.U., Amaechi, E.C., Elendu, T.C., Ezeh, C.P., et al., 2024. The impact of simulation-based training in medical education: a review. Medicine (United States) 103 (27), e38813. Available at https://doi.org/10.1097/MD.0000000000038813, https://journals.lww.com/md-journal/pages/default.aspx.

Gonzalo, J.D., Chuang, C.H., Huang, G., Smith, C., 2010. The return of bedside rounds: an educational intervention. Journal of General Internal Medicine: Official Journal of the Society for Research and Education in Primary Care Internal Medicine 25, 792–798.

Hulland, O., Farnan, J., Rabinowitz, R., Kearns, L., Long, M., Monash, B., et al., 2017. What's the purpose of rounds? A qualitative study examining the perceptions of faculty and students. Journal of Hospital Medicine 12 (11), 892–897. Available at https://doi.org/10.12788/jhm.2835.

Hsieh, M.C., Lee, J.J., 2018. Preliminary study of VR and AR applications in medical and healthcare education. Journal of Nursing and Health Studies 3 (1), 1.

McLean, S.F., 2016. Case-based learning and its application in medical and health-care fields: a review of worldwide literature. Journal of Medical Education and Curricular Development 3. Available at https://doi.org/10.4137/jmecd.s20377.

Onishi, H., 2008. Role of case presentation for teaching and learning activities. The Kaohsiung Journal of Medical Sciences 24 (7), 356–360. Available at https://doi.org/10.1016/s1607-551x(08)70132-3, PMID:18805750, PMCID:PMC11917907.

Peters, M., Ten Cate, O., 2022. Bedside teaching in medical education: a literature review. Perspectives on Medical Education 3 (2), 76–88. Available at https://doi.org/10.1007/s40037-013-0083-y.

Ramirez, J., Singh, J., Williams, A.A., 2016. Patient satisfaction with bedside teaching rounds compared with nonbedside rounds. Southern Medical Journal 109 (2), 112–115. Available at https://doi.org/10.14423/SMJ.0000000000000419, http://journals.lww.com/smajournalonline.

Shaik, T., Aggarwal, K., Singh, B., Sawhney, A., Naguluri, R., Jain, R., et al., 2023. A comprehensive analysis of different types of clinical rounds in hospital medicine. Baylor University Medical Center Proceedings 37, 1–7. Available at https://doi.org/10.1080/08998280.2023.2261086.

Spencer, J., 2003. ABC of learning and teaching in medicine: Learning and teaching in the clinical environment. BMJ (Clinical research ed.) 326, 591–594. Available at https://doi.org/10.1136/bmj.326.7389.591.

Wolpaw, Wolpaw, Papp, 2003. SNAPPS: a learner-centered model for outpatient education. Academic Medicine: Journal of the Association of American Medical Colleges 78, 893–898. Available at https://doi.org/10.1097/00001888-200309000-00010.

Zhao, Y, 2022. Contrast Media & Molecular Imaging 26.

Chapter 19

# Artificial intelligence in medical education

*"AI will not replace doctors but instead will augment them enabling physicians to practice better medicine with greater accuracy and increased efficacy."*

—Benjamin Bell

## 19.1 Introduction

Artificial intelligence (AI) has undoubtedly made a big impact on healthcare and is starting to have an influence on medical education. Traditional teaching approaches are no longer enough because future doctors must traverse a healthcare system powered by data, digital tools, and cutting-edge technology. AI-powered platforms are transforming the way medical students learn, practice, and refine their clinical skills by offering personalized learning experiences, realistic simulations, and instant feedback that increases competence and confidence. However, using AI also brings up significant issues around ethics, data literacy, and the changing role of the doctor. Understanding how AI affects medical education is essential to improving training paths and preparing the next generation of healthcare professionals for a future enabled by AI (Narang, 2023). One of the main things that ponders us is that AI entered the teaching and learning process later than it did in clinical medicine. Some of the reasons for this late entry include the lack of standardized curricula and faculty readiness, the emphasis on humanistic skills, the uncertainty surrounding the effectiveness of education, infrastructure, and financial constraints, the slow institutional change in academia, and the rapid advancements in AI surpassing curricula.

## 19.2 Current status of artificial intelligence in medical education

AI has transitioned from an experimental concept to a core strategic pillar in medical education, driven by the need to prepare graduates for an AI-enabled clinical environment.

### 19.2.1 Integration into curricula

#### *19.2.1.1 Mandatory artificial intelligence literacy*

AI literacy is becoming an essential component of medical education due to the quick integration of AI into clinical workflows. To guarantee that future doctors can navigate AI-driven healthcare in a safe and moral manner, leading organizations have devised frameworks and standards. In this sense, medical schools worldwide now mandate that all healthcare professionals finish basic AI training that covers ethical reasoning and practical applications.

#### *19.2.1.2 Formal competency frameworks*

Formal AI principles and competency frameworks, such the AAMC principles on AI Competencies for Medical Educators and the Global Framework for an AI Curriculum in Medicine, have evolved from theoretical suggestions into organized, practical manuals that are embraced by international health and education organizations. In an AI-integrated healthcare system, these frameworks specify the particular knowledge, abilities, and moral principles that educators and doctors must possess. In order to guarantee that graduates can critically assess and safely use AI tools in practice, medical schools are also setting minimal competency standards akin to evidence-based medicine.

#### *19.2.1.3 Specialty-specific training*

Specialty-specific training is replacing broad AI literacy in medical education. To guarantee that graduates can handle the specialized technical tools in their areas of expertise, various medical schools have started including specific AI

**The Principles and Practice of Medical Education. DOI:** https://doi.org/10.1016/B978-0-443-44197-4.00021-7

competencies into their residency programs. Furthermore, specific AI modules are becoming commonplace in disciplines where AI-assisted diagnosis is already deeply ingrained, such as radiology, pathology, and cardiology.

### 19.2.2 Practical educational tools

#### *19.2.2.1 Artificial intelligence–powered simulations*

AI-powered simulations are now vital teaching aids rather than experimental novelties. These simulations use physiology engines, virtual reality, and generative AI to build "scaffolded" learning environments that adjust to each student's unique performance and speed. With the help of these resources, students can develop their diagnostic reasoning, empathy, and history-taking skills in a low-risk setting (Garg, 2020).

#### *19.2.2.2 Personalized learning engines*

Due to the emergence of personalized learning engines, the "one-to-many" medical education approach has given way to the "one-to-one" model. These systems, which are frequently referred to as precision medical education, use AI to customize curriculum, content, and assessments in real time to each student's distinct learning style and performance. Additionally, in order to move away from a "one-to-many" lecture approach, these AI platforms can now analyze individual student performance, automatically identify knowledge gaps, and create tailored study aids or test questions.

## 19.3 Need for artificial intelligence in healthcare and medical education

AI is essential for precision, effectiveness, and customization in healthcare and education. It addresses problems including expanding medical knowledge and meeting the needs of diverse learners. Better patient monitoring, personalized treatment plans, early disease detection, and administrative automation are all made possible by AI in healthcare. AI in education provides platforms for adaptive learning, individualized feedback, and efficient assessment tools. Additionally, it prepares upcoming professionals to employ AI-driven technologies in increasingly challenging fields. AI is essential in medical education to enhance student learning through personalized, adaptive content, virtual patients for risk-free practice, and automated feedback, in addition to enhancing curriculum design and streamlining administrative tasks for teachers. Due to the rapidly evolving medical field and the growing number of tech-savvy students, future physicians need to be educated about AI applications and ethical issues in their practice (Paranjape et al., 2019). AI is revolutionizing medical education by providing personalized learning paths, realistic simulations, and automated assessment tools to improve student performance and engagement. Robotic simulators and AI-powered virtual patients offer safe environments for practicing challenging procedures and clinical skills. However, concerns including algorithmic bias, data privacy, ethical considerations, and the need for faculty training must be addressed for responsible and fair inclusion into medical courses (Fig. 19.1).

AI in healthcare and medical education is crucial for increasing training through simulations, lowering administrative burdens, boosting diagnostic accuracy, and personalizing patient care. With AI-driven tools that enable students to perform diagnostics in risk-free, simulated situations while providing individualized learning experiences, it closes the gap between theory and practice. The various needs of AI in medical education and healthcare are described below:

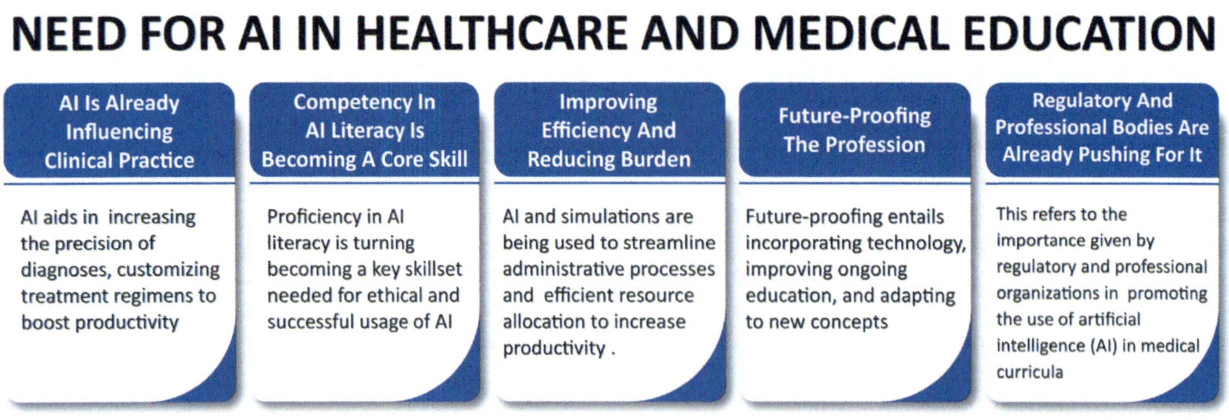


FIGURE 19.1 **Need for artificial intelligence (AI) in healthcare and medical education.** This figure shows the need for AI in healthcare and medical education.

### 19.3.1 Artificial intelligence is already influencing clinical practice

AI is transforming healthcare practice by improving diagnostic accuracy, personalizing treatment plans, and optimizing administrative procedures to increase efficiency. Among the various tasks that AI assists with are managing electronic health data, transcribing notes, analyzing medical imaging, and even predicting disease outbreaks. This allows medical personnel to focus more on patient care while potentially lowering costs and improving health outcomes.

### 19.3.2 Competency in artificial intelligence literacy is becoming a core skill

Proficiency in AI literacy is becoming a crucial skill set as it becomes necessary for the learner to interact with more prevalent AI systems in an ethical and successful manner. Beyond technical proficiency, this skill includes the ability to understand AI concepts, evaluate applications of intelligent teaching systems critically, and use AI tools in both personal and professional settings in an ethical manner. In order to navigate the developments in contemporary medicine and foster innovation, AI literacy today includes technical, ethical, critical thinking, and practical application characteristics. In this situation, it is necessary for the learner to go beyond merely learning about AI and work toward developing the skills that entails understanding AI concepts, interacting with tools effectively, critically assessing outputs for bias and accuracy, and using it in clinical practice and patient care in an ethical and responsible manner.

### 19.3.3 Improving efficiency and reducing burden

It is crucial to leverage AI and simulations, simplify administrative processes, and concentrate on efficient resource allocation in order to increase productivity and lessen the load in medical education. AI is helping us accomplish this aim and freeing up educators to concentrate on their primary teaching responsibilities by automating repetitive work, using simulation for safe and repeatable practice, and rethinking protocols to let allied health professionals handle administrative tasks (Davenport & Kalakota, 2019).

### 19.3.4 Future-proofing the profession

Future-proofing the medical field means integrating technology, enhancing continuing education, and adapting to new treatment models in order to stay up to date and improve patient outcomes. This means specializing in high-demand disciplines, embracing technology like AI and telemedicine, and concentrating on a multidisciplinary, patient-centered approach through ongoing training and a willingness to innovate.

### 19.3.5 Regulatory and professional bodies are already pushing for it

Regulatory and professional organizations are encouraging the use of AI in medical curriculum by defining fundamental competencies, providing guidance on ethical use and legal accountability, and developing helpful training resources. Their objective is to teach aspiring physicians how to use AI in clinical settings in a way that is ethical, effective, and safe.

## 19.4 Synergistic relationship of artificial intelligence and medical education

Medical education and AI collaborate to create a dynamic learning environment by augmenting human capabilities with AI-driven insights and tools. AI enables individualized training through interactive platforms and lifelike simulations while automating administrative tasks for educators. This collaboration fosters data-driven decision-making, which ultimately leads to more effective, patient-centered care by providing useful insights, improving skill development, and preparing aspiring medical professionals for difficult clinical scenarios. The various benefits from the synergy of AI and medical education include the following:

### 19.4.1 Augmented human capabilities

AI improves human intellect by handling data processing and pattern recognition, while humans provide empathy, common sense, and moral judgment. AI is improving human capabilities in medical education by providing customized learning, complex simulations, data-driven decision support, and repetitive tasks. Teachers can now focus on developing higher-order abilities like empathy and ethical reasoning (Chan & Zary, 2019).

### 19.4.2 Improved skill development

AI helps medical students build their skills through personalized learning, realistic simulations, and better diagnostic training. AI-powered platforms create personalized learning paths by assessing student performance, while virtual patients and simulations provide safe environments for honing clinical and diagnostic abilities. Additionally, AI tools increase learning efficiency by automating certain assessments and offering instant feedback.

### 19.4.3 More informed workforce

AI is creating a more informed medical staff by providing tailored learning, enhancing clinical skill development through simulation, automating administrative tasks, and providing robust decision support. This integration improves educational results and better prepares healthcare professionals for an AI-enhanced healthcare setting.

### 19.4.4 Driving innovation

AI is driving innovation in medical education by facilitating tailored learning, providing realistic simulations, and improving assessment methods. It allows practice on virtual patients and robotic simulators for skill development, allows automated scoring and exam creation to increase efficiency and accuracy, and tailors learning paths to individual needs. AI also increases the bar for transdisciplinary abilities since future healthcare settings will require students to understand how to collaborate with both human intelligence and AI.

## 19.5 Incorporating artificial intelligence into medical curriculum

Incorporating AI into the medical curriculum is crucial for contemporary medical education. This entails utilizing AI simulations and clinical decision support systems to create experiential learning opportunities, building specific AI-focused modules, and incorporating AI tools into already-existing courses. In order to prepare future doctors for AI-integrated healthcare settings, educators must address ethical issues, faculty training, and infrastructure while concentrating on utilizing AI to customize learning and increase administrative efficiency.

### 19.5.1 Benefit of integrating artificial intelligence into medical education

#### *19.5.1.1 Personalized learning*

Personalized learning is essential to medical education since it adapts to each student's individual strengths, weaknesses, and learning preferences. Additionally, it contributes to bettering student retention, engagement, and clinical preparedness. This approach moves away from a one-size-fits-all approach and develops self-directed, lifelong learners ready for the rapidly changing healthcare sector through the use of adaptive technologies, adaptable curricula, and customized feedback.

#### *19.5.1.2 Enhanced teaching and learning*

Teaching and learning in medical education can be improved through the use of technology-enhanced learning, active learning strategies including flipped classrooms, team-based learning, and precision education. While technology enables customized learning experiences, virtual simulations, and flexible access to content, active approaches involve students in problem-solving and collaborative learning to improve both knowledge acquisition and clinical skills. Precision education further enhances this by tailoring training to a student's current level and giving them specific future steps.

#### *19.5.1.3 Improved assessment*

AI's automatic scoring and real-time review capabilities have the ability to completely transform the evaluation process. AI is influencing medical school assessment by offering more automated and objective evaluations through technologies like AI-driven grading systems and simulations, which provide faster, more dependable feedback than traditional methodologies. Human oversight is necessary to maintain quality and handle limitations, but it also makes it possible to develop innovative assessment methods like procedural competency analysis and AI-generated test items.

#### *19.5.1.4 Administrative efficiency*

Owing to their commitments towards their administrative duties, faculty members may occasionally produce work of a poorer quality and experience burnout. Relieving the teaching staff of administrative duties became essential, especially in the clinical setting. For instance, entering patient data into electronic medical records is a very strenuous task for staff members. In this way, by automating administrative tasks with AI solutions like Abridge, educators and students can focus on higher-value tasks like healthcare and critical thinking.

#### *19.5.1.5 Preparation for the future*

As medical education continues to grow and evolve to keep up with the rapidly expanding technology, preparing future physicians with AI knowledge and abilities is crucial and is the need of the hour. To ensure that students are ready for AI-integrated healthcare environments and can advance medical innovations, it is imperative to integrate AI competencies into the curriculum at the educational level. Additionally, requiring undergraduates to gain knowledge about the fundamental ideas of AI and incorporating AI literacy into the curriculum will significantly boost the student's career in order to tackle the rapid advances in modern medical education.

## 19.6 Artificial intelligence tools used in medical education

The way medical education is taught, learned, and implemented in both academic and clinical settings is being redefined by the AI tools. By expanding medical education outside the boundaries of conventional medical education, AI has in fact altered the dynamics of medical education. The balance has moved in favor of AI-based education with the introduction of technologies like virtual learning paths, personalized learning systems, and simulation-based education to mention a few. Furthermore, the way medical education is approached has altered due to the impact on patient-centered care, system dynamics, and AI's involvement in shifting power relations. Some of the important tools used in the field of medical education are described below (Fig. 19.2).

### 19.6.1 Chatbots

Chatbots are AI systems designed to comprehend, analyze, and react to human inquiries regarding certain input data. They do this by extracting data from online sources and using an advanced natural language processing model to produce natural language. Since most medical students are tech-savvy, chatbot systems can be readily incorporated as online applications to support education. There has been a lot of interest in chatbots' ability to produce text that looks human and communicate with customers. In the context of medical education, chatbots can effectively automate and speed up time-consuming tasks like summarizing and assessing research and medical literature. Chatbots are mostly used as virtual teaching assistants to respond to questions from students and provide resources.

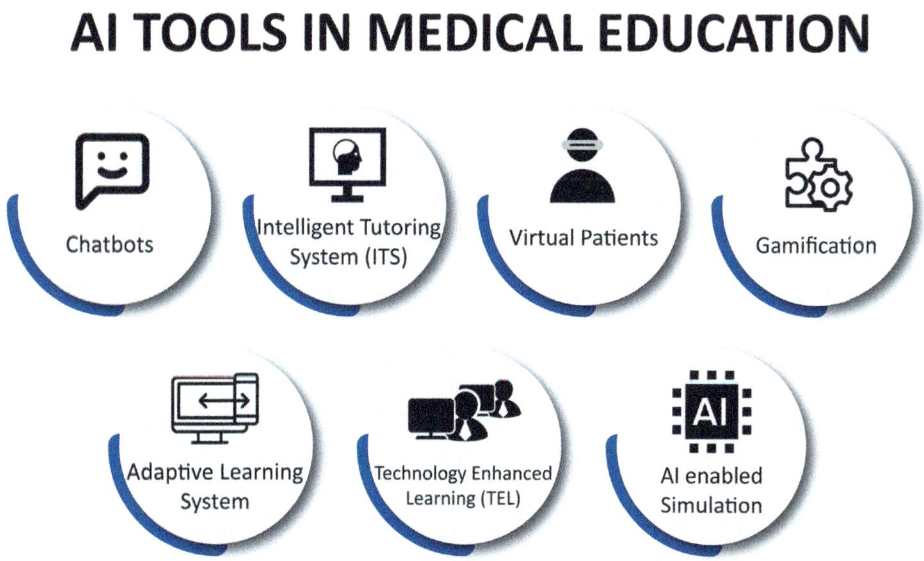


FIGURE 19.2 **Artificial intelligence (AI) tools in medical education.** This figure illustrates the various AI tools in medical education.

### 19.6.2 Intelligent tutoring system

Intelligent tutoring systems (ITSs) are becoming more and more potent instruments in the field of education by utilizing AI's ability to give individualized training. An intelligent teaching system is a cutting-edge platform that offers a highly customized learning experience by utilizing AI and natural language processing. These systems create an efficient and personalized learning environment by giving prompt feedback and taking into consideration each student's individual learning rate. In order to guarantee proper student progress, the ITS also frequently employs sophisticated algorithms to continuously evaluate and modify the content's difficulty (Sanae, 2024). ITS is in the vanguard of the educational revolution, revolutionizing how students acquire and assimilate knowledge by modifying instructional materials to accommodate various learning styles. It can modify the course material, adapt to different learning styles, and give fast feedback. The goal of ITSs is to improve our comprehension of the cognitive processes involved in teaching and learning. In order to provide educators control over what to teach, who to teach it to, and how to teach it, their design integrates AI technology. The algorithms use this data to pinpoint learning gaps and develop individualized lesson plans that can be modified to meet each student's needs and developmental stage. Through interpretation and adaptive learning, an ITS effectively replaces a human instructor. The ITS aims to address a number of common issues in medical education, including teaching course material, identifying students' knowledge gaps or strengths, providing automated feedback, choosing instructional materials based on students' needs, and promoting student collaboration. In order to help students acquire knowledge and skills pertinent to preclinical and clinical disciplines, ITSs are frequently used in medical education.

### 19.6.3 Virtual patients

Since virtual patients are thought to offer advantages over traditional teaching techniques, they are becoming more and more common in medical settings. They serve as an interactive tool that lets students hone their clinical abilities in a secure setting without endangering actual patients. Additionally, they offer a versatile, economical, and efficient method. Clinical reasoning has traditionally been taught to medical students through small group tutorials, problem-based learning, clinical communication skills, seminars, and clinical internships. However, in modern medical education, virtual patient teaching systems are gaining popularity, wherein they are used to explicitly teach clinical reasoning abilities. In order to train and educate medical professionals, virtual patients which are interactive computer simulations of real-world clinical environments can be of real help. The goal of the virtual patients is to mimic real patients, wherein they might be made to exhibit real ailments, respond to the students' treatments, and provide engaging clinical settings. Through these simulations, the students have an opportunity to experience real-world challenges. Medical students can improve their clinical reasoning and communication skills and build a compelling virtual environment that simulates real-life situations by interacting with virtual patients. This chance allows the student to assume the position of a healthcare professional by collecting data, formulating suggestions for medical therapy, differential diagnosis, and patient follow-up.

### 19.6.4 Gamification

The process of adding various aspects of gaming to a task or any activity is known as gamification. Since ITSs have been successful in boosting student enthusiasm and engagement, gamification in medical education is currently gaining a lot of attention. It enables students to interact with the subject in their own way, make connections with the learning resources, and delve deeper into the nuances of a certain issue, in contrast to knowledge that is given to them in a one-size-fits-all fashion. The focus may shift from memorizing to the development of critical thinking, problem-solving, leadership, teamwork, and communication skills as a result of gamification. It can therefore boost motivation and encourage imagination, creativity, and curiosity (Krishnamurthy et al., 2022). When game elements like leaderboards, medals, and points are added, learning becomes more engaging and enjoyable. More importantly, however, gamification helps students overcome their greatest fear by creating a secure and supportive learning environment and reducing the stress and cost of academic failure. Gamification may improve collaboration, learning, and engagement by facilitating real-world application. Additionally, they may facilitate risk-free healthcare decision-making, learning analytics, remote learning, and timely feedback. The benefits of gamification in medical education are clear since it enhances clinical performance, develops important soft skills like empathy and communication, and makes it easier for residents to engage with their study.

### 19.6.5 Adaptive learning system

Medical students today choose to study in a variety of methods, at various rates, and with a variety of instructional materials in order to acquire new knowledge and skills due to developments in the teaching and learning process. In its most basic form, adaptive learning is a method that uses technologies to identify a learner's strengths and weaknesses in order to create a tailored learning experience. This model provides an alternative approach to educational innovation for medical schools that seek to be sensitive to the distinct learning needs of every student with a range of backgrounds and skill sets (Patil et al., 2025). It looks into and comprehends the learner's preferred learning style, learning rate, and areas of knowledge shortage, much like a medical diagnostic instrument. These systems also offer customized therapies based on each student's unique needs and learning preferences. They are employed to enhance every student's capacity to gain from the educational system. For instance, medical students can study difficult medical topics at their own pace using websites like lecturio and osmosis, which offer personalized learning experiences.

### 19.6.6 Technology-enhanced learning

Technology that facilitates learning is called technology-enhanced learning. Although this term can be used to describe both digital and analog technologies, digital technology has recently had a significant impact on education, especially in the form of educational software. Technology-enhanced learning is surpassing our expectations in terms of enhancing and transforming education and educational establishments. This is especially true given the growing popularity of e-learning platforms, instructional software, and educational applications. In today's educational setting, technology is not only a helpful tool, but incorporating ITSs into the classroom is crucial to improving both teachers' and students' educational experiences.

### 19.6.7 Artificial intelligence–enabled simulation

Including simulation-based medical education in curricula is one of the most crucial elements in creating a novel approach to medical education. In the healthcare sector, simulation is growing in popularity as a teaching and training tool. Simulation is the technique of producing an experience without really going through the event. Simulation is a helpful modality to supplement training in actual clinical situations because it gives learners control over the tasks that are presented to them, prevents unsafe and dangerous situations, creates tasks that are uncommon in the real world, and offers opportunities to offer support and guidance. When compared to more conventional "chalk and talk" methods, these approaches have demonstrated the potential for increased efficiency in recent decades. They have also ushered in a new era in education by opening the door for the creation of ubiquitous learning environments.

#### *19.6.7.1 Simulation-based learning*

Simulated "real-life scenarios" and challenging problems are presented to students in controlled environments as a part of simulation-based learning, a form of experiential learning. Students can practice skills and absorb knowledge in a secure, realistic, yet virtual setting through simulation-based experiential learning. Simulation-based learning environments are dynamic, immersive virtual learning environments that imitate a series of problem-solving scenarios or an actual, realistic event.

#### *19.6.7.2 Simulation-based medical education (SBME)*

Internal medicine, anesthesia, emergency care, and surgery are just a few of the medical areas where simulation-based medical education is frequently used to teach technical or nontechnical abilities. Additionally, by enhancing patient or medical safety, simulation-based medical education has enhanced undergraduate and specialized education in the healthcare sector. These days, clinical training for a variety of medical specialties depends heavily on simulation-based medical education, particularly when paired with the development of instructional technology for medical education. In recent years, AI has had a big impact on healthcare and education technologies. Simulation-based medical education must therefore adapt to these developments in order to facilitate clinical learning in the AI era.

#### *19.6.7.3 Clinical simulation*

This innovative method of teaching medicine has been expanding rapidly in the last several years. Clinical simulation is defined as "a technique, not a technology, to replace or amplify real experiences with guided experiences" that replicates or evokes important aspects of the real world in a fully interactive manner. Simulation allows training through guided experiences in safe environments, enabling adequate learning and standardized assessment of the skills needed to deal with a changing reality. Healthcare professionals can learn and refine cognitive, technical, and behavioral skill sets using the innovative method of healthcare simulation, which offers well-crafted, cutting-edge experiences. Using this experiential learning approach, professionals and students can test and enhance their skills prior to working in real-world circumstances while increasing confidence in their talents. Additionally, medical simulation mimics a clinical environment without the potential for negative clinical outcomes. Medical, clinical, and occasionally surgical and nursing simulations are other names for healthcare simulators.

### 19.6.8 Artificial intelligence–based personalized learning

The days of "standardization" and "one-size-fits-all" are long gone in the current era of personalization. In the present world, when technology permeates every part of our lives, "personalization" is extremely important in all fields, including medicine. Personalization is particularly crucial for e-learning in order to make a learning experience more appealing to the target audience. Because of this, the idea of personal learning environments is becoming more and more significant in today's educational setting (Khan, 2025). A teaching strategy known as "personalized learning" modifies the curriculum to fit each student's unique needs, interests, and learning preferences. It is predicated on the notion that each student is unique and may realize their full potential if their needs are met by specialized training as opposed to a general approach. Even said, educational technology is just one of many instructional strategies that can be applied to individualized learning. Customized lesson planning and opportunities for experiential learning are the main goals of this method. It is impossible to overstate the significance of the learning process in medicine. In order to succeed in clinical practice, medical students must acquire skills and competences in addition to knowledge of facts and concepts. However, learning is an individual process that expands one's perspectives, knowledge, and useful skills. Even though all students must adhere to the same learning schedules, a significant disadvantage of traditional teaching is that students often share the same information, interests, learning objectives, or prerequisites. Students are urged to follow a series of learning sequences utilizing conventional learning resources in order to improve their academic performance. It is acknowledged that certain students are not suited for specific learning routes.

Modern course designers carefully consider individualized learning paths that take into consideration each student's needs, motivation, interests, behavior patterns, and talents. Personalized learning, on the other hand, is a methodical approach to education that customizes instruction to each student's unique requirements, preferences, strengths, and objectives. This results in a well-rounded educational experience by offering a variety of new courses and opportunities for skill development. The goal of personalized learning is a student-centered educational strategy that adjusts instruction to each student's particular requirements, interests, and learning preferences. It promotes self-directed learning, mastery, and involvement in addition to offering professional support and therapies. When creating individualized learning, it is also important to consider the unique circumstances, aptitudes, preferences, prior knowledge, interests, and objectives of each learner. It should also be flexible enough to evolve as the learner's knowledge and abilities do.

### 19.6.9 Continuous medical education

Continuous medical education is a crucial component of professional development since it keeps healthcare professionals up to date on the latest advancements in their industry. Traditional methods of continuing medical education, such as lectures, seminars, and conferences, have been demonstrated to be effective, but they are limited by factors such as time constraints, geographic restrictions, and personal preferences. AI has revolutionized medical education by offering fresh approaches to increase accessibility, customize content delivery, and improve educational opportunities. By using AI technology like VR/AR and online resources, continuous medical education can benefit a wider audience, including remote learners and those with limited access to traditional educational opportunities.

### 19.6.10 Telemedicine and remote patient monitoring

Telemedicine and remote patient monitoring are crucial elements of modern medical education because they increase access to clinical experiences, teach basic skills for digital healthcare delivery, and give aspiring physicians the knowledge and skills to deliver care in a technologically advanced setting. One of the primary purposes of telemedicine and remote patient monitoring is to increase access to healthcare. Telemedicine provides virtual consultations for patients who are elderly, live in rural areas, or have mobility issues. It also makes it possible to consult with experts remotely, which would previously require travel. Chronic illnesses like diabetes and heart failure, which can be continuously monitored utilizing data from home-based technology, can be better managed with remote patient monitoring. This enables medical professionals to monitor a patient's development in real time and promptly modify their treatment plan. Additionally, by delivering a constant flow of patient data, these technologies shift the focus from reactive to proactive treatment.

### 19.6.11 Bridging the gap between human and artificial intelligence: collaborative intelligence

The integration of AI technologies to enhance human expertise rather than replace it is required to close the gap between human and AI in medical education, with a focus on developing a curriculum that balances technological proficiency with essential human skills like empathy, critical thinking, and ethical reasoning. Among the methods that can be used to foster this collaboration between humans and AI are described below:

#### *19.6.11.1 Integrating the medical curricula with artificial intelligence tools*

This is achieved by developing AI-integrated curricula that enable medical schools to incorporate AI literacy teaching into their core curricula. This training should address the functioning of AI systems, their potential, limitations, and ethical implications, including data privacy, bias, and accountability (Abd-alrazaq et al., 2023). Additionally, by combining AI-powered learning tools with traditional human mentoring and face-to-face conversation, the learning process can be made more effective and interesting as the students get the benefit of the methodologies. The AI tools give the much needed flexibility to learning, and the face-to-face interaction and mentoring sessions help the students learn the soft skills mainly by interaction and observation. Since AI can handle administrative tasks, generate customized learning paths, and provide quick feedback, educators can focus on teaching complex skills, critical analysis, and the humanistic aspects of care.

#### *19.6.11.2 Fostering collaboration and trust*

To foster collaboration and trust between humans and AI, it is crucial to establish a foundation of transparency, reliability, and a clear division of roles as well as to improve user training and communication. Important strategies include designing AI for explainability, ensuring that AI performance is reliable and morally consistent, and educating people on how to interact and govern AI systems.

##### 19.6.11.2.1 Human-in-the-loop (HITL) and augmented intelligence

In AI-driven medical education, human educators actively participate in the training, validation, and ongoing improvement of AI systems through a pedagogical and technical method known as "human-in-the-loop" (HITL). AI serves as a tool, such as an automated grader or simulation engine, rather than taking the place of teachers, and specialists provide vital monitoring to guarantee educational correctness, safety, and moral standards. As part of HITL in medical education, human experts actively monitor and provide feedback to AI systems to guarantee accuracy, safety, and moral performance. This technique is used to improve AI systems, such those that generate test questions or provide diagnostic support, by allowing humans to correct errors, lessen bias, and give subtle understanding that AI would miss. This partnership enhances student learning while lowering the risks associated with relying solely on autonomous AI in a vital sector like healthcare. Through AI-driven diagnostic tools, simulation, and personalized learning, augmented intelligence in medical education improves training by complementing human clinical judgment rather than replacing it. It develops immersive, adaptive, and ethical learning using virtual and augmented reality, enhances diagnostic abilities with AI-powered analytics, and equips upcoming medical professionals for AI-integrated patient care.

##### 19.6.11.2.2 Explainable artificial intelligence

Explainable AI (XAI) in medical education helps teachers and students understand and trust AI-driven tools by making the AI's decision-making process clear. XAI is necessary in medical education in order to build trust, ensure safety, and foster critical thinking by allowing trainees to comprehend the reasoning behind AI-driven diagnoses, rather than relying on "black box" algorithms. It helps students make better clinical judgments by enabling them to validate AI, identify bias, and learn how to work with technology. This promotes trust in AI systems by showing how they arrive at a result using techniques like feature importance or counterfactual explanations. It also enables better learning, customized feedback, and curriculum refinement. It enhances learning because it provides students with unambiguous feedback, enabling them to take charge of their own education and understand the reasoning behind recommendations made by AI-powered tutoring systems.

##### 19.6.11.2.3 Seamless integration

The smooth integration of AI in medical education includes tailored learning platforms that adapt to each student's needs, virtual patients and simulations for practical instruction, and administrative support for instructors. The seamless integration of AI in medical education plays a vital role in promoting confidence as it changes AI from a being a black box into a dependable, transparent, and essential element for clinical decision-making. When AI tools are integrated directly into current workflows, such as diagnostic imaging software or electronic health records, students and practitioners can see firsthand how useful they are in low-risk, simulated, or supervised settings (Gordon et al., 2024).

### *19.6.11.3 Prioritizing human-centric skills and education*

A thorough reform of medical educational environments and organizational strategies is necessary to prioritize human-centric education and skills, with a focus on encouraging the development of unique human qualities like creativity, empathy, and critical thinking. This approach ensures that people can thrive alongside technological advancements such as AI.

#### 19.6.11.3.1 Reskilling and upskilling

While reskilling involves learning new skills for a different career, upskilling in medical education focuses on developing talents to improve performance in a present role. Both are crucial for improving patient care, handling staffing shortages, and adapting to the rapidly evolving healthcare sector. For instance, reskilling in medical education refers to transitioning to new roles such as from clinical to managerial roles, wherein the teacher or clinicians are trained to take on leadership, administrative, or hospital management roles. Similarly an example of upskilling involves enhancing the current roles such as training clinicians to interpret AI-driven diagnostics or operate advanced robotic surgical equipment. One of the many advantages of reskilling and upskilling is that it can help close skill gaps by equipping the current workforce with the abilities needed for future professional endeavors. Additionally, it helps improve patient care by training the healthcare professionals to develop new skills that allow them to provide the most cutting-edge and effective care possible, which can improve patient outcomes and reduce hospital readmission rates.

#### 19.6.11.3.2 Artificial intelligence literacy

In medical education, AI literacy refers to teaching aspiring medical professionals how to comprehend, assess, and use AI tools in clinical settings and education in an ethical manner. With an emphasis on algorithm openness, data bias, and enhancing patient care through individualized learning and simulations, it is evolving into a key skill. Understanding AI literacy has grown increasingly important in recent years. AI literacy in medical education refers to the capacity of future doctors and healthcare professionals to understand, evaluate, apply, and collaborate effectively with AI technologies. The curriculum's integration of AI literacy and its practical applications has gained importance, as it helps us to carry out various tasks including diagnosis support, administrative work, and research (Roveta et al., 2025). The goal is to ensure that medical professionals can use AI responsibly and competently within a framework of accountability for the decisions they make while utilizing new technology. By doing this, we can make sure that future educators are prepared to both profit from and overcome whatever challenges they may face.

### *19.6.11.4 Ethical development and governance*

Ethical development and governance refers to the ideas and practices that guide decision-making to ensure fair, responsible, and just outcomes, particularly in domains like public service and technology. It includes creating frameworks to prevent harm and promote trust among stakeholders, promoting transparency and accountability, and using moral principles to direct actions. This is crucial for both societal well-being and the moral advancement of technologies like AI (Salem et al., 2025).

#### 19.6.11.4.1 Inclusive development

In medical education, inclusive growth refers to ensuring that all students, especially those from diverse and historically impoverished backgrounds, feel appreciated, accepted, and given equal opportunities to succeed. In order to create a sense of community and improve learning outcomes for future physicians, it involves changing the curriculum, teaching methods, and institutional structures to be inclusive and accessible to all students. Teachers can be confident that fairness and equal opportunity have been attained by maintaining the inclusive development culture. A learning environment that concurrently cultivates a sense of inclusion, acceptance, and worth from peers and teachers is associated with increased academic performance and retention.

#### 19.6.11.4.2 Clear ethical frameworks

Clear ethical frameworks in medical education have many advantages, chief among them being that future healthcare professionals will be equipped to deal with challenging moral dilemmas, uphold professional standards, and prioritize patients' needs in an increasingly diverse and technologically advanced field. Ethical frameworks that ensure practitioners focus on what is best for the patient while avoiding damage results in a better, more thorough, and effective treatment programs that takes into account each patient's needs, goals, and experiences. However, when a whole healthcare facility functions within the same ethical framework, it also facilitates communication and mutual trust among coworkers, leading to a more cohesive and successful interdisciplinary team approach to patient care. The ethical development and governance in medical education also affects the teaching and learning process. For instance, it helps to develop medical students into capable, caring, and responsible professionals. It offers the essential, methodical foundation for handling difficult situations, guaranteeing patient safety, and building confidence in the healthcare system. Moreover by minimizing negligence, upholding patient autonomy, obtaining informed consent, and protecting patient privacy, teaching ethics directly improves patient results. It promotes a change from traditional system to a patient-centered approach.

## 19.7 Role of stakeholders in artificial intelligence for medical education

Stakeholders in AI for medical education, including institutional leaders, educators, and developers, are responsible for integrating, governing, and curating ethical AI tools within curricula to enhance training. Their roles focus on updating pedagogical methods, training staff, ensuring data privacy, and developing AI competencies to prepare future clinicians for technology-driven healthcare. The key stakeholder roles in this regard include the following:

### 19.7.1 Institutional leaders

Institutional leaders like the deans and program directors are in charge of incorporating AI into the curriculum, guaranteeing its ethical use, safeguarding students, and supplying the required resources. The success of AI in medical education is largely dependent on institutional leaders. In order to move AI initiatives from experimental pilot projects to integrated, standardized components of education and training, they are in charge of establishing the strategic vision, allocating resources, managing change, and guaranteeing ethical compliance.

### 19.7.2 Educators and faculty

Educators and professors play a crucial role in incorporating AI into medical education, as they play a vital role in promoting AI literacy, creating morally acceptable curricula, and assisting students in utilizing AI tools for clinical reasoning, personalized learning, and simulation-based training. They have to close the gap between humanistic, compassionate therapeutic practice and technological AI capabilities. They are also responsible for adopting AI tools for personalized learning, providing feedback, updating teaching methods, and mentoring students on AI applications.

### 19.7.3 Artificial intelligence developers and technologists

In order to create individualized, interactive, and effective teaching environments, AI developers and technologists in medical education develop, deploy, and manage cutting-edge digital technologies including AI-powered simulations, natural language processing based chatbots, and predictive analytics. To prepare future professionals, they support curriculum optimization, improve clinical skills training, and guarantee safe, ethical technology integration. They also ensure the validity, transparency, and fairness of AI datasets, mitigate algorithmic bias, and provide user-friendly tools tailored for educational use. In this regard, the success of AI in medical education depends on the effective collaboration between AI developers and educationists.

### 19.7.4 Medical students and learners

AI tools are being used more and more by medical students and learners to improve clinical reasoning, increase learning efficiency, and participate in tailored teaching. While negotiating ethical issues and becoming ready for a future AI-driven healthcare ecosystem, key tasks include using generative AI for research, researching, interacting with simulated virtual patients, and enhancing diagnostic abilities. Medical students are encouraged to actively engage with AI tools to build competence, provide feedback on usability, and develop critical thinking regarding AI-assisted clinical decisions.

### 19.7.5 Regulatory bodies and policymakers

Regulatory agencies and legislators play a crucial role in the use of AI in medical education, as they make sure that adequate measures have been taken to promote innovation while guaranteeing safety, ethics, and equity. In order to minimize algorithmic bias, ensure compliance with data protection acts such as Health Insurance Portability and Accountability Act or General Data Protection Regulation (GDPR), and uphold academic integrity, these organizations are moving from voluntary norms to more comprehensive, formal frameworks as AI tools are included into training ranging from administrative support to simulation. In addition to this, they also play a crucial role in developing guidelines, standards, and ethical frameworks for AI use in education and training to ensure safety and quality.

### 19.7.6 Patients and public

As codesigners, validators, and educators who guarantee AI tools are sympathetic, objective, and patient-centered, patients and the general public are crucial stakeholders in AI-driven medical education. They ensure that training focuses on moral, compassionate care by bridging the gap between technical AI capabilities and actual clinical needs. They also play an important part in providing insights on the human-centric aspects of AI, ensuring that patient safety and trust are central to the educational focus.

In essence, an effective, safe, and ethical AI integration requires collaboration among all these groups to align educational goals with technological advancements.

## 19.8 Impact of artificial intelligence in medical education

AI is having a huge impact on medical education, transforming how students and professionals learn, practice, and improve their skills. In terms of personalized learning, the goal of AI is to offer a customized learning experience, wherein the student can acquire knowledge and be provided with customized feedback while taking into consideration each student's unique learning needs. AI is revolutionizing medical education by making learning more efficient, accessible, and interactive. However, it is important that students and educators remain aware of potential risks and continue to focus on the human aspects of healthcare while integrating AI into their learning process (Fig. 19.3).

### 19.8.1 Increased engagement and motivation

AI in medical education significantly increases student motivation and engagement by providing personalized, interactive, and adaptable learning experiences that considers each student's needs and preferences. It is a well-known fact that increased motivation is associated with better learning results, more positive learning attitudes, and higher levels of enjoyment. AI systems assess a student's performance, identify knowledge gaps, and quickly adjust the curriculum and level of difficulty. This tailored method ensures that the material is continually relevant and appropriately challenging by fostering a sense of competence and autonomy, which boosts intrinsic motivation (Sriram et al., 2025).

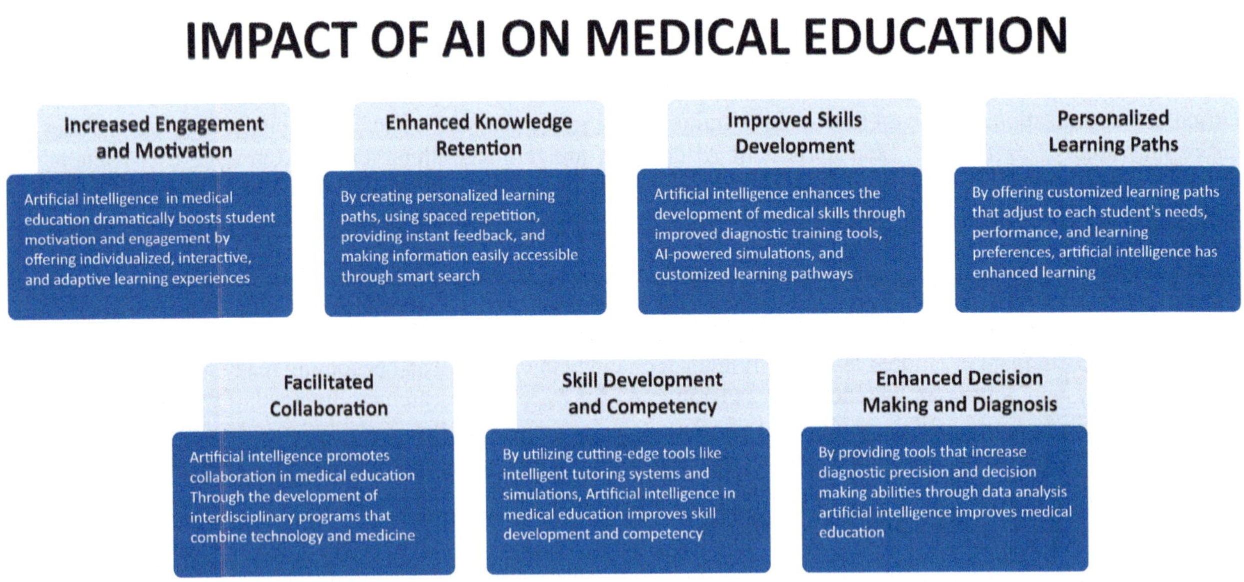


FIGURE 19.3 **Impact of artificial intelligence (AI) medical education.** This figure shows impact of AI medical education.

### 19.8.2 Enhanced knowledge retention

AI enhances knowledge retention by creating personalized learning pathways, using spaced repetition, providing instant feedback, and making information easily available through intelligent search. Additionally, it ensures that knowledge stays current and accessible, improves memory, and boosts engagement. It goes beyond one-size-fits-all training to improve comprehension and application by modifying difficulty, offering concentrated practice, and identifying at-risk learners.

### 19.8.3 Improved skills development

AI in medical education speeds up skill development since it facilitates immersive, individualized, and risk-free training and enables students to hone their diagnostic, procedural, and communication skills more quickly than they could with more conventional approaches. Important tools that improve clinical reasoning and competence include virtual patients, AI-driven simulations, and ITSs that offer instant feedback. AI evaluates student performance instantly and objectively, allowing for quick skill improvement and correction. These technologies improve readiness for real-world situations as they ensure that there is repetitive practice and honing of skills before a student can actually work in the clinical environment (Shankar, 2022).

### 19.8.4 Personalized learning paths

The emergence of personalized learning paths have brought in a fresh perspective to the teaching and learning process in medical education. AI-powered personalized learning paths, due to their ability to adjust the curriculum to each student's performance, increase diagnostic decision-making, speed up skill development, and improve information retention. For instance, real-time feedback, simulation-based training, and knowledge gap identification are offered by these systems which aid in maximizing learning, leading to quicker competency acquisition and better, uniform educational outcomes across institutions.

### 19.8.5 Facilitated collaboration

AI fosters collaboration in medical education through the creation of multidisciplinary programs that integrate technology and medicine, the use of collaborative learning platforms that enhance teamwork abilities, and immersive tools like virtual patient systems and simulations. AI promotes collaboration by initiating multidisciplinary programs because the incorporation of AI in medical education necessitates collaboration between healthcare institutions, medical schools, industry partners, and regulatory bodies. This leads to the development of new programs, such as those that combine computer science and medical curricula, where professionals from many fields work together to solve difficult clinical problems or develop new AI tools.

### 19.8.6 Skill development and competency

AI in medical education enhances skill development and proficiency while emphasizing the need for new abilities in data analysis, ethics, and AI fundamentals through the use of cutting-edge tools like ITSs and simulations. Additionally, as AI takes on more computational duties, the doctor's role will become more human interaction-focused, fostering soft skills like teamwork and communication. Improved communication skills are necessary to interact with interdisciplinary teams and share AI-generated insights with patients in a compassionate manner (Chatzichristos et al., 2025).

### 19.8.7 Enhanced decision-making and diagnosis

AI enhances medical education by offering tools that improve diagnostic accuracy and decision-making skills through data analysis, realistic simulations, and personalized learning. These technologies are meant to supplement human clinical judgment and empathy, not to replace it. Clinical decision support systems, which analyze vast amounts of patient data and medical literature to provide evidence-based recommendations, are one of the technologies that medical schools employ to teach students how to use. Although technology might not always much increase immediate diagnostic accuracy, teaching clinical decision support systems in medical education boosts student confidence, increases adherence to clinical recommendations, and encourages digital literacy. Students can better regulate their cognitive load and get ready for technologically advanced clinical settings by including clinical decision support systems into their coursework.

## 19.9 Challenges of artificial intelligence in medical education

### 19.9.1 Ethical challenge

The main ethical issues with AI in medical education are protecting data security and privacy, preventing algorithmic bias that could lead to discrimination, maintaining transparency in AI decision-making, and precisely defining culpability for errors. Other concerns include ensuring that students receive adequate training to understand the ethical implications of AI in practice and over-reliance on AI, which may impair clinical judgment.

### 19.9.2 Technological challenge

The numerous technological impediments of AI in medical education include infrastructure and costs, data problems, technical limitations like algorithmic failures, and significant pedagogical challenges like curriculum integration and faculty training. The knowledge gap between data scientists, engineers, and medical professionals is a significant barrier that makes it difficult to develop specialized AI tools for medical curriculum and requires a multidisciplinary team for successful deployment (Zarei et al., 2024).

### 19.9.3 Over reliance

Overuse of AI in medical education may cause students to rely more on automated responses than on in-depth research, which could undermine clinical judgment and critical thinking. Other issues that could arise from this dependence include a lack of preparedness for the ethical complexities of modern therapeutic treatment, which require human intuition, empathy, and experience that AI cannot fully replicate, and a susceptibility to AI-generated errors known as hallucinations.

### 19.9.4 Lazy doctors or deskilling

Deskilling in AI in medical education or "lazy doctors" carry a significant, established risk. Future medical professionals run the risk of losing their critical thinking and basic clinical skills if they rely too much on automated technologies. Concerns about the gradual loss of well-established skills, such as radiology pattern recognition, the ability to perform a thorough physical examination, and creating a broad differential diagnosis from first principles, have made the concept of deskilling more significant and urgent (AL-Qahtanee, 2025).

### 19.9.5 Factual hallucinations

Factual hallucinations in AI pose a serious threat to medical education and practice since AI models can deliver convincing but factually incorrect information that could mislead users and potentially lead to diagnostic or treatment errors. AI hallucinations are false or misleading results generated by AI models that appear certain and convincing but are not supported by the input data or real facts. In medical contexts, these can be very dangerous.

### 19.9.6 Digital divide

In medical education, the phrase "digital divide" refers to the unequal access to digital infrastructure and AI technologies, which can worsen health inequities and create educational gaps between students with and without sufficient access to technology and training. Disparities in digital literacy, institutional resources, and access to high-speed internet and devices are the main causes of this divide, which mostly impacts students in low-income countries or rural areas. To lessen this, policymakers, educators, and developers must collaborate to create inclusive AI tools, enhance digital literacy through faculty development, and offer equitable access to technology and training.

### 19.9.7 Variability in artificial intelligence adoption and regulatory frameworks

The implementation of AI in medical education and the associated regulatory frameworks vary widely due to a number of factors, including regional infrastructure, institutional agendas, and differing legal attitudes. Despite the enthusiasm of physicians and students, official participation in courses and consistent regulations remain uneven globally (Fig. 19.4).

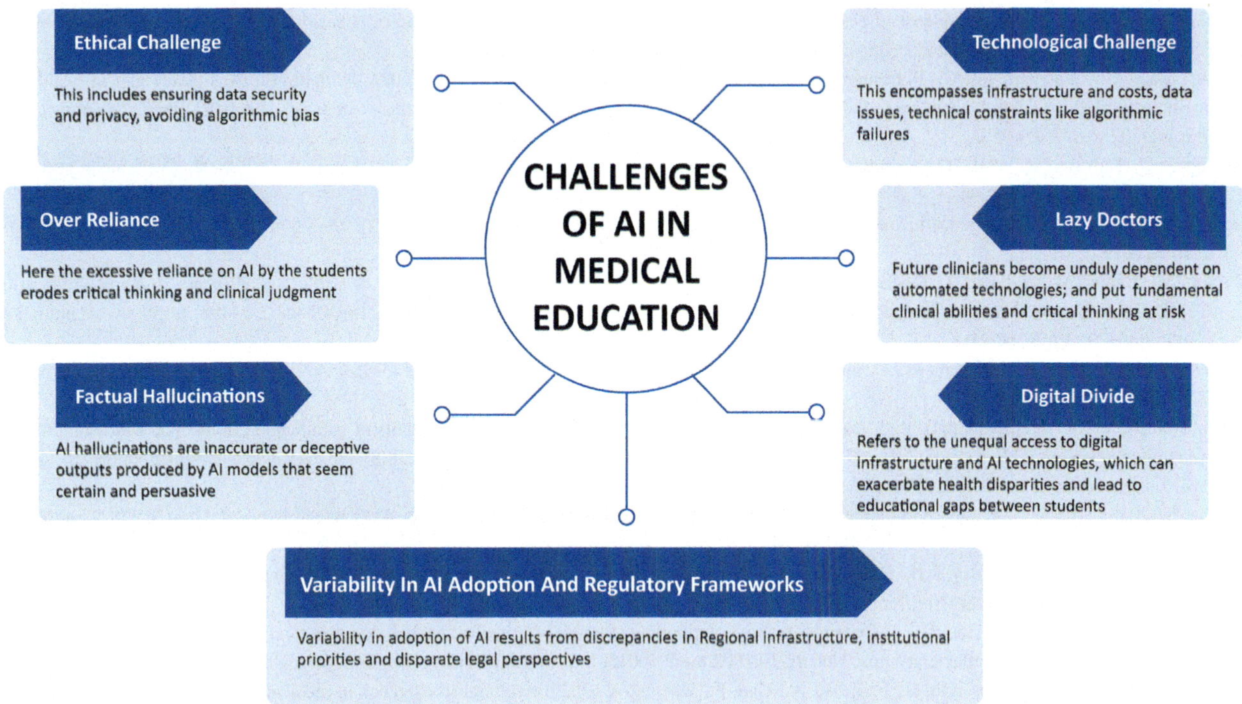


FIGURE 19.4 **Challenges of artificial intelligence (AI) in medical education.** This figure depicts the challenges of AI in medical education.

**Summary**

From basic training to lifetime professional development, artificial intelligence (AI) is changing the face of medical education. Adaptive learning platforms, virtual patients, simulation systems, predictive analytics, and generative models are examples of AI solutions that facilitate individualized learning, strengthen procedural training, and improve diagnostic reasoning. However, by automating processes like curriculum administration, assessment, and feedback production, AI can optimize educational workflows and free up teachers to concentrate more on mentoring and higher-order instruction. These prospects are accompanied by the difficulties that come with using AI, including algorithmic bias, data governance, legal issues, and the possible effects on students' clinical judgment and critical thinking. The necessity of openness, responsible data use, and teaching students how to use AI tools efficiently while retaining fundamental humanistic qualities are all highlighted in the discussion of ethical integration tactics.

**Key point**

**Future perspectives: a collaborative approach to healthcare or medical education**

Artificial intelligence (AI) in medical education has the potential to revolutionize student learning, faculty instruction, and clinician competency maintenance. While maintaining the importance of human judgment, empathy, and professionalism in medical education, the focus will be on individualized instruction, realistic training environments, ongoing evaluation, global accessibility, and ethical practice. Collaboration between AI and humans, where AI is anticipated to become a cognitive partner, will always be valued, highlighting how to incorporate AI insights into collaborative decision-making, when to depend on AI, and when to override it. The "human–AI team" model that results from all of this will influence clinical practice in the future.

## References

Abd-alrazaq, A., AlSaad, R., Alhuwail, D., Ahmed, A., Healy, P.M., Latifi, S., et al., 2023. Large language models in medical education: opportunities, challenges, and future directions. JMIR Medical Education 9, e48291. Available at https://doi.org/10.2196/48291.

AL-Qahtanee, H., 2025. Artificial intelligence in medical education: a cautionary advice for medical students. World Journal of Surgery and Surgical Research 8, 1599.

Chan, K.S., Zary, N., 2019. Applications and challenges of implementing artificial intelligence in medical education: integrative review. JMIR Medical Education 5 (1), e13930. Available at https://doi.org/10.2196/13930.

Chatzichristos, C., Chatzichristos, G., Borremans, I., Gruyaert, S., De Vos, I., De Vos, M., et al., 2025. Bridging the AI-literacy gap in health care: qualitative analysis of the flanders case study. Journal of Medical Internet Research 27. Available at https://doi.org/10.2196/76709, https://www.jmir.org/2025/1/e76709/PDF.

Davenport, T., Kalakota, R., 2019. The potential for artificial intelligence in healthcare. Future Healthcare Journal 6 (2), 94–98. Available at https://doi.org/10.7861/futurehosp.6-2-94.

Garg, T., 2020. Artificial intelligence in medical education. The American Journal of Medicine 133 (2), e68. Available at https://doi.org/10.1016/j.amjmed.2019.08.017.

Gordon, M., Daniel, M., Ajiboye, A., Uraiby, H., Xu, N.Y., Bartlett, R., et al., 2024. A scoping review of artificial intelligence in medical education: BEME Guide No. 84. Medical Teacher 46 (4), 446–470. Available at https://doi.org/10.1080/0142159X.2024.2314198, http://www.tandfonline.com/loi/imte20#.VufR2bdf2DY.

Khan, S.M., 2025. AI-based Personalized Learning in Medical Education. Elsevier, Amsterdam, pp. 145–166. Available at http://doi.org/10.1016/b978-0-443-33584-6.00007-4.

Krishnamurthy, K., Selvaraj, N., Gupta, P., Cyriac, B., Dhurairaj, P., Abdullah, A., et al., 2022. Benefits of gamification in medical education. Clinical Anatomy (New York, N.Y.) 35 (6), 795–807. Epub 2022 Jun 8. Erratum in: Clin Anat. 2025 Nov;38(8):886. Available at https://doi.org/10.1002/ca.24250, https://doi.org/10.1002/ca.23916, PMID:35637557.

Narang, S., 2023. Artificial intelligence in medical education. RUHS Journal of Health Science 8 (1). Available at https://doi.org/10.37821/ruhsjhs.8.1.2023.11.

Paranjape, K., Schinkel, M., Panday, R.N., Car, J., Nanayakkara, P., 2019. Introducing artificial intelligence training in medical education. JMIR Medical Education 5 (2). Available at https://doi.org/10.2196/16048, https://mededu.jmir.org/2019/2/e16048.

Patil, N.G., Kou, N.L., Baptista-Hon, D.T., Monteiro, O., 2025. Artificial intelligence in medical education: a practical guide for educators. MedComm - Future Medicine 4 (2). Available at https://doi.org/10.1002/mef2.70018, https://onlinelibrary.wiley.com/journal/27696456.

Roveta, A., Castello, L.M., Massarino, C., Francese, A., Ugo, F., Maconi, A., 2025. Artificial intelligence in medical education: a narrative review on implementation, evaluation, and methodological challenges. AI 6 (9), 227. Available at https://doi.org/10.3390/ai6090227.

Salem, M.A., Zakaria, O.M., Aldoughan, E.A., Khalil, Z.A., Zakaria, H.M., 2025. Bridging the AI gap in medical education: a study of competency, readiness, and ethical perspectives in developing nations. Computers 14 (6). Available at https://doi.org/10.3390/computers14060238, www.mdpi.com/journal/computers.

Sanae, L., 2024. Advances in Smart Medical, IoT & Artificial Intelligence. Springer Nature, Switzerland.

Shankar, P.R., 2022. Artificial intelligence in health professions education. Archives of Medicine and Health Sciences 10 (2), 256–261. Available at https://doi.org/10.4103/amhs.amhs_234_22.

Sriram, A., Ramachandran, K., Krishnamoorthy, S., 2025. Artificial intelligence in medical education: transforming learning and practice. Cureus 17 (3). Available at https://doi.org/10.7759/cureus.80852.

Zarei, M., Eftekhari Mamaghani, H., Abbasi, A., Hosseini, M.-S., 2024. Application of artificial intelligence in medical education: a review of benefits, challenges, and solutions. Medicina Clínica Práctica 7 (2), 100422. Available at https://doi.org/10.1016/j.mcpsp.2023.100422.

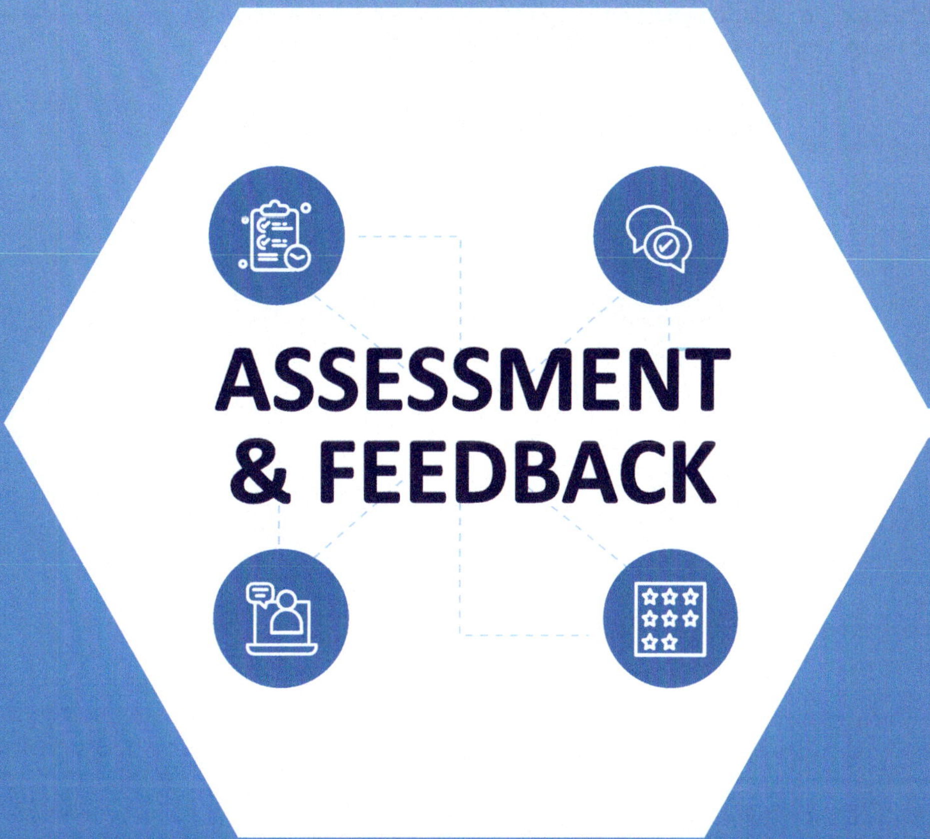


# PART 4

## ASSESSMENT AND FEEDBACK IN MEDICAL EDUCATION

# Chapter 20

# Introduction to assessment

*"Assessment is not about proving what students know, but about improving what they learn."*

—Dylan Wiliam

## 20.1 Introduction

In medical education, assessment is a dynamic, multidimensional process that evaluates a student's learning and growth in knowledge, abilities, and attitudes to make sure they meet learning goals and are prepared for professional practice. It includes summative tests to gauge proficiency at the conclusion of a learning session and formative tests to track learning and offer feedback along the way. A strong assessment system uses a variety of techniques, such as performance-based exams and portfolios, to direct improvements for students and the curriculum as a whole. It is built on the fundamental concepts of validity, reliability, fairness, and practicality (Cox et al., 2007).

One of the best definitions of what the assessment is in medical education can be obtained from the World Federation for Medical Education Global Standards for Quality Improvement 2020. It states that "Assessment assures, drives, guides, creates, and optimizes learning while providing feedback. In the context of a medical school, a system of assessment must exist, which incorporates multiple assessments that achieve the purposes of the school and its stakeholders." This statement explains well what is expected from the assessment and what medical schools should do to meet this expectation. Learning objectives and assessment are two sides of the same coin in education since assessment determines whether or not students have met the learning objectives. In order to guarantee a legitimate and successful learning experience, assessments are made to evaluate the particular information and abilities specified in the objectives (Beard et al., 2005). This alignment helps students understand what they need to learn and how their progress will be evaluated. It also directs the entire educational process, from instruction to assessment. Medical programs are structured according to predetermined goals and objectives. To put it another way, the objectives are the declarations outlining the qualities that students should acquire at the conclusion of the course. In order to create programs that help students acquire the knowledge, abilities, and attitudes necessary to ultimately achieve graduate competencies, medical schools set learning objectives.

## 20.2 Need for assessment in medical education

The "need" for assessment in medical education originates from its function in monitoring student development, accomplishing learning objectives, offering constructive criticism, and directing instruction. Assessments are essential for confirming whether students have acquired the required knowledge, abilities, and attitudes, as well as for assisting teachers in improving their teaching methods and guaranteeing alignment between intended and actual learning (Norcini et al., 2011). Some of the key reasons for assessment in medical education include the following:

### 20.2.1 Evidence of learning and achievement

Assessment offers verifiable proof that students have met learning goals and picked up the necessary knowledge, abilities, and behaviors. Assessment provides evidence of learning and achievement by gathering, analyzing, and interpreting data using a range of methods, including tests, projects, and observations. Teachers are able to direct instruction, give feedback, and report on progress since this evidence demonstrates what the students know and are capable of. Additionally, it fosters students' confidence, self-assessment abilities, and reflection on their own learning (Ferris and O'Flynn, 2015). For educators, assessment data helps teachers understand a student's current knowledge and skills, which allows them to adapt their teaching to better meet individual needs. Finding the areas in which students thrive and those in which they require

**The Principles and Practice of Medical Education. DOI:** https://doi.org/10.1016/B978-0-443-44197-4.00032-1

additional assistance is also beneficial. However, by recognizing their accomplishments, assessments help students do better by raising their self-esteem and motivation. Evaluation also helps students develop self-reflection since it enables them to set goals, monitor their own progress, and consider how they think and learn.

### 20.2.2 Feedback for improvement

Assessments provide students with useful feedback by pointing out mistakes and areas of weakness, allowing them to enhance their performance and successfully guide their learning. Assessment provides feedback for improvement by identifying areas that need further development, identifying strengths and shortcomings, and offering focused support to reduce the gap between current and intended performance. This continuous cycle of assessment and feedback helps students develop their skills, acquire confidence, and obtain a deeper understanding of the subject matter. Assessment and feedback support students' growth by highlighting misconceptions or knowledge gaps that they might not otherwise be aware of. Assessments also show a student's abilities, providing them with motivation and a springboard for growth.

### 20.2.3 Informing instruction

In order to make sure the educational program successfully achieves its objectives, educators use assessment findings to assess and modify their teaching strategies and course material. Assessment provides teachers with data on their students' development, enabling them to modify their instruction by pointing out areas of strength and weakness. This comprises future planning, curriculum development, and professional development at the individual and program levels, as well as formative evaluations to make adjustments to fit current student needs and summative assessments to evaluate the effectiveness of teaching practices. In the end, assessment ensures that instruction is effective, responsive, and in line with learning objectives.

### 20.2.4 Alignment with learning outcomes

Assessments confirm that the results attained by students correspond with the curriculum's intended learning objectives. Assessment supports alignment with learning objectives by verifying student learning through tasks that test the knowledge and skills outlined in the outcomes. It guides students by focusing their attention on what matters and providing a clear path for advancement from the first course admittance to the development of deeper understanding. Well-aligned exams provide a basis for giving students feedback, adjusting instruction, and ensuring the curriculum effectively accomplishes its educational goals.

### 20.2.5 Motivation and goal direction

One of the important features of doing an assessment on a regular basis is that it has the capacity to inspire students by providing students a clear picture of their progress and assisting them in maintaining focus on their learning objectives. Furthermore, assessment supports motivation and goal direction by fostering the establishment of specific objectives and establishing a cycle of success. It boosts motivation and promotes ownership of one's learning and growth by enabling people to recognize their strengths and limitations, match goals with reality, and track their progress. Through self-awareness, which enables people to comprehend their present abilities, values, and goals, assessment promotes motivation and goal orientation. This realization is essential for establishing attainable, realistic goals that support personal growth rather than ones that are overly simple or complex. As assessment establishes a baseline and aids in the definition of precise, quantifiable learning objectives, it also aids in the setting of particular targets. Setting clear objectives helps the learner realize what they need to improve, which is crucial for motivation and concentrated effort.

### 20.2.6 Quality assurance

A well-thought-out assessment procedure is essential to preserving the standard of medical education and guaranteeing the development of qualified future healthcare professionals. Assessment helps with quality assurance by providing a way to compare and improve products, services, and processes to predefined standards. It ensures uniformity, identifies errors and gaps, and provides data for continuous improvement, all of which lead to learners of higher quality. Assessment aids in quality assurance since it ensures consistency and equity. Assessment uses standards like rubrics to ensure that evaluation is consistent and fair across all the learners as well as teaching strategies. As a result, inconsistencies are discovered and corrected, yielding valid and reliable outcomes. It also encourages continuous improvement by identifying gaps or weaknesses and comparing performance to standards. This information is used to create improvement strategies that help educators adapt to changing needs of medical education.

## 20.3 Principles of assessment

Principles of assessment ensure evaluations are fair, accurate, and beneficial for learning, focusing on key concepts like validity, reliability, fairness, flexibility, clarity, and practicality (Fig. 20.1).

### 20.3.1 Validity

The assessment measures what it is intended to measure. In medical education, validity refers to the degree to which an assessment tool like an exam or clinical evaluation accurately measures the intended knowledge or skill. It is not about the tool itself but the evidence supporting the interpretation of the assessment results for a specific purpose and population. Validity is a hypothesis that requires ongoing data collection to support or refute the interpretation of assessment scores, ensuring assessments are meaningful and appropriate for making important decisions about a learner's competence.

### 20.3.2 Reliability

In medical education, reliability is defined as the consistency and repeatability of assessment results, guaranteeing that an assessment tool produces comparable results when used again or by various raters. It is an essential component of effective assessment as faulty tests yield uneven results that make it challenging to gauge a student's proficiency and result in incorrect learning decisions. There are several forms of dependability, including inter-rater consistency for performance-based assessments and internal consistency for written examinations, all of which seek to reduce measurement error and guarantee trustworthy evaluation.

### 20.3.3 Fairness

Creating equitable methods that eradicate prejudice and advance diversity, equity, and inclusion in assessment and learning environments is essential to achieving fairness in medical education. Important tactics include giving constructive and frequent feedback, putting in place dynamic and interesting learning activities, making sure assessment procedures are transparent, and encouraging staff and students to have a growth attitude. The aim here is to produce skilled doctors who are dedicated to social justice and health equity for all patients.

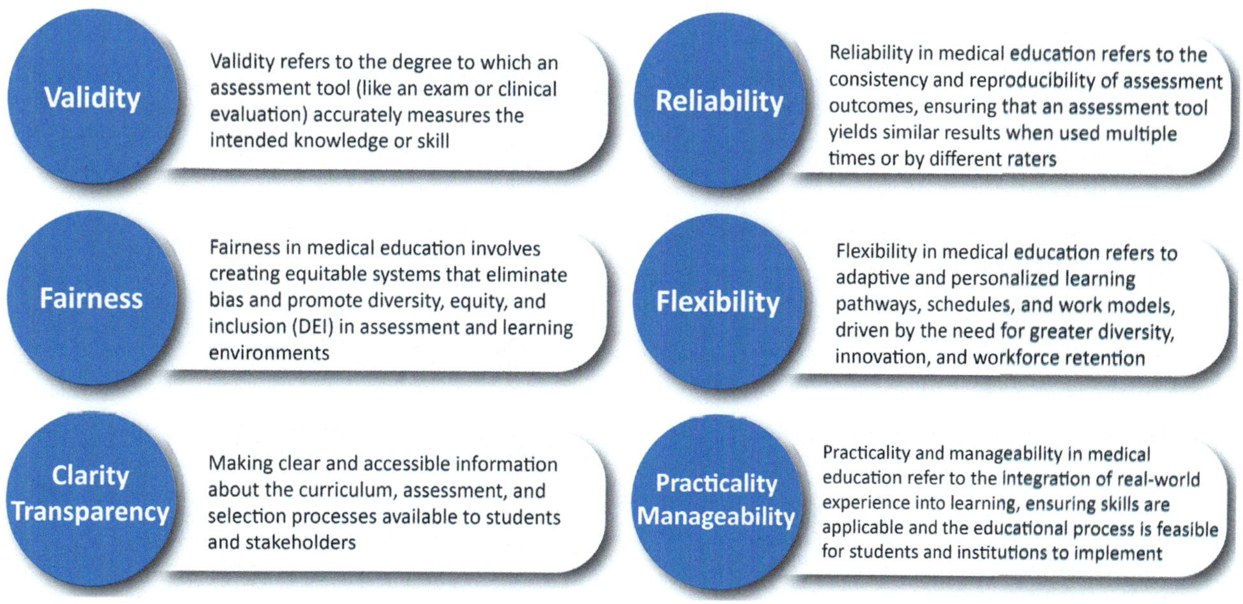


FIGURE 20.1 Principles of assessment.

### 20.3.4 Flexibility

Flexibility in medical education refers to customized and flexible schedules, work patterns, and learning pathways that are motivated by the need for increased workforce retention, diversity, and innovation. In terms of flexibility, the assessment can be modified to accommodate various learners, circumstances, and contexts. The use of online and technology-enhanced learning, flexible pathways that let students enter and exit multiple professions, and possibilities for customized timetables including less-than-full-time training are important ways in which flexibility in assessment can be brought about.

### 20.3.5 Clarity and transparency

It is very important to ensure that the learner is informed of the assessment's goal, criteria, and methodology. In medical education, transparency and clarity entail providing students and stakeholders with easily understandable information regarding the curriculum, evaluation, and selection procedures. This approach guarantees selection equity, promotes accountability in medical schools, helps students understand expectations, and helps match exams with clinical practice. Medical education may foster a varied and representative physician workforce, improve student advancement, and increase public confidence in the healthcare system by increasing transparency.

### 20.3.6 Practicality and manageability

In medical education, practicality and manageability refer to the incorporation of real-world experience into instruction, guaranteeing that skills are transferable and that the educational process is doable for both students and institutions. This entails giving patients and equipment access, developing practical clinical skills, encouraging critical decision-making and communication skills, and striking a balance between essential necessities and adaptable solutions to satisfy a range of needs and drive innovation.

## 20.4 Types of assessment

The types of assessment in medical education can be classified based on many factors such as purpose, method, and mode of delivery (Al-Wardy, 2010). Each of these classifications have a definitive aim to be achieved. The following are the various types of assessments used in medical education (Fig. 20.2).

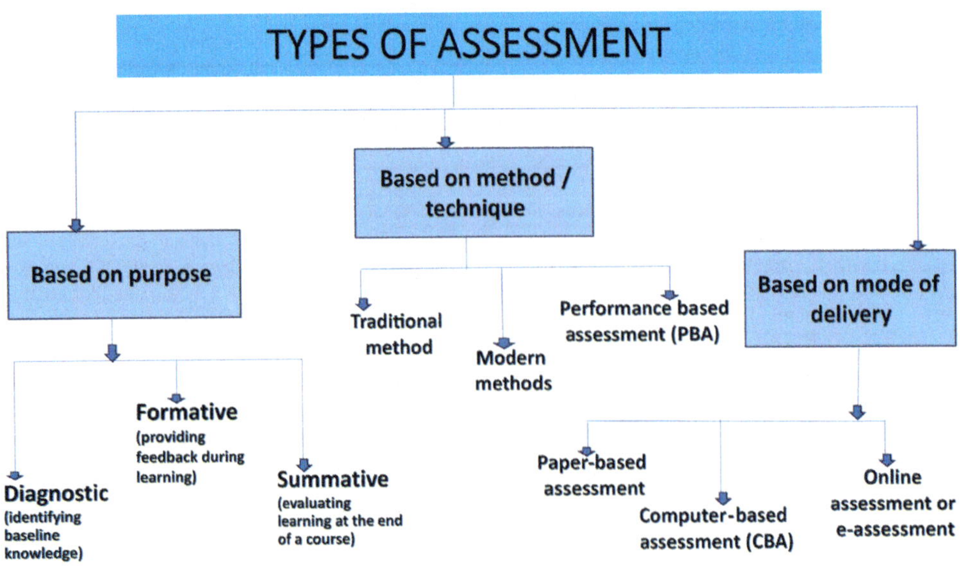


FIGURE 20.2 **Types of assessment.** This figure reflects the types of assessment.

## 20.4.1 Based on purpose

### 20.4.1.1 Diagnostic assessment

In medical education, diagnostic assessment is the process of evaluating a student's current knowledge and abilities in order to pinpoint their particular areas of strength and weakness either before or during training. Teachers are better able to build on a student's strengths and eliminate learning gaps with targeted training when they use this method to determine the student's current level of competence.

### 20.4.1.2 Formative assessment

Instead of giving a final grade, formative assessment in medical education is an ongoing process that evaluates student performance to provide prompt feedback for improvement. It emphasizes assessment for learning which includes techniques like self-evaluations, low-stakes tests, and simulations to help students pinpoint their strengths and shortcomings and direct their learning in real time. This method helps teachers improve their teaching techniques, promotes a deeper comprehension, and aids in the development of clinical reasoning (Sharma et al., 2015).

### 20.4.1.3 Summative assessment

In medical education, summative assessment assesses a student's mastery of a body of information and abilities, usually at the conclusion of a learning cycle, and establishes if they have fulfilled requirements for advancement or graduation. These are frequently high-stakes tests that evaluate general proficiency in areas like clinical problem-solving using a range of techniques, including multiple-choice questions, essays, and simulation (Kibble, 2017). Summative tests are used to make critical decisions like passing or failing, as opposed to formative tests, which concentrate on feedback for growth (Table 20.1).

## 20.4.2 Based on method/technique

Assessment types based on the technique or method used to gather evidence of learning include a variety of approaches, such as written assignments, oral questioning, observation, and practical skills tests. These techniques are distinct from the purpose or timing of the assessment

### 20.4.2.1 Traditional method

Oral tests like viva voce and written tests like multiple-choice questions and essays are the main components of traditional medical education assessment. Practical evaluations of clinical competence, such as long and short cases, are among the other conventional techniques (Epstein and Hundert, 2002). In contrast to more contemporary approaches like the Objective Structured Clinical Examination (OSCE), these techniques emphasize information and are simple to score.

**TABLE 20.1** Difference between summative and formative assessment.

| Sl no. | Formative assessment | Summative assessment |
|---|---|---|
| 1 | It is ongoing and occurs throughout the learning process | It occurs in the final stages of the course, or after the training is complete |
| 2 | It is an assessment for learning | It is an assessment of learning |
| 3 | It is of low stakes | It is of high stakes |
| 4 | The aim here is to assess the practice, preparation, and knowledge periodically | The aim here is to obtain a certification or a degree, usually at the end of the program or a course |
| 5 | It helps in understanding the gaps or deficiencies which can be worked upon | It helps in the process of grading and tells us how much the student has learned |
| 6 | It is in the form of quizzes, preassessments, or mock exams | It is in the form of end-of-course exams, final exams, certification tests |

However, they frequently have flaws, such as promoting rote memory over deeper understanding and being less successful at evaluating practical skills.

#### *20.4.2.2 Modern methods*

OSCEs, Mini-Clinical Evaluation Exercise (mini-CEX), and Direct Observation of Procedural Skills are only a few of the innovative techniques used in medical education assessment (Veloski et al., 2006). Programmatic assessment, which integrates several forms of evaluation for ongoing feedback, and simulation-based assessments, which use simulators and standardized patients to test skills in a controlled setting, are other important contemporary methods. Lastly, while online platforms increasingly provide automated and instantaneous feedback, portfolios and multisource feedback enable a more thorough perspective of a learner's competency.

#### *20.4.2.3 Performance-based assessment*

Beyond standard testing, performance-based evaluation in medical education assesses students' capacity to apply knowledge and abilities in practical, clinical settings. Written clinical simulations, computer-based clinical simulations, oral examinations, and standardized patients which evaluate skills, including clinical communication, physical examination, and patient interviews, are some of the methods used. The objective here is to assess a student's preparedness for independent practice by looking at their real performance.

### 20.4.3 Based on mode of delivery

#### *20.4.3.1 Paper-based assessment*

Both students and teachers have a high appreciation for paper-based assessment systems. Due to its familiarity, accessibility, and perceived dependability, many educators continue to use printed materials for evaluations. They believe that paper grading offers flexibility in grading and the capacity to provide individualized comments based on the needs of each student. From the standpoint of the students, handwritten assignments and tests can also provide a tangible picture of their work, encouraging a more in-depth interaction with the subject matter.

#### *20.4.3.2 Computer-based assessment*

Computer-based assessments (CBAs), which use computers for both exam administration and scoring in medical education, offer benefits including quick feedback, consistent scoring, flexibility, and the ability to include multimedia like videos and simulations. CBAs, which are used for a number of purposes, including formative and summative assessments, in-class tests, self-assessments, and professional competency testing, frequently include multiple-choice questions. Compared to traditional paper-based techniques, CBA in medical education offers substantial administrative and educational benefits, particularly in improving clinical reasoning and logistical efficiency.

#### *20.4.3.3 Online assessment/e-assessment*

The use of digital tools and platforms to assess medical students' knowledge, clinical reasoning, abilities, and professional behaviors is known as online assessment in medical education. Online assessment has also gained popularity with the emergence of competency-based medical education, remote learning, and blended learning. Automated scoring offers rapid results and feedback, which helps students uncover knowledge gaps early and enhance learning outcomes. This is one of the main benefits of the online assessment. In addition to ensuring consistent questions, marking schemes, and scoring across big student populations, computer-based examinations lessen examiner bias. Successful adoption requires using multimodal tools, aligning online tests with online training, and potentially adopting a hybrid method. The several forms of online evaluation are described below.

##### 20.4.3.3.1 Virtual objective structured clinical examination and objective structured video examination

Traditional OSCEs are performance-based assessments used to evaluate clinical skills through a series of stations with standardized patients or mannequins to measure students' clinical skills to assess their competencies. Virtual OSCE, on the other hand, is a modified version of traditional OSCEs, which use video conferencing to simulate patient interactions and a structured video examination with prescripted scenarios to assess clinical skills. Similarly, the Objective Structured Video Examination evaluates students' awareness of and comprehension of the implications of different communication abilities. Additionally, the evaluation measures how many different abilities the student thinks will help to enhance the patient–doctor relationship.

#### 20.4.3.3.2 Case-based scenarios

Students can go through clinical cases in an online format by using the case-based scenarios, which is helpful for evaluating clinical reasoning, problem-solving, and decision-making abilities. Online exams have several advantages, including accessibility and flexibility due to their remote delivery, which permits more customizable arrangements. Additionally, it makes data collecting easier and gives students quick feedback.

## 20.5 Qualities of a good assessment

### 20.5.1 Relevant to the curriculum

Relevance to the curriculum is a foundational characteristic of a high-quality assessment. This concept, often termed constructive alignment, ensures that what is being tested directly reflects the knowledge, skills, and learning outcomes defined in the intended curriculum.

### 20.5.2 Focus on the important skill

One of the hallmarks of effective assessment is its focus on important skills prioritizing the evaluation of core competencies that are essential for future success rather than merely testing rote memorization.

### 20.5.3 Promote learning of skills

A defining characteristic of good assessment is its ability to promote the learning of skills by serving as a "bridge" between theory and practice, rather than merely measuring a student's final state of knowledge. This shift reflects an educational landscape where the process of acquiring skills is valued as much as the skills themselves.

### 20.5.4 Defining the level of attainment

Defining the level of attainment is recognized as a critical characteristic of a good assessment. This practice involves explicitly stating the standards and performance descriptors that determine what students know and can do at various stages of their learning journey.

### 20.5.5 Discriminating good and bad student

The ability of an assessment to discriminate refers to its technical capacity to distinguish between learners who have mastered the material and those who have not. While the term "discriminate" has negative social connotations, in the context of educational psychometrics, it is a hallmark of a fair and valid test.

### 20.5.6 Providing feedback

Providing feedback is considered the most powerful characteristic of a good assessment, shifting the goal from merely grading to actively guiding student development. Effective feedback acts as a bridge that helps students understand where they are, where they need to be, and exactly how to close that gap.

## 20.6 Assessment drives learning

"Assessment drives learning" refers to the idea that how learning is measured affects what and how students learn since they will concentrate their efforts on assignments and material that will be evaluated. Assessments support learning by providing feedback, promoting progress, establishing clear learning goals, and focusing study habits on specific skills and knowledge (Wormald et al., 2009). On the other hand, poorly aligned tests may promote performance without true comprehension or rote memory. The following are some of the main advantages of this connection between learning and assessment:

### 20.6.1 Shaping study habit

Assessment acts as a "map" for students' educational journey and advances learning by essentially influencing what and how they choose to study. It shapes study habits by outlining necessary performance and knowledge in detail and encouraging certain actions prior to assessment. Here are some ways that assessments could influence study habits. First of all, it helps students prioritize the material and abilities they anticipate being tested on, directing their time and effort appropriately. While tests that emphasize application and critical thinking stimulate deeper, more thorough study, those that concentrate on factual recall encourage rote memorization, which contributes to surface learning. Simultaneously this encourages self-regulation because students who participate in peer and self-evaluation learn how to keep an eye on their own learning, evaluate their own performance, and take ownership of their academic advancement. This promotes critical lifelong learning skills, including goal-setting, time management, and self-directed remedial action.

### 20.6.2 Providing feedback and guidance

Assessment promotes a continuous cycle of growth that propels learning as it provides students feedback and direction regarding their strengths, shortcomings, and future steps,. Students are given clear information about where they are, where they need to go, and what they need to do to progress using both formative and summative approaches, which fosters a growth attitude and aids in skill development. For example, the formative approach gives students quick feedback to help them identify their areas of strength and weakness as well as what they need to do to get better. This information enables individuals to modify their study techniques, such as looking for more materials or practice, in order to close the gap between their present performance and the intended results.

### 20.6.3 Motivating progress

Assessment encourages progress as it provides performance evaluation, thereby pointing out the strengths and shortcomings, boosting confidence through observable progress, and involving students in the learning process. Summative tests offer a sense of achievement and motivate further work, while formative tests, like self-reflections, enable ongoing modification of study techniques. Additionally, by enabling the learner to monitor and document their development over time, it fosters self-assurance and a sense of achievement. Additionally, it aids in the reinforcement of accomplishments because each successfully finished assessment and comprehension of the success criteria using tools like rubrics creates a sense of satisfaction that spurs additional work.

### 20.6.4 Directs learning goals

Assessment advances learning objectives by using performance data to create a roadmap for instruction and student progress. Assessment can assist educators and learners in concentrating on certain abilities and knowledge, establishing precise objectives for the future, by highlighting the discrepancy between present knowledge and intended learning outcomes. Students can see where they are, where they need to go, and what should happen next thanks to this procedure. Assessment defines the learner's initial level of understanding, providing a baseline from which all subsequent learning may be measured and guided. This helps the student establish a starting point. Assessment also helps in setting the future goals by identifying the "gap" between what is known and what should be known. In this regard, it helps both teachers and students set clear and achievable learning goals for what comes next.

## 20.7 When, how, and why of assessment

Assessment is a process of evaluating student learning, conducted by keeping in mind the three important aspects which are when, how, and why. "When" refers to the fact that whether the assessment is conducted throughout the learning cycle, "how" referring to the various methods used in conduction like observation, tests, and portfolios assessment, and "why" refers to the purpose for which it is conducted such as to measure progress, inform instruction, or provide feedback for improvement. The goal is to use this information to guide both teaching and learning, making education a dynamic and responsive process.

### 20.7.1 When: at every stage of learning

#### *20.7.1.1 Diagnostic (before)*

This assessment is carried out at the beginning of a course or unit to ascertain a student's starting point and measure prior knowledge. This assessment seeks to identify a student's diagnostic strengths, weaknesses, knowledge, and skill levels in order to inform future instruction and remediation. The evaluation technique is designed to assess the multidimensional capacity of diagnostic thinking, a multistep metacognitive process. For example, pretests, which are brief quizzes or examinations administered at the start of a course or unit, are used to assess students' prior knowledge before instruction starts.

#### *20.7.1.2 Formative (during)*

This type of evaluation, which involves quick exams or observations, is used throughout the learning process to provide prompt feedback that directs instruction and helps students make adjustments. Feedback from formative assessments is timely, precise, and directly related to learning objectives. These evaluations are carried out on a regular basis. Here, the goal is to assist instructors and students in identifying knowledge and skill gaps that may be filled to enhance performance. Formative evaluation emphasizes learning and growth rather than just final results. It offers a secure setting where students can experiment and learn from errors without facing serious repercussions. For instance, the formative evaluation can be carried out by obtaining comprehensive remarks on assignments or performance through verbal and written feedback.

#### *20.7.1.3 Summative (after)*

As a final evaluation of a student's knowledge and abilities, the summative assessment aids in measuring achievement. Additionally, it aids the assessor in establishing competency because it is utilized to make critical judgments, such as whether a student is prepared for professional practice or the next phase of their study. Summative assessment includes, for example, final exams, licensure exams, and clinical performance evaluations (Lehl et al., 2022).

#### *20.7.1.4 Interim*

In medical education, an intermediate assessment is a periodic examination that is administered at predetermined intervals throughout a curriculum to determine how well students are doing in relation to particular learning objectives. They can enhance overall performance on formal exams and are intended to identify learning gaps and provide instructors and students with information for future instructional planning.

### 20.7.2 How: through diverse methods

#### *20.7.2.1 Informal*

Facilitating class discussions and monitoring students are examples of informal assessment techniques. Informal assessment can take several forms, such as watching students while they engage in activities and observing how they interact, use resources, and solve problems. Real-time progress tracking is facilitated by taking notes or anecdotal records. Another strategy is to encourage the students to give their opinions or suggestions while testing their understanding and critical thinking abilities in a casual setting by posing questions.

#### *20.7.2.2 Formal*

Formal evaluations, which frequently produce numerical scores or grades, employ standardized techniques like exams, criterion-referenced examinations, and surveys to evaluate performance against particular objectives. They are different from informal techniques like casual observation because they are organized and consistently used to enable outcomes to be compared against benchmarks. These tests guarantee that curriculum objectives are fulfilled by giving teachers and students crucial, data-driven feedback.

#### *20.7.2.3 Performance-based*

In order to assess the application of skills, it involves examining student work samples, portfolios, and projects. Beyond conventional multiple-choice exams, performance-based evaluations in medical education assess a student's capacity to carry out activities and apply information in a clinical situation. The OSCE for fundamental skills, the mini-CEX for routine interaction observation, standardized patients for realistic scenarios, and workplace-based assessments that may include rating scales, checklists, and 360-degree evaluations are some of the methods (Vageriya, 2018).

#### *20.7.2.4 Student-led*

Student-led assessments are a valuable pedagogical approach in medical education, primarily used for formative learning, enhancing engagement, and developing professional skills. They offer an authentic, low-stakes environment that helps students identify learning gaps and adjust their study strategies and can even predict performance in faculty-led exams. Peer-led assessments, student-generated test items, and curriculum feedback are examples of student-led evaluations in medical education that promote deep learning, boost engagement, and enhance clinical skills. By encouraging higher-order thinking and self-directed learning, these programs boost academic achievement and fill in curricular design gaps.

### 20.7.3 Why: to guide and improve learning

#### *20.7.3.1 To inform teaching*

Assessment supports medical education by providing data on student learning that can be used to improve teaching strategies, modify the curriculum, and identify areas where students require remediation. Teachers may learn more about how well they are teaching, make the necessary changes to improve learning outcomes, and make sure the curriculum is still relevant and challenging by using the findings of both kinds of assessments.

#### *20.7.3.2 To guide students*

Assessment guides medical students by providing regular feedback on their progress in knowledge, skills, and attitudes through both formative and final summative evaluations. Assessments also help to identify knowledge and skill gaps and highlight areas where a student needs more work, allowing for immediate correction. It also helps to drive self-reflection and metacognition, as self-assessment tools encourage students to reflect on their own performance, fostering the critical thinking and self-awareness necessary for continuous improvement and lifelong learning.

#### *20.7.3.3 To measure progress*

Assessment is done to document academic readiness, skill acquisition, and progress toward learning goals. It also measures progress in medical education by evaluating knowledge, skills, and behaviors through various methods like the OSCE and progress tests. These tools provide evidence of learning, guide instructional adjustments, and offer feedback to students, enabling them to improve performance before critical outcomes arise.

#### *20.7.3.4 To identify "trouble spots"*

Here, assessment is utilized to identify areas where students need more help and when they are struggling with the course material. Assessment is also used to identify problematic areas in medical education by systematically collecting and analyzing data on student performance, which in turn reflects the effectiveness of the curriculum and teaching practices. This process makes use of a variety of evaluations and data analysis methods.

#### *20.7.3.5 To enhance motivation*

Through formative approaches, which offer constructive feedback and promote self-reflection, and continuous assessment, which can cultivate a sense of progress and competition, assessment improves motivation in medical education by clearly communicating learning goals and criteria and acknowledging progress. Institutions may foster a more stimulating environment that encourages in-depth learning and student participation by moving away from only summative, high-stakes assessments.

## 20.8 Assessment cycle

Setting learning objectives, offering learning opportunities, gathering evidence of student learning, and utilizing that evidence to enhance future instruction and learning experiences are all part of the ongoing medical education assessment cycle. It is an essential component of medical education and includes planning, data gathering, and analysis to assess the success of the program as a whole as well as the progress of individual students (Gibbs et al., 2006). Using evaluation to motivate learning, mentor students, and eventually create qualified healthcare workers is one of the cycle's main objectives. The steps in the assessment cycle are illustrated in Fig. 20.3.

### 20.8.1 Articulate goals and objectives

In the medical education assessment cycle, defining goals and objectives is an essential initial step. They offer the essential foundation for all ensuing actions, directing the creation of educational exercises and evaluations and guaranteeing conformity with program objectives and accreditation requirements. Since goals and objectives specify the desired

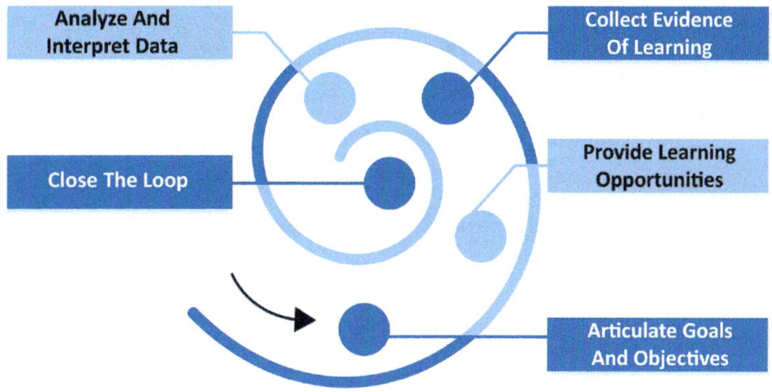


FIGURE 20.3 **Assessment cycle.** This figure shows the assessment cycle.

learning outcomes at the institutional, program, course, and activity levels, their articulation aids in establishing direction. Additionally, it aids in curriculum and assessment alignment since well-defined objectives guarantee that instructional strategies and test items are closely linked to the knowledge that students are supposed to acquire. Validity and reliability in evaluations depend on this alignment.

### 20.8.2 Provide learning opportunities

The assessment cycle offers several learning opportunities mainly through programmatic evaluation, self-reflection, and formative feedback. These components assist students in identifying knowledge gaps, honing abilities, and forming the self-directed learning habits necessary for long-term career advancement. The assessment cycle contributes to learning opportunities by promoting prompt, targeted feedback that helps students feel more confident and perform better.

### 20.8.3 Collect evidence of learning

Since assessing a student's real work or performance provides direct evidence of their mastery of particular abilities, gathering evidence of learning in medical education is prioritized in the assessment cycle. The collected data is utilized for summative judgments, like certification, as well as formative ones, such as giving comments for growth.

### 20.8.4 Analyze and interpret data

An essential part of the cycle of continuous medical education assessment is data analysis and interpretation. This procedure turns unprocessed assessment data into insightful information that can be utilized to pinpoint program advantages and disadvantages, guide enhancements, and encourage accountability.

### 20.8.5 Close the loop

"Closing the loop" refers to the crucial last phase of the medical education evaluation cycle, wherein the findings are used to guide and execute tangible enhancements in the curriculum, instructional strategies, or academic procedures. It changes the assessment process from one of gathering data to one of ongoing quality improvement. In order to create the action plan, the closing the loop phase is crucial. Particular action plans must be created if goals were not achieved. To address a particular skill gap found in the data, for instance, a program can develop a new course or activity. It is also essential for the implementation of changes, which may include updating the course material, modifying the teaching strategies, changing the prerequisites for the course, or modifying the admissions requirements.

## 20.9 Assessment versus evaluation

In medical education, assessment focuses on monitoring individual student learning and providing feedback to guide their progress, using methods like pretests and skill-based observations to identify strengths and weaknesses. In contrast, evaluation is a broader process of judging the overall quality, effectiveness, and value of a program, curriculum, or educational activity, using criteria to determine if objectives were met. Essentially, assessment is about the individual learner, while evaluation is about the program or system (Table 20.2).

**TABLE 20.2 Difference between assessment and evaluation.**

| Sl no. | Assessment | Evaluation |
|---|---|---|
| 1 | Ongoing, during the activity or process | It is done after the end of activity or process |
| 2 | It improves quality of the learner | It judges the quality of the learner |
| 3 | Diagnostic in nature as it provides specific strength and weakness | It is judgmental as it determines the extent to which objectives are achieved |
| 4 | It is completely process oriented | It is product oriented |
| 5 | The result is continuous development and adjustment | The result is in the form of final decision |

## 20.10 Process of assessment

A cyclical framework of planning, carrying out, assessing, and refining assessments is used in medical education to gauge students' advancement in their knowledge, abilities, and behaviors. Setting specific learning goals is the first step, which is followed by choosing suitable techniques like OSCEs or portfolios, gathering information on student performance, evaluating the findings, and giving instructors and students feedback. This ongoing process, commonly known as Plan–Do–Check–Act, guarantees that assessments are credible, valid, and utilized to improve the curriculum and direct student learning.

### 20.10.1 How to design a good assessment

A good assessment must be transparent to students and in line with the learning objectives of the course. Clear expectations, excellent feedback, and techniques that evaluate what matters most such as using real or alternative tests to gauge abilities beyond merely recall are all components of a good assessment. The following are essential guidelines for creating a successful assessment:

#### *20.10.1.1 Align with learning outcomes*

A key component of good instructional design is matching assessments to learning objectives. This guarantees that the evaluation examines the precise knowledge that students were supposed to acquire from the course or module, not other abilities. The knowledge and abilities you want learners to acquire should be directly measured by the assessment. In order for learners to practice the material being evaluated, it should also be in line with teaching and learning activities.

#### *20.10.1.2 Ensure validity, reliability, and fairness*

In high-stakes assessments, validity, reliability, and fairness are all important. This means that while the assessment should be inclusive and equitable, avoiding presumptions about prior knowledge and offering realistic challenges, it should also accurately measure the intended learning outcomes, be consistent and free from bias, and give every student an equal chance to succeed.

#### *20.10.1.3 Promote transparency*

Students should be made aware of the assessment goals, duties, and standards, so they may anticipate how their work will be evaluated. Encouraging openness is essential to creating an effective assessment since it improves student learning, validity, reliability, and fairness. Communicating expectations, standards, and procedures to all the stakeholders involved, especially students, is a key component of assessment transparency.

#### *20.10.1.4 Need to enhance learning*

In addition to measuring students' knowledge, effective evaluations should promote and improve learning. Additionally, it must give students the chance to use criticism to enhance their work. For example, rather than only measuring student achievement at the conclusion of a learning experience, formative assessment employs continuous feedback and reflection to enhance student performance.

#### *20.10.1.5 Use authentic or alternative methods*

Instead of just having students memorize facts, authentic or alternative assessment methods use real-world tasks to test students' application of knowledge. Creating assignments that are pertinent, encourage higher-order thinking, and provide a concrete proof of student's abilities like carrying out an experiment or making a presentation is necessary for effective evaluation.

#### *20.10.1.6 Provide clear feedback*

Clear feedback should be specific, timely, and actionable, focusing on the work itself rather than the person. It should include both strengths and weaknesses, provide clear examples, link to learning objectives or criteria, and offer actionable suggestions for improvement. Effective feedback also creates a supportive dialogue and encourages reflection and future growth.

#### *20.10.1.7 Give choices to student*

It is important to provide students with good assessment choices and offer options for topic or product and use a single, clear rubric to assess all choices fairly. This guide students by providing clear expectations, offering examples of good performance, and using formative assessments to practice skills with feedback.

#### *20.10.1.8 Plan strategically*

To plan strategically as part of student assessment, the process should be integrated into an ongoing cycle of defining goals, gathering evidence, analyzing data, and making improvements. This ensures that assessment is not just an endpoint but a continuous process that supports student success and institutional effectiveness. The key steps in the assessment process are described below.

#### *20.10.1.9 Planning*

In order to validate the assessment's aim, make sure it is in line with learning objectives and select suitable assessment techniques that accurately measure students' knowledge, abilities, and behaviors; planning is essential in medical assessment. In addition to facilitating the evaluation of curricular efficacy and resource allocation, planning produces a coherent assessment strategy that drives student learning and yields valuable feedback. The main actions in this respect are the following:

##### 20.10.1.9.1 Articulate learning objectives

Determining what students should know or be able to do at the conclusion of a learning experience is part of articulating learning objectives during the assessment preparation phase. This process is crucial for creating focused, measurable tests since it guarantees that assessments are in accordance with teaching strategies and the overall course objectives. In order to achieve this, objectives should be stated in accordance with the SMART (Specific, Measurable, Attainable, Relevant, Time-bound) criteria, utilizing action verbs that describe observable activities.

##### 20.10.1.9.2 Map curriculum and select methods

Mapping curriculum and selecting methods during the planning phase of assessment is a structured process that ensures alignment between what is taught, how it is taught, and how learning is measured. This alignment helps identify gaps or redundancies and ensures students have opportunities to achieve all intended outcomes.

#### *20.10.1.10 Role of blueprint*

A blueprint functions as a strategic "map" or "table of specifications" during the design stage of assessment in medical education, ensuring that tests are methodically matched with the learning objectives and competences of the curriculum. It is essential to creating legitimate, trustworthy, and high-quality assessments. During the planning stage, an assessment blueprint's main functions are to guarantee coverage and content authenticity. The blueprint outlines the subjects and skills that will be evaluated and offers an organized overview of the complete curriculum. This process guarantees a fair and objective evaluation that accurately reflects the variety of the course material by preventing the overrepresentation or underrepresentation of specific subject areas. Additionally, it helps us align with learning objectives since it makes a clear connection between the chosen assessment methods and tools and the intended learning outcomes and instructional activities. This promotes a coherent educational experience by guaranteeing that the assessment assesses what was really taught and what students are expected to know and be able to accomplish.

#### 20.10.1.10.1 Setting benchmarks

During the planning stage of assessment, setting benchmarks include a methodical process of creating precise standards and benchmarks that future performance can be evaluated against. This guarantees that the evaluation is focused, impartial, and in line with the project's or organization's overarching objectives. Key performance indicators must be identified in this step by choosing pertinent measurements that will accurately show progress toward the objectives. These key performance indicators must be pertinent, quantifiable, and in line with the learning objectives.

#### 20.10.1.10.2 Design assessment tools

During the planning stage, assessment instruments should be designed with defined objectives, suitable assessment kinds, and alignment with assessment principles. The setting and conditions, student assignments, an outline of the evidence to be obtained, and performance evaluation criteria are then produced as part of the tool. These will outline the assessment's specifics, including its location, time constraints, and accessible resources. Additionally, it will assist the tutor in creating the precise assignments that students will finish to show off their abilities or understanding. During the planning stage of medical education assessment, it is critical to determine what should be assessed in accordance with learning outcomes, when to assess by combining formative and summative assessments, and how to assess by selecting practical, valid, and reliable techniques. Along with the learning objectives, assessments should be prospective and intended to gauge performance as proof of learning. In order to control student burden and be appropriate for the duration of the learning experience, the assessments should be scheduled strategically. This decision will be influenced by the assessment method's timing. Lastly, in order to obtain a complete picture of a student's performance across all skill and knowledge levels, we must prepare to use a range of assessment instruments and sample techniques.

### *20.10.1.11 Implementation*

In order to ensure that efficient, evidence-based assessment techniques are included into regular teaching and learning, implementation is essential in medical education assessment. This will enhance student competency, foster deep learning, and eventually result in better healthcare outcomes. When evaluations are implemented correctly, they are guaranteed to be in line with learning objectives, provide insightful feedback, and promote the methodical development of the clinical knowledge and abilities needed for competent and safe patient care. It consists of two main steps:

#### 20.10.1.11.1 Conducting the assessment

A selected evaluation technique must be carefully planned, the criteria must be communicated clearly, the assessment must be carried out, and the findings must be used to guide future actions. The specifics of the assessment task should be communicated to the students well in advance if it is scheduled in between classes, such as a seminar or problem-based learning. This is typically done at the beginning of the course to give the student enough time to get ready and ask questions if they have any. It is the tutor's duty to guarantee that the assessment is impartial, trustworthy, and offers significant proof of the desired learning objectives or skills.

#### 20.10.1.11.2 Collecting data

A methodical, cyclical process of selecting suitable techniques, creating trustworthy tools, acquiring data, and arranging it for analysis is involved in data collection during the execution of an assessment. A combination of quantitative data, such as test and assignment scores, and qualitative data, such as portfolios and observations, can be utilized to gather information on student performance during assessment. It is crucial to remember that your assessment objectives should guide the data collection techniques that are being selected. Both forms of data are examined at the conclusion of the data gathering process in order to obtain a comprehensive picture of student development and guide instructional modifications.

### *20.10.1.12 Evaluation and analysis*

In order to direct changes, offer practical feedback, and ascertain the efficacy of training programs and initiatives, evaluation and analysis are essential components of medical education assessment. This process allows educators to enhance student learning outcomes, adjust instructional strategies, and ensure the development of essential knowledge, skills, and attitudes for effective patient care by identifying strengths and weaknesses in a student's learning and program's structure.

#### 20.10.1.12.1 Role of item analysis

Item analysis evaluates exams to ascertain the quality of both individual items and the exam as a whole by statistically evaluating student responses. It uses metrics like the difficulty index, discrimination index, and distractor analysis to determine whether items are too easy or hard, whether they differentiate between high and low performers, and whether the incorrect response choices, often known as distractors, are effective. This process helps to improve subsequent assessments, increase test validity and reliability, and detect learning gaps among students.

#### 20.10.1.12.2 Provide feedback

Creating a secure, nonthreatening environment, giving timely, targeted feedback, and concentrating on well-defined learning objectives are all necessary for providing successful feedback in medical education. Reinforcing good performance, offering constructive criticism with practical solutions, and promoting self-evaluation are some of the key goals of giving feedback. For the purpose of creating an action plan and tracking advancement, the learner and the assessor must establish a solid rapport.

#### 20.10.1.12.3 Evaluate the assessment

In medical education, evaluating an assessment entails assessing its effectiveness by looking at student performance as well as taking into account the assessment's wider influence on curriculum and learning. Determining the validity and reliability of the assessment, the fairness of its criteria, and the caliber and application of the feedback given to students to direct their future development are some of the crucial components of evaluation. A thorough evaluation makes use of student performance data to enhance both the individual assessment and the curriculum as a whole.

### *20.10.1.13 Improvement*

Enhancing medical professional competency, guaranteeing patient safety, cultivating a culture of quality, and encouraging ongoing learning all depend on improvements in medical education assessment. It enables the development of critical thinking abilities, the detection of learning gaps through formative feedback, and the modification of curriculum to satisfy changing patient and professional demands. Medical education can better match the intricacies of contemporary healthcare and educate graduates for their vital position on the healthcare team by consistently improving evaluation techniques. Among the crucial actions in this procedure are the following:

#### 20.10.1.13.1 Identify areas for improvement

Finding opportunities for improvement in medical education assessment entails assessing and matching tests to learning objectives, making sure they measure clinical competence effectively, and giving students quick, helpful feedback. Examining the validity and reliability of assessment instruments, ensuring the caliber of feedback channels, and evaluating the alignment between curriculum and assessment are all beneficial ways to accomplish this.

#### 20.10.1.13.2 Revise the curriculum

A curriculum can be revised for efficacy during assessment by using data to pinpoint areas that need work, ensuring that teaching, material, and assessment are all in line. In order to sustain constructive alignment, this entails collecting both quantitative and qualitative data, consulting stakeholders like teachers and students, and taking into account how changes in one area may affect others.

## 20.11 Assessment in the era of artificial intelligence

In the era of AI, assessment entails utilizing AI's potential for automated tasks and tailored learning while also modifying methods to control dangers like cheating and guarantee authenticity. Redesigning exams to be "AI-proof" through authentic, real-world tasks or integrating AI as a tool in the learning process itself are the main actions to be made in this regard. While addressing ethical issues like algorithmic bias and data protection, educational institutions are investigating the use of AI for assessment design, grading, and personalized learning. One of the most significant advantages of artificial intelligence in assessment is that it can automate the grading of some tests, saving teachers time and freeing them up to concentrate more on instruction.

## 20.12 Benefits of assessment in medical education

In medical education, assessment helps students learn better, become more motivated, and develop their capacity for introspection. It also benefits teachers and the program by providing evidence of learning, directing curriculum changes, and ensuring proficiency before certification. Some of the key benefits include identifying knowledge gaps, providing insightful feedback, promoting lifelong learning, and monitoring improvements in knowledge, skills, and behaviors.

### 20.12.1 For students

#### *20.12.1.1 Enhances learning and identifies gaps*

Assessment enhances learning and identifies gaps primarily through a feedback mechanism that informs both students and teachers, allowing for targeted adjustments to teaching and study strategies. This process is crucial for continuous improvement and achieving learning objectives. In this regard, assessments help pinpoint missing knowledge components, particularly diagnostic and formative ones, and serve as a "mirror" to reveal specific areas where a student lacks the necessary knowledge or skills to fully grasp a concept (Kordestani Moghaddam et al., 2019). It also helps in highlighting misconceptions by analyzing assessment results, in which the educators can spot common errors or patterns of misunderstanding across a class, which helps in understanding why students are struggling with a particular concept.

#### *20.12.1.2 Increases motivation*

Assessment in medical education improves motivation, as it guides students' learning attempts and encourages both intrinsic incentive through feedback and extrinsic drive through grades. It encourages participation, self-control, and the development of therapeutic skills, especially when formative techniques that offer feedback are used instead of only summative, high-stakes testing.

#### *20.12.1.3 Develops self-reflection*

Assessment, particularly self-assessment, is a powerful tool for developing self-reflection, which in turn helps individuals become more aware of their own learning processes, identify strengths and weaknesses, and take responsibility for their ongoing growth. It also enhances self-reflection and helps individuals gain insights into their own emotions, thoughts, behaviors, and learning preferences. This self-awareness is a foundational component of emotional intelligence and personal development. It also helps improve critical thinking, as the process of critically reflecting on one's work and experiences develops analytical skills, allowing individuals to make sense of information and draw informed conclusions.

#### *20.12.1.4 Builds practical competency*

Practical assessments are highly effective for developing a range of work-ready skills by closing the gap between theoretical knowledge and real-world application. Practical assessments foster skills crucial for professional success that often go unmeasured by traditional exams. One of the advantages is that learners apply knowledge to complex, real-world scenarios, which improves their ability to analyze situations and find effective solutions by fostering problem-solving skills (Preston et al., 2020).

### 20.12.2 For educators and the program

#### *20.12.2.1 Validates learning outcomes*

Validating learning outcomes is a fundamental benefit of assessment because assessment is the primary mechanism used to determine if students have achieved the intended knowledge, skills, and competencies. It also ensures the attainment of quality assurance and accountability as the data gathered from assessments helps educational institutions meet the requirements of accrediting bodies and funding agencies. It provides a solid basis for justifying institutional claims that graduates are prepared for future endeavors, thereby ensuring accountability and maintaining standards.

#### *20.12.2.2 Informs instructional design*

Assessment informs instructional design by providing feedback on learner progress; this in turn helps educators and designers adjust teaching strategies and refine course content. It helps determine if learning objectives have been met, identify areas where students struggle, and verify if the instructional materials are effective and engaging.

#### *20.12.2.3 Improves curriculum and teaching*

Assessment improves curriculum and teaching by providing data to identify and address student learning gaps, informing adjustments to instructional methods, and revealing the effectiveness of current curriculum. This allows teachers to provide targeted feedback and administrators to make evidence-based decisions for curriculum refinement and professional development.

#### *20.12.2.4 Ensures professional standards*

Assessment ensures professional standards by validating that individuals possess the necessary skills and knowledge for their field, providing a mechanism for quality assurance, and driving continuous improvement in practice. It is important to note that assessments are essential for awarding qualifications and certifications, which signal to employers and the public that an individual has met the minimum requirements for a profession.

#### *20.12.2.5 Supports quality improvement*

Assessment supports quality improvement by providing the data and insights needed to identify problems, measure performance, and track the effectiveness of changes made. It creates a feedback loop essential for continuous, systematic improvement. It identifies areas for improvement as assessment methods provide data collection and stakeholder feedback and highlight where current performance is below standard or where processes are failing, thus pinpointing specific areas that need attention.

## 20.13 Challenges of assessment in medical education

Challenges in medical education assessment include delays in feedback, the difficulty of reliably measuring complex competencies like clinical judgment and attitude, time constraints for educators, subjectivity in grading, the burden of managing data across multiple sources, lack of student engagement or skills in self-reflection, the need to shift from a "better than others" to a "better than yesterday" assessment culture, and the increasing complexity of adapting to online assessments. Addressing these issues requires careful planning, clear communication, and a collaborative effort between educators and learners to create a more informative and formative assessment environment.

### 20.13.1 Time and resource demands

Time and resource demands represent a significant challenge in assessment, affecting both educators and institutions by requiring substantial investments of time, money, and personnel. This challenge can impact the quality, frequency, and methods of assessment used. In addition, it leads to staff workload and burnout as educators often face immense time pressure and job demands. The additional workload associated with designing and marking assessments can strain psychological resources, impacting overall well-being and job satisfaction.

### 20.13.2 Measuring complex competencies

Measuring complex competencies is challenging due to the difficulty in creating valid, reliable, and practical assessment tools. Some of the challenges in this regard include the subjectivity and cost of scoring complex tasks, the inability of traditional tests to capture real-world performance, and a lack of expert consensus on competency definitions and measurement criteria.

### 20.13.3 Feedback delays

Delays in providing feedback can hinder student learning and morale, as students are left in uncertainty about their performance and progress. Feedback delays present a significant challenge in assessment, primarily because they can disrupt the learning loop by disconnecting the feedback from the original context of the work. However, some research also suggests that strategic delays can actually enhance long-term retention and deeper learning. The primary purpose of feedback is to look ahead into future assignments and help students improve their performance. Lengthy delays often mean students receive feedback after they have moved on to other topics or completed subsequent, related assignments, making it difficult to apply the insights gained.

### 20.13.4 Subjectivity in grading

Subjective elements in performance-based assessments can lead to inconsistency and may require clear rubrics to mitigate. Subjectivity in grading is a challenge because it can lead to inconsistent, unfair evaluations based on a grader's personal opinions rather than facts, potentially damaging the credibility of the assessment and causing student frustration. It is particularly prevalent in assessing complex or creative tasks, like essays or long-answer questions, where a rubric provides a guideline but human judgment is still required to evaluate understanding, reasoning, and expression.

### 20.13.5 Managing assessment data

Collecting, analyzing, and managing assessment data from various sources and formats can be an overwhelming and duplicative effort. Managing assessment data is a challenge due to issues like data quality, analysis, and organization. Key challenges include inaccurate or inconsistent data, difficulty in analyzing large datasets, and storing data across multiple sources and formats. Overcoming these requires clear data policies, robust quality control, and the use of appropriate technology for automation.

### 20.13.6 Adapting to online assessments

The shift to online platforms presents challenges related to maintaining reliability and validity, addressing technical issues, and preventing cheating. Adapting to online assessments presents significant challenges related to technology access and proficiency, academic integrity, and pedagogical shifts from traditional, in-person evaluation methods. These challenges affect both students and instructors and can impact the validity, reliability, and fairness of the assessment process.

**Summary**

Assessment is a core component of medical education, serving as both a tool for evaluating learner performance and an engine for driving learning itself. Assessment in medical education plays a crucial role in ensuring that learners develop the knowledge, clinical skills, professional attitudes, and competencies required for safe and effective patient care. Assessment in medical education is multifaceted, evolving, and central to producing competent healthcare professionals. Effective assessment requires alignment with educational goals, use of multiple complementary methods, commitment to fairness, and emphasis on feedback and continuous improvement. As medical education moves toward competency-based and programmatic models, the focus increasingly shifts toward authentic, longitudinal evaluation of learners' readiness for independent practice.

**Key point**

**Future perspectives: placing the emphasis on the competencies**

The landscape of medical education assessment is rapidly evolving in response to technological advances, shifts in healthcare systems, growing emphasis on competency, and calls for fairness and authenticity. Future assessment practices are expected to be more integrated, individualized, data-informed, and aligned with real-world clinical practice. The future of assessment in medical education is moving towards an integrated, technology-enhanced, individualized, and ethically grounded ecosystem. It aims to improve the accuracy of competency decisions, enhance learning through meaningful feedback, ensure fairness, and better prepare learners for the realities of modern healthcare. The ultimate goal is to create assessment systems that not only measure performance but actively shape adaptable, reflective, and patient-centered physicians.

## References

Ahmed, Y., Khayal, S., Wadi, M., Abuzaid, M., 2024. Evolving assessment in medical education: exploring the role of open-book examinations. Journal of Advances in Medical Education and Professionalism 12 (4), 215–225. Available at https://doi.org/10.30476/jamp.2024.102078.1955, https://jamp.sums.ac.ir/article_50395_b752dc8e4f16f7c68507cf9e2761ed9b.pdf.

Al-Wardy, N.M., 2010. Assessment methods in undergraduate medical education. Sultan Qaboos Univ Med J. 10 (2), 203–209. Epub 2010 Jul 19. PMID: 21509230; PMCID: PMC3074721.

Beard, J., Strachan, A., Davies, H., Patterson, F., Stark, P., Ball, S., et al., 2005. Developing an education and assessment framework for the Foundation Programme. Medical Education 39 (8), 841–851. Available at https://doi.org/10.1111/j.1365-2929.2005.02236.x.

Cox, M., Irby, D.M., Epstein, R.M., 2007. Assessment in medical education. New England Journal of Medicine 356 (4), 387–396. Available at https://doi.org/10.1056/NEJMra054784.

Epstein, R.M., Hundert, E.M., 2002. Defining and assessing clinical competence. JAMA the Journal of the American Medical Association 387, 226–235. Available at https://doi.org/10.1001/jama.287.2.226.

Ferris, H.A., O' Flynn, D., 2015. Assessment in medical education; what are we trying to achieve? International Journal of Higher Education 4 (2). Available at https://doi.org/10.5430/ijhe.v4n2p139.

Gibbs, T., Brigden, D., Hellenberg, D., 2006. Assessment and evaluation in medical education. South African Family Practice 48 (1), 5–7. Available at https://doi.org/10.1080/20786204.2006.10873331.

Kibble, J.D., 2017. Best practices in summative assessment. Advances in Physiology Education 41 (1), 110–119. Available at https://doi.org/10.1152/advan.00116.2016, http://advan.physiology.org/content/ajpadvan/41/1/110.full.pdf.

Kordestani Moghaddam, A., Khankeh, H.R., Shariati, M., Norcini, J., Jalili, M., 2019. Educational impact of assessment on medical students' learning at Tehran University of Medical Sciences: a qualitative study. BMJ Open 9 (7), e031014. Available at https://doi.org/10.1136/bmjopen-2019-031014.

Lehl, S.S., Gupta, M., Hande, S.H., Hande, M.H., 2022. The long and short of summative assessment in competency based medical education: time to raise the bar? The Journal of the Association of Physicians of India 70 (2), 11–12.

Norcini, J., Anderson, B., Bollela, V., Burch, V., Costa, M.J., Duvivier, R., et al., 2011. Criteria for good assessment: consensus statement and recommendations from the Ottawa 2010 conference. Medical Teacher 33 (3), 206–214. Available at https://doi.org/10.3109/0142159X.2011.551559.

Preston, R., Gratani, M., Owens, K., Roche, P., Zimanyi, M., Malau-Aduli, B., 2020. Exploring the impact of assessment on medical students' learning. Assessment and Evaluation in Higher Education 45 (1), 109–124. Available at https://doi.org/10.1080/02602938.2019.1614145, http://www.tandfonline.com/toc/caeh20/current.

Sharma, S., Sharma, V., Sharma, M., Awasthi, B., Chaudhary, S., 2015. Formative assessment in postgraduate medical education - perceptions of students and teachers. International Journal of Applied and Basic Medical Research 5 (4), 66. Available at https://doi.org/10.4103/2229-516X.162282.

Vageriya, V., 2018. Assessment and evaluation- in perspective of medical education. Nursing & Healthcare International Journal 2 (4). Available at https://doi.org/10.23880/nhij-16000154.

Veloski, J., Boex, J.R., Grasberger, M.J., Evans, A., Wolfson, D.W., 2006. Systematic review of the literature on assessment, feedback and physicians' clinical performance*: BEME Guide No. 7. Medical Teacher 28 (2), 117–128. Available at https://doi.org/10.1080/01421590600622665.

Wormald, B.W., Schoeman, S., Somasunderam, A., Penn, M., 2009. Assessment drives learning: an unavoidable truth? Anatomical Sciences Education 2 (5), 199–204. Available at https://doi.org/10.1002/ase.102, http://www3.interscience.wiley.com/cgi-bin/fulltext/122593991/PDFSTART.

2008. Assessment methods in medical education. International Journal of Health Sciences (Qassim) 2 (2), 3–7.

Chapter 21

# Pillars of assessment

*"Assessment is not about proving what students know but about improving what they learn."*

—Dylan Wiliam

## 21.1 Introduction

Medical schools should adopt a comprehensive, methodical, and strategic approach to assessment that complements the rest of the curriculum, provides valid, reliable, and otherwise appropriate assessment methods, and is overseen by assessment specialists with psychometric support and the required authority within the governance systems. Delegating decision-making to specific leads in specializations or years who are not involved in a coordinated and strategic approach carries dangers. It is indeed very important that the medical school should be capable of tracking learner development and guarantee that results are properly evaluated. This may sometimes be jeopardized by an unstructured and delegated approach. In this regard, the assessment pillars are essential since they guarantee that evaluations are significant, reliable, and helpful for decision-making and student development (Cox et al., 2007).

Moreover, the pillars of assessment offer a framework for creating tests that precisely measure the things they are supposed to measure, yield reliable findings, have a clear purpose, and enhance learning and growth. They also guarantee that tests are impartial, accurate, and useful for certifying qualified medical professionals, enhancing learning, and directing the curriculum. Following these guidelines enables meaningful assessments of knowledge, abilities, and behaviors, which improves student performance, enhances instructional strategies, and eventually results in safer patient care.

## 21.2 Key components and characteristics of a good assessment

### 21.2.1 Learning outcomes

Learning outcomes serve as the basis for assessment design and are quantifiable statements that specify what students will know or be able to perform at the end of a course. In order to make sure that learning activities and evaluation techniques correspond with what students are expected to accomplish, assessments are then designed to directly measure these results. The topic matter, a condition or context, and particular action verbs are all essential elements of a well-written learning outcome. Since they offer a clear framework for developing tests, projects, and quizzes that are intended to gauge students' accomplishment of particular objectives, the learning outcomes aid in guiding the assessment design. Additionally, by making results quantifiable, it guarantees alignment. Teachers can select assessment techniques that actually measure the needed knowledge and abilities rather than merely covering content (Norcini et al., 2011).

### 21.2.2 Assessment methods

The instruments and strategies used to assess student learning are known as assessment methods, and they are often divided into three categories: formative, summative, and diagnostic. Summative tests are used at the conclusion of a course to assess overall performance; formative tests are used throughout to track progress and direct instruction; and diagnostic tests are used at the start of a course to determine prior knowledge. The duties, the evidence collected, the assessment criteria, and the situation are additional elements.

### 21.2.3 Performance criteria

The precise standards, benchmarks, or observable behaviors used to gauge the caliber, efficacy, and competence displayed in a performance evaluation are known as performance criteria. They specify what is considered "good" work or a

**The Principles and Practice of Medical Education. DOI:** https://doi.org/10.1016/B978-0-443-44197-4.00005-9

successful performance in respect to predetermined learning objectives or job goals. The goal of the performance criterion is to make the evaluation procedure impartial, open, and equitable. They eliminate uncertainty from the assessment process by giving both the assessor and the individual being evaluated precise instructions. The fundamental elements of a rubric are performance criteria, which deconstruct a complicated activity into precise, quantifiable standards and offer descriptions like performance standards for various success levels, such as excellent, good, and requires improvement for each criterion.

### 21.2.4 Feasibility

As a part of assessment, feasibility is a methodical analysis of a project's viability to ascertain whether it can be carried out successfully and realistically. In order to evaluate possible success, pinpoint hazards, and direct decision-making, it entails examining a number of variables, including technical, operational, and economic viability. The purpose is to assess a project's likelihood of success objectively before committing substantial resources.

### 21.2.5 Simplicity

Simplicity is an essential feature of assessment design and practice as it improves usability, efficiency, and clarity for administrators and participants alike. Since it encourages use and interpretation, the principle of simplicity guarantees that the assessment procedure itself does not become a barrier to correct measurement or effective action. Assessment methods and instruments should also be simple to use and comprehend. This makes it possible for teachers and other stakeholders to swiftly and properly analyze the data, allowing for focused actions like adjusting instruction to meet a student's unique learning needs. When an examination is simple, it may be completed quickly and without needless complications. Simplified assessment approaches in a variety of disciplines, including education impact assessments. The use of simplified assessment also helps in saving the time and resources.

## 21.3 Frameworks in medical education assessment

While the pillars of assessment are the essential quality standards that any good assessment must meet, frameworks of assessment in medical education are the broad conceptual maps or models that organize a whole assessment system. The quality and efficacy of the framework's application are guaranteed by the pillars. A medical education assessment framework provides a structured conceptual model for evaluating learner competency and progress by clearly connecting learning outcomes, assessment methods, and performance criteria. These frameworks ensure the validity and reliability of assessments, give teachers and students a common vocabulary, guide the selection and design of assessment tools, and ultimately provide a road map for achieving competency by defining objectives and tracking progress across the curriculum. To put it simply, developing high-quality, trustworthy, and valid assessment programs that successfully direct learner growth and guarantee patient safety by generating qualified healthcare professionals requires a strong assessment framework in medical education.

## 21.4 Aims of an assessment framework

### 21.4.1 Guide for learners

Assessment serves to direct learners' progress by giving them feedback, assisting them in understanding learning objectives, and fostering the development of self-evaluation skills. The assessment process helps to improve learning throughout the course rather than just at the conclusion by identifying strengths and weaknesses to direct future actions and increase motivation. Assessments assist learners to identify where they are and what they need to change in order to accomplish learning objectives. It also enables the student to take control of their own education by aligning learning objectives with success criteria. All of this contributes to the learner's increased motivation as they see their progress and learn how to do better.

### 21.4.2 Reliability and validity

The fundamental goals of assessment are validity and reliability, which guarantee that tests are accurate, valid, and consistent, that is, that they measure what they are supposed to and produce steady findings. While validity ensures that the test evaluates the desired notion and nothing else, reliability concentrates on repeatability (Htay et al., 2023).

### 21.4.3 Competency-based evaluation

Competency-based evaluation instead of focusing solely on theoretical knowledge, measures a person's practical capacity to apply skills, information, and behaviors in real-world settings. Its main objective is to make sure students can do the tasks necessary for a position or learning objective. It goes beyond conventional assessments to find strengths, identify skill gaps, and direct development for better organizational performance and career readiness (Epstein & Hundert, 2002).

### 21.4.4 Window to learning

"Window to learning" as an aim of assessment highlights the use of assessment not only to evaluate a student's final knowledge but also to obtain a clear picture of their continuous learning process. It entails using assessment as a tool for continuous development rather than a definitive judgment in order to give feedback for improvement, guide instruction, and assist students in understanding their own learning. It gives students a better understanding of what they already know and where they need to grow, making their learning process more apparent. Students can improve their learning strategies and gain a greater understanding of their cognitive processes by taking part in self and peer assessments. Simultaneously, it enables educators to assess the efficacy of their pedagogical approaches and make prompt modifications as necessary. Additionally, teachers can use assessment to pinpoint particular challenges and offer prompt, targeted feedback or assistance to help students bridge the gap between their present comprehension and learning objectives (Fig. 21.1).

Some of the common frameworks of assessment are described below:

1. Miller's pyramid,
2. Reporter–interpreter–manager–educator (RIME) framework.

### 21.4.5 Millers pyramid

Miller's pyramid, which is a tiered paradigm for evaluating clinical competency, moves from fundamental understanding to real-world application. "Knows" (knowledge), "Knows How" (understanding), "Shows How" (demonstration), and "Does" (performance) are its four levels. The approach is used in health education to match learning objectives with anticipated competencies, with evaluation in practical settings being necessary at the highest levels (Witheridge et al., 2019).

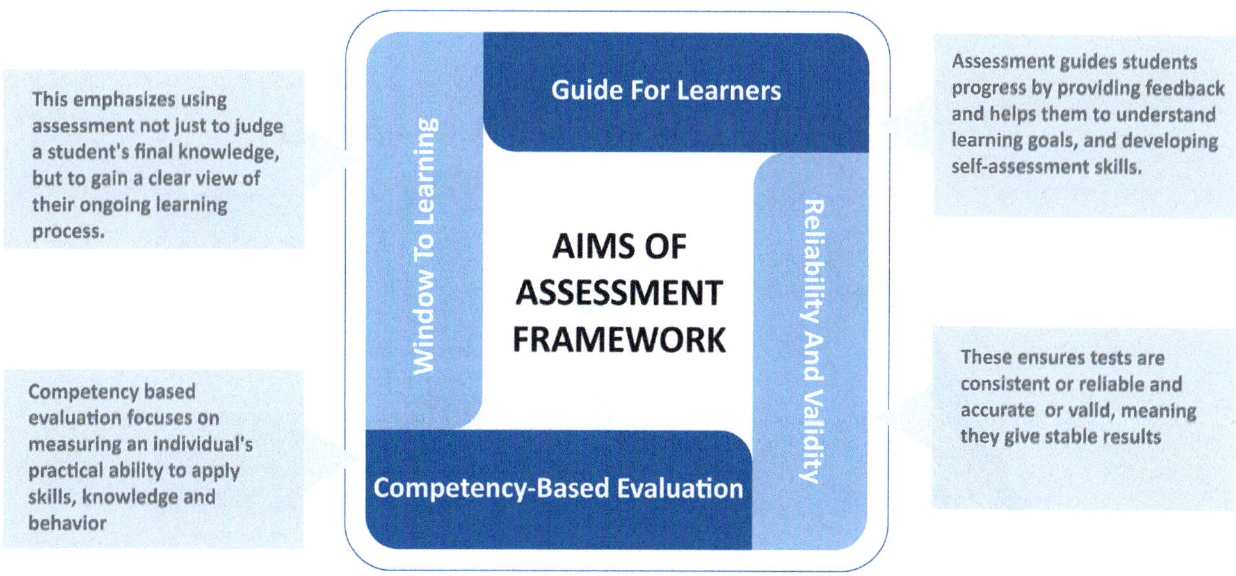


FIGURE 21.1 Aims of assessment framework.

## 21.5 The four levels of Miller's pyramid

### 21.5.1 Knows

The acquisition of factual and fundamental knowledge is represented by the "knows" phase, which is the lowest level of Miller's pyramid of clinical competence. The ability of a student to retain knowledge and comprehend fundamental ideas is the main focus of this stage. The possession of scientific ideas, hypotheses, and factual information about a certain topic or ability is a characteristic feature of this stage.

### 21.5.2 Knows how

This level assesses the capacity to use the information, for instance, in management plan synthesis or problem-solving activities. "Knows how" is the second level of Miller's pyramid, where a person can use fundamental information to finish a task or solve a problem, but they might not yet be able to do so on their own in a real-world situation. It symbolizes the capacity to use knowledge to comprehend how to do something, bridging the gap between merely remembering facts and practically exhibiting a talent. Instead of using a straightforward knowledge recall test, this level is evaluated using techniques that call for the application of information, including case studies or problem-based questions.

### 21.5.3 Shows how

In this phase, students use standardized patient activities or simulations to show that they can execute abilities in a clinical setting. According to Miller's pyramid, "shows how" refers to a learner's capacity to exhibit proficiency and apply information in a real-world, simulated situation, such conducting a physical examination or handling a clinical case during an Objective Structured Clinical Examination (OSCE). This goes beyond just knowing the facts and being able to describe how to do something. This step warrants that the student must be able to demonstrate a skill in a controlled environment.

### 21.5.4 Does

This stage is the highest level that uses direct observation to evaluate a learner's autonomous performance in actual, authentic practice environments. It represents a learner's capacity to operate autonomously in actual clinical practice and is the highest level of Miller's pyramid of clinical skill assessment in contrast to the lower levels of "knows," "knows how," and "shows how." The "does" level, which is the hardest to quantify, focuses on what an individual actually does in their regular work and is frequently evaluated using techniques like patient satisfaction surveys and in-person observation in a clinical context (Fig. 21.2).

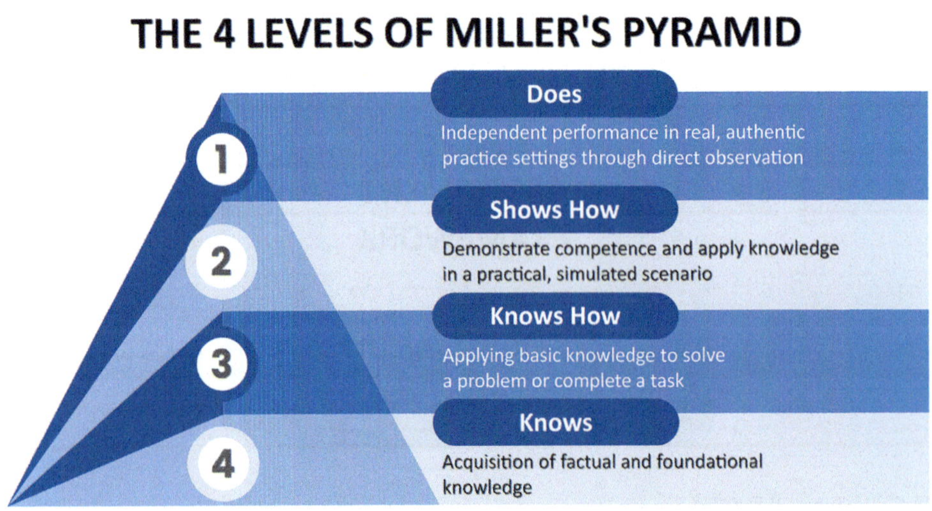


FIGURE 21.2 The four levels of Miller's pyramid.

## 21.6 Role of Miller's pyramid in assessment

The Miller's pyramid plays a role in assessment in the following ways:

Miller's pyramid, which offers a hierarchical structure that progresses from fundamental knowledge to practical application, is used in evaluation to evaluate clinical competency. The four levels namely, knows, knows how, shows how, and does correspond to various forms of evaluation, such as written tests for the base and performance-based assessments for the top. This approach transitions from theoretical knowledge to practical application, matching learning outcomes with a learner's developmental stage. In assessment, the Miller's pyramid is used to achieve the following goals:

1. It uses objective examinations, such as multiple-choice questions or written exams, to evaluate the learner's knowledge base.
2. Assesses knowledge application skills through case studies, essays, or problem-solving activities.
3. Assesses abilities in a regulated clinical setting, like in simulations or standardized patient activities.
4. Evaluates performance in a real-world clinical environment where the student operates on their own.
5. Moving beyond skills and knowledge to being a professional, the fifth level of Miller's pyramid evaluates a learner's professional identity, values, and attitudes.

In addition to the above-mentioned roles, Miller's pyramid, which moves from theoretical knowledge to practical application, is a highly relevant hierarchical assessment of clinical competence in medical education. It gives teachers a framework for assessing how well a student progresses from understanding fundamental concepts to demonstrating how to apply them, and finally completing the assignment on their own in an actual clinical situation. This is achieved in the following ways:

### 21.6.1 Alignment with learning outcomes

Miller's pyramid offers alignment with learning outcomes since it provides a hierarchical structure that charts the development of a learner's skills from fundamental knowledge to actual, autonomous performance in practice. In addition, Miller's pyramid helps in bridging the gap between theory and practical implementation, thereby assisting instructors to ensure that instructional strategies and evaluations change as students grow.

### 21.6.2 Diverse assessment methods

Miller's pyramid, which outlines a hierarchy of clinical competence with distinct evaluation tactics appropriate for each level, offers a framework for employing a variety of assessment techniques. In order to properly assess a learner's development, a range of instruments are required as this framework progresses from fundamental knowledge to practical performance.

### 21.6.3 Focus on practice

In order to close the gap between knowing and doing, the model highlights the significance of evaluating how students use their knowledge in practical situations. By creating a clear hierarchy that leads to the "Does" level, which evaluates a person's actual, unsupervised performance in a real-world clinical context, Miller's pyramid of clinical competence offers a focus on practice. It advances evaluation from simple knowledge recall to skill display in real-world contexts.

### 21.6.4 Evolution of competence

It highlights the necessity of repeated experience and a variety of assessment situations while offering a framework for monitoring the development of competence from novice to expert. Miller's pyramid provides a framework for the progressive development of competence, moving from basic knowledge to actual performance in a professional setting, by linking many learning levels to appropriate assessment methods. Moving up the pyramid represents increasing complexity, professional authenticity, and the integration of knowledge, abilities, and attitudes.

### 21.6.5 Reporter–interpreter–manager–educator (RIME) framework

The RIME framework is a developmental model used in medical education to evaluate and direct a learner's advancement through clinical competence by classifying their skills into four roles: Educator, who masters clinical skills and instructs others; Manager, who chooses and implements treatment plans; Interpreter, who identifies common diagnoses; and Reporter, who gathers and communicates information (Parmar, 2025). This framework helps educators and students comprehend and track their progress from novice to expert by offering a consistent language for evaluation and feedback.

### 21.6.6 Reporter

The first step in the RIME structure is the reporter component, where a learner concentrates on precisely and successfully obtaining and conveying clinical information about a patient. In order to paint a clear image, a reporter can provide answers regarding a patient's symptoms, medical history, and test findings while separating pertinent information from irrelevant data. For all clinical specialists, this phase is fundamental.

### 21.6.7 Interpreter

The RIME framework's interpreter component is where students go beyond just collecting patient data to independently evaluating and explaining it. They also are able to recognize and rank issues and create a differential diagnosis. Building on the reporter stage, which concentrates on data collection, this level focuses on "why" questions, such as why a patient has particular symptoms or test findings. When a student goes beyond merely collecting patient data and starts analyzing it to create a differential diagnosis, they enter the interpreter phase of the RIME system. At this point, students are able to prioritize a patient's difficulties, pinpoint the most crucial ones, and bolster their arguments in favor of or against various diagnoses. This stage comes before the "manager" phase and after the "reporter" phase. The ability of the learner at this level to take the unprocessed data from the "reporter" stage and start deciphering its meaning is one of the key characteristics of the interpretation phase. As interpreters, the students are also capable of setting priorities since they are able to determine which of the patient's problems are most important.

### 21.6.8 Manager

The learner's capacity to create and carry out a patient care plan, progressing from the interpretation stage of diagnosis to the "action" stage of therapy, is represented by the manager component of the RIME system. A manager may prioritize work for several patients, independently develop and defend a diagnostic and therapeutic plan, and weigh the advantages and disadvantages of various treatment alternatives. The RIME framework's manager phase is when a student can autonomously create and carry out a patient management plan, evaluate a plan's risk-benefit ratio, and set priorities for a group of patients. In therapeutic practice, this stage denotes the shift from comprehension and interpretation to action and decision-making. This phase is characterized by the students' ability to develop and defend a plan since they are able to develop a therapeutic and diagnostic plan for a patient and justify it. They can weigh the advantages and disadvantages of various interventions and use clinical judgment to decide whether action is required. This entails planning management for a group of patients rather than simply one and prioritizing which patients require care first. Additionally, it provides an opportunity for the learner to transition from theory to practice. In fact, the manager phase signifies a change from the interpreter phase's cognitive comprehension to management and treatment's active action.

### 21.6.9 Educator

The highest level of clinical development is represented by the Educator phase of the RIME system, when a learner may detect knowledge gaps and contribute to the larger medical community in addition to managing challenging situations. In addition to teaching and mentoring junior colleagues and educating patients and their families about care and discharge plans, an educator can identify key research issues. Moving from comprehension and action to becoming a source of knowledge for others is part of this stage. One of the key characteristics of being a student educator is that the student educates peers and junior professionals, exhibiting a mastery of clinical skills and a willingness to share knowledge. Additionally, because students at this level are capable of critically evaluating the data, identifying unresolved issues, and developing research or quality improvement questions to further the profession, they support reflective practice and research (Fig. 21.3).

## 21.7 Role of reporter–interpreter–manager–educator (RIME) in assessment

### 21.7.1 Describes learner progress

Learner progress is described by the RIME paradigm as a developmental continuum via four distinct stages of growing autonomy, responsibility, and knowledge, skill, and attitude synthesis. In medical education, it is frequently used to track professional development and offer controlled feedback. As they move from comprehending information to acting and instructing others, learners advance by continuously exhibiting dependability at every level.

FIGURE 21.3 Reporter–interpreter–manager–educator (RIME) framework.

### 21.7.2 Facilitates feedback

Educators can utilize RIME to give students focused feedback that identifies their current level and offers suggestions for moving on to the next. By offering an organized, progressive, and shared vocabulary for evaluating a learner's clinical performance, the RIME framework encourages feedback and enables educators to provide precise, practical advice for advancement to the next level of competence. In order to provide a common understanding of performance goals and reduce misunderstandings during feedback sessions, the RIME framework employs straightforward, descriptive categories that are simple for both educators and students to understand.

### 21.7.3 Sets expectations

At various stages of their training, the framework aids in setting clear expectations for both residents and students. Additionally, by offering a developmental continuum of professional competence that clearly outlines the required skills and responsibilities at various levels of a learner's training, this framework establishes clear expectations in evaluation. Preceptors and students can better grasp current performance levels and identify areas for development by using this common, nonjudgmental terminology.

### 21.7.4 Enhances reliability

RIME aids in standardizing assessment and enhancing the dependability of evaluations across many specialties and contexts by offering an organized method. The framework enhances assessment reliability by providing educators with an organized, standardized, and behaviorally specific set of criteria to monitor and evaluate student growth. This systematic approach reduces subjectivity and promotes consistency across numerous assessors and circumstances. Furthermore, this framework aids in the development of consistent standards by delineating discrete, increasing stages of clinical competence by way of just gathering information to teaching others and self-evaluation. This rigorous structure ensures that all faculty members are evaluating students using the same set of defined standards rather than their own implicit expectations.

### 21.7.5 Guides development

This paradigm directs advancements in assessment as it offers a structured, developmental continuum of observable clinical performance. It assists teachers in diagnosing the learner and provide focused feedback to support their development from fact-finding to self-directed, skilled practitioners.

## 21.8 Pillars of assessment

Purpose, validity, reliability, and value are usually regarded as the fundamental pillars of assessment in medical education. These pillars guarantee that tests are significant, measure what they are supposed to, yield consistent results, and support students' learning and growth. Fairness, feasibility, feedback, and authenticity are additional important characteristics that support these pillars and are essential for efficient evaluation in medical education (Fig. 21.4).

### 21.8.1 Purpose

In medical education, assessment serves a variety of purposes, including guiding and enhancing student learning through formative feedback, evaluating student performance for certification and licensing, ensuring and upholding academic

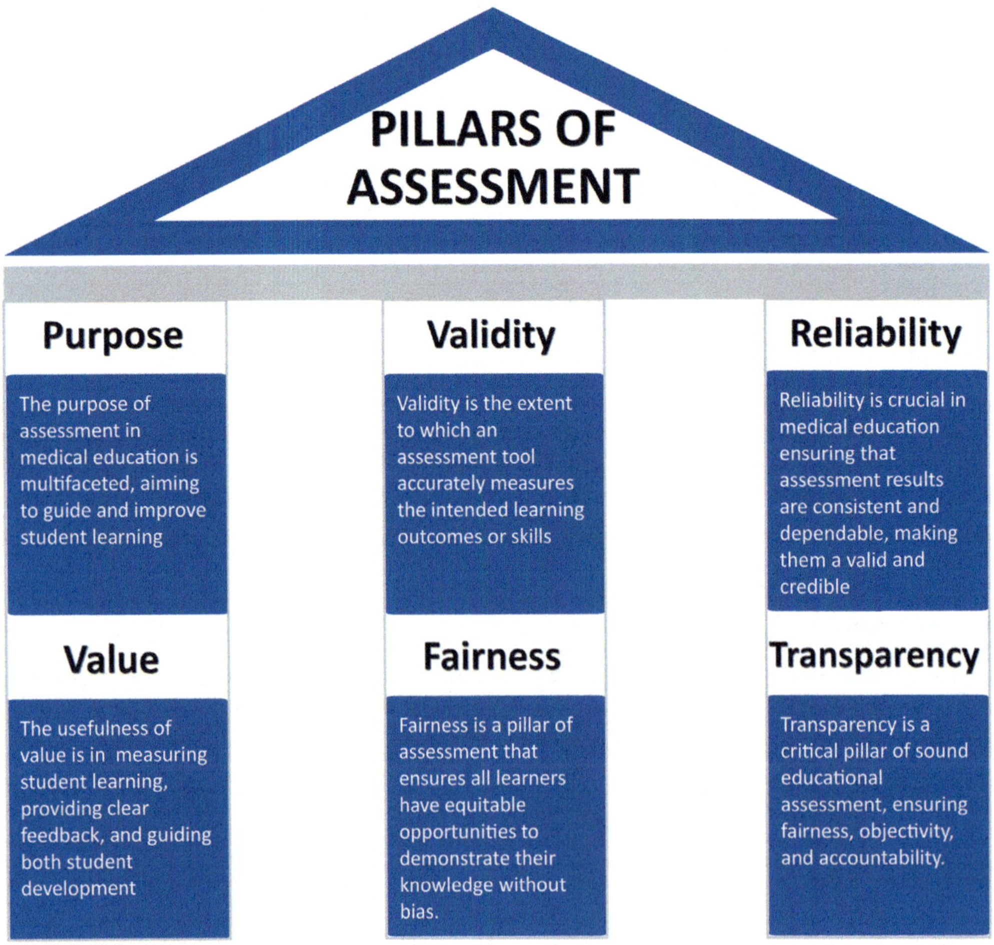


FIGURE 21.4 Pillars of assessment.

standards, and informing teachers about the efficacy of their instruction and the overall advancement of the curriculum. By giving both the student and the instructor information on their progress, strengths, and flaws, assessment serves to enhance student learning. Assessments assist in identifying learning gaps, tracking students' advancement toward learning goals, and provide feedback to direct future instruction and research. They also help decide whether learning objectives have been accomplished, provide final assessments of student performance, and inspire students to put in more effort. The goal is crucial because it tracks students' development and enables them to determine whether they are fulfilling course objectives and curriculum expectations. Additionally, it provides detailed information on students' strengths and shortcomings to assist them perform better. Additionally, the evaluation encourages students to keep getting better by demonstrating the results of their hard work. The test also helps uncover barriers by detecting learning issues, such as comprehending challenging medical topics or concepts, so that students can get the help they need. It aids teachers in assessing the efficacy of their teaching methods and making decisions about whether to repeat content, move on, or adopt new ways. In addition, it enables teachers to pinpoint the areas in which students are struggling to grasp the subject matter. It also helps the teacher monitor the student's development over time with respect to the syllabus. Along the same lines, determining whether students have fulfilled the learning objectives necessary for a degree or certificate is beneficial to the institutions. However, determining what, when, and how items are to be evaluated is another goal of the evaluation. In light of this, there are three categories of evaluation purposes:

#### *21.8.1.1 Formative assessment*

Formative assessment's goal is to track students' progress and offer continuous feedback that enables both students and teachers to make real-time instructional adjustments. By giving students immediate, targeted feedback, it helps to promote learning by helping them identify their areas of strength and weakness. The primary goal here is to direct instruction since it provides teachers with information about students' comprehension, enabling them to adjust activities and teaching strategies to clear up misconceptions and better serve students' needs (Bennett, 2011).

#### *21.8.1.2 Summative assessment*

The goal of summative assessment is to gauge students' learning at the conclusion of a unit or course by comparing their material and skill mastery to a standard. The findings of this kind of formal, high-stakes assessment are used to measure a student's overall performance in a course, inform stakeholders about the efficacy of the program, and direct future instruction. Summative assessment's primary goal is to assess students' knowledge and abilities at a single moment in time, giving us a full estimate of what they have learned throughout a unit, course, or program. Additionally, it enables the instructor to assess if learning objectives have been fulfilled by comparing student performance to a predetermined benchmark or standard.

#### *21.8.1.3 Certification and licensing*

Assessment for certification and licensing in medical school is essentially a summative procedure intended to safeguard the public by guaranteeing that all doctors fulfill a similar, uniform standard of competence and fitness to perform safely and effectively. Across many medical schools and training programs, these evaluations set a uniform and defined threshold for safe practice. They serve as a crucial screening tool to decide who is eligible to move on to the next training phase or start working independently. To get a license for full practice, a person usually needs to pass these important tests.

## 21.9 Advantages of the purpose of assessment

A clear purpose of assessment in medical education provides significant advantages, primarily by driving student learning, ensuring the validity and reliability of evaluation methods, and ultimately protecting the public by certifying competent physicians. The clear purpose benefits both the learners and the educators.

### 21.9.1 Benefits for students and learning

#### *21.9.1.1 Guides learning and effort*

A clear purpose of assessment in medical education is crucial because "assessment drives learning." Through this statement, it explicitly makes students' aware of what content and skills are important, thereby directing their study efforts and guiding them toward effective learning strategies and deeper understanding. A clear purpose guides learning and effort as it clarifies goals and objectives (Khalaf et al., 2025). When the purpose and criteria are transparent, students understand exactly what is expected of them, from foundational knowledge to higher-order clinical reasoning skills. This alignment ensures that students focus their time and energy on mastering relevant competencies rather than guessing what might be on a test. This also directs the student in the right direction.

#### *21.9.1.2 Enhances motivation and engagement*

Student interest and engagement are greatly increased in medical education when assessment has a clear purpose. Students can better concentrate their study efforts when they are aware of the purpose of the assessment, such as formative feedback to enhance skills or summative certification. The learning process feels less random and more intentional because to this clear direction, which offers a concrete objective. Students will have less worry and uncertainty as a result, which will increase their confidence and improve their performance. Additionally, it will allow the student to focus their efforts on learning instead of worrying about the future.

#### *21.9.1.3 Facilitates self-reflection and self-assessment*

By giving students clear criteria to evaluate their performance, a clear assessment purpose in medical education promotes self-reflection and self-assessment. This increases students' self-awareness, encourages self-directed learning, and fosters the development of critical professional competencies. Students can accurately evaluate their own work against the expected standards when the assessment's goal and criteria are made clear, frequently through the use of rubrics or guidelines. This improves the procedure' efficacy and lessens the subjectivity that comes with self-evaluation. Additionally, it aids in the identification of gaps since students are able to identify specific areas of strength and weakness in their knowledge, abilities, and behavior through the process of self-assessment against specific targets. The main benefit of this insight is that it helps them focus their future learning efforts where they are most required.

#### *21.9.1.4 Promotes critical thinking and reasoning*

Higher-order thinking, problem-solving, and clinical reasoning skills are tested in well-designed, purpose-driven tests that go beyond simple knowledge recall. In this instance, application rather than recollection will be the main focus. Students are encouraged to go beyond surface learning and acquire deeper analytical abilities when they are aware that exams will require them to apply knowledge to realistic patient settings, such as in OSCEs or case-based discussions.

#### *21.9.1.5 Provides effective feedback*

Students can effectively address knowledge or skill gaps and enhance performance in following tasks when feedback is timely, relevant, and specific, which is ensured by a clear purpose. In medical education, having a clear evaluation goal has numerous important benefits for providing useful feedback. It will make it possible to provide personalized and pertinent comments. Feedback can be precisely matched with the intended learning goals and assessment criteria when the goal is clear. This guarantees that the feedback is pertinent to both the curriculum's objectives and the learner's present requirements.

#### *21.9.1.6 Reduces anxiety*

Having a clear purpose of assessment in medical education can indeed reduce anxiety for both faculty and students. This is important to ensure clarity and predictability. Moreover, when the students understand why they are being assessed the process becomes predictable and transparent. It also encourages the focused preparation as a clear purpose defines the scope and standards required an also the students know exactly what to study and how to prioritize their efforts, which minimizes the anxiety associated with overwhelming workloads and uncertainty about exam content.

### 21.9.2 Benefits for educators and programs

#### *21.9.2.1 Ensures alignment with objectives*

A clear purpose of assessment in medical education ensures alignment with objectives by providing curriculum coherence, wherein the clearly defined assessment purposes serve as a constant compass, ensuring that all assessment tasks from written exams to practical skill evaluations directly measure the specific learning outcomes and competencies outlined in the medical school curriculum. This coherence prevents a mismatch between what is taught and what is tested. It also helps in effective resource allocation as when the objectives are clear, educators can design targeted assessments efficiently, optimizing the use of valuable resources like faculty time, standardized patients, and clinical settings.

#### 21.9.2.2 *Improves assessment quality*

Having a clear purpose of assessment significantly improves assessment quality in medical education by enhancing validity, reliability, and impact. This clarity ensures that assessment methods are appropriate, well-defined, and serve their intended function effectively. Some of the advantages offered here are in terms of enhanced validity as when the purpose is clear, assessments can be specifically designed to measure the intended knowledge, skills, or behaviors, ensuring they accurately reflect the curriculum's objectives and the competencies required of a medical professional. Moreover, increased reliability and fairness are achieved through a well-defined purpose that leads to standardized assessment construction, administration, and scoring. This consistency minimizes bias and error, ensuring that results are reliable and fair across all students and evaluators.

#### 21.9.2.3 *Allows for program evaluation and improvement*

Assessment data provides valuable information to determine the effectiveness of teaching methods and the overall program, allowing educators to make evidence-based revisions and improvements. When the purpose of an assessment is clearly defined, it establishes what knowledge, skills, or behaviors are being measured and why. This clarity helps align the assessment methods with the curriculum's learning objectives. A clear assessment data can also highlight the areas where students consistently struggle, indicating potential gaps or weaknesses in the curriculum, teaching methods, or resources. In addition, it allows educators to verify if the actual learning outcomes match the intended program goals.

#### 21.9.2.4 *Facilitates communication with stakeholders*

A clear purpose of assessment in medical education significantly facilitates communication with stakeholders by providing transparency, focus, and a shared understanding of goals. This clarity helps in aligning the expectations as a clear purpose ensures that everyone involved that is the learners, faculty, administrators, and accreditors understands exactly what is being assessed and why. This alignment prevents confusion about the expectations for performance and achievement. This will provide a clarity in the feedback as when the assessment objectives is explicit the feedback provided to learners is more targeted and actionable. Learners better understand the areas they need to improve to meet specific, established competencies.

#### 21.9.2.5 *Aids in standard setting*

It provides a clear basis for determining performance standards and making fair, evidence-based decisions about student progression and certification. A clear purpose of assessment in medical education aids in standard setting by establishing consistent and defensible benchmarks, whereby a defined purpose clarifies the specific knowledge, skills, or behaviors being measured, allowing educators to set consistent performance standards or benchmarks that are defensible and aligned with the learning objectives of the curriculum. Furthermore, it guarantees fairness and validity by clearly articulating assessment goals that help the educators to ensure that the standards set are appropriate for the level of training like for example to distinguish between a first-year student and a final-year resident and that the assessment tools used are valid and reliable for their intended use.

### 21.9.3 Validity

Validity in medical education assessment refers to how well an evaluation instrument captures the desired learning objectives or competencies. It requires developing an evidence-based case to support the interpretation of assessment results for a particular purpose, population, and time period rather than being a straightforward yes or no. Therefore, an evaluation that offers significant and convincing proof that the student has mastered the particular knowledge or skill it is intended to measure is considered valid (Gupta, 2023).

### 21.9.4 Key aspects of validity

#### 21.9.4.1 *Measurement of intended outcomes*

Measuring intended outcomes is a crucial component of validity in medical education. Making sure the assessment instruments truly measure the particular learning outcomes they are designed to evaluate is a crucial component of generating this validity evidence. There are two possible kinds of these:

#### 21.9.4.2 Construct validity

Construct validity, as used in medical education today, guarantees that the tests actually measure constructs like clinical reasoning or empathy, and not just surface-level knowledge, by collecting data demonstrating that scores are consistent with theory, correlate with other related measures like convergent validity, and differ from unrelated measures like discriminant validity. It is essential for correctly reading test results and making wise decisions, including determining whether a student passed or failed a given exam. To make sure that the tests aren't tainted by unrelated elements, it needs a variety of evidence, such as expert evaluations, item analysis, and evaluating real-world implications.

#### 21.9.4.3 Content validity

In medical education, content validity guarantees that assessment instruments, such as tests and portfolios, accurately and thoroughly cover the whole body of pertinent information or skills that are the "construct" required for a particular medical competency, avoiding gaps or extraneous content. It is created by expert panels assessing item comprehensiveness, relevance, and clarity using techniques like the content validity index to guarantee that important choices, such as competency or license, are reliable and significant.

#### 21.9.4.4 Evidence-based argument

The modern concept of validity in medical education is basically an argument backed by data. Instead of being a static aspect of an assessment tool itself, it is now recognized as a robust, ongoing process of collecting and applying many sources of data to support the intended interpretations and applications of assessment results. Medical educators often use frameworks like Kane's argument-based approach, which offers a framework for validating assessments by methodically constructing a logical argument for the intended interpretation and use of scores. This approach focuses on the interpretation-use argument and its supporting validity argument, going beyond simple psychometrics to take into account the complex assumptions in evaluating healthcare professional performance, particularly in real-world settings.

#### 21.9.4.5 Context-specific

It is a known fact that a physician's competency and an assessment tool's performance can differ greatly based on the particular scenario, setting, and resources available, thereby highlighting the importance of the context specificity as an essential component of both validity and medical education. This implies that proof of validity gathered in one situation does not necessarily translate to another instead, each new setting necessitates a critical evaluation. Validity is directly impacted by context specificity in a variety of ways. It means to say that the validity evidence is strongly limited to the particular purpose, context, and learner group in which it was gathered since it is context-bound. For instance, an assessment tool validated in an urban hospital environment with adequate resources might not be valid in a rural clinic with limited resources. This might be due to variations in available technology, support personnel, and patient demographics.

#### 21.9.4.6 Focus on interpretation

Interpretation is an essential part of validity in medical education since assessment data have no inherent value on their own; rather, their value is derived from the interpretation and application that are applied to them. Validity is not a feature of a test; rather, it is an evaluation of the acceptability and scientific soundness of these proposed interpretations and the conclusions derived from them. It should be noted that the modern approach to validity views it as a single, cohesive concept that requires evidence from multiple sources to support the intended score interpretation. Fundamentally, validation is the process of assembling reasoning and proof into a coherent argument to bolster the viability of a specific interpretation.

### 21.9.5 The advantages of validity in medical education assessment

Validity in medical education assessments has the many advantages as it guarantees that assessments measure what they are supposed to which results in a true portrayal of student performance, gives confidence in the results, and immediately affects patient care by enhancing learner competence. It aids guarantee that the evaluation is pertinent to the learning objectives of the program and can be utilized to produce just and convincing judgments regarding a student's aptitudes.

#### 21.9.5.1 Meaningful results

The assurance that the tests accurately measure the intended medical knowledge, clinical skills, and professional behaviors is the main benefit of validity in assessment for producing meaningful results in medical education. This is crucial for making wise decisions regarding a student's competence and preparedness for practice. Assessment data can be deceptive and have no intrinsic relevance in the absence of validity evidence. A legitimate assessment enables teachers to draw relevant and defensible conclusions about a student's performance, such as whether a student is capable of carrying out a certain clinical assignment.

#### 21.9.5.2 *Justifiable decisions*

The main benefit of assessment validity in medical education is that it offers the proof and rational justification required to support the appropriateness and defensibility of critical judgments regarding students' competency and future practice. This guarantees that the choices fulfill the profession's social responsibility to patients and are equitable to students. Since the ultimate goal of medical education assessments is to guarantee that trainees and graduates are capable of giving patients safe, high-quality care, it also helps to protect public safety. By successfully separating individuals who have fulfilled the necessary requirements from those who have not, valid assessments serve as a quality control mechanism that directly safeguards patient welfare.

#### 21.9.5.3 *Alignment with learning*

High assessment validity in medical education, which means that the test measures what it is supposed to measure, directly supports and improves the learning process by guaranteeing positive alignment with learning objectives, directing student effort, and encouraging deeper learning and critical clinical skills. Since a valid assessment is clearly connected to the desired learning outcomes and curriculum objectives, validity in assessment for learning takes the form of explicit learning goals. Students may focus their efforts on studying pertinent competencies rather than speculating about what might be on an exam which helps them grasp what is expected of them.

### 21.9.6 Tests to ensure validity of assessments

Numerous statistical tests and techniques, such as assessing item difficulty and consistency using Cronbach's alpha, assessing inter-rater reliability, and carrying out predictive validity checks, can be used to guarantee the validity of medical education assessments. Standardized techniques, such as the mini-CEX and the OSCE, are intended to increase validity through reliable, superior evaluation. The following statistical and psychometric tests are useful in this context:

#### 21.9.6.1 *Item analysis*

By statistically analyzing individual questions to make sure they accurately measure intended learning, identifying and correcting problematic items such as questions that are too easy or hard, ambiguous, or contain poor distractions that skew results and verifying that the test as a whole successfully distinguishes between knowledgeable and struggling students, item analysis increases assessment validity by boosting confidence that the test measures what it claims to measure.

#### 21.9.6.2 *Internal consistency*

Internal consistency ensures an assessment's items measure the same underlying concept, supporting validity by showing the test isn't measuring different things. A high consistency suggests that the test is accurately gauging the intended construct. A high internal consistency such as a high Cronbach's alpha indicates coherence.

#### 21.9.6.3 *Correlation coefficients (validity coefficients)*

Correlation coefficients such as Pearson's "$r$" quantify the linear relationship between variables and serve as important markers of validity by demonstrating how well a new test result corresponds with known measures provided by criterion validity or related constructs given by construct validity. Stronger validity is indicated by a coefficient near +1, whereas low validity is suggested by the coefficient which is near 0.

### 21.9.7 Reliability

Reliability in medical education assessment refers to the consistency and repeatability of evaluation outcomes. When an evaluation is trustworthy, it will yield comparable results under the same circumstances, guaranteeing that the scores represent the learner's consistent skill rather than chance or measurement mistake (Baniasadi et al., 2023).

## 21.10 Importance of reliability in assessment

In medical education, reliability is essential because it guarantees that test results are reliable and consistent, which makes them a legitimate and trustworthy indicator of a student's proficiency. Lack of reliability can result in injustice, uncertainty for students, and a decline in trust in the evaluation process since results might differ greatly based on the rater or the time of the assessment. Scores with high dependability are repeatable and significant, which is crucial for making wise judgments regarding a student's development and preparedness for clinical practice. Reliability becomes crucial for a number of reasons, including the following:

### 21.10.1 Ensures consistency and fairness

In medical education, reliability in evaluation offers a number of important benefits that guarantee uniformity and equity. Reproducibility, or consistency, is the ability of a trustworthy assessment to produce identical or nearly identical results when repeated by various assessors or under identical circumstances. Because of this consistency, a student's score accurately represents their level of knowledge or proficiency rather than the subjective bias of a particular examiner, the time of day the test was administered, or the particular collection of questions (Levy-Feldman, 2025). However, in terms of fairness or equitable grading, all students are assessed using the same stable standard when assessments are consistent. This ensures that results are determined by performance and merit rather than by chance or artificial unpredictability, creating a level playing field. For high-stakes decisions like certification, promotion, and licensure, where choices must be justifiable and fair to all applicants, this is essential.

### 21.10.2 Supports accurate decision-making

Accurate decision-making is supported by assessment reliability in medical education in a number of crucial ways. When given again under the same circumstances, a trustworthy assessment consistently yields results that are comparable. This guarantees that a student's score accurately reflects their knowledge or ability rather than being the product of chance, the particular test version, or the rater. For a trainee's competency to be fairly and accurately evaluated, this consistency is essential. By reducing measurement error, a high dependability guarantees that every student is assessed using the same uniform standard. In high-stakes decisions like moving on to the next training, certification, or licensure stage, this encourages equity and fairness.

### 21.10.3 Provides a basis for validity

Since an assessment cannot be deemed valid unless it is first dependable, dependability in medical education assessment serves as the essential basis for validity. For the outcomes of an evaluation to be interpreted as valid, reliability is a necessary criterion. Additionally, it contributes to the reduction of measurement error, as a high level of dependability implies a lower level of random error in the measuring process. Any claims to validity are undermined if an assessment has a lot of random errors because the scores will be unstable and cannot possibly reflect a true, accurate estimate of a student's actual skill.

### 21.10.4 Builds trust and credibility

By guaranteeing fairness, consistency, and dependability of outcomes for all stakeholders from students to accrediting agencies and the general public, reliability in assessment contributes to the development of trust and credibility in medical education. Reliability in assessment has the benefit of measuring the same item consistently across many situations and individuals, which reduces prejudice and fosters equity in decisions about promotion and evaluation. For students to believe that the system is just, it must be fair. This also increases the learner's engagement and trust. Learners are more likely to trust their supervisors, the feedback they get, and the program leadership as a whole when evaluations are reliable and repeatable. Instead of concentrating only on marks, this trusting environment promotes student engagement with the feedback and the learning process itself.

## 21.11 Types of reliability

### 21.11.1 Internal consistency

In medical education assessments, internal consistency is a type of dependability that gauges how well test or assessment items evaluate the same underlying concept or construct. A high internal consistency score, which is frequently determined using metrics like Cronbach's alpha, shows that the various components of an assessment are consistent and that random error has no effect on the results, guaranteeing a more reliable and accurate assessment of a student's proficiency. For instance, a student should continuously score highly on other relevant questions in the same assessment if they perform well on one question pertaining to a certain medical issue.

### 21.11.2 Inter-rater reliability

In medical education evaluation, inter-rater reliability gauges how consistently various raters evaluate the same performance, skill, or patient case. Using techniques like Cohen's Kappa or the intraclass correlation coefficient to measure agreement, it is an essential kind of reliability that guarantees assessments are objective and reliable by reducing subjective judgment errors. It is crucial in medical education because it offers the opportunity for objective evaluation, which helps to guarantee that a student's performance is assessed consistently across various assessors and keeps the bias of a single rater from influencing the result. However, it is crucial for standardizing subjective assessments, like grading a portfolio of work, evaluating a clinical presentation, or judging surgical skills on a video (De Beer et al., 2023).

### 21.11.3 Test–retest reliability

Test–retest reliability is a kind of dependability in medical education evaluation that evaluates a test's consistency by giving it to the same set of people twice. It evaluates a measure's stability over time, making sure that the findings should be comparable on both occasions if neither the test's environment nor the students' knowledge have changed. A high test–retest reliability shows that the test is a dependable instrument for frequent use and is unaffected by random errors. It has become more important in medical education since it helps to ensure that the test is assessing a stable characteristic, such as a student's comprehension of a fundamental medical idea, rather than being impacted by sporadic variations. In addition, it ensures that the outcomes of novel assessment instruments in medical education are repeatable, which is essential for their development and validation. In order to guarantee repeatable and trustworthy results, reliability in medical education assessment is attained by standardized instruments, consistent administration, qualified assessors, and a variety of statistical techniques such test–retest, internal consistency, and inter-rater reliability. Reducing random error and enhancing reliability also depend on using an adequate number of items or observations and maintaining a consistent environment. The following are a few methods for achieving reliability:

#### *21.11.3.1 Standardizing the evaluation*

To lessen variability, this is accomplished by making sure that every participant is in a uniform setting. To avoid mistakes, it's critical to give participants clear, consistent instructions and acquaint them with the user interface.

#### *21.11.3.2 Train and calibrate assessors*

Inter-rater reliability can be decreased by providing assessors with clear, comprehensive rubrics, using continuous calibration sessions with practice scoring, incorporating feedback loops, and monitoring scores using metrics like Cohen's Kappa. It is important to make sure that training is frequent, intensive, and employs consensus methods to align subjective judgments with objective standards.

#### *21.11.3.3 Increase the length and quality of assessments*

It is important to have a sufficient number of questions or tasks which helps to provide a more dependable measure of ability. Moreover, a well-constructed items lead to more consistent and reliable results.

#### *21.11.3.4 Employ statistical reliability methods*

For the test–retest reliability, it is important to administer the same assessment to the same group at different times to see if scores are consistent. For the internal consistency reliability, it is preferred to use the methods like Cronbach's alpha to measure how well different items or questions within a single assessment correlate with each other. Using the statistical models to account for multiple sources of error, for instance, different assessors and different tasks, in a single analysis will aid in assessing reliability more comprehensively.

#### 21.11.3.5 Incorporate multiple sources of feedback

For workplace-based assessments, a combination of different feedback sources such as a patient and colleague feedback can provide a more reliable picture of performance.

#### 21.11.3.6 Value

The "value" of an assessment in medical education is its ability to measure student learning, give explicit feedback, and direct curriculum reform and student progress. A valuable evaluation is one that is fair, legitimate, and reliable. It should also take into account the information, abilities, behaviors, and attitudes that are actually crucial for professional competence. In order to prepare well-rounded, lifelong learners for practice, assessments should not only evaluate competency but also promote ongoing learning and introspection.

### 21.11.4 Key aspects of value in assessment

#### 21.11.4.1 Guides teaching and learning

Values influence learning, teaching, and assessment by providing a basic framework for educational goals, influencing student motivation and behavior, and influencing pedagogical decisions and assessment practices. Students' approach to learning is greatly influenced by guiding learning, which helps to strengthen personal values like self-direction, accomplishment, and compassion. Students who prioritize self-direction, for example, might choose a deep learning method, aiming for a thorough knowledge, whereas those who are more concerned with external incentives, such as riches, might take a surface or achieving strategy, focused on grades. Additionally, it facilitates self-reflection because values assessment is a process that helps people become more self-aware and match their behaviors with the things that are most important to them. A more meaningful and satisfying educational path may result from this self-awareness.

#### 21.11.4.2 Provides feedback

An assessment's value is essential for giving feedback since it gives a clear picture of a learner's strengths and shortcomings, which can then be used to provide precise, practical advice for progress. Assessments find knowledge or skill gaps by comparing performance to predetermined criteria, and the feedback they provide helps students stay motivated, grow, and take responsibility for their education.

#### 21.11.4.3 Curriculum development

The value of assessment in curriculum development is crucial as it provides the necessary data and feedback to systematically improve the quality, relevance, and effectiveness of educational programs and teaching methods. It acts as a guide to ensure that the curriculum meets its intended goals and the evolving needs of students and society.

#### 21.11.4.4 Evaluating effectiveness and impact

Value is widely utilized in medical education as a metric to assess the impact (what changed?) and efficacy (did it work?) of educational interventions. Value-based evaluation goes beyond traditional criteria to assess the value of training programs in terms of improved clinical performance, patient outcomes, and long-term, high-value care by linking educational outcomes to patient care quality, efficiency, and costs.

#### 21.11.4.5 Ensuring fairness and accountability

In order to promote justice and accountability and guarantee that educational institutions and graduates satisfy the changing demands of society, values-based methods in medical education are crucial. Medical schools can encourage impartial curricula, fair admissions, and responsible professional conduct by emphasizing fundamental values, including equity, excellence, relevance, and cost-effectiveness.

#### 21.11.4.6 Fairness

Fairness is a pillar of assessment that ensures all learners have equitable opportunities to demonstrate their knowledge without bias. It involves providing appropriate support based on individual needs and designing assessments that are transparent, unbiased, and accurately measure the intended learning outcomes. Achieving fairness requires flexibility in methods and considering factors like developmental stage, culture, and learning styles.

### 21.11.5 Key components of fairness in assessment

#### 21.11.5.1 Equity and access

Equity and access are foundational components of fairness in assessment, acting as the mechanisms through which fair outcomes are achieved for all students, especially those from diverse or marginalized backgrounds. Fairness in assessment is a broad concept that essentially means all students have a just and impartial opportunity to demonstrate their learning. It stresses on the fact that all learners should have the resources and opportunities to learn and prepare for the assessment. This may also involve providing accommodations for students with special needs.

#### 21.11.5.2 Unbiased design

Unbiased design is a critical component of fairness in assessment, aiming to ensure all individuals have an equal opportunity to demonstrate their knowledge and skills without being advantaged or disadvantaged by irrelevant personal characteristics such as race, gender, or socioeconomic status. This also highlights the importance of creating assessments that are free from bias related to factors such as gender, culture, or socioeconomic status. This includes clear and culturally sensitive instructions and tasks.

#### 21.11.5.3 Accurate measurement

Accurate measurement, a core principle of psychometrics, is a fundamental requirement for achieving fairness in assessment. In this context, accuracy is typically understood through the technical concepts of validity and reliability, which work to ensure that an assessment measures only the intended knowledge or skills and does so consistently across all test-takers. In this regard, it should be noted that inaccurate measurement compromises fairness because it means that the assessment is measuring something other than the intended learning outcomes or is doing so inconsistently, leading to unreliable and unjust outcomes for students.

#### 21.11.5.4 Transparency

Transparency is a critical component of fairness in assessment, functioning as a moral precondition and an essential mechanism for ensuring equitable and just evaluation processes. By making the assessment process clear and open, institutions can build trust, minimize bias, and provide all students with an equal opportunity to succeed based on their abilities.

#### 21.11.5.5 Flexibility

Flexibility is essential to assessment fairness, mainly by fostering inclusion and educational justice. Flexible assessments offer a variety of alternatives to accommodate different student needs, histories, and circumstances, guaranteeing that every student has an equal opportunity to demonstrate their learning thereby providing equity, whereas traditional, "one-size-fits-all" exams treat all students equally, thereby upholding the principle of equality.

#### 21.11.5.6 Continuous improvement

Continuous improvement is a critical component of fairness in assessment, acting as a mechanism for identifying, correcting, and preventing biases and inconsistencies. It ensures that assessment processes, tools, and outcomes are systematically reviewed and refined to provide all individuals with an equitable opportunity to demonstrate their knowledge and skills.

### 21.11.6 Transparency

Transparency is a critical pillar of sound educational assessment, ensuring fairness, objectivity, and accountability. It goes beyond merely providing clear documentation. It involves active engagement and communication with all stakeholders to ensure a shared understanding of expectations and criteria. Transparency in assessment primarily means that the requirements, expectations, and grading criteria are made explicit and understandable to the students and other stakeholders. Some of the important aspects of transparency in assessment can be described as follows:

#### 21.11.6.1 Explicit requirements

Explicit requirements transparency is a foundational principle in educational assessment, widely considered essential for fairness and effective student learning by ensuring that students clearly understand the purpose, criteria, and standards of an assessment task. It involves moving assessment from a well-guarded secret to a public, shared understanding. The idea here is that the assessment tasks, intended learning outcomes, and grading criteria should be clearly stated and easily accessible.

#### *21.11.6.2 Clear communication*

In assessment, clear communication in transparency refers to openly and honestly sharing the criteria, processes, and results of evaluations to build trust and ensure fairness. It involves making sure that all stakeholders, such as students, employees, or the public, understand how decisions are made and what is expected of them. Transparency requires that the standards for evaluation are defined and communicated using clear, plain language, avoiding jargon. This ensures that individuals understand what they are being assessed on and how their performance is measured.

#### *21.11.6.3 Use of tools*

In educational assessment, various tools are used to operationalize transparency as a pillar, which involves ensuring that the purpose, criteria, and processes of assessment are clear and understandable to all stakeholders, especially students. These can be in the form of rubrics which are a fundamental tool for transparency, as they explicitly outline the quality criteria and performance levels for an assessment task. The benefit provided by using rubrics is that they help minimize subjectivity in grading, allow students to understand expectations, and facilitate self-assessment and peer-assessment, thereby promoting a shared understanding of the criteria. Since they provide capabilities that improve openness in the assessment process, learning management system platforms like Canvas, Moodle, and Google classroom are also utilized as assessment tools. The benefit is that they give teachers a single place to exchange comprehensive rubrics, assignment guidelines, and prompt, real-time feedback on student work. Additionally, they provide statistics to monitor student development.

#### *21.11.6.4 Opportunities for dialogue*

Transparency as a pillar of assessment creates significant opportunities for dialogue by establishing trust, ensuring mutual understanding of expectations, and fostering a collaborative learning environment. This shift moves assessment from a one-way, evaluative process to an interactive one that supports student self-regulation and improvement. By engaging stakeholders in discussions about assessment criteria, rationales, and expectations, it helps build a shared understanding and addresses potential ambiguities.

#### *21.11.6.5 Alignment*

In education, alignment and transparency are complementary principles that ensure assessments are fair, relevant, and effective for learning. Transparency is considered a pillar of sound assessment, working in concert with alignment to build trust and direct student effort.

## 21.12 Impact of pillars of assessment on the teaching and learning process

The pillars of assessment impact teaching and learning by making them more focused, effective, and meaningful. They guide teachers in selecting appropriate tasks and strategies, provide students with clear feedback for improvement, and ensure that assessment results are fair and useful for measuring progress and refining instruction. This framework helps to align assessment with the goals of learning, creating a feedback loop that enhances the overall educational process.

## 21.13 Benefits of pillars of assessment in medical education

The pillars of assessment in medical education provide a framework for developing high-quality assessment systems that offer numerous benefits to students, faculty, and ultimately, patient care. The benefits of utilizing these pillars include the following:

### 21.13.1 Ensuring competence and patient safety

The pillars of assessment in healthcare education and practice are critical for ensuring professional competence and consequently enhancing patient safety. These assessment frameworks help to identify and address weaknesses, ensure adherence to standards, and foster continuous improvement within the healthcare system. Some of the benefits of the pillars of assessment for competence and patient safety includes prevention of error as a rigorous, high-quality training and consistent adherence to established protocols, which are verified through comprehensive assessments. These protocols when applied effectively prove to be very crucial in minimizing preventable errors that can have severe consequences for patients. It also helps in identifying competency gaps as standardized competency assessments allow educators and leaders to identify specific areas where healthcare professionals or students may be lacking in knowledge, skills, or behavior. Addressing these gaps early, especially during training, is more effective than correcting them in active practice (Oyinloye & Imenda, 2019).

### 21.13.2 Driving student learning

Instead of focusing solely on rote memory, a well-designed exam promotes deep learning, critical thinking, and the application of knowledge to real-life circumstances. Essentially, by putting the principles of robust assessment pillars into practice, educators can establish a system in which assessment is an essential component of the learning process rather than merely a final judgment, enabling students to recognize their own strengths and weaknesses and know what they need to do to improve (Miller, 1990).

### 21.13.3 Providing meaningful feedback

A core benefit of a robust assessment system is the provision of timely and helpful feedback. This feedback helps students identify their strengths and weaknesses, enabling them to focus their learning and improve their performance over time. Providing meaningful feedback is a core benefit and indeed an essential component of the assessment process itself. Assessments provide the data necessary to deliver meaningful, actionable guidance that drives continuous improvement for both learners and educators. Feedback also offers an explanation of what students are doing well and where they need to improve, based on the assessment criteria. Most importantly it guides further action as it provides specific, actionable steps and suggestions for improvement, essentially acting as a map for the student to bridge the gap between their current performance and the desired learning outcomes.

### 21.13.4 Supporting program evaluation

Assessment data offers vital information for program evaluation, helping educators determine the effectiveness of the curriculum and identify areas for improvement in teaching methods and content delivery. The pillars of assessment support program evaluation by providing a structured, reliable framework for gathering and interpreting data, which ultimately informs decision-making and ensures accountability. The assessment pillars guide the collection of relevant and consistent data, a fundamental step in program evaluation. This ensures that the information gathered is meaningful and directly related to the program's objectives. By analyzing data collected through valid and reliable assessments, educators and program managers can provide concrete evidence of the program's effectiveness to stakeholders. This is essential for accountability and demonstrating that goals are being met.

### 21.13.5 Promoting lifelong learning

The pillars of assessment can be structured to specifically promote lifelong learning by moving beyond traditional testing and fostering self-directed, continuous engagement with knowledge. One of the ways in which lifelong learning can be promoted is to emphasize on reflective practice. The assessments that incorporate reflective practices, such as portfolios or postproject reviews, help individuals learn from their experiences and identify strategies for future improvement (Ben-David, 2009). This habit of reflection is a key component of self-improvement and lifelong growth. It also promotes collaboration and social learning. Assessments that involve group projects and collaboration help students develop vital social and communication skills. Learning from peers and working effectively in teams are valuable skills that extend into professional and personal life, facilitating continuous learning through social interaction.

### 21.13.6 Enhancing fairness and transparency

Fairness in assessment means that all individuals have an equal opportunity to demonstrate their knowledge and skills, taking their diverse backgrounds and needs into account. By enhancing fairness, assessments are rendered appropriate for diverse learners. This may involve providing reasonable accommodations or adjustments for individuals with special needs or different learning styles, rather than treating everyone exactly the same way. Fair assessments actively work to counter implicit biases related to an individual's background, culture, or learning style. By using objective marking guidelines and diverse assessment methods, personal prejudices are minimized.

### 21.13.7 Covering a wide range of competencies

Using a variety of assessment methods is a key benefit because it allows for the comprehensive coverage of a wide range of knowledge, skills, and behaviors that a single assessment method cannot capture. Different assessment types provide unique evidence to build a holistic picture of a learner's abilities. This approach ensures holistic evaluation competencies often integrate knowledge, skills, and attitudes. A single method, such as a written test, might only measure factual knowledge, whereas an observation or a simulation can assess practical application, problem-solving, and interpersonal skills like communication and teamwork. On the other hand, it enhances validity when evidence is gathered from multiple sources. This helps confirm that the inferences made about a learner's capabilities are accurate and appropriate for the intended purpose.

### 21.13.8 Meeting accreditation standards

Meeting accreditation standards is a primary benefit and a core function of the pillars of assessment. The assessment process, when structured around key pillars or standards, provides the robust, evidence-based documentation required for external peer review and validation. The key benefits include external validation and credibility. Accreditation signals all the stakeholders including students, employers, and the public that an institution or program meets established minimal quality requirements. It also ensures the structured quality assurance, which is guided by accreditation standards and forces institutions to implement robust internal quality assurance mechanisms. This systematic planning, development, and review of processes is central to demonstrating compliance.

**Summary**

This chapter highlights the significance of assessment in educational settings. It clarifies that assessment involves more than merely gauging student performance. It also involves directing instruction, offering feedback, and encouraging in-depth learning. Assessments assist teachers and students in identifying and filling in learning gaps. The entire educational process is improved, resulting in better outcomes for teachers and students, when the pillars of assessment purpose, validity, reliability, and value are properly incorporated into assessment methods. The chapter emphasizes that good assessment is complex and necessitates a thorough comprehension of the fundamental ideas that support it. Teachers can design assessment systems that accurately reflect students' abilities and promote a culture of ongoing learning and development by emphasizing validity, reliability, fairness, and transparency.

**Key point**

**Future perspective: ensuring professional competence and patient safety**

The foundations of assessment will change significantly as technology, pedagogy, and our understanding of human learning continue to progress education. Personalized, dynamic, and data-driven solutions that improve every student's educational experience will be the hallmark of assessment in the future. Teachers may create more thorough and inclusive tests that accurately reflect students' abilities and learning progress by accepting the changing viewpoints on validity, reliability, fairness, and transparency. The function of assessment will go beyond measurement as technology and educational philosophy grow further, becoming a crucial instrument for promoting skill development and lifetime learning.

## References

Baniasadi, A., Salehi, K., Khodaie, E., Bagheri Noaparast, K., Izanloo, B., 2023. Fairness in classroom assessment: a systematic review. The Asia-Pacific Education Researcher 32 (1), 91–109. Available at https://doi.org/10.1007/s40299-021-00636-z.

Ben-David, M.F., 2009. The role of assessment in expanding professional horizons. Medical Teacher 22 (5), 472–477. Available at https://doi.org/10.1080/01421590050110731.

Bennett, R.E., 2011. Formative assessment: a critical review. Assessment in Education: Principles, Policy and Practice 18 (1), 5–25. Available at https://doi.org/10.1080/0969594X.2010.513678.

Cox, M., Irby, D.M., Epstein, R.M., 2007. Assessment in medical education. New England Journal of Medicine 356 (4), 387–396. Available at https://doi.org/10.1056/NEJMra054784.

De Beer, V., Nel, J., Pieterse, F.P., Snyman, A., Joubert, G., Labuschagne, M.J., 2023. An evaluation of the inter-rater reliability in a clinical skills objective structured clinical examination. African Journal of Health Professions Education 13–17. Available at https://doi.org/10.7196/ajhpe.2023.v15i2.1574.

Epstein, R.M., Hundert, E.M., 2002. Defining and assessing professional competence. The Journal of the American Medical Association 287 (2), 226–235. Available at https://doi.org/10.1001/jama.287.2.226, http://jama.jamanetwork.com/journal.aspx.

Gupta, K., 2023. Validity and reliability of students' assessment: case for recognition as a unified concept of valid reliability. International Journal of Applied & Basic Medical Research 13 (3), 129–132. Available at https://doi.org/10.4103/ijabmr.ijabmr_382_23.

Htay, M.N.N., Kamath, M.G., Anand, K.M., Sugathan, S., Luyee, E.O., Sahoo, S., 2023. Overview of reliability and validity of assessments in medical education. International Journal of Transformative Health Professions Education 1 (1), 14–18. Available at https://doi.org/10.71354/tr76gs42.

Khalaf, E., Ennab, F., Otaki, F., Guraya, S.S., Amir-Rad, F., Khan, E., et al., 2025. Medical students' perception of assessment and its effects on their learning in Dubai: a convergent mixed methods study. Frontiers in Medicine 12. Available at https://doi.org/10.3389/fmed.2025.1620437, journal.frontiersin.org/journal/medicine.

Levy-Feldman, I., 2025. The role of assessment in improving education and promoting educational equity. Education Sciences 15 (2), 224. Available at https://doi.org/10.3390/educsci15020224.

Miller, G.E., 1990. The assessment of clinical skills/competence/performance. Academic Medicine 65 (9), S63. Available at https://doi.org/10.1097/00001888-199009000-00045.

Norcini, J., Anderson, B., Bollela, V., Burch, V., Costa, M.J., Duvivier, R., et al., 2011. Criteria for good assessment: consensus statement and recommendations from the Ottawa 2010 conference. Medical Teacher 33 (3), 206–214. Available at https://doi.org/10.3109/0142159x.2011.551559.

Oyinloye, O.M., Imenda, S.N., 2019. The impact of assessment for learning on learner performance in life science. EURASIA Journal of Mathematics, Science and Technology Education 15 (11). Available at https://doi.org/10.29333/ejmste/108689.

Parmar, P., 2025. Forensic Leg Med 115.

Witheridge, A., Ferns, G., Scott-Smith, W., 2019. Revisiting Miller's pyramid in medical education: the gap between traditional assessment and diagnostic reasoning. International Journal of Medical Education 10, 191–192. Available at https://doi.org/10.5116/ijme.5d9b.0c37.

Chapter 22

# Exam blueprinting and item analysis

*"Before anything else, preparation is the key to success."*

—Alexander Graham Bell

## 22.1 Introduction

Instruction and evaluation must adapt as the teaching and learning process in medical education changes. Tests with irrelevant variance or poor alignment pose a threat to the validity of an assessment. A well-crafted test plan ensures an accurate depiction of learning outcomes and domains by bridging gaps and integrating different education pillars.

Medical education has indeed had a long and transformative journey. Some of the notable advancements in the field of curriculum include it being built on solid pedagogical concepts, the emergence of innovative teaching methods like problem-based learning, and a greater emphasis on interaction and learner-centeredness in medical education. Teachers' roles have evolved from identifying problems to offering solutions. Assessment in medical education has shifted from a 20th-century focus on isolated, objective, knowledge-based testing to a modern, holistic approach focused on competency and continuous improvement. Effective and efficient healthcare delivery requires knowledge, technical, analytical, and communication skills, interdisciplinary care, counseling, and evidence- and system-based care. Our evaluation techniques should be thorough, trustworthy, and strong enough to examine the essential qualities in addition to knowledge and skills in order to accommodate this shift. Furthermore, evaluation is now a crucial part of all stages of professional development. An exam blueprint is a plan for an assessment that outlines content areas and cognitive levels to guarantee thorough and valid coverage, whereas an item analysis is the statistical assessment of individual test questions following the exam to determine their level of difficulty, efficacy in differentiating between students with high and low ability, and quality of distractions. Together, the item analysis offers input to enhance current questions and provide better assessments in the future, while the blueprint offers the structure for a high-quality test (Hamdy, 2006).

## 22.2 Exam blueprint

An exam blueprint, also known as a table of specifications, is a thorough method for creating an exam that connects learning objectives to the content and structure of the test. It serves as a framework for test designers, ensuring that the exam covers certain topics and skills, evaluates them at appropriate cognitive levels like recall, analysis, or evaluation, and has an equitable distribution of question types and difficulty levels. The blueprint promotes fairness and consistency in evaluation by aligning the test with the course objectives and specifying precisely what will be assessed and how (Gunathilake et al., 2024). For instance, specific learning objectives dictate the course design and assessment techniques in medical education. The ideal specified learning outcomes are defined by the SMART pattern (specific, measurable, achievable, relevant, and time-bound). On the other hand, students are occasionally asked to pick up a lot of knowledge in a short amount of time. Due to this, knowing where to focus their study efforts can be difficult and a daunting task for the students. Exam plans provide students with a thorough overview of the course material, which may help them concentrate on their preparation.

## 22.3 Need for an exam blueprint

An exam blueprint is necessary to ensure that exams cover certain topic areas and cognitive levels, align with course objectives, and provide students with fair and genuine assessments. It serves as a guide for test development, highlighting the significance of each subject and skill, encouraging consistency, enabling the production of multiple similar exam

**The Principles and Practice of Medical Education. DOI:** https://doi.org/10.1016/B978-0-443-44197-4.00023-0

versions, and guiding students' preparation. In the assessment process, an exam blueprint serves as a comprehensive plan or a strategic map. It is an essential tool for creating fair, legitimate, and dependable tests that complement curricular material and learning objectives. An exam blueprint serves many important purposes as listed below:

### 22.3.1 Aligns assessments with learning outcomes

Aligning assessments with learning outcomes as part of an exam blueprint is one technique to ensure that an exam's content and structure are closely aligned to the course's educational objectives. A test blueprint, which outlines the specific learning outcomes to be assessed, the cognitive level, the number and percentage of questions, and the weight allocated to each domain, acts as a road map to guarantee an accurate, fair, and comprehensive evaluation. This process improves student preparedness, strengthens the validity of tests, and makes clear the skills and knowledge that students need to demonstrate.

### 22.3.2 Ensures validity

An exam blueprint guarantees validity as it helps in developing a structured, content-valid, and aligned assessment that directly links test items to particular learning objectives and curricular requirements. In addition, risks such as construct underrepresentation and concept-irrelevant variance are minimized as the blueprint guarantees balanced content sampling, appropriate cognitive levels, and equitable weighting (Abdellatif et al., 2024).

### 22.3.3 Promotes fairness and consistency

Exam blueprints encourage consistency by offering a standardized, structured framework for exam creation and marking, as well as fairness by guaranteeing that all students are evaluated using the same standards and content. A test blueprint promotes fairness by providing each student with an equal opportunity to demonstrate their skills and knowledge without prejudice or unexpected questions. The blueprint ensures that the exam accurately reflects the entire spectrum of the course content and learning objectives by preventing the overrepresentation or underrepresentation of specific topics. This implies that students who have studied the full curriculum are not disadvantaged by an exam that is mostly focused on one subject. The exam blueprint serves as an organized, standardized road map that synchronizes testing with learning objectives, topic coverage, and difficulty levels. In this way an exam blueprint guarantees and encourages consistency in assessment. It ensures that every exam version is equal and fair to students, regardless of when they take it, by doing away with subjective grading and arbitrary content selection.

### 22.3.4 Supports effective student preparation

An exam blueprint enables students to focus their study efforts efficiently and ensures a fair evaluation process by providing them with a clear grasp of the course material and assessment requirements. The blueprint offers clarity on both scope and substance because it enumerates the major topics and concepts that will be covered on the test. Students are able to understand the breadth of the subject matter they are expected to learn when guesswork is eliminated. Prioritizing study efforts is also made easier by figuring out the proportionate weight or percentage of questions allocated to each topic or learning objective. This suggestion helps students prioritize their study time by focusing more on challenging or significant subjects rather than spreading their efforts too widely.

### 22.3.5 Guides test development

A test blueprint guides test development by creating a matrix that aligns test items with course learning objectives and topic areas. This development plan specifies the number of questions, point values, and cognitive difficulty levels for each goal in order to guarantee fairness, content validity, and a clear path for both test producers and students. The primary focus here is on alignment with learning goals to ensure that the test appropriately evaluates the knowledge and skills outlined in the course's learning objectives. Additionally, it guarantees topic validity by carefully tying assessment items to the curriculum. This procedure also helps us comprehend that the test appropriately covers the intended content.

### 22.3.6 Facilitates evaluation and improvement

By offering a precise, organized framework that enables methodical examination, data analysis, and focused modifications, a blueprint promotes assessment and progress. It serves as a single source of truth, guaranteeing that every endeavor is in line with precise, quantifiable goals. The plan clearly outlines the performance criteria, content categories, and desired results. This makes the evaluation process objective rather than subjective by enabling assessors to precisely ascertain if and to what degree objectives have been met. This offers precise standards for evaluation. Additionally, by comparing actual performance or results to the comprehensive plan described in the blueprint, it aids in the identification of gaps and inefficiencies.

### 22.3.7 Allows for creating multiple versions

By acting as a template that automatically chooses questions from a question bank based on particular, predetermined criteria, such as the learning objectives, difficulty, and content areas, the blueprint enables the creation of various test versions. This procedure guarantees that the structure and content of every version are similar. It aids in this process by enabling randomized selection, in which the system chooses a certain amount of items at random from a question bank for every subject area or learning goal indicated in the blueprint. This ensures uniform coverage and difficulty across all versions while producing distinct examinations for various test-takers. Additionally, each topic's weight, learning objective, and degree of cognitive difficulty are all specified in the design. By adhering to these guidelines, validity and fairness are maintained since every created version satisfies the same set of requirements.

## 22.4 Components of an exam blueprint

An exam blueprint is a comprehensive plan that links content to learning objectives and specifies the number of questions for each topic or skill, cognitive level, and weighting. This ensures that the test is in line with curriculum goals by covering key topics in an acceptable manner. Content areas, learning objectives, cognitive levels, question types/format, and weight or proportion for each are essential elements that form a matrix to direct the creation of a legitimate and trustworthy assessment (Fig. 22.1).

### 22.4.1 Learning objectives

A blueprint's basic element is learning objectives, which are the quantifiable, precise skills that students must acquire. They are explicitly mapped to activities, content, and assessments to guarantee positive alignment. In order to direct test design, weight, and topic coverage, they specify what students should know or be able to accomplish, usually classified by cognitive levels. The blueprint serves as a road map, connecting the learning objectives with the methods of assessment. An assessment blueprint enables teachers to balance cognitive complexity, assign appropriate weight, and guarantee thorough covering of the course material as it facilitates the mapping of objectives to particular exam items.

### 22.4.2 Cognitive levels

Cognitive levels are a crucial part of an educational or assessment plan that makes use of frameworks such as Bloom's taxonomy to make sure that tests are accurate, balanced, and measure more than just rote memory. The cognitive levels in a test plan classify learning objectives to guarantee a fair evaluation, ranging from basic memory to sophisticated critical thinking. Recalling, comprehending, applying, analyzing, evaluating, and producing are the six hierarchical levels. These levels guarantee that tests cover a range of cognitive abilities. In this sense, a test blueprint maps course subject regions against particular cognitive levels in the form of a two-dimensional matrix. In addition to guaranteeing balance and avoiding an excessive focus on low-level recall questions, this thoughtful design method guarantees coverage of higher-order thinking abilities like problem-solving and critical analysis. The blueprint also helps in matching assessment items to the intended learning outcomes and the cognitive demands that go along with them, thereby improving the test's content validity. It also guarantees that item writers create questions that focus on specific cognitive abilities in accordance with the curriculum.

### 22.4.3 Weighting

A crucial part of an assessment blueprint is the weighting or proportion of questions, which acts as a sampling strategy to guarantee the exam's fairness and content validity. In order to reflect their relative importance in the curriculum, it determines how many test items or how much of the total mark value is assigned to particular academic areas and learning objectives. This procedure is crucial to guarantee that every relevant facet of the topic is addressed and evaluated fairly. The percentage allotted to a topic usually indicates how much time is spent on it during teaching and learning process, how important it is in professional practice, how powerful is its impact on learning, and how important it is for future professional growth.

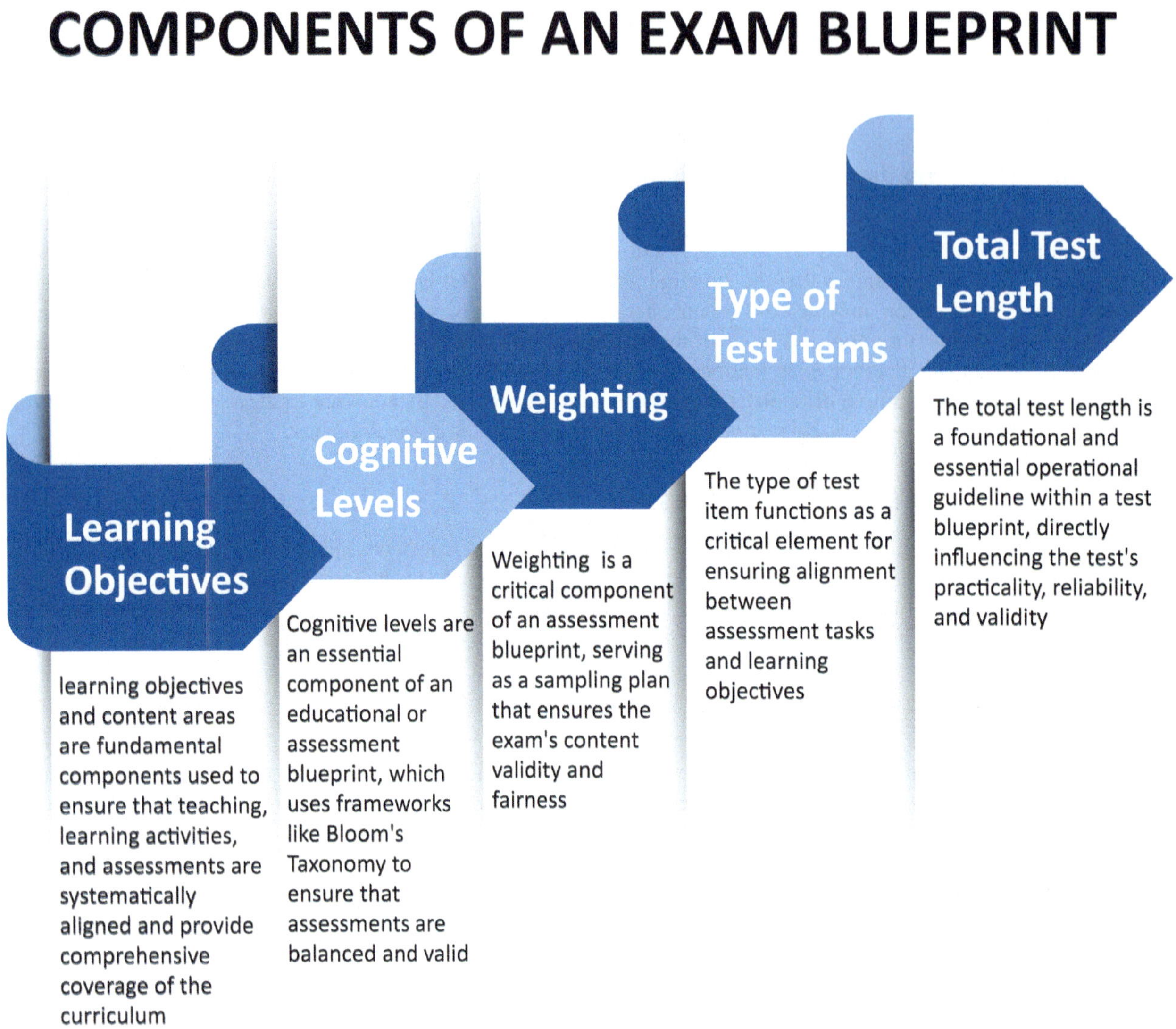


FIGURE 22.1 **Components of an exam blueprint.** This figure shows the components of an exam blueprint.

### 22.4.4 Type of test items

One of the major features of the blueprint is the different types of test items that it represents. Item types are also distributed based on the importance of the content, with specific numbers of questions allocated to each format. An important point to be noted here is that the test item in a blueprint guarantees alignment between learning objectives and assessment. Multiple-choice questions (MCQs), short-answer questions (SAQs), and questions that target the assessment of practical sessions are important item types. These elements guarantee a fair and reliable evaluation of knowledge, comprehension, application, and analysis. Based on the desired learning outcome, the blueprint directs the choice of item types. Each question type is also selected based on the action verb used in the specific learning outcome. For instance, SAQs, essays, or scenario-based questions that require students to generate answers and demonstrate higher-order thinking usually are aligned to the specific learning outcomes, wherein the action verbs like define, list, or mention are used.

### 22.4.5 Total test length

A test blueprint's overall test length is a fundamental and crucial operational guideline that directly affects the test's validity, reliability, and practicality. It acts as a fixed parameter that directs how each test item is distributed among different subject areas and cognitive levels. In order to determine how many questions should be assigned to each particular topic or learning target based on their proportionate weight in the curriculum, the total test length will describe the total number of items chosen. It will also support practicality at the same time. It should be remembered that the overall duration must be reasonable in light of the assessment's time constraints, the subject's nature, and the breadth of the material to be covered. This is how the blueprint determines that the majority of students can finish the test in the allocated time.

## 22.5 Construction of blueprint

Exam blueprints are fundamental documents that map exam questions to learning objectives in order to ensure content validity, topic balance, and appropriate cognitive levels. To produce a balanced assessment, the blueprint construction includes defining scope, assigning weights to content areas, choosing question types and levels of difficulty, and confirming alignment with course objectives.

### 22.5.1 Steps in blueprint construction

Determining the course's learning objectives, choosing the knowledge and abilities to evaluate, choosing suitable question forms such as multiple choice or open-ended, and creating a table that assigns weights to each content area and skill level are all crucial steps in creating an exam blueprint (Ismail et al., 2020). This guarantees that the test is in line with the objectives of the course, gives a clear framework for producing excellent test items that satisfy certain aims, and accurately samples the content. The steps involved in the construction of the blueprint are described as follows:

#### *22.5.1.1 Step 1: define the blueprinting purpose and scope*

Increasing content validity, enhancing fairness, lowering bias, and coordinating instruction with evaluation are the goals of defining the purpose and scope of blueprinting. For both formative and summative assessments, the scope includes mapping themes, cognitive levels, skills, and weighting. It acts as a comprehensive method to ensure that assessments are objective, comprehensive, and cover every crucial aspect of the learning objectives. By establishing a link between assessment items and learning objectives, the blueprint enhances content validity. Additionally, it makes the examination transparent and fair by giving students explicit expectations about what will be examined. It also serves as a teaching guide, guaranteeing that the material being taught corresponds with the material being evaluated. The blueprinting aids in defining the themes, subject matter, and learning objectives such as attitudes, abilities, and knowledge that will be covered in the assessment. Apart from this, it helps to determine how much of the evaluation will be devoted to each topic and cognitive level.

#### *22.5.1.2 Step 2: tabulate curricular content*

This stage involves organizing and listing the curriculum content, which typically takes the form of course themes, subjects, or particular learning outcomes that make up one axis of the blueprint matrix. A spreadsheet table containing all course learning objectives or subjects is made in order to tabulate curriculum content for a medical education blueprint. We can easily understand this by taking cardiovascular system as an example as given in Table 22.1.

The whole content for the cardiovascular system is broken down into eight themes in this table. The basic sciences are covered in themes one through four, while the clinical topics are covered in themes four through eight. The timeframe, the particular learning objectives, and the topics covered under each theme are further explained. It also explains the quantity of questions that will be utilized in each theme for the final exam, as well as the assessment techniques such as MCQs, SAQs, objective structured practical examination, and objective structured clinical examination. It also emphasizes how much weight is given to each topic based on its frequency and significance.

#### *22.5.1.3 Step 3: identify impact and frequency*

In medical education blueprint construction, "impact" refers to the significance or weight given to specific topics, skills, or learning domains in an exam, ensuring assessments accurately reflect clinical importance and curriculum priorities, thereby enhancing validity, fairness, and constructive alignment between teaching and testing. On the other hand, the term frequency is a key factor, scored by experts, indicating how often a topic or competency is encountered in clinical practice or asked in exams, guiding assessment weight, with higher scores for very common or frequently asked topics. It indicates that these topics will gain more coverage in the exam. For instance, topics like common health issues like life-threatening conditions or frequent presentations have a chance of more representation during exams rather than topics which are less practical or clinical significance.

**TABLE 22.1** Sample of an example blueprint.

| Theme wise weightage as per SLOs, duration, and importance | | | Theme 1 | Theme 2 | Theme 3 | Theme 4 | Theme 5 | Theme 6 | Theme 7 | Theme 8 | Total | |
|---|---|---|---|---|---|---|---|---|---|---|---|---|
| **Specific learning outcomes** | | | 40 | 32 | 26 | 35 | 32 | 29 | 22 | 18 | 233 | |
| **Duration (days)** | | | 5 | 7 | 4 | 5 | 4 | 4 | 3 | 2 | 30 | |
| **Total questions** | MCQs/SAQs | **Knowledge** | 10/1 | 8/2 | 7/1 | 8/1 | 5/1 | 8/0 | 5/1 | 5/1 | 56 | 165/33 |
| | | **Understanding** | 16/2 | 11/3 | 12/2 | 12/2 | 11/2 | 12/2 | 10/1 | 12/1 | 96 | |
| | | **Application** | 1 | 2 | 1 | 2 | 2 | 2 | 2 | 1 | 13 | |
| | | **Total** | **27/4** | **21/5** | **20/4** | **22/4** | **18/4** | **22/4** | **17/4** | **18/4** | **MCQs/SAQs165/33** | |
| | **OSPE** | | **3** | **3** | **2** | **2** | **2** | **1** | **1** | **2** | **16** | |
| | **OSCE** | | **0** | **0** | **0** | **0** | **1** | **2** | **1** | **1** | **5** | |
| **Final exam** | **Weight% MCQs** | | 8% | 7% | 7% | 7% | 18% | 20% | 17% | 17% | 100% | |
| | **No of MCQs** | | 5 | 4 | 4 | 4 | 11 | 12 | 10 | 10 | 60 | |
| | **Weightage % SAQs** | | 6% | 9% | 7% | 7% | 24% | 23% | 12% | 12% | 100% | |
| | **No of SAQs** | | 1 | 1 | 1 | 1 | 2 | 2 | 1 | 3 | 12 | |
| | **OSPE** | | 2 | 1 | 0 | 1 | 0 | 0 | 0 | 3 | 7 | |
| | **OSCE** | | | | | | 1 | 1 | | | 2 | |

Abbreviations: *SLOs*, Specific learning outcomes; *MCQs*, multiple-choice questions; *SAQs*, short-answer questions; *OSPE*, objective structured practical examination; *OSCE*, objective structured clinical examination.

### 22.5.2 Impact frequency score in exam blueprint

The impact–frequency model is an organized method for creating test blueprints, particularly in competency-based medical education, as it establishes appropriate weighting for various themes to ensure assessment validity. It determines the overall weight of each item in the test by combining two important metrics: the Impact (*I*) of a topic on patient safety or clinical practice and the Frequency (*F*) with which it is addressed. These two factors are multiplied to determine each topic's impact–frequency (*I*×*F*) score, which determines its weight and priority in assessments. A higher *I*×*F* score indicates that a topic is more important and should be reviewed more often, ensuring that the assessment aligns with the most important components of the curriculum.

#### 22.5.2.1 Step 4: categorize curricular content based on relative weightage

Based on the *I*×*F* as describe above, the curriculum elements are categorized as "must know," "should know," and "nice to know." Curriculum items with an *I*×*F* of 6 or 9 are categorized as "must know" information, those with an *I*×*F* of 3 or 4 are categorized as "should know" knowledge, and those with an *I*×*F* of 1 or 2 are categorized as "nice to know" knowledge. After that, each curriculum item will be categorized and organized within the Microsoft Excel file. An additional method for calculating weighting is to divide the *I*×*F* for each curriculum item. This yields a relative weighting for each curriculum item that matches the percentage of evaluation items. More questions should be created based on the curriculum content since bigger the quantity produced, the more significant it is. It is important to note that the first method works better for assessments with fewer questions than curriculum content items, while the second method works better for assessments with more curriculum content items than total questions.

#### 22.5.2.2 Step 5: decide on percentage of questions for each category

It is to be noted that the exam questions cannot be dispersed equitably across the course material. The percentage of questions in each area in the blueprint varies according to the particular course, learning objectives, and intended emphasis. Nonetheless, a typical method is assigning percentages to evaluation domains, including knowledge, comprehension, and application, as well as content areas like particular illnesses or systems. Deciding on the percentage of questions for each domain in an exam blueprint requires aligning assessment weight with educational priorities, such as topic importance, instructional time, and cognitive complexity. The weight assigned to a topic should generally reflect the amount of time spent instructing it. For instance, if the if topic takes 30% of class time, it should be 30% of the exam.

#### 22.5.2.3 Step 6: decide on number of items for each assessment task

There should be a set number of items in each assessment task. We can now determine how many questions should be created from each category using the data from the previous phase. The procedure entails determining the overall exam duration, giving themes and domains weight, and computing the number of items per task according to the difficulty and amount of time required for each question. You must first develop a table of specifications that includes learning objectives and content categories in order to determine the quantity of items for each assessment activity in a blueprint. Next, based on the importance of each objective and content area, establish the proportion of the test that should cover them. Lastly, multiply the total number of questions by the assigned percentage to determine the number of items for each cell in the table. Multiply the total number of test items by the proportion allotted to that particular content area and learning target combination for each cell in the table. For example, if the test consists of 50 questions and a certain content area and objective should account for 10% of the test, you would put $0.10 \times 50 = 5$ items for that cell.

#### 22.5.2.4 Step 7: assign questions to lecturers for items preparation

Assign questions to lecturers for items preparation is the last stage in the blueprint development process. Lecturers should first be given particular subject areas or learning objectives based on their areas of expertise and the blueprint's weightings before assigning questions. In accordance with the blueprint's distribution of content areas, question types like MCQs, and cognitive levels like knowledge, comprehension, and application, questions are assigned to lecturers who are actively involved in teaching the topic, such as in problem-based learning, or subject experts, such as in lectures, after the blueprint is finalized and specific content weights and desired cognitive levels are determined (Fig. 22.2).

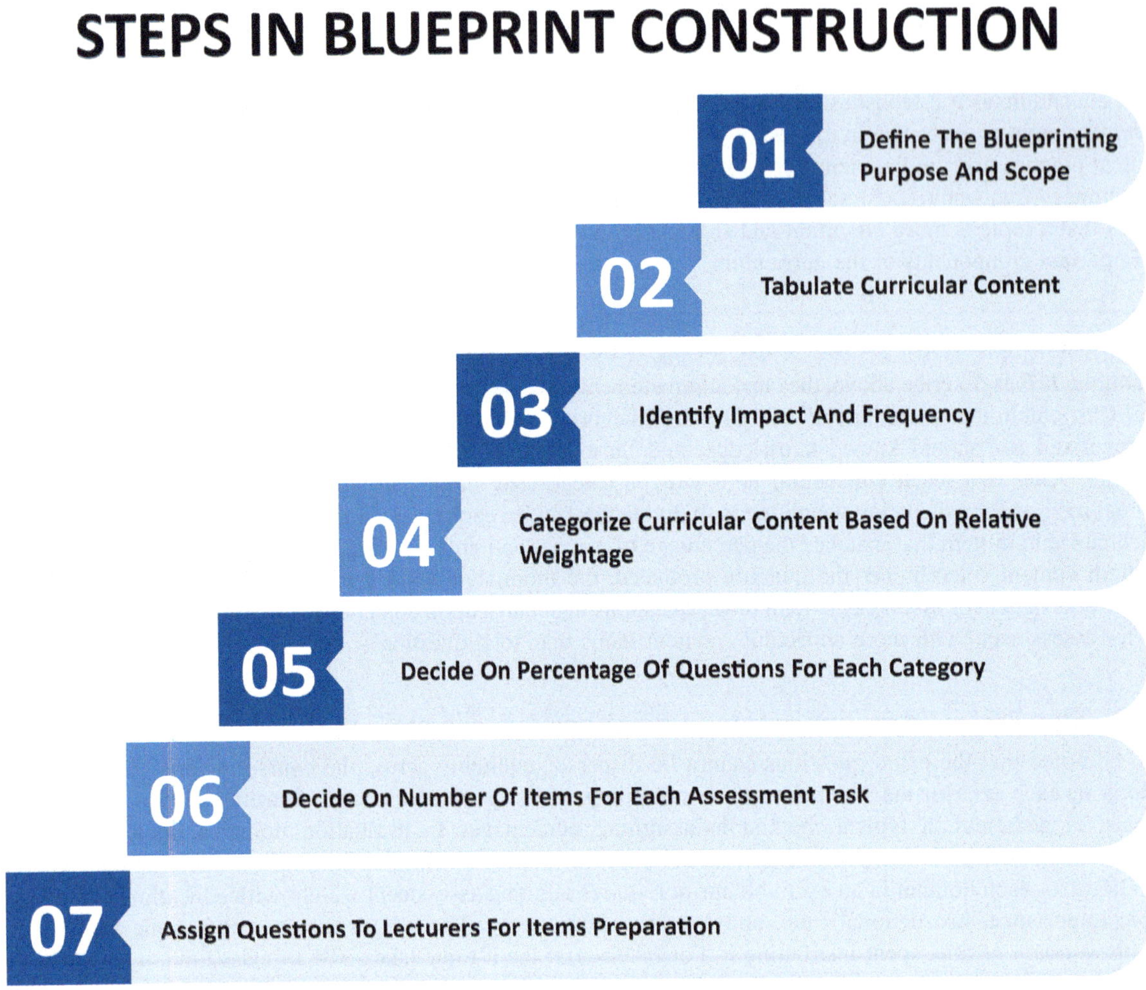


FIGURE 22.2 **Steps in blueprint construction.** This figure shows the steps in blueprint construction.

## 22.6 Item development and item analysis

In blueprint design, item development entails matching assessment questions to a test blueprint that outlines the relative importance of various subjects and learning goals. In order to ensure that the assessment appropriately reflects the curriculum and evaluation goals, item development is the process of producing particular questions, also known as items, that evaluate the targeted competencies. A combination of subject-matter competence and evaluation expertise is necessary for item production in medical education (Downing and Haladyna, 1997). In order to provide high-quality assessment items that are suitable, valid, and reliable for assessing medical students' knowledge and clinical reasoning abilities, this dual expertise is essential. The content specialists, who are usually faculty members or clinicians with specialized, current expertise in the pertinent medical field, often carry out this process. In this regard, the job of the expert is to make sure the item is medically accurate and concentrates on key, professionally authentic topics. Additionally, it incorporates the opinions of assessment or psychometric specialists who focus on measurement theory, test design, and psychometrics. They guarantee that the evaluation program as a whole satisfies the requirements for validity and reliability and that the items are technically sound and free of construction defects. The key steps in item development can be described as follows (Fig. 22.3):

1. Conceptualization

   Conceptualization is the first step in item development, where the abstract idea of what is being measured is precisely defined and broken down into observable components. This process directly informs the creation of the blueprint, which then serves as a detailed, structured guide for item writers. Conceptualization defines the purpose and scope of the test, including the knowledge, skills, or abilities to be assessed.

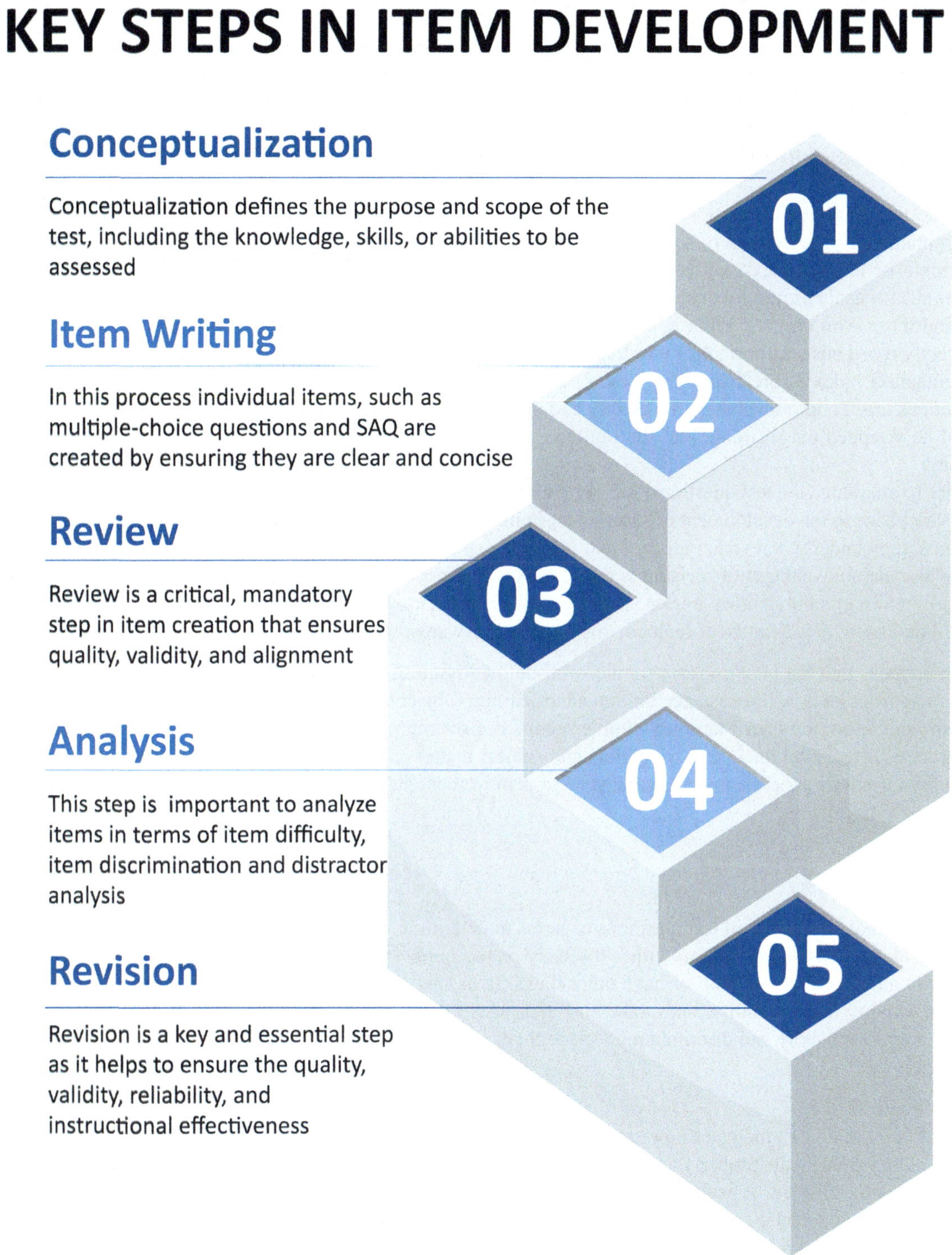


FIGURE 22.3 **Key steps in item development.** This figure depicts the key steps in item development.

2. Item writing
   Individual items, including MCQs and SAQs, are made throughout this procedure by making sure they are precise, succinct, and pertinent to the test's subject matter. As was previously noted, subject matter experts play a critical role in this respect because they are able to not only formulate sound questions but also review and validate them. The subject matter experts decide which particular material is crucial for the item and identify the core knowledge area (Haladyna and Downing, 1989).

3. Review
The test items are examined during this procedure. A preliminary review for clarity, correctness, and fundamental quality standards by colleagues or subject matter experts is undertaken. In order to guarantee quality, validity, and alignment with the initial goals and user requirements, review is an essential and required phase in the item production process. Prior to final approval or release, it assists in improving the quality of items created, identifying defects, and eventually increasing its efficacy.
4. Analysis
After the test is given, it's critical to examine the items in terms of distractor analysis, item complexity, and item discrimination. The percentage of students that properly answered an item is a measure of item difficulty. Items that are too simple or too complex may need to be revised since they have little discriminating power. Furthermore, item discrimination analysis provides us with an evaluation of an item's ability to distinguish between students who perform well and those who do not. While items with a low or negative discrimination index frequently have errors, such as incorrectly typed answers or unclear wording, they should be reviewed, updated, or deleted from the item bank. A high discrimination index is preferred. Conversely, the distractor analysis assesses how successful the wrong choices or distractions are. To improve the item's quality and difficulty, nonfunctional distractors like those that are never selected should be swapped out for more realistic options.
5. Revision
In order to guarantee the test questions' quality, validity, reliability, and instructional efficacy, revision is a crucial and necessary phase in the development of assessment items in medical education. Before the items are extensively utilized in exams, they undergo a revision process that includes expert assessment and feedback methods to fix potential faults. One of the main advantages of revision is that it makes it easier to guarantee clarity and quality. The revision process assists in removing ambiguities, unclear language, and item-writing errors that could otherwise jeopardize the validity of the item, such as grammatical faults or overlapping answer options.

This guarantees that the query poses a single, well-defined issue. Additionally, it assists the teacher in confirming that every item appropriately addresses specific, intended learning objectives and evaluates suitable cognitive levels, such as application and analysis rather than only simple recall. It also improves validity and reliability since a comprehensive review process, often involving a team of editors and subject matter experts, results in a test that is often of higher quality. High-quality questions generate more reliable student achievement statistics, which leads to more perceptive conclusions about the learning process.

### 22.6.1 Item analysis

The statistical study of individual test questions or items to determine their efficacy and quality is known as item analysis. It assists in identifying things that are either too easy or too hard, fail to accurately distinguish between students who perform well and those who do not, or have other flaws. Item analysis helps educators make future tests better, increase assessment validity and reliability, and make sure that tests effectively measure the desired knowledge or abilities by examining item complexity and discrimination (Sadeghi et al., 2025). Some of the important components of item analysis are listed below:

1. Item difficulty
Item difficulty index (*P*) indicates how simple or complex a test question is for the student. A question is deemed easy if the majority of students properly answer it. However, if the majority of students give a wrong response, the question is deemed to be difficult. A well-balanced test should ideally include questions with different levels of difficulty to evaluate both fundamental comprehension and advanced knowledge. A test item's difficulty is typically expressed as a difficulty index, or *P*-value, which is determined by calculating the percentage of students who correctly answer the question (D'Sa and Visbal-Dionaldo, 2017).
2. Difficulty index
The percentage of examinees who properly answered an item is the difficulty index. This number falls between 0 and 1, wherein as 90% of students successfully answered the question, a rating near 1 such as 0.90 indicates that it is very simple. A number near zero, such as 0.20, indicates that the question is difficult because only 20% of students answered it correctly. An easy question is indicated by a higher difficulty index, whilst a more difficult question is indicated by a lower value. Item difficulty is given by the following formulae (Table 22.2):

$$\text{Item difficulty} = \text{number of students with correct responses/total number of students}$$

**TABLE 22.2** Level of difficulty of an item.

| Range of difficulty index | Interpretation | Action |
|---|---|---|
| 0–0.25 | Difficult | Revise or discard |
| 0.25–0.75 | Slightly difficult | Retain |
| 0.75–above | Easy | Revise or discard |

Here are the key advantages of using a difficulty index in an exam blueprint:

1. Identifies learning gaps
A question with a low difficulty index reflects a learning gap since it indicates that the subject is too tough for the current cohort, as indicated by the low percentage of students who properly answered the question. Teachers can focus on the most difficult areas to improve student performance by identifying which questions have low difficulty indices. This allows them to pinpoint specific concepts that need to be clarified or taught again.
2. Ensures test balance
The difficulty index ensures test balance by helping teachers create a mix of simple, moderate, and hard questions rather than one that is exceptionally easy or challenging overall. In order to ensure that the exam effectively differentiates students of different skill levels and reliably measures a wide variety of knowledge, test designers can detect questions that are too simple with a high index value or too tough with a low index value by looking at each question's index (Rao et al., 2016).
3. Improves future tests
The difficulty index is an essential technique in item analysis that helps make future exams better by pointing out tough questions, raising the general caliber and validity of assessments, and offering insightful feedback for better teaching. It also aids in determining which questions are flawed. The difficulty index identifies things that are either too easy that is when almost everyone answers them correctly or too difficult is when almost everyone answers them incorrectly. These items frequently have low discriminating power and may be confusingly phrased, have technical errors, or have the wrong key.
4. Determines discrimination power
Since items with moderate difficulty levels are best at differentiating between high- and low-performing students, the difficulty index aids in determining an item's discrimination power. Excessively simple or complex items have a limited capacity to demonstrate variations in students' abilities.

### 22.6.2 Item discrimination

The ability of an assessment item to differentiate between students who perform well and those who do not is measured by item discrimination. It is computed by comparing the percentage of right answers between higher and lower score groups, or by employing techniques like the point-biserial correlation. A higher discrimination of index values above 0.2 are typically regarded as good, indicating that the item is successfully measuring the same construct as the remainder of the test. An item may be too simple, overly difficult, poorly constructed, or assessing a completely unrelated construct if it has low or negative discrimination. The item discrimination enables us to determine whether an item is too easy, too difficult, or not constructed well. Item discrimination also gauges how well a single test item distinguishes between high-performing and low-performing students. Item discrimination serves as a quality control measure to guarantee that the questions used to evaluate that material are operating effectively, hence enhancing the validity and reliability of the assessment as a whole, while a blueprint guarantees content coverage. It may need to be changed if it is extremely high, indicating that it is easy, or extremely low, suggesting that it is difficult. Moreover, test-takers may become confused and experience poor discrimination if the right answer is wrong or if there are several right answers. Inconsistent answers may result from unclear or badly written questions. If an item doesn't measure the same aptitude as the other objects, it will have low discrimination, which is another issue that may arise.

### 22.6.3 KR20 and Cronbach's alpha

Kuder Richardson Formula 20, also referred to as KR20, is a statistical metric used in item analysis to evaluate a test's internal consistency reliability, especially for items that have dichotomous choices, such as MCQs. It measures how well a test's different items capture the same underlying trait and how much score variability results from random error. High dependability is indicated by a KR20 value near 1.0, whereas low reliability is indicated by a number close to 0.0.

KR20 value is very significant part of the item analysis as it is used to confirm the dependability of the blueprint's implementation. Calculating the KR20 after an exam is given helps ascertain whether the test's real items function coherently to measure a single, common trait or construct as intended by the blueprint's design. As the KR20 is an essential part of the postexamination item analysis procedure, it also helps with quality assurance. The quality of the assessment created from the blueprint is validated if the KR20 value is high; usually, 0.70 or above is deemed acceptable for classroom tests and >0.80 for high-stakes exams. This indicates that the test items are consistent and interrelated. KR20 is essentially Cronbach's alpha for binary data and helps assess if a test is a consistent measure, especially with varying item difficulties. Item analysis with Cronbach's alpha is a statistical method for evaluating the internal consistency and reliability of a set of survey items, a scale designed to measure a single characteristic or latent variable. The values of Cronbach's alpha ensure reliability. A high Cronbach's alpha typically above 0.70 suggests that the items within a test that are intended to measure the same specific concept or domain are indeed correlated and consistently measuring that concept. This ensures that the test is a dependable measurement tool.

### 22.6.4 Point biserial

The correlation between students correct or incorrect scores on a particular item and their overall scores when adding up their results on the remaining items is known as the point-biserial correlation. It is a unique kind of correlation between a continuous variable and a dichotomous variable such as the multiple-choice item score, which is either right or wrong. The test's overall score ranges from 0 to the maximum number of MCQs. Point-biserial values fall between −1.0 and +1.0, just like in other correlations. A strong positive point-biserial value means that students who scored highly on the test as a whole are also answering the item correctly, whereas those who scored poorly are answering it incorrectly. Essentially, the percentage of students who correctly answer an item is indicated by its *P*-value. The percentage of students who correctly answered the question is the result of multiplying the *P*-value by 100. The *P*-value statistic has a range of 0 to 1.

### 22.6.5 Distractor analysis

A plausible but inaccurate answer choice is referred to as a distractor in item analysis for MCQs. Teachers can evaluate a student's comprehension by examining the incorrect answers they choose when a distractor diverts uninformed students from the correct response. A nonfunctional distractor is too obvious or is not selected by any students, but an effective distractor is chosen by some students, especially those who score lower overall (Tarrant et al., 2009). Distractor analysis compares the performance of incorrect answer alternatives, or distractors, in MCQs to the correct answer and assesses how well they differentiate between test-takers who perform well and those who do not. A functional distractor appears to be a feasible but inaccurate alternative because it is chosen by a reasonable percentage of test-takers, including some high-scoring students but mostly low-scoring ones. To improve the test's overall quality, validity, and reliability, nonfunctional distractors are found using statistical techniques, such as item difficulty and item discrimination. These distractors are then changed or eliminated.

#### *22.6.5.1 Importance of distractor analysis*

A crucial part of item analysis is distractor analysis, which assesses how well the wrong answers or distractors in MCQs work. It guarantees that the assessment maintains high reliability and validity, measures student knowledge accurately, and is in line with pedagogical objectives within the framework of an exam blueprint.

1. Identify student misconceptions
   Since frequently selected distractors often represent common faults in understanding, distractor analysis helps identify student misconceptions by examining which incorrect answer choices students choose. This study aids educators in improving the overall validity of exams, honing questions, and modifying instructional techniques to target particular challenges. Tutors can identify misconceptions and modify instruction to better correspond with learning objectives by looking at response patterns, particularly among lower-performing students.

2. Improve test item quality
   In order to improve the test item quality, it is vital to determine, edit, or eliminate nonfunctional distractors which are options selected by relatively few students to enhance test item quality using distractor analysis. At the same time, it is important to calculate the percentage of students who choose each distractor to assess its effectiveness. If in case a distractor is not selected by a sizable portion of students, it is deemed as ineffective. Enhancing distractions helps distinguish between students with varying levels of knowledge and increases the plausibility of test items.
3. Evaluate item discrimination
   In distractor analysis, item discrimination is assessed by determining how well each wrong choice distinguishes between students with high and low scores. A nonfunctional distractor is selected by more high-scoring students or by both groups equally, whereas a functional distractor is selected by a far greater percentage of low-scoring students than high-scoring students. In order to enhance an MCQ's overall quality, this analysis assists in locating and eliminating unproductive distractions.

## 22.7 Transparency of the exam blueprint

The transparency of an exam blueprint in medical education is considered a vital component of fair, valid, and reliable assessment practices. It involves clearly communicating the exam's structure, content areas, and cognitive expectations to students and faculty, which helps align teaching and learning with assessment objectives. Some of the important benefits of transparency of an exam blueprint can be listed as follows:

1. Builds student trust and reduces anxiety
   Since the exam blueprints clearly describe content domains, cognitive levels such as knowledge, application, and analysis and their corresponding weightings, their transparency helps to decrease ambiguity and foster student trust. Additionally, a clear blueprint acts as a guide, enabling students to prioritize important topics and devote their study efforts more effectively rather than speculating about what might be on the test, which helps to increase study focus.
2. Ensures fairness and objectivity
   Transparency reduces the possibility of bias in the questions chosen by paper setters and helps guarantee that no test-taker is taken aback by the format of the test. All students have an equal chance to successfully prepare and concentrate their study efforts on the pertinent content if the exam's scope and criteria are made clear beforehand. Additionally, because they know what to anticipate, students experience less tension and anxiety before exams due to the worry of being tested on unimportant or unmentioned topics.
3. Enhances content validity and reliability
   A blueprint helps teachers and students make sure that preparation and instruction match the actual assessment content by clearly breaking down themes, cognitive levels like memory, analysis, and synthesis, and their corresponding weightings. Additionally, a clear plan guarantees a representative and balanced sampling of the full curriculum.
4. Guides learning and teaching
   When released, the blueprint directs faculty instruction and student learning toward important learning goals and clinically pertinent subjects, creating a supportive, goal-driven learning environment. In order to ensure thorough coverage and lessen the likelihood of a hidden curriculum, teachers might utilize the blueprint to design their lessons in accordance with the curriculum rules and intended outcomes.
5. Promotes social accountability
   A key element of a medical school's social responsibility is blueprint's transparency, which guarantees that instructional materials are in line with community needs and societal expectations. Transparency makes it possible for outside review and criticism, which can assist in locating any potential biases or problems with the assessment's design. This public accountability system promotes ongoing development and aids in bringing the evaluation procedure into compliance with ethical norms and societal values.

## 22.8 Benefits of an exam blueprint

### 22.8.1 Ensures content validity

An exam blueprint by connecting assessment items to learning objectives, giving each topic or skill the proper weight, defining cognitive levels, making sure the test appropriately reflects the scope and importance of the curriculum, avoiding underrepresentation or irrelevant content, and coordinating instruction with testing objectives ensures content validity.

In order to ensure that the test measures what it is intended to assess, it serves as a master plan that directs the production of questions that cover important information in the right proportions.

### 22.8.2 Promotes fairness

A blueprint provides a rigorous, transparent, and objective framework for assessment design, thereby ensuring that all students are evaluated using the same criteria and relevant content. Additionally, by ensuring that every version of the test covers topics in the same proportions and employs uniform criteria for all students, the exam design supports fairness by offering standardized and consistent standards. Teachers are no longer able to make arbitrary decisions as a result.

### 22.8.3 Improves clarity for students

The blueprint specifies the relative importance or weight of each topic and the cognitive level such as basic recall, application, analysis, or problem-solving at which students are expected to perform. This transparency allows students to gauge their preparation level effectively.

### 22.8.4 Facilitates constructive alignment

A blueprint facilitates constructive alignment by providing a structured, two-dimensional matrix that explicitly links the three core pillars of an educational course, namely, the intended learning outcomes, teaching and learning activities, and assessment tasks.

### 22.8.5 Aids in designing a balanced assessment

A blueprint aids in designing a balanced assessment by providing a strategic road map that ensures alignment with learning objectives, comprehensive coverage of content, and a suitable balance of cognitive skill levels and assessment types. This structured approach enhances the validity, reliability, and fairness of the assessment.

## 22.9 Benefits of item analysis

### 22.9.1 Identifies flawed items

Item analysis helps identify defective items since it employs statistical indicators, like the item difficulty index and the item discrimination index, to identify questions that are not functioning as predicted. These metrics assist in identifying items that are too simple, too complex, unclear, incorrectly typed, or contain distracting or ineffective options (Siri and Freddano, 2011).

### 22.9.2 Highlights learning gaps

Item analysis contributes to reveal learning gaps by examining student performance on individual test questions to pinpoint particular areas of difficulty. Teachers can identify whether questions or subjects are confusing students by looking at elements like item difficulty, discrimination, and distractor effectiveness. In the end, this knowledge helps to close learning gaps by enabling focused interventions, modified instruction, and enhanced test design.

### 22.9.3 Boosts instructional efficacy

Item analysis boosts instructional efficacy by providing data-driven insights that allow educators to identify specific learning gaps, adjust teaching strategies, and improve the quality and relevance of assessment materials. This process informs better instructional decisions and leads to improved student outcomes.

### 22.9.4 Enhances exam reliability

Item analysis enhances exam reliability by identifying and correcting problematic questions through statistical evaluation of difficulty and discrimination. It helps ensure that each item is fair, has clear distractors for MCQs, and accurately differentiates between high- and low-performing students, leading to a more consistent and trustworthy measure of student knowledge (Zubairi et al., 2025).

### 22.9.5 Supports academic integrity

Item analysis supports academic integrity by revealing potential issues with an exam, such as confusing questions or patterns of incorrect answers, which can highlight areas where students may have cheated. It helps educators improve assessment quality by identifying difficult or biased questions, strengthening fairness, and informing future test design. Analysis also enables educators to identify specific student misconceptions and potential misconduct, leading to more accurate and reliable evaluations (Tejinder et al., 2009).

In summary, an exam blueprint ensures fairness, validity, and alignment with learning outcomes by detailing the test's content and difficulty, while item analysis identifies problematic questions and areas of confusion, ultimately leading to more effective teaching and better student assessment. Together, they enhance content validity, improve exam design, help pinpoint teaching gaps, and provide clearer study guidance for students.

## 22.10 Challenges in exam blueprinting

### 22.10.1 Lack of standardization and experience

Lack of expertise and standardization poses serious problems for item analysis, mainly because it introduces subjectivity and inconsistency that jeopardizes the validity and reliability of the findings. Due to the absence of a common set of rules for item development and analysis, alternative criteria may be used by teachers or test creators, resulting in differences in discrimination indices and item difficulty between tests or institutions.

### 22.10.2 Difficulty in content and cognitive alignment

The task of making sure test items effectively reflect the intended curriculum content and cognitive complexity without introducing unnecessary difficulty or misalignment is known as difficulty in content and cognitive alignment in item analysis. When an item does not appropriately reflect the learning objectives, it is said to be misaligned. This frequently leads to things that are inadvertently too simple, too challenging, or test the incorrect ability.

### 22.10.3 Item volume estimation

A crucial, frequently challenging component of item analysis that directly affects the validity and reliability of test results is item volume estimation, or the problem of figuring out the ideal amount of items to include in an assessment. The challenge is striking a balance between thorough topic coverage, test-taker tiredness, and practical constraints. The development of an item bank, which entails creating a sizable pool of validated items or an item bank that enables improved item selection where only the best things are used, hence boosting efficiency, is one method to address the item volume estimation.

### 22.10.4 Adapting to new assessment models

Adapting to new assessment models, such as competency-based education and adaptive testing, challenges traditional item analysis by shifting the focus from statistical measures of individual items to a holistic, programmatic evaluation of skills and behaviors across an entire curriculum.

### 22.10.5 Ensuring blueprint quality

Ensuring blueprint quality is a challenge in medical education because while item analysis provides quantitative data on question performance, it does not inherently verify the initial, planned alignment of the test content with the curriculum objectives. The primary challenge lies in bridging the gap between the intended test design and the actual test items developed by faculty.

## 22.11 Challenges in item analysis

### 22.11.1 Identifying poorly constructed items

It is difficult to identify badly written items because they generate errors and ambiguity that compromise validity and reliability, the two main objectives of assessment. Flawed items might evaluate other, irrelevant abilities or allow for right answers by chance or test-taking tactics, rather than accurately assessing an individual's genuine knowledge or skill.

### 22.11.2 Addressing ambiguous items

Addressing ambiguous items in medical education is a challenge because it can cause significant psychological distress in students, negatively impact clinical performance and patient care, and affect career choices. While clinical practice is inherently ambiguous, a lack of formal education on how to manage it means students struggle to develop tolerance for it.

### 22.11.3 Discarding low-reliability items

Discarding low-reliability items in medical education is a challenge because it can reduce the breadth of content coverage, necessitate increased time and resources for new item development, and potentially compromise the validity of the overall assessment.

### 22.11.4 Overthinking by high performers

Overthinking by high-performing medical students presents a significant challenge in medical education by leading to anxiety, burnout, impaired clinical decision-making, and an inability to adapt to uncertainty. This behavior stems from perfectionism and an intense pressure to perform flawlessly in an environment where mistakes can have serious consequences. In conclusion, nonstandardized design techniques, a lack of implementation experience, the challenge of accurately estimating item volume based on exam length and credit hours, and the difficulty of aligning the blueprint with diverse learning outcomes and adaptive testing are all obstacles in exam blueprinting. Finding poorly written items and differentiating between student errors are difficult tasks for item analysis.

## 22.12 Modernization of exam blueprinting

In order to improve the validity, reliability, and fairness of assessments, modernization efforts in test blueprinting and item analysis mainly focus on utilizing technology and implementing data-driven techniques. Efficiency, openness, and better alignment are the main goals of modernizing blueprinting. Additionally, paper-based matrices have given way to digital technologies and software with the introduction of electronic blueprinting (Owan et al., 2023). As a result of this, the process of creating, storing, and managing blueprints will be streamlined. A fair representation of subjects and cognitive abilities across several exam versions can be ensured with the aid of these technologies. Moreover, the modern blueprints will bring in the integration with curriculum management systems in which they are often integrated directly. This allows for seamless mapping of assessment items to specific learning outcomes and competencies, ensuring comprehensive coverage and reducing the risk of content underrepresentation (Samuel and Extension, 2024).

**Summary**

A strategic technique for creating assessments that are in line with curriculum objectives and competences is exam blueprinting. Exam blueprints guarantee that tests represent the breadth and depth of the material students are expected to master by providing a clear map of the subject areas, cognitive levels, and desired learning outcomes. However, item analysis is a crucial procedure that assesses each exam question's performance after it has been administered. Teachers can investigate important statistical measures, including item difficulty, discrimination index, and distractor analysis, using item analysis. This chapter emphasizes the importance of both exam blueprinting and item analysis in producing high-quality tests that assist students' learning and development in addition to accurately measuring their knowledge. The training of qualified healthcare workers is eventually aided by the efficient use of these technologies, which guarantee that medical education evaluations are reliable, equitable, and in line with learning objectives.

**Key point**

**Future perspectives: flexibility of blueprint**

A more dynamic, competency-based, and individualized approach, bolstered by technology and data analytics, will characterize test blueprinting and item analysis in medical education in the future. A more thorough and flexible assessment system that takes into consideration not only factual knowledge but also critical clinical reasoning, communication, teamwork, and cultural competency will result from this progression. A more comprehensive approach to medical education will be made possible by the incorporation of new assessment methods, such as formative evaluations and simulations, which will guarantee that aspiring medical professionals possess the entire spectrum of abilities necessary for providing efficient patient care.

## References


Abdellatif, H., Alsemeh, A.E., Khamis, T., Boulassel, M.-R., 2024. Exam blueprinting as a tool to overcome principal validity threats: a scoping review. Educación Médica 25 (3), 100906. Available at https://doi.org/10.1016/j.edumed.2024.100906.

D'Sa, J.L., Visbal-Dionaldo, M.L., 2017. Analysis of multiple choice questions: item difficulty, discrimination index and distractor efficiency. International Journal of Nursing Education 9 (3), 109. Available at https://doi.org/10.5958/0974-9357.2017.00079.4.

Downing, S.M., Haladyna, T.M., 1997. Test item development: validity evidence from quality assurance procedures. Applied Measurement in Education 10 (1), 61–82. Available at https://doi.org/10.1207/s15324818ame1001_4, http://www.tandf.co.uk/journals/titles/08957347.asp.

Gunathilake, S., Alvis, K., Gunawardana, K., Rajapaksha, S., Warnakulasooriya, A., Athulgama, P., et al., 2024. The Role of Test Blueprints in Medical Exam Design.

Haladyna, T.M., Downing, S.M., 1989. Validity of a taxonomy of multiple-choice item-writing rules. Applied Measurement in Education 2 (1), 51–78. Available at https://doi.org/10.1207/s15324818ame0201_4.

Hamdy, H., 2006. Blueprinting for the assessment of health care professionals. The Clinical Teacher 3 (3), 175–179. Available at https://doi.org/10.1111/j.1743-498x.2006.00101.x.

Ismail, M.A.-A., Mat Pa, M.N., Al-Muhammady Muhamad, J., Yusoff, Muhamad S.B., 2020. Seven steps to construct an assessment blueprint: a practical guide. Education in Medicine Journal 12 (1), 71–80. Available at https://doi.org/10.21315/eimj2020.12.1.8.

Owan, V.J., Abang, K.B., Idika, D.O., Etta, E.O., Bassey, B.A., 2023. Exploring the potential of artificial intelligence tools in educational measurement and assessment. Eurasia Journal of Mathematics, Science and Technology Education 19 (8). Available at https://doi.org/10.29333/ejmste/13428, http://www.ejmste.com/.

Rao, C., Kishan Prasad, H.L., Sajitha, K., Permi, H., Shetty, J., 2016. Item analysis of multiple choice questions: assessing an assessment tool in medical students. International Journal of Educational and Psychological Researches 2 (4), 201. Available at https://doi.org/10.4103/2395-2296.189670.

Sadeghi, P., Pourabbas, A., Dehghani, G., Katebi, K., 2025. Quantitative and qualitative item analysis of exams of basic medical sciences departments of Tabriz University of Medical Sciences in 2023. BMC Medical Education 25 (1). Available at https://doi.org/10.1186/s12909-025-07539-3.

Samuel, K., Extension, K., 2024. Publication. The Role of Artificial Intelligence in Educational Assessment 5, 44–48.

Siri, A., Freddano, M., 2011. The use of item analysis for the improvement of objective examinations. Procedia - Social and Behavioral Sciences 29, 188–197. Available at https://doi.org/10.1016/j.sbspro.2011.11.224.

Tarrant, M., Ware, J., Mohammed, A.M., 2009. An assessment of functioning and non-functioning distractors in multiple-choice questions: a descriptive analysis. BMC Medical Education 9 (1). Available at https://doi.org/10.1186/1472-6920-9-40.

Tejinder, S., Piyush, G., Daljit, S., 2009. Principles of Medical Education, 3rd ed. Jaypee Brothers Medical Publishers (P) Ltd, New Delhi, pp. 70–77.

Zubairi, N.A., Alahmadi, T.S., Ibrahim, M.H., Hegazi, M.A., Gadi, F.U., 2025. Effective use of item analysis to improve the reliability and validity of undergraduate medical examinations: evaluating the same exam over many years: a different approach. Journal of Medical Sciences 41 (3), 810–815. Available at https://doi.org/10.12669/pjms.41.3.10693, https://pjms.org.pk/index.php/pjms/article/download/10693/2663.

Chapter 23

# Written assessment

*"Assessment is not about proving what students know, but about improving what they learn."*

—Dylan William

## 23.1 Introduction

A written assessment is a procedure or test in which individuals demonstrate their knowledge, abilities, or aptitude by responding in writing to a prompt. These tests are used in education for a number of tasks, including class placement, grading, and assessing student learning. Written tests are used in professional settings, especially hiring, to examine a candidate's analytical thinking, problem-solving, and writing and communication skills.

## 23.2 Purpose of written assessment

A written assessment's main goal is to assess and show an individual's knowledge, comprehension, and abilities in a particular subject or competency by asking them to submit written answers. Assigning grades, identifying strengths and weaknesses, ensuring subject mastery, placing people in suitable courses or roles, and offering comments for growth are all possible with written evaluations. The primary goal of assessment in medical education should be to give educators and administrators interpretive data regarding the effects of teaching and learning process on students achievement in terms of gaining and application of knowledge. This will enable these educators to make the best decisions about how to modify and adapt their teaching methods (Epstein, 2007). In medical schools, assessments are frequently used to let students know about their development and accomplishments. In order to ensure competency for patient care and advancement to the next training level, written examinations in medical education assess students' cognitive knowledge, clinical reasoning, and application skills. These tests such as multiple-choice questions (MCQs), essays, and short answers influence learning habits, offer feedback, and aid in certification or licensing decisions (Fig. 23.1).

### 23.2.1 For students

#### *23.2.1.1 Demonstrate knowledge*

The written exams let students demonstrate their knowledge as it allows the student to develop their own responses, synthesize information, and exhibit higher-order thinking skills like critical analysis and problem-solving. They give students a way to connect concepts, explain their thinking, and apply knowledge in an organized manner, demonstrating a deeper comprehension of the subject matter that would not be seen through other forms of assessment. Additionally, it gives students the opportunity to go beyond simple recollection and show how they have synthesized material and formed connections between various concepts, allowing them to exhibit deeper understanding. Essays and long-answer questions are useful tools for assessing reasoning, critical thinking, and problem-solving abilities. The learner has the chance to exhibit higher-order thinking in this way. Written exams allow students to clarify things they have acquired, such as difficult-to-observe ethical reasoning or sophisticated decision-making (Cantillon, 2010).

#### *23.2.1.2 Provide feedback*

Written assessments let students provide feedback and encourage them to review, think about, and act upon the remarks at their own pace since they help in creating a permanent, referable record of their work. It can record progress over time,

**The Principles and Practice of Medical Education. DOI:** https://doi.org/10.1016/B978-0-443-44197-4.00037-0

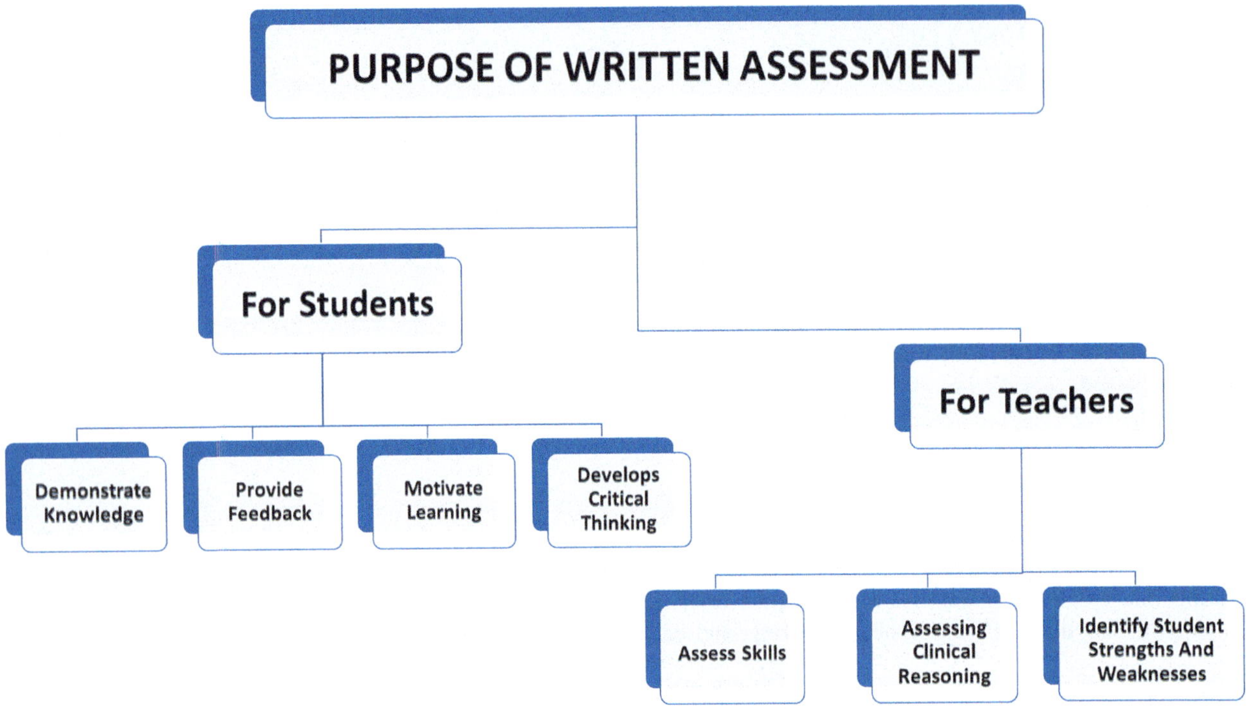


FIGURE 23.1 **Purpose of written assessment.** This figure reflects the purpose of written assessment.

provide more in-depth commentary than spoken feedback, and enable students to take greater responsibility for their education by participating in their own self-evaluation. Unlike verbal criticism, which can be forgotten, the written assessment serves as a permanent record and a lasting document that the student can read and refer to numerous times. Additionally, it helps students engage in self-paced reflection, allowing them to carefully read and consider the criticism without the pressure of an in-person discussion. Additionally, it facilitates a deeper investigation because written feedback is frequently more meticulously crafted and structured, offering students clear, cohesive input that they may thoroughly examine.

### *23.2.1.3 Motivate learning*

Students are motivated by written exams because they give them feedback, which enables them to set objectives, monitor their progress, and feel proud of themselves when they succeed. Assessments may foster student confidence and autonomy when they are straightforward, attainable, and offer precise guidance. These tests help students learn in the future by giving them precise, in-depth feedback that highlights their strengths and areas for improvement. Students are better able to work toward standards when clear learning objectives and assessment criteria are communicated up front. Additionally, it helps students assess their performance by serving as a tool for self-improvement and offering insightful feedback. The written assessment helps students set goals and can increase their motivation to learn by allowing them to track their progress toward those goals.

### *23.2.1.4 Develops critical thinking*

Written assessments, especially those that take the form of essays and open-ended questions, help students develop critical thinking by transforming students from passive to active learners who must examine, evaluate, and synthesize information in order to organize their thoughts and deliver a cohesive argument. Students go beyond merely memorizing knowledge through this approach. Students must deconstruct complex information, assess the credibility of various sources, discern between facts and opinions, and look for errors or biases in existing arguments when completing writing assignments like essays and reports. Additionally, it aids students in developing arguments and justifications in which they must decide what information is most crucial and how to persuade an audience of their point of view using facts, evidence, and observations. They improve their argumentation skills by expressing and defending their own well-reasoned findings.

### 23.2.2 For teachers

#### 23.2.2.1 Assess skills

Written tests give teachers insight into students' language use, critical thinking, and problem-solving abilities. Teachers can use them to discover areas of weakness, track development over time, and evaluate certain abilities like fluency, content, and conventions. Teachers can use this information to assess learning results, design lessons, and provide focused feedback. The capacity of the student to write clearly and effectively is one of the specific skills that the teacher can evaluate by concentrating on how the student can remember the information and clearly and succinctly convey the various concepts. Higher-order thinking, problem-solving, and reasoning abilities can be evaluated through written tests such as essay questions.

#### 23.2.2.2 Assessing clinical reasoning

Written assessments let teachers assess clinical reasoning by evaluating students' ability to analyze situations, provide differential diagnoses, explain their reasoning, and create treatment plans in a structured way. These assessments, which might be written notes, case-based essays, or case-based questions, are helpful for both formative and summative evaluation since they provide an insight into a student's mental processes.

#### 23.2.2.3 Identify students' strengths and weaknesses

Written evaluations let teachers evaluate their students by giving information on their knowledge and skills to guide future instruction. These assessments enable teachers to identify each student's strengths and shortcomings, which aids in lesson planning and teaching modification to meet the needs of both the class and the individual. This process allows teachers to track students' progress over time, provide targeted comments, and ensure the curriculum's effectiveness.

## 23.3 Types of written assessment

In medical education, written assessment is a technique that uses written answers to questions to evaluate students' knowledge, abilities, and application of medical concepts. MCQs, short-answer questions (SAQs), essays, and patient management issues are just a few of the forms that can be utilized to assess what a student has learned and accomplished. In an effort to evaluate knowledge in context more validly and reliably, contemporary methods prioritize therapeutically relevant scenarios above memorization of discrete information. The written evaluation exams fall into one of the following categories:

1. Traditional written assessment methods,
2. Innovative written assessment methods (Fig. 23.2).

### 23.3.1 Traditional written assessment methods

MCQs, essay questions, and patient management problems (PMPs) are examples of traditional written evaluation techniques in medical education that are mostly used to gauge a student's conceptual understanding and application. Essays and PMPs allow for a more thorough evaluation of comprehension, critical thinking, and clinical reasoning, but they are more subjective and time-consuming to grade. MCQs, on the other hand, are effective for testing factual recall across a broad curriculum and can be objectively graded.

### 23.3.2 Types of traditional written assessments

Some of the important types of traditional written assessments are described below:

1. Long-essay questions (LEQs),
2. Multiple-choice questions (MCQs),
3. Short-answer questions (SAQs),
4. Short-essay questions (SEQs),
5. Patient management problems (PMPs),
6. Short written responses.

### 23.3.3 Long-essay questions (LEQs)

Although they have drawbacks including time-consuming grading and the possibility of subjective marking, LEQs in medical education evaluate higher-order thinking, reasoning, and knowledge application. Modern modifications, such as

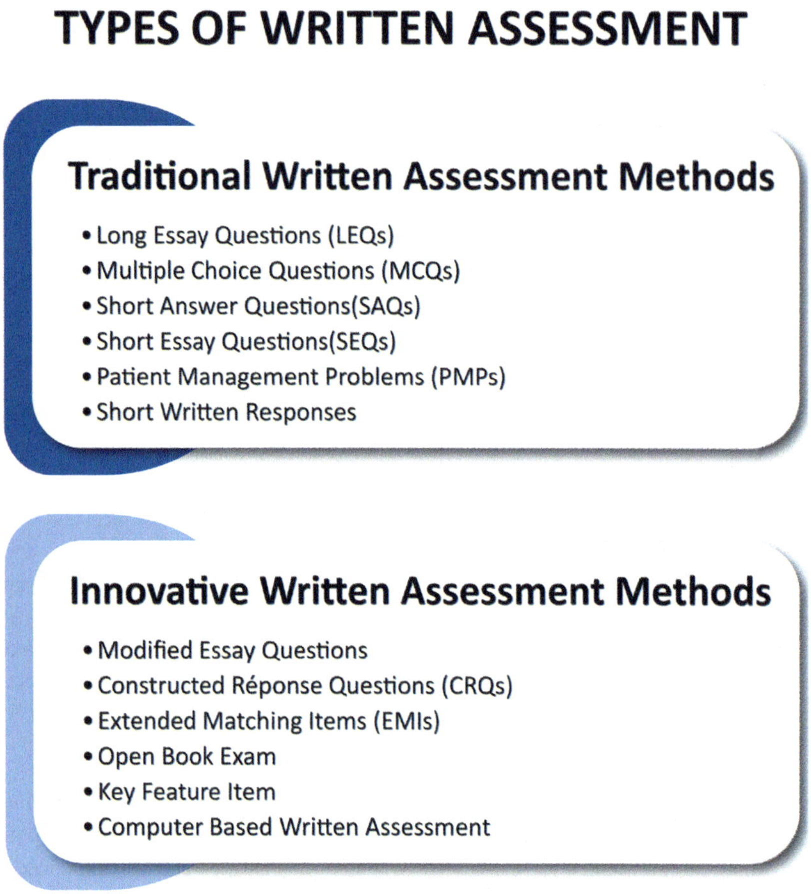


FIGURE 23.2 **Types of written assessment.** This figure represents the types of written assessment.

modified essay questions (MEQs) and SEQs, are used to assess clinical scenarios, factual recall, and reasoning skills more successfully, even though they have historically been difficult to grade consistently. LEQs are still thought to be useful for evaluating higher-order thinking abilities and clinical reasoning that predicts future competence, but their use has been questioned in favor of more objective tests like MCQs and MEQs.

### *23.3.3.1 Purpose of long essay questions*

#### 23.3.3.1.1 Higher-order thinking

LEQs enhance higher-order thinking in medical education by pushing students to synthesize information, evaluate evidence, and create solutions that closely mimic real-world clinical practice. This promotes deeper, more active learning strategies in contrast to mere recall. Additionally, it makes it easier for students to synthesize and integrate their knowledge. In order to create a cohesive response or solution, lengthy essays need students to take information from a variety of sources and combine numerous concepts, such as basic sciences, clinical expertise, and ethics. Students can evaluate validity, compare and contrast various treatment alternatives or diagnostic approaches, form opinions, and support their arguments with evidence with the aid of written assessments. All of these processes advance to the higher levels of Bloom's taxonomy, going beyond simple comprehension.

#### 23.3.3.1.2 Integration of knowledge

LEQs enhance knowledge integration in medical education by challenging students to synthesize information, relate basic science to clinical settings, participate in deep learning, and develop higher-order cognitive and reasoning skills. As it fosters deeper conceptual learning, the written assessment encourages knowledge integration. This is due to the fact that LEQs require the spontaneous creation and retrieval of information from memory, in contrast to MCQs, which frequently assess isolated facts or rely on recognition. Medical students exhibit noticeably better retention and comprehension of information they have actively created, produced, or constructed themselves when compared to information they have only read or passively received. This phenomenon is known as the "generation effect" in written assessment within medical education. Well-crafted LEQs require students to apply their knowledge to challenging, real-world patient problems. This is especially true with structured or MEQs that employ clinical vignettes. This method breaks down the traditional barriers between basic and clinical sciences by forcing students to draw connections between seemingly unrelated information to provide a complete approach to diagnosis and treatment.

#### 23.3.3.1.3 Enhancing ability to explain

Written exams improve students' capacity for explanation by challenging them to arrange their ideas, present arguments, and synthesize material in an organized manner. Students develop their ability to express themselves clearly, comprehend the expectations of their audience, and strengthen their communication skills by recognizing and fixing their shortcomings through the process of writing and getting feedback. Students must organize their ideas rationally, a process that is made tangible through writing. In written assessments, students must not only explain concepts but also make connections between them, combining data from multiple sources to create a logical argument. Additionally, it promotes self-reflection because writing assignments can incorporate self-evaluation, in which students consider their own writing process. They become more conscious of their own advantages and disadvantages as a result of this introspection, which is essential to improving their communication skills.

#### 23.3.3.1.4 Communication skills

LEQs help medical students strengthen their communication skills, which directly transfer to effective patient and interprofessional communication, by pushing them to arrange and accurately express complex medical data. Additionally, it encourages conciseness and clarity because writing essays forces students to communicate their thoughts clearly and succinctly, eliminating superfluous jargon that the reader might not understand. Giving patients clear directions, such as writing instructions to the nursing staff or discharge instructions, and successfully interacting with coworkers are all made possible by this approach.

### *23.3.3.2 Advantages of long essay questions*

#### 23.3.3.2.1 Assessing complex abilities

LEQs in medical education offer a significant advantage in assessing complex skills by requiring students to demonstrate critical thinking, clinical reasoning, and the synthesis of content in an authentic manner. LEQs are especially useful for assessing higher-order cognitive abilities like assessment, synthesis, and analysis, which are frequently challenging to gauge using other techniques like MCQs. Furthermore, the lengthy essays, especially the MEQs that frequently employ clinical vignettes, can evaluate students' ability to apply established practices to novel, challenging patient scenarios. This is similar to how diagnostic thinking and decision-making actually occur in a clinical context.

#### 23.3.3.2.2 Evaluates reasoning and synthesis

Since they force students to go beyond mere factual recall and exhibit higher-order cognitive abilities like reasoning, problem-solving, and the synthesis of knowledge from multiple sources into a coherent response, LEQs are useful in medical education. In order to replicate the process of actual medical practice, students are prompted to integrate basic science knowledge with clinical concepts by the lengthy essay questions, particularly those that are based on clinical scenarios like MEQs. Additionally, it enables the students to arrange their information and prepare a well-organized, rational answer that immediately responds to the issue. This demonstrates their capacity to organize a complicated argument or devise a management plan, which is an essential component of clinical thinking.

#### 23.3.3.2.3 Encourages deeper learning

In medical education, LEQs are well known for encouraging deeper learning by forcing students to exhibit higher-order cognitive abilities that go beyond simple factual memory. This type of evaluation promotes knowledge integration and application, which are crucial for clinical practice. LEQs promote deeper learning since they make it possible to analyze complex learning outcomes like analysis, synthesis, and evaluation which may be challenging to assess using other techniques like MCQs. Additionally, the lengthy essay questions foster critical thinking since they force students to consider a subject in-depth, evaluate data, formulate logical arguments, and provide evidence to back up their views.

### *23.3.3.3 Limitations of long essay questions*

The primary drawbacks of LEQs in medical education are their limited capacity to evaluate a wide range of content in comparison to other formats, their poor reliability due to subjective grading, and the substantial time and effort required for both students to write and faculty to grade. LEQs are frequently more challenging to create than other forms, and they can also evaluate writing abilities in an unfair manner due to ambiguity in the questions.

#### 23.3.3.3.1 Subjectivity in grading

A well-known drawback of extended essay questions in medical education is subjectivity in grading, mainly because of the possibility of prejudice and inconsistent assessment. Low reliability is a manifestation of this subjectivity. Lack of intra-rater reliability, or agreement by the same grader over time, and inter-rater reliability, or agreement across various graders, is the primary issue with lengthy essay questions. When assessing the same question, experienced markers differ significantly. However, essay responses are subject to interpretation, in contrast to objective question formats such as MCQs. This permits the score to be influenced not just by the content and clinical logic but also by the grader's own judgment, opinions, and even unrelated elements like handwriting or writing style.

#### 23.3.3.3.2 Time-intensive

LEQs take a lot of time, which is a major drawback for both students and teachers in medical education. From the standpoint of the students, they frequently find it difficult to formulate a comprehensive response within the constrained time frame of an exam, which may lead to incomplete responses that fall short of properly demonstrating their comprehension or potential. Furthermore, an exam can only cover a small portion of the syllabus because lengthy essays take a lot of time to prepare, possibly leaving a lot of material untested. Traditional essay grading is a labor-intensive and time-consuming process from the faculty's perspective, which is particularly troublesome in the fast-paced medical school setting. Furthermore, grading can be subjective even with a well-designed marking scheme, and various markers may give the same essay different ratings. Fairness and dependability may be impacted by this inconsistency, which is occasionally caused by halo effects in which a strong first response affects later grading.

#### 23.3.3.3.3 Focus on recall

The tendency of LEQs in medical education to place more emphasis on factual recollection than on evaluating higher-order cognitive abilities like problem-solving or clinical reasoning is one of its main drawbacks. The majority of lengthy essay questions only demand students to recall and replicate vast amounts of material without necessarily comprehending the context or application. This method encourages rote learning and other surface-level learning techniques that are less useful for long-term retention and practical application in a healthcare context.

### *23.3.3.4 Designing long essay questions*

It is important for a teacher to focus on intricacy and clarity when creating lengthy essay questions. It is good to use exact terminology and include complicated subjects that need for more than just explanation in the question. A strong thesis and a cohesive essay style with an introduction, body, and conclusion are necessary to support a nuanced argument. Providing context and employing particular action verbs like discuss, describe, or analyze will help the students to direct their approach towards the formulation of a good quality response.

### *23.3.3.5 Write clear and specific questions*

#### 23.3.3.5.1 Avoid ambiguity

When crafting lengthy essay questions, it is essential to avoid ambiguity by emphasizing clarity, specificity, and organization. This will ensure that the prompt inspires a targeted, well-reasoned response rather than an ambiguous or generic overview. The aim here is to make sure every student comprehends the question in the same way and is aware of exactly what is expected of them.

#### 23.3.3.5.2 Avoid questions with simple answers

It is vital to understand that while designing the LEQs, asking "how," "why," or "to what extent" type of questions that demand for analysis, comparison, and the synthesis of complex knowledge is always preferred rather than just memorization. It is good to use phrases like "analyze," "compare," "evaluate," or "discuss" to indicate the necessity for a thorough response when responding to questions that require an argument, interpretation, or assessment. It is also important to steer clear of queries that can be answered with a single fact, a "yes" or "no," or a brief account of what happened.

##### 23.3.3.5.3 Use action verbs

When writing LEQs, the use of action verbs, also known as directive or task verbs, to clarify the specific task or thinking process expected of the student is a vital part of designing the LEQ. Some of the common action verbs used are describe, discuss, and explain. These verbs act as instructions, guiding the focus and scope of the response.

#### *23.3.3.6 Make questions complex and multifaceted*

##### 23.3.3.6.1 Encourage deeper themes

Incorporate prompts that force students to go beyond simple summaries and participate in critical analysis, synthesis, and reflection in order to promote deeper topics when crafting lengthy essay questions. It is crucial to formulate questions that accommodate many viewpoints and interpretations. Students are encouraged to use sophisticated reasoning and analysis as a result.

##### 23.3.3.6.2 Require complex understanding

Writing lengthy essay questions necessitates a thorough comprehension of a subject that goes much beyond simple data recall. They are intended to evaluate higher-order cognitive abilities and ask for critical thinking, argument building, and information synthesis. In contrast to MCQs or SAQs, which frequently rely on recognition or easy recollection, essay questions require you to produce and present your own ideas in an organized way, which is one of the reasons deep comprehension is required. A excellent essay also requires a well-informed viewpoint or a succinct, persuasive thesis statement, which involves choosing a position and formulating a path of argument. This method necessitates a deep comprehension of the subject in order to produce a compelling argument.

#### *23.3.3.7 Provide context and stimulus*

##### 23.3.3.7.1 Add context

In order to help the tutor and the reader understand the breadth and complexity of the subject, a well-written LEQ naturally provides and necessitates context. The phrasing of a comprehensive essay question, such as using instructional verbs like analyze, evaluate, compare or discuss, defines the precise boundaries of the required response, preventing the tutor from merely listing all known facts about a subject. This is how LEQs establish scope and purpose. It provides context since many lengthy essay prompts start with a general remark or a brief paragraph that provides crucial background information. This aids the student in comprehending the particular context or situation that the question is requesting them to think about.

##### 23.3.3.7.2 Include stimulus materials

Stimulus materials can and frequently are used while writing lengthy essay questions. The use of such resources encourages critical thinking and greater comprehension by situating the concerns in real-world contexts. Texts, case studies, scenarios, and graphic components might be used to make the LEQs more interesting and challenging.

### 23.3.4 Multiple-choice questions (MCQs)

In medical education, MCQs are frequently used to evaluate knowledge, encourage active learning, and give effective feedback on a student's clinical competency. A relevant stem, which is typically the question or problem, and a collection of possibilities, such as the one right answer and plausible wrong answers, or distractors, make up a successful MCQ. Good MCQs should be well-written, test knowledge applicability to scenarios, and evaluate higher-order thinking abilities beyond mere recall (Salam et al., 2020). The validity and usefulness of the questions depend on their quality, which includes avoiding ambiguity, nonparallel alternatives, and typical errors like utilizing the statement all or none of the above.

#### *23.3.4.1 Components of a multiple-choice question*

##### 23.3.4.1.1 Stem

An MCQ's "stem" is the question or incomplete statement that serves as its foundation. The test taker must choose the right response from a list of possibilities that contains incorrect options known as "distractors" in order to solve the problem. The stem may consist of a formula, a chart, an image, or a direct query or an unfinished statement that needs to be completed.

#### 23.3.4.1.2 Options or alternatives

The list of possible responses to MCQs is known as the alternatives or options. It consists of a single right response known as the "key" and a number of wrong replies known as distractors. Well-written choices are consistent in both length and content, which makes them believable but not overtly clear. Since it can be difficult to create many convincing distractions, it is usual practice to limit the number of possibilities to three to five.

### *23.3.4.2 Types of multiple-choice questions*

MCQs in medical education can take many different forms, such as extended matching questions, true or false questions, standard multiple-choice with a single correct option, and the ubiquitous single best answer. To evaluate higher-order thinking abilities, MCQs might be organized around a clinical situation or vignette. Several frequently used MCQs in medical education include the following:

#### 23.3.4.2.1 Single best answer (SBA)

The single best answer format is often the most widely used format. In this type, there is only one best option that coincides with the intended response. The type A MCQ is a typical and commonly used single-best-answer MCQ used in medical education. It has a stem in the form of a question or problem and numerous alternatives, only one of which is the right or "key" answer, with the others acting as distractions. These questions are recognized for their impartiality, effectiveness, and capacity to give instructors and students diagnostic feedback. They evaluate a variety of cognitive levels, from simple recollection to application and analysis of medical knowledge (Moss, 2001).

#### 23.3.4.2.2 Extended matching questions (EMQ)

This format starts with a list of possibilities, which is then used to answer a number of questions or items. Extended matching questions are a type of MCQ used in medical education to evaluate higher-order thinking abilities, such as clinical decision-making and diagnostic reasoning. These questions present brief clinical scenarios that must be matched with the best response from a longer list of potential diagnoses, treatments, or management options. Extended matching questions, in contrast to standard MCQs, require a deeper, more applied comprehension of medical knowledge rather than the memorization of discrete facts because they include a single list of solutions that may be applied to various circumstances.

#### 23.3.4.2.3 Scenario-based or case-based multiple-choice questions

One or more MCQs pertaining to a clinical vignette or case scenario serve as the stem for these questions, which frequently evaluate comprehension and application. MCQs based on scenarios or cases unlike traditional recall questions are essential in medical education for evaluating clinical reasoning and fusing theoretical knowledge with real-world application. They encourage critical thinking, are more professionally authentic, and provide a more comprehensive assessment of cognitive abilities by mimicking real clinical settings (Coughlin and Featherstone, 2017). Even though they need to be properly crafted to be effective, both students and lecturers greatly value their applicability and ability to evaluate challenging medical concepts.

#### 23.3.4.2.4 True or false questions

These are the simplest MCQs in which the student must decide if a statement is true or false. Although they can be prone to guesswork and may only evaluate lower-level cognitive abilities, true or false questions are employed in medical education to examine factual knowledge and concept comprehension. Poorly written questions can lack clarity and discriminatory power, even though numerous true or false questions have more diagnostic potential than single true or false questions since they allow for the evaluation of several facets of a topic. Notwithstanding their drawbacks, well-crafted MCQs are regarded as a useful instrument for evaluating higher-level cognitive abilities, such as comprehension and concept mastery, and for offering a high level of test-retest reliability.

#### 23.3.4.2.5 Multiple-response multiple-choice questions

Multiple-response MCQs that let students choose one or more right answers are employed to successfully test higher-order cognitive abilities like clinical reasoning and complicated decision-making. These questions, which are frequently scenario-based, assess knowledge, application, and analysis while challenging students to distinguish between several legitimate therapeutic options. Instructors select these questions to assess multiple aspects of a topic simultaneously and to lessen the possibility of guesswork. Depending on the test's goals, these questions have either a partial or dichotomous scoring system. One of the distinguishing characteristics of these kinds of questions is that, in contrast to single-best-answer questions, multiple response items can assess intricate clinical situations in which several therapies, investigations, or symptoms are true at the same time. These are perfect for evaluating understanding of multiple-step care regimens, clinical diagnostic criteria, or various possible causes of a patient's disease. These questions have several benefits, including encouraging more in-depth analytical thinking, lowering the likelihood of making an accurate guess, and simulating clinical decision-making in the real world where a variety of considerations need to be taken into account.

#### 23.3.4.2.6 Multiple true/false

In this version, a statement is followed by a number of sub-statements that each student must determine whether or not is true. Compared to traditional true or false or single-best-answer MCQs, multiple true or false questions allow for a more thorough evaluation of knowledge and decision-making abilities by having students assess the veracity of several statements within a single question stem. Although they can reveal more nuanced understanding and provide better content validity by covering more topics, creating good multiple true or false items and developing efficient scoring methods that account for false options are major challenges, and some research suggests inherent flaws that may limit their efficacy.

#### 23.3.4.2.7 Problem-solving multiple-choice questions

In order to evaluate higher-order cognitive skills like diagnosis or management, these questions go beyond mere memory and call for the application of knowledge to solve a particular clinical situation. Clinical scenarios are used as a "stem" in problem-solving MCQs in medical education to evaluate students' abilities to diagnose, manage, and carry out treatment plans, necessitating higher-order thinking beyond mere knowledge recall. They usually include a clinical case, a question, and a list of possible responses. The objective is to find the one best response that exemplifies the use of medical knowledge in a practical setting.

#### 23.3.4.2.8 Recall versus application-based multiple-choice questions

The degree of cognitive ability being evaluated is a crucial distinction, even though it is not a format. While some MCQs are intended to assess the capacity to apply that knowledge in a clinical setting, others concentrate on retaining fundamental information. Recall-based MCQs in medical education evaluate a student's ability to retrieve facts and concepts from memory, while application-based MCQs gauge their ability to apply that knowledge in novel, frequently clinical, scenarios to solve problems or make sense of complex situations. Although they necessitate more intricate question design, application-based questions are typically seen as a more rigorous evaluation of deep learning and clinical competence since they gauge clinical reasoning and professional sincerity.

#### 23.3.4.2.9 Computer-based multiple-choice questions (CB-MCQs)

Computer-based MCQs (CB-MCQs), a popular assessment technique in medical education, include interactive forms such case-based questions, visual components, and automatic feedback to measure knowledge and clinical reasoning. For both formative and summative evaluations, these tests provide efficiency, uniformity, and objective scoring. Advanced CB-MCQs can evaluate higher-order thinking abilities through simulated clinical settings, even though they frequently test factual memory.

### *23.3.4.3 Benefits of computer-based multiple-choice questions*

CB-MCQs offer significant benefits for both educators and students, primarily in efficiency, objectivity, and data analysis. They streamline the assessment process while providing a consistent and scalable method for evaluating knowledge.

#### 23.3.4.3.1 Efficiency and coverage

Through automation and simplified administration, CB-MCQs significantly improve the effectiveness and coverage of educational examinations. CB-MCQs are very successful since they are digital and automate a large portion of the assessment process. These MCQs provide rapid, objective grading, which eliminates subjective bias and relieves teachers of the burden of manual marking. These assessment enhance the learning process as they provide the students immediate feedback. The administrative burden is also reduced because the digital platforms streamline the entire exam administration process, from scheduling and delivery to monitoring and results aggregation. Logistics such as printing, distribution, and physical venue management consequently require less time and effort.

#### 23.3.4.3.2 Objective scoring

One of the main advantages of CB-MCQs is objective scoring, which allows for automatic, impartial, and uniform evaluation. Based on a predetermined key, the computer can instantaneously identify whether an answer is correct or incorrect, doing away with the need for human judgment and the possibility of grader bias. The elimination of bias is one of the main advantages of objective scoring in CB-MCQs. It is beneficial since human graders can occasionally be swayed by elements unrelated to the content being evaluated, including the student's handwriting or their perception of them. Every test is assessed consistently due to automated scoring, giving each applicant a fair and trustworthy assessment.

#### 23.3.4.3.3 Diagnostic feedback

CB-MCQs are also beneficial as they facilitate diagnostic feedback, which offers quick, thorough, and customized insights into a learner's strengths, shortcomings, and misconceptions. This makes it possible for teachers and students to take prompt, focused action to enhance learning outcomes. They can automatically score a variety of question kinds, including multiple-choice, and deliver feedback immediately following completion, as opposed to days or weeks later as with traditional paper-based examinations. This keeps misconceptions from solidifying and enables students to respond to feedback while the assessment material is still fresh in their minds.

#### 23.3.4.3.4 Interactive and visual content

CB-MCQs with interactive and visual information offer several benefits, chief among them being increased engagement, enhanced comprehension and retention, and the ability to evaluate a greater variety of practical skills. It increases motivation and engagement. It should be mentioned that interactive features and images, like simulations or drag-and-drop interfaces, are more captivating than plain text, which helps draw in and retain students' attention while lowering exam anxiety. Incorporating videos, graphics, and interactive features into computer-based exams can enhance their engagement and enable the evaluation of complex information that is challenging to articulate verbally.

#### 23.3.4.3.5 Personalized learning

CB-MCQs provide tailored learning and maintain student interest by automatically adapting question complexity to a student's ability level. This adaptive exam ensures that students are challenged without being overwhelmed, in addition to utilizing interactive formats that support different learning styles. This personalized experience allows students to learn at their own speed, have a better understanding of their skills, and increase their motivation.

#### 23.3.4.3.6 Assessment of complex skills

Computer-based testing offers significant advantages for the assessment of complex abilities by using technology to use interactive exercises, simulations, and adaptive questions that are difficult or impossible to replicate with traditional paper-based procedures. With computer-based testing, questions can include multimedia elements such as videos, audio clips, images, and animations. In medical education, CB-MCQs have developed beyond basic factual memory to become a dependable, affordable instrument for evaluating sophisticated cognitive abilities like clinical reasoning, problem-solving, and decision-making.

### *23.3.4.4 Guidelines to create good multiple-choice questions*

#### 23.3.4.4.1 Blueprint

Having a blueprint or the table of specifications is the first step in writing any exam. Planning an exam in accordance with a course's learning objectives is one of the vital step in the creation of the MCQ. A list of the competencies and subjects that should be covered in an exam, along with a weight assigned to each, makes up a test blueprint, which serves as a guide for developing a balanced exam. The blueprint also gives an idea about the number of questions allocated to that particular activity based on the weight and importance of the topic (Holland, 2021).

#### 23.3.4.4.2 Writing the option first

It is advised to write the options first when creating MCQs with a single best option. The chosen topic and context are then used to build a list of potential homogenous options. The selections should be as brief and easy to understand as feasible.

Even though the final version often uses only four options, it is advisable to start with a list of more than four. It is crucial to observe that this list of possibilities is uniform. For instance, all of the options should be related to diagnosis, treatments, laboratory tests, complications, and one of the options chosen as the primary response to the query.

#### 23.3.4.4.3 Using a good distractor

A good distraction should be both convincing to an incompetent candidate and inferior to the right response. Every choice should be accurate and include information, that is, to varied degrees, acceptable. The most suitable, most prevalent, least hazardous, or any other feature that is at the top or bottom of a range would be requested by the examiner. The fact that there is just one right response must be made very apparent. If a candidate's choice corresponds with the examiner's key, their response is deemed accurate (Al-Rukban, 2006).

#### 23.3.4.4.4 Question stem

In this step, a question stem with a lead-in sentence based on the right choice is written. A well-crafted MCQ should assess medical knowledge application rather than merely memorization. Context-rich questions are believed to elicit thought processes that more accurately reflect problem-solving skills than context-free questions. Instead of evaluating the candidate's understanding of unimportant details or difficult issues that are rarely encountered, the emphasis should be on issues that would commonly arise in clinical practice. Questions might be developed using examples of the kinds of situations that are frequently faced in one's own profession. MCQs should be carefully employed to assess critical material and clinical competency in order to make testing both equitable and consequentially valid. The presentation of a problem should be the first step in the clinical case, followed by pertinent signs and symptoms, diagnostic study results, initial treatment, follow-up findings, etc. Essentially, the stem should contain all the information required for a qualified applicant to respond to the question. The lead-in question should specify exactly how the applicant is to respond to the question. It is best to avoid using ambiguous or imprecise terminology. Trick questions have no place in multiple-choice exams. Avoiding negative and double negative stems is a smart idea.

#### 23.3.4.4.5 Tailoring of multiple-choice questions

To guarantee that the assessment is legitimate, dependable, equitable, and successfully fosters meaningful learning, MCQs should be developed based on the students' proficiency level. It is important to keep in mind that different question designs are needed for different learning levels. In the medical field, for example, creating an MCQ for a second-year student differs from creating one for a fifth-year student due to the disparity in learning levels. For students in year two, the emphasis is largely on remembering and applying the fundamental knowledge, but for students in year five, analysis and critical thinking will be given more weight.

### *23.3.4.5 Assessment of multiple-choice questions*

Psychometric analysis and item analysis are used to evaluate the quality and efficacy of MCQs, taking into account variables such as discrimination, validity, reliability, and difficulty. While there are benefits to MCQs, such as objective grading, extensive subject coverage, and effective administration to large groups, there are drawbacks as well, such as the effort needed for appropriate formulation and the possibility that poorly designed questions will only evaluate lower-order thinking. Instructors should strive for well-designed questions that address learning objectives, evaluate higher-order thinking, and are checked for accuracy and clarity in order to provide high-quality MCQs.

### *23.3.4.6 Role of item analysis in assessing the multiple-choice questions*

After MCQs are administered, item analysis assesses their quality by providing information on distractor effectiveness, difficulty, and discrimination. It aids in the identification, revision, or elimination of subpar questions, resulting in more accurate and trustworthy evaluations, enhanced test-building abilities, and better instructional techniques. Some of the key roles of item analysis can be described as follows:

#### 23.3.4.6.1 Assesses question quality

In order to enhance assessments and instruction, item analysis evaluates the complexity, discrimination, and distractor efficacy of questions. It finds items that are too simple or too complex, those that don't distinguish between good and low performers, and problems with MCQs' distractions. In order to develop meaningful and trustworthy assessments that precisely gauge student learning and guide instructional modifications, this analysis is essential.

#### 23.3.4.6.2 Identifies poorly performing items

Item analysis determines if an item is too easy, too complicated, or fails to differentiate between students with high and low scores by examining item difficulty and discrimination. It also assesses the test's overall reliability to help enhance items for use in the future and looks at distractions to see if wrong responses are practical or nonfunctional.

#### 23.3.4.6.3 Evaluates distractors

Item analysis evaluates the effectiveness of distractions by examining how students chose incorrect answers in MCQs. It helps distinguish between high-performing and low-performing students by identifying diversions that are too evident or defective, making sure that each distraction is realistic and accomplishes its purpose of distracting. A good distractor has a negative discrimination score, draws more low-performing students, and is chosen by a respectable portion of test takers, according to analysis.

#### 23.3.4.6.4 Improves future tests

Item analysis improves subsequent exams by highlighting errors in test questions, such as being too simple, excessively complex, or poorly written. In order to more accurately assess student knowledge and differentiate between high and low performers, this approach helps to improve the validity and reliability of the questions. The method also helps highlight areas where students struggled, which can be useful for future education or content revision.

#### 23.3.4.6.5 Enhances teaching and learning

Item analysis helps teachers and students detect knowledge gaps and misconceptions, which enhances teaching and learning by supplying information for improved assessments and instructional strategies. It improves test quality by allowing teachers to modify or eliminate ineffective questions, and it benefits students by highlighting specific areas where they need to work harder. This process produces more reliable and accurate assessments of students comprehension.

#### 23.3.4.6.6 Increases test validity and reliability

Item analysis increases a test's validity and reliability by statistically examining each item to identify and correct poorly performing items. When item quality is enhanced, the test as a whole becomes more reliable and a more accurate evaluation of what it is intended to measure. This process helps remove flawed items that are too easy, too difficult, or incapable of differentiating between students who perform well and those who do not.

## 23.3.5 Short-answer questions (SAQs)

In medical education, SAQs are assessment techniques that test students' recall, interpretation, and application of knowledge by having them write a succinct response to a question. They can be used for both factual recall and problem-solving, and unlike MCQs, they require students to create their own answers, which can be more realistic to clinical practice (Sam et al., 2016). SAQs are a useful substitute or addition to other assessment techniques since they are more demanding than MCQs when it comes to evaluating depth of comprehension.

### *23.3.5.1 Characteristic features of short answer questions*

#### 23.3.5.1.1 Constructed response

A crucial feature of SAQs in medical education is a built response, which evaluates a student's capacity to come up with an answer as opposed to just choosing one from a list. This format assesses higher-order thinking abilities like analysis, synthesis, problem-solving, and clinical reasoning in addition to simple recall. Instead of choosing from a list of predetermined answers, students must create their own response from scratch while answering a constructed-response question (CRQ). The required response is purposefully limited to being extremely brief in very short answer questions (VSAQs), which are a highly particular kind of constructed response. One of the most important features of the CRQ is that it requires the student to come up with the answer without being given any options. VSAQs do away with the potential for guesswork or the use of recognition cues, which may be a problem with MCQs.

#### 23.3.5.1.2 Concise response

A concise response in a SAQ is characterized by being succinct, high-scoring, and brief—usually two to three lines or one paragraph—avoiding unnecessary introductions, directly addressing the question in the opening sentence, and providing supporting details or examples. It effectively demonstrates exact expertise by swiftly addressing the prompt, utilizing essential clinical phrases without needless explanation.

##### 23.3.5.1.3 Tests various skills

In medical education, short response questions are frequently utilized in the context of clinical scenarios to evaluate a variety of cognitive and practical skills beyond simple knowledge recall. In order to evaluate skills like recall of knowledge, students must be able to recollect factual information without the help of multiple-choice options. Additionally, it evaluates the capacity to apply fundamental clinical and biomedical knowledge to particular, real-world clinical situations. It assesses the student's clinical reasoning and problem-solving skills, simulating actual clinical practice by analyzing patient data, interpreting findings, making a diagnosis, and recommending a course of treatment.

##### 23.3.5.1.4 Ease of creation

When compared to other formats, such as MCQs, SAQs are thought to have a relative advantage in medical education due to their ease of development. Although it still takes deliberate work to write good questions, it is typically a property that makes them more appealing to assessors (Bala et al., 2023). As teachers simply need to define a clear, succinct question and a brief, precise answer key, SAQs are thought to be very efficient to create. They offer greater structural flexibility than true or false questions and may be drafted more quickly than MCQs, which necessitate the creation of distractors.

##### 23.3.5.1.5 Clinical scenario

In order to evaluate clinical reasoning and decision-making rather than just memory, a clinical scenario in a SAQ offers a realistic, concise, and focused patient case which includes presentation, examination, and pertinent investigation results. These questions are designed to test higher-order cognitive processes like analysis and diagnosis rather than just memorization.

##### 23.3.5.1.6 Brief and open-ended answer

A feature of VSAQs that permits a free-form response rather than a predetermined choice like "yes" or "no" is called an open-ended answer. Compared to closed-ended questions, this kind of inquiry is intended to generate a succinct, direct response in the respondent's own words, which may provide deeper insights. It also assesses the capacity to directly recall and articulate a particular fact or piece of information.

#### *23.3.5.2 Very short-answer questions (VSAQs)*

In medical education, VSAQs are CRQs that test higher-order cognitive processes by requiring a succinct free-text response, usually 1–5 words, to a clinical scenario. They are more similar to clinical practice than MCQs and increase assessment validity by eliminating cueing and minimizing guesswork.

#### *23.3.5.3 Advantages of very short answer question in medical education*

The benefits of VSAQs include encouraging real-world clinical thinking, lowering cueing from alternatives, and enabling thorough error analysis to inform future instruction. VSAQs, in contrast to MCQs, demand students to recall and produce answers, which results in a more accurate assessment of applied knowledge and promotes analytical reasoning rather than test-taking techniques. Furthermore, extremely these type of questions are a useful tool for both formative and summative evaluations and can offer insightful feedback on students' areas of difficulty. In medical education, VSAQs have several significant benefits, including the following:

##### 23.3.5.3.1 Promotes authentic clinical reasoning

Since they force students to come up with answers from memory, which is similar to real-life clinical practice where doctors must recall and apply knowledge without preprovided options, very short response questions encourage true clinical reasoning in medical education. Short response questions, which are frequently based on real-world clinical vignettes, more accurately reflect situations in which a medical practitioner must make a diagnosis, organize an investigation, or develop a management approach without a list of options.

#### 23.3.5.3.2 Minimizes cueing and test-taking behaviors

Very short response questions effectively reduce cueing and test-taking tendencies by requiring students to generate responses from memory rather than just choosing the correct response from a list. This method encourages deeper learning and improves the validity and authenticity of assessments. VSAQs drastically lower the likelihood that students will get the right response via recognition, word association, or guesswork by eliminating a list of possible replies (Desjardins et al., 2014).

#### 23.3.5.3.3 Enables detailed error analysis

In medical education, very short response questions are very beneficial for in-depth mistake analysis because they force students to come up with answers based on memory, exposing their true cognitive flaws and knowledge gaps without the "cueing effect" of MCQs. The benefit of a thorough error analysis is that, in contrast to MCQs, which allow students to guess the answer or employ test-taking techniques, VSAQs force students to provide a particular response.

#### 23.3.5.3.4 Provides valuable feedback

Both teachers and students receive rich, genuine feedback from VSAQs, which improves learning outcomes and clinical practice abilities. They accomplish this by making students come up with answers from memory, which reveals specific knowledge gaps and true comprehension. In contrast to MCQs, these questions provide insightful information on a student's cognitive blunders and mental processes.

#### 23.3.5.3.5 Encourages deeper learning

Very short response questions promote deeper learning, which enhances long-term memory retention and more closely matches real clinical practice by requiring students to develop knowledge rather than just recognize it. These kinds of questions promote deeper learning and application of knowledge because, unlike MCQs, they require students to construct their own answers, which stimulates higher-order cognitive processes like analysis and application of knowledge rather than superficial memorization or recognition of correct options.

#### 23.3.5.3.6 Offers a balanced assessment

Compared to MCQs, VSAQs provide a more authentic and rigorous evaluation of knowledge retention and application, making them a fair assessment approach in medical education. They provide rich feedback opportunities for teachers and students while mitigating the cueing and guessing problems that come with MCQs.

## 23.3.6 Short-essay questions (SEQs)

In medical education, SEQs evaluate higher-order thinking abilities like reasoning and problem-solving beyond mere recollection by presenting a clinical scenario and then requiring a written, structured answer. They are seen as a useful tool and are frequently employed in formats such as MEQs to assess a student's capacity to arrange information and apply it to a clinical issue. SEQs are thought to encourage deeper learning and higher retention since they challenge students to come up with their own responses, unlike MCQs.

### *23.3.6.1 Characteristic features of short-essay questions*

SEQs are designed to assess a student's comprehension, analytical skills, and ability to articulate ideas concisely within a limited scope. Some of the important features of the SEQs include the following:

#### 23.3.6.1.1 Conciseness

The length of the responses is restricted, usually falling between one paragraph and a maximum of 400–500 words. This calls for concentrated writing that stays away from repetition and superfluous elements. Here, precision and clarity without ambiguity are the main objectives. To guarantee that the reader grasps the intended meaning right away, this entails adopting precise, straightforward language and avoiding jargon unless it is topic-specific and defined.

#### 23.3.6.1.2 Structured response

One of the characteristics of an organized response to an SEQ is that it is concise, targeted, and well structured, directly addressing every aspect of the challenge in a constrained amount of space. In order to show higher-order thinking abilities like analysis and synthesis rather than merely memorization, the response nonetheless uses the essay's introduction, body, and conclusion.

#### 23.3.6.1.3 Higher-order thinking assessment

SEQs typically require students to develop a focused response that exhibits complicated cognitive processes beyond simple memory in order to test higher-order thinking through unique distinguishing qualities. Unlike multiple-choice or extremely SAQs, they go beyond mere knowledge recall to assess critical thinking abilities including synthesis, analysis, and evaluation.

#### 23.3.6.1.4 Specific focus

One of the main features of SEQs is specific emphasis, which calls for a succinct, focused response that directly addresses a particular facet of a subject within a constrained word count. An SEQ uses limiting terms to purposefully restrict the breadth of the required response, in contrast to a long essay that could delve deeply into a wide topic. Instead of covering a broad variety of basic material, the inquiry is intended to assess a student's ability to cover a specific topic in detail. Another crucial feature is that the questions frequently contain precise instructions, restricting terms, and directive words like "explain," "discuss," "compare," and "justify" that tell the learner exactly what to do.

#### 23.3.6.1.5 Argumentation and evidence

Evidence and argumentation are essential components of SEQs. Instead of just summarizing data, argumentation requires the learner to take an informed stance, assess contradicting evidence, and reason out the solution. Additionally, the students are able to demonstrate that their thoughts are supported by facts, statistics, examples, and expert opinions, which makes them more legitimate and earns the tutor's trust.

#### 23.3.6.1.6 Subjective grading

In SEQs, subjective grading refers to evaluating student answers based on an experienced grader's judgment, opinion, and interpretation rather than a single, objectively correct response. This approach is utilized to assess students' complicated thinking abilities, reasoning, and the caliber of insight they provide, which frequently have several acceptable modes of presentation.

#### 23.3.6.1.7 Open-ended, yet focused

SEQs are also characterized by being open ended and yet focussed. This means that the question allows for a thorough, unrestricted response while still giving the student precise guidelines or an assumption to clarify. These are open-ended inquiries that cannot be answered with a straightforward "yes" or "no" or a single fact. They usually start with phrases like "How," "Why," or "What," which encourage students to elaborate, think critically, and respond in their own words. SEQs help measure higher-order cognitive skills since they require students to analyze, synthesize, and assess material rather than just memorize facts. It also encourages original thought since the students must formulate their own arguments and integrate relevant facts and ideas into a coherent response.

## 23.3.7 Patient management problems (PMPs)

In medical education, a written assessment technique called "patient management problems" is used to mimic real-world clinical scenarios and gauge a student's capacity for clinical reasoning, problem-solving, and decision-making. In order to assess clinical reasoning, diagnostic skills, and treatment options, test takers can traverse patient scenarios from intake to management using the well-known, traditional PMP which is a type of written, simulated assessments. Using MCQs or SAQs, they evaluate a learner's ability to elicit clues, pick investigations, and choose suitable therapy. They are thought to be helpful for assessing how a physician would treat a patient at various stages of care, such as getting a patient's medical history and starting treatment. The following is a description of the typical characteristics of written PMPs:

### *23.3.7.1 Sequential presentation*

PMPs, which comprise simulation exercises designed to teach and evaluate clinical reasoning, are characterized by their sequential presentation. This approach mimics real medical practice by asking participants to make decisions at each stage of patient care before receiving feedback or moving on to the next. The purpose of PMPs is to mimic the decision-making process a doctor goes through when looking into and treating a patient, where information is provided gradually rather than all at once.

#### 23.3.7.2 Interactive format

A key component of PMPs utilized in clinical practice and medical education is their interactive nature, which goes beyond conventional, static case studies to offer a dynamic, decision-based experience. This style simulates real-world situations by requiring students or clinicians to actively make judgments and get feedback.

#### 23.3.7.3 Immediate feedback

As a high-impact, formative assessment method in medical education, immediate feedback in PMP type questions gives students quick, practical, and helpful answers to their diagnostic or therapeutic choices. In contrast to summative tests, this method greatly improves learning and retention by enabling students to correct mistakes before they become deeply embedded.

#### 23.3.7.4 Focus on process

Instead of focusing only on the right response or result, the process feature in PMPs refers to the evaluation of a clinician's clinical reasoning, decision-making, and the series of activities they take when managing a patient. It assesses how a medical practitioner makes a diagnosis and develops a treatment plan, reflecting actual practice.

#### 23.3.7.5 Written component

The written component of a PMP is a simulation exercise designed for teaching or assessment in a medical context, moving the user through the clinical decision-making process. The exercise is typically broken down into distinct phases that require the respondent to actively engage with the patient's case. The main phases that constitute the written component include Phase 1, also known as the Information Provision, which involves presenting preliminary patient data, such as demographics, primary complaints, and pertinent medical history. Phase 2, which is a request for action, is the next phase in which the respondent is asked to make choices, such as asking for a more detailed history, a physical examination, particular diagnostic tests, or preliminary treatment measures. Phase 3, often known as the feedback phase, comes next and presents the outcomes or repercussions of the activities that the respondent desired. The entire process ends with the conclusion of the problem which is the phase 4 where the final diagnosis, management plan, and expected outcomes are often discussed. This phase may cover aspects like patient education, advice to relatives, specific medications, or referral for specialist opinions.

#### 23.3.7.6 Benefits of patient management problems

PMPs in written assessments offer significant benefits, primarily by focusing learning and evaluation on real-world clinical skills, which traditional tests often fail to capture. Some of the key benefits include development of problem-solving skills in which the PMPs force learners to actively diagnose and manage a patient by presenting a realistic clinical scenario, frequently with information that is revealed dependent on the user's choices. This directly fosters problem-solving abilities in a secure setting. It also provides a vital link between theoretical understanding and clinical practice is provided by the use of virtual simulations, high-fidelity manikins, and standardized patients in PMP questions. These simulations help healthcare workers to improve clinical reasoning, practice decision-making, and improve patient safety without endangering real patients by simulating realistic, high-stakes, or unusual events. In addition to this, unlike passive learning such as like lectures, PMPs promote active learning by requiring participants to actively process information, make choices, and observe the consequences of their actions in a low-risk environment. They also allow the examiners to assess a student's ability to manage a patient over time, including their ability to respond to changes in the patient's condition, rather than just factual knowledge. When used in formal assessments, the written format provides a permanent, traceable record of the student's decisions and path through the problem, which is valuable for grading and feedback. Moreover, the format of PMPs encourages a systematic approach to patient assessment and management, which helps reduce the chance of missing important signs or symptoms and can improve overall patient outcomes. It also facilitates students and educators to identify specific areas where the student's knowledge or decision-making process needs improvement for the purpose of self-assessment or learning.

### 23.3.8 Short written responses

Short written responses, such as exit slips or extremely SAQs, are employed in medical education to gauge students' critical thinking and reasoning skills. While exit slips give teachers immediate feedback in a classroom context, VSAQs are a dependable substitute for MCQs by exploring deeper understanding. These techniques are useful for formative and summative evaluations, which aid in determining comprehension, exposing misunderstandings, and directing future instruction. Some of the important features of the short written responses is that the questions are based on realistic clinical scenarios, vignettes, or data, such as lab results and images, which reflects the real-world clinical practice more effectively than traditional MCQs. It is also characterized by concise and free-text answers, wherein the students are required to generate their own brief responses, rather than selecting from a list of options. This format removes cueing and the ability to guess the correct answer. Moreover, this format focuses on testing the application of knowledge, clinical reasoning, and problem-solving skills such as diagnosis, investigation, and management decisions. The free-text nature of responses allows educators to analyze student errors, identify common misconceptions, and provide targeted feedback to both students and teachers, which can inform and enhance future teaching.

#### *23.3.8.1 Benefits in medical education*

The benefits of the SAQs are that they provide a higher validity and reliability and encourage deep learning as these types of questions can promote deep learning by making the student to construct an answer from memory rather than recognizing one from a list. These questions can also be delivered and efficiently machine-marked online, with human review for alternative yet plausible answers, making them a practical alternative to purely manual assessment methods.

### 23.3.9 Innovative written assessment tests

Innovative written assessment tests employ technology and practical application to assess comprehension. They go beyond conventional approaches by incorporating digital tools like e-portfolios and simulations, as well as innovative task formats like creating visual artifacts, providing explanations for multiple-choice answers, or finishing project-based assessments. These tests are more interesting and successful at capturing a student's variety of skills since they can be adaptive, include multimedia components, and deliver real-time feedback.

#### *23.3.9.1 Types of innovative written assessments*

##### 23.3.9.1.1 Modified essay questions (MEQ)

An MEQ is an assessment technique used in medical education that presents a brief clinical scenario followed by a series of structured questions to gauge a student's ability to solve problems, reason, and apply knowledge in a way that is similar to a real-life clinical case. MEQs provide a useful, applicable format that goes beyond factual recollection to measure higher-order cognitive skills and can be used to evaluate attitudes and communication, in contrast to typical essay questions or even MCQs (Elstein, 1993).

##### 23.3.9.1.2 Difference between essay and modified essay question

An open-ended written response to a prompt is required for a classic essay question, which frequently assesses writing abilities and general knowledge. In order to assess higher-order cognitive abilities including problem-solving, clinical reasoning, and the capacity to integrate information in an organized, sequential manner, an MEQ presents a particular clinical situation or case study followed by a series of questions. Compared to a traditional essay, the MEQ's structured framework offers greater objectivity in grading and may be modified for a wider range of skills. The table below provides a summary of the key distinctions between an essay and an MEQ (Table 23.1).

**TABLE 23.1** Difference between essay and modified essay question.

| Sl no. | Feature | Traditional essay question | Modified essay question |
|---|---|---|---|
| 1 | Format | A general prompt or question. | A sequential, evolving case study presented in sections. |
| 2 | Purpose | Tests broad knowledge, argumentation, and writing skills. | Tests problem-solving, reasoning, and integration of knowledge in a specific context. |
| 3 | Structure | Open-ended, allowing for extensive discussion and argumentation. | Structured around a patient scenario, with decisions required at each step before moving to the next. |
| 4 | Skills assessed | To assess higher-order cognitive skills such as critical thinking, logical reasoning, and the ability to synthesize and communicate complex information. | To assess a student's clinical reasoning and problem-solving skills through a sequential, case-based format. |

Example of traditional essay question:

"Describe the events of an action potential, including the roles of specific ion channels and the concepts of absolute and relative refractory periods."

Example of modified essay question:

Mrs. X, a 32-year-old female patient, is your first patient with a 6-month history of loose bowel movements, about 8 per day. Blood has been present in many of them. She has lost 15 kg. On examination the patient looks ill. Her BP is 130/70 and pulse 108/mi. There is generalized rebound tenderness with no rebound. A sigmoidoscopy done reveals friable rectal mucosa with multiple bleeding points.

**Question 1.1.** What further history would you inquire about?

**Question 1.2.** List two differences between ulcerative colitis and Crohn's disease?

**Question 1.3.** What form of inflammatory bowel disease would this patient most likely have?

**Question 1.4.** The patient is concerned that she may have cancer, discuss the relationship between inflammatory bowel disease and the risk of acquiring carcinoma of the colon.

### *23.3.9.2 Key features of a modified essay question*

Higher-order cognitive abilities including problem-solving, reasoning, and decision-making in the context of real-world situations are commonly assessed using MEQs in medical and professional education. A highly controlled framework and objective evaluation set MEQ apart from traditional essays. Among the crucial characteristics of MEQs are the following:

#### 23.3.9.2.1 Case-based scenarios

A brief, realistic case or clinical scenario which typically concentrates on the patient's issue encountered in day-to-day practice forms the foundation of each question. They act as a means of moving evaluations away from straightforward factual recollection toward clinical thinking, problem-solving, and decision-making.

#### 23.3.9.2.2 Evolving information

The scenario is unfolded stage-by-stage, with new information provided in sequence. The examinee must respond to each stage before proceeding to the next, preventing them from previewing later parts of the case.

#### 23.3.9.2.3 Focuses on problem-solving

MEQs primarily assess the ability to apply knowledge and understanding in new circumstances rather than just factual recall. They can also assess professional skills like communication, ethics, and attitudes.

#### 23.3.9.2.4 Structured format

MEQs are short-answer, structured tests consisting of four to ten SAQs that are mutually exclusive after a brief, changing patient scenario.

#### 23.3.9.2.5 Objective and reliable scoring

A detailed, predetermined model answer and marking scheme is used by examiners to ensure consistency and reduce inter-rater variability. This increases the objectivity of marking compared to traditional, free-form essays.

#### 23.3.9.2.6 Defined scope

Each question or section of a question is framed to cover a specific, well-defined area of practical experience, allowing for a wider range of content to be tested across a single exam compared to conventional, broader essay questions.

#### 23.3.9.2.7 Assessment of writing skills

While structured, they still require students to compose their responses in their own words, allowing for an assessment of writing and organizational skills.

#### *23.3.9.3 Creating a modified essay question*

In an attempt to create an MEQ, it is always good to start with a central scenario and divide it into successive sections, each of which builds upon the one before it. In order to assess students' reasoning and problem-solving abilities, this format involves posing a problem and then disclosing fresh facts with each successive question. For consistent grading, it is essential to write questions that are straightforward, clear, and include a structured answer key. The steps to create an MEQ can be summarized as follows:

##### 23.3.9.3.1 Start with a scenario

For an MEQ to be valid, relevant, and capable of testing higher-order cognitive skills, an effective scenario must be chosen. The questions are based on the scenario, which should be carefully crafted. Aligning the scenario with the learning objectives or competences being evaluated is one of the most crucial factors to take into account while choosing the scenario. Simultaneously, attention should be taken to ensure that the scenario reflects actual circumstances, such as a patient case or a particular professional issue. A deeper participation and critical thinking are encouraged as a result of the assessment's increased authenticity and relevance. Additionally, common or possibly critical issues should be the main focus rather than unimportant or extremely uncommon subjects.

##### 23.3.9.3.2 Formulate the first question

Making a direct connection to the original scenario is crucial when crafting the first inquiry. The initial patient presentation or circumstance detailed in the MEQ stem should logically lead to the first question. "List," "Identify," "Outline," "Describe," and "What is the most likely cause?" are examples of explicit command language that specify the desired task.

##### 23.3.9.3.3 Reveal new information

An MEQ must be framed as a sequential, changing situation that requires students to use problem-solving techniques at each level in order to expose new information. New information that wasn't covered in the first section must now be provided, for instance, revealing about the findings of an investigation like an X-ray. For the "modified" part of the question, this is an essential step.

##### 23.3.9.3.4 Posing the next question

When posing the following question in a modified essay, make sure it follows logically from the previous one. This is typically accomplished by adding more sequential information to the scenario before posing a new question that calls for the application of extra knowledge or problem-solving techniques. To prevent punishing a student repeatedly for a single initial mistake, the next question should build on the previous scenario without requiring the response to be correct. For example, one could ask, "How do you manage this case?"

##### 23.3.9.3.5 Repeat as necessary

The stage of posing the question can be continued if necessary, with each new part of the question building on the previous answer and new information. Repeating the step of posing questions gives students additional, previously undisclosed facts to enable further, more in-depth questioning.

##### 23.3.9.3.6 Develop a scoring checklist

A methodical strategy that guarantees objectivity and is in line with learning objectives is necessary when creating a scoring checklist or rubric for an MEQ. Using an analytical rubric, which divides the response into distinct criteria and establishes performance levels for each, is the most efficient approach.

#### *23.3.9.4 Benefits of modified essay questions in medicine*

MEQs are useful in medical education because they assess clinical problem-solving abilities, provide a more comprehensive evaluation than standard essays, and improve objectivity and reliability. They are useful for evaluating a broader range of skills, such as thinking, attitudes, and factual memory, and they can even reveal gaps in the curriculum. Some of the main benefits of employing MEQs in the medical field include assessment of higher-order thinking in which the questions are designed to test a student's ability to think critically and solve problems in a clinical context by presenting a case in a chronological sequence. The modified essay questions provide a more comprehensive assessment as they are more flexible than MCQs and can cover more content and assess a wider range of abilities, including attitudes and ethical considerations, which are difficult to evaluate with other formats. Moreover, with a structured scoring system and predetermined answers, MEQs can reduce examiner unreliability and provide more objective and reliable marking compared to traditional essays. It also fosters deeper learning by helping students to better understand the application of basic science principles to new scenarios. It also has a role in enhancing curriculum development as it gives educators valuable insights into areas where students consistently struggle, highlighting weaknesses in the curriculum that need strengthening.

### 23.3.10 Constructed-response questions (CRQs)

In medical education, CRQs use styles, including essays, case studies, and brief replies, to test students' higher-order thinking abilities like critical thinking and problem-solving. They are used to assess a student's capacity to apply knowledge in a clinical setting, but in order to guarantee validity and reliability, they call for sophisticated marking schemes, such as level-based rubrics.

### 23.3.11 Extended matching items (EMIs)

EMIs are a kind of assessment used in medical education that tests clinical reasoning by combining a more applicable, contextual format with the structure of MCQs. They provide for a more thorough assessment of a student's capacity to diagnose, prioritize, or make judgments in clinical situations by presenting a series of related questions that relate to a single, lengthy list of possibilities (Fenderson et al., 1997). The following are some of the extended matching questions' salient characteristics:

#### *23.3.11.1 Single option list*

Often referred to as R-type EMQs or single best answer EMIs, EMIs with a single-option list are homogeneous, have a set number of alternatives usually 8–20, and are normally given in either alphabetical or numerical order. One of the primary characteristics is homogeneity, in which every item on the list falls into the same category, such as diagnoses, certain medications, laboratory tests, or management techniques. This compels applicants to choose between reasonable options in a particular therapeutic setting.

#### *23.3.11.2 Multiple stems*

The single option list serves as the basis for a number of linked clinical scenarios or vignettes. These vignettes are designed to be similar in structure and length. One of the main characteristics that sets EMIs apart from conventional single-best-answer MCQs is their multistem format, which makes it possible to evaluate a candidate's capacity to distinguish between comparable conditions or interventions across a range of clinical presentations, thereby more accurately testing clinical reasoning abilities.

#### *23.3.11.3 Themed content*

The term "theme" in EMIs refers to a broad topic area or clinical setting that connects a series of linked questions, each of which draws from a single, comprehensive list of potential replies. The theme offers a targeted framework for the evaluation, enabling teachers to assess knowledge in a practical and comprehensive way, frequently focusing on a particular set of differential diagnoses, management strategies, or investigations.

#### *23.3.11.4 Longer option lists*

EMIs' longer choice lists are a crucial component that reduces the impact of guesswork and efficiently evaluates higher-order cognitive abilities and clinical reasoning in medical education.

#### *23.3.11.5 Application-based*

The application-based aspect of EMIs in medical education refers to their capacity to evaluate higher-order cognitive abilities such as clinical reasoning, decision-making, and the application of theoretical information to actual clinical situations. This sets them apart from conventional MCQs, which frequently simply assess factual memory.

## 23.3.12 Open-book exam (OBEs)

In medical education, OBEs move the emphasis from rote memorizing to problem-solving employing extensive medical knowledge by providing access to resources like notes or texts to evaluate application, analysis, and critical thinking. This is similar to real clinical practice where information retrieval is crucial (Ahmed et al., 2024). They can lessen stress, foster deeper comprehension, and encourage self-directed learning. However, their effective design necessitates scenario-based questions that go beyond mere memory, which frequently results in more difficult tests that examine synthesis and evaluation abilities. Some of the key characteristics of the open-book assessment are as follows:

### *23.3.12.1 Resource access*

The ability for students to use their approved materials is one of the features of OBEs. It is important to remember that your access to resources during an open-book test is strictly restricted to the items your instructor has designated, which may include your prepared course notes, textbooks, and other reference materials. Since these tests are meant to assess your comprehension and ability to apply knowledge rather than merely memorize it, they typically do not allow free internet access or consulting others.

### *23.3.12.2 Focus on application*

Exams that are open-book do emphasize application over memory. They are made to assess the student's capacity for higher-order thinking skills, which include the ability to synthesize, analyze, and use knowledge to solve issues or react to situations. OBEs emphasize application by shifting the emphasis from recall to comprehension. Rather than assessing students' memorization skills, these tests demand that they have a thorough understanding of the course material in order to use it to answer questions. In order to examine the topic, choose pertinent details from their notes or textbook, and then develop a well-reasoned response, the student must also exercise critical thinking. Compared to rote memorization, this is a more advanced skill.

### *23.3.12.3 Higher-order thinking*

Higher-order thinking abilities like application, analysis, and evaluation can be helped by OBE, but only if the questions are structured to demand them and go beyond simple memorization. Instead of just finding and copying answers, a well-designed OBE pushes students to interpret, synthesize, and apply knowledge, reflecting real-world situations. Open books promote higher-order thinking when students move from memory to reasoning because they push them to use the knowledge to develop their responses. They imitate real-world difficulties because they are adept at assessing students' capacity to apply knowledge to unfamiliar situations.

### *23.3.12.4 Time limits*

Most OBEs have time limitations, even though they may be longer than closed-book exams. The time restrictions are in place to ensure that the student cannot look up every answer and still need to understand the material well. Time limits prevent students from using brute force study to demonstrate their knowledge. Instead, they must assess and apply concepts rather than just memorize them.

## 23.3.13 Key feature item

A format that examines clinical decision-making using a brief case scenario and questions that concentrate on the most crucial steps where mistakes are likely to happen is an important feature item in a written assessment. Two scholars, Georges Bordage and Gordon Page, first used the term "key feature." This method assesses the application of knowledge in a practical setting by concentrating on challenging or "key" parts of problem-solving rather than a thorough, linear explanation. In order to evaluate a candidate's capacity to make critical choices in areas like diagnosis and management, the questions may be given in multiple-choice or open-ended formats. A clinical case or vignette that gives the examinee a synopsis of the patient's case is a major feature item. Additionally, it includes one or more questions that specifically address the crucial phases in the clinical scenario. Candidates are more likely to make mistakes at these decision moments. Depending on what is being evaluated, the questions may be given in a variety of formats, such as MCQs or SAQs.

### *23.3.13.1 Designing a key feature item*

Writing a realistic case scenario based on clinical experience, defining the clinical problem, identifying the crucial "key features" of decision-making, and developing one question for each key feature are all vital steps in creating a key feature item in medical education. In order to improve the item before it is used in an exam, it is crucial to choose an appropriate question structure, create a scoring key, and conduct a pilot test. The following are some actions that should be taken in this regard:

#### 23.3.13.1.1 Define the problem and key features

The first and most important stage is to define the clinical problem since the clinical scenario that the item will evaluate must be precisely described. Since they are the "key features" the item will test, it is important to identify the crucial decisions or actions in diagnosing and treating the clinical problem.

#### 23.3.13.1.2 Write the case scenario

When developing a scenario, great effort must be given to ensure that it is realistic and accurately depicts a clinical circumstance that occurs in the real world. In order for the user to respond to the questions, it is important to remember that the case should include sufficient information, such as the patient's medical history, physical examination findings, and test results. The duration of the case should be modified because questions about investigations and management may require a longer case scenario, while questions about history and physical examination may require a shorter one.

#### 23.3.13.1.3 Develop the questions

When drafting the questions, it is important to keep in mind that each significant element should have one question. It is crucial to select the most appropriate format for each question after it has been written, such as multiple-choice, short response, or a combination of forms. Making a scoring key that lists the right responses to each question is also important at this stage.

#### 23.3.13.1.4 Validate and refine the item

To ensure that a key feature item in medical education accurately evaluates clinical decision-making abilities, it must undergo a rigorous, multistep validation and refinement procedure. Expert evaluation, alignment with learning objectives, piloting, and postadministration item analysis are all part of this procedure. Refinement is the process of enhancing an item's quality by utilizing data and feedback. Items with low clarity ratings or unfavorable comments from pilot testing should be altered for sentence structure, spelling, or format during this round. Additionally, this technique allows us to change the level of difficulty. If item analysis shows that a question is either too simple or too difficult, it may need to be changed, for example, by altering distractions or the clinical information. Additionally, it provides us the chance to get rid of things that have technical defects, including grammatical errors that give hints or unrealistic distractions.

## 23.3.14 Computer-based written assessment

A computer-based written assessment is one that uses electronic equipment in place of traditional pen and paper for both administration and marking. Both objective questions like multiple-choice and nonobjective questions with typed solutions, like essays or short answers, can be included (Cantillon et al., 2004).

### *23.3.14.1 Key features of computer-based written assessment*

#### 23.3.14.1.1 Delivery and marking

The delivery and marking of computer-based written assessments are expedited through the use of technology, frequently involving secure platforms for delivery and a mix of automated and on-screen systems for marking. Compared to conventional paper-based procedures, this procedure seeks efficiency, objectivity, and quicker outcomes. Conversely, tests are administered to students through a computer terminal with specialized software, either on standalone computers, via a local area network, or through a secure web browser. These tests are usually administered in a controlled setting, frequently with invigilation (Al-Amri and Ali, 2016). In order to preserve exam integrity, this can involve both in-person testing locations and remote proctoring via cameras and screen recording. In this assessment, the type of question determines the marking procedure. For example, the usage of automated marking is objective, automatic, and instantaneous. The student's response is compared by the system to a predetermined proper answer key.

#### 23.3.14.1.2 Question types

A range of question forms, from conventional formats to interactive, technologically enhanced items, are used in computer-based written assessments. Multiple-choice and nonobjective questions, such as essays or short replies, where responses are entered into the system, are the most prevalent kinds. The most common of these question kinds are MCQs, in which the test taker is given a question or statement and must choose one accurate response from a list of possibilities.

#### 23.3.14.1.3 Feedback

Instead of merely displaying correctness, feedback in computer-based written assessment should be detailed, goal-oriented, and customized. When used properly, it is a potent tool that promotes metacognition and self-regulated learning.

#### 23.3.14.1.4 Flexibility

In order to improve students' motivation and engagement during the assessment process, flexibility in computer-based written assessment entails giving them a variety of options and tailoring the test delivery to each student's performance. Within the parameters of security and invigilation, computer-based written assessments can provide more flexibility in terms of when and where examinations are administered than traditional paper-based exams. Additionally, it gives students a variety of evaluation methods and assignments to show what they have learned while letting them play to their strengths. In order to lessen students' anxiety about a single high-stakes exam, instructors might simultaneously provide them alternatives to modify the relative weighting of various assessment components toward their final mark.

#### 23.3.14.1.5 Security and randomization

In order to avoid cheating, guarantee test integrity, and preserve the validity and dependability of exam outcomes, security and randomization are crucial elements of computer-based assessments in medical education. In computer-based tests, security measures are intended to verify the examinee's identification and stop access to unapproved materials while the test is being administered. Medical facilities frequently use specialized computer-based assessment rooms with physical security measures, such as opaque screens separating candidates, to reduce potential for copying.

### *23.3.14.2 Benefits of computer-based written test*

Faster, automated results, more efficiency through auto-grading and less administrative labor, improved security, environmental friendliness by doing away with paper, and more flexibility and convenience for test takers are just a few advantages of computer-based written assessment. Additionally, they enable features like extensive statistics for both students and teachers, adaptive testing, and accessibility choices for students with impairments. The following are some of the advantages for test takers:

#### 23.3.14.2.1 Convenience and flexibility

Candidates can frequently select from a variety of test dates and times, and in certain situations, they can even take the test from the comfort of their own homes. This saves time and money by removing the need to travel to particular testing facilities at predetermined times. In terms of adaptability, it is feasible to adjust the level of difficulty of questions in real-time in response to a candidate's performance, offering a more accurate and customized evaluation of their skills.

#### 23.3.14.2.2 Improved experience

With greater accessibility, an easy-to-use interface, and instant feedback and results, computer-based testing offers test takers a far better experience. One of the key benefits is that the digital format frequently has clear, uncluttered screens with simple navigation tools, which makes the testing process less stressful and easier to manage. Features like time tracking and the ability to underline important wording allow students to focus solely on the questions.

#### 23.3.14.2.3 Accessibility

With built-in tools like text-to-speech, font resizing, and magnification, computer-based assessments provide substantial accessibility benefits for students with visual impairments or learning disabilities. Additionally, they provide flexibility and offer a more equal testing experience for a larger spectrum of students with different requirements by enabling remote test-taking and configurable modifications like extra time.

#### 23.3.14.2.4 Typing advantage

Strong typing abilities in medical education are very beneficial for computer-based written exams because they enable students to express their knowledge more effectively, generate more thorough answers, and concentrate more on the content of their responses than the writing mechanics. Additionally, the benefit of typing in computer-based assessments guarantees that a student's performance is a more accurate representation of their comprehension and clinical reasoning capacity, free from the physical constraints of handwriting legibility or speed.

## 23.4 Standard setting in written assessment

The process of determining a minimum passing score, or cut score, to ascertain whether a candidate is "just good enough" to be deemed competent is known as standard setting in medical education's written exams. In order to guarantee that graduates are safe to practice medicine, this is an essential, expert-driven procedure. Common techniques include the Ebel method, which is frequently applied to MCQs, and the Angoff approach, which entails a panel of experts determining the likelihood that a "borderline" candidate will properly answer an item (Ben-David, 2000).

### 23.4.1 Importance of standard setting in written assessment

Instead of merely comparing students to one another, standard setting is essential in written exams in order to produce a defensible, consistent, and equitable minimum passing score based on a predefined standard of performance. For high-stakes judgments like those in medical certification, it guarantees that the test is neither too simple nor too tough over time and that the cut score is transparent and well-supported. Additionally, this procedure synchronizes the content of the evaluation with the intended learning objectives.

### 23.4.2 Common methods for standard setting in written assessment

Some of the best-known standard setting methods which have a role to play in written assessments are as follows:

#### *23.4.2.1 Angoff method*

The Angoff method has endured over time, as have its revisions. They remain the most used techniques for MCQ standard setting. They enable the experts to evaluate the things using their anecdotal experience. The changes make it possible to make better decisions. One of the key components of the Angoff technique is that it involves a group of subject matter experts from different fields, often six to eight in number, who attempt to estimate the likely percentage of borderline examinees who would properly answer the MCQs. For every question that is included, the experts carry out this estimation procedure. After that, these ratios are totaled for each expert judge. The examination's cut score is determined by taking the median total of item proportions among judges. Angoff's function in written exams is significant since it establishes the minimal passing score for students, which is the cutoff on written examinations based on the opinions of subject matter experts.

#### *23.4.2.2 Ebel method*

Ebel method, which is a variant of the Angoff method, also depends on the expert opinion of the objects. In this method, a team of specialists evaluates each exam item according to its difficulty, such as easy, medium, and hard, and significance, which include essential, important, and acceptable. Each object is placed by experts into a grid that represents a 3×3 or 4×3 matrix. An algorithm that takes into consideration the difficulty and significance of the questions is then used to determine the final cut score, which provides a pass or fail result based on a predetermined standard rather than relative student performance.

### 23.4.3 Evaluation and quality assurance of written assessment

Through a process of ongoing review and improvement, evaluation and quality assurance for written assessments guarantee the outcomes are valid, accurate, and dependable. Clear standards, uniform grading criteria, the use of both internal and external review methods, process evaluation, and continual development are important components. All of these elements guarantee that the written evaluation will be advantageous in terms of credibility, guarantee fairness, and increase trust in the evaluation's results (Iroro, 2023).

### 23.4.4 Role of written assessment in teaching and learning process

The process of written assessment is essential to the teaching and learning process because it serves as a key instrument for assessing student learning, directing student development, and shaping instruction, as discussed in the many forms of written assessment. It assists in determining a student's present level of knowledge, monitors their development, identifies areas of weakness, and offers a final assessment of mastery. Written tests that are well designed can examine abilities like analysis and application in addition to factual knowledge.

## 23.5 Advantages and disadvantages of written assessment

Although written tests can evaluate higher-order thinking and be objectively assessed, they can take a long time to grade, may only measure surface-level information, and may not provide immediate feedback or be misinterpreted. The benefits of written assessments include the capacity to evaluate big groups, create a permanent record, and foster critical thinking and writing abilities. In addition, written evaluations, such as essays, allow students to convey their knowledge in a more comprehensive manner because they may articulate the subject matter in its entirety. Students can more effectively communicate their knowledge and comprehension of the subject by being encouraged to use a variety of techniques, such as creating tables or diagrams. However, the drawbacks of written tests include time-consuming marking, challenges in evaluating abilities like verbal communication, and the possibility that students will prioritize quantity over quality because of time constraints.

**Summary**

Written assessments are a reliable and scalable means of evaluating students' comprehension, reasoning, and application of biological and clinical concepts, making them a crucial component of medical education. They also provide an efficient, unbiased, and scalable way to measure the knowledge and understanding ability of the student pertaining to a particular topic. As they assess the information, reasoning, and communication skills that underpin safe and efficient clinical practice, their ability to assess much more than factual memory has become increasingly important. The usage of various types of written assessments in medical education helps the student to develop essential skills of remembering, recalling, understanding, and interpreting various facts and concepts in medicine. The development of these skills prove to be very beneficial in future clinical practice.

**Key point**

**Future perspective: moving towards technologically enhanced model**

Written assessments are rapidly evolving due to sociological, pedagogical, and technological developments. In the future, more authentic written exams in medical education will reward a variety of skills, including clinical reasoning, decision-making, evidence appraisal, professionalism, and communication. Future written tests will assess competencies rather than just factual memory. Written assessments will always be crucial, but they will evolve to better represent the mental work of real therapists. As written exams become more authentic, digital, and AI-supported, tutors' roles will shift from markers and content-transmitters to learning facilitators, coaches, and assessment designers. As their role becomes increasingly important, they take on greater responsibilities.

## References

Ahmed, Y., Khayal, S., Wadi, M., Abuzaid, M., 2024. Evolving assessment in medical education: exploring the role of open-book examinations. Journal of Advances in Medical Education and Professionalism 12 (4), 215–225. Available at https://doi.org/10.30476/jamp.2024.102078.1955, https://jamp.sums.ac.ir/article_50395_b752dc8e4f16f7c68507cf9e2761ed9b.pdf.

Al-Amri, S., Ali, Z., 2016. Systematic review of computer based assessments in medical education. Saudi Journal of Medicine & Medical Sciences 4 (2), 79–88. Available at https://doi.org/10.4103/1658-631X.178288. Epub 2016 Mar 9. PMID: 30787703; PMCID: PMC6298330.

Al-Rukban, M.O., 2006. Guidelines for the construction of multiple choice questions tests. Journal of Family and Community Medicine 13 (3), 125. Available at https://doi.org/10.4103/2230-8229.97543.

Bala, L., Westacott, R.J., Brown, C., Sam, A.H., 2023. Twelve tips for introducing very short answer questions (VSAQs) into your medical curriculum. Medical Teacher 45 (4), 360–367. Available at https://doi.org/10.1080/0142159x.2022.2093706.

Ben-David, M.F., 2000. AMEE guide no. 18: standard setting in student assessment. Medical Teacher 22 (2), 120–130. Available at https://doi.org/10.1080/01421590078526, http://www.tandfonline.com/loi/imte20#.VufR2bdf2DY.

Cantillon, P., Irish, B., Sales, D., 2004. Using computers for assessment in medicine. BMJ Clinical Research (ed.) 329 (7466), 606–609. Available at https://doi.org/10.1136/bmj.329.7466.606.

Cantillon, P.W.D., 2010. ABC of Learning and Teaching in Medicine, 2 nd ed. Blackwell Publishing Ltd, Malden (MA).

Coughlin, P.A., Featherstone, C.R., 2017. How to write a high quality multiple choice question (MCQ): a guide for clinicians. European Journal of Vascular and Endovascular Surgery 54 (5), 654–658. Available at https://doi.org/10.1016/j.ejvs.2017.07.012.

Desjardins, I., Touchie, C., Pugh, D., Wood, T.J., Humphrey-Murto, S., 2014. The impact of cueing on written examinations of clinical decision making: a case study. Medical Education 48 (3), 255–261. Available at https://doi.org/10.1111/medu.12296.

Elstein, A.S., 1993. Beyond multiple-choice questions and essays: the need for a new way to assess clinical competence. Academic Medicine 68 (4), 244–249. Available at https://doi.org/10.1097/00001888-199304000-00002.

Enameguolo-Yarhere, I., 2023. Quality assurance of assessment in medical education: A faculty workshop for facilitators by University of Port Harcourt's College of Health Sciences. IOSR Journal of Research & Method in Education 13 (2), 38–44.

Epstein, R.M., 2007. Assessment in medical education. New England Journal of Medicine 356 (4), 387–396. Available at https://doi.org/10.1056/NEJMra054784, http://content.nejm.org/cgi/reprint/356/4/387.pdf.

Fenderson, B.A., Damjanov, I., Robeson, M.R., Veloski, J.J., Rubin, E., 1997. The virtues of extended matching and uncued tests as alternatives to multiple choice questions. Human Pathology 28 (5), 526–532. Available at https://doi.org/10.1016/S0046-8177(97)90073-3, http://www.elsevier.com/inca/publications/store/6/2/3/1/3/9/index.htt.

Holland, J., & Stevens, N. (2021). Guidelines for the development of multiple-choice items and assessments. Royal College of Surgeons in Ireland.

Jane, H., Niall, S., 2021. Guidelines for the Development of Multiple Choice Items & Assessments. Royal College of Surgeons in Ireland. Available at https://doi.org/10.25419/rcsi.13947164.v1.

Moss, E., 2001. Multiple choice questions: their value as an assessment tool. Current Opinion in Anaesthesiology 14 (6), 661–666. Available at https://doi.org/10.1097/00001503-200112000-00011.

Salam, A., Yousuf, R., Bakar, S.M.A., 2020. Multiple choice questions in medical education: how to construct high quality questions. International Journal of Human and Health Sciences (IJHHS 4 (2), 79–88. Available at https://doi.org/10.31344/ijhhs.v4i2.180.

Sam, A.H., Hameed, S., Harris, J., Meeran, K., 2016. Validity of very short answer versus single best answer questions for undergraduate assessment. BMC Medical Education 16 (1). Available at https://doi.org/10.1186/s12909-016-0793-z, http://www.biomedcentral.com/bmcmededuc/.

Yarhere, I.E., & Akadiri, A.O. (2023). Quality assurance of assessment in medical education: A faculty workshop for facilitators by University of Port Harcourt's College of Health Sciences. IOSR Journal of Research & Method in Education (IOSR-JRME), 13 (2), 38–44.

Chapter 24

# Assessment of clinical skills

*"Medicine is learned by the bedside and not in the classroom. See and then reason and compare and control. But see first."*
—William Osler

## 24.1 Introduction

Assessment of clinical skills in medical education refers to the process of determining a student's ability to apply medical knowledge to patient care, including communication, history-taking, physical examination, and procedural skills (Cox et al., 2007). Some of the most important tools include methods, such as objective organized clinical assessments, direct observation of procedural skills (DOPS), and mini-clinical evaluation exercises (Mini-CEX), where students must perform tasks in a real or simulated clinical setting for feedback and observation. These tests are designed to be valid, reliable, and unbiased. In order to help students become competent and prepare for real clinical practice, they provide them with structured feedback and opportunities to improve their self-esteem.

## 24.2 Need for assessment of clinical skills

In order to guarantee competency, quality, and patient safety, clinical skill assessment is essential for medical education and professional development. This assessment of clinical skills ensures that graduates satisfy professional standards and confirms that they can use knowledge and judgment to patient care. It also provides crucial feedback for improvement (Miller, 1990). Modern methods, including OSCEs, are preferred over conventional lengthy and short case exams since they are more valid and accurate in measuring skills consistently across different circumstances. Evaluation of clinical skills is essential because it guarantees that practitioners and students have the abilities required to safely diagnose, treat, and care for patients. Additionally, it promotes professional development since students' professional development is greatly aided by clinical and soft skills that a student imbibes during the clinical years of their study. Additionally, it offers comprehensive feedback on areas in which a person needs to grow, which is essential for learning and skill development. In the end, it ensures that graduates fulfill the prerequisites before being granted a license or permitted to work independently.

## 24.3 Principles of clinical skills assessment

The principles of clinical skills assessment ensure that evaluations are meaningful, equitable, and effective for training and practice. These principles can be categorized into two main areas, namely, the core psychometric principles and ethical and practical principles. By adhering to these principles, educators and regulatory bodies can develop credible and defensible clinical skills assessments that protect the public and foster the development of competent clinicians. The accuracy, consistency, and fairness of assessments of are guaranteed by fundamental psychometric concepts—namely validity, reliability, and standardization—in clinical skill assessment. On the other hand, beyond psychometrics, several other principles like fairness, authenticity, sufficiency, and feedback of learning guide the practical and ethical application of clinical skills assessment.

### 24.3.1 Core psychometric principles

#### *24.3.1.1 Validity*

A fundamental psychometric principle of clinical skill assessment is validity, which guarantees that an assessment measures the precise clinical skills it purports to measure, thereby offering a reliable and significant foundation for critical

**The Principles and Practice of Medical Education. DOI:** https://doi.org/10.1016/B978-0-443-44197-4.00036-9

decisions regarding a clinician's competency and patient care. As it verifies that the assessment tool is actually testing the required skills, such as communication, physical examination, and clinical judgment, validity indicates the accuracy of measurement (J., & Ziv, 2003). Additionally, validity highlights how ensuring an evaluation is important and helps in minimizing bias and guarantees equity for all students. Using reliable assessments provides legal defensibility for decisions made in professional sectors.

#### *24.3.1.2 Reliability*

In clinical skill evaluation, reliability guarantees consistent, trustworthy outcomes over time and among various raters. Clinical skill assessment reliability measures whether an assessment produces comparable results under comparable circumstances and guarantees consistent, predictable, and repeatable results. It is maintained by standardized checklists, rater training, and the use of numerous observers or stations to minimize subjectivity. It is essential for evaluating actual proficiency as opposed to chance performance. Clear rubrics, extensive rater training, uniformity, and the use of several low-stakes tests are all necessary for making correct clinical judgments, particularly for high-stakes choices like graduation.

#### *24.3.1.3 Standardization*

In clinical skill assessment, standardization refers to the use of consistent techniques, such as standardized patients and OSCEs, to objectively measure learner competence in areas such as history taking, physical examinations, and communication. This ensures fairness, reliability, and comparability across various settings and learners by reducing variability in patient encounters and evaluation criteria, which is essential for high-stakes testing and certification.

### 24.3.2 Ethical and practical principles

#### *24.3.2.1 Fairness*

A fundamental ethical criterion in clinical skill assessment is fairness, which guarantees that assessments of competence are devoid of prejudice, discrimination, and unrelated elements, treating all students equally. It is a multifaceted concept that guarantees that tests appropriately gauge necessary clinical competencies while guarding against the unjust disadvantage of students due to characteristics like gender, color, or disability. In terms of fairness evaluations must not bias against any person or group on the basis of characteristics unrelated to the abilities being evaluated. Fairness is ensured by inclusive assessment techniques and plain, unambiguous standards.

#### *24.3.2.2 Authenticity*

Since it guarantees that assessments accurately represent actual clinical practice, authenticity in clinical skill evaluation is seen as an ethical principle that upholds the obligation to produce competent, safe professionals while honoring the patient–provider relationship. It promotes moral, patient-centered care by bridging the gap between theoretical understanding and real-world execution. The assessment tasks should reflect real-world clinical situations and demands. Assessing a skill in a context-relevant manner provides more meaningful evidence of a clinician's competence.

#### *24.3.2.3 Sufficiency*

In the context of ethical clinical skill assessment, the term sufficiency refers to the need for students to have the knowledge and abilities necessary to deliver safe, effective, and compassionate care while avoiding patient damage. It bridges the gap between theoretical knowledge and real-world patient care by acting as a threshold principle in competency-based education. The assessment must gather enough evidence to make an informed and confident decision about a learner's capabilities. This often involves using multiple assessment methods to obtain a comprehensive view of a student's skills.

#### *24.3.2.4 Feedback for learning*

A fundamental ethical principle in clinical skill evaluation is feedback for learning such as a formative feedback, which moves the emphasis from just evaluating performance to promoting ongoing professional development and guaranteeing patient safety. Based on the moral precepts of beneficence and nonmaleficence, it is regarded as a crucial component for learner development. Assessment should not only confirm standards but also provide timely, specific, and constructive feedback to help students improve their skills. This principle emphasizes the formative role of assessment in professional development.

#### 24.3.2.5 *Acceptability and cost-effectiveness*

Acceptability and cost-effectiveness are becoming more widely acknowledged as crucial, practical, and moral elements of contemporary clinical skill evaluation, going beyond just technical psychometrics to address sustainability, equity, and resource management. They serve as useful considerations and as guidelines that guarantee evaluations are equitable for students and sustainable for organizations (Fig. 24.1).

## 24.4 Clinical competencies

The information, abilities, judgment, and attitudes necessary to deliver safe, efficient, and moral patient care are known as clinical competencies in medicine. These competencies, which encompass patient care, medical knowledge, practice-based learning, systems-based practice (SBP), professionalism, and interpersonal and communication skills, are grouped under frameworks such as the ACGME core competencies. In order to enhance patient outcomes and professional satisfaction, maintaining clinical competence requires ongoing education, introspection, and keeping up with medical developments. Below is a description of some of the key competencies of clinical skills.

### 24.4.1 Patient care

The ability of a healthcare professional to use their knowledge, abilities, and professional behaviors to deliver safe, efficient, and patient-centered care is referred to as patient care as a competency. In order to enhance well-being and achieve favorable health outcomes, it entails a comprehensive strategy that includes professional skill, communication, compassion, ethical practice, and an awareness of patient requirements. In the intricate and ever-changing healthcare system, maintaining patient care competence requires constant self-evaluation, continuing education, and a dedication to quality improvement.

### 24.4.2 Medical knowledge

Medical knowledge is a combination of thorough comprehension of both basic and clinical sciences, including the application of scientific ideas to evidence-based practice and clinical decision-making. A fundamental prerequisite for healthcare workers is medical knowledge, which entails gathering and using up-to-date biological, clinical, social-behavioral, and epidemiological knowledge to patient care. In order to ensure efficient and secure patient care, it places a

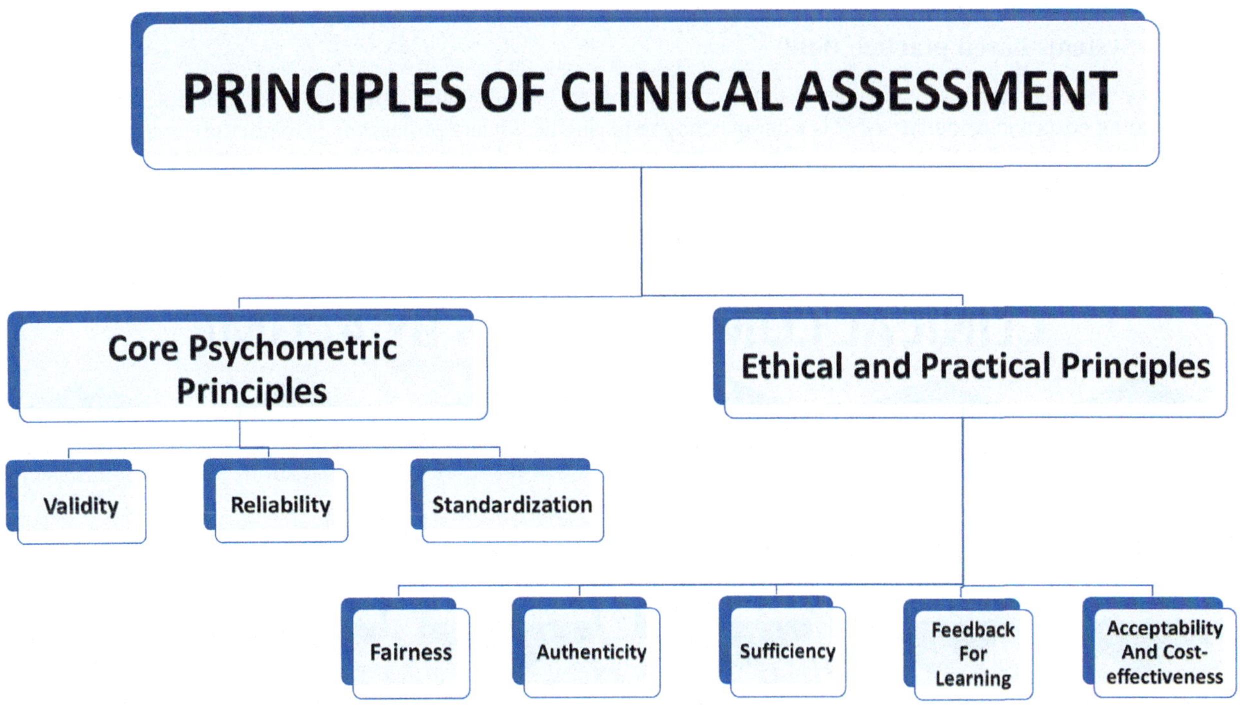


FIGURE 24.1 **Principles of clinical assessment.** This figure shows the principles of clinical assessment.

strong emphasis on using this knowledge to distinguish between health and illness and updating it frequently to reflect new developments in medical knowledge.

### 24.4.3 Interpersonal and communication skills

The capacity to build trust and cooperation via effective communication with patients, families, and other medical professionals is one of the important skills that the learner should acquire. As a competency, interpersonal and communication skills include the capacity to successfully engage with others in order to exchange information, establish connections, and work together to accomplish objectives. Active listening, effective verbal and nonverbal communication, empathy, and conflict resolution are all essential for creating a healthy work environment, increasing productivity, and developing careers.

### 24.4.4 Practice-based learning and improvement (PBLI)

A key skill in medicine is PBLI, which emphasizes lifelong learning and ongoing self-evaluation to enhance patient care. It is indeed a dedication to lifelong learning, which includes introspection, the pursuit of new information, and gradual practice improvement (Faraj Pour et al., 2022). It entails evaluating scientific information, spotting knowledge gaps, cultivating critical thinking skills, incorporating evidence into clinical judgments, and implementing a cycle of introspection and development. Accessing and evaluating medical literature, comprehending research design, using information technology, and supporting others' learning are important components.

### 24.4.5 Professionalism

Professionalism is characterized by adherence to ethical codes, accountability, compassion, and a commitment to patient well-being, privacy, and autonomy. Professionalism as a competency involves consistently displaying work ethic, integrity, and excellent work habits while behaving ethically and acting in the interest of your community or profession. This competency encompasses various attributes such as accountability, responsibility, reliability, and a commitment to excellence and continuous learning, demonstrating a high standard of personal and professional behavior to achieve quality results.

### 24.4.6 Systems-based practice (SBP)

The ability to understand and navigate healthcare systems to provide safe, effective, and efficient care, including understanding costs and resources. SBP is a competency requiring healthcare professionals to not only understand but also effectively use complex healthcare systems to provide optimal patient care (Colbert, 2011). It involves being aware of the larger healthcare context, including social determinants of health, to improve quality, safety, and cost-effectiveness. SBP also emphasizes collaborative, multidisciplinary teamwork and the ability to call on other resources within the system to advocate for patients (Fig. 24.2).

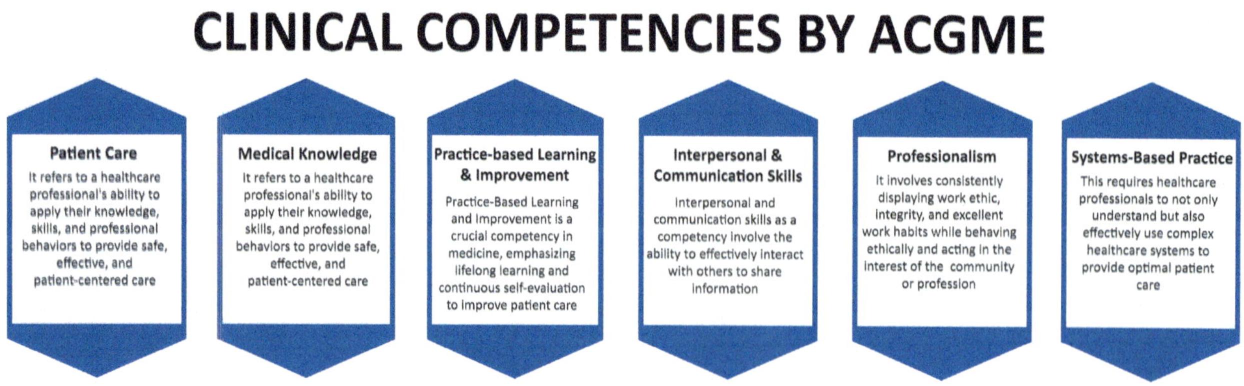


FIGURE 24.2 **Clinical competencies by Accreditation Council for Graduate Medical Education (ACGME).** This figure depicts the clinical competencies by ACGME.

## 24.5 Frameworks for clinical skill assessment

Frameworks for clinical skill evaluation provide a structured approach to evaluating a clinician's skills using methods such as clinical competency frameworks that outline specific skills and the utility formula that balances reliability, validity, and feasibility. Other frameworks include performance-based examinations, such as the clinical and professional skills assessment, and systematic methods, such as the airway, breathing, circulation, disability, and exposure (ABCDE) assessment, for critically ill patients. These frameworks are designed to guide the evaluation process, ensure quality, and aid in skill development.

### 24.5.1 Clinical competency frameworks

In medical education, clinical competency frameworks are outcome-based methods of training and evaluation that concentrate on the observable skills needed for professional practice. They change the focus from training time to the display of particular knowledge, skills, attitudes, and abilities. The American Medical Association's model, which covers technical skills like suturing, professionalism, and clinical decision-making, and CanMEDS, which defines roles like medical expert and communicator, are two instances of clinical competency frameworks in medical education (Frank and Danoff, 2009). Frameworks frequently classify competences into domains including knowledge, skills, and professional behaviors like communication and teamwork. They may also incorporate more recent topics like sustainable practice and climate change awareness. In clinical skill evaluation, a competency-based framework is used to specify the particular abilities, knowledge, and behaviors needed for a position, offering a uniform standard by which to evaluate a clinician's performance. By emphasizing real-world performance rather than merely theoretical knowledge and connecting advancement to mastery rather than time invested, it enables customized instruction. In the end, this strategy seeks to guarantee that clinicians are prepared for practice and capable of providing high-quality patient care by setting clear objectives for both students and teachers and facilitating focused feedback.

### 24.5.2 Millers pyramid

In 1990, George Miller introduced a new method for assessing clinical competency. Miller argued that the traditional approach to assessing medical students placed too much emphasis on their knowledge and not enough on how they would behave in an actual consultation. His "Miller pyramid" paradigm laid the groundwork for a dramatic change from the traditional Flexnerian medical school approach, which was mostly made up of assessments based on theoretical knowledge, to tests based on clinical performance (Hampton et al., 2024).

Each layer of Miller's pyramid, a hierarchical framework for clinical skill assessment, indicates the degree of a learner's skill development. Each tier establishes the foundation for subsequent tiers and indicates a higher level of clinical development. "Knows," the base of the pyramid and the most basic level of clinical evaluation, examines a learner's foundational knowledge of a particular subject and is often assessed using objective tests. The next level, "knows how," assesses how well a student applies their subject-matter knowledge through the collection, analysis, and synthesis of management plans. The third level, "shows how," evaluates a learner's performance in a clinical environment with a patient.

### 24.5.3 Advantages of Miller's pyramid in clinical assessment

Miller's pyramid provides several advantages in clinical assessment, including providing a clear framework for understanding skill growth, enabling evaluation that is realistic to real-world practice, and guiding learning by linking testing to specific developmental phases. Additionally, it facilitates the use of a range of assessment methods appropriate for different skill levels and promotes autonomous thought and practice. Among the main benefits are the following:

#### *24.5.3.1 Structured progression of competence*

Miller's pyramid's planned progression of competence, which provides teachers with a clear, hierarchical framework to connect learning objectives and evaluation methods with expected student advancement, is one of its primary benefits. This framework ensures a logical and thorough development of skills by guiding students from basic knowledge to independent practice. In this, competencies can be developed gradually through the pyramid's four levels, namely, knows, knows how, shows how, and does.

#### 24.5.3.2 Authentic assessment

Since its top level, the "does" concentrates on evaluating a learner's performance in actual clinical settings. Miller's pyramid provides the advantage of authentic evaluation. This emphasizes that practice-based evaluation is the cornerstone of authentic assessment. One of the primary benefits of the authentic assessment is that it provides a real-world validity. In contrast to artificial test scenarios, the "does" level assesses what trainees do on a daily basis, which is the most authentic and valid measure of their ability to successfully perform their professional role. It also allows for the analysis of several important abilities, including professionalism, communication, and teamwork, which are difficult to assess using either simulations or conventional written exams.

#### 24.5.3.3 Alignment of learning and assessment

The fact that Miller's pyramid provides a structured framework for linking learning objectives with appropriate evaluation methods at each stage of competency development is one of its key benefits. This alignment ensures that the learning objectives are appropriately defined since Miller's pyramid helps educators define specific, hierarchical learning outcomes, from basic knowledge to real performance in practice. It also guarantees that the methods are appropriate for the level by helping teachers select assessment strategies appropriate for each competency level.

#### 24.5.3.4 Guidance for learning

Miller's pyramid provides a structured framework that guides the entire learning process and helps to coordinate teaching and assessment methods by recognizing incremental levels of ability. This learning structure's primary advantage is that it ensures a clear, sequential progression from basic knowledge to useful performance. Miller's pyramid works well as a learning guide because it enables educators and learners to precisely specify learning goals and anticipated results at every stage of the learning process. Moreover, it ensures a structured process in which learning advances iteratively. This implies that the development of practical abilities and behavior at the higher levels is based on the foundational information at the lower levels.

#### 24.5.3.5 Flexibility in assessment methods

Miller's pyramid provides a particularly flexible framework since its hierarchical design easily accommodates a range of assessment methods, from traditional written exams to direct observation in real-world clinical settings. This flexibility is a major benefit since it enables teachers to match assessments to learning objectives by selecting the best assessment technique for the particular level of competency being assessed. Additionally, Miller's pyramid offers a thorough picture of a learner's abilities by encouraging the use of several evaluation techniques.

#### 24.5.3.6 Identifies gaps in performance

One of the important benefit and use of Miller's pyramid in clinical assessment is its ability to identify performance gaps, especially when employing assessment techniques aimed at the higher levels of the pyramid. This is made feasible by the pyramid's hierarchical structure, which progresses from knowledge to action and permits focused evaluation at every stage, making it simpler to identify certain areas in which a learner might be having difficulty.

#### 24.5.3.7 Promotes competency-based medical education

Miller's pyramid is a perfect fit for competency-based medical education since it offers an organized framework that is very helpful for gaining and utilizing competencies. In this sense, a clear, sequential pathway for learning objectives is created by the pyramid's four levels: knows, knows how, shows how, and does. The objectives of competency-based medical education are directly reflected in this path, which takes students from basic knowledge to real-world performance. Additionally, by assisting instructors in choosing suitable assessment instruments for every competency level, it harmonizes assessment techniques. For instance, the higher "shows how" and "does" levels, where observable behavior is crucial, use 360-degree evaluations or DOPS abilities, while multiple-choice questions are appropriate for the "knows" level.

### 24.5.4 Dreyfus developmental framework

The brothers, Stuart and Hubert Dreyfus, created the Dreyfus developmental framework, also known as the Dreyfus model of skill acquisition (Batalden et al., 2002). This model explains how learners advance through five stages of proficiency when learning a new skill. The five levels of advancement of a learner include the stages of novice, advanced beginner, competent, proficient, and expert. This paradigm, which shows a transition from rule-based, analytical thinking in the

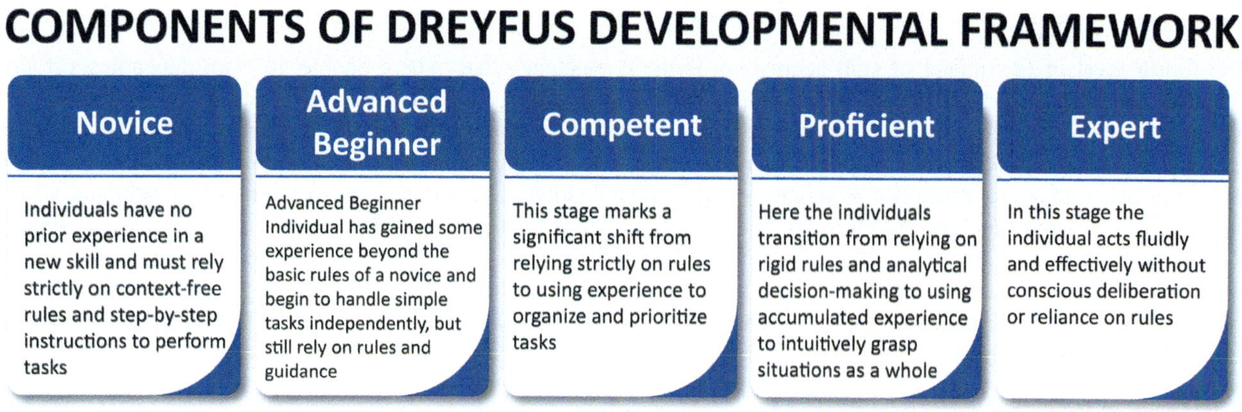


FIGURE 24.3 **Components of Dreyfus developmental framework.** This figure shows the components of Dreyfus developmental framework.

early stages to intuitive, experience-driven decision-making in the later stages, is a useful tool for educators to customize training and promote the growth of learner. The following is a description of the five stages (Fig. 24.3):

### *24.5.4.1 Novice*

The Dreyfus developmental framework's foundational level, known as the novice stage, requires people to rely solely on context-free rules and detailed instructions to complete tasks because they have no past experience with the new ability. At this point, students rely on rules and processes that are independent of context in order to complete activities. They don't comprehend the subtleties of various circumstances and simply follow directions. One of the key characteristics of this stage is that the novice exhibits rule-driven behavior, which involves operating by strictly adhering to a predetermined set of rules. They don't have the experience to use discretion or adjust when circumstances don't exactly follow directions.

Additionally, the learner views tasks in isolation and is unable to identify the "bigger picture" or comprehend the context in which material makes sense due to their inadequate contextual understanding. This phase also makes supervision necessary because a student with little expertise needs thorough guidance and constant oversight by a mentor or teacher. Students must make mistakes, get real-world experience, and start to identify situational nuances and recurring patterns beyond the strict principles they first acquired in order to develop from the novice stage to the advanced beginner level.

### *24.5.4.2 Advanced beginner*

According to the Dreyfus developmental framework, the second level of skill acquisition is the advanced beginner stage, where a student has acquired some experience beyond the fundamental guidelines of a novice. They are able to identify situational components and start managing basic tasks on their own, but they still depend on rules and might require assistance in more complicated circumstances. They advance beyond a novice's strictly rule-based approach at this phase, gaining expertise as they start to identify recurrent components and patterns in a skill. The advanced beginner begins to use situational and context-free cues, attempting to combine fresh situational knowledge with the rules they learnt as a novice to direct their behavior. At this point, students are able to recognize several elements of a situation that may call for a different set of rules to address the issue.

### *24.5.4.3 Competent*

The third stage of skill development, known as the competent stage, is when people start to accept accountability for their actions and results and develop their ability to plan and solve problems. At this point, there is a noticeable change from carefully following rules to organizing and prioritizing work based on experience. The learners now have a holistic recognition, which allows them to transition from viewing tasks as discrete processes to having a broader, holistic picture of the problem. This is one of the significant improvements observed throughout this stage. In this phase, the learner is able to efficiently manage their work, prioritize pertinent components, plan, and arrange. A student must take the risk of letting go of strict rules and start to trust their developing intuition in order to advance past the competent level. This shift to the skilled stage entails gaining a deeper comprehension and the capacity to recognize, without conscious thought, what a circumstance requires.

#### 24.5.4.4 *Proficient*

The fourth level of this model of skill acquisition is the competent stage, where people go from depending on strict guidelines and analytical judgment to employing collected experience to intuitively understand circumstances as a whole. Although they view the problem holistically and based on experience, proficient learners nevertheless utilize analytical thinking to determine a path of action. This stage is characterized by the learner's intuitive understanding of situations, when they use their prior experiences and profound situational involvement to intuitively understand what is happening in a given scenario. Additionally, they employ experience-based maxims or guiding concepts that adjust to various settings rather than rigidly adhering to context-free norms. Even if they have an intuitive grasp of a problem, they nonetheless choose a course of action after carefully considering all of their options. In order to get to expertise, people at the proficient level must continue to gain a great deal of experience and risk abandoning rule-based decision-making in favor of fully relying on their developing intuition. At this point, experienced mentoring can be especially helpful in guiding this change.

#### 24.5.4.5 *Expert*

The expert stage is the fifth and final stage of the Dreyfus developmental framework, where performers possess an intuitive and holistic grasp of situations, allowing them to act fluidly and effectively without conscious deliberation or reliance on rules. Experts are recognized as go-to resources by their peers and often innovate in their domain. At this stage, emphasis shifts from rules to intuition, as experts move beyond reliance on guidelines or maxims and make decisions based on deep, tacit understanding developed through extensive experience. Additionally, when students become experts, their awareness shifts to higher-level strategic issues and they get completely engrossed in the work. An experienced learner may apply analytical techniques to new or complicated situations, come up with creative solutions, coach others, and make a substantial contribution to their area. (Peña, 2010) Reaching this level necessitates a great deal of experience and practice in a range of contexts, as well as the readiness to abandon a deliberate, rule-based approach and have faith in one's developing intuition. The primary distinction between an expert and a proficient performer is that the expert's actions also become instinctive and fluid, eliminating the necessity for deliberate action selection.

## 24.6 Methods of clinical skills assessment

The methods of clinical skill assessment can be classified as "structural assessments" and "work-based and simulation methods." These are depicted in the following figure (Fig. 24.4).

### 24.6.1 Structured assessments

A structure-based clinical assessment is a methodical, standardized approach that uses a preset format to guarantee consistency and impartiality when assessing a person's clinical abilities, knowledge, and professional behaviors. The OSCE is the most widely used example in medical education.

### 24.6.2 Objective structured clinical examination (OSCE)

A standardized test used in healthcare education to gauge a student's clinical proficiency in a simulated setting is called an OSCE. With the assistance of actors or standardized patients, students move through a sequence of stations where they carry out particular activities, like obtaining a patient's history, conducting a physical examination, or interpreting test results. This method ensures impartiality and consistency in the assessment of a variety of competencies by employing rating scales or checklists to evaluate performance (Harden et al., 1975).

#### 24.6.2.1 *Key features of an objective structured clinical examination*

##### 24.6.2.1.1 Standardized stations

The OSCE's standardized stations, which require all applicants to complete similar, predetermined clinical activities in a set amount of time under the same conditions, are a key component that guarantees fair and comparable assessments. Examiners use standardized checklists to objectively assess performance at these stations, which may involve procedures or obtaining a history. This method shifts away from conventional, possibly subjective tests and concentrates on reliably and consistently evaluating clinical abilities and competencies (Hetaimish and Elbadawi, 2015).

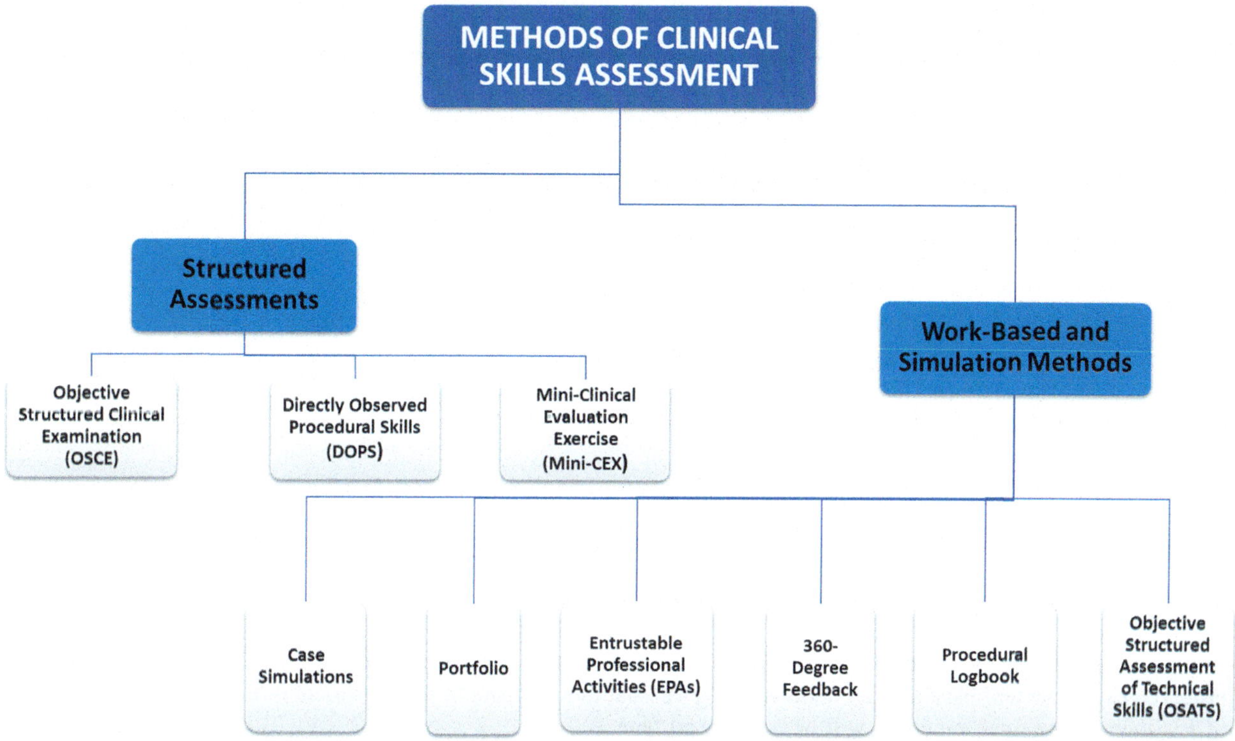


FIGURE 24.4 **Methods of clinical skills assessment.** This figure depicts the methods of clinical skills assessment.

#### 24.6.2.1.2 Simulated environment

One fundamental and crucial component of the OSCE is the utilization of a simulated environment. This setting is essential for offering a consistent, secure, and impartial way to evaluate clinical skills without endangering actual patients. Replicating real-life clinical scenarios in a controlled context is one of the most crucial roles of a simulated environment in an OSCE. Because of this, there is a benefit in terms of standardization and each candidate is exposed to the same situation, instructions, and patient reactions through manikins or standardized patients. This lessens examiner bias while testing with real patients by ensuring consistency and fairness in assessment. Additionally, it maintains safety since it gives the learner a risk-free environment to practice and make mistakes. It is safe to evaluate complex or urgent situations without jeopardizing patient safety, such as handling a cardiac arrest or breaking bad news.

#### 24.6.2.1.3 Objective assessment

In contrast to traditional exams where bias or variability could affect results, objective assessment in an OSCE uses standardized, timed stations with checklists and standardized patients to evaluate specific clinical skills like history taking, physical examinations, and communication for all candidates identically. This ensures fairness, reduces subjectivity, and allows precise, comparable scoring. The use of rating scales and specific, premade checklists created by the examiners to rate performance on actions, behaviors, and communication is another crucial component of the objective assessment. This ensures uniform grading for all candidates.

#### 24.6.2.1.4 Broad skill assessment

One important aspect of the OSCE is its broad skill assessment. The multistation style of the OSCE is intended to cover a wide variety of clinical abilities in a standardized and objective manner, in contrast to older clinical tests that frequently concentrated on a restricted range of skills with considerable variability. The OSCE's capacity to evaluate a learner's variety of skills is one of its qualities that has grown in significance. Each station in this format can be designed to evaluate a particular, targeted skill set, enabling the overall exam circuit to cover a wide range of talents. History taking, physical examination methods, counseling, giving bad news, getting consent, problem-solving, and clinical reasoning are some of the crucial abilities that are examined.

#### 24.6.2.1.5 Structured format

The structured format of an OSCE is a methodical and uniform organizational structure intended to guarantee impartiality and justice in the evaluation of clinical competency. This framework entails dissecting the wide range of clinical skills into distinct, manageable parts, which are then evaluated uniformly for each candidate. It is distinguished by the existence of several stations, with candidates often rotating from five to ten stations. Every station gives each applicant the same clinical task or scenario, which could include data interpretation, a real patient, a highly trained standardized patient, or a manikin.

### *24.6.2.2 Conduction of the objective structured clinical examination exam*

Using standardized patients, mannequins, or materials, candidates rotate through a circuit of timed stations during an OSCE. Each station tests a particular clinical skill, such as taking a history, performing physical examinations, or interpreting data. The various stages of conduction of the OSCE exam are described below:

#### 24.6.2.2.1 Planning and organization phase

An OSCE's planning and organization phase is a complicated, multistep procedure that guarantees the assessment's uniformity, dependability, and fairness. It covers a number of important topics and usually starts well before the exam date.

#### 24.6.2.2.2 Design a blueprint

Making a blueprint that links the capabilities to be tested such as taking a history, doing a physical examination, and communicating effectively to the course or module's learning objectives is the first stage in the planning and organization phase. The examiner will have a better understanding of what to anticipate from the student in this step. Additionally, the OSCE's themes and question count will be determined here.

#### 24.6.2.2.3 Form a committee

Creating a committee is one method of organizing the execution of an objective, systematic clinical examination. The entire design, validity, reliability, and execution of the OSCE fall within the purview of this committee, which is made up of the course coordinator, members of the student assessment committee, and topic experts.

#### 24.6.2.2.4 Develop stations

The next stage would be to ascertain the quantity and kind of stations required, including whether they should be unmanned and require a written response or manned with an examiner present. In an OSCE, stations are generally classified as either dynamic or static, designed to test different facets of a candidate's clinical competence and knowledge within a timed, structured circuit.

### *24.6.2.3 Dynamic objective structured clinical examination stations*

Dynamic stations are interactive, "manned" stations that evaluate behavioral competency and practical therapeutic abilities. It uses a manikin, simulated patient, or standardized patient. The evaluation of history-taking, physical examination, counseling, and patient communication abilities is the primary focus of this kind of station. In order to obtain data, complete tasks, and establish rapport, the student must actively interact with the patient. For instance, the dynamic OSCE station could be a 10-minute station in which a student takes a history from a patient complaining of chest pain and explains the diagnosis.

### *24.6.2.4 Static objective structured clinical examination stations*

Static stations evaluate knowledge and interpretive abilities without involving patients. It often entails reading questions, looking at images, or doing data analysis. Here, the interpretation of tests such as arterial blood gases, chest X-rays, and ECGs is the main focus. This method is quite similar to responding to a multiple-choice or short-answer question. Instead of testing physical technique, these stations are frequently used to test data interpretation. For example: A station where the student is shown an ECG and a blood report must write down the diagnosis and management plan.

#### 24.6.2.4.1 Set time limits

While individual station time restrictions normally range from 6 to 20 minutes, depending on the particular activity and the examining body, the overall duration of an OSCE is approximately three hours for all stations combined.

#### 24.6.2.4.2 Create scoring criteria

To guarantee that every student is assessed using the same precise, predetermined performance criteria, it is crucial to create a preestablished, standardized checklist or scoring guide. Subject-matter experts should create the checklist, and the examiners should discuss it in detail well in advance. Clinical knowledge, technical proficiency, interpersonal and communication skills, problem-solving and decision-making, professionalism, and ethical behavior should all be included in the checklists.

### *24.6.2.5 Implementation phase*

The carefully prepared exam is actively given to the applicants throughout the implementation phase of an objective structured clinical assessment. In this stage, the stations are physically carried out, candidates are moved, and their performance is assessed in real time by standardized patients and qualified assessors.

#### 24.6.2.5.1 Set up the environment

The skill laboratories used to set up the OSCE exam should be large enough to hold the number of students taking the test. During the test, students are typically split up into two or three groups. In order to prepare for an OSCE, it is necessary to create discrete, quiet stations that mimic clinical settings using partitions or in separate rooms; make sure that all equipment mannequins, vitals monitors, and drug trolleys are present and operational. In addition, it is important to prepare standardized patients with clear signage and logistics for candidate flow.

#### 24.6.2.5.2 Provide instructions

Instructions for the OSCE focus on organized clinical evaluation; for example, the instructions include informing the students of the exam structure, a dress code, and the need to attend early. It's crucial to be aware of each station's duration, thoroughly study the directions, adhere to time constraints, and act professionally.

#### 24.6.2.5.3 Candidate rotation

Candidates move through a series of independent "stations" that each concentrate on a distinct clinical assignment or scenario during an OSCE. This procedure guarantees that every student encounters the same obstacles under uniform circumstances, improving impartiality and fairness in evaluation. All candidates are signaled to go to the next station in the circuit in a predetermined, structured order at the conclusion of the allocated time, which is typically 6–8 minutes. Typically, a short movement or "changeover" period of 1–2 minutes is included. This period may also contain a "reading time" to go over the instructions for the subsequent station.

#### 24.6.2.5.4 Monitoring and assessment

In an OSCE, a candidate's clinical competencies are monitored and assessed using an organized, standardized approach with trained examiners and certain criteria. In this regard, checklists are considered as the main tools. The assessors directly watch the candidate's performance as they engage with a patient, mannequin, or data to make sure the monitoring process runs properly. Furthermore, every station has a rigorous time constraint that is the same for every candidate. The entire test environment is managed, with security measures in place to prevent applicants from communicating about particular station material, to guarantee that there is no setback during the procedure.

### *24.6.2.6 Feedback and evaluation phase*

An OSCE's feedback and evaluation phase, which focuses on organized assessment and focused feedback delivery, is an essential part of learning and professional development. In order to help students realize their strengths and flaws, this step entails giving them clear pass scores, comprehensive results, and a variety of feedback.

#### 24.6.2.6.1 Documentation

Examiners rate candidates' performance at each station using structured marking tools in order to document OSCE scores. The data are then centrally collated to produce an overall result. Using checklists, which include a comprehensive list of precise actions expected of the applicant, such as presenting yourself to the patient and verifying the patient's basic information, is the most popular approach. Using a binary or multipoint rating system, the examiner indicates whether or not each stage was completed.

##### 24.6.2.6.2 Analysis

In this stage, the validity and reliability of the findings are examined. A psychometric assessment of the exam data is part of the analytic process for validity and reliability in an OSCE. Reliability is measured using important metrics like Cronbach's alpha and generalizability theory, whereas validity is evaluated by compiling data from several sources.

##### 24.6.2.6.3 Feedback and debriefing

Feedback and debriefing are crucial educational components of an OSCE, especially in formative assessments, since they help students identify knowledge gaps, improve future performance, and develop clinical reasoning skills. Debriefing is a formal, bidirectional, and cooperative reflection process, whereas feedback is a more unidirectional information flow. The unpredictability frequently present in traditional clinical examinations is eliminated by this methodical procedure of conducting OSCEs, which guarantees that each candidate faces the same challenges and is scored using the same objective standards.

## 24.6.3 Role of the examiner during the objective structured clinical examination

In order to ensure fairness and a consistent assessment of competency for patient safety, an examiner's job in the OSCE exam is to objectively evaluate a candidate's clinical skills, knowledge, and professionalism in a standardized manner while acting as an impartial observer and assessor using checklists. They watch how candidates interact with mannequins or standardized patients, evaluate their performance according to preestablished standards, and only step in when a candidate misinterprets the purpose of the station or endangers others. Using particular checklists and marking schemes to maintain consistency and fairness and reduce prejudice, as well as closely monitoring the candidate's history taking, communication, physical examination, and clinical reasoning at each station, are some of the crucial duties that must be adhered to by the examiner.

## 24.6.4 Role of the student during the objective structured clinical examination

A student's responsibility during an OSCE is to actively demonstrate integrated clinical skills in brief, timed stations, acting as a capable healthcare professional by taking a patient's history, conducting focused physical examinations, effectively communicating with standardized patients, solving clinical problems, and documenting findings all while managing time and upholding professional standards like informed consent and hand hygiene.

## 24.6.5 Advantages of the objective structured clinical examination

The OSCE's primary benefits are its standardization and objectivity, which guarantee fairness by evaluating each candidate on the same tasks under the same conditions, minimize bias, and provide trustworthy, thorough feedback on a variety of clinical skills, including taking a history, performing a physical examination, and applying clinical reasoning (Lakshman, 2025). It provides a secure setting for practice with standardized patients or simulators, enabling the assessment of difficult skills without endangering patients. It also makes it easier to identify students' strengths and weaknesses for focused instruction and encourages introspection. Some of the prime advantages of the OSCE exam include the following:

### *24.6.5.1 Standardization and objectivity*

The OSCE's main benefits are in terms of standardization and objectivity. It maintains standardization through the presence of identical stations in which all candidates face the same, highly structured, time-bound stations, which can include simulation, history taking, or technical tasks. The OSCE also involves standardized patients ensuring that each candidate receives the same prompt and difficulty level. It upholds objectivity by making sure that a structured checklists is used which ensures that assessment is based on performance rather than the examiner's subjective impression which reduces examiner bias and boosts reliability and permits legitimate comparisons of clinical skills like communication, taking a history, and technical proficiency in a secure setting. Compared to typical exams, this methodical approach analyzes a wider variety of competencies, encourages self-reflection, and offers explicit feedback.

### *24.6.5.2 Comprehensive skill assessment*

The OSCE provides a comprehensive skill assessment by using a multistation approach to examine a wide range of talents that are difficult to evaluate using traditional approaches. An extensive assessment of numerous practical and nontechnical abilities necessary for clinical practice is made possible by this arrangement. In this sense, unlike traditional formats that might concentrate on a single case, a wide range of competences, including communication, clinical reasoning, technological skills, and ethics, are examined in a single exam.

#### 24.6.5.3 Safety and realism

The OSCE has significant advantages in terms of safety and realism in healthcare education and evaluation. It makes it possible to evaluate clinical skills in a secure, real-world environment, preparing trainees for real-world application without putting actual patients in danger. Moreover, it maintains safety by giving trainees a secure environment in which to carry out intricate procedures and interactions with the help of mannequins or standardized patients without endangering the lives of actual patients.

#### 24.6.5.4 Targeted feedback and learning

The OSCE format facilitates targeted feedback and learning through its structured, station-based approach and the potential for immediate, individualized feedback. This concrete feedback highlights the specific areas needing improvement, thereby fostering self-reflection and ultimately guiding ongoing professional development.

#### 24.6.5.5 Educational value

As a learning experience as well as an assessment tool, it offers both the teachers and students' substantial information which is of immense educational value. This enhances the instructional value by giving the instructors prompt, helpful feedback following each station, which is essential for seeing and fixing performance gaps in the teaching and learning process while the event is still fresh in their memories. Additionally, the methodical assessment procedure successfully identifies certain knowledge or skill gaps amongst the students necessitating focused interventions and customized learning programs at both the individual and curriculum levels.

#### 24.6.5.6 Improved reliability

One major benefit of the OSCE is increased dependability, which is made possible by standardization and the utilization of several brief stations. This strategy lessens the subjective variances included in conventional evaluation techniques. The consistency and reproducibility of an assessment's findings are referred to as its reliability. The exam results' overall validity and reliability are greatly increased by using a large sample of short stations to evaluate a wide variety of abilities and competences. Additionally, the thorough training of examiners and standardized patients contributes to uniformity in the presentation of scenarios and the assessment of each candidate's performance.

### 24.6.6 Challenges of objective structured clinical examination

The primary challenges of the OSCE stem from high stress and anxiety, immense time pressure, and the unfamiliar format of being observed while performing clinical tasks. The challenges of the OSCE conduction can affect both the students and the logistics involved in the exam.

#### 24.6.6.1 Challenges for students

##### 24.6.6.1.1 Exam stress

Exam stress is a significant problem for medical students undertaking OSCE due to the high stakes, timed, and observed nature of the stations, often creating higher anxiety than written tests. This severe real-life performance anxiety can cause panic attacks, hamper memory recall, and make it difficult to demonstrate actual clinical skills.

##### 24.6.6.1.2 Time management

Due to short station durations, time management is a significant difficulty in OSCEs. This can result in incomplete assignments, lost grades, and high levels of stress, which are frequently brought on by overthinking, inadequate preparation, distractions, and insufficient practice. Candidates find it difficult to strike a balance between patient-centered care, efficiency, and safety in a matter of minutes. This circumstance emphasizes the necessity of concentrated practice and wise time management in order to be successful.

##### 24.6.6.1.3 Unfamiliar format and environment

Students who are unfamiliar with the OSCE format and atmosphere face considerable problems, chief among them being increased worry and tension, which can impede their capacity to exhibit their actual clinical competence. Uncertainty about the procedure and expectations, rather than a lack of expertise or understanding, may also cause students to do less well. This difficulty may be made worse by a lack of experience with a simulated test setting or "mock" OSCE.

##### 24.6.6.1.4 Communication skills

Beyond clinical expertise, the test thoroughly evaluates soft qualities like professionalism, empathy, and communication, all of which can be challenging to develop and execute under examination stress. The applicants' capacity to establish rapport, completely sympathize, and offer sufficient explanations and clarification may be hampered by the tight time limit per station, which affects communication skills. Additionally, the high-stakes nature of the test, the presence of performance anxiety and tension, and being constantly watched by examiners or videotaped can all contribute to a great deal of stress and anxiety, making it challenging to remain composed and perform naturally.

##### 24.6.6.1.5 Fear of failure

A major psychological obstacle in the OSCE is fear of failing, which can cause high levels of stress and anxiety and have a detrimental effect on a student's real performance. Students develop a fear of failing due to a variety of factors, including stress and worry, time limits, the urge to do better, and the dread of making mistakes.

##### 24.6.6.1.6 Inconsistent examiner behavior

Since the evaluation format mostly depends on the subjective judgment of several examiners, inconsistent examiner conduct is a well-known and substantial difficulty for the OSCE. The fairness and dependability of the test results may be jeopardized by this fluctuation, which is sometimes referred to as examiner stringency or leniency.

#### *24.6.6.2 Institutional and logistical challenges*

A popular, effective, and trustworthy method for assessing clinical competency in healthcare fields is the OSCE. Despite its advantages, OSCE implementation is fraught with institutional and practical difficulties, especially in situations with low resources or developing nations. Institutions implementing the OSCE also face challenges, which can indirectly affect students.

##### 24.6.6.2.1 Resource intensity

Adequate facilities, equipment, faculty training, and the recruitment and training of standardized patients are all necessary resources for conducting an objective organized clinical examination. Logistical complexity, or the requirement for extensive preparation to organize several stations at once, is one of the key components of resource intensity. To ensure uniformity among all candidates, each station requires the proper supplies, tools, and a particular configuration.

##### 24.6.6.2.2 Organization and planning

Students may become difficult and stressed out as a result of poor organization, such as unclear instructions, insufficient time between tests, or lengthy wait periods. Planning also entails obtaining sufficient physical space, several rooms or cubicles for stations, a substantial quantity of tools and supplies, including manikins, instruments, and supplies, as well as a sizable number of skilled human resources, such as examiners, standardized patients, and support personnel.

##### 24.6.6.2.3 Standardization

The OSCE's main goal is standardization; however, putting it into practice is extremely difficult. The exam's dependability and impartiality may be compromised by bias and inconsistency brought about by variations in patients, examiners, and resources. Both students and teachers can enhance performance and results by comprehending and resolving these issues in order to better prepare for the OSCE.

### 24.6.7 Directly observed procedural skills (DOPS)

DOPS is a workplace-based assessment tool used in the healthcare industry. It involves a supervisor watching a trainee perform a clinical skill, such as a physical exam or a procedure, in a real-world setting. The supervisor provides prompt, constructive feedback and documents the performance using a structured checklist to ensure competency and direct future learning, bridging the gap between theoretical knowledge and practical application (Ahmad et al., 2024). It's a crucial formative assessment technique that uses organized observation, feedback, and improvement action planning to build practical competence.

## 24.6.8 Key features of directly observed procedural skills

### *24.6.8.1 Direct observation*

The DOPS framework requires a qualified assessor to physically see a trainee perform an entire practical operation in a real-world clinical setting. This allows the assessor to look at the performance's crucial intermediary processes, professionalism, communication skills, and adherence to safety protocols in addition to the final product. A skilled supervisor closely monitors a trainee completing a complete procedural skill in a real-world clinical situation, such as while doing a colonoscopy, drawing blood, or suturing a wound.

### *24.6.8.2 Structured assessment*

The assessor rates the trainee's performance on a set of criteria, including comprehension of the operation, getting consent, technical aptitude, communication skills, and postprocedure management, using a standardized checklist or form. One important and distinguishing characteristic of DOPS is structured assessment. It is essential to the tool's validity, reliability, and educational efficacy and sets DOPS apart from casual observation.

### *24.6.8.3 Immediate and structured feedback*

This is regarded as the most important aspect. The assessor and trainee have a confidential discussion about the performance right after the observation, usually within 5–10 minutes. This conversation analyzes strengths, points out weaknesses, makes recommendations for remedial actions, and settles on objectives for additional education.

### *24.6.8.4 Authentic context*

When it comes to DOPS, authentic context refers to the evaluation being conducted in a real-world clinical setting with genuine patients and equipment, as opposed to a simulated or artificial setting. This authenticity is an essential component that improves the assessment's validity and usefulness.

### *24.6.8.5 Conduction of directly observed procedural skills (DOPS) exam*

In medicine, the DOPS exam is an organized, workplace-based evaluation that is intended to assess a trainee's practical skills and offer prompt feedback for improvement. The conduction generally involves the following steps:

#### 24.6.8.5.1 Planning and preparation

The first phase in the planning and preparation process is the trainee's initiative in choosing a relevant clinical procedure to be evaluated and scheduling an appropriate time with a qualified and available assessor. Later on, choosing the technique is crucial. The method selected should be appropriate for the trainee's training level. The trainee is also in charge of setting up the relevant administrative activities in this step, such as making sure the assessor has the appropriate DOPS form or digital assessment instrument. Another crucial necessity at this point is getting the patient's genuine informed consent for both the procedure itself and the presence of an observer or assessor.

#### 24.6.8.5.2 Observation of the procedure

In this step, the assessor watches the trainee conduct the complete process in a real clinical setting such as a ward, clinic, or operating room. The accessor's lack of involvement is one of this step's key characteristics. During the observation, the assessor's role is often passive. Unless intervention is necessary for the patient's safety, they shouldn't take part in the procedure. In order to objectively record the trainee's performance across a variety of criteria, including professional approach and communication skills, applied knowledge, appropriate preparation, technical ability and procedural sequence, patient interaction and reassurance, and postprocedure management and documentation, a standardized, validated checklist tailored to the procedure is used.

#### 24.6.8.5.3 Feedback session

This feedback stage includes the assessor providing the trainee with structured, constructive feedback as soon as the procedure is finished. It also entails focusing on an organized conversation in which key topics are covered in detail, such as what was done successfully, suggestions for improvement, and a clear strategy for future development. It is crucial to involve the trainees throughout this period, have them reflect on their performance, and provide their own feedback and action plans. The observation and comment session is usually brief lasting 30–40 minutes.

#### 24.6.8.5.4 Assessment and documentation

In this stage, the assessor uses a predetermined scale to rate the trainee's performance for the pertinent criteria. The total comments and ratings are used to determine the outcome, which helps determine the trainee's level of competency. At the conclusion, the trainee and the assessor usually sign the DOPS form either electronically or physically to confirm the completion of the assessment and feedback. In order to demonstrate progress, the completed form is then turned in as part of the trainee's portfolio or logbook.

### *24.6.8.6 Advantages of direct observation of procedural skills*

Moving beyond conventional approaches to promote true skill acquisition through direct observation, formative evaluation, and targeted developmental plans, DOPS offers several benefits, including authentic assessment in real-time, immediate, targeted feedback, improved patient safety, learner confidence, and overall clinical competency. Several of the main benefits of this kind of evaluation include the following:

#### 24.6.8.6.1 Authentic and workplace-based

Authentic and workplace-based refers to the use of DOPS as an assessment method in which a supervisor observes a trainee carrying out a real clinical task that is authentic in a real workplace referring to workplace-based, offering immediate feedback for skill development rather than merely grading. Unlike high-stakes tests, it's about learning while working and concentrating on formative feedback to develop competency in technical and professional abilities.

#### 24.6.8.6.2 Immediate and specific feedback

The main benefit of immediate and targeted feedback via direct observation is that it enables students to clear up misunderstandings and improve their technique in real time, which immediately improves patient safety and skill development. Instead of focusing on broad impressions, direct observation enables the assessor to provide comments on particular, visible behaviors and activities. Since the learner is aware of precisely what needs to be reinforced or altered, the feedback becomes actionable.

#### 24.6.8.6.3 Enhanced patient safety

DOPS significantly increases patient safety by ensuring that medical trainees achieve and maintain high levels of clinical competency. This immediately leads to a decrease in medical errors and postprocedural sequelae. Furthermore, it has been repeatedly demonstrated that regular DOPS evaluations enhances trainees' overall procedural performance, information retention, and practical abilities. The likelihood of iatrogenic errors declines as trainees gain confidence and proficiency.

#### 24.6.8.6.4 Identifies strengths and weaknesses

A useful technique for evaluating and improving clinical competence is DOPS. Its primary benefit is the integration of instantaneous, organized feedback with real-time evaluation, which facilitates the efficient identification of strengths and flaws. In addition to technical abilities, DOPS offers a chance to evaluate and provide comments on critical nontechnical skills like professionalism, patient communication, and adherence to universal precautions.

#### 24.6.8.6.5 Fosters skill development

The main way that DOPS promotes skill development in a real-world clinical context is by providing prompt, organized, and behavior-based feedback. This method goes beyond theoretical understanding to evaluate and enhance real-world performance on the job, which boosts confidence and improves clinical competence.

#### 24.6.8.6.6 Motivates learners

Learners are motivated by DOPS in a number of ways, chief among them being the prompt, organized, and helpful feedback they receive in a real-world clinical context. The test has the benefit of guiding students toward particular, essential clinical skills because its content closely mirrors practical performance. Compared to conventional approaches, it also offers a clear route for progress, raising engagement and pleasure.

#### 24.6.8.6.7 Standardized and fair

When used with organized, prompt, and helpful feedback, DOPS contributes to a standardized and equitable approach to improving clinical abilities. For every procedure, DOPS uses particular, premade checklists that offer precise standards for what qualifies as competent performance. Regardless of the assessor or the particular clinical scenario, this structure guarantees that all trainees are evaluated using the same set of standards.

#### 24.6.8.6.8 Comprehensive skill assessment

DOPS is a comprehensive skill assessment that allows for the evaluation of multiple domains of competence and the rapid, focused feedback for development by comparing a trainee's performance in a real-world clinical environment to a standardized checklist. Unlike simulated or traditional written exams that might solely measure theoretical knowledge, direct observation of procedural skills evaluates performance in real clinical settings with real patients, providing an authentic depiction of a trainee's abilities.

### *24.6.8.7 Disadvantages of direct observation of procedural skills*

In addition to having problems with assessor bias, subjectivity in feedback, and limited scope, DOPS has disadvantages such as being time-consuming, resource-intensive, and stressful for learners, which may influence performance. Finding enough trained observers, integrating it seamlessly into hectic clinical days, and maintaining consistent grading standards, which frequently require further evaluations to get the whole picture, are challenges.

## 24.6.9 Mini-clinical evaluation exercise (Mini-CEX)

The Mini-CEX is a structured, short workplace assessment of a trainee's patient encounter. Its main features include a brief, focused interaction which usually lasts for 10–20 minutes, assessment across multiple core competencies such as communication, professionalism, judgment, and history taking, immediate formative feedback, and the use of multiple observers, settings, and occasions for a reliable, multifaceted evaluation of clinical skills and behaviors for improved learning (Nair et al., 2008). Through focused, evidence-based feedback in the hectic clinical setting, the Mini-CEX's main goal is to evaluate and enhance critical clinical skills and behaviors for high-quality patient care. Some of the important features of the Mini-CEX can be discussed below:

### *24.6.9.1 Brief and focused*

The term "mini" in Mini-CEX suggests that the interactions are short and targeted. It differs from the conventional, hours-long clinical evaluation exercise as here each observation of a doctor–patient encounter is intended to be brief, usually lasting 15 minutes, with an extra 5–10 minutes devoted to feedback. Additionally, the Mini-CEX concentrates on particular clinical skills or issues within a single encounter, such as history taking, physical examination, counseling, professionalism, or clinical judgment, rather than a thorough assessment of a single case. It is not necessary to evaluate every component in every session.

### *24.6.9.2 Direct observation*

A fundamental and important component of the Mini-CEX is direct observation. It is the fundamental mechanism that enables prompt, helpful feedback and the genuine evaluation of a trainee's clinical skills in a real-world situation. The Mini-CEX direct observation component necessitates that an assessor watches a trainee during a real-world patient encounter in a clinical setting, such as a ward or clinic. This guarantees that the performance being assessed accurately reflects the trainee's real-world professional practice.

#### 24.6.9.3 *Comprehensive competency assessment*

A crucial component of the Mini-CEX is a thorough competency assessment, which is intended to examine many fundamental clinical skills, attitudes, and behaviors necessary for providing patients with high-quality treatment. It assesses fundamental clinical abilities such as taking a history and doing a physical examination. Clinical judgment, decision-making, communication, professionalism, ethics, organization, and efficiency in counseling are some of the competencies being assessed here.

#### 24.6.9.4 *Immediate and formative feedback*

One of the Mini-CEX's main and most important features is immediate and formative feedback. Through organized, real-time observation and helpful feedback in a real-world clinical context, the Mini-CEX's primary goal is to foster learning and skill development. The assessor gives feedback right away following the observed patient interaction. The learning process is more effective because of this immediacy.

#### 24.6.9.5 *Structured criteria*

The Mini-CEX evaluates a trainee's performance in a number of fundamental clinical skills during an actual patient encounter using established criteria. The main component that directs observation, enables consistent scoring, and guarantees focused, prompt feedback is these set criteria. The Mini-CEX's structured criteria center on identifying and grading particular, crucial clinical competencies.

#### 24.6.9.6 *Multiple encounters*

Since a trainee's performance can differ greatly based on the particular patient and clinical issue encountered, many encounters are a crucial component of the Mini-CEX. A more accurate and thorough assessment of a trainee's overall clinical competence can be obtained by evaluating them over several, diverse interactions. Reliability is increased and the range of abilities is sampled by several observations across various patients, issues, and environments, such as a ward or clinic.

### 24.6.10 Conduction of the mini-clinical evaluation exercise

An organized, workplace-based assessment tool called the Mini-CEX is used to watch a trainee–patient interaction in a genuine clinical setting and give quick feedback. The trainee often leads the short procedure, which takes 10–20 minutes for observation and 5–10 minutes for feedback. Both the assessor and the trainee must follow these steps when conducting a Mini-CEX:

#### 24.6.10.1 *Preparation*

The trainee, assessor, and clinical setting work together to choose the encounter, patient, and time during the Mini-CEX preparation phase. To make sure the activity is a worthwhile learning and assessment opportunity, preparation is crucial. In this stage, the trainee is increasingly responsible for making plans, choosing an acceptable clinical encounter, and scheduling a time with a qualified assessor.

#### 24.6.10.2 *Consent*

A crucial and required component of the Mini-CEX procedure is consent. Participation in the observed interaction requires consent from both the patient and the trainee or student. The patient must give permission for the trainee and assessor to observe them during the engagement. The procedure is generally well received by the patients. It is crucial to remember that consent must be sought prior to the start of the Mini-CEX encounter, usually when the appointment is made or by the reception staff upon arrival. The trainee should then confirm this once more.

#### 24.6.10.3 *Observation*

The main observation of a Mini-CEX entails an expert, typically a faculty member, directly observing a trainee engage with a genuine patient in a professional context. In order to evaluate particular clinical skills and professional actions in a real-world setting, direct observation is crucial. The fact that observation takes place during regular clinical practice, such as in an emergency room, inpatient ward, or outpatient clinic, as opposed to an artificial, simulated setting, is one of its main characteristics. This guarantees the validity of the evaluation. During the patient contact, the assessor watches the trainee closely, concentrating on certain abilities such obtaining a patient's history, performing a physical examination, communicating effectively, being professional, and exercising clinical judgment.

#### 24.6.10.4 *Feedback*

The most important part of a Mini-CEX is effective feedback, which should be an organized, quick, and participatory discussion. By encouraging self-reflection and motivation, it seeks to enhance future performance. The assessor gives the trainee organized, helpful feedback right away following the observation. This is an important, formative portion of the exercise that is meant to draw attention to both areas of strength and regions that require work. Both particular areas of strength and constructive criticism that need improvement should be covered in the conversation. Avoiding giving the student too many corrections at once is crucial.

#### 24.6.10.5 *Assessment and documentation*

An organized procedure centered on direct observation, prompt feedback, and official documenting of the trainee's performance in critical clinical competencies is used to assess and document a Mini-CEX. Here, the assessor uses a standardized form to grade the trainee's performance. The ratings and feedback summary are recorded in a portfolio or on paper, which can be utilized at first if it is transferred later.

### 24.6.11 Work-based and simulation methods

#### 24.6.11.1 *Case simulations*

With evidence demonstrating improved performance in history-taking, examination, and management skills, case simulations are effective tools for clinical skill assessment. They provide a safe, realistic setting to assess healthcare learners' technical skills, decision-making, and communication without putting patients at risk. They bridge theory and practice through immersive scenarios that increase competence, confidence, and preparedness for actual clinical challenges (Zeng, 2026). They produce realistic, graded experiences for complicated or uncommon situations using high-fidelity mannequins, standardized patients, or virtual platforms. This allows for organized feedback and self-evaluation, which works well for evaluating both inexperienced and seasoned experts.

#### 24.6.11.2 *Key aspects of case simulations*

The capacity of case simulations to offer a secure, realistic, and standardized setting for assessing both technical and nontechnical skills is one of their key features as a clinical skill evaluation tool. In a controlled environment, they are crucial for bridging the gap between theoretical understanding and real-world application. It should be mentioned that the case simulations use a variety of modalities, including virtual reality systems, high-fidelity manikins, and standardized patients, to produce scenarios that closely resemble actual clinical settings. This authenticity enhances the validity of the exam by motivating students to apply their abilities in a relevant context. One of the most crucial aspects of simulations is that they allow trainees to make mistakes and learn from them without putting actual patients in danger. This safe environment makes trainees feel less nervous and more competent when they eventually transition into real clinical practice.

#### 24.6.11.3 *Conduction of the case simulations*

In order to examine a learner's skills, decision-making ability, and aptitude for problem-solving, a case simulation technique of assessment entails an organized process of creating, executing, and assessing a realistic scenario. This approach is frequently utilized to close the gap between theory and practice in industries like business and healthcare. Case simulation execution can be divided into multiple crucial phases:

##### 24.6.11.3.1 Define objectives and scope

This step clearly defines the learning objectives, and the particular characteristics the assessment seeks to measure such as clinical skills, leadership, and teamwork is one of the first tasks.

##### 24.6.11.3.2 Design the scenario

This step is essential to construct a consistent, realistic, and captivating case scenario that reflects actual circumstances. Creating precise rules and instructions, together with any necessary tools or resources, is one of the main features of this step. To guarantee the scenario's applicability and relevance, subject-matter experts should validate it.

#### 24.6.11.3.3 Develop assessment tools

This step involves establishing precise evaluation standards that are in line with the goals, such as global rating scales or standardized checklists. To develop assessment tools effectively and to guarantee consistency and dependability in evaluation, it is vital to train the assessors.

#### 24.6.11.3.4 Prebrief the participants

It is advisable to make sure that every participant is aware of the goal of the encounter, the ground rules, and the pertinent background information prior to the simulation. Additionally, prebriefing aids in creating a learning or assessment setting that is both psychologically secure and productive. It entails establishing expectations, acquainting participants with the logistics, and describing the goals of the scenario and the evaluation procedure without disclosing particular actions or results.

#### 24.6.11.3.5 Running the simulation

During the execution phase of a case simulation assessment, known as "running the simulation," participants actively participate in the realistic scenario to apply information, make decisions, and display abilities. These actions are then monitored and evaluated. Usually, careful planning comes before this stage, which is then followed by critical analysis and debriefing.

#### 24.6.11.3.6 Collect and analyze data

In order to assess a participant's capacity to apply knowledge and skills in a practical, problem-solving setting, data collection and analysis are incorporated into the scenario in a case simulation assessment. The information utilized may be generated by the participant's choices and activities throughout the simulation, or it may be predefined inside the case. During this procedure, observers gather information about the participant's performance using the predetermined assessment tools. After that, this data is examined to reveal their strengths and potential growth areas.

#### 24.6.11.3.7 Debriefing and feedback

Debriefing and feedback are essential, planned elements for fostering learning, skill development, and self-reflection while using simulation methods of evaluation. Despite their differences, both are crucial components that assist participants in assessing their performance and bridging the gap between their present activities and intended results. The debriefing session is an important step following the simulation. Participants can consider their choices, get feedback, and pinpoint opportunities for development during this moderated conversation. Learning and self-evaluation are improved by this method.

#### 24.6.11.3.8 Implementing findings

In order to incorporate results into a case simulation assessment technique, the simulation must be set up so that students must collect information, interpret it as findings, make judgments based on these discoveries, and then evaluate their decisions. Assessors can examine students' critical thinking, problem-solving, and decision-making abilities using this approach. Teachers and assessors can utilize case simulations to test complex skills that are difficult to measure with traditional methods by adhering to these principles.

### *24.6.11.4 Types of case simulations*

Medical case simulations fall into several categories, such as standardized patients for communication, mannequin-based for physical skills like cardiopulmonary resuscitation and various procedures, computer-based for clinical decision-making, and virtual reality or augmented reality for complex, immersive surgical or anatomical training, frequently combined in hybrid models. These categories range from straightforward text-based scenarios to immersive virtual reality. These provide secure, realistic settings for practicing everything from fundamental abilities to intricate team dynamics.

#### 24.6.11.4.1 High-fidelity mannequins

High-fidelity mannequins are realistic, computerized human simulators that are used in medical education to offer immersive, hands-on training for complex clinical skills. In addition to helping students practice critical interventions in a secure setting, they enhance cooperation and decision-making abilities by reacting dynamically to therapy with vital signs and physiological changes. Although they are occasionally used in conjunction with task trainers or standardized patients for a thorough skill evaluation, high-fidelity mannequins are widely used and highly effective for evaluating clinical skills because they provide realistic, hands-on practice for procedures, diagnosis, and communication in controlled settings, bridging theory and practice. Learners frequently find them more useful and fulfilling than lower-fidelity tools because they offer dynamic patient responses that aid in evaluating decision-making, cooperation, and procedural proficiency for uncommon occurrences or routine treatment.

#### 24.6.11.4.2 Computer-based simulations

In order to improve competency and patient safety, computer-based simulations in medical education replicate clinical scenarios using interactive digital tools like virtual patients, virtual or augmented reality, and high-fidelity mannequins. This provides a safe, controlled environment for students to learn, practice, and master complex skills like diagnosis, teamwork, and emergency response without endangering real patients. By enabling mastery of uncommon or challenging issues and promoting critical nontechnical abilities like leadership and decision-making, these approaches bridge gaps in traditional learning through actual experiences, instant feedback, and repeated practice. Computer-based simulations are widely used and effective for evaluating clinical skills, providing a safe, realistic way to assess decision-making, problem-solving, and technical abilities in healthcare professionals and students.

#### 24.6.11.4.3 Immersive environments

Immersive environments virtual or augmented reality are used in medical education to train future doctors with realistic, safe simulations for skills like surgery, anatomy for virtual dissection, complex procedures like cardiopulmonary resuscitation or intubation and improving clinical decision-making, communication, empathy, and spatial understanding, offering deep engagement, better retention, and remote access, thereby enhancing patient care readiness without real-world risks.

### *24.6.11.5 Benefits of case simulation in assessment of clinical skills*

In order to bridge the theory-practice gap before actual patients are involved, case simulation in clinical skills assessment provides a safe, realistic, and controlled environment for learners to hone technical skills, critical thinking, communication, and teamwork. This allows for repeated practice, instant feedback, and mastery in handling complex, high-stakes scenarios without risk. It improves decision-making, increases self-assurance, finds skill gaps, and increases general readiness for real-world clinical scenarios.

### *24.6.11.6 Disadvantages of case simulation in assessment of clinical skills*

High expenses, difficulties simulating real-world complexity, and the possibility of incorrect outcomes because of learner attitude and assessment subjectivity are the primary drawbacks of using case simulation to evaluate clinical abilities. One of the disadvantages is that implementing high-fidelity simulation programs necessitates a substantial financial investment for the acquisition and upkeep of cutting-edge equipment, including sophisticated mannequins and virtual systems, as well as specialized facilities and technical support personnel. Furthermore, the dynamic, complicated, and frequently chaotic character of real-life clinical scenarios including ambient stresses, patient unpredictability, and the emotional impact of treating actual humans is difficult for simulations, even high-fidelity ones, to accurately represent.

## 24.6.12 Portfolio

For competency-based assessment, self-directed learning, critical thinking, and ongoing professional development from undergraduate to graduate levels, medical education portfolios are dynamic collections of a learner's work in the form of evidence, reflections, and accomplishments, demonstrating growth in knowledge, skills, and professional values (Haldane, 2014). Portfolios facilitate lifelong learning and professional growth by encouraging self-reflection and providing a continuous record of a student's journey. Portfolios can be used for both formative and summative purposes by undergraduate, graduate, and practicing medical professionals. They can be used to monitor procedural experience and competencies.

### *24.6.12.1 Portfolios as an assessment tools*

In medical education, portfolios are dynamic collections of a student's work that are useful for both formative and summative assessment, particularly in competency-based models. They are effective tools for tracking competency development, encouraging self-reflection, and promoting ongoing professional growth beyond traditional exams. They capture skills like professionalism, critical thinking, and practical experience like procedures through evidence and reflective writing. The ability to examine knowledge, abilities, attitudes, and professional behaviors like ethics and communication all of which are frequently overlooked in exams is one of the main characteristics of portfolios as an evaluation instrument. Additionally, it encourages lifelong learning by pushing students to evaluate their experiences, recognize their needs, and plan their own growth. Procedural logs, feedback, leadership examples, and patient education materials are examples of meaningful work that can be gathered through the use of portfolios as a source of authentic evidence. The portfolio promotes formative evaluation and individualized learning by facilitating structured communication between students and supervisors.

#### 24.6.12.2 Benefits of using portfolios as clinical assessment tool

Portfolios as clinical assessment tools offer deep insights beyond individual exams, promoting self-directed learning, competency development, and introspection by showcasing growth over time through a range of data, including reflections, observations, and assignments. Although they can take a lot of time, they connect theory to practice, enhance professionalism, critical thinking, and self-awareness, and allow for ongoing feedback and individualized learning.

#### 24.6.12.3 Entrustable professional activities (EPAs)

EPAs are observable, fundamental medical tasks or responsibilities, such as taking a patient's history or treating a particular condition, that trainees can eventually carry out on their own, without direct supervision, once they show enough proficiency, connecting broad competencies with practical clinical practice. They function as hands-on units for competency-based medical education, enabling supervisors to evaluate trainees in day-to-day operations, provide focused feedback, and make judgments on entrustment based on decreasing levels of supervision and growing autonomy.

#### 24.6.12.4 Entrustable professional activities as an assessment tool

EPAs are essential clinical assessment tools that define fundamental, observable tasks, such as taking a patient's history or conducting an exam, that trainees must master to practice independently. They integrate professionalism, knowledge, and skills into real-world, workplace-based assessments that offer structured feedback for increasing autonomy, going beyond simple checklists to determine whether a trainee can be trusted with a task safely (ten Cate O 2024). Unlike fragmented testing, they are authentic and integrated since they evaluate actual clinical tasks, combining many qualities like professionalism, communication, and knowledge into a single assessment. Here, the emphasis is on entrustability, which is a crucial objective of competency-based medical education and aims to ascertain if a student can be trusted to carry out an activity without supervision. In order to use EPAs for clinical assessment, specific, observable tasks such as performing a basic history and physical examination are defined, standardized rubrics for supervision levels ranging from can do with direct supervision to can do independently are created, learners are observed performing these tasks, and structured feedback is provided to track progress toward full entrustment. This process bridges competencies like communication and knowledge to real-world performance in a competency-based system.

The first stage in this process is to identify and define core EPAs by choosing particular high-stakes clinical tasks that the trainees must complete, such as managing a patient with chest discomfort or prescribing a common drug. Every EPA must have a distinct start, finish, and quantifiable result that reflects real-world professional practice. Detailed rubrics which typically range from 1 to 5, and clearly defining the level of supervision required is vital. That is, whether the task calls for direct supervision, general supervision, or the trainee's ability to complete it on their own is the next crucial step in developing entrustment scales and rubrics. To lessen subjectivity, each level uses detailed rubrics with particular requirements. Subsequently, it is crucial to demonstrate how the abilities are integrated by connecting the EPAs to more general competencies. In this context, the supervisors' job is to watch the students do EPAs in actual clinical settings, such a clinic or bedside. Checklists, observation forms, portfolios, and technological platforms can be used to carry out this duty of supervision. It is crucial to give prompt, useful feedback that is directly related to the EPAs and entrustment level by emphasizing the steps the learner must take to advance to the next level of trust. Ultimately, all of the progress is monitored, and the data is used to make the decision.

#### 24.6.12.5 Benefits of using entrustable professional activities as an assessment tool

##### 24.6.12.5.1 Practical assessment

EPAs offer a defined framework for evaluating a learner's preparedness for autonomous, real-world professional practice, which is the main advantage of employing them for practical assessment. EPAs directly bridge the gap between theoretical knowledge and real clinical performance by converting abstract competencies into tangible, observable, and quantifiable tasks.

#### 24.6.12.5.2 Improved feedback

EPAs provide improved feedback by offering a structured framework that makes feedback more specific, timely, and actionable compared to traditional assessment methods. This enhances learner development and supports the progressive acquisition of autonomy.

#### 24.6.12.5.3 Contextualized evaluation

A key benefit of using EPAs as an assessment tool is that they facilitate contextualized evaluation by assessing a trainee's ability to perform real-world professional tasks in authentic clinical settings. This approach moves beyond theoretical knowledge evaluation and ensures that assessment is relevant to actual practice.

#### 24.6.12.5.4 Focus on professional duties

Using EPAs as an assessment tool provides substantial benefits for professional duties by bridging the gap between theoretical knowledge and real-world practice and by fostering accountability, reliability, and ethical behavior in trainees.

## 24.6.13 The 360-degree feedback

In medical education, 360-degree feedback, also known as multisource feedback, is a thorough assessment tool that gathers information from peers, teachers, patients, the healthcare team, and self-evaluation to provide a comprehensive picture of a learner's strengths and weaknesses in clinical skills, professionalism, communication, and leadership. By providing a variety of viewpoints, it promotes self-awareness, pinpoints areas for improvement, increases teaching efficacy, and strengthens interprofessional learning. However, there are obstacles to overcome, such as creating specialized tools and removing practical obstacles (Awdishu, 2016).

### *24.6.13.1 The 360-degree feedback as a clinical skill assessment tool*

In medical education, 360-degree feedback is frequently employed as a clinical skill assessment method, mainly for the evaluation of nontechnical qualities including professionalism, interpersonal skills, and communication. This approach gathers anonymous, private feedback from people who collaborate closely with the trainee in order to provide a thorough, all-encompassing assessment of performance. The clinician or student, superiors or faculty such as supervisors, clinical instructors, peers or colleagues such as other doctors, residents, nurses, allied health professionals, subordinates such as junior staff, administrative staff, and patients and their families are all likely to participate in this process. The feedback is usually collected via validated surveys or rating scales that focus on specific, observable behaviors and is compiled into a single report, which an accredited facilitator discusses with the individual to create a development plan.

### *24.6.13.2 Advantages in clinical assessment*

The main benefit of 360-degree feedback in a clinical context is that it offers a thorough and all-encompassing picture of a professional's performance, especially with regard to nonclinical abilities like teamwork, professionalism, and communication that are frequently overlooked by conventional evaluations. The main advantage of the 360-degree evaluation is that it offers a thorough assessment by collecting input from a variety of sources, such as peers, superiors, allied health professionals, and patients. This results in a comprehensive picture of performance that is not possible with evaluations from a single source. Additionally, it provides a chance for the doctors to pinpoint areas in which their own assessments of their abilities and behaviors diverge from those of others. This knowledge is a potent stimulant for both career and personal development. The interpersonal and communication skills that are sometimes overlooked in traditional evaluations are the primary emphasis of the 360-degree feedback. In particular, the 360-degree approach evaluates critical "soft skills" that have a direct bearing on patient satisfaction and care quality, such as empathy, listening, and explanation clarity.

### *24.6.13.3 Disadvantages of 360-degree feedback in clinical assessment*

The drawbacks of 360-degree feedback in clinical settings include the possibility of bias and personal grudges, the difficulty of getting honest input, particularly from juniors or patients who are afraid of the consequences, the lack of context due to different work environments, subjectivity, time cost, the risk of overwhelming data, poor implementation, and the potential to focus on personality rather than performance, which could impede genuine development if not carefully managed with clear goals and psychological safety.

### 24.6.14 Procedural logbook

In medical education, a procedural logbook is an essential tool for trainees to record every clinical procedure they perform. This allows them to track their skill development, demonstrate competency for assessments, receive feedback from trainers, direct self-directed learning, and provide institutions with high-quality data on service activity, all of which ultimately improve training and patient safety. It describes the type of procedure, the degree of involvement, the date, supervisor sign-offs, and the transition from paper-based systems to sophisticated digital platforms with multimedia.

#### *24.6.14.1 Procedural logbook as a clinical skill assessment tool*

In clinical skill assessment, a procedural logbook is an essential tool that serves as a structured record for tracking student performance, defining learning objectives, monitoring skill acquisition such as sutures or injections, highlighting gaps, and preparing for formal exams like OSCEs by providing a clear overview of performed procedures. However, its effectiveness depends on clear guidelines, consistent use, and integration with direct feedback, not just counting procedures.

#### *24.6.14.2 Advantages of procedural logbooks as a clinical assessment tool*

Procedural logbooks are an essential clinical evaluation tool that facilitates self-directed learning, performance monitoring, and targeted feedback by providing a structured, objective, and continuous record of a trainee's practical experience. In order to help standardize training across different clinical sites and ensure that trainees are exposed to a minimum number of procedures, procedural logbooks provide a clear set of learning objectives. This is one of the main advantages of using them as a clinical skill assessment tool. Additionally, they enable supervisors and trainees to keep an eye on the trainee's development, pinpoint their strengths and shortcomings, and make sure they are on track to attain the necessary competences. Additionally, logbooks provide a written framework for discussions between supervisors and trainees, fostering chances for critical self-reflection on clinical judgment and results as well as frequent, constructive feedback.

#### *24.6.14.3 Disadvantages of procedural logbooks as a clinical assessment tool*

The main drawbacks of procedural logbooks as a clinical assessment tool are their low validity and reliability, practical challenges, and ineffective feedback systems. Instead of focusing on actual expertise, they mainly record activity. The significant drawbacks are related to the fact that the logbooks record a learner's actions but not their decision-making process, the quality of their work, or how they carried out the method. Furthermore, when documentation is not personally controlled, it is frequently regarded as untrustworthy. Furthermore, conclusions concerning clinical competence based just on logbook entries are not supported by any data. They frequently overlook crucial information, such as the degree of supervision or help given by a supervisor.

### 24.6.15 Objective structured assessment of technical skills (OSATS)

Although it focuses on technical execution rather than surgical outcomes, the OSATS is a crucial tool in medical education, particularly in surgery, for objectively assessing a learner's performance in procedural skills using checklists and five-point Likert scales, providing structured feedback for assessment for learning and exhibiting high reliability and validity. It helps standardize training and monitor progress for procedures like suturing in both real-world clinical settings and simulations (Anderson, 2016).

## 24.7 Assessment of core clinical competencies

Using techniques like OSCEs, written tests, on-the-job observations, 360-degree feedback, simulations, self-assessments, and performance reviews, assessing core clinical competencies entails evaluating a healthcare professional's knowledge, skills, and attitudes across important areas like communication, clinical reasoning, patient care, and professionalism in order to guarantee safe, high-quality patient care. These evaluations monitor progress in critical areas for competency-based medical education, such as history taking, diagnosis, treatment, documentation, and patient education. Some of the major clinical competencies that are being assessed include the following:

### 24.7.1 Assessment of patient care

To guarantee patient safety, care quality, and operational efficiency, healthcare personnel's knowledge, skills, and attitudes are methodically evaluated through methods such as observation, self-evaluations, interviews, and portfolios.

In order to detect and manage competence-related concerns and improve patient outcomes, this assessment is essential for initial orientation, continued annual reviews, and professional development. In order to provide compassionate, appropriate, and successful patient care for the treatment of health issues and the health program, residents or fellows must show that they are capable of doing so. OSCEs, direct observation using checklists and rating scales, skills assessments, written or oral exams, and simulations are some of the techniques used to assess patient care competency. To present a complete picture of the patient caregiver's abilities and performance in areas including communication, technical proficiency, professionalism, and clinical decision-making, include self-assessments, peer reviews, and supervisor evaluations utilizing 360-degree feedback.

### 24.7.2 Assessment of medical knowledge

Assessing medical knowledge competency is determining how well and safely a healthcare practitioner can use their knowledge. Written exams and progress assessments are examples of theoretical approaches for evaluating medical knowledge. However, OSCE, objective structured practical examination, Mini-CEX, DOPS, OSATS, and portfolio and logbook assessment can all be used for the clinical and practical assessment.

### 24.7.3 Assessment of interpersonal skills

Self-evaluations, peer feedback, structured interviews, performance-based tasks, and standardized tests, including psychological tests like personality tests and scales intended to assess social and behavioral skills, are some of the techniques used to assess interpersonal competence. The selection of an evaluation method frequently relies on the situation, such as when evaluating new teachers or employees. It may include the use of observational instruments as well as the examination of relationship management and communication styles.

### 24.7.4 Assessment of practice-based learning and improvement

A trainee's capacity to examine patient care, locate and evaluate scientific evidence, and use this information to enhance practice through lifelong learning is assessed as part of the PBLI competence evaluation process. Assessment techniques include chart-stimulated recalls and record reviews to evaluate clinical decision-making, resident portfolios that record progress and reflections, self-assessments and 360-degree evaluations to measure personal development and teamwork, and simulation-based scenarios for crucial procedures. Evidence-based practice and reflective practice that promotes personal development are the two primary areas of focus for the evaluation.

### 24.7.5 Assessment of professionalism

In clinical medicine, evaluating professionalism necessitates a multifaceted approach that uses a variety of tools, such as 360-degree feedback, OSCEs, patient surveys, self-assessments, and critical incident reports. It focuses on observed behaviors, ethics, accountability, and patient-centered care, with ongoing, longitudinal feedback for improvement rather than a single test. Using frameworks that integrate virtue-based and behavior-based perspectives, key areas evaluated include ethical behavior, communication, teamwork, integrity, respect for patients, self-reflection, and stress management.

### 24.7.6 Assessment of system-based practice

Evaluating a healthcare professional's capacity to comprehend and engage with the larger healthcare system in order to deliver the best possible patient care is known as SBP competency assessment. Performance-based evaluations, such as the planning and execution of quality improvement initiatives, direct system usage observation, 360-degree feedback from colleagues and affiliated personnel, self-evaluations, and simulated case studies, are examples of assessment techniques. The need for more reliable, targeted, and multifaceted evaluation techniques is highlighted by the ongoing development of validated tools for thorough SBP assessment.

**Summary**

As a key component of medical education, clinical assessments are used to gauge medical professionals' and students' proficiency in actual healthcare environments. These tests are intended to evaluate practical abilities, judgment, and professional conduct in addition to theoretical knowledge. Good clinical evaluations guarantee that medical professionals are prepared to provide patients with safe, superior treatment. In order to guarantee that healthcare professionals possess the abilities, expertise,

**(cont'd)**

and professionalism required to provide patient care, clinical evaluations are an essential part of medical education. Even if there are difficulties, a strong system for producing qualified medical professionals is created via the use of a variety of organized assessment instruments together with frequent feedback and introspection. Clinical evaluations will change as medical education continues to adjust to the shifting healthcare environment in order to better serve patients and future healthcare professionals.

**Key point**

Future perspectives: mirroring the real-world practice

A number of developing trends, technological developments, and a greater comprehension of the skills necessary for contemporary healthcare practice will influence clinical evaluation in medical education in the future. Clinical evaluations will probably change significantly as medical education adjusts to the shifting needs of the healthcare system. The incorporation of technology and digital tools is one of the main viewpoints regarding the future of clinical evaluations in medical education. Furthermore, clinical evaluation in medical education is expected to evolve into a more dynamic, individualized, and technologically integrated process in the future, which will have the ability to mirror the real-world practice. The increasing use of digital technologies, artificial intelligence, and competency-based education models will improve the way clinical competence is evaluated, even though conventional techniques like objective structured clinical examinations and direct observation of procedural skills will still be important. This will result in assessments that are more precise, comprehensive, and formative, better preparing students for the intricacies of contemporary healthcare. Medical education may guarantee that future healthcare workers are not only clinically competent but also prepared to function well in varied, team-oriented, and patient-centered settings by embracing these developments.

# References


Ahmad, M.R., Abu, H., Helal, U., Ahmed, Hossin, M., Akm, K., et al., 2024. Direct observation of procedural skills (DOPS) in enhancing clinical competencies of postgraduate medical and dental education: a scoping review. IOSR Journal of Dental and Medical Sciences (IOSR-JDMS), 23, 59–70. Available at https://doi.org/10.9790/0853-2309085970.

Anderson, D.D., 2016. Clin Orthop Relat Res 474.

Awdishu, L., 2016. MedEdPublish 7.

Batalden, P., Leach, D., Swing, S., Dreyfus, H., Dreyfus, S., 2002. General competencies and accreditation in graduate medical education. Health Affairs 21 (5), 103–111. Available at https://doi.org/10.1377/hlthaff.21.5.103.

Colbert, C.Y., Ogden, P.E., Ownby, A.R., Bowe, C., 2011. Systems-based practice in graduate medical education: systems thinking as the missing foundational construct. Teaching and learning in medicine 23 (2), 179–185. Available at https://doi.org/10.1080/10401334.2011.561758.

Cox, M., Irby, D.M., Epstein, R.M., 2007. Assessment in medical education. New England Journal of Medicine 356 (4), 387–396. Available at https://doi.org/10.1056/NEJMra054784.

Faraj Pour, A., Yazdani, S., Farmad, S.A., 2022. A conceptual model for practice-based learning and improvement competency in medicine. Journal of Family Medicine and Primary Care 11 (6), 3230–3237. Available at https://doi.org/10.4103/jfmpc.jfmpc_1878_21 Epub 2022 Jun 30. PMID: 36119169; PMCID: PMC9480706.

Frank, J.R., Danoff, D., 2009. The CanMEDS initiative: implementing an outcomes-based framework of physician competencies. Medical Teacher 29 (7), 642–647. Available at https://doi.org/10.1080/01421590701746983.

Haldane, T., 2014. Gastroenterol Hepatol Bed Bench 7.

Hampton, D., Melander, S., Tovar, E., Falls, C., Makowski, A., Grubbs, A.B., et al., 2024. Value of miller's pyramid for clinical skills assessment in the evaluation of competency for nurse practitioner students. The Journal for Nurse Practitioners 20 (4), 104952. Available at https://doi.org/10.1016/j.nurpra.2024.104952.

Harden, R.M.G., Downie, W.W., Stevenson, M., Wilson, G.M., 1975. Assessment of clinical competence using objective structured examination. British Medical Journal 1 (5955), 447–451. Available at https://doi.org/10.1136/bmj.1.5955.447.

Hetaimish, B., Elbadawi, H., 2015. OSCE assessment for medical student clinical skills and performance, are we doing it the right way? International Journal of Science and Research (IJSR 4, 963.

J., Ziv, A., 2003. Reliability and validity of a simulation based acute care skills assessment for medical students and residents. Anesthesiology 99, 1270–1280.

Lakshman, K., 2025. Is OSCE the answer to proper clinical skills assessment? Karnataka Journal of Surgery 2, 4–5. Available at https://doi.org/10.25259/kjs_1_2025.

Miller, G.E., 1990. The assessment of clinical skills/competence/performance. Academic Medicine 65 (9), S63. Available at https://doi.org/10.1097/00001888-199009000-00045.

Nair, B.R., Alexander, H.G., McGrath, B.P., Parvathy, M.S., Kilsby, E.C., Wenzel, J., et al., 2008. The mini clinical evaluation exercise (mini-CEX) for assessing clinical performance of international medical graduates. Medical Journal of Australia 189 (3), 159–161. Available at https://doi.org/10.5694/j.1326-5377.2008.tb01951.x.

Peña, A., 2010. The Dreyfus model of clinical problem-solving skills acquisition: a critical perspective. Medical Education Online 15 (1), 4846. Available at https://doi.org/10.3402/meo.v15i0.4846.

ten, Cate O., 2024. Entrustable Professional Activities and Entrustment Decision-Making in Health Professions Education. Ubiquity Press, London.

Zeng, J, 2026. BMC Med Educ 26.

Chapter 25

# Workplace-based assessment

*"The two fulcra of medicine are reason and observation. Observation is the clue to guide the physician in his thinking."*
—Giorgio Baglivi

## 25.1 Introduction

Workplace-based assessment is a technique for determining a person's proficiency in professional practice by seeing and analyzing their performance, abilities, and attitudes in an actual work environment. Unlike traditional tests that evaluate "knows" rather than "does," workplace-based assessment are essential in medical education because they assess trainees' clinical abilities, knowledge application, and professional behaviors directly in real clinical settings. These genuine evaluations are essential to competency-based medical education for monitoring advancement, guaranteeing patient safety, and directing development toward independent practice. They also offer ongoing, formative feedback for learning. With a focus on patient safety, it encourages and promotes learning in critical areas of competency and offers helpful criticism on areas of strength and areas in need of improvement (Liu, 2012).

As it evaluates real performance in the workplace, it has the benefit of high content validity. In curriculum design and educational planning, where teaching, learning, assessment, and feedback are tightly interwoven, workplace-based assessments are encouraged as essential components. Workplace-based assessments are also a great source of information for educational supervision and feedback (Anderson et al., 2021). They are designed to show that students have made satisfactory progress and accomplishments, as well as to identify areas that still require improvement and to discuss and agree on ways to address those areas. The benchmark that trainees should have attained at the end of their current training phase is typically used to evaluate them. In essence, workplace-based assessments are a type of realistic evaluation that evaluates performance in the actual setting that physicians encounter on a daily basis.

## 25.2 Evolution of workplace-based assessments

In the late 2000s, workplace-based assessment was first used in medical education. The Postgraduate Medical Education and Training Board in the United Kingdom made significant advancements and publications around 2007, and Norcini and Burch formalized the idea that same year. Although Miller's pyramid from 1990 is where performance-based assessment got its start, workplace-based assessments became widely used and structured in the mid-to-late 2000s as a way to evaluate clinical performance in real-world settings.

## 25.3 Role of workplace-based assessment in medical education

Workplace-based assessment gained importance as it provided a way to evaluate a trainee's everyday performance and professional skills. They provide real-time feedback that traditional exams cannot, which shapes clinical learning, ensures patient safety, and guides a trainee's transition to independent practice. They provide an organized way to see abilities in practical contexts, pinpoint strengths and shortcomings, and connect instruction and evaluation, resulting in a whole portfolio that shows continuous competency development (Burch, 2019). Workplace-based assessments have a significant impact on teaching and learning because they provide real-time, authentic feedback in real-world work situations, shift education toward competency-based models, and improve professional abilities. They enhance learner growth through focused observations and portfolio monitoring, boosting performance and introspection (Miller, 1990).

**The Principles and Practice of Medical Education. DOI:** https://doi.org/10.1016/B978-0-443-44197-4.00028-X

## 25.4 Workplace-based assessment tools

Workplace-based assessment tools are methods used to evaluate a learner's performance and competency in a real-world work environment, commonly in fields like medicine and other professions.

### 25.4.1 Types of workplace-based assessment tools

Direct observation of procedural skills (DOPS) for practical skills, case-based discussions (CbD) for clinical reasoning and knowledge application, mini-clinical evaluation exercise (Mini-CEX) for observed clinical interactions, and multisource feedback (MSF) for patient and colleague perspectives are common workplace-based assessment tools (Fig. 25.1).

### 25.4.2 Direct observation of procedural skills (DOPS)

DOPS is a workplace-based assessment tool which assesses practical, hands-on skills and technical competence in a real-time setting. It is used to evaluate a trainee's competence in practical tasks through direct observation by a supervisor. It involves a structured checklist and aims to provide immediate, constructive feedback to facilitate skill development and improvement. Strengths include its ability to assess real-world skills, improve performance, and boost confidence, while potential limitations can involve time constraints, assessor bias, and the need for proper training for both assessors and trainees (Khan et al., 2013).

#### *25.4.2.1 Role of direct observation of procedural skills as workplace-based assessment*

DOPS is an essential workplace-based assessment tool in which a supervisor observes a trainee performing a practical skill, such as suturing or examining a patient, and provides prompt, structured feedback, emphasizing strengths and areas for improvement. This helps trainees become proficient in routine tasks, develop evidence for advancement, and promote self-reflective learning in actual clinical settings. In addition, unlike delayed feedback, it offers instant, focused feedback, enabling real-time learning correction and goal-setting. Additionally, it identifies areas that require work and highlights the trainee's strengths, directing future learning. It also acts as a tool to help supervisors evaluate the program's overall

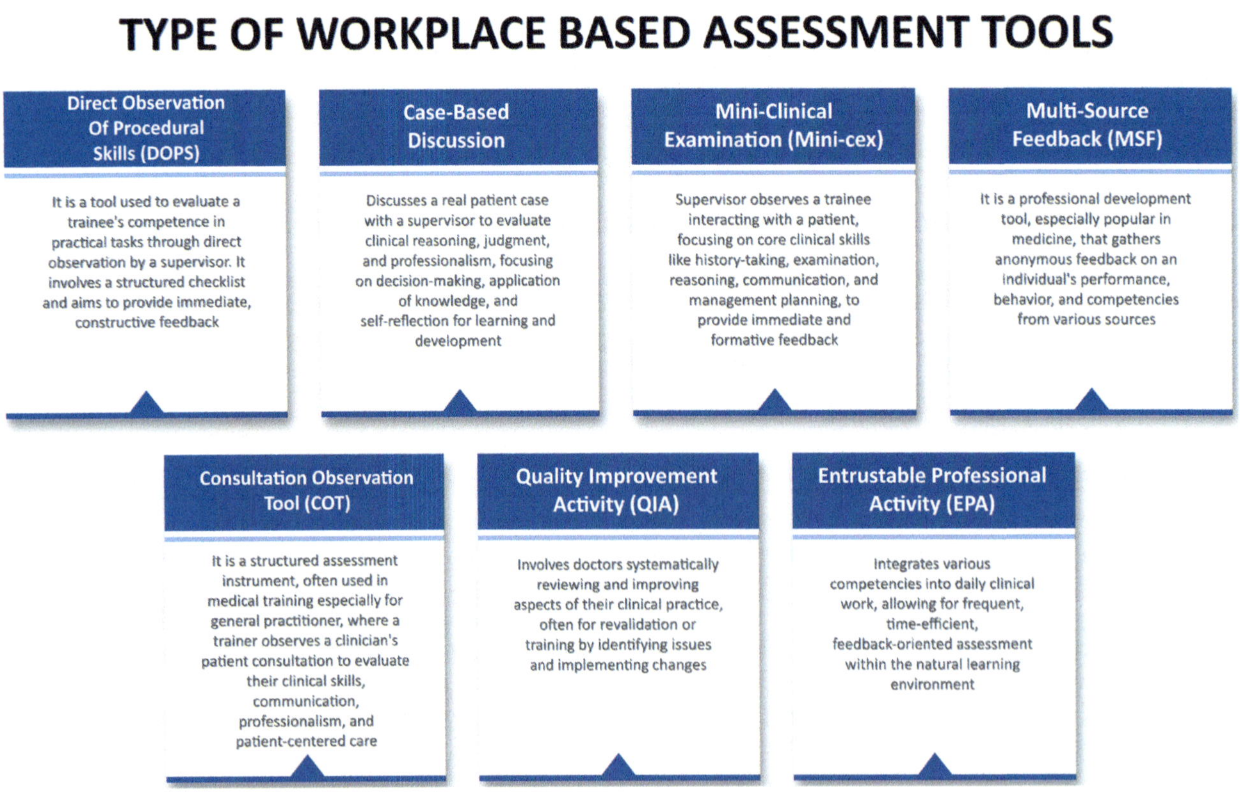


FIGURE 25.1 **Types of workplace-based assessment tools.** This figure depicts the types of workplace-based assessment tools.

progress or preparedness for increasingly challenging tasks. Ultimately, it fosters professional growth by motivating students to analyze their own learning needs, choose assessors, and evaluate their own performance.

### 25.4.3 Case-based discussion

A CbD is a structured oral assessment that involves a healthcare trainee discussing a real patient case with a supervisor in order to assess clinical reasoning, judgment, and professionalism. It focuses on decision-making, knowledge application, and self-reflection for learning and development, frequently involving feedback and action planning. This important workplace-based assessment method uses written records as a foundation to examine the trainee's problem-solving abilities, ethical considerations, and holistic approach to patient care. In order to demonstrate clinical judgment, decision-making, and knowledge application, a learner discusses a patient case with an assessor during this oral assessment. Some of the key features of a CbD are discussed below:

#### *25.4.3.1 Purpose*

A CbD's primary function is to act as an effective teaching and learning tool, as well as a formative assessment. It aims to promote learning and professional growth by providing structured feedback on a trainee's performance in an actual work environment. Evaluating the trainee's professional judgment, decision-making, and application of medical knowledge in a real-world patient scenario is the primary goal of the CbD. Instead of focusing only on factual recollection, it investigates the reasoning behind their decisions and behaviors. Additionally, it gives the trainee rapid, precise, and helpful feedback from a supervisor. This feedback highlights the trainee's strengths and areas for development, enabling them to consider and implement the suggestions made.

#### *25.4.3.2 Process*

In this step, the trainee discusses the case's background, management, outcomes, and thought process with an experienced assessor or supervisor after presenting a case they handled. The case selection procedure is one of the most important components. Usually, the assessor chooses a case that presents a challenge or opportunity for learning, or the trainee chooses one from their most recent practice. The case selection entails selecting two to four intricate, difficult, or fascinating patient cases that the trainee individually manages, with an emphasis on scenarios involving many encounters as opposed to one-time consultations. Cases should emphasize professional judgment, clinical reasoning, and decision-making, frequently addressing particular curricular competencies.

#### *25.4.3.3 Areas of focus*

The area of focus is a crucial component of a CbD because, as opposed to an open-ended discussion, the process purposefully focuses on particular elements of a single, real-life case to evaluate a trainee's clinical reasoning, decision-making, and application of knowledge. Since the entire discussion revolves around a single, particular patient case or professional event that the trainee has managed or seen, one of the most important topics to concentrate on is the individual case review. A thorough examination of decision-making and performance in the real world is made possible by this foundation in a realistic experience.

#### *25.4.3.4 Feedback*

The most crucial component of a CbD is feedback. It is primarily a formative assessment tool meant to guide the trainee's learning and growth rather than a strictly summative test. Feedback plays a crucial role in helping trainees gain insight into their performance, foster reflective learning, and eventually enhance their clinical decision-making and patient management abilities. The conversation also aids in pinpointing the trainee's particular strengths and, more crucially, any knowledge or skill gaps or places for improvement (Saedon et al., 2010).

#### *25.4.3.5 Documentation*

Since the evaluation employs the trainee's actual patient data, such as case history, as the foundation for a structured conversation, documentation is an essential component of a CbD. One key ability that is specifically evaluated is the written record's quality. The assessor is prompted to investigate the trainee's clinical reasoning and decision-making process via written notes. Additionally, it aids the assessor in precisely assessing the case notes' accuracy, completeness, conciseness, and clarity. This guarantees that the learner satisfies professional requirements for clinical documentation.

#### 25.4.3.6 *Environment*

Environment forms an essential component of a CbD, as it directly affects the quality of feedback, the transparency of communication, and the assessment's overall educational value. An appropriate location guarantees anonymity and seclusion, allowing both the trainee and the assessor to speak honestly and freely without worrying about being overheard. The case-based assessment necessitates a defined time slot, usually 20–30 minutes. This protected time is made possible by a decent setting, free from the pressures of a hectic workday, which can otherwise result in hurried and subpar feedback sessions.

### 25.4.4 Conduction of case-based discussion

The conduction of a CbD mainly involves three steps:

1. Preparation,
2. Case presentation,
3. Structured conversation,
4. Feedback session,
5. Documentation and follow-up.

#### 25.4.4.1 *Preparation*

In order to prepare for a CbD, the trainee must first choose a suitable case, go over all pertinent information, and create a brief case summary to provide to the assessor before the meeting. There are particular preparation steps for both the assessor and the trainee.

#### 25.4.4.2 *Trainee/presenter's steps*

##### 25.4.4.2.1 Select appropriate cases

It is essential that trainees select challenging, complex, or fascinating cases that they handled independently so that supervisors can assess clinical judgment and decision-making skills in real-world scenarios. These cases should demonstrate a range of competences, including data collection, management, referral, and uncertainty, in a number of contexts, including hospitals and home visits.

##### 25.4.4.2.2 Prepare case overview

A trainee should choose a difficult case and provide a concise synopsis of the patient's journey, emphasizing their role and choices, in order to create a case review for a CbD. This must then be translated into one to three key competences and prepared to use patient data to discuss their clinical reasoning, ethical issues, and self-reflection. In order to allow the conversation to concentrate on the trainee's actions and learning rather than just facts, the summary should be brief between one and three pages and submitted in advance.

##### 25.4.4.2.3 Submit documentation

A trainee submits documentation for a CbD by choosing two to four pertinent patient cases in which they were directly responsible, compiling their thorough case notes that include assessments, management, and results, mapping these cases to particular training competencies like clinical reasoning and communication, and creating a succinct summary for their supervisor to review and discuss in a focused, private session, concentrating on strengths and areas for improvement. The documentation aids in defining learning objectives and demonstrates clinical judgment rather than merely facts.

##### 25.4.4.2.4 Identify learning objectives

During a CbD, supervisors and trainees work together to create learning objectives. The trainee's presentation and reflection on a real-life situation determine the precise goals. Examining and evaluating the trainee's clinical judgment, reasoning, and knowledge application is the main objective. The application of knowledge, clinical reasoning, and decision-making are the primary areas of emphasis.

### 25.4.4.3 *Assessor/facilitator's steps*

#### 25.4.4.3.1 Review of the material

The facilitator's review of the content in a CbD includes leading the students through an organized discussion of a real-world situation, evaluating their clinical reasoning, and giving prompt, helpful feedback. The review typically occurs during and after the conversation rather than during a presession presentation. The assessor reviews the provided case notes and summary to learn more about the scenario and identify potential conversation topics.

#### 25.4.4.3.2 Preparing questions

Preparing effective questions for CbDs involves planning with clear learning objectives, using open-ended and varied question types, and organizing them to guide participants through the case analysis.

#### 25.4.4.3.3 Selecting a case

The facilitator regularly presents difficult scenarios with appropriate complexity and chooses examples that align with learning objectives to promote problem-solving and competency demonstration. The facilitator uses open-ended questions to guide the trainee through data collection, diagnosis, management, and ethical considerations while making sure the discussion stays focused on the trainee's actions rather than hypothetical situations to provide summative assessment and feedback.

### 25.4.4.4 *Case presentation*

#### 25.4.4.4.1 Introduction and scene setting

The role of this stage is to establish a private, protected environment and ensure that the discussion takes place in a quiet location where both participants are free from interruptions. In addition to that, it is important to clarify the purpose of the discussion where the main focus is to explore clinical reasoning and professional judgment, not just a test of factual knowledge. At the same time, in this stage, the various ground rules pertaining to the session are set such as the session's duration may be from 20 to 30 minutes and to inform the participants that the discussion will focus on the thinking process behind decisions made in a real patient case.

#### 25.4.4.4.2 Case presentation

In a CbD, a trainee presents a real patient case to an assessor in order to improve practice through structured, developmental feedback on history, examinations, investigations, and ethical aspects. This feedback is frequently given in chunks for better focus. Instead of examining speculative "what if" scenarios, the presentation should discuss what the trainee did, why, and what data supports it. It should also use the case as a starting point for evaluating skills like communication and problem-solving. Here, the form of the case and whether or not it is reasoning-focused are very important. Additionally, it is crucial to ensure that it is in line with the competencies that it aims to address.

### 25.4.4.5 *Structured conversation and assessor observation*

Using prepared questions, the assessor guides a structured discussion to examine the trainee's handling of the case. This covers topics including data collection, differential diagnosis, therapy planning, ethical issues, and collaboration. The student should evaluate their strategy and provide justification for their choices and actions. In a CbD assessment, the assessor uses a structured discussion about a patient case to evaluate a trainee's clinical reasoning, decision-making, and knowledge application. This kind of evaluation places more emphasis on real performance. In order to encourage the trainee to think critically about their activities and defend their capacity for problem-solving and decision-making, it is imperative that the assessors use open-ended questions rather than merely analyzing factual recollection. Additionally, in this instance, the assessor employs a particular CbD mark form or rating scale that lists a number of competences or domains that need to be evaluated, such as clinical judgment, professionalism, and recordkeeping. Each pertinent domain is graded during the assessment procedure in accordance with the anticipated performance standards for that training stage. The assessor records their reasoning for the marks given, along with performance evaluations and suggestions for further improvement.

### 25.4.4.6 Feedback session

The feedback session is essential for reflection in CbD. It focuses on what went well and areas for improvement, employs a positive approach with clear, doable goals for future learning, and results in development plans that are agreed upon and signed forms that demonstrate progress. It ensures that supervisors provide targeted guidance, trainees reflect on their clinical reasoning, and specific actions for improvement are formed by bridging the gap between clinical discussion and structured learning.

### 25.4.4.7 Documentation and follow-up

Documentation is essential for keeping the record of performance, feedback, and future learning goals during a postdiscussion phase in a CbD. The procedure usually entails the assessor filling out an electronic or paper-based structured assessment form while the learner is present. The agreement on a corrective or developmental action plan is an essential component of the documentation. This describes the actions the learner will take to address identified learning requirements, such as additional conversation and self-directed learning. Here, both the trainee and the assessor must sign and date the completed paper to attest to their participation in the conversation and their receipt of the feedback. Both parties usually retain copies, which are added to the trainee's portfolio. In a CbD follow-up, the emphasis switches from presentation to in-depth introspection and future development. Supervisors probe clinical reasoning, investigate alternative actions related to patient safety, medico-legal aspects, and nonwork factors, and jointly set development goals. This ensures that trainees identify learning gaps, comprehend missed opportunities, and commit to actions for improved practice, going beyond simple critique to actionable learning plans.

### 25.4.4.8 Advantages of the case-based discussion as a workplace-based assessment

Because they are realistic, evaluate higher-order thinking skills like judgment, ethics, and complex reasoning, thereby directly connecting assessment to learning through formative feedback which will also help in assessing a trainee's comprehension of holistic care, such as recordkeeping and communication, and offer a structured method of tracking competency development for supervisor review. CbDs offer significant advantages in workplace assessment. They examine how trainees apply knowledge in actual patient situations rather than merely memorize it, bridging theory and practice. Essentially, case-based conversations turn evaluation into an effective teaching opportunity that offers profound insights into a trainee's developing clinical competency. Some of the prime benefits obtained by case-based learning can be enumerated as follows:

#### 25.4.4.8.1 High validity and realism

Since case-based conversations assess real performance in the workplace using real patient circumstances, they are seen as having great validity and authenticity. It allows medical students to showcase their diagnostic reasoning and critical thinking skills, which can be difficult to assess with more traditional methods like multiple-choice exams. Additionally, by giving a glimpse of the skill being acquired by the trainee and clear evidence that the trainee is ready for independent practice, it offers reliable data for summative assessments.

#### 25.4.4.8.2 Tests higher-order skills

CbDs are a useful teaching and evaluation technique since they concentrate on higher-order thinking skills like clinical reasoning and problem-solving, which are difficult to test using traditional approaches. This approach aids in bridging the knowledge gap between theory and practice. In this sense, critical thinking, analysis, synthesis, evaluation, and clinical judgment are all actively encouraged by CbD. Traditional lectures, on the other hand, frequently concentrate only on lower-order rote memory. Another benefit of the CbD is that since it tests for higher-order skills, it enables the student to prepare for professional practice by applying skills like clinical reasoning and problem-solving to challenging real-world situations.

#### 25.4.4.8.3 Formative and reflective learning

Case-based talks offer numerous advantages for formative and reflective learning by promoting critical thinking, offering prompt and tailored feedback, and fostering the development of essential professional skills including teamwork and self-evaluation. Formative assessment, which focuses on providing feedback to improve future performance rather than just assessing results, is a crucial component of case-based talks. CbDs are concentrated, frequently one-on-one conversations in which a teacher can offer prompt, helpful, and thorough feedback on a student's decision-making process. This enables students to quickly relate the feedback to their performance, filling in knowledge and skill deficiencies while the experience is still new. In this context, the student and the supervisor can identify areas of strength, weakness, and ongoing shortcomings that need to be addressed, allowing for targeted support and intervention. This is possible due to the development of effective communication between the two.

#### 25.4.4.8.4 Evaluation of broad range of skills

Case-based conversations are a helpful evaluation method as they use real-world scenarios to assess a variety of skills that are difficult to measure with regular tests, like professional judgment, critical thinking, and communication. Assessors can examine higher-order thinking skills such as analysis, synthesis, problem-solving, and decision-making by having students defend their choices in difficult situations. CbDs enable the examination of professional and ethical aspects of practice, such as patient care, recordkeeping, and confidentiality, which are challenging to assess using other techniques.

#### 25.4.4.8.5 Evidence of competence

A CbD provides a direct method for gathering evidence of competence by assessing an individual's professional judgment and application of knowledge in real-world scenarios. This approach moves beyond theoretical knowledge to evaluate performance in actual practice, offering several advantages for learning and assessment. In this regard, the assessor can probe the rationale behind investigations, referrals, treatment plans, and overall patient management. In addition, it ensures that medical or professional knowledge is effectively integrated into the actual case management, bridging the gap between theory and practice.

### *25.4.4.9 Challenges of case-based discussion*

Assessment variability, which includes inconsistent training and standards leading to subjectivity, learner issues like uncertainty, lack of confidence, poor communication, dominance by a small number of people, logistical challenges like time-consuming, connecting theory to practice, and assessment validity, which includes generalizability, potential bias, and proving reliability, are all obstacles in CbDs. To promote efficient learning and equitable assessment, resolving issues frequently requires improved assessor training, well-defined frameworks, and organized facilitation.

## 25.4.5 Mini-clinical examination (mini-CEX)

In order to provide quick, formative feedback for professional development in real-world clinical settings, a supervisor observes a doctor or trainee interacting with a patient during a Mini-CEX, also known as mini-clinical examination, which is a brief, practical assessment that focuses on fundamental clinical skills like history taking, examination, reasoning, communication, and management planning (Norcini J J 2003). Unlike standardized tests, it provides an organized method of evaluating a trainee's skills in a real-world setting, making it a crucial tool in workplace-based assessment for medical training.

### *25.4.5.1 Role of mini-clinical examination as workplace-based assessment*

The Mini-CEX is an essential workplace-based assessment for witnessing actual patient interactions. It offers quick, structured feedback on fundamental clinical skills like history, examination, reasoning, and professionalism to promote formative learning and show supervisors that trainees are prepared for unsupervised practice, not just a checklist. Its function is to provide prompt, situation-specific performance coaching in real-world clinical situations, promoting ongoing patient care improvement (Hill et al., 2009). Some of the main benefits of Mini-CEX as a workplace-based assessment tool include its encouragement of real-time, realistic evaluation, which evaluates abilities during real-world patient contacts and provides context and validity that are frequently lacking in simulated scenarios. It also focuses on the targeted skill evaluation, emphasizing core competencies such as history taking, physical examination, clinical reasoning, professionalism, and communication.

## 25.4.6 Multisource feedback (MSF)

In order to provide a comprehensive view for self-reflection, learning, and improvement, MSF, also known as 360-degree feedback, is a professional development tool that is particularly popular in the medical field. It collects anonymous feedback on an individual's performance, behavior, and competencies from a variety of sources, including peers, supervisors, subordinates, and occasionally patients. It frequently combines ratings with narrative comments for better understanding (Björklund, 2022).

### 25.4.7 Consultation observation tool (COT)

A COT is a structured assessment tool that is frequently used in medical education, particularly for general practitioners. It involves a trainer watching a clinician consult with a patient in order to assess the clinician's clinical skills, professionalism, communication, and patient-centered care while offering feedback for improvement. It aids in evaluating skills in examination, shared decision-making, obtaining a patient's history, and providing holistic care. It uses video, audio, or in-person observations to provide feedback on how successfully the practitioner conducts a consultation, addressing topics such patient contact, clinical reasoning, and collaborative decision-making. The COT provides proof of professional growth and competence.

#### *25.4.7.1 Key features of the consultation observation tool*

In terms of evaluation methodology, it facilitates direct observation, which enables a trained reviewer or trainer to see a clinical consultation in real time. In terms of structured feedback, the observer grades particular criteria using a defined form and gives the trainee prompt, helpful comments. In terms of performance, each item is usually scored on a scale. In this sense, the COT is a type of workplace-based evaluation, which takes place as part of continuing education and professional development in a real-world clinical situation. Numerous talents, which are usually divided into distinct categories, are evaluated by the instrument. The COT assesses a variety of skills that contribute to high-quality care, such as opening and gathering information, building rapport and introducing oneself clearly, using open questions appropriately and actively listening, clarifying, and exploring cues, both verbal and nonverbal, discovering the patient's perspective in terms of ideas, concerns, and expectations, obtaining pertinent history, and identifying potential red flags.

#### *25.4.7.2 Conduction of the consultation observation tool*

Consultation abilities are assessed in real-time or through recordings using the consultation observation instrument, a standardized evaluation used in healthcare education, especially for advanced practitioners and trainees in general practice. It offers a structure for professional growth and structured feedback. The following steps are involved in the conduction process:

#### *25.4.7.3 Preparation*

##### 25.4.7.3.1 Patient consent

In a medical COT, patient consent is a crucial ethical and legal step. The patient must willingly agree to the observation or review of their consultation for training, whether it be audio, video, or live. Before, during, or after the session, this guarantees that the patient is aware of the procedure, risks, advantages, and options. This is often accomplished through a particular form or recorded verbal consent. It protects both patient autonomy and doctor liability by requiring explicit disclosure about the observer's role and potential use of the recording, along with express agreement for various purposes.

##### 25.4.7.3.2 Selecting the consultation

Choosing a consultation for the COT entails picking a video or audio recording of a patient interaction that highlights the trainee's abilities in a variety of patient categories, including children, the elderly, mental health, and settings like home visits or surgery. In order to achieve a representative sample for the practice, the case should ideally be complex for deeper learning. Aside from that, the primary emphasis will be on the crucial phases, such as establishing rapport, obtaining information, analyzing, planning, and closing in order to obtain insightful input for improvement.

##### 25.4.7.3.3 Gathering materials

When discussing clinical education and assessment tools such as the COT, the term gathering materials refers to the method by which a medical professional obtains the information required from a patient in order to make a diagnosis and develop a treatment plan. Understanding patients' viewpoints and expectations regarding their symptoms and the entire consultation process is also highly helpful.

##### 25.4.7.3.4 Trainee preparation

The trainee can prepare in a variety of ways, such as by engaging in the process of self-reflection, which allows the trainee to review either through audio or video by focusing on diverse, complex cases across various scenarios, such as face-to-face, video, and phone, understanding key competencies, and working closely with supervisors for feedback, with the goal of improving skills for holistic patient care.

### *25.4.7.4 The observation or review phase*

#### 25.4.7.4.1 Observation

In a COT, the term "observation phase" describes the time during the patient encounter when an observer, usually a supervisor or trainer, methodically observes and listens to the practitioner's interaction with the patient. It is crucial that the observer pays attention by concentrating on abilities like rapport-building, communication, and nonverbal clues, as well as abilities unique to certain stages of the consultation, including taking a history, providing an explanation, and making plans. In order to stimulate remembering during the feedback session, the observer records certain conversational excerpts and evaluative notes in this step. One way to differentiate between the two is to use different colored pens or highlighters.

#### 25.4.7.4.2 Note taking

The purpose of taking notes in a COT is to record the observer's assessments, offer concrete proof for comments, and assist the observer in remembering significant aspects of the exchange. Both the conversation's content and the observer's assessment of the skills displayed should be included in the notes. Additionally, the notes serve as proof of feedback since they offer specific instances to back up the comments made to the trainee following the consultation.

#### 25.4.7.4.3 Timing

Although there isn't a set minimum or maximum time for COTs, face-to-face or phone consultations often last 8–15 minutes, with the goal of achieving sufficient complexity without going overboard. Consultations of any length can be used, but those that are too short or too long may not demonstrate enough. The tool assists in evaluating abilities in a variety of consulting contexts.

### *25.4.7.5 After the consultation*

#### 25.4.7.5.1 Joint review

The crucial conversation and feedback session that takes place between the supervisor and the trainee following the observation phase is known as the joint review step in the context of the COT. The goal of this joint review is for the supervisor to give formal, organized comments and for both parties to discuss the observed consultation, connecting the observations to applicable capability frameworks or defined performance objectives.

#### 25.4.7.5.2 Grading

Following the consultation, the supervisor assesses each part according to predetermined standards and offers structured feedback while the trainee and supervisor jointly review the observed or recorded session. The supervisor rates each particular area of the consultation in accordance with the predetermined performance criteria during this phase, which comes after the review and discussion.

#### 25.4.7.5.3 Feedback and action plan

The feedback and action plan phase is an essential part of the COT process, which assesses and improves healthcare professionals' consultation skills. The primary goals of this phase are to provide organized input and collaborate to design a strategy for future development. The main goal of this stage is to identify strengths when the assessor acknowledges good work and offers supportive comments. This helps create a safe learning atmosphere in addition to preventing defensiveness. The trainee's recommendations for areas that they think require improvement may subsequently be expanded upon by the assessor. The assessor provides thorough, beneficial feedback on areas that require improvement by linking observations to the performance criteria and the broader capability framework.

### *25.4.7.6 Advantages of consultation observation tool*

The COT assesses a practitioner's clinical and communication abilities during patient consultations. The COT's benefits are intended for both patients and trainees.

### *25.4.7.7 For the trainees*

#### 25.4.7.7.1 Skill development

Since it allows for organized observation and focused feedback on critical clinical, communicative, and professional skills, the COT is a useful for trainee skill development. This procedure increases self-awareness, pinpoints certain learning requirements, and offers doable suggestions for development. Additionally, by watching their own recorded consultations, trainees can witness their performance directly, which is a potent motivator for self-awareness and behavioral change that is frequently more effective than simply receiving advice on how to get better. Additionally, by using particular performance criteria, the COT enables trainers to offer precise feedback on observable behaviors rather than hazy opinions. This focused feedback is essential for skill improvement.

#### 25.4.7.7.2 Targeted, real-time feedback

A COT's main benefit is that it allows a supervisor to give a trainee quick, targeted, and organized feedback—a critical component for quick skill improvement and enhanced performance in clinical practice. Instead of offering broad remarks, this personalized feedback concentrates on the particular, visible actions and performance standards. Feedback might therefore focus on specific aspects, such as communication of a management plan, the use of open-ended questions, or active listening techniques.

#### 25.4.7.7.3 Holistic evaluation

The COT provides trainees with a comprehensive assessment advantage and a detailed grasp of their ability to apply abilities in real-world clinical settings by evaluating a wide range of competencies beyond basic knowledge testing. The COT's ability to assess a variety of abilities required for successful clinical practice, such as professionalism, communication, interpersonal skills, and clinical skills, is one of its key advantages. Aspects of performance that are frequently overlooked by conventional, single-focus evaluations are captured by this integrated method. Additionally, the evaluation is carried out through direct observation or recording review during actual patient appointments. This guarantees that the assessment is grounded on real performance in an authentic clinical setting, capturing the subtleties and complexity of real patient interactions.

#### 25.4.7.7.4 Self-reflection

The COT has developed into a helpful tool for promoting self-reflection by providing trainees with a structured framework for assessing, evaluating, and learning from their interactions with patients. In this context, the trainees obtain an objective appraisal of their performance, including verbal and nonverbal abilities, by recording and evaluating their own consultations. This can be challenging to accomplish through recollection alone. However, the COT offers particular performance standards, such as active listening and the use of open-ended questions, which direct the trainee's introspection. Using this framework, they are able to move from a general impression of an interaction to a critical, in-depth analysis of specific parts.

#### 25.4.7.7.5 Professional confidence

A COT significantly increases a trainee's professional confidence by providing a methodical framework for direct observation, comprehensive feedback, and self-reflection. Through objective, constructive evaluation, this technique lowers performance uncertainty and fosters competence. Positive feedback in areas where a trainee does well confirm their current skills and promotes excellent practices, which immediately increases the learner's confidence and self-efficacy. Although using observation tools for the first time can be unsettling, trainees can get over this discomfort as they become more comfortable with the procedure and the feedback loop. Their confidence in managing different consultation circumstances, such as challenging cases or telephone consultations, grows as they gain experience and notice improvements.

### *25.4.7.8 For patients and healthcare systems*

#### 25.4.7.8.1 Improved patient care and safety

The COT improves patient care and safety by raising the caliber of clinical consultations, which reduces misdiagnosis, promotes team decision-making, and ensures comprehensive care delivery. This tool focuses on improving diagnostic accuracy by highlighting meticulous history taking, attentive listening, and sound clinical reasoning, all of which are crucial for reaching the correct diagnosis. The tool assists in preventing diagnostic errors that could result in inappropriate therapy and a delayed recovery by guaranteeing a thorough information-gathering procedure. Additionally, this tool offers feedback on communication skills, such as using language that is patient-friendly, properly describing diagnosis, and making sure patients understand. Misunderstandings regarding treatment plans and follow-up schedules, which are frequent causes of medical errors, are lessened by effective communication.

##### 25.4.7.8.2 Enhanced patient experience

The COT enhances the patient experience by improving the quality of the consultation process, primarily through better communication, shared decision-making, and a more patient-centered approach to care. A key component of effective and compassionate care is making the patient feel heard, respected, and valued by emphasizing their personalized experience of their sickness rather than just the condition. In view of this, the COT evaluates the clinician's capacity to investigate the patient's viewpoint, including their thoughts, worries, and expectations. The end result of all these initiatives is improved patient care. Additionally, the COT lowers these risks by enhancing the quality of consultations, resulting in a safer and more efficient overall care experience.

##### 25.4.7.8.3 Shared decision-making

The use of a COT fosters shared decision-making by providing a structured framework for observation and feedback on the key communication skills required for the process. This structured observation helps to translate the theoretical concepts of shared decision-making into actual clinical practice.

##### 25.4.7.8.4 Efficiency

The primary way in which a COT, particularly in a digital or virtual version, increases productivity is by optimizing time for patients and healthcare providers. This technique can be useful since it allows the healthcare professional to respond to inquiries in batches when they have time rather than during a specific, scheduled appointment time. This cuts down on waiting room time as well as the total amount of time spent on each case. Additionally, it aids in better preparation by making it easier for the patient to assess their symptoms or take a preconsultation history. This makes it possible for the physician to go over pertinent information in advance, allowing them to concentrate the actual consultation time on addressing certain issues and coming to well-informed conclusions, resulting in a more efficient dialogue.

#### *25.4.7.9 Quality improvement activity (QIA)*

In healthcare and other fields, a quality improvement activity is a methodical process that involves audits, case reviews, or tracking personal outcome data. It involves professionals reviewing, evaluating, and improving the quality, safety, and effectiveness of their work with the goal of identifying strengths, weaknesses, and areas for positive change, often for revalidation or accreditation. It involves deliberate, data-driven adjustments to enhance system performance and patient care (Clement 2023). As a workplace-based assessment, a quality improvement activity entails physicians methodically examining and enhancing elements of their clinical practice, frequently for revalidation or training. This is done by identifying problems, putting new protocols into place, creating templates, conducting audits, reflecting on results, and showcasing improved patient care or efficiency. This is a formative assessment for growth. It's a methodical approach to demonstrate continuous professional development through practical improvements in areas such as prescription, consultation skills, or service delivery, extending beyond straightforward audits to a complete cycle of improvement. The primary goals of the quality improvement activity are to show ongoing professional development, increase clinical efficacy in the workplace, and improve patient safety.

#### *25.4.7.10 Key features of the quality improvement activity*

When utilized as a workplace-based assessment tool, a quality improvement activity is a methodical procedure that motivates physicians to examine, appraise, and enhance the efficacy, safety, or quality of their work or clinical setting. The following are a few traits of the quality improvement activity:

##### 25.4.7.10.1 Relevance to practice

A quality improvement activity is relevant to practice because it involves methodically reviewing the work of the clinician and identifying areas for improvement, such as patient safety and practice efficiency, making changes, and reflecting on the results to demonstrate that an improvement has occurred often by using a data-driven, cyclical approach. Relevance, a methodical approach, evidence-based, reflection and action, and demonstrable improvement or upholding high standards which are demonstrated by audits, case reviews, or process adjustments are some of the most crucial characteristics in this regard.

#### 25.4.7.10.2 Systematic and robust approach

It should be a systematic, evidence-based process that often follows methods such as the clinical audit cycle or the plan–do–study–act cycle. It involves establishing clear, quantifiable, attainable, pertinent, and time-bound objectives in addition to gathering baseline data to comprehend the scope of the issue and existing practices. Additionally, it also looks into the process of improving the standard of care in response to the results or facts, assessing the action's outcomes, typically by data collection or another audit, to see whether improvements have been made or whether best practices have been upheld.

#### 25.4.7.10.3 Reflection and learning

Reflection and learning are essential elements of quality improvement programs because they support continuous professional development, make it easier to apply theory to practice, and are required for making educated, data-driven adjustments to improve processes and outcomes. Here, the purpose of learning and reflection is to pinpoint areas that require development. Professionals and teams can critically examine and assess the caliber of their work through reflection, determining what functions effectively and where adjustments are required. The essential initial step in any cycle of progress is this self-evaluation. Healthcare workers can also improve their critical thinking abilities, see any biases, and create more efficient problem-solving techniques for challenging circumstances by thinking back on prior experiences and the logic behind the choices made.

#### 25.4.7.10.4 Demonstration of change or outcome

A crucial component of a quality improvement activity is the demonstration of change or outcome, which offers the crucial proof required to verify that the modifications implemented have, in fact, produced a quantifiable improvement. It is impossible to determine whether the steps made were successful without this demonstration. Demonstration plays a crucial part in quality improvement activities by offering proof of efficacy. Improving the standard of care or services is the main objective of a quality improvement endeavor. The initiative's success in closing the gap between actual performance and the desirable outcome can be objectively demonstrated by showcasing a change or a favorable result.

#### 25.4.7.10.5 Discussion and collaboration

Collaboration and discussion are usually regarded as crucial components of successful quality improvement initiatives. These are essential ideas that propel learning, creativity, and the effective execution of long-lasting improvements. Since it frequently deals with complicated, systemic issues that benefit from multidisciplinary input, one of the main roles of debate and collaboration here is to provide viewpoints. Collaboration allows for the creation of creative, comprehensive solutions that an individual might miss as well as a deeper understanding of the problem by bringing together individuals with diverse backgrounds and skills. Additionally, because it explicitly uses shared learning sessions when teams from various settings exchange their experiences, achievements, and failures, it also improves collaboration.

As a workplace-based assessment, the quality improvement activities are essentially a useful, introspective instrument used to guarantee lifetime competence and a dedication to ongoing improvement in the clinical setting. The main benefit of quality improvement initiatives in workplace-based evaluation is that they directly connect individual professional growth to observable enhancements in clinical practice and patient care. These exercises are a very successful way to build professional skills and conduct formative assessments. Additionally, the trainee's involvement in these activities fosters the development of transferable abilities, such as teamwork, communication, leadership, and project management, that go beyond clinical knowledge. Furthermore, it increases motivation and engagement because trainees who work on projects that solve real-world problems they have identified are frequently more genuinely interested and driven since they can understand how their effort will directly benefit their team and patients. It encourages direct practice improvement for the workplace or organization, where the primary goal of quality improvement initiatives is to attain measurable improvements to specific aspects of healthcare delivery, safety, efficacy, and efficiency in the community. It also fosters an organizational culture of problem-solving, continuous improvement, and a proactive approach to error prevention by involving employees at all levels in quality improvement initiatives.

## 25.4.8 Entrustable professional activity (EPA)

EPA is a workplace-based assessment method that incorporates a variety of abilities into routine clinical practice, enabling frequent, efficient, feedback-oriented evaluation in the natural learning context. It consists of specific learning and evaluation objectives for classifying and evaluating fundamental tasks within a field. In this, the trainees must complete defined, observable professional tasks known as EPAs in order to prove their competence and be given unsupervised practice. They complement competency-based medical education by reorienting the emphasis from disjointed skills to comprehensive, practical professional responsibilities (Pinilla, 2021).

### *25.4.8.1 Role of entrustable professional activity in the workplace-based assessment*

EPAs are crucial to workplace-based assessment because they provide visible, unitized tasks like "managing a patient's initial presentation" that link curriculum to practice. Instead of only grading skills, this allows supervisors to make entrustment decisions like "can I trust this trainee to do this safely?" EPAs are essential to the organization of workplace-based assessment because they define core competencies, guide formative feedback, monitor progressive independence from direct supervision to independent practice, advise competency committees on important decisions, and shift assessment from time-based to competency-based education.

### *25.4.8.2 Role of trainee and assessor in workplace-based assessment*

By actively engaging in their everyday responsibilities, demonstrating behaviors and abilities, assessing their performance, and providing evidence of their learning, trainees take part in workplace-based evaluations. Assessors must observe trainees in real-world work settings, provide timely, constructive feedback on performance, identify areas of strength and weakness, compare the trainee's performance to expected standards, and raise concerns about poor practice in order to support the trainee's development or ensure patient safety.

### *25.4.8.3 Role of the trainee*

Trainees participate in workplace-based evaluations by actively carrying out their daily tasks, exhibiting behaviors and skills, evaluating their performance, and presenting proof of their learning. Assessors must observe trainees in real-world work settings, provide timely, constructive feedback on performance, identify areas of strength and weakness, compare the trainee's performance to expected standards, and raise concerns about poor practice in order to support the trainee's development or ensure patient safety.

#### 25.4.8.3.1 Active participation

In order to truly engage with feedback, set goals, and promote learning, active trainee participation in workplace-based assessments necessitates trust in supervisors, an understanding of the purpose of the assessment, supportive systems like time allocation, and basic tools to overcome obstacles like power dynamics and anxiety, shifting the focus from high-stakes judgment to formative growth. Workplace-based assessments become genuine learning opportunities rather than merely performance evaluations when trainees voluntarily select examples, debate them, and use expert feedback to identify strengths and shortcomings and develop learning plans for growth.

#### 25.4.8.3.2 Demonstrate skills

Using tools like DOPS or mini-CEXs to evaluate technical and nontechnical skills like communication, decision-making in real-time, focusing on strengths and development areas, and developing action plans for continuous improvement, trainees can demonstrate their skills in workplace-based assessments through direct observation, targeted feedback, and reflection. This goes beyond a score to promote true competence.

#### 25.4.8.3.3 Evidence for learning

The evidence for trainee learning in workplace-based assessments consists of a combination of structured tools such as mini-CEXs, CbDs or DOPS, and naturally occurring clinical encounters that are recorded in a portfolio. It focuses on both assessment for learning for competence and assessment for learning for development, driven by direct observation, MSF, reflection, and dialogue between trainee and supervisor to demonstrate growth over time.

#### 25.4.8.3.4 Seeking feedback

In order to obtain feedback for workplace-based assessments, trainees must proactively plan, ask targeted questions about observed performance that focus on both strengths and weaknesses, actively listen, take notes, and make a commitment to using the feedback to guide learning. This process creates a cycle of self-reflection, improvement, and closing the feedback loop with trainers for successful skill development in real-world work settings.

#### 25.4.8.3.5 Meeting requirements

In order to fulfill the requirements of workplace-based assessment, trainees must actively participate by completing a variety of assessments, gathering evidence for their e-Portfolio, asking assessors for constructive feedback on a regular basis, discussing their progress with their supervisor, and exhibiting competence across professional capabilities, ultimately demonstrating readiness for independent practice. Timely feedback, connecting evidence to capabilities, correcting subpar performance, and making sure tests encompass clinical skills, professionalism, and decision-making in real-world scenarios are some of the crucial components.

### *25.4.8.4 Role of the assessor*

#### 25.4.8.4.1 Observe and judge

In a workplace-based assessment, the assessor serves as both a judge who renders unbiased, fact-based judgments regarding competence based on predetermined criteria and an observer who methodically gathers evidence of performance in a genuine work environment. Here, the assessor's primary responsibility as an observer is to collect genuine proof of a candidate's abilities, knowledge, and behaviors during regular work or particular assessment activities, such as a DOPS. To analyze the evidence, the assessor also takes on the role of a judge. In this sense, assessors employ their subject-matter expertise to compare all gathered evidence such as work samples, verbal interrogations, and observations against predetermined assessment criteria and occupational standards.

#### 25.4.8.4.2 Provide feedback

In workplace-based assessment, the assessor's role is to provide timely, specific, and constructive feedback based on objective observation to guide learning and improve performance, not just to grade. This involves highlighting strengths, identifying development areas, encouraging reflection, and creating a safe space for open dialogue about the trainee's progress towards competency.

#### 25.4.8.4.3 Evaluate performance

As an expert observer, coach, and judge, the assessor's role in workplace-based assessment is to systematically evaluate a learner's practical performance and competence against predetermined standards in real-world work settings, provide timely, targeted feedback, connect theory to practice, and make well-informed judgments to guide development and progression, bridging the gap between learning and doing.

#### 25.4.8.4.4 Identify skill gaps

In addition to offering constructive criticism, an assessor in workplace-based assessment is also in charge of identifying skill gaps, maintaining integrity, and promoting ongoing learning. Important concerns frequently include validity and reliability, time constraints, administrative burden, psychological safety, objectivity, consistency among assessors, and obtaining timely, high-quality feedback for improvement rather than merely grading.

#### 25.4.8.4.5 Professional responsibility

The assessor's role in providing professional responsibility in workplace-based assessment involves upholding ethical standards, ensuring fairness and impartiality, maintaining occupational competence, and role-modeling professional conduct. These actions ensure the credibility of the assessment process and foster a culture of professionalism in the trainee.

#### 25.4.8.4.6 Recommending further action

In a workplace-based assessment, the assessor's role in recommending further action is to provide objective, specific, and forward-looking guidance to support the individual's continued development and improvement. This is a crucial part of the feedback loop, helping the learner bridge the gap between their current performance and the required standards of competence.

### 25.4.9 Role of learning environment in workplace-based assessment

The learning environment is essential to workplace-based assessment because it offers the real-world, authentic context for skill development, allowing for direct observation and hands-on experience. It also shapes learning through social or cultural factors, trust, and power dynamics and serves as a framework for assessment to drive formative learning via goal-setting, self-regulation, and structured feedback, moving learners towards mastery. However, challenges include unpredictability and ensuring consistent, high-quality feedback opportunities. Additionally, it promotes learning since the environment's tests offer useful feedback that helps students produce, comprehend, and apply knowledge, ultimately enhancing performance.

## 25.5 Workplace-based assessments and competency-based medical education

In competency-based medical education, workplace-based assessments have gained importance as they provide real-time, authentic evaluation of trainees' clinical skills in real-world settings and provide quick, useful feedback for improvement rather than merely grading. Workplace-based assessments offer direct observation and the quick, helpful feedback required to satisfy predefined competency criteria (Begum et al., 2023). They serve as the foundation of programmatic assessment, employing a variety of low-stakes observations such as Mini-CEX, DOPS, and CbD to provide a thorough picture of a learner's competency, directing their path toward independent practice and facilitating entrustment decisions.

### 25.5.1 Alignment of workplace-based assessments with competency-based medical education

In competency-based medical education, aligning workplace-based assessment entails connecting real-world clinical observations, feedback, and performance data to particular competencies within a standardized framework. This creates a continuous feedback loop that allows trainees to develop skills for predetermined outcomes instead of merely evaluating knowledge (Booth et al., 2009). In order to guide learning and certify practice readiness, key alignment strategies include using structured tools such as in communication or procedures, recording real-world encounters in portfolios, ensuring multiple assessors for validity, and feeding this data to competence committees for progress decisions. Workplace-based assessment evaluates a wide range of professional skills, including communication and consultation skills, data collection and interpretation, clinical examination and procedural skills, clinical judgment and decision-making, professionalism, and clinical management, which includes managing both common and complex medical conditions, including comorbidities and uncertainty, as well as working with colleagues.

## 25.6 Factors affecting workplace-based assessment

Workplace-based assessment is impacted by a number of factors, such as trainer or assessor issues like training, bias, engagement, trainee factors like learning style, engagement, and feedback literacy, organizational context in terms of time, workload, and culture, psychosocial elements like communication, support, and work-life balance, and assessment design elements like validity, reliability, and tools. In order to provide genuine, excellent feedback for ongoing learning and competency development rather than merely evaluation, effective workplace-based assessment strikes a balance between these factors. The following categories apply to the factors (Fig. 25.2):

### 25.6.1 Assessor and trainer factors

#### *25.6.1.1 Training and development*

For workplace-based assessment to be implemented effectively, faculty development and training are essential because they address issues like faculty capacity, time restrictions, and the requirement for regular, high-quality feedback. The quality of feedback and the general perception of workplace-based assessment can both be greatly enhanced by faculty training. On the other hand, inconsistent application of criteria may result from a lack of trainer training.

#### *25.6.1.2 Cognitive bias*

When assessors see and evaluate a trainee's performance in an actual clinical context, they frequently make systematic mistakes in judgment and decision-making. This is known as cognitive bias in workplace-based assessment. These biases

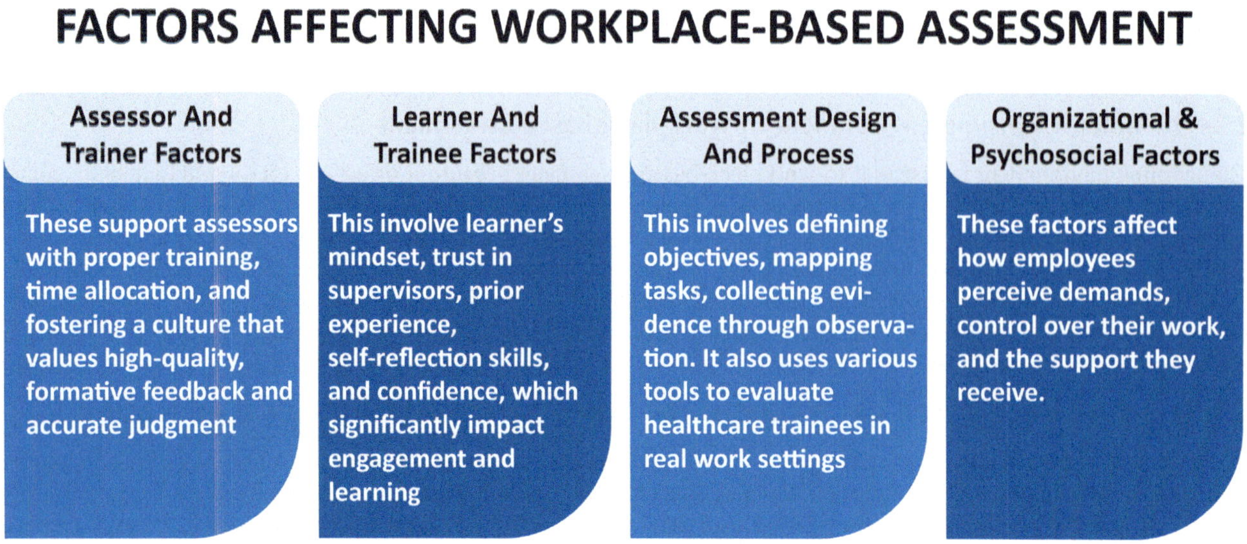


FIGURE 25.2 **Factors affecting workplace-based assessment.** This figure shows the factors affecting workplace-based assessment.

result from mental shortcuts that our brains employ to quickly handle complicated information, but they can cause erroneous and subjective evaluations that may jeopardize truth and fairness.

#### *25.6.1.3 Assessor engagement*

In workplace-based assessment, faculty members actively participate as observers, assessors, and feedback providers in actual clinical settings. They guide trainees' development by providing immediate verbal and written feedback, identifying strengths, and pinpointing areas for improvement. This is essential for ensuring competence despite obstacles like time constraints and connecting daily work with learning goals.

### 25.6.2 Learner and trainee factors

In workplace-based assessment, learner and trainee factors include mindset, trust in supervisors, prior experience, self-reflection skills, and confidence, all of which have a significant impact on engagement and learning. Positive outcomes depend on trainees feeling safe, getting timely and relevant feedback, comprehending the assessment goals, and viewing supervisors as reliable mentors who support their development rather than merely evaluate them.

#### *25.6.2.1 Trust and power dynamics*

In workplace-based assessments, trust and power dynamics refer to the important, frequently unequal relationship between the supervisor and the trainee. The supervisor has formal power, such as job title and assessment authority, but the trainee's willingness to be honest and learn depends on the supervisor's fairness, knowledge, and commitment to the trainee. Since trainees estimate vulnerability dependent on perceived supervisor trustworthiness and prospective negative power, this interaction decides whether an evaluation is a meaningful learning opportunity or a tense, guarded event.

#### *25.6.2.2 Motivation and mindset*

Motivation in the context of workplace-based evaluation refers to a person's desire and drive to do well, whereas a growth mindset is the conviction that one's skills can be improved with commitment and hard work. Both are important determinants of engagement, performance, and growth potential.

#### *25.6.2.3 Self-reflection*

Self-reflection in workplace-based assessment is the critical process where trainees think deeply about their own performance, behaviors, strengths, weaknesses, and learning experiences on the job, comparing their self-perception with actual outcomes and feedback to foster growth, improve decision-making, and align personal development with organizational goals. It involves asking "what, why, and how" about your work to find better ways to do things, making it a core part of continuous professional development.

### 25.6.3 Assessment design and process

The process involves defining objectives, mapping tasks, collecting evidence through observation, providing timely, specific feedback, and tracking progress to ensure trainees meet standards, enabling entrustment decisions for independent practice.

#### *25.6.3.1 Validity and reliability*

Despite implementation challenges like time constraints, workplace-based assessments offer high validity by evaluating actual clinical performance in context. However, their reliability can be questioned due to assessor variability, though this is improved with multiple encounters or assessors, leading to strong educational impact, particularly for formative feedback.

#### *25.6.3.2 Tools and methods*

Workplace-based assessment employs a variety of techniques to assess healthcare trainees in actual work environments. It focuses on abilities that are challenging to test elsewhere, such as communication, professionalism, and clinical judgment. It does this through structured discussions like CbDs or direct observation like Mini-CEX. It is enhanced by MSF and evaluated through portfolios for formative and summative feedback to ensure competency and guide learning.

#### *25.6.3.3 Authenticity*

Authenticity in workplace-based assessment means evaluating trainees' clinical skills and knowledge through real-world tasks in their actual work setting, moving beyond rote memorization to judge practical application. Key aspects include using diverse tools, multiple assessors, longitudinal records, and real-time feedback to ensure assessments reflect actual professional performance and foster growth in authentic clinical scenarios, enhancing validity and educational impact.

### 25.6.4 Organizational and psychosocial factors

Organizational and psychosocial factors significantly impact the effectiveness and outcomes of workplace-based assessment by influencing employee stress, performance, and overall well-being. These factors affect how employees perceive demands, control over their work, and the support they receive. Organizational factors are the underlying elements within an organization's structure and management that shape the work environment and influence employee responses.

#### *25.6.4.1 Clear leadership and expectations*

In the context of workplace assessment, organizational factors like clear leadership and expectations play a critical role. A lack of clarity regarding roles, responsibilities, and expected performance can create significant psychological distress and confusion during assessments.

#### *25.6.4.2 Organizational culture*

In addition, a supportive, people-oriented culture with a focus on psychological safety is an important variable in employee motivation and performance.

## 25.7 Significance of workplace-based assessment

Workplace-based assessment is important because it assesses professional competency by seeing people in their real-world workplace, connecting theoretical understanding with real-world application, and offering crucial, ongoing feedback for professional growth. Workplace-based assessment by way of recognizing both strengths and chances for improvement offers targeted support to guarantee that the individuals complete professional standards and advance successfully in their professions (Dong et al., 2024). The workplace-based evaluation is important because it affects both the assessors and the trainees (Fig. 25.3).

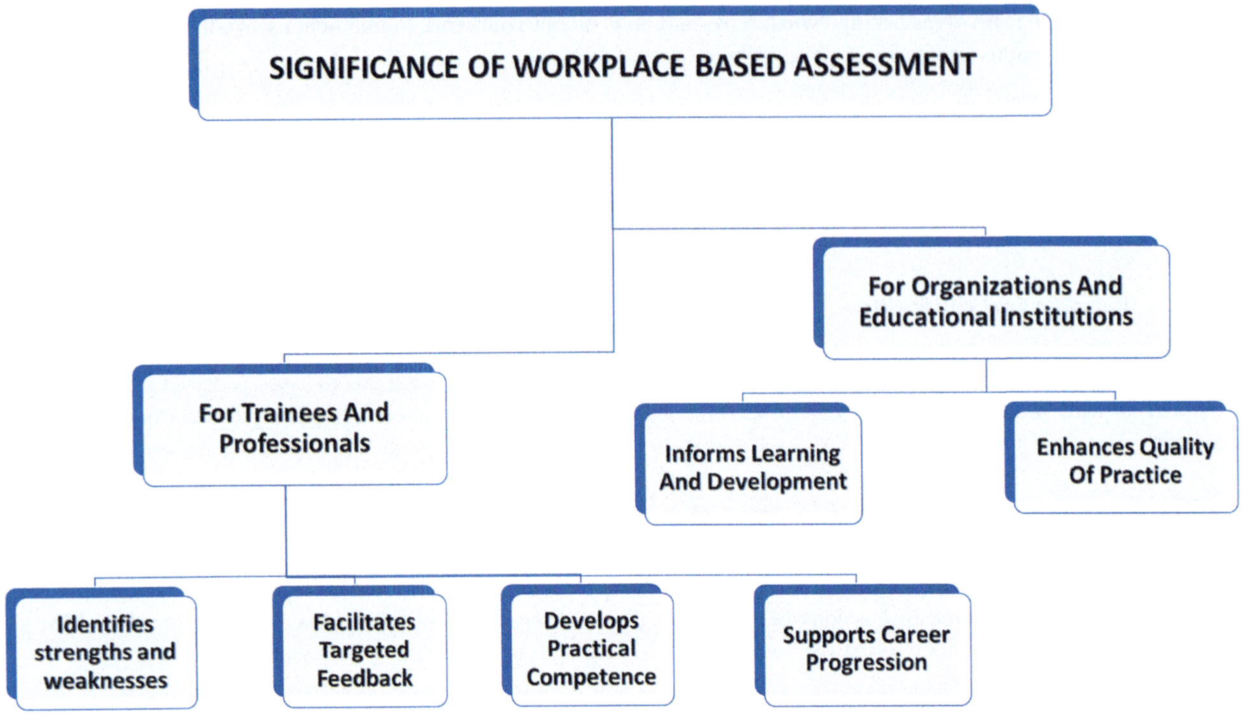


FIGURE 25.3 **Significance of workplace-based assessment.** This figure shows the significance of workplace-based assessment.

## 25.7.1 For trainees and professionals

### *25.7.1.1 Identifies strengths and weaknesses*

One of the main benefits and goals of workplace-based assessment is identifying trainees' strengths and limitations. Instead of merely certifying competence, it is intended as a formative evaluation approach to offer insightful, structured feedback and direct learning. One of the main benefits is that the trainee is observed in the real world, as assessments take place while they work with genuine patients on a daily basis. This allows for the evaluation of actual performance and professional behaviors in a natural setting. This makes it easier to see how trainees use their knowledge and abilities in real-world situations (Nair et al., 2012).

### *25.7.1.2 Facilitates targeted feedback*

By evaluating trainees' performance in actual clinical settings and offering prompt, precise, and helpful feedback on their daily practice, workplace-based assessment enables trainees to receive focused feedback. In contrast, traditional exams may yield just general marks. The crucial element in this situation is the assessors' direct observation of trainees carrying out actual activities with real patients and coworkers, which enables comments to be rooted in particular, real-world occurrences rather than made-up scenarios. Additionally, the observation approach helps pinpoint a trainee's unique strengths, shortcomings, and opportunities for improvement in a variety of abilities, including as communication, clinical skills, and teamwork, making the feedback extremely pertinent and useful.

### *25.7.1.3 Develops practical competence*

Workplace-based assessment evaluates trainees in actual clinical settings to help them gain practical competence. Compared to fake exams, workplace-based assessment delivers a more accurate assessment of trainees' abilities by using actual patients and clinical circumstances (Erumeda et al., 2024). This guarantees that the abilities acquired will be immediately useful in their future work. Supervisors can give quick, detailed, and useful comments about the trainee's performance in that specific setting because the assessment takes place in the real workplace. For quick skill development, this direct application of feedback is essential.

#### 25.7.1.4 Supports career progression

Workplace-based evaluation significantly promotes career advancement by monitoring professional development, making sure clinical competencies are fulfilled, and offering organized, useful feedback. In the field of medical education, workplace-based evaluation is a crucial component of formal training and credentialing procedures (Saedon et al., 2010). The findings are used to determine if a student is prepared to move on to the next phase of their training or obtain full certification or registration. The fact that these tests assess abilities, professional conduct, clinical reasoning, and communication in authentic contexts, all of which are challenging to grade in conventional exams, is one of their key features. In many organized training programs, showing these competencies successfully is a prerequisite for development.

### 25.7.2 For organizations and educational institutions

#### 25.7.2.1 Informs learning and development

By offering real-time, constructive feedback in real-world work situations, workplace-based assessment is essentially intended to guide learning and development by assisting trainees in developing self-awareness, enhancing skills, tracking progress, and addressing developmental needs for improved performance and patient safety. By evaluating "does" level performance rather than just "knows" or "knows how," it goes beyond traditional tests, making learning instantly applicable and promoting ongoing progress.

#### 25.7.2.2 Enhances quality of practice

By integrating assessment with clinical work and giving healthcare professionals rapid, precise, and organized feedback, workplace-based assessment can improve practice quality. By assessing performance in real-world work environments, workplace-based evaluation helps close the gap between formal education and practical clinical practice. This guarantees that experts fulfill pertinent, practical competency requirements. It must be mentioned that the primary advantage of workplace-based assessment is the chance for prompt, helpful comments based on firsthand performance observation. Trainees are more likely to perform well when they receive feedback that is clear, reliable, and connected to a plan of action.

## 25.8 Challenges of workplace-based assessment

Time constraints for busy assessors, a lack of assessor buy-in or training, problems with standardization and reliability, challenges integrating teaching in assessment roles, the possibility of bias, administrative burden, and making sure assessments are genuinely formative rather than high-stakes that trainees perceive as "tick-box" exercises are some of the challenges associated with workplace-based assessment. The following list includes some of the main obstacles to workplace-based learning (Fig. 25.4):

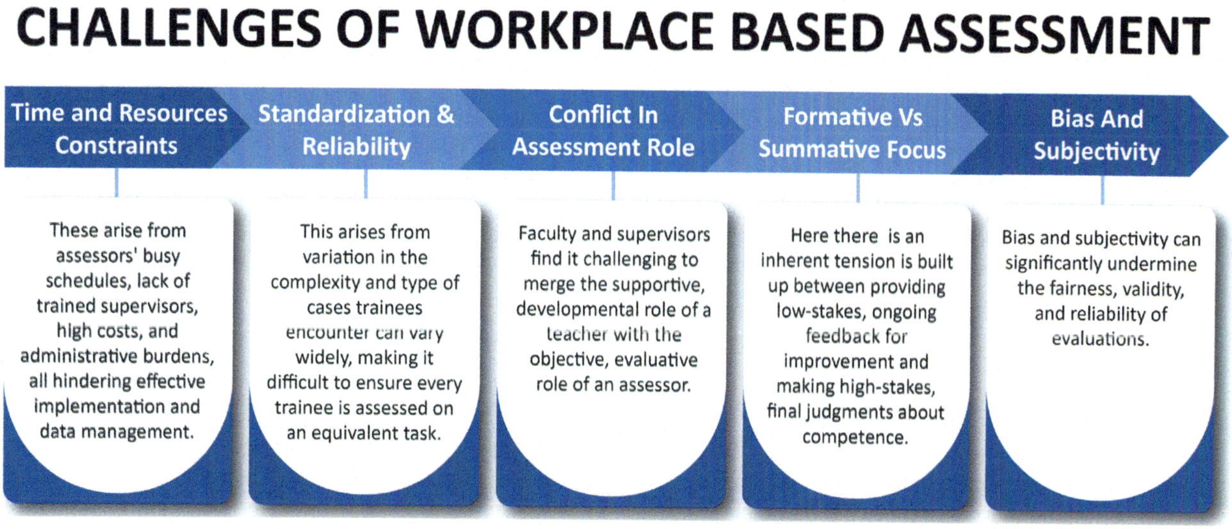


FIGURE 25.4 **Challenges of workplace-based assessment.** This figure shows the challenges of workplace-based assessment.

### 25.8.1 Time and resources

Workplace-based assessments face significant time and resource constraints due to assessors' hectic schedules, a lack of supervisors with the necessary training, high costs, and administrative burdens. These factors all impede effective implementation and data management, necessitating strategies like technology and process simplification to make assessments more effective and integrated into daily work.

### 25.8.2 Standardization and reliability

In terms of standardization, it is a reality that assessment tasks in the real workplace are not prescripted, which means that the complexity and kinds of situations that trainees meet might vary greatly, making it challenging to guarantee that every trainee is evaluated on an identical assignment. Furthermore, there is frequently disagreement among assessors on the required standards and how to interpret the assessment criteria.

### 25.8.3 Conflict in assessment role

One major issue with workplace-based assessment is the conflict in the assessor job, particularly the combined responsibilities of serving as an examiner and an educator/supervisor. Combining the objective, evaluating job of an assessor with the supporting, developmental role of a teacher is difficult for supervisors and faculty. When they judge a student, assessors may feel as though they are evaluating their own instruction, which can cause discomfort.

### 25.8.4 Formative versus summative focus

The inherent conflict between giving high-stakes, final assessments of competence and offering low-stakes, continuous feedback for development makes it difficult to strike a balance between the formative and summative focus in workplace-based assessment. These difficulties may result from the trainee's perspective. For example, when the evaluation has summative repercussions, trainees may be reluctant to discuss difficult situations, acknowledge their shortcomings, or participate completely in the feedback process out of concern that it will negatively affect their final score. Instead of having a sincere learning dialog, this could result in a careless attitude of simply fulfilling the evaluation requirements.

### 25.8.5 Bias and subjectivity

Bias and subjectivity pose serious problems for workplace-based assessment because they can seriously compromise the validity, reliability, and fairness of assessments. The intrinsic dependence on human judgment and the nonstandardized character of real-world work situations are the root causes of these problems. Biased evaluations can exacerbate already-existing disparities within an organization by resulting in biased decisions about promotions, pay, and development opportunities.

**Summary**

The self-directed learning environment that workplace-based evaluations foster is crucial for ongoing professional growth. In the workplace, a wide range of clinical encounters that are highly relevant to the trainees' entire curriculum can be evaluated, and the interaction between trainees and their assessors offers a priceless learning opportunity. Workplace-based exams assist in bridging the gap between theoretical learning and practical application by focusing on how students apply clinical knowledge, engage with patients and colleagues, and demonstrate professional attitudes during normal patient care. Workplace-based assessments offer ongoing, formative insights into learner performance across a range of competencies through instruments like the mini-clinical evaluation exercise, direct observation of procedural skills, case-based discussion, and multisource feedback. In order to assist students in improving their clinical reasoning, procedural abilities, and interpersonal competencies, it highlights the significance of providing high-quality, useful feedback.

**Key point**

Future perspectives: moving towards excellence in patient-centered care

The continual advancements in healthcare, educational technologies, and the changing needs for competency-based training all influence the future prospects of workplace-based assessment in medical education. Assessment of students is evolving as a result of the shift to competency-based medical education. Workplace-based evaluation methods will become even more important as medical education shifts away from time-based training and toward acquiring specified competencies. Future outlooks point to a

**(cont'd)**

stronger focus on individualized learning through workplace-based assessments, which are customized to each learner's needs and measure their progress in real time across several domains of competency. The future of workplace-based evaluation in medical education is expected to be marked by more technological integration, personalized learning, and a broader emphasis on professional abilities beyond technical proficiency. The methods and instruments used to evaluate aspiring medical professionals will also grow more sophisticated as healthcare demands do. Workplace-based evaluation will continue to be a crucial tool for guaranteeing that students acquire the skills required to provide excellent, patient-centered care while simultaneously encouraging a culture of ongoing professional development throughout their careers.

# References


Anderson, H.L., Kurtz, J., West, D.C., 2021. Implementation and use of workplace-based assessment in clinical learning environments: a scoping review. Academic Medicine 96 (11), S164. Available at https://doi.org/10.1097/ACM.0000000000004366, https://journals.lww.com/academicmedicine/pages/default.aspx.

Begum, J., Sidhu, T.K., Mahajan, A., Dutta, S., Singhal, A., Kwatra, G., 2023. Workplace-based assessment in the era of competency based medical education: perception and experiences as a medical teacher. Medical Journal of Dr. D.Y. Patil Vidyapeeth 16 (2), 255–259. Available at https://doi.org/10.4103/mjdrdypu.mjdrdypu_79_22.

Björklund, K., 2022. BMC Med Educ 1.

Booth, J., Johnson, G., Wade, W., 2009. Workplace-based assessment pilot. Report of Findings of a Pilot Study Royal College of Physicians.

Burch, V.C., 2019. The changing landscape of workplace-based assessment. jatt 20, 37–59.

Clement, T, 2023. Professional Development in Education 50.

Dong, C., Low, S.C., Yan, C.C., 2024. Workplace-based assessment in healthcare: key concepts and best practices. Proceedings of Singapore Healthcare 33. Available at https://doi.org/10.1177/20101058241310618, https://journals.sagepub.com/home/PSH.

Erumeda, N.J., George, A.Z., Jenkins, L.S., 2024. Evidence of learning in workplace-based assessments in a family medicine training programme. South African Family Practice 66 (1). Available at https://doi.org/10.4102/safp.v66i1.5850.

Hill, F., Kendall, K., Galbraith, K., Crossley, J., 2009. Implementing the undergraduate mini-CEX: a tailored approach at Southampton University. Medical Education 43 (4), 326–334. Available at https://doi.org/10.1111/j.1365-2923.2008.03275.x.

Khan, Adil, M., Gorman, Mark, Gwozdziewicz, Lukasz, et al., 2013. Direct observation of procedural skills (DOPS) as an assessment tool. Journal of Pakistan Medical Students 3, 137–140.

Liu, C., 2012. An introduction to workplace-based assessments. Gastroenterology and Hepatology From Bed to Bench 5 (1), 24–28 PMID: 24834194; PMCID: PMC4017451.

Miller, G.E., 1990. The assessment of clinical skills/competence/performance. Academic Medicine 65 (9), S63. Available at https://doi.org/10.1097/00001888-199009000-00045.

Nair, B.R., Hensley, M.J., Parvathy, M.S., Lloyd, D.M., Murphy, B., Ingham, K., et al., 2012. A systematic approach to workplace-based assessment for international medical graduates. Medical Journal of Australia 196 (6), 399–402. Available at https://doi.org/10.5694/mja11.10709Australia, https://www.mja.com.au/system/files/issues/196_06_020412/nai10709_fm.pdf.

Norcini, J.J., 2003. ABC of learning and teaching in medicine: work based assessment. British Medical Journal 326, 753–755. Available at https://doi.org/10.1136/bmj.326.7392.753.

Pinilla, S, 2021. BMC Medical Education 21.

Saedon, H., Saedon, M.H.M., Aggarwal, S.P., 2010. Workplace-based assessment as an educational tool: guide supplement 31.3 – viewpoint. Medical Teacher 32 (9), e369. Available at https://doi.org/10.3109/01421590903548547.

Chapter 26

# Feedback in medical education

*"The purpose of feedback in the learning process is to improve a student's performance—definitely not put a damper on it. The ultimate goal of feedback is to provide students with an 'I can do this attitude.'"*

—Laura Reynolds

## 26.1 Introduction

Feedback in health professional education refers to particular information given to a student regarding their observed performance in relation to a standard or goal, with an aim of improving future performance and learning outcomes. It plays a vital role in skill acquisition and professional growth by emphasizing strengths and indicating opportunities for improvement in knowledge, abilities, and professional behavior. One of the most significant ways that the "teacher" and the "student" connect is through feedback. However, it has been well documented that during their clinical rotations, medical and other health professional students are hardly ever personally watched and provided feedback. Giving feedback is a crucial step in the learning process since it helps close the performance gap between desired and actual performance (Hewson and Little, 1998).

Learning tactics are improved as a result of the feedback process, which informs the learner about the caliber of their performance. Feedback enhances learning outcomes and helps students make wise decisions. It also acts as an effective tool that helps students advance academically as through feedback students are able to identify their strengths and can work upon their weaknesses. Feedback in medical education is precise, information-sharing that evaluates a student's performance against a benchmark in order to enhance their professional behaviors, knowledge, and abilities. Instead of assessing the individual, effective feedback is quick, clear, actionable, and concentrates on observable behaviors to direct future performance. It is also an essential tool for skill development, offering helpful advice that enables students to recognize their areas of strength and growth in clinical settings.

## 26.2 Purpose of feedback

Feedback helps guide learners' progress in medical education by providing clear, accurate information about their performance, bridging the gap between current and expected competence, promoting self-reflection, and fostering the development of critical clinical, professional, and communication skills. These skills will eventually ensure better patient care. It is crucial for enhancing self-control, adjusting abilities, and transforming vague observations into specific development objectives that go beyond assessment to real learning.

### 26.2.1 Performance calibration

The main goal of performance calibration through feedback in medical education is to improve calibration accuracy, which is the process of reducing the discrepancy between a learner's self-assessed perceived performance and their actual performance. Increased self-awareness is essential for developing clinical reasoning, which in turn improves patient care and diagnostic precision. One of the key functions of feedback in performance calibration is to allow students to compare their personal assessments of high-quality work to anticipated professional standards. As a result, they are more equipped to track and assess their own performance throughout their careers.

### 26.2.2 Skill development

The main goal of feedback in medical education is skill development. Feedback helps reduce the gap between current and desired performance by fostering self-regulated learning, generating patient-centered, compassionate physicians, and

**The Principles and Practice of Medical Education. DOI:** https://doi.org/10.1016/B978-0-443-44197-4.00027-8

enhancing clinical, communication, decision-making, and professionalism competencies through accurate, practical insights. It is essential for formative learning, assisting trainees in developing reflective practice, calibrating performance, and achieving professional growth. It goes beyond mere evaluation to promote behavioral change (Ende, 1983).

### 26.2.3 Promotes reflection

Feedback's primary function in medical education changes from evaluation to reflection, turning students' experiences into profound, integrated knowledge for better clinical skills, patient care, self-regulated learning, and professional identity formation. This helps students analyze their behavior, comprehend context, and create practical plans for future development.

### 26.2.4 Guides future performance

The primary goal of feedback in medical education is to specifically direct future performance. It is important to keep in mind that the feedback is a formative evaluation tool, which means it is an essential component of the continuous learning process rather than a definitive assessment. It is intended to support students in staying on course to meet their learning objectives. Effective feedback's primary objective is to assist the learner in recognizing and reducing the discrepancy between their actual performance and the intended or expected standard of performance, which is one of its key components.

### 26.2.5 Enhances self-regulation

Feedback in medical education attempts to enhance self-regulation by empowering students to become autonomous, goal-directed learners who can organize, track, and modify their own learning and clinical performance. It goes beyond merely fixing mistakes to develop metacognitive skills like self-assessment and strategic adjustment, ultimately promoting lifelong learning and professional competence. Feedback makes students more aware of their strengths, weaknesses, and current skills by contrasting their self-analysis with external analysis. Students can establish clear, attainable learning objectives for next assignments by comprehending performance gaps (Ramani and Krackov, 2012).

### 26.2.6 Drives motivation and direction

Feedback provides motivation and clarity by emphasizing areas that require attention and places that have made progress. Additionally, it preserves motivation and self-efficacy by recognizing and rewarding exceptional performance, which enhances a learner's confidence and motivates them to continue positive actions. Additionally, students are more likely to actively participate in the learning process and seek out additional feedback when it is positive, explicit, and given in a supportive setting.

## 26.3 Components of feedback

Appropriate feedback must be based on an accurate assessment and given to the learner in a supportive, self-directed setting. The next stage or level to reach should then be discussed and guided by the feedback source. The facilitation of how to get there helps with this process (Fig. 26.1).

### 26.3.1 Specificity and descriptiveness

The fact that specific feedback focuses on specific behaviors, actions, or results rather than making general remarks like "good job" or "bad job" is one of its key characteristics. When feedback is given in a certain manner, the learner can clearly see their strengths and weaknesses as well as the specific behaviors that need to be changed. Additionally, it establishes credibility and trust by demonstrating that the person giving the criticism has listened and is concerned enough to make insightful commentary. In terms of descriptiveness, observations of the activity and its effects should be the main focus of the feedback. Here it is always better to avoid personal assessment or judgment. Additionally, by placing more emphasis on effort and the process of improvement than on natural aptitude, this method promotes a growth mentality. It enables the recipient to take charge of their own growth and turns feedback into a two-way learning process.

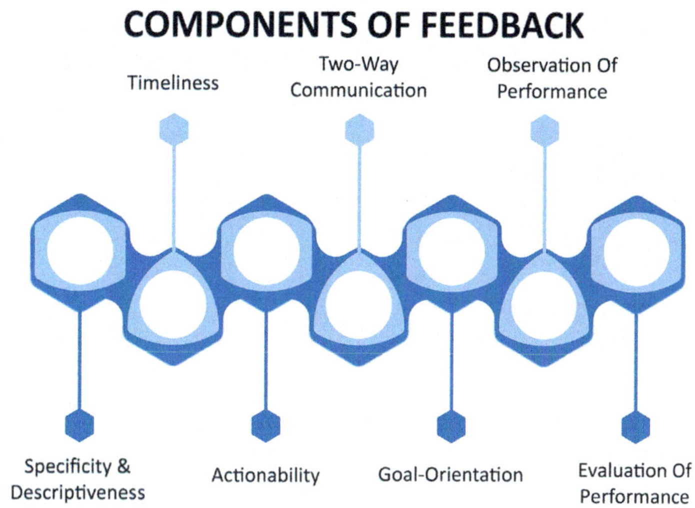


FIGURE 26.1 **Components of feedback.** This figure shows the components of feedback.

## 26.3.2 Timeliness

Giving feedback as soon as possible after an activity makes it more memorable, relevant, and actionable, which helps recipients change their behavior, develop trust, and speed up learning. Delivering feedback later loses its impact and may lead to confusion or resentment. Ideally, the feedback should be given quickly but carefully, striking a balance between immediacy and the recipient's readiness to receive it. This guarantees that the feedback is still applicable and useful. Giving feedback within a day following the activity is one of the best practices. This keeps the information current and enables the student to fully comprehend the tutor's point of view.

## 26.3.3 Actionability

The effectiveness of actionability in medical education feedback depends on learner receptivity, educator skill, and a supportive culture. Action plans and follow-up are frequently necessary to bridge the gap between receiving feedback and actually changing performance. Actionability in medical education feedback refers to specific, clear, nonjudgmental guidance focused on behaviors, allowing learners to create concrete improvement plans and demonstrate progress.

## 26.3.4 Two-way communication

Two-way communication in feedback is a cooperative, dialogic process in which the teacher and the student actively exchange observations, evaluate performance, and jointly develop a plan for progress. This tactic fosters a supportive learning environment and builds trust by transforming feedback from a one-way to a reciprocal process. This establishes a shared responsibility because both the teacher and the student are actively involved. While the teacher offers their insights and views, the student is encouraged to evaluate their own performance. This two-way dialogue improves the impact of subsequent feedback interactions by fostering a strong educational relationship and incorporating empathy and thoughtful listening.

## 26.3.5 Goal orientation

In medical education, goal-oriented feedback goes beyond mere evaluation to actively direct skill development and future professional progress by establishing clear, explicit learning objectives, offering concrete steps, and promoting self-reflection. It places a strong emphasis on using descriptive, nonjudgmental language, connecting feedback to specific objectives such as enhancing clinical or communication skills, and encouraging learners to evaluate themselves and design their own learning trajectory. This method increases the effectiveness of feedback, enabling students to become proactive learners who can make adjustments and produce higher results.

### 26.3.6 Observation of performance

The fundamental element of feedback is performance observation, which acts as the objective data collection stage that comes before communication intended to direct improvement. It is impossible to provide insightful and useful feedback without precise, methodical, and unambiguous observation. A comprehensive and extensive observation is one of the key components of an accurate evaluation and feedback. In order to develop their assessment, feedback providers must critically observe the learners' performance. Crucial observation is an educational talent that requires ongoing development.

### 26.3.7 Evaluation of performance

The foundation of effective feedback in medical education is performance evaluation, which provides the objective data in the form of clinical skill observation and knowledge application which are needed to identify strengths and weaknesses, direct students toward mastery, and close gaps between present and desired performance levels. This makes feedback specific, reliable, and useful for professional development. Without thorough, direct observation and assessment, feedback remains vague, failing to motivate or facilitate the deep learning required for competent clinical practice, though challenges exist in consistent, high-quality implementation.

## 26.4 Types of feedback in medical education

The feedback in medical education is broadly classified into two types based on purpose and focus and based on settings and delivery. These types are depicted in the below diagram (Fig. 26.2).

### 26.4.1 By purpose and focus

#### *26.4.1.1 Corrective or negative feedback*

In medical education, corrective or negative feedback is a crucial part of formative assessment that aims to bridge the gap between a trainee's present performance and expected standards. Corrective feedback is essential for enhancing clinical skills, communication, and professionalism when given constructively, despite the fact that it is frequently feared and occasionally seen as detrimental to self-esteem. In this sense, in a growth-oriented setting, successful techniques like ask–tell–ask, Pendleton's model, and developing rapport are essential for encouraging progress rather than just criticism (Velou and Ahila, 2022).

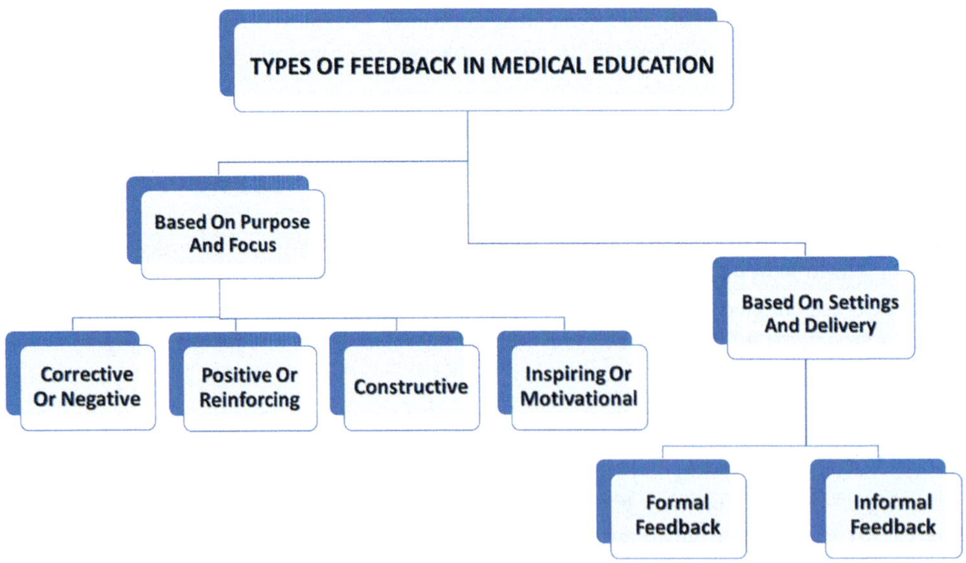


FIGURE 26.2 **Types of feedback in medical education.** This figure shows the types of feedback in medical education.

#### 26.4.1.2 *Positive or reinforcing feedback*

Positive or reinforcing feedback is an essential communication method in medical education that recognizes excellent performance and conduct, inspiring the student and promoting best practices. It helps students discover their skills, boosts their self-esteem, and directs them toward reaching their academic and career objectives. It gives the learner confidence that they are on the right track and encourages them to keep up those useful skills and behaviors. Constructive criticism is said to have an impact on overall learning since it acknowledges that successful acts are highly motivating and boost a trainee's self-efficacy.

#### 26.4.1.3 *Constructive feedback*

By concentrating on what was done or said and how to improve, constructive feedback in medical education helps learners identify strengths and weaknesses in a safe manner, bridging the gap between current performance and desired clinical standards and promoting self-awareness, motivation, and better patient care. It's a crucial component of formative evaluation, going beyond simply highlighting mistakes to offer concrete recommendations for improvement and proficiency.

#### 26.4.1.4 *Inspiring or motivational feedback*

In medical education, motivating or inspiring feedback is a constructive, nonjudgmental discussion that centers on a person's performance and is intended to boost self-efficacy, promote introspection, and direct the learner towards development and career advancement. In order to avoid defensiveness and humiliation, the emphasis here is on performance rather than personality, with feedback focused on the observable behavior and actions rather than the person's character. Based on firsthand observation of a task, motivational feedback employs precise, descriptive language to assist students pinpoint their strengths and areas for improvement (Gigante et al., 2011).

### 26.4.2 Based on settings and delivery

#### 26.4.2.1 *Formal feedback*

Formal feedback is a crucial component of formative assessment for the development of capable, thoughtful clinicians. It is organized, planned, and scheduled information provided to trainees about their performance with the goal of bridging the gap between actual and desired skills for future improvement. It focuses on observable behaviors rather than the individual. One of the key characteristics of formal feedback is that, in contrast to informal feedback, it is planned and scheduled, taking place at particular times, such as midway through the course or following any noteworthy events. Furthermore, it is performance-based since it outlines behaviors and activities to direct future performance. Closing the gaps and minimizing the disparity between present performance and the necessary criteria is one of the goals of the formal feedback. Bidirectional communication between the trainee and the supervisor is another aspect of this kind of feedback. Instead of the objective summative assessment, it is mostly formative and focused on learning and progress. For instance, structured models such as SNAPPS, R2C2, and Pendleton's model and instruments such as OSCEs and peer assessment tool like the Mini-PATs are used in formal feedback in medical education to offer documented, precise instruction on core competencies, concentrating on observable behaviors.

#### 26.4.2.2 *Informal feedback*

In contrast to planned formal assessments, informal feedback in medical education is spontaneous, daily guidance which is often verbal and given by any member of a multidisciplinary team that focuses on immediate performance improvement for skills like communication and procedures. It is essential for ongoing learning and should ideally be specific, timely, nonjudgmental, and supportive of self-reflection. Through frequent engagement and observation, it assists students in identifying their strengths, addressing their deficiencies, and honing their abilities outside of official examinations. One of the main characteristics of informal feedback is that, unlike scheduled meetings, it is impromptu and occurs in real time, such as "hallway consultations" or sticky notes. Feedback of this type is common, multidirectional, and may originate from colleagues, superiors, or even patients. The informal feedback is skill-focused and focuses on professional behavior, communication, and practical skills.

## 26.4.3 Effective feedback

In medical education, effective feedback is precise, timely, actionable information that compares observed performance to standards. It is intended to improve future clinical skills, knowledge, and behavior by encouraging self-reflection and bridging the gap between current and desired competence (Huma et al., 2025). It is given in a supportive setting with clear improvement steps. It is more than simply an assessment, wherein it serves as a developmental tool that emphasizes observable behaviors and uses organized models, such as Pendleton's or R2C2, to help learners understand and use it.

## 26.4.4 Giving effective feedback

In medical education, providing effective feedback entails being prompt, precise, learner-centered, concentrating on observable behaviors, beginning with the learner's self-evaluation, striking a balance between constructive criticism and positive reinforcement, setting up a private space for challenging feedback, and concluding with a specific action plan for improvement often by using models like the one-minute preceptor. Feedback can cause performance to decline if it is not communicated appropriately. The learner may become defensive and embarrassed as a result of poor feedback. Feedback needs to be descriptive and nonjudgmental. When providing constructive criticism, some important guidelines to take into account are as follows:

### *26.4.4.1 Planning effective feedback*

In order to create a descriptive, two-way dialogue that promotes reflection and skill development rather than merely summative judgment, medical education feedback planning entails clearly communicating goals, concentrating on particular, observable behaviors rather than the person, beginning with learner self-assessment, being timely, and concluding with specific action plans for improvement. Using models like Pendleton or SET-UP, making sure settings are private and suitable, and cultivating an environment where students actively seek out and consider feedback are all important tactics (Boud and Molloy, 2013).

### *26.4.4.2 Explicit feedback*

In medical education, explicit feedback is precise, direct information about a trainee's performance that clearly identifies what was done well and what needs improvement to guide future actions. It goes beyond implicit corrections, such as simply restating something correctly, to a detailed "why" and "how," frequently using structured methods like ask–tell–ask for better learning and skill development. It is an essential teaching tool, but good delivery requires context such as the difficulties of a clinical environment and making sure the student actively participates with the material for genuine improvement.

### *26.4.4.3 Descriptive feedback*

In medical education, descriptive feedback is specific, nonjudgmental information about a trainee's observable behaviors that focuses on what they did and its impact rather than who they are. It goes beyond simple evaluation to encourage reflection and improvement in areas like clinical skills and communication. Different from grading, it is an important teaching technique that aids residents and students in comprehending their behavior, learning from errors, and developing competence.

### *26.4.4.4 Behavior-focused feedback*

Using structured models such as R2C2, SNAPPS for dialogue, self-assessment, and the creation of clear action plans for improvement, behavior-focused feedback in medical education focuses on specific, observable actions rather than personality that learners can change (Stowers et al., 2024). It goes beyond simple praise or criticism to promote real skill acquisition and professionalism. It links directly to professional competencies like CanMEDS by highlighting the learner's actions such as communication, clinical reasoning, and empathy and how they affect the patient or team. It works best when it is timely, detailed, and a part of a supportive learning environment.

### *26.4.4.5 Specific feedback*

In medical education, specific feedback concentrates on observable behaviors and uses structured models, such as Pendleton's or SNAPPS, to provide timely, practical, and descriptive guidance on what was done well and areas that require improvement. This approach promotes professional development, competence, and self-reflection rather than merely evaluation. Setting specific objectives, giving specific examples from firsthand observation, maintaining a secure environment, and developing an action plan are all essential components. For instance, in order to change a behavior in a student, it is necessary that the feedback given is positive and constructive criticism and at the same time is directed towards specific measures to be taken by the student to bring about the change.

#### 26.4.4.6 *Concise feedback*

Feedback on professional behavior in medical education must be brief, timely, and concentrated on observable behaviors rather than personal characteristics. The objective is to direct future performance and encourage introspection and a plan of action for improvement. Effective feedback in medical education is by its very nature brief, precise, and behavior-focused. Instead of bombarding the learner with information, it focuses on a few key ideas that are immediately applicable and pertinent to the learning goals. One of the key features of conciseness is that it only covers a small number of the most crucial areas for improvement in a single session. This keeps the learner from experiencing cognitive overload and raises the possibility that they will absorb and apply the knowledge.

#### 26.4.4.7 *Honesty in feedback*

In medical education, honest feedback is essential for skill development. It focuses on particular behaviors using descriptive, nonjudgmental language, guides students to the next level, and requires both giving and receiving skills. It is frequently organized to be timely, credible, and supportive, creating a culture where students feel safe to grow and become independent practitioners. In this case, the criticism ought to be based on firsthand observation, emphasizing particular behaviors or acts rather than conjecture, rumors, or personality assessments. This lends credibility to the input and offers specific instances for discussion.

## 26.5 Feedback as an integral part of curriculum

As a vital instrument for enhancing students' knowledge, abilities, professional performance, and attitudes, feedback is in fact a fundamental and crucial part of medical education. In the process of becoming competent doctors, effective, frequent feedback helps learners comprehend their progress, promotes good practices, inspires them toward desired objectives, and eventually reduces the gap between actual and intended performance. The contribution of effective feedback towards curriculum may be summarized as follows:

### 26.5.1 Drives improvement

Good feedback acts as a constant coach that speeds up learning, increases engagement, clarifies expectations, and fosters trust for improved individual and organizational performance. It does this by giving clear direction, improving self-awareness, correcting mistakes, and rewarding positive behaviors. It increases motivation, skill development, and goal achievement by assisting individuals in realizing their impact, identifying their strengths, and concentrating their efforts on improvement. Additionally, it gives precise details regarding how a trainee's observed performance compares to a standard, giving them guidance on how to get better.

### 26.5.2 Supports goal attainment

Effective feedback encourages goal attainment by providing a clear route for improvement, motivating action, and making the necessary adjustments to reduce the gap between current performance and expected results. Additionally, goal achievement is supported by appropriate feedback since it gives people direction and clarity, enabling them to comprehend the objective, where they are headed, and the particular success criteria. In order to determine what needs to change, it sheds light on the discrepancy between present performance and the intended goal state.

### 26.5.3 Fosters self-directed learning

Effective feedback changes learners from passive recipients to active managers of their own development by encouraging self-awareness, setting goals, and guiding future actions. It accomplishes this by offering specific insights into strengths and weaknesses, promoting autonomy through coaching questions and clear criteria, and empowering students to take charge of setting goals and locating resources. Additionally, it empowers students to actively seek out criticism, enhancing their capacity to track and assume responsibility for their professional development.

### 26.5.4 Enhances clinical skills

Effective feedback enhances clinical skills by narrowing the gap between actual and desired performance, promoting self-reflection, boosting confidence, and guiding future actions through precise, timely, and constructive guidance—all of which turn experiences into learning opportunities for competence and expertise. It enables students to recognize their strengths, address their shortcomings, and hone their professional judgment which in turn contribute to a safer, better patient care (Veloski et al., 2006).

## 26.6 Models for delivering feedback

It is beneficial for the clinical educator to investigate various feedback models and methods that have been documented in the literature during the teaching and learning process. Important feedback models in medical education include learner-centered models like R2C2 and agenda-led outcome–based analysis (ALOBA) that encourage deeper reflection, as well as structured approaches like the feedback sandwich, Pendleton's rules, and one-minute preceptor for rapid clinical feedback. Newer models like BOOST, which emphasize clarity and actionability, all aim to direct improvement in clinical skills and reasoning.

Before we dive into individual models used in the delivery of feedback, it has to be kept in mind that the success of a feedback session is dependent on three broad areas as outlined below. These being structure, format, and content.

### 26.6.1 Structure

In medical education, structuring feedback is essential because it guarantees that it is precise, useful, and successfully encourages introspection and performance enhancement in a nurturing setting. Feedback that lacks structure is frequently ambiguous, easily disregarded, or even demoralizing to the student. Since general remarks like "good job" are useless, structuring the feedback improves clarity and specificity. Feedback is more comprehensible and useful when it is linked to particular, observed behaviors and actions rather than personal traits, which is ensured by a systematic method. In order to provide both parties enough time to prepare, the timing of the feedback must also be taken into account. It could be essential to make sure the feedback is provided in a private setting and that everyone is aware of the meeting's goal. In order to avoid frightening the learner, the environment of the room should also be taken into account. Feedback should be centered on the attitudes, behaviors, and knowledge that were noticed. It is also necessary that a descriptive language be utilized to make the feedback easier to understand. Establishing mutual respect and trust with the common objective of enhancing the learner's performance is important.

### 26.6.2 Format

In medical education, formatting feedback is essential because it guarantees that the information is precise, timely, actionable, and unambiguous. This helps students comprehend performance standards, fill in knowledge gaps, and eventually enhance clinical competence and patient safety. Effective feedback formatting makes expectations and standards obvious, ensuring that students understand the expected level of performance and that learning objectives and benchmarks are clearly established. Simultaneously, it directs progress since detailed, descriptive feedback that emphasizes observed behaviors rather than personal characteristics offers useful ideas for improvement, assisting students in fixing errors and reiterating best habits.

### 26.6.3 Content

Both the teacher and the student must have enough time to be ready for the meeting. When evaluating what has been learned, both formal and personal learning objectives must be taken into account. When a teacher watches a student in action, they might identify areas for development as well as specific instances of excellent performance. A single feedback session should only target a small number of particular areas for improvement, such as two or three of the most important ones. Some of the most common and accepted feedback models are described below. The seven models have similarities and differences in their structure and objectives for the feedback encounter, from the simplest and most educator-centered, such as the feedback sandwich, to the most complex and learner-centered models, such as the ALOBA. Several aspects must be considered as part of the decision-making process when choosing the ideal model for a feedback encounter:

1. Feedback sandwich model,
2. The Pendleton's rules,

3. The one-minute preceptor model,
4. The SET-GO model,
5. The R2C2 model,
6. The ALOBA model,
7. BOOST model.

### 26.6.4 Feedback sandwich

A popular strategy for providing constructive criticism in medical education is the feedback sandwich, often known as the praise-improve-praise method (Milan et al., 2006). In order to make criticism more digestible and less likely to damage the learner's ego, the method "sandwiches" critical or corrective remarks between two layers of positive feedback. Schwenk and Whitman in 1987 are frequently credited with highlighting the feedback sandwich as a specific teaching strategy to improve medical education, making it a significant model for doctors to give constructive feedback, even though the general idea of sandwiching criticism between praise emerged earlier (1940s) and was popularized in business by Mary Kay Ash (Fig. 26.3).

### 26.6.5 Structure of the feedback sandwich

The structure of the feedback sandwich resembles a sandwich in which the upper and lower parts focus on the positive opening and closing stages while the middle part represents the constructive criticism.

#### *26.6.5.1 Positive opening (the first "slice of bread")*

The first step in this process is to recognize particular areas of the learner's performance that were completed successfully. Building rapport, lowering defensiveness, and increasing the learner's openness to the impending criticism are the objectives of this part. In this step, the students feel more accepted as the positive opening can break down boundaries and foster trust as their qualities or efforts are recognized and appreciated by the teacher.

#### *26.6.5.2 Constructive criticism (the "meat")*

In the feedback sandwich method, the constructive criticism step is the middle layer (the "meat"), sandwiched between two positive comments (the "bread"). This involves clearly and specifically addressing the area for improvement, focusing on behavior rather than personality, and offering actionable suggestions to help the person grow, all delivered with a consistent, caring tone.

#### *26.6.5.3 Positive closing (the second "slice of bread")*

In order to ensure that the person leaves the conversation feeling inspired and supported rather than just criticized, the positive closing step of the feedback sandwich method entails concluding the conversation with another compliment or encouraging statement, reiterating the person's value, and expressing confidence in their future improvement. Due to the positive closing part of the feedback, the students feel better and encouraged. This also has a positive influence on their improvement process, as this part highlights their strengths and lessens the impact of the criticism.

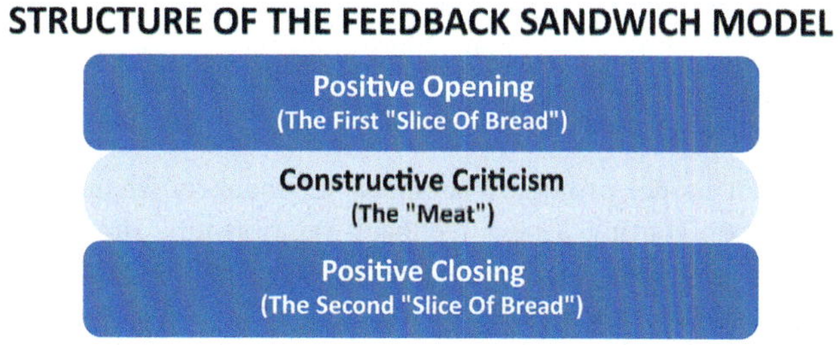


FIGURE 26.3 **Feedback sandwich model.** This figure shows the feedback sandwich model.

### 26.6.6 Advantages of the feedback sandwich model

The benefits of the feedback sandwich approach include softening criticism, making it simpler to give and receive, raising morale by concluding on a good note, fostering a supportive culture, and enhancing receptiveness to improvement, particularly for sensitive persons. It ensures that staff feel appreciated and are inspired rather than demoralized by carefully framing challenging topics. Some of the important advantages of this model are as follows:

#### *26.6.6.1 Softens criticism*

The negative feedback is less harsh and simpler for the recipient to process because of the encouraging opening and closing remarks. The feedback sandwich model's main advantage is that it lessens the impact of criticism, making it simpler for the recipient to hear and comprehend the unfavorable comments. One clever method to lessen defensiveness and anxiety in the individual receiving the feedback is to frame criticism between two layers of compliments. The learner is more receptive to feedback when criticism is softened. When the talk starts with praise, the recipient may feel more relaxed and open-minded, which increases the likelihood that they will listen to the areas that need work rather than getting defensive right away.

#### *26.6.6.2 Increases receptiveness*

Feedback sandwiches make people more open to criticism. The strategy seeks to create a supportive environment where the recipient is more receptive to receiving and acting upon feedback by sandwiching constructive criticism between two complimentary remarks. Instead of becoming disheartened, the learner is able to maintain their enthusiasm and confidence by ending the discussion with a second positive statement. They are reminded of their advantages and areas for development.

#### *26.6.6.3 Boosts morale and confidence*

Increasing the recipient's confidence and morale is one of the main goals of the feedback sandwich approach. The purpose of the positive feedback layers is to increase the person's sense of worth, motivation, and openness to the constructive criticism in the middle. The concept of beginning with sincere optimism enables the tutor to recognize the strengths and contributions of the individual, reinforcing their importance to the team and preventing the criticism from damaging their self-esteem.

#### *26.6.6.4 Eases delivery for tutors*

The feedback sandwich model's ability to facilitate the delivery of challenging or unfavorable remarks is one of its primary benefits for tutors. The method seeks to ease the discomfort of the feedback-giver, particularly for individuals who find criticism difficult or are afraid of conflict. The straightforward positive-negative-positive style provides teachers with a clear, methodical approach that can help them stay focused and lessen anxiety during a potentially unpleasant conversation.

#### *26.6.6.5 Ends on a positive note*

The feedback sandwich model wraps off the discussion with either a constructive summary that focuses on the future or the last layer of positive reinforcement. This approach provides a significant advantage by preserving the learner's sense of self-worth and ensuring that they are receptive to additional learning and development. Another advantage is that closing with praise ensures the student leaves feeling good about their total contribution, preventing lingering bad feelings.

#### *26.6.6.6 Encourages balanced feedback*

The idea that the feedback sandwich model promotes balanced feedback is frequently mentioned as a major benefit, especially for novice teachers or when starting a new feedback relationship. However, by beginning with praise, the approach seeks to safeguard the recipient's self-worth and foster an atmosphere that is more open to the critical criticism, which should result in a more equitable reception of all views.

#### *26.6.6.7 Good for sensitive individuals*

The feedback sandwich model's ability to provide constructive criticism to people who are sensitive is usually regarded as a benefit. The encouraging remarks at the start and finish are meant to lessen the impact of the criticism, lessen defensiveness, and support the recipient's motivation and sense of self. It lessens the impact since the compliments serve as a buffer against the criticism, making it seem less like a personal assault. Additionally, it lessens fear since a good introduction establishes a nonthreatening, encouraging tone that can assist control the recipient's emotional reaction to challenging remarks.

## 26.6.7 Disadvantages of the feedback sandwich model

The feedback sandwich model's drawbacks include diluting criticism, fostering mistrust and misunderstanding, making praise appear insincere, and possibly impeding genuine development by hiding serious problems because recipients might only pay attention to the positive aspects or reject the entire message as deceptive. Some of the important disadvantages of feedback sandwich model are described below:

### *26.6.7.1 Perceived insincerity*

One well-known drawback of the feedback sandwich paradigm in medical education is perceived insincerity. This issue arises because the person receiving the input, such as a medical student or trainee, may rapidly see through the strategy of expressing praise only to lessen the impact of criticism. When compliments are used as a buffer for unfavorable comments, it might give the impression that the person giving the criticism is being dishonest or manipulative.

### *26.6.7.2 Confusion*

One well-known drawback of the feedback sandwich strategy in medical education is confusion. Learners may become confused about the true priorities or important lessons learned from the feedback session as a result of the conflicting information. The strategy conveys contradictory messages by sandwiching criticism between two affirmative comments. It could be challenging for the learner to get the true message if they are unsure whether to concentrate on the compliments or the criticism.

### *26.6.7.3 Lack of dialogue*

One of the main drawbacks is the absence of dialogue, which is frequently acknowledged as a one-way information transfer from teacher to student with no organized chance. The feedback sandwich model's main drawback in medical education is that it promotes a one-way, educator-centered information transfer and inhibits discussion. Deep learning in therapeutic contexts requires natural, two-way interaction and self-directed reflection, which are often inhibited by the sandwich's strict, predetermined format. Additionally, there is a lack of learner engagement because the model does not ask students to evaluate themselves or contribute to their own learning objectives, which restricts their ability to actively participate in the feedback process. A more dialogic approach is necessary since contemporary approaches to feedback highlight the learner as the most active agent.

### *26.6.7.4 Minimal performance impact*

The feedback sandwich model's main drawback in medical education is that it promotes a one-way, educator-centered information transfer and inhibits discussion. This restriction results from a number of related problems, such as the sandwich's strict, predetermined format, which tends to inhibit the organic, two-way dialogue and self-directed reflection required for deep learning in therapeutic settings. The learner's ability to actively participate in the feedback process is limited because the model frequently does not ask them to self-evaluate or contribute to their own learning objectives. Many instructors are urged to investigate different, more participatory feedback models that emphasize communication, self-evaluation, and cooperative action planning for development in light of these constraints.

## 26.6.8 Pendleton's rules

In medical education, Pendleton's rules provide a structured, learner-centered approach to providing balanced feedback. The learner first identifies what went well, the educator reinforces strengths, the learner identifies areas for improvement, the educator adds theirs, and the learner summarizes with an action plan to promote self-reflection and lessen defensiveness (van de Ridder, 2023). This five-step dialogue, which was created by David Pendleton, is perfect for new teachers and students who need to develop reflective abilities in areas like patient consultations since it fosters ownership, self-evaluation, and a safe setting. Using five essential phases, the Pendleton's model in medical education offers a fair, learner-centered framework for feedback:

1. The learner states what went well,
2. The observer (educator) states what went well,
3. The learner states what could be improved,
4. The observer states what could be improved,
5. The learner summarizes and plans for the future.

This model promotes self-reflection, ownership, and a constructive dialogue starting with the learner's perspective, ensuring a balanced, specific, and actionable feedback session. The five steps are described in detail in the below paragraphs:

#### 26.6.8.1 *Learner's perspective*

The learner's perspective is at the center of Pendleton's model, which empowers them through structured self-reflection and active participation. They begin by identifying what they did well and what they believe needs improvement. This reduces defensiveness, builds ownership, and fosters a learner-led, balanced dialogue for creating concrete action plans, transforming them from passive recipients to engaged problem-solvers in their own development. The instructor asks the student to list the elements of their performance that they felt went well.

#### 26.6.8.2 *Educator's reinforcement*

This step of educator reinforcement entails a structured dialogue in which the teacher strategically employs positive feedback or reinforcement by first asking students to identify their strengths before adding their own. This creates a safe environment for constructive criticism and group action planning, which encourages self-evaluation and learner engagement. The teacher concurs with the student and goes into further detail about additional positive observations they made.

#### 26.6.8.3 *Learner's self-assessment (what could be improved?)*

As a fundamental component of Pendleton's learner self-assessment model, students take the lead in providing feedback by first identifying what went well and what could be improved. This encourages ownership and introspection before the educator adds their balanced viewpoint, leading to a shared action plan for improvement and improved performance in clinical settings. Here, the teacher basically encourages self-reflection by asking the student to pinpoint areas in which they might develop.

#### 26.6.8.4 *Educator's guidance*

The Pendleton's model in medical education directs teachers to support learner-centered feedback by beginning with the student's self-evaluation, concentrating first on what went well, then asking the student what could be improved, and working together to create an action plan. This promotes reflection and lessens defensiveness by building on strengths before addressing weaknesses. In order to make it a supportive, two-way conversation rather than a lecture, the educator's role is to set the scene, use open-ended questions like "what went well" or "what could be different," which would reinforce positives, gently guide improvement areas, and ensure a concrete, agreed-upon plan for future action.

#### 26.6.8.5 *Summarize and agree on action plan*

In order to ensure that the feedback results in progress, the teacher and student agree on a plan for future action after summarizing the most crucial points. After a fair debate of the observed performance, the final stages involve summarizing and deciding on an action plan. The learner or observer highlights the key points of conversation in this step, emphasizing both the areas for improvement and the qualities that have been highlighted and should be maintained.

After that, the learner and the observer agree on the precise actions that need to be done. This guarantees that the student accepts responsibility for their learning and development objectives. This shared understanding results in the creation of a precise, detailed, and workable development strategy. This action will specify the actions the learner will take, the resources such as particular reading, practice opportunities, or mentorship, and the schedule for evaluation or reassessment (Fig. 26.4).

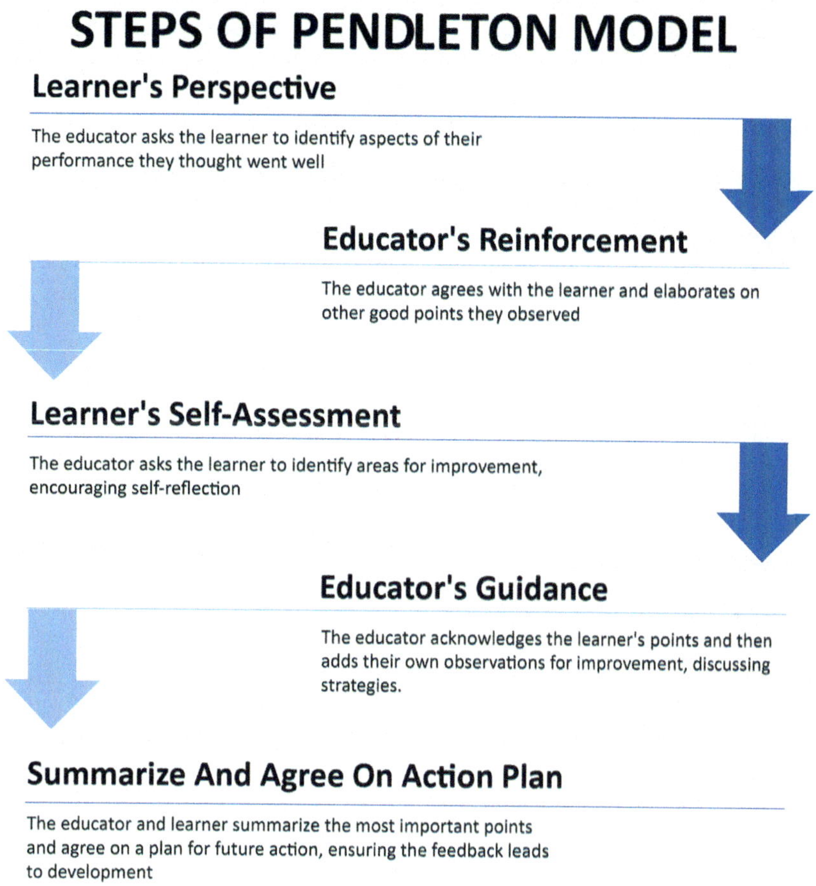


FIGURE 26.4 **Pendleton's model.** This figure shows the Pendleton's model.

### 26.6.9 Benefits of Pendleton's model

In medical education, the Pendleton's feedback model establishes a safe, learner-centered dialogue by beginning with the learner's self-assessment of strengths, adding positives, investigating areas for improvement through learner-led questions, and developing an agreed-upon action plan. It also encourages self-reflection, lessens defensiveness, and promotes improved performance and self-assessment skills for future practice. Some of the key benefits of the Pendleton's feedback model in medical education are as follows:

#### *26.6.9.1 Promotes reflection*

The Pendleton's approach mainly promotes reflection in medical education by placing the student at the center of the feedback process and asking them to actively participate in self-evaluation. This framework transforms the learner from a passive recipient to an active participant, which is essential for the development of professional knowledge and lifelong learning skills. Self-assessment is required by the Pendleton's model since the student must first determine their own areas of strength and growth. Before the observer offers feedback, this first phase compels people to reflect critically on their performance, remember their behaviors and mental processes, and express their self-awareness.

#### *26.6.9.2 Reduces defensiveness*

One well-known benefit of the Pendleton's feedback approach in medical education is its ability to lessen defensiveness. The framework of the model aids in establishing a "safe learning environment" for the student. The Pendleton's technique aims to provide constructive feedback by following a specific, nonjudgmental sequence that actively involves the learner. It supports the positive first approach, which starts with the student and the teacher determining what aspects of the observed performance went well. This makes the learner more open to subsequent feedback by highlighting strengths and reinforcing best practices.

#### 26.6.9.3 *Fosters dialogue*

The Pendleton model's capacity to foster discussion is one of its recognized advantages in medical education. A two-way communication between the feedback provider and the feedback recipient is naturally encouraged by the model's structured, conversation-based approach. Learner-centeredness, which starts the process by asking the learner to first reflect on their own performance, is one of the major processes inside the model's framework that fosters the dialogue. This puts the learner's viewpoint first, transforming them from passive information consumers into active participants.

#### 26.6.9.4 *Supports novices*

For beginners who might be more sensitive to criticism, the model helps reduce defensiveness and creates a less daunting, more positive environment by having the learner state what went well at the beginning of the feedback session. Additionally, it encourages self-evaluation because the structured format necessitates self-reflection and self-evaluation, which are fundamental skills that students must acquire early in their study. Instead of encouraging passive information reception, this promotes active learning. However, the Pendleton model's drawbacks in medical education include its strict structure, which may prevent in-depth discussion, the possibility of diluting critical feedback, its time-consuming nature for busy clinicians, and its reliance on students having strong self-assessment skills, which aren't always present and could result in superficiality if not applied carefully.

### 26.6.10 One-minute preceptor model

The one-minute preceptor model, commonly referred to as the five-step "micro-skills" model, is one paradigm that is especially helpful in hectic clinical settings and micro-feedback sessions. It offers a succinct and simple structure for instructing and giving feedback while providing patient care. After getting a commitment from the student on a particular topic, like the diagnosis or treatment plan, the educator looks for evidence to support the student's reasoning, teaches general guidelines if needed, and then has a quick conversation to reinforce the positive aspects and correct any errors (Ravichandran et al., 2019). This just-in-time feedback model supports the development of clinical reasoning and decision-making abilities, ideally on an individual basis. It requires medium learner reflection and self-assessment skills as well as medium feedback-giving expertise from the educator to examine a single aspect and provide balanced feedback.

### 26.6.11 Steps in one-minute preceptor model

The one-minute preceptor model uses five quick steps for effective clinical teaching. This framework efficiently guides learners to deeper thinking and skill development in busy settings. The five steps of one-minute preceptor model are as follows:

#### 26.6.11.1 *Get a commitment*

In the one-minute preceptor paradigm, obtaining a commitment entails asking the learner to articulate their precise diagnosis, plan, or interpretation of a situation, going beyond merely accepting and disclosing their thought process. Use questions like "What do you think is going on with this patient?" or "What would you like to accomplish in this visit?" instead of just asking, "What do you think?" to help them take ownership of a decision. This will allow the teacher to probe their reasons and offer specific feedback.

#### 26.6.11.2 *Probe for supporting evidence*

The second important step in the one-minute preceptor model is to probe for supporting evidence. After the learner has committed to a diagnosis or plan, the preceptor uses questions like "What led to that conclusion?" or "What else did you consider?" to assess their clinical reasoning rather than just the answer. This allows for targeted teaching by exposing the learner's knowledge gaps and reasoning.

#### 26.6.11.3 *Teach a general principle*

The one-minute preceptor model's teaching a general principle approach aims to assist learners in extrapolating a particular case experience to other contexts. The key is to keep the principal brief, focused, and generalizable, ideally taking 30 seconds or less. The goal is to make the lesson more memorable and transferable by offering a takeaway message or a guiding principle that may be used in other, comparable clinical interactions. A knowledge gap found in previous steps of the one-minute preceptor model, such as when the preceptor probed for supporting evidence and found an area for development, is frequently the focus of this step.

#### *26.6.11.4 Reinforce what was done well*

This step of reinforcing what was done well in the one-minute preceptor model entails providing precise, encouraging feedback about the learner's strong actions, abilities, or thought processes to promote their repetition. This goes beyond simple compliments like "good job" to emphasize the significance of the learner's positive behavior, such as "Your organized presentation clearly linked the patient's history to your assessment, making it easy to follow." This boosts self-assurance and reinforces important lessons for upcoming interactions.

#### *26.6.11.5 Correcting mistakes or giving guidance*

Finding areas where the learner's performance, knowledge, or clinical reasoning needs to be improved is the main goal of the correct mistakes stage. Instead of making the learner feel condemned or judged, the objective is to provide them targeted, behavior-focused feedback that will help them get better in the future. Additionally, by addressing the problem as soon as it happens and concentrating on particular actions and behaviors rather than broad negative assessments, it guarantees that the students receive quick and precise feedback. It also encourages self-evaluation since it allows the learner to assess their own performance first. This makes them more receptive to your feedback and promotes self-monitoring (Fig. 26.5).

### 26.6.12 Advantages of one-minute preceptor model

Through its five essential steps, which include getting commitment, probing for supporting evidence, teaching a general principle, reinforcing good performance, and correcting mistakes, the one-minute preceptor model offers a structured, effective way to teach in busy clinical settings, boost learner confidence, improve diagnostic skills, and enhance feedback quality. This makes learning more active and effective for both students and preceptors. Both the learner and the preceptor stand to gain from this paradigm.

#### *26.6.12.1 Advantages for learners*

##### 26.6.12.1.1 Enhanced clinical reasoning

The one-minute preceptor model significantly enhances clinical reasoning in medical education by providing preceptors with an organized, time-efficient framework for guiding students through cases. It focuses on "probing for evidence" and teaching underlying principles, which goes beyond simple fact recall to deep analysis, improves diagnostic skills, and fosters confident clinical decision-making by identifying gaps and guiding focused study. It is very successful for building strong reasoning skills because its five micro-skills methodically engage students, encouraging active problem-solving and critical thinking in hectic clinical environments.

##### 26.6.12.1.2 Increased engagement

One well-known benefit and essential component of the one-minute preceptor paradigm in medical education is increased participation. A more active and individualized learning experience is promoted by the model's structure, which actively encourages the student into a teaching conversation and decision-making process. In the end, this dynamic engagement ensures that students are active participants in their own education by bridging the gap between academic knowledge and practical application.

##### 26.6.12.1.3 Targeted feedback

One of the main benefits of the one-minute preceptor model in medical education is that its methodical approach makes it easier to provide precise, timely, and high-quality feedback that is focused on the learner's actions and clinical reasoning. Since the feedback is targeted towards a precise component of the topic or a particular quality of the student, it proves to be of immense values in the process of improvement.

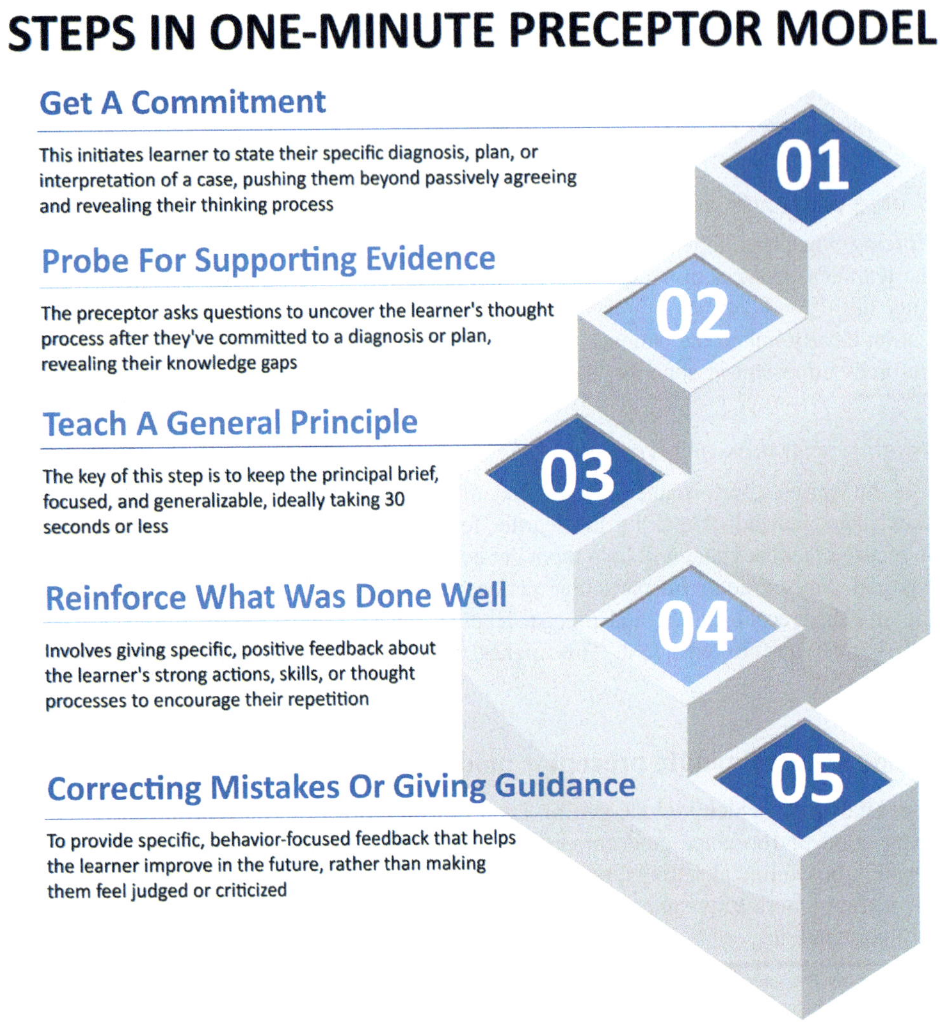


FIGURE 26.5 **Steps in one-minute preceptor model.** This figure depicts the steps in one-minute preceptor model.

#### 26.6.12.1.4 Improved confidence

One well-established benefit of the one-minute preceptor paradigm is increased confidence for both the preceptors, that is, faculty and the students. It is observed that following one-minute preceptor sessions, the learners in this approach show increased confidence in their clinical skills, capacity to evaluate patients, and general performance. Additionally, approaches such as obtaining a commitment and looking for evidence, which would improve a student's capacity to express their ideas and recognize their own knowledge gaps, boosts their self-confidence.

### *26.6.12.2 Advantages for preceptors*

#### 26.6.12.2.1 Efficiency and structure

This model's main benefit is that it saves a lot of time because it was created especially for clinical settings with limited time. The one-minute preceptor model effectively provides excellent instruction during quick exchanges. Additionally, compared to traditional techniques, it results in greater learning outcomes in less time by concentrating instructional efforts and feedback, which contributes to high yield feedback. This model's simplicity and structure make it easier to learn and use, which lessens the mental strain on preceptors. This is another benefit of the model's efficiency and structure.

#### 26.6.12.2.2 Increased confidence

One of the main benefits of the one-minute preceptor model for preceptors is enhanced confidence. Preceptors are more confident in their ability to deliver improved learning experiences, give targeted feedback, and develop their own clinical reasoning skills because of the successful method that makes teaching manageable and effective. It offers a rational framework for systematically providing feedback, which lowers concerns about undermining confidence and improves the general standard of clinical coaching.

#### 26.6.12.2.3 Identifies learning gaps

The one-minute preceptor model's organized approach, especially the probing stage, which effectively identifies the learner's knowledge gaps or clinical reasoning by posing open-ended questions "What led you to that conclusion?" is a major benefit for preceptors. This enables preceptors to directly adapt instruction to the deficiencies that have been detected, increasing relevance and efficiency by targeting particular needs rather than providing general knowledge, encouraging learner ownership, and boosting feedback quality in a nonthreatening way.

#### 26.6.12.2.4 Positive learning environment

The one-minute preceptor model fosters a positive learning environment for preceptors by providing an effective way to provide high-quality, learner-centered feedback, boosting preceptor confidence, making teaching manageable in hectic settings, and increasing learner engagement through active participation and respect. This results in improved clinical reasoning and a more fulfilling teaching experience for both parties.

### 26.6.13 SET-GO model

In medical education, the SET-GO model is a structured, descriptive feedback system that emphasizes learner-driven reflection and problem-solving through the following steps: See (explain what happened), Else (what else you saw), Think (learner's perspective), Goal (what to achieve), and Offers (how to get there). Although it takes time and expert facilitation, it encourages nonjudgmental feedback, peer learning, and skill rehearsal, making the learner involved in creating development goals (Fig. 26.6).

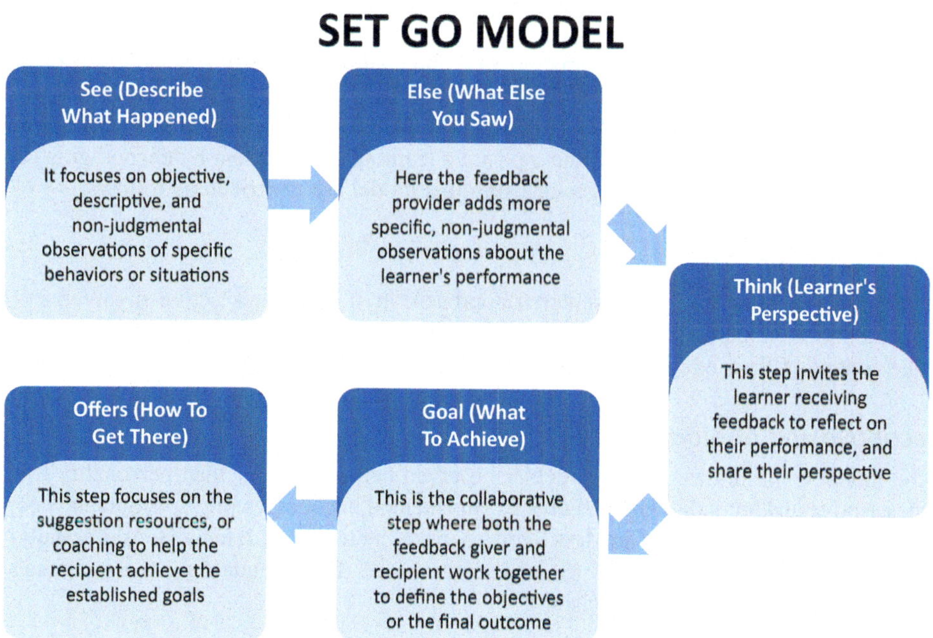


FIGURE 26.6 **Components of the SET-GO (see, else, think, goal, and offers) model.** This figure depicts the SET-GO model.

## 26.6.14 Components of the SET-GO model

### 26.6.14.1 See

S stands for "what I Saw" in the SET-GO feedback model, which emphasizes objective, descriptive, and nonjudgmental observations of certain behaviors or circumstances. This is essential for initiating a productive feedback discussion, particularly in medical training. Trainers and trainees are guided through feedback by the complete model.

### 26.6.14.2 Else

"E" stands for "what Else I saw" (or "what Else did I see"), which denotes the second descriptive step in which the feedback provider builds on the first "S," that is, "what I Saw," adding more detailed, nonjudgmental observations about the learner's performance to paint a more complete picture before going on to reflection and goal setting.

### 26.6.14.3 Think

The "T" stands for "what the recipient Thinks" in the SET-GO feedback approach, which makes it a conversation rather than a monologue by encouraging the receiver to consider their performance, express their viewpoint, and actively engage in identifying areas for improvement.

### 26.6.14.4 Go

"Goals" (the "G" in SET-GO) is the step in the SET-GO feedback paradigm when participants specify what constitutes effective performance or behavior going forward. In order to achieve improvement, both the feedback provider and the recipient collaborate to create goals or the ultimate result they agree upon. Clarifying expectations is aided by this step.

### 26.6.14.5 Offers

The "O" in the SET-GO feedback model stands for "Offers" or "Ongoing support," which refers to recommendations, tools, or guidance to assist the recipient in reaching the predetermined objectives. It's about offering practical assistance to close the gap between present performance and the intended future result, making sure that feedback results in progress rather than just criticism.

## 26.6.15 Advantages of the SET-GO model

The SET-GO model's structured, descriptive, nonjudgmental approach makes feedback more receptive and effective by concentrating on observed behaviors and empowering learners to identify solutions for specific goals. It also fosters learner autonomy, improves self-reflection, builds trust, and promotes collaborative goal setting. It moves away from top-down criticism and produces better performance and a positive learning environment by encouraging communication and learner-driven improvement. The main advantages of using this model are described as follows:

### 26.6.15.1 Learner-centered

The SET-GO paradigm promotes the learner-centered approach, as it stimulates active involvement and compels the student to actively define their performance and development rather than passively taking instructions. Students are empowered to take control of their education as they are allowed to select their own courses and objectives.

### 26.6.15.2 Descriptive and nonjudgmental

Since the SET-GO model concentrates on the particular observed behavior rather than personality traits or presumptive intents, it offers descriptive and nonjudgmental feedback. This method encourages introspection and cooperative problem-solving by making feedback more objective and less likely to make the recipient defensive. Additionally, it defines actions objectively by concentrating on "what I Saw" (S) and "what Else" (E), safeguarding the recipient's self-esteem and encouraging them to be more receptive to criticism.

### 26.6.15.3 Promotes self-reflection

The main way that the SET-GO approach promotes self-reflection is by asking students to evaluate their own performance and offer solutions before getting advice from outside sources. The third phase of the five-stage feedback model incorporates this procedure. Additionally, by encouraging students to examine their ideas and choices, the "think" stage in the model promotes metacognition and deeper comprehension.

#### 26.6.15.4 Clear goal setting

The SET-GO model's clear goal setting offers a number of benefits, chief among them being focus and direction, which boosts performance, accountability, and motivation. Additionally, it helps with direction and attention because the precise and well-defined goals provide teams and people with a road map, enabling them to comprehend precisely what must be accomplished and the necessary measures to get there. This removes uncertainty and guarantees that all efforts are directed toward shared, significant goals. However, it also boosts engagement and motivation since a clear goal gives one a feeling of direction and a standard for achievement, both of which are strong motivators. People are more likely to be involved and provide their best effort when they are aware of the goals they are pursuing.

#### 26.6.15.5 Builds trust and rapport

The cooperative, two-way communication approach fosters trust and a solid educational relationship between the teacher and the student. Additionally, it creates organized, transparent, and user-focused experiences that foster trust and support. It fosters confidence by synchronizing activities with stakeholder expectations and providing consistent, dependable support along the route. This expression of sincere attention and compassion strengthens the bond and builds trust.

#### 26.6.15.6 Facilitates vicarious learning

The SET-GO paradigm allows participants to learn from the viewpoints and experiences of others by encouraging peer feedback and cooperative problem-solving in a group setting. The practice of shared observation, which starts with the teacher asking both the observed learner and the group members to report what they witnessed during a performance or scenario, is one way to improve vicarious learning. This typical observation serves as a starting point for conversation. Additionally, by promoting contributions from all group members, the model creates a discussion and fosters reflection via other people's experiences. Students are exposed to a variety of viewpoints on a same event, which can extend their comprehension beyond their immediate performance.

#### 26.6.15.7 Actionable and concrete

The SET-GO approach supports an actionable and tangible strategy by using a systematic, collaborative feedback process that explicitly moves from shared observation and contemplation to defining specific goals and, most importantly, identifying the practical "offers" or proposals for achieving those goals. It goes beyond observation to specific planning, offering doable strategies for achieving objectives that result in actual behavioral change.

### 26.6.16 R2C2 model

The R2C2 model, which stands for Relationship, Reactions, Content, and Coaching, is a four-phase, empirically supported paradigm for coaching and feedback. It focuses on encouraging an interactive dialogue between the giver and recipient of feedback to improve performance by building rapport, analyzing the recipient's response, clarifying the meaning of the criticism, and collaborating to develop an action plan for improvement. This iterative paradigm is intended to improve the acceptance and application of feedback, foster introspection and self-evaluation, and support advancement in a variety of educational and professional contexts. The R2C2 approach was created to deliver feedback based on performance and assessment rather than on daily practice or specific rotation times.

The method generates a conversation by looking at an assessment result, its importance, and the viewpoint and reaction of the students. The teacher first builds a relationship with the student in order to investigate the learners' reactions to the assessment, encourage introspection and self-evaluation, and create a respected and reliable atmosphere. After assessing the students' comprehension of the assessment's contents and findings, the teacher takes a coaching approach, reaching a consensus on solutions and a plan of action. The R2C2 model offers a learner-centered framework that makes it easier for students to accept the assessment and the feedback they receive. This requires them to look beyond the assessment's outcome, which calls for medium to high levels of reflection and self-evaluation abilities. The teacher must have strong feedback-giving abilities since they should be ready to deal with unfavorable responses and well comprehend the goal of the exam and the material to be reviewed. The following are the components of R2C2 model (Fig. 26.7).

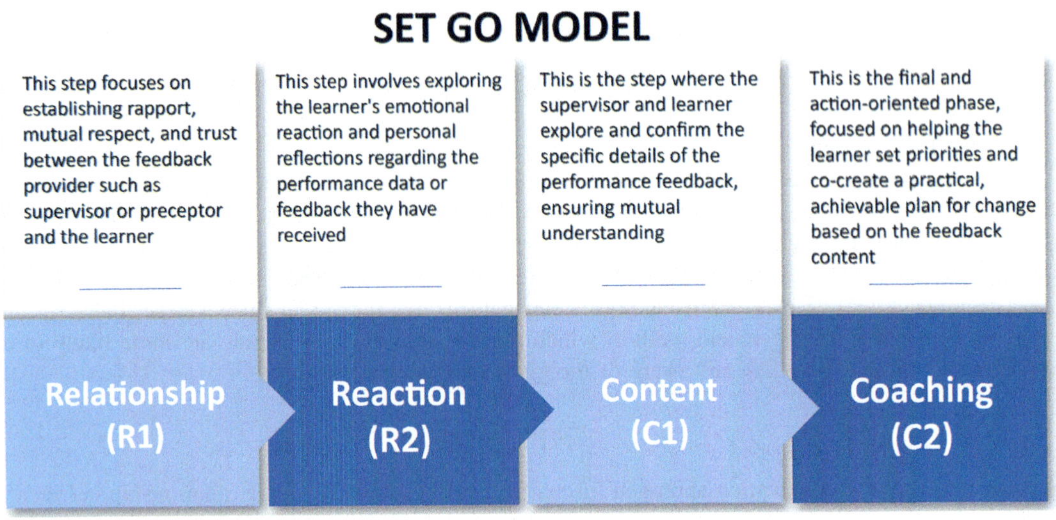


FIGURE 26.7 **Components of R2C2 (relationship, reactions, content, and coaching) model.** This figure shows the components of R2C2 model.

#### *26.6.16.1 Relationship (R1)*

The essential first stage of the R2C2 model is the relationship component, which focuses on building rapport, respect, and trust between the learner and the feedback giver, such as a supervisor or preceptor. This stage lays the groundwork for establishing a secure setting that encourages candid, nonjudgmental, and reciprocal communication, increasing the learner's openness to coaching and feedback. Engaging the learner in the feedback dialogue as an active participant rather than a passive recipient is the main objective here.

#### *26.6.16.2 Reaction (R2)*

The R2C2 feedback model's reaction component entails investigating the learner's feelings and introspection in relation to the performance information or feedback they have been given. This stage is essential for controlling the learner's feelings and getting them involved in the feedback process. The main goal here is to give the learner a secure environment to share their initial thoughts about their performance statistics, which could be both favorable and unfavorable. This aids the supervisor in comprehending the learner's viewpoint and any obstacles that might prevent them from embracing the criticism.

#### *26.6.16.3 Content (C1)*

In order to ensure mutual understanding, the supervisor and learner investigate and validate the specifics of the performance feedback during the content component of the R2C2 feedback model. This phase avoids making assumptions about the learner's behavior in favor of concentrating on the data, observations, and facts. The preceptor and student go over the actual feedback data, observations, or assessment reports together during this phase to make sure everyone understands the information.

#### *26.6.16.4 Coaching (C2)*

The last and action-oriented stage of the R2C2 paradigm is the coaching component, which focuses on assisting the learner in setting priorities and cocreating a workable plan for change based on the feedback content. One of the most important features of the coaching phase is that it facilitates goal setting, wherein the learner and supervisor or preceptor collaborate to establish clear, doable objectives for progress.

### 26.6.17 ALOBA model

Agenda-led outcome–based analysis, or ALOBA, is a learner-centered feedback model that prioritizes the needs of the recipient by first examining their agenda and intended results for the feedback session. According to the paradigm, the student sets goals, the teacher analyzes them, and then provides structured, descriptive, and nonjudgmental feedback that is in line with those predetermined results.

With the help of the educators' perspective and the learners' agenda and learning needs, the ALOBA approach seeks to create a learner-centered dialogue or interview-style feedback. Prior to the feedback interaction, the learner is invited to consider and determine their requirements and agenda. After that, the instructor supports theory-practice connections, promotes self-evaluation and problem-solving abilities, and offers fair criticism. After discussing ideas and options to meet the learner's goal and learning requirements, the teacher verifies the learner's approval, summarizes the interaction, and decides on an action plan. The ALOBA model is regarded as an extension of the Pendleton's rules because it adds learner-centeredness and flexibility to the feedback encounter where the learner is an active participant throughout rather than a passive recipient of suggestions. In order to guide the conversation and determine their requirements and objectives, the learner needs to possess strong insight, introspection, and self-evaluation abilities. In this model, strong feedback-giving skills and judgment are necessary for the educator to lead the debate, provide fair criticism, and draw links between theory and practice. The following are the components of ALOBA model (Fig. 26.8).

### *26.6.17.1 Agenda-setting*

The first step in this method is setting the agenda, which involves asking the student to identify their own learning needs, problems, and specific objectives or areas in which they would like to have an input from the facilitator. This open recognition of challenges lessens defensiveness and turns criticism into something useful.

### *26.6.17.2 Observation and data gathering*

The facilitator's direct observation of the learner's performance and the learner's subsequent self-evaluation serve as the foundation for observation and data collection in the ALOBA approach. In this step, the learner's performance in a real or simulated clinical scenario is observed directly by the facilitator. In order to provide objective and nonjudgmental

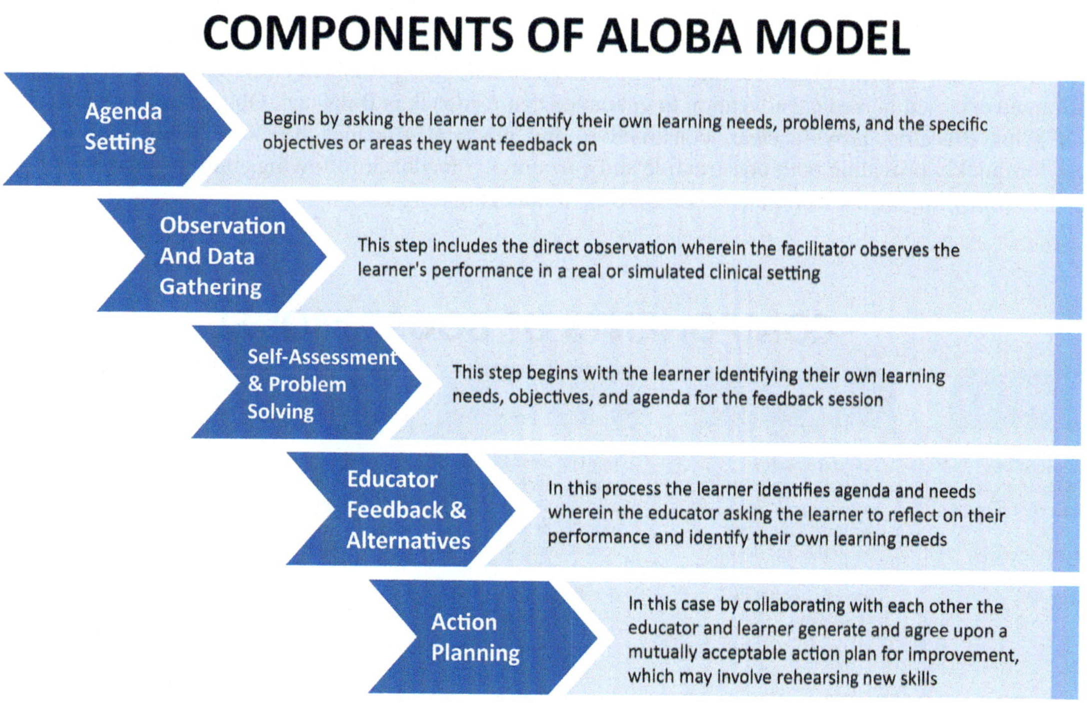


FIGURE 26.8 **Components of ALOBA (agenda-led outcome–based analysis) model.** This figure shows the components of ALOBA model.

feedback, this observation should be centered on particular, behaviorally focused acts rather than assessments of the learner's character.

#### 26.6.17.3 *Self-assessment and problem solving*

Since the process starts with the learner determining their own learning needs, objectives, and agenda for the feedback session, the ALOBA model's role in self-assessment and problem-solving has grown in importance. Before the teacher gives official evaluation, the student must assess their own performance, pinpoint their areas of strength and weakness, and recommend areas for improvement. This calls for a high level of insight and introspection skills. The instructor encourages this self-evaluation by asking open-ended questions such as "what do you think?" or "what went well?"

#### 26.6.17.4 *Educator feedback and alternatives*

An organized, five-step procedure that promotes an engaging, nonjudgmental atmosphere is usually part of the ALOBA paradigm. In this procedure, the student determines their agenda and needs. The teacher then asks the student to evaluate their performance and determine their own learning needs, goals, and particular topics they wish to talk about during the feedback session. This prioritizes the viewpoint of the learner. Additionally, the instructor promotes problem-solving and self-evaluation in which instead of only pointing out errors, the teacher and student talk about alternative ways to solve problems.

#### 26.6.17.5 *Action planning*

The last and most important phase in converting feedback into future performance improvement is action planning. It is defined as a cooperative effort between the teacher and the student to specify precise, doable actions. In this instance, by working together, the teacher and student create and approve a mutually beneficial improvement plan that might include practicing new abilities.

### 26.6.18 BOOST model

The BOOST feedback model, which focuses on behaviors rather than personality for better learning outcomes, is a checklist framework used in medical education to guarantee that feedback is Balanced, Objective, Observed, Specific, and Timely. It helps educators provide clear, constructive, and practical guidance for student improvement. It offers a framework for quickly providing both constructive and encouraging feedback following an event, making it pertinent and inspiring. The following are the components of BOOST model (Fig. 26.9).

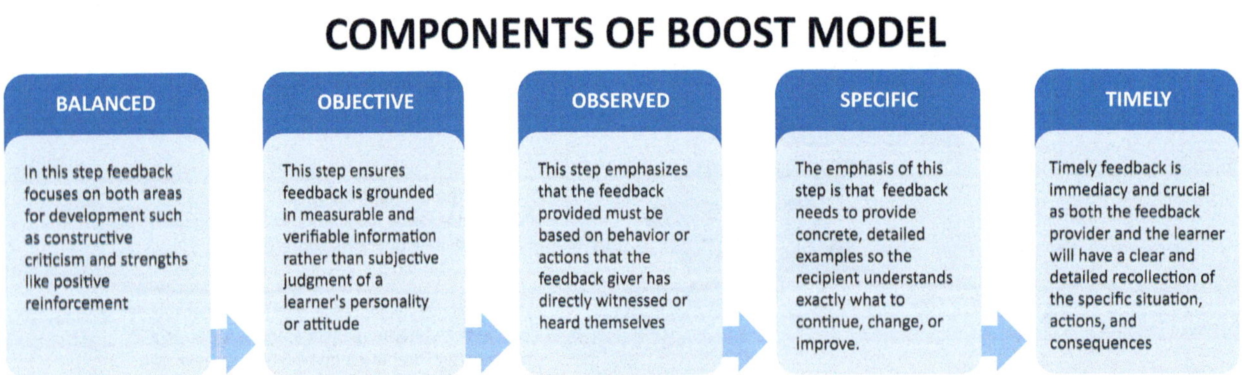


FIGURE 26.9 **Components of BOOST (balanced, objective, observed, specific, and timely) model.** This figure shows the components of BOOST model.

#### 26.6.18.1 *Balanced*

Feedback in medical education should emphasize both strengths in terms of positive reinforcement and areas for improvement by constructive criticism according to the balanced component of the BOOST feedback model. The objective is to offer a comprehensive viewpoint that encourages rather than demotivates the student, rewarding positive conduct while directing progress.

#### 26.6.18.2 *Objective*

The objective component guarantees that feedback is based on quantifiable and verifiable data rather than a subjective assessment of a learner's disposition or personality. This approach is crucial in medical education because impartiality and evidence-based assessment are necessary to prevent bias and preserve credibility.

#### 26.6.18.3 *Observed*

The observed component of the BOOST feedback model highlights that the feedback must be based on behavior or activities that the person providing the feedback has personally seen or heard, rather than on hearsay or assumptions from third parties. Ensuring fairness, accuracy, and accountability in the feedback process is a crucial principle in medical education.

#### 26.6.18.4 *Specific*

It is always good to avoid making generalizations. In order for the recipient to know exactly what needs to be improved, changed, or continued, the feedback must provide specific, in-depth examples. It is important to remember that utilizing specific examples instead of ambiguous phrases like "good job" will help to make concepts obvious and applicable.

#### 26.6.18.5 *Timely*

Giving a timely feedback, especially immediately after an activity or task, is of immense importance as both the learner and the feedback provider will have a clear and thorough memory of the particular circumstance, actions, and outcomes, making the feedback more pertinent and significant. This is crucial because it reinforces learning since timely feedback enables the learner to instantly relate the instruction to their performance, allowing them to make quick modifications and implement remedial actions in subsequent tasks or patient contacts.

## 26.7 Role of feedback in medical education

By offering precise, practical insights into clinical performance, self-reflection, and professional development, feedback is essential in medical education for directing skill acquisition, rewarding positive behaviors, and closing performance gaps. It goes beyond simple assessment to actual learning (Chowdhury and Kalu, 2004). In order to support self-regulated learning and guarantee that trainees comprehend and can use the guidance for future patient care, it must be timely, focused, and structured, frequently utilizing models such as Pendleton's.

### 26.7.1 Role of feedback process in teaching and learning process

The feedback process is essential to teaching and learning because it helps students self-regulate, identify their strengths and weaknesses, and modify their learning strategies for mastery. It also helps teachers improve their instruction for better outcomes. It transforms education from a one-way process to a dynamic, learner-centered cycle of comprehension, advancement, and ongoing growth. From the learner's point of view, it is beneficial since it improves performance by pointing out strengths and deficiencies and directing students to develop their clinical knowledge, abilities, and professional behaviors. Additionally, by encouraging students to assess their own performance and take responsibility for their education, it develops self-reflection and metacognition and helps them become self-regulated learners (Van Der Leeuw and Slootweg, 2013).

### 26.7.2 Role of feedback process in assessment process

In medical education, feedback is more than just a summative evaluation. It is in fact an integral part of professional development and formative learning. It helps bridge the gap between actual and expected performance, direct future learning, and promote professional development by offering precise, practical data about a learner's skills, knowledge, and behaviors (Devi et al., 2012). It identifies learning needs, promotes successful practice, and helps students make the transition to self-reflection and ongoing development in clinical settings.

## 26.8 Challenges in providing feedback

From understanding, reading, and analyzing feedback to observing and measuring the educational consequences, using feedback at different levels has numerous obstacles. The following list includes some of the difficulties in giving feedback:

### 26.8.1 Feedback underuse

Feedback is frequently neglected in educational activities, which is a widespread problem. There are several reasons why this omission occurs. These factors include hectic clinical schedules, unforeseen circumstances, needless delays, underappreciated significance, and inadequate feedback-giving abilities. Raising awareness of the value of feedback in medical education, improving institutional and cultural feedback practices, conducting frequent faculty development activities, planning feedback appropriately and in advance, incorporating feedback into the curriculum, and routinely reviewing the progress made in improving the practice of feedback are some of the ways to increase the level of feedback use.

### 26.8.2 Generality and brevity

Brief and general feedback is used by many feedback sources. Subpar evaluation frequently uses terms like "good job," "perfect," "well done," and other such short phrases. Other than being motivating and uplifting, these terms and descriptions have no educational value. The feedback given by the teachers must enable the students to understand the high and poor points of their performance as well as the reasons behind them. They must understand how to use guidance and improvement initiatives to maximize high performance and enhance low performance. In this regard, giving a thorough feedback is a crucial part of teaching.

### 26.8.3 Delayed feedback

Immediate feedback is considerably more valuable than feedback received far from the performance time. Feedback given shortly after a performance has greater educational value, depending on the style and goal. It has a strong corrective influence, corresponds well to the events, and includes sufficient particular details. Immediate debriefing has a great educational value when it comes to group performance and feedback. Summative feedback for term training or prolonged courses is an exception to this norm, as sufficient observation and evaluation are required before relevant feedback can be offered.

### 26.8.4 Misperception

One of the hardest problems when giving comments is misperception. When they are told about their shortcomings, learners frequently misinterpret their criticism and become depressed, which makes them avoid it or defend their performance. Every effort should be made to help students see feedback as a chance to learn, much like one-on-one coaching. Giving the learner a leadership role in their feedback, utilizing dialogue rather than monologue, offering feedback rather than forcing it, and selecting the appropriate location for feedback are all beneficial factors for the right impression of feedback.

### 26.8.5 Improper setting

Unfortunately, feedback is frequently offered in a hurry, while on the go, in front of others, or while the learner is unprepared or unaware. These situations are all not conducive to providing insightful comments. Such circumstances are frequently dictated by a hectic clinical schedule or academic involvement of the facilitator. It is strongly advised that feedback be arranged ahead of time and that a brief debriefing session be conducted on a regular basis following a significant therapeutic activity that includes instruction.

### 26.8.6 Acceptance or rejection

Learners' acceptance of feedback is one of the most difficult and restrictive variables. Learners find it difficult to accept their shortcomings and are curious about their performance. This is particularly true when they are forced to give feedback. When students ask for and seek for their own evaluation, it becomes much simpler.

**Summary**

Effective medical education is built on feedback, which is essential for producing skilled, thoughtful, and secure healthcare workers. This chapter describes the goals, tenets, modes of delivery, and difficulties of providing feedback in academic and clinical training settings. Timely, precise, and useful feedback is essential. It works best when it is connected to clear standards or learning objectives and concentrates on observable behaviors rather than personal characteristics. The various models described in the chapter showcase the various advantages of feedback in terms of improving learner achievement, clinical performance, skill acquisition, and self-regulated learning.

**Key point**

**Future perspectives: shifting towards a collaborative process**

Technology advancements, competency-based teaching, and a better knowledge of how students acquire expertise are driving the rapid evolution of feedback in medical education. Feedback will probably become a continuous, data-rich, collaborative and a learner-owned process in the coming years. As a result, medical programs are shifting away from "feedback as correction" and toward a feedback culture that emphasizes psychological safety. Additionally, the students will receive clear instruction on how to seek for, analyze, and apply criticism rather than only accepting it. Faculty development, on the other hand, will emphasize relationship-centered feedback, coaching, and reflective inquiry. The development of digital and artificial intelligence (AI)-supported feedback systems, which use AI to evaluate clinical paperwork, procedural films, and communication performance like simulated interactions, will improve the feedback process. The result will offer prompt, tailored, and useful feedback. Additionally, performance measurements for tasks like physical examinations and surgeries may be obtained through the use of wearables and smart devices.

## References

Boud, D., Molloy, E., 2013. Feedback in Higher and Professional Education: Understanding It and Doing It Well. Routledge, New York, NY.

Chowdhury, R.R., Kalu, G., 2004. Learning to give feedback in medical education. The Obstetrician & Gynaecologist 6 (4), 243–247. Available at https://doi.org/10.1576/toag.6.4.243.27023.

Devi, V., Mandal, T., Kodidela, S., Pallath, V., 2012. Integrating students' reflection-in-learning and examination performance as a method for providing educational feedback. Journal of Postgraduate Medicine 58 (4), 270–274. Available at https://doi.org/10.4103/0022-3859.105447.

Ende, J., 1983. Feedback in clinical medical education. JAMA: The Journal of the American Medical Association 250 (6), 777–781. Available at https://doi.org/10.1001/jama.250.6.777.

Gigante, J., Dell, M., Sharkey, A., 2011. Getting beyond "Good Job": how to give effective feedback. Pediatrics 127 (2), 205–207. Available at https://doi.org/10.1542/peds.2010-3351.

Hewson, M.G., Little, M.L., 1998. Giving feedback in medical education: verification of recommended techniques. Journal of General Internal Medicine 13 (2), 111–116. Available at https://doi.org/10.1046/j.1525-1497.1998.00027.x.

Huma, S., Ullah, H., Sarfraz, J., 2025. *Strategies to develop*, "an efficient feedback system in medical education and its outcome – an interventional study" Medical Teacher 47 (7), 1152–1158. Available at https://doi.org/10.1080/0142159X.2024.2418586.

Milan, F.B., Parish, S.J., Reichgott, M.J., 2006. A model for educational feedback based on clinical communication skills strategies: beyond the "feedback sandwich" Teaching and Learning in Medicine 18 (1), 42–47. Available at https://doi.org/10.1207/s15328015tlm1801_9.

Ramani, S., Krackov, S.K., 2012. Twelve tips for giving feedback effectively in the clinical environment. Medical Teacher 34 (10), 787–791. Available at https://doi.org/10.3109/0142159X.2012.684916.

Ravichandran, L., Sivaprakasam, E., Balaji, S., Swaminathan, N., 2019. Effectiveness of the 1-Minute preceptor on feedback to pediatric residents in a busy ambulatory setting. Journal of graduate medical education 11 (4), 204–206. Available at https://doi.org/10.4300/JGME-D-19-00440.

Stowers, K., Timme, K., Giardino, A., 2024. Feedback in clinical medical education: then vs. now. Medical Research Archives 12 (12). Available at https://doi.org/10.18103/mra.v12i12.5946.

van de Ridder, J.M.M., 2023. Pendleton's rules: a mini review of a feedback method. American Journal of Biomedical Science & Research 19 (1), 19–21. Available at https://doi.org/10.34297/ajbsr.2023.19.002542.

Van Der Leeuw, R.M., Slootweg, I.A., 2013. Twelve tips for making the best use of feedback. Medical Teacher 35 (5), 348–351. Available at https://doi.org/10.3109/0142159X.2013.769676.

Veloski, J., Boex, J.R., Grasberger, M.J., et al., 2006. Systematic review of the, 'literature on assessment, feedback and physicians' clinical performance: BEME Guide No. 7. Medical Teacher 28 (2), 117–128. Available at https://doi.org/10.1080/01421590600622665.

Velou, S., Ahila, E., 2022. Types of feedback in medical education -A new hypothesis in alignment with competency-based medical education.

Chapter 27

# Program evaluation

*"Without proper self-evaluation, failure is inevitable."*

—John Wooden

## 27.1 Introduction

The methodical assessment of educational programs' impact, efficacy, and quality in order to promote advancements in patient care and training is known as program evaluation in medical education. It entails determining the needs of stakeholders, establishing success criteria, gathering and evaluating data on program procedures and results, and applying the results to support ongoing curriculum and training activities, guarantee accountability, and inform decision-making (Ullah, 2024). There are various evaluation models and techniques, and a thorough and successful evaluation process depends on the participation of all stakeholders. For anyone in charge of a medical education program, program evaluation is a crucial duty.

A "program" can be as tiny as a single class, a course, or a medical school clerkship rotation, or it can be as extensive as an entire educational program. The "program" could take place during postgraduate training, continuing professional development, or medical school. A solid evaluation strategy is necessary for all such programs. Evaluation, at its most basic level, entails determining the worth of the information at hand. Therefore, information is used in educational program evaluation to determine an educational program's value or worth. The process of educational program evaluation is more formally defined by ACGME as the "systematic collection and analysis of information related to the design, implementation, and outcomes of a program, for the purpose of monitoring and improving the quality and effectiveness of the program." Program evaluation, as this definition makes clear, is about comprehending the program through a regular, methodical, intentional collection of data in order to discover or identify what contributes to the program's "success" and what steps need to be taken in order to address the evaluation process' findings (Durning et al., 2007).

## 27.2 Need for program evaluation

In order to provide high-quality, patient-centered care in medical education, program evaluation is essential. This is done by systematically evaluating programs, offering evidence for improvement, and guaranteeing accountability to stakeholders and accreditation agencies. It promotes program enhancement, maximizes the use of resources, and aids in upholding quality assurance, all of which are necessary for accreditation and proving the program's worth to society (Shrivastava and Shrivastava, 2023). The following justifies the necessity of program evaluation in medical education:

### 27.2.1 Quality improvement

Program evaluation aims to ensure that the graduates satisfy professional standards and community needs. Program evaluation is also crucial for improving the quality of medical education. It provides a systematic way to gather information about the effectiveness, applicability, and results of programs. It ensures that programs stay in line with accrediting requirements and changing healthcare demands by offering proof of responsibility, assisting in the identification of gaps in teaching, curriculum, or faculty training, and promoting ongoing improvement through iterative cycles.

### 27.2.2 Accountability

Program evaluation is essential for accountability in medical education because it systematically determines whether programs address community health needs, produce qualified physicians, and accomplish their goals. Additionally, it

**The Principles and Practice of Medical Education. DOI:** https://doi.org/10.1016/B978-0-443-44197-4.00019-9

fulfills accreditation criteria, encourages ongoing development, and provides information for decision-making. It ensures graduates are truly prepared and provides statistics to show the value to stakeholders, including students, teachers, and the community, by directly linking educational quality to societal responsibility and efficient resource usage.

### 27.2.3 Evidence-based decisions

Program evaluation is essential to move educational practices away from presumptions and toward ongoing, accountable improvement. Program assessment is essential to medical education because it provides the systematically collected data needed to enable informed, evidence-driven decisions. Additionally, program evaluation generates trustworthy evidence because it employs dependable research methods to collect both qualitative and quantitative data about the effectiveness, caliber, and impact of an educational program. This data serves as the evidence for decisions.

### 27.2.4 Patient safety

Program evaluation is essential for patient safety because it determines whether safety training actually alters behavior, finds curriculum gaps such as in teaching and learning process or insufficient training, assures accountability, and promotes ongoing improvement in the production of qualified, safety-conscious professionals, ultimately preventing harm and improving patient outcomes. It increases the effectiveness of education by measuring attitudes, actions, and practical application in addition to basic knowledge assessments.

### 27.2.5 Resource optimization

Program evaluation is crucial for medical education resource optimization because it methodically evaluates the efficacy and efficiency of programs. When a systematic program evaluation is executed, it provides the faculty and the institution a much needed systematic and evidence-based information about "what works" and "what does not." Program evaluation also aids in resource optimization by ensuring that financial, human, and physical resources are allocated to the most productive endeavors. It changes program management from a reactive to a proactive, data-driven process that maximizes effect and reduces waste. It aids medical schools in defending their choices on how to allocate resources and in coordinating their curricula with institutional and societal demands.

### 27.2.6 Meeting global standards

Program assessment is essential for medical education institutions to meet and maintain global standards since it provides evidence of quality, ensures accountability, encourages continuous improvement, and facilitates international recognition for graduates. Since a strong program evaluation system is a fundamental requirement for accreditation by major international organizations like the World Federation for Medical Education (WFME), meeting global standards has become a critical requirement for accreditation. The Educational Commission for Foreign Medical Graduates, for example, has linked its certification requirements to WFME-recognized accreditation, making evaluation a de facto requirement for graduates hoping to practice abroad, especially in the United States (Fig. 27.1).

## 27.3 Program evaluation models

Using frameworks such as Kirkpatrick's four levels (reaction, learning, behavior, results), the CIPP model (context, input, process, product), and logic models, medical education program evaluation models systematically evaluate programs with an emphasis on improvement rather than just judgment (Balmer et al., 2019). Evidence-based choices for curriculum improvement are led by statistics on satisfaction, skill acquisition, practical application, and final impact on patient care. These models offer organized methods for determining whether a program is effective, for whom, and how it might be improved.

### 27.3.1 Kirkpatrick's model

Donald Kirkpatrick created the four-level Kirkpatrick model of program evaluation in the 1950s to evaluate the efficacy of training and educational initiatives. It gauges the effectiveness of a program by looking at how participants responded to the instruction, what they learned, if they changed their behavior at work, and the organization's overall outcomes (Masood and Usmani, 2015).

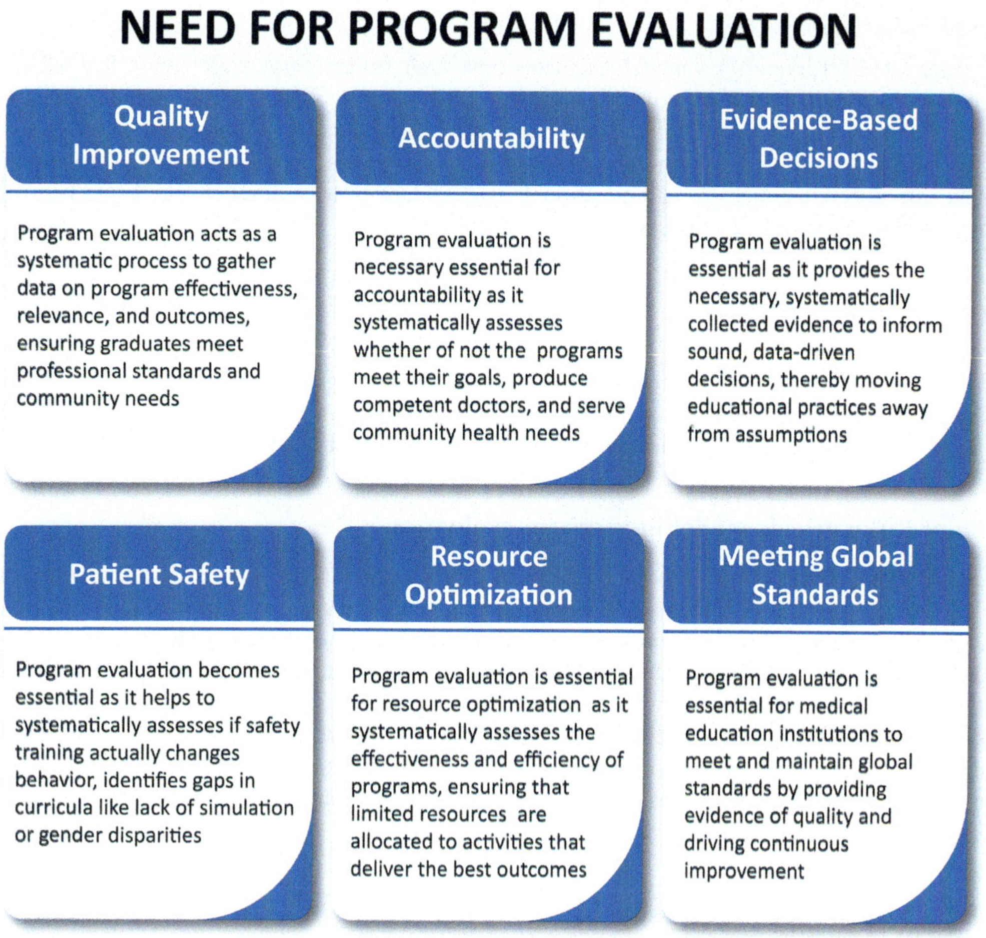


FIGURE 27.1 Need for program evaluation.

### 27.3.2 Levels of Kirkpatrick's model

The Kirkpatrick model's four training evaluation levels, which include reaction, learning, behavior, and results, assess participant satisfaction, knowledge acquisition, workplace application, and ultimate organizational influence.

#### *27.3.2.1 Level 1: reaction*

Although it doesn't ensure actual learning or behavior change, the reaction stage of the Kirkpatrick's model in medical education assesses trainees' immediate satisfaction and engagement with a training by focusing on their feelings about the content, instructors, materials, and environment. The data on the learner satisfaction is usually done through surveys or interviews. It is the first stage in determining whether or not participants enjoyed the training and thought it would be helpful for future practice, which influences their willingness to participate at higher levels.

#### *27.3.2.2 Level 2: learning*

The learning stage of the Kirkpatrick's model forms the vital link between initial response and actual on-the-job application. It also helps the faculty and the administration to understand how well participants in a training program pick up the desired knowledge, skills, attitude, confidence, and dedication from training using tests, simulations, or observations. It extends beyond participant satisfaction, as depicted by Level 1 of the model, to determine whether real learning took place, usually through evaluations, simulations, and pretests and posttests.

#### 27.3.2.3 Level 3: behavior

The behavior stage of the Kirkpatrick's model evaluates how well students use their newly learned information and abilities in a real-world clinical setting or professional practice. Transferring knowledge from the training environment to the workplace is the main goal of Level 3. It goes beyond merely being aware of the material to really applying it in the workplace. This assessment aids in determining whether the training has improved medical professionals' everyday performance and activities in a noticeable and quantifiable way. A lack of behavioral change could be a sign of problem with the training or the support of new practices in the workplace.

#### 27.3.2.4 Level 4: results

The results stage of this model assesses how training ultimately affects organizational objectives and patient outcomes. It looks beyond individual learning to determine whether increased knowledge leads to better clinical performance, fewer medical errors, increased patient satisfaction, increased efficiency, or even improved institutional reputation, directly linking training to observable benefits like learner efficiency and quality of healthcare (Fig. 27.2).

### 27.3.3 Role of Kirkpatrick's model in program evaluation

Systematic improvement, which offers an organized method for identifying strengths and shortcomings and promoting curriculum enhancement, is one of the main advantages of this model in program evaluation. Additionally, it goes beyond level 1 of reaction to comprehend the more profound effects of practice and competency (Rouse, 2011). It also involves a variety of stakeholders in the evaluation process, including students, instructors, leaders, and patients, thereby helping the faculty to gather their opinions and suggestions regarding the program. This methodology encourages accountability by showcasing the program's value and effectiveness to organizations and accreditation agencies. It also helps teachers improve their professional development programs and teaching methods. With the help of this paradigm, medical educators can make sure that their programs are successful in developing trained healthcare professionals that enhance patient care, in addition to being well received.

### 27.3.4 CIPP (context, input, process, product) model

Daniel Stufflebeam created the CIPP (context, input, process, product) model, which is a thorough framework for program assessment that offers data to enhance programs rather than only assess them. It assesses a program from four perspectives, namely, the context referring to the needs, issues, and objectives in the learning environment, input pertaining to the various strategies and resources, process indicating towards various activities and execution, and finally product represented by impact and results. This decision-focused strategy gives different stakeholders both formative assessment targeting towards improvement during implementation and summative assessment which is needed for final

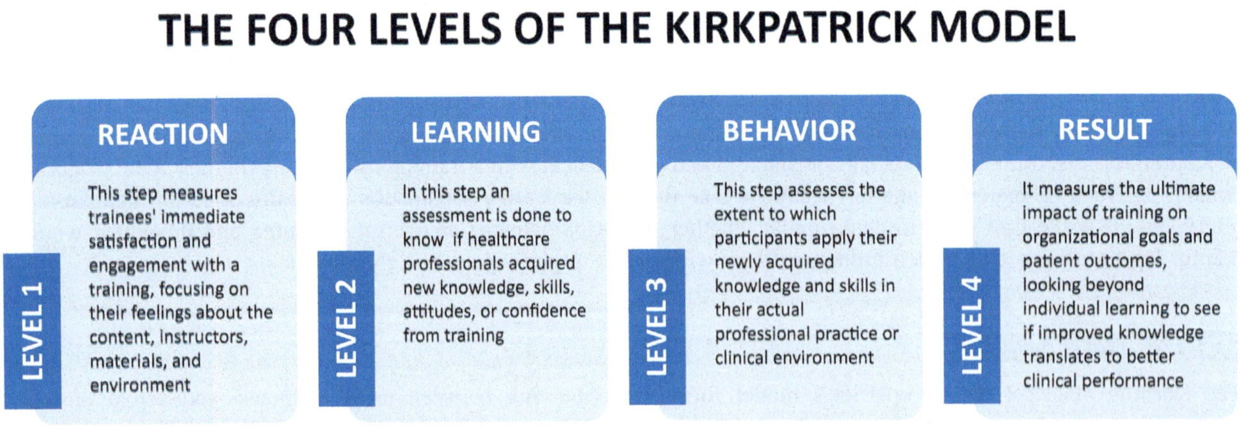


FIGURE 27.2 The four levels of the Kirkpatrick's model.

judgment (Warju, 2016). As mentioned above, the CIPP model consists of four distinct evaluation types which are described as follows:

#### *27.3.4.1 Context evaluation*

The CIPP model's context evaluation stage in medical education identifies needs, resources, opportunities, and issues in the learning environment, such as gaps in the current curriculum, learner difficulties, and societal health needs. This helps to justify a new or updated program and directs decision-making regarding what needs to be done to enhance physician training and meet demands on the health system. It serves as a thorough needs analysis, assisting in the alignment of learning objectives with practical needs and guaranteeing efficacy and relevance from the outset.

#### *27.3.4.2 Input evaluation*

Before a program begins in medical education, the input evaluation stage of the CIPP model evaluates the necessary resources, strategies, and plans, such as budget, faculty, curriculum design, facilities, and policies, to determine the best approach for meeting identified student or community needs, ensuring feasibility, effectiveness, and alignment with goals for improving decision-making and program structure. "What resources are needed, and what is the best plan to achieve the objectives?" is one of the problems it addresses.

#### *27.3.4.3 Process evaluation*

The process evaluation stage of the CIPP model is a formative, continuous quality control step in medical education that examines how a program is carried out, verifying fidelity to the plan, identifying operational issues like teaching delivery and resource use, providing real-time feedback to guide necessary mid-course revisions, guaranteeing efficient resource use, and influencing decisions about program continuation or modification. Before evaluating the program's final results, it focuses on activities, implementation fidelity, resources, and practical problems encountered throughout execution to ensure that the program operates as intended.

#### *27.3.4.4 Product evaluation*

The CIPP model's product evaluation stage in medical education evaluates a program's effectiveness, sustainability, and wider influence on students and the healthcare system, as well as if the program's objectives were reached. It provides vital information for program improvement and decisions on continuation or modification. It focuses on intended or unintended results, impact on patient care, long-term skill retention, and if the graduates can adapt to the gained skills (Fig. 27.3).

### 27.3.5 Role of CIPP model in program evaluation

For thorough, improvement-focused program evaluation in medical education, the CIPP model is essential. It assists educators in evaluating needs, methods, implementation, and results in order to make well-informed judgments regarding curriculum development, from design to revision (Toosi, 2021). In order to ensure relevance, efficacy, and ongoing quality improvement for stakeholders, it provides a structured, cyclical strategy to systematically examine all program elements. This approach helps determine whether to continue, alter, terminate, or disperse programs. The CIPP model is essentially a methodical structure that guarantees medical education programs are pertinent, adequately funded, efficiently administered, and produce the intended educational and professional results.

## 27.4 Program evaluation designs

In medical education, program evaluation designs are frameworks for assessing educational programs, such courses and residencies, to ascertain their worth, improve quality, ensure accountability, and guide choices. With a focus on stakeholder demands for practical change, they employ a range of methods to address problems about design, implementation, and outcomes, from simple pretests or posttests to sophisticated randomized trials. They fall into three primary categories:

1. Experimental designs,
2. Quasi-experimental designs,
3. Nonexperimental designs.

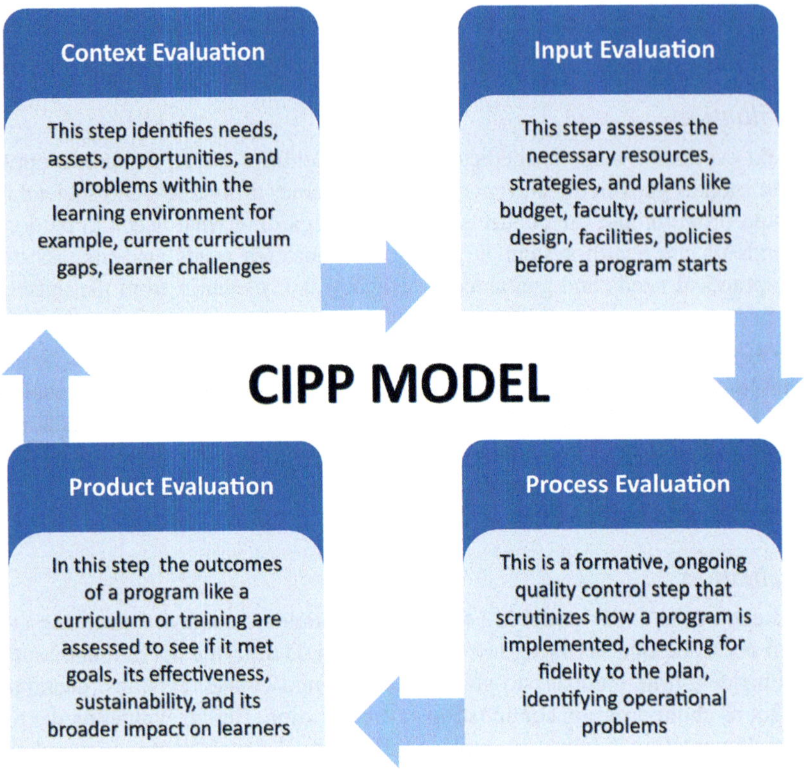


FIGURE 27.3 CIPP (context, input, process, product) model.

### 27.4.1 Experimental designs

Experimental designs in medical education program assessment systematically test educational interventions such as a new curriculum or simulation training by comparing outcomes between groups. Pertaining to the experimental design randomized controlled trials (RCTs) are considered as the gold standard for evaluating medical education programs. Experimental designs establish cause-and-effect by randomly allocating students to intervention and control groups. In this design, the participants, such as students or residents, are randomly assigned to either an intervention group that receives new training or a control group that receives conventional training or no new training.

### 27.4.2 Role of randomized controlled trials in program evaluation

As RCTs offer the most reliable way to show a cause-and-effect relationship between an educational intervention and its outcomes, RCTs are the "gold standard" for program evaluation in medical education. They provide high internal validity by reducing bias, guaranteeing that observed results are actually attributable to the program itself. The primary duty is to ascertain whether a certain educational program which might be in the form of a new curriculum, teaching method, or training really improves outcomes as opposed to changes caused by other causes. Additionally, it helps minimize bias because RCTs guarantee that both known and unknown confounding factors are, on average, equally distributed between the groups at the beginning of the study by randomly assigning participants such as students, residents, clinics and hospitals. Additionally, it makes it easier to provide high-quality information since well-conducted RCTs rank highest in the hierarchy of evidence, which is essential for evidence-based clinical practice and well-informed policy-making in medical education. RCTs go beyond merely asking "what works?" to provide quantifiable evidence of how much an intervention works in comparison to standard practice or an alternative approach. Furthermore, the outcomes of thorough RCTs assist educators and policy makers in making well-informed choices about which initiatives to scale up or implement, guaranteeing that funds are directed toward successful treatments. RCTs can assist understand why and for whom an intervention works by combining with process assessments and subgroup analyses.

### 27.4.3 Quasi-experimental designs

When random assignment is not practical, quasi-experimental designs in medical education assess interventions like a new teaching techniques or simulation training using nonrandom groups like current classes or departments. In this type of design, various techniques like nonequivalent control groups, interrupted time series, or regression discontinuity are employed to estimate effectiveness by comparing outcomes before and after or between groups. These designs provide useful insights despite having weaker internal validity than RCTs because of potential selection bias. They are essential for practical applications, evaluating technologies such as simulation-based learning by employing robust designs and statistical techniques to control for validity concerns (Capili and Anastasi, 2024).

#### *27.4.3.1 Nonequivalent control groups*

In program assessment, nonequivalent control group designs are quasi-experimental techniques that compare a treatment group with a comparable. In real-world situations when randomization is impracticable or unethical, this design evaluates program impact utilizing pretest and posttest data to determine changes between groups. Here the students are placed in preexisting groups, such as different classes or cohorts, that are fundamentally different or nonequivalent prior to the intervention rather than being randomly assigned. In order to determine whether the new approach was successful, the researchers compare outcomes such as test scores or skill performance between the group receiving the new training, which is the treatment group, and the group not receiving the training, which is the control group. However, they acknowledge that potential preexisting differences, such as selection bias, may affect the results.

#### *27.4.3.2 Role of quasi-experimental designs in program evaluation*

Quasi-experimental designs have proved to be beneficial to evaluate the impact of public policies, educational, and social programs in conditions where the random assignment is not feasible. They provide reliable causal inference in real-world settings by establishing or identifying comparison groups, such as regression discontinuity, analyzing prepost data, and using statistical techniques to estimate counterfactuals statements. For instance, this involves estimating what would have happened in the absence of the program (Song et al., 2016). Assessing the impact without randomization is one of the most important aspects of the quasi-experimental design since it enables evaluators to examine program effects in real-world situations when random assignment to treatment or control groups is impractical due to ethical or practical concerns. This design are also quite successful with preexisting data, which makes them perfect for assessing programs or policies that are already in place.

### 27.4.4 Nonexperimental designs

In medical education, nonexperimental designs are perfect for evaluating prevalence, characterizing trends, or investigating correlations in situations when experiments are unethical or impossible since they monitor naturally occurring occurrences without changing factors. Descriptive studies such as surveys, case studies, correlational designs, and retrospective investigations are common types. They are great for real-world insights and patterns, but because they don't have control over variables, they mostly demonstrate associations rather than direct cause-and-effect, which makes them more biased than actual studies (Vassar, 2010).

#### *27.4.4.1 Descriptive type of nonexperimental design*

Descriptive nonexperimental designs in medical education focus on "what is" by using techniques such as case studies, case series, cross-sectional surveys—for instance, measuring student stress levels at one point—and observational studies such as tracking common learning difficulties to map prevalence, patterns, and demographics in order to form hypotheses rather than prove cause-and-effect, which is essential for comprehending educational phenomena in real-world settings.

#### *27.4.4.2 Role of nonexperimental designs in program evaluation*

In program evaluation, nonexperimental designs are essential, particularly in situations where random assignment is not feasible. They offer insights into "how" and "why" programs function by observing natural occurrences, spotting trends, and investigating relationships between variables, such as correlations or differences between existing groups, using techniques like pretests or posttests, case studies, or realist evaluation to comprehend program context and outcomes, even without establishing strict causality as experiments do (Gibson et al., 2008). With an emphasis on relational or descriptive findings rather than conclusive cause-and-effect, they aid in describing how programs operate, spot trends, and provide useful qualitative data for enhancing efficacy.

## 27.5 Conduction of program evaluation

In order to assess a program's efficacy, value, and potential for improvement, a systematic process of data collection and analysis is required. This process usually entails the following steps, namely, defining goals, designing the study, gathering and analyzing data, and reporting findings to inform decisions, improve outcomes, and ensure accountability to stakeholders. It is an iterative cycle that begins with specific queries and employs data to direct the creation of subsequent programs (Gursel, 2008).

### 27.5.1 Internal stakeholders

Students, residents, faculty, program directors, coordinators, and institutional leadership such as committee heads and deans are examples of internal stakeholders in program evaluation. These individuals directly participate in or are impacted by the program and offer vital perspectives on curriculum, learning, resources, and outcomes for decision-making and improvement. Internal stakeholders are also essential to the evaluation of medical education programs because they offer distinctive viewpoints, define pertinent questions, guarantee data validity, drive curriculum improvement, boost buy-in, and ensure that the program meets competency goals. The input drawn from these stakeholders in terms of curriculum, training needs and program efficiency ultimately lead to better-trained physicians and quality healthcare.

### 27.5.2 External stakeholders

Patients and the community, healthcare facilities, professional associations, government or funders, licensing or accreditation bodies, alumni, and industry partners are examples of external stakeholders in program evaluation. These groups have a vested interest and offer vital perspectives on relevance, quality, and real-world impact for accountability and improvement. By offering real-world perspectives, guaranteeing program relevance, validating outcomes, identifying societal needs, and promoting accountability, external stakeholders contribute to program evaluation in medical education. The input given by the external stakeholders helps the faculty and institution connect educational objectives to practical healthcare quality and public trust. As a result of all these, there is an improvement in terms of graduate quality, program quality, and societal effect. This guarantees that the outcoming graduates are ready for a variety of patient demands and community health concerns.

### 27.5.3 Expertise involved

#### *27.5.3.1 Evaluation experts*

In medical education, evaluation specialists are essential for creating thorough evaluation plans, gathering and evaluating qualitative and quantitative data, and offering unbiased, evidence-based feedback to improve program quality, curriculum, and outcomes. This ensures that programs fulfill objectives, adhere to accreditation requirements, and ultimately improve patient care. They serve as negotiators and adapters, concentrating on the demands of stakeholders and applying methodical techniques to assess the value and usefulness of a program.

#### *27.5.3.2 Data analysts*

In order to provide actionable insights for curriculum improvement, personalized learning, and better student or patient outcomes, data analysts are essential to the evaluation of medical education programs. They do this by gathering, cleaning, and interpreting a variety of data, including assessments, electronic health records, and surveys. This allows for evidence-based decisions and adaptive strategies, which not only measure outcomes but also improve the entire educational cycle.

## 27.6 Key components of program evaluation

### 27.6.1 Defining the program

Some of the essential features of program evaluation lie in clearly defining the scope, aims, and objectives of the educational program under assessment. While the scope establishes the parameters for the evaluation itself by deciding which particular areas, timeframes, and target groups will be examined, goals specify the anticipated results and purpose of the program being assessed. Additionally, it should also be evident as to exactly what is being assessed. For example, whether the program is focused on evaluating the teaching methods to determine the value of new simulation techniques, or whether it aims to evaluate the curriculum to determine the effectiveness of content delivery, alignment with goals, or the outcomes to determine how well graduates perform their impact on patient care.

### 27.6.2 Stakeholder involvement

Involving important stakeholders like students, faculty, patients, and administrators is beneficial as they influence evaluation questions, data collection, and interpretation to meet social accountability goals and improve healthcare outcomes. Additionally, by providing a range of perspectives, ensuring relevance, increasing buy-in, and enhancing utility, their participation provides value. In the past, stakeholders were often passive, but modern initiatives like collaborative evaluation aim for deeper, participative engagement, even though more patient and community involvement is still needed (Musick, 2006).

### 27.6.3 Establishing criteria

In order to assess a program's performance, establishing criteria entails defining precise, quantifiable characteristics, including relevance, effectiveness, efficiency, impact, and sustainability. Determining the evaluation's goal, involving stakeholders, establishing precise performance standards, and choosing suitable data collection techniques are some of the crucial tasks in this respect. These criteria should be practical, transparent, and aligned with the program's objectives. Moreover, in order to guarantee that the criteria are properly related to the program's particular goals, learning objectives, and anticipated outcomes, it is essential to develop criteria for program evaluation. This procedure guarantees that the assessment is methodical, impartial, and defendable.

#### *27.6.3.1 Data collection*

In medical education, data collection is an essential and fundamental part of program evaluation since it systematically accumulates information to evaluate program efficacy, guide decision-making, and promote ongoing quality improvement. Every evaluation step, from setting goals to assessing results, is intimately related to the data gathering stage. In this context, the main goal of data collecting is to obtain timely, accurate, and pertinent information, so that decisions regarding the worth of an educational program can be made.

#### *27.6.3.2 Data analysis and interpretation*

In order to make data-informed decisions, improve curricula, and raise the caliber of healthcare practitioners, data analysis and interpretation are essential components of medical education program assessment. This procedure entails a methodical strategy to evaluating the program's efficacy, pinpointing areas in need of development, and guaranteeing responsibility. This data analysis helps program directors make evidence-based decisions, such as changing teaching tactics or allocating resources, by providing them with a clear picture of the program's benefits and drawbacks. Additionally, it guarantees accountability because program efficacy is proved to funding sources, accrediting agencies, and stakeholders through the use of aggregated data and its interpretation.

#### *27.6.3.3 Reporting and utilization*

Reporting and utilization play a vital role as they facilitate the sharing of results of evaluation with stakeholders and actively applied to support decision-making, enhance program quality, guarantee accountability, and promote organizational learning. This reporting and usage method also contributes to capacity building by encouraging a sense of ownership over the process of improvement by involving stakeholders in the interpretation of results and the development of action plans. Nonetheless, the program becomes more flexible and self-regulating as a result of the reporting and implementation of results, which promotes a culture of continuous quality improvement (Fig. 27.4).

## 27.7 Domains of program evaluation

The domains of program evaluation in medical education evaluate different phases and elements by concentrating on needs, design, implementation, outcomes, and efficiency in order to improve curriculum, teaching, environment, and student performance and eventually guarantee high-quality medical training. These areas offer a thorough framework for comprehending and assessing an educational program by looking at its necessity, the reasoning behind its creation, how

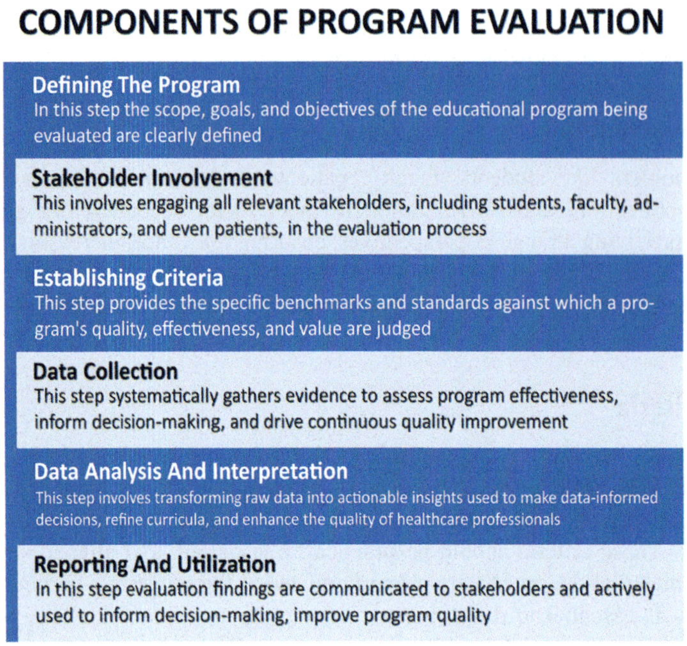


FIGURE 27.4 Components of program evaluation.

well it was implemented, its final results, and if the process was a cost-effective in terms of use of resources. The following is a description of the five domains:

### 27.7.1 Needs assessment

Needs assessment is a key component of medical education program evaluation, including methods such as surveys, interviews, and literature reviews to guide curriculum design and quality improvement. It serves as the essential beginning point to identify gaps between current learner performance or knowledge and desired competencies, ensuring programs are relevant, effective, and address actual needs of students, faculty, and the healthcare system.

### 27.7.2 Theory assessment

As a domain in medical education program evaluation, theory assessment focuses on inputs, processes, outcomes, and context to ensure quality, identify strengths or weaknesses, and drive continuous improvement in preparing competent physicians. It does this by using established evaluation models and educational theories to systematically judge a program's worth. In order to create strong, efficient medical training, it looks at elements like learner-environment interaction, context, and stakeholder needs in addition to outcomes to understand why things work or don't.

### 27.7.3 Implementation assessment

Implementation assessment, a critical area of program evaluation in medical education, focuses on determining whether a new curriculum, program, or educational reform is being delivered as intended, to whom, how, and at what cost. Different from evaluating the end learning results, this type of process evaluation aids in understanding the reality and possible obstacles of the implementation process.

### 27.7.4 Impact assessment

A crucial component of evaluating medical education programs is impact assessment, which focuses on identifying the program's long-term, persistent, and frequently wider consequences on individuals, the healthcare system, and the community at large rather than merely short-term results like knowledge acquisition. The primary purpose of impact assessment is to establish a causal relationship between the educational intervention and observable changes, whether positive or negative, and whether purposeful or unintended.

## 27.7.5 Efficiency assessment

Efficiency assessment, which focuses on the connection between resource inputs and educational outputs or results, is a well-known and important area of program evaluation in medical education. It helps determine whether a program is delivering value by optimizing resource use to achieve desired goals. Efficiency in this context refers to using the fewest medical resources possible while optimizing desired outputs and objectives, such as the number of graduates, knowledge advancement, and improved patient care. Ensuring responsibility, directing resource allocation, identifying areas that need improvement, and guaranteeing program sustainability are the main goals. Program evaluation domains in medical education are crucial for ensuring quality, accountability, and relevance because they systematically assess needs, design, implementation, outcomes, and costs. This results in data-driven decisions for ongoing improvement, improved graduate readiness, and alignment with societal needs. They offer an organized framework, like CIPP, to comprehend the program's advantages, disadvantages and effects, directing adaptation and guaranteeing that graduates satisfy changing healthcare needs.

## 27.7.6 Timing and types of program evaluation

In order to guarantee accountability, accreditation, and ongoing quality improvement for better patient care, program evaluation in medical education should begin in the planning stage prior to launch and take place in formal cycles on a regular basis, such as annually or every 3–5 years, as well as whenever major changes are made or problems emerge. Program evaluation in medical education can be conducted at the following junctures:

### *27.7.6.1 Before program launch*

Prelaunch or formative evaluation's primary objective is to evaluate the program's viability, suitability, and acceptability so that any necessary modifications and enhancements can be made before it is completely implemented. It raises the program's chances of success and helps guarantee that it begins with a strong foundation. The needs assessment, which determines whether the program is necessary in the first place and whether the suggested objectives fill recognized gaps, is one of the crucial factors to be taken into account in the prelaunch review. Assessing the underlying data, available resources like the budget, personnel, time, and any potential issues with the program's operating environment are also crucial. Reviewing the suggested tactics, instruments, and resources is impertinent in this regard to make sure they are suitable and adequate to achieve the program's objectives.

### *27.7.6.2 During the program*

Program evaluation that offers continuous, timely input to enhance a program's design, execution, and general efficacy is known as formative evaluation. Rather than evaluating the final result, it is carried out during the development and administration of a program to identify achievements, problems, and places for improvement. Formative evaluation aids in directing modifications and improving the program while it is still in process by looking at the inner workings of the program and the environment in which it functions. One of the most important aspects of the formative assessment is that it is carried out either at different stages of the program's ongoing development or throughout its planning, development, and implementation phases. Rather than determining success or failure, the main goal of this evaluation is to improve the program by providing data for adjustments. The formative assessment's objective is to analyze the "how" of the program's operations in terms of achievements, flaws, challenges, and modifications. This type of evaluation uses both quantitative and qualitative data from a range of sources, including students, employees, and other stakeholders. The findings are meant to ensure the program's applicability and effectiveness and help the decision-makers make the necessary modifications. Some of the advantages of the formative assessment are focused on program modification, where staff members can use the evaluation's comments to quickly alter the program's delivery or content. Additionally, it aids in planning since the knowledge acquired can lead future planning and program management as a whole. It can also help employees reflect on their job and provide them constructive criticism.

### *27.7.6.3 Summative evaluation*

It is done at the conclusion of a program to assess its overall impact and efficacy. Summative evaluation is a kind of program assessment that is carried out at the conclusion of a program to evaluate its overall efficacy, impact, and value by ascertaining whether its planned goals and objectives were fulfilled. It offers a comprehensive evaluation of the program, which frequently results in choices on its continuation, modification, or termination. Outcome and impact evaluations are common techniques that measure results using data from final exams, performance evaluations, surveys, and interviews. A summative assessment's main objective is to evaluate the program's overall efficacy in accomplishing its objectives. It offers a thorough picture of the program's execution, activities, results, and effects. Determining the program's total value and if it produced the desired result is the main goal. Additionally, it directs the decision-makers about the continuation, expansion, or termination of a program. Summative evaluation collects data from a variety of sources over time to demonstrate the program's efficacy. The summative evaluation's retroactive nature is one of its main features. It provides an overview of the program's performance by looking back at it. It is also evidence-based since findings on the program's effectiveness are supported by gathered data and evidence.

#### *27.7.6.4 Periodic program evaluation*

In medical education, periodic program evaluation is a methodical, continuous process that uses data from students, outcomes, and accreditation to assess a program's efficacy, guarantee quality, and make evidence-based improvements for better patient care. It involves steps like planning, data collection, analysis, and feedback to stakeholders. It is frequently guided by frameworks from center for disease control and prevention or Kirkpatrick's model and is essential for fulfilling standards and social obligations. Ensuring quality, meeting accreditation requirements, informing decision-making, and improving educational efficacy for high-quality patient care are the primary goals here. It also includes setting objectives, gathering information such as student input, evaluating results in relation to standards, and giving feedback for ongoing development. It is linked to certification cycles and is carried out frequently, for example, every 3–5 years with the possibility of interim evaluations. Statistics from several sources, including student evaluations, course or clerkship comments, performance statistics, and outcomes, are used in this procedure. Faculty, staff, students, advisers, and accrediting organizations are all involved in this process.

#### *27.7.6.5 Process evaluation*

It focuses on how the program is carried out and whether or not it is being delivered according to schedule. Program evaluation that focuses on evaluating a program's execution, activities, and quality rather than its final results is known as process evaluation. It looks at whether a program is carried out as planned, looks at difficulties and changes from the original plan, and gives early alerts about any issues. In order to comprehend the implementation process, offer insights into program reach and integrity, and guide program improvement, process evaluations employ a variety of techniques, including qualitative and quantitative data collection. Finding out how closely program actions match the original plan is made easier with the help of the process evaluation. It also aids in monitoring the volume, caliber, and nature of operations carried out. Additionally, by collecting input on client satisfaction and staff experiences with the program, it aids in understanding participant and staff perceptions. Apart from this, it also offers early warning indicators, such as seeing possible obstacles, difficulties, or inefficiencies before they have a major negative influence on program progress.

Some of the key questions addressed by this evaluation are as follows:

1. How well is the program being implemented?
2. Are the program's activities occurring as planned?
3. What are the challenges and facilitators of implementation?
4. How satisfied are participants and staff with the program?

#### *27.7.6.6 Outcome evaluation*

It gauges the program's effects and outcomes, including how learners' knowledge, abilities, and patient care have changed. Measuring a program's long-term efficacy in accomplishing its stated goals and objectives is the focus of outcome evaluation, a subset of program assessment. In order to ascertain whether the program had an impact and to guide future funding and improvement decisions, it evaluates the extent to which the target population's behaviors, attitudes, or situations have changed. Although results are frequently measured after a certain amount of time, such as 6 months to a year or more, following program implementation, outcome evaluation is usually carried out at or close to the program's conclusion. The anticipated outcomes are measured by the outcome evaluation. It also aids in determining whether the program's objectives have been fulfilled. It will also highlight any noteworthy shifts brought about by the program, whether good and bad. Outcome evaluation is crucial because it shows how successful the program's tactics and activities were in bringing about the desired change. Making educated judgments about whether to maintain, enhance, or grow the program is aided by the results of outcome evaluation. The outcome evaluation method helps prove responsibility by allowing programs to demonstrate their impact and contribution to funders and other stakeholders. What happened is more important to the outcome evaluation than how it happened. It often occurs after the program has been implemented and enough time has elapsed for the results to show. In conclusion, outcome evaluation is a summative form of assessment that delivers a final judgment on how well a program has achieved in accomplishing its ultimate goals and objectives.

## 27.8 Steps in the conduction of program evaluation

Program evaluation involves key steps like planning, designing, collecting data, analyzing findings, and reporting using results to inform decisions and improve the program, focusing on clear outcomes and stakeholder engagement throughout. The key steps in program evaluation can be described as follows:

### 27.8.1 Assessing context and needs

Using techniques like needs assessments to ascertain "what needs to be done" before proceeding to design and data collection, assessing context and needs is a fundamental, first step in program evaluation. It is essential for comprehending the environment, identifying gaps, setting pertinent questions, and designing effective evaluations, which frequently involve community history, power dynamics, and existing assets to ensure relevance and success. Understanding the program's background, objectives, structure, and stakeholders is crucial in this situation in order to decide what needs to be evaluated and why.

### 27.8.2 Describe the program and define scope

Before organizing data collection and analysis, it is important to define the program's essential components which include needs, goals, activities, resources, and intended outcomes using tools like a logic model or theory of change to create a clear roadmap. This is often the second step in program evaluation. This foundation guarantees the accuracy and relevance of evaluation data.

### 27.8.3 Focus evaluation questions

Focus evaluation questions is a crucial first step in program evaluation. It directs the entire process by defining what to look into, that is, the relevance, outcomes, implementation, who to ask in terms of stakeholders, and what data is required. It moves from general concepts to specific, prioritized questions about program quality, impact, and improvement, ensuring that the evaluation remains manageable and relevant within resource constraints.

### 27.8.4 Design the evaluation

A key planning step in program evaluation is designing the evaluation, which entails defining the program, identifying stakeholders, defining goals and questions, choosing a design or methods, organizing data collection and analysis, and outlining reporting. All of these steps are necessary to guarantee that the evaluation is accurate, practical, useful, and suitable for guiding decisions and enhancing the program. In this sense, defining the program's rationale and results, concentrating the questions, selecting data sources and analytic techniques, and establishing deadlines and roles prior to the start of data collecting are all crucial components.

### 27.8.5 Data collection

In order to answer particular evaluation questions, evaluate results, and guide decisions, data collection is a crucial, methodical step in program evaluation. It involves obtaining reliable evidence through techniques like surveys, interviews, observations, and document reviews. Careful planning of sources, tools, and quality is necessary to guarantee trustworthy, useful insights for stakeholders.

### 27.8.6 Analyze and interpret data

A crucial and difficult stage in program evaluation is data analysis and interpretation, which acts as a link between unprocessed data and useful insights. It converts gathered data into insightful conclusions regarding the efficacy, efficiency, and applicability of a program. In order to answer the particular evaluation questions presented at the beginning of the program's assessment, statistical and qualitative methodologies are applied methodically throughout the entire process.

### 27.8.7 Report findings

In the program evaluation process, reporting findings is an essential stage that usually comes after data analysis and before acting upon or utilizing the findings for program improvement. It is an essential part of the last phases of the evaluation framework in many standard models. Creating concise, visually appealing reports that share insights and lessons learned with various audiences is crucial in this step.

### 27.8.8 Application of the findings

In program assessment, the application of the findings is an essential final stage. By applying the findings, conclusions on the program's worth, merit, or importance can be bolstered by combining and analyzing data in light of the examined evidence. This entails evaluating results by contrasting them with predetermined standards of success or criteria. As these results culminate in precise, doable advice for stakeholders, it also helps with the formulation of recommendations. Program managers, legislators, and providers are advised on what to do in light of the evidence by these recommended courses of action.

## 27.9 Impact on teaching and learning

Program evaluation has a big impact on teaching and learning because it provides information to improve curriculum design, instructional strategies, and policies. This results in better student outcomes in terms of engagement and achievement, as well as stronger accountability for all stakeholders and the efficient use of resources to achieve educational objectives (Bukhari, 2021). Determining "what works" and "what doesn't" offers feedback to improve teaching strategies, customize education, and create more effective, entertaining, and creative learning environments. Additionally, it enables educators to make data-driven adjustments by assisting assessors in identifying the advantages and disadvantages of instructional strategies. Additionally, it promotes professional development by pointing out the areas in which teachers require additional training, directing successful professional development programs.

## 27.10 Accreditation and program evaluation

Program evaluation is essential for medical education accreditation since it ensures that programs meet accreditation standards through ongoing monitoring, stakeholder feedback, outcome assessment such as board exams, and identifying areas for improvement, all of which ultimately support accountability and evidence-based development. It provides systematic data to demonstrate quality, effectiveness, and fitness for purpose. It also serves as the catalyst for self-evaluation, assisting institutions in proving that they meet strict external standards established by accrediting organizations in order to get or keep their authorized status.

**Summary**

The methodical process of determining the efficacy and influence of educational programs is known as program evaluation in medical education. It seeks to assess how well an educational program accomplishes its stated objectives and results, pinpoint areas in need of development, and hold stakeholders accountable. A crucial part of medical education is program evaluation, which offers insightful information that can raise the caliber and efficacy of educational initiatives. Medical educators can evaluate their programs in a methodical and significant fashion by utilizing a range of approaches and frameworks. Despite obstacles, review is essential to guaranteeing that medical education programs continue to generate skilled, caring healthcare workers and are in line with the changing demands of the healthcare system.

**Key point**

Future perspectives: emphasizing on the role of stakeholder

Program evaluation in medical education is changing due to changes in healthcare delivery, educational approaches, and technological improvements. Future viewpoints on program evaluation will probably concentrate on enhancing the caliber, involving the stakeholder, scope, and impact of assessments as medical education grows more intricate and varied. Technological developments, changing educational frameworks, and an increasing focus on competency-based and outcome-focused approaches will all influence program assessment in medical education in the future. In order to guarantee that medical education programs not only fulfill academic requirements but also enhance healthcare delivery worldwide, evaluators will need to implement more dynamic, individualized, and evidence-based methodologies.

## References


Balmer, D.F., Rama, J.A., Simpson, D., 2019. Program evaluation models: evaluating processes and outcomes in graduate medical education. Journal of Graduate Medical Education 11 (1), 99–100. Available at https://doi.org/10.4300/JGME-D-18-01084.1.

Bukhari, N., 2021. Twenty-first century program evaluation: a comparative analysis of three frameworks-based approaches. The International Journal of Assessment and Evaluation 28 (1), 51–66. Available at https://doi.org/10.18848/2327-7920/cgp/v28i01/51-66.

Capili, B., Anastasi, J.K., 2024. An introduction to types of quasi-experimental designs. American Journal of Nursing 124 (11), 50–52. Available at https://doi.org/10.1097/01.NAJ.0001081740.74815.20, http://www.lww.com/product/?0002-936X.

Durning, S.J., Hemmer, P., Pangaro, L.N., 2007. The structure of program evaluation: an approach for evaluating a course, clerkship, or components of a residency or fellowship training program. Teaching and Learning in Medicine 19 (3), 308–318. Available at https://doi.org/10.1080/10401330701366796, http://www.tandfonline.com/toc/htlm20/current.

Gibson, K.A., Boyle, P., Black, D.A., Cunningham, M., Grimm, M.C., McNeil, H.P., 2008. Enhancing evaluation in an undergraduate medical education program. Academic Medicine 83 (8), 787–793. Available at https://doi.org/10.1097/ACM.0b013e31817eb8ab, http://www.lww.com/product/?1040-2446.

Gursel, Y., 2008. An Example of Program Evaluation Project in Undergraduate Medical Education. Education for Health, Abingdon, England.

Masood, R., Usmani, M., 2015. A study for program evaluation through kirkpatrick's model. Khyber Medical University Journal 7, 2015.

Musick, D.W., 2006. A conceptual model for program evaluation in graduate medical education. Academic Medicine 81 (8), 759–765. Available at https://doi.org/10.1097/00001888-200608000-00015.

Rouse, D.N., 2011. Employing Kirkpatrick's evaluation framework to determine the effectiveness of health information management courses and programs. Perspectives in Health Information Management / AHIMA, American Health Information Management Association 8 (Spring), 1c. Available at 21464860. PMCID: PMC3070232.

Song, D.J., Choi, J.W., Kim, K., Kim, M.S., Moon, J.M., 2016. Quasi-experiment study on effectiveness evaluation of health communication strategies. Journal of Korean Medical Science 31 (7), 1027. Available at https://doi.org/10.3346/jkms.2016.31.7.1027.

Shrivastava, S.R.B.L., Shrivastava, P.S., 2023. Program evaluation in the field of medical education. Current Medical Issues 21 (3), 181–182. Available at https://doi.org/10.4103/cmi.cmi_31_23.

Toosi, M., Modarres, M., Amini, M., Geranmayeh, M., 2021. Context, input, process, and product evaluation model in medical education: a systematic review. Journal of Education and Health Promotion 10 (1). Available at https://doi.org/10.4103/jehp.jehp_1115_20, www.jehp.net.

Ullah, H., Huma, S., Yasin, G., Ashraf, M., Tahir-ud-Din, Q., Shabana, H., et al., 2024. Curriculum and program evaluation in medical education: a short systematic literature review. Annals of Medicine & Surgery 86 (10), 5988–5994. Available at https://doi.org/10.1097/ms9.0000000000002518.

Vassar, M., Wheeler, D.L., Davison, M., Franklin, J., 2010. Program evaluation in medical education: an overview of the utilization-focused approach. Journal of Educational Evaluation for Health Professions 7, 1. Available at https://doi.org/10.3352/jeehp.2010.7.1.

Warju, W., 2016. Educational program evaluation using CIPP Model. Innovation of Vocational Technology Education 12 (1). Available at https://doi.org/10.17509/invotec.v12i1.4502.

# PART 5

## QUALITY IN MEDICAL EDUCATION

Chapter 28

# Quality improvement

*"Quality is never an accident. It is always the result of intelligent effort."*

—John Ruskin

## 28.1 Introduction

In the context of healthcare, quality is defined as delivering timely, efficient, equitable, patient-centered, safe, and effective care to achieve desired health outcomes. In order to get the optimal outcomes, it entails acting appropriately for the appropriate patient at the appropriate time. Evidence-based procedures and patient feedback are used to measure and constantly improve this approach. The IOM defined healthcare quality in 2001 as "the extent to which health care services for individuals and populations increase the likelihood of desired health outcomes and are consistent with current professional knowledge."

A planned, methodical, and data-driven approach to continuously improve the safety, efficacy, efficiency, patient experience, and overall results of healthcare delivery is known as quality improvement (QI). Healthcare organizations may make quantifiable progress and give patients better, more consistent care by identifying areas for improvement, applying scientific methods and tools, and implementing improvements in processes and systems (Misra, 2019). This entire concept, which includes pertinent curricula, efficient teaching strategies, qualified instructors, ongoing development, and student-centered services, is focused on developing competent doctors who satisfy social demands and established norms. It is a dynamic, multifaceted idea that calls for constant assurance and assessment, concentrating on both the institution's internal procedures and the exterior results in terms of graduate readiness and community involvement (Buttell et al., 2007).

### 28.1.1 Quality in medical education

Quality in medical education focuses on improving health outcomes since it guarantees that the medical programs successfully train qualified and caring doctors through well-organized quality assurance (QA) and QI procedures. To satisfy high standards and stakeholder needs, this entails establishing precise program objectives, a well-designed and executed curriculum, efficient assessment techniques, suitable resources, ongoing review, and accreditation by professional authorities (Walshe, 2007). QI in medical education also plays a critical role in developing students into proactive problem-solvers and improving both the educational process and future patient care by integrating systematic enhancement into curricula and cultivating skills in efficiency, safety, and patient-centeredness. This results in better outcomes, fewer errors, and a more prepared physician workforce ready to continuously improve healthcare delivery.

### 28.1.2 Quality in healthcare

QI practices significantly improve healthcare by enhancing patient safety, efficacy, and experience. As a result, there are fewer errors, cheaper costs, better results in terms of fewer infections or readmissions, improved staff, and patient satisfaction and more productivity due to streamlined workflows. In order to create a more dependable, patient-centered, and long-lasting health system, it does this by methodically identifying issues, putting improvements into place such as improved communication and standardized procedures, and assessing outcomes (Wong et al., 2012).

### 28.1.3 Terminologies associated with quality

A wide range of terms are associated quality in medical education, with the main goals being to guarantee that medical education satisfies established criteria, generates qualified practitioners, and promotes ongoing development. It is

**The Principles and Practice of Medical Education. DOI:** https://doi.org/10.1016/B978-0-443-44197-4.00014-X

imperative that we understand these terms in order to appreciate the roles each of these terms play in upholding quality medical education and healthcare.

### 28.1.4 Quality improvement (QI)

QI in medical education is the ongoing process of improving curricula, teaching, and learning to better prepare future doctors to improve healthcare systems by using structured techniques and frameworks. Through process reduction, improvement, or standardization, QI is a methodical, data-driven strategy to improving health services, safety, and efficiency. It focuses on making continuous, quantifiable improvements to close the gap between evidence-based care and existing practice, frequently through quick, small-scale testing cycles. It entails incorporating QI training into core teaching, utilizing experiential learning, and cultivating a culture of continuous improvement for better patient outcomes.

### 28.1.5 Quality assurance (QA)

In general terms, QA is considered as a strategy for proactively preventing flaws and mistakes to guarantee that goods, services, or procedures constantly satisfy specifications and client expectations. To increase confidence that quality standards will be fulfilled, QA entails putting in place methodical processes, procedures, and activities throughout all phases of development and delivery. Instead of only identifying errors after they happen, it aims to stop errors before they start (Tabish, 1990). In medical education, QA entails well laid down procedures to guarantee that curricula fulfill strict requirements for generating skilled healthcare workers, thereby improving patient safety and care. It involves establishing and keeping an eye on standards through external accreditation by organizations such as the General Medical Council (GMC) and internal quality management. Continuous self-evaluation, feedback systems, and adherence to national and international standards are some of the essential elements of QA, which aims to provide reliable, consistent, and high-quality medical education and training.

### 28.1.6 Quality control (QC)

Generally speaking, QC is the process of examining goods and services to make sure they fulfill consumer demands and established standards. It involves testing and methodical activities at different stages of production or delivery to find and fix flaws before they are delivered to the client. It emphasizes on finding and fixing problems after they arise in order to preserve constant quality, avoid mistakes, and cut expenses, all of which eventually foster client loyalty and confidence (ALBayati, 2025). Establishing and upholding high standards for medical programs through monitoring, evaluation, and ongoing improvement is known as QC in medical education, and it is frequently directed by accrediting organizations such as the World Federation for Medical Education (WFME) and the GMC. Implementing strong feedback mechanisms from graduates and hospitals, assessing curriculum effectiveness through student pass rates and graduate performance, and cultivating a culture of QA through procedures like peer review and frequent institutional visits are some of the crucial elements here. The objective is to guarantee that medical schools generate skilled, prepared physicians who satisfy community demands and uphold international standards (Savsar, 2017).

### 28.1.7 Quality management

In medical education, quality management is a methodical technique that makes use of concepts like accreditation and continual development to guarantee that programs fulfill requirements for producing qualified physicians, which benefits students, institutions, and patient care. Peer review, student feedback, and self-evaluation are used to monitor instruction, services, and results against benchmarks such as WFME standards. The goal is to provide relevant, high-quality training that equips globally competitive physicians for societal demands, with an emphasis on structures, processes, and outcomes.

## 28.2 Purpose of quality improvement

In addition to ensuring safety, patient-centeredness, and lower costs in healthcare settings, the main goal of QI is to systematically improve processes and systems to generate better results, efficiency, and well-being for people and organizations. QI is an organized method to learning and development in a variety of industries, including healthcare, manufacturing, and services. It entails an ongoing cycle of measurement, analysis, and implementation of changes to enhance performance. It is about a dedication to continuous development, utilizing feedback and measurement to comprehend performance and make adjustments that increase both person well-being and organizational success. A complex web of care channels and processes forms the foundation of any health care system. The effectiveness of this network and the collaboration of those who administer and provide care are key factors in determining the quality of care provided by the system. The essence of QI lies in providing the time, authority, resources, and skills with an aim to improvise the quality of medical education and healthcare. It entails a methodical and coordinated approach to problem-solving with particular techniques and instruments in order to achieve a quantifiable improvement. Some of the important purposes of QI in medical education can be summarized as follows:

### 28.2.1 Enhance outcomes

One of the main goals of QI in medical education is to enhance patient outcomes by producing highly qualified doctors who can provide safe and effective care. The ultimate objective of incorporating QI into medical education is to equip aspiring physicians to recognize and resolve systemic problems that impact patient care. It also emphasizes how understanding QI principles helps medical students and trainees adopt a risk-identification and error-prevention mindset, which immediately improves patient safety and lowers adverse occurrences.

### 28.2.2 Improve processes and systems

One of the main goals and tenets of QI in medical education and healthcare in general is process and system improvement. QI aims at analyzing and redesigning work processes and systems to enhance results for patients, students and healthcare organizations. Additionally, in order to minimize undesired variation and produce more consistent quality and predictable outcomes, the QI process aims to standardize procedures and structures. It is crucial to remember that defects in intricate systems and procedures are addressed by the QI process. This guarantees that medical errors are addressed.

### 28.2.3 Reducing the costs

Although cutting costs is a big advantage and a major motivator for QI projects in medical education and healthcare systems, it is typically seen as a byproduct of improvement rather than the main goal. Improving patient outcomes, safety, and the efficacy and efficiency of care delivery are the main goals. Enhancing healthcare and education delivery to reach high standards of excellence is the main objective of QI. The quality of care can be adversely affected by an exclusive focus on cost-reduction measures without an integrated quality strategy, even if financial pressures frequently require efforts to increase quality while decreasing costs. In order to benefit patients and the system overall, true QI incorporates a mindset of ongoing learning and data-driven decision-making.

### 28.2.4 Increase efficiency and productivity

Increasing productivity and efficiency is a major goal of QI in medical education, which is in line with more general healthcare objectives of optimizing resources and reducing waste. The ultimate goal is to work smarter, not simply harder, so that medical personnel may devote more time to direct patient care and enhance system performance. Standardizing workflows and clinical recommendations are only two of the many ways that QI aims to boost productivity and efficiency, which is one of its main advantages. This minimizes variation and yields more predictable outcomes, which aids in eliminating pointless work and optimizing resource use. It should be noted that as data collection and analysis help administrators and educators anticipate staffing needs, identify bottlenecks, and monitor performance indicators to guide decisions, they are crucial to QI.

### 28.2.5 Ensuring safety

One of the main goals of QI in medical education is to provide future healthcare workers with the abilities, information, and perspective necessary to safeguard patients. Long before graduates start working in clinical settings, institutions can foster a culture of safety by incorporating QI concepts into their curricula. This is accomplished by introducing official patient safety and QI training into the core curriculum, going beyond traditional clinical skills to include mistake prevention and systems thinking. Teaching trainees to see medical errors as malfunctions in intricate healthcare systems rather than as personal failings is a major goal of QI education. This strategy promotes the creation of system-based remedies to stop harm in the future.

### 28.2.6 Promoting patient-centeredness

Encouraging patient-centeredness as the primary goal of QI in medical education guarantees that aspiring doctors will be prepared to deliver treatment that respects and responds to each patient's unique preferences, requirements, and values.

This strategy transcends a conventional biomedical model in favor of a holistic one, which eventually results in better health outcomes, more patient satisfaction, and more effective healthcare systems. Patient-centered care has been shown to improve disease management, enhance treatment plan adherence, and lower medical errors and readmissions to hospitals. Additionally, stressing the patient's viewpoint, building trust, and encouraging candid conversation are all facilitated by QI. These procedures turn out to be crucial for precise diagnosis and collaborative decision-making. Furthermore, this improves communication and trust between the doctor and the patient.

### 28.2.7 Facilitate learning and innovation

As it transcends academic knowledge to experiential, action-oriented learning where students and trainees identify, evaluate, and resolve real-world healthcare issues, QI fosters learning and innovation. Aspiring doctors learn how to use data-driven strategies to improve patient safety, efficacy, and care quality through the QI process. Focusing on particular, locally relevant issues, like cutting back on pointless testing and simplifying discharge summaries, is one way the QI approach fosters creativity. In this case, the process of QI promotes original, innovative solutions that are adapted to particular situations.

### 28.2.8 Promote standardization

Standardization promotion is a key element of QI in medical education since it guarantees consistency, reliability, and equity in the training of future healthcare professionals. Reducing unwanted variability in learning outcomes, assessments, and educational procedures is the aim of this technique. Regardless of their training location or specific faculty preceptor, standardization helps ensure that all students are exposed to the same basic curriculum, learning objectives, and clinical experiences. This is one of the primary objectives of encouraging standardization in medical education. This reduces disparities in educational opportunities and guarantees that all graduates have a basic capability.

## 28.3 Basic concepts of quality improvement

In medical education, QI refers to the methodical use of techniques to improve the effectiveness, safety, and quality of training programs and curricula with the goal of producing healthcare professionals who are better equipped. Since it emphasizes ongoing, formative, and frequently student-driven improvement, it differs from QA, which is summative. The QI process is grounded in the following basic concepts:

### 28.3.1 Establishing a culture of quality in clinical practice

A culture of quality is being established in clinical practice in a number of ways, including incorporating QI principles into the curriculum, fostering leadership, involving all stakeholders which include the students, faculty, staff, and patients upholding their viewpoint, and creating supportive procedures like teams, meetings, and a clear vision to drive ongoing learning and improvement in patient care and education.

### 28.3.2 Determining and prioritizing potential areas for improvement

In medical education, as well as in clinical practice, it is critical to recognize and comprehend areas where things could be improved. While prioritization uses data to rank the needs based on impact such as patient safety, learning outcomes, efficiency, and feasibility, frequently focusing on key areas like curriculum, communication, and emerging health challenges, QI in medical education identifies improvement areas by gathering stakeholder feedback, analyzing performance data using frameworks like plan–do–study–act (PDSA) cycle and benchmarking against standards.

### 28.3.3 Collecting and analyzing data

Collecting and analyzing data in medical education is crucial for QI because it provides objective evidence to identify problems, measure progress, tailor teaching, personalize patient care, reduce costs, and foster a culture of data-driven decision-making, ultimately leading to better learner outcomes and enhanced patient safety. It transforms education from subjective to evidence-based, ensuring curricula and training meet evolving healthcare needs and improve overall quality. The importance of collecting and analyzing data is to identifying gaps as the data reveals where learners struggle, such as pertaining to the specific clinical skills, knowledge areas, or where curricula fall short, pinpointing areas needing improvement. Additionally, it aids in tracking progress because it offers specific measures, such as test scores and clinical competency rates, to monitor the effectiveness of interventions like new teaching strategies and simulation training.

### 28.3.4 Communicate the results

As it promotes long-lasting change, improves learning, fosters a culture of safety and cooperation, and directly connects educational initiatives to better patient outcomes, communicating QI findings in medical education is essential. It converts individual achievements into systemic advancements. One of the most important features of sharing the results is that it speeds up the dissemination of best practices since sharing discoveries outside of the immediate project team makes it easier for other departments and institutions to adopt promising local improvements. This keeps successful initiatives from being "isolated secrets" and increases their influence throughout the healthcare system. Additionally, it promotes a culture of continuous learning, which is a fundamental component of a QI mindset in medical education. The practice of sharing outcomes stimulates continual reflection, thinking, and acting in continuous cycles of development.

### 28.3.5 Commit to ongoing evaluation

As it establishes feedback loops, finds learning gaps, guarantees that training satisfies practical requirements like patient safety, improves patient outcomes by producing skilled, adaptable clinicians, and fosters a culture of continuous learning, a commitment to continuous evaluation in medical education is essential for QI. It goes beyond simply ticking boxes to actively create safer care and better physicians. It is important to be kept in mind that the process of improvement is not a one-time event but rather feeds the cycle of experiencing, reflecting, and acting, incorporating a QI viewpoint into training also encourages continuous learning. It also helps discover gaps and ineffectiveness by determining what training isn't working or what new abilities are missing, from small operations to large systems.

## 28.4 Frameworks of quality improvement

The PDSA cycle, the Donabedian model, and other well-known approaches like Lean and Six Sigma are examples of frameworks for QI in medical education. Additionally, some of the institutions examine and enhance their curricula using standards from groups like the World Federation for Medical Education, with an emphasis on topics like mission, curriculum design, assessment, and resources. The main objective is an ongoing process to evaluate and improve the caliber of medical education programs and graduates' competencies, frequently by incorporating QI concepts into the actual educational process. Some of the common frameworks are described below:

### 28.4.1 Plan–do–study–act (PDSA) cycle

A key component of QI, the PDSA cycle was first created by physicist Walter Shewhart in the 1920s as the plan–do–check–act (PDCA) cycle. W. Edwards Deming popularized it and changed it for management and ongoing improvement. In order to learn more thoroughly, he eventually concentrated on the "Study" phase. The PDSA cycle is now widely used in medical education as a structured, iterative method for continuous QI (Rose et al., 2021). It enables teams and educators to test minor adjustments to curricula, teaching strategies, or procedures like assessment or simulation to see if they enhance learning, progressively accumulating knowledge for wider application. The PDSA cycle is applied in medical education with the goal of addressing key issues about what we are attempting to achieve, how we will be able to determine whether the change is an improvement, and what changes we can do that will lead to an improvement (Cleghorn and Headrick, 1996).

### 28.4.2 Stages of the plan–do–study–act cycle

In medical education, the PDSA cycle is a fast, iterative method for testing and putting small changes into practice to improve processes, such as improving a teaching module or clinical workflow. It consists of four stages which are listed below:

#### *28.4.2.1 Stage 1: plan stage*

In medical education, the plan stage of the PDSA cycle is an essential first step where teams identify an area for improvement, establish clear goals, create a specific test plan, forecast results, and choose what data to gather, concentrating on small, manageable changes to optimize learning or procedures. In order to lay the groundwork for systematic testing, this step entails group brainstorming, defining "who, what, when, where," connecting to a theory of change, and planning for qualitative or quantitative feedback. Among the crucial actions carried out in this stage are as follows:

#### 28.4.2.1.1 Identifying the problem or opportunity

The first step in every QI project is to identify the issue or opportunity in the plan stage of the PDSA cycle. It entails a well-organized approach to precisely identifying the current state of affairs, examining the root causes of a problem, and establishing precise objectives for development. Gathering information about the current state of affairs through the collection of baseline data or the creation of a strategy to acquire fresh data in order to comprehend the extent and size of the issue or opportunity is one of the crucial components of the identification process. This information offers a dispassionate view of the existing procedure and the locations of malfunctions. At this point, it's also critical to characterize the issue precisely. The clearer the definition, the more targeted and effective the potential solutions will be.

#### 28.4.2.1.2 Set goals and questions

Setting goals and formulating questions are one of the initial actions in the "Plan" stage of the PDSA cycle. They outline the objectives and constraints of the improvement project, guiding the subsequent analysis and activities. It is essential to keep in mind that the goals ought to be specific, measurable, reachable, relevant, and time-bound. They must clearly state the objectives for the cycle of improvement. Additionally, the desired outcome must be clearly stated. After this, it's critical to establish metrics that clearly indicate how performance will be evaluated.

#### 28.4.2.1.3 Develop a hypothesis

A hypothesis in the plan stage of the PDSA cycle should specify the anticipated result of a particular, small-scale change and the reasoning behind it. It is not merely a guess rather it is a prediction based on evidence and current knowledge. When developing a hypothesis, it is important to specify exactly what the educator hopes to accomplish, select the specific intervention that will be tested to do so, clearly and quantitatively describe the expected outcomes, and, when practical, use SMART principles. In addition to this, documenting the hypothesis and plan and describing the data gathering are essential elements of this stage.

#### 28.4.2.1.4 Design the test

In this phase of the PDSA cycle, we design the test by establishing goals, formulating hypotheses or predictions, organizing the small-scale implementation in terms of who, what, where, and when, and choosing what data to gather to gauge success. The most important details to provide are "Who," which relates to identifying the particular individuals engaged, such as any department or field from which the participants would be drawn. "What" referring to the exact change to be tested for instance using the new checklist or a template in clinical examination, "Where" specifying the location such as in the outpatient department or the ward and "When" specifying the time frame as to when exactly the test will be carried out.

#### 28.4.2.1.5 Plan data collection

Planning data collection entails deciding what will be measured and what information is required to address the listed concerns at the Plan stage of the PDSA cycle. It also involves determining who will gather the information and when and where it will be collected to test a particular change, answer important questions about its impact, make predictions, and set up logistics such as forms, deadlines, and small-scale testing to enable effective learning before taking action. Determining how to measure performance is also crucial; this includes quantitative information such as test results and observation checklists, as well as qualitative data from comments or surveys. Teachers can ensure that the subsequent Do, Study, and Act phases are effective and lead to meaningful progress if the data collection process is well planned.

### *28.4.2.2 Stage 2: do*

The "Do" step of the PDSA cycle involves implementing the plan on a small, manageable scale, gathering data, and documenting findings. Before assessing the outcomes in the "Study" phase, the execution phase is concentrated on conducting the test and obtaining evidence on issues, achievements, and unforeseen effects. Implementation of the plan, which calls for carrying out the precise actions specified in the "Plan" stage, frequently in a short trial or pilot, is one of the crucial aspects to be considered in this stage. Additionally, it's critical to gather information about the change's effects by tracking occurrences with tools like checklists or charts. The observations are then recorded, and any issues or unexpected results are highlighted. Instead of doing a complete check of the system, the objective here is to conduct a small-scale test to learn rapidly and acquire knowledge.

### *28.4.2.3 Stage 3: study*

In order to determine whether to adopt, adapt, or abandon the change in the "Act" stage, the PDSA's study stage involves analyzing the results of the change that was tested in the "Do" phase, comparing outcomes to the predictions to learn what happened, identifying successes, failures, or unintended consequences, and summarizing lessons learned. Data analysis, which involves looking at the data gathered during the Do phase, is one of the main activities that take place during the study stage. In order to determine whether the outcomes align with the hypotheses and expectations established during the "Plan" stage, this is subsequently compared to the predictions. At this point, it's also critical to determine whether the adjustment has produced the desired improvement and evaluate its efficacy. This phase also provides an opportunity to record any issues, surprising outcomes, or new perspectives, including the reasons why things succeeded or failed. Documenting what was learned to guide the subsequent actions in the "Act" phase is essential to making this stage more genuine. Essentially, the study phase involves introspection and using the test results to inform future decisions, guaranteeing ongoing progress as opposed to merely making changes at random.

### *28.4.2.4 Stage 4: act*

The outcomes of the study phase are analyzed in the "Act" stage of the PDSA cycle to determine the next course of action. This includes either standardizing the successful change, making adjustments and conducting another PDSA cycle, or abandoning it if it failed, promoting continuous improvement by learning from both successes and failures. During this phase, some significant events take place in the area of decision-making, whereby the researcher decides what to do next based on the information gathered during the "Study" step. Regarding the anticipated shift, this is one of the most important things to note. If the modification was successful, the process is standardize as the new procedure or it can be used more widely. However, if the outcomes were inconsistent or not what was anticipated, we make changes and create a new, somewhat different PDSA. The entire PDSA cycle is restarted if, for whatever reason, the modification was ineffective. In such case, we learn from the failure and move on to a new idea. In order to acquire knowledge gradually, this stage also aids in determining whether to conduct another cycle to test the modification on a wider scale, over a longer period of time, or with more people. The Act stage, which is a crucial component of continuous improvement initiatives, essentially closes the loop by incorporating learning to promote either broader acceptance or additional refining. The PDSA cycle is applied in medical education in the following areas:

#### 28.4.2.4.1 Curriculum refinement

The PDSA cycle offers an organized, iterative framework for testing minor modifications, evaluating their effects using data, and implementing evidence-based enhancements, all of which aid in curricular improvement. It moves curriculum development away from sudden large scale implementation towards a controlled, ongoing process of improvement. The curriculum refinement process is also based on student input using which the new active learning sessions are tested, instructional materials are modified, and session organization is improved.

#### 28.4.2.4.2 Training through simulation

Training through simulation is improved by the PDSA cycle, which turns it from a one-time event into an ongoing, data-driven improvement process. Additionally, it assures that training immediately results in improved patient safety and team performance, enhances the quality of debriefings, helps develop simulation situations, and safely evaluates new clinical practices.

#### 28.4.2.4.3 Professional development

The PDSA cycle fosters professional development as it provides a structured, data-driven framework for action-oriented learning and continuous, incremental improvement. Moreover, it allows professionals to test changes on a small scale, learn from the results, and refine their skills or processes before full implementation, minimizing risk and fostering a proactive, rather than reactive, approach to career growth.

#### 28.4.2.4.4 Team-based learning

The PDSA cycle drives continuous improvement in team-based learning, as it gives an opportunity for the teams to quickly test, evaluate, and refine their processes in real-time. Thereby it helps move from theoretical knowledge to practical, action-oriented learning (Fig. 28.1).

### 28.4.3 Donabedian model

In his seminal 1966 work "Evaluating the Quality of Medical Care," physician and healthcare researcher, Avedis Donabedian, first presented the Donabedian model. By evaluating the educational context in terms of structure, teaching or learning activities, which comprise the process, and results, which are the outcomes like student knowledge or performance, the Donabedian model, also known as the structure–process–outcome model, offers a framework for QI in medical education. In order to use it, first the educational environment, including the faculty, resources, and curriculum are

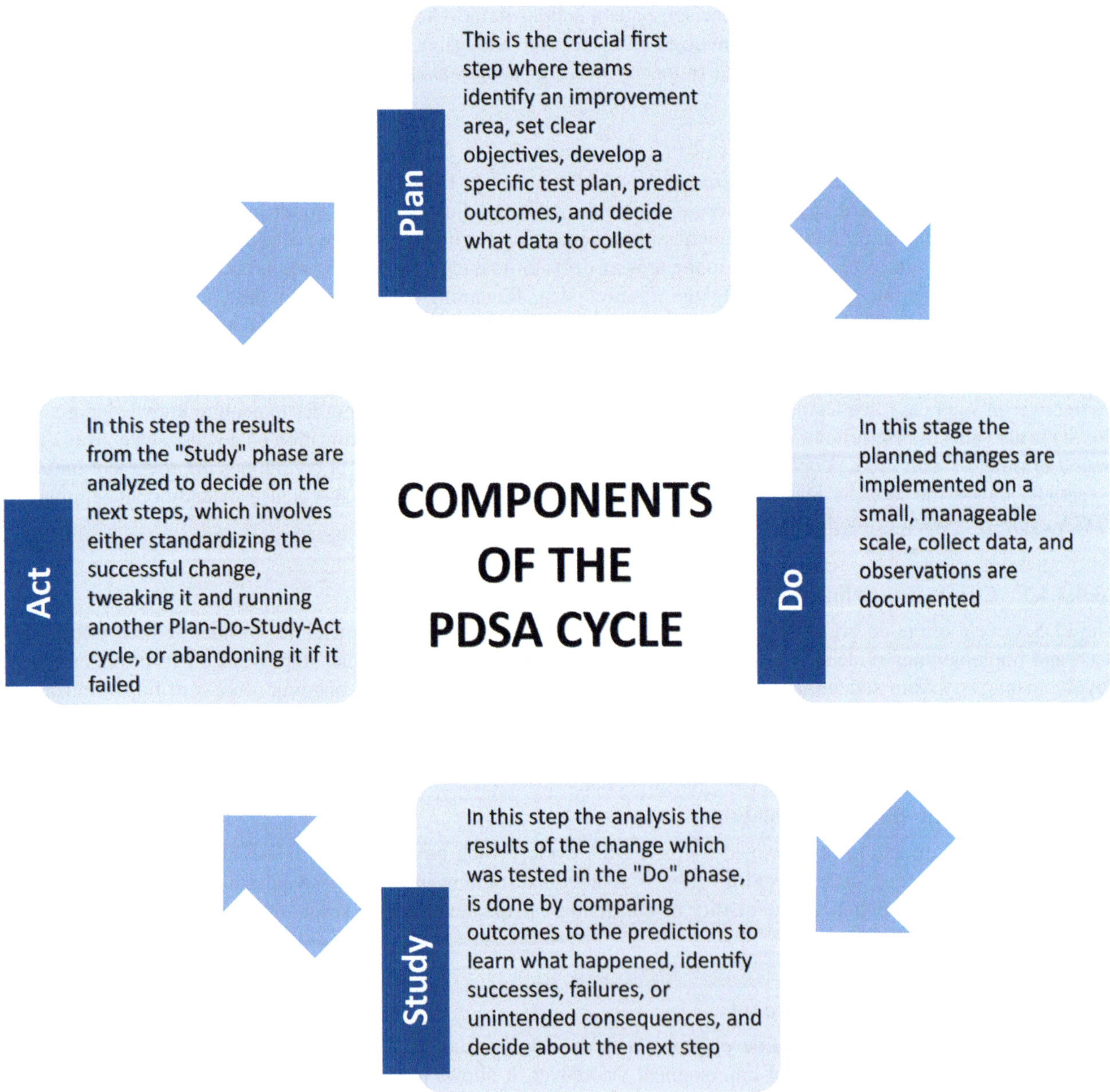


FIGURE 28.1 **Components of the plan–do–study–act (PDSA) cycle.** This figure shows the components of the PDSA cycle.

identified. Then, it is necessary to examine instructional strategies in terms of evaluation and communication. Lastly the outcomes, such as exam scores, clinical skills, and graduate success, are measured (Robby et al., 2025). This all-encompassing perspective aids in identifying weak aspects, such as inadequate structure that could result in subpar procedures, and implementing focused adjustments for improved learning and qualified future physicians.

### 28.4.4 Structure

The Donabedian structure component of medical education includes the fundamental components where learning and care take place, such as physical facilities like classrooms and labs, resources like technology and libraries, budgets, human resources like qualified faculty and other staff, organizational policies like the curriculum and accreditation, patient-to-provider ratios, and the availability of a variety of cases for training, all of which set the stage for high-quality instruction and efficient patient care. It concerns the "where and who" of instruction, making sure the system has the necessary components to generate effective procedures and results.

### 28.4.5 Process

The Donabedian model's process component in medical education concentrates on the actual teaching or learning activities, such as teacher–student communication, curriculum implementation, feedback, evaluation techniques, and adherence to educational guidelines. It measures how education is delivered to ensure quality, not just what is taught or the outcomes. It's about how successfully evidence-based teaching methods are used to promote learning, as well as transactional steps like workshops and student-staff interactions. The process stage has a variety of effects on medical education, including serving as a direct quality indicator since it shows whether best practices are being followed, which makes it essential for spotting quality gaps. Furthermore, it has a direct impact on student outcomes including knowledge and abilities, which in turn affects patient outcomes.

### 28.4.6 Outcomes

The Donabedian model's outcome component in medical education focuses on the final results of the educational process, measuring effects on students and the healthcare system, such as learner satisfaction, patient outcomes like improved health or safety due to better-trained doctors, graduate success in terms of competence or licensure, and broader impacts like improved professional reputation or health system efficiency. By evaluating shifts in knowledge, abilities, attitudes, and practical performance, it provides a response to the question, "Did the education work?" The outcome stage shows if the structural and procedural modifications have improved patient experience and health as expected. Furthermore, it assesses actual efficacy by demonstrating whether the organization and procedures genuinely enhance healthcare delivery by producing better doctors. The approach also helps detect possible quality issues by looking at the relationships between structure, process, and outcomes (Opele and Adepoju, 2024). Additionally, because results from process and structure measures can explain why a particular result was reached, it provides a framework for evidence-based review. Some of the examples of how the Donabedian model is used in medical education are in assessing a medical program curriculum, wherein there is an evaluation of the structure in terms of facilities, faculty, process such as simulation labs or skill labs, and outcome such as student performance in exams or clinical skills (Fig. 28.2).

### 28.4.7 World Federation for Medical Education standards

The World Federation for Medical Education is a global organization committed to raising the standard of medical education by setting standards, promoting accreditation for medical schools and accrediting bodies, and maintaining the World Directory of Medical Schools to ensure high scientific and ethical standards for physicians worldwide. WFME establishes best practices in basic, postgraduate, and ongoing medical education in collaboration with educators and academic institutions. With a focus on eight major areas which include mission/values, curriculum, assessment, students or doctors, staff, resources, QA or improvement, and governance the WFME standards for QI in medical education offer a framework for basic, postgraduate, and continuing professional development stages. These areas are intended to guide program design, evaluation, and enhancement for better physician training and outcomes globally, with an emphasis on local adaptation. The World Federation for Medical Education publishes comprehensive Global Standards for medical education that cover basic training, postgraduate education, and continuing professional development and serve as a guide for self-evaluation and improvement. This is just one of the ways the organization contributes to QI. In a variety of contexts, including hospitals, communities, and primary care, integrated curricula and training are emphasized. Additionally, it assesses regional and national accrediting organizations to make sure they adhere to strict quality

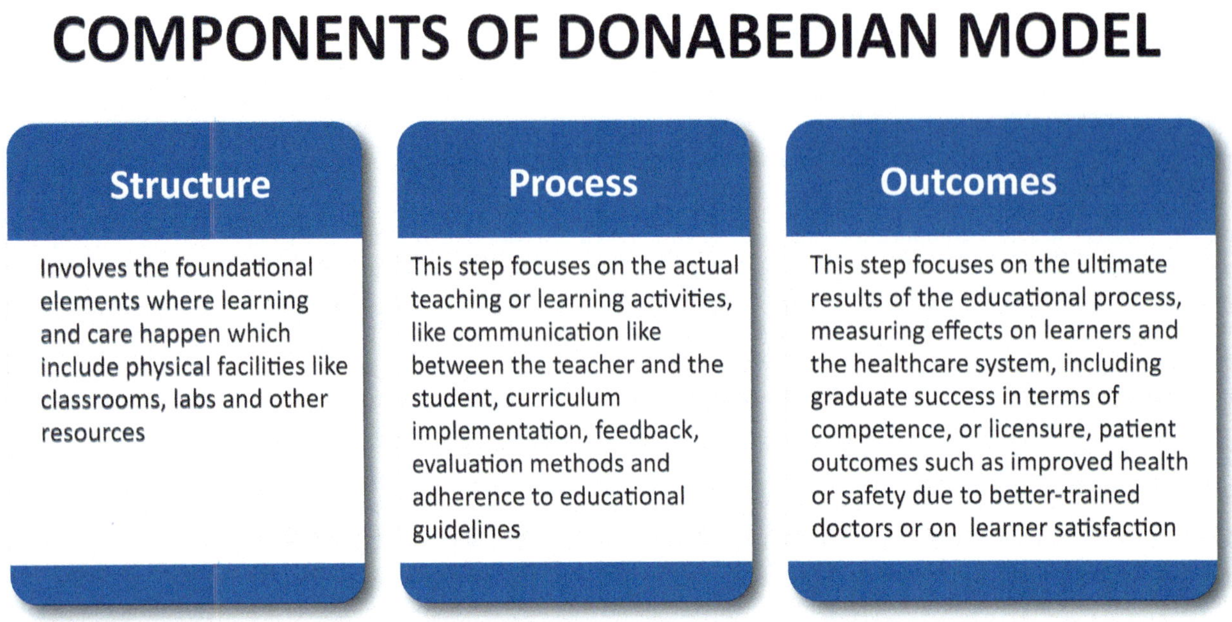


FIGURE 28.2 **Components of Donabedian model.** This figure shows the components of Donabedian model.

standards, increasing trust in recognized educational institutions. It aids in QA by demonstrating to students, employers, and regulators that a medical program meets strict requirements and facilitates graduates' international recognition. The World Federation for Medical Education promotes ongoing QI by sharing best practices and offering feedback through the recognition process. Furthermore, in order to prepare physicians for complicated healthcare requirements, the World Federation for Medical Education advocates a change from straightforward learning objectives to integrated competences.

## 28.5 Conduction of the quality improvement process

QI is an ongoing endeavor for sustained excellence rather than a one-time fix. It can be carried out continuously, beginning whenever a problem or opportunity for better performance is identified, and repeating the PDSA cycle to refine changes until goals are met. It is preferable to begin by identifying objectives and issues, then test minor adjustments, track outcomes, and, if successful, implement permanently, always engaging teams and patients. There are three occasions at which this process can begin. For instance, it can start when any concerns shows up and needs to be solved. For example, it involves addressing problems like lengthy wait times, mistakes, waste, or low patient satisfaction. It can also be employed to enhance current procedures as in identifying methods to improve procedures so they are more effective, efficient, or patient-centered, even in the absence of a crisis. For example, it involves better managing the patients in the outpatient department. Lastly they can also be used when introducing new services, such as the launch of a course, to adhere to quality requirements from the outset. In all of these occasions, the conduction of QI process follows certain key steps as described below (Fig. 28.3).

### 28.5.1 Identification of a problem or gap

A first and fundamental and first step in any strategy for improving quality is identifying a problem or gap. The discrepancy between the desired state or standard of what should be happening and the current reality, or what is happening, is referred to as a problem or gap. This stage is critical because it establishes the initiative's overall direction, aids in setting priorities, and clearly explains why a change is required. QI initiatives may be diverted toward symptoms rather than underlying causes if the problem is not well defined, which could result in ineffective solutions.

### 28.5.2 Identifying the cause and confirm data

Fishbone diagrams and root-cause analysis are useful methods for determining the source of QI in medical education. A methodical procedure combining QA, continuous monitoring, and best practices for data management is needed to confirm

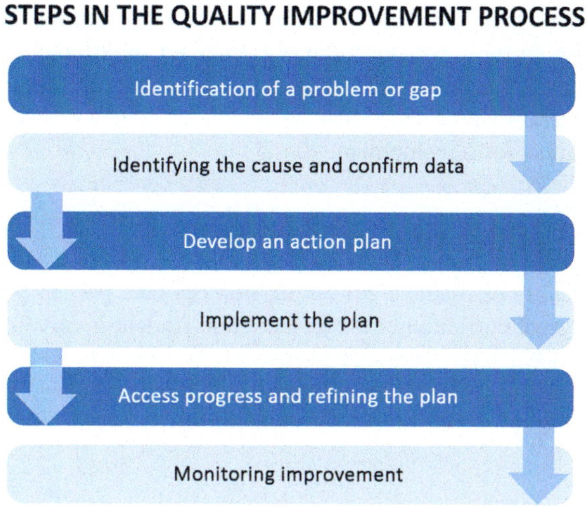


FIGURE 28.3 **Steps in the quality improvement process.** This figure shows the steps in the quality improvement process.

data conduction. Valid QI results depend on precise, comprehensive, and trustworthy data collection. This entails keeping an eye on data quality all the way through the project lifecycle.

### 28.5.3 Develop an action plan

Creating a QI action plan entails a sequential procedure to identify the issue, make adjustments, and track results. Making a list of the precise actions or tasks required to accomplish the objectives is one of the crucial factors to take into account while creating an action plan. Prioritizing jobs is crucial in this situation, and it's necessary to note any dependencies that require one work to be finished before another can begin. A key component of creating an action plan is assigning resources, such as personnel, funds, equipment, and training, to each task. Assigning accountability for each action item is crucial, either by designating a team or an individual. A good action plan will also set clear deadlines and milestone for the overall plan and individual steps. It is important to identify potential barriers and develop contingency plans.

### 28.5.4 Implement the plan

This step entails carrying out the particular interventions and tactics described in the planning phase in order to execute the intended changes. Before a larger rollout, it is advisable to test the modifications inside a particular team, department, or restricted process area. This aids in managing any hazards and improving the plan. In order to be ready for possible challenges and have backup plans in place to deal with them when they come up, it's also critical to be watchful and anticipate hurdles.

### 28.5.5 Access progress and refining the plan

The main methods for accessing the progress are systematic data collecting and monitoring. Continuous data collection is crucial. All team members should have easy access to this data, which is frequently presented at regular meetings or shown on performance dashboards. The "Study" and "Act" phases of the PDSA cycle usually serve as the framework for the process of refining the plan, which is a fundamental component of the continuous QI mindset. Analyzing the feedback and data gathered is one of the first things to do in order to ascertain whether the modifications made during the "Do" phase have resulted in the anticipated improvements. It is vital to examine what succeeded, what failed, and why. It is critical to identify particular areas that fell short of expectations or where new problems have surfaced based on the observations made during data analysis.

### 28.5.6 Monitoring improvement

Monitoring improvement in quality requires a consistent, data-driven, and systematic approach to evaluating training, curricula, and student achievement. Internal benchmarking, tracking metrics through stakeholder involvement, and formal, ongoing evaluation of clinical training are important tactics for achieving long-lasting changes. This guarantees that everyone continuously adheres to the best practices. Additionally, it's critical to make sure that every employee has received training on the new standard processes and has access to the tools and resources needed to sustain the change. Assigning distinct roles for overseeing the new procedure and tracking its effectiveness over time is also crucial. It is feasible to increase motivation and strengthen the quality culture by routinely sharing results and recognizing accomplishments. Conducting routine evaluations, such as audits or reviews, guarantees compliance with the new standards and identifies potential areas for development.

## 28.6 Roles and responsibilities of participants

In the process of improving the quality of medical education, trainees take part in projects to learn and apply changes, professors or educators are in charge of curriculum development and student involvement, and leadership offers strategic direction. While healthcare quality specialists make sure the process is efficient and the care is equal, patients serve as the final goal and offer insightful feedback. In order to increase safety, efficacy, and patient-centeredness, QI is a collaborative, iterative process that requires everyone involved to accept change, promote communication, and understand this.

### 28.6.1 Leadership

As the catalyst that propels an organization from merely managing quality to actively pursuing continuous improvement by empowering employees, encouraging open communication, and guaranteeing accountability for better results, leadership is essential to QI. An effective leadership pertaining to QI is considered to be of vital importance as the leader sets vision, creates a quality culture, aligns goals, inspires teams, and provides resources. In order to make it more than simply a project but an entrenched value, leaders must inspire buy-in, take on difficult problems, and incorporate quality into the core strategy.

### 28.6.2 Faculty or medical educators

As primary agents for improving student success, program relevance, and overall academic excellence through ongoing pedagogical development and engagement in continuous assessment and feedback loops, faculty play a critical role in QI through developing and delivering effective teaching, designing curricula, mentoring students in QI projects, taking part in institutional assessments, and inventing teaching methods.

### 28.6.3 Trainees

As frontline observers, active participants in change initiatives, and future leaders of a quality-focused culture, trainees who are medical students play a crucial and diverse role in the process of QI. Trainees, whether medical students or residents, are in the best position to identify opportunities for improvement because they are regularly exposed to system inefficiencies and safety problems at the front lines of operations. They can uncover valuable areas where present methods might be changed to improve patient care, safety, efficiency, and patient experience, thanks to their distinct viewpoint.

### 28.6.4 Patients

Moving beyond simple satisfaction to true cocreation and partnership in healthcare, patients are crucial partners in QI, acting as experts in their own experience to help identify needs, design better services, test solutions, and evaluate outcomes in order to guarantee that care is truly patient-centered, safer, and more effective. They contribute to better health outcomes and more responsive care by defining quality, evaluating treatment, providing direct feedback in a number of ways, and even participating in training and the development of educational materials.

### 28.6.5 Healthcare quality professionals

In order to improve patient outcomes and organizational efficiency, healthcare quality professionals use data analytics to identify problems, design and implement solutions such as root-cause analysis, ensure standard compliance, train staff, and cultivate a culture of safety and collaboration. They serve as change agents, connecting useful advancements in care delivery with data insights.

## 28.7 Assessment of quality improvement

In order to improve the quality of future healthcare delivery, the assessment of QI in medical education is an ongoing process that entails comparing the curriculum and learner competencies to particular standards. This procedure employs a variety of approaches to gather and examine data from several stakeholders and is directed by well-established frameworks. Stakeholder feedback is one of the key elements of the QI process since it greatly depends on input from a variety of stakeholders, such as patients, teachers, alumni, and medical students. Surveys, focus groups, and curriculum committees are frequently used to gather this information. Furthermore, a combination of qualitative and quantitative methodologies is used. Even though learner surveys are frequently used, a comprehensive picture requires other data gathering methods like action research initiatives, just-in-time feedback, and free-text responses.

### 28.7.1 Assessment frameworks and models

#### *28.7.1.1 Plan–do–study–act cycle*

The PDSA cycle is a fundamental, iterative framework for small-scale QI testing. It functions as a scientific method for action-oriented learning, building knowledge about possible improvements, identifying obstacles, and refining ideas prior to full implementation, making changes safer, less disruptive, and more effective. Its function is to methodically lead teams from a suggested idea to a tested solution by organizing the test, carrying it out, analyzing the findings, and acting on the lessons learned to continue the cycle until improvement is attained, guaranteeing that modifications are effective in real-world situations. This cycle helps different teams understand what works, why, and under what circumstances, which aids in the process of QI. It enables teachers and staff to make ongoing program improvements based on student performance and feedback. Additionally, it assists instructors in identifying obstacles and unearthing latent issues that require modifications throughout implementation.

#### *28.7.1.2 DMAIC (define–measure–analyze–improve–control) methodology*

The DMAIC (define, measure, analyze, improve, control) methodology, a structured, data-driven framework from Six Sigma, is used in medical education for systematic QI. It focuses on problem-solving by identifying problems like subpar student outcomes, assessing present performance, identifying underlying causes like curriculum gaps, putting new teaching strategies into practice, and controlling to maintain gains like tracking exam results with control charts. It frequently builds on the more straightforward PDSA cycle by reducing variety, mistakes, and inefficiencies, thereby improving health care and educational efficacy (Fig. 28.4).

#### *28.7.1.3 Steps of DMAIC methodology*

##### 28.7.1.3.1 Define

The DMAIC (define, measure, analyze, improve, control) methodology's fundamental first stage in a medical education QI process is the define phase. Before any solutions are taken into consideration, its main objective is to clearly define the problem, project goals, and scope and to unite all stakeholders around a shared goal (Monday, 2022). This stage helps identify the issue by having the concerned team identify a particular problem that needs to be fixed in the medical education process. The problem statement ought to be precise and measurable. For instance, "Medical students in their third year report an average of 4 hours of self-directed learning time per week, exceeding the 1-hour target, which may impact their final exam scores."

##### 28.7.1.3.2 Measure

In order to create a baseline, the measure phase of the DMAIC technique in medical education QI is a data-gathering stage that focuses on quantifying a process's present performance. This stage lays the groundwork for further investigation and development by offering unbiased proof of the issue found in the define phase. Applying the measure phase in medical education entails particular tasks that are related to the administrative or learning process that is being enhanced, such as determining essential metrics that are pertinent to the identified problem. For instance, it involves resident mistake rates in clinical documentation, curriculum delivery time, faculty satisfaction, and student performance. Additionally, it aids in the development of a structured strategy that outlines the types of data to be collected, how they will be collected, where they will come from, and how often they will be collected. This guarantees the consistency and dependability of the data.

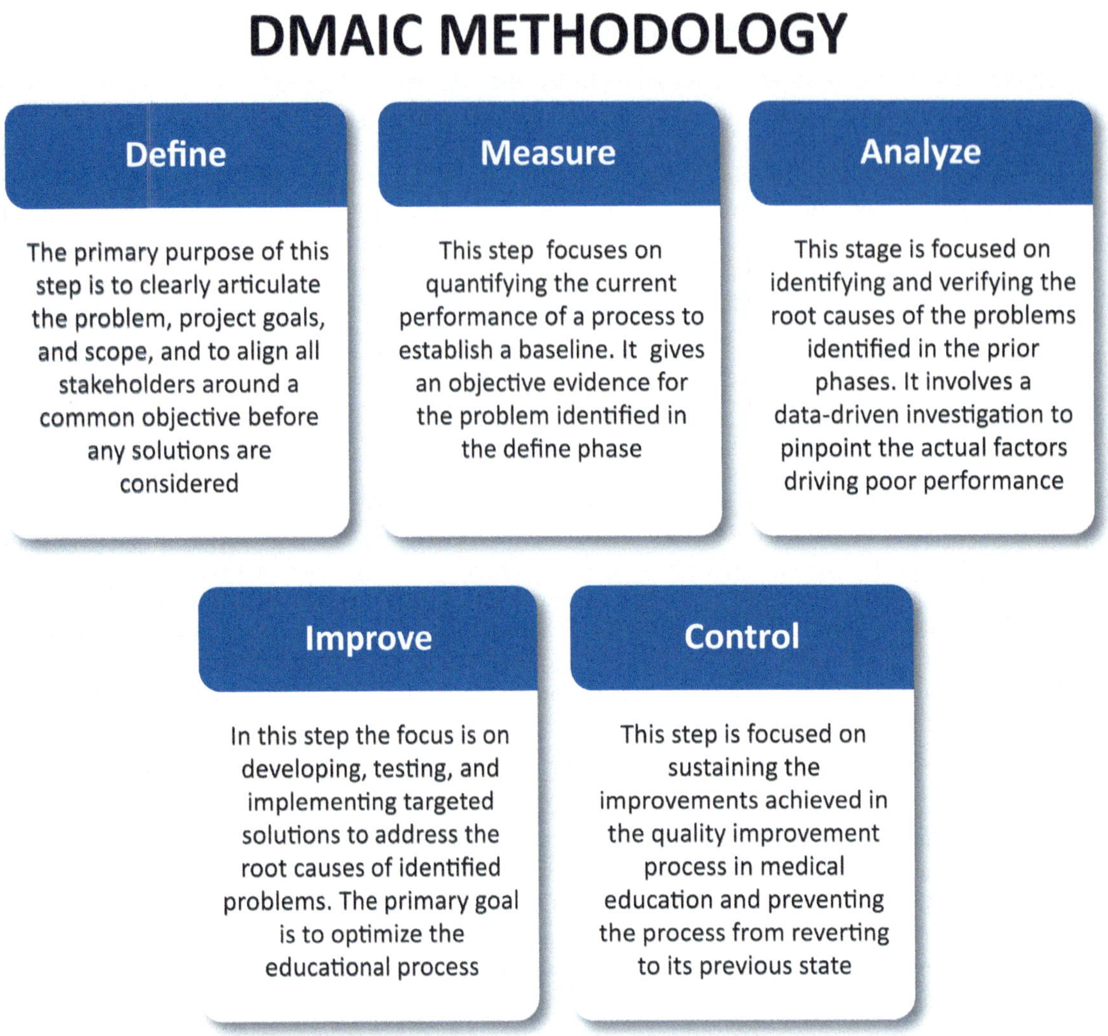


FIGURE 28.4 **DMAIC (define–measure–analyze–improve–control) methodology.** This figure depicts the DMAIC methodology.

#### 28.7.1.3.3 Analyze

In a medical education QI process, the analyze phase of the DMAIC approach is a crucial step that focuses on locating and confirming the underlying causes of the issues or inefficiencies found in the earlier phases. It entails a thorough, data-driven examination to identify the real causes of subpar performance rather than just symptoms. It assists in identifying the fundamental elements causing the issue that is defined in the "Define" phase and measured in the "Measure" phase. However, this stage also avoids making assumptions about the reasons by using data and statistical techniques to objectively confirm the relationship between inputs and outcomes.

#### 28.7.1.3.4 Improve

The improve phase of the DMAIC technique in medical education is on creating, evaluating, and putting into practice focused solutions to deal with the underlying causes of issues found in current procedures. Optimizing the educational process to attain quantifiable gains in efficacy and quality is the main objective. The improve phase entails a methodical approach to implementing change in order to produce potential solutions. The project team employs methods such as brainstorming or the nominal group technique to generate a variety of innovative ideas in order to eliminate the verified root causes of a problem, such as subpar exam performance, low module satisfaction, or ineffective administrative procedures. In order to find the optimal solution, concepts are also assessed during this phase based on their viability, cost, impact, and available resources.

#### 28.7.1.3.5 Control

The last and most important stage of the DMAIC technique is the control phase, which aims to maintain the gains made in the medical education QI process and keep it from going back to its initial state. Long-term success is ensured by incorporating the new process into routine practice. Documentation and standardization are some of the most critical duties at this phase. All updated processes and new workflows need to be thoroughly documented as standard operating procedures or work instructions. These materials should be written with end users in mind, be easily accessible through the electronic health record or quick-reference guides, and should be easy to understand. Additionally, a system for continuous monitoring is set up during this phase to track the effectiveness of the new process using key metrics or key performance indicators.

## 28.7.2 Ps of quality

A comprehensive assessment approach that assesses quality in four distinct dimensions which include Product, Process, People, and Propellers is known as the "4 Ps" paradigm for QI in medical education. Medical facilities can organize and concentrate their QI and innovation activities with the help of this framework. The 4 Ps of quality framework are shown in Fig. 28.5.

### *28.7.2.1 Product*

In essence, the product in the 4 Ps model of quality assessment include the medical professionals themselves and the services they are taught to provide which are the result of the educational system. The technical solution's quality, adherence to professional and educational standards, and student happiness are important factors in this regard.

### *28.7.2.2 Process*

The mechanisms and techniques used to offer education are the main emphasis of this component. This covers instructional strategies like online modules, flipped classrooms, didactic lectures, assessment techniques, and verification and validation procedures. The PDSA cycle is a popular approach for making incremental changes in this field.

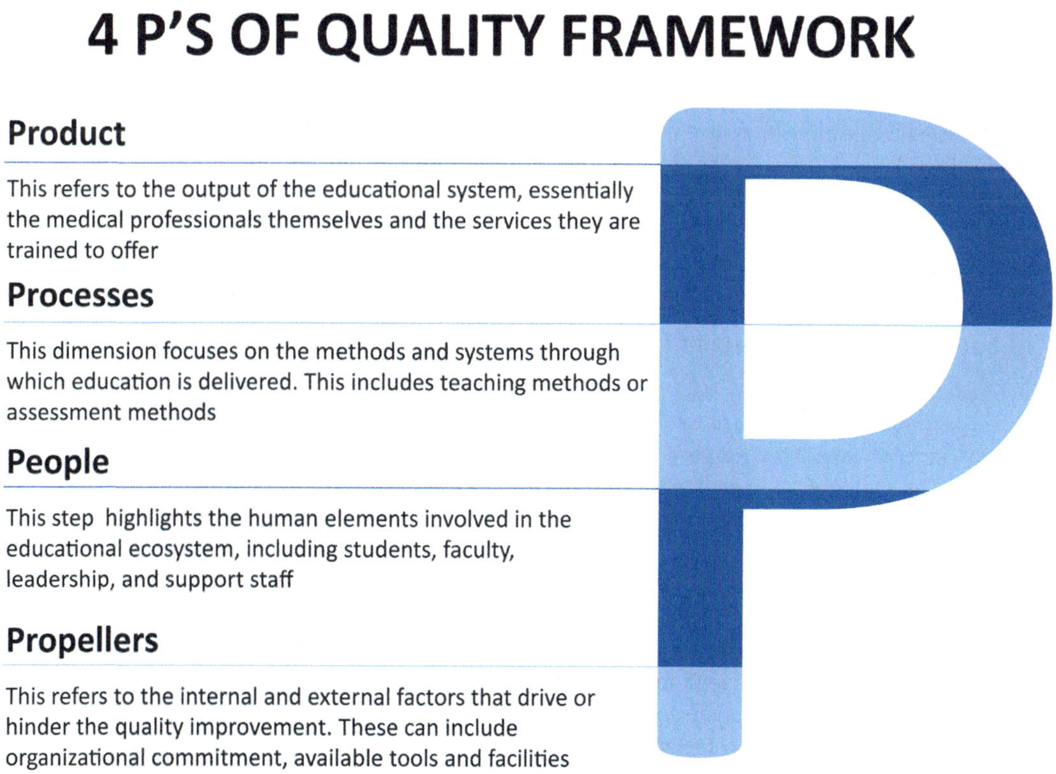


FIGURE 28.5 **The 4 Ps of quality.** This figure depicts the 4 Ps of quality.

#### 28.7.2.3 People

This dimension draws attention to the human components of the educational ecosystem, such as students, teachers, administrators, and support personnel. Recruitment, training, workplace culture, leadership vision, and faculty development initiatives are all components of quality.

#### 28.7.2.4 Propellers

This alludes to the internal and external factors that either strengthen or weaken the QI initiative. These can include organizational dedication, leadership support, easily accessible facilities and tools, and a positive work environment, all of which contribute to the transition. Institutions can methodically address different aspects of their educational programs to guarantee ongoing progress and satisfy accreditation requirements by employing an all-encompassing strategy.

### 28.7.3 Benchmarking and quality improvement process

Benchmarking significantly improves QI in medical education by providing clear performance goals, pointing out strengths and problems, and encouraging evidence-based changes to curricula and outcomes like test pass rates. Benchmarking enhances student performance, contentment, and professional growth by comparing data to top performers. It helps uncover gaps, reduces needless variance, and motivates faculty and institutions to utilize targeted strategies like audit-feedback loops to enhance patient care and system efficiency (Willmington et al., 2022).

## 28.8 Benefits of quality improvement in medical education

QI in medical education equips future doctors to enhance patient safety, care delivery, and system efficiency by promoting a mindset for continuous learning, data analysis, and collaboration. Better patient outcomes, more satisfaction, reduced expenses, and a more adaptable, evidence-based healthcare workforce are the outcomes of this. The QI process guarantees that the graduates can establish a culture of excellence and creativity as it goes beyond theory to practical abilities in problem identification and solution implementation (Arja et al., 2024). The following lists some of the main advantages of QI:

### 28.8.1 Improved learner competencies

QI programs in medical education directly improved learner competencies by providing the frameworks and practical experiences needed to develop important knowledge, skills, and professional behaviors. Learners can get specialized knowledge of QI principles and methodology, such as data analysis, system mapping, and change implementation, by integrating QI components into the curriculum. This enhances learner competencies. This provides them with a systematic approach to evaluate and improve healthcare delivery (Nair et al., 2016).

### 28.8.2 Enhanced clinical environment

Through structured feedback, simulation, and practical application of improvement principles, QI in medical education improves clinical settings by developing capable, self-assured learners, fostering positive learning environments, and promoting patient-centered care. This results in better patient outcomes and more engaged future physicians. Students, teachers, and patients all gain from the transformation of learning from passive observation to active, system-level change (Gould et al., 2002).

### 28.8.3 Community relevance

There are many advantages to incorporating community relevance into medical education QI projects, chief among them being the alignment of educational goals with real population health needs and the provision of practical, real-world skills to future physicians. The QI strategy satisfies social accountability standards set forth by organizations like the World Health Organization by directly linking the medical school's mission to societal expectations. Additionally, it gives students a sense of purpose and meaning while working on projects that actually enhance local communities, which boosts motivation and engagement for learning QI skills.

### 28.8.4 Efficiency and resource optimization

One of the main ways that medical education can increase its effectiveness and maximize resource allocation is through QI, which is a crucial part of value in healthcare. QI programs make sure that resources which include people, equipment, money, and time are gripped to optimize health and educational outcomes by systematically locating and eliminating waste and bottlenecks. Additionally, it lowers waste and inefficiencies, lowers operating expenses, and improves the sustainability of medical education and healthcare delivery.

### 28.8.5 Focus on key principles

Incorporating QI into medical education provides significant benefits by promoting a culture of lifelong learning and providing future doctors with the resources they need to provide safer and more effective care. One of the key concepts that the QI process benefits from is data-driven decision-making, which use regular and systematic data collection and analysis to quantify the impact of interventions, assess performance fluctuations, and drive decision-making for improvement. In order to ensure that care is thoughtful and sensitive to the patient's preferences, it also ensures that all efforts are ultimately centered on the patient's experience, needs, and values.

### 28.8.6 Improved patient outcomes

QI in medical education directly improves patient outcomes by training future doctors how to deliver safer, more efficient, and patient-centered treatment. Through enhanced communication and specialized training, this results in higher adherence, fewer errors, fewer readmissions, and more health literacy, eventually fostering a culture of continuous improvement for better patient health and satisfaction.

## 28.9 Challenges of quality improvement in medical education

It is difficult to implement significant changes beyond basic curriculum updates to actual practice improvement in medical education due to obstacles like resource scarcity, including time, money, organizational inertia, lack of faculty or resident training, poor data systems, resistance to change, weak leadership, and fragmented cultures. Overcoming professional barriers, gaining support from busy employees, incorporating QI into hectic schedules, and converting learning into long-lasting system-level improvements are some of the major obstacles.

### 28.9.1 Lack of faculty or resident training

Lack of QI-trained faculty to mentor residents is a significant obstacle to improving the quality of medical education. This problem is exacerbated by residents' conflicting clinical demands, a lack of time set aside for projects, and a disconnect between theoretical instruction and real-world application, which frustrates and burns out both parties and prevents them from developing meaningful skills for future practice. The solution in this situation includes identifying mentors, frequently from outside the department, incorporating QI into the curriculum, allocating resident time, and organizing faculty development.

### 28.9.2 Resource and time constraints

Time and resource limitations affect almost every facet of the process, from planning and execution to evaluation and sustainability, making them major and ubiquitous obstacles to QI in medical education. One of the biggest issues is the workload and staffing shortages that plague medical education and healthcare systems, especially in environments with limited resources. As a result, current staff members have enormous workloads and large clinical volumes, leaving little to no time or energy for the extra work required for QI projects. It is also true that staff members involved in QI projects find it difficult to find time for crucial tasks like goal-setting, data collection and analysis, intervention implementation, and evaluation due to conflicting clinical, educational, and administrative priorities.

### 28.9.3 Organizational rigidity and resistance

Organizational rigidity and resistance provide significant barriers to QI in medical education by hindering the adoption of new curricula, limiting the resources available for QI projects, and fostering a culture that is hostile to change. Fear of the unknown, a lack of supportive networks, and deeply ingrained professional conventions are often the causes of these challenges. Medical institutions may have cultures that are reluctant to alter the status quo because of their lengthy history and deeply rooted customs. Clinicians may be less amenable to changes brought about by those outside of their professional group, such as managers or persons from various disciplines, and they may be defensive of their professional autonomy.

### 28.9.4 Data collection and privacy issues

QI initiatives in medical education are severely hampered by data collection and privacy issues, mainly because they create a challenging balance between the need to protect sensitive student and patient information and the need for detailed data to identify areas for improvement.

### 28.9.5 Leadership and engagement

Initiatives for QI in medical education and healthcare settings face major obstacles from leadership and engagement. These difficulties are frequently caused by preexisting hierarchies, conflicting priorities, and a lack of reliable, committed assistance. Regarding leadership and involvement, medical professionals frequently oppose modifications to long-standing procedures because of their professional autonomy and mistrust of initiatives spearheaded by outside parties. Collaboration is hampered by the lack of a common vision across various professional groups, such as doctors, managers, and nurses, who may have different priorities and speak different languages.

**Summary**

A continual, data-driven approach to improving curricula, instructional strategies, assessment procedures, and learning environments is known as quality improvement (QI) in medical education. In order to produce qualified doctors who can provide safe and efficient patient care, the chapter highlights the importance of high-quality medical education. The methodical, iterative process of QI aims to find gaps, carry out focused actions, and assess their results. Important frameworks like root-cause analysis and plan–do–study–act cycles are utilized as useful methods for identifying educational issues and testing remedies on a limited scale prior to wider adoption.

**Key point**

**Future perspectives: focusing on learning by doing**

The future of quality improvement (QI) in medical education will be technology-driven, learner-centered, equity-focused, and closely linked to clinical quality and safety. It will change medical schools and training programs into dynamic, data-driven learning environments that better equip aspiring doctors for the difficulties of contemporary healthcare. In order to ensure that trainees are active contributors to improving patient care and safety rather than only observers, QI will go beyond academic knowledge to "learning by doing" through practical initiatives. The future of QI in medical education will be largely dependent on advanced analytics, learning management system data, AI-based assessment tools, and real-time feedback dashboards in relation to the advent of artificial intelligence in medicine. Compared to traditional evaluations, these technologies will be faster in identifying performance trends, identifying curriculum gaps, and customizing learning support.

## References

ALBayati, M.T.A., 2025. Quality control in the healthcare sector: a comprehensive literature review. American Journal of Economics and Business Innovation 4 (3), 152–163. Available at https://doi.org/10.54536/ajebi.v4i3.5964.

Arja, S.B., White, B.A., Fayyaz, J., Thompson, A., 2024. The impact of accreditation on continuous quality improvement process in undergraduate medical education programs: a scoping review [version 2; peer review: 2 approved, 1approved with reservations]. MedEdPublish 14, 13. Available at https://doi.org/10.12688/mep.20142.2.

Buttell, P., Hendler, R., Daley, J., 2007. Quality in healthcare: concepts and practice. United States The Business of Healthcare: [3 volumes] 1-3. Bloomsbury Publishing Plc, pp. 61–94. Available at http://www.bloomsbury.com/9780313084669.

Cleghorn, G.D., Headrick, L.A., 1996. The PDSA cycle at the core of learning in health professions education. The Joint Commission Journal on Quality Improvement 22 (3), 206–212. Available at https://doi.org/10.1016/S1070-3241(16)30223-1.

Gould, B.E., Grey, M.R., Huntington, C.G., Gruman, C., Rosen, J.H., Storey, E., et al., 2002. Improving patient care outcomes by teaching quality improvement to medical students in community-based practices. Academic Medicine 77 (10), 1011–1018. Available at https://doi.org/10.1097/00001888-200210000-00014, http://www.lww.com/product/?1040-2446.

Misra, S., 2019. Quality improvement in medical education. Annals of the National Academy of Medical Sciences (India) 55, 117–118. Available at https://doi.org/10.1055/s-0039-3402363.

Monday, L.M., 2022. Define, measure, analyze, improve, control (DMAIC) methodology as a roadmap in quality improvement. Global Journal on Quality and Safety in Healthcare 5 (2), 44–46. Available at https://doi.org/10.36401/JQSH-22-X2, PMID:37260837, PMCID:PMC10229001.

Nair, P., Barai, I., Prasad, S., Gadhvi, K., 2016. Quality improvement teaching at medical school: a student perspective. Advances in Medical Education and Practice 7, 171–172. Mar 18. Available at https://doi.org/10.2147/AMEP.S101395, PMID:27051330, PMCID:PMC4807889.

Opele, J.K., Adepoju, K.O., 2024. Validation of the donabedian model of health service quality in selected States in Nigeria. Nigerian Journal of Clinical Practice 27 (2), 167–173. Available at https://doi.org/10.4103/njcp.njcp_220_22, http://www.njcponline.com/backissues.asp.

Robby, K.N.A., Nurjazuli, N., Izzah, L., Tallapessy, A., Astuti, I.S.W., 2025. From structure to outcome: the role of the donabedian model in global health service management research (1987-2024). E3S Web of Conferences 650. Available at https://doi.org/10.1051/e3sconf/202565002025, www.e3s-conferences.org/.

Rose, S., Hamill, R., Caruso, A., Appelbaum, N.P., 2021. Applying the plan-do-study-act cycle in medical education to refine an antibiotics therapy active learning session. BMC Medical Education 21 (1). Available at https://doi.org/10.1186/s12909-021-02886-3.

Savsar, M., 2017. Quality control application in healthcare management. Trauma Acute Care 2, 30.

Tabish, S., 1990. Quality assurance in health care. Journal (Academy of Hospital Administration (India)) 2, 2–5.

Walshe, K., 2007. Understanding what works—and why—in quality improvement: the need for theory-driven evaluation. International Journal for Quality in Health Care 19 (2), 57–59. Available at https://doi.org/10.1093/intqhc/mzm004.

Willmington, C., Belardi, P., Murante, A.M., Vainieri, M., 2022. The contribution of benchmarking to quality improvement in healthcare. A systematic literature review. BMC Health Services Research 22 (1). Available at https://doi.org/10.1186/s12913-022-07467-8.

Wong, B.M., Levinson, W., Shojania, K.G., 2012. Quality improvement in medical education: current state and future directions. Medical Education 46 (1), 107–119. Available at https://doi.org/10.1111/j.1365-2923.2011.04154.x.

Chapter 29

# Quality assurance

*"Quality is everyone's responsibility."*

—W. Edwards Deming

## 29.1 Introduction

In order to generate qualified healthcare professionals, quality assurance (QA) in medical education entails methodically guaranteeing and raising the standards of medical training through procedures including establishing criteria, keeping an eye on compliance, gathering feedback, and making the required adjustments. Here, the main emphasis is on structure, process, and results through self-evaluation, peer review, and external accreditation such as by the World Federation for Medical Education (WFME) (Joshi, 2012). In the end, patient care is improved by tracking curricula, faculty, resources, student performance, and well-being through ongoing cycles of feedback and improvement, connecting theory and practice for global preparedness.

### 29.1.1 Definition of quality assurance

Defining "quality" is difficult because it is subjective and dynamic. It can have different meanings for different stakeholders. Quality in medical education can be defined relatively as a condition of meeting the required standards as prescribed by external agencies and it meets those standards again and again. In medical education, QA is a methodical process that ensures high standards by assessing curriculum, instruction, assessment, and results. It moves toward continuous improvement through feedback, self-evaluation, and external reviews to produce qualified, moral physicians for improved healthcare. It entails establishing standards, keeping an eye on procedures such as faculty development and student evaluations, and utilizing information from colleagues, alumni, and students for improvement and responsibility. Curriculum design, the learning environment, student advancement, and governance are important aspects that are all documented and evaluated to satisfy changing global needs. Some of the key elements of QA include systematic evaluation, which includes the continuous, thorough assessment of educational programs against predetermined standards, and continuous improvement, which is a cycle that goes from merely meeting standards to improving them for better educational outcomes.

## 29.2 Need for quality assurance in medical education

Driven by globalization, shifting healthcare needs, and the need for public trust, QA in medical education is essential to ensuring that future physicians are competent, safe, and meet international standards. This requires systematic monitoring, accreditation, and ongoing curriculum, teaching, and outcome improvement for the benefit of patients and the health system. In order to develop better-prepared doctors for any place in the globe, it moves the emphasis from simply reaching minimums to a dynamic cycle of setting criteria, monitoring performance, gathering input, particularly from students, and innovating. The key reasons for QA in medical education are shown in Fig. 29.1.

**The Principles and Practice of Medical Education. DOI:** https://doi.org/10.1016/B978-0-443-44197-4.00020-5

FIGURE 29.1 Need for quality assurance in medical education.

### 29.2.1 Patient safety and healthcare quality

QA is crucial for patient safety and healthcare quality because it provides procedures like audits, training, and standards to avoid errors, reduce harm, ensure compliance, and boost good outcomes. Additionally, it promotes a culture that prioritizes safety and consistently provides high-quality treatment, which enhances patient experiences and organizational efficacy. It is the framework that changes healthcare from only healing illness to consistently delivering patient-centered, safe, and effective care.

### 29.2.2 Regulatory compliance

As it offers the proactive processes, documentation, and monitoring required to achieve the norms and legal requirements, prevent costly failures, protect public trust, and uphold professional standards, QA is crucial for regulatory compliance. Regulatory compliance in medical education guarantees patient safety and competent healthcare providers by ensuring that training programs follow established national and international standards. In order to comply with regulations, curriculum, assessment techniques, and institutional resources are continuously monitored, evaluated, and improved through QA. QA promotes compliance through the evaluation and feedback. The feedback provided by the faculty and the students help the educators to evaluate the efficacy of instruction, guaranteeing consistency with learning objectives. Additionally, the method guarantees that quality criteria are always met.

### 29.2.3 Credibility and trust

In order to build confidence and trust in medical education, QA is essential. It guarantees that the training satisfies established criteria of excellence and generates qualified and ethical healthcare professionals. It bridges the gap between public trust and educational institutional procedures by providing evidence-based training standard verification. Assuring the validity, reliability, and fairness of high-stakes tests, including written exams, OSCEs, and workplace-based assessments, is one of QA's key functions (Nawaz et al., 2025). This gives the public and regulatory agencies convincing evidence that a graduate is qualified to practice. Additionally, institutions reduce prejudice and guarantee consistent graduate quality across various places and examiners by including standardized assessment forms and rigorous validation.

### 29.2.4 Continuous improvement

QA is essential for continuous improvement in medical education because it provides structured procedures to create standards, evaluate outcomes including student competency and patient safety, gather stakeholder feedback from students, staff, and the public, and apply data-driven adjustments. This guarantees that graduates fulfill evolving healthcare demands and that universities continue to be relevant and well-known throughout the world. It shifts the focus from simply meeting standards to a proactive cycle of feedback, analysis, and improvement in order to generate trained physicians and provide high-quality patient care (Dolmans et al., 2003). QA provides the standards for structure, process, and outcomes. The quality improvement process leverages these elements of QA to drive continuous improvement through feedback loops and systemic modifications, turning occurrences that are only adequate into the best outcomes.

### 29.2.5 Accountability and transparency

QA is crucial for encouraging accountability and transparency in order to guarantee that medical education programs adhere to established standards and produce competent, trustworthy healthcare providers. This is essential for both public safety and maintaining public trust in the medical sector. These positions in modern medical education are becoming more and more defined by digital transformation and a shift toward continuous, real-time monitoring. Medical schools and clinical facilities are held accountable by quality assurance procedures for the quality of doctors they create and the resources they use.

QA process by ensuring that medical curricula focuses on public health and community needs makes sure that educational outputs are in line with societal healthcare priorities, thereby supporting the idea of social accountability. Furthermore, strict quality assessment programs like OSCEs and workplace evaluations ensure that graduates fulfill fundamental safety and ethical requirements prior to starting their careers. This also makes sure that the student has attained professional competence.

### 29.2.6 Stakeholder involvement

QA serves medical education stakeholders, including students, teachers, institutions, patients, and society, by ensuring competent graduates, building trust, meeting standards, improving patient care, and encouraging continual development. In the end, this produces a trained labor force ready to safely and successfully meet community demands. Through QA processes, advisory boards can integrate a range of stakeholder perspectives, transforming suggestions into useful adjustments for better healthcare and educational outcomes (Attree, 2025).

### 29.2.7 Standardization and benchmarking

Since it offers the fundamental framework for creating, putting into practice, and overseeing the standards that allow for efficient standardization and benchmarking, QA is vital in medical education. All graduates are guaranteed to possess the minimal skills required to provide safe and efficient healthcare due to this dynamic approach (Amaral and Norcini, 2023). Benchmarking helps identify performance gaps and encourage improvement by comparing an institution's performance with "best-in-class" peers using QA data. In addition, using QA metrics like graduation rates, faculty qualifications, and student satisfaction, medical schools can identify performance gaps in relation to external benchmarks. Furthermore, the QA systems produce standardized KPIs, like as residency match rates or board exam scores, that allow direct comparison between different medical programs.

## 29.3 Frameworks for quality assurance in medical education

These offer organized procedures and principles to guarantee excellent standards in medical education, including methodical program and practice monitoring, assessment, and ongoing improvement. Important examples include frameworks from international organizations like the World Health Organization, which offers recommendations for regional and national QA systems in basic medical education, and regulatory organizations like the General Medical Council, which employs proactive and reactive QA to guarantee standards are met. In order to ensure alignment with stated educational objectives and comply with regulatory requirements, these frameworks are customized to match particular needs and frequently contain elements such as quality standards, audit procedures, stakeholder feedback, and risk management. Some of the important frameworks and their components are described below:

### 29.3.1 General Medical Council (GMC) framework

Using tools like self-assessments, data analysis, and quality visits to medical schools and postgraduate bodies, the GMC's Quality Assurance Framework guarantees high standards in medical education and training through approval, proactive, and reactive processes that emphasize standards, collaboration, and continuous improvement. It is an adaptable, data-driven system based on openness and equity that promotes excellence by finding and disseminating best practices, addresses deficiencies with targeted measures, and ultimately safeguards the public by making sure physicians adhere to standards (Crampton, 2019). In the end, the GMC's Quality Assurance Framework protects the public by guaranteeing high-quality care through proactive monitoring and reactive interventions across the United Kingdom's multitiered training system. It does this by regulating physicians and medical education, establishing standards like good medical practice, holding training bodies accountable, and looking into concerns. For the purpose of patient health and safety, the GMC's framework essentially serves as a system to ensure that all medical professionals adhere to strict criteria from their initial training to continued practice.

### 29.3.2 World Federation for Medical Education (WFME)

The WFME guarantees high scientific and ethical standards across basic, postgraduate, and continuing professional development by promoting QA in medical education globally through the establishment of standards, requirements for medical schools and accrediting organizations, and a recognition program to validate national accreditation systems. Their responsibilities include assessing accreditation organizations, encouraging ongoing development, and collaborating with the WHO and governments to create robust quality frameworks. The WFME contributes to QA by establishing standards for medical education that cover curriculum, outcomes, evaluation, and institutional operation at every stage of a doctor's training (Tackett, 2015). As part of the accreditation process, it also evaluates national and regional accrediting bodies to ensure that their practices follow international standards, which is an essential step for global mobility and qualification recognition. Additionally, it guarantees the preservation of rigor and openness by offering an impartial, open process for evaluating accreditation systems and disclosing the outcomes. Additionally, it works with nations and institutions such as the World Health Organization to support the development and improvement of national medical education QA systems.

## 29.4 Quality assurance cycle

The QA cycle in medical education, a continuous loop of plan–do–check–act or comparable models, aims to improve program excellence by establishing standards, putting procedures into place, keeping an eye on results through data, audits, and feedback, and making methodical adjustments that are in line with mission, vision, and graduate needs (Patel and Deshpande, 2017). In order to ensure that graduates satisfy competency standards for global practice, it incorporates stakeholders such as teachers, students, and administrators in data gathering, including assessments and feedback, analysis, planning corrective actions, and ongoing monitoring. The key stages of a typical QA cycle in medical education are as depicted in the below diagram (Fig. 29.2

### 29.4.1 Plan (setting standards and objectives)

The plan phase is the cornerstone of the QA cycle in medical education. Establishing the standards and procedures necessary to fulfill regulatory obligations and educational goals is its main focus. Determining the precise educational objectives, graduate competences, and community health needs that the program must meet is one of the essential tasks in the plan phase of the QA cycle. Furthermore, the WFME and other local and worldwide benchmarks are adopted or modified by the institutions as part of the establishment of standards. Another crucial part of this phase is identifying the

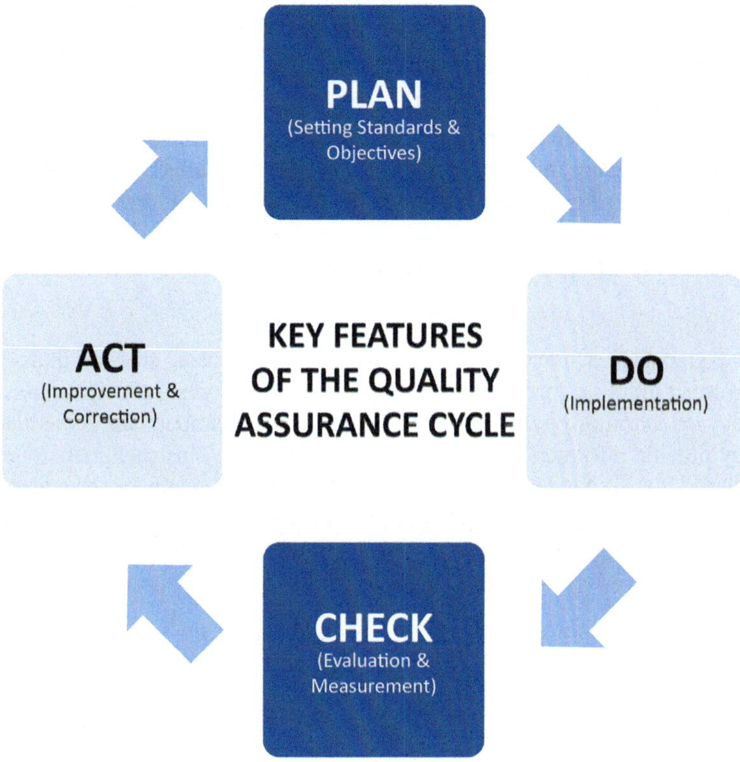


**FIGURE 29.2** Quality assurance cycle.

internal and external stakeholders, such as students, professors, healthcare providers, and the general public, to ascertain their requirements and expectations. Additionally, it aids in risk assessment, which makes it feasible to pinpoint possible obstacles to quality, like curriculum gaps or resource constraints, and create mitigation plans.

### 29.4.2 Do (implementation)

The "Do" phase in QA cycle's is the stage of execution. Here the plans and tactics created during the "Plan" stage are put into action during this phase in order to test modifications or implement the educational program. This phase focuses on carrying out procedures like conducting lectures, clinical rotations, or simulation-based training in accordance with the daily schedule of the educational program. Apart from this, collecting data and observations in a systematic way during implementation is beneficial. This covers performance indicators, attendance trends, and student comments. Curriculum reform is one of the main uses of the "Do" phase in medical education. If the medical school intends to switch to competency-based teaching, the "Do" phase entails implementing a pilot module utilizing the new framework.

### 29.4.3 Check (evaluation and measurement)

The third phase of the QA cycle in medical education is called the check phase. Its main goal is to keep an eye on, assess, and examine the outcomes of educational initiatives that have been put into practice to make sure they adhere to set criteria. The check phase contributes to performance monitoring by routinely assessing clinical training and courses using suitable tools, like program reports and student questionnaires. The check phase also contributes to gap identification by assisting in the detection of differences between the actual accomplishments of medical graduates and the anticipated learning outcomes.

### 29.4.4 Act (improvement and correction)

The act phase of the QA cycle is the last stage where institution put long-term fixes into place based on assessed outcomes. This stage in medical education makes sure that the advancements in training, curriculum, or clinical simulation are maintained and standardized. Corrective measures, process updates, and employee training are all part of it. After that, the cycle is restarted with the updated baseline.

## 29.5 Role of participants in quality assurance process

Stakeholders, including students, teachers, patients, employers, and regulatory agencies, are crucial to the QA of medical education in order to improve graduate competency and better address community health needs. They offer constructive criticism, mold courses, guarantee applicability, promote accountability, and encourage ongoing development (K Raman, 2025). Their diverse perspectives ensure that programs stay current and produce qualified physicians, bridging the gap between theory and practice, by offering feedback on teaching, curriculum, and graduation outcomes.

### 29.5.1 Roles of stakeholders of quality assurance

#### *29.5.1.1 Students*

In order to guarantee that graduates achieve high standards for patient care, students play a critical role in medical education QA by offering direct input on curriculum, instruction, and resources, participating as partners in improvement, and evaluating the efficacy of learning environments. They have a big impact on accreditation and ongoing quality improvement because they provide information about curriculum relevancy, instructional strategies, fair assessment, and facility sufficiency (Ulewicz, 2017). Students who take on the role of feedback providers often provide thorough quantitative and qualitative feedback on courses, instruction, clinical placements, resources, and even assessment techniques. However, as curriculum reviewers, students contribute to making sure the curriculum is current, structured, and in line with international standards and modern medical practice.

#### *29.5.1.2 Faculty*

Faculty members are essential to the implementation, monitoring, and enhancement of medical education QA because they deliver curriculum, evaluate students, provide feedback, develop programs, ensure that resources are used, and participate in accreditation. Ultimately, they influence the educational setting and ensure that graduates fulfill the requirements for providing competent healthcare. Faculty members are essential to QA in the areas of curriculum delivery and innovation, including curriculum implementation, technology supervision, and suggestions for improved teaching methods and content. Faculty members also contribute significantly to the monitoring and evaluation process, which includes assessing student performance, monitoring the effectiveness of courses, providing feedback on learning objectives, and producing internal reports.

#### *29.5.1.3 Patients*

Patients play a critical role in medical education QA by offering practical feedback on the quality of care, bridging the gap between academic instruction and clinical practice, and assisting in the development of patient-centered curricula. This ensures that graduates have critical communication and empathy skills, which improves outcomes and satisfaction. Through surveys and direct interaction, their input assesses the efficacy of providers, identifies areas that require development, and guarantees that education meets the requirements and expectations of patients. Patients play a crucial role in the QA process because they provide firsthand information on a trainee's professionalism, communication, clinical abilities, and empathy which are difficult to evaluate through tests alone. Additionally, their experiences influence curriculum design, which shifts the emphasis from merely biomedical information to essential interpersonal skills.

#### *29.5.1.4 Regulatory and accrediting bodies*

Regulatory and accrediting bodies like the WFME are essential to ensuring the quality of medical education because they set standards, ensure accountability, encourage ongoing improvement, and maintain public trust. Through self-evaluation, peer review, and adherence to stringent, dynamic standards, they also act as external evaluators, making sure that programs produce competent, safe-to-practice graduates. They provide a standard for quality, prevent poor practices, and foster public confidence in the training of the profession by clearly linking education to patient safety and effective healthcare delivery.

#### *29.5.1.5 Institutions*

The QA units and the administrators which are a part of the institution also play a vital role in QA. In order to satisfy local and international accreditation criteria, the QA unit in medical education guarantees, oversees, and enhances academic standards, curricula, and training. It promotes ongoing advancements in research, education, and administrative procedures to generate skilled, secure medical practitioners. Managing staff and student feedback, performing internal audits, and promoting a culture of excellence are important responsibilities. On the other hand, the administrator by creating, overseeing, and improving academic standards to satisfy accreditation requirements serve as key pillars in QA. Moreover administrators also promote continuous improvement, as they are actively involved in evaluating curricula, overseeing assessment systems, encouraging a culture of excellence, assisting faculty development, and utilizing data to make sure educational outcomes are in line with institutional objectives and legal requirements.

## 29.6 Tools used in quality assurance process

Fishbone (Ishikawa) diagrams and SWOT analysis for root cause analysis and strategic planning, as well as assessment-specific tools like the objective structured practical examination and medical appraisal feedback questionnaire to assess learning outcomes and staff performance, are examples of traditional data analysis tools for QA in medical education. Comprehensive quality manuals, accreditation procedures, and established QA committees that direct the overall framework are additional crucial tools.

### 29.6.1 Data analysis and improvement tools

#### *29.6.1.1 Check sheets*

In order to enable data-driven quality improvement, process monitoring, and the identification of root causes for proactive solutions, check sheets are essential QA tools for systematic data collecting that reveal patterns, trends, and problem areas. They also serve as input for other tools like Pareto charts or histograms. They make complicated problems more comprehensible for improved decision-making by converting unstructured observations into structured and visual data. In clinical skills assessment, check sheets are frequently used to confirm that trainees have finished all required stages during procedures, guaranteeing conformity to standards. Additionally, it is utilized in the data collection and monitoring process to track the frequency of particular events, like prescription errors or patient safety problems(Chance, 2024).

### 29.6.2 Assessment and feedback tools

The objective structured clinical examination is a crucial evaluation and feedback tool in medical education. Although it requires a lot of resources, the objective structured clinical examination is a potent QA tool in medical education that guarantees graduates meet clinical standards by objectively evaluating competencies like history-taking, physical examination, communication, and professionalism through standardized, timed stations, providing valid feedback for continuous improvement, and mapping curriculum to practice (Vimpani et al., 2020).

### 29.6.3 Framework and management tools

QA frameworks, such as the WFME Global Standards and the WHO Academy Quality Management Framework, which offer the standards and structural principles for medical programs, are essential to the process of QA because they set strict guidelines and a benchmark that medical institutions must follow. Medical institutions also use several strategies, like the plan–do–study–act cycle, which is used to repeatedly evaluate educational reforms on a small scale before implementing them fully. However, in order to guarantee adherence to standards, management tools such as the curriculum Mapping Matrix are utilized to connect curriculum content, learning objectives, and assessment techniques. Even the key performance indicators also known as KPIs which are used to gauge how well the educational system is working, including student pass rates, faculty development involvement, or student feedback ratings, can be used as a useful management tool (Asiyai, 2022).

### 29.6.4 Monitoring and control tools

In order to uphold academic standards and guarantee that graduates possess professional competencies, monitoring and control technologies are crucial in medical education. These tools track performance, spot standard deviations, and carry out corrective measures as part of a larger QA system. Control charts, which measure graduation rates or student assessment scores against upper and lower bounds, are frequently used to evaluate the type of variation in processes over time. Additionally, educators can analyze the frequency distribution of data and use the 80/20 rule to highlight the most important quality issues by utilizing Pareto charts and histograms.

## 29.7 Determinants of the quality assurance process in medical education

The determinants of the QA process in medical education include external factors like regulatory bodies, accreditation standards, public accountability demands, and societal needs for safe and competent practitioners, as well as internal factors like faculty competency, curriculum design, student feedback, and institutional commitment to continuous improvement. To guarantee that medical graduates fulfill high standards of patient care, effective QA systems necessitate precise quality standards, stringent monitoring and evaluation, data-driven decision-making, and cooperation among all stakeholders (Fig. 29.3.

### 29.7.1 Internal determinants

Internal determinants of QA in medical education are the institution-driven, systematic, and continuing activities aimed at maintaining, evaluating, and upgrading the quality of medical training, rather than only complying with external accreditation. These factors are ingrained in the institutional culture and are crucial for raising educational standards over time.

#### *29.7.1.1 Institutional commitment and culture*

The mobilization of internal resources and management support to maintain high educational standards is referred to as institutional commitment. Support from the institution's leaders and management is one of the key elements that characterizes an institution's commitment to quality. Support from the leadership is crucial for directing the application of quality policies and enhancing the internal QA system's credibility. The establishment of permanent accreditation secretariats, funding for quality development, and the provision of high-quality infrastructure are examples of the institution's dedication to resource allocation.

The institutions must not only supply the resources but also formally document their commitment through quality manuals that outline protocols and guarantee that ongoing development is in line with the strategic direction of the college. The institution's strong culture of quality, which is motivated by leaders, is what makes QA successful. All members of the institution, including the academic and administrative personnel, should agree on the institution's values and beliefs. This greatly helps to improve quality over time.

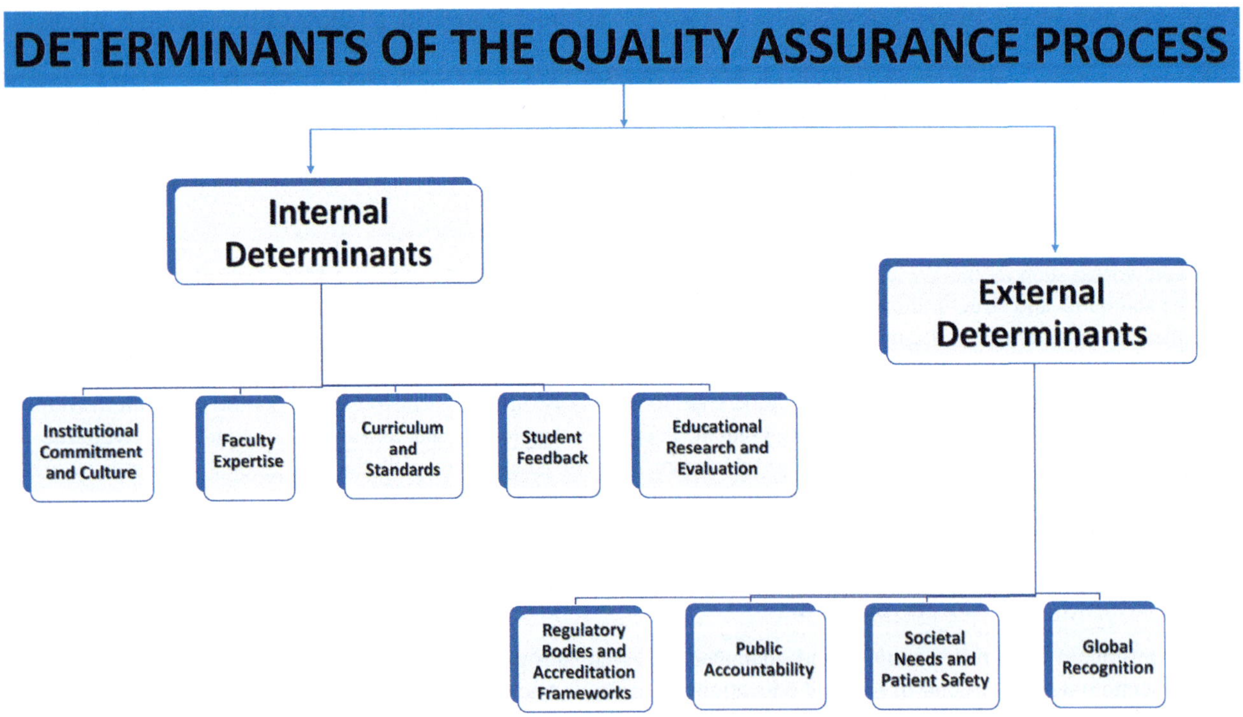


FIGURE 29.3 Determinants of the quality assurance process.

#### 29.7.1.2 *Faculty expertise*

Faculty knowledge serves as both a main input and a catalyst for institutional performance, making it a crucial internal determinant of QA in medical education. It has an impact on a number of important areas, including the provision of high-quality instruction, where faculty knowledge directly affects how well students are taught and learn. In order to improve student comprehension, knowledgeable instructors are expected to break down intricate medical procedures into manageable parts. Similarly, faculty experience is crucial for developing and evaluating curriculum to make sure they continue to be relevant to contemporary healthcare demands and worldwide trends, such as competency-focused outcomes. Additionally, the faculty makes a contribution by demonstrating their proficiency in creating valid and trustworthy evaluation instruments to precisely determine students' competencies. Additionally, the faculty's adoption of international bioethics and professionalism standards, which serve as a standard for student learning, is impertinent.

#### 29.7.1.3 *Curriculum and standards*

As the main guide and standard for institutional excellence, curriculum and standards are important internal determinants of internal QA in medical education. As the structural basis for instruction, the curriculum is regarded as the "heart and soul" of a medical school. Additionally, it serves as a competency blueprint by outlining the format, subject matter, and teaching strategies needed to develop moral and flexible healthcare workers. It is also important that the curriculum content must closely match the program's intended goals, objectives, and graduate attributes in order to maintain quality.

#### 29.7.1.4 *Student feedback*

Since students are the "pivot stakeholders" who have firsthand experience with the curriculum and teaching strategies, student feedback is a key internal predictor of QA in medical education. Institutions go beyond simple standard compliance to ongoing quality improvement by incorporating their insights. Students are always in the ideal position to observe the medical program and provide feedback on the caliber of several aspects of medical education, including the curriculum, teaching strategies, evaluation, and even logistics. A student's feedback also plays a crucial role in order to ensure that the medical program provides high-quality medical education and healthcare. Having regular, cooperative conversations with student is one of the finest methods to get their input. Focus groups and in-person consultative discussions offer a true picture of administrative deficiencies and institutional culture that anonymous surveys could overlook. Additionally, activities like course evaluation surveys and mentorship offer a chance to make sure the QA process is headed in the proper direction.

#### 29.7.1.5 *Educational research and evaluation*

As they offer an evidence-based framework for institutional accountability and ongoing improvement, educational research and evaluation are important internal determinants of QA. The theoretical and empirical underpinnings of a top-notch medical curriculum are established by educational research. It helps administrators make well-informed judgments about curriculum changes and instructional techniques by substituting systematic research for anecdotal information. Additionally, educational research keeps the curriculum current and adaptable to the demands of contemporary healthcare by assisting us in identifying new worldwide trends and technology developments. Evaluation functions as applied research that uses this data to decide on program efficacy and essential adjustments, whereas research concentrates on producing new knowledge and profound insights into the educational process.

### 29.7.2 External determinants

The various standards and agencies outside of a medical school's direct control that drive it to uphold, assess, and enhance its curricula are referred to as external determinants of QA in medical education. These elements are intended to guarantee standardization, public accountability, and the development of qualified, secure doctors.

#### 29.7.2.1 *Regulatory bodies and accreditation frameworks*

Regulatory bodies and accreditation frameworks are essential to the QA of medical education because they set standards, carry out external program evaluations, ensure safe and competent graduates, foster public trust, and encourage continuous improvement through defined metrics for curriculum, faculty, resources, and outcomes. They not only improve global mobility and allow graduates to practice, but they also provide external validation that a medical school meets criteria, adapting to healthcare advancements and protecting patients by ensuring excellent training. Regulatory agencies are also essential in setting standards because they develop national and worldwide benchmarks for curriculum, instruction, assessment, and resources. Requiring or recognizing accreditation agencies to carry out formal program reviews is also advantageous.

#### *29.7.2.2 Public accountability*

Public accountability, also known as social accountability, sets the bar for quality in medical education by ensuring that institutions align their teaching, research, and services with society's most pressing health needs. It acts as a crucial driver for QA by integrating public accountability into mandatory certification processes at all levels. Additionally, the statutory authorities require institutions to show how their programs address societal health priorities through the use of accreditation as a QA instrument. Medical schools are compelled by accountability to change their goals from ones set by the institution to ones based on actual health facts. Additionally, public accountability highlights the need for institutions to collaborate with governments, healthcare groups, and the general public to identify priority health issues. Furthermore, the QA procedures assess whether curriculum incorporates contextualized learning, such as exposure to underprivileged groups and community health issues (Upadhyay et al., 2020).

#### *29.7.2.3 Societal needs and patient safety*

In medical education, QA is a proactive, methodical way to make sure that future medical professionals have the clinical and ethical skills needed to address the many requirements of society and deliver safe care. In order to create an institution's goal and the particular competences graduates must possess, QA combines feedback from patients, society, and medical experts. This is one of the key parts of QA's role in relation to societal demands and patient safety. Additionally, it ensures labor preparation by monitoring and updating courses. QA guarantees that students are taught topics that are essential to the contemporary healthcare environment, such as ethical behavior and evidence-based decision-making.

#### *29.7.2.4 Global recognition*

In order to stay competitive and credible, universities must match their internal standards with international benchmarks, which makes global recognition a potent external predictor of QA in higher education. Validation through international accreditation is one of the key ways that global recognition functions as an external determinant. To obtain an international reputation for excellence, institutions apply for accreditation from international organizations. They are compelled by this process to embrace strict, predetermined standards for student outcomes, faculty development, and curriculum. Additionally, the desire for international recognition encourages an institutional culture of excellence. Feedback from external reviews promotes ongoing program updates and strategic alignment with global best practices.

## 29.8 Benefits of quality assurance

QA in medical education ensures high standards and produces qualified, prepared physicians who improve patient care, reduce errors, and meet international standards by standardizing curricula, improving instruction, promoting continuous improvement, and increasing graduate employability (Sethuraman, 2025). Among the key benefits are better patient outcomes, data-driven program improvement, enhanced faculty development, and meeting accreditation standards for a globally competitive medical profession (Fig. 29.4

### 29.8.1 For students and graduates

#### *29.8.1.1 Competency and readiness*

QA plays a vital role in bridging the gap between clinical practice and academic instruction in medical education. Two main advantages of a strong QA system are competency and readiness, which guarantee that graduates can provide safe, superior patient care when they enter the field. As a method for creating, monitoring, and enforcing strict standards across curricula, examinations, and training environments, QA in medical education guarantees competency and preparedness for practice. Additionally, it converts academic objectives into quantifiable results, guaranteeing that graduates have the professional behaviors, knowledge, and abilities needed to deliver safe patient care. This guarantees that every graduate has attained a high level of technical and clinical proficiency (Skeff and Chandramohan, 2014).

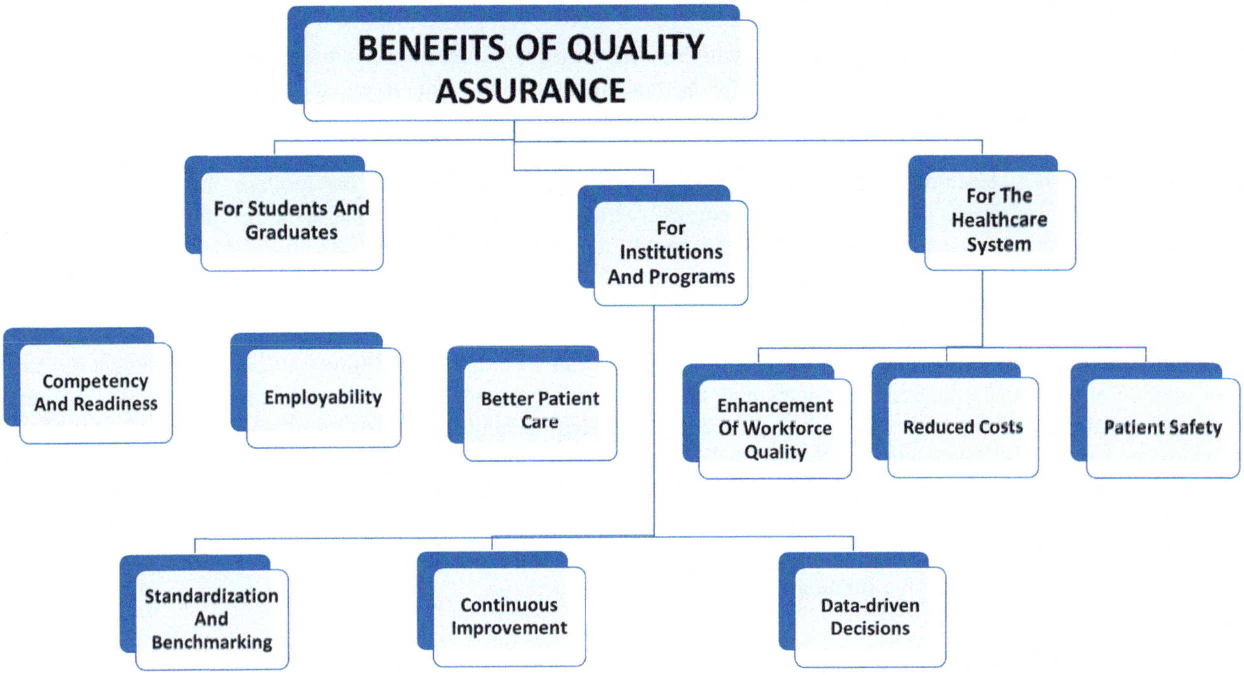


FIGURE 29.4 **Benefits of quality assurance.**

#### 29.8.1.2 *Employability*

QA in medical education directly increases graduate employability by aligning academic instruction with the evolving demands of the healthcare labor market. By standardizing competences and taking stakeholder feedback into account, QA ensures that graduates are not only ready for the workforce but also possess the essential credentials required for both local and international practice. Moreover the QA program of the WFME confirms that graduates have the fundamental knowledge, clinical skills, and professional behaviors such as ethics, teamwork, and communication. Furthermore with the help of robust and trustworthy assessment methods like objective structured clinical examinations and workplace-based evaluations, it makes sure that only competent graduates enter the profession.

#### 29.8.1.3 *Better patient care*

QA in medical education directly improves patient care by creating competent, qualified doctors who adhere to best practices, reducing errors, increasing safety, and promoting a patient-centered approach. Better, more reliable findings and increased trust in the healthcare system follow from this. It ensures that the curriculum meets high standards, promotes evidence-based practice, and cultivates a culture of continuous improvement in future providers in order to make care safer and more effective from the outset (Burdick and Dhillon, 2020).

### 29.8.2 For institutions and programs

#### 29.8.2.1 *Standardization and benchmarking*

Standardization and benchmarking are crucial parts of a strong QA framework in medical education. These procedures help organizations develop a data-driven culture of continuous improvement rather than just self-evaluation. In order to guarantee consistency in student learning outcomes, program quality, and conformity to international standards, standardization and benchmarking are essential, interdependent components of QA in medical education. These strategies enable medical schools transition from mere compliance to continuous quality improvement. Standardization offers QA in medical education by setting standard, evidence-based norms for curriculum, evaluation, and faculty development, ensuring graduates reach uniform competence levels. It promotes responsibility, facilitates ongoing development through accreditation, and lessens institutional variation in training quality. Benchmarking, on the other hand, offers a systematic, evidence-based approach to compare educational programs, procedures, and results against healthcare standards and accepted criteria, thereby guaranteeing QA in medical education.

#### 29.8.2.2 *Continuous improvement*

Continuous improvement in medical education, which is driven by QA, creates a cycle of learning that leads to better-prepared doctors, better patient care, and more efficient systems by utilizing data to enhance curricula, teaching methods, and resource allocation. In the end, this guarantees that graduates fulfill evolving healthcare demands and get better patient results. Additionally, the information obtained from student performance and feedback encourages gradual improvements in curriculum design, teaching methods, and evaluation to guarantee relevance to modern medicine. Continuous improvement also contributes to better patient outcomes by training future physicians in a culture that values excellence. Better, safer patient care, fewer mistakes, and better health outcomes are the immediate effects of this.

#### 29.8.2.3 *Data-driven decisions*

QA in medical education helps make data-driven decisions by transforming raw institutional data into relevant evidence. This method ensures that educational policies and student support systems are efficient and measurable by substituting systematic analysis for intuition. Data-driven decisions influence curriculum reform by allowing educators to identify training gaps or out-of-date course materials through continuous monitoring of student outcomes and performance measures. Then, decisions on curriculum changes are based on recognized learning trends rather than anecdotal feedback. These data-driven judgments help to optimize resource allocation since real-time data is used to make well-informed decisions about staffing, facility utilization, and clinical placement rotations. This ensures that the areas with the greatest impact on student progress receive funding.

### 29.8.3 For the healthcare system

#### 29.8.3.1 *Enhancement of workforce quality*

QA in medical education immediately raises the caliber of the workforce by guaranteeing that graduate professionals meet stringent, consistent requirements for competence and practice. QA future-proofs the healthcare workforce by integrating ongoing monitoring and assessment into training programs. Furthermore, QA procedures like accreditation and licensure guarantee that graduates have the technical expertise and soft skills like empathy and communication which are vital to provide safe and efficient care. It directly raises clinical learning levels by switching from time-based to outcome-based modes of instruction. Comprehensive QA auditing also assesses if training programs are current and comply with changing industry standards and technology developments. This guarantees that staff members are prepared for the use of digital technologies and simulation in contemporary healthcare settings.

#### 29.8.3.2 *Reduced costs*

QA in medical education provides significant cost benefits by enhancing long-term institutional stability, reducing clinical errors, and streamlining administrative processes. QA helps prevent expensive, time-consuming changes or the need to retrain graduates later by spotting problems early in the educational cycle. Additionally, by guaranteeing that graduates are prepared to perform QA, it reduces the significant human and financial costs linked to medical errors, morbidity, and mortality.

#### 29.8.3.3 *Patient safety*

QA in medical education directly enhances patient safety by ensuring that future doctors are competent, risk-aware, and skilled in error prevention. Due to these characteristics possessed by the doctors, patient safety is upheld, as it leads to reduction in medical errors, improved care coordination, and improved patient outcomes from the start. QA ensures that trainees learn basic safety protocols, such as infection control, reporting errors, and understanding human fallibility, which will enable them to deliver a high-quality healthcare and also provide the highest quality of patient safety.

## 29.9 Challenges of quality assurance

Political meddling, commercialization, lax accreditation, and a lack of finance, infrastructure, and qualified personnel are some of the obstacles that medical education's QA must overcome. Inconsistent standards, inadequate faculty training, and challenges in monitoring outcomes such as patient health and faculty reluctance to change are some of the major obstacles that necessitate integrated frameworks for true quality improvement.

### 29.9.1 Institutional and administrative hurdles

#### 29.9.1.1 Resource constraints

Strategic prioritization, automation, and effective management are necessary to minimize effects on scope and results. Resource constraints are a significant QA challenge that limit skilled personnel, time, money, and tools, resulting in incomplete testing, delayed releases, compromised quality, and increased risks. These restrictions compel concessions, such as employing fewer specialists or next-best substitutes, which may worsen healthcare patient outcomes or raise software flaws.

#### 29.9.1.2 Political and commercial pressures

Political and commercial influences provide significant challenges to QA in medical education by influencing accrediting decisions, focusing on financial metrics rather than educational quality, and creating conflicts of interest. Due to these limitations, the focus often shifts from true advancement and academic standards to external conformity and financial survival. However, the increasing commercialization of medical education sometimes transforms educational institutions into money-making business opportunities, which contradicts the core objectives of public service and high-quality training.

#### 29.9.1.3 Documentation overload

When the amount and complexity of records become so great that they impede rather than aid the quality process, this is known as documentation overload in QA. Transparency and traceability depend on documentation, yet an imbalance can result in a number of serious problems. Compared to actual testing or process improvement, QA teams frequently spend more effort developing, updating, and sorting through documentation.

#### 29.9.1.4 Lack of ownership

A major problem in QA is lack of ownership, which happens when people or groups don't feel accountable for the end product's quality and frequently see it as someone else's job. One of the problems here is that choices are carried out more slowly since nobody feels entirely accountable for advancing the project. Faculty members may also view QA as a bureaucratic burden unrelated to their primary teaching and clinical responsibilities, a phenomenon known as faculty indifference. The belief that QA is a top-down administrative necessity rather than a tool for their own professional growth or student achievement is the root of this lack of perceived ownership.

### 29.9.2 Accreditation and regulatory issues

#### 29.9.2.1 Weak regulatory bodies

Weak regulatory bodies substantially jeopardize QA in medical education by not enforcing standards, which could lead to the production of inferior healthcare professionals. It may be difficult to ensure that medical schools follow staffing and teaching infrastructure requirements due to the rapid expansion of medical schools, which could overwhelm the regulatory bodies with inadequate compliance monitoring.

#### 29.9.2.2 Inconsistent standards

Effective QA in medical education is severely hampered by inconsistent standards, mainly because they leave gaps between institutional procedures and global standards. Variable accreditation benchmarks are the cause of this issue. It should be mentioned that because different countries and regions have different accrediting criteria, some organizations place more emphasis on student outcomes, while others place more emphasis on structural inputs, such as lecture hall dimensions or library capacity.

#### 29.9.2.3 Box-ticking

Box-ticking is the term used to describe a compliance-driven strategy in medical education when QA is reduced to a superficial administrative chore that is more concerned with satisfying minimum accreditation standards than with real institutional improvement. This is the outcome of putting compliance ahead of quality. The institutions here ignore their own strategic objectives and ongoing quality improvement in favor of carefully adhering to the standards established by outside authorities in order to preserve accreditation.

### 29.9.3 Curriculum and teaching challenges

#### *29.9.3.1 Linking the outcomes*

Since traditional QA frequently concentrates on short-term student satisfaction or knowledge gain rather than long-term patient outcomes, connecting educational outcomes to clinical performance and patient health is a major problem in QA for medical education. The underutilization of data is one of the problems that contribute to this obstacle. Even though institutions get a lot of data, they frequently don't use it to guide curricular modifications that directly increase clinical efficacy.

#### *29.9.3.2 Knowledge gaps*

Knowledge gaps, which are defined as the difference between current learner or faculty knowledge and required professional standards, are one of the primary obstacles to effective QA in medical education. One reason for this is that traditional didactic curricula may lag behind the rapid advancements in healthcare, leading to disjointed education that overlooks the complexity of the modern healthcare system. Additionally, there may be a know–do gap since students frequently have academic knowledge but lack the cognitive depth or practical experience to apply it in real-world situations.

#### *29.9.3.3 Faculty and interprofessional friction*

QA in medical education is greatly hampered by faculty and interprofessional disagreement, which interferes with curriculum delivery, assessment, and standardization. These disputes are often rooted in deeply embedded professional hierarchies, conflicting educational philosophies, and institutional silos. Faculty members usually prioritize their own specialized domains and consider integrated or interprofessional education as a danger to their discipline's territory or curriculum time, which is why they reject integrated curricula. Additionally, when collaborative teaching methods that challenge established power dynamics are demanded by QA systems, there is resistance to change because outdated hierarchical models are still extensively used.

**Summary**

In order to preserve and improve educational standards, this chapter has looked at quality assurance (QA) in medical education as a methodical and ongoing process. It underlines that QA includes continuous improvement in curriculum design, teaching and learning strategies, assessment procedures, faculty development, and institutional governance in addition to adhering to accreditation standards. The World Federation for Medical Education and the General Medical Council Framework are two significant topics covered in this chapter, with an emphasis on how they affect QA. Alignment between educational objectives, assessment methodologies, and clinical training environments was emphasized, as was the responsibility of faculty and leadership in cultivating a culture of excellence, accountability, and reflective practice.

**Key point**

Future perspectives: upholding the global standards

Rapid changes in healthcare delivery, educational theory, technology, and societal expectations will determine the future of quality assurance (QA) in medical education. It is anticipated that QA systems will become more dynamic, learner-centered, and outcome-driven in the future, with a significant focus on ongoing improvement as opposed to recurring inspection. In order to guarantee excellence and relevance in healthcare training, the future of QA in medical education depends on dynamic, learner-centered, and outcome-focused systems that incorporate technology, international standards, and social responsibility. International accreditation and harmonizing standards for interprofessional, team-based treatment are receiving more attention. This guarantees that training incorporates allied health professions and is in line with the demands of a global healthcare environment.

## References

Amaral, E., Norcini, J., 2023. Quality assurance in health professions education: role of accreditation and licensure. Medical Education 57 (1), 40–48. Available at https://doi.org/10.1111/medu.14880.

Asiyai, R.I., 2022. Best practices for quality assurance in higher education: implications for educational administration. International Journal of Leadership in Education 25 (5), 843–854. Available from. Available at https://doi.org/10.1080/13603124.2019.1710569, http://www.tandf.co.uk/journals/titles/13603124.asp.

Attree, K.J., 2025. Stakeholder engagement in accreditation and quality assurance via the program advisory board: contributions and benefits. Quality Assurance in Education Available from. Available at https://doi.org/10.1108/QAE-04-2025-0112, http://www.emeraldinsight.com/info/journals/qae/qae.jsp.

Burdick, W., Dhillon, I., 2020. Ensuring quality of health workforce education and practice: strengthening roles of accreditation and regulatory systems. Human Resources for Health 18 (1). Available at https://doi.org/10.1186/s12960-020-00517-4.

Crampton, P., 2019. BMJ Open 9.

Chance, E.A., 2024. Int J Nurs Sci 11.

Dolmans, D.H.J.M., Wolfhagen, H.A.P., Scherpbier, A.J.J.A., 2003. From quality assurance to total quality management: how can quality assurance result in continuous improvement in health professions education? Education for Health 16 (2), 210–217. Available at https://doi.org/10.1080/1357628031000116899.

Joshi, M.A., 2012. Quality assurance in medical education. Indian Journal of Pharmacology 44 (3), 285–287. Available at https://doi.org/10.4103/0253-7613.96295. PMID: 22701233.

Raman, K.S., 2025. How to use the pdca cycle and 'quality pillars' for quality improvement in health professions education (with a worked-out example on improving the communication skills of students). 10.13140/RG.2.2.23018.79042.

Nawaz, Q., Ghaffar, S.A., Shafiq, S., Tufail, S., Rehman, M., Khan, J.S., 2025. Ensuring competency: quality assurance in high-stakes assessments in medical education. Biological and Clinical Sciences Research Journal 6 (5), 172–174. Available at https://doi.org/10.54112/bcsrj.v6i5.1761.

Patel, P., Deshpande, V., 2017. Application of plan-do-check-act cycle for quality and productivity improvement-a review. International Journal for Research in Applied Science & Engineering Technology 5, 197–201.

Ulewicz, R., 2017. The role of stakeholders in quality assurance in higher education. Human Resources Management and Ergonomics XI, 93–107.

Upadhyay, P.K., Tiwary, G., Pant, I., 2020. Quality medical service for quality assurance in neurosciences and medical sciences using international standard organization, quality standards and concepts of total quality management (TQM). Acta Scientific Medical Sciences 4 (12), 44–52. Available at https://doi.org/10.31080/asms.2020.04.0794.

Vimpani, J.W., Wright, S.M., Subramaniam, M., 2020. The why and how of quality assurance in high-stakes medical education assessments. Postgraduate Medical Journal 96 (1139), 544–551.

Sethuraman, K.R., 2025. Quality assurance in medical education: are we "ticking the boxes" or "making a difference"? Indian Pediatrics Available from. Available at https://doi.org/10.1007/s13312-025-00222-2, https://www.springer.com/journal/13312.

Skeff, K.M., Chandramohan, B., 2014. Quality assurance in high-stakes medical education examinations. Medical Education 48 (8), 753–763.

Tackett, S., 2015. Medical teacher 38.

Chapter 30

# Quality control

*"Quality control starts and ends with training."*

—Kaoru Ishikawa

## 30.1 Introduction

In medical education, quality control refers to the systematic observation and assessment of healthcare procedures in order to satisfy legal and standard requirements. Quality control in medical education ensures that medical training programs and services meet established criteria of excellence. It is a crucial component of the broader concept of quality management and is often contrasted to quality assurance. In order to make sure that standards are being met, quality control in medical education looks at the "end products" or direct results of a process. These procedures often involve analyzing the generated data and comparing it to preestablished criteria. Quality control forms a vital part of healthcare, as it plays an essential role in preventing errors, enhancing treatment outcomes, and optimizing healthcare services. In contrast to quality assurance, which is a preventive activity centered on the process itself, it is essentially a detection activity focused on examining the final product to confirm if it complies with the requirements. Quality management include both quality assurance and quality control (Elshama, 2022).

Exam result analysis to guarantee validity, consistency, and reliability, for example, by determining whether pass rates meet expectations, is one of the most well-known instances of quality control procedures in medical education. One of the most important areas where quality control is used is the monitoring of trainee performance, which is done to make sure trainees are fulfilling anticipated annual milestones. Additionally, in order to find areas that fell short of quality standards, educational program audits involve examining course evaluations and student feedback following a course.

## 30.2 Need for quality control

In order to guarantee that graduates are capable, secure, and satisfy the needs of patients and society all of which eventually result in improved healthcare outcomes, quality control is crucial in medical education. It is a continuous process of monitoring, evaluating, and improving educational programs, curricula, teaching methods, and assessment standards to bring them into accordance with national and worldwide quality benchmarks. The ultimate objective is to find and fix flaws, encourage a culture of excellence, and guarantee responsibility in order to safeguard patients and maintain public confidence in the medical field (Teigland et al., 2013). Some of the main justifications for the necessity of quality control in medical education are shown in Fig. 30.1.

### 30.2.1 Patient safety

Through ongoing monitoring, risk management, and adherence to regulations like the International Organization for Standardization and the Health Insurance Portability and Accountability Act, quality control in healthcare plays a critical role in patient safety by acting as a systematic process to prevent errors, ensure accurate diagnostics, guarantee safe medical devices or drugs, and maintain high care standards. Ultimately, this reduces harm, improves outcomes, and builds

**The Principles and Practice of Medical Education. DOI:** https://doi.org/10.1016/B978-0-443-44197-4.00035-7

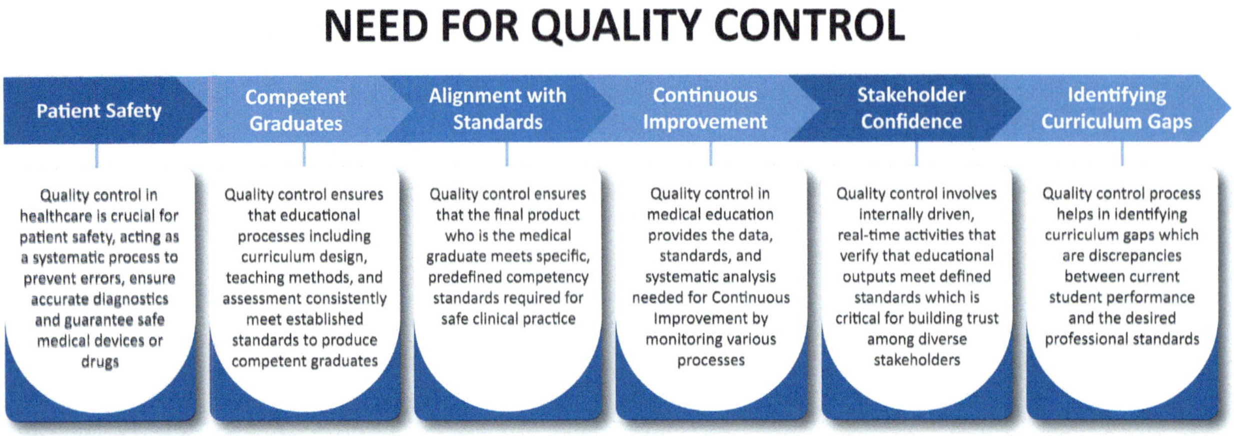


FIGURE 30.1 Need for quality control.

patient trust. From lab findings to treatment administration, it protects patients by identifying problems before they worsen and makes sure that the right care is provided at the right time. Error prevention is one of quality control's most important functions in patient safety. It finds and fixes process faults to prevent medical errors, such as incorrect diagnoses or prescription errors, before they endanger patients. Additionally, it helps with diagnostic accuracy by guaranteeing accurate and dependable laboratory testing, avoiding incorrect diagnoses, and guaranteeing successful treatment strategies. Furthermore, it assists healthcare professionals in confirming that drugs and medical equipment adhere to stringent safety regulations, avoiding negative reactions or malfunctions (Anugrahsari et al., 2022).

### 30.2.2 Competent graduates

In order to generate competent graduates prepared for the workforce, quality control in medical education acts as a mechanism to guarantee that educational activities, including curriculum design, teaching strategies, and assessment, consistently fulfill specified criteria. One of the key roles of the quality control process in ensuring the competency of graduates in medical education is to test and measure student outcomes against predefined performance indicators to ensure they have the technical aptitude and higher level academic skills required for their degree (Zaini et al., 2011). Additionally, it guarantees that educational programs are updated on a regular basis and in line with the demands of the healthcare industry, emphasizing both specialized technical abilities and general qualities like problem-solving and cooperation. Mechanisms for assessment and verification, such as internal audits and external exams, confirm that the academic norm satisfies national or international standards and that grading is consistent.

### 30.2.3 Alignment with standards

Quality control is used in medical education as a verification method to make sure that the medical graduate meets certain, predetermined competency requirements necessary for safe clinical practice. Before entering residency, quality control procedures such as comprehensive final exams and objective structured clinical examinations serve as gatekeepers to ensure that students have acquired the clinical skills and knowledge required by national and international standards. In order to ascertain if the educational program effectively prepares students for the demands of the workplace, the quality control process also monitors outcome metrics by examining graduation rates and performance on standardized licensure examinations like the The United States Medical Licensing Examination (USMLE).

### 30.2.4 Continuous improvement

Quality control in medical education, which forms a key component of accountability and excellence, provides the data, standards, and analysis required for continuous improvement by monitoring procedures, spotting gaps like curriculum relevance or teaching methods, and guiding data-driven decisions to improve learning environments, train better doctors, and enhance patient care (Vroeijenstijn, 1995). It also involves establishing what appears to be good, such as curriculum outcomes, staff development, student achievement, and setting a baseline for progress. In order to identify areas that require attention, the quality control method routinely collects data on student performance, teacher efficacy, and program alignment with societal demands. Quality control identifies shortcomings in curriculum, instructional strategies, or administrative services by comparing current performance to standards.

### 30.2.5 Stakeholder confidence

In medical education, quality control refers to internally motivated, real-time activities that track performance and confirm that learning outcomes adhere to predetermined criteria. These procedures are essential for fostering confidence among various stakeholders, such as the public, employers, regulatory agencies, and students. Quality control provides objective proof that graduates have the requisite clinical skills through performance-based tests like OSCEs. It thereby offers a confirmation of graduate proficiency. These processes boost employers' confidence and reduce the possibility of medical errors by assuring that recent graduates are work-ready and capable of providing safe, high-quality care. Additionally, students feel more confident in the value of their own professional training when fairness and reliability in high-stakes exams are ensured through visible quality control.

### 30.2.6 Identifying curriculum gaps

In medical education, quality control plays a crucial role in guaranteeing that curriculum stay current, therapeutically applicable, and grounded in research. Curriculum gaps, which represent the discrepancies between current student performance and the anticipated professional requirements, can be found by systematic monitoring and data-driven analysis. In quality control methods, curriculum mapping is used to identify the exact places and frequency of education of specific knowledge, abilities, and attitudes. This helps in pinpointing the blind spots, where vital competencies like medical ethics and health systems science which concentrates on the delivery of healthcare, professional collaboration, and ways to enhance patient outcomes and system effectiveness are lacking. Additionally, by analyzing data from formative and summative assessments, quality control identifies recurrent areas of student weakness. The CIPP model and other standardized tools help determine whether educational processes are producing the required professional behaviors.

## 30.3 Components of quality control

Establishing clear standards, putting in place stringent evaluation and feedback procedures for both students and instructors, making sure that the learning environment is encouraging and well-resourced, giving faculty members continual training and development and developing a system for ongoing monitoring and improvement are all components of quality control in medical education. These elements, which are frequently organized under quality management frameworks, seek to maintain academic standards, raise graduates' general caliber, and improve patient safety and healthcare results. Some of the key components of the quality control in medical education include the following:

### 30.3.1 Clear standards and frameworks

#### *30.3.1.1 Defined quality standards*

Medical education quality standards, which are frequently guided, guarantee that academic programs fulfill, uphold, and surpass established criteria. For instance the competency of medical graduates is assessed in medical education using established quality standards as the necessary benchmark. These standards offer an impartial foundation for the physical examination and testing of educational outputs as part of the quality control procedure. Establishing baseline performance, which refers to the lowest acceptable level of performance, results, or graduate outcomes necessary to guarantee patient safety and professional competence, is one of the functions of set standards in quality control. Additionally, these standards aid in objective measurement by giving supervisors and stakeholders an authoritative foundation for communication and evaluation by providing clear definitions and vocabulary.

#### *30.3.1.2 Quality management frameworks*

Quality management frameworks are structured systems used in medical education that combine continuous quality improvement, quality assurance, and quality control to guarantee that educational programs fulfill professional and national standards. Quality control is at the forefront of these frameworks, concentrating on sustaining steady, predictable performance through prompt remedial action and real-time monitoring. These frameworks frequently classify requirements into distinct categories, offering a checklist that is used to evaluate educational institutions. The quality control process also facilitates operational monitoring by conducting daily or periodic inspections, which include keeping an eye on trainee feedback, auditing training sessions, and convening team meetings to address critical delivery gaps.

## 30.3.2 Assessment and feedback

### 30.3.2.1 Student performance assessment

Student performance assessment is a crucial part of educational quality control because it provides data to measure learning effectiveness, identifies problems like teaching gaps or student needs, guarantees standards are met, and promotes continuous improvement. All these aspects closely link student performance to institutional excellence. It goes beyond simple grading to assist teachers, motivate students, and match programs with objectives. Assuring stakeholders, including employers and regulators, that standards are always maintained is also crucial. Additionally, the student outcomes demonstrate the effectiveness of key program quality indicators such resource allocation, instructional strategies, and curriculum design. In order to enable early intervention, the quality control process also assists in identifying problems such as poor performance, which highlights particular difficulties like inadequate support, poor instruction, or problems with admissions.

### 30.3.2.2 Trainee feedback

As a formative assessment method that offers precise, useful information to enhance a trainee's abilities and knowledge, trainee feedback is an essential component of quality control in medical education. When given well, it encourages self-reflection, inspires students, reinforces best practices, and eventually improves patient safety by guaranteeing excellent medical care and professional development. Feedback must be timely, precise, nonjudgmental, centered on observable performance rather than the individual, and include concrete plans for improvement in order to be effective. Trainee feedback is an essential auto-regulatory mechanism for quality control in medical education since it is the primary tool for minimizing the gap between a learner's actual performance and predefined educational requirements. From a simple process, it has developed into a dynamic, interpersonal interchange focused on professional development and competency-based milestones. Performance calibration, which enables trainees to compare their clinical performance to necessary competences like the ACGME milestones, is one of the crucial roles of training feedback in quality control. It also helps with error reduction, as it is simple to address clinical errors and significantly lower error rates in practice by providing systematic and targeted feedback.

### 30.3.2.3 Program evaluation

Assessing the curriculum and learning environment's efficacy on a regular basis is crucial in order to highlight issue areas and fill in any gaps. Because it provides a systematic way to collect and analyze data on programs to ascertain their effectiveness and direct future improvements, program evaluation is crucial for quality control in medical education. Medical education institutions can find areas for development, maximize resources, and guarantee graduates achieve the necessary requirements by utilizing criteria to evaluate the value and impact of programs. This will ultimately improve patient care and satisfy societal obligations. It is a fundamental component of quality control that is used to monitor results, find gaps, and promote ongoing quality improvement.

## 30.3.3 Learning environment and resources

### 30.3.3.1 Well-equipped facilities

One of the key component of quality control in medical education is well-equipped facilities, which guarantee that the teaching environment satisfies established standards for clinical competence and safety. To preserve accreditation and safeguard public safety, universities must integrate cutting-edge technology with strict infrastructure upkeep ensuring that students have access to these facilities, which ultimately improve the general caliber and efficacy of medical education, including crucial infrastructure like auditoriums, labs with modern equipment, libraries with extensive reference materials, and technology for both staff and students. Having well-equipped facilities also guarantee that students receive sufficient instruction in contemporary, useful settings for lectures, hands-on learning, and research.

### 30.3.3.2 Supportive environment

A supportive environment is a crucial component of quality control in medical education because it directly links student well-being and psychological safety to academic success and patient safety. It creates a warm, safe space where students feel comfortable voicing concerns and where procedures are in place to address them. Since it preserves the learner's psychological safety by fostering an environment where students can ask questions and make mistakes without fear of judgment, a supportive environment is crucial to quality control in medical education. It also highlights how crucial it is for educators to serve as role models and friends in addition to being instructors. Maintaining adequate physical space, security, and protected learning time are also necessary to uphold organizational norms. It also promotes the development of a peer community and lessens the risk of burnout by creating a community of practice where students collaborate rather than compete. A supportive environment is also necessary to help students manage stress, acquire vital skills, and eventually satisfy the demanding requirements of the medical profession.

## 30.3.4 Faculty and staff development

### *30.3.4.1 Trainer standards*

In order to carry out the process of quality control, it is crucial to establish clear standards for trainers and ensure that they have the skills and knowledge necessary to instruct and supervise trainees. Trainer standards are an essential part of quality control in medical education because they set requirements for teacher performance and competency, which directly impacts the standard of patient care. This monitoring of the training standards guarantees that instructors have the pedagogical abilities, clinical knowledge, and professional values needed to create competent and safe healthcare workers. In order to safeguard patient safety, these requirements are still developing in the direction of integrated credentialing and improved monitoring. One of the responsibilities of the trainers in quality control is that they put patient safety first by combining clinical treatment with chances for supervised learning. Furthermore, regulatory authorities guarantee a consistent educational experience across various training sites by setting common criteria like the General Medical Council (GMC)'s domains. These criteria guarantee that trainers have the professional values, expertise, and abilities needed to supervise, present curriculum, and assist trainees' growth. Organizations can ensure that medical professionals are taught to fulfill high clinical standards, protecting patient safety and enhancing overall healthcare results, by establishing clear expectations and putting appraisal mechanisms in place.

### *30.3.4.2 Continuous professional development (CPD)*

The faculty are given the chance to refresh their knowledge and teaching techniques through the process of CPD, which is essential for upholding high standards in medical education. CPD is a crucial component of quality control in medical education because it guarantees that healthcare professionals maintain and grow their knowledge, abilities, and professional integrity to provide high-quality, evidence-based therapy. It serves as a regulatory framework to ensure that healthcare professionals maintain and update the skills, knowledge, and attitudes required for safe and effective practice throughout their careers. CPD minimizes knowledge decay and maintains competency by keeping practitioners informed about the latest scientific discoveries, treatments, and therapeutic recommendations. It also shifts the focus from simply attending lectures to demonstrating behavioral changes in clinical contexts. This is often monitored through personal learning plans and portfolios that record learning application and reflection.

## 30.3.5 Monitoring and improvement systems

### *30.3.5.1 Audits and reviews*

Carrying out the process of routine internal and external audits ensures that the quality standards are being followed. Audits and reviews are essential quality control elements in medical education that systematically evaluate performance against predetermined standards to find gaps and promote process improvement. Moreover, they bridge the gap between educational theory and practical application by highlighting discrepancies and providing a strategy for corrective action. Audits ensure that medical training programs meet strict national and international standards like WFME or GMC in order to maintain accreditation and legal standing. This helps maintain compliance and adherence to regulations. They also serve as mirrors that enable educational institutions to find, examine, and address differences between intended curricular objectives and actual learning outcomes.

### *30.3.5.2 Exception reporting*

Exception reporting, a vital quality control technique in medical education, is primarily used to monitor and resolve discrepancies between a trainee's agreed-upon work plan and their actual working environment. It offers real-time data to identify systemic weaknesses that can compromise patient safety and academic standards. The purpose of exception reporting is to verify that safety and educational requirements are upheld while addressing deviations from planned activities, such as missing training sessions, working excessive hours, or having insufficient supervision. One of the main purposes of exception reporting in quality control is training safeguard, which allows us to identify situations in which trainees are unable to receive the appropriate supervision or training opportunities because of staff shortages or excessive workloads.

#### *30.3.5.3 Systemic monitoring*

The methodical and ongoing process of gathering, evaluating, and applying data to monitor a program's advancement toward fulfilling predetermined goals and standards, thereby directing management choices and guaranteeing continuous quality improvement, is known as systemic monitoring in quality control in medical education. This is a continuous, real-time monitoring method as opposed to a one-time evaluation, which sets it apart from periodic evaluation. It is a structured, continuous process created to guarantee that training, educational initiatives, and student evaluations fulfill predetermined, exacting standards. To find, stop, or fix flaws in real time, it entails examining data at every stage of the educational process, from student selection to curriculum delivery to graduate competency results. This strategy also focuses on tracking program inputs, such as funding, staff credentials, and facilities, and outputs, such as the quantity of trained staff, the availability of learning resources, and student progress statistics, to ensure they meet preset requirements.

#### *30.3.5.4 Documentation and record-keeping*

Documentation and record-keeping are the cornerstones of quality control in medical education. They offer a reliable record of instructional activities, student achievement, and institutional compliance with legal requirements. Documentation is a fundamental pillar of operational efficiency and educational integrity in the context of quality control, rather than just an administrative activity. It is important to remember that institutional accreditation requires proof of conformance to national and international standards. The documented records show that instruction is carried out according to schedule and that curricula fulfill the required standards. Furthermore, systematic clinical audits and internal evaluations to assess the efficacy of educational initiatives are made possible by the comprehensive data. Additionally, structured documentation guarantees a uniform training experience for students at various locations.

## 30.4 Frameworks for quality control process

The frameworks for quality control process provide structured guidance for ensuring quality in educational programs. These frameworks emphasize key elements like establishing clear learning outcomes, implementing robust monitoring systems, ensuring consistent standards, and fostering continuous improvement through stakeholder involvement and feedback (Karle, 2002) .Some of the important frameworks for quality control in medical education include the following:

1. World Federation for Medical Education (WFME) standards,
2. UK GMC standard for postgraduate education.

### 30.4.1 World Federation for Medical Education (WFME) standards

To improve the caliber of medical education around the globe, the WFME establishes international standards for medical education, accreditation, and ongoing professional development. Its standards support uniformity and rigor in curriculum, faculty, resources, and evaluation procedures by serving as a thorough quality benchmark for medical schools and accrediting bodies. The WFME plays a crucial role in quality control in medical education, ensuring high quality and international consistency across basic, postgraduate, and continuing medical education through the establishment of global standards, evaluation of accreditation organizations through its recognition program, and maintenance of the world directory of medical schools. It acts as a benchmark-setter, promoting continuous improvement and helping to maintain standards for a globally mobile medical workforce.

### 30.4.2 UK General Medical Council standard for postgraduate education

The GMC of the United Kingdom establishes national standards for postgraduate medical education. The general medical council propels quality control by establishing desired results, authorizing training providers, and overseeing a quality assurance structure that ensures consistent high standards across the United Kingdom. Ultimately, the GMC safeguards patients by verifying that training equips physicians with the knowledge and abilities necessary for safe, ethical practice through approval, oversight, and prompt action.

## 30.5 Steps in quality control

Quality control in medical education involves continuous cycles of standard setting, internal self-assessment, external review, and improvement implementation to ensure high-quality education. The key steps in medical education quality control are depicted in Fig. 30.2.

### 30.5.1 Establish clear standards and objectives

To ensure that medical education programs create qualified, socially responsible doctors, the first step in quality control is to establish defined standards and objectives. One of the important aspects in this regard is the adoption of global models, in which the institutions conform to international frameworks such as the WFME worldwide standards. To ensure that students are truly meeting the necessary competency levels, assessment methods must be strictly compatible with the established learning objectives. Clearly defining the educational objectives and allowing for the alignment of assessment are also important steps in this process (Walter et al., 2025).

### 30.5.2 Develop a framework and assign roles

Developing a framework for quality control necessitates a proper strategy that incorporates stakeholder demands, competency-based outcomes, and continuous improvement processes. A strong framework ensures that educational programs generate qualified healthcare workers by shifting the emphasis from simple compliance to cultivating a culture of excellence. Assigning roles for quality control, on the other hand, guarantees that academic activities fulfill the necessary requirements and generate qualified healthcare workers. Usually, this structure functions at the programmatic, local, and national levels. It is vital to ensure that all the stakeholders have clearly defined roles and responsibilities, and their duties are outlined properly. For example, in medical education, the quality assurance unit serves as the focal point for upholding strict standards, promoting ongoing development, and guaranteeing accreditation compliance. In order to close performance gaps, they are also responsible for managing quality improvement projects, conducting internal audits, and obtaining feedback through course evaluations.

### 30.5.3 Implement and integrate quality control

Medical education ensures that student outcomes, instructional standards, and institutional services consistently meet predefined benchmarks by implementing a quality control method. This process is essential to maintaining certification and producing highly skilled healthcare workers. A change from passive learning to a methodical, data-driven strategy focused on continuous quality improvement is necessary for incorporating quality control into medical education. Instead of being taught as a stand-alone course, the principles of quality control can be integrated longitudinally throughout all training years. For example, the preclinical years can offer fundamental techniques like the plan–do–study–act cycle, while the clinical years encourage students to lead or participate in real-world quality improvement projects during their rotations in order to address genuine patient safety issues.

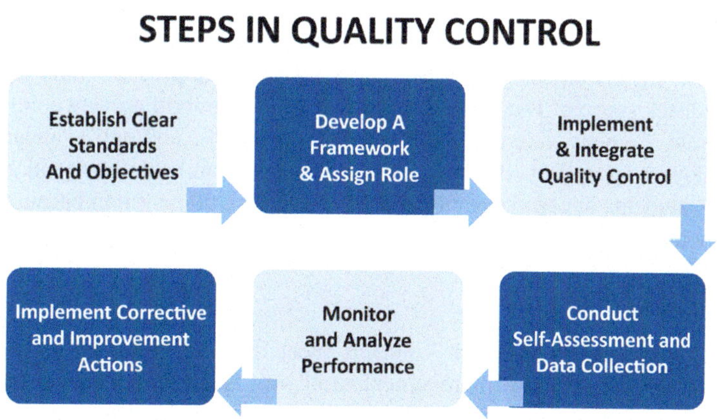


FIGURE 30.2 Key steps in medical education quality control.

### 30.5.4 Conduct self-assessment and data collection

Integrating self-evaluation and data gathering is a continuous process in medical education that guarantees training programs fulfill predetermined standards of quality. The WFME global standards are one example of an international benchmark that is increasingly being used to structure this process. Internal reviews of policies and procedures in comparison to standards can be carried out utilizing self-study methods. Another method is external review and peer assessment, in which external teams, such as accreditation agencies, visit the location and examine the evidence to produce unbiased reports. When it comes to data collection, it can be accomplished by asking faculty members about their teaching and learning resources, as well as students' training, evaluation, and surroundings.

### 30.5.5 Monitor and analyze performance

Performance monitoring and analysis in the medical education quality control process ensure that educational outputs meet preestablished standards for student competency and patient safety. Two methods to accomplish this are curriculum mapping and audits, which meticulously align learning objectives with assessment methods, and regular internal audits to ensure compliance with global standards such as those of the WFME. This approach can also be aided by continuously obtaining feedback from teachers and students through digital media. Metrics like student satisfaction scores are also becoming more widely acknowledged as crucial quality indicators (Jette et al., 2014).

### 30.5.6 Implement corrective and improvement actions

Implementing corrective and improvement actions into practice in quality control is essential in order to satisfy certification and training requirements. Implementing corrective measures and improvement actions are highly dependent on the results of performance analysis. For example, the decision may be in the form of improving resources, providing faculty development, or updating the curriculum. Additionally, the process of continuous quality improvement ensures that procedures are transparent and flexible by fostering a culture in which employees and students embrace continual development.

## 30.6 Roles and responsibilities of participants in quality control

Students, instructors, administrators, regulatory agencies, and outside quality assurance organizations are important players in quality control for medical education. Each has specific responsibilities, such as giving feedback and guaranteeing compliance, delivering high-quality instruction, supervising educational programs, establishing standards, and carrying out impartial evaluations. All of these roles contribute to ongoing improvement and upholding high standards of patient safety and healthcare quality.

### 30.6.1 Students or learners

Students are essential to quality control in education because they are the primary consumers and provide vital feedback on curriculum, instruction, and services through surveys, committees, and direct participation. From the perspective of the user, this serves to guarantee standards, guide improvement, and validate institutional quality. In this context, some of the vital roles undertaken by the students is in serving on quality assurance committees, participating in evaluations, giving course feedback, and actively working to enhance the entire educational process. Through satisfaction surveys, course evaluations, and program assessments on curriculum relevance, instructional practices, and resource efficacy, students or learners largely serve as feedback sources. The students play a key role by providing viewpoints on a given issue by acting as representatives on program councils, internal quality assurance boards, and self-evaluation teams. Additionally, it involves working with employees to create better programs, providing helpful solutions, and confirming accurate data for outside audits. Additionally, students are an essential part of the early warning system because they improve institutional responsiveness by promptly alerting others about possible issues or areas that require attention.

### 30.6.2 Faculty or educators

In order to ensure that instruction meets standards, students meet learning objectives, and programs are in line with institutional goals through self-evaluation and improvement strategies, faculty and educators must actively participate in curriculum design, effective teaching, student assessment, feedback mechanisms, and ongoing professional development. Faculty members or educators form a vital component of the quality control process, as they apply quality through training, skill assessment, and technique improvement to close gaps. They take an active part in developing and improving courses. In addition to this, they are also responsible for delivering effective training, utilizing state-of-the-art methods and tools. They also foster a pleasant learning environment by serving as role models and motivators. In terms of assessment and feedback, educators provide constructive criticism to students in order to bridge any gaps in their knowledge.

### 30.6.3 Administrators and institutional leaders

Administrators and institutional leaders play a critical role in quality control in medical education by promoting a culture of continuous improvement, ensuring regulatory compliance, managing resources, and providing the necessary support for instructors and staff. Additionally, while institutional leaders provide vision, culture, and strategic direction, promoting change and growth, administrators focus on managing daily operations, resources, and processes, ensuring effectiveness and goal attainment. They often combine these duties in complex institutions such as universities, directing both teaching and learning as well as operational aspects to foster innovation and acceptability. In order to ensure that quality goals are in line with professional standards and society health needs, administrators and leaders actively participate in the alignment of mission and vision. Furthermore, to preserve a coherent learning environment, administrators oversee cooperation and communication between teachers, students, and support personnel.

### 30.6.4 Regulatory and external bodies

To guarantee quality control in medical education, international, national, and regional regulatory and external accreditation agencies collaborate. These organizations set standards, evaluate courses, and ensure that graduates are competent to practice safely and effectively. They are essential to the accrediting process because they formally recognize that a medical program meets performance standards, which are sometimes required for graduates to be eligible for registration or residency. One of the key roles of the regulatory and external bodies is in standard setting, which creates technical, ethical, or safety standards. They also have a part in granting licenses for medical practices, university course accreditation, and other activities that allow people or organizations to operate. Additionally, these organizations actively participate in external audits and assessments by offering unbiased feedback and analysis to gauge institutional performance and compliance.

### 30.6.5 Quality assurance managers or teams

Quality assurance teams and managers serve as the strategic architects in medical education, creating the mechanisms that quality control subsequently oversees. In order to prepare students for competent patient care, quality assurance teams make sure that medical education programs adhere to strict criteria and incorporate the most recent best practices. They create the quality management approach, which includes standard operating processes for hiring professors, delivering curricula, and evaluating students. Additionally, they are crucial in serving as a bridge between students, professors, and leadership, enabling feedback loops to find and fix systemic problems before they affect the caliber of graduates.

## 30.7 Total quality management (TQM)

In medical education, TQM is a strategic philosophy that applies continuous improvement to all facets of training future physicians. It focuses on patient-centered outcomes, staff involvement, leadership, and data-driven processes with the goal of improving educational quality, efficiency, and relevance to societal needs, ensuring graduates meet high standards of care and practice (Yousuf, 2011). It integrates quality in teaching, research, administration, and services like libraries and hospitals, encouraging long-term success and ultimately benefiting the patient and the community by making everyone accountable for exceeding expectations rather than just meeting them. Quality control is a part of TQM, a comprehensive philosophy for excellence in medical education, which aims to provide high-quality, patient-centered treatment and effective learning. It focuses on analyzing and fulfilling certain requirements for procedures, courses, and results, such as graduate competency (Coate, 1993). TQM provides the strategic framework of continuous improvement, employee involvement, leadership for quality controls' tactical inspections, such as evaluation, accreditation, and the integration patient outcomes with education quality, in order to achieve overall institutional excellence.

### 30.7.1 Difference between quality assurance and quality control

The difference between quality assurance and quality control is shown in Table 30.1

**TABLE 30.1** Difference between quality assurance and quality control.

| | Quality assurance | Quality control |
|---|---|---|
| 1 | Quality assurance (QA) in medical education is a systematic process ensuring high standards in training competent doctors, focusing on curriculum, teaching, assessment, and outcomes to meet societal needs. | Quality control (QC) in medical education involves systematic processes to ensure training meets standards, focusing on checking the end product through activities like assessment, evaluation, and monitoring teaching/curriculum, to guarantee high-quality, safe patient care and professional proficiency. |
| 2 | It is process oriented. | Quality control is product oriented. |
| 3 | It is a proactive process aimed to prevent errors. | Quality control is a reactive process to identify and correct defects. |
| 4 | The goal here is to improve development and test processes to reduce defects. | The goal here is to identify defects in a developed product before it is released. |
| 5 | Preventing errors through staff training in good clinical practice is an example of quality assurance. | Detecting errors through sampling and measurements is an example of quality control. |

## 30.8 Role of modern technologies in shaping the quality control

Modern technologies are fundamentally transforming quality control in medical education by shifting assessment from traditional, subjective methods to objective, data-driven, and standardized systems. Technologies such as artificial intelligence (AI), virtual or augmented reality, and advanced simulation allow institutions to monitor student performance in real-time, identify knowledge gaps, and ensure consistent training standards that directly impact patient safety. Some of the key technologies shaping quality control are described below.

### 30.8.1 Artificial intelligence (AI) and machine learning

AI is used to analyze vast datasets, including electronic medical records and simulation performance, to identify trends and areas for improvement in both student performance and curriculum design.

### 30.8.2 Virtual and augmented reality

Since they offer regulated, objective, and repeatable training settings that guarantee consistent, high-fidelity skill development, virtual reality and augmented reality are essential tools for quality control in medical education. They make it possible for educational institutions to provide all students with a consistent, standardized curriculum.

### 30.8.3 Simulation-based medical education

Simulation-based medical education serves as a vital quality control tool by offering a safe, standardized, and repeatable environment to teach and assess healthcare personnel, ultimately minimizing clinical errors and enhancing patient safety. It permits the evaluation of competency in technical processes and nontechnical abilities like communication and teamwork, ensuring that practitioners achieve standardized benchmarks before treating real patients.

### 30.8.4 Learning management systems and digital analytics

These platforms, such as Moodle or specific data repositories like REDEI, track performance data, enabling educators to monitor curriculum effectiveness, identify struggling students early, and tailor support.

### 30.8.5 Computer-based testing and e-objective structured clinical examinations

The transition to electronic assessments increases efficiency, flexibility, and security, allowing for on-demand testing and immediate, objective analysis of student abilities. These above-mentioned key technologies shaping quality control have a profound impact on quality control and assessment, as they help enable objective and real-time evaluation, wherein the

AI-driven tools can evaluate technical skills such as surgical knot-tying with high accuracy, often exceeding human observers. It also facilitates the personalized learning pathways, as AI tailors educational content to individual student needs, providing targeted resources to address specific deficiencies rather than a "one-size-fits-all" approach. Technologies also ensure that all trainees meet specific, measurable criteria before moving to clinical rotations, addressing variability in traditional apprenticeship models. In addition, technologies like simulation allows for the practice of rare or critical events, reducing errors in clinical settings.

## 30.9 Benefits of quality control

In medical education, quality control guarantees the competence of future physicians, which improves patient care, lowers errors, and boosts public confidence. Aligning curriculum with society demands, encouraging a culture of continuous development through feedback, boosting professionalism, guaranteeing international standards, and making education more relevant and effective for generating safe, qualified healthcare workers are some of its advantages. Some of the key benefits of quality control are shown in Fig. 30.3.

### 30.9.1 Improved patient outcomes

In medical education, quality control guarantees that graduates have the standardized skills required to provide safe, efficient care. Quality control greatly reduces avoidable injury, such as medication errors and diagnostic inaccuracy, by standardizing tests and rigorously validating student performance to guarantee that graduates fulfill critical safety benchmarks. Measurable gains in the management of chronic diseases and survival rates are achieved by systematic quality control procedures in clinical learning environments (Tariq and Ali, 2014).

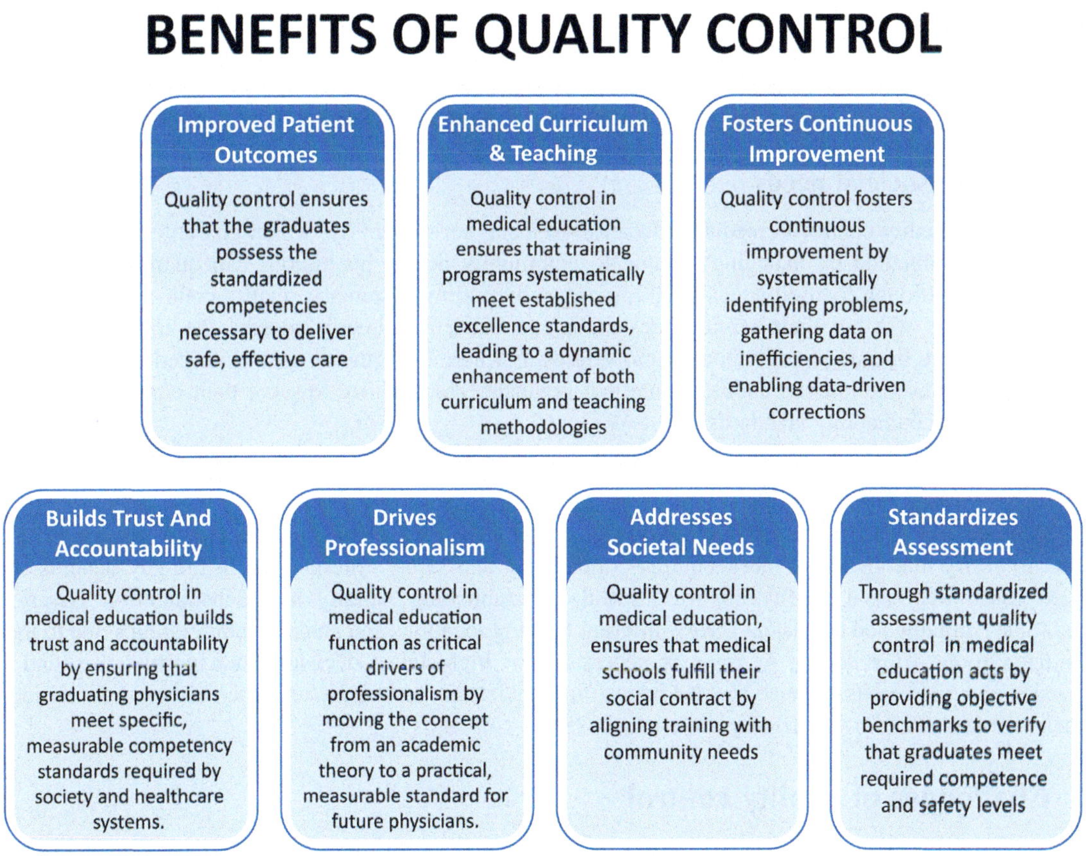


FIGURE 30.3 Benefits of quality control.

### 30.9.2 Enhanced curriculum and teaching

In medical education, quality control guarantees that training programs consistently fulfill predetermined excellence requirements, resulting in a dynamic improvement of curriculum and instructional strategies. Additionally, quality control serves as a formative process that uses feedback from stakeholders to detect shortcomings, enabling curriculum teams to implement remedial actions and reassess their efficacy. Teachers can switch from traditional lectures to active learning models like flipped classrooms and simulation-based training in real time due to the feedback provided by the quality control process (Alotiby et al., 2021).

### 30.9.3 Fosters continuous improvement

A key advantage in medical education is the promotion of continuous quality improvement through quality control, which turns routine assessments into a meticulous, iterative engine for excellence. Shifting medical schools from a tick-box accreditation approach to an intentional framework of discovering, implementing, and assessing program enhancements is one of quality control's main functions. This leads to quantifiable gains in the caliber and results (Berwick, 1989).

### 30.9.4 Builds trust and accountability

Quality control in medical education promotes confidence and accountability by guaranteeing that graduating physicians meet exact, measurable competency standards required by society and healthcare systems. Quality control systems foster trust amongst various stakeholders by confirming the graduate. Strict quality control measures like OSCEs and licensure tests lower the likelihood of medical errors by assuring the public that education is safe and that trainees are qualified to practice. Furthermore, reputable and trustworthy ratings endure scrutiny from outside organizations, sustaining the medical profession's reputation as a whole.

### 30.9.5 Drives professionalism

Quality control in medical education is essential to promoting professionalism because it turns the concept from an academic theory into a practical, measurable standard for future physicians. Medical schools make sure that graduates have the technical expertise and moral principles required for safe patient care by integrating quality control methods into their courses. Furthermore, quality control provides a concrete framework for professional behavior, changing professionalism from a challenging notion into a set of requirements that students must consistently meet.

### 30.9.6 Addresses societal needs

A key advantage of quality control in medical education is social accountability, which guarantees that medical schools uphold their social obligation by matching training to community needs. Institutions can quantifiably enhance health outcomes and equity through focused educational initiatives by putting systematic quality control mechanisms in place. Institutions can assess whether their services, research, and curricula directly address the top health issues of the communities they serve by using quality control procedures. Furthermore, quality control procedures, including external accreditation evaluations and internal audits, ensure that graduates are qualified to serve their communities by enforcing adherence to social accountability standards.

### 30.9.7 Standardizes assessment

Standardized evaluation acts as the primary quality control mechanism in medical education by providing objective benchmarks to verify that graduates meet required competency and safety criteria. One of the key benefits of adopting standardized assessment as a quality control method is establishing validity and dependability. Assessments are guaranteed to be equitable and consistent across different training locations and student demographics due to the presence of standardized tools. This degree of rigor is necessary for high-stakes decisions like certification and licensure. Standardized evaluation results can also be used by institutional leaders to identify underperforming training facilities and modify their courses accordingly (Srijampana et al., 2025).

## 30.10 Challenges of quality control

Resource limitations like in terms of funding, infrastructure, faculty, out-of-date curricula that struggle to keep up with technology and practice, lax regulation pertaining to varying standards, untrained inspectors, administrative gaps, commercialization, and bridging the theory-practice gap are some of the challenges that quality control in medical education faces, all of which have an impact on patient safety and student preparedness. Inconsistent accreditation, a lack of faculty growth, adjusting to digital health, and guaranteeing relevance to changing healthcare demands are the main obstacles.

### 30.10.1 Resource constraints

Resource constraints in medical education significantly hinder quality control by lowering the material, human, and financial resources needed to maintain high standards of instruction and evaluation. In many medical schools, there is a severe instructor shortage. Additionally, a shortage of modern equipment and essential medical supplies like gloves or reagents limits opportunities for practical education.

### 30.10.2 Added costs and workload

There are many operational difficulties in implementing quality control in medical education, particularly with relation to staff workload and budgetary expenses. These elements frequently serve as obstacles to obtaining certification and upholding academic quality. The fact that establishing strong quality management systems necessitates significant financing for new technology, software for monitoring performance data, and physical training facilities are some of the crucial elements in this respect. Additionally, the lack of protected time for educators and physicians adds to their workload, making these duties seem like extra labor that interferes with patient contact time.

### 30.10.3 Faculty development needs

Faculty development is a key concern for quality control in medical education since the professional competency of medical educators directly impacts the quality of healthcare systems. Faculty development requirements must be identified and addressed in order to guarantee that educational goals are met and institutional standards are maintained. In addition, many medical faculty members identify more as doctors or subject matter experts than as teachers. It can be detrimental to their professional identity and hinder the implementation of good teaching practices to force them into new educational roles, such as coach or navigator.

### 30.10.4 Poor infrastructure

Inadequate infrastructure makes it difficult to maintain quality control in medical education, particularly in low- and middle-income nations. For instance, many institutions lack the robust information technology required to collect, store, and assess data on clinical and educational outcomes. However, real-time monitoring of quality metrics is time-consuming and error-prone due to reliance on paper-based records or archaic technological forms.

### 30.10.5 External factors

Quality control in medical education is being increasingly challenged by external variables, which put pressure on educational institutions that are not directly under their control. These elements have the potential to interfere with standardized tests, put a burden on resources, and compel curriculum changes that are challenging to handle within conventional quality assurance systems. The availability of qualified academics and clinical preceptors is impacted by external variables, such as the worldwide scarcity of medical professionals. It becomes challenging to maintain a constant level of clinical competence in the absence of enough outside experience to supervise student training.

**Summary**

To guarantee that medical personnel are trained to satisfy set standards of competence, ethics, and patient safety, quality control in medical education is crucial. The systems, procedures, and parties involved in preserving and raising the standard of medical education at the undergraduate, graduate, and continuing professional development levels are crucial for quality control in medical education. Regular programmatic and institutional accreditation guarantees responsibility, encourages openness, and propels ongoing development. Peer review, academic audits, and student feedback systems are examples of internal quality control methods that support external supervision. Effective quality control is shown to be greatly aided by faculty development, stakeholder engagement, leadership commitment, and data-driven decision-making. In the end, quality control in medical education helps to produce skilled, moral, and patient-centered healthcare workers by combining systematic monitoring with ongoing improvement initiatives.

**Key point**

Future perspectives: the importance of simulation-based training

Rapid technological advancements, changing healthcare requirements, and rising standards for patient safety and responsibility will all influence quality control in medical education in the future. Periodic inspections are giving way to continuous, learner-centered, data-driven methods in quality control. Future quality control in medical education will be learner-centered, technology-enabled, dynamic, and intimately related to healthcare results. The goal of these advancements is to guarantee that medical graduates are adequately equipped to handle the intricate and constantly changing requirements of contemporary healthcare systems. As quality control will increasingly concentrate on demonstrated competencies rather than time-based training, one of the crucial factors to be considered in the future is the change to competency-based and outcome-oriented approaches. Measurable results including clinical reasoning, professionalism, communication, teamwork, and patient safety will be prioritized. The core of quality monitoring will be the ongoing evaluation of competencies along the training continuum.

# References


Alotiby, A., Almaghrabi, M., Alosaimy, R., Alharthi, A., Khawandanah, B., Alansari, R., et al., 2021. Learning environment quality for medical students at Umm Al-Qura university: a comprehensive study on stressors, sources, and solutions after introduction of a new bachelor of medicine and bachelor of surgery (MBBS) curriculum. Advances in Medical Education and Practice 12, 1487–1497. Available at https://doi.org/10.2147/AMEP.S343792, https://www.dovepress.com/getfile.php?fileID=76796.

Anugrahsari, S., Chaeruman, U.A., Abbas, H., Suryadi, S., 2022. Patient safety education for clinical students: a systematic literature review. Open Access Macedonian Journal of Medical Sciences 10 (F), 208–214. Available at https://doi.org/10.3889/oamjms.2022.8345.

Berwick, D.M., 1989. Continuous improvement as an ideal in health care. New England Journal of Medicine 320 (1), 53–56. Available at https://doi.org/10.1056/NEJM198901053200110.

Coate, E., 1993. The introduction of total quality management at oregon State University. Higher Education 25 (3), 303–320. Available at https://doi.org/10.1007/bf01383856.

Elshama, S.S., 2022. Quality management in medical education between theory and application: paradigm shift or falsification of reality. Journal of Clinical Case Reports and Studies 3 (3), 01–05. Available at https://doi.org/10.31579/2690-8808/109.

Karle, H., 2002. Global standards in medical education – an instrument in quality improvement. Medical Education 36 (7), 604–605. Available at https://doi.org/10.1046/j.1365-2923.2002.01221_2.x.

Jette, D.U., Nelson, L., Palaima, M., Wetherbee, E., 2014. How do we improve quality in clinical education? Examination of structures, processes, and outcomes. Journal of Physical Therapy Education 28, 6–12. Available at https://doi.org/10.1097/00001416-201400001-00004.

Srijampana, V.V.G.R., Vennam, B.S., Dulipala, P., Potti, R., 2025. Impact of "assessment and accreditation" process on quality of higher and medical education in India: a systematic review and meta-analysis. Archives of Medicine and Health Sciences 13 (1), 139–144. Available at https://doi.org/10.4103/amhs.amhs_342_24.

Tariq, M., Ali, S.A., 2014. Quality assurance and its application in medical education. College of Physicians and Surgeons Pakistan. Pakistan Journal of the College of Physicians and Surgeons Pakistan. 24 (3), 151–152. Available at http://www.jcpsp.pk/archive/2014/Mar2014/01.pdf.

Teigland, C.L., Blasiak, R.C., Wilson, L.A., Hines, R.E., Meyerhoff, K.L., Viera, A.J., 2013. Patient safety and quality improvement education: a cross-sectional study of medical students' preferences and attitudes. BMC Medical Education 13 (1). Available at https://doi.org/10.1186/1472-6920-13-16.

Vroeijenstijn, A.I., 1995. Quality assurance in medical education. Academic Medicine: Journal of the Association of American Medical Colleges 70 (7 Suppl), S59–S67. Available at https://doi.org/10.1097/00001888-199507000-00021. PMID: 7626161.

Walter, M.M., Rogan, S., Schurz, A.P., Zinzen, E., 2025. How to evaluate the quality of the clinical learning environment in health professions education? Protocol of a systematic review. PLoS One 20 (4). Available at https://doi.org/10.1371/journal.pone.0293773, https://journals.plos.org/plosone/article?id=10.1371/journal.pone.0293773.

Yousuf, N., 2011. Total quality management: do health profession educators need to be educated? Education in Medicine Journal 3 (2), e65. Available at https://doi.org/10.5959/eimj.3.2.2011.c1.

Zaini, R.G., Bin Abdulrahman, K.A., Al-Khotani, A.A., Al-Hayani, A.M.A., Al-Alwan, I.A., Jastaniah, S.D., 2011. Saudi Meds: a competence specification for Saudi medical graduates. Medical Teacher 33 (7), 582–584. Available at https://doi.org/10.3109/0142159X.2011.578180.

Chapter 31

# Benchmarking

*"If you can't measure it, you can't improve it."*

—Peter Drucker

## 31.1 Introduction

In medical education, benchmarking is the methodical process of contrasting recognized "best practices" or peer performance with instructional strategies, student results, or institutional norms. It is a fundamental element of ongoing quality development, going beyond straightforward internal evaluation to offer objective performance goals and external context. The primary goal of benchmarking is quality assurance (QA), which guarantees that programs fulfill national or international accreditation and registration requirements. Additionally, it aids in performance tracking, which measures the student's advancement toward mastering particular learning objectives at regular intervals (Wilkinson et al., 2015). It also helps identify gaps by highlighting areas where an organization or individual may be falling short of required standards. It aids in enhancing stakeholder responsibility by proving to educators, students, and regulatory bodies that educational standards are being met.

### 31.1.1 Definition

World Federation for Medical Education (WFME) defines benchmarking in medical education as comparing standards and practices with external criteria, often using its own principles-based global standards, to drive quality improvement, ensure global relevance, and assure stakeholders that high standards are met, adapting universal principles to specific regional and institutional contexts for better outcomes rather than prescribing rigid methods.

## 31.2 Need for benchmarking in medical education

Benchmarking plays a crucial role in medical education to ensure consistent quality, promote improvement, and ensure competency by comparing programs and student outcomes to internal or external standards, identifying best practices, guaranteeing accountability, and aligning curricula with evolving healthcare needs. In the end, this gets graduates ready for productive practice (Pantall, 2001). It goes beyond mere accreditation to promote creativity and evidence-based learning by standardizing professionalism, clinical skills, and informatics expertise. Several significant justifications for the necessity of benchmarking in medical education are described in Fig. 31.1.

### 31.2.1 Quality improvement

In order to identify strengths, weaknesses, and best practices, quality improvement in medical education requires benchmarking either internally or externally as a methodical tool to compare practices, processes, and outcomes. This will ultimately lead to data-driven decisions for improved curriculum, student performance, and relevance to healthcare needs. In addition to promoting a culture of continual improvement and guaranteeing that educational standards satisfy social and professional expectations, benchmarking offers objective data to compare actual performance versus goals (Ettorchi-Tardy et al., 2012).

### 31.2.2 Identification of best practices

Benchmarking also plays a role in identifying, disseminating, and modifying best practices to enhance academic results and institutional performance. It functions as a tool for continuous quality improvement by comparing internal processes

**The Principles and Practice of Medical Education. DOI:** https://doi.org/10.1016/B978-0-443-44197-4.00009-6

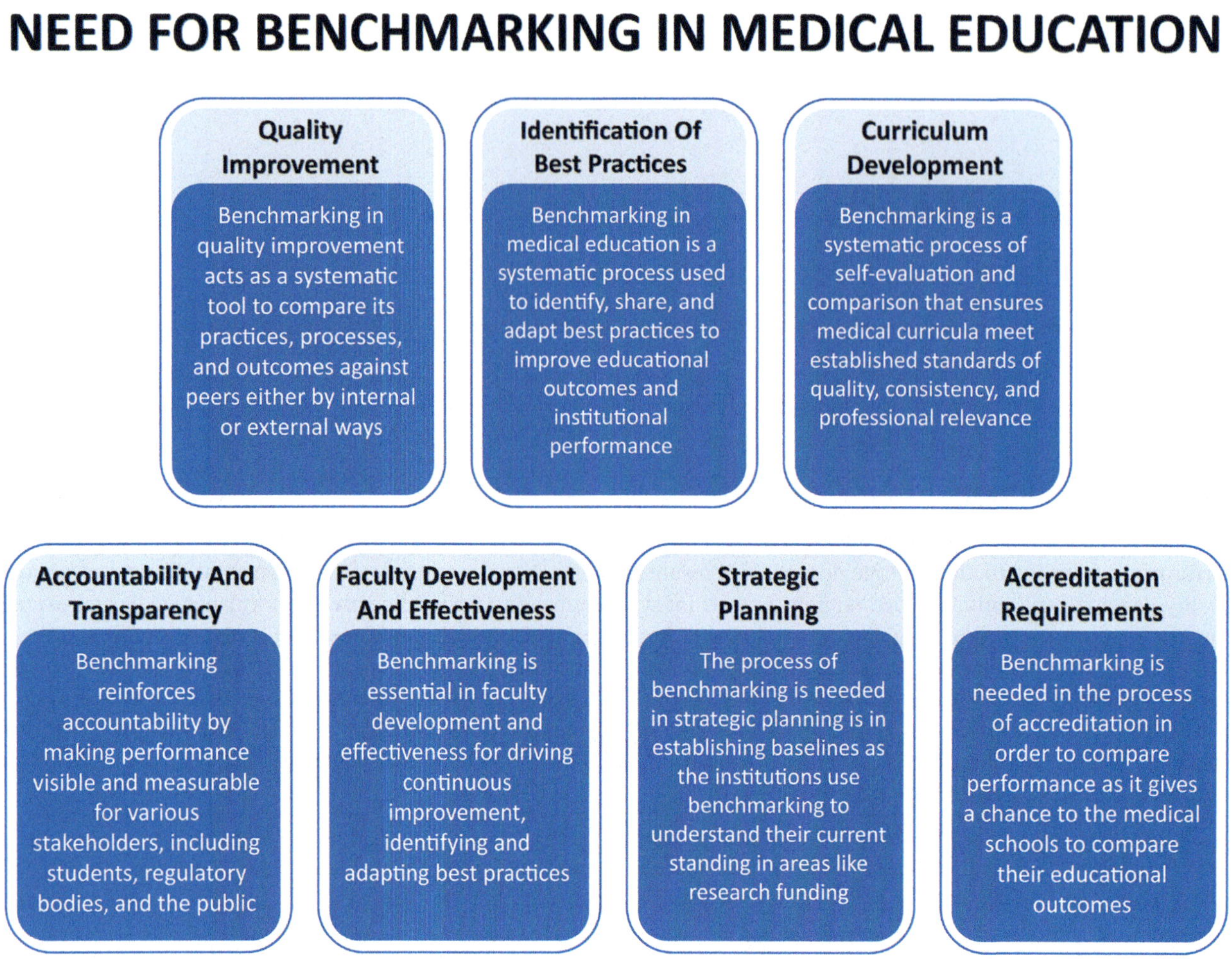


FIGURE 31.1 Need for benchmarking in medical education.

with set benchmarks or high-achieving peers. It helps institutions to pinpoint specific educational shortcomings and best practices by comparing internal data, such as student board scores and clinical competency ratings, with regional or worldwide counterparts.

### 31.2.3 Curriculum development

Benchmarking in medical education curriculum development is a vital procedure for ensuring that training programs satisfy defined standards of quality, relevance, and in some circumstances, international comparison. It provides a data-driven framework for planning, assessing, and revising courses to guarantee graduates are sufficiently equipped for the workforce.

### 31.2.4 Accountability and transparency

Benchmarking is a strategic method used in medical education to encourage continuous quality improvement by comparing institutional performance to peers, industry leaders, or established standards. Benchmarking is essential for accountability and transparency because it provides an unbiased, data-driven framework for comparing performance to peers or predetermined criteria. This strategy moves performance measurement from just an intuitive feeling to reality and holds organizations accountable for their actions and outcomes. Benchmarking improves accountability by making performance measurable and apparent to a range of stakeholders, including the public, regulatory bodies, and students. Establishing accountability to society and promoting transparency in training standards require benchmarking in medical education. In order to promote ongoing quality improvement, it entails evaluating institutional, program-level, and student performance measures against "best-in-class" peers or recognized international norms.

### 31.2.5 Faculty development and effectiveness

In medical education, benchmarking's function has grown from that of a diagnostic tool to that of a prescriptive force behind ongoing accountability and quality improvement. In order to drive, measure, and maintain quality improvements in teaching, research, and clinical care, benchmarking in faculty development and effectiveness within medical education is crucial. It offers a methodical, data-driven way to find gaps, implement best practices, and match departmental, institutional, and individual objectives with national or worldwide standards. In addition to its role in promoting ongoing development, identifying and implementing best practices, and ensuring that programs fulfill accreditation requirements, benchmarking also helps institutions maintain faculty development's relevance by monitoring trends like the shift towards digital technology, online learning, and competency-based assessment.

### 31.2.6 Strategic planning

Benchmarking is a data-driven methodology used in medical education strategic planning to assess institutional performance, set long-term objectives, and guarantee educational quality in comparison to local or international standards. Establishing baselines through benchmarking is essential for strategic planning because it allows institutions to assess their existing performance in areas such as research funding, faculty-to-student ratios, and graduation rates, which serve as the basis for well-informed planning. It also helps define appropriate goals by comparing metrics with peer institutions of similar size. This comparative process gives the institutional leaders the opportunity to set challenging but realistic goals that align with healthcare standards.

### 31.2.7 Accreditation requirements

Benchmarking is the practical technique used in medical education to convert abstract accrediting standards into quantifiable performance and quality indicators. In order to evaluate a medical school's performance, curriculum, and results to those of peer institutions or international norms, benchmarking is an essential, methodical process in medical education certification. It is necessary to go beyond simply fulfilling the basic standards in order to encourage ongoing quality improvement and guarantee that graduates have competencies that are globally competitive. As it allows medical schools to compare their educational outcomes, such as graduation rates, national exam scores, and student satisfaction, against peer institutions or national and international standards, benchmarking is essential to the accreditation process in order to compare performance.

## 31.3 Types of benchmarking

Benchmarking in medical education, used to improve curriculum, trainee performance, and institutional quality, consists of two primary types, namely, benchmarking by comparison and benchmarking by focus. These types mainly give an idea of what is compared or focused in the process of benchmarking. The main types and their components are described in Fig. 31.2.

### 31.3.1 By comparison

In order to identify strengths and weaknesses and implement best practices for improved quality, consistency, and student preparation, medical education institutions, programs, or individuals compare their performance metrics, such as student outcomes, curriculum, or assessment strategies, against peers or established standards, such as WFME. This process is known as benchmarking by comparison. To ensure graduates achieve high standards, it entails the methodical study of data, such as progress test scores or clinical reasoning abilities, frequently utilizing longitudinal or direct and indirect approaches to analyze changes and drive curriculum development.

#### *31.3.1.1 Internal benchmarking*

In medical education, internal benchmarking is the process of comparing performance metrics, procedures, and results within a single institution or network, such as various departments, cohorts, or years, in order to find internal best practices, identify variances, and promote ongoing quality improvement in administration, teaching, and assessment while promoting a culture of shared learning rather than rivalry. This process helps to standardize high-quality practices like assessment fairness or student support by learning from high-performing areas, staying consistent, and meeting accreditation standards (Altan et al., 2022). In order to increase validity and reliability, some of the most important

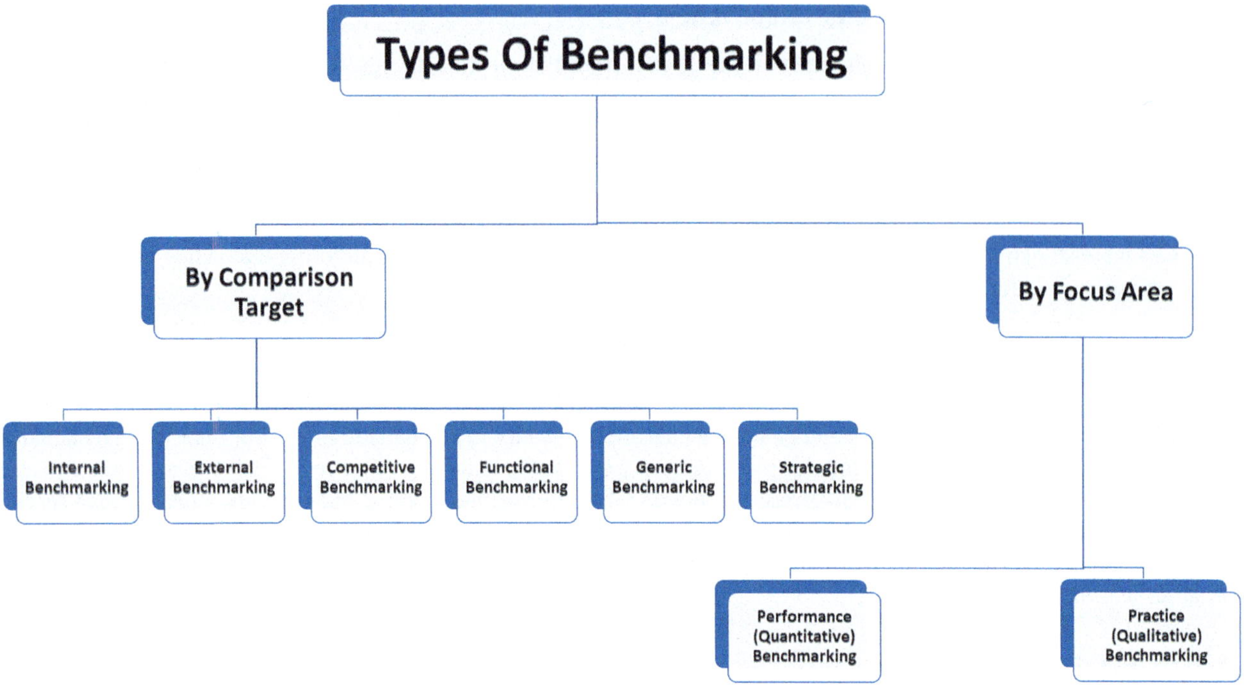


FIGURE 31.2 Types of benchmarking.

uses of this kind of benchmarking in medical education include quality assessment, which compares exam question difficulty, discrimination indices, or grading consistency across departments or faculty. In order to identify areas that require curricular improvement, it is also utilized to analyze student performance in various clinical rotations or preclinical disciplines. Comparing student satisfaction, remediation rates, or advising efficacy across student cohorts or support services is another benefit of this kind of benchmarking.

### *31.3.1.2 External benchmarking*

The systematic process of comparing an institution's performance, curriculum, and student results to those of peer organizations or national or international standards is known as external benchmarking in medical education. It functions as a tool for quality improvement to make sure that educational standards satisfy international standards and to pinpoint areas that need pedagogical development. Different kinds of comparable frameworks are used in external benchmarking. Process benchmarking, which examines specific teaching techniques, course completion cycle lengths, and evaluation efficacy, is a type of external benchmarking (Halm, 2025). This strategy identifies bottlenecks in curriculum design, testing, or teaching, allowing programs to implement best practices that increase student results and operational efficiency.

### *31.3.1.3 Competitive benchmarking*

Medical schools and teaching hospitals assess their training programs, results, and procedures by contrasting them with those of peer institutions or industry leaders through a strategic process known as competitive benchmarking. Here, finding performance gaps and implementing best practices to promote ongoing quality improvement are the main objectives. This strategy goes beyond internal evaluation to incorporate best practices from high-achieving institutions in an effort to improve graduation outcomes, strengthen academic standing, and guarantee adherence to accrediting requirements.

### *31.3.1.4 Functional benchmarking*

Medical education institutions use functional benchmarking which is a strategic approach that compares their own internal functions or processes to high-performing companies with the goal of identifying and implementing best practices. They frequently look outside of the immediate medical sector to find best practices. One of the key characteristics of this kind of benchmarking is that it involves cross-industry comparison. Functional benchmarking in medical education involves comparing specific processes or functions within medical institutions to identical or similar practices in organizations outside the healthcare industry to identify and adopt best practices for quality improvement. Here, the primary emphasis is on the execution of a certain activity rather than the end result, like test scores for students. It looks at systems, workflows, and management strategies. In order to guarantee a smooth "customer" experience for potential students, medical schools may compare their application and enrollment procedures to those of elite business schools or the hospitality sector. This is one way that functional benchmarking is applied in medical education.

#### *31.3.1.5 Generic benchmarking*

Comparing generic workflows and processes in one healthcare facility to generic practices in other industries is known as generic benchmarking. For instance, here the medical institution can compare its methods for gathering patient input with customer feedback systems and strategies used in the food service sector. In medical education, scheduling and resource management are two areas where generic benchmarking can be applied. In this situation, a commercial airline's logistics for handling frequent personnel and equipment rotations could serve as a baseline for a medical simulation center's scheduling system.

#### *31.3.1.6 Strategic benchmarking*

Strategic benchmarking in medical education is a voluntary process where institutions compare their curricula, assessment techniques, student results, and educational procedures to top-performing peers or recognized norms. Unlike basic data comparison, this strategy is oriented toward long-term quality improvement, identifying best practices, and creating cultural change. It assists medical schools in maintaining the competitiveness of their programs, ensuring that they meet worldwide standards, and producing graduates of the highest caliber. Strategic benchmarking's emphasis on macro-level concerns including resource allocation, long-term planning techniques, and institutional reputation rather than operational chores is one of its key characteristics. Program development is one of the most prevalent instances of strategic benchmarking, where medical schools utilize benchmarks to make sure their curricula continue to be on par with those of elite universities in terms of quality and equivalency.

### 31.3.2 By focus area

In order to drive quality improvement, identify best practices, and set goals for programs or institutions, benchmarking in medical education by focus area entails comparing performance, practices, and outcomes in particular domains, such as student assessment, professionalism, curriculum, or simulation, against established standards or top performers. Concentrating on certain areas will give an opportunity to medical education institutions to go beyond overall rankings to identify practical enhancements for improved quality, efficiency, and student outcomes which are in line with more general healthcare benchmarking objectives.

#### *31.3.2.1 Performance (quantitative) benchmarking*

In order to promote ongoing quality improvement, quantitative benchmarking in medical education entails systematically comparing educational procedures and results to predetermined benchmarks or peer performance. This approach translates qualitative goals into quantifiable data to facilitate objective comparison. The key performance indicators (KPIs) are the foundation of this kind of benchmarking. The performance of students in national tests, where medical schools regularly compare their students' scores to national averages to assess the efficacy of their curriculum, is one of the main instances of quantitative benchmarks in medical education.

#### *31.3.2.2 Practice (qualitative) benchmarking*

Instead of depending just on quantitative measurements, qualitative benchmarking in medical education aims to raise institutional standards and teaching quality by comparison with best practices. In order to promote ongoing development, it pinpoints the "how" and "why" underlying excellent educational results. Finding the best practices by comparing institutional procedures to recognized international standards, like those set by the WFME, is one of the key elements of qualitative benchmarking. In addition, by pointing out differences between intended standards and present educational practices in areas like professionalism, curriculum design, and assessment, it also aids in gap analysis. Determining and evaluating professionalism by comparing student and faculty expectations to regulatory guidelines is a typical example of qualitative benchmarking in medical education.

## 31.4 Tools for benchmarking

In medical education, benchmarking tools include data-driven platforms for gathering and comparing performance metrics, such as the CHKS data analysis toolkit (DAT) and MGMA data for practice operations, as well as standardized assessment tools, such as the objective structured clinical examination (OSCE), mini-clinical evaluation exercise (Mini-CEX), and directly observed procedural skills (DOPS) to assess clinical competence. While acute physiology and chronic health evaluation (APACHE) and simplified acute physiology score (SAPS) models are utilized for critical care outcomes, specialized tools like the World Health Organization (WHO)'s global benchmarking tool (GBT) concentrate on regulatory systems and health emergency preparedness (Fig. 31.3).

### 31.4.1 Standardized assessment instruments

#### *31.4.1.1 Objective structured clinical examination (OSCE)*

The OSCE is a standardized instrument with rating scales and checklists to guarantee dependability that evaluates a variety of clinical and communication skills using several stations. In medical and healthcare education, OSCEs are an essential benchmarking tool that offer a systematic framework for assessing clinical competency. Institutions can objectively evaluate practical skills, including communication, physical examination, and clinical reasoning, against predetermined performance goals by rotating candidates through timed, task-oriented stations. Moreover, in order to assess whether students are attaining the clinical skill levels required for professional licensing, educational institutions also use OSCE data to evaluate their curricula against national standards (Artyukhov et al., 2023).

#### *31.4.1.2 Mini-clinical evaluation exercise (Mini-CEX)*

This is a technique for assessing clinical competency in particular areas. Mini-CEX makes benchmarking easier by providing systematic, objective, and consistent performance statistics of trainees in real-world clinical contacts. This allows for comparison against predefined standards and the identification of program, institutional, and individual strengths and weaknesses. Bias is reduced by the use of many observers and a standardized grading system, and progress is tracked over time and targeted improvement is made possible by the quick, targeted feedback.

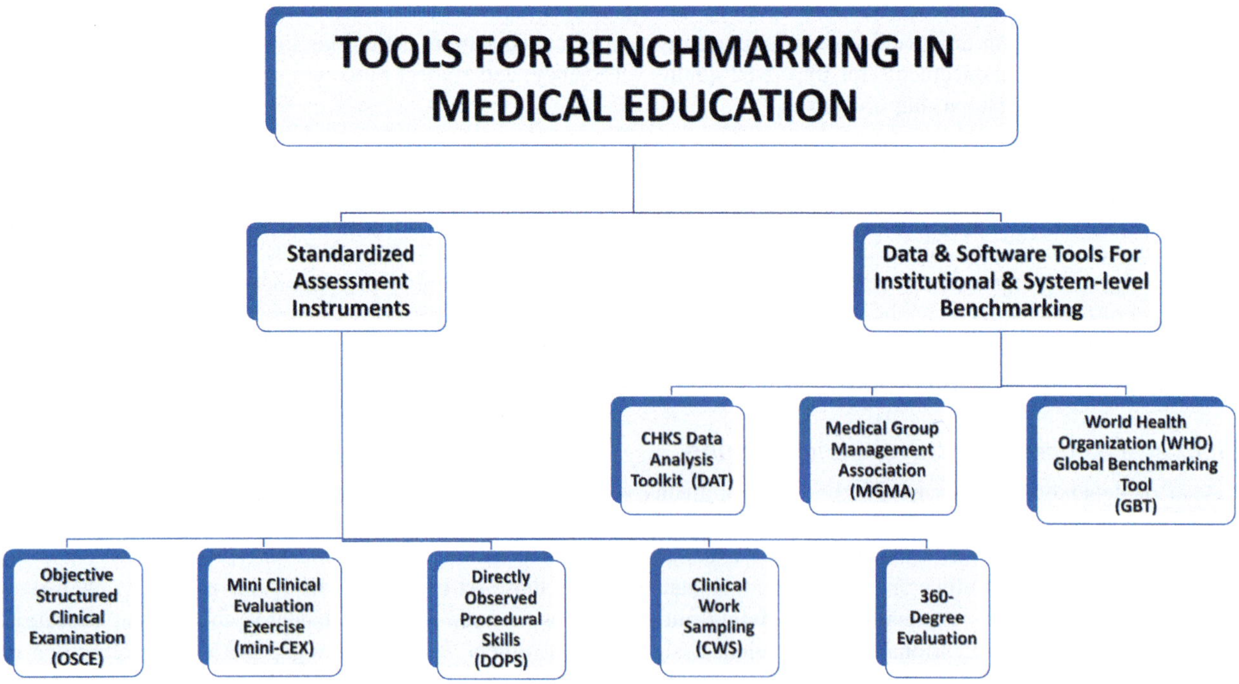


FIGURE 31.3 Tools for benchmarking in medical education.

#### 31.4.1.3 Directly observed procedural skills (DOPS)

DOPS serves as an instrument for evaluating the effectiveness of particular technical processes. DOPS are useful for benchmarking because they provide an organized, objective method of assessing and comparing learners' performance in practical activities against predefined criteria. DOPS makes it possible to assess individual growth and compare competency levels among various trainees, programs, or institutions by recording performance with thorough, consistent feedback. The process of DOPS ensures that trainees satisfy predetermined competency criteria prior to advancing in their training. Furthermore, this workplace-based evaluation makes it easier to identify skill gaps and offers a quantitative foundation for quality improvement (Collier et al., 2006).

#### 31.4.1.4 Clinical work sampling (CWS)

Another recognized technique for collecting clinical competence is CWS. It is an observational time-study methodology used in medical education to examine how clinicians and trainees allocate their time among different tasks. It provides a reliable, impartial, and practical tool for comparing efficiency, performance, and educational experience. CWS can be used to compare the clinical exposure of trainees at different sites, ensuring consistent educational experiences One use in healthcare benchmarking is CWS, which calculates the proportion of time spent on direct patient care, such as bedside operations, compared to administrative or unit-related tasks. Additionally, it can assist in comparing the amount of time a doctor spends on direct care versus administrative duties. In order to find performance gaps, hospitals also utilize it to compare their clinical workflows with those of best-in-class companies. Instead of using subjective summaries, CWS is utilized in educational evaluation to compare student performance to established behavioral objectives.

#### 31.4.1.5 The 360-degree evaluation

This is a technique that collects input from various sources to offer a thorough evaluation of a person's performance. A 360-degree evaluation is an effective benchmarking technique that collects anonymous feedback on an individual's performance, skills, and behaviors from a range of sources, including colleagues, supervisors, reports, clients, and others. It creates a comprehensive, multiperspective view that reveals strengths, weaknesses, and blind spots, enabling customized development plans that compare current performance to organizational needs and best practices for growth. It goes beyond standard top-down appraisals by helping people evaluate how they are regarded versus how they see themselves, identifying leadership potential, and providing data for trends and strategic workforce planning (Berk, 2009).

### 31.4.2 Data and software tools for institutional and system-level benchmarking

#### 31.4.2.1 CHKS data analysis toolkit (DAT)

The CHKS DAT makes medical education benchmarking easier by allowing institutions to undertake custom data analysis to assess performance against peer groups or national standards, identify areas for improvement, and encourage data-driven decisions for enhanced quality and efficiency. It allows administrators, doctors, and performance specialists to conduct in-depth studies of operational and clinical performance, going beyond conventional dashboards.

#### 31.4.2.2 Medical Group Management Association (MGMA) data

The MGMA data offers fundamental standards and KPIs for the operations of medical practices, encompassing measurements such as patient flow, productivity, and satisfaction. It is useful for benchmarking in medical education because it provides industry norms for physician compensation, productivity, and practice operations. This enables institutions to compare their own academic faculty and programs to top-performing groups and the general market. Academic institutions can find inequalities, improve pay and incentive programs to draw and keep faculty, maximize productivity in teaching and research, and assess the financial sustainability of their curricula in the context of actual medical practice by examining these metrics.

#### 31.4.2.3 World Health Organization (WHO)'s global benchmarking tool (GBT)

It is used to evaluate health emergency preparedness capabilities and national medicinal product regulation systems. The WHO's GBT is useful for benchmarking medical education since it provides a platform for assessing and improving the regulatory frameworks controlling medicinal products. This indirectly improves medical education outcomes by promoting a strong, open, and efficient public health regulatory environment. Using maturity levels to identify strengths and weaknesses, the GBT assesses both broad regulatory frameworks and particular functions, such as clinical trial oversight. It also guides the creation of institutional development plans to increase capacity and encourages good practices and transparency within national regulatory authorities. The GBT guarantees the quality and safety of medical products by creating a consistent global standard for regulatory systems, which is essential for experts to utilize them sensibly and for the general public to effectively access them.

### 31.4.3 Frameworks for benchmarking

Benchmarking frameworks for medical education provide structured ways to compare learning activities and outcomes to predefined standards or best practices. In order to improve quality, medical education benchmarking frameworks like the CANMEDS and NCAAA standards are used to compare curriculum, assessment techniques, and institutional performance. In order to guarantee consistent trainee preparation and academic achievement, these tools frequently concentrate on competency-based outcomes that are in line with worldwide standards. For example, competency-based frameworks guarantee accurate evaluation of abilities, knowledge, and attitudes while defining, measuring, and benchmarking trainee readiness for progress.

## 31.5 Roles and responsibilities of stakeholders in benchmarking

With roles ranging from executives setting vision and allocating resources to employees or teams providing operational details and customers or partners providing external perspectives, stakeholders in benchmarking define goals, provide critical insights, validate data, and drive implementation. Together, they work to identify gaps, foster innovation, and ensure actionable, relevant improvements. Their combined participation guarantees support and effective change.

### 31.5.1 Internal stakeholders

Internal stakeholders in medical education, comprising faculty, administrators, students, and clinical staff, execute a crucial role in benchmarking since they help in establishing performance criteria, gathering data, pinpointing areas for quality improvement and putting changes into place. In addition, the internal stakeholders play a very crucial role in promoting advancements in curriculum, evaluation, and educational results since they compare internal metrics to peers or national standards (Mohajeri et al., 2009).

#### *31.5.1.1 Institutional leadership*

Using data-driven criteria to enhance educational quality, social accountability and operational efficiency is the primary focus of institutional leadership, such as deans and designated institutional officials. Acting as catalysts, leaders go above their administrative responsibilities to establish goals that match institutional success with the demands of community health. In the domain of strategic visioning, one of the key responsibilities of institutional leadership is defining institutional components to be benchmarked, such as faculty productivity and student achievement, based on similarities with peer institutions and the most recent international trends. Additionally, they support QA by ensuring that benchmarking is used for both compliance and ongoing research and instructional development (Alosani et al., 2016).

#### *31.5.1.2 Faculty and educators*

As they move from traditional teaching to data-informed positions that guarantee institutional quality and learner competency, faculty and educators are the main forces behind benchmarking in medical education. As educators use big data and learning analytics to find performance gaps in both individual students and entire cohorts, they mostly serve as diagnostic assessors. This enables them to assess if the current curriculum satisfies national or international standards and customize training. They also serve as standard-setters and curators of content. In this sense, educators choose and arrange excellent, evidence-based resources that conform to accepted educational standards rather than merely producing content. They work together across schools to establish what constitutes appropriate performance at particular benchmarks, including professionalism upon graduation.

#### *31.5.1.3 Students*

Students are an essential component in medical education benchmarking because they provide the end-user perspective, encourage institutional quality improvement through performance statistics, and actively engage in self-directed learning. In the end, they reflect the efficacy of programs against established standards and best practices to guarantee excellence and comparability. Students participate in the benchmarking process by contributing data that medical schools use to compare themselves to internal objectives or external institutions. These metrics include academic achievement, skill acquisition, and attainment levels. Additionally, interns and students recognize their own needs and pursue learning through practice-based learning and reflection, which is a sign of a robust educational process.

#### *31.5.1.4 Quality assurance (QA) teams*

In medical education, QA teams are strategic facilitators that go beyond basic compliance to promote excellence through methodical benchmarking. They are responsible for creating, keeping an eye on, and assessing standards in relation to both domestic and global benchmarks. Establishing standards and framework development are two of the main tasks of QA teams in benchmarking. These teams help create academic standards based on subject benchmark statements, which serves as a benchmark for QA and accreditation in medical education by outlining the fundamental knowledge, abilities, and competences that graduates must have. They guarantee that these frameworks are in line with global norms, like those set forth by the WFME.

### 31.5.2 External stakeholders

External stakeholders are crucial to benchmarking in medical education because they provide an objective, useful perspective on the quality, applicability, and outcomes of training programs. Medical schools can better align their standards and curricula with the actual needs and expectations of society and the broader healthcare system with the help of this input. When it comes to QA and accreditation, external stakeholders like professional associations are crucial since they use data from official quality reviews and compliance requirements. Their assessments aid in ensuring that the educational and clinical requirements are being met by the institutions. Curriculum designers can also analyze current market trends, necessary graduate skills, and performance gaps with the support of external stakeholders such as local companies, professional groups, and practicing physicians. This guarantees that graduates of medical school are qualified to address the demands of modern society and healthcare.

#### *31.5.2.1 Patients and the community*

Patients and the community play a critical role in medical education benchmarking by providing a sincere, practical perspective that ensures medical education is socially responsible, patient-centered, and relevant to actual healthcare requirements. Their participation aids in establishing standards that go beyond academic requirements and encompass humanistic traits, communication abilities, and an awareness of various social circumstances. Establishing curriculum standards and objectives, wherein patients and community members assist in identifying the fundamental abilities and attributes that future physicians should have, is one of the crucial responsibilities in benchmarking and quality improvement. Patients and community members help choose graduation skills by participating in steering committees and consultative workshops to establish desired qualities and competencies in graduates, such as empathy, respect, and cultural awareness.

#### *31.5.2.2 Health and medical organizations*

Health and medical organizations, which are the main employers, play a crucial role in benchmarking in medical education by highlighting the practical outcomes, quality, and applicability of graduate talents to actual healthcare needs. This entails ensuring that training improves patient outcomes, successfully controls healthcare costs, and complies with industry standards. Establishing the standards and competencies that certified physicians and other healthcare professionals need to have is the primary duty of employers in benchmarking. Their involvement guarantees that medical education programs generate graduates with the abilities and information needed for successful practice in contemporary healthcare environments. They also help by offering comments on the performance of graduates. In this regard, medical organizations give medical schools vital input on the efficacy of their curricula by assessing the performance of recent graduates. Medical schools can find gaps and opportunities for improvement with the use of this feedback loop.

#### *31.5.2.3 Government and policymakers*

In medical education benchmarking, the government and policymakers play a crucial role in providing leadership, establishing regulatory standards, guaranteeing accountability, and utilizing comparable data to promote ongoing quality improvement. They use their special position as financiers, regulators, and frequently healthcare providers to influence the direction and caliber of medical education in order to address the health requirements of the general public. Some of the most important roles that the government and policymakers play in the context of benchmarking are setting quality standards and governance, whereby governments establish the minimal requirements for medical education through regulatory and accreditation procedures and working with independent accrediting bodies. This ensures that all medical programs meet a minimal standard of quality and that graduates are qualified to practice medicine. Policymakers create the governance structures in educational institutions to provide transparency and accountability in decision-making.

## 31.6 Process of benchmarking

In order to find gaps and promote ongoing quality improvement, benchmarking in medical education is a methodical process that compares your program's performance in terms of curriculum, outcomes, and teaching methods against established standards or top-performing peers. This process usually entails planning, data collection and analysis, identifying best practices, and putting changes into place to ensure better educational quality and trainee preparedness. These components are built on the core principles of improving quality, patient safety, and overall performance through data-driven insights and continuous improvement cycles. The key steps in benchmarking is depicted in Fig. 31.4.

Benchmarking involves the following key steps:

1. Planning,
2. Data collection,
3. Data analysis,
4. Implementation,
5. Monitoring.

### 31.6.1 Planning

As a tool for ongoing quality improvement, benchmarking in medical education is planned using a meticulous, multistep process to find, implement, and modify best practices. The initial step in this approach is identifying the specific areas that need improvement, such as curriculum design, student evaluation methods like OSCEs or portfolio utilization, or specialized clinical skills training. The scope of the clarification of what needs to be benchmarked will be established by identifying specific, high-priority areas of medical education that need improvement, such as faculty performance, clinical simulation, or student development. It is also essential to assemble a team to lead the process, which should include important stakeholders including academic leaders, educators, and, in certain situations, students and clinical professionals, and to secure executive support in order to ensure that the necessary resources are available. In order to assure useful data comparison, it is essential to establish peer groups that identify whom to benchmark. This is accomplished by choosing suitable comparable institutions, such as teaching hospitals or medical schools. Setting relevant and trustworthy key performance measures, like as graduation rates, faculty-to-student ratios, or curriculum mapping metrics, is also essential.

### 31.6.2 Data collection

Data collection in medical education benchmarking refers to a structured process used to compare institutional performance against peers or national standards to drive quality improvement. It involves gathering both quantitative and qualitative data on curriculum, student performance, and faculty, enabling institutions to identify strengths and weaknesses to inform evidence-based, strategic decisions. Determining the pertinent KPIs and the particular data needed after identifying the precise procedures or results to benchmark, such as patient wait times, death rates, and curriculum delivery efficiency, is also vital in this stage. It is also important to ascertain whether the data is obtained from internal or external sources. When it comes to internal data, it is essential to gather performance measurements and information from within the institution, frequently from administrative databases, incident reports, and electronic health records. However, in terms of the external data, it must be collected from designated benchmarking partners, which may include national databases or other organizations. Additionally, it is advisable to employ techniques that guarantee consistency and accuracy, such as surveys, interviews, standardized forms, or automated data extraction using software and preexisting databases.

### 31.6.3 Data analysis

In order to find performance gaps and opportunities for development, the data analysis step in medical education benchmarking entails a methodical process of comparing gathered performance data against a predetermined standard or

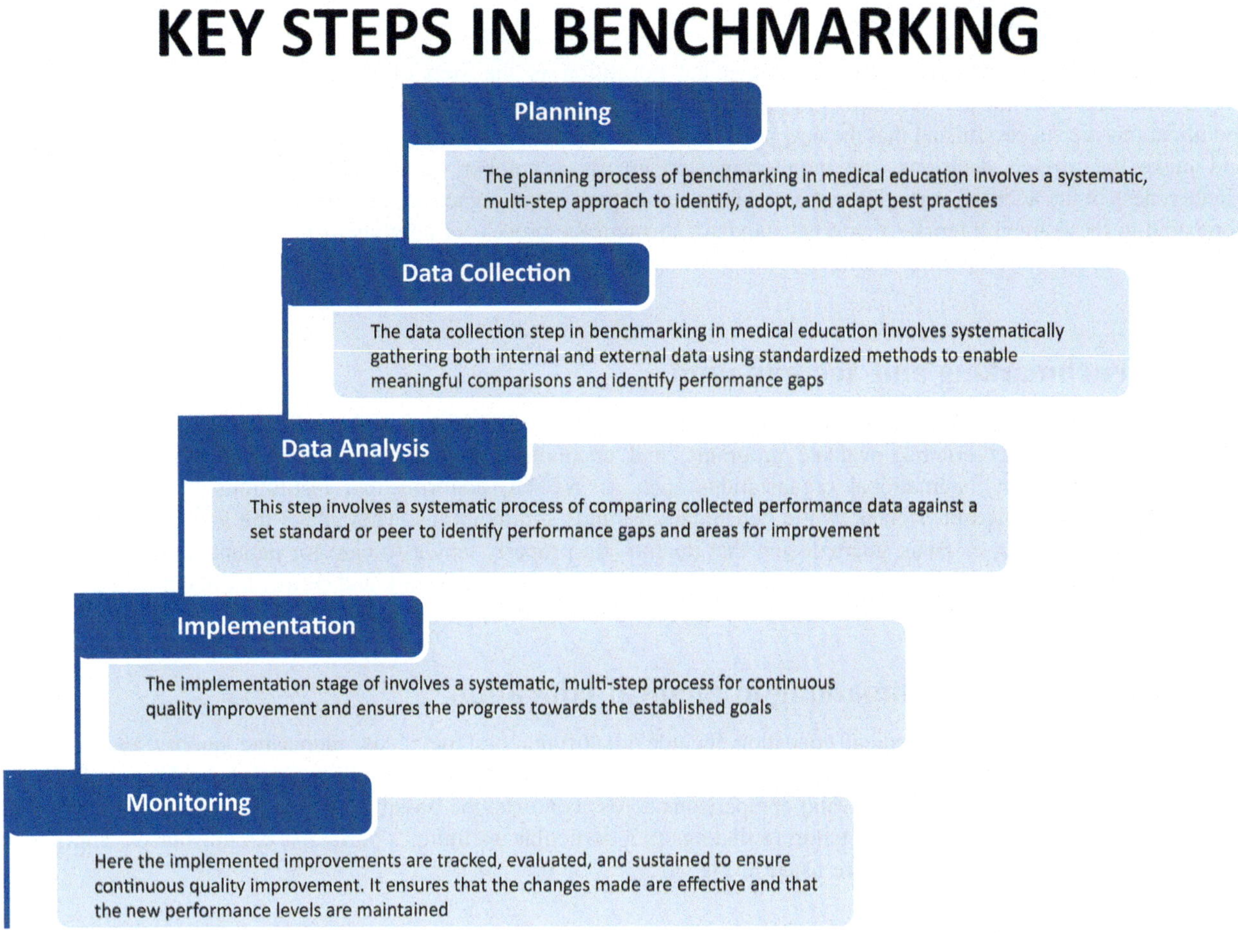


FIGURE 31.4 Key steps in benchmarking.

peer. Data analysis is a critical, proactive component of benchmarking in medical education, transforming raw institutional data into actionable insights for QA, curriculum development, and student support. It involves a structured process of collecting, cleaning, and modeling data such as board scores, admission metrics, and student feedback. Additionally, it enables a performance comparison between the institution's measurements and the predetermined benchmarks. This comparison aids in determining how performance compares to the best in the company as well as comparing current performance to a certain reference point. For instance, one of the uses of data analysis in medical education is in curriculum and performance, wherein a comparison is made of student performance on licensing exams against national averages to identify gaps in educational delivery.

### 31.6.4 Implementation

Implementing benchmarking in medical education is a continuous, structured process designed to improve curriculum, assessment quality, and student learning outcomes by comparing performance against peer institutions, national standards, or best practices. Successful implementation requires a shift from passive data collection to active, evidence-based improvements in educational programs. Analyzing performance gaps in terms of identifying areas of strength and weakness and their underlying causes is crucial before beginning the implementation process. This can be accomplished by contrasting benchmark data from peer businesses with the collected internal data. To assure widespread support and a consistent vision for quality improvement, the results and action plans must subsequently be shared with all essential stakeholders, including academics and management. Linking the benchmarking data to action plans is also crucial for the implementation strategy to be successful. This entails determining gaps, creating interventions, and keeping track of developments.

### 31.6.5 Monitoring

The critical last stage of benchmarking in medical education is the step of monitoring, where the implemented improvements are monitored, assessed, and maintained to guarantee ongoing quality improvement. It guarantees that the modifications are successful and that the new performance standards are upheld or surpassed. In order to measure progress and make data-driven decisions, the monitoring step entails a number of crucial tasks. In order to evaluate the effectiveness of the adopted strategies and ascertain whether performance gaps are decreasing, continuing data is routinely compared to the defined internal or external standards through the process of continuous monitoring. Additionally, if the intended outcomes are not being attained, it directs data-driven decisions and the necessary modifications to the action plans.

## 31.7 Benchmarking and accreditation

Benchmarking and accreditation in medical education are crucial methods for ensuring quality, comparability, and patient safety because they set criteria, evaluate programs, and encourage ongoing development. Accreditation, which is sometimes overseen by international organizations such as WFME, confirms that institutions fulfill the necessary requirements for qualified physicians, while benchmarking finds shortcomings. They make the shift from basic QA to data-driven quality improvement, guaranteeing that doctors are properly taught to care for patients and assist a skilled, adaptable worldwide workforce. While accreditation provides the broad framework and certification, benchmarking helps medical schools satisfy accreditation standards by educating and supporting the process.

## 31.8 Benefits of benchmarking in medical education

Benefits of benchmarking in medical education include identifying best practices, increasing quality and efficiency, encouraging reflective practice, and providing data for strategic decisions. However, there are drawbacks as well, such as implementation complexity, ensuring valid comparisons across contexts, the possibility that interventions won't result in better outcomes, and the challenge of tailoring findings to a particular institution's particular circumstances. Some of the important benefits of benchmarking are listed in Fig. 31.5.

### 31.8.1 Quality improvement

Benchmarking in medical education fosters quality improvement and ultimately results in a culture of continuous learning and better-prepared physicians by exposing best practices, identifying curriculum or competency gaps, encouraging innovation through comparison with leaders, setting reasonable goals, standardizing procedures, and improving patient outcomes by matching training with high-performing peers. Benchmarking allows educators to identify best practices by comparing educational programs, instructional strategies, and outcomes such as resident performance, exam scores, and clinical competencies to top-performing institutions. Setting realistic objectives for program quality, student competency, and operational efficacy is another benefit of the data collected throughout the benchmarking process (Brah et al., 2000).

### 31.8.2 Identification of best practices

The systematic discovery, comprehension, and adoption of best practices from high-performing peers or industry leaders is one of the main advantages of benchmarking in medical education. This procedure enables focused quality

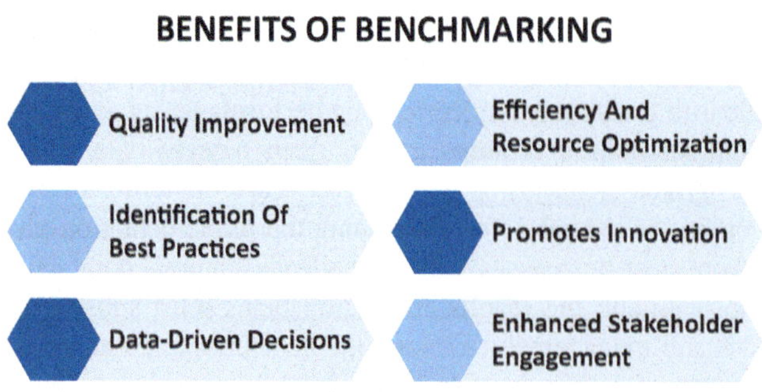


FIGURE 31.5 Benefits of benchmarking.

improvement and innovation by going beyond mere comparison. Benchmarking offers a data-driven way to evaluate an institution's performance metrics such as student results, the effectiveness of educational procedures, and the adoption of new teaching technologies along with those of other programs that have been successful. This comparison identifies particular areas in which an institution might be performing well or poorly. Additionally, medical educators might research the particular tactics, procedures, and treatments that produce better outcomes by identifying best-in-class healthcare institutions. Understanding the fundamental reasons for high performance rather than merely the performance metric itself is a key component of this learning process.

### 31.8.3 Data-driven decisions

Benchmarking assists medical education in making data-driven decisions by providing a systematic, evidence-based framework for identifying performance gaps, understanding best practices, and efficiently allocating resources. Data-driven decisions are made possible through benchmarking. Teachers can objectively identify areas of excellence and areas that need improvement by allowing the identification of strengths and weaknesses in which an institution's performance indicators are compared against internal, national, or peer standards. As a result, rather than depending only on intuition or speculation, decisions are made utilizing a more exacting, intentional process. Additionally, benchmarking transforms raw data into useful and relevant knowledge. For example, if data indicates that a specific teaching strategy raises board test scores in peer institutions, educators can use that information to guide curricular changes.

### 31.8.4 Efficiency and resource optimization

Benchmarking boosts productivity and optimizes resources in medical education, as it identifies best practices, optimizes operational workflows, and guarantees that resources are used wisely. Additionally, it aids in rerouting funds to projects that provide greater returns and increased productivity. Benchmarking enhances operational efficiency, as it identifies areas where procedures can be enhanced, automated, or streamlined, benchmarking. Adopting best practices, which is made easier by the benchmarking process and allows educators to learn from top performers and comprehend how they obtain superior results, is one of the main advantages in this respect. The process of benchmarking is also beneficial as it allows an organization to use tried-and-true strategies instead of starting from scratch, thereby reducing unnecessary work and wasted resources (Housawi et al., 2020).

### 31.8.5 Promotes innovation

Benchmarking encourages innovation in medical education by drawing attention to performance gaps, encouraging the adoption of best practices, and fostering a culture of continuous improvement. Benchmarking promotes transformation and opens the door to new tactics. It breaks organizational complacency by enabling instructors to observe how peers or great achievers attain higher results. It also encourages teachers and administrators to come up with original ideas and implement cutting-edge instructional techniques or technical innovations. Employee participation in the benchmarking process fosters collaboration, communication, and a common understanding of anticipated standards, all of which contribute to the development of a culture that is open to fresh concepts and creative behaviors. Consequently, it fosters an inventive culture.

### 31.8.6 Enhanced stakeholder engagement

Increased stakeholder engagement is a significant advantage of benchmarking in medical education since it leads to more relevant, comprehensive, and ultimately effective quality improvement measures. Institutions can make sure that their programs meet real needs and get the support required for change by integrating a variety of viewpoints. Students, teachers, patients, and community members are stakeholders with distinct perspectives on what makes a reliable, excellent, and beneficial educational or medical outcome. Their contributions aid in the development of evaluation criteria that take into account real-world issues and patient outcomes in addition to academic metrics. Additionally, stakeholders are more likely to trust and apply the results when they feel heard and respected throughout the review process. As a result, best practices found through benchmarking are adopted more widely.

## 31.9 Challenges of benchmarking in medical education

### 31.9.1 Methodology and implementation

The complexity and variety of healthcare systems, problems with data collection, accuracy, and sharing, the absence of defined measures, and contextual considerations are the main obstacles to the methodology and application of benchmarking in medical education. It is well established that medical education and healthcare systems are intrinsically complicated and context-dependent, which makes it challenging to establish performance measures that are comparable and generally applicable. The notion of excellent practice could range greatly between departments or even organizations. Additionally, it is difficult to undertake systematic comparisons or meta-analyses due to the prevalence of heterogeneity in study design and reporting, which typically results in a wide range of study designs and reporting standards in the body of existing literature.

### 31.9.2 Contextual differences

Contextual considerations present a significant challenge for benchmarking in medical education since shifts in local settings may impact performance outcomes and the implementation of best practices across institutions. It is important to understand that what is advantageous in one situation may not be in another for a number of reasons. One of the primary contextual issues is diverse learning environments. It is true that medical education takes place in a variety of environments, ranging from major university hospitals to tiny, neighborhood clinics, each with unique organizational cultures, patient populations, and resources. Inaccurate inferences about institutional performance may result from benchmarking results without taking this natural diversity into consideration.

### 31.9.3 Measuring impact

Long time lags, a lack of standardized measurements that fully capture the scope of patient care and societal impact, and the difficulty of attributing specific outcomes to educational initiatives are the main reasons why measuring impact poses substantial obstacles for benchmarking in medical education. In this sense, defining and assessing impact presents a significant challenge. Since there isn't a single, accepted definition of impact in this setting, the emphasis is on easily quantifiable but less significant metrics like skill retention or confidence rather than observable effects on patient outcomes like morbidity or complication rates.

### 31.9.4 Adapting findings

Adapting results from benchmarking exercises is a significant challenge in medical education because of significant factors such as cultural and contextual differences, the difficulty of obtaining and comparing consistent data, and the inherent complexity of connecting educational outcomes to long-term patient care improvement. Medical education systems and institutional cultures are very different in terms of structure, governance, and ideologies. A best practice that works in one institution could not work in another due to differences in local clinical norms, faculty knowledge, and available resources. Finding similar and trustworthy data across several institutions is likewise exceedingly challenging. If not appropriately taken into consideration, differences in data gathering techniques, metrics definitions, and reporting time gaps may result in findings that are not accurate.

**Summary**

Benchmarking is a crucial management and quality-improvement technique used in the field of medical education. Benchmarking is the meticulous comparison of an organization's performance measurements, procedures, and results to industry best practices, peer organizations, internal standards, or recognized international standards such as those set forth by the World Federation for Medical Education. Ensuring ongoing quality improvement, establishing program comparability, and satisfying stakeholder expectations for educational standards are the main objectives. Establishing benchmarking is crucial for leaders in medical education to make sure their programs are competitive, of the highest caliber, and in line with the needs of the contemporary healthcare system. It advocates for a comprehensive, cooperative, and data-driven approach of ongoing performance improvement, going beyond mere comparison.

(cont'd)

**Key point**

Future perspective: fostering professional growth

Benchmarking in medical education is evolving from a static, annual evaluation to a dynamic, data-driven, longitudinal assessment that incorporates artificial intelligence. In an effort to close the gap between classroom knowledge and clinical proficiency, key trends show a shift away from basic ranking and toward offering formative, real-time information for ongoing professional growth. With a focus on competency and professionalism, it is indeed fast progressing towards positive behavior mapping. As a result, benchmarking will transition from straightforward retrospective reporting to real-time and predictive analytics on program and student success. Essentially, benchmarking in medical education will be more data-rich in the future, more concerned with the how and why of teaching, and deeply entwined with the objective of enhancing patient care and public health outcomes.

## References

Alosani, M.S., Al-Dhaafri, H.S., Yusoff, R.Z., 2016. Mechanism of Benchmarking and Its Impact on Organizational Performance. International Journal of Business and Management. 11, 172–172. Available at https://doi.org/10.5539/ijbm.v11n10p172.

Altan, D., Ahuja, V., Kelleher, C.M., Chang, D.C., 2022. Look in the mirror, not out the window: in favor of internal benchmarking. Annals of Surgery Open 3 (3), e184. Available at https://doi.org/10.1097/AS9.0000000000000184. PMID: 36199485.

Artyukhov, A., Gavurova, B., Volk, I., Bilan, S., Lyeonov, S., Mudarri, T., 2023. External assessment of medical education quality: indicative model development considering paradox of skill. Frontiers in Public Health 11. Available at https://doi.org/10.3389/fpubh.2023.1184861.

Berk R. A., 2009. Using the 360° multisource feedback model to evaluate teaching and professionalism. Medical Teacher 31 (12), 1073–1080. Available at https://doi.org/10.3109/01421590802572775.

Brah, S. A., Lin Ong, A., Madhu Rao, B., 2000. Understanding the benchmarking process in Singapore. International Journal of Quality & Reliability Management 17 (3), 259–275. Available at https://doi.org/10.1108/02656710010306157.

Collier, D.A., Collier, C.E., Kelly, T.M., 2006. Benchmarking physician performance, Part 2. Journal of Medical Practice Management 21 (5), 273–279.

Ettorchi-Tardy, A., Levif, M., Michel, P., 2012. Benchmarking: a method for continuous quality improvement in health. Healthcare Policy 7 (4), e101. Available at http://www.longwoods.com/product/download/code/22872.

Halm, M., 2025. External or internal benchmarking? AJN, American Journal of Nursing 125 (9), 59. Available at https://doi.org/10.1097/ajn.0000000000000144.

Housawi, A., Al Amoudi, A., Alsaywid, B., Lytras, M., bin Moreba, Y.H., Abuznadah, W., et al., 2020. A progressive model for quality benchmarks of trainees' satisfaction in medical education: towards strategic enhancement of residency training programs at saudi commission for health specialties (SCFHS). Sustainability 12 (23), 10186. Available at https://doi.org/10.3390/su122310186.

Mohajeri, K., Nayeri, M.D., Mashhadi, M.M., 2009. A Model for Stakeholder-Oriented Benchmarking Process. Journal of Applied Sciences 9. Available at https://doi.org/10.3923/jas.2009.237.247.

Pantall, J., 2001. Benchmarking in healthcare. NT Research 6 (2), 568–580. Available at https://doi.org/10.1177/136140960100600203.

Wilkinson, T.J., Hudson, J.N., Mccoll, G.J., Hu, W.C.Y., Jolly, B.C., Schuwirth, L.W.T., 2015. Medical school benchmarking – from tools to programmes. Medical Teacher 37 (2), 146–152. Available at https://doi.org/10.3109/0142159X.2014.932902.

Chapter 32

# Accreditation

*"The higher the standard of education in a profession, the less marked will be the charlatanism."*

—William Osler

## 32.1 Introduction

In medical education, accreditation is a formal quality assurance process conducted by independent bodies with an aim to evaluate curriculum, faculty, resources, and student assistance. It also ensures that medical institutions and programs provide high-quality training, produce competent and safe doctors, protect patients, and ensure graduates meet professional and international standards for practice and further training, such as residency (Abdalla, 2012). Quality assurance, a procedure that allows us to confirm that a medical school's curriculum, faculty, learning environment, and resources satisfy set educational standards, is one of the main components of accreditation. Additionally, it helps us maintain patient safety and guarantees that graduates are adequately equipped to practice medicine in a safe and efficient manner, thereby safeguarding the general public. It also entails creating and implementing standards towards infrastructure, faculty qualifications, curriculum, and assessment.

### 32.1.1 Definition

In medical education terminology, accreditation has numerous definitions. However, according to the World Health Organization, accreditation is "a voluntary peer review process designed to test the educational quality of new and established medical programs." In other words, accreditation is the certification of the suitability of medical education programs and of the competence of medical institutions in the delivery of medical education.

## 32.2 Need for accreditation

In medical education, accreditation is essential for maintaining patient safety, ensuring qualified, skilled physicians, and standardizing quality across institutions. It also ensures that graduates meet requirements for licensure, international mobility, and efficient care, all while promoting ongoing program quality improvement. It serves as a seal of approval, indicating that a medical school satisfies societal health needs and offers a strong basis for lifelong learning (Van Zanten et al., 2008). In terms of quality assurance, accreditation in medical education is essential for students because it ensures a standardized, high-quality education, ensures that their degrees and qualifications are recognized nationally and internationally for licensing and residency, protects their investment, and opens doors for international mobility and additional training (Blouin and Tekian, 2018). It encourages medical institutions to keep improving while reassuring the public that graduates are equipped to deliver safe, efficient care. Additionally, as many institutions favor degrees from reputable universities, graduates from accredited institutions are guaranteed good employment chances due to their greater legitimacy and improved job prospects.

From the medical institutions' point of view, accreditation is crucial for guaranteeing patient safety, raising the standard of care, winning over the public, and frequently for monetary and legal reasons including government funding and insurance reimbursement. It is a methodical, external peer-review procedure that assesses an organization's performance in relation to a set of predetermined criteria. Furthermore, accreditation gives healthcare companies a framework for meeting strict performance requirements, lowering medical errors, preventing infections, controlling hazards, and safeguarding patient safety.

**The Principles and Practice of Medical Education. DOI:** https://doi.org/10.1016/B978-0-443-44197-4.00024-2

## 32.3 Components of accreditation

A predetermined set of standards or criteria, an independent body's external peer-review procedure to evaluate adherence to these standards, and a system for ongoing improvement and correction based on the results are the essential elements of accreditation. Together, these elements foster organizational development, guarantee quality, and build trust across a range of industries, including healthcare and higher education. A few of the primary elements of accreditation are shown in Fig. 32.1.

### 32.3.1 Published standards

A standardized framework for assessing medical facilities is provided by published accreditation criteria, which act as standards to guarantee a minimal degree of safety and quality in the provision of healthcare. They promote ongoing quality improvement, foster public confidence, and provide a framework for management and organizational structure. The process of quality assurance and improvement is one of the main justifications for the stated standards since these criteria set predetermined standards and best practices that institutions must adhere to. The process of assessment against these criteria fosters a culture of continuous quality improvement that extends beyond the survey period by pointing out deficiencies and encouraging institutions to make adjustments. Following established recommendations also helps to ensure patient competence and safety by ensuring that the institution has effective systems for risk management, infection control, and appropriate medication management. It assists academic institutions in ensuring that graduates can practice safely and effectively (Boulet and van Zanten, 2014).

### 32.3.2 External review process

The external review procedure, which provides unbiased, independent assurance that an institution meets set standards for performance, safety, and quality, is a crucial part of medical institution accreditation. It serves as a solid foundation for continuous improvement, quality assurance, and public accountability. Since external reviewers, who are often experts in the field, and peer reviewers from other reputable organizations provide a fresh, unbiased perspective on the institution's operations and patient care procedures, the objective evaluation is one of the primary purposes of the external review

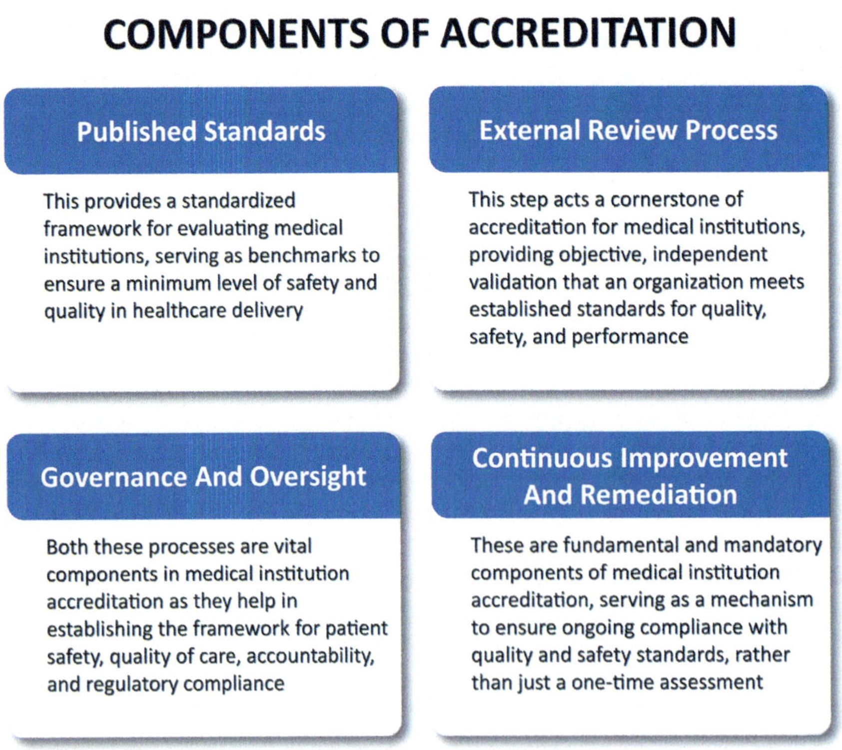


FIGURE 32.1 Components of accreditation.

process. This aids in identifying advantages and disadvantages that internal employees might miss because of internal prejudice or familiarity. Additionally, the procedure confirms that the medical facility complies with a predetermined set of strict, evidence-based national or international standards and regulatory criteria, including facility management, infection control, and patient safety procedures (Davis and Ringsted, 2006).

### 32.3.3 Governance and oversight

As they set the foundation for patient safety, care quality, accountability, and regulatory compliance, governance and oversight are essential to the certification of medical facilities. In the end, the governing board is in charge of establishing the organization's strategic direction and making sure it satisfies and surpasses accreditation requirements. In addition, the governing body establishes the organization's mission, vision, and values and makes sure that strategic plans and resources are in place to achieve high-quality, safe patient care, which is a fundamental requirement for accreditation. Good governance and oversight also lay the groundwork for strategic direction. Additionally, governance makes all employees from executives to frontline clinicians accountable for their contributions to quality and safety by establishing clear lines of authority and decision-making procedures at every level.

### 32.3.4 Continuous improvement and remediation

As the means to guarantee continuous adherence to quality and safety requirements rather than only a one-time evaluation, continuous improvement and remediation are essential and required elements of medical institution accreditation. They promote a culture of excellence and a proactive approach to both operational effectiveness and patient care. Remediation is also a vital component of accreditation which serves as a required quality-assurance method in health professions education and other regulated domains to guarantee that practitioners or trainees fulfill minimum competency requirements. While accreditation establishes benchmarks for excellence, remediation offers the means to assist those who have lagged behind in meeting those benchmarks, safeguarding the public and upholding professional integrity.

## 32.4 Frameworks of accreditation

A medical facility's accreditation entails a number of crucial frameworks and elements that guarantee quality control and ongoing development. Organizational structure, standards and criteria, and an evaluation procedure are usually included in these frameworks. In order to ensure the competence of medical professionals and patient safety, these frameworks specify particular requirements, create a legal and operational framework for the accrediting organization, and lay out a procedure for peer review, site visits, and continuous monitoring. The World Federation for Medical Education (WFME) and the National Accreditation Board for Hospitals and Healthcare Providers (NABH) have developed some of the most well-known frameworks of accreditation in medical education. These frameworks are organized procedures for assessing and certifying medical education programs to make sure they adhere to strict quality standards. The following are some of the fundamental ideas and functional components of accreditation frameworks:

### 32.4.1 Legal or regulatory framework

An essential and fundamental component of any accrediting structure is the legal or regulatory framework. To guarantee that recognized firms function lawfully, morally, and professionally, accreditation criteria are built upon, incorporate, and frequently surpass the minimal requirements specified by applicable laws and regulations. Legal and regulatory frameworks in medical education are crucial tools for preserving ethical norms, safeguarding public health, and ensuring high-quality care.

### 32.4.2 Accrediting bodies

As the highest level of an accreditation framework, accrediting bodies are autonomous, powerful organizations that formally acknowledge the ability and objectivity of other organizations, such as certification bodies, labs, or educational institutions, to carry out particular tasks or adhere to predetermined standards. Accrediting bodies in the field of medical education function as independent, approved organizations that guarantee institutional quality and safety. As a result, they increase patient safety by promoting ongoing development, building public trust, facilitating worldwide recognition, and guaranteeing that graduates are qualified to practice.

### 32.4.3 Standards and criteria

The broad, high-level principles of quality, such as mission, governance, and student support, are known as standards in an accreditation framework. The specific, quantifiable evidence or indicators that are used to assess how well a program or institution satisfies those standards, creating a roadmap for quality assurance and improvement, are known as criteria. Criteria specify the "how" to prove competence and obtain accreditation, whereas standards establish the "what." The standards and criteria act as benchmarks for evaluating the caliber, organization, and results of medical programs. They serve as a catalyst for continuous quality improvement and guarantee that organizations fulfill the prerequisites for producing physicians who are qualified, safe, and professional.

### 32.4.4 Process of accreditation

As part of a larger framework, the accreditation process usually entails steps like self-evaluation, external evaluation, and a final decision followed by ongoing monitoring. These steps are intended to confirm that an organization satisfies particular requirements for quality, capacity, and compliance, guaranteeing accountability and improvement through documentation and peer review.

### 32.4.5 Quality improvement

The primary driver of accreditation is quality improvement, which takes organizations beyond simple compliance to achieve sustained excellence by integrating cycles of assessment, change, and monitoring into their core operations. This ensures that standards aren't only met once but are continuously improved in areas like patient safety, processes, and outcomes. The systematic standards and evaluation that propel this quality improvement are provided by accreditation frameworks, which promote a culture of learning, more efficient use of resources, and better services.

### 32.4.6 Transparency and accountability

In accrediting systems, transparency and accountability are essential, integrated principles that promote public trust, guarantee quality, and promote ongoing development in a variety of fields, particularly healthcare and education. Accountability and transparency offer an organized way to assess how organizations use their resources and achieve their objectives, guaranteeing that they follow minimal quality requirements and maintaining quality assurance.

### 32.4.7 World Federation for Medical Education (WFME)

Instead of directly accrediting medical schools, the WFME creates standards and recognizes accrediting organizations, which improves healthcare quality globally and gives students options for international mobility. In order to ensure excellent standards for basic, graduate, and continuing medical education globally, WFME's primary responsibility is to assess the evaluators. Instead of reviewing medical institutions directly, WFME's function in accreditation is to evaluate agencies by comparing national and regional accreditation organizations to worldwide standards (Alsheikh Mustafa, 2004). Furthermore, it supports international standards for basic medical education that address mission, curriculum, assessment, resources, quality assurance, and governance. Additionally, the WFME's recognition program bestows status to organizations whose certification procedures satisfy its standards, indicating that their medical schools offer top-notch instruction.

### 32.4.8 National Accreditation Board for Hospitals and Healthcare Providers (NABH)

As a mark of excellence for ongoing improvement, public trust, and international recognition, particularly for government empanelment and medical tourism, the NABH establishes quality standards and evaluates healthcare organizations, such as hospitals and labs, for compliance, guaranteeing patient safety, high-quality care, and ethical practices (David and Valas, 2017). As it creates thorough, evidence-based standards and objective components spanning clinical, operational, and governance concerns such as patient rights, infection control, infrastructure, and staff training, NABH plays a standard-setting role in accreditation. Additionally, NABH employs skilled auditors to carry out thorough evaluations in order to verify conformity to standards through peer, documentation, and self-evaluation.

## 32.5 Purpose of accreditation frameworks

Accreditation frameworks offer an organized method to guarantee and improve quality by establishing standards that are met by medical institutions, thereby assuring consistent, high-quality services, boosting employability and international recognition, and promoting continuous improvement through peer review and accountability (Taber et al., 2020). Their main goal is to verify proficiency and dependability so that stakeholders can trust certificates, reports, and services (Fig. 32.2).

### 32.5.1 Quality assurance and standards

The fundamental goal of accreditation is quality assurance and standards. Accreditation is the formal process by which an independent body confirms that an institution satisfies predetermined quality benchmarks, guaranteeing educational excellence, accountability, continuous improvement, student confidence, and employability by comparing performance to national or international standards. In essence, accreditation verifies that students receive a worthwhile, acknowledged education by offering external confirmation that internal quality standards are operating.

### 32.5.2 Public trust and confidence

One of the main goals of accreditation in medical education is to foster public trust and confidence. A medical institution's competence, integrity, adherence to best practices, and dedication to ongoing progress are all validated objectively by a third party through accreditation. Accreditation fosters confidence amongst the patients and the general public, as the independent, objective verification is conducted by an unbiased, external agency which offers a more reliable assessment than an institution's self-proclaimed declarations of excellence. Stakeholders are reassured that criteria are being met by this third-party validation.

### 32.5.3 Continuous improvement

One of the main goals of accreditation is continuous improvement, which transforms it from a one-time check to a continuous cycle of self-evaluation, data-driven action, and improved quality. This ensures that institutions continuously meet and surpass standards for improved outcomes, stakeholder trust, efficiency, and innovation, ultimately fostering institutional growth and relevance. Using accrediting frameworks to identify problems, make changes, and monitor results leads to incorporating quality into the institution rather than just getting a stamp of approval.

### 32.5.4 Accountability

One of the main goals of accreditation is accountability. Educational institutions are held responsible for their performance, quality, and transparency to students, the public, and governments by mandating adherence to standards, public reporting, and continuous improvement. This justifies funding and fosters trust. Using peer review and frequent audits to confirm operations and improve outcomes, it goes beyond simply guaranteeing quality to actively showing efficacy in areas like curriculum, teaching, and learning process and administration.

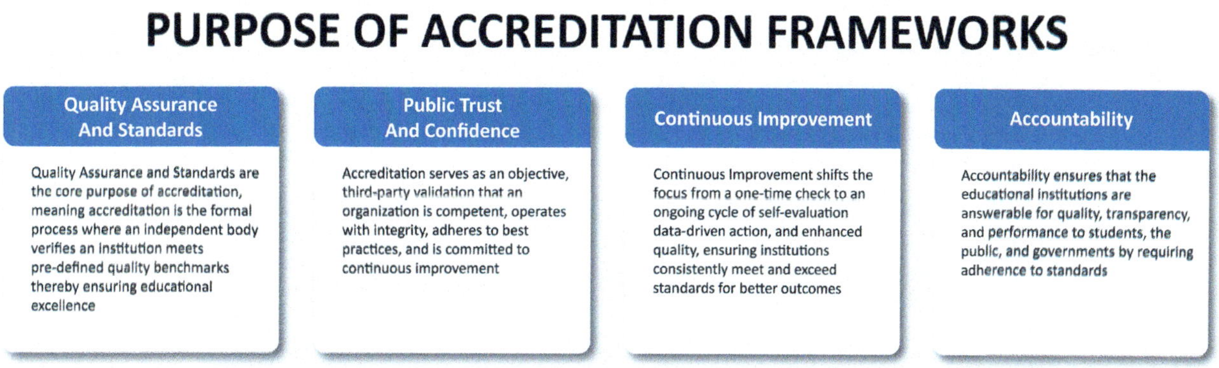


FIGURE 32.2 Purpose of accreditation frameworks.

## 32.6 Process of accreditation

In order to ensure that medical schools provide high-quality training for qualified physicians, the accreditation process in medical education is a strict, cyclical quality assurance system that involves self-evaluation, application, document review, a critical on-site visit by external reviewers, and a final accreditation decision regarding full, conditional, or no accreditation. Fig. 32.3 depicts the various steps in the accreditation process.

### 32.6.1 Eligibility and application

A medical program must normally be a legally recognized curriculum that regularly offers a physician education program in order to be eligible for medical education accreditation. When a medical facility satisfies certain requirements, such as possessing a final operating license, adhering to safety regulations, and going through a self-evaluation process, it can submit an application for accreditation. Normally an institution can apply for accreditation once it has implemented the required quality standards for at least 3 months. Some accrediting bodies typically call for the program to be running in order for at least one student cohort to have finished or to be about to finish. The application is the first stage in a series of steps that culminate in a thorough evaluation of quality. The preapplication or initial inquiry phase of the application process begins with the institution reviewing the specific guidelines provided by the accrediting body and discussing eligibility with their team. The formal application, which takes the form of an official letter requesting accreditation and is typically signed by a recognized institutional authority, is then sent along with the completed application form.

FIGURE 32.3 Process of accreditation.

### 32.6.2 Self-assessment

A vital and fundamental part of the accreditation process for medical education is self-assessment. It acts as a thorough internal evaluation that helps organizations assess their curricula in comparison to recognized national or worldwide standards, pinpoint areas in need of development, and promote a culture of ongoing quality improvement. Beyond only meeting an accreditation requirement, the self-assessment process offers many advantages. For example, it compels a critical and analytical examination of the program's curriculum, instructors, resources, and student services in order to pinpoint both its strong points and its needs for improvement. However, as a fundamental component of the ongoing renewal process required by accrediting bodies, the results of the self-assessment directly influence the development of action plans for resolving inadequacies.

### 32.6.3 Peer review and site visit

Peer review and the site visit are essential, required steps in the accreditation process for medical education, acting as quality control methods to assess a program's adherence to national requirements. Peer evaluation is at the core of the accreditation process which includes that a group of seasoned experts, including teachers, deans, and occasionally students from other universities, review the educational institution or program. One benefit is that peers in the area have the knowledge and experience needed to apply certification requirements effectively and fairly in order to assess whether a high-quality learning environment is offered. The critical on-site component is the site visit, during which the peer-review team verifies and elucidates the data supplied by the institution during its comprehensive self-study phase. Verifying that the data and self-analysis provided in the institution's data collecting tool and self-study report (SSR) appropriately reflect the program's actual operations and resources is one of the crucial elements in this process. The site visit essentially operationalizes the peer-review process by verifying that the institution's self-description is correct and that the requirements are being followed in practice, eventually reassuring the public and students about the caliber of the medical education program.

### 32.6.4 Review and decision

The medical school's self-evaluation is usually followed by an external review with a thorough report and an on-site visit, which results in a final decision regarding accreditation, probation, or denial and continuous quality improvement monitoring. The medical education accreditation review process is a multistage, peer-review system that ensures quality. The review procedure can be carried out in several phases, such as the following:

#### *32.6.4.1 Self-evaluation or internal review*

As the main, evidence-based method by which professional programs and higher education institutions evaluate their own performance in relation to predetermined standards, self-evaluation, and internal review are fundamental elements of the accreditation process. This internal process is a crucial, continual cycle of self-reflection intended to identify strengths, flaws, and places for ongoing progress rather than just a compliance exercise. In addition the entire accreditation decision is based on the self-evaluation report (SER), which is the primary document utilized by external peer-review teams. Self-evaluation and internal review as a part of decision-making in accreditation is important, as the weaknesses identified during self-evaluation are expected to lead to a concrete improvement plan, which is reviewed by accreditors.

#### *32.6.4.2 External review preparation*

In medical school accreditation, preparation for external reviews is a demanding, multiyear process that serves as a foundation for monitoring, quality control, and ultimate decision-making. It turns accreditation from a passive audit into an active cycle of continuous quality improvement, allowing institutions to assess their own performance in relation to national or international standards prior to a site visit by an independent panel. The accrediting organization appoints a team of specialists to conduct the on-site evaluation after all required documentation has been submitted. The main document is the self-study report, or SSR, which is a self-analysis of the institution's performance in comparison to accreditation requirements. In order to gather information, evaluate performance, and pinpoint shortcomings, it brings together the administration, teachers, students, and other relevant parties.

#### *32.6.4.3 Site visit (on-site evaluation)*

As an essential, on-site, peer-reviewed assessment to confirm that a medical school or program complies with set quality criteria, a site visit is a fundamental component of the accreditation process in medical education. It goes beyond documentation to actively monitor learning settings, enabling accrediting bodies to make well-informed, crucial decisions on accreditation status. The SER's correctness is confirmed by site teams. The report that is created from the data collected during the site visit directly influences the accreditation council's or committee's ultimate decision.

### 32.6.5 Monitoring and reaccreditation

Medical education accreditation's monitoring and reaccreditation procedures are cyclical mechanisms intended to guarantee ongoing quality improvement and conformity to set standards. In order to assure continued adherence to standards, accrediting authorities may monitor institutions in between official reaccreditation cycles. Additionally, the accredited institutions are required to produce annual data reports that include information on student data, faculty changes, financial records, and curriculum updates as part of the monitoring process. An interim report or mid-cycle inspection may be necessary to show remediation if any particular areas of noncompliance were identified during a prior assessment or if the institution undergoes major changes. Reaccreditation is a thorough reevaluation that usually takes place on a set cycle, such as every 4, 6, or 10 years, depending on the accrediting organization and previous compliance status.

## 32.7 Role of the stakeholders in the accreditation process

In order to ensure that standards are met, programs are relevant, and quality improves, stakeholders—including leadership, staff, students, patients, employers, and industry partners—provide a variety of perspectives, feedback, and expertise. This promotes buy-in, accountability, and alignment with real-world needs. Their participation guarantees that accreditation is a cooperative pursuit of excellence, linking theory and practice and fostering confidence in the recognized organization rather than merely a compliance exercise.

### 32.7.1 Leadership or administration

Throughout the accreditation process, the administration and leadership play a crucial, strategic role by offering the resources, accountability culture, and vision required to fulfill and maintain quality standards. Setting goals and a clear vision for the organization that is in line with the norms of the accrediting body and the institution's mission is one of their primary responsibilities. Additionally, they play a crucial role in ensuring that quality assurance is a continuous priority rather than merely a sporadic activity by integrating accrediting criteria into the institution's long-term strategic plans. In addition, effective leadership guarantees that sufficient financial, human, and material resources are allotted to support the accreditation self-study process and required enhancements. Additionally, the administration guarantees that the employees have the time and training necessary to contribute to the process in terms of data collecting and documentation in an efficient manner.

### 32.7.2 Faculty or staff

In order to ensure that the institution satisfies standards through peer review and ongoing quality improvement for improved student outcomes, faculty and staff play a critical role in accreditation. They drive the crucial self-study process by gathering data, evaluating program quality, creating reports, and implementing improvements. The creation of committees to review performance, gather information on student success, assess program quality, and evaluate advising are some of the major ways that faculty members contribute to accreditation. These committees directly impact the SSR. Additionally, the faculty supervises the entire process, organizing assignments, compiling data, and creating the final report. Essentially, teachers and staff are the primary contributors that evaluate, document, and eventually propel the advancements that accreditation aims to accomplish.

### 32.7.3 Students or patients

As the main stakeholders in the accreditation process, students and patients are essential because they offer input and take part in self-evaluation exercises to ensure and enhance the caliber of medical or educational services. Their opinions are crucial to a thorough and impartial assessment. As they participate in self-assessment, students play a crucial role by serving on internal committees and task groups for self-study and offering insightful feedback on issues impacting their education and student life. Additionally, they offer feedback through focus groups, interviews, and formal satisfaction surveys when external expert commissions visit the site. This input aids in assessing the program's results and learning environment. Through satisfaction surveys, feedback procedures, and in-person interviews with accrediting team during on-site inspections, patients play a crucial role in offering input on the quality of care. Additionally, they actively participate in quality improvement since health institutions may invite patients to serve as consumer representatives on quality improvement committees that examine continuing safety and quality procedures.

### 32.7.4 Employers or industry partners

Employers are essential to the medical school accreditation process because they make sure that graduates are prepared for the workforce, that training is still relevant to today's healthcare demands, and that the caliber of future medical professionals is maintained. Their participation aids in bridging the gap between academic norms and the real-world requirements of the healthcare sector. Employers also contribute to the definition of standards and competences by offering crucial feedback on the professional behaviors, practical skills, and core competencies that medical graduates require to thrive in the contemporary healthcare setting. The standards established by accrediting organizations are directly influenced by this input.

### 32.7.5 Accrediting bodies

With an aim to foster public trust and facilitate licensure, accrediting bodies develop standards, examine medical facilities through peer-review methods like self-evaluation, site inspections, and reports, and at the same time guarantee that programs produce qualified, safe physicians. They act as quality guardians, fostering ongoing development, by regularly assessing standards, stimulating innovation, and ensuring graduates attain professional competency for safe patient care. The accrediting agencies contribute to the process of quality assurance and standard setting by establishing and enforcing rules for curriculum, faculty, resources, and student outcomes that medical schools must satisfy. This ensures a minimal level of educational excellence. In addition, they oversee the accreditation procedure, which offers an unbiased evaluation of program quality and involves site inspections, self-studies by medical institutions, and public reporting.

### 32.7.6 Governing bodies or policymakers

Legislators and governing bodies play a crucial, fundamental role in medical education accreditation by providing the legal framework and regulatory authority required for accrediting systems to operate and be recognized. This guarantees both a trained health workforce and public safety. Setting the legislative mandate that requires accrediting systems to operate within a legal framework in compliance with a government law or decree is the primary duty of the governing bodies. Accrediting bodies have the power to set standards, conduct evaluations, and award or revoke accreditation, thanks to this legal support. Additionally, governments cooperate with or via independent accrediting agencies to guarantee that medical education institutions adhere to a predetermined set of basic criteria. This contributes to ensuring that the education offered is of the highest caliber and compliant with national and industry demands. Several internationally renowned medical education accrediting organizations include the following:

#### *32.7.6.1 Liaison Committee on Medical Education (LCME)*

The US Department of Education and WFME acknowledge the LCME as the premier authority for accrediting medical education schools that lead to MD degrees. Establishing standards, such as those for a medical school's operations and structure, conducting surveys on a regular basis which is typically every eight years, and making accreditation decisions are all part of its function in accreditation. Additionally, it authorizes government funding, facilitates licensure, and guarantees that programs generate qualified physicians.

#### *32.7.6.2 Foundation for Advancement of International Medical Education and Research (FAIMER)*

In order to help regulators and medical institutions meet international standards for physician competency and licensure, the FAIMER conducts research, provides data training accreditors in WFME standards, and facilitates international partnerships to improve quality.

#### 32.7.6.3 World Federation of Medical Education (WFME)

Instead of directly accrediting, the WFME enhances quality by evaluating and certifying national accrediting organizations through its recognition program and ensuring they follow international standards. This international certification denotes excellence, shields students from subpar medical schools, and permits graduates to practice abroad by meeting requirements set by organizations such as the ECFMG. WFME creates international standards, oversees the World Directory of Medical Institutions, and strives to enhance medical education across the globe.

#### 32.7.6.4 National Medical Commission (NMC)

The NMC of India regulates and accredits medical education by setting standards, assessing institutions, and ensuring quality through a robust framework for infrastructure, faculty, curriculum, and research in order to provide equitable access to qualified physicians. Indian medical graduates are now more well-known worldwide due to their new WFME accreditation. Establishing ethical standards, monitoring practice, giving recognition, and promoting continuous improvement in medical schools across the nation are some of its duties.

## 32.8 Emerging trends in accreditation

Competency-based medical education (CBME), data-driven quality improvement, globalization through cross-border accreditation, digital competencies like artificial intelligence (AI), social accountability, and interdisciplinary or team-based learning, as well as a shift from merely quality assurance to ongoing improvement of learning and practice readiness are emerging trends in medical education accreditation. Important topics include integrating technology while maintaining humanistic abilities, guaranteeing cultural flexibility in global standards, and connecting education to patient results. The following are some of the major new developments in medical education accreditation:

### 32.8.1 Shift to competency-based medical education (CBME)

In response to new accrediting trends, switching to CBME necessitates a methodical, multifaceted approach focused on establishing defined outcomes, changing the curriculum and assessment techniques, and cultivating a supportive culture. Redesigning the curriculum and structuring the learning process around competencies rather than merely traditional disciplines or time-based rotations is crucial in this situation (Dalseg et al., 2024). Incorporating entrustable professional activities, which specify the fundamental professional tasks that graduates should be able to complete without direct supervision upon commencing residency, is also crucial. Accreditation systems may facilitate this transition by promoting quality improvement, being flexible, and recognizing innovative methods of implementing CBME.

### 32.8.2 Data-driven quality improvement

A notable and well-established trend in accrediting procedures in healthcare and education is data-driven quality improvement. This change entails switching from conventional, episodic quality assurance techniques that frequently largely rely on recurring site visits to ongoing, technologically enabled cycles of review and improvement that make use of both quantitative and qualitative data. Since the employment of technology is essential to this trend, the integration of technology and analytics is crucial. Digital platforms, data analytics, artificial intelligence, and machine learning are being used by accrediting organizations and institutions to gather, examine, and present enormous volumes of data on performance indicators, results, and efficiency. This enables more thorough and precise assessments.

### 32.8.3 Global harmonization of standards

The WFME is the main force behind the notable and continuing trend of global standardization in medical education accreditation. This movement seeks to guarantee a basic standard of medical education across the globe, make it easier for doctors to go abroad, and eventually raise the standard of healthcare. The globalization of the medical field, which emphasizes the need for a common concept of what qualifies as high-quality medical education, is one of the main forces behind accreditation. This is because doctors are migrating more frequently, and medical students are exchanging ideas across national boundaries. Furthermore, it guarantees quality control because international standards give national and regional accreditation bodies a framework or yardstick to assess medical facilities and make sure they adhere to widely recognized fundamental criteria.

### 32.8.4 Digital competencies and artificial intelligence integration

AI can significantly enhance and speed up the medical education certification process by automating administrative procedures, improving data analysis for quality assurance, and enabling real-time standard compliance monitoring. This change enables human evaluators to concentrate on crucial regions and qualitative assessments. AI systems can automatically collect, arrange, and format the substantial documentation needed for accreditation from a variety of sources, including student records, curriculum metrics, and faculty credentials. This is one of the main areas in which AI can be useful. This lessens human error and the administrative load. Additionally, it plays a part in curriculum mapping and alignment, whereby AI-powered systems may evaluate course material and match it with certain certification requirements and competences. This keeps the curriculum up to date with changing medical knowledge and best practices by assisting instructors in identifying and addressing knowledge gaps or redundancies in real time.

### 32.8.5 Social accountability

Beyond standard quality assurance, social responsibility is a major and expanding trend in medical education accreditation that specifically requires medical institutions to match their mission, education, research, and service activities with the community's top health needs. Due to the need for global health, which is fueled by enduring health disparities and changing global health dynamics, social accountability has emerged as a major trend. Healthcare workers that are prepared to handle a variety of societal issues are also desperately needed. Furthermore, accrediting organizations are realizing that, in addition to individual clinical abilities, high-quality medical education must also be effective in enhancing population health outcomes and pertinent to the local environment.

## 32.9 Benefits of accreditation

Accreditation improves medical education by increasing and standardizing quality, ensuring graduates meet worldwide standards, promoting continuous improvement, and ultimately safeguarding the medical industry and patient safety. Obstacles include high costs, limited resources for smaller institutions, the heavy administrative burden of the process, potential resistance to change, and the need to keep up the momentum of continuous improvement following accreditation. The following are some of the primary benefits of accreditation:

### 32.9.1 Ensures quality and standards

In medical education, accreditation ensures quality and standards by establishing benchmarks for curriculum, faculty, resources, and student outcomes, ensuring graduates have the skills required for safe patient care, promoting continuous improvement through frequent reviews, and increasing public trust by ensuring programs meet strict criteria for efficacy, patient-centeredness, and equity. It creates a framework for responsibility, ensures that education is in accordance with societal demands, allows medical professionals' greater mobility, and improves the standing of institutions.

### 32.9.2 Promotes continuous improvement

Accreditation in medical education promotes ongoing advancement since it aids in establishing a framework for regular assessment, self-evaluation, and the implementation of improvements to educational programs and procedures. To be recognized, institutions must do in-depth self-studies to identify their own benefits, drawbacks, and areas for improvement. This introspective approach often ignites change and quality-improvement initiatives. However, regular evaluations and site visits by outside peer assessors provide objective feedback and recommendations for enhancements. This external perspective helps firms focus their efforts on meeting and exceeding quality requirements.

### 32.9.3 Enhances patient safety

Patient safety is enhanced by accreditation in medical education through strict requirements for certified graduates, tough training, and ethical professionalism. As a result, the facilities where they train and work experience fewer errors, better risk management, and stronger safety cultures. In addition, as it guarantees a continuous focus on excellent, patient-centered care and evidence-based practices, it ultimately enhances clinical outcomes and boosts public trust in healthcare providers.

FIGURE 32.4 Benefits of accreditation.

### 32.9.4 Supports staff development

Accreditation in medical education significantly aids staff growth by mandating continuous professional development, promoting a culture of quality improvement, and providing a framework for teachers to increase their skills. The accreditation process itself serves as a catalyst for continuous quality improvement initiatives. Staff and faculty members engage in self-assessment, data analysis, and program evaluation to enhance their skills in educational research and program evaluation. This enables them to identify areas that require improvement and implement the required changes.

### 32.9.5 Improves resource management

Accreditation in medical education improves resource management by promoting strategic planning, ensuring efficient resource allocation, and cultivating a culture of continuous quality improvement. The process provides a systematic framework for effectively managing an organization's resources. Accreditation standards sometimes address the adequacy of a range of resources, including personnel, financial resources, physical buildings, equipment, and clinical training sites. Medical facilities can guarantee that money, staff time, and infrastructure are distributed effectively to fulfill the primary educational objective and satisfy student requests by distributing resources in compliance with the specified standards (Fig. 32.4).

## 32.10 Challenges of accreditation

Some of the major challenges of accreditation hindering quality improvement despite accreditation's goal to ensure standards and accountability are listed below:

### 32.10.1 Resource limitations

Resource limitations present significant challenges to attaining and maintaining accreditation in medical education because they directly impact an institution's ability to meet stringent requirements pertaining to faculty, buildings, equipment, and cash. One of the biggest obstacles in this case may be financial constraints. Lack of funding may make it difficult for institutions, especially smaller ones, to make the required investments. Budget deficits make it challenging to purchase modern medical equipment and maintain an adequate educational infrastructure, both of which are occasionally required by certification standards. Institutions' incapacity to offer competitive compensation and incentives owing to financial strain may be the cause of faculty shortages and high turnover rates. This puts at risk the degree of oversight and training required for accreditation.

### 32.10.2 Administrative burden

A well-known issue with medical education accreditation is administrative burden, which includes a lot of paperwork, high time and resource requirements, and more work for staff and faculty. The substantial paperwork and record-keeping needed to prove adherence to certification requirements is one of the main causes of load. For self-evaluation study reports, administrative workers in particular frequently point to content duplication and the sheer amount of material as a significant challenge. Additionally, it has been noted that the accrediting process takes a significant amount of time and human resources from all parties involved, including academics, administrative personnel, and members of the self-evaluation committee. When institutional systems are not in line with accreditation requirements, this frequently entails laborious manual data collecting.

### 32.10.3 Resistance to change

One major and frequent obstacle in the certification process for medical education is resistance to change. Both organizational and interpersonal reasons contribute to this reluctance, which frequently results from a conservative academic atmosphere, an increasing workload, and a fear of the unknown. Faculty and staff may feel anxious, disoriented, and unsure about how the accreditation process will impact their responsibilities, status, and job security in this situation. Additionally, the accreditation process is time-consuming and requires a significant amount of extra work, including data collecting, paperwork, and curriculum modifications, which adds to the academic staff's already heavy workload from teaching, research, and clinical responsibilities. Additionally, staff members may not completely comprehend the goals, advantages, or processes of certification, which could result in a lack of interest and buy-in.

### 32.10.4 Sustaining improvement

As the process is frequently seen as a recurring, compliance-driven event rather than a continuing culture of quality improvement, sustaining progress is a major problem of accreditation in medical education. Resource limitations, cultural resistance, and the absence of integrated mechanisms for ongoing monitoring are the main obstacles. The propensity for quality procedures to deteriorate once the immediate pressure of an accreditation visit has passed is one of the main obstacles. Until the next accrediting cycle comes around, the emphasis frequently moves away from sustaining the gains. A lack of efficient, continuous processes for tracking, assessing, and offering feedback on quality procedures may be the cause of this. Effectively identifying and resolving persistent problems is challenging without thorough data collection and analysis to quantify certain results.

### 32.10.5 Complexity and expertise

Accreditation in medical education is severely hampered by complexity and a lack of pertinent knowledge, which affects the evaluation procedure, the clarity of criteria, and the capacity to measure important results. Faculty and administrative staff who are in charge of SERs may not fully comprehend accreditation standards due to their complexity and breadth. Additionally, accreditation requirements need to be revised on a regular basis due to the rapid evolution of medicine and the ongoing advances in medical knowledge, technology, and care delivery. This makes it difficult for accrediting agencies to uphold current and pertinent standards and for universities to stay up to date. In addition, a major obstacle is frequently the lack of qualified and experienced people to act as reviewers or evaluators who have a thorough understanding of clinical practice and educational philosophy across a variety of specializations.

**Summary**

In medical education, accreditation is an essential quality control method that guarantees medical facilities adhere to strict, peer-reviewed guidelines. It protects the public by guaranteeing that graduates possess the skills and information required to provide safe and effective patient care. Clearly defined standards, institutional self-evaluation, external peer review, site visits, and continuous monitoring are essential elements of accreditation systems. Curriculum design, faculty qualifications, student assessment, learning settings, governance, and resources are all commonly covered by these standards. Accreditation frameworks increasingly incorporate competency-based education, outcome-based measurements, and alignment with national and international healthcare agendas. New developments in medical education accreditation, such as the use of digital technology, international standardization, risk-based accrediting models, and increased focus on patient-centered care and social accountability are gaining a lot of attention in modern medical education accreditation process. In general, accreditation is portrayed as a dynamic and crucial procedure that supports the integrity and caliber of medical education systems across the globe.

**Key point**

Future perspective: upholding continuous quality improvement

Due in part to initiatives by organizations like the World Federation for Medical Education, accreditation in medical education is gradually shifting toward globally linked norms. In order to improve the legitimacy and comparability of medical programs around the globe, institutions and accrediting bodies strive to seek recognition from international quality assurance frameworks. The future of medical education accreditation is turning toward outcome-based, continuous quality improvement rather than static, process-oriented checklists. More internationally standardized criteria and recognized systems are probably in store for medical education certification in the future, which will increase mobility and transparency. Results, continuous improvement, and graduates' preparedness for the actual world will be given more weight. It is urged that digital tools and analytics be integrated to assist data-informed appraisal. All of these developments point to an accreditation environment that is increasingly technologically advanced, outcomes-oriented, and integrated, supporting high-quality medical education globally.

# References

Abdalla, M.E., 2012. Accreditation in Medical Education: Concepts and Practice.

Alsheikh Mustafa, G.Y., 2004. Accreditation of Medical Education: Viewpoints from the Regions II: Eastern Mediterranean WHO Regional Office with notes. Educacion Medica 9. Available at https://doi.org/10.13140/RG.2.1.2537.4484.

Blouin, D., Tekian, A., 2018. Accreditation of medical education programs: moving from student outcomes to continuous quality improvement measures. Academic Medicine 93 (3), 377–383. Available at https://doi.org/10.1097/acm.0000000000001835.

Boulet, J., van Zanten, M., 2014. Ensuring high-quality patient care: the role of accreditation, licensure, specialty certification and revalidation in medicine. Medical Education 48 (1), 75–86. Available at https://doi.org/10.1111/medu.12286.

Dalseg, T.R., Thoma, B., Wycliffe-Jones, K., Frank, J.R., Taber, S., 2024. Enabling implementation of competency based medical education through an outcomes-focused accreditation system. Perspectives on Medical Education 13 (1), 75–84. Available at https://doi.org/10.5334/pme.963, https://pmejournal.org/.

David, S.N.J., Valas, S., 2017. National accreditation board for hospitals and healthcare providers (NABH) standards: a review. Current Medical Issues 15 (3), 231. Available at https://doi.org/10.4103/cmi.cmi_51_17.

Davis, D.J., Ringsted, C., 2006. Accreditation of undergraduate and graduate medical education: how do the standards contribute to quality? Advances in Health Sciences Education 11 (3), 305–313. Available at https://doi.org/10.1007/s10459-005-8555-4.

Taber, S., Akdemir, N., Gorman, L., Van Zanten, M., Frank, J.R., 2020. A "fit for purpose" framework for medical education accreditation system design. BMC Medical Education 20. Available at https://doi.org/10.1186/s12909-020-02122-4, http://www.biomedcentral.com/bmcmededuc/.

Van Zanten, M., Norcini, J.J., Boulet, J.R., Simon, F., 2008. Overview of accreditation of undergraduate medical education programmes worldwide. Medical Education 42 (9), 930–937. Available at https://doi.org/10.1111/j.1365-2923.2008.03092.x.

# PART 6

## RESEARCH IN MEDICAL EDUCATION

Chapter 33

# Introduction to medical research

*"We look to medical research to discover remedial measures to insure better health and more happiness for mankind."*

—Thomas Hunt Morgan

## 33.1 Introduction

With the ultimate goal of enhancing human health and quality of life, medical research is the methodical study of diseases, their prevention and treatment, and health promotion by scientific methods. It includes both preclinical research, which is conducted in laboratories, and clinical research, which tests interventions on humans, such as through clinical trials, to ascertain their safety and efficacy. Research aims to solve scientific issues and benefit society, not necessarily the individual participants, in contrast to individual medical care (Norman, 2002). Medical research is crucial to improve healthcare by identifying novel medicines and enhancing patient care. It is essential for developing students' critical thinking, evidence-based practice, and scientific inquiry in medical education. It prepares aspiring doctors for changing medical difficulties by bridging the knowledge gap between theory and practice. Early curricular integration of research through projects and mentorship prepares future physicians to advance science and enhance patient outcomes.

## 33.2 Characteristics of good educational research

Good educational research is methodical, objective, empirical, and repeatable. It is based on well-defined questions, employs reliable techniques for verifiable data analysis and makes precise and logical contributions to theory or practice that close the gap between educational concepts and practical advancements. It seeks to solve issues, promote comprehension, and create practical solutions, frequently utilizing a variety of disciplines. The characteristics of good educational research are depicted in Fig. 33.1.

### 33.2.1 Systematic and scientific approach

A systematic methodology in medical research guarantees an orderly, comprehensive, and bias-free research process. One of the key components of a systematic methods is that the study must begin with a clear, testable research topic or hypothesis. The usage of structured methodology should also be prioritized since the design, data collection techniques and analytic strategy should all be well defined and strictly adhered to. Focusing on the thorough data collection which is a vital component of research is also essential. The data must be collected in an orderly and consistent manner, using validated instruments if feasible.

Using scientific criteria ensures that the findings are authentic, reliable, and significantly advance the area. In this case, researchers should try to remain impartial and keep their own biases from influencing the design, execution, or interpretation of the study's results. A scientific approach to research places a strong emphasis on adhering to an organized procedure that includes problem identification, hypothesis or research question formulation, methodical data gathering and analysis, and drawing conclusions based on evidence. Following these processes will enable the research to target specific problems by assessing variables, identifying, and synthesizing evidence to improve educational results.

### 33.2.2 Objectivity and ethical conduct

In order to ensure that studies are trustworthy, equitable, and protective of participants' rights and welfare, objectivity and ethical behavior are essential components of quality medical education research. In terms of objectivity, it guarantees that the

The Principles and Practice of Medical Education. DOI: https://doi.org/10.1016/B978-0-443-44197-4.00022-9

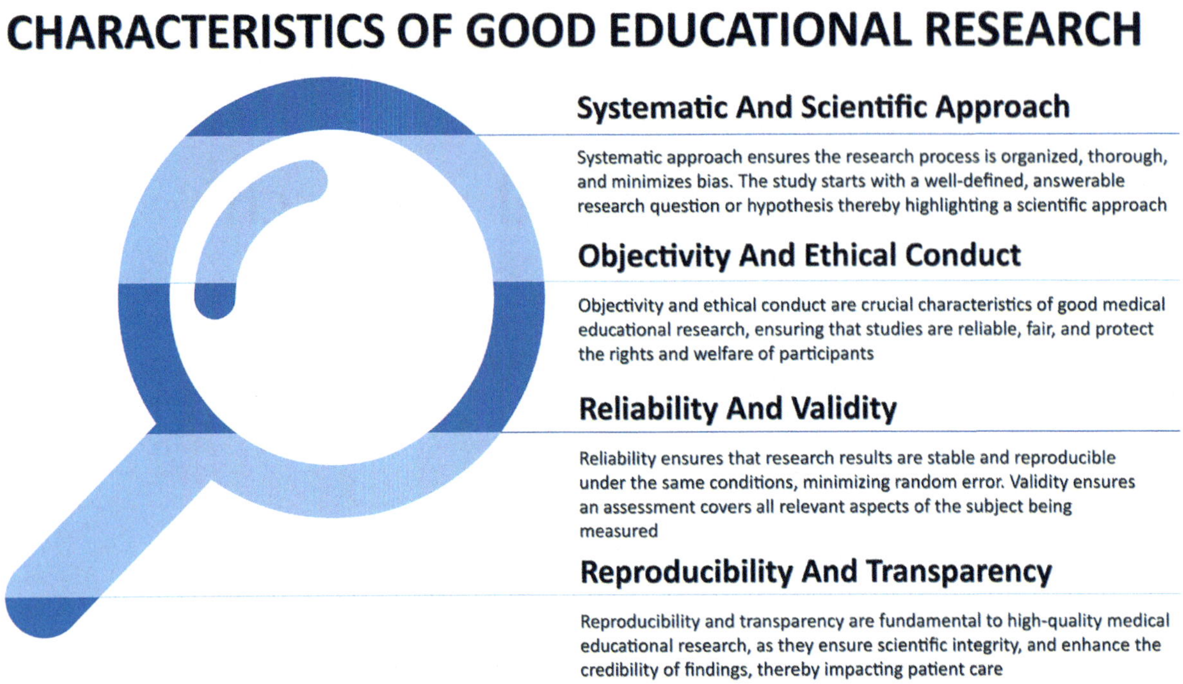


FIGURE 33.1 **Characteristics of good educational research.** This figure depicts the characteristics of good educational research.

study is devoid of prejudice, subjective beliefs, or other influences, making the results reliable and repeatable. It is impertinent in this regard to use strong procedures that reduce systematic error and guarantee uniform and unbiased data collecting.

However, as medical research frequently uses human beings, sensitive data, and professional practices, ethical behavior is crucial. Institutional review boards (IRBs) or ethics committees are usually in charge of monitoring adherence to moral standards. Placing a strong emphasis on informed consent will ensure that participants voluntarily agree to participate in the study after being fully informed of its objectives, risks, and benefits. Data must be anonymized and stored securely in order to protect participants' identities and private information.

### 33.2.3 Reliability and validity

Validity and reliability are essential ideas for assessing the caliber and reliability of medical education research. Validity is the degree to which a measurement accurately measures what it is supposed to measure, whereas reliability is the consistency of a measurement. Eliminating random error will guarantee that study outcomes are consistent and repeatable under the same circumstances. This is crucial in medical education to ensure that assessments and data collection methods yield accurate results for a range of situations and observers. The degree to which a study or evaluation instrument truly evaluates the fundamental idea it purports to measure is known as validity. Validity guarantees that an evaluation encompasses all pertinent facets of the topic being assessed. It also assesses how well a measurement tool aligns with existing knowledge and conceptual ideas.

### 33.2.4 Reproducibility and transparency

Transparency and reproducibility are essential to high-quality medical education research because they guarantee scientific integrity, improve the validity and dependability of results, and ultimately affect patient treatment and public confidence in the evidence. The ability of independent researchers to produce consistent results using the same data, materials, and analytical techniques as the original author is known as reproducibility. Reproducibility verifies that the initial findings accurately represent the study's design and data rather than being the result of bias, error, or chance. All technical and methodological aspects of the research process, including data, code, and study protocols, must be fully and adequately disclosed in order to be considered transparent. Policymakers, practitioners, and the general public can trust the research process when it is transparent.

## 33.3 Need for research in medical education

In order to improve healthcare teaching, learning, and assessment, as well as to promote innovation and evidence-based decision-making in medical practice and training, research in medical education is essential. It enhances educational programs, offers a path for academic careers, and gives present and future healthcare professionals essential skills. The various needs for research in medical education are shown in Fig. 33.2.

### 33.3.1 To improve teaching and learning

In order to improve teaching and learning in the healthcare, medical research is essential because it ensures that future clinicians have the skills necessary for evidence-based practice and lifelong learning, provides the evidence base for educational practices, and propels the development of creative pedagogical approaches. One of the most important benefits of medical education research is that it creates evidence-based practices. This is because research in medical education changes instruction from depending only on tradition to employing data-supported strategies like problem-based learning or simulation training, which have been shown to improve student performance and creativity. In addition, it has been noted that students who participate in research activities are better able to develop the critical appraisal and analytical abilities necessary for assessing the extensive and continuously changing medical literature (Ofei-Dodoo, 2025).

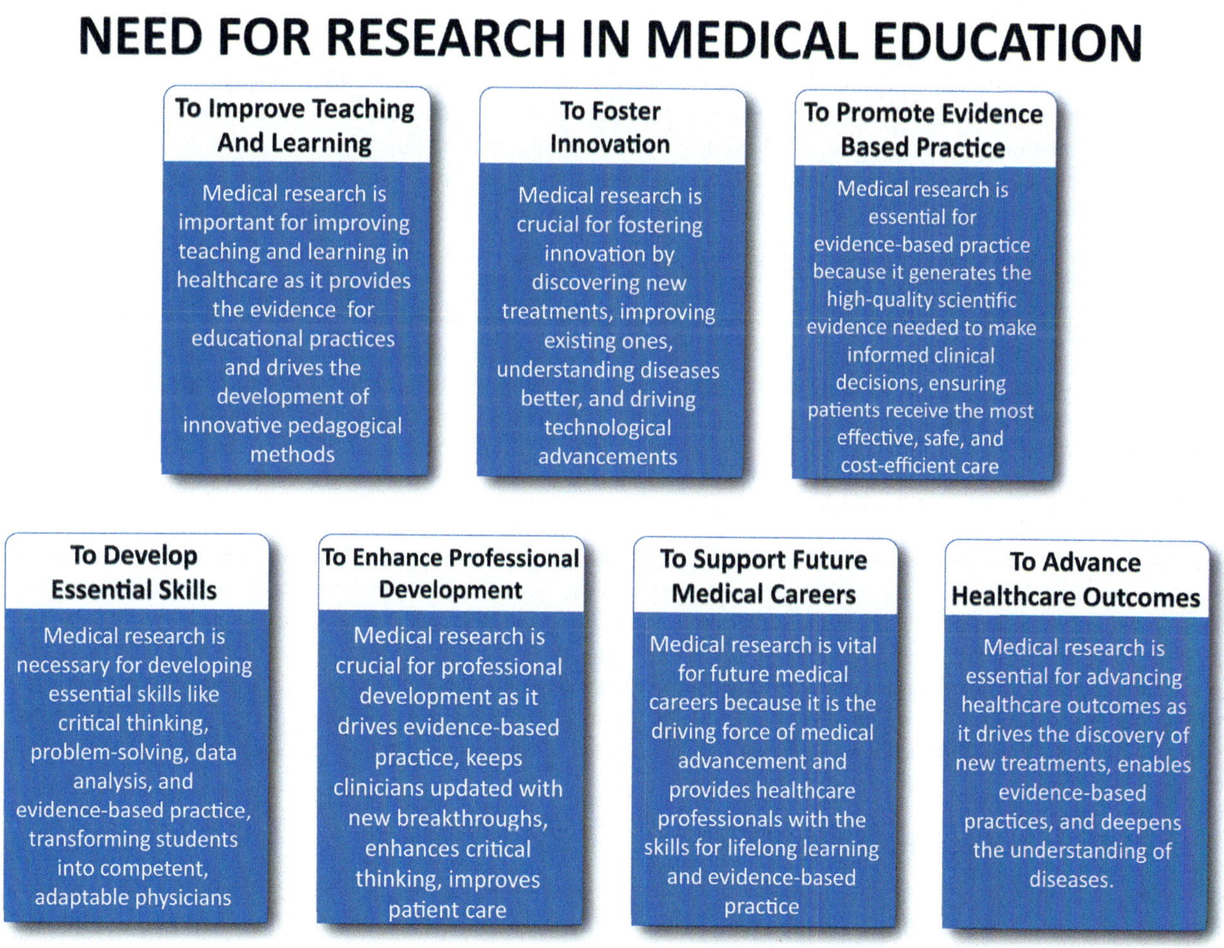


FIGURE 33.2 **Need for research in medical education.** This figure shows the need for research in medical education.

### 33.3.2 To foster innovation

In order to improve patient outcomes, safety, and the efficiency of healthcare systems through evidence-based practices and cooperative efforts between academia, industry, and public health, medical research is essential for promoting innovation through the discovery of new treatments, the improvement of existing ones, a better understanding of diseases, and the advancement of technologies like artificial intelligence (AI) in health. Translational research, which moves from basic science to clinical application, serves as the basis for new medications, vaccines, technologies, and diagnostic tools.

### 33.3.3 To promote evidence-based practice

Medical research is essential to evidence-based practice because it generates the high-quality scientific information needed to make educated clinical decisions. This guarantees that patients receive the most efficient, secure, and affordable treatment possible, enhancing results, reducing mistakes, and consistently increasing the bar for healthcare. Medical research is essential to evidence-based practice because well-designed studies, which provide scientifically proved methodologies over anecdotal evidence or tradition, form the foundation of evidence-based practice. Additionally, it enhances patient outcomes by directly implementing research findings, which results in improved outcomes, lower mortality, and quicker patient recovery.

### 33.3.4 To develop essential skills

The practice of undertaking medical research must be encouraged amongst both the teaching faculty and the students, as it proves to be beneficial for both. For the teaching faculty, in addition to promoting analytical thinking and research methodology, medical education research builds important abilities in learner evaluation, evidence-based teaching, and curriculum design. In order to promote clinical competency, it strengthens educators' leadership, data analysis, and simulation-based training skills. From the perspective of the student, medical research proves to be beneficial, as it educates students for evidence-based practice and helps them gain important skills, such as advanced data analysis, literature interpretation, and study design. In addition to enhancing critical thinking and academic proficiency, these activities promote vital soft skills like collaboration, leadership, and communication. This makes them better candidates for residency and better equipped for the changing medical landscape (Lee et al., 2021).

### 33.3.5 To enhance professional development

Medical research advances professional development, as it promotes critical thinking, evidence-based decision-making, and specialized knowledge. Additionally, it enhances patient care and helps healthcare professionals develop in their careers. Engaging in active medical research also offers opportunities for scholarly recognition, networking, and mentoring, particularly in academic medicine. Furthermore, research facilitates ongoing professional growth by keeping medical personnel abreast of the most recent developments, breakthroughs, and treatment approaches.

### 33.3.6 To support future medical careers

Since medical research is the engine of medical progress and equips healthcare workers with the knowledge and skills necessary for lifelong learning and evidence-based practice, it is essential for future careers in medicine. It ensures that future physicians can adjust to a changing healthcare environment by influencing how they diagnose, treat, and prevent illnesses. Healthcare practitioners can improve the lives of thousands of patients, not just the ones they treat individually, by contributing to the development of novel therapies, diagnostic tools like enhanced imaging or genetic testing, and medical gadgets. Furthermore, it equips the healthcare practitioner for emerging developments such as big data analytics, biotechnology, AI, precision medicine, and tailored treatments. Clinicians who have a background in research are more equipped to apply and contribute to these developments.

### 33.3.7 To advance healthcare outcomes

As medical research propels the development of novel therapies, permits evidence-based practices, and expands our knowledge of illnesses, it is critical to improving healthcare outcomes. These days, practically every drug, diagnostic test, and medical procedure is a direct result of past research and clinical trials. Medical research enables new strategies for controlling, curing, and treating diseases. This leads to the creation of innovative medications, therapies, and medical devices, such as life-saving vaccines, state-of-the-art cancer treatments like immunotherapy and targeted therapy, and long-term illness management tools like continuous glucose monitors for diabetes. Before any new approach can be widely adopted, extensive clinical study is required to demonstrate its efficacy and safety. For this reason, medical research is crucial. This systematic approach ensures that treatments provide real benefits with controllable side effects while protecting patients.

## 33.4 Common research methodologies in medical education

Primary research, which gathers fresh, unprocessed data, and secondary research, which analyzes previously collected data, are the two main categories of medical research. Primary research is further divided into basic research which is frequently lab-based, clinical research which is human-focused, and epidemiological research which is mainly population-based investigations, extending from experimental trials to observational studies that improve patient care (Fig. 33.3.

### 33.4.1 Primary research

With the ultimate goal of producing more skilled, flexible, and creative doctors through evidence-based practices, primary research in medical education entails gathering fresh data to comprehend and enhance how doctors are trained. It focuses on curriculum effectiveness, teaching strategies, assessment, and professional development using techniques like surveys, interviews, or observations. It includes basic science, clinical, and educational studies to solve practical issues in healthcare training and is different from secondary research. Basic research, clinical trials, and epidemiological trials are other classifications for primary research (Boet et al., 2012).

#### *33.4.1.1 Basic research*

Basic research in medical education refers to basic, curiosity-driven scientific investigation that is often lab-based and investigates basic biological processes such as DNA, cells, and molecules in order to develop the fundamental knowledge needed to comprehend disease. The goal of basic research is to understand how life works, which will drive future breakthroughs and cures rather than immediate application. It is the cornerstone of medical research and lays the groundwork for later clinical discoveries in disciplines including physiology, biochemistry, microbiology, and genetics.

#### *33.4.1.2 Clinical trials*

Clinical trials are practical research studies used in medical education that teach aspiring physicians and healthcare workers how to test novel medications, therapies, devices, or patient behavioral changes. They also teach them how to assess safety, efficacy, and patient outcomes in order to improve care and advance medical knowledge. They are essential for teaching evidence-based medicine, demonstrating how therapies go from the lab to the patient's bedside, and

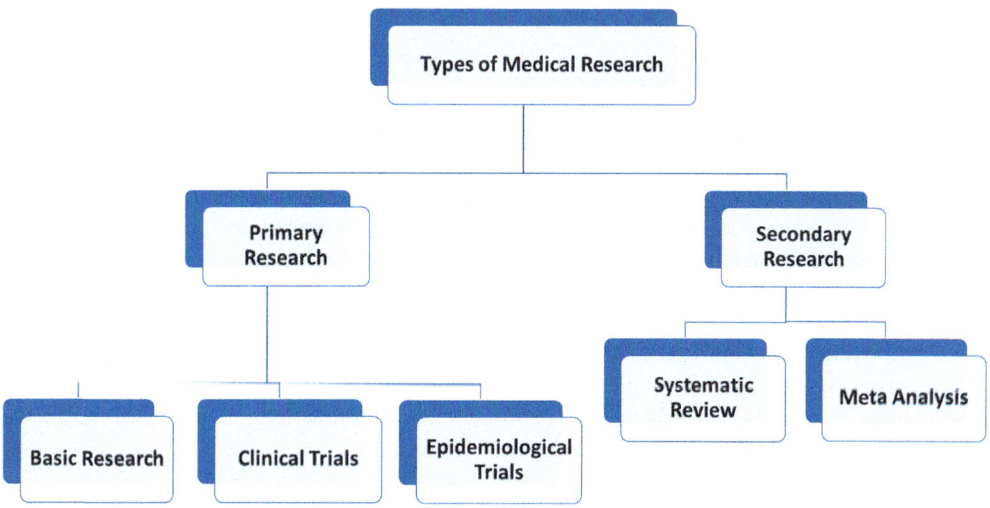


FIGURE 33.3 Types of medical research.

incorporating stringent standards to safeguard volunteers. They also teach aspiring physicians how to analyze data and make defensible conclusions (Röhrig et al., 2009).

#### 33.4.1.3 *Epidemiological trials*

A fundamental kind of medical research, epidemiological trials examine the prevalence, causes, and consequences of diseases in communities in order to guide clinical practice, public health, and policy. These studies can be broadly categorized as either interventional or observational, such as cohort and case–control studies. They are crucial for assessing risk factors and evaluating the effectiveness of interventions in specific groups.

### 33.4.2 Secondary research

In medical education, secondary research makes use of already-existing data and literature, such as studies, reviews, and datasets, to identify trends, gaps, and evidence for better treatment planning, curriculum development, or understanding learning outcomes. This is frequently accomplished through systematic reviews or meta-analyses, which save time and money while utilizing large-scale findings for better patient care and instructional strategies (Ratajczak et al., 2025).

#### 33.4.2.1 *Systematic review*

In medical education, systematic reviews are rigorous research summaries that employ clear, transparent methods to locate, evaluate, and synthesize all available evidence on a particular question, such as the efficacy of a teaching method to provide trustworthy answers, guide best practices in training future doctors, and improve educational strategies by lowering bias and summarizing complex data. To create a solid body of evidence, they entail a precise query, thorough searches, evaluation of the caliber of studies, and data synthesis. In medical education, systematic reviews offer thorough, objective, and evidence-based summaries of pedagogical literature that are crucial for guiding curriculum development, instructional strategies, and educational policy. To assess the efficacy of various initiatives, pinpoint research gaps, and assist educators and policymakers in making evidence-based, data-driven decisions, they synthesize a variety of studies (Garg et al., 2008).

#### 33.4.2.2 *Meta analysis*

A meta-analysis is a potent statistical technique used in medical education that quantitatively combines findings from several related studies, frequently found through a systematic review, to obtain a single, more accurate estimate of an effect. This technique helps resolve conflicting findings, boost statistical power, and direct best practices for teaching or interventions, such as comparing active versus passive learning for medical students. Meta-analysis is widely used in medical education to identify the most effective teaching strategies. Finally, meta-analysis goes beyond grades to look at the long-term effects of educational programs. It can also be used to support the integration of digital resources (Simmonds et al., 2005).

### 33.4.3 Qualitative versus quantitative research

Using techniques like ethnography, grounded theory, phenomenology, narrative inquiry, and case studies to comprehend learner or teacher experiences, professionalism, and clinical contexts, qualitative research in medical education investigates the "how" and "why" of human experiences, behaviors, and complex social issues. Through focus groups, interviews, and observation, it helps develop theory and enhance practice. From patient care to curriculum design, it offers context, subtlety, and depth to comprehend the human aspects of medicine while guaranteeing rigor through challenging questions and open-minded approaches. In order to make data-driven decisions and enhance healthcare training, quantitative research in medical education uses numerical data and statistics to measure learning, teaching, and outcomes. It also assists in evaluating curricula, teaching methods, and assessment tools by identifying patterns, comparing interventions such as new versus old teaching methods, and evaluating trends in student performance test scores and graduation rates. To discover unbiased, trustworthy evidence for best practices, it frequently employs designs like descriptive, correlational, quasi-experimental, and experimental studies.

## 33.5 Research process

Determining the problem, examining the literature, developing hypotheses or questions, designing the study, gathering data, analyzing it, interpreting the results, and reporting the findings are all part of the methodical processes that make up the research process. This methodical approach guarantees validity, rigor, and the systematic and repeatable creation of new information (Abramson et al., 2018). The steps in the research process can be described as in Fig. 33.4.

### 33.5.1 Identifying a research problem

The critical first step in the research process is identifying a research problem, which entails identifying a knowledge gap, issue, or area that requires investigation. This process establishes the focus, direction, and objectives of the study by turning a general interest into a specific, important, and researchable question through literature review, expert consultation, and observation, ultimately serving as the study's foundation.

### 33.5.2 Conducting a literature review

A vital research phase is conducting a literature review, which entails framing the issue of interest, finding databases, critically evaluating and synthesizing previous research to identify trends, gaps, and controversies, and then writing to place your work within the scholarly discourse. Identifying keywords, selecting databases, establishing criteria, reading, assessing sources, synthesizing findings in terms of citing, comparing, critiquing, and writing are some of the tasks that go into creating the basis and contribution of the study.

### 33.5.3 Formulating the research questions or hypotheses

Research questions ask what you want to know, and hypotheses offer specific predictions like if-then statements about variable relationships. Both must be feasible, clear, relevant, and informed by preliminary research. Formulating research questions and hypotheses is an essential early step in research, providing direction by transforming broad interests into focused, testable inquiries that guide methodology, data collection, and analysis (Eva and Lingard, 2008).

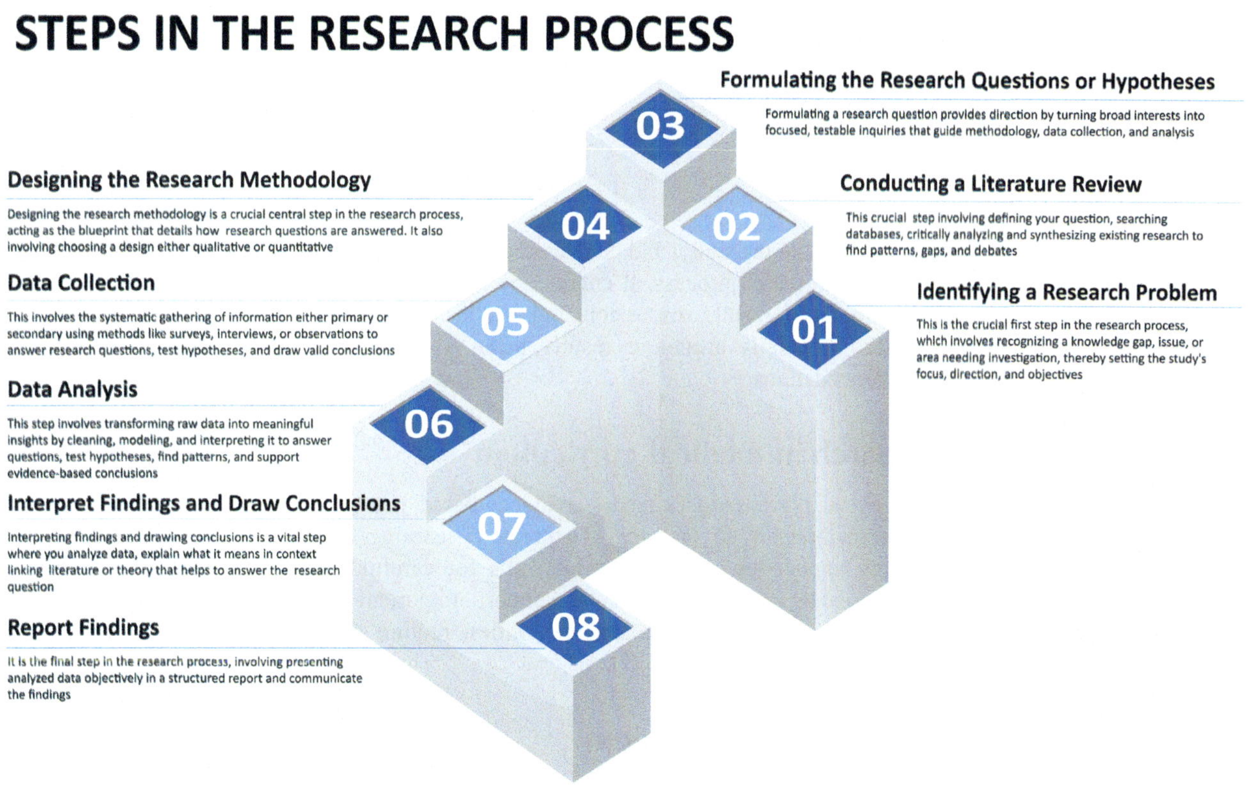


FIGURE 33.4 Steps in the research process.

### 33.5.4 Designing the research methodology

In order to ensure valid, dependable, and targeted results, designing the research methodology is an essential, central step in the research process. It serves as the blueprint that outlines how the research questions will be addressed. This includes selecting a qualitative or quantitative design, identifying the population and sample, choosing data collection tools, and planning analysis. It guides all subsequent efforts to maintain the study's integrity by bridging the gap between the broad research topic and the actual data collecting. Following the finalization of the research methodology, it is important to secure approval from the IRB to ensure the ethical treatment of participants (Chen and Reeves, 2020).

### 33.5.5 Data collection

In order to answer research questions, test hypotheses, and draw valid conclusions that serve as the foundation for trustworthy study outcomes, data collection is an essential step in the research process. It entails the methodical gathering of information, either primary or secondary, using techniques like surveys, interviews, or observations. This step also includes implementing the study design, gathering data through tools like questionnaires, assessments, or interviews.

### 33.5.6 Data analysis

In order to answer questions, test hypotheses, identify trends, and support evidence-based conclusions, data analysis is an essential and crucial stage in the research process. In order to identify patterns and trends and make well-informed judgments or gain a deeper understanding of the issue, it entails methodical activities including as identifying objectives, gathering, cleaning, transforming, modeling, and displaying data. In order to assess curricula and student performance, data analysis in medical education research uses quantitative methods like descriptive or inferential statistics, regression, qualitative methods like thematic analysis, grounded theory, and mixed-methods approaches.

### 33.5.7 Interpret findings and draw conclusions

Interpreting results and making conclusions are crucial research steps. This entails evaluating the data, elucidating its significance in light of literature or theory, answering the research questions, endorsing or disproving the hypotheses, talking about the implications, admitting limitations, and suggesting further studies that will eventually increase knowledge and guide action. It entails moving from what the data indicates to what it signifies for the study and the field at large.

### 33.5.8 Report findings

In order to communicate discoveries, validate the study, and enable others to learn from the research, reporting findings is an essential, frequently last, step in the research process. It entails objectively presenting analyzed data in a structured report that includes an introduction, methods, results, discussion, and conclusion. In order to bridge the gap between data collection and practical application, it entails arranging results using visualizations like charts, evaluating their significance, and formulating recommendations.

## 33.6 Integration of research in medical curriculum

Incorporating research into the medical curriculum enhances evidence-based practice, critical thinking, and lifelong learning by teaching aspiring doctors how to do research. To create a learning environment that links theoretical knowledge with real-world research application, this integration calls for careful curriculum design, cultivating an institutional culture that supports research, and gaining faculty support. Important tactics include using longitudinal courses that gradually develop research concepts and practical abilities, require autonomous research projects, and integrate current research into coursework (Ahmed et al., 2024).

## 33.7 Research-oriented medical curriculum

In order to develop clinician-researchers and improve students' critical thinking, evidence-based practice, and professional skills, a research-oriented medical curriculum incorporates instruction in research methodology, critical evaluation of literature, and scientific inquiry into the medical education process. Research ethics, biostatistics, study designs, and practical skills like protocol preparation and data analysis are frequently taught in this curriculum. These methods of teaching might differ among medical schools to meet their particular settings and learning objectives (Ahmed et al., 2016). The following are some of the essential elements of a medical curriculum focused on research:

## 33.7.1 Core research skills

### *33.7.1.1 Formulating research questions*

Using frameworks like PICOT (population, intervention/exposure, comparator, outcome, time) for clinical question criteria, medical students must be able to formulate research questions in order to transform their clinical curiosities into targeted, answerable inquiries that will ultimately guide the literature review, study design, and significant contributions to evidence-based medicine. This ability, which requires curiosity, mentorship, and methodical improvement, connects clinical observation with scientific investigation.

### *33.7.1.2 Knowledge about study designs*

Teaching medical students about study designs entails categorizing them into observational and interventional or experimental types. An organized method is used in effective teaching, which begins with fundamental ideas, progresses to practical application and integrates research with clinical practice often with the use of real-world examples and guidelines. In order to ensure that medical students get the "why" and "how" of designing studies, a combined strategy of planned didactic sessions, practical application, and critical assessment is the most effective way to teach research design (Cole, 2025).

### *33.7.1.3 Understanding research ethics and governance*

A thorough research curriculum must include research ethics and governance since they offer the necessary foundations for carrying out ethical, secure, and reliable research. They guarantee that research adheres to legal and regulatory obligations, upholds scientific integrity, and safeguards the rights and welfare of participants. In this sense, it is crucial that medical students learn about the components of research ethics and governance, including autonomy, beneficence, nonmaleficence justice, fundamental ethical principles, regulatory guidelines, research integrity, and the function of ethics review boards. In addition to these, a medical research ethics and governance program should emphasize fundamental ideas and how they are applied in real-world situations inside a formal regulatory framework.

### *33.7.1.4 Appraising scientific literature critically*

Focusing on study design, bias, data analysis, and applicability, critical appraisal of scientific literature in medical education teaches aspiring physicians to assess research validity, reliability, and relevance to enhance patient care. Early in the curriculum, structured tools such as Critical Appraisal Skills Program checklists are frequently used, progressing from basic validity checks to clinical application to develop critical evidence-based medicine skills. To address typical issues such student disinterest or perceived lack of relevance to patient management, curricula should incorporate this training early and progressively, possibly through interactive sessions and actual journal articles.

### *33.7.1.5 Biostatistics and data analysis*

Medical students receive biostatistics instruction that combines basic statistics, such as probability, hypothesis testing, and data kinds, with practical skills, such as using SPSS software, excel study design, clinical trials, epidemiology, data analysis, and communication. Additionally, through specialized modules, short courses, or online platforms, students should be taught how to use real-world health data to develop basic abilities for evaluating research and developing studies.

### *33.7.1.6 Writing research proposals and manuscripts*

A medical research curriculum should include knowledge of the standard structures, key components, and writing practices for both research proposals and manuscripts. Students need to understand how to articulate a research question, select appropriate methodology, and present findings clearly and concisely.

### *33.7.1.7 Communication and teamwork skills*

In order to improve patient outcomes, safety, and satisfaction while lowering errors, medical research curricula integrate communication and teamwork by teaching critical skills like active listening, empathy, shared decision-making, and clear information exchange for patients, families, and the healthcare team. This is frequently accomplished through long-term training and simulation. Beyond clinical treatment, these abilities are essential for collaborating on research projects, engaging the public, resolving patient complaints regarding communication, and promoting improved teamwork.

### 33.7.2 Integration into the curriculum

Using a scaffolded approach from fundamental concepts like epidemiology and ethics to designing studies, analyzing data, and presenting findings, integrating research skills into medical curricula entails required projects, long-term mentorship, dedicated research time, and integrating statistics into core subjects. This fosters a culture where all future physicians, not just researchers can apply evidence-based care. To incorporate research into the medical curriculum, it is best to begin with basic knowledge in the first year and develop skills gradually over the subsequent years, preferably culminating in a research project. To ensure that every student gain experience, it is also crucial to incorporate modest research projects or literature evaluations into courses like public health and basic epidemiology and research. Scaffolded learning, which builds skills one after the other, is another way to present the fundamentals of research. For example, the primary emphasis in years 1 and 2 will be on the fundamentals of research, such as literature searches, critical evaluation, workshops on basic statistics, and summer research projects. However, mastering data analysis and producing reports or abstracts should be promoted during the sporadic years of small group research in lab or lecture classes. In order for the students to actively participate in the research process, it is also crucial to guarantee that they have regular time set aside for research.

#### *33.7.2.1 Research-based learning*

Through techniques like team-based learning and case-based learning that emphasize active student participation and lifelong scholarship, research-based learning in medical education incorporates scientific inquiry into the curriculum, going beyond traditional lectures to foster critical appraisal, problem-solving, and innovation. This prepares flexible physicians who can practice evidence-based medicine, contribute to advancements, and better engage with patients seeking health information.

#### *33.7.2.2 Skill development*

A research-oriented medical curriculum fosters critical thinking, evidence evaluation, data analysis, and scientific literacy by gradually incorporating research activities such as projects or workshops. It also improves evidence-based medicine and future responsibilities for researchers and clinicians by shifting the focus from facts to lifelong learning skills. To avoid bad experiences, however, good mentoring and curriculum design are crucial. Some of the key competencies that should be taught to students in order to produce creative and analytical doctors are literature review, research design, statistics, ethics, and writing.

#### *33.7.2.3 Fostering a research culture*

Incorporating practical research, mentoring, and ethical training into the curriculum, together with institutional policies that acknowledge and reward research and encourage teamwork, open communication, and openness, are all necessary to cultivate a research culture among students. Project-based learning, putting students in touch with working researchers, teaching ethics, offering rewards like research credits, and creating a welcoming atmosphere that encourages inquiry, critical thinking, and a variety of contributions are all important tactics.

### 33.7.3 Teaching research skills in medical education

In order to develop critical thinking, literature review, data analysis, and writing skills, medical educators integrate practical projects, mentorship, and structured curricula. They also emphasize teamwork, ethical considerations, and practical application to promote lifelong learning and evidence-based practice. Small group exercises, mentored projects within real studies, and making sure faculty members support the development of a robust research culture are important strategies.

## 33.8 Impact of research on medical education

Medical research has a big influence on medical education since it drives the curriculum's content, provides a foundation for evidence-based medicine, develops critical thinking and analytical skills in students, and gives aspiring doctors special career development options. Additionally, it strengthens the medical school's capacity to develop novel therapies and methods, which enhances patient care and public health outcomes.

### 33.8.1 Medical students

Medical research significantly enhances medical students' education by promoting a deeper comprehension of complex medicine, supporting evidence-based practice, enhancing communication, and preparing them for future academic or clinical careers by combining theory with real-world application and advancing professional development.

#### *33.8.1.1 Critical thinking and analysis*

Medical research fosters critical thinking in students by providing them with the means to assess literature, interpret complicated data, locate trustworthy evidence, and reach well-reasoned conclusions. Students can go beyond rote memorization to active problem-solving and application of evidence-based medicine in real-world situations through active involvement and collaborative learning, enhancing reasoning, self-correction, and objective evaluation skills.

#### *33.8.1.2 Information gathering*

Through practical experience in systematic data collection, critical evaluation of information sources, and problem-solving in real-world medical contexts, medical research helps medical students enhance their information gathering skills. Whether it is secondary data from databases already in existence and published literature, or primary data via experiments, surveys, or interviews, students learn the organized, rigorous process of gathering data. Due to this practical experience, they learn about a variety of processes, including qualitative and quantitative approaches, as well as which instruments are appropriate for specific research challenges.

#### *33.8.1.3 Literature review and appraisal*

Medical research helps students develop their critical appraisal and literature review skills by turning them from passive learners into active, problem-based judges of evidence and by combining theoretical knowledge with real-world application. This hands-on method helps students understand the entire cycle of scientific inquiry. Here, active problem-solving abilities will be emphasized. Students undertaking research will be expected to analyze and present research papers related to a specific clinical condition or project instead of receiving didactic instruction. This active method deepens understanding while fostering an interesting learning atmosphere.

#### *33.8.1.4 Research design*

Through practical application of theoretical information and hands-on, experiential learning, medical research helps medical students expand their grasp of study design. Because it enables students to move from theory to real-world circumstances and directly engage with the process and challenges of study design, this method is more effective than standard classroom training alone. Encouraging students to apply their theoretical knowledge of various study types, such as descriptive, analytical, experimental, methodologies and to real projects, is crucial because it helps them understand when to use a particular design to address a given research question (Kapoor, 2016).

#### *33.8.1.5 Data analysis and statistics*

Medical research helps medical students improve their data analysis and statistical skills by providing them with hands-on experience with real-world data, bridging the gap between abstract theory and practical application, and fostering critical thinking that is essential for evidence-based care. One advantage of this skill is that rather than only learning abstract statistical formulas, medical research students may apply concepts like hypothesis testing, $P$-values, and confidence intervals to actual patient data. As a result, the ideas become clearer and more tangible.

#### *33.8.1.6 Research ethics*

Through integrated curricula, historical case studies, simulated IRB reviews, and hands-on participation in ethical review procedures, medical research develops students' knowledge of research ethics. It emphasizes fundamental concepts like autonomy, beneficence, nonmaleficence, and justice while teaching them skills in informed consent, privacy, risk assessment, and conflict of interest management, preparing them to ethically advocate for patients in research. These aforementioned abilities are crucial for fostering a lifetime learning habit because they help students develop into autonomous, lifelong learners who are able to assess new material critically and adjust to medical advancements. Medical practitioners can also incorporate evidence into their practice and make more informed judgments for patient care by comprehending research ideas.

### 33.8.2 Teaching and learning process

Medical research has a significant influence on the teaching and learning process in healthcare education by encouraging the adoption of evidence-based practices, fostering critical thinking and analytical abilities, and bridging the gap between theoretical knowledge and real-world clinical application. Some of the most significant ways that medical research has impacted the teaching and learning process are listed below:

#### *33.8.2.1 Foundation for evidence-based medicine*

Evidence-based medicine, which promotes the careful, transparent, and prudent application of the best available data when making decisions about patient treatment, is based on medical research. Students learn to critically evaluate scientific material and apply it in practice by incorporating research findings into the curriculum which improves patient outcomes.

#### *33.8.2.2 Development of critical thinking and analytical skills*

Research activities assist students to improve on their critical thinking, problem-solving, and analytical skills. In order to prepare them for challenging clinical situations, students also have the chance to learn how to assess evidence, recognize issues, and create solutions.

#### *33.8.2.3 Innovation in teaching methods*

New teaching and learning approaches are developed and evaluated as a result of research in medical education. Effective techniques like simulation-based training have been adopted as a result, enabling students to learn difficult operations and decision-making techniques in a secure setting with high-fidelity mannequins or standardized patients without endangering real patients. Students are successfully steered toward the art of connecting theory to practice through cutting-edge teaching techniques like problem-based or case-based learning, which employ actual clinical examples to foster active, inquiry-based learning. These approaches also provide the student with an opportunity to collaborate with a multidisciplinary team.

#### *33.8.2.4 Curriculum development and integration*

Research results have an impact on curriculum design and enable medical curricula to be updated on a regular basis. Integrating research technique both vertically and horizontally throughout the program to increase its effectiveness is one of the crucial steps in this regard. Medical research also shapes educational goals to generate skilled physician-scientists, incorporates critical thinking and evidence-based medicine, and provides the most recent evidence for clinical practice, all of which have an impact on curriculum creation.

#### *33.8.2.5 Professional growth of educators*

Participating in research will give an opportunity for the medical educators to enhance their own pedagogical expertise, ensuring they are knowledgeable about the most recent developments and capable of employing the most effective teaching strategies. This also helps close the knowledge gap between educational research and practical teaching.

#### *33.8.2.6 Promotion of lifelong learning*

Medical research fosters lifelong learning since it encourages a culture of ongoing inquiry, requiring the regular updating of information, and cultivating the critical evaluation skills required to adjust to the quick changes in healthcare. It compels professionals to actively participate in changing evidence-based practices and go beyond static, textbook knowledge.

## 33.9 Driving forces of modern medical research

Modern medical research is driven by a complex interplay of technological advancements, evolving healthcare needs and cost pressures, and the increasing demand for personalized care. Some of the key driving forces of modern medical research are illustrated in Fig. 33.5.

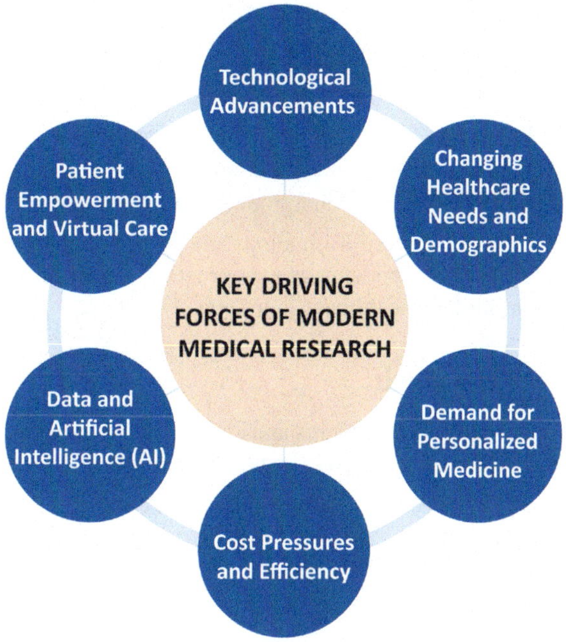


FIGURE 33.5 Key driving forces of modern medical research.

### 33.9.1 Technological advancements

New research continues to be sparked by radical advances resulting from developments in computer technology, biomedical technologies, and molecular and cell biology. Advanced imaging such as MRI, CT scans, AI, and nanotechnology are examples of specific technologies that allow for more precise diagnosis, focused medication distribution, and better patient outcomes.

### 33.9.2 Changing healthcare needs and demographics

The need for novel cures, treatments, and technologies is fueled by the world's aging population and the incidence of chronic illnesses. Global pandemics and other public health problems call for quick research and development of innovative treatments and preventative measures.

### 33.9.3 Demand for personalized medicine

Due to their increased knowledge, empowerment, and access to medical information, patients are actively searching for tailored treatments. As a result of developments in genomics, proteomics, and metabolomics, scientists may now develop medications and lifestyle recommendations that are specific to each person's unique genetic makeup in order to maximize effectiveness and minimize side effects.

### 33.9.4 Cost pressures and efficiency

Modern medical research is heavily influenced by economic concerns and the need for efficiency, which forces a change in focus from just finding novel treatments to finding ones that are affordable. Research is increasingly concentrated on optimizing health outcomes (quality-adjusted life years, or QALYs) within set budgetary restrictions, as healthcare costs rise as a result of aging populations and chronic illnesses.

### 33.9.5 Data and artificial intelligence

The development of AI and machine learning for predictive analytics, clinical decision support systems, and seeing minute patterns that human eyes would overlook is fueled by the exponential expansion of medical data, including genomes, patient records, and remote monitoring data.

### 33.9.6 Patient empowerment and virtual care

Patients can now monitor their own health, obtain information, and receive remote consultations due to the growth of social media, telemedicine, and mobile health. This change encourages research into user-centric technology and virtual care models while empowering people in their healthcare journeys.

## 33.10 Benefits of medical research

Medical research offers benefits like advances in treatments and cures, leading to better disease prediction, prevention, and control, ultimately improving patient outcomes and public health. Some of the key benefits of medical research are listed below:

### 33.10.1 Develops new treatments and cures

In order to improve health, prolong life, and improve quality of life by making healthcare safer, more efficient, and more individualized, medical research is essential as it contributes in creating new cures and treatments, better diagnostics, medications like vaccines and antibiotics, surgical methods, and technologies like gene editing. In order to create long-term value for patients and healthcare systems, this approach entails rigorous clinical trials to demonstrate safety and efficacy. This process drives innovation from basic discoveries to patient-ready medicines for ailments ranging from cancer to rare genetic disorders.

### 33.10.2 Improves disease management

Medical research significantly improves disease management by finding new treatments, refining existing methods, and enabling personalized, proactive care through understanding disease biology, developing diagnostics, and integrating technology like AI for early prediction and monitoring. This ultimately improves patient's quality of life and lowers healthcare costs. Medical research contributes to the control of disease by helping us discover new drugs, treatments, and methods for better management or cures. It also aids in our understanding of risk factors and disease mechanisms, leading to more targeted and effective interventions.

### 33.10.3 Enhances patient care

Medical research enhances patient care by creating novel medicines, improving existing therapies, establishing evidence-based standards, and promoting a culture of continuous learning. Through advancements in drugs, equipment, and treatment methods, this leads to improved results, more patient confidence, and more efficient, individualized therapy. Medical research also enables healthcare personnel to deliver better care by establishing a learning health system and training them to be leaders who identify gaps and seek solutions.

### 33.10.4 Informs public health policy

Medical research is crucial to public health policy because it provides the evidence needed to create effective, data-driven programs for illness prevention, health promotion, and resource allocation. In order to improve population outcomes, it also aids in identifying health trends, validating interventions like screens or vaccinations, and improving healthcare delivery. It serves as the basis for evidence-based policy and guarantees that decisions are based on scientific findings rather than just anecdotal.

### 33.10.5 Contributes to knowledge

Medical research advances knowledge and ultimately improves our ability to prevent, diagnose, and manage health conditions for a healthier society by understanding population health needs, creating new theories, developing novel drugs, vaccines, and medical devices, and informing evidence-based practices for better patient care.

### 33.10.6 Staying updated

Keeping up with medical research is essential because it promotes better patient care through evidence-based practices, guarantees that healthcare professionals maintain their credibility and expertise, encourages innovation, and increases the effectiveness of the health system, all of which eventually lead to better health outcomes and new treatments. In order to keep up with the rapid advancements, new methods of learning beyond textbooks are even needed.

## 33.11 Challenges of medical research

Securing funding, navigating complicated regulations like IRB approvals and agreements, recruiting and retaining diverse patients, managing ethical dilemmas like informed consent, data privacy, and AI use, ensuring data quality, dealing with workforce shortages, high costs, and time constraints, and combating bias and errors that can skew results and impede the development of new treatments are just a few of the many challenges that medical research faces. Some of the vital challenges in medical research are listed below:

### 33.11.1 High costs

Complex regulations, patient recruitment or retention, specialized staff, manufacturing, and data management all contribute to high costs in medical research, particularly for clinical trials. These factors ultimately limit innovation, slow drug development, raise final product prices, and may even result in underfunded, unsuccessful trials that compromise patient access and research ethics.

### 33.11.2 Time-consuming

Due to complicated methodologies, drawn-out procedures like grant reviews and institution board review approvals, sluggish patient recruitment, longitudinal study designs and the sheer amount of data or information to process, medical research is intrinsically time-consuming. This results in high costs, protracted treatment wait times, and the requirement for ongoing funding and qualified researchers.

### 33.11.3 Potential risks for participants

In addition to psychological distress like anxiety from traumatic recall and social or economic impacts like breach of confidentiality, lost wages, and damaged reputation, risks for participants in medical research can range from minor discomforts like pain, fatigue, and time commitment to serious physical harm like adverse drug reactions, injury, or even death in rare cases. Other challenges include controlling these hazards, getting informed consent, and handling trial-related issues like travel or complex procedures while upholding ethics and minimizing burdens.

### 33.11.4 Resource limitations

Resource constraints are significant obstacles to medical research, affecting funding, personnel, infrastructure, and data. They can cause projects to stall, data to be of lower quality, and capacity to be reduced, particularly in low-income settings. However, these issues can be lessened through cooperation, capacity building, and improved systems. Inadequate funding, a lack of qualified staff, equipment shortages, inadequate IT infrastructure, time restrictions, difficulties integrating or accessing data, bureaucracy, and difficulties recruiting or retaining patients are some of the other major problems.

### 33.11.5 Ethical considerations

Protecting human subjects through informed consent, maintaining privacy and confidentiality, weighing risks and benefits, and upholding scientific integrity, particularly with vulnerable populations, as well as data security and equitable compensation, are all important ethical issues in medical research that call for strict oversight to avoid exploitation, coercion, and misuse of data for the benefit of society.

### 33.11.6 Academic overload

Due to the enormous quantity of material to cover, extreme time constraints, and extremely rigorous schedules, academic overload is a major problem in medical education and research. Chronic stress, burnout, and detrimental effects on students' performance and well-being can result from this. In this setting, it is noted that medical researchers and students must deal with an overwhelming amount of information and a quick pace of new publications, which makes it challenging to absorb, comprehend, and apply all pertinent knowledge. In addition to time management challenges brought on by hectic schedules, there is little time for relaxation, personal life, or learning efficient time management techniques due to back-to-back lectures, clinical rotations, research obligations, and exams.

**Summary**

In order to guarantee that healthcare workers receive effective, pertinent, and evidence-based training, research in medical education is essential. Educational practices must be regularly assessed and enhanced through methodical study as medical knowledge, technology, and healthcare systems continue to advance. Improving teaching and learning strategies is one of the main benefits of research in medical education. The best teaching methods, curriculum, and evaluation instruments for fostering students' critical thinking, clinical skills, professionalism, and knowledge retention are determined by educational research. Education becomes more effective and learner-centered as a result. Lastly, research improves accountability and policymaking in medical education. Research-produced evidence influences resource allocation, educational policy, and accrediting requirements. In the end, superior medical education research results in more skilled medical personnel and better patient outcomes.

**Key point**

**Future perspectives: the emergence of precision education**

Rapid developments in healthcare, technology, and educational philosophy will influence medical education research in the future. Medical education research will become more and more important in producing skilled, sympathetic, and flexible practitioners as the demands on healthcare workers continue to rise. Competency-based medical education is another crucial viewpoint. The goal of future research is to investigate ways to identify, evaluate, and monitor competencies throughout the medical training continuum. Workplace-based assessment, trustworthy professional activities, and long-term learner performance evaluation will be prioritized. In order to create a "precision education" model that is customized for each student, medical research in education is moving toward digital integration with artificial intelligence, precision medicine, and simulation. Lastly, outcomes-based research that connects educational interventions to actual clinical performance and patient care results will receive more attention. Future medical education research will directly improve healthcare quality and safety by bridging the knowledge gap between education and practice.

## References

Abramson, E.L., Paul, C.R., Petershack, J., Serwint, J., Fischel, J.E., Rocha, M., Treitz, M., McPhillips, H., Lockspeiser, T., Hicks, P., Tewksbury, L., Vasquez, M., Tancredi, D.J., Li, S.T.T., 2018. Conducting quantitative medical education research: from design to dissemination. Academic Pediatrics 18 (2), 129–139. Available at https://doi.org/10.1016/j.acap.2017.10.008, http://www.elsevier.com/wps/find/authored_newsitem.cws_home/companynews05_01099.

Ahmed, Y., Taha, M.H., Khayal, S., 2024. Integrating research and teaching in medical education: challenges, strategies, and implications for healthcare. Journal of Advances in Medical Education and Professionalism 12 (1), 1–7. Available at https://doi.org/10.30476/jamp.2023.99751.1854, https://jamp.sums.ac.ir/article_49775_386c8f64e10895b88bf5080321a4d561.pdf.

Ahmed, R., Farooq, A., Storie, D., Hartling, L., Oswald, A., 2016. Building capacity for education research among clinical educators in the health professions: a BEME (Best Evidence Medical Education) systematic review of the outcomes of interventions: BEME Guide No. 34. Medical Teacher 38 (2), 123–136. Available at https://doi.org/10.3109/0142159X.2015.1112893.

Boet, S., Sharma, S., Goldman, J., Reeves, S., 2012. Review article: medical education research: an overview of methods. Canadian Journal of Anesthesia/Journal canadien d'anesthésie 59 (2), 159–170. Available at https://doi.org/10.1007/s12630-011-9635-y.

Chen, W., Reeves, T.C., 2020. Twelve tips for conducting educational design research in medical education. Medical Teacher 42 (9), 980–986. Available at https://doi.org/10.1080/0142159x.2019.1657231.

Cole, R., 2025. Building a program of research in medical education: recommendations from the professional literature. Medical Education Online 30 (1), 2546869. Available at https://doi.org/10.1080/10872981.2025.2546869 Epub 2025 Aug 12. PMID: 40798939; PMCID: PMC12351726.

Eva, K.W., Lingard, L., 2008. What's next? A guiding question for educators engaged in educational research. Medical Education 42 (8), 752–754. Available at https://doi.org/10.1111/j.1365-2923.2008.03135.x.

Garg, A.X., Hackam, D., Tonelli, M., 2008. Systematic review and meta-analysis: when one study is just not enough. Clinical Journal of the American Society of Nephrology 3 (1), 253–260. Available at https://doi.org/10.2215/CJN.01430307.

Kapoor, M.C., 2016. Types of studies and research design. Indian Journals Anaesth 60 (9), 626–630. Available at https://doi.org/10.4103/0019-5049.190616 PMID: 27729687; PMCID: PMC5037941.

Lee, G.S.J., Chin, Y.H., Jiang, A.A., Mg, C.H., Nistala, K.R.Y., Iyer, S.G., Lee, S.S., Chong, C.S., Samarasekera, D.D., 2021. Teaching medical research to medical students: a systematic review. Medical Science Educator 31 (2), 945–962. Available at https://doi.org/10.1007/s40670-020-01183-w, springer.com/education+&+language/journal/40670.

Norman, G., 2002. Research in medical education: three decades of progress. BMJ (Clinical Research ed.) 324 (7353), 1560–1562. Available at https://doi.org/10.1136/bmj.324.7353.1560.

Ofei-Dodoo, S., 2025. The importance of basic research knowledge in undergraduate medical education. Kansas Journal of Medicine 18 (1), 28–30. Available at https://doi.org/10.17161/kjm.vol18.22599.

Röhrig, B., Wachtlin, D., Blettner, M., 2009. Types of study in medical research - Part 3 of a series on evaluation of scientific publications. Deutsches Ärzteblatt international 106, 262–268. Available at https://doi.org/10.3238/arztebl.2009.0262.

Ratajczak, P., Oziewicz, K., Sommer, I., Kopciuch, D., Paczkowska, A., Zaprutko, T., Kus, K., 2025. The perspective on secondary research practices: a cross-sectional analysis. Healthcare 13 (8), 927. Available at https://doi.org/10.3390/healthcare13080927.

Simmonds, M.C., Higgins, J.P.T., Stewart, L.A., Tierney, J.F., Clarke, M.J., Thompson, S.G., 2005. Meta-analysis of individual patient data from randomized trials: a review of methods used in practice. Clinical Trials 2 (3), 209–217. Available at https://doi.org/10.1191/1740774505cn087oa.

Chapter 34

# Quantitative research methods in medical education

*"Without data, you're just another person with an opinion."*

—W. Edwards Deming

## 34.1 Introduction

In order to test hypotheses, spot trends, and make evidence-based decisions about curriculum, instruction, and assessment effectiveness, quantitative research in medical education uses numerical data and statistical analysis. It frequently employs techniques like surveys, experiments, and standardized tests on large samples. There are numerous uses for quantitative research methods in medical education. For example, experimental methods like randomized controlled trials (RCTs) are essential to establish cause-and-effect links between interventions and outcomes in medical learners. These techniques are used to evaluate causal linkages, quantify and characterize educational events, and look into correlations between variables.

In order to analyze and create medical education research studies, one must comprehend the fundamentals of quantitative research. Understanding quantitative study design, technique, and results is crucial, as evidenced by the growing emphasis in medical education on evidence-based decisions based on thorough, high-quality research. The findings of research studies on medical education can offer instructors important information about excellent practices (Ratelle et al., 2019). Essentially, quantitative research involves the systematic investigation of phenomena through statistical, mathematical, or computational techniques. It aims to quantify data and generalize results from a sample to a larger population.

## 34.2 Characteristic features of quantitative research

Quantitative research, which explores deeper understanding, primarily focuses on providing answers to the questions of how many and how much, frequently employing statistical analysis to produce results that are reliable, repeatable, and predictable (Tavakol and Sandars, 2014). In order to uncover trends, test theories, and generalize results, it gathers objective, quantifiable data from sizable groups using organized instruments like surveys. Some of the important features of the quantitative research are described in Fig. 34.1.

### 34.2.1 Numerical data and statistics

The essential components of quantitative research in medical education are numerical data and statistics, which offer the objective, quantifiable proof required for hypothesis testing, trend identification, and evidence-based decision-making. They make it possible for researchers to accurately, methodically, and broadly measure educational processes and effects (Slater and Hasson, 2025). The premise of quantitative research is that the variables being examined may be quantified and expressed numerically. Exam results, GPAs, clinical rotation hours, patient satisfaction scores on a scale, and illness prevalence rates are a few examples of data gathered in medical school.

**The Principles and Practice of Medical Education. DOI:** https://doi.org/10.1016/B978-0-443-44197-4.00016-3

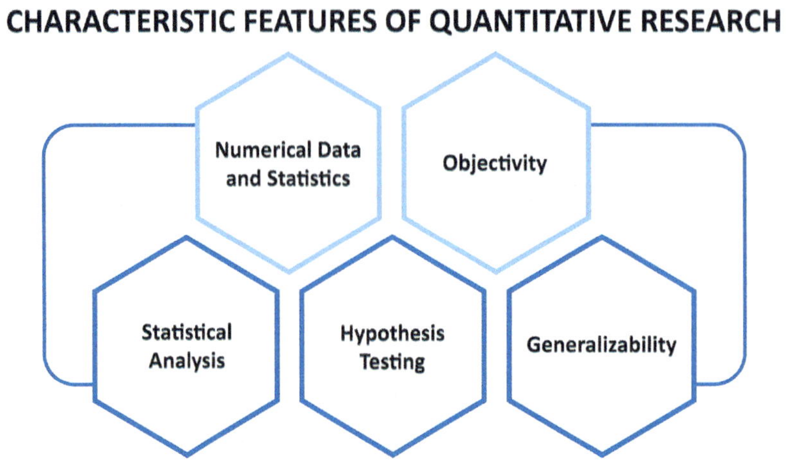


FIGURE 34.1 Characteristic features of quantitative research.

### 34.2.2 Objectivity

Based on the positivist paradigm, objectivity is a fundamental and essential component of quantitative research in medical education. In order to guarantee that conclusions are supported by empirical data and are verifiable by all, it refers to reducing researcher bias and subjectivity. Objectivity guarantees that the researcher's personal convictions, opinions, and biases do not skew the procedures of gathering and analyzing data. Quantitative research in medical education mostly uses structured and standardized tools, such as validated surveys, experiments like RCTs, and physiological measures, to attain impartiality. These techniques guarantee that data is gathered uniformly among various individuals and environments.

### 34.2.3 Generalizability

One essential quality of quantitative research in the realm of medical education is generalizability. In order to guarantee that results may be transferred or expanded beyond the particular study group to larger populations, settings, and contexts this, method makes use of statistical analysis and high sample sizes. External validity, or the degree to which the study's findings may be applied to circumstances outside of the study itself, is one of the most important components of generalizability in medical education research. In the context of medical education, this could refer to whether the results of a novel teaching approach that was tested at one university can be applied to other medical schools.

### 34.2.4 Hypothesis testing

One distinguishing feature and fundamental component of quantitative research techniques used in medical education is hypothesis testing. In quantitative research, a hypothesis is a tested, tentative statement that is based on current theory and literature and predicts the relationship between independent and dependent variables. It offers a precise, quantifiable, and verifiable prediction regarding a study's result, enabling researchers to use statistical analysis to accept or reject it. It gives academics a methodical framework for making unbiased, data-driven choices on educational interventions, curriculum, and results. One of the main characteristics of hypothesis testing in medical education is that it advances objectivity and empirical research since it goes beyond subjective observations by evaluating a proposed theory or assumption using empirical data, such as test results, surveys with numerical scales, and graduation rates. This increases the validity and dependability of the study's conclusions. Moreover, the research design is guided by the hypothesis, which establishes the sampling strategy, instruments for gathering data, and analytical methods required to verify the claim.

### 34.2.5 Statistical analysis

One of the main and distinguishing characteristics of quantitative research in medical education is statistical analysis. It gives researchers the necessary tools to organize, compile, and analyze the numerical data gathered during studies, enabling them to make unbiased, broadly applicable judgments on educational interventions, policies, and practices. There are two forms of statistical analysis used in medical education research, namely, descriptive and inferential. Researchers use descriptive statistics to summarize and characterize the fundamental characteristics of the data using metrics like means, medians, modes, and standard deviations. This gives the sample and important variables a clear image, for example, survey results, student test results, and demographics. Conversely, inferential statistics like regression analysis, ANOVA, and t-tests are used to test hypotheses and make inferences about a broader population from a smaller sample. This could entail evaluating if a novel simulation curriculum considerably enhances clinical abilities in comparison to a conventional approach in medical education.

## 34.3 Quantitative research designs

Experimental, quasi-experimental, and descriptive correlational research are examples of quantitative research approaches used in medical education. The study issue determines the approach to be used. While observational studies, such as surveys and correlational studies, are frequently used to describe trends and relationships, experimental designs are most suited for inferring cause-and-effect relationships.

### 34.3.1 Descriptive research

Descriptive quantitative research in medicine uses surveys, observations, and case series to measure traits, behaviors, or health statuses in populations, such as disease prevalence and demographics without changing variables. The goal is to create a comprehensive overview to identify trends and direct future research but not causality. Case reports or series, surveillance, and cross-sectional or longitudinal surveys are some of the significant forms of descriptive research that concentrate on the "what,“ "where,“ and "when“ of health occurrences (Deckert and Wilson, 2023). One of the main features of descriptive research is its observational nature, in which the researchers act as observers, seeing and measuring variables as they naturally occur. Furthermore, it explains relationships but is unable to establish cause and effect. Since descriptive research offers baseline data in all these areas, it becomes crucial for comprehending health trends, prevalence, and demography. In medicine, case reports and series are common quantitative descriptive approaches that are effective for generating hypotheses because they provide in-depth descriptions of individual cases or groups of patients with distinct illnesses (Aggarwal and Ranganathan, 2019). Another type of descriptive method is surveillance studies, which involve the ongoing tracking of health habits, risk factors, or disease incidence in communities across time. In medical education, surveillance studies serve as a descriptive research method that entails the systematic, ongoing gathering, analysis, and interpretation of data pertaining to clinical outcomes, learner performance, or educational processes. In contrast to experimental study, surveillance observes the educational environment in order to spot trends, patterns, and areas that require improvement. This method is essential for tracking training quality, clinical learning environment safety, and trainee competency over time in medical education. Some of the important application of descriptive research is in evaluating training programs, determining student requirements and competencies, seeing trends in medical education, and creating baseline data for upcoming, more intricate educational interventions.

### 34.3.2 Correlational research

In medical research, correlational research is a quantitative, nonexperimental methodology that finds statistical relationships between two or more variables, such as exercise and cardiac health or diet and blood pressure, without changing them. This method can help generate hypotheses for future studies but it can only demonstrate association in terms of positive, negative, or none at all. Its application has proven crucial in medical domains like epidemiology, such as cohort, cross-sectional, and case-control studies when ethical or pragmatic considerations prohibit experimentation, enabling the identification of patterns in patient data already in existence (Devi et al., 2023). In correlational research, variables are measured while being passively observed. It is distinguished by being nonexperimental since the researcher does not intervene or change any factors. Determining whether the variables move together, indicating a positive correlation, or move in opposite directions, indicating a negative correlation, or they do not move at all, indicating a zero correlation, is the strength of correlational research. Correlational research in medical education establishes connections between educational parameters, such as learning strategies, test scores, or student well-being, without changing the variables. Important applications include forecasting performance, validating assessment tools, and evaluating the effectiveness of courses. Teachers can identify areas for improvement by, for instance, comparing test scores to student motivation levels or comparing workplace burnout rates.

### 34.3.3 Causal-comparative or quasi-experimental research

A crucial quantitative technique in medicine is causal-comparative or quasi-experimental research which compares existing groups, such as patients with various conditions or exposures, to determine how an independent factor, such as lifestyle, affects an outcome, such as disease progression, when true experiments are unethical or impractical. Using designs like as nonequivalent groups, regression discontinuity, or natural experiments to infer causality after an event or exposure, it is effective for real-world health concerns by simulating trials by examining preexisting differences without random assignment (Schenker and Rumrill, 2004). Comparing entire, naturally occurring groups, such as smokers versus nonsmokers or patients from various clinics, is what distinguishes casual comparative research. There is no randomization in this kind of study because of logistical or ethical issues. In order to identify the sources of current consequences, this research frequently looks backward. When random student assignment is unfeasible or unethical, quasi-experimental research in medical education assesses the effects of educational interventions, such as new curriculum, simulation training, or teaching techniques. It is a useful substitute for RCTs, providing useful information about the efficacy of education in practical contexts.

### 34.3.4 Experimental research

Experimental research, a type of basic quantitative methodology in medicine, actively manipulates variables, such as a new drug or therapy, in controlled settings using randomization and control groups as in an RCT in order to establish definitive cause-and-effect relationships between interventions and health outcomes. This offers compelling proof of the effectiveness of interventions, illness processes, and treatments. The introduction or modification of an independent variable, such as dosage or treatment type, is one of the main features of experimental research. Dependent variables like symptom reduction and survival rates both before and after the intervention are measured in experimental research. Assigning participants to groups, balancing unknown influences, and bolstering internal validity are all made possible by this kind of randomization. The RCT is the best illustration of an experimental study design. experimental research in medical education assesses the efficacy of instructional strategies, curricula, and assessment instruments, as it can compare interventions versus control groups. RCTs are frequently used to assure evidence-based improvements. Applications include evaluating new, technology-enhanced curricula, testing virtual simulations, and determining how instructional strategies affect student results.

### 34.3.5 Randomized control trial (RCT)

An RCT is considered the gold standard for assessing the effectiveness of interventions because it uses random assignment to treatment and control groups to minimize bias and demonstrate cause-and-effect relationships. This real experiment allows researchers to ascertain whether a new drug, therapy, or program actually works by comparing outcomes between groups (Bhatia and Grewal, 2021). One of the key components of an RCT as an experimental design is the random assignment of participants to either the experimental group or the control group. In this case, the researcher actively modifies the independent variable. In this instance, the control group acts as a baseline, allowing researchers to distinguish the intervention's effect from other factors. Blinding, in which participants, researchers, or analysts are not aware of who is in which group, is often used to prevent bias. In public health, medicine, and other fields, RCTs are basically the strictest way to generate high-quality data for interventions (Sullivan, 2011). RCTs in medical education, by randomly allocating students or residents to an intervention or control group, assess the efficacy of novel teaching strategies, curricula, and technological advancements. When evaluating instructional tactics, such as contrasting simulation-based training with traditional lectures, they offer robust, causal evidence to lessen prejudice.

### 34.3.6 Survey method

Although it can also include qualitative components with open-ended questions, the survey method is a fundamental and popular quantitative research technique used widely in medical education to collect numerical data such as ratings or frequencies from numerous learners about their knowledge, attitudes, behaviors, or experiences. This allows for statistical analysis to describe trends or test hypotheses (Gürbüz, 2017). Likert scales and multiple-choice questions, which produce numerical data, are common closed-ended survey questions. In order to identify trends or test theories, the gathered data are statistically analyzed using means, correlations, or t-tests. This approach makes it possible to effectively reach a bigger population in order to obtain comprehensive insights. Assessing student satisfaction with a curriculum or gauging attitudes toward novel teaching strategies are two significant applications of surveys in medical education. In medical education, surveys are crucial instruments for collecting information from students, instructors, and practitioners regarding attitudes, beliefs, behaviors, and educational results. They are widely used in curriculum review, competency assessment, educational needs assessment, and program enhancement, especially in graduate medical education.

## 34.4 Steps in quantitative research

Identifying a research problem and question, analyzing the literature to develop hypotheses, designing a study with particular variables and methods, gathering and preparing numerical data, analyzing the data using statistical techniques, and interpreting and sharing the results are all steps in quantitative research in medical education. Ensuring the validity and reliability of measurement instruments and carefully choosing a representative sample to improve the results' generalizability are important factors to take into account (Abramson et al., 2017). The steps in quantitative research are as follows (Fig. 34.2):

1. Identifying a research problem,
2. Reviewing the literature,
3. Specifying a research purpose,
4. Designing a study,
5. Analyzing and interpreting the data,
6. Reporting research.

FIGURE 34.2 Steps in quantitative research.

### 34.4.1 Identifying a research problem

The fundamental first step in quantitative research is identifying a research problem, which entails identifying a quantifiable gap, contradiction, or unanswered question in the body of knowledge. This is usually done through literature review and observation in order to establish a focused, pertinent, and feasible study that directs the development of variables, hypotheses, and data collection techniques for empirical investigation. One of the primary features of the research problem is that it allows the researcher to identify gaps since the research can concentrate on areas where existing theories are inadequate, inconsistent, or poorly researched, particularly in recent literature. To enable quantitative testing, it is also important to keep in mind that the problem must be specific, quantifiable, and not ambiguous. Additionally, it is crucial to make sure the problem can be studied with the resources at hand and makes a significant contribution to theory or practice. Evidence supporting the problem's significance should be included when identifying a research problem (Kittur, 2023).

### 34.4.2 Reviewing the literature

A critical step in quantitative research is reviewing the literature which helps researchers define the problem, find gaps, develop theoretical frameworks, select methodologies, avoid duplication, and build credibility by synthesizing existing knowledge. This results in refined research questions and hypotheses that ensure relevance and build upon previous findings. The background, history, and present state of the subject are some of the main goals of a literature review. It also assists us in identifying areas that require further investigation or unresolved questions. It is crucial for defining variables, focusing the issue and developing precise questions or hypotheses. It is also helpful in exposing other people's successful or failing strategies, plans, and ways of analysis. The literature review contributes to credibility since it shows that the research is grounded on and builds upon current knowledge.

### 34.4.3 Specifying a purpose and research questions

A fundamental step in quantitative research is defining the purpose and research questions. This establishes the focus of the study by defining what the researcher wants to know and precisely how the researcher will find out through measurable variables. This leads to specific hypotheses that direct data collection and analysis, ultimately clarifying the problem, variables, and expected outcomes. It involves breaking down a broad problem into precise, verifiable statements that guide the entire study design, guaranteeing feasibility and emphasis. From design to data analysis, the entire research process is guided by the process of defining the research question. Additionally, it guarantees measurability by concentrating on factors that can be measured, which is crucial for numerical data. It also clarifies for the researcher the purpose of the investigation. Completing this stage will enable the researcher to translate a general interest into precise, measurable objectives, ensuring that the quantitative study is viable, logical, and focused.

### 34.4.4 Designing a study

Designing a study is a crucial phase where the researcher plans how to address the research issues in order to guarantee validity and reliability for testing hypotheses with numerical data. Choosing a design, such as descriptive, correlational, or experimental, determining the population or sample, selecting data gathering techniques, such as surveys and observations, defining variables, and detailing analysis procedures are all part of this process. The study design serves as the basis for the ensuing data collection, analysis, and interpretation procedures, ensuring that the results are logical, trustworthy, and appropriately address the research issue. Quantitative data collecting techniques concentrate on using organized, impartial, and repeatable ways to collect numerical data for statistical analysis. Surveys and questionnaires, experiments, organized observation, and secondary data analysis are important techniques that usually use closed-ended questions (such multiple-choice and Likert scales) to measure trends.

### 34.4.5 Analyzing and interpreting data

The researcher should have the data ready for analysis when it has been gathered. A critical step in quantitative research is data analysis and interpretation which involves cleaning, statistically testing, visualizing patterns, and elucidating findings to address research questions. This process moves from descriptive summaries to inferential conclusions about populations, all of which support evidence-based conclusions and decisions. Finding patterns, trends, and connections is the primary goal of data analysis and interpretation. It can also be used to verify assumptions and test hypotheses. This procedure is also used to support decisions and conclusions with empirical data.

### 34.4.6 Reporting research

In order to ensure transparency, replicability, and real-world impact, reporting research is an essential last step in the quantitative process. It involves organizing and clearly communicating the analyzed results, interpretations, limitations, and implications to an audience, frequently using tables and graphs in accordance with the standard IMRaD (introduction, methods, results, discussion) format. Reporting research is crucial because it enables the researcher to communicate findings to stakeholders, policymakers, and other researchers. It also encourages transparency and reproducibility by allowing others to understand and confirm the methods used in the particular study. By making it simpler to translate complex numbers into understandable concepts, it influences practice or further research. Even though reporting is typically the last major stage after data collecting and processing, it involves organizing the entire research process, from the initial problem to the final conclusions, ensuring the findings appropriately address the initial concerns.

## 34.5 Applications of quantitative research in medical education

In order to improve teaching, assessment, and curriculum design for better clinical preparedness, quantitative research in medical education uses surveys, experiments, and correlational studies to measure learning outcomes, test interventions such as new curricula or teaching methods for effectiveness, track student progress, evaluate satisfaction, and identify learning patterns. Going beyond simply why and how to offer data-driven insights, it helps address concerns like how many, how much, and how effective. The essential applications in medical education are shown in Fig. 34.3.

### 34.5.1 Curriculum evaluation and effectiveness

By gathering and quantitatively analyzing numerical data to produce unbiased, broadly applicable insights, quantitative research is used in medical education to assess curricula and gauge their efficacy. This method aids in assessing program results and the effects of educational interventions. The main purposes of quantitative approaches are to examine cause-and-effect links between educational interventions and learning outcomes and to provide answers to queries regarding the magnitude of an impact. It also helps to monitor the learning outcomes by using standardized, quantitative assessments to assess changes in knowledge, skills, and attitudes following a curriculum revision. Quantitative research methods also aid in evaluating program satisfaction through the use of surveys and questionnaires containing numerical or scaled responses. This approach gathers information on how satisfied participants are with the curriculum's resources, content, and delivery.

### 34.5.2 Assessment validation

For assessment validation in medical education, which entails gathering empirical data to support the proper interpretation and application of assessment data, quantitative research is crucial. The validity and reliability of evaluation instruments are assessed using quantitative techniques. This involves calculating inter-rater or test–retest reliability using correlation coefficients and utilizing statistics such as Cronbach's alpha for internal consistency reliability. In addition, to make sure the instrument sufficiently samples the topic, researchers employ quantitative survey methodologies with content specialists to systematically review how effectively assessment items cover the pertinent learning domain or curriculum.

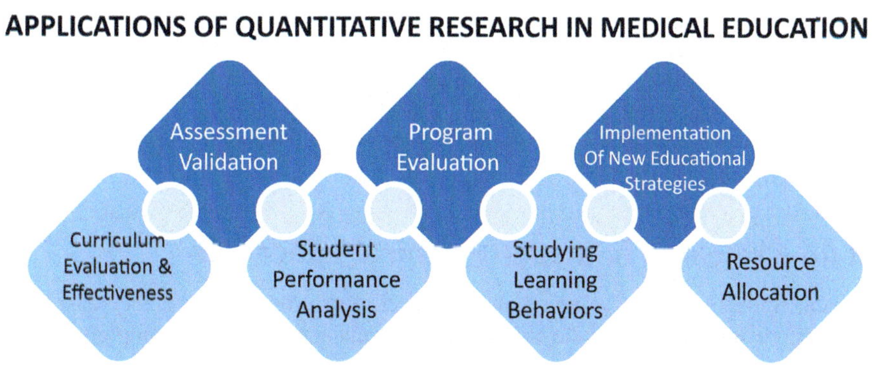


FIGURE 34.3 Applications of quantitative research in medical education.

### 34.5.3 Student performance analysis

Quantitative research is frequently used in medical education to assess student performance since it can provide objective, data-driven insights that can be applied to larger populations. Applications include evaluating teaching methods and predicting future academic and clinical success. Standardized tests, exams, and GPA data are utilized in relation to student performance analysis to assess the efficacy of various curricula or teaching strategies, such as contrasting e-learning with conventional teaching methods. Significant performance disparities between student groups can also be found using statistical techniques like ANOVA and t-tests.

### 34.5.4 Program evaluation

In order to give objective, quantifiable, and broadly applicable statistics on the efficacy, efficiency, and impact of medical education programs, quantitative research is used. It assists in determining if educational interventions are effective in altering learners' behaviors, knowledge, and abilities, which in turn affects patient outcomes. Clinical effectiveness can be ascertained by carrying out thorough research such as RCTs to examine the impact of various educational interventions on real patient outcomes such as lower readmission rates and improved health indicators. Determining clinical efficacy provides insight into the results and influence of medical programs. Conversely, evaluating program satisfaction and perception through the use of structured surveys and questionnaires with quantitative scales to get input from faculty and students regarding their satisfaction, opinions of the learning environment, and the caliber of courses or clerkships provides insight into program implementation and procedures.

### 34.5.5 Learning behaviors

Measuring the frequency and duration of learning activities, assessing the efficacy of teaching interventions, examining factors influencing student performance, and monitoring learning outcomes over time are just a few of the ways that quantitative research in medical education is used to investigate learning behaviors. To find out how much time medical students spend on different learning tasks such as patient engagement, literature search, formal instruction, etc., researchers employ quantitative techniques like log diaries and surveys. This makes it easier to compare the effects of various learning environments, like rotations in a hospital versus a general practice.

### 34.5.6 Implementation of new educational strategies

In order to give objective, data-driven proof of the efficacy, efficiency, and generalizability of novel medical education initiatives, quantitative research is widely used in their implementation. It enables teachers to assess results, test theories, and make well-informed, fact-based choices on curricula and instructional strategies. Large-scale surveys and correlational studies can reveal patterns that can guide program modifications, such as the relationship between study habits and academic achievement or the variables affecting students' participation in online learning.

### 34.5.7 Resource allocation

By offering an evidence-based, data-driven strategy to maximize the use of scarce resources including staff, buildings, equipment, and financing for optimal efficiency and equity, quantitative research in medical education is used to resource allocation. For instance, by examining several inputs and outputs, the Data Envelopment Analysis model is used to assess and compare the effectiveness of various units, such as departments or hospitals. In order to improve service quality, this aids in locating inefficiencies and optimizing resource distribution. In essence, medical education research uses quantitative methodologies to produce objective, broadly applicable evidence to improve the quality and results of training future healthcare workers.

## 34.6 Quantitative research in modern medical education

In contemporary medical education, quantitative research is crucial for promoting evidence-based practice, assessing the success of educational initiatives, and guiding the creation of curricula and policies. It complements qualitative methods to provide a thorough grasp of intricate educational dynamics by offering objective, quantifiable insights that can be extrapolated to broader populations. Quantitative approaches are used to assess educational programs, including curriculum changes, new training programs, and the effectiveness of instructional methodologies, by assessing outcomes such as student test scores, knowledge acquisition, and skill development. Additionally, student performance is evaluated using the quantitative approach. In this sense, the numerical data produced by objective assessments and standardized examinations to evaluate student and teacher performance aids in identifying areas that require more assistance or modifications to teaching strategies. Furthermore, in order to critically analyze clinical research and apply findings to patient care, supporting evidence-based medicine, medical students, and residents must have a solid grasp of statistics and quantitative research principles.

## 34.7 Role of technology in quantitative research in medical education

In order to objectively measure outcomes like skill acquisition, knowledge retention, and decision-making, technology in quantitative medical education research offers powerful tools for data collection like virtual reality, apps, or e-learning. In terms of analysis, technologies like artificial intelligence and learning analytics play a key role and in terms of delivery of immersive, scalable learning simulations, and virtual patients have a key contribution. This allows for data-driven curriculum improvements and individualized training for future doctors. It offers profound insights through measurable outcomes by converting conventional learning into quantified, interactive experiences.

## 34.8 Benefits of quantitative research in medical education

Benefits of quantitative research in medical education include data-driven evaluation, the ability to evaluate novel curricula and teaching strategies like simulations for efficacy, the identification of learning gaps, and the assurance that results are quantifiable and transferable to larger student populations. It offers unbiased, statistical proof to enhance curriculum design, gauge student performance indicators, identify the best dosage and training regimens, and produce trustworthy, repeatable results for improved clinical and educational decision-making. Among the main advantages are the following:

### 34.8.1 Evidence-based curriculum

Quantitative research is crucial for evidence-based medical curricula because it provides objective data to test student skill acquisition, such as statistical literacy and critical appraisal, identify gaps, and drive data-informed curriculum improvements for better patient outcomes. It illustrates advantages including better problem-solving, improved clinical reasoning, and more engaged learning using methods like surveys and trials.

### 34.8.2 Objective measurement

As the quantitative research in medical education employs standardized instruments and numerical data to minimize bias and enable direct comparisons and statistical analysis, objective measurement is a major advantage. When assessing educational initiatives or results, this method improves validity and reliability. Multiple-choice questions with standardized scoring, simulated checklists, and validated scales are examples of objective measurements that guarantee consistent data gathering throughout time and among various participants. This lessens the effect of individual grader bias, which might occur with more subjective assessments like essays or observational ratings.

### 34.8.3 Generalizability

One of the main advantages of quantitative research in medical education is generalizability, which enables conclusions from a particular, researched group to be securely extended to a larger population such as the students, instructors, or institutions within that environment. Researchers can choose a subset of participants that closely resembles the traits of the target population by employing random or representative sampling techniques. Additionally, inferential statistics can be used to interpret quantitative data, which is frequently gathered from large samples. This gives researchers a quantifiable level of confidence when estimating and drawing conclusions about the larger population.

### 34.8.4 Pattern identification

The statistical basis for evidence-based decision-making is provided by quantitative research in medical education, which is very successful in finding common, objective patterns and trends within huge datasets. For instance, educators can make well-informed, evidence-based decisions about curriculum design, assessment techniques, and resource allocation by identifying quantifiable patterns such as the relationship between particular teaching methods and student performance outcomes. Data patterns also aid in assessing the efficacy of clinical or educational interventions. For instance, following the introduction of a novel surgical procedure, researchers can monitor patterns in patient recovery rates, enabling a precise evaluation of its effects.

### 34.8.5 Efficiency and scalability

Quantitative research in medical education offers significant advantages in terms of speed and scalability by providing an organized technique to generate reliable, widely applicable data quickly and across large populations. The data collected in quantitative research is often numerical and may be quickly processed and analyzed using statistical methods. This expedites the process of drawing conclusions from data collection, which is crucial for making prompt judgments when creating medical curricula. Because quantitative instruments are standardized, they may be easily deployed across multiple institutions, regions, or even countries, making large-scale comparison studies easier.

### 34.8.6 Testing theories

One of the primary benefits of quantitative research in medical education is its capacity to provide the comprehensive testing of widely accepted theories and hypotheses. This methodical deductive approach allows researchers to apply well-established psychological or educational theories to novel contexts and use empirical evidence to support, enhance, or contradict them. Quantitative research typically starts with a theory in a deductive framework and builds particular, verified hypotheses from it. These concepts are then examined using numerical data and statistical analysis.

### 34.8.7 Reliability and replicability

The results of quantitative research are more credible and scientifically rigorous due to their greater dependability and reproducibility, which has a significant positive effect on medical education. The consistency of study findings over time and under various circumstances is referred to as reliability. Similar to a trustworthy medical test, quantitative approaches guarantee that a study produces consistent results if repeated by using standardized tools and procedures. Replicability is essential for confirming outcomes and demonstrating the validity of conclusions in other contexts or demographics. Confidence in the original claim's scientific validity is increased when a different researcher can repeat a study and obtain a consistent outcome.

## 34.9 Challenges of quantitative research in medical education

High costs, time constraints, data scarcity, complicated ethical approvals, and a lack of qualified researchers or faculty are some of the obstacles that quantitative research in medical education must overcome. These obstacles are exacerbated by difficulties in connecting education to patient outcomes, which makes it challenging to evaluate teaching effectiveness and necessitates substantial infrastructure, funding, and statistical expertise for reliable generalizable findings. The following is a list of some of the main difficulties with quantitative research.

### 34.9.1 Resource intensive

Large sample sizes, strict data collection procedures, and substantial statistical infrastructure and experience are the main reasons why quantitative research in medical education requires a lot of resources. These elements need large time, financial, and staff requirements. One of the key elements is that the studies are often longitudinal, monitoring patient outcomes and educational interventions over lengthy stretches of time necessitating ongoing financing and resources; because of this, the research takes a long period. However, quantitative studies frequently need data from a significant section of the target population to guarantee the generalizability and statistical power of findings, which can be costly and time-consuming to gather particularly across several institutions.

### 34.9.2 Data issues

The validity and dependability of results in quantitative research in medical education are greatly impacted by data-related issues. These problems affect every step of the research process, including data collection, processing, and interpretation. First, it is evident that errors, inconsistencies, and inaccuracies are frequently present in raw data used in healthcare and medical education which leads to disparities in data quality and accuracy. Inconsistent data entry, when data for a same variable may be entered in different formats, can also be a barrier to coding and analysis. Incomplete datasets are also a frequent issue, usually brought about by participants missing measurement points, skipping questions, or quitting a research. A significant number of missing data points might cause the dataset to become unbalanced, which can affect the data's validity and conclusions' accuracy.

### 34.9.3 Methodological complexity

For quantitative research, methodological complexity is a major obstacle because the method's emphasis on numerical data and statistical analysis frequently makes it difficult to fully capture the complexity, context, and dynamic nature of complex events. In this regard, methodological complexity causes phenomena to be oversimplified. This is due to the fact that by reducing intricate processes to quantifiable variables and figures, quantitative research may oversimplify the rich contextual character of medical education, possibly losing significant insights and depth. Furthermore, contextual factors, including individual patient variance, institutional culture, and local policies, have a significant impact on educational achievements. These crucial subtleties are difficult for quantitative approaches to capture, which might result in conclusions that are difficult to apply in different contexts.

### 34.9.4 Ethical and regulatory hurdles

Informed consent is one of the main ethical challenges. Obtaining appropriately informed consent from medical trainees in this context may be difficult due to the inherent power dynamics between faculty researchers and students. Because they fear that their evaluations or future job may suffer if they refuse to engage, trainees may feel under pressure to do so. Ensuring that trial participants are fully informed that their data is being utilized for research or education rather than for direct therapeutic intervention may be challenging. Apart from this, negotiating the institutional board review procedure or similar ethics committees is a significant regulatory barrier to institutional board review approval and compliance. Participants' vulnerability and the sensitive nature of the data frequently result in increased scrutiny and convoluted approval procedures.

### 34.9.5 Lack of expertise

A major obstacle to quantitative research in medical education is a lack of knowledge, which affects study design, data analysis, and result interpretation. One of the causes of the lack of expertise is that inexperienced researchers may find it difficult to create sound quantitative studies, which includes coming up with precise research questions, choosing suitable study designs like experimental or quasi-experimental, figuring out appropriate sample sizes, and putting strict data collection procedures into practice. Studies with low statistical power, selection bias, or confounding variables may result from this, ultimately jeopardizing the validity of findings.

### 34.9.6 Infrastructure gaps

Infrastructure gaps present significant challenges to quantitative research in medical education and can compromise the validity and scope of studies by limiting data availability, technological capabilities, and physical resources. For instance, it can be challenging to use online platforms, data collection tools, and digital resources that are crucial for modern quantitative research due to inadequate technology and connectivity, poor or inconsistent internet access, and unstable power supplies in some institutions, particularly in low- and middle-income countries. It can also be challenging to conduct research efficiently in addition to clinical duties in the absence of specialist research facilities, such as consultation offices lacking the necessary computer equipment.

### 34.9.7 Linking education to outcomes

Connecting education to outcomes is a significant barrier in quantitative research in medical education due to the complexity of outcomes, the longitudinal nature of medical training, and methodological issues. Research on medical education aims to produce a range of outcomes, from expanded clinical skills and professionalism to information gathering and, ultimately, improved patient outcomes. One of the components that poses a challenge for quantitative research is longitudinal follow-up. It is well established that long-term patient well-being, practice habits, and career choices are often the first signs of the true impact of medical education years after training ends. Since tracking pupils over such long periods of time costs a lot of resources and presents practical challenges, many studies focus on short-term results.

### 34.9.8 Contextual factors

Contextual factors are a significant barrier to the application of quantitative research in medical education because they introduce variability that is difficult to measure, regulate, or standardize across numerous contexts. It is more challenging to separate certain variables and produce broadly applicable conclusions because of this variety. One of the key factors affecting quantitative research is the difficulty of standardization. As everyone knows, medical education takes place in a range of clinical settings, from big university hospitals to small community clinics, with different patient populations, resources, and institutional cultures. Since it is very hard to standardize study interventions or assessment tools across these disparate contexts, direct comparisons are challenging.

### 34.9.9 Student engagement

Due to their demanding coursework and time constraints, medical students find it difficult to participate in extra research activities or finish lengthy research-related questionnaires, which may result in lower response rates and selection bias in quantitative studies. Additionally, students frequently express a lack of statistics and research methodological abilities, which prevents them from actively participating in research-related learning opportunities and, consequently, the process of generating data for pertinent investigations.

**Summary**

As they allow researchers and educators to systematically measure, analyze, and interpret educational phenomena, quantitative research methods are essential to the advancement of medical education. In the context of medical education, this chapter presents the fundamental ideas and uses of quantitative research, highlighting its importance in producing unbiased, broadly applicable data to guide instructional strategies, curriculum development, and educational policy. The main quantitative research designs that are frequently used in medical education include observational studies, cross-sectional surveys, cohort studies, experimental and quasi-experimental studies, and randomized controlled trials.

**Key point**

**Future perspectives: incorporating the big data**

As educational settings, technologies, and healthcare systems become increasingly sophisticated and data-driven, quantitative research in medical education is predicted to undergo substantial change. Its future role is anticipated to be shaped by a number of important trends and directions. For example, the breadth of quantitative research will be expanded by the growing use of big data and learning analytics. Large amounts of learner data are produced via electronic portfolios, learning management systems, simulation systems, and digital learning platforms. These data will be used more and more in quantitative research in the future to analyze learning trajectories, forecast performance, identify learners who are at risk, and assess the efficacy of educational interventions in real time. More learner-centered and equity-focused quantitative research in medical education will be conducted in the future. In order to assist in the creation of more inclusive curricula and equitable assessment systems, researchers will use quantitative data to investigate differences in educational access, assessment, and results.

## References

Abramson., E., Paul, C., Petershack, J., Serwint, J., Fischel, J., Rocha, M., et al., 2017. From design to dissemination: conducting quantitative medical education research. Academic Pediatrics 18. Available at https://doi.org/10.1016/j.acap.2017.10.008.

Aggarwal, R., Ranganathan, P., 2019. Study designs: Part 2 - descriptive studies. Perspectives in Clinical Research 10 (1), 34–36. PMID. Available at https://doi.org/10.4103/picr.PICR_154_18, 30834206.

Bhatia, N., Grewal, A., 2021. Randomized controlled trials. International Journal of Regional Anaesthesia 2 (2). Available at https://doi.org/10.13107/ijra.2021.v02i02.037.

Deckert, J., Wilson, M., 2023. Descriptive Research Methods. University Press of Florida, pp. 153–165. Available at http://doi.org/10.5744/florida/9780813069548.003.0011.

Devi, B., Lepcha, M., Basnet, S., 2023. Application Of correlational research design in nursing and medical research. Xi'an Shiyou Daxue Xuebao (Ziran Kexue Ban)/Journal of Xi'an Shiyou University 65, 60–69. Available at https://doi.org/10.17605/OSF.IO/YRZ68.

Gürbüz, S., 2017. Survey as a quantitative research method. Research Methods and Techniques in Public Relations and Advertising. Peter Lang AG, pp. 141–162. Available at http://doi.org/10.3726/b10899, https://www.peterlang.com/view/product/79203.

Kittur, J., 2023. Conducting quantitative research study: a step-by-step process. Journal of Engineering Education Transformations 36 (4), 100–112. Available at https://doi.org/10.16920/jeet/2023/v36i4/23120, https://sciresol.s3.us-east-2.amazonaws.com/srs-j/jeet/pdf/Volume-36/Issue-4/Final%20Mrge%20JEET%20April%202023-101-113.pdf.

Ratelle, J.T., Sawatsky, A.P., Beckman, T.J., 2019. Quantitative research methods in medical education. Anesthesiology 131 (1), 23–35. Available at https://doi.org/10.1097/aln.0000000000002727.

Schenker, J.D., Rumrill Jr, P.D., 2004. Causal-comparative research designs. Journal of Vocational Rehabilitation 21 (3), 117–121. Available at https://doi.org/10.3233/jvr-2004-00260.

Slater, P., Hasson, F., 2025. Quantitative research designs, hierarchy of evidence and validity. Journal of Psychiatric and Mental Health Nursing 32 (3), 656–660. Epub 2024 Dec 2. PMID. Available at https://doi.org/10.1111/jpm.13135, 39620883.

Sullivan, G.M., 2011. Getting off the "gold standard": randomized controlled trials and education research. J Grad Med Educ 3 (3), 285–289. PMID. Available at https://doi.org/10.4300/JGME-D-11-00147.1, 22942950.

Tavakol, M., Sandars, J., 2014. Quantitative and qualitative methods in medical education research: AMEE Guide No 90: part II. Medical Teacher 36 (10), 838–848. Available at https://doi.org/10.3109/0142159x.2014.915297.

Chapter 35

# Qualitative research methods in medical education

*"Research is formalized curiosity. It is poking and prying with a purpose."*

—Zora Neale Hurston

## 35.1 Introduction

In order to get profound insights and comprehend complicated phenomena, qualitative approaches have become essential instruments. This chapter attempts to demystify the qualitative research process by offering a comprehensive introduction and useful techniques for navigating its numerous parts (Kuper et al., 2008). While quantitative research provides numerical data and statistical analysis, qualitative research explores the varied textures of human experience and viewpoint, revealing situations and nuances that are sometimes lost in numerical translation. In addition to being a methodological choice, qualitative research entails a commitment to exploring social issues in depth, which enables researchers to relate to the subjective experiences of their subjects.

Qualitative research is a methodology for scientific inquiry that emphasizes the depth and richness of context and voice in understanding social phenomena. This methodology is constructive or interpretive aiming to unveil the "what," "why," "when," "where," "who," and "how" behind social behaviors and interactions rather than merely quantifying occurrences.

## 35.2 Characteristics of qualitative research

Instead of quantifying variables, qualitative research in medical education aims to comprehend human experiences in their natural environments. A naturalistic approach, interpretive and inductive reasoning, an emphasis on context and meaning, data collection through techniques like observations and interviews, the use of words and images in the data, and the production of rich descriptive findings rather than generalizations are some of the key traits (Cristancho et al., 2018). Some of the key characteristics of qualitative research are illustrated in Fig. 35.1.

### 35.2.1 Naturalistic approach

As it requires examining phenomena in their actual, natural environments and seeing how individuals act when engrossed in their authentic, everyday life experiences, the naturalistic approach is a fundamental part of qualitative research. In order to obtain a genuine, in-depth grasp of how people view their own realities, researchers take on the role of observers or participants. The fundamental tenet of qualitative research is that social processes and human behavior cannot be comprehended in isolation from their particular circumstances. In order to witness interactions and events as they naturally occur, this kind of research enables the researchers to fully immerse themselves in the area, such as a classroom, a community, or an organization.

### 35.2.2 Interpretive and inductive

As the qualitative research enables researchers to investigate complicated phenomena, develop novel ideas from data, and comprehend the lived experiences and viewpoints of participants, interpretive and inductive procedures are crucial elements of qualitative research in medical education. Understanding how people perceive their experiences is the main goal of the interpretative approach in medical education research. This method highlights how reality is context-dependent and socially manufactured. Researchers seek to understand how medical students, teachers, and clinicians interpret their

**The Principles and Practice of Medical Education. DOI:** https://doi.org/10.1016/B978-0-443-44197-4.00001-1

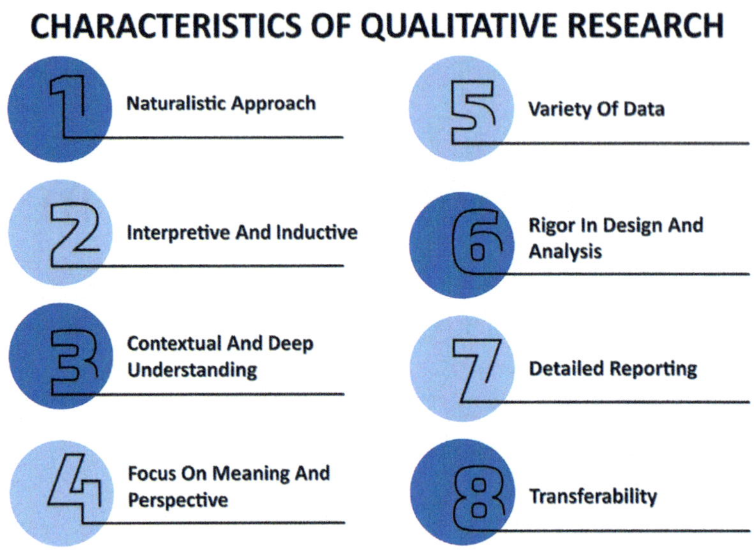


FIGURE 35.1 **Characteristics of qualitative research.** This figure depicts the characteristics of qualitative research.

training environments, clinical experiences, and professional identities by using an interpretive lens. An inductive approach is a bottom-up research methodology in which conclusions are drawn directly from the data, free from the limitations of a predetermined theory or hypothesis. These researchers start with particular observations and data points, which are subsequently examined to find trends, themes, and eventually more general theoretical ideas. When combined, these methods assist medical educators in gaining a more comprehensive grasp of the many human and social elements that impact clinical practice, teaching, and learning (Denzin and Lincoln, 2005).

### 35.2.3 Contextual and deep understanding

In order to thoroughly examine the intricacies of human experience, interactions, and systems that statistical approaches alone are unable to fully capture, qualitative research in medical education relies on the pursuit of contextual and deep understanding. This approach is crucial because medical education is a dynamic field that is deeply interwoven with specific social, cultural, and institutional contexts. Beyond simply identifying a problem like high student burnout rates, qualitative research looks at human perspectives, motivations, and experiences to understand why students feel burnout and how institutional cultures contribute to it. The particular context of a program, including its resources, local patient population, and institutional history, may have an impact on its performance. To correctly interpret results and assess whether a program or intervention can be applied in different contexts, a thorough grasp of these contextual elements is required.

### 35.2.4 Focus on meaning and perspective

Qualitative research in medical education, which aims to comprehend complicated human experiences rather than quantify them, relies heavily on meaning and viewpoint. Understanding the lived experience, which gives medical students, educators, and patients a forum to express their individual experiences, interpretations, and understandings of phenomena like burnout, professional identity formation, or illness, is one of the key reasons this approach is essential in qualitative research.

### 35.2.5 Variety of data

In order to improve the rigor, validity, and depth of the study findings through a process known as triangulation, a variety of data is an essential part of qualitative research in medical education. This method goes beyond depending just on one viewpoint, offering a more thorough and solid comprehension of complicated phenomena in medical education and practice. Researchers can corroborate findings by gathering and evaluating a variety of data sources. This increases the credibility of the conclusions by allowing insights from one data source to be verified or refuted by others.

### 35.2.6 Rigor in design and analysis

A crucial element of qualitative research in medical education is rigor, which guarantees the reliability, caliber, and integrity of the study's conclusions. It creates criteria by which the scientific community evaluates and critiques research and contributes to the development of trust that the results are reliable and well-supported. In qualitative research, rigor is essential because it guarantees quality and credibility by laying the groundwork for results that are seen as trustworthy and significant in the field of health professions education. Additionally, it reduces prejudice through methodical use of techniques by assisting researchers in recognizing and reducing their own subjective influences and biases.

### 35.2.7 Detailed reporting

As the qualitative research is capable of establishing reliability, openness, and rigor, detailed reporting is an essential part of qualitative research in medical education. In contrast to quantitative research, which aims for objectivity, qualitative research uses rich, descriptive data to comprehend complicated phenomena. Hence, a detailed account of the research process is crucial for readers to evaluate the reliability and relevance of the results. Detailed reporting enhances trustworthiness and credibility by addressing qualitative research quality criteria such as credibility, transferability, dependability, and confirmability. Researchers make sure that the results are based on the data and not just their personal bias or interpretation by clearly outlining every stage of the investigation.

### 35.2.8 Transferability

As the qualitative counterpart of generalizability in quantitative research, transferability is an essential element of qualitative research in medical education. It describes the extent to which a particular study's conclusions can be applied or transferred to other comparable situations, environments, or people. Transferability aids in determining the practical utility of research findings in medical education, where learning settings and interventions are frequently complex and context-dependent. Researchers must give thorough, in-depth explanations of the study site, participants, and procedures in order to promote transferability. This enables readers, such as other educators or curriculum designers, to evaluate how comparable the study's environment is to their own and determine whether the results are applicable to them.

## 35.3 Qualitative research designs

Using techniques, including phenomenology, grounded theory, ethnography, narrative inquiry, action research, and case studies, frequent qualitative research types in medical education investigate experiences and difficult situations (Atkinson and Pugsley, 2005). Discourse analysis, consensual qualitative research, and photovoice are additional useful methods for gaining a thorough understanding of patient interactions, teaching, and learning (Fig. 35.2).

### 35.3.1 Phenomenological research

Using interviews and narrative to uncover universal themes by suspending researcher biases, phenomenological research in medical education is a qualitative approach that investigates the lived experiences of students, physicians, or patients to understand the essence or meaning behind their perceptions of particular educational or clinical phenomena, such as stress, challenging diagnoses, or learning a new skill. It offers important insights by understanding the "what" and "how" of human experience rather than merely objective facts, often through interpretive or descriptive approaches that improve practice and education (Laverty, 2003). This approach's fundamental tenets include its emphasis on lived experience, which aims to comprehend how people view and interpret their surroundings by going beyond objective facts to subjective meaning. Furthermore, it analyzes words and themes rather than numbers through in-depth interviews, narratives, and observations. Additionally, it makes use of the bracketing technique, in which researchers set aside their preconceptions in order to see participant experiences in a novel way. It is used in medical education to comprehend student deal via examinations of stress, burnout, high-stakes tests, or entering clinical settings. Gaining insight into patients' experiences with illness also improves the practice and empathy of healthcare professionals. It also aids in curriculum building by offering advice on how to teach complex skills while considering the difficulties and perspectives of students. It is considered helpful in medical education since it provides a deeper, humanistic understanding that quantitative data often

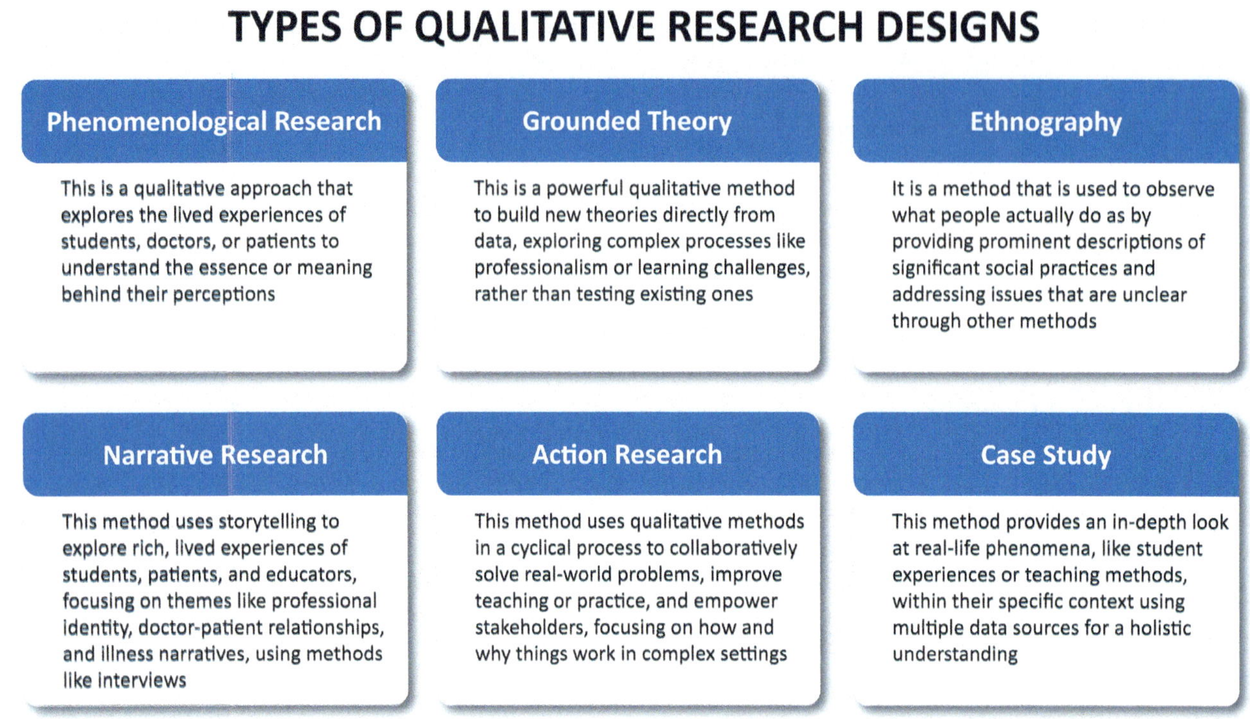


FIGURE 35.2 **Types of qualitative research designs.** This figure shows the types of qualitative research designs.

ignores. It also helps close the gap between abstract theory and real-world application in the domains of education and medicine.

### 35.3.2 Grounded theory

Instead of evaluating preexisting ideas, grounded theory is a potent qualitative technique used in medical education research to develop new theories directly from data, examining intricate processes like professionalism or learning difficulties. In order to uncover underlying social processes and produce rich, context-specific understandings that can enhance curriculum and practice, it employs repetitive cycles of data collecting and analysis, theoretical sampling, and inductive coding. This theory's fundamental tenets include its inductive methodology, which begins with observations and progresses toward more general concepts and theories rather than preconceived notions. Additionally, it entails the simultaneous collection and analysis of data, whereby data is collected and processed in cycles, with analysis guiding subsequent data collecting procedures. To improve comprehension, fresh data is continuously compared with preexisting codes, concepts, and categories. Participants in these studies are not chosen at random, rather they are carefully selected as trends appear to investigate or refute new concepts. Grounded theory is used in medical education to investigate the unknown because it is perfect for comprehending phenomena that are not well understood, like team dynamics, clinical reasoning, or student burnout. Additionally, it assists researchers in producing new, data-driven insights that are essential for creative curriculum design. More significantly, it draws attention to the "how" and "why" of actions and experiences in intricate healthcare environments. Grounded theory is useful because it is realistic, contains conclusions that are profoundly contextualized, and is flexible enough to produce new, useful information.

### 35.3.3 Ethnography

Researchers in medical education employ ethnography, a qualitative technique that goes beyond surveys to see what individuals actually do as opposed to what they claim to do, to provide rich, insider insights on student behaviors, cultures, and underlying issues. It is ideal for understanding complex issues like professionalism, curriculum impact, or learning cultures through in-depth observation and interviews, helping educators to customize interventions and comprehend lived experiences by offering prominent descriptions of important social practices and addressing issues that are unclear through other methods (Reeves et al., 2008). One of the key components of ethnography is immersion, where researchers spend a substantial period of time in the topic setting, such as a medical school or clinical rotation. It also uses observations, interviews, and document analysis to gather detailed accounts. It also provides thorough, in-depth explanations that provide cultural context in addition to basic information. Hidden curriculum is one example of how qualitative research is used in medical education to help identify unwritten rules, norms, and values that affect students' learning. It also promotes professionalism by helping us look at how professionalism is acquired and used in real-world situations. It also helps assess how curriculum modifications are viewed and their actual impacts. It promotes an understanding of student–teacher dynamics by assisting the researcher in comprehending behaviors, performance traits, and difficulties in teaching or learning. Ethnography documents people's actual behaviors as opposed to just their opinions or claims. It is helpful in medical education because it addresses issues that other approaches ignore that are ingrained in complex social systems, making it easier to handle challenging problems. It also provides the comprehensive information needed to develop more successful educational policies and practices (McLeod, 2024).

### 35.3.4 Narrative research

Using techniques like interviews, diaries, and documents to understand meaning-making and context, narrative research in medical education employs storytelling to explore rich, lived experiences of students, patients, and educators (Gill et al., 2008). It focuses on issues like sickness narratives, doctor–patient relationships, and professional identity, offering deep insights that are often missed by statistics but require careful examination. It provides a powerful qualitative lens to understand the human side of medicine, often complementing other methods like ethnography or phenomenology. The fact that people use tales to make sense of their lives and experiences is one of the key aspects of narrative research. Narrative inquiry records these firsthand perspectives in order to comprehend deeper meanings. Through in-depth interviews, field notes, letters, diaries, photographs, and already-existing materials, it assists us in gathering stories (Parks, 2023). Some of the uses in medical education include examining how students form their professional identities or how doctors view power relationships and patients' experiences with disease. As it goes beyond surface-level information to comprehend the intricacies, feelings, and context of medical training and practice, narrative research is thought to be beneficial in medical education. It also aids in our comprehension of the difficulties, development, and changing self-perception that come with being a doctor. It also helps close the gap between narrative-based medicine and evidence-based medicine by recognizing the particular circumstances and patient experience.

### 35.3.5 Action research

In contrast to strictly quantitative studies, action research in medical education empowers educators and stakeholders by applying qualitative approaches in a cyclical process to address real-world problems collaboratively, enhance teaching and practice, and focus on how and why things function in complicated environments. It frequently involves students, teachers, and patients as coresearchers and combines research with direct action to provide useful knowledge for improved clinical training and healthcare delivery (Carter and Little, 2007). The fact that medical education tackles particular difficulties in curricula, clinical skills, patient communication, or organizational issues within healthcare settings is one of its most important features. Additionally, it fosters ownership and relevance by involving all stakeholders as active participants rather than merely subjects. In order to refine strategies and effect significant change, action research also uses an ongoing cycle of planning, acting, assessing results, and reflecting. It also offers "why" and "how" answers that quantitative data does not by using techniques like focus groups, observations, and interviews to comprehend context, meaning, and experiences (Hamada, 2019).

### 35.3.6 Case study

In qualitative medical education research, a case study offers an in-depth examination of real-life phenomena, such as student experiences or teaching methods, within their particular context using multiple data sources for a holistic understanding. It differs from brief clinical case reports by providing deep contextual analysis to explain complex issues, such as learning barriers or factors influencing the development of clinical skills. A case study's use of in-depth investigation, where the emphasis is on comprehending a case thoroughly rather than merely summarizing it is one of its unique characteristics. It examines occurrences in their natural settings while offering thorough, contextualized data. It also uses records, surveys, interviews, and observations to provide a full picture. In terms of application, it helps in understanding student experiences, evaluating teaching methods, and analyzing learning environments.

## 35.4 Tools used in qualitative research

Focus groups, participant observation, in-depth or semi-structured interviews, phenomenology, ethnography, grounded theory, and narrative inquiry are some of the methods used in qualitative research in medical education to collect rich data. In order to gain a thorough understanding of experiences, actions, and learning processes, researchers employ theme analysis, which is frequently supported by CAQDAS software, document analysis, and visual approaches like photovoice. Interviews, which can take the shape of unstructured, semi-structured, or in-depth discussions to examine individual experiences and viewpoints, are among the frequently used data collection methods. Focus groups are another method used in the qualitative study to investigate social dynamics and common understandings. In order to comprehend contexts, researchers can also use observation as a method of gathering data by seeing real activities in various circumstances. Thematic analysis, on the other hand, is one of the analytical tools that aids in finding, examining, and summarizing data patterns.

## 35.5 Steps in qualitative research

Framing a compelling question, selecting an appropriate methodology such as phenomenology or grounded theory, purposeful sampling, data collection such as focusses groups and interviews, rigorous analysis such as coding or thematic analysis, reporting rich narratives, frequently in tandem with analysis, and concentrating on deep understanding of experiences such as student learning or clinical reasoning are all steps in qualitative research in medical education (Clarke et al., 2021). Some of the vital steps in qualitative research are described in Fig. 35.3.

### 35.5.1 Developing the research question

Instead of measuring associations, the process of developing a qualitative research question in medical education entails starting broad, reviewing literature, consulting stakeholders, and refining to specific, open-ended questions using verbs like explore or describe to uncover meanings, processes, and experiences. It also ensures that the question is focused, researchable, and in line with qualitative goals of depth over breadth. One of the characteristics of a good qualitative research question is that it is open-ended, allowing for rich, nuanced responses rather than just yes or no answers. The

FIGURE 35.3 **Steps in qualitative research.** This figure depicts the steps in qualitative research.

questions should be both broad enough to allow for the emergence of new themes and precise enough to direct the investigation. A major problem in medical education should be addressed by the research question which should also be simple to comprehend and clearly relate to the goals of the study.

### 35.5.2 Choosing a methodology

In medical education, selecting a qualitative methodology depends on the purpose of the research issue, such as whether to produce theory that is consistent with your beliefs or explore experiences. It also entails matching established methods and a particular philosophical framework with the study objectives, problems, and personal interests. Providing a deep, comprehensive understanding of intricate social issues is the aim. In order to do rigorous, reliable research, it is important to define the research topic, comprehend paradigms, choose a framework, decide on data collecting, and make sure that the research question, paradigm and procedures are all consistent.

### 35.5.3 Choosing the study design

Choosing a qualitative research study design in medical education means matching a specific methodological approach, such as phenomenology, grounded theory, ethnography, narrative, or case study, with the research question, focusing on the "why" and "how," considering the underlying beliefs about knowledge and ensuring feasibility all while acknowledging that, unlike more rigid quantitative designs, the design is iterative and can change. The approach starts with a clearly stated research question and moves on to choosing a plan that guides data collection and analysis in order to produce profound, meaningful insights about medical education phenomena.

### 35.5.4 Data collection

In order to gather rich, descriptive data in the form of observations, interviews, or documents to comprehend experiences, viewpoints, and processes, qualitative medical education research use techniques like focus groups, in-depth interviews, and observation. These approaches, which are driven by particular research concerns and aim for rigor and credibility, sometimes involve transcribing and methodical analysis to capture subtleties that go beyond straightforward facts. Important steps in the data collection process include establishing goals, selecting appropriate methods such as focus groups or interviews, setting up consent or recording procedures, collecting data, cleaning or organizing it for analysis, and ensuring that the process captures real participant voices.

### 35.5.5 Data analysis

Data analysis in qualitative medical education research is an iterative process that uses coding, classification, and theme identification to make sense of rich text or narrative data from observations and interviews in order to find patterns, meanings, and insights into intricate human experiences. In order to develop knowledge for interventions or programs, this method sometimes employs software such as NVivo and progresses from particular "codes" to more general ideas that are backed by statements from participants. This crucial phase, which starts early and occurs concurrently with data collection, transforms raw data into actionable conclusions.

### 35.5.6 Ensuring rigor

Moving beyond quantitative validity or reliability, ensuring rigor in qualitative research entails using methodical, transparent techniques to ensure the findings' credibility, dependability, confirmability, and transferability. To make sure the research process is thorough, consistent, and credible, this entails techniques like triangulation, member checking, reflexivity, extended involvement, thick description, and keeping an audit trail (Busetto et al., 2020). Rigor can be ensured at every stage of qualitative research. For example, creating a research topic that is precise, targeted, and backed by a solid conceptual framework consistent with the selected technique ensures rigor during the design process. This guarantees that every facet of the research functions consistently. In the same way, employing triangulation and using several data sources, including observations, interviews, documents, and methodologies, to cross-verify findings during the data gathering process strengthens credibility and offers a more thorough understanding of the phenomenon.

### 35.5.7 Write and report

In qualitative research writing and reporting, rich participant narratives are woven with transparent methods. In this, rich descriptions, direct quotes, and contextual details are emphasized to make the findings credible and understandable. Here the report is frequently structured in accordance with standards like Standards for Reporting Qualitative Research for clarity. A solid introduction, a literature review, a thorough methodology, findings supported by evidence, and a compelling discussion or conclusion that emphasizes implications and future study are essential components of writing and reporting qualitative research.

## 35.6 Impact of qualitative research in medical education

Medical education is greatly impacted by qualitative research because it offers a thorough, contextual understanding of why and how learning and professional development occur, explores complex human experiences like professionalism, stress, and patient-provider interactions, and reveals subtleties that are overlooked by statistics, resulting in deeper insights and more successful, person-centered teaching methods. In the end, it improves curriculum design and supports trainees' holistic development by helping to understand student or physician behaviors, difficulties, and social dynamics within learning contexts (Brundha, 2023). Some of the key impacts of qualitative research in medical education are described below.

### 35.6.1 Uncovers "why" and "how"

Investigating and comprehending the "why" and "how" behind human behavior, experiences, and social phenomena is the fundamental goal of qualitative research. In contrast to quantitative approaches that concentrate on "how much" or "how many," the qualitative technique offers nuanced, contextual, and comprehensive insights into complicated issues and underlying motivations. Qualitative approaches allow participants to express their opinions in their own words and prioritize depth over breadth. It promotes in-depth interviews by encouraging one-on-one discussions with open-ended questions, which are essential for thoroughly examining delicate subjects or unique, personal experiences. This approach is perfect for immediately eliciting the "why" behind participants' activities by asking them to describe their motivations, difficulties and unfulfilled needs.

### 35.6.2 Explores lived experiences

Qualitative research on lived experiences in medical education offers significant insights into students' entire journeys by highlighting subtle aspects like peer support, mental health, and learning environments that quantitative data misses. In the end, this data influences patient care by understanding student–patient connections and empowering diverse learners, as well as curriculum, teaching, and support systems for improved training. Some of the major effects include exposing hidden obstacles like uncertainty, facilitators like strong mentorship, supportive environments, emphasizing the needs of a variety of students, and directing more compassionate, successful teaching methods, frequently employing phenomenology to understand the essence of being a medical student.

### 35.6.3 Understands context and culture

Because it offers a thorough grasp of complicated topics that immediately affects the appreciation of context and culture in healthcare and educational settings, qualitative research is crucial to medical education. This research reveals how social, cultural, and situational aspects influence students', teachers' and patients' perceptions, behaviors, and learning processes by delving into the subtleties of human experience. Additionally, people from many backgrounds participate in medical education. These varied viewpoints can be captured by researchers using qualitative techniques like ethnography and interviews which reveal cultural ideas about health, sickness, and the educational setting that standardized quantitative metrics frequently overlook. This knowledge is essential for creating practices and courses that are culturally appropriate.

### 35.6.4 Informs curriculum and pedagogy

Because qualitative research offers a deep understanding of complicated human experiences, interactions, and situations that quantitative methods frequently overlook, it greatly influences curriculum and pedagogy in medical education. This depth makes it possible to identify the complex requirements and difficulties that patients, teachers, and students encounter, which results in focused and successful educational interventions. Curriculum development is one area where qualitative research has a significant impact since it helps create curricula that are more applicable to actual clinical practice. Researchers can identify important subjects like cultural competency, communication skills, and empathy by examining the lived experiences of patients and healthcare professionals. This ensures that these subjects are incorporated into core learning objectives rather than being considered as ancillary subjects. It aids in the creation of curricula that tackle the hidden curriculum, which consists of social messages and unwritten conventions in educational settings that promote more moral and competent learning environments. Innovative teaching strategies that go beyond conventional lectures are informed by qualitative research in the field of pedagogy. For instance, research on how medical students acquire professional identities has led to the adoption of reflective practice and narrative medicine in the classroom, which encourages students to examine their values and beliefs.

### 35.6.5 Enhances understanding of patient care

Qualitative research significantly increases our understanding of patient care in medical education and promotes more thorough, patient-centered, and culturally sensitive processes by looking at subjective experiences and real clinical settings. One of the primary outcomes is the appreciation of patient experiences. In this sense, qualitative research gathers patient perspectives on illness, treatment, and healthcare delivery in addition to clinical data. This immediately affects medical education and leads to more compassionate care by teaching future medical professionals to recognize and value the patient's experience. Qualitative research also aids in the development of communication skills by analyzing real patient-provider encounters and identifying both effective and ineffective communication strategies.

### 35.6.6 Identifies barriers and solutions

Qualitative research in medical education identifies key barriers like lack of training/mentorship, time constraints, weak institutional support, funding issues, and resistance to new methods, hindering deeper understanding of complex health issues. Solutions involve targeted training, dedicated time, strong leadership, cultural shifts, integrated mixed-methods, and improved recognition for qualitative work, enabling better teaching, research, and problem-solving in healthcare. Some of the important barriers identified are in relation to training and skills such as insufficient practical experience, methodological knowledge, and skills in qualitative methods. In addition, the qualitative research helps the researcher to identify time-intensive nature conflicts with clinical duties, lack of funding, and human and physical resources.

### 35.6.7 Impact of qualitative research on teaching and learning process

Qualitative research has a significant influence on teaching and learning because it provides deeper insights into the "how" and "why" of educational experiences and goes beyond statistics to understand students' feelings, behaviors, and social-emotional development. It assists educators in tailoring approaches, improving surroundings, and fostering holistic development by looking at a range of perspectives and real-world situations. It transforms instruction into a more introspective and adaptable process that allows teachers to change strategies, increase student engagement, and create stronger relationships. All these aspects will lead to more effective and student-centered learning. It improves students' understanding of the learning process by assisting them in expressing their experiences and the causes of their achievements or failures (Brown et al., 2023). When teachers are aware of each student's particular circumstances and experiences, they are better equipped to offer more pertinent help. Qualitative research fundamentally adds the human element to education by offering the nuanced knowledge required for teaching and learning that is genuinely effective, responsive, and comprehensive.

## 35.7 Qualitative research in modern medical education

Modern medical education is significantly influencing qualitative research by accepting it as an essential tool for understanding the complex social, emotional, and lived experiences inside healthcare rather than relying solely on quantitative approaches. This shift promotes more context-aware and human-centered research and practice (Schneider et al., 2017). The idea that qualitative research is less valuable than quantitative methods is being refuted by its resurgence in health professions education. It is now widely acknowledged as a crucial method for generating data that informs clinical practice and policy decisions.

### 35.7.1 Mixed-methods research (MMR)

MMR, which is becoming more and more common, integrates the collecting and analysis of both qualitative and quantitative data into single research. The "why" behind quantitative results can be explained by qualitative data. For example, interviews can be used to determine the causes of particular survey patient satisfaction levels.

### 35.7.2 Technology integration

Although it takes some time to learn, CAQDAS (Computer-Assisted Qualitative Data Analysis Software) assists medical education researchers in effectively managing, coding, and analyzing qualitative data such as interviews and videos, supporting a variety of approaches such as grounded theory and phenomenology by organizing text, connecting concepts, and visualizing patterns. This increases transparency and depth in studies on student experiences, curriculum effectiveness, or patient perceptions.

### 35.7.3 Digital data collection

Digital data collection in qualitative research uses online tools for gathering rich insights through methods like video interviews, online focus groups, social media analysis, and analyzing digital texts or images, extending reach, saving time, and capturing multimodal data, such as text, audio, visual while requiring careful consideration of ethics, digital literacy, and unique analytical skills to manage the complex data streams. It enhances traditional methods like interviews and observations by overcoming geographical barriers and allowing for longitudinal studies providing deeper context and diverse participant voices.

### 35.7.4 Creative data elicitation method

Innovative techniques that can uncover emotional and organizational aspects that traditional interviews might overlook include asking participants to draw complex experiences, using guided walks or walking interviews through particular settings, or using photovoice in which participants take pictures to depict their experiences.

### 35.7.5 Emphasis on rigor and reporting

The 32-item COREQ (Consolidated Criteria for Reporting Qualitative Research) checklist was created to enhance the reporting quality of qualitative studies that include focus groups and interviews. It focuses on the research team, study design, and data analysis or reporting to guarantee transparency and rigor. Interviewer characteristics, participant knowledge, sample strategies, data collecting settings, analytical approaches (coding, theme derivation, software), and reporting of findings, including utilizing participant quotes and ensuring data consistency, are all covered.

### 35.7.6 Role of artificial intelligence

While issues like data privacy and maintaining human oversight remain critical, artificial intelligence (AI) is revolutionizing qualitative research in medical education by simplifying laborious tasks like coding and thematic analysis with natural language processing, improving data exploration, finding patterns in massive datasets like student engagement, feedback, and enabling deeper understanding of learner experiences and curriculum effectiveness. In order to produce deeper insights into student opinions, training requirements, and curriculum design, AI serves as a potent aid for the analysis of focus groups, interviews, and open-ended survey replies.

## 35.8 Benefits of qualitative research in medical education

Through rich, descriptive data, qualitative research in medical education provides profound insights into the "how" and "why" of learning. It explores complex human experiences, behaviors, and social contexts that numbers cannot capture, aids in understanding student challenges, develops better teaching theories, enhances curriculum, and identifies obstacles to new programs, ultimately improving educational quality and practice (Fig. 35.4).

### 35.8.1 Understanding learner and faculty experiences

Qualitative research offers important benefits for medical education because it provides a comprehensive understanding of the intricate and subjective experiences of teachers and students. This research goes beyond quantitative measurements and yields meaningful results that can direct educational practices and policies by looking at the "how" and "why" of their perspectives. Examples of qualitative research methods that are unique in capturing lived experiences are focus groups and interviews, which allow participants to articulate their authentic, real-world experiences in their own words. This offers comprehensive, context-specific data that sheds light on the intricacies of medical education and training settings. Furthermore, the qualitative research approach is quite successful in identifying the hidden curriculum, which consists of

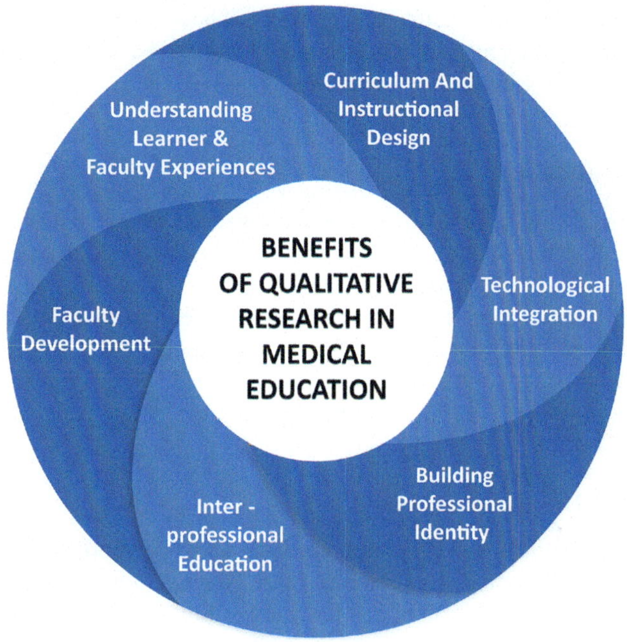


FIGURE 35.4 **Benefits of qualitative research in medical education.** This figure depicts the benefits of qualitative research in medical education.

implicit rules, values, and behaviors that are taught in medical settings. It assists in locating unforeseen effects of educational initiatives that standardized assessments could overlook.

### 35.8.2 Curriculum and instructional design

Qualitative research greatly enhances curriculum and instructional design in medical education because it provides a thorough, contextual understanding of learning experiences that quantitative methods sometimes overlook. With this information, teachers can more effectively modify curricula to meet the needs and challenges of the real world. It improves understanding of educational experiences by capturing the subtleties of how medical students and teachers evaluate their learning environment, the curriculum, and instructional practices through qualitative techniques like observations and interviews. This profound understanding aids in recognizing unspoken needs, difficulties, and successes. Qualitative research also explains why certain teaching methods are ineffective and offers useful teaching suggestions. This instant feedback loop helps to improve instructional strategies.

### 35.8.3 Technological integration

According to qualitative study, technology integration in medical education boosts engagement, confidence, and abilities by offering safe, immersive learning, enhancing access to resources, and boosting comprehension of complicated subjects. However, curriculum change, staff training, and addressing digital inequities are required for genuinely effective adoption. One significant benefit that qualitative research reveals is that technology provides a means of moving beyond "see one, do one" to experiential learning for complex skills and rare events. Additionally, technology improves decision-making and helps visualize challenging concepts like anatomy through virtual reality by enabling students to replay events and choose their own learning speed.

### 35.8.4 Building professional identity

As it thoroughly examines the intricate, subjective path of becoming a doctor and reveals how students integrate personal identity with professional values through lived experiences like clinical encounters, mentorship, and cultural context to build self-awareness, purpose, resilience and a strong sense of self as a physician which goes beyond mere skill acquisition to foster genuine professional commitment and well-being. In this regard, qualitative research is essential in medical education for professional identity formation. Additionally, it reveals how students find coherence, meaning, and purpose in their work, thus connecting the development of professional identities to resilience and well-being which is a crucial component that quantitative approaches frequently overlook.

### 35.8.5 Interprofessional education

Since it delves into the "how" and "why" of students' experiences, qualitative research is essential to medical education for interprofessional education. Unlike quantitative data, which only assesses effectiveness, qualitative research reveals profound insights into diminished hierarchies, shared knowledge, and increased empathy, resulting in improved teamwork and patient care. Studying their lived experiences in interprofessional settings reveals how students understand complex patient needs, dispel myths, and develop respect for different professions. These will contribute to better curriculum design. For instance, students share discipline-specific knowledge and have a better understanding of each other's responsibilities (Almoghirah et al., 2023).

### 35.8.6 Faculty development

Since it reveals profound insights into faculty experiences, challenges, and perceptions, qualitative research is a potent tool for faculty development in medical education. This leads to more effective, customized programs that improve teaching abilities, curriculum, leadership, faculty satisfaction, and retention, ultimately strengthening the entire medical education system by creating a culture of growth and innovation. In qualitative research, faculty members' life experiences are captured using qualitative methods like interviews, which reveal subtleties and obstacles overlooked by quantitative surveys, such as the reasons faculty members find it difficult to adopt new teaching strategies like critical thinking. Simultaneously, qualitative findings assist create interventions that are relevant, engaging, and address real-world difficulties rather than assumptions by facilitating the understanding of faculty needs, which may include career advancement, managing burnout, and strengthening leadership.

## 35.9 Challenges of qualitative research in medical education

Time-consuming data collecting and analysis, problems with generalizability, a lack of training or experience, a lack of funding, the possibility of researcher bias, handling large datasets, and institutional obstacles including low recognition are all obstacles that qualitative research in medical education must overcome.

### 35.9.1 High cost and demanding workload

High expenses and a taxing workload are major obstacles to qualitative research in medical education, which affects researchers' time, institutional resources, and overall research output. These difficulties are frequently caused by the methodology's intrinsic time-consuming character as well as the demanding atmosphere of medical practice. The main workload issues stem from the substantial effort needed at various phases of the research process, which frequently clashes with medical academics' clinical and instructional obligations. Additionally, recruiting, scheduling, and carrying out the data collecting process with each participant take a substantial amount of time. This is especially challenging in clinical settings when patient care takes precedence over research efforts due to high patient demands and staff shortages. Organizing, transcribing, and methodically analyzing vast amounts of textual or audio data is frequently one of the most time-consuming parts of qualitative research. Compared to quantitative analysis, this approach might take a lot more time because it requires great focus and complete immersion in the data in terms of its coding and interpretation.

### 35.9.2 Technological advancement and e-learning

Technological developments, such as AI, VR, telemedicine, and e-learning, present challenges to qualitative research in medical education because they change data sources from in-person to online, call for new skills such as digital pedagogy, raise ethical and privacy concerns in terms of data security, affect clinical skill development, and highlight issues with access, faculty training, and adapting to new learning or teaching styles. To fully understand their true impact on learning, new research methodologies and strong support are needed.

### 35.9.3 Ensuring equity and access

A number of practical, methodological, and systemic problems make it difficult for qualitative research in medical education to guarantee equity and access. Ensuring a varied representation of participants can be challenging, which is one of the main obstacles of qualitative research. Convenience sampling, which may involve inviting students from a particular institution or those who are easily accessible, is frequently used in recruitment, which may exclude voices from underrepresented minority groups, people from diverse socioeconomic backgrounds, or people pursuing nontraditional educational pathways. This restricts the findings' inclusivity and generalizability.

### 35.9.4 Adapting to evolving curricula

Due to time restrictions, the necessity of quickly adapting study designs, and a widespread lack of institutional support and faculty experience in qualitative methodology, evolving medical curriculum pose serious problems for qualitative research. For example, medical academics generally lack mentorship in qualitative research methods, focused training programs, and practical experience. Researchers may find it difficult to plan and carry out excellent qualitative investigations if they lack the necessary experience, which could result in methodological flaws and findings with little significance.

### 35.9.5 Generalizability and bias

The underlying goal and methodology of qualitative research differ from the statistical generalizability sought in quantitative studies, which is the main obstacle to generalizability. Instead of using large, random samples intended to be statistically representative of a larger population, qualitative studies usually utilize small, carefully chosen samples to investigate phenomena in depth. The results of qualitative research are frequently firmly rooted in a particular context, such as a medical school or a particular student body, and are therefore not meant to be widely applicable without taking into account the similarities of different contexts.

### 35.9.6 Maintaining rigor

It is difficult to maintain rigor in qualitative research since it necessitates going beyond quantitative validity or dependability to guarantee reliability through openness, methodical procedures, and critical reflexivity to control innate subjectivity and researcher bias. The main challenges in this area include striking a balance between flexibility and systematicity, demonstrating rigor without strict templates, and responding to criticisms of subjectivity by employing techniques like member verification, triangulation, reflexivity, and thorough audit trails.

**Summary**

The function, tenets, and use of qualitative research approaches in the field of medical education were examined in this chapter. Qualitative research has gained a lot of importance in medical education, as it allows for in-depth investigation of complex phenomena like learning experiences, professional identity formation, curriculum implementation, clinical reasoning, and educational culture areas that are difficult to capture through quantitative measures alone.

Popular qualitative research designs, such as case study research, grounded theory, ethnography, phenomenology, and narrative inquiry, showcase the importance of recurrent cycles of data gathering and ongoing comparison to find new social processes and patterns. Moreover, in terms of technological developments and the application of artificial intelligence, the impact of contemporary medical education on qualitative research is gaining interest, as it gives a new perspective on how the research is conducted and addresses various aspects such as ethics and digital literacy in qualitative research.

**Key point**

**Future perspectives: comprehending the human experience**

As the profession continues to tackle difficult, human-centered issues that go beyond what quantitative approaches alone can handle, qualitative research in medical education has a bright and significant future. The potential of qualitative research to enhance comprehension of human experiences, guide innovative teaching methods, and supplement quantitative approaches is what will determine its future in medical education. Technological developments will expand the sources of qualitative data, such as online learning environments, digital ethnography, virtual interviews, learning analytics in conjunction with qualitative research, and analysis of reflective e-portfolios. Researchers will be able to examine educational processes in real time and across geographically separated environments, thanks to these developments. Qualitative research will continue to be crucial for understanding the subtleties of learning, teaching, and professional growth in healthcare settings as medical education develops.

## References


Almoghirah, H., Illing, J., Nazar, H., 2023. A qualitative study to explore student learning and development of interprofessional collaboration during an online interprofessional education intervention. BMC Medical Education 23 (1). Available at https://doi.org/10.1186/s12909-023-04885-y.

Atkinson, P., Pugsley, L., 2005. Making sense of ethnography and medical education. Medical Education 39 (2), 228–234. Available at https://doi.org/10.1111/j.1365-2929.2004.02070.x.

Brundha, M.P., 2023. Qualitative research and its significance in medical education – a review. International Journal of Head and Neck Pathology 6 (1), 18–24. Available at https://doi.org/10.56501/intjheadneckpathol.v6i1.918.

Busetto, L., Wick, W., Gumbinger, C., 2020. How to use and assess qualitative research methods. Neurological Research and Practice 2 (1). Available at https://doi.org/10.1186/s42466-020-00059-z.

Brown, M.E.L., Ahuja, N., Sivam, V., Khanna, A., Parekh, R., 2023. A qualitative evaluation of the impact of a medical student school outreach project on both medical students and school pupils. BMJ Open 13 (11), e070508. Available at https://doi.org/10.1136/bmjopen-2022-070508.

Carter, S.M., Little, M., 2007. Justifying knowledge, justifying method, taking action: epistemologies, methodologies, and methods in qualitative research. Qualitative Health Research 17 (10), 1316–1328. Available at https://doi.org/10.1177/1049732307306927.

Cristancho, S.M., Goldszmidt, M., Lingard, L., Watling, C., 2018. Qualitative research essentials for medical education. Singapore Medical Journal 59 (12), 622–627. Available at https://doi.org/10.11622/smedj.2018093, http://www.smj.org.sg/sites/default/files/SMJ-59-622.pdf.

Clarke, S.O., Coates, W.C., Jordan, J., 2021. A practical guide for conducting qualitative research in medical education: part 3—using software for qualitative analysis. AEM Education and Training 5 (4). Available at https://doi.org/10.1002/aet2.10644, http://onlinelibrary.wiley.com/journal/24725390.

Denzin, N.K., Lincoln, Y.S., 2005. Introduction: the discipline and practice of qualitative research. In: Denzin, N.K., Lincoln, Y.S. (Eds.), The Sage Handbook of Qualitative Research, 3rd ed. Sage Publications, Thousand Oaks, CA, pp. 1–33.

Gill, P., Stewart, K., Treasure, E., Chadwick, B., 2008. Methods of data collection in qualitative research: interviews and focus groups. British Dental Journal 204 (6), 291–295. Available at https://doi.org/10.1038/bdj.2008.192.

Hamada, H., 2019. Action research to enhance quality teaching. Arab World English Journal (1), 4–12. Available at https://doi.org/10.24093/awej/chief1.1.

Kuper, A., Reeves, S., Levinson, W., 2008. An introduction to reading and appraising qualitative research. BMJ (Clinical Research ed.) 337, a288. Available at https://doi.org/10.1136/bmj.a288.

Laverty, S.M., 2003. Hermeneutic phenomenology and phenomenology: a comparison of historical and methodological considerations. International Journal of Qualitative Methods 2 (3), 21–35. Available at https://doi.org/10.1177/160940690300200303, https://journals.sagepub.com/home/IJQ.

McLeod, S., 2024. Ethnography In Qualitative Research. https://doi.org/10.13140/RG.2.2.27135.62884.

Parks, P., 2023. Story Circles: A New Method of Narrative Research 7, 58–72. Available at https://doi.org/10.29333/ajqr/12844.

Reeves, S., Kuper, A., Hodges, B.D., 2008. Qualitative research methodologies: ethnography. BMJ (Clinical Research ed.) 337, a1020. Available at https://doi.org/10.1136/bmj.a1020.

Schneider, N.C., Coates, W.C., Yarris, L.M., 2017. Taking your qualitative research to the next level: a guide for the medical educator. AEM Education and Training 1 (4), 368–378. Available at https://doi.org/10.1002/aet2.10065, http://onlinelibrary.wiley.com/journal/24725390.

Chapter 36

# Explorative studies and remodeling

*"We look to medical research to discover remedial measures to insure better health and more happiness for mankind."*
—Thomas Hunt Morgan

## 36.1 Introduction

Before delving into large-scale studies, exploratory studies in medical education research seek to obtain preliminary understanding, identify factors, or develop frameworks for complex learning or teaching issues. They frequently use qualitative methods like focus groups and interviews to explore feedback or advising. These studies provide foundational insights for future, more thorough research on student experiences, curriculum development, or interventions like simulation. However, in medical education research, "remodeling" refers to the fundamental restructuring or reimagining of how medicine is taught and learned. This includes integrating mentorship or early clinical exposure, changing the focus from just facts to skills or competencies, adapting to new needs like technology or evidence, and essentially changing curricula and training paradigms for contemporary healthcare challenges rather than merely repairing existing structures.

The chapter is divided into two portions for easier understanding. The importance of exploratory research in medical education will be covered in Section A. However, Section B will address remodeling and how it affects the medical education.

## 36.2 Section A: exploratory studies in medical education

In medical education, exploratory studies are preliminary research projects that look into novel or poorly understood subjects with the goal of gaining understanding, generating hypotheses, and testing viability before larger, more definitive studies. They frequently use qualitative techniques like observations or interviews to understand complex educational phenomena, teaching strategies like problem-based learning, or student experiences like academic stress. Rather than offering definitive conclusions, they clarify research topics, identify important variables, and assist in designing future robust evaluations (Godage et al., 2025).

## 36.3 Need for exploratory studies

In medical education research, exploratory studies are essential for identifying new issues, elucidating difficult concepts, testing the viability of interventions such as simulation or CBL, generating hypotheses, and comprehending particular contexts like cultural or linguistic barriers. These studies serve as a crucial basis for larger, more comprehensive studies, particularly in areas that are understudied, such as emergency medicine training or particular student needs (Szabelska et al., 2021). Before making significant investments, they assist researchers in understanding why and how things work or don't function, ensuring that treatments are appropriate and successful for a variety of learners. The following are some important areas where the exploratory research have been helpful:

### 36.3.1 Identify knowledge gaps

The quick changes in medical practice and pedagogy frequently result in knowledge gaps in medical education research. These gaps highlight the necessity for exploratory investigations, which aid in problem definition, the identification of pertinent variables, and the generation of research hypotheses. The way medical education was previously taught and learned has altered due to the introduction of new technologies like virtual reality and augmented reality. In this sense, it is necessary to determine the most effective ways to include new technologies into curricula and evaluate their long-term

**The Principles and Practice of Medical Education. DOI:** https://doi.org/10.1016/B978-0-443-44197-4.00006-0

efficacy in developing clinical skills in comparison to conventional approaches. In medical education research, exploratory investigations are crucial for navigating unexplored or complicated areas where current knowledge is insufficient to draw firm conclusions (Chellamuthu et al., 2025). These investigations serve as a first step in identifying important knowledge gaps and refining future research topics by employing adaptable, open-ended methodologies. To put it simply, exploratory investigations are essential in these fields to go beyond conjecture, offer rich descriptive data, and lay the groundwork for future, more targeted hypothesis-driven research.

### 36.3.2 Develop and refine interventions

For complex interventions, developing and improving them through exploratory studies in medical education research requires a methodical, iterative process. Prior to a comprehensive assessment, this exploratory study aims to make sure the intervention is well-thought-out, workable, and acceptable. Pilot and feasibility studies are frequently included in exploratory research, which is a fundamental stage in ensuring that an intervention is workable and reliable (Liu et al., 2021). Through the use of exploratory research techniques, focus groups or small-scale pilots can be used to test early intervention concepts. In order to maximize engagement and relevance for the target audience, feedback obtained during these stages is used to improve the materials, delivery method, and content.

### 36.3.3 Understand student experiences

In medical education research, exploratory investigations are crucial because they aid in comprehending complicated student experiences that frequently lack preceding study or standardized measurement. To obtain in-depth information, spot developing themes, and lay the groundwork for more organized study in the future, these investigations employ qualitative approaches. New clinical settings, technologies, and curriculum are all part of the ongoing evolution of medical education (Walton and Oswald, 2024). In these novel or understudied fields, such as the effects of virtual reality training or modifications to clinical rotations because of global health emergencies, exploratory studies might reveal student perspectives and difficulties. Additionally, exploratory research aids in the researchers' comprehension of the subtleties of student experiences and aids in the creation of more accurate and trustworthy surveys and assessment instruments. The development of survey items that appeal to the target group is directly influenced by the themes found in an exploratory study.

### 36.3.4 Contextualize learning

According to contextualized learning, learning works best when it takes place in the particular setting or circumstance in which the information will be used. Exploratory studies are particularly well suited to investigate this. For instance, researchers can observe how medical students learn in real-world settings by immersing themselves in clinical settings using techniques like ethnography and observation, which reveal the unspoken social and cultural factors influencing education. Additionally, focus groups and qualitative interviews enable teachers and students to discuss their experiences in-depth, yielding valuable, contextual information on the learning process.

### 36.3.5 Assess feasibility

For medical education exploratory studies, feasibility assessment is essential because it establishes whether a promising idea, such as a new curriculum or technological tool, can actually function in the real world. It also identifies obstacles, such as time, money, and acceptance, prior to a large, expensive main study, ensuring the full research is practical, effective, and efficient by testing recruitment, procedures, and outcomes on a small scale, ultimately saving resources and increasing the likelihood of success for major interventions. Exploratory studies also help to determine whether larger study is viable by evaluating enrollment rates, data collection methods, resource availability, resource allocation, protocol clarity, and acceptability of novel methodologies.

## 36.4 Characteristics of explorative studies

Exploratory studies in medical education are characterized by their flexibility, focus on uncovering insights into poorly understood problems, and function as a foundational step for more structured research. Some of the vital characteristics of explorative studies are depicted in Fig. 36.1.

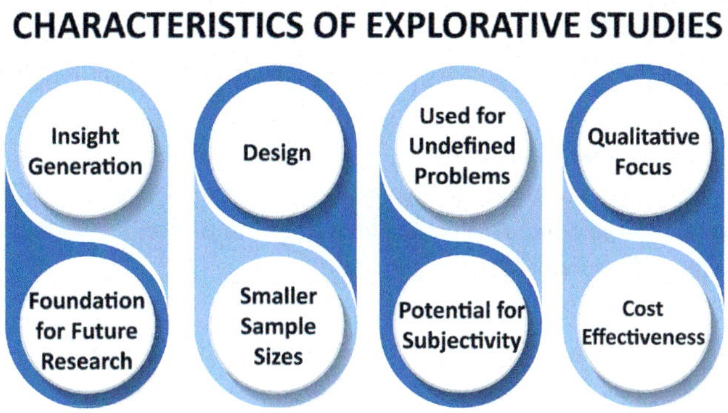


FIGURE 36.1 Characteristics of explorative studies.

### 36.4.1 Insight generation

One of the main and distinguishing features of exploratory research is the development of insights. The main goal of this kind of research is to explore novel or poorly understood subjects in order to obtain preliminary insight and develop new concepts, rather than to offer conclusive answers or test accepted theories. Understanding novel phenomena is the primary goal of exploratory research. When there isn't much previous research on a subject, exploratory research serves as a first step.

It enables scholars to explore "uncharted areas" and gain a better understanding of the topic. Furthermore, the knowledge gathered from exploratory work is essential for formulating precise, verifiable hypotheses. Later on, more organized, definitive study is guided by these hypotheses. An exploratory study's primary benefit is essentially its capacity to produce novel ideas that are then utilized to guide and organize more thorough studies later.

### 36.4.2 Flexibility in design

Flexibility in design is a defining and crucial trait since exploratory investigations are conducted when a problem is not clearly defined or understood. This flexibility allows researchers to adjust their methodology when new insights and unexpected findings arise, rather than being constrained by a rigid, predetermined strategy. The exploratory investigations are special because they allow for design freedom. In response to the data they acquire, researchers can use exploratory investigations to adjust their study topics and data gathering techniques, such as altering interview questions and adding new observations and emphasis areas. A deeper and more thorough comprehension of the subject is made possible by this iterative process.

### 36.4.3 Used for undefined problems

Exploratory research is employed for situations that are not clearly characterized. Relying on the data itself to develop new ideas or frameworks instead of testing preestablished assumptions is one of the main characteristics of exploratory studies. When the issue is complicated, involves a novel field of study, or lacks preexisting theoretical frameworks, this method is especially helpful. Gaining initial insights, seeing trends, and coming up with ideas for more organized, future study are the primary objectives.

### 36.4.4 Qualitative focus

The hallmark of exploratory studies is a qualitative focus, which uses techniques like focus groups and interviews for rich, nonnumerical data rather than testing hypotheses in order to gain deep understanding, uncover new ideas, and investigate complex "why" and "how" questions when little is known about a topic. This approach provides basic insights for future, more targeted research by prioritizing depth over breadth. Seeking depth and nuance which is a rich, thorough account of feelings, attitudes, and actions that captures intricacies ignored by numbers is one of the main components of qualitative focus in exploratory research. Exploratory research employs a qualitative approach to investigate the unknown, emphasizing context and meaning to develop comprehension from the ground up.

### 36.4.5 Foundation for future research

As the main objective of exploratory research is to find preliminary insights and patterns in areas with little current information, exploratory studies serve as a foundation for future research. They invariably discover gaps and generate suggestions that can be investigated further in subsequent, more focused research since they are venturing into uncharted territory. Exploratory research highlights the need for more study by identifying facets of a topic that have not yet been examined. These studies yield speculative conclusions that form the foundation for tested hypotheses in subsequent stages of research, rather than testing predetermined hypotheses.

### 36.4.6 Smaller sample sizes

As exploratory studies' main objectives are hypothesis generation, first insight finding, and methodological development rather than definitive hypothesis testing or generalizability to a wider population, they frequently have smaller sample sizes. Rich, in-depth knowledge about a complicated or understudied subject is prioritized in exploratory investigations, which are frequently better accomplished through rigorous data collecting from fewer participants. Additionally, smaller samples make this preliminary research more manageable and economical because exploring new areas might require a significant investment of time, money, and manpower.

### 36.4.7 Potential for subjectivity

Exploratory studies are subjective due to their dependence on qualitative methods and focus on obtaining profound understanding of new or obscure occurrences. Unlike quantitative research, which strives for complete objectivity and generalizable, numerical truths, exploratory research actively engages with individual perspectives and interpretations. To collect rich, detailed data, exploratory research frequently use qualitative techniques, including open-ended interviews, observations, and case studies. By its very nature, this data is based on the researcher's and the participants' subjective experiences and views.

### 36.4.8 Cost-effectiveness

Cost-effectiveness is a crucial feature and important benefit of exploratory investigations, not because the study is inherently inexpensive but rather because it helps avoid the enormous financial losses that come with pursuing failing goods or tactics without first conducting fundamental research. Before devoting significant resources to extensive development or assessment, an organization might test the waters and obtain fundamental insights through exploratory research. Early detection of possible issues or low demand helps prevent the significantly greater expenses associated with creating and introducing a failed product or solution.

## 36.5 Exploratory research design

In order to generate insights, find patterns, and formulate hypotheses for future, more structured research using both primary and secondary data sources, exploratory research design employs flexible methods to investigate problems with little prior knowledge. It mainly relies on literature reviews, in-depth interviews, focus groups, case studies, and observations, which are frequently qualitative. Both primary and secondary research methodologies are used in exploratory research designs (Fig. 36.2).

### 36.5.1 Primary research methods

Information obtained directly from the topic is referred to as primary research. It might happen through a single person or a group of people. The researcher may choose to carry out this kind of research themselves or by hiring a third party to do it for them. Primary research is specifically done to investigate a particular issue that calls for a thorough investigation. In order to understand a problem, primary research methods for exploratory design concentrate on direct data collection using qualitative techniques such as surveys or polls, in-depth interviews, focus groups, and observational studies, as well as some quantitative tools such as to obtain preliminary insights, identify patterns and develop hypotheses for additional research. Researchers can obtain first-hand, context-specific data on attitudes, actions, and opinions using these techniques.

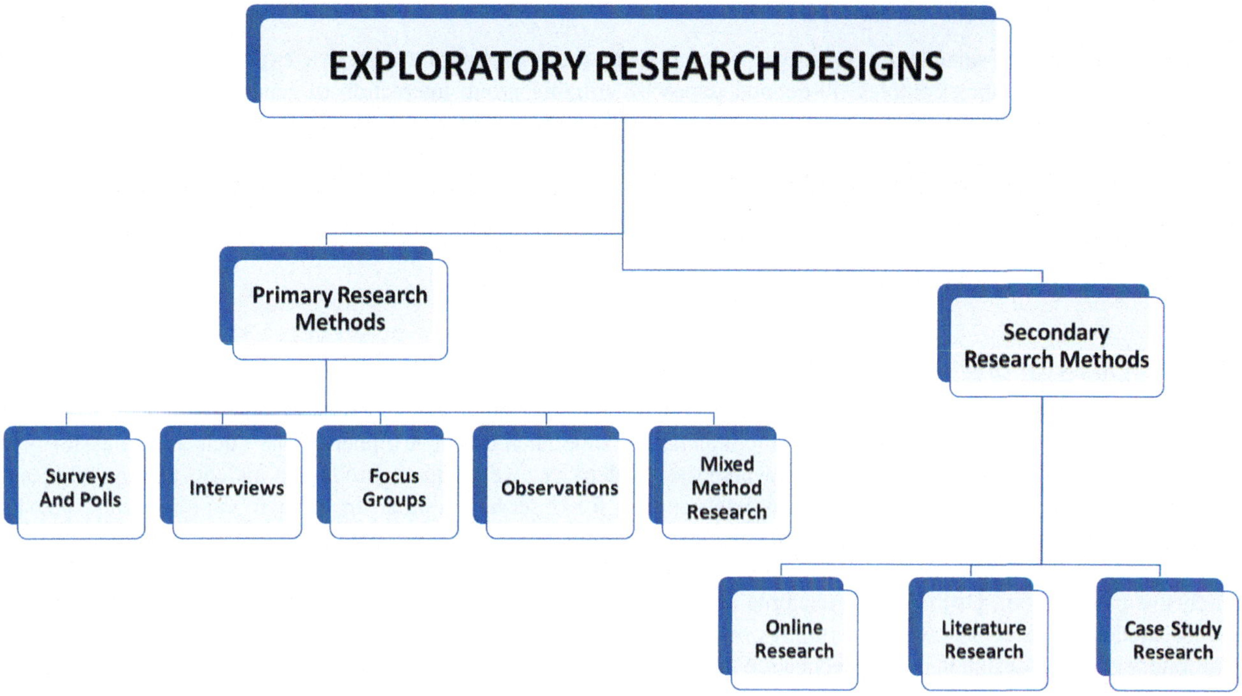


FIGURE 36.2 Exploratory research designs.

### 36.5.1.1 Surveys and polls

Information is gathered from a predetermined group of respondents through surveys and polls. Although they can also be used for broad trend-spotting, surveys and polls are excellent primary research methods for exploratory research, particularly when they are designed with open-ended questions to uncover new ideas, themes, and deeper insights into an under-researched topic. This allows researchers to identify patterns and formulate stronger research questions for later studies. They provide flexibility, are an economical means of obtaining direct input, may effectively collect data from huge populations, and can incorporate qualitative data for a deeper understanding. In exploratory research, surveys and polls are used for exploration, discovery, and preliminary understanding rather than for statistical verification of facts. Surveys and polls are useful main instruments for exploratory designs in the field of exploratory studies. They bridge the gap between initial curiosity and more structured research by gathering fundamental qualitative and quantitative data. For example, researchers survey students' perceptions of learning, their teachers, the academic atmosphere, and their social self-perceptions within a medical school using validated instruments such as the Dundee Ready Education Environment Measure (DREEM) inventory (Alshahrani, 2025).

### 36.5.1.2 Interviews

The foundation of exploratory research is interviews, which are a key technique for obtaining in-depth, qualitative insights into complicated subjects, behaviors, and experiences. Open-ended questions with small, focused samples are used to define issues, unearth new concepts, and produce hypotheses for additional research rather than conclusive answers. Through one-on-one discussions and the use of semi-structured guides for flexibility, they enable researchers to investigate viewpoints, comprehend underlying causes, and gather rich, nuanced data. Researchers can learn about participants' subjective opinions, emotions, and experiences through interviews, which can highlight problems the researcher may not have thought of. As a method of exploratory research, interviews are crucial for developing hypotheses since they provide a thorough examination of complicated occurrences and aid in the formulation of future study topics or possible answers. For instance, the use of interview in exploratory research design in medical education is in gaining deep understanding of students' or professionals' experiences, perceptions, and motivations.

#### 36.5.1.3 *Focus groups*

Although they are less suitable for sensitive personal data, focus groups are effective exploratory tools in medical education for understanding complex, ill-defined issues by utilizing group interaction to generate rich qualitative data, uncovering new perspectives on curriculum, professional challenges like ethics or new policies and patient experiences. This lays the groundwork for larger studies like surveys. They support the investigation of learner needs, the evaluation of innovative teaching techniques, the understanding of difficult practice areas, and even the enhancement of survey questions by providing context that is occasionally missed in quantitative approaches. Focus group discussions, for instance, are essential to curriculum development because they provide feedback on how to create or improve courses, comprehend the needs of students, and identify learning obstacles.

#### 36.5.1.4 *Observations*

Using observation as an exploratory research design in medical education entails investigating teaching or learning in natural environments, such as classrooms or clinicals, to comprehend complex phenomena such as professionalism, communication, skill acquisition in situations where the problem isn't well defined, producing insights and hypotheses, or guiding future, more rigorous studies such as interventions for observation skills. In order to capture rich, contextual nuances that could be overlooked in structured surveys, researchers frequently use qualitative methods like ethnography or field notes to observe relationships, behaviors, and processes. This approach provides deep knowledge rather than conclusive causal proof. One excellent example of using observations as an exploratory research strategy is seeing how medical students learn to deal with uncertainty or medical jargon. Some of the key benefits of using primary research as an exploratory research design in medical education include direct relevance to a particular context, the generation of original hypotheses and customized insights, and the ability to gain a thorough understanding of complex educational phenomena that may not be covered in existing literature. This strategy is essential for tackling certain issues in educational programs and fostering creativity.

### 36.5.2 Mixed method research

As it combines qualitative depth facilitating the understanding of "why" and "how" with quantitative breadth measuring "what" and "how many" for a holistic view, mixed methods research is essential in exploratory research. This allows researchers to first explore themes and then test those themes broadly or vice versa to gain richer more validated insights into complex phenomena that single methods miss. It provides answers to issues that neither strategy could adequately handle on its own, strengthening theories and guiding better choices in the real world (Rana, 2025).

### 36.5.3 Secondary research methods

In an exploratory study design in medical education, secondary research methodologies are extremely useful for understanding the current state of knowledge, identifying research gaps, and guiding future primary research paths. The goal of exploratory research is to look into issues that are not well defined or have not been thoroughly examined. Secondary research aids in this early stage by swiftly and affordably offering fundamental insights. When using secondary research methods, researchers can identify areas that require primary investigation and ascertain what is already known about a topic in medical education, such as teaching strategies for a certain subject. Furthermore, knowledge gleaned from current data aids in narrowing down general subjects into precise, verifiable research questions or hypotheses for subsequent investigations. Some of the often-employed secondary research techniques are as follows:

#### 36.5.3.1 *Online research*

In exploratory studies, online research is a potent secondary method that provides rapid, affordable access to a wealth of existing data reports, journals, and social media to find patterns, test preliminary ideas, and identify knowledge gaps for further investigation. However, in order to establish a strong foundation for new research, careful source vetting for credibility and potential bias is necessary. Online research turns out to be the initial stage in exploratory investigations, offering context, background, and current knowledge about a subject before primary research starts. Academic databases, government websites, medical publications, and international standards from medical education regulatory authorities are some of the main sources of internet research.

#### 36.5.3.2 Literature research

By examining published sources like journals, books, and reports, literature research is a critical secondary method which is essential in exploratory studies to map existing knowledge, identify gaps, define concepts, and formulate hypotheses. It also provides an affordable foundation prior to possible primary data collection. It is an effective place to start when investigating new research issues since it helps researchers comprehend what is previously known, find discrepancies, and identify areas that require more research. One of the benefits of literature research is its authenticity, since any information presented in a book or journal is carefully examined by subject matter specialists before being published. Instead of merely summarizing these sources, rigorous literature study can assist in critically combining, assessing, and interpreting them to provide a whole picture.

#### 36.5.3.3 Case study research

Unlike short clinical case reports, case study research is a potent qualitative approach in medical education that focuses on in-depth, real-life contextual investigation of phenomena like teaching, learning, or treatments. Its strength is in describing complex situations where experiments aren't practical, encouraging critical thinking, and connecting theory to practice. It goes beyond surface-level data to offer rich, detailed insights about a particular individual, group, event, or phenomena (Cheek, 2018). Additionally, by spotting trends in unusual or understudied fields, it aids in the creation of new hypotheses or the improvement of preexisting ones. For instance, in medical education, deep, detailed awareness of intricate clinical or educational settings enables in-depth insights (Anaf, 2007).

## 36.6 Steps of exploratory studies

Instead of offering definitive solutions, exploratory investigations look into, define, and comprehend vaguely defined situations. Identifying the study topic, reading the body of literature, developing flexible hypotheses or questions, creating a qualitative approach, gathering data, evaluating the results, and recommending more research are all important processes. The various steps of the exploratory research are described below:

### 36.6.1 Defining a research problem

In exploratory design, defining a research problem entail starting broad, refining it through literature review and observation to identify knowledge gaps, developing open-ended research questions and potential hypotheses, and then choosing adaptable techniques like focus groups or interviews to obtain preliminary insights, laying the groundwork for more focused future research issues.

### 36.6.2 Conduct literature review

In exploratory research, conducting a literature review entails a thorough search of the body of knowledge to identify gaps, themes, and context, guiding open-ended questions to find what is unknown rather than to test hypotheses, identifying important concepts and laying the groundwork for further investigation, which frequently results in the development of preliminary hypotheses for later stages. Determining the general topic, conducting extensive database searches, spotting trends and gaps, synthesizing themes, and then developing flexible research questions are some of the crucial tasks in this respect.

### 36.6.3 Formulate a hypothesis

Instead of testing a predetermined hypothesis as in confirmatory research, which focuses on observation, literature review, variable identification, open-ended questions, and then generating hypotheses to test later, exploratory research formulates a hypothesis by investigating an unknown area to obtain comprehensive understanding, employing adaptable, unrestricted techniques such as focus groups, in-depth interviews, and context-aware observation. This method, which frequently uses purposeful sampling to choose important informants, puts comprehension, pattern recognition, and in-depth interpretation ahead of numerical data., leading to initial, tentative statements or hypotheses that guide deeper confirmatory research. Through these methods, it creates a qualitative approach.

### 36.6.4 Data collection

In order to understand a broad topic, identify variables, and formulate clearer questions for future, more structured research rather than definitive answers, data collection in exploratory research is a flexible, iterative step focused on gaining initial insights. It frequently uses qualitative methods like focus groups, interviews, and observations in addition to secondary sources. It entails picking primary or secondary methods, selecting instruments like interview guides, sampling, collecting rich data, and then iteratively evaluating patterns to improve hypotheses.

### 36.6.5 Data analysis

The iterative, early-stage, discovery-focused process of data analysis in exploratory research design includes understanding the problem, cleansing data, summarizing significant features like distributions, outliers, visualizing patterns, and formulating hypotheses. Rather than validating current research, the objective is to gather preliminary ideas and guide future, more formal studies. It starts with gathering data, which is followed by the refinement of research questions and the discovery of hidden structures using techniques like constant comparison and memoing a critical process in which researchers write memos to record ideas, interpretations, patterns, and connections as they analyze data progressing from raw data to theoretical concepts.

### 36.6.6 Drawing conclusions

Rather than offering conclusive answers, generating conclusions in exploratory research entails synthesizing information to produce fresh insights and viable hypotheses for future, more formal studies. In order to find patterns, themes, and trends in the qualitative data gathered from focus groups, interviews, and observations, this procedure adheres to the data analysis phases. Understanding the underlying problems or occurrences under investigation is the main goal of this step. Subsequently, the results are connected to the research questions, and the interpreted data is analyzed in relation to the original, open-ended research questions to determine what has been discovered about the subject. The researcher recommends certain avenues or useful actions for next research based on the knowledge acquired. This frequently entails suggesting a change to more formal study designs that can yield outcomes that are more conclusive (Fig. 36.3.

## 36.7 Application of explorative studies in medical education

In medical education, exploratory research is used to look into complicated or poorly understood educational phenomena, like how students view learning tactics, the effects of particular teaching techniques, or the elements that affect professional development like empathy. Before more extensive interventions are planned and carried out, they help identify important variables, present preliminary data on prospective advantages or issues, and provide fundamental insights for future, more structured study. These studies frequently use qualitative techniques, such as observations or interviews, to reveal complex experiences and guide the creation of more focused study topics and successful teaching strategies. A few of the most important uses in medical education are depicted in the following figure (Fig. 36.4.

### 36.7.1 Curriculum development and evaluation

Since the early phases of curriculum development and evaluation involve investigating educational requirements, gaps in knowledge, and potential innovative approaches, it can be seen as an application of exploratory investigations. Before creating and assessing a curriculum, this exploratory method aids in defining the goals and scope. It also helps in determining educational needs when it comes to curriculum development. It is crucial that educators and curriculum developers consider the needs of students, the demands of the labor market, or emerging social issues before developing any curricula. This entails using techniques like focus groups, interviews, and surveys to collect qualitative information about what needs to be taught and learned. Exploratory research also helps evaluate existing gaps by helping the researcher find the shortcomings in the current curricula. developers can pinpoint specific areas that need improvement as they can look into current student performance data and feedback from educators and administrators.

### 36.7.2 Assessing new educational tools and methods

Exploratory studies are frequently and directly used to the assessment of new educational tools and procedures, especially when the tools, methods, or context in which they are employed are novel, unclear, or lack substantial preceding study. Gaining preliminary understanding, identifying important factors, and formulating hypotheses for further, more organized

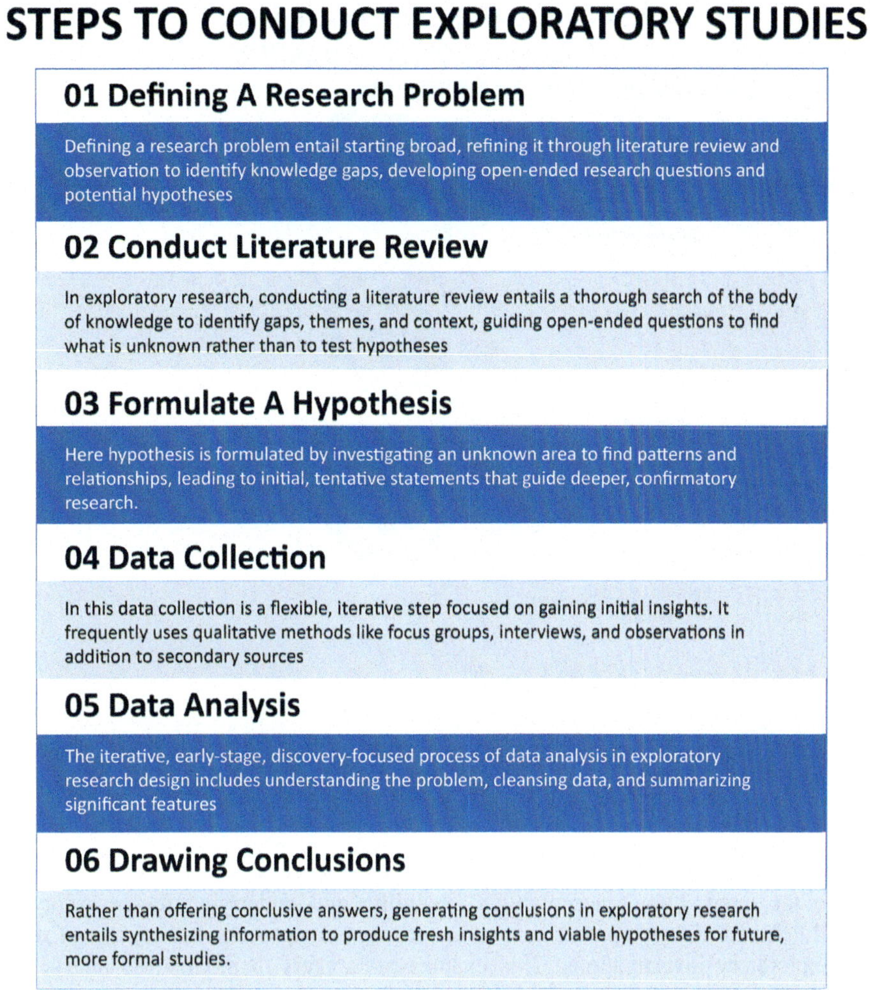


FIGURE 36.3 Steps to conduct exploratory studies.

research are the main objectives. Exploratory research is crucial for examining novelty. This is advantageous when a new digital learning tool, such an AI-driven platform or a new augmented reality application, is presented because there is frequently no prior knowledge regarding its impact or ideal implementation. Researchers and educators can better grasp "what is happening" and "how" the technology is being utilized in a real-world context by conducting exploratory studies. Additionally, they assist in uncovering potential like higher engagement and self-efficacy that might not be visible with more rigorous study designs, as well as unanticipated obstacles like teacher skill gaps, the digital divide, and student distraction.

### 36.7.3 Understanding student and educator experiences

In order to gain a deeper understanding of student and educator experiences in exploratory research, qualitative methods such as focus groups, interviews, and case studies are used to uncover deep insights into active, student-centered learning, challenges with new methods like online or gamified learning, skill development in research, and factors influencing engagement. These methods reveal how students navigate inquiry-based learning and how educators adapt, resulting in better support systems and teaching strategies for deeper understanding and critical thinking.

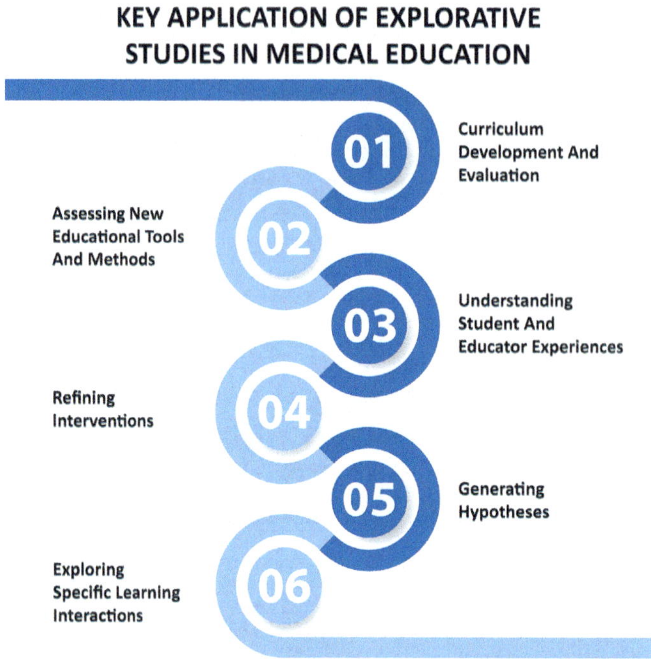


FIGURE 36.4 Key application of explorative studies in medical education.

### 36.7.4 Refining interventions

In exploratory research, refining interventions is a flexible and iterative process that mostly employs qualitative and mixed approaches to improve the intervention's acceptability, viability, and underlying theory prior to a more extensive, conclusive evaluation. Intervention development is seen as a continuous process with cycles of development, feedback, and improvement in exploratory investigations. This enables researchers to modify the intervention when fresh data becomes available. Exploratory research helps establish or enhance the underlying theory or logic model by outlining the components, activities, and expected outcomes of the intervention. As a result, the study progresses from a purely structural framework to a theory that clarifies the connections between various phenomena.

### 36.7.5 Generating hypotheses

In exploratory research, hypotheses are not predetermined but rather arise from an open examination of evidence, serving as a bridge to more formal testing in the future. The process of developing hypotheses begins with broad interests, employs techniques such as observations and interviews to identify patterns, and then develops initial concepts such as exploratory hypotheses about possible relationships. These are then refined into testable statements for subsequent confirmatory studies with an initial focus on understanding rather than proving causation.

### 36.7.6 Exploring specific learning interactions

Particular learning interactions in exploratory research frequently revolve around open-ended, inquiry-based, social procedures intended to produce new understanding and insights in domains with limited prior knowledge. These exchanges embrace experimentation, dialogue, and feedback loops in addition to traditional training. Learner content interaction, in which students connect with the course material and content, is one of the fundamental forms of interactions. Learner-to-learner interaction is one of the various forms of interaction. It allows peers to address cognitive disagreement and bring about conceptual transformation. The instructor's position in learner-instructor interaction changes from that of a typical "sage on the stage" to that of a "guide by the side" in exploratory learning.

## 36.8 Advantages of exploratory research

### 36.8.1 Flexibility and adaptability

Numerous professions can considerably benefit from the substantial advantages that exploratory research brings. Its adaptability is one of its main advantages. Based on preliminary results, researchers can modify their strategy to further explore unanticipated areas of interest. Instead of following strict frameworks, this flexibility promotes a more thorough understanding of complicated problems. An essential part of exploratory research is flexibility and adaptability, which enable researchers to react to new information and evolving circumstances. This dynamic approach makes it easy to delve deeply into uncharted territory by allowing the investigation of various techniques and methodologies. The ability to change course and modify research designs in response to participant input and preliminary results is one of the main benefits of flexibility. Being flexible allows scholars to investigate different viewpoints without being restricted by a strict framework.

### 36.8.2 Uncovering new insights

Finding fresh perspectives is another benefit of exploratory research. Rich qualitative data is frequently revealed by exploratory research, offering insightful context regarding behaviors, motivations, and difficulties. This aids in the creation of hypotheses that can then be investigated using more methodical research techniques. Organizations might find knowledge gaps or emerging trends that might otherwise go undiscovered by embracing the open-ended character of exploratory research. All things considered, these benefits lead to comprehensive understanding, which makes exploratory research a crucial instrument for well-informed decision-making.

### 36.8.3 Foundation for future research

In order to establish the groundwork for subsequent study, exploratory research is essential. This kind of research is essential to medical education because it establishes the foundation for subjects that are novel, ambiguous, or have not been thoroughly studied before, directing the creation of later, more targeted in-depth studies. The impact of new teaching technology is one example of an ambiguous or broad subject in medical education that exploratory research helps to define into more manageable and focused research concerns. Researchers can develop formal hypotheses that can be extensively verified in later, more structured investigations by identifying patterns and possible links through the collection of preliminary findings.

### 36.8.4 Deeper understanding of complex issues

Exploratory research in medical education offers the advantage of providing a better knowledge of complex difficulties by enabling researchers to investigate innovative or understudied subjects in a flexible manner. This method is essential when there is a paucity of starting evidence or when current ideas are not adequate to properly explain a phenomenon. Researchers can identify underlying causes, settings, and subtleties of problems rather than just their outward manifestations by employing qualitative and adaptable techniques like observations and interviews. For instance, research might look at systemic or cultural factors that contribute to resident burnout in addition to work hours. However, this study approach makes it possible to find unanticipated variables or interactions that organized research would overlook in the intricate and dynamic field of medical education. This all-encompassing perspective is essential for creating comprehensive successful solutions.

### 36.8.5 Informed decision-making

Exploratory research is crucial to medical education because it provides the basic data and insights needed to make evidence-based decisions on curriculum design, program implementation, and policy development. It helps educators go beyond assumptions by clarifying the exact nature of a problem before investing significant resources. Exploratory research aids in the identification and definition of issues that are still unclear such as the reasons behind a specific teaching strategy's poor performance or how students see a novel clinical encounter. Decision-makers can concentrate their attention on the most pertinent topics due to this preliminary understanding. Researchers can find new information and potential causal aspects that were not previously thought of by employing flexible, open-ended techniques like observations and interviews. In order to ensure that future research is well-targeted, these first findings are then used to develop precise, testable hypotheses for more structured studies in the future.

## 36.9 Disadvantages of exploratory research

Exploratory research has its downsides, which are crucial to consider alongside its advantages.

### 36.9.1 Lack of structure

As it might result in a number of difficulties, a lack of structure is regarded as a drawback of exploratory research in medical education. Data collection and interpretation can be uneven and challenging to standardize because research questions are frequently broad and study strategies are flexible. Furthermore, it is challenging for other researchers to accurately duplicate the study and validate the results due to the subjective and open-ended character of the research.

### 36.9.2 Bias during data collection

The possibility of bias during data gathering is another issue. Because of their personal experiences or expectations, researchers may inadvertently distort the results. Furthermore, lower sample sizes are frequently used in exploratory research, which may not accurately reflect the larger population. As a result, conclusions from exploratory research are often not generalizable, which limits their applicability. Finally, because stakeholders may prefer more structured research approaches that yield quicker, measurable results, it might be challenging to obtain adequate funding for exploratory research. Understanding these disadvantages is necessary to make informed decisions about the role of exploratory research in larger investigations.

### 36.9.3 Lack of conclusive results

The absence of definitive findings is one drawback of exploratory research. This kind of research seeks to find trends, produce insights, and lay the groundwork for more investigation. However, because of this intrinsic exploratory aspect, outcomes are frequently unclear and subject to interpretation. Researchers may find insightful information when collecting qualitative data, such as through client interviews or open-ended survey replies, but they sometimes encounter difficulties drawing firm conclusions. There may be advantages and disadvantages to these conflicting findings. On the one hand, they provide freedom in comprehending intricate issues without being constrained by predetermined measures. On the other hand, decision-making processes may be hampered by the difficulty to reach definite, practical judgments. In order to ensure that everyone is aware of the significance of additional research to validate findings, it is critical that research teams inform stakeholders of these limitations. As a result, even while exploratory research provides insightful preliminary findings, it requires additional research to produce more precise and definitive results.

### 36.9.4 Limited generalizability

In actuality, the restricted generalizability of exploratory research is one of its primary disadvantages in medical education. The results are typically only applicable to the particular conditions or populations researched because this research methodology frequently uses small, targeted samples and context-specific case studies, making it challenging to generalize the findings to other institutions, demographics, or educational contexts. Small participant groups are often used in exploratory research to obtain precise insights, which unavoidably reduces the statistical power and external validity needed for generalization. Medical education settings vary widely in terms of curriculum design, student demographics, and institutional culture. Research done in one setting may not be transferable to another without further confirmation.

## 36.10 Section B: remodeling in medical education

A long-standing effort in medical education is remodeling, which aims to modify curricula and instructional strategies to address changing healthcare demands. This entails exploring novel strategies and recognizing present issues in an open-ended way in the framework of exploratory research, frequently employing qualitative techniques to comprehend various viewpoints and circumstances. In general, remodeling refers to a paradigm change in curriculum design and teaching strategies to address the healthcare demands of the twenty-first century. This entails switching from traditional knowledge transfer to competency-based training, which emphasizes clinical reasoning, practical skills, and the incorporation of new technology and interdisciplinary approaches (Brenner and Ertl, 2012).

### 36.10.1 Definition

In medical education, "remodeling studies" refers to the deliberate reorganization, modernization, or reform of medical curricula and training programs in response to evolving healthcare needs, technology breakthroughs, and pedagogical innovations. Instead of focusing on a particular kind of scientific experiment, this research aim to change the way medicine is taught, frequently moving from teacher-centered to learner-centered methods like competency-based education, AI integration, and simulation-based training. Remodeling medical education is not a standard medical term but rather a conceptual phrase referring to the comprehensive review, redesign, and updating of the structure, style, and content of medical training programs to better meet the demands of modern healthcare and society.

## 36.11 Key areas of remodeling in medical education

### 36.11.1 Curriculum modernization

Changing the focus from rote memorization of facts to a learner-centered approach is one of the main components of curriculum redesign. It also makes it easier to comprehend the doctor's position in the larger healthcare system, including teamwork, leadership, and communication abilities. In order to satisfy the needs of the community and students in contemporary healthcare, the curriculum will be reviewed and redesigned. The structure, style, and substance of medical training programs can be thoroughly reviewed, redesigned, and updated due to curriculum modernization, which makes it easier for students to assume the obligations and difficulties they may encounter in the future.

### 36.11.2 Technological integration

Technology has undoubtedly altered how current medical education is taught and practiced. In this context, it is crucial to remember that technology is an essential component of medical education. Medical instructors now have the chance to incorporate technology because of remodeling. The key technological areas that are changing medical education are data analytics, telemedicine training, online learning environments, virtual simulation and extended reality, and artificial intelligence (AI). These technologies are transforming traditional education by making it more accessible, customizable, and interactive.

### 36.11.3 Competency-based medical education

As a revolutionary approach to medical education, competency-based medical education (CBME) replaces the conventional time-based model with one that is focused on attaining certain, observable skills and results necessary for efficient medical practice. Since CBME has replaced the conventional time-based learning model with a system centered on attaining precise, quantifiable results, it has become a true game-changer (Penrabel et al., 2022). This method guarantees that all medical graduates have the abilities and information needed to practice safely and successfully after completing their training. The need for increased public accountability and to make sure doctors are prepared to handle changing healthcare needs is what is driving this global movement.

### 36.11.4 Active learning and mentorship

The participation of students is one of the key components of renovation. It is a reality that a lot of contemporary teaching techniques have changed the focus of instruction from the teacher to the student. In this regard, putting into practice cutting-edge student-centered learning techniques like problem-based learning helps with the remodeling because these approaches are essential in changing the emphasis of medical education from traditional, passive learning to active, application-based learning. This approach also prepares future physicians for the complexities of modern healthcare by promoting critical thinking, clinical skills, and teamwork.

### 36.11.5 Addressing community needs

A key component of modernizing medical education is addressing community needs, which guarantees that instruction is applicable, fair, and efficient in enhancing public health. Medical schools produce doctors who are better prepared to meet the requirements of society by concentrating their curricula on local, real-world health issues. It is important to remember that program learning outcomes should be the first step in addressing community needs. However, it is crucial to engage in the process of curriculum revision in order to highlight the significance of primary care and prevention by incorporating thorough instruction in these areas throughout the curriculum. In order to extend learning environments outside of traditional hospitals and into community settings, it is also crucial to take community participation into account as a fundamental component of medical education. As partners in the educational process rather than merely passive users, this calls for cooperation with local health professionals, community-based groups, and community stakeholders.

### 36.11.6 Student well-being and professional identity

In order to produce resilient, humanistic, and capable doctors, medical education is being redesigned with a greater emphasis on combining student well-being and structured professional identity construction. In the end, this integration enhances both individual well-being and the standard of patient care by assisting students in managing stressors and internalizing professional values. In medical education, the development of a professional identity and well-being are closely related processes. In this regard, a trainee's feeling of purpose in their job can be strengthened by having a strong sense of professional identity, which also protects them from stress and burnout. Additionally, encouraging self-care and student wellness helps students deal with the rigors of medical school and think about their future roles, which is crucial for creating a strong professional identity. Medical education reforms aim to produce physicians who are not only clinically competent but also self-aware, resilient, and better equipped to meet the changing challenges and societal expectations of the medical profession by proactively addressing the complex interplay of student well-being and professional identity (Fig. 36.5.

## 36.12 Factors affecting remodeling in medical education

Rapid technological advancements, an increasing global illness load, changing healthcare needs, and a transition from knowledge-based to competency-based training all call for changes in medical education. The integration of technology like AI and virtual reality, a focus on multidisciplinary collaboration and lifelong learning, the desire for greater educational justice, and a renewed emphasis on skills like critical thinking and leadership are some of the key elements influencing renovation.

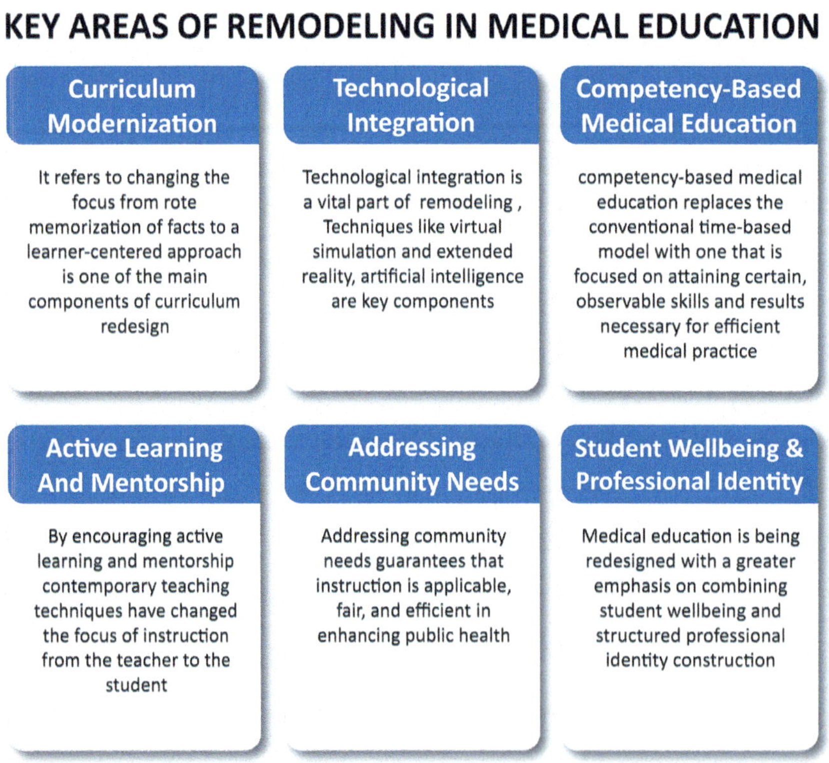


FIGURE 36.5 **Key areas of remodeling in medical education.** This figure shows the key areas of remodeling in medical education.

### 36.12.1 Technological advancements

Medical education is undergoing a significant transformation due to technological breakthroughs, which are replacing traditional lecture-based models with hybrid, individualized, and immersive learning experiences. Personalized medicine and immersive learning have been made possible by significant developments like AI and machine learning, which have completely changed medical education. This will facilitate the development of new skills, increase the accuracy of diagnoses, and increase access to care and training. Similarly, technologies such as telemedicine and remote patient monitoring have dismantled international borders and expanded the scope of clinical practice and medical education by enabling virtual consultations and remote learning. These developments are compelling medical education to embrace a hybrid approach that combines traditional, human-centered instruction with digital resources.

### 36.12.2 Shifting healthcare demands

Medical education is being completely redesigned to address physician shortages, emphasize public health and social determinants, and promote interdisciplinary collaboration in response to changing healthcare demands. The aging population and lack of skilled physicians are two of the main challenges facing modern clinical practice. There is a severe physician shortage due to the rising demand for healthcare worldwide, which is partly caused by an aging population. Medical schools around the world are eager to increase enrollment in order to effectively address these demands by producing more "practice-ready" professionals. Additionally, because of contemporary medical education, a doctor's responsibilities are expanding beyond treating specific patients to include spearheading changes to the larger healthcare system. This calls for new abilities in population health, system reform, and leadership.

### 36.12.3 Competency-based education

CBME fundamentally alters traditional medical education by shifting the focus from time-based training and knowledge acquisition to the display of observable talents and outcomes relevant to real-world patient requirements. This paradigm shift necessitates significant changes to curriculum design, faculty responsibilities, and assessment methods. Because CBME embodies the outcomes-based approach, curricula here are based on a set of predefined core competencies required of a medical graduate, ensuring alignment with societal and patient expectations. Additionally, CBME places a strong emphasis on early clinical experience and hands-on, active learning to assist students in bridging the gap between theoretical knowledge and clinical application, hence increasing the relevance of basic sciences.

### 36.12.4 Educational equity

In order to diversify the physician workforce and give all doctors the tools they need to address health inequities, educational equality is a major driving force behind changes to medical education. The objective is to genuinely change the institutional conditions that allow all students to succeed, going beyond merely expanding access. For instance, by modernizing the teaching strategies to incorporate case studies and different patient examples that speak to students from a range of backgrounds, all students will feel included and involved (Khan, 2012)

### 36.12.5 Alignment with practice

As it guarantees that instruction is applicable to actual healthcare demands and equips students with the practical skills, critical thinking, and professional identity necessary for efficient patient care, alignment with practice is a key factor in the transformation of medical education. Students' learning experiences become more relevant and engaging when the curriculum is in line with clinical practice. Students gain a better understanding of how to apply knowledge in practical settings when basic sciences are integrated into a clinical setting.

## 36.13 Impact of remodeling in medical education

By improving clinical skills, incorporating technology, and encouraging holistic patient care, medical education remodeling entails changing curricula and teaching strategies to generate more capable, patient-centered, and responsive healthcare providers. A better mix of clinical experiences, more multidisciplinary training, an emphasis on soft skills like communication, the integration of patient feedback, more research possibilities, and the creation of more effective and compassionate future physicians are some of the major effects. The following is a list of some of the main effects of remodeling on medical education:

### 36.13.1 Improved clinical skills and competence

Medical education reform significantly enhances clinical skills and competency by bridging the gap between theory and practice and integrating modern, evidence-based teaching methods into the curriculum. For instance, the transition from a time-based to a CBME guarantees that all students meet precise, quantifiable clinical milestones prior to progressing, ensuring that instruction is in line with real patient-centered demands and professional practice criteria (Todres et al., 2012). In addition, rather than waiting for traditional clinical years, exposing students to patient interactions and clinical settings early in their education fosters the development of critical communication skills, professional identity and confidence, and the application of theoretical knowledge in a practical setting. In essence, these changes make education more interesting, useful, and goal-oriented, better preparing graduates for the complexity and demands of contemporary healthcare and improving patient care and safety outcomes.

### 36.13.2 Curricular structure and balance

There are many important advantages to redesigning medical education for better curricular balance and structure, especially when integrated and competency-based approaches are used. This reorganization improves student learning and clinical skills while bringing the curriculum into line with the demands of modern healthcare. The combination of basic and clinical sciences is one of the main features of contemporary medical education. In this sense, remodeling has the benefit of dismantling conventional boundaries between preclinical and clinical years as well as between disciplines. This promotes a deeper, more meaningful comprehension and improved knowledge retention by assisting students in making connections between fundamental science concepts and actual clinical situations. Furthermore, a revamped curriculum usually ensures a natural flow of knowledge and eliminates needless duplication between classes. This well-organized framework guarantees that the content is provided in a manner consistent with the practical application of knowledge.

### 36.13.3 Professional development and research

By encouraging lifelong learning, critical thinking, and research-oriented abilities in both students and professors, medical education reform greatly advances professional development and research. Throughout their careers, medical professionals will be able to stay up to date with the rapid changes in medical knowledge, technology, and treatment procedures due to revised curriculum that support self-directed learning skills. Furthermore, cutting-edge teaching strategies like problem-based learning, team-based learning, and case-based training enhance clinical reasoning, problem-solving, communication, and general clinical competence, thereby contributing immensely to patient care.

### 36.13.4 Alignment with community needs

One of the main advantages of redesigning medical education is that it will better equip the healthcare workforce to handle the unique health issues, cultural circumstances, and goals of the communities they will serve. This strategy goes beyond conventional, standardized methods to produce healthcare service that is more pertinent and efficient. Graduates who get education focused on common local health challenges, such as managing chronic diseases in metropolitan areas or maternity and child health in rural areas, are immediately prepared to have a real influence on community health statistics. Additionally, students who receive training in the places where they are needed are more likely to open their practices there once they graduate, which helps to alleviate the physician shortage in underprivileged areas.

## 36.14 Interplay of exploratory studies and remodeling

The relationship between exploratory research and remodeling in medical education explains how instructors and educational institutions actively investigate new pedagogical strategies and technological advancements, then modify curricula, instructional strategies and physical spaces to incorporate these innovations into productive, student-centered learning environments. The ensuing process of curriculum remodeling in medical education is directly informed and driven by exploratory investigations which offer the evidence basis and fresh insights required to uncover gaps or weaknesses in current curricula. One way to conceptualize the relationship is as an evidence-based cyclical process. In this context, remodeling is the next intervention or simulation, informed by the results of exploratory research, which serve as the first diagnostic, collecting data to detect trends.

### 36.14.1 Identifying the need for change

Exploratory research investigates topics that have not previously been completely studied or assesses phenomena from a novel angle in order to ascertain "what is happening" in modern educational practices. For example, a qualitative exploratory study might demonstrate that students are not adequately prepared for clinical practice by the current curriculum or that the "hidden curriculum" negatively impacts students' well-being or professionalism.

### 36.14.2 Informing redesign

The findings of these exploratory studies highlight specific issues, such as the need to improve communication skills instruction, incorporate empathy training, or better integrate research and teaching. This data provides precise direction, ensuring that renovation projects are targeted and relevant to the actual requirements of students and the healthcare environment.

### 36.14.3 Developing new approaches

Exploratory research can also assess the feasibility and value of innovative pedagogical interventions, such as a specific combination of learning theories or teaching modes like case-based learning before they are completely adopted. This initial study ensures that the proposed changes will be successful.

### 36.14.4 Guiding implementation

The insights gained from exploratory work serve as a guide for the actual remodeling processes such as faculty development programs, the creation of new modules, or the alignment of formal and informal curricula.

### 36.14.5 Continuous improvement

The method is iterative, using the outcomes of a reformed curriculum as subjects for additional assessment and exploratory study to guarantee continuous improvement in medical education. In essence, exploratory research acts as the diagnostic stage that directs and confirms the therapeutic phase, that is, remodeling of the medical education curriculum, ensuring that the changes are based on empirical data and address real-world problems.

**Summary**

The function of exploratory investigations as a fundamental strategy for comprehending and enhancing medical education has gained a lot of importance in medical education. In complicated educational settings where issues are not well defined and results are impacted by a variety of institutional, cultural, and environmental factors, exploratory investigations are especially beneficial. These studies help identify gaps in medical education curriculum, teaching methodologies, assessment methods, and learning environments by gathering qualitative and quantitative data from students, teachers, and clinical settings. The chapter explores the idea of educational renovation, building on the results of exploratory study. In order to better match medical education with changing healthcare demands, remodeling refers to the methodical reform of curricula, teaching strategies, and assessment frameworks. This entails incorporating learner-centered approaches, technology-enhanced instruction, interprofessional learning, and competency-based education. In order to ensure that reforms are both contextually appropriate and long-lasting, exploratory studies provide evidence on learner experiences, faculty viewpoints, and institutional restrictions.

**Key point**

**Future perspectives: towards an evidence-based transformation**

As healthcare systems and learning environments continue to change, exploratory research and remodeling in medical education are anticipated to become more important, strategic, and data-driven in the future. In order to find trends in student performance, engagement, and clinical competency, data-exploratory research will make more use of artificial intelligence, learning analytics, and huge educational datasets. All things considered, iterative and evidence-based transformation is the key to the future of exploratory research and remodeling in medical education. Medical education may continue to be responsive, creative, and capable of educating healthcare professionals for an increasingly complex and dynamic healthcare landscape by consistently investigating educational obstacles and adapting programs accordingly.

## References

Alshahrani, B, 2025. Journal of Pioneering Medical Sciences. 14.

Anaf, S., 2007. Qualitative Health Research. 17.

Brenner, S., Ertl, G., 2012. Remodelling and adverse remodelling in CAD. Herz 37 (6), 590–597. Available at https://doi.org/10.1007/s00059-012-3660-7.

Cheek, C, 2018. Medical Education. 52.

Chellamuthu, L., Vasudevan, J., Subhaashini, T.D., Govindan, D., Palanichamy, S., 2025. An exploratory study among medical students and faculty on competency-based medical education and family adoption programs. Journal of Dr. NTR University of Health Sciences. 14 (2), 252–258. Available at https://doi.org/10.4103/jdrntruhs.jdrntruhs_90_24.

Godage, S., Pn, R., Vidanagama, U., Fernando, C., Ekanayake, U., Hapuarachchi, T., et al., 2025. The Importance of Exploratory Data Analysis in Clinical Studies. Available at https://doi.org/10.13140/RG.2.2.28599.64166.

Khan, A.S., 2012. Equity or equality in medical education. Journal of Educational Evaluation for Health Professions 9, 3. Available at https://doi.org/10.3352/jeehp.2012.9.3 Epub 2012 Jan 30. PMID: 22323930; PMCID: PMC3274562.

Liu, W., Huang, W., Liu, C., Li, P., Chen, J., 2021. An exploratory study on needs for clinical research training: data from Chinese hospitals. BMC Medical Education 21 (1). Available at https://doi.org/10.1186/s12909-021-02993-1.

Penrabel, R.P.M., de Oliveira Bastos, P.R.H., Biberg-Salum, T.G., 2022. The perspectives and challenges of the competency-based curriculum in medical education: a literature review. Creative Education 13 (10), 3191–3203. Available at https://doi.org/10.4236/ce.2022.1310203.

Rana, K., 2025. MDPI encyclopedia. 5, .

Szabelska, A., Pollet, T., Dujols, O., Klein, R., IJzerman, H., 2021. A Tutorial for Exploratory Research: An Eight-Step Approach. Available at https://doi.org/10.31234/osf.io/cy9mz.

Todres, M., Tsimtsiou, Z., Sidhu, K., Stephenson, A., Jones, R., 2012. Medical students' perceptions of the factors influencing their academic performance: an exploratory interview study with high-achieving and re-sitting medical students. Medical Teacher 34 (5), e325. Available at https://doi.org/10.3109/0142159X.2012.668626.

Walton, J.M., Oswald, A.E., 2024. An exploratory study of small-group learning interactions in pre-clerkship medical education: uncovering a mismatch between student perceptions and real-time observations. International Medical Education 3 (4), 449–460. Available at https://doi.org/10.3390/ime3040034, https://www.mdpi.com/journal/ime.

# Index

*Note*: Page numbers followed by "*f*," "*t*," and "*b*" refer to figures, tables, and boxes, respectively.

## B

## C

## D



## E

Printed and bound by CPI Group (UK) Ltd, Croydon, CR0 4YY
25/09/2026
02222747-0014